THE CELL SURFACE
IN ANIMAL EMBRYOGENESIS AND DEVELOPMENT

1

CELL SURFACE REVIEWS

VOLUME 1

NORTH-HOLLAND PUBLISHING COMPANY
AMSTERDAM · NEW YORK · OXFORD

THE CELL SURFACE
IN ANIMAL EMBRYOGENESIS AND DEVELOPMENT

Edited by

GEORGE POSTE

Roswell Park Memorial Institute, Buffalo, N.Y.

and

GARTH L. NICOLSON

University of California, Irvine, CA

and

Salk Institute for Biological Studies, San Diego, CA

1976

NORTH-HOLLAND PUBLISHING COMPANY
AMSTERDAM · NEW YORK · OXFORD

ISBN 0 7204 0597 1

PUBLISHED BY:
Elsevier/North-Holland Biomedical Press
335 Jan van Galenstraat, P.O. Box 211
Amsterdam, The Netherlands

SOLE DISTRIBUTORS FOR THE U.S.A. AND CANADA:
Elsevier/North-Holland Inc.
52 Vanderbilt Avenue
New York, N.Y. 10017

Library of Congress Cataloging in Publication Data

Main entry under title:

The cell surface in animal embryogenesis and development.

(Cell surface reviews; v. 1)
Bibliography: p.
Includes index.
1. Developmental cytology. 2. Plasma membranes.
3. Embryology. I. Poste, G. II. Nicolson, Garth L.
III. Series.
QL963.5.C44 591.3'33 76-39792
ISBN 0-7204-0597-1

PRINTED IN THE NETHERLANDS

General preface

Research on membranes and cell surfaces today occupies center stage in many areas of biology and medicine. This dominant position reflects the growing awareness that many important biological processes in animal and plant cells and in microorganisms are mediated by these structures. The extraordinary and unprecedented expansion of knowledge in molecular biology, genetics, biochemistry, cell biology, microbiology and immunology over the last fifteen years has resulted in dramatic advances in our understanding of the properties of the cell surface and heightened our appreciation of the subtle, yet complex, nature of cell surface organization.

The rapid growth of interest in all facets of research on cell membranes and surfaces owes much to the convergence of ideas and results from seemingly disparate disciplines. This, together with the recognition of common patterns of biological organization in membranes from highly different forms of life, has led to a situation in which the sharp boundaries between the classical biological disciplines are rapidly disappearing. The investigator interested in cell surfaces must be at home in many fields, ranging from the detailed biochemical and biophysical properties of the molecules and macromolecules found in membranes to morphological and phenomenological descriptions of cellular structure and cell-to-cell interactions. Given the broad front on which research on cell surfaces is being pursued, it is not surprising that the relevant literature is scattered in a diverse range of journals and books, making it increasingly difficult for the active investigator to collate material from several areas of research. Thus, while scientists are becoming increasingly specialized in their techniques, and in the nature of the problems they study, they must interpret their results against an intellectual and conceptual background of rapidly expanding dimension. It is with these conflicting demands and needs in mind that this series, to be known under the collective title of CELL SURFACE REVIEWS, was conceived.

CELL SURFACE REVIEWS will present up-to-date surveys of recent advances in our understanding of membranes and cell surfaces. Each volume will contain authoritative and topical reviews by investigators who have contributed to progress in their respective research fields. While individual reviews will provide comprehensive coverage of specialized topics, all of the reviews published within each volume will be related to an overall common theme. This format represents a departure from that adopted by most of the existing series of "review" publications which usually provide heterogeneous collections of reviews on unrelated topics. While this latter format is considerably more convenient from an editorial standpoint, we feel that publication together of a number of related reviews will better serve the stated aims of this series – to bridge the information and specialization "gap" among investigators in related areas. Each volume will therefore present a fairly complete and critical survey of the more important and recent advances in well defined topics in biology and medicine. The level will be advanced, directed primarily to the needs of the research worker and graduate students.

Editorial policy will be to impose as few restrictions as possible on contributors. This is appropriate since the volumes published in this series will represent collections of review articles and will not be definitive monographs dealing with all aspects of the selected subject. Contributors will be encouraged, however, to provide comprehensive, critical reviews that attempt to integrate the available data into a broad conceptual framework. Emphasis will also be given to identification of major problems demanding further study and the possible avenues by which these might be investigated. Scope will also be offered for the presentation of new and challenging ideas and hypotheses for which complete evidence is still lacking.

The first four volumes of this series will be published within one year, after which volumes will appear at approximately one year intervals.

George Poste
Garth L. Nicolson
Editors

Contents of forthcoming volumes

Volume 2
VIRUS INFECTION AND THE CELL SURFACE
Host and Tissue Specificities in Virus Infections of Animals – H. Smith (Birmingham)
Structure and Assembly of Enveloped Viruses – R. Rott and H.D. Klenk (Giessen)
Antigenic Alterations in Virus-Infected Cells – A.C. Allison and W. Burns (London)
Virus-Erythrocyte Interactions – T. Bächi (Zurich), J.E. Deas and C. Howe (New Orleans)
Virus-Host Cell Interactions in "Slow" Virus Diseases – R.T. Johnson and L.P. Weiner (Baltimore)
T-Cell Recognition of Virus-Infected Cells – R.V. Blanden (Canberra)
Host Antigens in Viruses – J. Lindenmann (Basel)
Interaction of Viruses with Model Membranes – J. Tiffany (Oxford)
Cell Fusion by Sendai Virus – Y. Hosaka (Philadelphia) and F. Shimizu (Sendai)

Volume 3
DYNAMIC ASPECTS OF CELL SURFACE ORGANIZATION
Dynamic Aspects of Membrane Organization – G.L. Nicolson (San Diego), G. Poste (Buffalo) and T. Ji (Laramie)
Freeze-Fracture Techniques in the Analysis of Membrane Structure – N.S. McNutt (San Francisco)
Manipulation of Membrane Lipid Composition – A.F. Horwitz (Philadelphia)
Exchange of Phospholipids between Membranes – J.C. Kader (Paris)
Membrane Fluidity and the Activity of Membrane-Associated Enzymes – H. Kimelberg (Albany)
Growth-Dependent Changes in Cell Surface Proteins – C. Gahmberg (Helsinki)
Growth-Dependent Changes in Cell Surface Glycolipids – D. Critchley (Leicester)

Preface

Over the past two decades the study of growth, differentiation and morphogenesis in the developing embryo has shifted in emphasis from preoccupation with descriptions of changing morphology at the tissue and cellular level to concern with the subcellular events and molecular mechanisms underlying these complex processes. Elusive as the final answers remain, there is now a general consensus that the cell surface is of major importance in regulating many aspects of differentiation and morphogenesis. The surface regions of cells, by their influence on cell growth, division, adhesion, movement and recognition, assume a central regulatory role in determining the complicated series of cellular interactions involved in the orderly development of tissue systems within the embryo.

The capacity of the cell surface to discharge these regulatory functions reflects the high degree of specificity and functional diversity displayed by this structure in different cells. These properties, in turn, reflect the fact that the plasma membrane is a dynamic structure which is able to undergo structural and functional changes in response to both genetic regulation and environmental stimuli. The modulation of cell surface organization by both intra- and extracellular factors not only provides a mechanism whereby highly specific patterns of surface organization can be generated in different cell types (i.e., differentiation), but also enables differentiated cells to modulate their surface properties in response to specific physiological stimuli.

In discussing the role of the cell surface in development, we cannot treat it as an isolated entity. It is a structure that is highly responsive to events occurring in both the intra- and extracellular compartments. In addition to its unique role in regulating the flow of substances between the intra- and extracellular environments, it is becoming clear that subtle changes in the topographic arrangement of macromolecules in the plasma membrane and its associated structures can serve as a powerful mechanism for transmission of "information" between the outside and inside of the cell. Reaction of a cell with extracellular factors, including other cells, may generate structural rearrangements of membrane components which, in turn, may result in alterations in intracellular metabolism, including possible changes in gene expression. The altered intracellular organization produced by an event occurring at the cell surface may then act as a feedback mechanism to stabilize existing surface organization or, alternatively, may impose entirely new arrangements of surface determinants with resulting change in cell surface properties. Is it these sequences of reciprocal information transfer between the intra- and extracellular compartments that drive cells along the pathways resulting in cytodifferentiation and the emergence of histotypic organization? Although the answer to this question is not known at present, the role of the cell surface as an exquisite mechanism for coupling the intra- and extracellular compartments is made possible by trans-membrane functional linkage between the various components of the plasma membrane and membrane-associated structures. This feature of membrane organization enables information to be transmitted in both directions across the membrane via various combinations of cooperative, allosteric and transductive coupling mechanisms.

The impressive range of physical, chemical, cytological, ultrastructural, immunological and genetic techniques which have been introduced in recent years to study membrane organization have also been embraced by developmental biologists. This has resulted in significant progress in our efforts to understand how changes in cell surface properties influence the complex processes of differentiation and morphogenesis in the developing embryo. This progress is reflected in the thirteen reviews presented in this volume which examine various aspects of cell surface function and cellular interactions in development.

We recognize that a subject as diverse as cellular interactions in embryogenesis cannot be treated adequately in a single volume. We have therefore included a number of reviews dealing with general aspects of the important topics of fertilization, cleavage, implantation, placentation, the immunobiology of development, cell communication and specification of cell position, morphogenetic cell movements and inductive tissue interactions. These articles discuss data obtained from a wide range of cell, tissue and organ systems, and each article offers an extensive bibliography. The few remaining reviews are devoted to discussion of development in three organ systems, the heart, the limb and the immune apparatus. These systems were selected largely because they represent areas of research in which considerable progress has been made recently. Equally detailed and impressive information on the importance of the cell surface in differentiation and morphogenesis are available for such tissues as muscle, cartilage, bone, connective tissue and for nerve cells and blood cells. These topics have been reviewed in detail in many publications in the last few years, and it was considered appropriate to confine discussion of these systems to the more general articles contained in this volume. An attempt has been made throughout this volume to achieve a balance between discussion of events occurring at the tissue and cell level and those at the subcellular and molecular level. This provides a blend between the older, but nonetheless elegant, descriptive data from classical embryology and data obtained in the newer era of developmental biology.

The contributors to this volume most receive ultimate credit for whatever success it achieves. To them we express our thanks and our appreciation of their willingness to accept editorial suggestions. We are also grateful to the many other individuals who contributed to the production of this volume. Particular thanks go to Adele Brodginski, Molly Terhaar, Judy Morey, Alice MacKearnin and Shirley Guagliardi for their assistance in preparing the edited manuscripts. Finally, we thank Dr. Jack Franklin of North-Holland for his encouragement and help in starting this new series.

George Poste
Buffalo, New York
December, 1975

Garth L. Nicolson
Irvine and San Diego, California
December, 1975

List of contributors

The numbers in parentheses indicate the page on which the author's contributions begin.

John M. ARNOLD (55), Kewalo Marine Laboratory, Pacific Biomedical Research Center, University of Hawaii, Honolulu, Hawaii 96813, U.S.A.

Randall W. BARTON (601), Department of Pathology, University of Connecticut Health Center, Farmington, Connecticut, 06032, U.S.A.

Robert L. BRENT (145), The Stein Research Center, Departments of Pediatrics and Radiology, Jefferson Medical College of the Thomas Jefferson University, Philadelphia, Pennsylvania 19107, U.S.A.

Donald A. EDE (495), Department of Zoology, University of Glasgow, Glasgow G12 9LU, Scotland.

Michael EDIDIN (127), The Mergenthaler Laboratory for Biology, The Johns Hopkins University, Baltimore, Maryland 21218, U.S.A.

Irving GOLDSCHNEIDER (601), Department of Pathology, The University of Connecticut Health Center, Farmington, Connecticut 06032, U.S.A.

Ralph B. L. GWATKIN (1), Merck Institute for Therapeutic Research, Rahway, New Jersey 07065, U.S.A.

Marketta KARKINEN-JÄÄSKELÄINEN (331), III Department of Pathology, University of Helsinki, SF-00290 Helsinki 29, Finland.

Thomas R. KOSZALKA (145), The Stein Research Center, Departments of Radiology and Biochemistry, Jefferson Medical College of the Thomas Jefferson University, Philadelphia, Pennsylvania 19107, U.S.A.

Eero LEHTONEN (331), III Department of Pathology, University of Helsinki, SF-00290 Helsinki 29, Finland.

Francis J. MANASEK (545), Department of Anatomy, University of Chicago, Chicago, Illinois 60637, U.S.A.

David MASLOW (697), Department of Experimental Pathology, Roswell Park Memorial Institute, Buffalo, New York 14263, U.S.A.

Daniel McMAHON (449), Division of Biology, California Institute of Technology, Pasadena, California 91109, U.S.A.

Richard K. MILLER (145), The Perinatal Center, Departments of Obstetrics-Gynecology and Pharmacology, University of Rochester School of Medicine and Dentistry, Rochester, New York 14642, U.S.A.

Stig NORDLING (331), III Department of Pathology, University of Helsinki, SF-00290 Helsinki 29, Finland.

Lauri SAXÉN (331), III Department of Pathology, University of Helsinki, SF-00290 Helsinki 29, Finland.

Judson D. SHERIDAN (409), Department of Zoology, University of Minneapolis, Minneapolis, Minnesota 55455, U.S.A.

Michael I. SHERMAN (81), Department of Cell Biology, Roche Institute of Molecular Biology, Nutley, New Jersey 07110, U.S.A.

J.P. TRINKAUS (225), Department of Biology, Yale University, New Haven, Connecticut 06520, U.S.A.

Jorma WARTIOVAARA (331), III Department of Pathology, University of Helsinki, SF-00290 Helsinki 29, Finland.

Christopher WEST (449), Division of Biology, California Institute of Technology, Pasadena, California 91109, U.S.A.

Linda R. WUDL (81), Department of Cell Biology, Roche Institute of Molecular Biology, Nutley, New Jersey 07110, U.S.A.

Contents

3 *The implanting mouse blastocyst, by M.I. Sherman and L.R. Wudl*

4 *Cell surface antigens in mammalian development, by M. Edidin*

5 *The transport of molecules across placental membranes, by R.K. Miller, T. R. Koszalka and R.L. Brent*

6 On the mechanism of metazoan cell movements, by J.P. Trinkaus

7 *Inductive tissue interactions, by L. Saxén, M. Karkinen-Jääskeläinen, E. Lehtonen, S. Nordling and J. Wartiovaara*

8 *Cell coupling and cell communication during embryogenesis, by J.D. Sheridan*

9 *Transduction of positional information during development, by D. McMahon and C. West*

10 *Cell interactions in vertebrate limb development, by D.A. Ede*

11 Heart development: interactions involved in cardiac morphogenesis, by F.J. Manasek

12 Development and differentiation of lymphocytes, by I. Goldschneider and R.W. Barton

13 In vitro analysis of surface specificity in embryonic cells, by David E. Maslow

Fertilization 1

R.B.L. GWATKIN

1. Introduction

The purpose of fertilization is to transmit and recombine genes. To realize these goals nature has devised specialized transports, the sperm and the egg that unite by a complex series of membrane interactions. Until comparatively recently, the study of fertilization in mammals had not attracted the same amount of research interest as fertilization in invertebrates. The experimental difficulties posed by the internal site of fertilization in mammals, the small number of mammalian eggs that can be obtained for experimentation (even from superovulated animals) and the rigorous environmental requirements for consistent success in achieving fertilization in vitro have all been at least partly responsible for this situation. Over the last 15 years, however, considerable effort has been devoted to the analysis of the mechanisms of mammalian fertilization, stimulated in large part by the economic incentive of developing methods for selective breeding and artificial insemination in farm animals and also by increasing awareness of the social importance of fertility regulation in human populations. The development of improved techniques for handling mammalian gametes in vitro, better recognition of the environmental requirements for fertilization of mammalian eggs in vitro and the availability of increasingly sophisticated techniques for ultrastructural and biochemical analysis have all served to greatly increase our knowledge of mammalian fertilization over the past few years.

In addition to analysis of the events occurring during actual fusion of the gametes, complete discussion of the subject of fertilization must also consider the changes undergone by gametes before and after they fuse and the role of various factors in the reproductive tract in influencing the interaction between sperm and egg. It is clearly beyond the scope of the present chapter to adequately cover the entire subject of fertilization and I have therefore limited this review to a discussion of the organization of the cell surface in mammalian gametes and the various membrane interactions involved in achieving successful fusion between gametes. For a general background on the wider aspects of fertilization the reader is referred to the books by Austin (1965, 1968), Austin and Short (1972), Coutinho and Fuchs (1974), Metz and Monroy (1969), and Zamboni (1971a)

G. Poste & G.L. Nicolson (eds.) The Cell Surface in Animal Embryogenesis and Development

and the recent reviews by Longo (1973) and McRorie and Williams (1974).

Fertilization, in its most restricted sense, involves fusion between the plasma membranes of male and female gametes with resulting introduction of the sperm nucleus into the cytoplasm of the egg followed by incorporation of the male and female genomes into a single nucleus which marks the end of fertilization. It is clear, however, that the fusion of gametes and the subsequent development of syngamy are but isolated parts in a highly organized multistep process in which the interacting gametes must first undergo necessary preparatory changes to enable them to fuse, while once fusion has taken place further changes must occur in the surface properties of the newly fertilized egg to allow its orderly development to the zygote stage and beyond.

Before formal union of the gametes can occur the sperm must first attach to the surface of the egg and penetrate the various investments surrounding the egg. This requires release of lytic enzymes carried in the acrosome of the sperm which are able to digest the egg investments. The acrosome reaction involves fusion between the acrosomal membrane and the overlying plasma membrane. The acrosome reaction must itself be preceded by the process of capacitation in which the surface properties of the sperm also change. Capacitation, in turn, is preceded by yet other changes in the sperm surface that occur during sperm maturation in the epididymis before ejaculation. A further series of important membrane interactions occur after successful fusion of the gametes. Once a spermatozoon has fused with the egg plasma membrane the egg is protected against fusion of additional spermatozoa. This block to polyspermy is mediated by alterations in the egg plasma membrane and, in certain species, also by the surrounding zona pellucida. The development of the block to polyspermy is itself temporally, and probably causally, related to the breakdown of the cortical granules of the egg which undergo massive synchronized exocytotic fusion with the egg plasma membrane immediately after fusion of the first spermatozoon. The block to polyspermy appears to be imposed by the direct modification of the egg surface by the membranes of the cortical granules. From this brief outline it is clear that cell surface organization is of fundamental importance in fertilization and that successful fertilization requires a wide variety of interactions between different cellular membrane systems. The aim of this review is to provide a mid-1975 picture of these membrane interactions and their importance in mammalian fertilization.

2. *The egg*

The structures of the mammalian egg involved in fertilization are shown in Fig. 1. In most mammals, the egg undergoes the first meiotic division with extrusion of the first polar body shortly before ovulation and remains arrested at the second meiotic metaphase (Fig. 1) until sperm penetration occurs. Because of the experimental difficulties involved in obtaining large numbers of mammalian eggs, little is known about the molecular structure of the egg plasma membrane and its

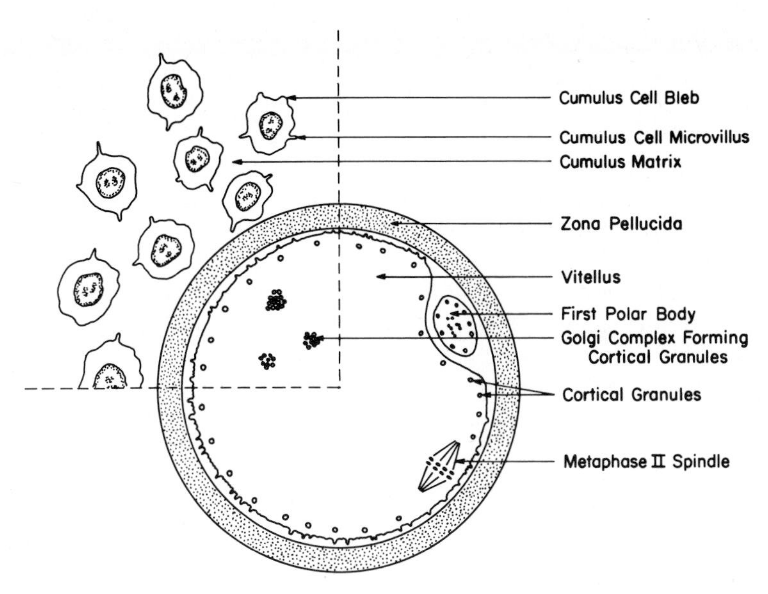

Fig. 1. Diagrammatic representation of the ovulated mammalian egg with associated cumulus oophorus (the latter is shown as present only in the upper left quadrant).

surrounding investments. Consequently, our present knowledge of the organization of the mammalian egg is limited largely to morphological description at both the light microscope and electron microscope levels. Numerous descriptions of the morphology of egg formation and maturation in various species of mammals are available (see, Austin, 1961, 1968; Hadek, 1965; Pikó, 1969; Zamboni, 1970, 1971a; Baker, 1972) and only sufficient discussion will be offered here to outline the importance of particular regions of the egg in different stages of the fertilization process.

Typically, the cytoplasm of the freshly ovulated egg is surrounded by a plasma membrane and two surrounding layers of investment, the zona pellucida and the cumulus oophorus (Fig. 1). The plasma membrane of the egg is also called the vitelline membrane and these two terms will be used interchangeably throughout this chapter. The general function of the egg plasma membrane corresponds to that of the plasma membrane in somatic cells, controlling the entry of material into the vitellus both by diffusion and by active transport (review, Hafez and Sugawara, 1969). The plasma membrane is surrounded by the translucent acellular zona pellucida, though the two structures are physically separated by the fluid-filled perivitelline space (Fig. 1). The thickness of the zona pellucida varies between 3 and 15 μ in different species of mammal (see, Austin, 1961). The fragmentary evidence available indicates that the zona pellucida is composed of proteins and carbohydrates (Braden, 1952; Stegner and Wartenberg, 1961, Jacoby, 1962; Loewenstein and Cohen, 1964; Austin, 1968; Pikó, 1969), possibly in the form of glycopeptide units stabilized by disulphide bonds (Gould et al., 1971, Inoué and Wolf, 1974; Oikawa et al., 1974), and possibly hydrophobic interactions or salt linkages (Nicolson et al., 1975). In view of the significant contribution of glycopeptides to the composition of the zona pellucida, it is perhaps not surprising that this structure is strongly antigenic (see, Glass and Hanson, 1974; Shivers and Dudkiewicz, 1974) and is also able to bind a variety of plant lectins (Oikawa et al., 1973, 1974; Nicolson et al., 1975). The zona is freely permeable to large molecules such as ferritin and horseradish peroxidase (Hastings et al., 1972), immunoglobulin M (Sellens and Jenkinson, 1975) and viruses (Gwatkin, 1967).

In the newly ovulated egg, the vitellus and the surrounding zona are surrounded by the cumulus oophorus consisting of several thousand granulosa cells, which are relatively widely separated from each other and interspersed in a gelatinous matrix (Fig. 1). Information on the chemical composition of the cumulus matrix is limited but hyaluronic acid is considered to be the major component (see, Pikó, 1969).

It is clear from this brief description that the investments surrounding the egg constitute major mechanical barriers which the sperm must pass through in order to achieve direct contact with the egg. Thus, spermatozoa must first interact with the cumulus cells (section 5), then attach and bind to the exterior of the zona pellucida (section 7) before subsequently penetrating the zona to interact directly with the egg plasma membrane (section 9). This is then followed by fusion between the plasma membrane of the penetrating spermatozoon and the egg plasma membrane (section 9). Once fusion has occurred, penetration of additional spermatozoa is blocked. This block to polyspermy is produced by discharge of the numerous cortical granules present in the egg cytoplasm (Fig. 1) which, following fusion of the fertilizing sperm with the egg, fuse with the overlying plasma membrane to release their contents into the perivitelline space (section 10). The material(s) released from these granules act on both the egg plasma membrane and the surrounding zona pellucida to block attachment and penetration of additional spermatozoa into the newly fertilized egg.

3. The sperm

Although there is a voluminous literature on the behavior of spermatozoa and the diverse factors that affect their ability to fertilize, our knowledge of membrane organization in mammalian spermatozoa is still limited. It is clear, however, that the functional properties of the sperm, including its surface characteristics, vary significantly at different stages in the life of the sperm. Experimental characterization of the surface properties of the sperm is probably needed at a minimum of six stages in the life of the sperm, each of which is associated with recognizable differences in sperm behaviour. These include the pre-ejaculation stages of development in the testis and subsequent maturation in the epididymis, and post-ejaculation stages involving capacitation, the acrosome reaction and final attachment to, and fusion with, the egg.

Unlike the situation in many invertebrates and certain non-mammalian vertebrates, the mammalian spermatozoon is not functionally competent when released from its association with the Sertoli cell in the germinal epithelium of the testis, and must undergo further maturation in the epididymis where the ability to fertilize is acquired (reviews, Orgebin-Crist, 1969; Hamilton, 1973a,b). Inherent in much of the work that has been done on sperm maturation, though infrequently stated as such, is the concept that the three gross anatomical segments of the epididymis, the caput, corpus and cauda, also constitute distinct physiological compartments in which specific maturational changes occur in the sperm. It is now clear from both ultrastructural and biochemical observations that this is an oversimplification (see, Hamilton, 1973a). For descriptive purposes, however, the three anatomical divisions of the epididymis are of some value and are still widely used in the literature to define the source of spermatozoa and will be used as such throughout this chapter.

The maturation of spermatozoa within the epididymis and the development of competence to penetrate the egg involves changes in several different organelles of the spermatozoon as well as changes in the behaviour of the intact cell. A number of phenomenological observations have been made concerning alterations in spermatozoa during their transport through the epididymis. These include: caudal migration and loss of the cytoplasmic (kinoplasmic) droplet; changes in the response of spermatozoa to heat, cold and alkaloids, increased specific gravity in mature sperm; and increased capacity of mature sperm for sustained motility (review, Bedford, 1972). Apart from changes in sperm motility which are of importance for subsequent penetration of the egg, it is difficult to define meaningful functional correlations between the above alterations and development of fertilization capacity in mature sperm.

In focusing on the various membrane interactions involved in fertilization, the sperm plasma membrane, the acrosome and the nucleus are of major interest and the remaining discussion in this chapter will therefore concentrate on these structures, all of which are located in the head region of the spermatozoon (Fig. 2).

3.1. The plasma membrane

Indirect evidence for changes in the plasma membrane of the spermatozoon during passage through the epididymis into the cauda region is provided by observations on changes in the light scattering properties of the sperm surface (Lindahl and Kihlstrom, 1952), changes in the affinity of the sperm surface for histochemical dyes (Ortavant, 1953; Brochart and Debatene, 1953; Glover, 1961) and fixatives (Fawcett and Phillips, 1969), changes in surface adhesiveness (Bedford, 1972) and changes in the electrophoretic mobility of the intact sperm (Bedford, 1963).

More direct evidence that the properties of the sperm plasma membrane change during maturation of sperm in the epididymis has been obtained from biochemical and ultrastructural studies.

Analysis of the phospholipids present in spermatozoa has established that in

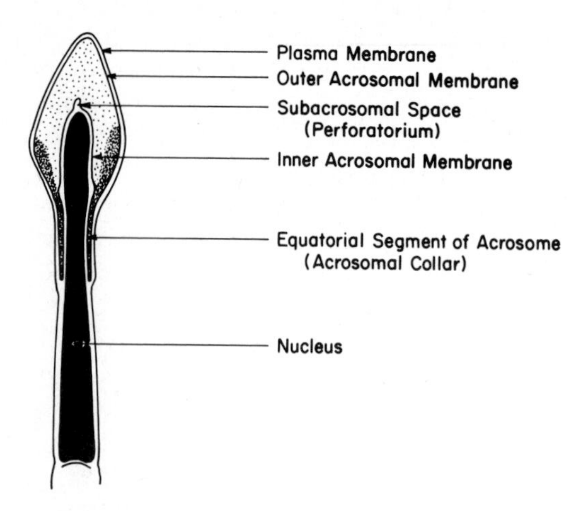

Fig. 2. Diagrammatic representation of a sagittal section of the head region of a hamster spermatozoon (redrawn from Yanagimachi and Noda, 1970d).

several species significant changes in lipid composition occur as the spermatozoa pass through the epididymis (Dawson and Scott, 1964; Crogan et al., 1966; Quinn and White, 1967; Scott et al., 1967; White, 1973; Terner et al., 1975). In the ram, the bull and the boar the membranes of testicular spermatozoa have a higher lecithin and cholesterol content and a higher ratio of saturated to unsaturated fatty acids than spermatozoa from the cauda epididymis or ejaculated spermatozoa. These findings suggest that the membrane lipids in mature spermatozoa would be more "fluid" than those of testicular spermatozoa. This data refers, however, to the bulk lipid composition in spermatozoa and localisation of different classes of phospholipid to different regions of the sperm has not yet been attempted.

Nonetheless, in most general terms, these differences in membrane lipid composition may be important in determining the ability of the sperm plasma membrane to fuse with other membranes during the acrosome reaction and subsequent fusion with the eggs. Recent studies on the in vitro fusion of model membranes of defined composition have shown that an essential requirement for fusion is that the lipids in the interacting membranes be "fluid", and that increases in membrane lecithin and cholesterol content reduce the capacity of membranes to fuse (Papahadjopoulos et al., 1974). If similar conditions apply to fusion involving the sperm plasma membrane, then the above changes in lipid composition accompanying sperm maturation in the epididymis would be expected to exert an important influence on the fusion capacity of the sperm.

The most convincing evidence for changes in plasma membrane organization during sperm maturation has been provided by a number of recent ultrastructural studies. Changes in the dimensions and staining affinity of the inner and outer electron-dense leaflets of the plasma membrane have been detected in rabbit and human spermatozoa during their passage through the epididymis (Bedford et al., 1972) but the significance of these changes for membrane function remains to be established.

A more profitable approach in defining plasma membrane alterations during sperm maturation has come from electronhistochemical observations on the binding of specific ligands to the plasma membrane. Changes in the chemical nature of the sperm surface during epididymal maturation have been demonstrated by Cooper and Bedford (1971) and by Bedford et al. (1973) who showed that the ultrastructural binding pattern of colloidal iron hydroxide (CIH) particles to the plasma membrane of rabbit, monkey and human sperm changed with transport of sperm from the caput to the cauda epididymis. The CIH-binding technique not only provides information on the distribution of negative charges on the cell surface but also affords insight into the chemical nature of the charged moieties. At low pH (1.8) binding of CIH particles indicates the presence of ionized radicals of carboxy sugars, primary phosphate and sulphate, while binding at higher pH (3.0–4.6) indicates the presence of amino acid carboxy groups. The work of Bedford and his colleagues has demonstrated that not only are there marked differences in the chemistry of the sperm surface in different species but also in epididymal sperm of differing maturity within a species. For example, the plasma membrane over the head region of rhesus monkey and human spermatozoa obtained from the caput and upper corpus epididymis fails to bind CIH particles at pH 1.8 whereas the same membrane region in sperm obtained from the lower corpus and cauda epidymis displays a high affinity for particle binding at low pH.

These studies have also established that passage of sperm through the epididymis is associated with changes in the affinity of different regions of the sperm for CIH particles. In monkey sperm, a higher charge density is detected by particle binding over the post-acrosomal region than on the acrosomal region, whereas in human spermatozoa both regions show equal charge density. In complete contrast, the head region of both immature and mature rabbit spermatozoa does not bind CIH particles to any significant extent at pH 1.8 but at pH 4.6 significant binding occurs to the head in sperm from the lower corpus and cauda epididymis but immature sperm remain unreactive (Cooper and Bedford, 1971; Bedford et al., 1973). Distinct CIH-particle binding patterns have also been detected in the plasma membrane overlying the midpiece and tail in spermatozoa from these species (Cooper and Bedford, 1971; Bedford et al., 1973).

Regional distribution of anionic residues has also been detected in hamster spermatozoa using similar CIH-binding techniques (Yanagimachi et al., 1972, 1973). In this species, mature spermatozoa show a dense concentration of residues over the tail region but binding is practically absent over the post-acrosomal plasma membrane and is scant over the acrosomal region.

These changes in the nature and the distribution of surface charges on the plasma membrane of the maturing sperm detected at the ultrastructural level are probably correlated with the differences in the electrokinetic mobility and head-tail orientation in electric fields displayed by sperm populations taken from various levels of the epididymis (Bedford, 1963).

Bedford et al. (1973) investigated the question of whether the changes in sperm surface charge density and distribution accompanying epididymal maturation resulted from an intrinsic mechanism associated with the age(ing) of the sperm or whether extrinsic epididymal factors were also involved. By ligating the lower corpus epididymis, these investigators found that spermatozoa retained in the caput epididymis never developed the characteristic charge pattern found in mature sperm in the cauda epididymis. Similar ligation experiments have also demonstrated that sperm artificially retained in the caput fail to develop the normal auto-agglutination behaviour displayed by mature sperm (Bedford, 1967).

These experiments indicate that the epididymis makes an important contribution to the plasma membrane changes that occur during transit of spermatozoa through the epididymis. The important question of whether the membrane changes acquired in the distal regions of the epididymis result from absorption and/or insertion of secreted epididymal components into the sperm plasma membrane, or from unmasking or chemical alteration of existing plasma membrane components by conditions in the lower epididymis, remains to be answered.

Specialization in the organization of the plasma membrane in different regions of mammalian spermatozoa has also been demonstrated in ultrastructural studies on the binding of plant lectins to the sperm surface. Plasma membrane receptors for the plant lectin, concanavalin A (Con A), appear to be distributed primarily over the acrosomal membrane in sperm from several species (Edelman and Millette, 1971; Nicolson and Yanigamachi, 1972; Johnson, 1975). Characteristic but different binding patterns have been identified for other lectins and for the binding of virus particles to the sperm surface (Ericsson et al., 1971; Nicolson and Yanigamachi, 1972; Gall et al., 1974). The existence of region-specific plasma membrane antigens has also been documented (Henle et al., 1938; Hjort and Brogaard, 1971; Johnson and Edidin, 1972) though certain antigens appear to be distributed over the entire plasma membrane (see, Erickson, 1972).

Striking regional specialization in the organization of the sperm plasma membrane has also been demonstrated recently by freeze-fracture electronmicroscopy (Friend and Fawcett, 1974; Fawcett, 1975; Koehler and Gaddum-Rosse, 1975). This technique reveals aspects of the internal organization of the plasma membrane by means of a fracture within the membrane. Using this method, significant differences have been noted in the number and distribution of intramembranous particles in different regions of the plasma membrane in guinea pig and rat spermatozoa. Over the acrosome the plasma membrane exhibits a periodic structure with highly ordered arrangement of particles into crystalline arrays. Over the midpiece the particles exist in long rows oriented circumferentially and which also appear to be concentrated over the underlying mitochondria, while in the midpiece plasma membrane the particles are arranged randomly.

In addition to variation in the type of structural determinants present in different regions of the sperm plasma membrane, the macromolecular organization of similar membrane components may differ in specific regions of the sperm surface. For example, Nicolson and Yanagimachi (1974) have shown that the translational mobility of D-galactose residues varies significantly within different regions of the plasma membrane in rabbit epididymal spermatozoa. When spermatozoa were labeled with *Ricinus communis* agglutinin (which binds specifically to D-galactose residues) at 0°C continuous labeling occurred over the whole sperm surface. However, when spermatozoa were labeled with the lectin at 0°C and then incubated at 37°C, clustering and patching of lectin-receptor complexes was observed in the post-acrosomal plasma membrane but not elsewhere. These observations suggest that the lateral translational mobility of integral membrane glycoproteins carrying D-galactose residues is significantly higher in the post-acrosomal plasma membrane. Restriction of the lateral mobility of membrane components in the sperm plasma membrane over the acrosomal region has also been identified for Con A receptors (Johnson, 1975) and heteroantigens (Koehler and Perkins, 1974).

These observations suggest that the organization of the plasma membrane in the mammalian spermatozoon conforms in general principle to the fluid mosaic model for membrane structure (Singer and Nicolson, 1972) in which different integral membrane components display varying degrees of translational mobility within the membrane. It is not yet clear, however, whether the differing mobilities of membrane components in specific regions of the sperm plasma membrane reflect differences in the fluidity of the membrane matrix in these regions, differences in the linkage of membrane receptors to membrane-associated cytoskeletal elements, or both.

The greater mobility of membrane components in the post-acrosomal region may be of particular significance since the sperm fuses with the egg plasma membrane at the post-acrosomal region. As will be discussed later the redistribution of integral membrane proteins may be an important event in membrane fusion, and it is tempting to speculate that the greater mobility of components in the post-acrosomal membrane is related to the important role of this region in achieving final fusion of the sperm with the egg.

There is now substantial evidence to indicate that the seminal plasma and epididymal secretions contain macromolecules, so called coating antigens, that bind to the surface of spermatozoa during their maturation in the epididymis (Chang, 1957; Bedford and Chang, 1962; Weil and Rodenburg, 1962; Weinman and Williams, 1964; Dukelow et al., 1966, 1967; Pinsker and Williams, 1967; Hunter and Nornes, 1969; Johnson and Hunter, 1972; Killian and Amann, 1973). These components appear to bind to the entire cell surface and affinities for specific regions of the sperm plasma membrane have not been identified. As we shall see later the removal of these coating factors may be an important event in the capacitation of spermatozoa within the female tract.

Although the ultrastructural and biochemical studies reviewed in this section have substantially enlarged our understanding of the structural organization of the mammalian spermatozoon, detailed characterization of the macromolecular organization of the plasma membrane and other subcellular structures in spermatozoa still lags far behind our understanding of these structures in somatic cells. The general lack of methods for the subcellular fractionation of spermatozoa has been the major contributory factor in this deficiency. Some progress has been reported recently, however, in the chemical dissection, separation and purification of spermatozoal subcellular components (Millette et al., 1973; Gall et al., 1974). The further development and application of these (and perhaps other) fractionation procedures is an obvious and important area for future research.

3.2. The acrosome

The acrosome in mammalian spermatozoa contains the lytic enzymes responsible for digestion of the egg investments. The shape and size of the acrosome varies markedly in different species. Variation exists mainly in the anterior portion of the acrosome which in some species is relatively small (man, monkey, boar, ram, rabbit) while in others it is much larger and elaborate (guinea pig, chinchilla). The formation and structural characteristics of the acrosome have been reviewed by Burgos (1974) and the properties of the various acrosomal enzymes by Fritz et al. (1973) and by McRorie and Williams (1974).

Three portions of the acrosome are recognized. The most anterior portion is termed the acrosome cap or apical segment (Fig. 2) and this extends beyond the anterior limit of the nucleus. The main segment of the acrosome extends back over the anterior half of the nucleus and then abruptly decreases in thickness to form the so called equatorial segment or acrosomal collar (Fig. 2). This latter region has a greater electron density than the rest of the acrosome. The apical and main segments are believed to be involved in the dispersal of the cumulus while enzymes associated with the equatorial segment and inner acrosomal membrane are suggested to be involved in lysis of the zona pellucida.

Little is known concerning the factors that dictate the shape and volume of the acrosome, though its development in the guinea pig is seemingly androgen-dependent (Blaquier et al., 1972). After leaving the germinal epithelium of the testis the acrosome in some species such as man may show little or no apparent morphological change, in species such as the rabbit, pig and the sheep limited changes occur, while in the guinea pig dramatic enlargement takes place during transit through the epididymis (see, Burgos, 1974).

3.3. The nucleus

The nucleus is the major organelle in the head region of the sperm. It is characteristically extremely electron-dense and devoid of obvious ultrastructure other than a variable number of small randomly dispersed cavities in the chromatin (so called nuclear vacuoles). For information on the detailed ultrastructure, development and maturation of the sperm nucleus the reader is refered to the reviews by Fawcett (1970) and Burgos (1974).

Despite the overall emphasis on membranes in this chapter, it is perhaps pertinent to briefly mention one feature of nuclear maturation in mammalian spermatozoa that may be important in protecting the nucleus from damage during penetration of the sperm through the egg investments. Calvin and Bedford (1971) and Bedford and Calvin (1974) have shown that the nucleus in spermatozoa of eutherian mammals is stabilized during passage of sperm through the epididymis by the formation of disulphide bridges in the nuclear chromatin. Similar cross-linking does not occur in the nuclei of spermatozoa from invertebrates, fish or chickens or even marsupial mammals. Calvin and Bedford propose that formation of disulphide bridges creates the necessary degree of rigidity in the chromatin needed to protect it from mechanical shear during passage of the spermatozoon through the thick resilent zona pellucida of the egg which, in this form, is peculiar to Eutheria. However, the observation that newly mature epididymal sperm are often unable to support normal embryological development after their entry into the egg (Orgebin-Crist, 1967) suggests that nuclear maturation may involve additional non-mechanical changes.

3.4. Cytoplasmic components

Three cytoplasmic regions are recognized in the head region of the mammalian spermatozoon. One region exists between the acrosome and the anterior pole of the nucleus in the so called subacrosomal space (Fig. 2). The shape of this layer of cytoplasm varies markedly in sperm from different species, being limited to a very thin film in human spermatozoa but is well developed in rodent sperm (see, Burgos, 1974). Another thin layer of cytoplasm exists over the anterior portion of the sperm head between the outer acrosomal membrane and the overlying plasma membrane (Fig. 2). The third cytoplasmic region in the sperm head is the so called post-acrosomal layer or post-nuclear cap. This region is continuous with the other two cytoplasmic layers, which thus form a continuous perinuclear layer of cytoplasm (Fig. 2).

The function of the perinuclear cytoplasm is unknown. It has been shown to possess enzymatic activity (Teichmann and Bernstein, 1969), and Bedford (1970) has proposed that it serves to activate the mechanism responsible for the rapid dissolution of the compact sperm nucleus within the ooplasm of the newly fertilized egg but definitive information on this point is not available.

4. Transport of gametes to the site of fertilization

The site of semen deposition in mammals is the vagina, the cervix or the lumen of the uterus depending on the species (reviews, Bishop, 1969; Blandau, 1969; Hafez, 1973). Following deposition within the female tract, transport of spermatozoa to the site of fertilization results from pulsatile contractile activity in the wall of the uterus and the oviduct and also from fluid currents created by ciliated cells lining the oviduct. The motility of spermatozoa appears to be of minimal importance in transport within the female tract, though it may play some role in enabling sperm to pass through cervical mucus and to traverse the uterotubal junction. The major role for sperm motility comes later in enabling sperm to penetrate the investments of the egg (see p. 30). The rate of transport of spermatozoa to the site of fertilization is very rapid, in many species only as little as 5 to 10 minutes elapsing before sperm arrive in the oviduct (see, Blandau, 1969).

The egg(s) after release from the ovary enter the oviduct as a result of fluid currents created by the beating of cilia lining the inner surface of the oviduct infundibulum. The cilia may also exert direct traction on the granulosa-cell mass surrounding the egg. Both of these mechanisms serve to transport the egg(s) into the ampulla of the oviduct. In the majority of mammals the various steps in the fertilization process described in the later sections of this article take place within the oviduct and the newly fertilized egg develops to a 8 cell-to-morula stage embryo before finally passing to the uterus. In some species, however, the eggs pass from the oviduct and enter the uterus during fertilization (see, Austin, 1961).

Wastage of male gametes is very high in mammals. Of the millions of sperm present in the ejaculate very few ever reach the egg(s) in the oviduct. At most a few hundred or, in many cases, only a few dozen spermatozoa are ever likely to have the chance to interact with the egg. This marked dilution effect, together with the very small number of female gametes produced by mammals and the general lack of prolonged viability in both male and female mammalian gametes, dictates that sperm must be introduced into the female tract at or very close to the time of ovulation for successful fertilization to occur. This synchrony is achieved in most species by a well-defined oestrus period which restricts mating to the period of ovulation. Detailed information on the fertilizable life of mammalian gametes, the duration and frequency of oestrus cycles in different species, and the pertinent aspects of mating behaviour that affect the success of fertilization can be obtained from the publications of Austin (1961), Bedford (1970, 1972), Bishop (1969), Blandau (1969) and Croxatto (1974).

5. *Sperm capacitation*

For many years it was assumed that the function of the female genital tract in fertilization was merely to serve as a mechanical vehicle for transporting sperm to the site of fertilization. It is now recognized, however, that spermatozoa must reside in the female tract for several hours before they acquire the capacity to penetrate the zona pellucida and fertilize the ovum.

The need for a physiological change in mammalian spermatozoa resulting from their residence in the female tract as a prerequisite for successful fertilization was first recognized by Austin (1951) and Chang (1951) and this change was termed "capacitation" by Austin (1952). Since capacitation was first recognized an enormous amount of research has been done on this phenomenon with the aim of elucidating the factors involved in the control of capacitation (reviews, Austin, 1968; Pikó, 1969; Bedford, 1970, 1972; Barros, 1974).

The time required for capacitation varies significantly between species (Table 1). This variation may well reflect differences in the type and/or availability of the female factors responsible for capacitation. Variation in the time required for capacitation may also in part reflect the marked variation between species in the site of deposition of sperm in the female tract (vagina or uterus) and the timing of insemination in relation to ovulation.

Table 1
Time required for capacitation of spermatozoa in vivo.

Species	Time (hrs)
Mouse	<1
Sheep	1.5
Rat	2–3
Hamster	2–4
Pig	3–6
Ferret	3.5–11.5
Rabbit	5
Rhesus monkey	5 or 6
Man	5 or 6

Austin (1974).

The site(s) of physiological capacitation of sperm within the female tract is uncertain. At the present time, it is difficult to define a coherent pattern in the different species. In particular, more information is needed on the respective capacitation potential of the uterus and the oviduct in different species. The initial experiments of Chang (1951, 1955) established that capacitation of rabbit spermatozoa can take place within the uterus and this observation has been confirmed for a number of other mammalian species (see, Bedford, 1970, 1972; Barros, 1974). However, there is an increasing amount of evidence which suggests that the spermatozoa of some species do not complete their capacitation in the uterus, the process being completed only when the sperm pass into the fallopian tube (see, Bedford, 1972; Barros, 1974).

Important information on the conditions for capacitation has been obtained from experiments on sperm capacitation in vitro.

Postovulatory oviduct contents containing eggs in cumulus will capacitate sperm in vitro (Barros, 1968) but lose this capacity when the cellular components, including the cumulus oophorus are removed by centrifugation (Gwatkin et al. 1972). The work of Gwatkin et al. (1972) has established that the cumulus oophorus by itself can act as a reliable inducer for capacitation of hamster spermatozoa in vitro. Spermatozoa attach to cumulus cells, remain associated with them for 2–3 hours and are then released in a capacitated state. A dialyzable factor present in the cumulus matrix is essential for capacitation to take place. The reaction can occur in a modified medium 199 but not in Tyrode's solution (Yanagimachi, 1969a; Gwatkin – unpublished observations). The cumulus cells would appear to alter the plasma membranes of adherent spermatozoa by secreting glycosidases, since the capacitation potential is blocked by glucaro (1→4) lactone, a specific inhibitor of β-glucuronidase, and by 2-acetamido-2-deoxygluconolactone, a specific inhibitor of β-N-acetylglucosaminidase (Gwatkin and Anderson, 1973). Gwatkin et al. (1974) have also shown that mouse spermatozoa can be capacitated in vitro by the cumulus oophorus. Cumulus-induced capacitation of mammalian spermatozoa may be analogous to the fertility-inducing interaction of epithelial cells with spermatozoa found in hydroids (O'Rand, 1972) and to the action of sperm transit cells in sponges, bed bugs and leaches (Lord Rothschild, 1956a,b).

Hamster epididymal spermatozoa can also be capacitated in vitro by both oviductal and follicular fluids, but the degree of success is highly variable (see, Mahi and Yanagimachi, 1973). Capacitation-inducing factors for hamster spermatozoa have been detected in follicular fluid from the hamster (Barros and Austin, 1967), the cow (Yanagimachi, 1969b; Gwatkin and Andersen, 1969; Barros and Garavagno, 1970), the mouse and the rat (Yanagimachi, 1969a) and in oviductal fluids from the hamster (Barros and Austin, 1967), the rat (Barros, 1968) and the rabbit (Iwamatsu and Chang, 1972). Capacitation of hamster spermatozoa has also been achieved by incubation with both homologous and heterologous blood sera (Barros and Garavagno, 1970; Yanagimachi, 1970a). It is possible that all of these fluids share common chemical properties which are responsible for the induction of capacitation and that variation in the ability of fluids from different species to induce capacitation might merely reflect quantitative variation in the inducing component(s).

In view of the above observation that the capacitation potential of the cumulus oophorus is blocked by glycosidase inhibitors, and the finding that direct treatment of uncapacitated hamster sperm with crude β-glucuronidase will capacitate them (Gwatkin and Hutchinson, 1971), it might be profitable to examine the role of similar enzymes in capacitation induced by the various body fluids listed above. Similarly, the reported capacitation of hamster (Toyoda et al., 1971; Bavister, 1973) and mouse spermatozoa (Gwatkin et al., 1974) after incubation with epididymal secretions might reflect the high content of glycosidases present in these secretions (Conchie and Mann, 1957).

Apart from the so called "decapacitation factors" discussed later in this section, little attention has been given to the question of experimental blocking of capacitation by exogenous factors. Gwatkin and Williams (1970) and Briggs (1973) have shown that the in vitro capacitation of hamster spermatozoa by the cumulus oophorus could be inhibited by steroids. Sterol sulphates, which are known to be much more effective than steroids in stabilizing cellular membranes (Bleau et al., 1974), were likewise more effective in blocking cumulus-induced hamster sperm capacitation in vitro (Bleau et al., 1975).

Apart from its fundamental role in altering the spermatozoon to allow it to fertilize the ovum, it was at one time considered that capacitation might also act as a mechanism for ensuring species specificity in the fertilization reaction. Although the female tract of certain species is hostile to foreign sperm (see, Adams, 1974), there does not appear to be any exacting species-specificity expressed in capacitation. Reciprocal capacitation has been demonstrated between related species such as sheep and goat, rabbit and hare, cotton tail rabbit and domestic rabbit and between mink and ferret (Chang and Hancock, 1967). Furthermore, as outlined above, hamster spermatozoa can be capacitated in vitro by various body fluids from heterologous species, though not as efficiently as with homologous fluids. There is some evidence, however, to suggest that capacitation of rabbit spermatozoa has a more rigorous species-specific requirement (see, Bedford, 1972).

Capacitation is not an irreversible process, and an extensive literature has developed on the subject of so called "decapacitation factors" in seminal plasma. These "factors" are assayed by their ability to block the fertilization potential of fully capacitated spermatozoa. The existence of decapacitation factors was first reported by Chang (1957) who found that capacitated sperm taken from the uterus 6 hours after insemination and then resuspended in the supernatant fluid from centrifuged semen lost their ability to fertilize when reintroduced into the oviduct. The fertilizing capacity of the decapacitated spermatozoa was regained, however, following a further period of residence in the female tract. These findings were interpreted as evidence that an inhibitory substance (the decapacitation factor) was present in seminal plasma and was able to interact with spermatozoa and prevent their immediate participation in fertilization. The existence of decapacitation factors has since been demonstrated in seminal plasma from man, the rabbit, the bull, the stallion and the boar (see, Davis, 1971; Bedford, 1972).

The physico-chemical nature of decapacitating factors remains ill-defined. Bedford and Chang (1962) and Dukelow et al. (1967) showed that decapacitation factor activity could be removed from rabbit seminal plasma by ultracentrifugation, was non-dialyzable and was resistant to mild heat and freezing. Information on the mode of action of decapacitation factors is also very limited. These factors do not affect sperm motility, and current evidence, albeit limited, indicates that they affect spermatozoa by binding directly to the surface of the sperm and "stabilize" the plasma membrane over the acrosome and thus hinder onset of the acrosome reaction (see, Bedford, 1970, 1972). Perhaps the most important general conclusion to emerge from studies on decapacitation is that the process of capacitation does not involve irreversible structural alteration of the spermatozoon since capacitation is readily reversed by decapacitation factors. Additional comments on the possible role of decapacitation factors in altering the plasma membrane properties of the spermatozoon will be presented later in this section.

Although capacitation represents a distinct physiological event there is only scant information available on the underlying changes in the organization of the spermatozoon and its constituent organelles. Experimental attempts to correlate the changes in sperm behaviour in capacitation with specific alterations in the sperm have logically focussed on the overall morphology of the spermatozoon, its surface properties, and its metabolic activity.

Studies with both the light and electron microscope have failed to reveal any significant structural alterations in the gross morphology of capacitated spermatozoa (see, Bedford, 1972). In making this statement, however, a semantic distinction has been made between capacitation and the acrosome reaction which will be discussed later. In some species, changes in acrosome morphology are considered as occurring during capacitation. However, since spermatozoa cannot undergo the acrosome reaction without a prior period of capacitation, the use of the term capacitation will be restricted in this article to the events which prepare the sperm for, and take place before, the acrosome reaction.

The general failure to observe changes in the gross morphology of capacitated spermatozoa is paralleled by the general absence of changes in the various spermatozoon organelles, including the plasma membrane, when examined by conventional transmission electronmicroscopy (see, Bedford, 1972). However, very recent freeze-fracture ultrastructural studies by Koehler and Gaddum-Rosse (1975) have shown that the characteristic longitudinal "strands" of intramembranous particles present in fractures of the plasma membrane over the mid-piece of the guinea pig spermatozoon undergo dissociation during in vitro incubation of spermatozoa in physiological media that promote capacitation. Temporally correlated with these structural changes in membrane organization is a change in the flagellar beat. Clearly, more work will be required before a causal relationship can be established. Nonetheless, these initial findings suggest that by using ultrastructural techniques with increasingly powerful resolution it may eventually prove possible to identify specific plasma membrane alterations in capacitation (also see below).

The functional role of capacitation in enabling spermatozoa to subsequently undergo the acrosome reaction raises the obvious question of whether the plasma membrane over the acrosome is modified during capacitation to permit the membrane fusion events involved in the acrosome reaction to take place. In this respect, it is of interest that Koehler (1975) has recently reported that in vitro incubation of rabbit sperm in physiological media that promote capacitation is accompanied by changes in the distribution of intramembranous particles in the acrosomal plasma membrane. In uncapacitated spermatozoa the particles are arranged uniformly but after only 30 minutes incubation in the capacitating medium the particles become aggregated into patches. It will be of considerable interest to see if the addition of seminal or epididymal decapacitation factors to capacitated sperm restores the original pattern of particle distribution within the acrosomal plasma membrane.

In addition to changes in the acrosomal plasma membrane, the finding that hamster spermatozoa can fuse with zonaless ova only after capacitation (Yanagimachi and Noda, 1970c), implies that at least in this species capacitation also changes the properties of the post-acrosomal plasma membrane involved in final fusion with the egg plasma membrane.

The most convincing evidence for alteration of the sperm plasma membrane during capacitation concerns the loss or removal of components from the surface of the sperm (Kirton and Hafs, 1965; Ericsson, 1967; Hunter and Nornes, 1969; Vaidya et al., 1969, Johnson and Hunter, 1972b; Aonuma et al., 1973; Gordon et al., 1974, 1975a,b; Oliphant and Brackett, 1973a,b; Rosado et al., 1973). The material(s) removed from the sperm surface are considered to correspond to the so called "coating" proteins from the epididymal secretions and the seminal plasma discussed earlier (p. 9) which attach to spermatozoa during their passage through the epididymis.

Oliphant and Brackett (1973a) prepared ^{14}C-labeled antibodies against rabbit seminal plasma and showed that rabbit spermatozoa lose their ability to bind the antibodies under capacitating conditions. Loss of surface material from the surface of capacitated spermatozoa might also explain the reported reduction in the binding of concanavalin A to capacitated spermatozoa (Gordon et al., 1974, 1975a,b). Similarly, the increase in the net negative charge of the sperm acquired during epididymal maturation is reversed in capacitated spermatozoa (Vaidya et al., 1971) and this would again be consistent with a loss of surface materials, in this case sialic acid residues.

The mechanism(s) responsible for the loss or removal of surface components from capacitated spermatozoa is presently unclear. Treatment of mouse and rabbit spermatozoa with hypertonic salt solutions, a procedure that induces elution of proteins from the sperm surface (Oliphant and Brackett, 1973b) also induces them to fertilize eggs in cumulus (Brackett and Oliphant, 1975). Similar capacitation of spermatozoa after incubation in various physiological salines devoid of any material from the female tract has also been reported for the hamster (Bavister, 1973; Austin et al., 1973) the guinea pig (Barros et al., 1973a; Yanagimachi, 1972) and the mouse (Miyamoto and Chang, 1973a), though no information was given in these studies on the release of surface-associated materials from capacitated spermatozoa. The ability to induce capacitation in vitro in high ionic strength tissue culture media and physiological salines raises the possibility that increases in the osmolarity of the secretions in the female genital tract might induce elution of coating material from the cell surface. There is presently no information available to directly support or refute this possibility.

The other possible mechanism for removal of coating material from the sperm surface involves direct digestion by enzymes present in female tract secretions. For example, treatment of spermatozoa in vitro with pronase, trypsin, β-glucuronidase, β-amylase or neuraminidase has been shown to induce capacitation (Kirton and Hafs, 1965; Gwatkin and Hutchinson, 1971; Johnson and Hunter, 1972b). It is tempting to speculate that the capacitating action of these enzymes is mediated via their ability to remove surface-associated materials from the sperm. The high levels of proteolytic activity found in the oestrus uterus (Joshi et al., 1970; Joshi and Murray, 1974) compared with the low enzyme activity and high protease-inhibitor content of the progestational uterus (Noyes et al., 1958; Soupart, 1967; Hamner et al., 1968) provides additional circumstantial evidence compatible with this interpretation. Conditions in the oestrus uterus would thus be expected to favour enzymatic removal of surface materials from spermatozoa, but definitive evidence on this point is still lacking.

The most straightforward interpretation of the experimental data reviewed above and in the preceding section is that as spermatozoa mature in the epididymis macromolecular material(s) ("coating factor(s)"; decapacitation factor(s)") present in the secretions of the male tract bind to the surface of the sperm. Following mating and introduction of spermatozoa into the female tract capacitation takes place which involves the removal of these coating materials from the sperm surface.

As mentioned earlier (p. 6), there is some evidence to suggest that during epididymal maturation the lipids of the sperm plasma membrane become more "fluid", perhaps rendering the membrane highly susceptible to fusion with other membranes. The binding of extrinsic coating factors from epididymal secretions and seminal plasma to the sperm surface might therefore conceivably play an important role in "stabilizing" the sperm plasma membrane to reduce its capacity to fuse. Membrane stabilization of this kind would facilitate storage of the sperm and prevent premature activation and acrosome release. However, once spermatozoa were exposed to appropriate capacitating conditions in the female tract (high osmolarity?; enzymes?; or both?) then capacitation would proceed. The process of capacitation can therefore be considered as involving loss or removal of the membrane "stabilizing" or "coating" factors added in the male tract. Once the stabilizing factors are removed, the sperm plasma membrane is again capable of undergoing fusion and this is soon manifest in the acrosome reaction. This apparent need for removal of surface coat materials from spermatozoa is reminiscent of the situation in virus-induced fusion of somatic cells where loss of cell surface glycoproteins is a necessary prerequisite for the fusion reaction (reviews, Poste, 1970, 1972).

In view of the documented importance of the plasma membrane in determining the responsiveness of various somatic cells to hormones, it is not unreasonable to suggest that the plasma membrane of the sperm might similarly function as a target site for the action of various hormones present in the male and female tracts. More specifically, hormone-induced changes in sperm plasma membrane function could be important in capacitation, particularly since a number of steroid hormones have been shown to affect membrane stability (see, Seeman, 1972). Recent studies on the binding of radiolabeled hormones to spermatozoa from several mammalian species have established that sperm can bind both testosterones and oestrogens and a variety of other hormones such as thyroxine and insulin (review, Hoskins and Casillas, 1975). Surprisingly, however, there is little experimental information available on the effects of exogenous hormones on the fertilizing ability of spermatozoa.

Circumstantial evidence to suggest that hormone-induced changes in the sperm plasma membrane might be involved in capacitation is provided by the observation that in some species the capacitation potential of the female tract is influenced by the endocrine status of the female. Progesterone almost completely inhibits the capacitation activity of the rabbit uterus (Chang, 1958; Soupart, 1967; Hamner et al., 1968; Bedford, 1970). Sperm capacitation is achieved in the oestrus rabbit uterus in about 10 hours, but in ovariectomized animals the spermatozoa are only partly capacitated within this time and up to 26 hours are required to achieve full capacitation (Bedford, 1970). The low capacitation potential of the oestrogen-deficient rabbit uterus can, however, be fully restored by exogenous oestrogen (Soupart, 1967). In contrast to the sensitivity of uterine capacitation to the host's endocrine status, the fallopian tube environment in the rabbit remains competent to capacitate sperm regardless of the endocrine status of the animal. After ovariectomy there is a reduction in the overall level of capacitation activity within the oviduct but the capacitation potential of the tubal environment is still maintained at a level adequate to capacitate larger numbers of spermatozoa than would ordinarily reach this site after normal mating (Bedford, 1970). In addition, progesterone does not inhibit the capacitation potential of the fallopian tube. The oviduct, at least in the rabbit, therefore appears to possess an inherent ability for inducing capacitation which can be enhanced by oestrogen but which cannot be suppressed by progesterone, even in the absence of oestrogen.

Another potentially important aspect of the action of hormones on spermatozoa during capacitation concerns hormone-induced changes in sperm motility and sperm metabolism (review, Hoskins and Casillas, 1975). Spermatozoa undergoing capacitation show a distinct change in their pattern of motility (Gwatkin and Andersen, 1969; Iwamatsu and Chang, 1969; Yanagimachi, 1969a,b; Barros et al., 1973). The beating of the flagellum in capacitated sperm is considerably greater than that of non-capacitated sperm and there is a general consensus that this enhanced motility would facilitate sperm penetration through the investments of the egg. Although exogenous progesterone and oestrogen appear to have little effect on sperm motility in vitro (for refs. see, Hoskins and Casillas, 1975), sperm motility in vitro can be significantly enhanced by cyclic nucleotides and/or phosphodiesterase inhibitors (see, Hoskins and Casillas, 1975). The biochemical basis for this response is presently unknown, but an attractive possibility is that cyclic nucleotides activate a cyclic AMP-dependent protein kinase which, in turn, activates a Ca^{2+}-controlled motility-regulating protein. Such a mechanism is not without precedent in other cells (see, Hoskins and Casillas, 1975).

Capacitation of mammalian spermatozoa is also accompanied by changes in metabolic activity, though it is not yet clear whether the increased respiratory and glycolytic activities found in capacitated sperm are causally related to their fertilizing capacity. This point is probably academic, however, since it is likely that the enhanced motility of capacitated sperm would require increased cellular energy production in order to sustain it. Data from in vitro studies suggest that cyclic nucleotides may again be important in regulating the metabolic activity of spermatozoa. Addition of exogenous cyclic nucleotides to uncapacitated spermatozoa in vitro has been shown to increase their metabolic activity to levels characteristic of capacitated spermatozoa (see, Hoskins and Casillas, 1975). These observations, together with the above data on the effects of cyclic nucleotides on sperm motility, have prompted numerous speculations on the role of cyclic nucleotides in both sperm maturation and subsequent capacitation. Indeed, Hicks et al. (1972) have even proposed that cyclic AMP acts as a *specific* inducer of capacitation for human spermatozoa.

Although the circumstantial evidence for the involvement of cyclic nucleotides in sperm capacitation is certainly compelling, it should be noted that intracellular levels of cyclic AMP in capacitated and non-capacitated spermatozoa have yet to be measured.

It is also necessary to consider the role of steroid hormones in the metabolic changes found in capacitated spermatozoa. Of particular interest is the finding that testerone and related steroids found in the secretions of the male tract suppress oxidative metabolism in spermatozoa (see, Hoskins and Casillas, 1975). This finding has prompted the obvious proposal that male sex hormone-induced metabolic suppression may be an important mechanism in conserving the endogenous lipids of the sperm during the comparatively long period of transport of sperm through the epididymis so that sufficient lipid will be available for use as an energy source by capacitated spermatozoa. In contrast, oestrogen and oestradiol have been shown to stimulate sperm metabolism (see, Hoskins and Casillas, 1975). If similar responses operate in vivo then exposure of spermatozoa to different male and female hormones at appropriate concentrations in the male and the female tracts would be expected to exert a significant effect on sperm metabolism, with activation being favored in the female tract environment.

However, irrespective of the exact mechanisms involved it seems likely that the plasma membrane of the sperm serves as the initial target site for hormone binding and that subsequent changes in the intracellular metabolic machinery following hormone binding will probably involve plasma membrane-associated enzymes. There is presently no information available on the rate kinetics of such membrane-associated enzymes as Na^+-K^+-ATPase or adenyl cyclase in capacitated and uncapacitated spermatozoa. This would appear to be an obvious area for future research in relation to the metabolic activation of capacitated spermatozoa.

6. *The acrosome reaction*

As a result of changes induced in the sperm plasma membrane by capacitation, the spermatozoon begins to develop multiple fusions between the outer acrosomal membrane and the overlying plasma membrane. Initiation of fusion between these two membranes marks the start of the acrosome reaction. The morphology of the acrosome reaction has been studied in several species by electron-microscopy (Barros et al., 1967; Bedford, 1968; Franklin et al., 1970;

Yanagimachi and Noda, 1970a,b; Bedford, 1972; Thompson et al., 1974). These studies have shown that the acrosome reaction is initiated as a series of focal fusion points between the outer acrosomal membrane and the plasma membrane, finally creating a vesiculated acrosome-plasma membrane complex over the head of the sperm, except over the equatorial segment which is unaffected at this time. The fusion between these two membranes creates connecting ports which presumably allow the release of the acrosomal contents, though the complex question of whether different acrosomal enzymes are released sequentially (cf. Srivastava et al., 1974) remains to be solved.

The major functional importance of the acrosome reaction appears to be in terms of the release of the acrosomal enzymes which are believed to facilitate passage of spermatozoa through the various investments of the egg. The release of hyaluronidase, which is thought to be present in the anterior portion of the acrosome (Gould and Bernstein, 1973), is considered to facilitate transport of spermatozoa through the cumulus matrix (Srivastava et al., 1965) since the zona pellucida is resistant to digestion by this enzyme. Digestion of the zona pellucida is believed to primarily involve the action of acrosin released from the acrosome. Aspects of sperm penetration through the zona will be discussed in more detail later (p. 30).

Despite the simplistic and attractive nature of the concept of digestion of the egg investments by acrosomal enzymes, certain reservations have been expressed recently concerning the role of hyaluronidase. Although the release of hyaluronidase from the acrosome has been utilized in several laboratories as a chemical assay for the acrosome reaction, recent observations suggest that at least in guinea pig spermatozoa the release of hyaluronidase is poorly correlated with the morphological acrosome reaction, significant enzyme release often being detected in the absence of the classical acrosome reaction (Rogers and Morton, 1973; Talbot and Franklin, 1974). These observations point to a need for further investigation of the question of whether the full acrosome reaction involving extensive membrane fusion and vesiculation is required for the release of this particular enzyme. The role of hyaluronidase in enabling spermatozoa to pass through the cumulus has also been questioned in the light of observations that not all spermatozoa that are able to pass through the cumulus have undergone the acrosome reaction (Bedford, 1968, 1970). If, however, as suggested above, release of hyaluronidase can precede the classical acrosome reaction, passage of spermatozoa with intact acrosomes through the cumulus does not provide evidence either for or against the role of this enzyme in altering the cumulus. Metz (1972, 1973), however, has reported that antibodies raised specifically against hyaluronidase can inhibit fertilization of rabbit eggs in vitro.

Although the ultrastructural basis of the classical acrosome reaction is well documented, both the timing of the reaction in relation to sperm passage through the various investments of the egg and the "trigger" stimulus for the acrosome reaction remain to be clarified.

It is now accepted that spermatozoa must first be prepared by capacitation before the acrosome reaction can occur (induction of 'false' acrosome reactions by membrane-active agents such as lysolecithin and detergents has been excluded from consideration here). It has not been established with any degree of certainty whether the acrosome reaction can simply proceed in a physically appropriate environment or whether a specific induction stimulus operating only at the site of fertilization is involved. There is some precedent for the latter type of mechanism in the "fertilizins" of the egg coat which trigger the acrosome reaction in echinoderm spermatozoa (see, Dan, 1967). However, as with so many aspects of the fertilization process in mammals, there appears to be significant species variation in the conditions required for triggering the acrosome reaction. The ability to induce the acrosome reaction in capacitated spermatozoa is shown by homologous follicular fluids in the hamster (Barros and Austin, 1967; Yanagimachi, 1969a) and the rabbit (Bedford, 1969) and also by tubal fluids in the hamster (Barros and Garavagno, 1970; Iwamatsu and Chang, 1972). In the rabbit, however, capacitated spermatozoa within the Fallopian tube do not generally show an acrosome reaction in the absence of eggs (Bedford, 1969). Yanagimachi (1969b) has isolated a heat-stable dialyzable factor from bovine follicular fluid capable of inducing the acrosome reaction in hamster spermatozoa in vitro. Whether this factor corresponds in any way to the induction stimulus for the acrosome reaction in vivo is not known. As reported in more detail below, the acrosome reaction in guinea pig spermatozoa can also take place in simple physiological salines (Barros et al., 1973a,b; Barros, 1974; Yanagimachi and Usui, 1974; Rogers and Yanagimachi, 1975).

Insight into the mechanism of membrane fusion in the acrosome reaction is also limited. Recent work by Yanagimachi and his colleagues (Yanagimachi and Usui, 1974; Rogers and Yanagimachi, 1975) has established an important role for Ca^{2+} in controlling the acrosome reaction. They found that incubation of guinea pig spermatozoa in a simple modified Krebs–Ringer or Tyrodes solution containing 0.1% bovine serum albumin resulted in capacitation and subsequent acrosome reactions in about two thirds of the spermatozoa after 10–12 hours. If Ca^{2+} was omitted from the incubation medium the acrosome reaction did not take place. The acrosome reaction could, however, be induced within minutes by the addition of Ca^{2+} to the preincubated spermatozoa. A full preincubation period was necessary to obtain this rapid response to Ca^{2+}, indicating that capacitation, presumably involving loss of surface coating materials, was an essential prerequisite for the acrosome reaction. Ca^{2+} has also been shown to be essential for the acrosome reaction in several marine invertebrates (Dan, 1967).

It does not seem unreasonable to suggest that the Ca^{2+}-dependence of the acrosome reaction results from the fact that Ca^{2+} is required for membrane fusion. Ca^{2+} has been shown to play a central role in controlling numerous examples of membrane fusion in somatic cells (review, Poste and Allison, 1973). Indeed, the rapid triggering of the acrosome reaction in "primed" capacitated sperm by the simple addition of Ca^{2+} described above is reminiscent of previous reports in which provision of Ca^{2+} has been shown to trigger rapid membrane fusion in both germ cells (Steinhardt and Epel, 1974; Vacquier, 1975b) and in a diverse range of somatic cells (see, Rubin, 1970; Poste and Allison, 1973; Douglas, 1974; Thorn and Petersen, 1975; Williams and Chandler, 1975) and also to induce rapid and extensive fusion in model membranes of defined composition (see, Papahadjopoulos and Poste, 1975; Papahadjopoulos et al., 1974, 1975).

In view of the parallels between membrane fusion occurring in the acrosome reaction and membrane fusion in other cell types it would be of interest to examine the response of the acrosome reaction to agents that have been shown to inhibit or enhance other types of membrane fusion (see, Poste and Allison, 1973).

If the above comments on the role of Ca^{2+} in "triggering" membrane fusion in the acrosome reaction are valid, it would appear that the natural inducer(s) of the acrosome reaction might act by altering the permeability of the plasma membrane to Ca^{2+}. Consequently, it is perhaps significant that recent work by Poste (unpublished observations) has shown that the acrosome reaction in guinea pig spermatozoa can be rapidly induced by the carboxylic acid ionophore A-23187 (Fig. 3) which facilitates Ca^{2+} transport across the plasma membrane.

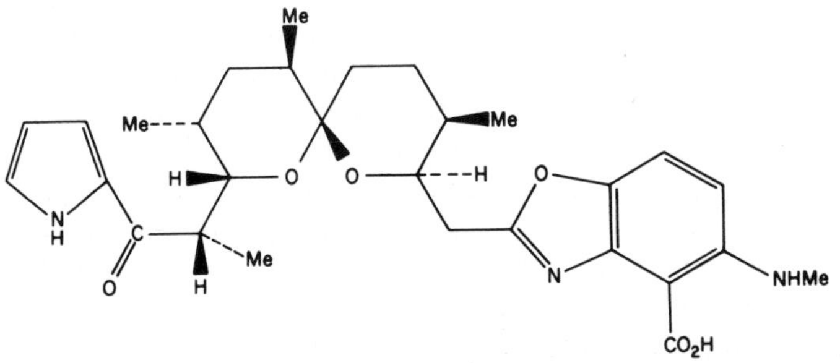

Fig. 3. Structure of the carboxylic acid ionophore A23187 (from Chaney et al., 1974).

Induction of the acrosome reaction by this Ca^{2+}-ionophore shares obvious similarities with other reports of ionophore-induced triggering of exocytosis in somatic cells in which secretory granules undergo rapid fusion with the plasma membrane after treatment with this compound (see, Cochrane and Douglas, 1974; Kagayama and Douglas, 1974; Eimerl et al., 1974; Garcia et al., 1975; Nordmann and Currell, 1975; Williams and Chandler, 1975). Further support for the role of an influx of Ca^{2+} as a "trigger" in the acrosome reaction is provided by the finding that calcium antagonists such as lanthanum, compound D600 (α-isopropyl-α[N-methyl-N-homoveratryl)-γ-aminopropyl]-3,4,5-trimethoxyphenylacetonitrile HCl) and Verapamil HCl (Isoptin) inhibit the acrosome reaction in guinea pig spermatozoa (Poste, unpublished observations).

By analogy with the documented effects of Ca^{2+} in stimulating contractile processes and motility in somatic cells, an enhanced rate of uptake of Ca^{2+} into the spermatozoon might also account for the marked "whiplash" movement developed by spermatozoa that have undergone the acrosome reaction.

As stressed already, the acrosome reaction shares obvious similarities with exocytotic secretion in various somatic cells where membrane-bound secretory granules fuse with the plasma membrane after binding of the appropriate secretagogue to the cell surface. In many cells of this type, microfilamentous contractile systems appear to be involved in the release reaction, since release can be blocked by cytochalasin B, a fungal antibiotic that disrupts microfilaments (see, Poste and Allison, 1973; Allison and Davies, 1974). The inhibitory effect of cytochalasin B on secretion has in general been interpreted as indicating that microfilaments associated with the membrane of the secretory granule(s) function to trans-locate the granule(s) to sites on the plasma membrane where membrane fusion takes place. Ca^{2+} again occupies a central position in this scheme, since the various secretagogues are believed to act by increasing the permeability of the plasma membrane to Ca^{2+}, thus raising the concentration of intracellular Ca^{2+} which, in turn, induces contraction of the actin-like microfilaments with resulting translocation of secretory granules to the plasma membrane (review, Allison and Davies, 1974). Actin-like microfilaments have been identified in association with the acrosome in echinoderm spermatozoa (Jessen et al., 1973; Tilney et al., 1973; Summers, 1975) raising the intriguing possibility that they might play a role in bringing the acrosomal and plasma membranes into close apposition during the acrosome reaction. No information appears to be available on the presence of similar structures in mammalian spermatozoa. However, Johnson (1975) has reported preliminary findings that cytochalasin B does not affect the acrosome reaction in mammalian spermatozoa (even at very high drug concentrations) suggesting that cytochalasin-sensitive microfilaments are probably not involved.

An ATPase has been identified on the outer surface of the acrosomal membrane and the inner surface of the acrosomal plasma membrane in mammalian spermatozoa (Gordon, 1973; Yanagimachi and Usui, 1974), prompting several authors to suggest that this enzyme might play some role in the acrosome reaction (Barros, 1974; Bedford, 1974; Johnson, 1975). However, meaningful data on this question are not yet available.

Completion of the acrosome reaction is indicated by sloughing of the vesiculated fused acrosome-plasma membrane complex from the head of the sperm. In most species the vesiculated products of membrane fusion are lost from the sperm head before penetration of the zona begins (Bedford, 1972; Thomson et al., 1974). Following loss of the vesiculated membrane elements continuity of the surface membrane over the sperm head is maintained by fusion at the anterior limit of the equatorial region between the plasma membrane and the remaining fragment of the outer acrosomal membrane. In some species such as the hamster further vesiculation of the equatorial segment will take place during penetration of the sperm through the zona (Yanagimachi and Noda, 1970b). Thus at the end of the acrosome reaction the sperm head is covered by the inner acrosomal membrane anteriorly and the original plasma membrane posteriorly. It is therefore the properties of these two membrane regions that will influence subsequent steps in the sperm-egg interaction.

7. Attachment and binding of spermatozoa to the zona pellucida

The region of the spermatozoon that first associates with the zona pellucida is not known. Hamster (Yanagimachi and Noda, 1970a,b) and rabbit (Bedford,

1968) spermatozoa have been observed at some distance from the egg, lacking both the plasma membrane and the outer acrosomal region over their anterior region. This suggests that the inner acrosomal membrane, the equatorial segment or the post-acrosomal region of the spermatozoon may be involved. However, as Zamboni (1971b) has pointed out, these may be supernumerary, not fertilizing, spermatozoa. Franklin et al. (1970) have suggested that the acrosome normally serves to attach spermatozoa to the zona surface. They consistently observed hamster spermatozoa associated with the zona by means of the vesiculated acrosome-plasma membrane complex. Thus, in our current state of knowledge we still need to answer the question whether attachment and binding of sperm to the zona involves the intact plasma membrane, the vesiculated acrosome-plasma membrane complex, the surface of the inner acrosomal membrane or even possibly all three in a sequential process if the acrosome reaction were to occur at the zona surface.

Recent observations on the in vitro interaction of hamster spermatozoa with the zona pellucida suggests that there are two successive steps, *attachment* and *binding*, involved and that the presence of the vitellus somehow influences these initial sperm-zona interactions (Hartmann et al., 1972; Hartmann and Hutchinson, 1974a,b,c).

The *attachment* of hamster spermatozoa to eggs in vitro involves loose association of sperm with the zona, since the sperm can be readily dislodged by pipeting. Attachment can occur equally well at both 2°C and 37°C and appears to be non-species specific (Hartmann et al. 1972). After 30–40 minutes, however, attachment is followed by *binding*, which represents a more tenacious union between sperm and the zona. In contrast to the attachment phase, binding is not disturbed by pipeting (Fig. 4) and, unlike attachment, binding does not occur at 2°C and is also species-specific. Approximately 20 minutes after the onset of binding the spermatozoa will have traversed the zona and entered the vitellus (Hartmann and Hutchison, 1974c).

During the initial attachment stage hamster spermatozoa undergo a modification, possibly involving acrosin activation. Hartmann et al. (1972) showed that when hamster eggs with sperm attached were transferred to a second microculture and the attached sperm dislodged by pipeting, the released spermatozoa were found to bind to eggs more rapidly and also to penetrate the egg earlier than untreated sperm. This modification appears to be species-specific since prior attachment of hamster spermatozoa to mouse eggs did not accelerate their subsequent binding to hamster eggs (Hartmann and Hutchison, 1974a). Attachment also appears to modify the surface of the zona since eggs previously subjected to attachment acquire the capacity to bind sperm rapidly (Hartmann and Hutchison, 1974c). Only 10 minutes of sperm attachment are necessary to produce detectable acceleration of sperm binding and the effect continues to develop after the original attached spermatozoa have been removed. When hamster zonae, but not those of mouse, are included in the same microculture as the hamster gametes binding is delayed. Hartmann and Hutchison (1974b) have interpreted this as indicating that a soluble material, which they designated *SI* factor, is released by sperm attached to the zonae. It is possible that this could

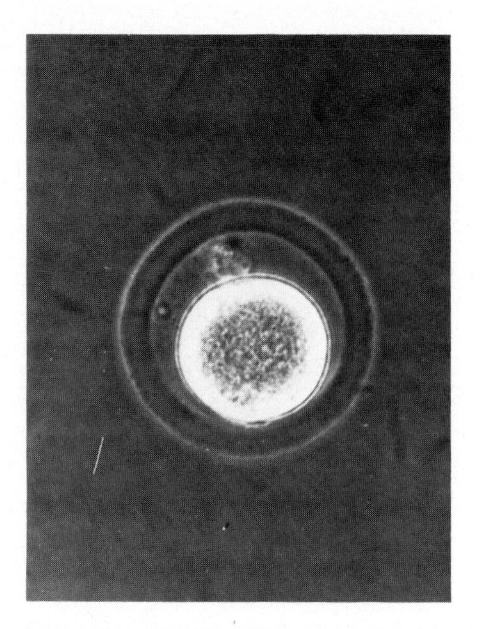

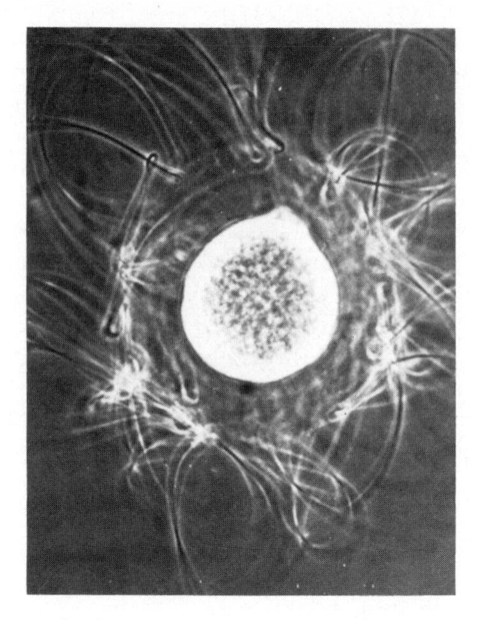

Fig. 4. Photomicrographs of hamster eggs exposed to capacitated spermatozoa showing the effect of washing on sperm attachment and binding. Left: egg incubated with spermatozoa for 15 minutes before washing showing complete removal of spermatozoa from the egg surface. Right: egg incubated with spermatozoa for 40 minutes before washing showing failure of washing to remove the now firmly bound spermatozoa (reproduced with permission from Hartmann et al., 1972).

be solubilized acrosin.

As mentioned earlier, binding of hamster spermatozoa to intact eggs requires 30–40 minutes but similar binding to isolated zonae pellucidae requires only 5–10 minutes (Fig. 5). This suggests that the vitellus must in some way control the interaction between the gametes at the zona surface, and also somehow prevent the type of rapid sperm binding found with isolated zonae. A specific vitelline factor (as yet undefined) has been postulated to explain this phenomenon (Hartmann and Hutchison, 1974a). Binding of spermatozoa to isolated zonae also seems to differ from binding to intact eggs in other ways. While pretreatment of hamster spermatozoa with the trypsin-acrosin inhibitor *p*-aminobenzamidine inhibits binding to intact eggs, this treatment causes only a slight delay in the binding of spermatozoa to isolated zonae (Hartmann and Hutchison, 1974a). Furthermore, binding of hamster spermatozoa to isolated zonae as a function of sperm concentration follows a straight line relationship whereas binding to intact eggs displays a sigmoidal curve (Fig. 6). The latter response was interpreted by Hartmann and Hutchison as suggesting a cooperative effect consistent with the proposed involvement of a factor from the vitellus.

On the basis of the findings discussed so far Hartmann and Hutchison concluded that normal binding of spermatozoa to intact eggs might occur as a

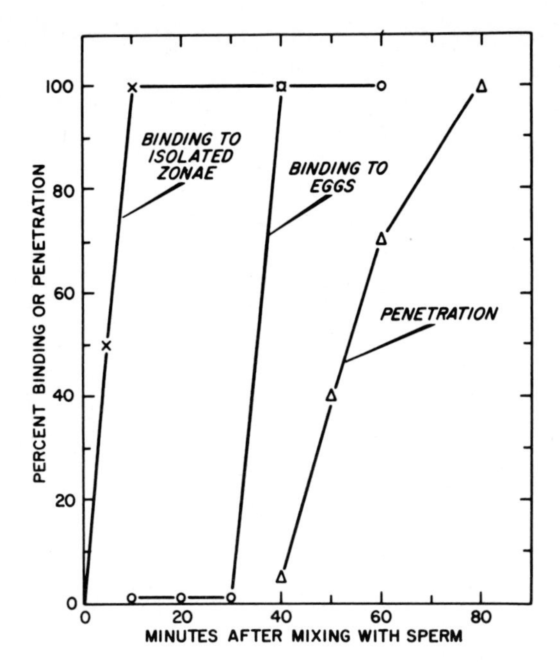

Fig. 5. Binding of hamster spermatozoa to isolated zonae pellucidae and intact eggs and penetration of sperm into intact eggs.

two-step process, the first step being analogous to the rapid binding found with isolated zonae. Some support for this possibility was offered by the observation that some sperm became bound to intact eggs after only 2–3 minutes incubation but not immediately thereafter (Hartmann and Hutchison, 1974a). However, later work by the same authors has shown that the number of sperm bound in this way is too low to account for all of those which are bound at 30–40 minutes and early binding in fact occurs only at very high sperm concentrations (Hartmann and Hutchison, 1975). As a result of these data Hartmann and Hutchison have now proposed that early binding of sperm to the egg, such as occurs with isolated zonae, is normally completely blocked but this block can be partially overcome when eggs are exposed to an excessively high sperm concentration.

The species-specific nature of sperm binding to the zona identified in the above experiments suggests that the zona pellucida may be a major factor in preventing fertilization by heterologous sperm (see also, Hanada and Chang, 1972; Yanagimachi, 1972b; Bedford, 1974).

Several attempts have been made to obtain information on the nature of the sperm "receptor" in the zona pellucida. In the hamster, the sperm-binding component of the zona is sensitive to both trypsin and chymotrypsin (Hartmann and Gwatkin, 1971). Binding of hamster spermatozoa is not affected, however, by pretreatment of eggs with neuraminidase, lysozyme, α- and β-amylase, glucoamylase, β-glucosidase, β-glucuronidase, β-galactosidase or phospholip-

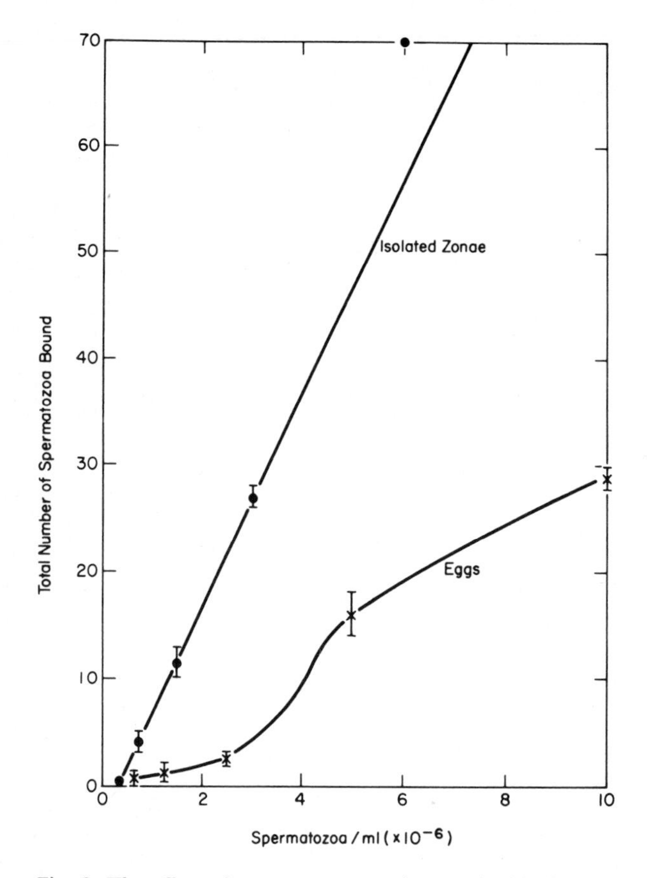

Fig. 6. The effect of sperm concentration on the binding of capacitated hamster spermatozoa to hamster eggs and isolated zona pellucidae from hamster eggs. The indicated values were obtained from counts on 10 eggs and 10 isolated zonae. (Reproduced with permission from Hartmann and Hutchison, 1975).

ases C and D (Gwatkin et al., 1973a). In contrast, sperm binding and penetration are both unaffected by trypsin treatment of the rabbit zona pellucida (Bedford, 1974), again emphasizing the role of species variation among the eutherians in the various stages of fertilization. Soupart and Clewe (1965) have reported that pretreatment of rabbit eggs with a partially purified neuraminidase reduced the number of eggs that were subsequently penetrated by sperm. However, the enzyme preparation used by these investigators may have contained contaminants since it also dissolved the cumulus. More recent observations reported by Bedford (1974) using a more rigorously purified neuraminidase have failed to find any effect of this enzyme on either sperm binding or penetration.

To date, a receptor for sperm binding has yet to be isolated from the mammalian zona pellucida. However, material has been extracted from the vitelline layer of sea urchin eggs that binds to spermatozoa and prevents them from attaching to eggs or fertilizing them (Aketa, 1967, 1973; Aketa et al., 1968, 1972).

The antigenicity of the zona pellucida has been demonstrated in the hamster, the rabbit, the mouse and the guinea pig by producing antiserum against saline extracts of homologous ovarian tissues (see, Shivers and Dudkiewicz, 1974). Shivers and his colleagues attempted to exploit this property of the zona to study the nature of the zona-sperm receptor in hamster and rabbit eggs (reviews, Shivers, 1974; Shivers and Dudkiewicz, 1974). These investigators demonstrated that coating of the zona with specific antibodies blocked both the binding and penetration of spermatozoa. Unfortunately, such experiments employing multivalent ligands to cross-link surface components presently offer little insight into the nature of the sperm binding site, since binding of antibodies to receptors adjacent to the actual sperm receptors on the zona surface would probably cover the specific receptor site and thus prevent binding of spermatozoa.

Plant lectins which bind to specific saccharides have also been used to explore the contribution of specific sugars on the zona surface as possible sperm binding sites (Oikawa et al., 1973, 1974). Wheat germ agglutinin (binds N-acetyl-D-glucosamine), *Ricinus communis* agglutinin (binds D-galactose) and *Dolichos biflorus* agglutinin (binds N-acetyl-α-D-galactosamine) were each found to bind to the zona of hamster eggs causing a change in the light-scattering properties of the zona and rendering the zona more resistant to digestion by proteases. The lectin concanavalin A (binds α-D-mannose) did not produce these changes in the zona. Distinct differences were also found in the ability of these various lectins to modify sperm binding and subsequent penetration through the zona. Capacitated hamster spermatozoa failed to bind to or penetrate the zona in eggs treated with wheat germ agglutinin, while treatment with *Ricinus communis* and *Dolichos biflorus* agglutinins as well as concanavalin A prevented sperm penetration but did not inhibit binding, suggesting that the sperm receptor does not involve terminal galactose, N-acetyl-galactosamine or mannose residues (Oikawa et al., 1973, 1974).

The use of plant lectins to study sperm binding to the zona pellucida suffers, however, from the same interpretational difficulties as the earlier work with antibodies. Although the chemical nature of the reactive moieties on the zona surface is better known for lectins than for antibodies, the multivalent nature of the lectins dictates that they induce substantial cross-linking of surface determinants and in so doing could produce significant steric hindrance of access of spermatozoa to their receptors. Thus, lectin-induced inhibition of sperm binding could equally well be interpreted as a steric hindrance phenomenon rather than revealing the chemical nature of the terminal saccharides of a sperm receptor. A further unresolved problem associated with the use of lectins for experiments of this type is that reactive saccharide residues on the zona surface need not necessarily be true structural components of the zona but might merely be associated with adsorbed glycoproteins derived from the cumulus. Finally, the possibility must be considered that the block to sperm penetration found in lectin-treated unfertilized eggs may in fact be a block to polyspermy resulting from premature cortical granule discharge. Indeed, concanavalin A has been shown to induce cortical granule discharge (Gwatkin, unpublished observations), perhaps by altering the permeability of the plasma membrane (cf. Aull and

Nachbar, 1974; Konig et al., 1973). Although Oikawa et al. (1974) were aware of this potential problem and stated that no cortical granule release was observed, a low level of release sufficient to cause alteration in the zona pellucida, as reported for example in parthenogenones by Mintz and Gearhart (1973), would be extremely difficult to detect.

8. Penetration of the zona pellucida

The next step in the fertilization process involves penetration of the spermatozoon through the zona pellucida. Mammalian spermatozoa characteristically pass through the zona on a slightly curving path at a tangent to the circumference of the eggs rather than at 90° or at random angles to the egg surface. A satisfactory explanation has yet to be offered for this unusual but consistent angle of entry. The path of the penetrating spermatozoon within the zona can be detected as a narrow slit in the substance of the zona and persists for a few hours after penetration.

The nature of the factors that facilitate penetration of the sperm head through the resilient substance of the zona is still not established with any certainty. While there is a general consensus that the tail of the spermatozoon provides an important motive force, there is less agreement over the role of digestion of the zona by acrosomal enzymes (so called zona lysins). It is considered important to stress the uncertainties still surrounding the zona lysin concept, since the casual reader of the literature in this area might well draw the conclusion that this concept had achieved the status of accepted dogma.

Most of the work on the role of enzymes associated with the sperm head in altering the zona substance to facilitate sperm penetration has focussed on the trypsin-like enzyme, acrosin (review, McRorie and Williams, 1974). Support for the role of acrosin, and perhaps other acrosomal proteases, in penetration of the zona rests largely on the observation that acrosomal extracts are able to digest the substance of the zona (Srivastava et al., 1965; Stambaugh and Buckley, 1969, 1972; Multamaki and Niemi, 1969) and that a variety of naturally occurring or synthetic trypsin inhibitors inhibit fertilization both in vitro (Stambaugh et al., 1969; Suominen et al., 1973) and in vivo (Zaneveld et al., 1971).

Reservations must be expressed, however, concerning the interpretation of the above enzyme inhibitor experiments. For example, in none of these studies has it been shown that the acrosome reaction was able to take place in the presence of the inhibitors. The observed failure of inhibitor-treated spermatozoa to penetrate and fertilize eggs could simply reflect inhibition of the acrosome reaction rather than direct inhibition of the enzyme itself. In addition, the very high concentration of enzyme inhibitors employed raises doubts regarding their specificity of action, and also introduces the additional possibility of functional inactivation of a wide variety of amino acids with nucleophilic groups (see, Bedford, 1974) which could lead to inhibition of fertilization at a step unrelated to zona penetration. Finally, the inhibition of fertilization by naturally occurring trypsin inhibitors is not always reproducible (see, Miyamoto and Chang, 1973b) while the inhibitory effects of synthetic inhibitors have been shown to be accompanied by a significant reduction in sperm motility (Miyamoto and Chang, 1973b).

The morphology of zona penetration also poses a number of questions concerning the distribution of acrosin and other enzymes within the sperm head. As mentioned earlier, in those species for which reliable electronmicroscopic data is available, the vesiculated acrosome-plasma membrane complex appears to be shed before the sperm head begins to penetrate the zona pellucida. This means that the most anterior portion of the sperm head is covered anteriorly only by the inner acrosomal membrane and posteriorly by the equatorial segment of the acrosome. Concomitant with shedding of the vesiculated membranes the visible electron-dense content of the acrosome is also dissipated. Thus if acrosin and/or other proteases play a role in digesting the zona they must be closely associated with either of these two membranes.

Bedford (1968) and Pikó (1969) have proposed that acrosin is bound to the inner acrosomal membrane. There is at least evidence to indicate that acrosin is more tightly bound within the acrosome than hyaluronidase (Fritz et al., 1974) and might be more likely to remain associated with the sperm head following loss of the vesiculated acrosome-plasma membrane complex. More direct evidence to support this conclusion comes from the work of Brown and Hartree (1974) who were able to solubilize a trypsin-like enzyme from acrosome-depleted ram spermatozoa indicating that the enzyme was closely associated with the inner acrosomal membrane. In contrast, Yanagimachi and Noda (1970a,b) have suggested that in hamster spermatozoa acrosin is associated with the equatorial segment of the acrosome. Their proposal was based on the finding that the equatorial segment undergoes vesiculation during penetration within the zona (but see Franklin et al., 1970) and thus might serve to release the enzyme directly into the substance of the zona. Species variation must again be acknowledged, however, since in rabbit spermatozoa vesiculation of the equatorial segment does not occur during transit through the zona (Bedford, 1968, 1972). Unfortunately, further meaningful comment on this question cannot be offered at the present time other than to draw the obvious conclusion that further work is needed.

9. *Fusion of the sperm with the vitellus*

Once through the zona pellucida the head of the spermatozoon makes immediate contact with the surface of the vitellus and becomes adherent to it. Phase contrast microscopic observations (Austin and Walton, 1960 and others) have shown that the motility of the spermatozoon rapidly ceases on reaching the vitelline surface. Recent electron-microscopic observations of these non-motile spermatozoa have revealed that fusion between sperm and egg membranes has already taken place at this stage (Austin 1975 citing personal communication from R. Yanagimachi).

Fusion between the plasma membrane of the egg and the penetrating spermatozoon was first described in mammals by Szollosi and Ris (1961) in rat gametes. This study was important in that it established that the plasma membranes of the two gametes did in fact fuse and that the spermatozoon was not phagocytosed by the vitellus. Electron-microscopic evidence of fusion between gamete plasma membranes has since been obtained for the hamster (Barros and Franklin, 1968; Yanagimachi and Noda, 1970a,c), the mouse (Stefani et al., 1969) the rat (Pikó, 1969) and the rabbit (Bedford, 1970, 1972). Yanagimachi et al. (1973) have also obtained electron-histochemical evidence for intermixing of plasma membrane components from male and female cells following gamete fusion.

The above ultrastructural studies have also established that the spermatozoon makes initial contact with the egg plasma membrane at the post-acrosomal region and it is this region that is involved in creating the initial fusion "bridge" with the egg. The role of the post acrosomal plasma membrane in initiating fusion between mammalian gametes is in striking contrast to gamete fusion in marine invertebrates in which initial contact with the egg membrane is made by the apical region of the sperm head (see, Colwin and Colwin, 1967; Epel, 1975; Summers et al., 1975).

Fusion between mammalian gametes is initiated by the development of point fusions between the post acrosomal plasma membrane of the sperm and microvillous-like processes on the vitelline surface (Fig. 7). Scanning electron-microscopy of this process shows that the vitelline microvilli wrap around the sperm head (Yanagimachi and Noda, 1972). It is of interest to note that microvilli are also prominent in fusion occurring between various somatic cells (see, Poste, 1970). The low radius of curvature of such structures may be of value in achieving close contact between interacting cells, since microvilli would encounter significantly less electrostatic repulsion to contact than larger cellular projections (see, Pethica, 1961; Poste, 1970). The plasma membrane around the post acrosomal region of the sperm head next disappears and this region of the nucleus then becomes enveloped by ooplasm (Fig. 7). Incorporation of the most caudal part of the sperm head, the neck and the sperm tail into the ooplasm are accompanied by loss of the overlying plasma membrane (see Bedford, 1972) but it is not yet clear whether this membrane becomes incorporated onto the egg plasma membrane. In contrast, the region of the sperm head covered by the inner acrosomal membrane and the equatorial segment (if it persists) do not fuse with the egg plasma membrane and are instead phagocytosed as a separate vesicular structure which gradually disintegrates within the egg (Barros and Franklin, 1968; Pikó, 1969; Yanagimachi and Noda, 1970b). The factors that dictate different uptake mechanisms for specific regions of the spermatozoon are presently unknown.

The induction stimulus for initial fusion between the post acrosomal plasma membrane and the egg is also unknown. The specific nature of this interaction is indicated by the fact that capacitation of the spermatozoon is a prerequisite for fusion and uncapacitated spermatozoa are unable to penetrate zona-free eggs (Yanagimachi and Noda, 1970c). The maturity of the oocyte can also affect the

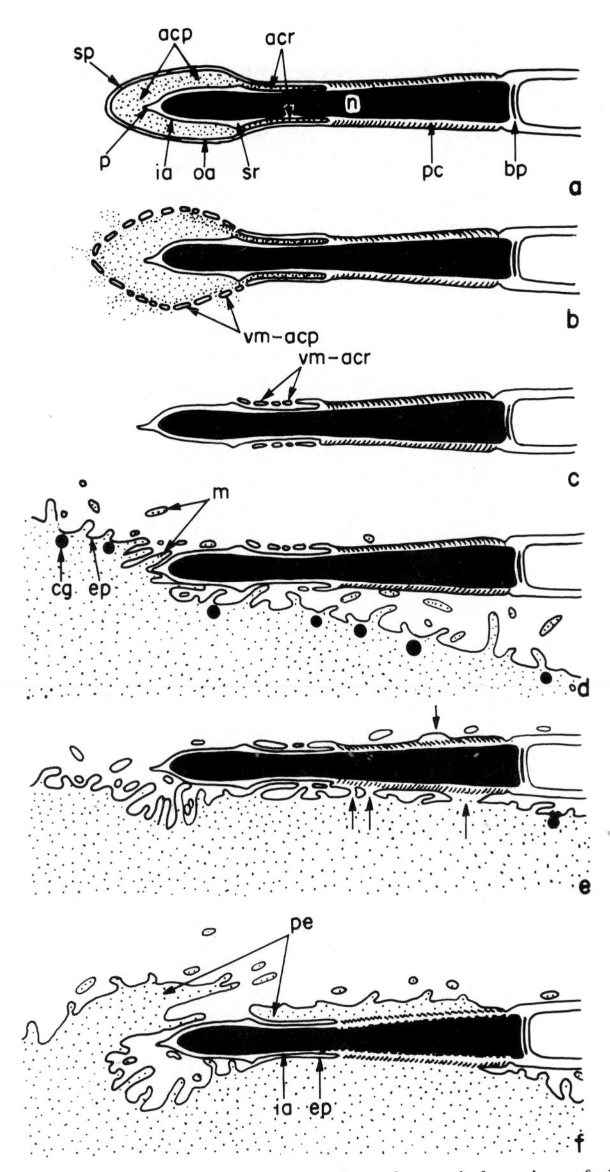

Fig. 7. Diagrammatic representation of a sagital section of the head region of a hamster spermatozoon showing the sequential changes involved in the acrosome interaction of the spermatozoon with the egg. (Redrawn from Yanagimachi and Noda, 1970a,b): (a) In the oviduct. No obvious morphological change in sperm morphology is observed at this stage; (b) In the cumulus oophorus showing vesiculation of the outer acrosomal membrane and the overlying plasma membrane (start of the acrosome reaction) and proposed release of hyaluronidase from the acrosome; (c) Penetration of the spermatozoon through the zona pellucida (substance of the zona not shown); (d) After penetration of the zona showing entrappment of the head of the spermatozoon by microvillous projections on the surface of the vitellus. (e and f) Fusion between the spermatozoon and the egg. Note that fusion occurs between the post-acrosomal region of the sperm plasma membrane and the closely apposed plasma membrane of the vitellus (shown by arrows). Key: acp = acrosomal cap (anterior portion of acrosome); acr = acrosomal collar (equatorial segment of the acrosome); bp = base plate; cg = cortical granule; ep = egg plasma membrane; ia = inner acrosomal membrane; mv = microvilli of vitellus; n = nucleus; oa = outer acrosomal membrane; pc = postacrosomal region of the plasma membrane; sp = sperm plasma membrane; vm = vesiculated membrane.

fusion reaction, sperm being unable to penetrate immature oocytes (Barros and Munoz, 1973; Overstreet and Bedford, 1974). The importance of species specificity in determining the ability of sperm and egg membranes to fuse has not been investigated in detail, though there is some evidence to suggest that it is not a major limiting factor. Hanada and Chang (1972) found that capacitated mouse spermatozoa readily entered zona-free eggs of both homologous and heterologous species in vitro. This apparent lack of species-specificity in sperm-egg fusion is perhaps not surprising in the light of experiments with somatic cells in which interspecific hybrid cells have been created by fusion of cells from diverse species and even different taxonomic orders.

The present lack of information on the biochemical events accompanying the fusion of sperm and egg reflects our general ignorance of the macromolecular organization of the membranes involved. However, as discussed earlier (p. 9), there is some evidence that the structural organization of the post acrosomal plasma membrane may differ from other regions of the sperm plasma membrane. The translational mobility of lectin receptors within this region of the plasma membrane appears to be higher than elsewhere on the sperm and this may be important in the involvement of the post acrosomal region in initial fusion with the egg (Nicolson and Yanagimachi, 1974). Clustering of integral membrane proteins has been proposed as a necessary event in membrane fusion (see, Poste and Allison, 1973; Satir, 1974). Consequently, the higher mobility of components in the post acrosomal membrane dictates that this region may be more appropriately organized to allow redistribution and clustering of membrane proteins prior to fusion with the egg plasma membrane.

Nicolson et al. (1975) have used similar lectin-binding techniques to monitor the translational mobility of components within the plasma membrane of hamster eggs. In contrast to the sperm plasma membrane, a uniformly high degree of lectin receptor mobility was found throughout the egg membrane. This observation suggests that fusion could probably occur at any point on the egg surface, assuming of course that correct apposition with the post-acrosomal region of the sperm had first occurred. However, recent observations by Johnson et al. (1975) suggest that in the mouse egg attachment of sperm in vitro rarely occurs to the membrane region overlying the second metaphase spindle. The functional significance of this finding has yet to be established. However, it may well be a protective mechanism to ensure that extrusion of the second polar body can take place unimpaired or, alternatively, it may serve to prevent sperm entry into a region in which the sperm head itself might be subsequently extruded.

10. The block to polyspermy and the cortical granule reaction

In mammals, and indeed most animals, fertilization is normally monospermic, that is, only one spermatozoon gains entry into the egg cytoplasm. The entry of two or more spermatozoa (polyspermy) leads to abnormal development and early embryonic death (review, Austin, 1969). The block to polyspermy in mammalian fertilization operates both at the zona pellucida and at the egg plasma membrane, though significant species variation exists concerning which of these two sites is more important. In the hamster, dog, pig and the sheep entry of the fertilizing spermatozoon has been shown to result in changes in the zona pellucida, the so-called zona reaction, that render it refractile to penetration by further spermatozoa (Braden et al., 1954; Austin and Braden, 1956; Barros and Yanagimachi, 1971; Gwatkin and Williams, 1974). Although the zona reaction may constitute the major block to polyspermy in these species, an additional block may also operate at the vitelline membrane. In the rabbit, however, the zona reaction does not appear to develop and the block to polyspermy is mediated solely by changes in the properties of the vitelline membrane (Braden et al., 1954).

It is now generally accepted that the block to polyspermy in fertilized mammalian eggs results from breakdown of the cortical granules of the egg and subsequent modification of the zona and/or the vitelline membrane by material(s) released from these granules. Direct evidence that the zona reaction is produced by material(s) from the cortical granules has been obtained by Barros and Yanagimachi (1971). They collected material discharged from fertilized hamster vitelli and found that in vitro treatment of unfertilized eggs with the harvested material rendered them infertile. Similar results were obtained by Gwatkin et al. (1973b) who found that treatment of hamster eggs with cortical granule material, collected either from fertilized or electrically pulsed vitelli, completely abolished the ability of the zona pellucida to bind spermatozoa.

Attachment and fusion of the fertilizing spermatozoon to the vitellus triggers rapid fusion of the cortical granules with the overlying plasma membrane with resulting discharge of the granule contents into the perivitelline space (see, Szollosi, 1967; Pikó, 1969). Cortical granule discharge occurs over the whole circumference of the egg, starting at the point of entry of the fertilizing spermatozoon and progressing as a propagated wave around the entire egg (Braden et al., 1954). The nature of the trigger stimulus for cortical granule discharge is still unknown. Cortical granule breakdown and development of a block to polyspermy have been induced artificially in mammalian eggs by cold-shock (Austin, 1956; Flechon and Thibault, 1964) and by electrical stimulation with a square-wave pulse of 150V for 1 msec (Gwatkin et al., 1973b; Gwatkin and Williams, 1974). More recently, the divalent cation ionophore A-23187 has been shown to be highly effective in inducing the cortical granule reaction in eggs or oocytes from sea urchins, starfish, tunicates, amphibians and hamsters (Steinhardt and Epel, 1974; Steinhardt et al., 1974). Exposure of hamster eggs to A-23187 (3 μM) for 2 minutes was sufficient to induce cortical granule discharge and impose a block to polyspermy.

The ability of A-23187 to trigger rapid discharge of cortical granules in the absence of sperm may offer the first clue to the trigger stimulus operating in the physiological situation. The rapid and extensive fusion of egg cortical granules with the plasma membrane induced by A-23187 is very similar to the action of this compound in promoting fusion of secretory granules with the plasma membrane in various somatic cells discussed earlier in this chapter (p. 23). This raises the obvious question of whether the physiological cortical granule reaction is triggered by a rise in the intracellular Ca^{2+} concentration which then acts to promote membrane fusion along the lines mentioned earlier (p. 23). For example, alteration in the permeability of the egg plasma membrane to Ca^{2+} with an enhanced influx of Ca^{2+} into the cell could occur as a consequence of fusion of the fertilizing spermatozoon with the vitelline membrane. A physico-chemical scheme for alterations in membrane permeability to Ca^{2+} resulting from close contact and fusion of biological membranes has been proposed by Poste and Allison (1973). Changes in intracellular free Ca^{2+} levels could also occur as a result of mobilization of Ca^{2+} ordinarily bound to intracellular membranes and to specific Ca^{2+}-binding proteins (cf. Steinhardt and Epel, 1974; Nakumaru and Yasumasu, 1974). At the present time, there is no information available on the permeability of the mammalian egg to Ca^{2+} at different stages in the fertilization process. However, an increase in free intracellular Ca^{2+} following fertilization has been known for considerable time in sea urchin eggs (Mazia, 1937 and others) and there is now a convincing body of evidence, reviewed recently by Epel (1975), to support the view that increases in intracellular free Ca^{2+} act not only as the primary trigger for cortical granule discharge and the block to polyspermy, but also serve to activate a variety of metabolic pathways in the newly fertilized egg.

The chemical nature of the cortical granule material(s) responsible for altering the zona pellucida and/or the vitellus in the block to polyspermy is still under investigation. Most interest has focused on the possible role of cortical granule enzymes, notably proteases, in modifying the egg surface. Gwatkin et al. (1973b) and Gwatkin and Williams (1974) have shown that hamster cortical granules release a heat-labile trypsin-like enzyme that prevents binding of spermatozoa to the zona pellucida. Interestingly, the action of this enzyme was not species-specific and it was equally effective in producing a zona reaction in both hamster and mouse eggs (Gwatkin and Williams, 1974). The ability of this enzyme to prevent sperm binding to the zona pellucida is consistent with the sensitivity of the sperm receptor on the normal zona to digestion by proteases (Hartman and Gwatkin, 1971; Oikawa et al., 1975).

A trypsin-like protease has also been implicated in the block to polyspermy in sea urchin eggs (Vacquier et al., 1972a,b; Tegner and Epel, 1972; Schuel et al., 1973; Vacquier, 1975b; Longo et al., 1975). This enzyme inhibits binding of sperm to the egg and also appears to be involved in the elevation of the vitelline layer immediately after fertilization. Carrol and Epel (1975) have recently identified two protease components, one of which breaks attachments between the vitelline layer and the plasma membrane and another which appears to alter the egg surface so that sperm no longer bind to it. In addition to the trypsin-like protease the cortical granules of sea urchin eggs also contain β-1,3-glucanohydrolase (Epel et al., 1969). The role of this enzyme is unknown but glycosidases of this type could be involved in dispersing the discharged cortical granule contents, which in both sea urchins and mammals are known to be rich in sulphated mucopolysaccharides (Schuel et al., 1974).

In addition to removal of sperm binding components on the surface of the zona the cortical granule contents may also alter the general mechanical properties of the zona so as to hinder sperm penetration (Wyrick et al., 1974). In this sense, the zona reaction might be viewed as analogous to the block to sperm penetration produced experimentally by treatment of the zona pellucida with plant lectins (Oikawa et al., 1973, 1974) or antibodies (Shivers and Dudkiewicz, 1974). These agents cause precipitation and/or cross-linking of zona components which probably prevents destabilization or depolymerization of the zona membrane by sperm-associated lysins. In this respect it is also of interest to note that the resistance of the zona pellucida to dissolution by mercaptoethanol (Inoué and Wolf, 1974) and by exogenous proteases increases following fertilization (Smithberg, 1953; Chang and Hunt, 1956; Gwatkin, 1964; Conrad et al., 1971; Moore and Cragle, 1971). However, some caution must be exhibited in interpreting the functional significance of these alterations since increased zona resistance to proteases occurs in fertilized rabbit eggs which lack a zona reaction but is not found in hamster eggs despite the strong zona reaction in this species (Chang and Hunt, 1956). In addition, the development of protease-resistance has been recognized in mouse (Krzanowska, 1972) and sheep eggs (Trounson and Moore, 1974) in the absence of sperm and as a result only of ovulation.

As mentioned earlier, in rabbit eggs the block to polyspermy, while still correlated with cortical granule discharge, is mediated at the plasma membrane and a zona reaction does not appear to occur. The plasma membrane in fertilized rabbit eggs is no longer able to bind spermatozoa. Following fertilization the vitelline membrane in the rabbit egg has been shown to acquire an increased number of sialic acid residues (Cooper and Bedford, 1971) and concanavalin A receptors (Gordon et al., 1975a,b). Fertilized rabbit vitelli also become susceptible to agglutination by low doses of concanavalin A (Pienkowski, 1974).

The onset of the block to polyspermy in relation to the time of penetration of the fertilizing spermatozoon varies significantly. In marine invertebrates the block to polyspermy occurs very rapidly, comprising an initial "fast incomplete" block imposed within the first few seconds and a "slow complete" block requiring several minutes. The cortical granule reaction is temporally correlated with the latter. In mammals, although the block to polyspermy is again correlated with cortical granule discharge, the development of a full block occurs more slowly. Barros and Yanagimachi (1972) have estimated that in the fertilized hamster egg 2 to $3\frac{1}{2}$ hours are needed for the vitelline block to develop while the zona reaction requires less than 15 minutes. The slow onset of the vitelline block is consistent with a partial reformation of the egg membrane from the membranes of the cortical granules.

The faster onset of the block to polyspermy in eggs of marine invertebrates compared with the mammals probably reflects the difference in the number of spermatozoa to which eggs from these groups are exposed during normal fertilization. In sea urchins the unfertilized egg is exposed to extremely large numbers of competent sperm while in the mammalian oviduct only a few spermatozoa actually establish contact with the zona. This difference has probably dictated that the selective pressures for development of a rapid block to polyspermy have been high in sea urchins and less intense in the mammals (Bedford, 1972). However, a significant degree of polyspermy has been detected during in vitro fertilization of various mammalian eggs (Yanagimachi and Chang, 1964; Barros and Austin, 1967; Austin, 1969; Iwamatsu and Chang, 1969), even though the phenomenon is a rarity in vivo (Austin and Walton, 1960). This increase in the incidence of polyspermy in vitro probably results from the fact that eggs in vitro are exposed to much higher numbers of capacitated spermatozoa than would ever occur in vivo.

11. Pronucleus formation and development of syngamy

The final stages in fertilization involve transformation of the sperm nucleus within the egg cytoplasm to form the male pronucleus and subsequent interaction of male and female pronuclei to achieve mixing of the male and female genomes and establishment of syngamy. These processes have been reviewed recently by Bedford (1972) and by Longo (1973) and only a brief summary will be presented.

As the sperm nucleus becomes enveloped by the egg cytoplasm its nuclear envelope breaks down into a number of small vesicles that become dispersed in the cytoplasm (Fig. 8). This change does not necessarily involve the whole nuclear membrane which may remain intact over the anterior and posterior poles of the nucleus (Fig. 8b). The central region of the nucleus which is now exposed directly to the ooplasm begins to expand and fine irregular chromatin strands move out into the ooplasm (Fig. 8b). This process continues until the whole of the nuclear chromatin is converted into a much less dense intricate matrix of fine fibrils and granules. These changes resemble those produced in rabbit sperm nuclei by treatment in vitro with dithiothreitol or sodium dodecyl sulphate which disrupt disulphide bonds in the chromatin (Calvin and Bedford, 1971; Bedford et al., 1973). The obvious implication is that changes of a similar chemical nature may be produced by components in the egg cytoplasm.

The properties of the cytoplasm undoubtedly affect the ability of the sperm nucleus to undergo swelling, chromatin disaggregation and initiation of DNA synthesis. Yanagimachi and Usui (1972) found that the chromatin of hamster sperm nuclei does not disperse in oocytes while they are at the germinal vesicle stage. Dispersion occurs with increasing rapidity as the oocyte matures, but again fails to take place at the two-cell stage. Thibault (1973) observed that rabbit oocytes matured in vitro, unlike those which have matured in vivo, may be incapable of supporting dispersal of the sperm nucleus and subsequent

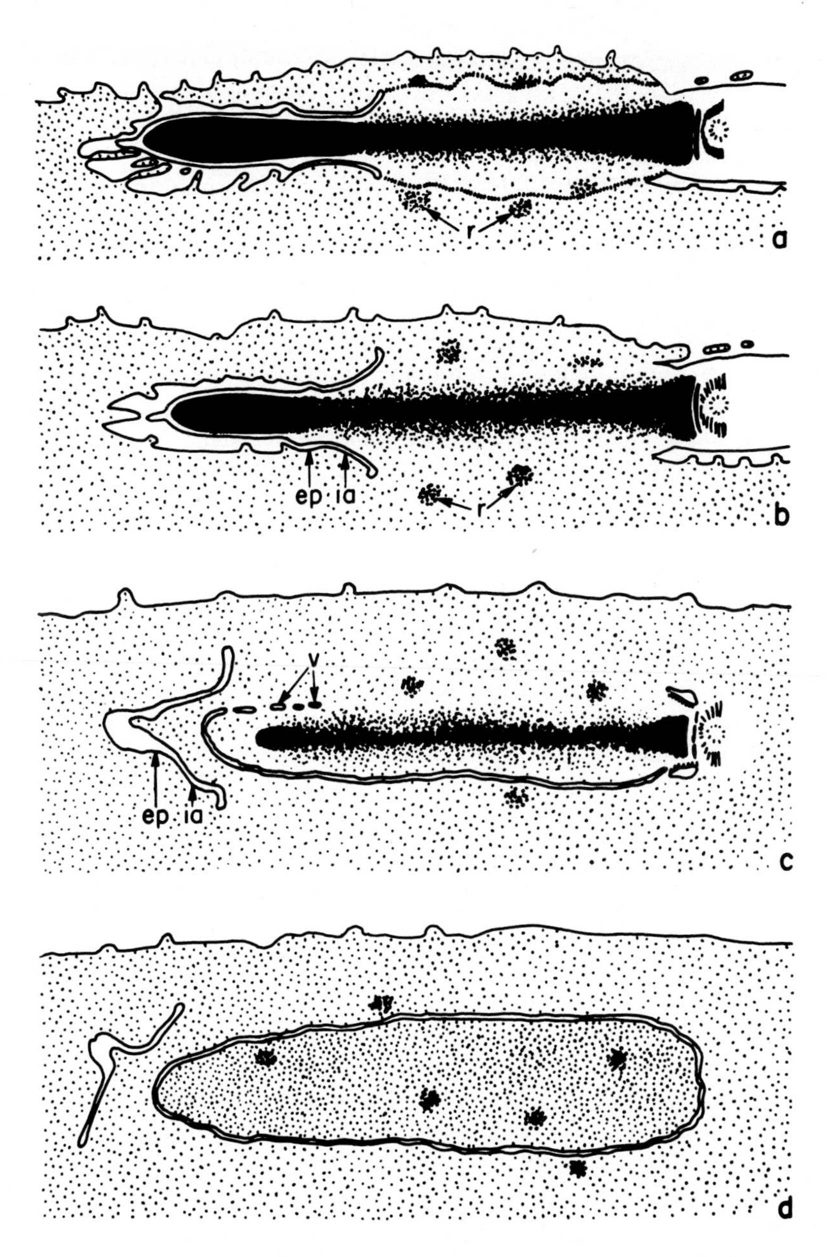

Fig. 8. Diagrammatic representation of formation of male pronucleus within the ooplasm (redrawn from Yanagimachi and Noda, 1970a). (a) Breakdown of nuclear membrane in the newly incorporated spermatozoon and start of chromatin dispersal. (b) Separation of the remnants of the inner acrosomal membrane from the dispersing chromatin. (c) Initiation of pronucleus envelope formation by coalescence of membranous vesicles. (d) Completion of pronuclear envelope formation and chromatin dispersal to produce the male pronucleus. Key: ep = egg plasma membrane; ia = inner acrosomal membrane; r = aggregates of particles in the ooplasm; v = membranous vesicles involved in formation of pronuclear envelope.

pronucleus formation. Taken together, these observations suggest that a male pronucleus growth factor may be elaborated by the egg during normal maturation and later disappears. In view of the suggested importance of disulphide-cross links in maintaining the highly condensed state of sperm chromatin in eutherian mammals, it is tempting to speculate that this factor may be a disulphide-reducing agent.

Soon after the major expansion of the sperm chromatin the pronuclear envelope begins to form along the periphery of the dispersed chromatin. The new nuclear envelope develops by union of a series of vesicular membranes that have accumulated around the expanded nucleus. This process is identical to nuclear envelope formation in meiotic and mitotic cells. One or several spherical nucleoli subsequently develop within the male pronucleus.

In the meantime, the female pronucleus develops in much the same manner. Both the male and the female pronuclei then increase greatly in size and move together in the center of the egg. In many species true pronuclear fusion does not take place. The two pronuclei instead associate via complex interdigitating membrane-bound projections. At this point the pronuclear chromatin begins to condense and the pronuclear envelopes break down. The chromosomes then intermix and recondense in association with microtubules to form a metaphase arrangement for the first mitotic division. In mammals a zygote nuclear envelope involving true pronuclear fusion (such as occurs in sea urchins) does not develop. The first time that the diploid chromosome complement will be surrounded by a single nuclear envelope is at the two-cell embryo stage.

12. Fate of non-fertilizing spermatozoa and interaction of spermatozoa with somatic cells

Of the many millions of mammalian spermatozoa deposited in the female genital tract, few ever reach the egg at the site of fertilization. In most species none or only very few of these non-fertilizing spermatozoa persist for any length of time in the female tract. The major mechanism for the removal of non-fertilizing spermatozoa appears to involve their phagocytosis (and presumably breakdown) by leukocytes that invade the tissues of the female tract after copulation (Austin, 1960; Bedford, 1965; Howe, 1967). Phagocytosis of non-fertilizing spermatozoa has been identified for leukocytes in the uterine cavity (Austin, 1960; Yanagimachi and Chang, 1963; Bedford, 1965; Mahajan and Menge, 1966; Moyer et al., 1967) and the uterine cervix (Moyer et al., 1970). Phagocytosis of spermatozoa by epithelial cells in the endometrium (Moyer et al., 1967), endometrial gland (Austin, 1960) and oviductal mucosa (Austin, 1960; Zamboni, 1971a) has also been described. Finally, non-fertilizing spermatozoa persisting in the female tract may on occasion be phagocytosed by embryos at the two-cell stage (Thompson and Zamboni, 1974) and at the blastocyst stage (Tachi and Kraicer, 1967; McReynolds and Hadek, 1971).

In all of these reports, cellular phagocytosis of sperm, rather than fusion of sperm with somatic cells, appears to be involved. The various structural elements of the sperm identified in somatic cells were all surrounded by membranes suggesting that they had been incorporated by phagocytosis. If fusion had taken place then some elements of the sperm, notably the nucleus, should be free within the cytoplasm. This apparent lack of fusion between sperm and somatic cells in vivo reinforces the view expressed earlier that sperm-egg fusion involves a highly specific membrane interaction (p. 32). Although fusion between gametes from heterologous species can occur, the overall fusion response displayed by germ cells nonetheless appears to be restricted to ensure that fusion with gametes rather than fusion with somatic cells takes place.

In rodents and lagomorphs where clearance of spermatozoa from the female tract has been studied in detail, non-fertilizing spermatozoa are rapidly removed and disappear within 24–36 hours of copulation (Howe, 1967; Bishop, 1969). The survival time (expressed in terms of fertilizing capacity) of sperm from these species within the female tract is, however, relatively short and little is known concerning clearance from the female tract of spermatozoa in species such as the dog, horse and the ferret in which sperm survive much longer (a property possibly correlated with the longer oestrus period in these species). Perhaps in these species leukocyte invasion is simply delayed or it may fail to occur altogether as has been suggested for the domestic fowl (see, Bishop, 1969).

Interest in the interaction of spermatozoa with somatic cells has increased recently in the light of recent experimental work which suggests that spermatozoa might be able to act as vectors for the transfer of genetic information to somatic cells both in vivo and in vitro.

Coppelson and Reid (1967, 1974) proposed that the uptake of non-fertilizing spermatozoa by epithelial cells in the cervix and the uterus might result in transfer of genetic material from the sperm to somatic tissue cells with possible risk of neoplastic transformation. Although there is presently no firm evidence to support this hypothesis, the successful transfer of sperm genetic material to somatic cells in vitro (vide infra) indicates that the hypothesis cannot be completely disregarded.

Several studies have been made over the past few years on the interaction of mammalian spermatozoa with mammalian somatic cells cultured in vitro. In several studies cell-fusing viruses (Sawicki and Koprowski, 1971; Zelenin et al., 1974) or chemicals (Croce et al., 1972) were used to introduce spermatozoa into somatic cells, but it is now clear that spontaneous uptake of spermatozoa by somatic cells in vitro can also occur (Higgins et al., 1975; Lau, 1975) though the question of whether sperm are incorporated by fusion or endocytosis has not been resolved.

Despite initial negative reports (Sawicki et al., 1971), evidence has been obtained to show that sperm nuclei can undergo limited activation, including initiation of DNA synthesis, within the cytoplasm of somatic cells (Johnson et al., 1970; Croce et al., 1972; Zelenin et al., 1974). However, the equivalent of syngamy, namely, the formation of a viable hybrid cell combining both sperm and somatic cell genomes within a single nucleus has yet to be achieved.

Data presented in a very recent paper suggests that despite the failure to achieve true hybridization sperm can nonetheless act as vectors for the transfer of genetic information to somatic cells (Higgins et al., 1975). These investigators showed that treatment of Chinese hamster DON cells with rat sperm resulted in subsequent expression of rat foetal antigens by the DON cells. Karyotypic analysis of these cells failed, however, to reveal the presence of any rat chromosomes, indicating that an extremely limited amount of genetic material had been transferred. Similar transfer of microscopically undetectable amounts of foreign genetic material and its subsequent expression have been reported during interaction between two types of somatic cell (Boyd and Harris, 1973; Klinger and Shin, 1974).

Apart from its obvious importance in relation to the role of transfer of genetic information in oncogenic transformation, the work of Higgins et al. (1975) may open the way for the experimental use of mammalian sperm as tools in somatic cell genetics. The experimental transfer of eukaryotic genetic information by nuclear transplantation and virus-induced cell fusion techniques has already emerged as a powerful approach in the genetic analysis of mammalian somatic cells in vitro, and a similarly bright future might well await a method for transferring very small amounts of genetic material. A particular attraction in transferring genetic information to somatic cells from spermatozoa, as opposed to any other cell, is that if the technique could be refined it would enable direct mapping of genes in a haploid genome and would also offer novel opportunities for genetic and biochemical analysis of a genome which has not yet been modified by the process of tissue differentiation.

References

Adams, C.E. (1974) Species specificity in fertilization. In: Physiology and Genetics of Reproduction. (Coutinho, E.M. and Fuchs, F., eds.) Part B, pp. 69–79, Plenum Press, New York.

Aketa, K. (1967) On sperm-egg binding as the initial step of fertilization in the sea urchin. Embryologia, 9, 238–245.

Aketa, K. (1973) Physiological studies on the egg surface component responsible for sperm-egg bonding in sea urchin fertilization. I. Effect of sperm-binding protein on the fertilizing capacity of sperm. Exp. Cell Res. 80, 439–441.

Aketa, J., Onitake, K. and Tsuzuki, H. (1972) Tryptic disruption of sperm-binding site of sea urchin egg surface. Exp. Cell Res. 71, 27–32.

Aketa, K., Tsuzuki, H. and Onitake, K. (1968) Characterization of the sperm-binding protein from sea urchin egg surface. Exp. Cell Res. 50, 676–679.

Allison, A.C. and Davies, P. (1974) Mechanisms of endocytosis and exocytosis. In: Transport at the Cellular Level (Sleigh, M.A. and Jennings, D.H., eds.) pp. 419–446, Cambridge University Press, Cambridge.

Aonuma, S., Mayumi, T., Suzuki, K., Noguchi, T., Iwari, M. and Okabe, M. (1973) Studies on sperm capacitation. I. Relationship between a guinea pig sperm-coating antigen and a sperm capacitation phenomenon. J. Reprod. Fertil. 35, 425–432.

Aull, F. and Nachbar, M.S. (1974) Concanavalin A-induced alteration in sodium and potassium content of Ehrlich ascites tumor cells. J. Cell Physiol. 83, 243–250.

Austin, C.R. (1951) Observations on the penetration of sperm into the mammalian egg. Aust. J. Sci. Res. (B) 4, 581–596.

Austin, C.R. (1952) The capacitation of mammalian sperm. Nature, 170, 326.

Austin, C.R. (1956) Cortical granules in hamster eggs. Exp. Cell Res. 10, 533–540.

Austin, C.R. (1960) Fate of spermatozoa in the female genital tract. J. Reprod. Fertil. 1, 151–156.

Austin, C.R. (1961) The Mammalian Egg, Charles C. Thomas, Springfield, Illinois.

Austin, C.R. (1965) Fertilization. Foundations of Developmental Biology Series, Prentice-Hall Inc., Englewood Cliffs, N.J.

Austin, C.R. (1968) Ultrastructure of Fertilization. Holt, Rhinehart and Winston, New York.

Austin, C.R. (1969) Variations and anomalies in fertilization In: Fertilization, Comparative Morphology, Biochemistry and Immunology (Metz, C.B. and Monroy, A., eds.) Vol. II, pp. 437–466, Academic Press, New York.

Austin, C.R. (1974) Principles of fertilization. Proc. Royal Soc. Med. 67, 925–927.

Austin, C.R. (1975) Membrane fusion events in fertilization. J. Reprod. Fertil. 44, 155–166.

Austin, C.R. and Bishop, M.W.H. (1958) Some features of the acrosome and perforatorium in mammalian spermatozoa. Proc. Roy. Soc. Series B, 149, 234–240.

Austin, C.R. and Braden, A.W.H. (1956) Early reactions of the rodent egg to spermatozoan penetration. J. Exp. Biol. 33, 358–365.

Austin, C.R. and Walton, A. (1960) Fertilization. In: Marshall's Physiology of Reproduction (Parkes, A.S., ed.) Vol. 1, part II, pp. 310–416, Longmans, Green, New York.

Austin, C.R. and Short, R.V., Eds. (1972) Germ Cells and Fertilization, Book 1, Reproduction in Mammals, Cambridge Univ. Press. Cambridge.

Austin, C.R., Bavister, B.D. and Edwards, R.G. (1973) Components of capacitation. In: The Regulation of Mammalian Reproduction. (Segal, S.J., Crozier, R., Corfman, P.A. and Condliffe, P.E., eds.) pp. 247–254, Charles C. Thomas, Springfield, Illinois.

Baker, T.G. (1972) Oogenesis and ovarian development. In: Reproductive Biology (Balin, H. and Glasser, S., eds.) pp. 398–437, Excerpta Medica, Amsterdam.

Barros, C. (1968) In vitro capacitation of golden hamster spermatozoa with fallopian tube fluid of the mouse and rat. J. Reprod. Fertil. 17, 203–206.

Barros, C. (1974) Capacitation of mammalian spermatozoa. In: Physiology and Genetics of Reproduction. (Coutinho, E.M. and Fuchs, F., eds.) Part B, pp. 3–24, Plenum Press, New York.

Barros, C. and Austin, C.R. (1967) In vitro fertilization and the sperm acrosome reaction in the hamster. J. Exp. Zool. 166, 317–323.

Barros, C. and Franklin, L.E. (1968) Behavior of the gamete membranes during sperm entry into the mammalian egg. J. Cell Biol. 37, C13.

Barros, C. and Garavagno, A. (1970) Capacitation of hamster spermatozoa with blood sera. J. Reprod. Fertil. 22, 381–384.

Barros, C. and Munoz, G. (1973) Sperm-egg interaction in immature hamster oocytes. J. Exp. Zool. 186, 73–78.

Barros, C. and Yanagimachi, R. (1971) Induction of zona reaction in golden hamster eggs by cortical granule material. Nature, 233, 268–269.

Barros, C. and Yanagimachi, R. (1972) Polyspermy-preventing mechanisms in the golden hamster egg. J. Exp. Zool. 180, 251–266.

Barros, C., Bedford, J.M., Franklin, L.E. and Austin, C.R. (1967) Membrane vesiculation as a feature of the mammalian acrosome reaction. J. Cell Biol. 34, C1–5.

Barros, C., Berrios, M. and Herrerg, E. (1973a) Capacitation in vitro of guinea pig spermatozoa in a saline solution. J. Reprod. Fertil. 34, 547–549.

Barros, C., Fujimoto, M. and Yanagimachi, R. (1973b) Failure of zona penetration of hamster spermatozoa after prolonged preincubation in a blood serum fraction. J. Reprod. Fertil. 35, 89–95.

Barros, C., Vliegenthart, A.M. and Franklin, L.E. (1972) Polyspermic fertilization of hamster eggs in vitro. J. Reprod. Fertil. 28, 117–120.

Bavister, B.D. (1973) Capacitation of golden hamster spermatozoa during incubation in culture medium. J. Reprod. Fertil. 35, 161–163.

Bavister, B.D. and Morton, D.B. (1974) Separation of human serum components capable of inducing the acrosome reaction in hamster spermatozoa. J. Reprod. Fertil. 40, 495–498.

Bedford, J.M. (1963) Changes in the electrophoretic properties of rabbit spermatozoa during passage through the epididymis. Nature, 200, 1178–1180.

Bedford, J.M. (1965) Effect of environment on phagocytosis of rabbit spermatozoa. J. Reprod. Fertil. 9, 249–256.

Bedford, J.M. (1967) Effect of duct ligation on the fertilizing ability of spermatozoa from different regions of the rabbit epididymis. J. Exp. Zool. 166, 271–282.

Bedford, J.M. (1968) Ultrastructural changes in the sperm head during fertilization in the rabbit. Am. J. Anat. 123, 329–358.

Bedford, J.M. (1969) Morphological aspects of capacitation. In: Schering Symposium on Mechanisms Involved in Conception (Raspe, G., ed.) Advances in the Biosciences, Vol. 4, pp. 35–50. Pergamon Press, Viewig.

Bedford, J.M. (1970) Sperm capacitation and fertilization in mammals. Biol. Reprod., Suppl. 2, 128–158.

Bedford, J.M. (1972) Sperm transport, capacitation and fertilization. In: Reproductive Biology (Balin, H. and Glasser, S., eds.) pp. 338–392. Excerpta Medica, Amsterdam.

Bedford, J.M. (1974) Mechanisms involved in penetration of spermatozoa through the vestments of the mammalian egg. In: Physiology and Genetics of Reproduction (Coutinho, E.M. and Fuchs, F., eds.), Part B, pp. 55–68, Plenum Press, New York.

Bedford, J.M. and Calvin, H.I. (1974) The occurrence and possible functional significance of -s-s- crosslinks in sperm heads, with particular reference to eutherian mammals. J. Exp. Zool. 188, 137–156.

Bedford, J.M. and Chang, M.C. (1962) Removal of decapacitation factor from seminal plasma by high-speed centrifugation. Am. J. Physiol. 202, 179–181.

Bedford, J.M., Cooper, G.W. and Calvin, H.I. (1973) Post-meiotic changes in the nucleus and membranes of mammalian spermatozoa. In: The Genetics of the Spermatozoon (Beatty, R.A. and Gluecksohn-Waelsch, S., eds.) pp. 69–89, Department of Genetics, University of Edinburgh and Albert Einstein College of Medicine.

Bishop, D.W. (1969) Sperm physiology in relation to the oviduct. In: The Mammalian Oviduct (Hafez, E.S.E. and Blandau, R.J., eds.) pp. 231–250, University of Chicago Press, Chicago.

Blandau, R.J. (1969) Gamete transport-comparative aspects. In: The Mammalian Oviduct (Hafez, E.S.E. and Blandau, R.J., eds.) pp. 129–162. University of Chicago Press, Chicago.

Blaquier, J.A., Cameo, M.S. and Burgos, M.H. (1972) The role of androgens in the maturation of epididymal spermatozoa in the guinea pig. Endocrinology, 90, 839–842.

Bleau, G., Bodley, F., Longpre, J., Capdelaine, A. and Roberts, K.D. (1974) Cholesterol sulfate. I. Occurrence and possible biological function as an amphipathic lipid in the membrane of the human erythrocyte, Biochim. Biophys. Acta 352, 1–8.

Bleau, G., VandenHeuvel, W.J.A., Andersen, O.F. and Gwatkin, R.B.L. (1975) Desmosteryl sulphate of hamster spermatozoa, a potent inhibitor of capacitation in vitro. J. Reprod. Fertil. 43, 175–178.

Boyd, Y.L. and Harris, H. (1973) Correction of genetic defects in mammalian cells by the input of small amounts of foreign genetic material. J. Cell Sci. 13, 841–846.

Brackett, B.G. and Oliphant, G. (1975) Capacitation of rabbit spermatozoa in vitro. Biol. Reprod. 12, 260–274.

Braden, A.W.H. (1952) Properties of the membranes of rat and rabbit eggs. Aust. J. Sci. Res. (B), 5, 460–471.

Braden, A.W.H., Austin, C.R. and David, H.A. (1954) The reaction of the zona pellucida to sperm penetration. Austral. J. Biol. Sci. 7, 391–409.

Briggs, M.H. (1973) Steroid hormones and the fertilizing capacity of spermatozoa. Steroids, 22, 547–553.

Brochart, M. and Debatene, D. (1953) Diminution de la permeabilitie a leosine de la capsule lipoidique de la tête des spermatozoides des ruminants domestiques après l'ejaculation. Comp. R. Seanc. Soc. Biol. 145, 20–25.

Brown, C.R. and Hartree, E.F. (1974) Distribution of a trypsin-like proteinase in the ram spermatozoon. J. Reprod. Fertil. 36, 195–202.

Burgos, M.H. (1974) Ultrastructure of the mammalian sperm head during differentiation and maturation. In: Physiology and Genetics of Reproduction (Coutinho, E.M. and Fuchs, F., eds.) Part A, pp. 209–225, Plenum Press, New York.

Calvin, H.I. and Bedford, J.M. (1971) Formation of disulphide bonds in the nucleus and accessory structures of mammalian spermatozoa during maturation in the epididymis. J. Reprod. Fertil. Suppl. 13, 65–75.

Carroll, E.J. and Epel, D. (1975) Isolation and biological activity of the proteases released by sea urchin eggs following fertilization. Dev. Biol. 44, 22–32.

Chaney, M.O., Demarco, P.V., Jones, N.D. and Occolowitz, J.L. (1974) The structure of A23187, a divalent cation ionophore. J. Am. Chem. Soc. 96, 1932–1933.

Chang, M.C. (1951) The fertilizing capacity of spermatozoa deposited into the Fallopian tubes. Nature, 168, 697–698.

Chang, M.C. (1955) Development of fertilizing capacity of rabbit spermatozoa in the uterus. Nature, 175, 1036–1037.

Chang, M.C. (1957) A detrimental effect of seminal plasma on the fertilizing capacity of sperm. Nature, 179, 258–259.

Chang, M.C. (1958) Capacitation of rabbit spermatozoa in the uterus with special reference to the reproductive phases of the female. Endocrinology, 83, 619–623.

Chang, M.C. and Hancock, J.L. (1967) Experimental hybridization. In: Comparative Aspects of Reproductive Failure. (Benirschke, K., ed.) pp. 206–217, Springer Verlag, New York.

Chang, M.C. and Hunt, D.M. (1956) Effects of proteolytic enzymes on the zona pellucida of fertilized and unfertilized mammalian eggs. Exp. Cell Res. 11, 497–499.

Cochrane, D.E. and Douglas, W.W. (1974) Calcium-induced extrusion of secretory granules (exocytosis) in mast cells exposed to 48/80 or the ionophores A-23187 and X-537A. Proc. Natl. Acad. Sci. U.S.A., 71, 408–412.

Colwin, L.H. and Colwin, A.L. (1967) Membrane fusion in relation to sperm-egg association. In: Fertilization (Metz, C.B. and Monroy, A., eds.) Vol. 1, pp. 295–367. Academic Press, New York.

Conchie, J. and Mann, T. (1957) Glycosidases in mammalian sperm and seminal plasma. Nature, 179, 1190–1191.

Conrad, K., Buckley, J. and Stambaugh, R. (1971) Studies on the nature of the block to polyspermy in rabbit ova. J. Reprod. Fertil. 27, 133–135.

Cooper, G.W. and Bedford, J.M. (1971) Charge density change in the vitelline surface following fertilization of the rabbit egg. J. Reprod. Fertil. 25, 431–436.

Coppelson, M. and Reid, B.L. (1967) Preclinical Carcinoma of the Cervix Uteri. Pergamon Press, Oxford.

Coppelson, M. and Reid, B.L. (1974) Interaction of sperm with somatic cells. Science, 185, 239.

Coutinho, E.M. and Fuchs, F. (1974) Eds. Physiology and Genetics of Reproduction, Parts A and B. Plenum Press, New York.

Croce, C.M., Gledhill, B.L., Gabara, B., Sawicki, W. and Koprowski, H. (1972) Lysolecithin-induced fusion of rabbit spermatozoa with hamster somatic cells. In: Workshop on Mechanisms and Prospects of Genetic Exchange (Raspe, G. ed.) Advances in the Biosciences Vol. 8, pp. 187–200. Pergamon Press, Vieweg.

Crogan, D.E., Mayer, D.T. and Sikes, J.D. (1966) Quantitative differences in phospholipids of ejaculated spermatozoa and spermatozoa from three levels of the epididymis of the boar. J. Reprod. Fertil. 12, 431–436.

Croxatto, H.B. (1974) The duration of egg transport and its regulation in mammals. In: Physiology and Genetics of Reproduction (Coutinho, E.M. and Fuchs, F., eds.) Part B, pp. 159–166, Plenum Press, New York.

Dan, J.C. (1967) Acrosome reaction and lysins. In: Fertilization, (Metz, C.B. and Monroy, A., eds.) Vol. 1, pp. 237–293, Academic Press, New York.

Davis, B.K. (1971) Macromolecular inhibitor of fertilization in rabbit seminal plasma. Proc. Nat. Acad. Sci. U.S.A. 68, 951–955.

Dawson, R.M.C. and Scott, T.W. (1964) Phospholipid composition of epididymal spermatozoa prepared by density gradient centrifugation. Nature, 202, 292–293.

Douglas, W.W. (1974) Mechanisms of release of neurohypophysial hormones: stimulus-secretion

coupling. In: Handbook of Physiology, Vol. IV, section 7, pp. 191–224. American Physiological Society Washington.
Dukelow, W.R., Chernoff, H.N. and Williams, W.L. (1966) Enzymatic characterization of decapacitation factor. Proc. Soc. Exp. Biol. Med. 121, 396–398.
Dukelow, W.R., Chernoff, H.N. and Williams, W.L. (1967) Properties of decapacitation factor and presence in various species. J. Reprod. Fertil. 14, 393–399.
Edelman, G.M. and Millette, C.F. (1971) Molecular probes of spermatozoan structures. Proc. Nat. Acad. Sci. U.S.A. 68, 2436–2440.
Eimerl, S., Savion, N., Heichal, O. and Selinger, Z. (1974) Induction of enzyme secretion in rat pancreatic slices using the ionophore A23187 and calcium. J. Biol. Chem. 249, 3991–3993.
Epel, D. (1975) The program of and mechanisms of fertilization in the echinoderm egg. Am. Zool., 15, 507–522.
Epel, D., Weaver, A.M., Muchmore, A.V. and Schimke, R.T. (1969) β-1,3-Glucanase of sea urchin eggs: release from particles at fertilization. Science, 163, 294–296.
Ericsson, R.J. (1967) A fluorometric method for measurement of sperm capacitation. Proc. Soc. Exp. Biol. Med. 125, 1115–1118.
Ericsson, R.J., Buthala, D.A. and Norland, J.F. (1971) Fertilization of rabbit ova in vitro by sperm with adsorbed Sendai virus. Nature, 173, 54–55.
Erickson, R.P. (1972) Alternative modes of detection of H-2 antigens on mouse spermatozoa. In: The Genetics of the Spermatozoon (Beatty, R.A. and Gluecksohn-Waelsch, S., eds.) pp. 191–202, Department of Genetics, University of Edinburgh and Albert Einstein College of Medicine.
Fawcett, D.W. (1970) A comparative view of sperm ultrastructure. Biol. Reprod. Suppl. 2, 90–127.
Fawcett, D.W. (1975) Gametogenesis in the male: Prospects for its control. In: The Developmental Biology of Reproduction (Markert, C.L. and Papaconstantinou, J., eds.) pp. 25–53, Academic Press, New York.
Fawcett, D.W. and Phillips, D.M. (1969) Observations on the release of spermatozoa and on changes in the head during passage through the epididymis. J. Reprod. Fertil. Suppl. 6, 405–418.
Flechon, J.E. and Thibault, C. (1964) Modifications ultrastructurales de l'ovocyte de la lapine an cours de l'activation. J. Microsc., Paris. 3, 34–41.
Franklin, L.E., Barros, C. and Fussell, E.N. (1970) The acrosomal region and the acrosome reaction in sperm of the golden hamster. Biol. Reprod. 3, 180–200.
Friend, D.S. and Fawcett, D.W. (1974) Membrane differentiations in freeze-fractured mammalian spermatozoa. J. Cell Biol. 63, 641–652.
Fritz, H., Schiessler, H., Schleuning, W-D, Fink, E., Forg-Brey, B., Arnhold, M., Meier, M., Werle, E. and Schirren, C. (1973) Proteinases and proteinase inhibitors in the fertilization process: new concepts of control. In: Schering Workshop on Contraception: The Masculine Gender (Raspe, G., ed.) Advances in the Biosciences Vol. 10, pp. 271–286, Pergamon Press, Vieweg.
Fritz, H., Schleuning, W.D. and Schill, W.-B. (1974) Biochemistry and clinical significance of the trypsin-like proteinase, acrosin, from boar and human spermatozoa. In: Proteinase Inhibitors (Fritz, H., Tschesche, H., Greene, L.J. and Truscheit, E., eds.) pp. 118–127. Springer, Berlin.
Gall, W.E., Millette, C.F. and Edelman, G.M. (1974) Chemical and structural analysis of mammalian spermatozoa. In: Physiology and Genetics of Reproduction (Coutinho, E.M. and Fuchs, F., eds.) Part A, pp. 241–257, Plenum Press, New York.
Garavagno, A., Posada, J., Barros, C. and Shivers, C.A. (1974) Some characteristics of the zona pellucida antigen. J. Exp. Zool. 189, 37–50.
Garcia, A.G., Kirpekar, S.M., Prat, J.C. (1975) A calcium ionophore stimulating the secretion of catecholamines from the cat adrenal. J. Physiol. 244, 253–262.
Garner, D.L., Easton, M.P., Munson, M.E. and Doane, M.A. (1975) Immunofluorescent localization of bovine acrosin. J. Exp. Zool. 191, 127–131.
Glass, L.E. and Hanson, J.E. (1974) An immunologic approach to contraception: Localization of antiembryo and antizona pellucida serum during mouse preimplantation development. Fertil. Steril. 25, 484–493.
Glover, J.D. (1961) Disintegrated spermatozoa from the epididymis. Nature, 190, 185–186.
Gordon, M. (1973) Localization of phosphatase activity on the membranes of the mammalian sperm head. J. Exp. Zool. 185, 111–123.

Gordon, M., Dandekar, P.V. and Bartoszewicz, W. (1974) Ultrastructural localization of surface receptors for Concanavalin A on rabbit spermatozoa. J. Reprod. Fertil. 36, 211–214.

Gordon, M., Dandekar, P.V. and Bartoszewicz, W. (1975a) The surface coat of epididymal, ejaculated and capacitated sperm. J. Ultrastruct. Res. 50, 199–207.

Gordon, M., Fraser, L.R. and Dandekar, P.V. (1975b) The effect of ruthenium red and Concanavalin A on the vitelline surface of fertilized and unfertilized rabbit ova. Anat. Rec. 181, 95–112.

Gould, S.F. and Bernstein, M.H. (1973) Localization of bull sperm hyaluronidase. J. Cell Biol. 59, 119a.

Gould, K., Zaneveld, L.J.D., Srivastava, P.N. and Williams, W.L. (1971) Biochemical changes in the zona pellucida of rabbit ova induced by fertilization and sperm enzymes. Proc. Soc. Exp. Biol. Med. 136, 6–10.

Gwatkin, R.B.L. (1964) Effect of enzymes and acidity on the zona pellucida of the mouse egg before and after fertilization. J. Reprod. Fertil. 7, 99–105.

Gwatkin, R.B.L. (1967) Passage of mengovirus through the zona pellucida of the mouse morula. J. Reprod. Fertil. 13, 577–578.

Gwatkin, R.B.L. and Andersen, O.F. (1969) Capacitation of hamster spermatozoa by bovine follicular fluid. Nature, 224, 1111–1112.

Gwatkin, R.B.L. and Andersen, O.F. (1973) Effect of glycosidase inhibitors on the capacitation of hamster spermatozoa by cumulus cells in vitro. J. Reprod. Fertil. 35, 565–567.

Gwatkin, R.B.L., Andersen, O.F. and Hutchison, C.F. (1972) Capacitation of hamster spermatozoa in vitro: The role of cumulus components. J. Reprod. Fertil. 30, 389–394.

Gwatkin, R.B.L., Andersen, O.F. and Williams, D.T. (1974) Capacitation of mouse spermatozoa in vitro: Involvement of epididymal secretions and cumulus oophorus. J. Reprod. Fertil. 41, 253–256.

Gwatkin, R.B.L. and Carter, H.E. (1975) Chapter 23, Cumulus Oophorus. In: Atlas of Mammalian Reproduction. (Hafez, E.S.E., ed.) Igaku Shoin, Tokyo.

Gwatkin, R.B.L. and Hutchison, C.F. (1971) Capacitation of hamster spermatozoa by β-glucuronidase. Nature, 229, 343–344.

Gwatkin, R.B.L. and Williams, D.T. (1970) Inhibition of sperm capacitation in vitro by contraceptive steroids. Nature, 227, 182–183.

Gwatkin, R.B.L. and Williams, D.T. (1974) Heat sensitivity of the cortical granule protease from hamster eggs. J. Reprod. Fertil. 39, 153–155.

Gwatkin, R.B.L., Williams, D.T. and Andersen, O.F. (1973a) Zona reaction of mammalian eggs: Properties of the cortical granule protease (Cortin) and its receptor substrate in hamster eggs. J. Cell Biol. 59, 128a.

Gwatkin, R.B.L., Williams, D.T., Hartmann, J.F. and Kniazuk, M. (1973b) The zona reaction of hamster and mouse eggs: Production in vitro by a trypsin-like protease from cortical granules. J. Reprod. Fertil. 32, 259–265.

Hadek, R. (1965) The structure of the mammalian egg. Int. Rev. Cytol. 18, 29–71.

Hafez, E.S.E. (1973) Gamete transport. In: Human Reproduction, Conception and Contraception (Hafez, E.S.E. and Evans, T.N. eds.) pp. 85–118. Harper and Row, Hagerstown, Maryland.

Hafez, E.S.E. and Sugawara, S. (1969) Biochemistry of oviductal eggs in mammals. In: The Mammalian Oviduct. (Hafez, E.S.E. and Blandau, R.J., eds.) pp. 373–385, University of Chicago Press, Chicago.

Hamilton, D.W. (1973a) The mammalian epididymis. In: Reproductive Biology (Balin, H. and Glasser, S., eds.) pp. 268–337, Excerpta Medica, Amsterdam.

Hamilton, D.W. (1973b) The epididymis as a possible site for control of fertility in the male. In: Schering Workshop on Contraception: The Masculine Gender (Raspe, G., ed.) Advances in the Biosciences, Vol. 10, pp. 128–144, Pergamon Press, Vieweg.

Hamner, C.E., Jones, J.P. and Sojka, N.J. (1968) Influence of the hormonal state of the female on the fertilizing capacity of rabbit spermatozoa. Fertil. Steril., 19, 137–143.

Hanada, A. and Chang, M.C. (1972) Penetration of zona-free eggs by spermatozoa of different species. Biol. Reprod. 6, 300–309.

Hartmann, J.F. and Gwatkin, R.B.L. (1971) Alteration of sites on the mammalian sperm surface following capacitation. Nature, 234, 479–481.

Hartmann, J.F., Gwatkin, R.B.L. and Hutchison, C.F. (1972) Early contact interactions between mammalian gametes in vitro: Evidence that the vitellus influences adherence between sperm and zona pellucida. Proc. Nat. Acad. Sci. U.S.A. 69, 2767–2769.

Hartmann, J.F. and Hutchison, C.F. (1974a) Nature of the prepenetration contact interactions between hamster gametes in vitro. J. Reprod. Fertil. 36, 49–57.

Hartmann, J.F. and Hutchison, C.F. (1974b) Contact between hamster spermatozoa and the zona pellucida releases a factor which influences early binding stages. J. Reprod. Fertil. 37, 61–66.

Hartmann, J.F. and Hutchison, C.F. (1974c) Mammalian fertilization in vitro: Sperm induced preparation of the zona pellucida of golden hamster ova for final binding. J. Reprod. Fertil. 37, 443–445.

Hartmann, J.F. and Hutchison, C.F. (1975) The effect of sperm concentration on binding to and penetration of hamster eggs in vitro. Abstract, 8th Ann. Meeting Soc. Study of Reprod., Fort Collins, Colorado.

Hastings, R.A., Enders, A.C. and Schlafke, S. (1972) Permeability of the zona pellucida to protein tracers. Biol. Reprod. 7, 288–296.

Henle, W., Henle, G. and Chambers, L.A. (1938) Studies on antigenic structure of some mammalian spermatozoa. J. Exp. Med. 68, 335–352.

Hicks, J.J., Pedron, N. and Rosado, A. (1972) Modifications of human spermatozoa glycolysis by cyclic adenosine monophosphate (cAMP), estrogens and follicular fluid. Fertil. Steril. 23, 886–893.

Higgins, P.J., Borenfreund, E. and Bendich, A. (1975) Appearance of foetal antigens in somatic cells after interaction with heterologous sperm. Nature, 257, 488–489.

Hjort, T. and Brogaard, K. (1971) The detection of different spermatozoal antibodies and their occurrence in normal and infertile women. Clin. Exp. Immunol. 8, 9–23.

Hoskins, D.D. and Casillas, E.R. (1975) Hormones, second messengers and the mammalian spermatozoon. In: Molecular Mechanisms of Gonadal Hormone Action. (Thomas, J.A. and Singhal, R.L., eds.) Vol. 1, pp. 283–324, University Park Press, Baltimore.

Howe, G.R. (1967) Leucocytic response to spermatozoa in ligated segments of the rabbit vagina, uterus and oviduct. J. Reprod. Fertil. 13, 563–566.

Hunter, A.G. and Nornes, H.O. (1969) Characterization and isolation of a sperm-coating antigen from rabbit seminal plasma with capacity to block fertilization. J. Reprod. Fertil. 20, 419–427.

Inoué, M. and Wolf, D.P. (1974) Comparative solubility properties of the zonae pellucidae of unfertilized and fertilized mouse ova. Biol. Reprod. 11, 558–565.

Iwamatsu, T. and Chang, M.C. (1969) In vitro fertilization of mouse eggs in the presence of bovine follicular fluid. Nature, 224, 919–920.

Iwamatsu, T. and Chang, M.C. (1972) Capacitation of hamster spermatozoa treated with rabbit tubal fluid or rabbit and bovine cystic fluids. J. Exp. Zool. 182, 211–219.

Jacoby, F. (1962) Ovarian histochemistry. In: The Ovary (Zuckerman, S., ed.) Vol. 1, pp. 189–245, Academic Press, New York.

Jessen, H., Behnke, O., Wingstrand, K.G. and Rostgaard, J. (1973) Actin-like filaments in the acrosomal apparatus of spermatozoa of a sea urchin. Exp. Cell Res. 80, 47–55.

Johnson, M.H. (1975) The macromolecular organization of membranes and its bearing on events leading up to fertilization. J. Reprod. Fertil. 44, 167–184.

Johnson, M.H. and Edidin, M. (1972) H-2 antigens on mouse spermatozoa. Transplantation, 14, 781–786.

Johnson, M.H., Eager, D., Muggleton-Harris, A. and Grave, H.M. (1975) Mosaicism in organization of concanavalin A receptors on surface membrane of mouse egg. Nature, 257, 320–322.

Johnson, R.T., Rao, P.N. and Hughes, S.D. (1970) Mammalian cell fusion. III. A HeLa cell inducer of premature chromosome condensation active in cells from a variety of species. J. Cell Physiol. 77, 151–158.

Johnson, W.L. and Hunter, A.G. (1972a) Immunofluorescent evaluation of the male rabbit reproductive tract for sites of secretion and absorption of seminal antigens. Biol. Reprod. 6, 13–22.

Johnson, W.L. and Hunter, A.G. (1972b) Seminal antigens: their alteration in the genital tract of female rabbits and during partial in vitro capacitation with β-amylase and β-glucuronidase. Biol. Reprod. 7, 332–340.

Joshi, M.S. and Murray, I.M. (1974) Immunological studies of the rat uterine fluid peptidase. J. Reprod. Fertil. 37, 361–365.
Joshi, M.S., Yaron, A. and Lindner, H.R. (1970) An endopeptidase in the uterine secretion of the preoestrus rat and its relation to a sperm decapacitating factor. Biochem. Biophys. Res. Commun. 38, 52–57.
Kagayama, M. and Douglas, W.W. (1974) Electron microscope evidence of calcium-induced exocytosis in mast cells treated with 48/80 or the ionophores A23187 and X537-A. J. Cell. Biol. 62, 519–526.
Killian, G.J. and Amann, R.P. (1973) Immunoelectrophoretic characterization of fluid and sperm entering and leaving the bovine epididymis. Biol. Reprod. 9, 489–499.
Klinger, H.P. and Shin, S-I. (1974) Modulation of the activity of an avian gene transferred into a mammalian cell by cell fusion. Proc. Nat. Acad. Sci. U.S.A. 71, 1398–1402.
Kirton, K.T. and Hafs, H.D. (1965) Sperm capacitation by uterine fluid or β-amylase in vitro. Science, 150, 618–619.
Koehler, J.K. (1975) Changes in anti-rabbit sperm antibody labeling patterns following "capacitation". J. Cell Biol. 67, 219A.
Koehler, J.K. and Gaddum-Rosse, P. (1975) Media induced alterations of the membrane associated particles of the guinea pig sperm tail. J. Ultrastruct. Res. – in press.
Koehler, J.K. and Perkins, W.D. (1974) Fine structure observations on the distribution of antigenic sites on guinea pig spermatozoa. J. Cell Biol. 60, 789–795.
König, E., Brittinger, G. and Cohnen, G. (1973) Relation of lysosomal fragility in CLL lymphocytes to PHA reactivity. Nature, New Biol. 244, 247–248.
Koo, G.C., Stackpole, C.W., Boyse, E.A., Hämmerling, V. and Lardis, M.P. (1973) Topographical location of H-Y antigen on mouse spermatozoa by immunoelectronhistochemistry. Proc. Nat. Acad. Sci. U.S.A. 70, 1502–1505.
Krzanowska, H. (1972) Rapidity of removal in vitro of the cumulus oophorus and the zona pellucida in different strains of mice. J. Reprod. Fertil., 31, 7–14.
Lau, L.C.T. (1975) Production of globules in mouse L cells penetrated with hamster sperms. Science, 190, 684–686.
Lewis, B.K. and Ketchel, M.M. (1972) Effects of female reproductive tract secretions on rabbit sperm. I. Release of hyaluronidase in vitro. Proc. Soc. Exp. Biol. Med. 141, 712–718.
Lindahl, P.E. and Kihlstrom, J.E. (1952) Alterations in specific gravity during ripening of bull spermatozoa. J. Dairy Sci. 35, 393–402.
Loewenstein, J.E. and Cohen, A.I. (1964) Dry mass, lipid content and protein content of the intact and zona-free mouse zona. J. Embryol. Exp. Morphol. 12, 113–121.
Longo, F.J. (1973) Fertilization: A comparative ultrastructural review. Biol. Reprod. 9, 149–215.
Longo, F.J., Schuel, H. and Wilson, W.L. (1975) Mechanism of soybean trypsin inhibitor induced polyspermy as determined by an analysis of refertilized sea urchin (*Arbacia punctulata*) eggs. Dev. Biol. – in press.
Lord Rothschild (1956a) Fertilization. Methuen & Co., Ltd., London.
Lord Rothschild (1956b) Unorthodox methods of sperm transfer. Sci. Amer. 195, 121–132.
Mahajan, S.C. and Menge, A.C. (1966) Factors influencing the disposal of sperm and the leukocytic response in the rabbit uterus. Int. J. Fertil. 11, 373–380.
Mahi, C.A. and Yanagimachi, R. (1973) The effects of temperature, osmolarity and hydrogen ion concentration on the activation and acrosome reaction of golden hamster spermatozoa. J. Reprod. Fertil. 35, 55–56.
Mazia, D. (1937) The release of calcium in *Arbacia* eggs upon fertilization. J. Cell. Comp. Physiol. 10, 291–304.
McReynolds, H.D. and Hadek, R. (1971) A study on sperm tail elements in mouse blastocysts. J. Reprod. Fertil. 24, 291–294.
McRorie, R.A. and Williams, W.L. (1974) Biochemistry of mammalian fertilization. Ann. Rev. Biochem. 43, 777–803.
Metz, C.B. (1972) Effect of antibodies on gametes and fertilization. Biol. Reprod. 6, 358–383.
Metz, C.B. (1973) Role of specific sperm antigens in fertilization. Fed. Proc. 32, 2057–2064.

Metz, C.B. and Monroy, A. (1969) Eds. Fertilization, Comparative Morphology, Biochemistry and Immunology. Academic Press, New York.
Millette, C.F., Spear, P.G., Gall, W.E. and Edelman, G.M. (1973) Chemical dissection of mammalian spermatozoa. J. Cell Biol. 58, 662–675.
Mintz, B. and Gearhart, J.D. (1973) Subnormal zona pellucida changes in parthenogenetic mouse embryos. Dev. Biol. 31, 178–184.
Miyamoto, H. and Chang, M.C. (1973a) The importance of serum albumin and metabolic intermediates for capacitation of spermatozoa and fertilization of mouse eggs in vitro. J. Reprod. Fertil. 32, 193–198.
Miyamoto, H. and Chang, M.C. (1973b) Effects of protease inhibitors on the fertilizing capacity of hamster spermatozoa. Biol. Reprod. 9, 533–537.
Miyamoto, H., Toyoda, Y. and Chang, M.C. (1974) Effect of hydrogen ion concentration on in vitro fertilization of mouse, golden hamster and rat eggs. Biol. Reprod. 10, 487–493.
Monroy, A. (1965) Chemistry and Physiology of Fertilization. Holt, Rhinehart & Winston, New York.
Monroy, A., Ortolani, G., O'Dell, D. and Millonig, G. (1973) Binding of Concanavalin A to the surface of unfertilized and fertilized Ascidian eggs. Nature, 42, 409–410.
Moore, R.M. and Cragle, R.G. (1971) The sheep egg: enzymatic removal of the zona pellucida and culture of eggs in vitro. J. Reprod. Fertil. 27, 401–409.
Morton, D.B. and Bavister, B.D. (1974) Fractionation of hamster serum-capacitated components from human serum by gel filtration. J. Reprod. Fertil. 40, 491–493.
Moyer, D.L., Legorreta, G., Maruta, H. and Henderson, V. (1967) Elimination of homologous spermatozoa in the female genital tract of the rabbit: a light and electron-microscope study. J. Path. Bact. 94, 345–350.
Moyer, D.L., Rimdusit, S. and Mishell, D.R., Jr. (1970) Sperm distribution and degradation in the human female reproductive tract. Obstet. Gynecol., 35, 831–840.
Multamaki, S. and Niemi, M. (1969) Zona pellucida dissolving protease in an acrosomal preparation of bull spermatozoa. Scand. J. Clin. Lab. Invest. 23, 108–115.
Nakamura, M. and Yasumasu, I. (1974) Mechanism for increase in intracellular concentration of free calcium in fertilized sea urchin egg. A method for estimating intracellular concentration of free calcium. J. Gen. Physiol. 63, 374–388.
Nicolson, G.L. and Yanagimachi, R. (1972) Terminal saccharides on sperm plasma membranes: Identification by specific agglutinins. Science, 177, 276–279.
Nicolson, G.L. and Yanagimachi, R. (1974) Mobility and restriction of mobility of plasma membrane lectin-binding components. Science, 184, 1294–1296.
Nicolson, G.L., Yanagimachi, R. and Yanagimachi, H. (1975) Ultrastructural localization of lectin-binding sites on the zona pellucida and plasma membranes of mammalian eggs. J. Cell. Biol. 66, 263–274.
Nordmann, J.J. and Currell, G.A. (1975) The mechanism of calcium ionophore-induced secretion from the rat neurohypophysis. Nature, 253, 646–647.
Noyes, R.W., Walton, A. and Adams, C.E. (1958) Capacitation of rabbit spermatozoa. J. Endocrinol. 7, 374–380.
Oikawa, T., Nicolson, G.L. and Yanagimachi, R. (1974) Inhibition of hamster fertilization by phytagglutinins. Exp. Cell Res. 83, 239–246.
Oikawa, T., Nicolson, G.L. and Yanagimachi, R. (1975) Trypsin-mediated modifications of the zona pellucida glycopeptide structure of hamster eggs. J. Reprod. Fertil. 43, 133–136.
Oikawa, T., Yanagimachi, R. and Nicolson, G.L. (1973) Wheat germ agglutinin blocks mammalian fertilization. Nature, 241, 256–259.
Oliphant, G. and Brackett, B.G. (1973a) Immunological assessment of surface changes of rabbit sperm undergoing capacitation. Biol. Reprod. 9, 404–414.
Oliphant, G. and Brackett, B.G. (1973b) Capacitation of mouse spermatozoa in media with elevated ionic strength and reversible decapacitation with epididymal extracts. Fertil. Steril. 24, 948–955.
O'Rand, M.G. (1972) In vitro fertilization and capacitation-like interaction in the hydroid *Campanularia flexuosa*. J. Exp. Zool. 182, 299–305.
Ortavant, R. (1953) Existence d'une phase critique dans la maturation épididymaire des spermatozoides de bélier et de taureau. Comp. R. Séanc. Soc. Biol. 147, 1552–1556.

Orgebin-Crist, M.C. (1967) Maturation of spermatozoa in the rabbit epididymis. Fertilizing ability and embryonic mortality in does inseminated with epididymal spermatozoa. Ann. Biol. Anim. Biochem. Biophys. 7, 373–389.

Orgebin-Crist, M.C. (1969) Studies on the function of the epididymis. Biol. Reprod., Suppl. 1, 155–175.

Overstreet, J.W. and Bedford, J.M. (1974) Comparison of the permeability of the egg vestments in follicular oocytes, unfertilized and fertilized ova of the rabbit. Dev. Biol. 41, 185–192.

Papahadjopoulos, D. and Poste, G. (1975) Calcium-induced phase separation and fusion in phospholipid membranes. Biophys. J. 15, 945–948.

Papahadjopoulos, D., Poste, G., Schaeffer, B.E. and Vail, W.J. (1974) Membrane fusion and molecular segregation in phospholipid vesicles. Biochim. Biophys. Acta 352, 10–28.

Papahadjopoulos, D., Vail, W.J., Jacobson, K. and Poste, G. (1975) Cochleate lipid cylinders produced by fusion of unilamellar lipid vesicles. Biochim. Biophys. Acta 394, 483–491.

Pethica, B.A. (1961) The physical chemistry of cell division. Exp. Cell. Res. Suppl. 8, 123–140.

Pienkowski, M. (1974) Study of the growth regulation of preimplantation mouse embryos using Concanavalin A. Proc. Soc. Exp. Biol. Med. 145, 464–469.

Pikó, L. (1961) Repeated fertilization of fertilized rat eggs after treatment with versene (EDTA). Am. Zoologist, 1, 467–468.

Pikó, L. (1969) Gamete structure and sperm entry in mammals. In: Fertilization, (Metz, C. and Monroy, A., eds.) Vol. 2, pp. 325–403, Academic Press, New York.

Pinsker, M.C. and Williams, W.L. (1967) Properties of a spermatozoa anti-fertility factor. Arch. Biochem. Biophys. 122, 111–117.

Poste, G. (1970) Virus-induced polykaryocytosis and the mechanism of cell fusion. Adv. Virus Res. 16, 303–356.

Poste, G. (1972) Mechanisms of virus-induced cell fusion. Int. Rev. Cytol. 33, 157–252.

Poste, G. and Allison, A.C. (1973) Membrane fusion. Biochim. Biophys. Acta 300, 421–465.

Quinn, P.J. and White, I.G. (1967) The phospholipid and cholesterol content of epididymal and ejaculated ram spermatozoa and seminal plasma in relation to cold shock. Austr. J. Biol. Sci. 20, 1205–1215.

Rogers, B.J. and Morton, B.E. (1973) The release of hyaluronidase from capacitating hamster spermatozoa. J. Reprod. Fertil. 35, 477–487.

Rogers, J. and Yanagimachi, R. (1975) Release of hyaluronidase from guinea-pig spermatozoa through an acrosome reaction initiated by calcium. J. Reprod. Fertil. 44, 135–138.

Rosado, A., Velazquez, A. and Lara-Ricalde, R. (1973) Cell polarography. II. Effect of neuraminidase and follicular fluid upon the surface characteristics of human spermatozoa. Fertil. Steril. 24, 349–354.

Rubin, R.P. (1970) The role of calcium in the release of neurotransmitter substances and hormones. Pharmacol. Rev. 22, 389–418.

Satir, B. (1974) Membrane events during the secretory process. In: Transport at the Cellular Level (Sleigh, M.A. and Jennings, D.H., eds.) pp. 399–418, Cambridge University Press, Cambridge.

Sawicki, W. and Koprowski, H. (1971) Fusion of rabbit spermatozoa with somatic cells cultivated in vitro. Exp. Cell Res. 66, 145–151.

Seeman, P. (1972) The membrane actions of anesthetics and tranquilizers. Pharmacol. Rev. 24, 583–655.

Schuel, H., Kelly, J.W., Berger, E.R. and Wilson, W.L. (1974) Sulfated acid mucopolysaccharides in the cortical granules of eggs. Exp. Cell Res. 88, 24–30.

Schuel, H., Wilson, W.L., Chen, K. and Lorand, L. (1973) A trypsin-like proteinase localized in cortical granules isolated from unfertilized sea urchin eggs by zonal centrifugation. Role of the enzyme in fertilization. Dev. Biol. 34, 175–186.

Scott, T.W., Voglmayr, J.K. and Setchell, B.P. (1967) Lipid composition and metabolism in testicular and ejaculated ram spermatozoa. Biochem. J. 102, 456–461.

Sellens, M.H. and Jenkison, E.J. (1975) Permeability of the mouse zona pellucida to immunoglobulin. J. Reprod. Fertil. 42, 153–157.

Shivers, C.A. (1974) Immunological interference with fertilization. In: Immunological Approaches to Fertility Control (Diczfalusy, E., ed.) pp. 59–74. Karolinska Institutet, Stockholm.

Shivers, C.A. and Dudkiewicz, A.B. (1974) Inhibition of fertilization with specific antibodies. In: Physiology and Genetics of Reproduction (Coutinho, E.M. and Fuchs, F., eds.) Part B, pp. 81–96, Plenum Press, New York.

Singer, S.J. and Nicolson, G.L. (1972) The fluid mosaic model of the structure of cell membranes. Science, 175, 720–731.

Smithberg, M. (1953) The effect of different proteolytic enzymes on the zona pellucida of mouse ova. Anat. Rec. 117, 554.

Soupart, P. (1967) Studies on the hormonal control of rabbit sperm capacitation. J. Reprod. Fertil. Suppl. 2, 49–64.

Soupart, P. and Clewe, T.H. (1965) Sperm penetration of rabbit zona pellucida inhibited by treatment of ova with neuraminidase. Fertil. Steril. 16, 677–689.

Srivastava, P.N., Adams, C.E. and Hartree, E.F. (1965) Enzymic action of acrosomal preparation on the rabbit ovum in vitro. J. Reprod. Fertil. 10, 61–67.

Srivastava, P.N., Munnell, J.F., Yang, C.H. and Foley, C.W. (1974) Sequential release of acrosomal membrane and acrosomal enzymes of ram spermatozoa. J. Reprod. Fertil. 36, 363–368.

Stambaugh, R., Brackett, B.G. and Mastroianni, L. (1969) Inhibition of in vitro fertilization of rabbit ova by trypsin inhibitors. Biol. Reprod. 1, 223–227.

Stambaugh, R. and Buckley, J. (1969) Identification and subcellular localization of the enzymes effecting penetration of the zona pellucida by rabbit spermatozoa. J. Reprod. Fertil. 19, 423–432.

Stambaugh, R. and Buckley, J. (1972) Histochemical subcellular localization of the acrosomal proteinase effecting dissolution of the zona pellucida using fluorescein-labeled inhibitors. Fertil. Steril. 23, 348–352.

Stefanini, M., Oura, C. and Zamboni, L. (1969) Ultrastructure of fertilization in the mouse. II. Penetration of sperm into the ovum. J. Submicrosc. Cytol. 1, 1–23.

Stegner, H.E. and Wartenberg, H. (1961) Elektronenmikroskopische und histochemische Untersuchungen über Struktur und Bildung der Zona pellucida menschlicher Eizellen. Z. Zellforsch. 53, 702–713.

Steinhardt, R.A. and Epel, D. (1974) Activation of sea urchin eggs by a calcium ionophore. Proc. Nat. Acad. Sci. U.S.A. 71, 1915–1919.

Steinhardt, R.A., Epel, D., Carroll, E.J. and Yanagimachi, R. (1974) Is calcium ionophore a universal activator for unfertilized eggs. Nature, 252, 41–43.

Summers, R.G., Hylander, B.L., Colwin, L.H. and Colwin, A.L. (1975) The functional anatomy of the echinoderm spermatozoon and its interaction with the egg at fertilization. Amer. Zool. 15, 523–551.

Suominen, J., Kaufman, M.H. and Setchell, B.P. (1973) Prevention of fertilization in vitro by an acrosome inhibitor from rete testis fluid of the ram. J. Reprod. Fertil. 34, 385–394.

Szollosi, D. (1966) Time and duration of DNA synthesis in rabbit eggs after sperm penetration. Anat. Rec. 154, 209–212.

Szollosi, D. (1967) Development of cortical granules and the cortical reaction in rat and hamster eggs. Anat. Rec. 159, 431–446.

Szollosi, D.G. and Ris, H. (1961) Observations on sperm penetration in the rat. J. Biophys. Biochem. Cytol. 10, 275–283.

Tachi, S. and Kraicer, P.F. (1967) Studies on the mechanism of nidation. XXVII. Sperm-derived inclusions in the rat blastocyst. J. Reprod. Fertil. 14, 401–405.

Talbot, P. and Franklin, L.E. (1974) The release of hyaluronidase from guinea pig spermatozoa during the course of the normal acrosome reaction in vitro. J. Reprod. Fertil. 39, 429–432.

Tegner, M.J. and Epel, D. (1972) Sea urchin sperm-egg interactions studied with the scanning electron microscope. Science, 179, 685–688.

Teichmann, R.J. and Bernstein, M.H. (1969) Regional differentiation in the head of human and rabbit spermatozoa. Anat. Rec. 163, 343A.

Terner, C., Maclaughlin, J. and Smith, B.R. (1975) Changes in lipase and phosphatase activities of rat spermatozoa in transit from the caput to the cauda epididymis. J. Reprod. Fertil. 45, 1–8.

Thibault, C. (1973) In vitro maturation and fertilization of rabbit and cattle oocytes. In: The Regulation of Mammalian Reproduction (Segal, S.J., Crozier, R., Corfman, P.A. and Condliffe, P.G., eds.) pp. 231–240, C.C. Thomas, Springfield, Illinois.

Thompson, R.S. and Zamboni, L. (1974) Phagocytosis of supernumery spermatozoa by two-cell mouse embryos. Anat. Rec. 178, 3–14.

Thompson, R.S., Moore Smith, D. and Zamboni, L. (1974) Fertilization of mouse ova in vitro: An electron microscopic study. Fertil. Steril. 25, 222–249.

Thorn, N.A. and Petersen, O.H. (1975) Eds. Secretory Mechanisms of Exocrine Glands. Academic Press, New York.

Tilney, L.G., Hatano, S., Ishikawa, H. and Mooseker, M.S. (1973) The polymerization of actin: its role in the generation of the acrosomal process of certain echinodermal sperm. J. Cell Biol. 59, 109–126.

Toyoda, Y., Yokoyama, M. and Hosi, T. (1971) Studies on the fertilization of mouse eggs in vitro. Jap. J. Reprod. 16, 147–151 and 152–157.

Trounson, A.O. and Moore, N.W. (1974) The survival and development of sheep eggs following complete or partial removal of the zona pellucida. J. Reprod. Fertil. 41, 97–105.

Vacquier, V.D. (1975a) Calcium activation of esteroproteolytic activity obtained from sea urchin egg cortical granules. Exp. Cell. Res. 90, 454–456.

Vacquier, V.D. (1975b) The isolation of intact cortical granules from sea urchin eggs: Calcium ions trigger granule discharge. Dev. Biol. 43, 62–74.

Vacquier, V.D., Epel, D. and Douglas, L.A. (1972a) Sea urchin eggs release protease activity at fertilization. Nature, 237, 34–36.

Vacquier, V.D., Tegner, M.J. and Epel, D. (1972b) Protease activity establishes the block against polyspermy in sea urchin eggs. Nature, 240, 352–353.

Vaidya, R.A., Bedford, J.M., Glass, R.H. and Morris, J. (1969) Evaluation of the removal of tetracycline fluorescence from spermatozoa as a test for capacitation in the rabbit. J. Reprod. Fertil. 19, 483–489.

Vaidya, R.A., Glass, R.H., Dandekar, P. and Johnson, K. (1971) Decrease in the electrophoretic mobility of rabbit spermatozoa following intrauterine incubation. J. Reprod. Fertil. 24, 299–301.

Von der Borch, S.M. (1967) Abnormal fertilization of rat eggs after injection of substances in the ampullae of the fallopian tubes. J. Reprod. Fertil. 14, 465–468.

Weil, A.J. and Rodenburg, J.M. (1962) The seminal vesicle as the source of the spermatozoa-coating antigen of seminal plasma. Proc. Soc. Exp. Biol. Med. 109, 567–570.

Weinman, D.E. and Williams, W.L. (1964) Mechanism of capacitation of rabbit spermatozoa. Nature, 203, 423–424.

White, I.G. (1973) Metabolism of spermatozoa with particular relation to the epididymis. In: Schering Workshop on Contraception: The Masculine Gender (Raspe, G., ed.) Advances in the Biosciences, Vol. 10, pp. 157–168. Pergamon Press, Viewig.

Williams, J.A. and Chandler, D. (1975) Ca^{2+} and pancreatic amylase release. Fed. Proc. 228, 1729–1732.

Williams, W.L., Abney, T.O., Chernoff, H.H., Dukelow, W.R. and Pinsker, M.C. (1967) Biochemistry and physiology of decapacitation factor. J. Reprod. Fertil. Suppl. 2, 11–21.

Wyrick, R.E., Nishihara, T. and Hedrick, J.L. (1974) Agglutination of jelly coat and cortical granule components and the block to polyspermy in the amphibian *Xenopus laevis.* Proc. Nat. Acad. Sci. U.S.A. 71, 2067–2071.

Yanagimachi, R. (1969a) In vitro capacitation of hamster spermatozoa by follicular fluid. J. Reprod. Fertil. 18, 275–286.

Yanagimachi, R. (1969b) In vitro acrosome reaction and capacitation of golden hamster spermatozoa by bovine follicular fluid and its fractions. J. Exp. Zool. 170, 269–280.

Yanagimachi, R. (1970a) In vitro capacitation of golden hamster spermatozoa by homologous and heterologous blood sera. Biol. Reprod. 3, 147–153.

Yanagimachi, R. (1970b) The movement of golden hamster spermatozoa before and after capacitation. J. Reprod. Fertil. 23, 193–196.

Yanagimachi, R. (1972a) Penetration of guinea pig spermatozoa into hamster eggs in vitro. J. Reprod. Fertil. 28, 477–480.

Yanagimachi, R. (1972b) Fertilization of guinea pig eggs in vitro. Anat. Rec. 174, 9–20.

Yanagimachi, R. and Chang, M.C. (1963) Infiltration of leucocytes into the uterine lumen of the

golden hamster during the oestrus cycle and following mating. J. Reprod. Fertil. 5, 389–396.
Yanagimachi, R. and Chang, M.C. (1964) In vitro fertilization of golden hamster ova. J. Exp. Zool. 156, 361–376.
Yanagimachi, R. and Noda, Y.D. (1970a) Electron microscopic studies of sperm incorporation into the golden hamster egg. Amer. J. Anat. 128, 429–462.
Yanagimachi, R. and Noda, Y.D. (1970b) Ultrastructural changes in the hamster sperm head during fertilization. J. Ultrastruct. Res. 31, 465–485.
Yanagimachi, R. and Noda, Y.D. (1970c) Physiological changes in the postnuclear cap region of the mammalian sperm: A necessary preliminary to the membrane fusion between sperm and egg cells. J. Ultrastruct. Res. 31, 486–493.
Yanagimachi, R. and Noda, Y.D. (1970d) Fine structure of the hamster sperm head. Am. J. Anat. 128, 367–388.
Yanagimachi, R. and Noda, Y.D. (1972) Scanning electron microscopy of golden hamster spermatozoa before and during fertilization. Experientia, 28, 69–72.
Yanagimachi, R., Nicolson, G.L., Noda, Y.D. and Fujimoto, M. (1973) Electron microscopic observations of the distribution of acidic anionic residues on hamster spermatozoa and eggs before and during fertilization. J. Ultrastruct. Res. 43, 344–353.
Yanagimachi, R. and Usui, N. (1972) The appearance and disappearance of factors involved in sperm chromatin decondensation in the hamster egg. J. Cell. Biol. 55, 293a.
Yanagimachi, R. and Usui, N. (1974) Calcium dependence of the acrosome reaction and activation of guinea pig spermatozoa. Exp. Cell Res. 89, 161–174.
Zamboni, L. (1970) Ultrastructure of mammalian oocytes and ova. Biol. Reprod. Suppl. 2, 44–63.
Zamboni, L. (1971a) Fine Morphology of Mammalian Fertilization, Harper and Row, New York.
Zamboni, L. (1971b) Acrosome loss in fertilizing mammalian spermatozoa: A clarification. J. Ultrastruct. Res. 34, 401–405.
Zamboni, L. (1972) Fertilization in the mouse. In: Biology of Mammalian Fertilization and Implantation (Moghissi, K.S. and Hafez, E.S.E., eds.) pp. 213–262, C.C. Thomas, Springfield, Illinois.
Zamboni, L., Moore-Smith, D. and Thompson, R.S. (1972) Migration of follicle cells through the zona pellucida and their sequestration by human oocytes in vitro. J. Exp. Zool. 181, 319–340.
Zamboni, L., Zemjanis, R. and Stefanini, M. (1971) The fine structure of monkey and human spermatozoa. Anat. Rec. 169, 129–153.
Zaneveld, L.J.D., Robertson, R.T., Kessler, M. and Williams, W.L. (1971) Inhibition of fertilization in vivo by pancreatic and seminal plasma trypsin inhibitors. J. Reprod. Fertil. 25, 387–391.
Zelenin, A.V., Shapiro, I.M., Kolesnikov, V.A. and Senin, V.M. (1974) Physico-chemical properties of chromatin of mouse sperm nuclei in heterokaryons with Chinese hamster cells. Cell Differ. 3, 95–101.

Cytokinesis in animal cells: new answers to old questions 2

John M. ARNOLD

1. Introduction

In the history of microscopic biology, possibly no other topic has been the subject of more curiosity, investigation, speculation and misinterpretation than that of cell division. The developing egg provides a particularly convenient system for the experimental study of cell division and much of our present insight into this process stems from observations on the initial series of cleavage divisions by which the zygote is converted into the population of cells that constitute the early stage embryo. The task of this paper is to review recent work in this area and to attempt to correlate some of the emerging data into an integrated scheme of cellular dynamics. Because the literature is vast, complex and sometimes contradictory it will be necessary to limit the scope of this article. Discussion has therefore been restricted to consideration of metazoan animals and, in particular, the mechanism of cytoplasmic division in invertebrate eggs. This is not to deny that much elegant research has been done on plants (review, Pickett–Heaps, 1972) and on the mechanism of chromosomal condensation, alignment and movement (review, Nicklas, 1971) but rather to concentrate the scope of the present article to make it manageable and, hopefully, comprehensible.

Classically, the topic of cell division has involved consideration of both mitosis (nuclear division involving chromosomal condensation and migration on the spindles) and cytokinesis (division of the extranuclear components of the cell). The former has received much more attention, possibly because of the complexity of mitosis and perhaps also because of its intrinsic importance in properly segregating the genetic material. However, cytokinesis is of indispensable importance in metazoan animals because without it differentiation at both the tissue and cellular level would be impossible, to say nothing of the maintenance of critical surface to volume ratios. There have been literally hundreds, if not thousands of papers dealing with various aspects of cytokinesis and I have purposely chosen to limit the area covered by this article. The topic of cytokinesis has been reviewed in detail by Rappaport (1971) and I will thus concentrate on more recent work and only bring in earlier work where necessary for clarification, background or, on occasion, for the sake of completeness. Hopefully, this

G. Poste & G.L. Nicolson (eds.) The Cell Surface in Animal Embryogenesis and Development

approach will avoid redundancy and promote brevity, while avoiding some of the interesting, but now discredited, theories, hypotheses and extreme speculations concerning cytokinesis that abound in the literature. At various times, mechanisms of cytokinesis in animal cells have been proposed that include polar expansion, polar relaxation (and hence relative equilateral constriction), inward growth of the plasma membrane, fusion of aligned vesicles in the future cleavage plane (the undisputed case in the majority of plant cells), separation by cytoplasmic streaming or migration and, finally, by the currently accepted idea of active contraction of the furrow base itself. If the reader is interested in the debunking of many of these proposed models of cytokinesis (s)he is referred to Rappaport's comprehensive review (1971) and to his many elegant papers (see particularly, Rappaport, 1961, 1965, 1969, 1973a).

2. *General aspects of cell division in the developing egg*

The series of mitotic cell divisions, widely referred to as cleavage divisions that convert the zygote into a population of cells, are fundamentally similar to those mitotic divisions that take place later in various somatic cells during both embryogenesis and post-natal life. The initial cleavage divisions differ, however, in the relation between cell growth and cell division, the rate of cell division, and the size and proportion of the mitotic apparatus (see, Rappaport, 1971).

Although the initial series of cleavage divisions occur in rapid succession, the total volume of cytoplasm involved remains constant. The progeny cells thus become progressively smaller in size and the ratio of nuclear to cytoplasmic volume changes. For example, in sea urchin eggs the ratio of nuclear to cytoplasmic volume changes from 1/550 to 1/6 between fertilization and the blastula stage (Brachet, 1950). The progressive reduction in cell size continues until the cell size that is characteristic of the particular species is reached. Once this is achieved, all subsequent cell divisions are followed by a period of cell growth before division again takes place. Embryos developing from eggs with experimentally reduced cytoplasmic volumes also undergo cleavage divisions to produce cells with a volume that is normal for the species. Such embryos are therefore composed of fewer cells than normal (Kühn, 1971).

The amount and distribution of yolk within the egg, together with the extent of separation or mixing of yolk with active cytoplasm, appear to be important factors in determining the pattern and rate of cell division. In eggs where the yolk is scant or distributed homogeneously, as in sea urchin and mammalian eggs, the rate of cleavage division is nearly uniform throughout the egg. In contrast, in amphibian eggs where the yolk is concentrated in the vegetal region, the division rate of the vegetal region is slower than that of cells in the animal region. In even more extreme cases such as the eggs of birds and certain teleost fishes where the yolk and the active cytoplasm are discrete, the yolk does not divide, being assimilated later by the developing embryo.

Most experimental work on cell division in the developing egg has been done with the eggs of marine invertebrates, notably those of the Echinodermata. As in the case of fertilization discussed in the preceding chapter, the large size and durability of invertebrate eggs, together with their availability in large numbers, dictate that they are a considerably more convenient experimental system to work with than comparable cells from mammals.

From the standpoint of cytokinesis, there are a number of central questions that have attracted the attention of researchers. These fall into five interrelated areas: (1) how does mitosis relate to cytokinesis?; (2) how does the contractile ring function in cytokinesis?; (3) how is the contractile ring controlled?; (4) what is the origin of new surface membrane?; and (5) how are the new daughter cells separated? In the following sections of this article we will examine each of these questions, and offer some discussion and speculation on certain of the still outstanding problems in cytokinesis. Since research in these areas is currently quite active and the available information on certain questions is in a state of flux, the present discussion cannot provide definitive answers to all of the questions discussed. Rather this article represents the current status of a very fluid field.

3. Mitosis in relationship to cytokinesis: positioning of the cleavage furrow

The relationship of the mitotic apparatus to the future cytokinetic (cleavage) furrow has been known for a considerable time and has been formalized as a set of "rules of cleavage". Hofmeister and Sachs both set forth cleavage rules, but somehow these rules in a slightly different form are also known as "Balfour's rules" (cited in Wilson, 1925). When all the redundancy is eliminated, two basic rules remain. These state: (1) the plane of cleavage is perpendicular to the plane of the previous cleavage; and (more interestingly), (2) the plane of cleavage is perpendicular to the long axis of the spindle. Although this relationship has been known for nine decades, only recently has much been done to demonstrate the mechanism by which the mitotic apparatus determines the position of the furrow.

There have been numerous studies on the effect of the mitotic apparatus on the egg surface. For example, Hiramoto (1956) displaced the mitotic apparatus in sea urchin eggs at various times during mitosis and found that the furrow formed opposite the metaphase plate until anaphase, or in some cases late metaphase. Subsequently, the mitotic apparatus could be removed, displaced, or replaced with an injected oil droplet without affecting either the position of the furrow or its formation. This also implies that once the furrow is induced it continues at the surface despite rearrangements of the internal cytoplasm. Rappaport's elegant experiments provide the most complete analysis of the contribution of various cellular components to cytokinesis (see 1971 review). He demonstrated that the nucleus was not essential for cytokinesis (Rappaport, 1961) and that the internal cytoplasm could be actively stirred during furrowing without effecting cytokinesis once it has begun at the surface (Rappaport, 1965, 1969). If, however, the mitotic apparatus was displaced before or during the "positioning interaction" the newly formed furrow followed the position of the metaphase plate (Rappaport, 1961). He concluded that the stimulus of the surface by components of the mitotic apparatus was rapid and once established was permanent. The possibility of the furrow being determined by a polar stimulation or a lack of stimulation has been ruled out by Rappaport's detailed analysis (see especially Rappaport, 1965). For example, when precleavage eggs are stretched by weighing them with small glass beads so the poles are far removed from the asters, the furrows still appear opposite the metaphase plate. Furthermore the zone of astral confluence corresponds to the position of the future furrow. By pressing down on the center of sand dollar eggs, Rappaport was able to convert the egg into a torus shape. The first cleavage caused the ring-shaped egg to form a horseshoe configuration. When the second cleavage occurred, the innermost astral rays overlapped although *there was no metaphase plate between them.* In the region where these achromatic asters overlapped, the surface was nonetheless stimulated and underwent furrow formation so that at second cleavage four cells were formed rather than the three that would be expected if the furrowing stimulus had originated in the chromosomes, nucleus, or closely associated structures.

Having unquestionably demonstrated the importance of astral confluence, Rappaport then went on to study several important characteristics of the aster-surface interaction. By compressing or otherwise distorting sea urchin eggs before cleavage, Rappaport (1969) found the maximum spindle to surface distance that permitted furrow stimulation to be 62.5 μ for first cleavage. If the interastral distance was increased to more than 35 μ furrowing was infrequent. Observations on eggs in which various combinations of increased interastral distance with reduced aster-to-surface-distance were produced, showed that compensation for weakened astral confluence could be made. The rate of cleavage stimulation movement was studied by displacing the mitotic apparatus so that it occupied an asymmetrical position (Rappaport, 1973b). In this system it was possible to demonstrate that the stimulus moved at a rate of $6.3 \pm 1.8\ \mu$ per minute. This rate is slower than diffusion and cytoplasmic streaming but is approximately that of microtubule outgrowth.

It would appear from the above data that the stimulus for localization of the furrow emanates from the astral rays and not from the metaphase plate chromosomes, the spindle microtubules, or other components of the cytoplasm or mitotic apparatus. Although this would seem to be the case where asters were present, most cell division is anastral. Rappaport and Rappaport (1974) compared the division of cultured newt kidney cells with that of sea urchin eggs. The cultured kidney cells are much smaller than sea urchin eggs and are anastral. However, when the surface of such cells was pushed in to displace the metaphase chromosomes, the cell surface in contact with the spindle was transformed into a cytokinetic furrow. This suggests that in smaller anastral cells the furrow stimulus originates from the spindles, and that the shorter surface to spindle distance obviates the necessity of astral confluence. Possibly the discontinuous spindle microtubules influence the surface. Thus the ratio of cytoplasmic volume to mitotic apparatus size could determine whether the asters or the spindle function as the primary organelle in determining furrow position.

The biochemical nature of the furrow stimulus is elusive and poorly understood. Rappaport (personal communication) has repeatedly emphasized that it probably involves the outgrowth of microtubules to the egg surface or to the subsurface egg cortex. Weisenberg (1972) has demonstrated that the in vitro assembly of microtubules isolated from rat brain requires the proper (low) level of ionized calcium in the medium. If the intact egg is able to regionally reduce Ca^{2+} in the region of the forming aster, it might be possible that there would be a temporary absolute *or* relative increase of Ca^{2+} in the future furrow region thus eliciting a contraction of the subsurface microfilaments (see below). The growth of the asters might therefore "sweep" Ca^{2+} to the site of the future furrow.

The above discussion pertains primarily to spherical eggs or cells dividing without the restrictions of surrounding cells or tissues. Obviously, this represents the minority of cases and can serve only as a model to be modified to fit a variety of cases. The question then becomes how do asymmetrical divisions occur and how does furrowing proceed in cells where the mitotic apparatus is unequally positioned? Within the limits of the present article, it would be impractical to discuss all of the possible cases, so a few examples will be mentioned as illustrative examples. Czihak (1973) studied the formation of the vegetal micromeres in sea urchin embryos and hypothesized that the vegetal egg cortex had a role in determining asymmetrical cleavage. According to Czihak, the mitotic apparatus is originally placed in a central position but as the astral rays contact the vegetal cortex the microtubules disaggregate and the nucleus and spindles are then pushed by the unaffected animal pole asters to a subequatorial position. Subsequently, the stimulus of the overlapping asters is vegetally displaced and causes unequal cleavage. This occurs despite stratification of the cytoplasm by centrifugation, hence the supposed cortical influence. Removal of larger and larger parts of the vegetal cortex caused more and more symmetrical cleavages.

In the case where the nucleus is displaced to the animal pole by large concentrations of yolk, the furrow is first initiated as an animal pole line which grows or self-assembles at either end. For example, in the first cleavage in *Rana* embryos, Kubota (1969) experimentally displaced the subcortical cytoplasm in advance of the furrow and caused the furrow to deviate toward the new position of the displaced subcortical cytoplasm. This would indicate that there is an "organizational" area at the advancing tip of the furrow. In cephalopod eggs, cleavage occurs in a blastodisc of clear cytoplasm and initially begins between the two telophase nuclei above the former position of the metaphase plate. The ends of the furrow each serve as the site of furrow "assembly" and progress toward the edges of the blastodisc (Arnold, 1969). It has been possible to surgically demonstrate that the actual furrow assembly precedes the apparent end of the furrow by at least 100 μ (Arnold, 1971). By cutting the surface in advance of an established furrow, the furrow can be made to shorten, thus demonstrating that the actual contractile tension extends beyond the area of actual surface ingression. Scanning electron micrographs show the area beyond the furrow to be covered with oriented microvilli (Arnold, 1974).

In summary, one can only admire the insight of E. B. Wilson when he wrote in 1925. "It nevertheless remains possible that the astral rays, when present, may be concerned in causing an equatorial increase in surface tension." In present day sophistication, we can only admire his perception and wonder how so many others missed the point for so many decades.

4. *The contractile ring*

4.1. Formation of the microfilamentous band

The concept of cytokinesis by contraction is an old one. However, it is only relatively recently that sufficient data has accumulated to adequately support this concept and to effectively rule out other hypotheses. Marsland and Landau (1954) more or less formalized the contractile ring hypothesis with their studies on the effect of high hydrostatic pressure and centrifugation on cytokinesis in marine eggs. They postulated that a contractile gel formed in the plane of the cleavage furrow and that the gel had many characteristics similar to those found in muscle contraction. Landau et al. (1955) demonstrated the dependency of furrowing on ATP in *Arbacia* and *Chaetopterous* cleavage. Mercer and Wolpert (1958) observed a dense layer about 0.1 μ thick situated beneath the cleavage furrow of dividing sea urchin eggs but the electron microscope techniques of the day did not allow them to adequately resolve the fine structure of that subcortical band. In the following decade dense layers or bands below cleavage furrows were occasionally remarked upon (e.g., Weinstein and Herbert, 1964) but the details of the fine structure of the cleavage furrow attracted little interest until the late nineteen-sixties. In retrospect this is somewhat surprising since Hoffman–Berling (1954) had studied glycerol/water extracted cells and found the cleavage furrows to be contractile and had postulated that the furrowing process was basically similar to that of muscular contraction. During the early nineteen sixties Rappaport accumulated evidence for a contractile furrow (see above) but it wasn't until the later part of that decade that suddenly, several independent investigations were published on the ultrastructure of various cleavage furrows. Baker (1965) found a dense fibrillar layer in the apex of cells involved in gastrulation of amphibian embryos and postulated that this dense layer was contractile and Cloney (1966) demonstrated that fine cytoplasmic filaments were associated with contraction of the epithelial cells during ascidian metamorphosis. With this background, it seemed natural to do high resolution electron microscopy on the cytokinetic furrow. Between 1968 and 1970 a number of very similar papers emerged from several separate laboratories, all with basically the same finding, namely, that a band of electron dense filaments existed beneath the cleavage furrow and was closely associated with the plasma membrane. These filaments were found in eggs or cells from a wide range of animals (Schroeder, in sea urchins, 1968; Goodenough, et al., in sea urchins, 1968; Szollosi, in a coelenterate and polychaete, 1968a,b, 1970; Arnold, in cephalopods, 1968a,b, 1969; Tilney and Marsland, in sea urchin, 1969; Selman and Perry, in newt eggs, 1970; Bluemink, in *Ambystoma*, 1970; Schroeder, in HeLa cells, 1970a,b; Scott and Daniel in mouse cells, 1970; etc.). Schroeder coined the term microfilaments to apply to these 4–7 nm filaments and the analogy to a primitive muscle system suddenly seemed much more obvious.

The microfilaments described in these studies measured between 4 and 7 nm in diameter and were of an indeterminate length. Occasionally in profile the individual filaments appeared to have a "hollow" appearance or were somewhat less electron dense in the center. Frequently, the filaments displayed a beaded appearance with the "beads" being the approximated diameter of the filament. Collectively, the microfilaments formed a band or layer approximately 0.1–0.2 μ thick and approximately 5–6 μ wide in fully formed furrows, though in the early stage of furrow formation the bands may be wider and more diffuse. Individual microfilaments within the band are spaced at 10 to 15 nm apart, though no obvious pattern exists within the band. Cytoplasmic organelles are excluded from the band region and it is noticeably more dense than other regions of the egg cortex. The filaments have a strong tendency to follow a parallel orientation along the axis of the cytokinetic furrow. In symmetrical holoblastic cleavage, the microfilamentous band continues completely around the dividing cell (Schroeder, 1968, 1972). In meroblastic cleavage (Fig. 1) the furrows first appear above the position of the former metaphase plate and extend in either direction, thus the ends of the extending furrows are younger. Arnold (1969) showed that the furrow in the squid egg first covered a rather broad area (16.75 μ) and seemed to contract isometrically at first. This contraction "gathered" the surface into many microvilli which subsequently became organized into longitudinal folds which paralleled the axis of the furrow (see also, Arnold, 1974). The contractile stress along the furrow was considered to cause the orientation of the longitudinal folds. As furrowing advanced the band of microfilaments increased in density and thickness until it reached a minimum width of approximately 6 μ and a thickness of 0.1–0.2 μ with a uniform density. In embryos from other species the condensation of the microfilamentous band tends to "unfold" the surrounding microvillous surface and form a region of smooth membrane (Szollosi, 1970).

The origin of the microfilaments in the band is presently unresolved. Unless the microfilaments are organized and oriented into some aggregate structure it is unlikely they will be noticed. Szollosi (1970) has suggested that the microfilaments found in the surface microvilli are recruited to form the cytokinetic microfilamentous band. Schroeder (1972) could not identify a source of precursor microfilaments for the contractile ring and he argued for the in situ assembly

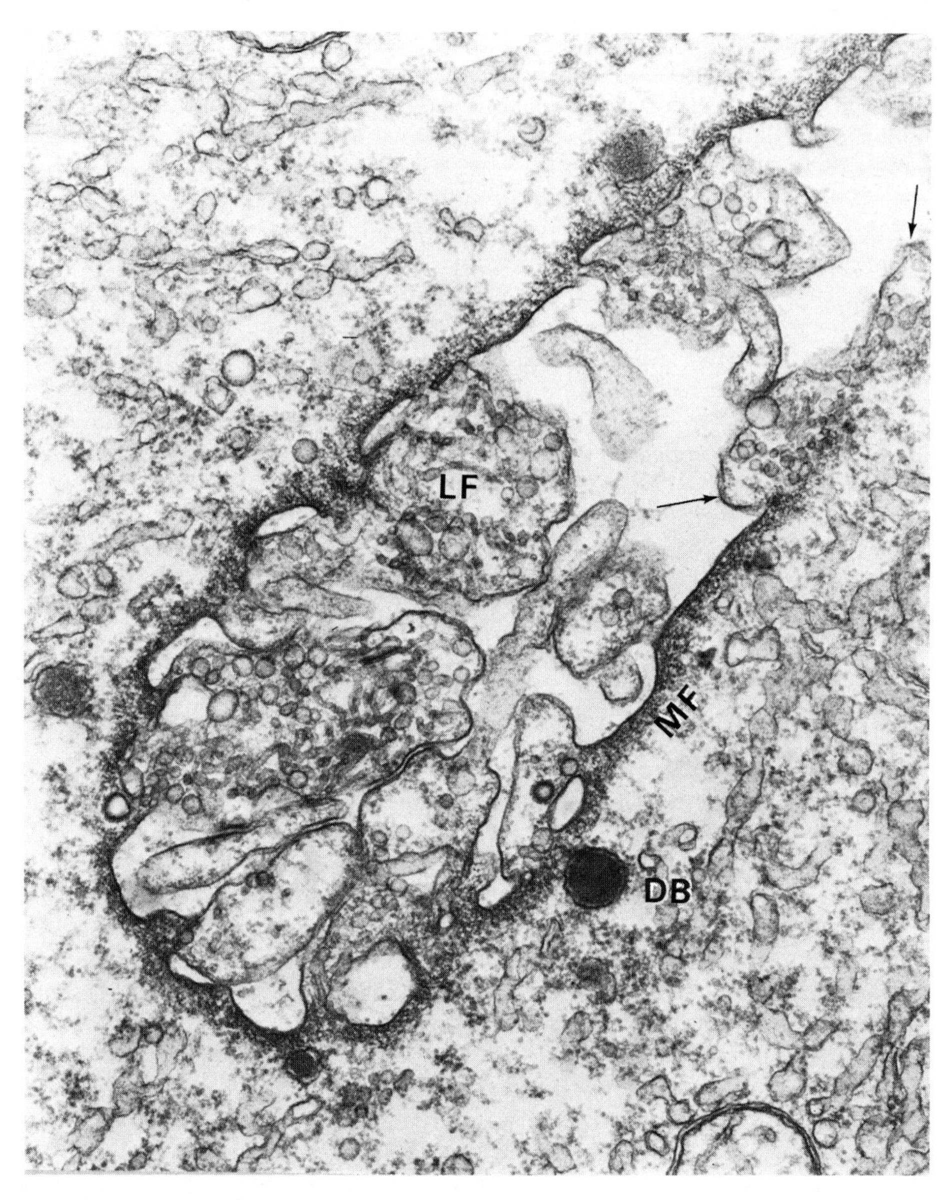

Fig. 1. Leading edge of the first cleavage furrow of *Loligo pealei.* The band of microfilaments (MF) is evident as a dark punctate layer beneath the plasma membrane. The longitudinal folds (LF) are seen in cross section and one appears to be "unfolding" (arrows). Dense bodies (DB) are evident beside the furrow. Approximately 35,500 X.

of the contractile ring in conjunction with the plasma membrane. Bluemink (1970) found randomly oriented filaments in the cortex of *Abystoma* but in the furrow region these filaments were organized into bundles, suggesting that a localized reorganization occurs at furrow initiation. This might account for the initial isometric contraction found in early furrowing which is then followed by anisometric contraction (e.g., Hiramoto, 1958; Scott, 1960). The suggestion of Arnold (1969) that the microfilaments might arise from membrane bounded vesicles produced by the Golgi apparatus is now more amusing than accurate, though such vesicles may well make a contribution to the "new surface" of the daughter cells (see below). It seems possible that the cell has a reserve of microfilaments diffusively distributed in the cortex or in other forms of aggregation elsewhere in the cell and these microfilaments or their precursors can be "recycled" into the furrow.

Relatively little work has been done on the intracellular origin and turnover of microfilaments. It is clear, however, from recent observations on the behavior of actin-containing microfilament bundles in cultured mammalian fibroblasts that they are dynamic structures whose organization can change rapidly in response to growth conditions (see, Goldman, 1975; Lazarides, 1975a). Further insight into the dynamic nature and interconversion of the various organizational states of actin-like microfilaments will probably contribute significantly not only to our understanding of the formation and functional properties of the cleavage furrow but also enlarge our appreciation of other cellular functions in which microfilaments appear to be important, such as cell locomotion and spreading, membrane ruffling endocytosis and the topographic arrangement of cell surface components.

When the microfilamentous band was found in the base on the cleavage furrow, it was immediately assumed that the filaments were the cause of contraction rather than the product of some yet unknown subsurface interaction. Because of the strength of the actomyosin analogy, it was assumed (with valid reason) that the microfilaments caused contraction. Rappaport's (1967) measurement of the contractile tension in the furrow (2.5×10^{-5} dynes/cm^2) of a sea urchin egg proved to be comparable to the force generated by an actomyosin thread of similar size ($0.1\ \mu \times 5\ \mu$). Yoneda and Dan (1972) estimated the equatorial constriction at 6×10^3 dynes in another sea urchin. Arnold (1971) microsurgically destroyed the furrow base and found that the cleavage furrows of squid eggs disappeared and the plasma membrane returned to the position of the uncleaved surface. Electron micrographs of the cut furrows revealed that the microfilamentous band was missing in the surgically ablated furrows. The same paper presented additional direct evidence for a contractile tension in the base of the cleavage furrow based on time-lapse cinematography, compression, and micromanipulation.

4.2. The actomyosin model: single use muscle

The hypothesis that cleavage is caused by contraction of a band of microfilaments attached to the plasma membrane is now fairly well accepted. However, the evidence for the mechanism of furrowing being a "primitive" actomyosin contraction deserves examination. Since the ultrastructural appearance of the cleavage furrow does not strongly resemble the classical picture of muscle, it is necessary to characterize the cleavage furrow by several criteria. First, what is the evidence for actin and/or myosin being associated with the cleavage furrow? Second, is the control mechanism for contraction analogous to muscle?

A definitive demonstration of the presence of actin in the microfilaments of the cleavage furrow has yet to be completed. Actin-like proteins have been isolated from early sea urchin embryos, constituting about 1% of the total soluble protein (Miki-Noumura, 1969; Miki-Noumura and Oosawa, 1969; Kane, 1975). This is obviously much more than the amount of actin needed to form a cleavage furrow. Perry et al. (1971) using the heavy meromyosin labeling technique of Ishikawa et al. (1969) "decorated" the microfilaments at the base of the cleavage furrow in an amphibian egg after it had been extracted in glycerine. Schroeder (1973) similarly labeled the contractile ring in dividing HeLa cells and found the microfilaments were decorated with the typical polarized, repeating arrowhead structures spaced at 27–35 nm (Fig. 2). He also observed thin unlabeled fibrils situated between the decorated microfilaments and speculated that these strands might possibly represent the myosin complement. Although reaction with heavy meromyosin cannot be considered a definitive identification of actin (see discussion by Pollard and Weihing, 1974) it can be regarded as a rather strong implication.

Sanger (1975) has used the fluorescent labelled heavy meromyosin staining technique of Aronson (1965) to follow the changing patterns of actin distribution during the division cycle of cultured chick fibroblasts. He found that during interphase when cells were flattened against the substrate, prominent long actin fibers were present which more or less paralleled the longitudinal axis of the cell. When the cell began to round up preparatory to karyokinesis and cytokinesis, the fluorescence became generally diffuse over the surface of the whole cell. The small spikes which attached the rounded cell to the surface remained quite fluorescent. In telophase, when the cells were beginning to separate and their furrow bases could be easily observed, intense fluorescence was detectable at the cleavage furrow region, but the mid-body was only slightly fluorescent. As cytokinesis was completed the fluorescence shifted from the furrow region to the distal poles of the daughter cells. Pseudopodia appeared at the distal poles at this time and Sanger (1975) suggests that the traction exerted by these pseudopodia enables the two sister cells to migrate apart.

Schroeder (1972) made careful volumetric measurements of the contractile ring in cleaving *Arbacia* eggs. The contractile ring in these eggs is a transitory structure, existing for only 6 and 7 minutes at 20°C. As contraction proceeds the contractile ring decreases in volume. Hence the strictest interpretation of a simple sliding of the filaments past each other cannot be immediately accepted as providing the contractile force. Instead, the interaction is more complex and may well involve the disassembly of some of the components of the contractile ring. One possibility is that actin-containing filaments from the contractile ring are recruited for utilization elsewhere in the cell, possibly, as is the case in fibroblasts, at the soon-to-migrate distal pole.

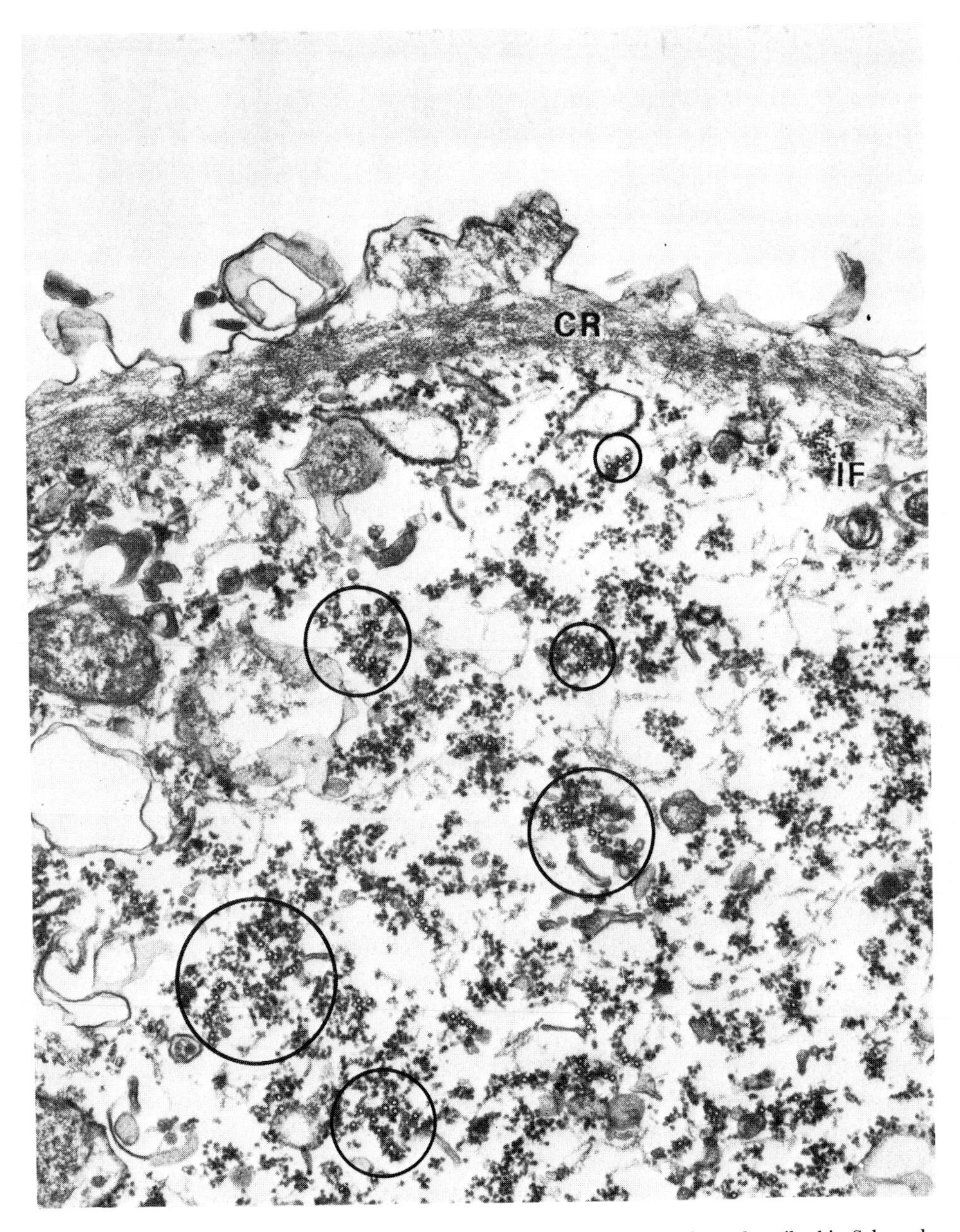

Fig. 2. Contractile ring microfilaments decorated with heavy mermoyosin as described in Schroeder (1973). Note the characteristic arrowhead appearance of the heavy meromyosin in this equatorial section of a HeLa cell in mid-cleavage. CR = contractile ring; circles indicate microtubules in the stem bodies; IF = intermediate filaments (100 Å). Approximately 14,900 X (courtesy of Dr. Thomas Schroeder).

It would thus appear that the cell has a reserve of actin-like protein which it can mobilize and utilize at various locales in the cell for various functions. Although the in situ demonstration of actin is not completely perfect because it has not yet been biochemically characterized, there seems to be no contrary evidence to argue in favor of any other contractile mechanism.

The question of myosin is not as easily approached and little if anything is known about the myosin component of the cleavage furrow (see, Pollard and Weihing, 1974). It is generally accepted that actin and/or myosin cannot generate movement by themselves hence the necessity of localizing them together in the furrow.

The identification of myosin is dependent largely on its ability to interact with actin. Two criteria must be met by any material suspected of being myosin: (1) it must be able to reversibly interact with actin; and (2) it must have ATPase activity in the presence of actin. Various myosins have been isolated from a range of non-muscle cells including brain, fibroblasts, platelets, and granulocytes, as well as *Physarum* and *Acanthamoeba*. The general conclusion is that myosin and myosin-like molecules from non-muscle cells are extremely variable in shape and enzymatic properties, though all are large asymmetrical molecules (Pollard and Weihing, 1974). Little if anything is known about the intracellular distribution of myosin to date. In some Sarcodina, thick filaments are seen with the electron microscope in association with actin filaments (the latter identified by heavy meromyosin binding) (Nachmias, 1968; Pollard and Ito, 1970), but comparable structures have not been seen in the cleavage furrow yet. As noted above, Schroeder (1972) did see thin non-heavy meromyosin decorated filaments in the cleavage furrow and suggested they might be myosin, but direct proof is still lacking.

A myosin-like protein has been isolated from the cortical layer of sea urchin eggs by Mabuchi (1973). When added to rabbit skeletal actin, this material produced an increase in viscosity as well as a significant ATP response. This myosin-like protein, or ovomyosin, as Mabuchi named it, has a molecular weight of approximately 200,000 and forms bipolar aggregations of low ionic strength as does muscle myosin. Although Mabuchi did not attempt to localize ovomyosin in the cleavage furrow, it did appear to be in the cortical layer. More recently, immunofluorescence studies using antibodies against native chicken gizzard myosin have shown that a myosin-like protein is associated with actin-containing microfilament bundles in various mammalian fibroblasts cultured in vitro (Weber and Groeschel-Stewart, 1974; Pollack et al., 1975).

The possibility exists that the myosin component might be incorporated into the plasma membrane itself. Gwynn et al. (1974) and Wallingham et al. (1974) have presented immunological evidence for the presence of myosin or myosin-like material at the surface of avian and mammalian fibroblasts cultured in vitro. The idea of myosin being a plasma membrane component is attractive because it could provide a means of anchoring actin-containing microfilaments to the plasma membrane. Such a linkage is necessary if the plasma membrane is to follow the ingression of the contractile ring.

Some evidence of connections between microfilaments and the plasma membrane in non-dividing cells has been obtained in ultrastructural studies. Perdue (1973) found actin-like microfilaments apparently inserting into the plasma membrane of cultured chick fibroblast and making contact with the external substrate. Yamada et al. (1971) have shown microfilaments apparently attached to the ends of microspikes growing out of cultured nerve cells. Less direct, but compelling, evidence for an association between microfilaments and plasma membrane components has also been provided by recent work on factors controlling the translational mobility of various integral proteins within the plasma membrane of mammalian cells cultured in vitro. There is now a large body of evidence which suggests that certain integral plasma membrane proteins are "linked" to microfilament systems and that this linkage is a major factor in controlling the lateral mobility of such proteins within the membrane (see, Edelman, 1974; Nicolson, 1975).

Comparable evidence for a direct interaction of membrane-associated microfilaments with plasma membrane components in the cleavage furrow has yet to be obtained. It seems likely, however, that such "linkages" do exist and this point represents an obvious area for further study.

5. *Control of contraction in the cleavage furrow*

If the actomyosin model of cytokinesis is valid, it is logical to look for regulatory factors in cleavage similar to those found in muscle contraction. In many types of muscle, there are regulatory proteins, tropomyosin and troponin, which act with myosin to split ATP. These, in turn, seem to be under the control of ionized calcium.

Although tropomyosin-like components have been isolated from non-muscle cells (Cohen and Cohen, 1972; Fine et al., 1973) and have been shown recently to be associated with the actin-containing microfilament bundles in such cells (see, Lazarides, 1975b), the presence of tropomyosin or troponin in the cleavage furrow has yet to be demonstrated.

Somewhat more is known about the role of calcium as a regulatory factor in cytokinesis. Tilney and Marsland (1969) speculated that Ca^{2+} might be a factor in initiating furrowing but thought it might be released by breakdown of the nuclear envelope. Since Rappaport (1971) has convincingly shown that the nucleus is not necessary for the induction of furrowing, this proposed nuclear stimulus now seems unlikely. Gingell (1970) was able to induce localized contractions in the cortex of *Xenopus* eggs with polycations and found that the plasma membrane was thrown into extensive folds as a result. Ionized calcium, strontium, or barium produced similar results but magnesium was ineffective. Schroeder and Strickland (1974) produced similar results using the divalent cation ionophore A23187 and were able to simulate "furrows" by streaming A23187 along the surface of *Rana* eggs. These "furrows" did not, however, extend beyond the treated area. Baker and Warner (1972) treated *Xenopus* eggs with EGTA and produced relaxation of the cleavage furrow and concluded that Ca^{2+} was necessary for cleavage but at a low level. Timourian, Clothier and Watchmaker (1972) were able to prevent cleavage in sea urchin eggs by treatment with EDTA before mid-metaphase. If the eggs were treated after the furrow had appeared, the EDTA treatment had no effect. In this latter case, however, it was not possible to distinguish between an EDTA effect on the mitotic apparatus (which disappeared) and, in turn, a lack of furrow initiation or a direct effect on the furrow itself. Clothier and Timourian (1972) studied the flux of ^{45}Ca in sea urchin eggs and found peaks of efflux at prophase, at metaphase, and when 50% of the eggs were in mid-cleavage. These results do not correlate with cytokinetic events but it must be kept in mind that this was a measurement of unbound Ca^{2+}.

In my laboratory we have been working with the role of Ca^{2+} in cytokinesis of the squid embryo (Arnold, 1975, and unpublished observations). If the fertilized eggs are soaked in Ca^{2+} free medium before first division for two hours or longer, the first cleavage is inhibited and only a shallow depression is found where the first furrow would normally appear. If these eggs are then transferred into normal sea water (Ca^{2+} available in the external medium) the furrows suddenly form and cut into the cytoplasm. Treatment with ionophore A23187 also had an interesting effect on cytokinesis. If the cleaving eggs were treated with ionophore and then transferred to normal sea water (Ca^{2+} available) the furrows initially diminished and then reappeared and deepened, lengthened, and contracted so strongly that individual cells rounded up and the blastoderm was eventually destroyed. Cytokinesis ceased after a few divisions but karyokinesis continued. If, however, cleaving eggs were treated with ionophore and then kept in a Ca^{2+} free medium, cleavage continued with only a slight delay. Apparently the internal Ca^{2+} concentration was sufficient to allow cytokinesis to continue. However, if these ionophore treated, Ca^{2+} deprived eggs were then transferred into normal sea water (Ca^{2+} available) the bases of the cleavage furrow underwent a sudden sharp contraction and the cells rounded up (Fig. 3).

A simple model to explain these results can be postulated. Suppose that cells maintain Ca^{2+} at a proper internal concentration by "pumps" at the plasma membrane; a high concentration of Ca^{2+} outside the membrane would cause Ca^{2+} to continually enter the cell. If no Ca^{2+} is available outside the cell, the pumps continue to remove the Ca^{2+} (or it is reduced by diffusion). When the cells are treated with ionophore, the pumps are "bypassed" and the Ca^{2+} level equilibrates with the outside medium. If under these conditions the external Ca^{2+} is low or non-existent, then cytokinesis temporarily continues because the level of internal Ca^{2+} still approximates that of the original untreated cell until it is leached out and cytokinesis is inhibited. If, however, the Ca^{2+} concentration in the external medium is high, Ca^{2+} will enter the cell and reach abnormally high levels leading to overstimulation of the contractile system, and cytokinesis ceases, probably because all the precursors are bound. This model is clearly simplistic and in need of further testing. Also, it is not applicable to all organisms. Cephalopods have evolved in the sea and have always had Ca^{2+} available to them

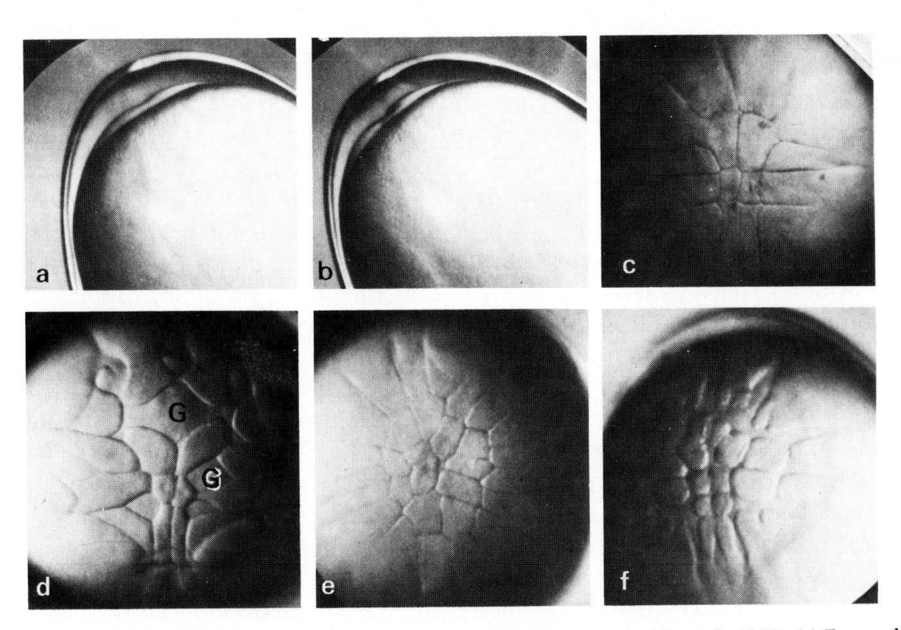

Fig. 3. Effect of Ca^{2+} changes on cytokinesis in the squid egg (approximately 50 X). (a) Egg soaked in Ca^{2+} free medium two hours prior to cleavage twelve minutes after cleavage had begun in the control. In the region where the furrow normally would be seen there is a shallow depression. (b) Same egg shown in (a) 60 seconds after transfer to Ca^{2+}-containing medium. Note the furrow has depth. (c) Normal sixteen cell stage. Control for (d). (d) Blastoderm of squid embryo 80 minutes after ionophore treatment. The furrows have contracted so strongly that large gaps (G) have appeared between the cells. The undercutting furrows are quite distinct. (e) Blastoderm of embryo treated with ionophore and incubated in Ca^{2+} free medium. The cleavage pattern is slightly distorted but is otherwise normal. (f) Same embryo as (e) six minutes after transfer into Ca^{2+} available medium. Note that large gaps have appeared in the furrows and that the cells have rounded up.

at a fairly uniform level in their external medium while fresh water organisms have had to evolve more elaborate mechanisms of regulating their internal Ca^{2+} at a proper level.

Because cytochalasins have been used extensively in the study of primitive motile systems and cytokinesis, their effects will be briefly discussed here. The interested reader is referred to the review of Copeland (1974) for an introduction to the vast literature on cytochalasin B.

Carter (1967) was the first to describe the effect of a series of mold metabolites called cytochalasins on cytokinesis. He demonstrated that when cultured mammalian cells were exposed to these compounds, karyokinesis would proceed but cytokinesis was blocked. Schroeder (1969) used cytochalasin B to inhibit the division of sea urchin eggs and demonstrated that the contractile ring was disrupted by this drug. This interesting observation precipitated a number of papers all of which demonstrated that microfilament-related phenomena (e.g., cytokinesis, nerve outgrowth, cytoplasmic streaming, cell locomotion, exocytosis etc.) could be inhibited by treatment with relatively low dosages of cytochalasin B (see Wessels et al., 1971a,b, for review). Unfortunately, cytochalasin B turned out *not* to be a specific "magic bullet" aimed only at microfilaments; rather it seems to react both at the plasma membrane and in the subplasmalemmal microfilament region (Mayhew et al. 1974; Tannenbaum et al., 1975). Sanger and Holtzer (1972) found that cytochalasin B had a broader effect than just interrupting primitive motile systems and found it affected cellular adhesion as well. Zigmond and Hirsch (1972) found cytochalasin B depressed glucose transport across the plasma membrane and Copeland (1974) in reviewing the many papers in this area concluded that cytochalasin B also alters the synthesis of mucopolysaccharides. There is also some evidence that at high concentrations cytochalasin B decreases the viscosity of actomyosin complexes. When cytochalasin B is added to actin-heavy meromyosin mixtures, the ATPase activity is depressed by up to 60% (Spudich and Lin, 1972). From this small representative sampling of the papers available, it would seem that the action of the cytochalasins remains to be completely elucidated but that these compounds still have interesting cytological effects on living cells. The fact that the cleavage furrow disappears concomitant with the disruption of the contractile ring when cells are treated with cytochalasin B (e.g., Schroeder, 1969) strongly implicates the microfilamentous contractile ring as instrumental in cytokinesis despite the poorly understood mode of action of cytochalasin B.

6. *The origin of new surface membrane*

The question of how dividing animal cells increase their surface area is not only interesting but also confusing, and there is considerable conflicting evidence on the origin of "new" plasma membrane during cytokinesis. Since the existing surface layers cannot be infinitely expandable without some new input, it would seem that some cellular mechanism would have appeared early in evolution for increasing cell surface area in dividing cells and that such a mechanism would have been conservatively maintained. Unfortunately, there is little general agreement concerning observations made on different species in different laboratories. Of course, there is no reason to believe that each possible source of new cell surface excludes any other source and two (or more) separate mechanisms might well be operating at once. It would appear that in many cells preexisting surface membrane is gathered together into microvilli or folds which are redistributed during the early phases of cytokinesis. This, in turn, is supplemented by a second source of intracellular membrane which is transported to the surface where it is incorporated into the plasma membrane, possibly by fusion of membrane-bound vesicles with the plasma membrane.

Using mammalian tissue culture cells, Erickson and Trinkaus (1976) have studied the source of membrane throughout the cell cycle as cells changed from a highly flattened state to a rounded pre-division shape, through division and back to the spread state again. When the cells were attached to the substrate and well spread, the surface appeared smooth and regular but in rounded cells (either pre-division or shaken free), the surface of the cells were covered with numerous microvilli and folds. Estimates of the total cell surface area in these two vastly different cell shapes were approximately equal, suggesting that the cell used the microvilli as a membrane reserve for shape changes and presumably for cytokinesis.

There is considerable evidence for similar utilization of gathered pre-existing surface membrane during cytokinesis in eggs. Szollosi (1970) showed that the microvilli of *Aequorea* eggs disappeared in the region of the furrow and speculated that this was not only due to the "recycling" of microfilaments but also to provide additional membrane to support an increase in surface area. Wolpert (1960) has calculated that the change from one sphere to two of equal volume requires a 26% increase in surface area. Similarly in the squid embryo the surface membrane at the end of the advancing furrow is gathered into a series of microvilli which change into longitudinal folds which parallel the future axis of the forming furrow (Arnold, 1974). As the microfilamentous band contracts and pulls the attached plasma membrane down to displace the cytoplasm to either side, the longitudinal folds unfold and are apparently incorporated into the "new" surface which separates the sister blastomeres. Not only does the decrease in number of folds support this idea (average of 16 folds when cleavage is just beginning versus average of 2 when the furrow base reaches the yolk) but actual measurements of membrane length in thin sections indicate that approximately 75% of the "new" membrane comes from unfolding of the original longitudinal folds (Arnold, unpublished observations). A similar situation exists during cytokinesis in the insect blastoderm. In *Drosophila* embryogenesis, cytokinesis follows after several free nuclear divisions so that new surface for about 3500 cylindrical cells must be provided at one time. Fullilove and Jacobson (1971) have made an elegant ultrastructural analysis of this striking example of membrane formation. As the final mitosis which precedes cytokinesis occurs, the formerly smooth surface becomes covered with microvilli. Fullilove and Jacobson estimated that these microvilli represented a five-fold increase in the surface of the plasma membrane. As cytokinesis begins, criss-crossing bands of microfilaments form between the future cells to produce a mesh-like array around the middle of the embryo. The cytokinetic bands contract and cut between the nuclei and pull the microvillous surface behind them as the depth of the cleavage furrow increases. Fullilove and Jacobson calculated that by "unfolding" the microvilli the cytokinetic furrows could extend the plasma membrane to the base of the elongating nuclei. Obviously a second source of membrane becomes necessary at this point. In the case of *Drosophila,* Fullilove and Jacobson found paired membranes in the cytoplasm which appeared parallel to the cleavage furrow paths below the level of the nuclei. As the nuclei elongate these paired membranes apparently become incorporated into the plasma membrane. Often these paired membranes had ribosomes attached to them. Sanders (1975) described intracellular lamellar bodies (also called myelin figures in the literature) and considered these to represent pools of phospholipid precursor for membrane synthesis in *Drosophila* blastoderm formation.

The situation seems to be comparable in other organisms. In *Loligo*, the cleavage furrow is preceded by an accumulation of dense bodies which originate in the Golgi apparatus (Arnold, 1969). These dense bodies come into actual contact with the down-cutting plasma membrane and their contents are released into the furrow space. In so doing their membrane becomes incorporated into the plasma membrane which now separates the new sister blastomeres. Unfortunately, it has not been possible to accurately measure the amount of membrane contributed by these vesicles to see if they could account for the missing 25% unaccounted for by the longitudinal folds.

As cleavage proceeds in *Loligo*, the blastodisc becomes separated from the underlying yolk by the formation of the undercutting furrows. These undercutting furrows are derived from the fusion of the first and second cleavage furrow (Arnold, 1971). Since there is no source of gathered membrane, the new membrane must come from internally derived sources just behind the microfilamentous band. There is a region of active fusion of small empty vesicles which apparently are derived from the Golgi apparatus (Fig. 4).

In amphibians the membrane that comes to separate the new blastomeres seems to be derived primarily from internal sources. Selman and Perry (1970) studied the differences between the pigment cell membrane on the outer surface of the new embryo during cleavage and the unpigmented surface between the blastomeres. In addition to the surface being unpigmented, the inner membrane was smoother and there was an abrupt junction between the pigmented and unpigmented membrane. A distinct groove was prominent at the junction of the pigmented and unpigmented membrane and Selman and Perry postulated this

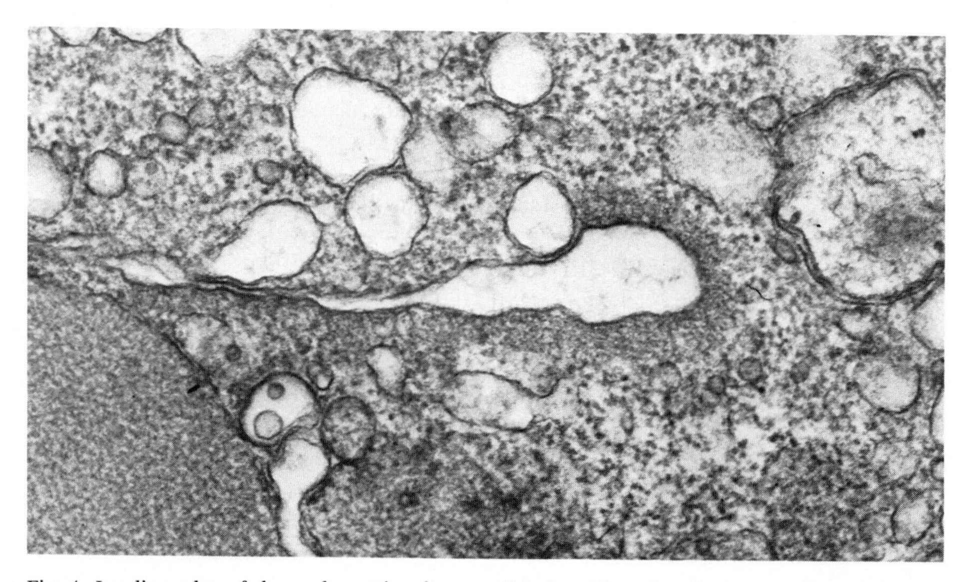

Fig. 4. Leading edge of the undercutting furrow of *Loligo*. The microfilamentous band is evident and just behind it a vesicle appears about to fuse with the furrow. Approximately 46,800 X.

might be a site of membrane growth. Bluemink and DeLaat (1973) studying cytokinesis in *Xenopus* eggs used ruthenium red surface marking and cytochalasin B treatments to argue for the de novo synthesis of new unpigmented membrane as opposed to expansion of the pre-existing pigmented membrane. Singal and Sanders (1974) also studied *Xenopus* eggs and found that microvilli followed the leading edge of the cleavage furrow only during the very early stage

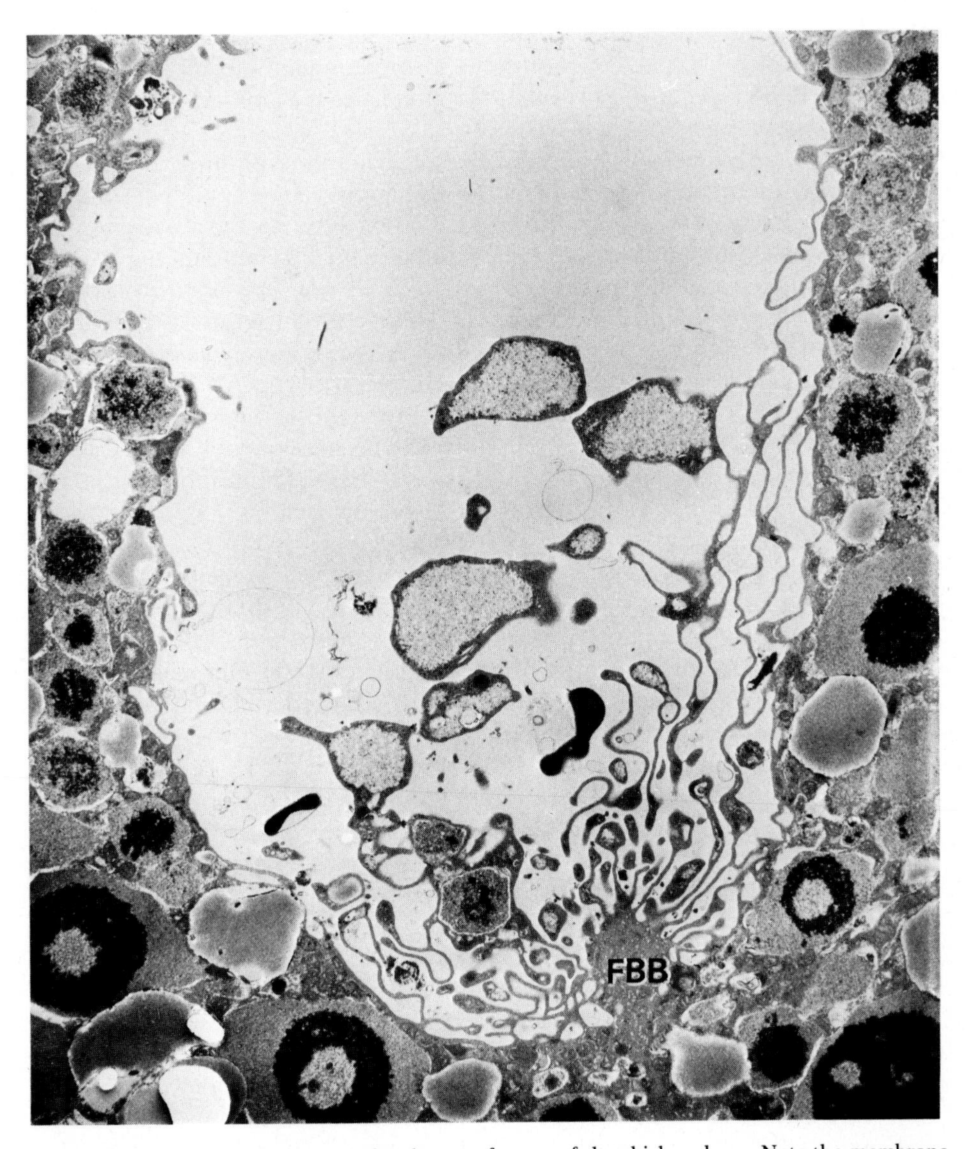

Fig. 5. Furrow base body (FBB) in the cleavage furrow of the chick embryo. Note the membrane-covered protrusions from the furrow base body to the furrow walls. Approximately 4,400 X (courtesy of Dr. I. Gipson and the Journal of Ultrastruct. Research).

of cytokinesis and thereafter they remained stationary. This would again agree with a program of de novo membrane synthesis deeper in the furrow. Singal and Sanders also described two morphologically distinct types of vesicles that were frequently seen in association with the furrow. The first comprised large vesicles with a fibrous content which fused with the furrow at a distance behind the leading edge, while the second type comprised smaller vesicles (40–60 nm) which were found in association with the leading edge of the furrow. Bluemink and DeLaat (1973) and Singal and Sanders (1974) suggested that dense lamellar bodies may be possible precursors of the new unpigmented membrane. Sanders and Singal (1975) have also shown that Golgi vesicles may be incorporated into the surface membrane of *Xenopus* furrows.

Cleavage of the chick blastoderm offers further insight into the possible mechanism(s) of membrane synthesis. In this highly telolecithal egg, Gipson (1974) has described two unique structures associated with the furrow. In the base of the well established furrow, a filamentous mass of cytoplasm, the "furrow base body" bulges into the furrow cavity (Fig. 5). There are extensive connections between the furrow wall and the furrow base body which ramify throughout the furrow cavity. In the walls of the furrow itself are extensive areas of randomly arranged membranes interspersed between glycogen particles. Gipson (1974) suggests this "membranous vesiculate reticulum" represents a site of synthesis for "preformed membrane" which will be subsequently utilized in the furrow. This membrane "factory" would seem to have significant potential not only for the study of cleavage furrow formation but also as a model of membrane biogenesis.

It would therefore seem that there is some tentative agreement that new membrane can arise in two stages during cleavage. The first step involves a "gathering and redistribution" of existing plasma membrane. The second step involves a true new synthesis of membrane, either by conventional organelles (e.g., Golgi or endoplasmic reticulum) or by a specialized and unique organelle such as the furrow base body found in the chick egg. The relative contribution of either step is possibly dependent on the absolute size of the cell undergoing cleavage and the area of new surface membrane required.

7. *Final separation of the daughter cells*

A question that is frequently ignored in discussions of cytokinesis is how the cells finally separate. Since the microfilamentous band cannot contract to infinity, it seems logical that another mechanism must be responsible for ultimately breaking the final link between the cells. It is usually assumed that "surface tension" breaks the final connection but this argument is sufficiently tenuous to merit close question. Sanger (1975) has postulated that the cultured fibroblasts examined in his study "crawled" apart by virtue of traction exerted by pseudo-podia on the distal poles of the daughter cells. Such a mechanism would only be feasible in cells in tissue culture attached to a substrate. However, most dividing cells are confined within a framework of tissue or, in embryos, encased in fertilization membrane or other accessory coat. In many cell types, including some molluscan embryos, midbodies are found in the final stages of daughter cell separation. Although midbodies have been known for a long time and many observations have been made on them (e.g., Slautterback and Fawcett, 1959; Skalko et al., 1972), little is known about their eventual fate. What I am going to present here is a brief suggestion that the midbodies may be the ultimate cause of cellular separation in at least some types of cells.

At first the midbodies contain the remnants of the spindle microtubules upon which electron dense material accumulates. In the blastoderm cells of *Loligo*, midbodies are present which may connect two or more cells (Arnold, 1974). When the remnant of the spindle disappears, a continuous cytoplasmic bridge remains through which organelles such as mitochondria can pass. Apparently this intercellular cytoplasm is eventually occluded by the appearance of cytoplasmic membranes which "plug" the cytoplasmic connection (Fig. 6). It is assumed that the stage of occluding membranes represents the final separation of the former daughter cells. It may be possible that the occluding membrane expands and forms new surface of the new sister cells. Admittedly, this is a speculative argument based on scant data but no other better data are available concerning the final separation of animal cells. Further work in this area is clearly needed.

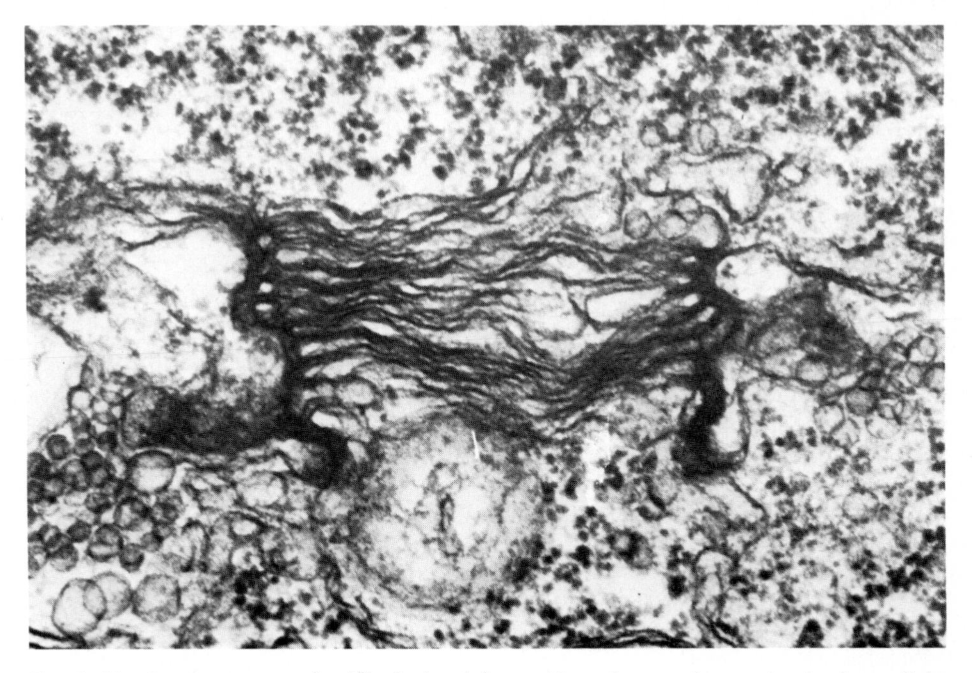

Fig. 6. Membranous stage of midbody breakdown. Note the membranes in the intercellular connection are continuous with the cytoplasm and the cell surface. Approximately 62,700 X (courtesy of J. Cartwright, Jr.).

8. *Concluding remarks*

In the last 10 to 15 years there has been a considerable change in our thinking about the underlying mechanisms of cytokinesis, primarily due to the incisive experiments of Rappaport (see 1965 and 1971) and more recently due to detailed fine structural analysis and comparisons with other studies on intercellular motility. The older ideas of animal cytokinesis being caused by expansion of the polar surface, fusion of aligned membranous vesicles without surface ingression, or protoplasmic streaming "pushing" the daughter cells apart have been relegated to historically interesting but experimentally unsubstantiated hypotheses. What has emerged is a concept of a primitive muscle-like contraction occurring which separates the daughter cells at a site determined by the position of the mitotic apparatus. The analogy with an actomyosin-like contraction has become stronger and stronger as more evidence has accumulated. The mechanism of cytokinesis appears to be a further example of the role of intracellular contractile proteins involving interaction of actin-like and myosin-like components in the presence of ATP and controlled by the intracellular calcium concentration (see Durham, 1974; Pollard and Weihing, 1974). Many unanswered questions remain. At this time much of the evidence concerning the contractile process is circumstantial and such problems as the localization of myosin in the furrow, the regulation of regional calcium concentration in the cell and the demonstration of regulatory proteins still remain to be satisfactorily elucidated. What is now needed is the ability to carefully measure the quantitative presence of these various components *in situ*. Hopefully the next ten years will be as productive as the past decade in increasing our knowledge of the mechanism of cytokinesis.

Acknowledgements

I would like to express my appreciation to Lois D. Williams-Arnold for her help not only with the preparation of this manuscript but also for her technical assistance with many of the experiments. Dr. Robert Kane read and improved the manuscript and Ms. Frances Horiuchi competently prepared it for publication. Much of the original work on squid was supported by grants from National Science Foundation and the National Institutes of Health.

References

Arnold, J.M. (1968a) Formation of the first cleavage furrow in a telolecithal egg (*Loligo pealii*). Biol. Bull. 135, 408.

Arnold, J.M. (1968b) An analysis of cleavage furrow formation in the egg of *Loligo pealii*. Biol. Bull. 135, 413.

Arnold, J.M. (1969) Cleavage furrow formation in a telolecithal egg (*Loligo pealii*). I. Filaments in early furrow formation. J. Cell Biol. 41, 894–904.

Arnold, J.M. (1971) Cleavage furrow formation in a telolecithal egg (*Loligo pealei*). II. Direct evidence for a contraction of the cleavage furrow base. J. Exp. Zool. 176, 73–86.
Arnold, J.M. (1974) Cleavage furrow formation in a telolecithal egg (*Loligo pealei*). III. Cell surface changes during cytokinesis as observed by scanning microscopy. Dev. Biol. 40, 225–232.
Arnold, J.M. (1975) An effect of calcium in cytokinesis as demonstrated with ionophore A23187. Cytobiologie 11, 1–9.
Aronson, J.F. (1965) The use of fluorescein-labeled heavy meromyosin for the cytological demonstration of actin. J. Cell Biol. 26, 293–298.
Baker, P.C. (1965) Fine structure and morphogenic movements in the gastrula of the treefrog, *Hyla regilla*. J. Cell Biol. 24, 95–117.
Baker, P.F. and Warner, A.E. (1972) Intracellular calcium and cell cleavage in early embryos of *Xenopus laevis*. J. Cell Biol. 53, 579–581.
Bluemink, J.G. (1970) The first cleavage of the amphibian egg: An electron microscope study of the onset of cytokinesis in the egg of *Ambystoma mexicanum*. J. Ultrastruc. Res. 32, 142–166.
Bluemink, J.G. and DeLaat, S.W. (1973) New membrane formation during cytokinesis in normal and cytochalasin B-treated eggs of *Xenopus laevis*. I. Electron microscope observations. J. Cell Biol. 59, 89–108.
Brachet, J. (1950) Chemical Embryology. Wiley – Interscience, New York.
Carter, S.B. (1967) Effects of cytochalasins on mammalian cells. Nature 213, 261–264.
Cloney, R.A. (1966) Cytoplasmic filaments and cell movements: Epidermal cells during ascidian metamorphosis. J. Ultrastruct. Res. 14, 300–328.
Clothier, G. and Timourian, H. (1972) Calcium uptake and release by dividing sea urchin eggs. Exp. Cell Res. 75, 105–110.
Cohen, I. and Cohen, C. (1972) A tropomyosin-like protein from human platelets. J. Mol. Biol. 68, 383–387.
Copeland, M. (1974) The cellular response to cytochalasin B: A critical overview. Cytologia 39, 709–727.
Czihak, G. (1973) The role of astral ray in early cleavage of sea urchin eggs. Exp. Cell Res. 83, 424–426.
Durham, A.C.H. (1974) A unified theory of the control of actin and myosin in non-muscle movements. Cell 2, 123–136.
Edelman, G.M. (1974) Origins and mechanisms of specificity in clonal selection. In: Cellular Selection and Regulation in the Immune Response (Edelman G.M., ed.) pp. 1–38. Raven Press, New York.
Erickson, C.A. and Trinkaus, J.P. (1976) Microvilli and blebs as sources of reserve surface membrane during cell spreading. Exp. Cell Res. 99, 375–384.
Fine, R.E., Blitz, A.L., Hitchcock, S.E., Kaminer, B. (1973) Tropomyosin in brain and growing neurones. Nature New Biology 245, 182–186.
Fullilove, S.L. and Jacobson, A.G. (1971) Nuclear elongation and cytokinesis in *Drosophila montana*. Dev. Biol. 26, 560–577.
Gingell, D. (1970) Contractile responses at the surface of an amphibian egg. J. Embryol. Exp. Morph. 23, 583–609.
Gipson, I. (1974) Electron microscopy of early cleavage furrows in the chick blastodisc. J. Ultrastruct. Res. 49, 331–347.
Goldman, R.D. (1975) The use of heavy meromyosin binding as an ultrastructural cytochemical method for localizing and determining the possible functions of actin-like microfilaments in nonmuscle cells. J. Histochem. Cytochem., 23, 529–542.
Goodenough, D., Ito, S., and Revel, J.-P. (1968) Electron microscopy of early cleavage stages in *Arbacia punctulata*. Biol. Bull. 135, 420.
Gwynn, I., Kemp, R.B., and Jones, B.M. (1974) Ultrastructural evidence for myosin of the smooth muscle type at the surface of trypsin-dissociated embryonic chick cells. J. Cell Sci. 15, 279–389.
Hiramoto, Y. (1956) Cell division without mitotic apparatus in sea urchin eggs. Exp. Cell Res. 11, 630–638.
Hiramoto, Y. (1958) A quantitative description of protoplasmic movement during cleavage in the sea urchin egg. J. Exp. Biol. 35, 407–424.

Wallingham, M.C., Ostlund, R.E. and Pastan, I. (1974) Myosin is a component of the cell surface of cultured cells. Proc. Nat. Acad. Sci. U.S.A. 71, 4144–4148.

Weber, K. and Groeschel-Stewart, U. (1974) Myosin antibody: the specific visualization of myosin containing filaments in non-muscle cells. Proc. Nat. Acad. Sci. U.S.A. 71, 4561–4564.

Weinstein, R.S. and Hebert, R.R. (1964) Electron microscopy of cleavage furrows in sea urchin blastomeres. J. Cell Biol. 23, 101A.

Weisenberg, R.C. (1972) Microtubule formation in vitro in solutions containing low calcium concentrations. Science 177, 1104–1105.

Wessells, N.K., Spooner, B.S., Ash, J.F., Bradley, M.O., Luduena, M.A., Taylor, E.L., Wrenn, J.T. and Yamada, K.M. (1971a) Microfilaments in cellular and developmental processes. Science 171, 135–143.

Wessells, N.K., Spooner, B.S., Ash, J.F., Luduena, M.A., and Wrenn, J.T. (1971b) Cytochalasin B: Microfilaments and "contractile" processes. Science 173, 356–359.

Wilson, E.B. (1925) The Cell in Development and Heredity. MacMillan and Company, New York and London.

Wolpert, L. (1960) The mechanics and mechanism of cleavage. Int. Rev. Cytol. 10, 163–216.

Yamada, K.M., Spooner, B.S. and Wessells, N.K. (1971) Ultrastructure and function of growth cones and axons of cultured nerve cells. J. Cell Biol. 49, 614–635.

Yoneda, M. and Dan, K. (1972) Tension at the surface of the dividing sea urchin egg. J. Exp. Biol. 57, 575–587.

Zigmond, S.H. and Hirsch, J.E. (1972) Effects of cytochalasin B on polymorphonuclear leukocyte locomotion, phagocytosis and glycolysis. Exp. Cell Res. 73, 383–393.

The implanting mouse blastocyst 3

Michael I. SHERMAN and Linda R. WUDL

1. Introduction

The implantation of the mammalian embryo in the uterus of the mother is a unique interaction between two organisms. This event has been the subject of an enormous number of papers, and has been exhaustively reviewed. Rather than recapitulating these summaries here with an occasional interspersion of data published since the last survey, we have endeavored to approach the subject from a slightly unorthodox aspect: although the implantation event is closely coordinated between uterus and the blastocyst, we shall attempt to discern specifically the role played by the latter. Thus, while implantation is subject to a number of controls and can be broken down into a series of phases, we shall examine the evidence that these controls are exerted on the blastocyst rather than on the uterus, and concern ourselves with the role of the blastocyst in each of these phases. We shall also consider how its previous developmental history has prepared the blastocyst for implantation. We shall attempt, wherever possible, to stress data which may relate to properties and reactivities at the cell surfaces of the blastocyst during implantation. Finally, in this review, we shall restrict ourselves to a consideration of implantation of the mouse blastocyst; other species will be referred to only for comparative purposes, or in cases where equivalent studies have not yet been carried out in the mouse.

2. Definitions and terminology

For the purposes of this review, the implantation event, as well as the periods directly preceding and following it, are divided into five phases, and these are shown schematically in Fig. 1.

The mouse embryo enters the uterus, at the morula stage, on the third day of pregnancy (the day of observation of the sperm plug is considered the first day of pregnancy). Shortly thereafter, by the 32-cell stage, blastocoelation has usually taken place. In preparation for implantation per se, during the *hatching phase*, the blastocyst must shed its zona pellucida. Subsequently, during the *apposition*

G. Poste & G.L. Nicolson (eds) The Cell Surface in Animal Embryogenesis and Development

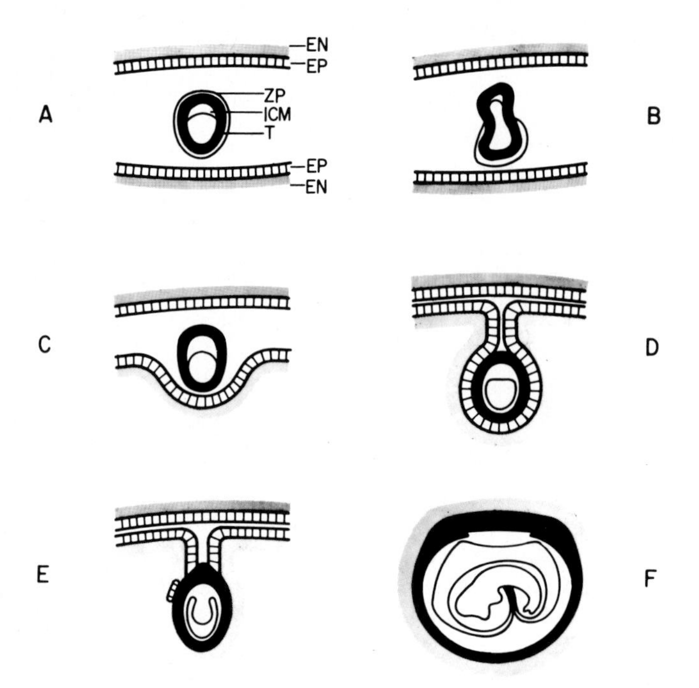

Fig. 1. Schematic representation of the phases of implantation. Figures are not drawn to scale: (A) Fourth day blastocyst in uterus; (B) Hatching phase, during which the zona pellucida is shed; (C) Apposition phase; the blastocyst comes to lie against the antimesometrial wall of the uterus with the ICM oriented as the mesometrial pole; (D) Attachment phase: the surface membranes of the trophoblast and uterine epithelial cells are in intimate contact; (E) Invasion phase: the uterine epithelial layer degenerates and trophoblast cells come into direct contact with the decidua; and (F) Postimplantation phase: trophoblast cells lose their invasiveness. EN: endometrium, later decidua; EP: uterine epithelium; ZP: zona pellucida; ICM: inner cell mass; T: trophectoderm, later trophoblast. The mesometrium is towards the top in all cases.

phase, the naked blastocyst must align and orient itself in preparation for attachment to the uterine wall. This occurs during the *adhesion phase*. As the blastocyst becomes firmly anchored to the uterus, the trophoblast cells begin to penetrate the uterine epithelial layer, characterizing the *invasion phase* of implantation. Obviously, invasiveness must be controlled so that the function of the maternal moiety of the placenta is not adversely affected. Consequently, during the *postimplantation phase*, trophoblast cells cease invasive activity and the maternal-foetal placental relationship is stabilized. The role of the conceptus in each of these phases will be considered in subsequent sections.

Because of the variations in the development of embryos from a single mother, not to mention differences between individual mothers or especially mouse strains, it is difficult to definitively schedule each of the phases to be discussed. Consequently, more stress will be placed upon the relative duration of each of the implantation-related events, rather than the gestation age in hours at which each might be expected to occur.

Prior to implantation, the blastocyst contains two cell types, the outer trophectoderm layer and the enclosed inner cell mass (ICM). This review will largely concentrate upon derivatives of the former group of cells (though it should be borne in mind that the ICM undergoes considerable proliferation and differentiation during the time period under discussion). The cells of the trophectoderm layer are often referred to as trophoblast cells. This creates some confusion since the term is also applied to the highly differentiated and specialized cells of the placenta which are derived from trophectoderm cells. We propose that the term "trophectoderm" be used to describe the layer of cells on the outside of the preimplantation blastocyst, while the term "trophoblast" be reserved for trophectoderm cells and their derivatives after the adhesion phase of implantation has been reached. Although this phase probably does not coincide with the time at which the differentiation processes are complete, the terminology is convenient because the transition point can be clearly discerned both in vivo and in vitro.

3. *Hatching phase*

Before the mouse blastocyst can implant in the uterus it must shed its outer protective coat, the zona pellucida. The mechanism by which it accomplishes this task has been studied for very many years and, until recently, was a topic for discussion and controversy but relatively poorly understood. The questions concerned with blastocyst hatching are: (a) is hatching a mechanical phenomenon or is it accomplished by means of an enzyme or enzymes, or both?, and (b) is there any contribution of the uterus to either the mechanism or timing of hatching?

In the absence of in vitro culture conditions which would support the growth of early mammalian embryos, these questions had to be attacked by means of available "in vivo" techniques. In early attempts to separate embryonic from maternal functions with respect to preimplantation development, fertilized mouse ova were transplanted to several extrauterine sites, including the anterior chamber of the eye (Runner, 1947; Fawcett, Wislocki and Waldo, 1947), abdominal cavity (Fawcett et al., 1947), kidney (Kirby, 1962a), testis (Kirby, 1963a), and spleen (Kirby, 1963b). In all of these ectopic sites, the mouse ovum was capable of developing to the blastocyst stage and hatching. Furthermore, equivalent results were obtained from transplants to various sites in pregnant, nonpregnant, or immature females, or to intact, castrated or immature males. From these experiments it was concluded that the mouse preimplantation embryo is capable of hatching in the absence of oviducal or uterine influence or of sex hormones. But there still remained the question as to whether or not the embryo could develop in the total absence of hormones or growth factors supplied by an adult host.

Whitten (1957) succeeded in growing mouse embryos in a chemically defined medium and found that in the absence of hormones, mouse embryos could develop from the 8-cell stage to blastocyst and hatch with a survival rate of approximately 50%. With present in vitro culture techniques, embryos can be grown from the one-cell stage to blastocyst at frequencies approaching 100% (Whitten, 1971). These observations demonstrate that the mouse embryo is capable of directing its own preimplantation development and hatching from the zona pellucida in the absence of external signals and controls.

Hatching of the blastocyst in vitro appears to be accomplished by means of mechanical rupture of the zona. As early as 1935, Lewis and Wright noted the ability of the blastocyst to expand and contract. Later, cinemicrography was used to study the regular pulsating movements of the blastocyst which result in rupture of the zona and subsequent escape of the embryo (Borghese and Cassini, 1963; Cole and Paul, 1965; Cole, 1967). Cole (1967) described the expansion and contraction cycles of a single blastocyst and estimated that the amount of time required for hatching in vitro is approximately 17 hours. Figure 5 illustrates several blastocysts in the process of hatching in culture.

These in vitro observations did not, however, settle the question of whether hatching is strictly mechanical in vivo as well, or whether the uterus plays a role in this process. In fact, it has become apparent that the uterus itself *is* involved in the hatching event. In the absence of uterine function, hatching is delayed by at least 24 hrs. This is observed during lactational delay (Rumery and Blandau, 1966; McLaren, 1967), ovariectomy delay (McLaren, 1968a), development of blastocysts retained in the oviduct (Orsini and McLaren, 1967), development in ectopic sites (Fawcett et al., 1947) and in vitro development (Whitten, 1957). Furthermore, under normal conditions: (a) blastocysts are seldom seen in the process of hatching, indicating that they require far less than the 17 hours reported by Cole (1967) for mechanical rupture; (b) those few blastocyst which are found partially hatched appear to have thinner, "stickier" zonae (McLaren, 1970a,b; Potts and Wilson, 1967; Mintz, 1971), unlike blastocysts hatching in vitro; and (c) rarely are zonae or fragments of zonae found in uterine flushings during or after hatching, whereas intact zonae are easily recovered from the uterus or oviduct after delayed hatching (Orsini and McLaren, 1967; McLaren, 1968a).

Conclusive evidence that the uterus participates in blastocyst hatching comes from two sets of experiments. McLaren (1970a) observed that when unfertilized eggs are present in the uterus of a pregnant mouse, they lose their zona, although slightly more slowly than normally developing blastocysts. In order to show this was not a secondary effect of the decidualization process, she transplanted unfertilized eggs into a pseudopregnant host and observed similar results. If unfertilized eggs are placed in an ovariectomized pseudopregnant host, on the other hand, they retain their zonae even after the vitelli have disintegrated. This suggests that zona lysis is due to a factor in the primed uterus and not a result of the release of intracellular degradative enzymes from the dead vitelli. Since there does appear to be a slight difference in timing between hatching of normal blastocysts and unfertilized eggs, McLaren concluded that hatching is a result of both zona lysis by the uterus and mechanical stress resulting from the expansion and contraction of the blastocyst.

Similar results have been obtained by Mintz (1971) using the homozyogous lethal mutant t^{12}. The t^{12}/t^{12} embryo is arrested at the morula (or early blastocyst) stage of development and is not capable of hatching in vitro (Mintz, 1963). However, when embryos from a $+/t^{12} \times +/t^{12}$ mating are recovered from the uterus on the fourth day, not only are normal ($+/+$ and $+/t^{12}$) blastocysts hatched, but so too are the t^{12}/t^{12} (dead) morulae. Once hatched, the dead morulae quickly disappear from the uterus. If embryos from the above mating are retained in the oviduct (by ligation of the uterotubal junction on the third day), normal blastocysts hatch (after some delay) but homozygous mutant morulae retain their zonae just as they do in vitro. Finally, if immature females are used in a similar mating and the uterus flushed on the fifth day of pregnancy, again only the normal blastocysts are hatched, while dead morulae and unfertilized eggs retain their zonae.

On the basis of their results, both Mintz and McLaren have postulated a uterine zonalytic factor which is at least partially responsible for hatching in vivo. The activity of this factor is directly or indirectly controlled by ovarian hormones (estrogen in the mouse, McLaren, 1968a). The question as to which of the two mechanisms, zona lysis or mechanical stress, is primarily responsible for hatching in vivo becomes purely academic at this point because it appears quite probable that both contribute toward the final result. Since, in the absence of the uterine factor, hatching is delayed approximately 24 hours, mechanical stress alone must not be the normal means of escape from the zona pellucida. On the other hand, the observation that unfertilized eggs and dead morulae (Mintz, 1971) appear to hatch slightly more slowly than normal blastocysts in the same uterine milieu indicates some active involvement of the embryo in this process.

Finally, it is probably erroneous to consider both blastocyst contraction and expansion as well as the zona lysin having as their only function the hatching of the blastocyst. Blastocysts continue to expand and contract even after hatching, and, considering the substantial increase in embryonic metabolism and macromolecular synthesis just prior to implantation, these motions may be the means, or the result, of an increased uptake of ions or solutes from the environment, for metabolic purposes (Cole, 1967). On the other hand, Mintz (1970; Pinsker, Sacco and Mintz, 1974) has postulated that the primary function of the uterine lytic factor is the initiation of implantation (see section 5.2.2.).

4. *Apposition phase*

During the apposition phase of implantation, the blastocyst, having hatched from its zona pellucida, must align itself correctly within the uterine lumen in preparation for the subsequent adhesion phase. Although it is difficult to fix the duration of the apposition phase with certainty, it is clear that this period is quite short. In vivo studies suggest a duration of as little as two hours in some cases (Kirby et al., 1967; Bergström and Nilsson, 1971). Experiments in vitro indicate that, on average, about ten hours elapse from the time that the blastocyst completely sheds its zona pellucida until it adheres to a monolayer of uterine cells (Sherman and Salomon, 1975 and unpublished results).

4.1. Morphological studies

The ultrastructure of the mouse blastocyst and the adjacent uterine epithelium just prior to, and during, implantation has been described by Potts and coworkers (1966, 1968; Potts and Wilson, 1967; Kirby, Potts and Wilson, 1967), Reinius (1967), Nilsson (1967) and Smith and Wilson (1974), and has been reviewed by Potts (1969). Detailed morphological studies on implantation in other mammals have been carried out as well (e.g., rat: Enders and Schlafke, 1967; Tachi, Tachi and Lindner, 1970; rabbit: Larsen, 1961, 1970; Enders and Schlafke, 1971). Comparative studies have also been published (e.g., Enders and Schlafke, 1969; Nilsson, 1970; Schlafke and Enders, 1975).

During the apposition phase of implantation in the mouse, the luminal surface of the uterine epithelial cells is characterized by large numbers of microvilli which are up to 1 μm in length (Potts, 1966, 1969; Reinius, 1967; Nilsson, 1967; Smith and Wilson, 1974). Although microvilli are present on the surface of trophectoderm cells in the early unhatched blastocyst (Potts, 1969; Nadijcka and Hillman, 1974), the luminal surfaces of the trophectoderm cells are relatively quiescent during the hatching and apposition phases (Reinius, 1967, Potts, 1969; Smith and Wilson, 1974; Nadijcka and Hillman, 1974). During apposition, all trophectoderm cells are greatly flattened and elongated, so much so that a single trophectoderm cell may overlie up to eight or ten uterine epithelial cells (Potts, 1968). It is at this time that the cells are undergoing the so-called "trophoblast giant cell transformation" (Dickson, 1963, 1966). This terminology is unfortunate, since it implies a great increase in the size of the trophectoderm cells. In the face of the dramatic flattening of these cells (to less than 1 μm over some parts of their surface), it is not clear if there is actually any increase in the *volume* of these cells. The phenomenon may be due only to a change in shape, and, as such, would be quite different from the "trophoblast giant cell transformation" that takes place in the postimplantation mouse placenta when trophoblast cells undergo cycles of endoreduplication and the nuclei alone can assume diameters greater than 100 μm (see Barlow and Sherman, 1972).

4.2. Positioning of the blastocyst

Orientation and attachment sites vary among mammals (Mossman, 1971). In the mouse, the first blastocyst attachment is antimesometrial, and the ICM is oriented mesometrially. Spacing of blastocysts along the uterine horn appears to be largely uterine-dependent in the rabbit and the rat (Boving, 1971). Even melanoma cells injected into the mouse uterus form clumps which assume proper spacing (Wilson and Potts, 1970). Since the blastocyst does not appear to have an active role in this phenomenon, and since the subject has been recently reviewed in detail (Wimsatt, 1975), it will not be discussed further here.

It has been noted that the uterine lumen is tightly occluded by the time of implantation (Potts and Wilson, 1967; Nilsson, 1967). Potts and Wilson (1967) have proposed that it is this closure of the uterine lumen which moves the blastocyst to its antimesometrial site in preparation for attachment. Reversal of the mesometrial-antimesometrial axis of the uterus does not prevent attachment to the antimesometrial wall, at least in the rat (Alden, 1945). Cowell (1969) noted, however, that mouse blastocysts can implant mesometrially in the uterus of a cyclic or ovariectomized mouse when an area on that wall is denuded of uterine epithelium. Kirby (1971) proposed that the blastocyst might be drawn to the normal antimesometrial attachment site or the induced mesometrial site by a differential in oxygen tension. In any case, it would appear as though the uterus, and not the blastocyst, is responsible for positioning the latter in preparation for implantation. This point was strongly made by Wilson (1963a) and Wilson and Potts (1970) who found in their studies that melanoma cells introduced into the uterine lumen become aligned against the antimesometrial wall.

Kirby et al. (1967; Kirby, 1971) proposed that orientation was a somewhat different story in that the blastocyst participated actively in this process. According to these investigators, there is a brief (two hour) period at the end of the preattachment phase during which the blastocyst is randomly oriented. Yet, the blastocyst is so intimately apposed to the uterine epithelium that the microvilli of epithelial and trophectoderm cells interlock. On the other hand, Kirby et al. (1967) found that trophectoderm – ICM junctions are only very tenuous at this time. In fact, trophectoderm and ICM cells actually appear to be separated by an extracellular deposit. These authors, therefore, maintained that the ICM is randomly oriented during initial attachment of the blastocyst to the uterine wall, but that subsequently, the ICM rotates on the inner surface of the trophectoderm cells so as to eventually assume a mesometrial location (Fig. 2). Kirby et al. (1967; Kirby, 1971) speculated that the same differential in oxygen tension that might lead to an antimesometrial attachment site could account for the final positioning of the ICM within the trophectoderm.

There has been some support for the ICM rotation hypothesis. The recent studies of Nadijcka and Hillman (1974) have confirmed that trophectoderm cells, although tightly connected to ICM cells by both desmosomes and focal tight junctions prior to hatching, subsequently lose these connections, and, as Kirby et al. (1967) reported, become separated by basement membrane material at the time of implantation. In microsurgical experiments, Gardner (1971) obtained isolated ICMs. He placed these inside trophoblastic vesicles consisting of trophectoderm cells which had previously constituted only the abembryonic and lateral walls of blastocysts. The ICMs were nevertheless able to appose themselves along these trophectoderm cells, and when the reconstructed blastocysts

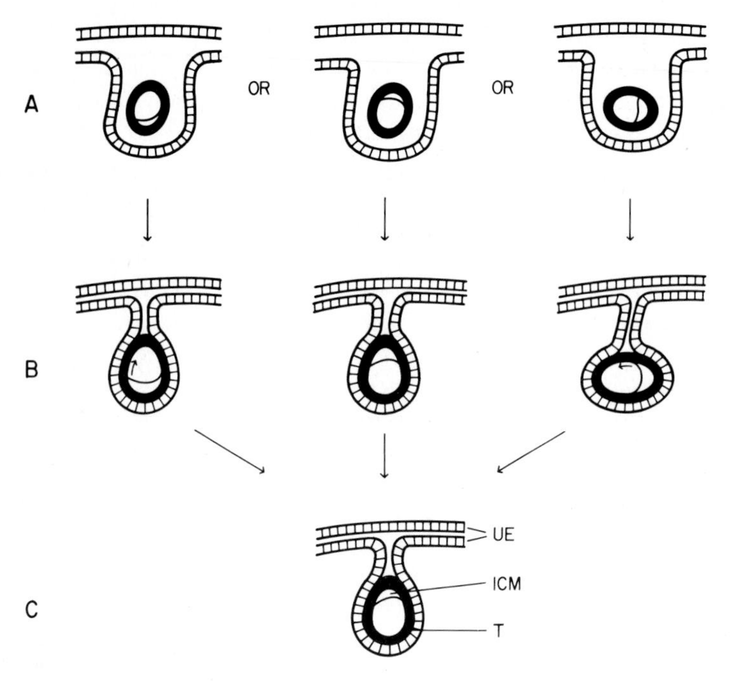

Fig. 2. Proposed scheme of blastocyst orientation. According to the proposal of Kirby et al. (1967), the ICM of the blastocyst is randomly oriented with respect to the mesometrial-antimesometrial axis during the apposition phase (A) and during the initial part of the adhesion phase (B). Thereafter, the ICM rotates about the inside of the trophoblast layer until it occupies a mesometrial location (towards the top of the figure) as in (C). UE: uterine epithelium; T: trophectoderm (later trophoblast); ICM: inner cell mass.

were placed in foster mothers, they presumably oriented themselves properly in the uterus, since they gave rise to live young. Kirby et al. (1967) reported that blastocysts transplanted to the anterior chamber of the eye inevitably align themselves so that the ICM is opposite the point of attachment. Finally, Jenkinson and Wilson (1970), during cinemicrographic observations of blastocysts implanting into lens fiber material in vitro, claimed to have actually observed the ICM of blastocysts rotating as much as 90° within the trophectoderm shell. The latter experiments would presumably preclude a specific role of the uterine environment in orientation of the blastocyst.

On the other hand, some recent observations may be inconsistent with the ICM rotation hypothesis. Ducibella et al. (1975) have found that prior to implantation, trophectoderm cellular projections appears to affix the ICM at one of the poles of the blastocyst. As these authors mention, however, it is quite possible for these interactions to decrease sufficiently to allow ICM rotation by the time implantation occurs. The following experiments by Gardner (1975) more seriously challenge the idea of ICM migration. Gardner marked individual trophectoderm cells with melanin granules. His procedure allowed him to mark cells at the embryonic or abembryonic poles, or on the lateral walls, of the blastocysts. He then placed these blastocysts in foster mothers and removed uteri for sectioning at a time when all but one of fifteen blastocysts were properly oriented. In no case was there any suggestion that the position of the ICM had changed relative to that of the marked trophectoderm cells.

Although Kirby and coworkers suggested that ICM rotation might explain blastocyst orientation in species other than the mouse, this hypothesis, even if it were correct, could not have universal applicability. For instance, in the guinea pig blastocyst, orientation must take place before hatching, since a multilayer "attachment cone" of trophectoderm cells forms at the abembryonic pole. It is these cells that project through the zona pellucida and anchor the blastocyst to the uterine wall (see Blandau, 1971, for a description).

The question of how the blastocyst orients itself in the uterus is therefore presently unresolved. Although the proper orientation of blastocysts transplanted to the anterior chamber of the eye certainly suggests that the blastocyst plays an active part in its orientation, it should be noted that blastocysts cultured in vitro do not appear always to attach to the substratum at the abembryonic pole, although they often do so (Cole and Paul, 1965; Sherman and Salomon, 1975). Enders (1971) has noted that the mouse blastocyst is larger, more flattened and more compressible at its abembryonic pole, and he has suggested that these factors may be involved in orientation. It is possible that the entire blastocyst may be rotated by uterine movements until the shape of the blastocyst caused it to settle in its proper orientation. In fact, the interdigitations between trophectoderm and uterine epithelial cells noted by Kirby et al. (1967) may suggest that a ratchet-like series of movements effects this rotation. Interdigitations between blastocyst and uterine surfaces, although frozen by fixation for electron microscopy, might nevertheless be fluid and temporary. If this is the case, then the role of the blastocyst in its orientation would be a passive one.

5. *Adhesion phase*

After the blastocyst has assumed its antimesometrial location and become properly aligned, it begins to implant by attachment to the uterine wall at its abembryonic pole. It should be noted that Potts (1969; Wilson and Potts, 1970) has considered "attachment" and "adhesion" as two separate phases of implantation. Since it is only the surface of the uterus, not the blastocyst, which distinguishes the two phases (an alteration from serpentine to a completely flat surface), we shall consider both of Potts' phases in a single classification. Also, Potts (1969; Wilson and Potts, 1970) suggests that part of the adhesion phase should include the insinuation of trophoblast cytoplasmic processes between groups of epithelial cells. We shall consider this event as the onset of the invasion phase, and it will be described later in the appropriate section. Because of differences in definitions and terminology, it becomes difficult to determine the length of time that the blastocyst spends adhering to the uterine wall prior to the beginning of the invasive phase. From the descriptions provided by Potts (1968), it is estimated that the adhesion phase persists for 10–15 hours. Certainly in vitro, blastocysts spend several hours anchored to a uterine monolayer before they begin to displace underlying uterine cells (see Fig. 2b of Salomon and Sherman, 1975).

The adhesion phase in the mouse differs notably from that in the rabbit. Attachment of the abembryonic trophoblast cells of the rabbit blastocyst to the uterine epithelium precedes attachment at the embryonic pole by a full day (Enders and Schlafke, 1971). Although the mouse blastocyst is first anchored to the uterus at the abembryonic pole, the membranes of the uterine epithelium appear to become juxtaposed to those on the remainder of the luminal surface of the blastocyst shortly thereafter (Potts, 1966, 1969; Nilsson, 1974).

5.1. Morphological studies

The luminal surface of the trophectoderm cell does not undergo any marked ultrastructural changes when it attaches to the uterine epithelium (Fig. 3). The surface of the uterine epithelial cell is, however, dramatically altered during adhesion: the regular pattern of microvilli disappears and is replaced with small, irregular projections (Potts, 1966, 1968, 1969; Reinius, 1967; Nilsson, 1967, 1974). Reinius (1967) and, in his earlier studies, Potts (1966, 1968) indicated that retraction of the microvilli from the uterine epithelial surface does not take place until after attachment has occurred, while Nilsson (1966, 1967) and Potts (1969), in a later description, suggest that disappearance of the microvilli increased the

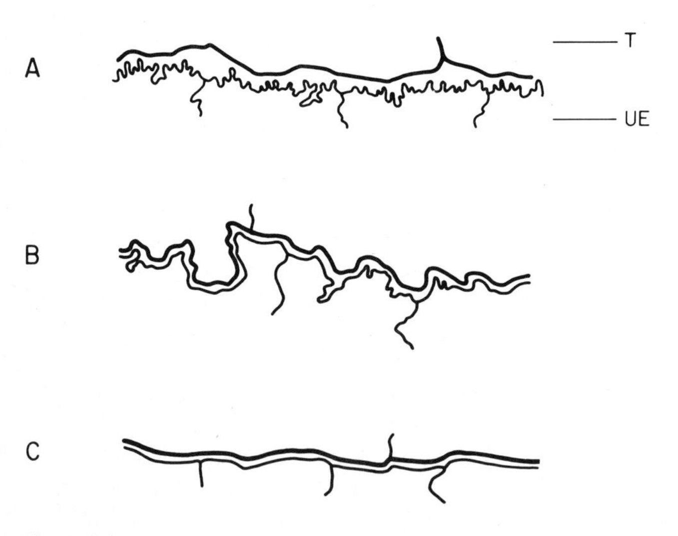

Fig. 3. Schematic representation of contacting embryonic and uterine surface membranes prior to, and during, implantation. During the apposition phase, the trophectoderm cell surfaces are smooth, while those of the uterine epithelium are microvillous (A). At the onset of the adhesion phase, the microvilli of the uterine epithelial surface membranes largely disappear and the trophoblast and uterine epithelial membranes are closely apposed and undulating (B). Finally, both surfaces flatten, forming a smooth border (C). T: trophectoderm (subsequently trophoblast); UE: uterine epithelium.

adhesiveness of the uterine surface and coincides with the onset of attachment. In fact, Nilsson (1967) proposed that it is these changes on the uterine epithelial surface which lead to increased adhesiveness and are therefore responsible for blastocyst attachment. To a large degree, the discrepancy arises because one can only guess when the blastocyst is actually first *attached* to the uterine wall, rather than being closely apposed, and the issue is further clouded by differences in terminology. In any case, it seems sensible that the smoothing out of the uterine epithelial cell surface, whether responsible for the initiation of attachment or not, provides a more compatible and complementary surface onto which the blastocyst can attach intimately and securely. However, if Nilsson is correct, this would suggest that the blastocyst has only a passive role during the beginning of the adhesion period, just as appears to be the case during most, or all, of the apposition phase.

During the initial part of the attachment phase (Fig. 3B), as the plasma membrane of the uterine epithelial cells produces a serpentine, gently undulating border, trophoblast cells become closely applied to the epithelium so that a gap of only 150–200 Å separates maternal and fetal cells over most of the contacting surfaces (Potts, 1966, 1968; Reinius, 1967; Smith and Wilson, 1974). Subsequently, the plasma membrane of uterine cells flatten out completely against the smooth trophoblast membranes (Fig. 3C). A homogeneous deposit which stains heavily with electron-dense stains soon appears in the extracellular spaces (Potts, 1966, 1968, 1969). The source of this material is not clear from ultrastructural studies. The material may be related to changes which occur in the surface properties of the implanting blastocyst (see section 5.4). On the other hand, Nilsson (1967) has noted, at least in the rat, that the closely apposed luminal surfaces not in contact with the blastocyst also have an electron-dense material between them. It may then be that both trophoblast and uterine cells secrete this material.

In certain areas, the plasma membranes of uterine epithelial and trophoblast cells seem to interconnect; both triple layered and septate desmosomes between maternal and fetal cells are observed (Potts, 1966, 1968, 1969). Occasionally, cytoplasmic continuity is established between cells. It would appear, however, that such occurrences are not as common in the mouse as they are in other species (Enders and Schlafke, 1969). Furthermore, neither Nilsson (1970) nor Smith and Wilson (1974) were able to find evidence of desmosomes or cytoplasmic continuity between trophoblast and uterine epithelial cells, and, as Potts (1969) has indicated, he may have been observing aberrant events associated with abnormal embryos. Certainly, fusion between trophoblast and uterine epithelial cells does not take place as it does during the adhesion phase in the rabbit, where there is widespread and complete fusion between maternal and embryonic cells (Larsen, 1961, 1970; Enders and Schlafke, 1971).

5.2. Hormone effects – implantation delay

Implantation delay is a naturally-occurring phenomenon in a number of mammals (see Enders, 1963 and Lanman, 1970). Lanman (1970) notes that

Lataste first recognized in 1891 that pregnant mice suckling a previous litter suffered prolonged pregnancies, and it was subsequently realized that lactational delay was due to a postponement of implantation through hormone imbalance. Implantation delay can also be induced in rodents by bilateral ovariectomy (Chambon, 1949). In the absence of any hormone treatment, unimplanted, but viable, mouse blastocysts can be removed from the uterus several days after ovariectomy (Weitlauf and Greenwald, 1968; Sherman and Salomon, 1975). Appropriate therapy with progesterone and estrogen, on the other hand, led to normal implantation and a resumption of gestational processes (Yoshinaga and Adams, 1966). Consequently, ovariectomy-induced implantation delay has been used effectively in studies designed to reveal the specific roles of steroid hormones in the various aspects of implantation.

5.2.1. Morphology of the blastocyst during delayed implantation

Ultrastructural studies on mouse blastocysts during, and after reversal of, implantation delay have been carried out by Nilsson and Bergström (Nilsson, 1967, 1974; Bergström and Nilsson, 1970, 1971, 1972, 1973, 1975; Bergström, 1972a,b) and by Potts and Psychoyos (1967; Potts, 1969).

Bergström and Nilsson (1971) have indicated that the blastocyst passes through three phases during and after implantation delay (Fig. 4). Following ovariectomy, the blastocyst hatches from its zona pellucida and is said to be in the "dormant" stage. This is so whether exogenous progesterone is supplied or whether there is no therapy at all, although there are some minor morphological differences in the two cases (Bergström, 1972b; Bergström and Nilsson, 1972). During the dormant stage, the uterine lumen begins to close about the blastocysts; although the blastocysts are free-floating two days after ovariectomy and in the presence of progesterone, by the seventh postoperative day, the uterine lumen is tightly closed and reminiscent of the apposition phase of normal pregnancy (according to Kirby, 1971) and the uterine lumen does not close unless progesterone is administered. Many microvilli are present on the surface of uterine epithelial cells, as are a number of apical protrusions (Bergström and Nilsson, 1972; Nilsson, 1974). The trophectoderm surface is also similar to that found in the apposition phase, except that it does have numerous microvilli, more so than in normal pregnancy (Potts and Psychoyos, 1967; Potts, 1969; Bergström, 1972b). Scanning electron microscopy reveals that the blastocyst surface is pocked with craters of varying size, presumably as a result of close contact with the protrusions on the surface of the uterine epithelial cells, as evidenced by conventional electron microscopy (Bergström, 1972b, Bergström and Nilsson, 1972, 1973; Nilsson, 1974). It is remarkable that the blastocyst is so closely apposed to the uterine surface, and yet attachment has not taken place.

Shortly after the administration of estrogen (ca. four hours), the blastocyst enters the "activated" stage (Fig. 4B; Bergström and Nilsson, 1971). The trophectoderm cell surface begins to bulge noticeably, particularly at the abembryonic pole, and microvillous projections become more numerous (Bergström and Nilsson, 1970, 1971, 1975; Bergström, 1972a), although probably not so

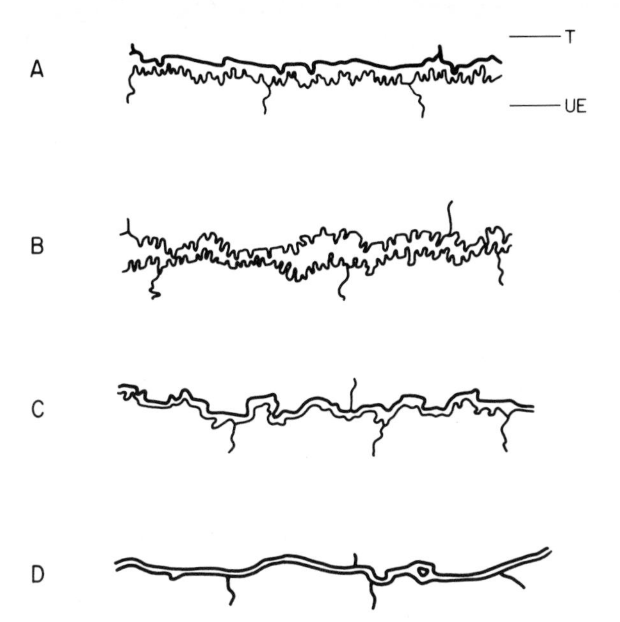

Fig. 4. Schematic representation of contacting embryonic and uterine surface membranes during, and following reversal of, delayed implantation. The sequence is derived largely from the experiments of Bergström and Nilsson (1971). During ovariectomy, in the "dormant" stage, the trophectoderm surface membrane is relatively smooth, while that of the uterine epithelium is microvillous (A). Following activation by administration of estrogen, both surfaces are initially microvillous (B). This is followed by the attachment stages, wherein the trophoblast and uterine epithelial surfaces are first smooth but undulating (C), and subsequently smooth and flattened (D). T: trophectoderm (subsequently trophoblast); UE: uterine epithelium.

numerous as those on the uterine epithelial surface (Nilsson, 1974). This stage appears to be uncharacteristic of the normal implantation process, since in the latter case, trophectoderm and trophoblast cells have relatively few microvillous projections from the time of hatching through the adhesion phase (see Fig. 3 and sections 4.1 and 5.1). Of course, it may be that this stage occurs very quickly in normal implantation and is difficult to detect. A further characteristic of the activation phase blastocyst is that its trophectoderm cells are no longer cratered (Bergström, 1972a, Nilsson, 1974; Bergström and Nilsson, 1975). According to Bergström and Nilsson, this is due to the secretion by uterine cells of a material which disrupts blastocyst-endometrial contacts (Nilsson, 1974; Bergström and Nilsson, 1975). If such secreted material exists, it is presumably not electron dense, since it is not evident in electron micrographs. In any case, the available evidence does suggest that following estrogen administration, the uterus loosens its grip somewhat upon the blastocyst. This phenomenon is also not observed in normal implantation.

The "activated" phase persists for several hours, since the third stage, "attachment" (Bergström and Nilsson, 1971), does not take place until about 16–24 hours after estrogen administration. This stage is very similar to the adhesion phase of normal implantation (Figs. 3B and 4C). The trophoblast cells flatten and increase in outer surface area. The surface membrane becomes quiescent; the few remaining microvilli have ridge-like processes (Bergström, 1972a; Bergström and Nilsson, 1970, 1971). Once again, the uterine epithelial cells undergo the same surface changes as they do in normal pregnancy (Figs. 4C,D); the attachment stage is characterized by the conversion of the uterine surface from microvillous to smooth and undulating (Potts, 1969; Nilsson, 1974). Finally, after attachment has occurred, an electron dense substance can be detected at the trophoblast-uterine epithelial interface (Nilsson, 1974), probably the same material that is deposited there during normal implantation.

In summary, there is much more evidence of changes in the blastocyst surface in the reversal of implantation delay than in normal implantation. It is, however, unclear whether those changes are a primary response to the administration of hormones, or whether the uterus responds initially to hormone therapy, and induces subsequent alterations on the part of the blastocyst. The mechanism of hormonal control over blastocyst implantation will be considered more fully in the section which follows.

5.2.2. Hormonal control of delayed implantation

There is ample evidence indicating that progesterone and estrogen initiate changes in the uterus at the time of implantation. On the other hand, there is little data to support the view that these hormones act in any direct way upon the implanting blastocyst, either in vivo or in vitro (see section 5.3.2). Smith (1968) has reported that blastocysts, after treatment with estrogen in vitro and transplantation to a progesterone-maintained uterus, were capable of implanting. These experiments were not conclusive, however, because small, but sufficient, amounts of hormone could have been carried over into the uterus, leading to local receptivity. In later studies, Smith and Smith (1971) reported that cultured blastocysts, but not morulae, showed transient increases in uptake and, at some dosages, incorporation into protein of amino acids in response to estrogen. However, the effect was only temporary (40 minutes or less), and studies were not carried out to demonstrate that other steroids did not also cause a stimulation. Furthermore, Weitlauf (1972) was unable to find an effect of estrogen in similar experiments. Finally, many other in vitro studies have failed to demonstrate a dependence on estrogen for proper blastocyst function and development (see section 5.3). While there is no solid evidence to rule out the possibility that steroid hormones act both on the uterus *and* the blastocyst in vivo, it is more likely that the effects of hormones and hormone withdrawal on blastocyst surface properties and metabolism are indirect, being mediated through uterine changes.

During implantation delay, metabolism of the blastocyst slows dramatically (for a recent review, see McLaren, 1973). Although cell division may continue for a day or two following the onset of delay, it eventually ceases with all cells in the G1 phase of the cell cycle (Sherman and Barlow, 1972). Consequently, any hypothesis put forward to explain how hormone deprivation blocks blastocyst attachment should also take into account the interruption of blastocyst metabolism and further development. Conversely, it must be decided whether or not the direct effect of hormone withdrawal on the blastocyst is to slow metabolism, and that this in turn leaves the blastocyst incapable of attaching. For instance, at the onset of implantation, the blastocyst has reached a stage in its development in which it is undergoing a marked transition in its intermediary metabolism and an acceleration of macromolecular synthetic processes (Biggers and Stern, 1973; Manes, 1975). The blastocyst at this time may thus require a very complex array of nutrients (Pincus, 1941; Gwatkin, 1966a,b; Hsu et al., 1974) and a high oxygen tension for carrying out energy dependent processes (Yochim, 1971). The embryo might then be forced into delay by the failure of the unprimed uterus to provide these essential nutritional or environmental requirements. This in turn may prevent synthesis by the blastocyst of surface components necessary for attachment (for a discussion of possible correlation between cell surface changes and metabolism, see Manes, 1975). On the other hand, the nonreceptivity of the uterine surface during delay (see previous section), and not interruption of normal blastocyst metabolism, may be responsible for the failure of the embryo to implant.

The observation that delayed blastocysts, when transferred to a suitable culture medium or to an ectopic site, can resume growth after a short lag (e.g., Gwatkin, 1966b) is consistent with the metabolic block hypothesis. But it also suggests another hypothesis, namely that the uterus produces an inhibitory substance which arrests blastocyst metabolism, growth and attachment. Estrogen administration would then promote blastocyst growth by neutralizing the inhibitor; at the same time the steroid could initiate surface changes in uterine epithelial cells, leading to attachment. A zona lysin (see below) could easily be a candidate for an estrogen-induced intermediate capable of inactivating an inhibitor and promoting blastocyst attachment.

Evidence for a uterine inhibitor in the rat and rabbit has been presented by Psychoyos (for a review, see Psychoyos, 1973) who found that uterine flushings from ovariectomized animals reduced the uptake and incorporation of radioactive uridine by blastocysts growing in vitro. On the other hand, in vitro observations cannot always be extrapolated to in vivo phenomena with respect to implantation delay (see section 5.3.2). Also, Cowell (1969) found that when blastocysts are transplanted to uteri of ovariectomized mice which had been previously traumatized with a glass scraper they were able to implant and grow at sites where damage had occurred to the epithelial surface. Although these results do not eliminate the possibility of a specific localized inhibitor being somehow removed as a result of the damage, such a case does seem unlikely.

Mintz (1971) has proposed that the zonalytic substance, produced in the uterus in response to estrogen administration (see section 3) is also directly responsible for initiating implantation. This "implantation-initiating factor" would presumably mediate attachment by altering blastocyst cell surface glycoproteins. This theory alone does not account for the decrease in blastocyst cell metabolism during delay nor does it explain the ability of blastocysts to attach in vitro or in ectopic sites in the absence of implantation-initiating factor.

In summary, during implantation delay, blastocyst metabolism is severely interrupted, as is its ability to attach to the uterine surface. The three most popular hypotheses put forward to explain the events which occur during implantation delay involve: (a) a metabolic block due to inadequacies in the unprimed uterine milieu; (b) a uterine inhibitor, inactivated by estrogen, which blocks implantation by interfering with normal blastocyst functions; and (c) an esterogen-induced activator, e.g. zona lysin, whose presence is necessary to initiate implantation. Many experiments, including the ones described above, have been carried out to help determine the possible role of these factors during delay and its reversal. The data obtained have often been contradictory or conflicting and, in our opinion, none of the hypotheses have been clearly proven or disproven. More conclusive experiments are necessary before the mechanism of hormonal control over the implanting blastocyst can be understood.

5.3. *In vitro studies*

While implantation is an interaction involving constant morphological changes, both on the part of the blastocyst and the uterus, the histological techniques used to study this phenomenon are static. The result can be compared to one's efforts to understand an intricate plot of a movie by assembling a large number of isolated frames. This situation could be remedied by the development of an in vitro system in which implantation took place just as it did in vivo. There have been relatively few efforts to develop such a system, and, unfortunately, none can yet faithfully reproduce the event as it takes place in the animal (see Sherman and Salomon, 1975 for a review). However, a number of analogies do exist between some in vitro studies and implantation in vivo, and some interesting findings have emerged.

The first efforts to study implantation in vitro were carried out by Glenister (1961). He followed the attachment of rabbit blastocysts onto isolated strips of endometrial tissue. However, parallel studies with mouse blastocysts were unsuccessful (Glenister, 1967). The simplest in vitro technique which provides information about the blastocyst when it is prepared for implantation is to allow it to hatch from its zona pellucida and attach to the surface of a culture dish (see, e.g., Fig. 5). Such studies with mouse blastocysts were first carried out by Mintz (1964), Cole and Paul (1965) and Gwatkin (1966a,b). Collagen has also been used as a substratum for blastocyst attachment (Hsu, 1971; Jenkinson and Wilson, 1973; Spindle and Pedersen, 1973). There are two critical ways in which this system deviates from implantation in vivo. Firstly, the substratum presents only two dimensions for attachment, and secondly unlike the uterine wall, a plastic or collagen substratum is an inert surface. Jenkinson and Wilson (1970) have attempted to overcome the former objection by studying blastocyst implantation into lens explants. An approach to the latter problem has been to allow blastocysts to implant on monolayers of cells (Cole and Paul, 1965; Salomon and Sherman, 1975; Sherman and Salomon, 1975; Sherman, 1975a). Grant (1973) has

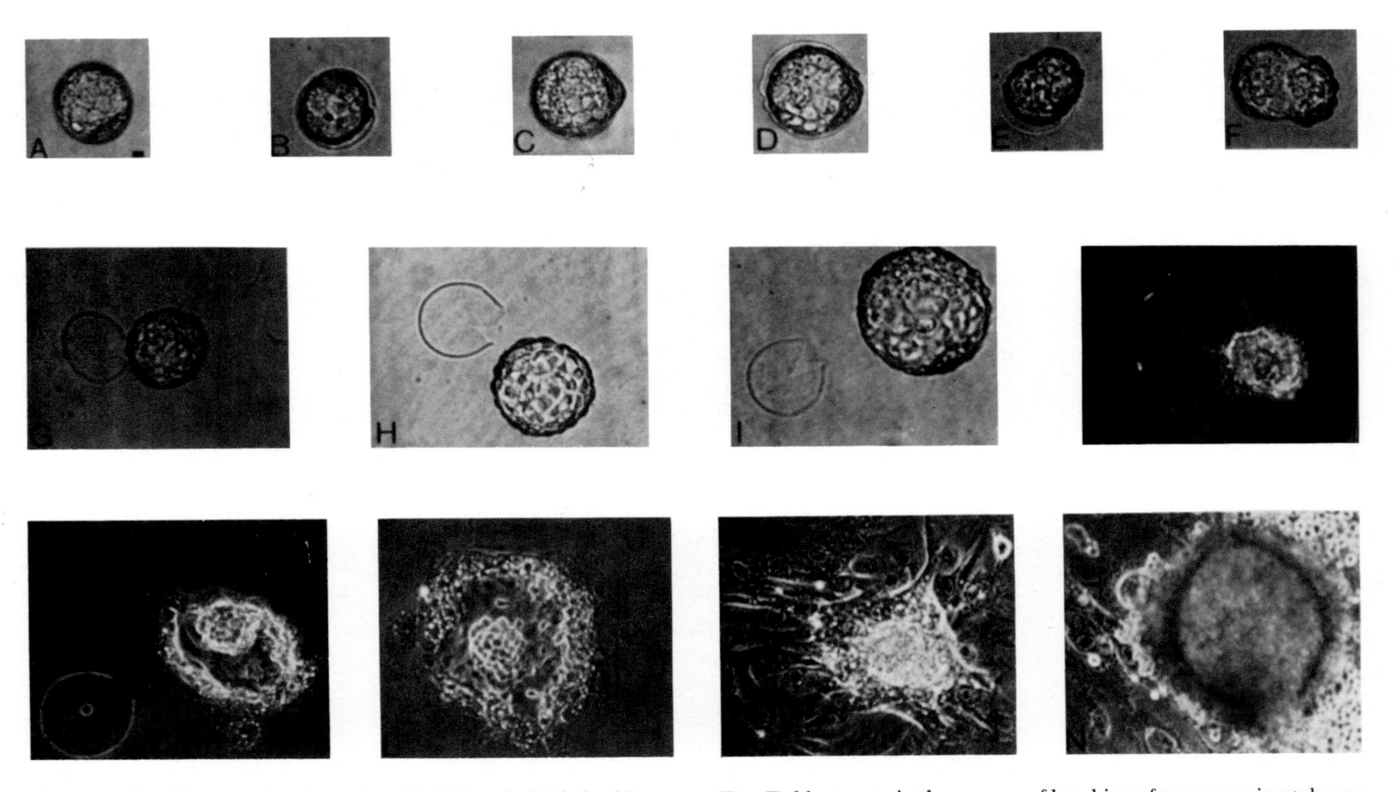

Fig. 5. In vitro blastocyst implantation. (A) Unhatched 4th day blastocyst; (B to F), blastocysts in the process of hatching after approximately one day of culture; (G–I), blastocysts, having hatched, attach to the surface of the culture dish and continue to expand (during second day of culture); (J), beginning of outgrowth of trophoblast cells along the culture dish (on second day of culture); (K–M), continuation of outgrowth of trophoblast cells and further development of inner cell mass (from 3 to 5 days of culture). Note that the nuclei of the cells in the trophoblast monolayer enlarge dramatically during polyploidization while the inner cell mass, which can be seen initially as a small ball of cells on the trophoblast monolayer (in K and L) forms a two-layered structure (in M) consisting of ectodermal cells surrounded by endoderm. Magnification is the same in all photos. Scale marker in A = 20 μm.

attempted to surmount both shortcomings by studying the implantation of mouse blastocysts in isolated, intact uteri. However, the result in the latter case is that, at least for morphological studies, the advantage of the in vitro system is lost, since the blastocysts can only be observed by fixation, sectioning and staining.

Before the above studies are considered further, it should be mentioned that a liberty is being taken here by the use of the word "implantation" to describe in vitro events. In the final section of this article, we shall consider, on the basis of available evidence, the extent to which this terminology is justified.

5.3.1. Requirements for blastocyst attachment in vitro

Standard cell culture media usually contain a balanced salt solution as well as glucose for an energy source, amino acids as a nitrogen source, some vitamins and serum, often fetal calf serum. All of these components are necessary for *optimal* development of mouse blastocysts in vitro. Although a balanced salt solution is required for embryo viability, experiments have been carried out to determine which, if any, of the other components are essential for blastocyst attachment to the substratum. According to Naeslund (unpublished work referred to by Nilsson, 1974), blastocysts are kept in a state of "delay" in vitro in the absence of glucose. However, no details were given as to the other components of the medium or to the nature of the substratum. Gwatkin (1966a,b) and Spindle and Pedersen (1973) have found that in vitro delay also takes place in the absence of amino acids. Gwatkin (1966a,b) reported that if arginine and leucine were omitted from the culture medium, blastocysts would fail to attach to the surface of the dish, at least for several days. These findings were confirmed by Sherman and Barlow (1972) and by Spindle and Pedersen (1973), though, interestingly, omission of either of these amino acids alone had little or no effect upon blastocyst attachment (Spindle and Pedersen, 1973). Omission of some other essential amino acids individually did, however, interfere somewhat with blastocyst attachment (Spindle and Pedersen, 1973).

Gwatkin (1966a) was unable to demonstrate any requirement for vitamins during blastocyst attachment to a culture dish. However, as Gwatkin has noted, these may be necessary only in trace amounts, so that endogenous levels are adequate.

A number of studies have been carried out on the necessity of serum or serum proteins for attachment of blastocysts in vitro. Gwatkin (1966b) found that whole serum or some serum fractions promoted blastocyst attachment to a culture dish, while other serum proteins (e.g., bovine serum albumin) could not. On the other hand, Menke and McLaren (1970) did observe attachment in vitro when bovine serum albumin served as the only macromolecular source. Sherman and Salomon (1975) reported that attachment of blastocysts to the surface of a culture dish could take place in the absence of any macromolecular source, although the percentage of blastocysts capable of attaching was less than that in the presence of serum, and the time required was also greater. If collagen is used as a substratum, attachment can take place either in the absence of serum proteins if all essential amino acids are present (Spindle and Pedersen, 1973), or, conversely, in the absence of amino acids in the presence of bovine serum albumin (Jenkinson and Wilson, 1973). Finally, Sherman and Salomon (1975) could find no effect of serum proteins on the number or rate of blastocysts attaching to a uterine monolayer. These observations, taken together, can probably best be explained by assuming that the role of serum proteins in blastocyst attachment in vitro is mainly to render an otherwise inhospitable culture surface receptive to blastocyst attachment. This has been proposed previously by Jenkinson and Wilson (1973; Wilson and Jenkinson, 1974). Thus, while a collagen substratum, uterine monolayer, or even some lots of plastic culture dishes might support blastocyst attachment, other lots of culture dishes might not until their surface had been coated with serum proteins.

The aforementioned observations on glucose and amino acid deprivation provide information which is probably more pertinent to implantation in vivo than are the studies on macromolecular requirements. The former experiments suggest that blastocysts must be actively metabolizing and synthesizing protein in order to become adhesive. The data is also consistent with the idea that the synthesis of new surface proteins by the blastocyst coincides with the implantation event (see section 5.4). If so, then ultrastructural studies suggesting that the trophoblast surface changes very little during the adhesion phase (see section 5.4) are misleading.

5.3.2. Hormonal aspects of blastocyst attachment in vitro

Studies described above have indicated that blastocysts remain closely apposed, but unattached, to the uterine epithelium during implantation delay. It is only with administration of steroid hormones that attachment takes place. If progesterone and estrogen were acting directly upon the blastocyst, then it might be the case that these hormones would also be required for attachment in vitro. The earliest studies directed to this question were by Glenister (1965). In his investigations on the implantation of rabbit blastocysts onto endometrial strips, Glenister found no stimulation of blastocyst attachment by the addition of progesterone and/or estrogen. The following year, Gwatkin (1966b) removed delayed blastocysts from uteri of ovariectomy-delayed mice and placed them in culture. These blastocysts experienced no difficulty in attaching to the surface of a culture dish in the absence of added steroid. Salomon and Sherman (1975; Sherman and Salomon, 1975) repeated Gwatkin's experiments using serum treated with dextran-norit so that neither progesterone nor estrogen could be detected and found that blastocysts were still able to attach to the culture dish at the same rate and frequency as blastocysts in medium supplemented with control serum. Nor were steroids required to effect attachment of blastocysts to a uterine cell monolayer under the same conditions.

It should be mentioned here that Dickmann and coworkers (Dickmann and Dey, 1974; Dey and Dickmann, 1974a,b) have claimed on the basis of histochemical studies that rodent blastocysts are capable of synthesizing both estrogen and progesterone, and proposed that this ability is necessary for implantation. This could also be imagined to affect attachment in vitro, i.e. blastocysts might produce their own steroid supply which allows them to attach in the absence of exogenous progesterone or estrogen. This is very unlikely to be the case. First, the existence of a putative estradiol dehydrogenase in blastocyst cells in no way provides evidence that blastocysts are capable of *synthesizing* estrogen as Dey and Dickmann (1974b) have suggested. Furthermore, the presence of a putative Δ^5,3β-hydroxysteroid dehydrogenase has only been demonstrated by Dickmann and coworkers with dehydroepiandrosterone as substrate, not with pregnenolone, the substrate that this enzyme converts to progesterone. In fact, extremely sensitive biochemical studies have been carried out which indicate that neither the preimplantation, nor the implanting, blastocyst can produce more than trace amounts of progesterone, if any. Significant progesterone synthesizing ability appears only after implantation, and is restricted to trophoblast cells (Chew and Sherman, 1975; Sherman and Atienza, 1975, 1976). Finally, blastocyst attachment in vitro is unaffected by cyanoketone, a potent inhibitor of Δ^5,3β-hydroxysteroid dehydrogenase activity (D.S. Salomon and M.I. Sherman, unpublished observations). Taken together, in vitro implantation studies strongly support the conclusion that the uterus and not the blastocyst is the target of steroid hormones during the attachment phase. In other words, implantation delay due to lactation or ovariectomy occurs primarily because the uterus is unreceptive to blastocyst attachment in the absence of steroid therapy. The lack of uterine receptivity in the absence of hormone priming may be due to properties of uterine epithelial cells. Cowell (1969) has observed that blastocysts are unable to attach to the uterine wall of cyclic or ovariectomized mice unless the uterine epithelium is scraped away. In the studies carried out by Sherman and Salomon (1975) and Salomon and Sherman (1975), the morphology of the uterine cell monolayer suggested that it consisted largely of stromal cells. It would be interesting to see if blastocysts are capable of attaching to monolayers of uterine epithelial cells with equal facility.

Although it has not yet proved possible to induce implantation delay in vitro by hormonal means, the studies reviewed above suggest that a similar phenomenon can be achieved by omission of essential metabolites from the culture medium. As with delay in vivo, the in vitro delay produced by amino acid deprivation can be reversed, in this case by the restoration of the missing amino acids to the culture medium. Blastocysts delayed as long as seven days in vitro can subsequently attach to the culture dish and continue to develop normally (Gwatkin, 1966b) and blastocysts delayed for four days in vitro can subsequently implant and develop normally in utero when transferred to foster mothers (Sherman and Barlow, 1972). To search for further similarities between ovariectomy-induced implantation delay in vivo and amino acid deprivation-induced delay in vitro, Gwatkin (1969) measured amino acid levels in uterine fluid from ovariectomized, pregnant mice to determine whether it might be low in levels of amino acids essential for attachment in vitro. This was found not to be the case (Gwatkin, 1969). Subsequently, cell cycle studies revealed a further difference between in vivo and in vitro delayed blastocysts (Sherman and Barlow,

1972). It would therefore appear as though delay induced in vitro by metabolite deprivation is not a satisfactory model for in vivo implantation delay, and the search is continuing for a closer analogy.

5.3.3. Surface of the blastocyst during attachment in vitro

Ultrastructural studies have not yet been carried out on the surface of mouse blastocysts undergoing implantation in vitro. However, there are some experimental results which offer clues about the changing nature of the blastocyst surface. For instance, it has been mentioned that the requirement by the attaching blastocyst for certain amino acids supports the idea that new surface proteins must be synthesized. Furthermore, the finite time interval between blastocyst hatching and attachment in vitro (Sherman and Salomon, 1975) also suggests that some alteration in the blastocyst surface must occur, and that shedding of the zona pellucida does not in itself prepare the blastocyst for attachment. In fact, premature removal of the zona pellucida by pronase does not shorten the time required before attachment can take place in vitro (Wilson and Jenkinson, 1974; Sherman and Salomon, 1975), indicating that the change in adhesiveness of the blastocyst surface is a temporally programmed event.

When blastocysts are removed from ovariectomy delay and placed in culture, their behavior is rather unusual. Initially, they are very sticky. After three hours in vitro, most are firmly attached to the culture dish or the uterine monolayer. After twenty hours in vitro, however, almost all the blastocysts have become disengaged from the cell surface and float freely once again. Thereafter, permanent attachment begins to take place, and subsequent development is the same as for control blastocysts (Sherman and Salomon, unpublished observations). It would therefore appear as though the surface of the delayed blastocyst undergoes an even more complex series of changes on its way to attachment than does a normal blastocyst. The adhesive→non-adhesive→adhesive sequence that the delayed blastocyst passes through may correspond, respectively, to the dormant, activated and attachment phases that Bergström and Nilsson (1971) have described (see section 5.1.1). The numerous microvilli on the blastocyst surface during the activated phase may signal a series of disruptions on the cell membrane that lead to the disengagement of the blastocyst from the culture surface.

5.4. Cell surface studies

Sophisticated genetic and immunological techniques have been utilized to describe the appearance and disappearance of specific surface antigens (e.g., F9 antigen, also found on teratoma cells and H-2, or major histocompatibility antigens) during preimplantation and early postimplantation stages. These surface antigen studies, which are discussed elsewhere in this volume (Edidin, 1976), have not been convincingly implicated in implantation, and will not be considered further here.

While morphological studies (sections 5.1 and 5.2.1) indicate that the uterine epithelial surface undergoes notable changes which could lead to attachment of the blastocyst, in vitro experiments (section 5.3) suggest that the blastocyst surface is also altered at the time of attachment. Recently, a series of studies have been initiated to characterize these presumptive implantation-related surface changes directly, especially those occurring on the blastocyst. To date, two basic approaches have been used. In the first, the surface charge has been studied on blastocysts prior to, and during, attachment. Histochemical methods have been used most often in these experiments. The second series of experiments have been designed to characterize biochemically the nature of surface components. Although these studies are mostly preliminary in nature and often open to criticism, they are described below.

5.4.1. Surface charge

Holmes and Dickson (1973) used colloidal iron-Prussian blue stain to demonstrate the presence of negatively charged molecules on the blastocyst surface. Since unhatched and delayed blastocysts did not stain, and the pretreatment of positively staining (fifth day) blastocysts with neuraminidase resulted in loss of staining, these authors concluded that estrogen stimulation of the blastocyst prior to attachment induces the synthesis and secretion (or insertion into the outer membrane surface) of negatively charged, sialic acid-containing, moieties. They further postulate that the temporal correlation of surface adhesiveness and an increase in tyrosine aminotransferase (TAT) activity with the estrogen surge indicates that blastocyst attachment and TAT activity are under hormonal control similar to that described by Ballard and Tomkins (1969) for the induction of cell adhesiveness and TAT activity in hepatoma cells. In the latter system, however, hormonal induction of cell adhesiveness results in a *decrease* of net negative charge on the cell surface. Contrary to the findings of Holmes and Dickson, other studies by Nilsson et al. (1973) on delayed and activated delayed blastocysts using the colloidal iron technique found that the surface of the blastocyst becomes *less* negatively charged at the time of implantation. This discrepancy, which is at present unresolved, could be due either to a difference in timing, i.e. a decreased surface charge just prior to attachment that is not observed with normal, unattached, blastocysts or to a difference in staining procedure. Nilsson et al. further suggested that the decrease in negative charge could be due to loss of sialic acid residues which, in turn, would increase the flexibility of the cell surface, and hence might account for the observed decrease in surface microvilli just prior to attachment (see section 5.2.1).

Clementson and coworkers (Clementson, Moshfeghi and Mallikarjuneswara, 1971; Clementson et al., 1972) have proposed that the electrostatic repulsion between negatively charged blastocyst and uterine surfaces can be neutralized without overt changes on the part of the cells. These investigators reported that the rat blastocyst has a net negative charge after hatching as, presumably, does the uterine surface. They suggest that the estrogen surge results in an increase in potassium ion concentration in the uterine fluid which lowers the membrane potential of the blastocyst and uterine cell surfaces. This, they propose, would allow surface interactions and attachment to take place. Progesterone treatment also increases the uterine potassium ion concentration, presumably by decreasing the net amount of uterine fluid. This latter observation may or may not have bearing on the ability of progesterone to help maintain blastocysts in delay and even induce implantation in some cases.

The above observations and hypotheses notwithstanding, it has been pointed out by Jones and Kemp (1969) that it is unlikely that the ability of cells to adhere depends purely on electrostatic forces. Their studies with tumor cells show little correlation between surface charge and adhesiveness. Still, at least in the case of hepatoma cells, steroid induction does result in a decreased surface charge and an increase in the ability of the cells to adhere to a given substratum. Nevertheless, before the role of surface charge in blastocyst implantation can be properly assessed, the discrepancies described above must be resolved. Then it must be decided whether any observed changes in surface charge are the cause or the effect of implantation.

5.4.2. Surface glycoproteins

Biochemical studies by Pinsker and Mintz (1973) suggest that changes in the glycoproteins on the surface of the embryo take place prior to implantation. These investigators cultured preimplantation embryos in the presence of radioactive D-glucosamine and measured incorporation of the label into cell surface glycoproteins, both quantitatively and qualitatively. They found an increase in incorporation from the 4- to 8-cell stage to the early blastocyst and a shift in the labeled components toward higher molecular weights. The increase in high molecular weight components in the proteolytic digests of morulae and blastocysts over those of 4- to 8-cell embryos could mean that new components are synthesized at the later developmental stages, or that there has been a shift in the ratio of large to small molecular weight components, or that a change in the membrane results in increased vulnerability of high molecular weight components to trypsinization, or possibly a combination of these mechanisms.

Enders and Schlafke (1974) used concanavalin A, ruthenium red and colloidal thorium dioxide to study surface properties of mature blastocyst and uterine epithelial cells. They concluded from these studies that both types of cells possess a negatively charged glycoprotein coat which differs very little in staining properties from those of blastocysts prior to hatching or during delayed implantation. The uterine surface coat is very thick compared to that of the blastocyst and does not change appreciably in preparation for attachment. However, these authors later suggested the possibility of a localized change in thickness at the attachment site (Schlafke and Enders, 1975).

The blastocyst and/or the uterine surface may be altered in preparation for implantation not only by synthesis of new moieties, but also by degradation of existing ones. Pinsker, Sacco and Mintz (1974) have postulated that the uterine zonalytic factor is a protease which alters the composition of glycoproteins on the blastocyst surface, thus making it possible for the blastocyst to adhere to the uterine wall. The observation that morulae will not attach (even after the zona is removed by the zona lysin) suggests that their previously observed change in the cell surface (Pinsker and Mintz, 1973; see above) might be necessary for the function of the protease in initiating attachment.

While much of the above data implicate hormonal controls and a maternally-produced protease in controlling possible changes on the blastocyst surface, data from in vitro (section 5.3) and ectopic (section 6.4) studies cannot be ignored. There is ample evidence to suggest that even outside of the uterus, the blastocyst passes from a stage during which it is unable to attach, to a phase in which it can do so.

6. *Invasive phase*

Perhaps the most fascinating aspect of implantation is the invasion of the uterine wall by trophoblast cells. The invasive phase takes place over a five day span (Kirby, 1965a). Although the subject of trophoblast invasiveness is of interest to embryologists, reproductive physiologists and oncologists, surprisingly little is known about several aspects of invasion of these cells during normal pregnancy. In fact, much of what we do know about trophoblast invasiveness is derived from ectopic or in vitro studies.

There are at least two reasons for trophoblastic cells invading the uterine wall during pregnancy: to firmly anchor the embryo in its proper position, and to acquire nutrients from the mother and pass them to the fetus. According to Boyd and Hamilton (1952), the latter can be achieved either by ingestion of dead cells which are victim to the invasion process or by coming into direct contact with the maternal blood supply. In fact, Böving (1966) has provided evidence that at least in the rabbit, maternal blood vessels underlying the uterine epithelium determine the initial sites and direction of trophoblast invasion. However, it should be stressed that there are vast differences among species in the extent of invasiveness. Invasiveness in the mouse is such that the placenta is of the hemochorial type where trophoblast cells are in direct contact with maternal blood. At the other extreme, e.g., in the horse or the pig, the placenta is epitheliochorial and the non-invasive trophoblast cells do not even penetrate the uterine epithelium (see Amoroso, 1952). Even among hemochorial types, e.g., mouse vs rabbit, there are marked differences in the patterns of invasiveness. These have been considered in detail by Schlafke and Enders (1975).

6.1. *Morphological studies*

The steps involved in the invasion process in the mouse are illustrated schematically in Fig. 6. According to Potts (1969), there is a pause between the time that the blastocyst becomes firmly adherent to the uterine epithelium and the onset of trophoblast invasiveness. Invasion in the loosest sense is probably initiated when narrow processes of attached trophoblast cells insinuate themselves between uterine epithelial cells, and reach almost down to the basement

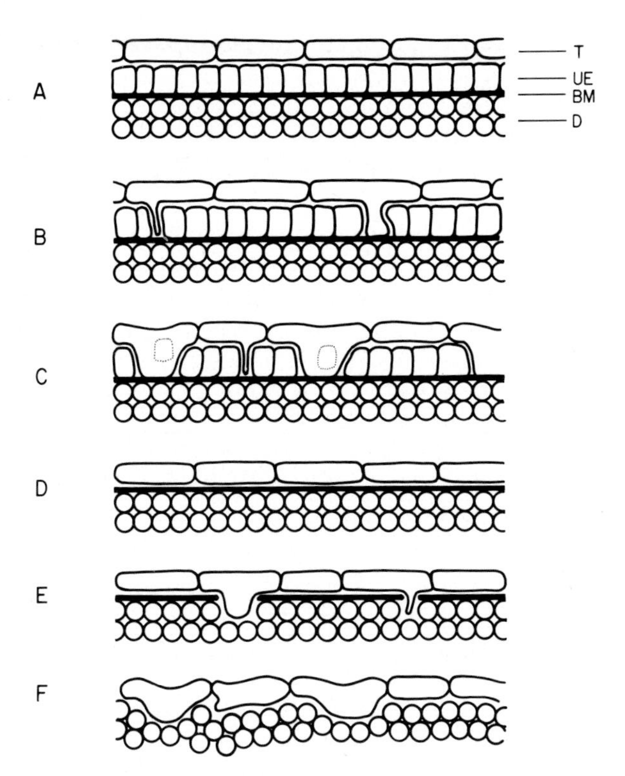

Fig. 6. Schematic representation of trophoblast invasion during implantation: (A), adhesion phase, prior to invasion; (B), insinuation of trophoblast cells into the uterine epithelial layer; (C) displacement of uterine epithelial cells by trophoblast (note phagocytosed uterine epithelial cells in trophoblast); (D) displacement has taken place, and trophoblast cells come to lie against basement membrane of the uterine epithelium; (E) penetration of basement membrane by trophoblast cells; (F) the basement membrane has been penetrated, and trophoblast cells come to lie in direct contact with decidual cells. T, trophoblast cells: UE, uterine epithelial cells; BM, basement membrane of uterine epithelial cells; D, decidual cells.

membrane of the latter (Potts, 1968; Finn and Lawn, 1968; Smith and Wilson, 1974). Although this usually takes place after attachment has occurred, it should be noted that Kirby (1971) claimed to have observed a similar phenomenon by light microscopy as unattached blastocysts in implantation delay became closely apposed to the luminal wall. Consequently, the significance of this phenomenon and its relationship to the actual invasion phase are open to some question.

Schlafke and Enders (1975) have described three ways in which trophoblast invasiveness can begin. In the first (intrusive implantation), trophoblast cells penetrate between intact uterine epithelial cells and come to rest upon the basal lamina of the epithelium. In the second (fusion implantation), the trophoblast as a syncytium actually fuses with uterine epithelial cells as the initial step in the invasive process. Finally, in displacement implantation, the uterine epithelium dislodges from the underlying basal lamina, and is replaced by trophoblast cells.

Although the initial infiltration of trophoblast cytoplasm suggests that intrusive implantation may be the mechanism by which mouse trophoblast cells invade the uterine epithelial layer, the bulk of the evidence supports the view that displacement best describes the process. According to Schlafke and Enders (1975), uterine epithelial cells are easily displaced from the underlying basal lamina in the mouse, and this displacement appears to begin before trophoblast invasion. In fact, according to El-Shershaby and Hinchliffe (1974), the first sign of degeneration of uterine epithelial cells can be detected *prior* to the breakdown of the zona pellucida. Even after the blastocyst has hatched, ultrastructural studies suggest that breakdown of the uterine epithelium is autolytic, rather than being due to contact with trophoblast cells (Smith and Wilson, 1974; El-Shershaby and Hinchliffe, 1975). Furthermore, earlier studies by Finn and Hinchliffe (1964, 1965) indicated that in deciduomas, uterine epithelial cells situated on the antimesometrial wall of the crypt begin to degenerate at the same time as they do in normal pregnancy. In both cases, mesometrially situated epithelial cells remain intact during this period. Finally, studies by Finn and Bredl (1973) with actinomycin D indicate that this antimetabolite prevents breakdown of the uterine epithelium, yet has no adverse effect upon the blastocyst at the doses used. This observation supports the idea that the degeneration of uterine epithelium is intrinsically programmed, and not due to trophoblast invasiveness.

Observations such as those described above lead to the idea that although trophoblast cells may be very migratory, insinuating themselves into spaces vacated by dead or dying epithelial cells, they are not themselves cytolytic. In fact, Finn and Lawn (1968) and Mulnard (1970) have proposed that those cells in the uterine epithelial layer which are the first to die serve as markers for the initial sites of penetration of invading trophoblast cells. Cowell's (1969) observation that trophoblast cells are unable to penetrate the intact uterine epithelial layer in unprimed (cyclic or ovariectomized) mice supports this view. In contrast to Cowell's observations, Finn and Bredl (1973) reported that trophoblast cells could invade the uterine epithelial layer after actinomycin D was administered to prevent its breakdown. However, it is not clear from the latter study whether or not trophoblast cells were merely penetrating discontinuities in the uterine epithelial layer where the antimetabolite had not been completely effective in preventing degeneration. In any case, the uterine epithelium becomes very abnormal in the presence of actinomycin D, forming a stratified multilayer, and it is therefore difficult to interpret the significance of trophoblast penetration of this unusual structure.

In contradistinction to the autolysis hypothesis, Wilson and Smith (1970) reiterated the early proposal of Duval (1891) that trophoblast cells might, at least initially, be active in the cytolysis of uterine epithelial cells. In support of their proposal, they subsequently presented data from histochemical and electron microscopic studies to show that a "border of lysosomes" was built up in trophoblast cells at the maternal junction at the time of blastocyst attachment to the epithelium (Smith and Wilson, 1971; Smith and Wilson, 1974). However, other electronmicroscopic studies have not been able to confirm these observations (El-Shershaby and Hinchliffe, 1975). The fact that dead uterine epithelial cells can be observed prior to hatching of the blastocyst is not quite such damaging evidence against a role for trophoblast participation in breakdown of uterine epithelium when it is considered that only very few (four or less) dead epithelial cells can be seen at this time, and that similar numbers of dead cells, presumbly from the ICM, can be seen within the zona-enclosed blastocyst itself (Wilson and Smith, 1970; El-Shershaby and Hinchliffe, 1975). The cause of death of these two classes of cells, which together constitute the so-called W-bodies (Wilson, 1963b; Finn and McLaren, 1967) is not clear, but there is no good reason to believe that it is in any way related, other than temporally, to implantation.

It should be pointed out, in the face of the above discussion, that breakdown of the uterine epithelium is not a *sine qua non* of invasiveness into the uterine wall: Wilson and Potts (1970) have found that melanoma cells invade through the layer of uterine epithelium without ever damaging individual epithelial cells.

Although some disagreement may exist about the cytolytic nature of mouse trophoblast cells, there is no argument that these cells are phagocytic. This fact has been recognized for many years (see Amoroso, 1952). Light microscopic and ultrastuctural studies have shown that invading trophoblast cells can isolate and phagocytose not only dead and degenerating uterine epithelial cells, but also apparently "healthy" cytoplasm from maternal cells and even intact uterine epithelial cells. In many cases, several (up to six) phagocytosed cells can be seen within the cytoplasm of single trophoblast cells (see Finn and Lawn, 1968; Mulnard, 1970; Nilsson, 1974; Smith and Wilson, 1974 and El-Shershaby and Hinchliffe, 1974 for studies in the mouse; and Tachi et al., 1970 for parallel work in the rat). Trophoblast cells begin to phagocytose degenerate uterine epithelial cells some ten hours after hatching from the zona pellucida, which is at the beginning of, or even prior to, the actual invasion by trophoblast into the uterine epithelium. Smith and Wilson (1974) have claimed, however, that there is a distinction between the phagocytic activities of trophoblast cells prior to, as opposed to during, the invasion phase. In the former case, only dead cells are phagocytosed, while in the latter case, both live and dead cells are engulfed. About ten hours after phagocytosis has begun, increased numbers of Golgi vesicles appear within the giant cells as they begin to digest cells which have been phagocytosed (El-Shershaby and Hinchliffe, 1975). If uterine epithelial cells do, in fact, die without trophoblast intervention, thereby allowing trophoblast cells to invade into the uterine stroma without resistance, then there would appear not to be any good reason for trophoblast cells to phagocytose degenerating epithelial cells other than for nutritive purposes, as Grosser first suggested (see Boyd and Hamilton, 1952).

It is interesting that trophoblast are not the only cells of the blastocyst capable of phagocytic activity. According to El-Shershaby and Hinchliffe (1974), dead cells present in the blastocyst prior to hatching are engulfed both by trophoblast and by ICM cells. Furthermore, Smith and Wilson (1974) have reported that dead uterine epithelial cells can be phagocytosed by other intact epithelial cells.

Although the advancing trophoblast layer is temporarily halted in its invasiveness by the basal lamina of the uterine epithelium (Potts, 1969; Smith and Wilson, 1974), it subsequently displaces this layer as well as it continues to migrate into the uterine stroma, particularly at the periphery of the ectoplacental cone. Although ultrastructural studies of mouse trophoblast cells at this stage do not appear to have been published, a thorough light microscopic description of trophoblast activity between the seventh and tenth days of gestation has been given by Amoroso (1952). As Amoroso indicated, invading trophoblast cells all around the embryo burrow into the decidual tissue and displace the endothelial cells lining the included maternal blood vessels. The end result is that the maternal circulating blood comes into close contact with fetal capillaries so that exchange of nutrients can take place with facility.

6.2. Decidualization

During implantation of the blastocyst, the uterine stroma undergoes a transformation, as the proliferation of connective tissue cells greatly enlarge the stromal area and displace the uterine glands. This phenomenon, known as decidualization, begins at the time of blastocyst attachment (Finn, 1972). Insofar as decidual cells are of maternal origin, the subject of decidualization lies outside of the scope of this review except for the questions of the role that the blastocyst might have in stimulating the phenomenon, and how the decidua might act to reverse or limit trophoblast invasiveness. A consideration of the first question follows, while the second is discussed in section 7.

The available evidence makes it quite clear that decidualization requires the correct hormonal conditions in order to occur at the time of blastocyst attachment to the uterine wall (Glasser, 1972; Finn, 1972; Psychoyos, 1973). During normal pregnancy, the other half of the necessary stimulus comes from the attachment of the blastocyst. However, although the blastocyst alone in a non-hormonally-primed uterus is inadequate to induce decidualization (Cowell, 1969), a variety of living and inert materials can replace the blastocyst in inducing the decidual response in a hormonally primed uterus. Besides blastocysts from heterologous species (Potts, 1969; Kirby, 1970), ectoplacental cones (Kirby, 1965a; Kirby and Cowell, 1968), sarcoma cells (Smith and Hartman, 1974), arachis oil (Finn and Hinchliffe, 1964) and even air bubbles (McLaren, 1970b) are all competent. On the other hand, neither cleavage stage embryos (Kirby, 1970), melanoma cells (Wilson and Potts, 1970) nor plastic beads the size of blastocysts (McLaren, 1968b) will produce a decidual response. Therefore, while the inducer of decidualization cannot be completely nonspecific, the unusual array of substitutes for the blastocyst suggests that the normal inducer is not unique in possessing the necessary qualifications. A consideration of the relationships among the various types of artificial inducers (McLaren, 1970b; Finn, 1972) has not yet led to the identity of the actual effector.

6.3. *Biochemical studies*

The morphological evidence cited above suggests that uterine epithelial cells at the time and site of implantation either die and undergo autolysis or are engulfed by invading trophoblast cells. Both of these mechanisms would involve catabolic enzymes such as the lysosomal acid hydrolases and, consequently, some of these enzymes have been studied in rabbit, rat and mouse implantation systems. Acid phosphatase is a commonly used lysosomal "marker" enzyme because it is easily localized in the cell by histochemical staining. Smith and Wilson (1971) utilized this enzyme to localize lysosomal-like organelles in mouse trophoblast and uterine epithelial cells during the invasive phase (see section 6.1). Since enzyme activity appears to be concentrated in a few lysosome-like particles at the fetal-maternal border of the trophoblast cells, these authors suggest that trophoblast lysosomal activity may be primarily responsible for membrane changes which facilitate closer contact between embryonic and maternal cells and, at the same time, aid in phagocytosis of dead uterine epithelial cells during the early phase of implantation. Furthermore, Smith and Wilson observed a much greater concentration of enzyme-rich particles in the uterine epithelial cells and attributed the general epithelial breakdown to cell death and autolysis. It was later suggested (Smith and Wilson, 1974), that phagocytosis by trophoblast of living and dead epithelial cells was also involved, and that epithelial cell death might be a result of the encroaching blastocyst and decidua cutting off or exhausting nutritive supplies.

Christie (1967) assayed a variety of enzymes in rat and rabbit blastocysts and uterine epithelial cells. He concluded that while various hydrolase activities are increasing in the blastocyst during implantation, none are exceedingly high. On the other hand, degeneration of the uterine epithelium is associated with a much greater amount of hydrolase activity in those cells. Furthermore, Abraham et al. (1970) found acid phosphatase activity and lysosome number in rabbit blastocysts to be constant during implantation while the activity of the enzyme not only increased in uterine epithelial cells, but was accompanied by a distinct series of changes in lysosome morphology which indicated that autolysis was taking place.

Hall (1971) described what might be lysosomal acid phosphatase activity in mouse uterine stromal cells in the vicinity of the blastocyst at the time of implantation. This activity appeared to spread out during the following 24 hour period. Smith (1972) obtained similar results with alkaline phosphatase and concluded that some factor at the onset of decidualization triggers synthetic activity in surrounding cells. Indeed, Finn and Hinchliffe (1964) considered alkaline phosphatase activity the "index of decidualization".

Although acid phosphatase may be a suitable marker for lysosomes, it is not likely to be the enzyme responsible for the breakdown of the uterine epithelium during implantation. Protease activity has also been measured in mouse uteri during this time. At least three different proteases have been observed to date. Pinsker, Sacco and Mintz (1974) described a protease in mouse uterine fluid which degrades casein at neutral pH and peaks on the fourth day, apparently in response to the estrogen surge. Bergström (1972a) has reported two acid aminoacylnaphthylamidase activities in the uterine epithelium at implantation with maximum activity appearing on the fifth day. These enzymes are present in uteri of ovariectomized mice maintained on progesterone and do not appear to be stimulated by estrogen. Combined with the earlier observation that proteolysis occurs at implantation sites and adjacent epithelium (Bergström, 1970), these results indicate that the mouse uterine epithelium does indeed have autolytic capabilities at the time of implantation. On the other hand, uterine protease activity can also be correlated with zona lysis (Pinsker et al., 1974) and perhaps with the breakdown of cell surface coating (see section 5.4) to facilitate attachment. The number and specificities of maternal proteases present in the uterus at the time of implantation are questions still to be answered.

Using the gelatin film technique, Denker (1972) reported the presence of a protease in rabbit trophoblast at the time of implantation. Owens and Blandau (1971) found proteolytic activity in guinea pig, but not in rat, trophoblast. Andary, Dabich and Van Winkle (1972) reported trypsin- and chymotrypsin-like enzyme activities in whole mouse blastocysts.

Taken together, all these experiments suggest that there is an array of proteases and other degradative enzymes at the implantation site during the invasive phase. Identification and characterization of these enzymes will require extensive and systematic studies with a variety of assay procedures and substrates. While the aforementioned production of proteases by uterine cells would be consistent with morphological studies suggesting that uterine epithelial cell breakdown is autolytic, the detection of proteases in trophoblast cells in the mouse and in other species should not be ignored. On the other hand, the presence of a protease activity in either trophoblast or uterine epithelial cells does not necessarily implicate that enzyme as being involved in cell degeneration. Perhaps enzyme inhibitor studies would shed some light on this problem.

In addition to producing catabolic enzymes, Christie (1967) pointed out that the implanting blastocyst is also characterized by elevated levels of enzymes which are necessary to meet increased energy requirements. Wong and Dickson (1969) similarly concluded that the change in dehydrogenase profiles in the fifth and sixth day mouse blastocyst suggests a decreased dependency on anaerobic respiration and an increase in oxidative phosphorylation. This conclusion was also reached by Ginsberg and Hillman (1975) based on the levels of, and effects of various metabolic inhibitors on, ATP synthesis in blastocysts.

In summary, combined morphological observations and biochemical data suggest, at this point, that trophoblast invasion and uterine epithelial cell degeneration are a result of a combination of events including the increased growth and metabolism of the blastocyst, death and autolysis of the epithelial cells, and phagocytosis by trophoblast of both dead and living epithelial cells.

6.4. Ectopic studies

As mentioned in section 3, ectopic studies clearly demonstrated that hatching of mouse blastocysts could take place independently of maternal factors. These same studies provided a good deal of information about the autonomy of blastocyst implantation, and particularly about the role of trophoblast cells in the process of invasion. From the earliest of these experiments (Nicholas, 1942), it was established that giant trophoblast cells, previously thought by a number of anatomists to be of maternal origin, could develop from cleavage stage rodent embryo transplants. The giant trophoblast cells so formed not only resemble those in normal pregnancy morphologically (Runner, 1947; Fawcett et al., 1947), but also biochemically (Barlow and Sherman, 1972; Sherman, 1972a,b; Chew and Sherman, 1975). Furthermore, both Runner (1947) and Fawcett et al. (1947) realized from their observations on preimplantation embryos placed in the anterior chamber of the eye that the developing trophoblast cells not only proceeded to become firmly anchored in the wall of the chamber, but actually invaded it in a process similar to implantation in utero. In fact, Kirby (1965b) noted that in some cases, blastocysts transplanted under the kidney capsule could give rise to trophoblast cells organized into structures remarkably similar to the fetal placenta in utero. It became patently clear that trophoblast differentiation (morphological, biochemical and functional) could take place independently of the uterine milieu or, for that matter, of any specifically maternal factors, since ectopic studies gave equivalent results in females and males. Furthermore, it also appeared that the presence of embryo proper was not necessary for this development of trophoblast, because in many ectopic implants derivatives of the ICM failed to proliferate.

The subject of experimental ectopic pregnancy and its bearing on the implantation event has been ably reviewed previously by Kirby (1965a) and by Billington (1971). Relatively little work in this field has been done since the latter of the two reviews. Consequently, rather than providing a detailed description of the experiments which have been carried out, we shall deal instead with the general concepts which have emerged from these studies, and consider how they are related to the invasion phase of implantation.

The most striking fact to have been revealed from ectopic studies is the degree of invasiveness of trophoblast cells. Trophoblast developing from mouse blastocysts can implant into, and invade, not only the wall of the anterior chamber of the eye (Runner, 1947; Fawcett et al., 1947) the abdomen (Fawcett et al., 1947) and the kidney (Fawcett, 1950; Kirby, 1960), but also the testis, spleen, liver and brain (see Kirby, 1965a). The invasive activities of implants of ectoplacental cone cells are even more remarkable. They will implant into all the above tissues, and also into the normally unreceptive unprimed uterine wall (Kirby, 1965a). In fact, Kirby (1962b) reported that trophoblast cells developing from ectoplacental cones can invade into mammary carcinomas and destroy many of the tumor cells therein. These observations have led Solomon (1966) to state that trophoblast cells are the most highly invasive cells known. Kirby (1965a) further maintained that ectopic trophoblast is far more destructive than the same cells in their normal uterine milieu. This may be due to the resistance of decidual cells to attack by trophoblast (see section 7).

Ectopic studies have not resolved the issue of whether trophoblast cells are cytolytic. In fact, they have further confused it. Trophoblast cells appear to invade all ectopic sites except for the testis by phagocytosis (Kirby, 1965a; Kirby and Cowell, 1968; Billington, 1971). In kidney implants, entire uriniferous tubules appear to be engulfed by trophoblast cells (Kirby, 1965a), as are apparently healthy tumor cells when ectoplacental cones and tumors are cotransplanted to the testis (Solomon, 1966). On the other hand, several laboratories (Kirby, 1963a; Billington, 1965; Solomon, 1966; James et al., 1972) have reported that trophoblast invasion into testis tissue is a cytolytic event. Kirby (1963a, 1965a) has described a border of lysed testicular tissue between advancing trophoblast cells and intact seminiferous tubules, as though the trophoblast were secreting cytolytic enzymes to break down the cells in front of it. This phenomenon is perhaps the strongest evidence in support of a cytolytic role for trophoblast cells in implantation and invasion. The paradox, however, is that there is little evidence of cytolysis in ectopic pregnancies in organs other than the testis, and trophoblast cells do not appear to be able to phagocytose live testis cells, again in contrast to their behavior in other sites. Consequently, trophoblast cells seem to possess the ability to destroy the host tissue in at least two different ways, the mode of action presumably depending upon the characteristics of the host tissue. As discussed above (section 6.1), just which of these modes is predominant in trophoblast invasion in utero is still open to some question.

Data from ectopic studies have also been considered with respect to the loss of invasiveness of trophoblast cells. This subject will be discussed in section 7.

6.5. *In vitro studies*

Ectopic studies have provided a good deal of evidence on the autonomous nature of trophoblast cells in the implantation event. However, with the development of appropriate conditions for the cultivation of blastocysts in vitro through the stages corresponding to implantation, it is now possible to test for trophoblast function under much better defined and controlled conditions. Furthermore, as mentioned previously, in vitro studies permit constant inspection of ongoing processes (see Figure 5). Finally, the role of cell contact can be assessed by comparing the behavior of the blastocyst in the presence or absence of other cells in the culture.

After blastocysts have attached to a substratum in vitro, the trophoblast cells begin to outgrow (Fig. 5). This occurs irrespective of whether the culture surface is inert, i.e., plastic, glass or collagen (Mintz, 1964; Cole and Paul, 1965; Gwatkin, 1966a,b; Barlow and Sherman, 1972; Spindle and Pedersen, 1973; Sherman, 1975b) or cellular (Cole and Paul, 1965; Sherman and Salomon, 1975; Salomon and Sherman, 1975; Sherman, 1975a). Gwatkin (1966b) has suggested that trophoblast outgrowth is analogous to the early stages of implantation. Studies of implantation on cellular monolayers indicate that trophoblast outgrowth is quite reminiscent of trophoblast invasiveness in utero or in ectopic sites, since, just as in these cases, outgrowing trophoblast cells can actively displace uterine cells (Sherman and Salomon, 1975; Salomon and Sherman, 1975; Sherman, 1975a) and a variety of cultured cell lines, including some that are normally highly invasive themselves (Cole and Paul, 1965; Salomon and Sherman, 1975; Sherman and Salomon, 1975). In their studies, Cole and Paul used cinemicrographic techniques to analyze trophoblast outgrowth into cell monolayers. They concluded from these unpublished observations that the formation of a plaque in the underlying monolayer was entirely by mechanical means, and that neither phagocytosis nor lysis played any part. This would be consistent with the proposal that trophoblast invasiveness in utero is mainly a migratory, rather than a cytolytic, phenomenon (see section 6.1). On the other hand, Salomon and Sherman (1975; Sherman and Salomon, 1975) have noted that a clear halo often forms ahead of advancing trophoblast cells as they invade the monolayer, reminiscent of Kirby's (1965a) observations with trophoblast cells growing under the testis capsule (see section 6.4), and suggestive of the secretion of cytolytic enzymes by the trophoblast. Furthermore, it has been reported that cultured mouse trophoblast cells are able to phagocytose red blood cells (Koren and Behrman, 1968) and thymus cells, provided that the latter are isoantibody-coated (Schlesinger and Koren, 1967). Also, Grant (1973) found that blastocysts implanting into uterine horns in vitro could phagocytose both uterine epithelial and stromal cells; in some cases, the engulfed cells still appeared to be healthy. To date, then, culture studies have failed to resolve the issue of the lack or presence of a cytolytic capability in trophoblast cells.

The requirements for trophoblast outgrowth are as stringent as, or more demanding than, those for blastocyst attachment. Spindle and Pedersen (1973) reported that whereas only some amino acids are necessary for blastocyst attachment, all essential amino acids except isoleucine are required for outgrowth on a collagen monolayer. Similarly, while some laboratories have found that blastocyst attachment can take place in the absence of serum or serum fractions, outgrowth does not take place on a plastic substratum under these conditions (Gwatkin, 1966a; Menke and McLaren, 1970; Jenkinson and Wilson, 1973; Sherman and Salomon, 1975), though marginal outgrowth can occur in the absence of macromolecules if a collagen substratum is used (Spindle and Pedersen, 1973). Normal extents of trophoblast development are observed with a collagen substratum plus bovine serum albumin (Jenkinson and Wilson, 1973), in lens explants (Jenkinson and Wilson, 1970) and on uterine monolayers (Sherman and Salomon, 1975). Since outgrowth did not take place when blastocysts were cultured in the presence of a collagen substratum, but not in contact with it, Jenkinson and Wilson (1973) concluded that, as is the case for attachment, protein is only necessary to coat the culture dish surface, making it receptive to cell adhesion. On the other hand, it is equally likely that trophoblast cells must be in direct contact with the collagen in order to catabolize it for subsequent use. In fact, Wilson and Jenkinson (1974) have subsequently proposed that part of the supportive role of lens explants may be related to the phagocytosis of lens fiber material by the trophoblast. More direct methods must be used to determine whether the required macromolecular sources are metabolized or merely used as a "basement membrane".

In studies using serum treated to remove endogenous steroids, Salomon and

Sherman (1975; Sherman and Salomon, 1975) could not demonstrate a requirement for progesterone or estrogen in trophoblast outgrowth or in the displacement of uterine, or other cell, monolayers by trophoblast. The outgrowths of trophoblast cells on collagen in serum free medium is also consistent with a lack of steroid dependence (Jenkinson and Wilson, 1973). On the other hand, Grant (1973) reported that trophoblast cells could not invade the luminal epithelium of isolated uteri in culture unless nonphysiologically high doses of progesterone were included in the culture medium. Estradiol neither increased the ability of blastocysts to implant and invade when added in conjunction with progesterone, nor stimulated trophoblast invasion when added alone. However, very large doses (5 μg/ml) of estradiol were used; in vivo, these concentrations of estrogen are antagonistic to blastocyst survival and implantation (Bergström, 1972a). The small amounts of estrogen present in the fetal calf serum may have been adequate to generate the desired effect in the presence of progesterone without further addition of estradiol. The nature of this steroid hormone dependence is not altogether clear, although the fact that progesterone administration was found to induce closure of the uterine lumen suggests that the added progesterone was necessary to render the isolated uteri receptive to blastocyst implantation. In fact, Grant (1973) did not indicate whether non-invasive blastocysts were at least able to attach to the uterine wall in the absence of progesterone. It is probable that even attachment does not take place in the non-primed uteri. In all, the in vitro data would appear to indicate that trophoblast cells can assume their invasive properties in the absence of any steroid hormone stimuli.

In all the studies discussed to this point, we have been concerned with the interaction of trophectoderm or trophoblast cells with the uterus. Studies have also been carried out to determine whether the derivatives of the ICM are in any way involved in the control of implantation. A number of ectopic studies seemed to rule out a role for the ICM cells (see section 6.4) but it could not be ascertained that *no* ICM cells were present in predominantly trophoblastic growths, nor was it possible to exclude the possibility that ICM cells might have triggered implantation in ectopic sites before they began to die. Subsequently, a number of in vitro studies have been carried out which unequivocally answer the question. By taking advantage of the fact that the ICM of the blastocyst is much more sensitive (for reasons as yet unknown) to X-irradiation and to a variety of antimetabolites than is the trophectoderm complement, Goldstein, Spindle and Pedersen (1975) (X-irradiation), Rowinski, Solter and Koprowski (1975) (actinomycin D, cordycepin and cycloheximide) and Sherman and Atienza (1975) (bromodeoxyuridine, cytosine arabinoside and Colcemid) have selectively killed all ICM cells shortly after blastocysts are placed in culture. Trophoblast cells not only survive these treatments, but they proceed to outgrow in all cases just as they do in the presence of the ICM. A second, and even more definitive, type of experiment involves the use of so-called "trophoblastic vesicles", which, according to the conventions used in this paper, should actually be referred to as "trophectodermal vesicles". These are embryonic structures containing only a layer of trophectoderm cells and no ICM. Trophectodermal vesicles can be generated by culturing disaggregated, cleavage-stage embryos (Tarkowski and Wroblewska, 1967) or by microsurgery (Gardner, 1971) or by culturing cleavage stage embryos in the presence of radioactive thymidine (Snow, 1973a,b). These structures can give rise only to giant trophoblast cells (Gardner, 1971, 1972; Snow, 1973a,b; Sherman, 1975a). Nevertheless, trophectodermal vesicles can induce decidualization and implant normally when transferred to pseudopregnant foster mothers (Gardner, 1971, 1972; Snow, 1973b). In vitro, they can attach and outgrow (Ansell and Snow, 1975; Sherman, 1975a), or displace uterine monolayer cells (Sherman, 1975a) just as do intact blastocysts.

All the in vitro studies described above involve the invasiveness of giant trophoblast cells. Similar studies on invasiveness have not been carried out with, e.g., the cells of the ectoplacental cone. The problem is that although ectoplacental cone cells do appear in blastocyst cultures under optimal conditions (Sherman, 1975b), they do not develop when blastocysts are cultured on uterine monolayers (Sherman and Salomon, 1975) or in intact uteri (Grant, 1973). To our knowledge, intact ectoplacental cones have not been removed from uteri and tested for invasive properties in vitro as they have been in ectopic sites (see section 6.4).

7. Postimplantation phase – Loss of trophoblast invasiveness

The phenomenon of implantation, which is so often compared with the invasive properties of malignant cells (Beard, 1902; Krebs, Krebs and Beard, 1950; Solomon, 1966 among others) is, however, almost always a benign event, although malignant trophoblastic tumors do occur in some species, notably humans (see Billington, 1971). In other words, trophoblast cells, although initially invasive, normally lose this property before permanent damage is done to the host. According to Kirby (1965a), trophoblast invasiveness ends on about the tenth day of pregnancy. Since the conceptus at this time has long since passed the stage of blastocyst, the subject of loss of trophoblast invasiveness is properly outside the scope of this review. However, since this phenomenon so definitively marks the end of the implantation event, it will be considered briefly. Only one issue will be discussed here. Is the loss of invasiveness of trophoblast cells a direct result of changes in the cells themselves, or are external factors operating to bring it about? Unfortunately, as is the case with most of the questions to which we have addressed ourselves, a conclusive answer cannot yet be given.

Kirby and coworkers (Kirby, 1965a; Kirby and Cowell, 1968; Billington, 1971) have provided attractive evidence in favor of an external control for the reversal of trophoblast invasiveness. The essential components of their argument are as follows: (a) the invasiveness of trophoblast cells in ectopic sites persists longer than it does in utero; (b) trophoblast cells are much more invasive in non-decidualized uteri than they are in uteri containing normal decidual growth. In fact, ectoplacental cone cells implanted into uteri of cyclic hosts can invade right through the uterine wall; (c) when deciduoma are induced and analyzed, it is found that necrosis takes place from the center of the growth such that the thickness of the deciduomal layer is comparable to that of the decidual layer of the same age in normal pregnancy. From these facts, Kirby and coworkers have proposed that trophoblast invasion is completely under the control of the decidual layer. Trophoblast cells can only invade through the centrally necrotic area of the decidua, and upon making contact with healthy decidual cells, invasion abruptly ends. It should be noted, however, that blastocysts are capable of implanting on, and invading into, monolayers of healthy decidual cells in vitro, just as they are for any other cell monolayer (Sherman and Salomon, unpublished observations). Other factors may, therefore, come into play.

Immunologic processes may be involved in the control of trophoblast invasiveness. Although paternal antigens on the trophoblast cell surface do not seem to depress trophoblast invasiveness (Billington, 1965; James et al., 1972) trophoblast-specific antigens (e.g., Beer, Billingham and Yang, 1972) might be involved. Koren, Abrams and Behrman (1968) have claimed that trophoblast cells show greater proliferation when they develop in kidneys of X-irradiated mice compared to immunologically normal hosts, irrespective of the genetic graft-host relationships.

It has been argued that trophoblast cells have a fixed lifespan, whether in utero, in ectopic sites (see Kirby, 1965 and Billington, 1971) or in vitro (Dorgan and Schultz, 1971). Consequently, it may be that the loss of invasiveness is related to the onset of death. However, about three days elapse between the loss of invasiveness and the time at which generalized death of trophoblast giant cells begins in utero or in ectopic sites. Although it is possible that trophoblast cells are in the process of dying for three days or more, it is not very likely. Recent in vitro work reduces this likelihood still more. Sherman and Salomon (1975; Salomon and Sherman, 1975) have observed that loss of trophoblast invasiveness into uterine monolayers takes place after about six days in vitro. This reversal, therefore, has taken place on the tenth equivalent gestation day, very close to the time at which it is estimated (Kirby, 1965a) to occur in utero. The noninvasive trophoblast cells appear morphologically healthy, and can, in fact, survive under these culture conditions for up to five weeks (Sherman and Salomon, 1975; Sherman, 1975b). These are perhaps the most compelling results which lead to the proposal that loss of trophoblast invasiveness may be only partly due to external influences, and that programming on the part of the trophoblast cells themselves is also involved.

8. *Concluding remarks*

As we indicated at the outset of this review, our objective has been to dissect the role of the blastocyst in each phase of implantation from that of the uterus, and to stress the various aspects of the former. This has proven to be a difficult task.

Although the blastocyst clearly participates actively in preparing itself for, and undergoing, implantation, its functions and activities during each stage are often so intimately interwoven with those of the uterus that the distinction between the two is not always clear. For example, we have concluded that during the hatching phase, a maternally-produced zona lysin plays a key role in denuding the blastocyst of the zona pellucida. On the other hand, in vitro studies make it clear that the blastocyst is capable of hatching through a series of expansions and contractions, without any maternal intervention. By the same token, the zona of abnormal embryos can be dissolved in vivo without any mechanical contributions from the embryo. So, we are faced with the question of the importance of expansion and contraction cycles for hatching of the blastocyst in vivo. There are several other unresolved issues concerning blastocyst activities during each phase of implantation. For example, does the ICM rotate within the trophectoderm during the apposition phase? Are the alterations of the uterine epithelial surface adequate to induce adhesion, as is suggested by morphological studies, or must alterations in the blastocyst surface also occur, as would be consistent with in vitro and biochemical observations? Does trophoblast invasiveness play any role in the breakdown of the uterine epithelial layer, or is the process strictly autolytic? Finally, does the decidua control the extent of trophoblast invasiveness, or is the loss of invasiveness programmed into the trophoblast cells themselves?

On the basis of a survey of the literature presented in this review, we propose that the conceptus does play an active role in each of the above processes, with the possible exception of orienting itself properly during the apposition phase. Much of the strength upon which we base our conclusion is drawn from results of in vitro studies. This raises the question of whether blastocyst behavior in vitro legitimately reflects its activities in vivo. Evidence has been provided in previous sections that cultured blastocysts undergo both morphological and biochemical transitions which are characteristic of their development in vivo. Furthermore, the blastocyst passes through each of the stages involved in implantation in utero in the correct sequence with only a slight initial delay due to an extended hatching time. After hatching, the blastocyst is directly apposed to the culture surface, but exhibits a transient period during which it will not attach. This is followed by adhesion of the blastocyst to the substratum. If the substratum is cellular, invasiveness ensues, and this phase is in turn followed by a loss of invasiveness. The fidelity with which these events occur lead us to contend that implantation in vitro is a reliable indicator of blastocyst actions in utero.

Suboptimal culture conditions may cause the blastocyst to depart from its normal behavior in vitro in ways which may lead to incorrect conclusions about in vivo development. For example, in vitro blastocyst delay can be induced by omitting certain amino acids from the culture medium, but the phenomenon does not strictly correlate with in vivo delay, which is apparently not caused by amino acid deprivation. Nevertheless, even these studies have been valuable in demonstrating how other nutritional deficiencies might conceivably block both normal metabolism of the blastocyst as well as blastocyst implantation during delay in vivo. In fact, we believe that adequate evidence has not been presented to support the idea that steroid hormones might control implantation, even in part, by acting directly upon the blastocyst. In our opinion, it is much more likely that these hormones act only upon the uterus to condition it properly for implantation. The control upon the blastocyst would then be indirect, i.e. absence of hormone conditioning of the uterus would lead to a surface nonreceptive for blastocyst adhesion as well as a uterine environment failing to provide the blastocyst with the metabolites necessary to synthesize macromolecules, including those which alter its own surface in preparation for adhesion.

Classical studies of implantation have leaned heavily upon conclusions based on morphological assessments. As we have mentioned above, however, morphological studies have an inherent shortcoming in that they freeze the implantation event at a particular instant, and because of variations in timing from one conceptus to another, the exact staging of the sectioned material is not always possible. Furthermore, we have also pointed out that even ultrastructural studies have failed to detect changes in the blastocyst surface which have been proposed to occur on the basis of histochemical, biochemical and in vitro studies. With the availability of improved systems for in vitro study, as well as relatively simple, yet often ultrasensitive, biochemical techniques, it is hoped that these and other procedures will be employed by investigators in their attempts to further clarify the many uncertainties that still remain with respect to the implantation process.

Finally, the questions that we have posed above concerning the participation of the blastocyst in implantation make it clear that both the blastocyst and the uterus contribute to each step of the process. In some of these instances, it might appear that actions by the blastocyst or the uterus are superfluous. For example, if blastocyst expansion and contraction did not take place in vivo, would the zona lysin be sufficient to properly hatch the blastocyst on schedule? If the uterine surface were not unreceptive prior to the adhesion phase, would the embryo nevertheless be prevented from precocious attachment because it had not yet synthesized the necessary surface components, even if it hatched prematurely? Would not the invasive properties of trophoblast cells be adequate to implant the embryo in the absence of autolysis of the uterine epithelium? We certainly cannot rule out the possibility that implantation could occur normally in vivo in the absence of these apparent double-checks, which might have arisen throughout the evolution of the implantation process. On the other hand, there is abundant evidence that nature is not given to encouraging redundancy, and it may be that each action of the blastocyst and the uterus during implantation is indispensable for the proper end result. Only carefully designed experiments with specific inhibitors or with strains of mice carrying appropriate mutations can provide us with a clear answer to such questions. Unfortunately, at the present time most or all of the necessary inhibitors and mutant strains are unavailable. It may well be profitable to search for them.

References

Abraham, R., Hendy, R., Dougherty, W.J., Fulfs, J.C., and Goldberg, L. (1970) Participation of lysosomes in early implantation in the rabbit. Exp. Molec. Path. 13, 329–345.

Alden, R.H. (1945) Implantation of the rat egg. I. Experimental alteration of uterine polarity. J. Exp. Zool. 100, 229–235.

Amoroso, E.C. (1952) Placentation. In: Marshall's Physiology of Reproduction (Parkes, A.S., ed.) vol. II, pp. 127–311. Longmans, London.

Andary, T.J., Dabich, D. and Van Winkle, L.J. (1972) Changes in proteinase activity in early vs. late mouse blastocysts. J. Cell Biol. 55, 3a.

Ansell, J.D. and Snow, M.H.L. (1975) The development of trophoblast in vitro from blastocysts containing varying amounts of inner cell mass. J. Embryol. Exp. Morph. 33, 177–185.

Ballard, P.L. and Tomkins, G.M. (1969) Hormone induced modification of the cell surface. Nature, 224, 344–345.

Barlow, P.W. and Sherman, M.I. (1972) The biochemistry of differentiation of mouse trophoblast: studies on polyploidy. J. Embryol. Exp. Morphol. 27, 447–465.

Beard, J. (1902) Embryological aspects and etiology of carcinoma. Lancet i, 1758–1761.

Beer, A.E., Billingham, R.E. and Yang, S.L. (1972) Further evidence concerning the autoantigenic status of the trophoblast. J. Exp. Med. 135, 1177–1184.

Bergström, S. (1970) Estimation of proteolytic activity at mouse implantation sites by the gelatin digestion method. J. Reprod. Fertil. 23, 481–485.

Bergström, S. (1972a) Scanning electron microscopy of ovoimplantation. Arch. Gynak. 212, 285–307.

Bergström, S. (1972b) Delay of implantation in the mouse by ovariectomy or lactation. A SEM Study. Fertil. Steril. 23, 548–561.

Bergström, S. (1972c) Histochemical localization of acid uterine aminoacylnaphthylamidases in early pregnancy and in different hormonal states of the mouse, J. Reprod. Fertil. 30, 177–183.

Bergström, S. and Nilsson, O. (1970) Morphological changes of the trophoblast surface at implantation in the mouse. J. Reprod. Fertil. 23, 339–340.

Bergström, S. and Nilsson, O. (1971) Scanning electron microscopy of mouse blastocysts before and at implantation. In: Current Problems in Fertility (Ingelman-Sundberg, A. and Lunell, N.-O., eds.) pp. 118–123, Plenum Press, New York.

Bergström, S. and Nilsson, O. (1972) Ultrastructural response of blastocysts and uterine epithelium to progesterone deprivation during delayed implantation in mice. J. Endocrinol. 55, 217–218.

Bergström, S. and Nilsson, O. (1973) Various types of embryo-endometrial contacts during delay of implantation in the mouse. J. Reprod. Fertil. 32, 531–533.

Bergström, S. and Nilsson, O. (1975) Embryo-endometrial relationship in the mouse during activation of the blastocyst by oestradiol. J. Reprod. Fertil. 44, 117–120.

Biggers, J. D., and Stern, S. (1973) Metabolism of the preimplantation mammalian embryo. Adv. Reprod. Physiol. 6, 1–59.

Billington, W.D. (1965) The invasiveness of transplanted mouse trophoblast and the influence of immunological factors. J. Reprod. Fertil. 10, 343–352.

Billington, W.D. (1971) Biology of the trophoblast. Adv. Reprod. Physiol. 5, 27–66.

Blandau, R.J. (1971) Culture of guinea pig blastocyst. In: The Biology of the Blastocyst (Blandau, R.J., ed.) pp. 59–69, University of Chicago Press, Chicago.

Borghese, E. and Cassini, A. (1963) Cleavage of mouse egg. In: Cinemicrography in Cell Biology (Rose, G.G., ed.) pp. 263–277. Academic Press, New York.

Böving, B.G. (1966) Some mechanical aspects of trophoblast penetration of the uterine epithelium in the rabbit. In: Egg Implantation (Wolstenholme, G.E.W. and O'Connor, M., eds.) pp. 72–82. Little, Brown and Co., Boston.

Böving, B.G. (1971) Biomechanics of implantation. In: The Biology of the Blastocyst (Blandau, R.J., ed.) pp. 423–442, University of Chicago Press, Chicago.

Boyd, J.D. and Hamilton, W.J. (1952) Cleavage, early development and implantation of the egg. In: Marshall's Physiology of Reproduction (Parkes, A.S., ed.) vol. II, pp. 1–126, Longmans, London.

Chambon, Y. (1949) Realization de retard de l'implantation par les faibles doses de progesterone chez la ratte. C. R. Soc. Biol. 143, 756–758.

Chew, N.J. and Sherman, M.I. (1975) Biochemistry of differentiation of mouse trophoblast: Δ^5,3β-hydroxysteroid dehydrogenase. Biol. Reprod. 12, 351–359.

Christie, G.A. (1967) Histochemistry of implantation in the rabbit. Histochemie 9, 13–29.

Clemetson, C.A.B., Moshfeghi, M.M. and Mallikarjuneswara, V.R. (1971) The surface charge on the five-day rat blastocyst. In: The Biology of the Blastocyst (Blandau, R.J., ed.) pp. 193–206. University of Chicago Press, Chicago.

Clemetson, C.A.B., Kim, J.K., Mallikarjuneswara, V.R. and Wilds, J.H. (1972) The sodium and potassium concentrations in the uterine fluid of the rat at the time of implantation. J. Endocrin. 54, 417–423.

Cole, R. J. (1967) Cinemicrographic observations on the trophoblast and zona pellucida of the mouse blastocyst. J. Embryol. Exp. Morph. 17, 481–490.

Cole, R.J. and Paul, J. (1965) Properties of cultured preimplantation mouse and rabbit embryos, and cell strains derived from them. In: Preimplantation Stages of Pregnancy (Wolstenholme, G.E.W. and O'Conner, M., eds.) pp. 82–112, Academic Press, New York.

Cowell, T.P. (1969) Implantation and development of mouse eggs transferred to the uteri of non-progestational mice. J. Reprod. Fertil. 19, 239–245.

Denker, H.W. (1972) Blastocyst protease and implantation: Effect of ovariectomy and progesterone substitution in the rabbit. Acta Endocrinol. 70, 591–602.

Dey, S.K. and Dickmann, Z. (1974a) Δ^5,3β-Hydroxysteroid dehydrogenase activity in mouse morulae and blastocysts. 7th Ann. Meeting Soc. Study Reprod., Abstract No. 150.

Dey, S.K. and Dickmann, Z. (1974b) Estradiol-17β-hydroxysteroid dehydrogenase activity in preimplantation rat embryos. Steroids 24, 57–62.

Dickmann, Z. and Dey, S.K. (1974) Steroidogenesis in the preimplantation rat embryo and its possible influence on morula-blastocyst transformation and implantation. J. Reprod. Fertil. 91–93.

Dickson, A.D. (1963) Trophoblastic giant cell transformation of mouse blastocysts, J. Reprod. Fertil. 6, 465–466.

Dickson, A.D. (1966) The form of the mouse blastocyst. J. Anat. 100, 335–348.

Dorgan, W.J. and Schultz, R.L. (1971) An in vitro study of programmed death in rat placental giant cells. J. Exp. Zool. 178, 497–512.

Ducibella, T., Albertini, D.F., Anderson, E. and Biggers, J.D. (1975) The preimplantation mammalian embryo: Characterization of intercellular junctions and their appearance during development. Dev. Biol. 45, 231–250.

Duval, M. (1891) Le placenta des rongeurs: Le placenta de la souris et du rat. J. Anat. Physiol. Paris 27, 24–106.

Edidin, M. (1976) Cell surface antigens in mammalian development. In: The Cell Surface in Animal Embryogenesis and Development (Poste, G. and Nicolson, G.L., eds.) Ch. 4, pp. 127–143, North-Holland, Amsterdam.

El-Shershaby, A.M. and Hinchliffe, J.R. (1974) Cell redundancy in the zona-intact preimplantation mouse blastocyst: a light and electron microscope study of dead cells and their fate. J. Embryol. Exp. Morph. 31, 643–654.

El-Shershaby, A.M. and Hinchliffe, J.R. (1975) Epithelial autolysis during implantation of the mouse blastocyst: an ultrastructural study. J. Embryol. Exp. Morph. 33, 1067–1080.

Enders, A.C. (1963) Ed. Delayed Implantation, University of Chicago Press, Chicago.

Enders, A.C. (1971) The fine structure of the blastocyst. In: The Biology of the Blastocyst (Blandau, R.J., ed.) pp. 71–94, University of Chicago Press, Chicago.

Enders, A.C. and Schlafke, S. (1967) A morphological analysis of the early implantation stages in the rat. Amer. J. Anat. 120, 185–226.

Enders, A.C. and Schlafke, S. (1969) Cytological aspects of trophoblast-uterine interaction in early implantation. Amer. J. Anat. 125, 1–30.

Enders, A.C. and Schlafke, S. (1971) Penetration of the uterine epithelium during implantation in the rabbit. Amer. J. Anat. 132, 219–240.

Enders, A.C. and Schlafke, S. (1974) Surface coats of the mouse blastocyst and uterus during the preimplantation period. Anat. Rec. 180, 31–46.

Fawcett, D.W. (1950) The development of mouse ova under the capsule of the kidney. Anat. Rec. 108, 71–91.

Fawcett, D.W., Wislocki, G.B. and Waldo, C.M. (1947) The development of mouse ova in the anterior chamber of the eye and in the abdominal cavity. Amer. J. Anat. 81, 413–443.

Finn, C.A. (1972) The biology of decidual cells. Adv. Reprod. Physiol. 5, 1–26.

Finn, C.A. and Bredl, J.C.S. (1973) Studies on the development of the implantation reaction in the mouse uterus: Influence of actinomycin D. J. Reprod. Fertil. 34, 247–253.

Finn, C.A. and Hinchliffe, J.R. (1964) Reaction of the mouse uterus during implantation and deciduoma formation as demonstrated by changes in the distribution of alkaline phosphatase. J. Reprod. Fertil, 8, 331–338.

Finn, C.A. and Hinchliffe, J.R. (1965) Histological and histochemical analysis of the formation of implantation chambers in the mouse uterus. J. Reprod. Fertil, 9, 301–309.

Finn, C.A. and Lawn, A.M. (1968) Transfer of cellular material between the uterine epithelium and trophoblast during the early stages of implantation. J. Reprod. Fertil. 15, 333–336.

Finn, C.A. and McLaren, A. (1967) A study of the early stages of implantation in mice. J. Reprod. Fertil. 13, 259–267.

Gardner, R.L. (1971) Manipulations on the blastocyst. Adv. Biosciences 6, 279–299.

Gardner, R.L. (1972) An investigation of inner cell mass and trophoblast tissues following their isolation from the mouse blastocyst. J. Embryol. Exp. Morph. 28, 279–312.

Gardner, R.L. (1975) Analysis of determination and differentiation in the early mammalian embryo using intra- and interspecific chimeras. In: The Developmental Biology of Reproduction (Markert, C.L. and Papaconstantinou, J., eds.) pp. 207–236, Academic Press, New York.

Ginsberg, L. and Hillman, N. (1975) Shifts in ATP synthesis during preimplantation stages of mouse embryos. J. Reprod. Fertil. 43, 83–90.

Glasser, S.R. (1972) The uterine environment in implantation and decidualization. In: Reproductive Biology (Balin, H. and Glasser, S.R., eds.) pp. 776–833, Excerpta Medica, Amsterdam.

Glenister, T.W. (1961) Observations on the behavior in organ culture of rabbit trophoblast from implanting blastocysts and early placentae. J. Anat. 95, 474–484.

Glenister, T.W. (1965) The behavior of trophoblast when blastocysts effect nidation in culture. In: The Early Conceptus, Normal and Abnormal (Park, W.W., ed.) pp. 24–26, University of St. Andrews Press, Edinburgh.

Glenister, T.W. (1967) Organ culture and its combination with electron microscopy in the study of nidation processes. Excerpta Medica Intern. Cong. Ser. 133, 385–394.

Goldstein, L.S., Spindle, A.I. and Pedersen, R.A. (1975) X-ray sensitivity of the preimplantation mouse embryo in vitro. Radiation Res. 62, 276–287.

Grant, P.S. (1973) The effect of progesterone and oestradiol on blastocysts cultured within the lumina of immature mouse uteri. J. Embryol. Exp. Morph. 29, 617–638.

Gwatkin, R.B.L. (1966a) Defined media and development of mammalian eggs in vitro. Ann. N. Y. Acad. Sci. 139, 79–90.

Gwatkin, R.B.L. (1966b) Amino acid requirements for attachment and outgrowth of the mouse blastocyst in vitro. J. Cell Physiol. 68, 335–344.

Gwatkin, R.B.L. (1969) Nutritional requirements for post-blastocyst development in the mouse. Int. J. Fertil. 14, 101–105.

Hall, K. (1971) 5-Nucleotidase, acid phosphatase and phosphorylase during normal, delayed and induced implantation of blastocysts in mice: a histochemical study. J. Endocrin. 51, 291–301.

Holmes, P.V. and Dickson, A.D. (1973) Estrogen-induced surface coat and enzyme changes in the implanting mouse blastocyst. J. Embryol. Exp. Morph. 29, 639–645.

Hsu, Y.-C. (1971) Post-blastocyst differentiation in vitro. Nature 231, 100–102.

Hsu, Y.-C., Baskar, J., Stevens, L.C., and Rash, J.E. (1974) Development in vitro of mouse embryos from the two-cell egg stage to the early somite stage. J. Embryol. Exp. Morph. 31, 235–245.

James, D.A., Acierto, S. and Murphy, B.D. (1972) Growth of mouse trophoblast transplanted to syngeneic and allogeneic testes. J. Exp. Zool. 180, 209–216.

Jenkinson, E.J. and Wilson, I.B. (1970) In vitro support system for the study of blastocyst differentiation in the mouse. Nature, 228, 776–778.

Jenkinson, E.J. and Wilson, I.B. (1973) In vitro studies on the control of trophoblast outgrowth in the mouse. J. Embryol. Exp. Morph. 30, 21–30.

Jones, B.M. and Kemp, R.B. (1969) Self-isolation of the foetal trophoblast. Nature, 221, 829–831.

Kirby, D.R.S. (1960) The development of mouse eggs beneath the kidney capsule. Nature, 187, 707–708.

Kirby, D.R.S. (1962a) The influence of the uterine environment on the development of mouse eggs. J. Embryol. Exp. Morph. 10, 496–506.

Kirby, D.R.S. (1962b) Ability of the trophoblast to destroy cancer tissue. Nature, 194, 696–697.

Kirby, D.R.S. (1963a) The development of mouse blastocysts transplanted to the scrotal and cryptochid testis. J. Anat. 97, 119–130.

Kirby, D.R.S. (1963b) Development of the mouse blastocyst transplanted to the spleen. J. Reprod. Fertil. 5, 1–12.

Kirby, D.R.S. (1965a) The "invasiveness" of the trophoblast. In: The Early Conceptus, Normal and Abnormal (Park W.W., ed.) pp. 68–74, University of St. Andrews Press, Edinburgh.

Kirby, D.R.S. (1965b) Endocrinological effects of experimentally induced extra-uterine pregnancies in virgin mice. J. Reprod. Fertil. 10, 403–412.

Kirby, D.R.S. (1970) Immunological aspects of implantation. In: Ovo-Implantation. Human Gonadotropins and Prolactin. (Hubinont, P.O., Leroy, F., Robyn, C. and Leleux, P., eds.) pp. 86–100, Karger, Basel.

Kirby, D.R.S. (1971) Blastocyst-uterine relationship before and during implantation. In: The Biology of the Blastocyst (Blandau, R.J., ed.) pp. 393–411, University of Chicago Press, Chicago.

Kirby, D.R.S. and Cowell, T.P. (1968) Trophoblast-host interactions. In: Epithelial-Mesenchymal Interactions (Fleischmajer, R. and Billingham, R.E., eds.) pp. 64–77, Williams and Wilkins, Baltimore.

Kirby, D.R.S., Potts, D.M. and Wilson, I.B. (1967) On the orientation of the implanting blastocyst. J. Embryol. Exp. Morph. 17, 527–532.

Koren, Z., Abrams, G. and Behrman, S.J. (1968) The role of host factors in mouse trophoblastic tissue growth. Amer. J. Obstet. Gynec. 100, 570–575.

Koren, Z. and Behrman, S.J. (1968) Organ culture of pure mouse trophoblast. Amer. J. Obstet. Gynec. 100, 576–581.

Krebs, E.T., Jr., Krebs, E.T., Sr. and Beard, H.H. (1950) The unitarian or trophoblastic thesis of cancer. Med. Rec. 163, 149–174.

Lanman, J.T. (1970) Delayed implantation of the blastocyst: An exploration of its effects on the developing embryo. Amer. J. Obstet. Gynec. 106, 463–468.

Larsen, J.F. (1961) Electron microscopy of the implantation site in the rabbit. Amer. J. Anat. 109, 319–334.

Larsen, J.F. (1970) Electron microscopy of nidation in the rabbit and observations on the human trophoblastic invasion. In: Ovo-Implantation. Human Gonadotropins and Prolactin (Hubinont, P.O., Leroy, F., Robyn, C. and Leleux, P., eds.) pp. 38–51, Karger, Basel.

Lewis, W.H. and Wright, E.S. (1935) On the early development of the mouse egg. Carnegie Inst. Contrib. Embryol. 25, 115–143.

McLaren, A. (1967) Delayed loss of the zona pellucida from blastocysts of suckling mice. J. Reprod. Fertil. 14, 159–162.

McLaren, A. (1968a) A study of blastocysts during delay and subsequent implantation in lactating mice. J. Endocrin. 42, 453–463.

McLaren, A. (1968b) Can beads stimulate a decidual response in the mouse uterus? J. Reprod. Fertil. 15, 313–315.

McLaren, A. (1970a) The fate of the zona pellucida in mice. J. Embryol. Exp. Morph. 23, 1–19.

McLaren, A. (1970b) Early embryo-endometrial relationships. In: Ovo-Implantation. Human Gonadotropins and Prolactin. (Hubinont, P.O., Leroy, F., Robyn, C. and Leleux, P., eds.) pp. 18–37, Karger, Basel.

McLaren, A. (1973) Blastocyst activation. In: The Regulation of Mammalian Reproduction (Segal, S.J., Crozier, R., Corfman, P.A. and Condliffe, eds.) pp. 321–334, Thomas, Springfield, Ill.

Manes, C. (1975) Genetic and biochemical activities in preimplantation embryos. In: The Develop-

mental Biology of Reproduction (Markert, C.L., and Papaconstantinou, J. eds.) pp. 133–163, Academic Press, New York.
Menke, T.M. and McLaren, A. (1970) Mouse blastocysts grown in vivo and in vitro: Carbon dioxide production and trophoblast outgrowth. J. Reprod. Fertil. 23, 117–127.
Mintz, B. (1963) Growth in vitro of t^{12}/t^{12} lethal mutant mouse eggs. Amer. Zool. 3, 550–551.
Mintz, B. (1964) Formation of genetically mosaic embryos and early development of lethal (t^{12}/t^{12})-normal mosaics. J. Exp. Zool. 157, 273–292.
Mintz, B. (1971) Control of embryo implantation and survival. Adv. Biosciences 6, 317–342.
Mossman, H.W. (1971) Orientation and site of attachment of the blastocyst: a comparative study. In: The Biology of the Blastocyst (Blandau, R.J., ed) pp. 49–57, University of Chicago Press, Chicago.
Mulnard, J.G. (1970) Aspects de l'activité phagocytaire du trophoblaste de la souris au début de l'ovo-implantation. In: Ovo-Implantation. Human Gonadotropins and Prolactin. (Hubinont, P.O., Leroy, F., Robyn, C. and Leleux, P., eds.) pp. 9–17, Karger, Basel.
Nadijcka, M. and Hillman, N. (1974) Ultrastructural studies of mouse blastocyst substages. J. Embryol. Exp. Morph. 32, 675–695.
Nicholas, J.S. (1942) Experiments on developing rats. IV. The growth and differentiation of eggs and egg-cylinders when transplanted under the kidney capsule. J. Exp. Zool. 90, 41–71.
Nilsson, O. (1966) Estrogen-induced increase of adhesiveness in uterine epithelium of mouse and rat. Exp. Cell Res. 43, 239–241.
Nilsson, O. (1967) Attachment of rat and mouse blastocysts onto uterine epithelium. Int. J. Fertil. 12, 5–13.
Nilsson, O. (1970) Some ultrastructural aspects of ovo-implantation. In: Ovo-Implantation. Human Gonadotropins and Prolactin. (Hubinont, P.O., Leroy, F., Robyn, C. and Leleux, P., eds.) pp. 52–72, Karger, Basel.
Nilsson, O. (1974) The morphology of blastocyst implantation. J. Reprod. Fertil. 39, 187–194.
Nilsson, O., Lindqvist, I., and Ronquist, G. (1973) Decreased surface charge of mouse blastocysts at implantation. Exp. Cell Res. 83, 421–423.
Orsini, M.W. and McLaren, A. (1967) Loss of the zona pellucida in mice, and the effect of tubal ligation and ovariectomy. J. Reprod. Fertil. 13, 485–499.
Owers, N.O. and Blandau, R.J. (1971) Proteolytic activity of the rat and guinea pig blastocyst in vitro. In: The Biology of the Blastocyst (Blandau, R.J. ed.) pp. 207–224, University of Chicago Press, Chicago.
Pincus, G. (1941) The control of ovum growth. Science 93, 438–439.
Pinsker, M.C. and Mintz, B. (1973). Change in cell-surface glycoproteins of mouse embryos before implantation. Proc. Nat. Acad. Sci. U.S.A. 70, 1645–1648.
Pinsker, M.C., Sacco, A.G., and Mintz, B. (1974) Implantation-associated proteinase in mouse uterine fluid. Dev. Biol. 38, 285–290.
Potts, M. (1966) The attachment phase of ovoimplantation. Amer. J. Obstet. Gynec. 96, 1122–1128.
Potts, M. (1968) The ultrastructure of implantation in the mouse. J. Anat. 103, 77–90.
Potts, M. (1969) The ultrastructure of egg implantation. Adv. Reprod. Physiol. 4, 241–267.
Potts, M. and Psychoyos, A. (1967) L'ultrastructure des relations ovoendométriales au cours du retard expérimental de nidation chez la souris. C. R. Acad. Sci. D 264, 956–958.
Potts, M. and Wilson, I.B. (1967) The preimplantation conceptus of the mouse at 90 hours post coitum. J. Anat. 102, 1–11.
Psychoyos, A. (1973) Hormonal control of ovoimplantation. Vitamins and Hormones 31, 201–256.
Reinius, S. (1967) Ultrastructure of blastocyst attachment in the mouse. Z. Zellforsch. Mikr. Anat. 77, 257–266.
Rowinski, J., Solter, D. and Koprowski, H. (1975) Mouse embryo development in vitro: effects of inhibitors of RNA and protein synthesis on blastocyst and post-blastocyst embryos. J. Exp. Zool. 192, 133–142.
Rumery, R.E. and Blandau, R.J. (1966) The loss of the zona pellucida in delayed implantation. Anat. Rec. 154, 485–486.
Runner, M.N. (1947) Development of mouse eggs in the anterior chamber of the eye. Anat. Rec. 98, 1–17.

Salomon, D.S. and Sherman, M.I. (1975) Implantation and invasiveness of mouse blastocysts on uterine monolayers. Exp. Cell Res. 90, 261–268.

Schlafke, S. and Enders, A.C. (1975) Cellular basis of interaction between trophoblast and uterus at implantation. Biol. Reprod. 12, 41–65.

Schlesinger, M. and Koren, Z. (1967) Mouse trophoblastic cells in tissue culture. Fertil. Steril. 18, 95–101.

Sherman, M.I. (1972a) The biochemistry of differentiation of mouse trophoblast: Alkaline phosphatase. Dev. Biol. 27, 337–349.

Sherman, M.I. (1972b) The biochemistry of differentiation of mouse trophoblast: Esterase. Exp. Cell Res. 75, 449–459.

Sherman, M.I. (1975a) The role of cell-cell interaction during early mouse embryogenesis. In: The Early Development of Mammals (Balls, M. and Wild, A.E., eds.) pp. 145–165, Cambridge University Press, London.

Sherman, M.I. (1975b) Long term culture of cells derived from mouse blastocysts. Differentiation 3, 51–67.

Sherman, M.I. and Atienza, S. B. (1975) Effects of bromodeoxyuridine, cytosine arabinoside and colcemid upon in vitro development of mouse blastocysts. J. Embryol. Exp. Morph. 34, 467–484.

Sherman, M.I. and Atienza, S.B. (1976) Production and utilization of progesterone and androstenedione by cultured mouse blastocysts. Submitted.

Sherman, M.I. and Barlow, P.W. (1972) Deoxyribonucleic acid content in delayed mouse blastocysts. J. Reprod. Fertil. 29, 123–126.

Sherman, M.I. and Salomon, D.S. (1975) The relationships between the early mouse embryo and its environment. In: The Developmental Biology of Reproduction (Markert, C.L. and Papaconstantinou, J., eds.) pp. 277–309, Academic Press, New York.

Smith, A.F. and Wilson, I.B. (1974) Cell interaction at the maternal-embryonic interface during implantation in the mouse. Cell Tiss. Res. 152, 525–542.

Smith, D.M. (1968) The effect on implantation of treating cultured mouse blastocysts with estrogen in vitro and the uptake of H^3-estradiol by blastocysts. J. Endocrinol. 41, 17–29.

Smith, D.M. and Smith, A.E. (1971) Uptake and incorporation of amino acids by cultured mouse embryos: estrogen stimulation. Biol. Reprod. 4, 66–73.

Smith, M.S.R. (1972) Changes in distribution of alkaline phosphatase during early implantation and development of the mouse. Aust. J. Biol. Sci. 26, 209–217.

Smith, M.S.R. and Hartman, S. (1974) Sarcoma cells as a blastocyst analogue in the mouse uterus. J. Reprod. Fertil. 36, 465.

Smith, M.S.R. and Wilson, I.B. (1971) Histochemical observations on early implantation in the mouse. J. Embryol. Exp. Morph. 25, 165–174.

Snow, M.H.L. (1973a) Abnormal development of preimplantation mouse embryos grown in vitro with [^{3}H]thymidine. J. Embryol. Exp. Morph. 29, 601–615.

Snow, M.H.L. (1973b) The differential effect of [^{3}H]thymidine upon two populations of cells in preimplantation mouse embryos. In: The Cell Cycle in Development and Differentiation (Balls, M. and Billett, F.S., eds.) pp. 311–324, Cambridge University Press, London.

Solomon, J.B. (1966) Relative growth of trophoblast and tumor cells coimplanted into isogenic mouse testes and the inhibitory action of 'methotrexate'. Nature, 210, 716–718.

Spindle, A.I. and Pedersen, R.A. (1973) Hatching, attachment, and outgrowth of mouse blastocysts in vitro: fixed nitrogen requirements. J. Exp. Zool. 186, 305–318.

Tachi, S., Tachi, C. and Lindner, H.R. (1970) Ultrastructural features of blastocyst attachment and trophoblastic invasion in the rat. J. Reprod. Fertil. 21, 37–56.

Tarkowski, A.K. and Wroblewska, J. (1967) Development of blastomeres of mouse eggs isolated at the 4- and 8-cell stage. J. Embryol. Exp. Morph. 18, 155–180.

Weitlauf, A.M. (1972) In vitro uptake and incorporation of amino acids by blastocysts from intact and ovariectomized mice. J. Exp. Zool. 183, 303–308.

Weitlauf, H.M. and Greenwald, G.S. (1968) Survival of blastocysts in the uteri of ovariectomized mice. J. Reprod. Fertil. 17, 515–520.

Whitten, W.K. (1957) The effect of progesterone on the development of mouse eggs in vitro. J. Endocrinol. 16, 80–85.

Whitten, W.K. (1971) Nutrient requirements for the culture of preimplantation embryos in vitro. Adv. Biosciences 6, 129–142.

Wilson, I.B. (1963a) A tumour tissue analogue of the implanting mouse embryo. Proc. Zool. Soc. London 141, 137–151.

Wilson, I.B. (1963b) A new factor associated with the implantation of the mouse egg. J. Reprod. Fertil. 5, 281–282.

Wilson, I.B. and Jenkinson, E.J. (1974) Blastocyst differentiation in vitro. J. Reprod. Fertil. 39, 243–249.

Wilson, I.B. and Potts, D.M. (1970) Melanoma invasion in the mouse uterus. J. Reprod. Fertil. 22, 429–434.

Wilson, I.B. and Smith, M.S. (1970) Primary trophoblast invasion at the time of nidation. In: Ovo-Implantation. Human Gonadotropins and Prolactin. (Hubinont, P.O., Leroy, F., Robyn, C. and Leleux, P., eds.) pp. 1–8, Karger, Basel.

Wimsatt, W.W. (1975) Some comparative aspects of implantation. Biol. Reprod., 12, 1–40.

Wong, Y.C. and Dickson, A.D. (1969) A histochemical study of ovo-implantation in the mouse. J. Anat. 105, 547–555.

Yochim, J.M. (1971) Intrauterine oxygen tension and metabolism of the endometrium during the preimplantation period. In: The Biology of the Blastocyst (Blandau, R.J., ed.) pp. 363–382, University of Chicago Press, Chicago.

Yoshinaga, K. and Adams, C. E. (1966) Delayed implantation in the spayed, progesterone-treated adult mouse, J. Reprod. Fertil. 12, 593–595.

Cell surface antigens in mammalian development 4

Michael EDIDIN

1. Introduction

For immunologists, a central problem in mammalian development is the immunologic relationship of a fetus to its mother. The mammalian fetus is a graft of foreign tissue within a species, an allograft, and such grafts are usually rejected. A discussion of cell surface differentiation in mammalian development clarifies to some extent the reasons for the success of the fetal graft. In the present review I will present such a discussion, but, at least in passing, will extend this traditional approach to summarize ideas about roles for cell surface antigens in determining cell association and position in the developing embryo. Though such ideas are still speculative and little developed, they lead us to approach development in vertebrates other than mammals in terms of cell surface antigens. Strong transplantation antigens are found throughout the vertebrate subphylum; though they raise an immunologic problem only in mammalian development.

I will mainly discuss experiments with inbred mice. However, much of the data apply to the development of humans and other placental mammals as well. For full details of mouse development, see Snell and Stevens (1966) and Rugh (1968).

2. Strong transplantation antigens and graft rejection

Early experiments on the survival of tumors grafted between inbred mouse strains indicated that there was a genetic basis to graft survival. By analyzing the very complicated data on growth of tumors from one inbred strain in the F2 generation of a cross between the tumor donor and a second strain, C.C. Little was able to make a formal demonstration that many genetic loci affected graft survival (summarized by Snell and Stimpfling, 1966). Only grafts sharing alleles at all loci with their recipients were accepted; all others were rejected. The alleles appeared to be co-dominant, since F1 hybrids accepted grafts from both parent inbred strains.

G. Poste & G.L. Nicolson (eds.) The Cell Surface in Animal Embryogenesis and Development

Independently of this work, following a suggestion by Haldane, Peter Gorer showed that an erythrocyte antigen of mice was associated with graft rejection (Gorer, 1938). Strain A animals, after rejecting strain B grafts, produced antibodies that agglutinated strain B erythrocytes. The erythrocyte antigen so defined, antigen II, was later linked genetically to a major histocompatibility locus, and the antigen locus was termed *Histocompatibility-2* (Gorer et al., 1948) or *H-2*. Compatibility at *H-2* alone was sufficient to ensure prolonged survival of an allograft, though many other genetic loci were mismatched between donor and recipient.

Parallel to the genetic work, Medawar and his colleagues showed the immunologic nature of graft rejection, initially studying the fate of skin grafts in humans, and then working with model systems, first random bred rabbits, then inbred mice (reviewed in Medawar, 1958). Disappointingly, it appeared that, though graft rejection was due to an immune response, it was not caused by antibodies even though antibodies to cell surface antigens appeared in serum after graft rejection. Passive immunity to grafts could not be procured by transfer of immune serum; only injection of spleen cells or lymphocytes from an immune animal accelerated or caused rejection of an otherwise tolerated graft (Mitchison, 1954). Thus it was not clear if the antigens causing graft rejection and those provoking anti-*H-2* antibodies were identical, or if they were merely closely linked. Initially the first appeared to be the case; typing mouse red cells with anti-*H-2* sera led to accurate predictions as to the survival of grafts between typed animals. Again, *H-2* locus products seemed to dominate graft rejection reactions. Experiments in which single histocompatibility loci were isolated in common genetic backgrounds also point up the strength of reactions caused by *H-2* locus products as opposed to those of other H-loci (Graff et al., 1966; Klein, 1975). For these reasons the *H-2* locus is said to determine strong transplantation antigens.

It further appears that the mouse *H-2* locus is an example of a genetic system found in all mammals, as well as in birds and probably fish. Each species examined has a single genetic locus, whose products make a dominant contribution to graft rejection, as well as stimulating formation of circulating antibody (Snell, 1968; Ivanyi, 1970). From work with mouse and humans further characteristics shared by these loci have become apparent.

Each major histocompatibility "locus" is properly termed a gene complex. In the case of *H-2*, five tightly-linked regions can be defined, each of which may, in turn, contain more than a single gene (Fig. 1) (for detailed discussion see, Klein et al., 1974; Shreffler and David, 1975). The products that can be associated with the region are: (1) antigens stimulating formation of serum antibodies; (2) antigens stimulating cellular metabolic responses, blast transformation of lymphocytes for example; and (3) factors affecting cell reactions and interactions, notably those of lymphocytes. The last region was initially defined in terms of control of the immune response to synthetic antigens (Benacerraf and McDevitt, 1972). Despite all this definition and subdivision, the genes and their products responsible for in vivo graft rejection remain unknown and it is likely that

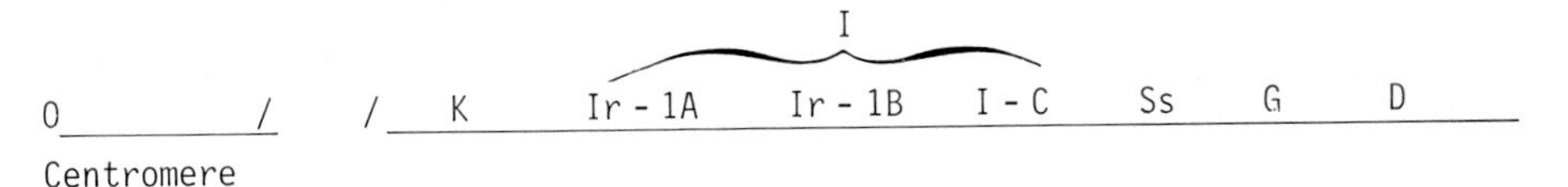

Fig. 1. Map of *H-2* gene complex in the mouse.

mismatch for more than one region is required for rejection (Alter et al., 1973; Edidin and Henney, 1973).

All the products of the *H-2* complex appear to be related to the cell surface, either as antigens (which have been shown to be integral membrane proteins; (Nathenson and Cullen, 1974), or in affecting reactions at the surface (the Ir genes). All are highly polymorphic, there are many alleles known for the D and K regions, the regions determining classic *H-2* antigens (detected with anti-sera) and further diversity of the complex is achieved due to the number of permutations possible between alleles, 7 closely linked genetic regions and subregions. These permutations are termed haplotypes.

A similar situation has been defined for the homologous locus in man, *HL-A* (reviewed in Bodmer, 1973; Amos, 1974). *HL-A* not only consists of a group of tightly linked genes affecting cell surface phenotypes, but its antigens also cross react with mouse *H-2* antigens (Klein, 1975, p. 532). The two complexes would appear to be evolutionary homologues. Other homologous complexes are found in non-human primates, dogs, and even chickens (Pazderka et al., 1975). All seem to affect both immune responses and graft survival, as well as specifying antigens detected with alloantibodies formed after graft rejection.

In discussing expression of these complex regions in development it is well to remember that typing for the products of one part of the complex does not indicate anything about expression of other parts of the complex. For example, while typing *H-2* haplotypes defined by some common laboratory strains often allows prediction of genotype and phenotype in another region, this may not be true for recombitant haplotypes (see, McDevitt and Chinitz, 1969, for one example involving immune response genes) and is usually not true for human *HL-A* haplotypes (Terasaki et al., 1974). Thus, in cataloging expression of *H-2* complex genes in development we can only note that the positive findings should perhaps be given more weight than negative findings. Until we know how to assay for all functions associated with the *H-2* complex we will not be certain as to the role its products may play in development.

3. *Transplantation antigen expression in adults and embryos*

3.1. *Adults*

Specific examples were not used in our discussion of tissue graft rejection. In fact, it appears that any tissue graft, whether of skin, or connective tissue, or brain, contains sufficient transplantation antigens, including *H-2* antigens, to cause its rejection (Edidin, 1972a). While there is some question as to the presence of *H-2* antigens in muscle or brain cells, the antigens of the complex detected by anti D or anti K antisera are well represented on lymphoid cells, on liver and kidney parenchymal cells (Graziano and Edidin, 1971), and on many established cell lines in tissue culture. In general, cells lacking *H-2* antigens cannot be selected from populations of tumor or cultured cells bearing *H-2*, whether selection is done in vivo or in vitro. It seems then that at least some part of the *H-2* complex is necessary for normal cell function (summary in Klein, 1975, chapter 13).

As mentioned, these antigens appear to be integral membrane proteins. They can be cleaved from the membrane with papain, but not with trypsin, and the cleavage products are of lower molecular weight than the antigens extracted with detergent. *H-2*D and *H-2*K molecules extracted with NP-40 detergent have apparent molecular weights on SDS acrylamide gels of around 44,000 (Nathenson and Cullen, 1974). *Ia* antigens, more restricted in their tissue distribution (Delovitch and McDevitt, 1975; Hämmerling et al., 1975), appear to be around 35,000 molecular weight. Generally the *H-2* antigens detected with anti-*H-2* sera are the 44,000 molecular weight antigens referred to. However, many anti-*H-2*D and K antisera contain antibodies to *Ia* antigens as well and these may complicate studies of *H-2* in gametes and embryos.

3.2. H-2 *complex products in embryos and gametes*

We begin this section with the observation made earlier. Mammalian embryos are not rejected by their mothers, though the fetal allograft is almost certain to differ in surface antigen makeup from its mother because of the high polymorphism of strong transplantation antigens. Why are these allografts not rejected? Three main mechanisms are conceivable: (1) The embryo is antigenically immature; histocompatibility antigens do not appear until late in term, or even after birth. (2) The pregnant female is immunologically suppressed, unable to respond to the antigens of the embryo. (3) All or part of the placenta constitutes a barrier between mother and fetus.

Mechanism 2 does not fall within the scope of this review. We may briefly summarize it by stating that pregnant females are generally immunologically responsive (Medawar and Sparrow, 1956; Woodruff, 1958) and that most theories of immunosuppression of pregnant females focus on specific unresponsiveness (Breyere and Barrett, 1961) to either transplantation or embryonic antigens involving suppressor cells or antibodies to fetal antigens (Hellström et al., 1971; Brawn, 1970; Rees et al., 1975), or on nonspecific immunosuppression which is due to hormones localized to the placenta (Caldwell et al., 1975). Beyond this, we need only point out that the mechanism takes for granted a high degree of antigenicity in the fetus.

The two remaining mechanisms are related to the extent that there is evidence showing the trophoblast to lack alloantigens, a condition necessary for barrier function. Hence, a catalog of cell surface antigen differentiation in embryos should lead us to a branch point at which trophoblast cells diverge from the cells of the embryo proper.

When considering antigens of the embryo we must remember that the portion of the *H-2* complex detected depends upon the method used to assay antigens. Either cell-mediated immunity or humoral antibody can be used to prove for *H-2* antigens in embryos. If some portions of the *H-2* complex affect all surfaces but are not antigenic, they will be overlooked in these assays. The two main experimental designs probing *H-2* expression in mouse embryos are: (1) measurement of in vivo cellular responses to an embryo tissue graft; and (2) measurement in vitro of anti-*H-2* antibody interaction with embryo cells. Since few histocompatibility antigens other than those of the strong locus provoke formation of humoral antibody, most probes of the weaker transplantation antigens have relied soley on host response to grafts as a measure of antigen expression, though a few exceptional antigens, *H-3* and *H-6*, have also been detected serologically. Throughout the discussion we will emphasize the strong transplantation antigens, since, because of the polymorphism of the complex determining them there is the greatest probability of a mismatch between fetus and mother at this locus. Also, many of the other histocompatibility antigens provoke such feeble immunity that responses are not liable to occur during the term of a pregnancy (for other recent reviews see Billington and Jenkinson, 1975; Johnson, 1975).

3.2.1. Surface antigens of gametes

Surprisingly little has been learned about the alloantigens of gametes, though there is a large literature on anti-sperm antibodies, especially as they affect fertility (Jones, 1975; Voisin et al., 1975). Analysis of alloantigens on gametes, both sperm and eggs, has been hampered by the presence of these anti-tissue antibodies, as well as by a virtual restriction to serological methods for an assay. Both male and female gametes have a limited lifespan. This makes interpretation of grafting experiments extremely difficult. Also, sperm preparations, even from ejaculates, are contaminated with non-sperm cells and hence any in vivo sensitization by these preparations may be due to contaminating cells (Vojtišková et al., 1969). Finally, eggs present still another problem; it is not easy to procure sufficient eggs to obtain in vivo responses, though a few experiments have been done in this way (Baranska et al., 1970).

Despite these difficulties, a small amount of work has been done on transplantation antigens of sperm. The results are still tentative, but encouraging.

First, some portion of the *H-2* complex is expressed on the surface of sperm. This has been demonstrated by fluorescence (Vojtišková et al., 1969; Erickson, 1972), and by cytotoxicity experiments, with reasonably careful controls (Goldberg et al., 1970; Johnson and Edidin, 1972). In the latter, sperm could be lysed by anti-*H-2* serum made against the appropriate mouse strain, even when such antisera had been absorbed on sperm of the serum donor strain to remove tissue-specific (anti-sperm) antibodies (Johnson and Edidin, 1972). This crucial control is not evident in some other work, but the titers of alloantisera against sperm expected to lyse and those not expected to lyse suggest some *H-2* associated specificity for the reaction. Similar experiments with the human *HL-A* system also suggest that some antigens of the *HL-A* complex are expressed on sperm (Fellous and Dausset, 1970). For both mouse and human, we cannot be sure just what parts of the major histocompatibility complex are expressed. Recently, antisera against *Ia* antigens have been clearly shown to react with sperm (Hämmerling et al., 1975). Hence, all of the papers to date detecting "*H-2*" antigens with antiserum on sperm may be reporting on *Ia* antigens, and we are left with no clear evidence for expression of D and K products on male gametes.

Given that the particular product of the *H-2* complex expressed in sperm is unknown, and conceding that some surface antigen specified by the major histocompatibility locus is expressed, we may ask if expression of the antigens is post-meiotic or not. Though in general post-meiotic expression of genome is not detected (Beatty, 1970), several reports suggest that haploid expression of *HL-A* or *H-2* occurs (Fellous and Dausset, 1970; Goldberg et al., 1970). These depend upon the differences in toxicity of antiserum for sperm from homozygotes, compared to those from F1. However, the values for kill of target sperm do not show clear-cut differences. Even sperm from homozygotes are not uniformly killed by anti-*H*-2 serum and may yield as low as 60% damaged cells, a level approaching that expected for the same antiserum reacting with cells from heterozygotes.

Two other antigen systems have been detected on mouse sperm surfaces; one of these is a transplantation antigen, the so-called male-specific or *H-Y* antigen. Existence of this antigen was first detected when females of an inbred strain were found to reject skin grafts from males of the same strain (Eichwald and Silmser, 1955; Hauschka, 1955). Rejection times due to this antigen were twice to three times as long as those across an *H-2* barrier, but this is still a relatively rapid rejection. Antibody is also produced to the *H-Y* antigen, as a result of graft rejection, and this antibody has been used to show the presence of *H-Y* on sperm (Goldberg et al., 1971). Again, cytotoxicity testing of sperm suggests that there is a differential in expression of *H-Y*, but again, the values for lysis were not clearly 50%; it may well be that the lytic assay is not sensitive enough to detect varying levels of *H-Y* on all sperm (Koo et al., 1973). Further evidence against post-meiotic expression of *H-Y* is that treatment of sperm with anti-*H-Y* does not alter the sex ratio of offspring produced from them. In another approach, Koo and co-workers showed by electron microscopy that as few as 20% of sperm fail to label with anti-*H-Y* and that there is a 10-fold range in intensity of labeling in the others (Koo et al., 1973). Thus, while there appears to be a quantitative variation in *H-Y* per sperm there is no evidence for haploid expression of *H-Y*.

Whatever the chemical nature of *H-Y*, it seems to be a structure highly conserved in evolution, always associated with the heterogametic sex. Thus, anti-mouse *Y* antigen sera react with other mammalian cells, including those of humans (Wachtel et al., 1974), while the same sort of antibody preparation also reacts with the cells of female birds and some female amphibia (Wachtel et al.,

1975a). *H-Y*, though serologically defined by reactions against sperm, is also found on other cells, notably epidermal epithelia and splenic lymphocytes. Recently, it has been suggested that *H-Y* antigen functions in the determination of sex rather than appearing on cells as a result of sex differentiation (Bennett et al., 1975; Wachtel et al., 1975b) and this antigen may be an example of a surface determinant whose appearance early in development leads cells down a particular pathway of differentiation. It would be of great interest to follow *H-Y* through embryonic development to see if it is continually expressed, and to note its fate during organogenesis.

A third antigen system has been described on sperm. Its antigens are products of a complex genetic locus linked to *H-2* in mouse, the *T* locus (Dunn, 1964; Bennett, 1964; Bennett, 1975; Hillman, 1975). Like *H-2*, the locus appears to consist of a series of tightly linked genes, each having a wild type and at least one other allele, usually recessive. Homozygotes for any of the alleles die at various stages of development, and it has been suggested that the *T* antigens, defined by anti-sperm antisera (Yanagisawa et al., 1974a, 1974b) or by antisera to a teratocarcinoma (Artzt et al., 1973), function as differentiation antigens, affecting cell surface interactions during development (Bennett et al., 1971; Gluecksohn-Waelsch and Erickson, 1971; Artzt et al., 1974). In this system there is again a hint that post-meiotic expression of the locus does occur, since, for example, sperm of $T/+$ animals are not lysed to the same extent as t^{w}/T sperm, when treated with an anti-t^{w}/T antiserum.

Again, one must bear in mind the differentials in antigen reactivity that seem to occur in most cell populations. Using low titered antisera or insensitive assay methods, heterozygous cells may appear to be unreactive with a particular serum.

Some quite informative experiments have also been done on transplantation antigens of zygotes and early embryos. Techniques of transplantation to ectopic sites are more than adequate to allow grafting experiments as well as serological measurements of zygote antigens. In transplantation experiments, zygotes or later stages are transferred to specifically immunized recipients, typically to the kidney capsule. Accelerated rejection of such grafts, compared to embryo grafts in unimmunized recipients provides evidence for transplantation antigens on the stages grafted. Such experiments clearly indicate that some transplantation antigens are present on zygotes, since inter-strain grafts are rapidly destroyed (Simmons and Russell, 1966). (Earlier experiments in which embryo grafts were transferred to unimmunized recipients cannot readily be taken as evidence for transplantation antigens on the stage grafted; while such grafts are rejected, their rejection could be due to differentiation of the graft and expression of antigens at a later stage of development.) However, it does not appear that *H-2* antigens are expressed, since when embryos of up to 6 days of age are grafted between cogenic resistant strains, differing genetically only at the *H-2* complex, no accelerated rejection is observed (Patthey and Edidin, 1973; Searle et al., 1974). In vitro model systems, using lymphocytes from immune animals to attack and destroy embryo tissue targets also give evidence that *H-2* is lacking in blastocytes (Jenkinson and Billington, 1974). The earliest stage of development that is rejected as if it bore *H-2* antigens is around 7 days of development (Patthey and Edidin, 1973).

Serological studies parallel and extend the data obtained in grafting experiments. Anti-sera to two moderately strong *H* antigens, *H-3* and *H-6*, react well with fertilized eggs and cleavage stages of mouse development. However, purely anti-*H-2* antisera do not react (Palm et al., 1971). Studies of cultured blastocysts which have shed trophoblast and hence present a naked inner cell mass for serum binding also fail to detect *H-2* (Heyner, 1973). However, culture of these blastocysts for a number of days results in expression of *H-2*, together with other developmental changes. Cultured embryos expressing *H-2*, as detected with fluorescent antibodies, appeared equivalent to normal $6\frac{1}{2}$-day embryos, around the stage of development determined to be *H-2* positive by grafting.

Later embryos appear to develop levels of strong transplantation antigens to some extent proportional to the levels that will be found in adult tissues (Graziano and Edidin, 1971). However, the *H-2* content of some organs, notably liver, has been shown to fluctuate and this may well reflect the traffic of various cell populations, especially erythroid cell precursors through the liver at this stage of mouse development. Liver in culture, like early embryos, exhibits an inherent timetable for *H-2* development (Klein, 1965). Cultured liver cells gradually increase in detectable *H-2* through a period equivalent to the normal gestation of their donor.

Even at birth, *H-2* antigen levels on a per cell or per unit protein basis are low compared to adult levels, the maximal values for lymphocytes and spleen appear to be around 10% of adult values; this estimate is based on quantitative absorption of antisera (Graziano and Edidin, 1971). The rate of post-natal maturation of erythrocyte *H-2* and perhaps of other cells appear to be under genetic control (Möller and Möller, 1962; Boubelik and Lengerova, 1971).

To summarize thus far, the antigens determined by some regions of the *H-2* complex appear after approximately one-third of development is completed and do not reach adult levels in various tissues until after birth. Thus it may be that one part of the protection afforded the embryo allograft is by the absence of strong alloantigens on the embryo cells.

It should be recalled that absence of antigens determined by *H-2* does not imply absence of other products in this region. If some *H-2* genes determine surface receptors, or otherwise modify cell interactions in development, they could be expressed, but not detected if their products fail to provoke immune responses or to react with antisera. Several authors have speculated that *H-2* complex genes might affect cell differentiation by modifying and affecting cell to cell interactions and by creating surface phenotypes, leading to particular paths of development and differentiation (Bennett et al., 1971; Burnet, 1971; Bodmer, 1972). It may be that in fact other portions of the complex contribute to such surface changes in differentiation, and that those *H-2* antigens serologically recognizable in adults represent only mature or modified forms of earlier surface features (Bodmer, 1973). Indeed, there is one suggestion that epigenetic variation of *H-2* and other surface antigens is the rule in development, with local environments ultimately determining the details of surface structures, and with them the fate of the cells bearing such structures (cf. Edelman, 1974). Less speculatively, we may suggest that *H-2* antigens, if they do determine or affect cell interactions, the degree of cell adhesion and/or the path of cell differentiation do so in the latter parts of development, during organogenesis.

Another class of cell surface antigens is also likely to affect cell association and differentiation. Stage specific antigens, appearing transiently in the embryo have been described on many occasions. Some of these are associated temporally or spatially with *H-2* and could function as the equivalents of strong transplantation antigens in early development. Two such antigen systems have been defined by antisera to cultured teratocarcinoma lines.

One of these, the F9 antigen, is detected with mouse antiserum against a cultured mouse teratocarcinoma line, F9, which has apparently lost all ability to differentiate in vivo or in vitro (Artzt et al., 1973). F9 antigen is detected in cleavage stages of normal mouse development, and on mouse and human sperm as well (Buc-Caron et al., 1974). Studies with sperm suggest that the antigen is determined by the wild-type allele of a *t* mutant, t^{12} (Artzt et al., 1974). Homozygotes for this mutant die at the morula stage, the latest stage shown to contain F9 antigens. Other studies with this system show that F9 cells lack *H-2* antigens (Artzt and Jacob, 1974), and that other lines of undifferentiated teratocarcinoma cells which can be induced to differentiate in culture, lose F9 antigen as they gain *H-2* (Nicolas et al., 1975). This reciprocal relationship between the two antigens in time suggests that the function of F9 antigens in cell interactions is taken over in development by components of the *H-2* locus.

A second antigen system defined by a teratoma is far more common than F9, appearing in many, though not all, transformed cultured cells tested, as well as the cultured teratocarcinoma, 402AX (Gooding and Edidin, 1974). This line does differentiate in vitro, and so cells used for immunization were a mixture of types. Despite this, a rabbit antiserum against 402AX reacts with a limited range of mouse tumors (excluding all leukemias tested), and reacts with only portions of normal embryos. One of the antigens defined by this serum, antigen I, is present in unfertilized eggs, zygotes, cleavage stages and blastulae (Edidin and Gooding, 1975; Gooding et al., 1976). It is not present on trophoblast cells derived by culturing blastocysts on a collagen substrate, and disappears from mouse embryos at around 8 days of development, which is one time at which *H-2* is first reliably detected. Like F9, 402AX cells do not bear *H-2* as judged by grafting as well as serological experiments (Edidin et al., 1975). Also, the temporal relationship between 402AX antigen I and *H-2* indicates the possibility of a complementary functional role as suggested above. Still further associations between *H-2* and the tumor defined antigen are found when the antigen is studied on cultured malignant murine fibroblasts, C1 1d cells. Here, the two antigens appear to be physically associated, since capping of one antigen system with aggregation of all of the surface antigens into one small area of the cell, brings antigens of the second system into the cap as well. Thus, while in embryos *H-2* antigens succeed teratoma-defined antigens, in some tumor cells the two are co-expressed and physically associated.

Whatever their function, at least some stage-specific antigens provoke immunity in pregnant females, to the extent that cells from such females, as well as serum, will react with tumors (Brawn, 1970; Girardi et al., 1973; Baldwin, et al., 1975; Edidin et al., 1975). It is not yet known at what stage in development immunization occurs, or whether or not deliberate immunization to embryonic antigens shared with tumors can suppress pregnancy.

In summary, the hypothesis of antigenic neutrality is not sufficient to explain embryonic survival in a host potentially able to reject it. Weak and strong histocompatibility antigens are expressed in the course of development, as well as differentiation antigens, peculiar to particular stages of development, which are foreign to the maternal host. However, considering the timing of expression of various antigens, and the magnitude of the responses that they provoke, it may well be that early stages of embryos do evade rejection by being poorly antigenic, and are able to move on to a further stage of development, implanted and separated from maternal circulation by a placenta, before the maternal immune response can get underway.

4. *Placenta and trophoblast as barriers to embryo rejection*

The placenta lies as a barrier between fetus and mother in all mammals. This complex organ varies in the number of cell layers between maternal and fetal circulations, but even in the most efficient placentae, those with the fewest layers between the circulations, there appears to be a complete cellular barrier between the two (Mossman, 1973; Billington, 1975). Low molecular weight nutrients and gasses pass this barrier by diffusion, higher molecular weight substances such as immunoglobulins are selectively transported (Wild, 1974), but there is no good evidence that cells can pass the barrier at all (Billington et al., 1969; Selier, 1970; but see Tuffrey et al., 1969). The fetal cells forming this barrier are trophoblast, derived from extraembryonic portions of the conceptus. It is trophoblast cells that invade the uterine linings in the course of embryonic implantation, and which later contribute to formation of the definitive placenta (for details of implantation see Wimsatt, 1975; Schlafke and Enders, 1975; and chapter 3 of this volume by Sherman and Wudl). Thus throughout the embryo's life in utero, there is close contact between trophoblast cells and maternal cells and circulation. The immunological implications of this contact have been discussed by Bardawil and Toy (1959) and later studies (McCormick et al., 1971) indicate that masses of immunoglobulin and complement components are deposited at the fetal-maternal interface in the placenta, again suggestive of an immune reaction going on continuously during pregnancy.

Pure trophoblast may be obtained by grafting cleavage mouse embryos to allogenic adults. The embryonic portions of these grafts are rejected, but they persist long enough to allow differentiation of trophoblast, which persists for weeks without promoting a host cellular reaction (Simmons and Russell, 1966; Searle et al., 1974). In mouse this differentiation depends upon the position of cells in the cleavage embryo and requires the persistence of inner cell mass cells (Herbert and Graham, 1974; Gardner, 1975). Hence presence of trophoblast is also an indication of the low level of alloantigenicity of the entire embryo at this stage. Another approach to procuring trophoblast uses grafts of the $6\frac{1}{2}$-day mouse embryo ectoplacental cone, the forerunner of the definitive placenta. Grafts of such rudiments grow even in specifically immunized adults (Kirby et al., 1966; Simmons and Russell, 1966). Again, no sign of cellular infiltration is seen around the graft. Similarly, cultured trophoblast is not destroyed by immune lymphocytes in culture, though embryo-derived portions of the culture are destroyed (Jenkinson and Billington, 1974). Thus, the trophoblast seems not only to resist killing by alloimmune cells, but it also fails to express the antigens that would attract such cells. The cells of trophoblast are sensitive to immune attack, since mouse to rat grafts are destroyed in vivo (Simmons and Russell, 1967), and human trophoblast cells in culture are sensitive to attack by anti-human chorionic gonadotropin (an antibody against a secretion product of trophoblast) and complement (Currie and Bagshawe 1967).

Are the alloantigens of trophoblast masked, or are they not expressed at all? If they are not expressed, are parts of the major histocompatibility complex expressed, or is trophoblast entirely free of all products of this complex?

Whole placentae are always seen to contain a mass of amorphous material, "fibrinoid." This material is a gross deposit that contains fibrin, as well as various immunoglobulins and complement components, and it is this deposit that was taken as evidence for a truce in the immunological "battlefield" of the placenta (Douglas, 1959). The term fibrinoid has also been used to describe a more discrete deposit of negatively charged material coating trophoblast cells (references are summarized in Edidin, 1972b). The latter fibrinoid seems to be a glycoprotein deposit, with a high content of sialic and hyaluronic acids. The appearance and chemistry of this deposit led Kirby (Kirby et al., 1964; Kirby, 1968) to suggest that it masked surface antigens on trophoblast and prevented killing of these cells. To test that proposal, Currie and coworkers treated mouse trophoblast (ectoplacental cones) with neuraminidase, and then used the treated materials to immunize adult allogeneic recipients, who were later challenged with test skin grafts from adults of the trophoblast donor strain (Currie et al., 1968). Ectoplacental cone cells which had been treated with neuraminidase were as effective as spleen cells (expressing their *H-2* antigens) in provoking accelerated graft rejection. Hence it was argued that alloantigens were present on trophoblast cells, but were masked by the highly charged cell surface coat. However, later experiments, both using the same grafting design and also probing for alloantigens with antiserum failed to detect any alloantigens on neuraminidase treated cells (Simmons et al., 1971; Searle et al., 1975). It may be that bacterial products, components of the neuraminidase used, adsorbing to the treated cells and, being cross-reactive with alloantigens, provoked accelerated graft rejection. Whatever the explanation, it now seems clear that neuraminidase treatment of mouse trophoblast does *not* lead to expression of alloantigens on these cells.

There is at present then no evidence for masking of trophoblast alloantigens. However, we also have no data on loss of antigen from the surface of differentiating trophoblast. If zygotes express non-*H-2* antigens then these must be cleared from the trophoblast if it is to be non-immunogenic as described. When this clearance begins, is uncertain. However, it is interesting to note that in a study of a teratoma defined, *H-2*-associated antigen on cleavage mouse embryos it appeared that some blastomeres lost their reactivity with the anti-teratoma serum while others retained this activity (Gooding et al., 1976). Trophoblast of early blastocyst stages was still somewhat reactive with the antiserum, but trophoblast from late blastocysts did not react at all with any of the antibodies in an anti-teratoma serum. This description hints at the stages during which alloantigens and early stage-specific antigens are lost from trophoblast, but much more work is needed to enlarge on these hints.

It should be noted that several cell lines have been established from the malignant trophoblast cells of human choriocarcinoma, and that these lines may offer opportunity for experimental manipulation and testing of the reasons for lack of expression of alloantigens on trophoblast.

5. *Two paradoxes*

The trophoblast appears to be a good candidate for an antigenically neutral barrier between fetus and mother. However, we are still not clear as to its expression of any part of the major histocompatibility complex. Several sets of observations suggest that though surface antigens of the complex may not be expressed in trophoblast, features affecting cell interactions are expressed.

First, in malignant trophoblast, such as human choriocarcinoma, it appears that *HL-A* compatibility allows greater invasiveness of trophoblast; metastases are more common in cases where husband and wife are *HL-A* compatible, in terms of reaction with typing sera, than when they are *HL-A* incompatible (the clinical papers are summarized in Edidin, 1972b). If some genes in the *HL-A* complex have effects on non-immunologic cell interactions, matching for alleles of these genes, indicated by the matching for closely linked antigens, could lead to more ready invasiveness of non-uterine tissues by the tumor cells.

A second case involves placental size in pregnancies between strains. Many observations have been made to suggest that the weight of the entire placenta is greater in such allogeneic pregnancies than in pregnancies within an inbred strain (Billington, 1965; McLaren, 1965; James, 1967). Recently, Beer et al. (1975) have shown convincingly that this is true for placentae of three rodent species, mouse, hamster and rat. In all cases, preimmunization of the mother against parental antigens led to even heavier placentae than when pregnancies were simply interstrain. Of course, the larger size of the placenta results in improved nutrient supply to the fetus and hence to heavier fetuses.

Beer and coworkers noted that all elements of the placenta appeared to have increased in quantity, not simply the trophoblast. However, since the trophoblast is a major fetal contribution to the placenta, whatever else has occurred, the trophoblast cells must have responded in some way to allogeneic immunity. It was early suggested that cell surface alloantigens were evolutionarily homologous to receptors required in cell adhesion – receptors sensing the associations of a single cell with its neighbors (Burnet, 1971, Bodmer, 1972). Perhaps in the placenta, and in trophoblast, this function once again becomes paramount, and so modified alloantigens, encountering an immune response or simply a display of other alloantigen stimulate the cells bearing them to proliferation. Here, as in the case of choriocarcinoma it appears than an aspect of histocompatibility is related to a more general cellular response: adhesion, growth, migration. It may well be that the details filling in the sketch given here will lead us to realize that portions of the major histocompatibility complex are expressed throughout development, but that these portions are involved in far more than the rejection of grafts by recipients, serving instead to co-ordinate, if not to control cell proliferation and differentiation in response to a host of external stimuli.

References

Alter, B.J., Schendel, D.J., Bach, M.L., Bach, F.H., Klein, J. and Stimpfling, J.H. (1973) Cell-mediated lympholysis: Importance of serologically defined H-2 regions. J. Exp. Med. 137, 1303–1309.

Amos, D.B. (1974) Genetics of the human histocompatibility system HL-1. Transplantation Proceedings, 6, 27–32.

Artzt, K. and Jacob, F. (1974) Absence of serologically detectable H-2 on primitive teratocarcinoma cells in culture. Transplantation 17, 633–634.

Artzt, K., Bennett, D. and Jacob, F. (1974) Primitive teratocarcinoma cells express a differentiation antigen specified by a gene at the T-locus in the mouse. Proc. Nat. Acad. Sci. U.S.A. 71, 811–814.

Artzt, K., Dubois, P., Bennett, D., Condamine, H., Babinet, C., and Jacob, F. (1973) Surface antigens common to mouse cleavage embryos and primitive teratocarcinoma cells in culture. Proc. Nat. Acad. Sci. U.S.A. 70, 2988–2992.

Baldwin, R.W., Embleton, M.J., Price, M.R., and Vose, B.M. (1975) Embryonic antigen expression on experimental rat tumors. Transplant Rev. 20, 78–99.

Baranska, W., Koldovsky, P., and Koprowski, H. (1970) Antigenic study of unfertilized mouse eggs: cross-reactivity with SV-40-induced antigens. Proc. Nat. Acad. Sci. U.S.A. 67, 193–197.

Bardawil, W. A. and Toy, B. L. (1959) The natural history of choriocarcinoma: problems of immunity and spontaneous regression. Ann. N.Y. Acad. Sci. 80, 197–257.

Beatty, R.A. (1970) The genetics of the mammalian gamete. Biol. Rev. 45, 73–119.

Beer, A.E., Scott, J.R., and Billingham, R.E. (1975) Histoincompatibility and maternal immunological status in determination of fetoplacental weight and litter size in rodents. J. Exp. Med. 142, 180–196.

Benacerraf, B. and McDevitt, H.O. (1972) Histocompatibility-linked immune response genes. Science, 175, 273–279.

Bennett, D. (1964) Abnormalities associated with a chromosome region in the mouse II. The embryological effects of lethal alleles at the *t*-region. Science, 144, 263–267.

Bennett, D. (1975) T-locus mutants: suggestions for control of early embryonic organization through cell surface components. In: The Early Development Of Mammals (Balls, M. and Wild, A.E., eds.) pp. 207–218, Cambridge University Press, Cambridge.

Bennett, D., Boyse, E.A., and Old, L.J. (1971) Cell surface immunogenetics in the study of morphogenesis. In: Cell Interactions, (Silvestri, L.G., ed.) pp. 247–263, North Holland, Amsterdam.

Bennett, D., Boyse, E.A., Lyon, M.F., Mathieson, B.J., Scheid, M. and Yanagisawa, K. (1975) Expression of H-Y (male) antigens in phenotypically female *Tfm/Y* mice. Nature 257, 236–238.

Billington, W.D. (1965) The invasiveness of transplanted mouse trophoblast and the influence of immunological factors. J. Reprod. Fertil. 10, 343–352.

Billington, W.D. (1975) Organization, ultrastructure and histocompatibility of the placenta: immunological considerations. In: Immunobiology of Trophoblast (Edwards, R.G., Howe, C.W.S., and Johnson, M.H., eds.), pp. 67–85, Cambridge University Press, Cambridge.

Billington, W.D. and Jenkinson, E.J. (1975) Antigen expression during early mouse development. In: The Early Development of Mammals (Balls, M. and Wild, A.E., eds.) pp. 219–232, Cambridge University Press, Cambridge.

Billington, W.D., Kirby, D.R.S., Owen, J.J.T., Ritter, M.A., Burtonshawe, M.D., Evans, E.P., Ford, C.E., Gauld, I.K., and McLaren, A. (1969) Placental barrier to maternal cells. Nature, 224, 704–706.

Bodmer, W.F. (1972) Evolutionary significance of the HL-A system. Nature, 237, 139–145.

Bodmer, W.F. (1973) Genetics of the HL-A and H-2 major histocompatibility systems. In: Defense and Recognition (Porter, R.R. ed.) pp. 295–328, Butterworths, London.

Boubelik, M. and Lengerová (1971) Genetic control of the developmental expression of H-2 antigens. In: Immunogenetics of the H-2 System (Lengerová, A. and Vojtišková, M. eds.) pp. 85–89, S. Karger, Basel.

Brawn, R.J. (1970) Possible association of embryonal antigen(s) with several primary 3-methylcholanthrene-induced murine sarcomas. Int. J. Cancer 6, 245–256.

Breyere, E.J. and Barrett, M.K. (1966) Tolerance induced by parity in mice incompatible at the H-2 locus. J. Nat. Cancer Inst. 27, 409–417.

Buc-Caron, M.H., Gachelin, G., Hofnung, M., and Jacob, F. (1974) Presence of a mouse embryonic antigen on human spermatozoa. Proc. Nat. Acad. Sci. U.S.A. 71, 1730–1733.

Burnet, F.M. (1971) "Self-recognition" in colonial marine forms and flowering plants in relation to the evolution of immunity. Nature, 232, 230–235.

Burnet, F.M. (1973) Multiple polymorphism in relation to histocompatibility antigents. Nature, 245, 359–361.

Caldwell, J.L., Sikes, D.P., and Fudenberg, H. (1975) Human chorionic gonadotropin: effects of crude and purified preparations of lymphocyte responses to phytohemagglutin and allogeneic stimulation. J. Immunol. 115, 1249–1253.

Currie, G.A. and Bagshawe, K.D. (1967) The masking of antigens on trophoblast and cancer cells. Lancet ii, 708–771.

Currie, G.A., van Doorninck, W., and Bagshawe, K.D. (1968) Effect of neuraminidase on the immunogenicity of early mouse trophoblast. Nature, 219, 191–192.

Delovitch, T.L. and McDevitt, H.O. (1975) Isolation and characterization of murine *Ia* antigens. Immunogenetics, 2, 39–52.

Douglas, G.W. (1959) Discussion following Bardawil and Toy (1959). Ann. N.Y. Acad. Sci. 80, 260–261.

Dunn, L.C. (1964) Abnormalities associated with a chromosome region in the mouse. I. Transmission and population genetics of the t-region. Science, 144, 260–263.

Edelman, G.M. (1974) Perspective, In: The Cell Surface (Kahan, B.D. and Reisfeld, R.A. eds.) pp. 257–266, Plenum Press, New York.

Edidin, M. (1972a) The tissue distribution and cellular location of transplantation antigens. In: Transplantation Antigens (Kahan, B.D. and Reisfeld, R. eds.) pp. 125–140, Academic Press, New York.

Edidin, M. (1972b) Histocompatibility genes, transplantation antigens and pregnancy. In: Transplantation Antigens (Kahan, B.D. and Reisfeld, R. eds.) pp. 75–114, Academic Press, New York.

Edidin, M. and Henney, C.S. (1973) The effect of capping H-2 antigens on the susceptibility of target cells to humoral and T cell-mediated lysis. Nature, 246, 47–49.

Edidin, M. and Gooding, L.R. (1975) Teratoma-defined and transplantation antigens in early mouse embryos. In: Teratomas and Differentiation (Sherman, M. and Solter, D. eds.) pp. 109–121, Academic Press, New York.

Edidin, M., Gooding, L.R., and Johnson, M. (1975) Surface antigens of normal early embryos and a tumor model system useful for their further study. Acta Endocrinol. 78, Supplement 194, 336–356.

Eichwald, E.J. and Silmser, C.R. (1955) Transplant. Bull. 2, 148–149.

Erickson, R.P. (1972) Alternative modes of detection of H-2 antigens on mouse spermatozoa. In: Genetics of the Spermatozoan (Beatty, R.A. and Gluecksohn-Waelsch, S. eds.) pp. 191–202, Edinburgh University Press, New York.

Fellous, M. and Dausset, J. (1970) Probable haploid expression of HL-A antigens on human spermatozoa. Nature, 225, 191–193.

Gardner, R.C. (1975) Origins and properties of trophoblast. In: Immunobiology of Trophoblast (Edwards, R.G., Howe, C.W.S. and Johnson, M.H. eds.) pp. 43–65, Cambridge University Press Cambridge.

Girardi, A.J., Repucci, P., Dierlam, P., Rutala, W. and Coggin, J.H. (1973) Prevention of simian virus 40 tumors by hamster fetal tissue: influence of partity status of donor females on immunogenicity of fetal tissue and an immune cell cytotoxicity. Proc. Nat. Acad. Sci. U.S.A. 70, 183–186.

Gluecksohn-Waelsch, S. and Erickson, R.P. (1971) A possible link between H-2 and T-locus effects. In: Immunogenetics of the H-2 System (Lengerová, A. and Vojtišková, M. eds.) pp. 120–122, S. Karger, Basel.

Goldberg, E.H., Aoki, T., Boyse, E.J., and Bennett, D. (1970) Detection of H-2 antigens on mouse spermatozoa by cytotoxicity tests. Nature, 228, 570–572.

Goldberg, E.H., Boyse, E.A., Bennett, D., Scheid, M., and Carswell, E.A. (1971) Serological demonstration of H-Y (Male) antigen in mouse sperm. Nature, 232, 478–480.

Gooding, L. and Edidin, M. (1974) Cell surface antigens of a mouse testicular teratoma. Identification of an antigen physically associated with H-2 antigens on tumor cells. J. Exp. Med. 140, 61–78.

Gooding, L., Hsu, Y.-C., and Edidin, M. (1976) Expression of teratoma-associated antigens on murine ova and early embryos. Dev. Biol. 49, 479–486.

Gorer, P.A. (1938) The antigenic basis of tumor transplantation. J. Pathol. Bacteriol. 47, 231–252.

Gorer, P.A., Lyman, S., and Snell, G.D. (1948) Studies on the genetic and antigenic basis of tumor transplantation. Linkage between a histocompatibility gene and "fused" in mice. Proc. Roy. Soc. B 135, 499–505.

Graff, R.J., Silvers, W.K., Billingham, R.E., Hildemann, W.H., and Snell, G.D. (1966) The cumulative effect of histocompatibility antigens. Transplantation, 4, 605–617.

Graziano, K.D. and Edidin, M. (1971) Serological quantitation of histocompatibility-2 antigens and the determination of H-2 in adult and fetal organs. In: Immunogenetics of the H-2 system (Lengerová, A. and Vojtišková, M. eds.) pp. 251–256, S. Karger, Basel.

Hämmerling, G.J., Mauve, G., Goldberg, E. and McDevitt, H.O. (1975) Tissue distribution of Ia antigens. Immunogenetics, 1, 428–437.

Hauschka, T.S. (1955) Probable Y-linkage of a histocompatibility gene. Transplant. Bull. 2, 154–155.

Hellström, I., Hellström, K.E. and Allison, A.C. (1970) Neonatally induced allograft tolerance may be mediated by serum-born factors. Nature, 230, 49–50.

Herbert, M.C. and Graham, C. (1974) Cell determination and biochemical differentiation of the early mammalian embryo. Curr. Top. Dev. Biol. 8, 151–178.

Heyner, S. (1973) Detection of H-2 antigens on cells of the early mouse embryo. Transplantation, 16, 675–678.

Hillman, N. (1975) Studies of the T-locus. In: The Early Development of Mammals (Balls, M.E. and Wild, A.E. eds.) pp. 189–206, Cambridge University Press, Cambridge.

Ivanyi, P. (1970) The major histocompatibility antigens in various species. Curr. Topics Microbiol. Immunol. 53, 1–90.

James, D.A. (1967) Some effects of immunological factors on gestation in mice. J. Reprod. Fertil. 14, 265–275.

Jenkinson, E.J. and Billington, W.D. (1974) Differential susceptibility of mouse trophoblast and embryonic tissue to immune cell lysis. Transplantation, 18, 286–289.

Johnson, M.H. (1975) Antigens of the peri-implantation trophoblast. In: Immunobiology of Trophoblast (Edwards, R.G., Howe, C.W.S., and Johnson, M.H. eds.) pp. 87–112, Cambridge University Press, Cambridge.

Johnson, M.H. and Edidin, M. (1972) H-2 antigens on mouse spermatozoa. Transplantation, 14, 781–786.

Jones, W.P. (1975) The use of antibodies developed by infertile women to identify relevant antigens. Acta Endrocinologica 78, Supplement 194, 376–404.

Kirby, D.R.S. (1968) Immunological aspects of pregnancy. Adv. Reprod. Physiol. 3, 33–79.

Kirby, D.R.S., Billington, W.D., Bradbury, S., and Goldstein, D.J. (1964) Antigen barrier of the mouse placenta. Nature, 204, 548–549.

Kirby, D.R.S., Billington, W.D. and James, D.A. (1966) Transplantation of eggs to the kidney and uterus of immunized mice. Transplantation, 4, 713–718.

Klein, J. (1965) The ontogenetic development of H-2 antigens in vivo and in vitro. In: Blood Groups of Animals (Matoušek, J. ed.) pp. 405–414, Junk, The Hague.

Klein, J. (1975) Biology of the Mouse Histocompatibility-2 Complex. Springer-Verlag, New York.

Klein, J., Bach, F.H., Festenstein, H., McDevitt, H.O., Shreffler, D.C., Snell, G.D., and Stimpfling, J.H. (1974) Genetic nomenclature for the H-2 complex of the mouse. Immunogenetics, 1, 184–188.

Koo, G.C., Stackpole, C.W., Boyse, E.A., Hämmerling, U. and Lardis, M.P. (1973) Topographical location of H-Y antigen on mouse spermatozoa by immunoelectronmicroscopy. Proc. Nat. Acad. Sci. U.S.A. 70, 1502–1505.

McCormick, J.W., Faulk, W.P., Fox, H., and Fudenberg, H.H. (1971) Immunohistological and elution studies of human placenta. J. Exp. Med. 133, 1–18.

McDevitt, H.O. and Chinitz, A. (1969) Genetic control of the antibody response: relationship between immune response and histocompatibility (H-2) type. Science, 163, 1207–1208.

McLaren, A. (1965) Genetic and environmental effects of fetal and placenta growth in mice. J. Reprod. Fertil. 9, 79–98.

Medawar, P.B. (1958) The immunology of transplantation. The Harvey Lectures, 52, 144–176.

Medawar, P.B. and Sparrow, E.M. (1956) The effects of adrenal cortical hormones, adrenocorticotrophic hormone and pregnancy on skin transplantation immunity in mice. J. Endocrinol. 14, 240–256.

Mitchison, N.A. (1954) Passive transfer of transplantation immunity. Proc. Roy. Soc. B 142, 72–87.

Möller, G. and Möller, E. (1962) Phenotypic expression of mouse isoantigens. J. Cell Comp. Physiol. 60 (suppl.) 107–128.

Mossman, H.W. (1937) Comparative morphogenesis of the fetal membranes and accessory uterine structures. Carnegie Contrib. Embryol. 24, 129–246.

Natheson, S.G. and Cullen, S.E. (1974) Biochemical properties and immunochemical-genetic relationships of mouse H-2 alloantigens. Biochim. Biophys. Acta 344, 1–25.

Nicolas, J.F., Dubois, P., Jakob, H., Gaillard, J. and Jacob, F. (1975) Tétratocarcinoma de la souris: Différenciation en culture d'une lignée de cellules primitini a potentialiter multiples. Annales de Microbiologie 126A, 1–22.

Padzerka, F., Longnecker, B.M., Law, G.R.J., and Ruth, R.F. (1975) The major histocompatibility complex of the chicken. Immunogenetics, 2, 101–130.

Palm, J., Heyner, S., and Brinster, R. L. (1971) Differential immunofluorescence of fertilized mouse eggs with H-2 and non-H-2 antibody. J. Exp. Med. 133, 1282–1293.

Patthey, H.L. and Edidin, M. (1973) Evidence for the time of appearance of H-2 antigens in mouse development. Transplantation, 15, 211–214.

Rees, R.C., Bray, J., Robins, R.A. and Baldwin, R.W. (1975) Subpopulations of multiparous rat lymph node cells cytotoxic for rat tumor cells and capable of suppressing cytotoxicity in vitro. Int. J. Cancer 15, 762–772.

Rugh, R. (1968) The mouse. Its Reproduction and Development. Burgess, Minneapolis.

Schlafke, S. and Enders, A.C. (1975) Cellular basis of interaction between trophoblast and uterus at implantation. Biol. Reprod. 12, 41–65.

Searle, R.F., Johnson, M.H., Billington, W.D., Elson, J. and Clutterbuck-Jackson, S. (1974) Investigation of *H-2* and non-*H-2* antigens on the mouse blastocyst. Transplantation, 18, 136–141.
Searle, R.F., Jenkinson, E.J. and Johnson, M.H. (1975) Immunogenicity of mouse trophoblast and embryonic sac. Nature, 255, 719–720.
Selier, M.J. (1970) Lack of porosity of the mouse placenta to maternal cells. Nature, 225, 1254–1255.
Shreffler, D.C. and David, C.S. (1975) The *H-2* major histocompatibility complex and the *I* immune response region: Genetic variation, function and organization. Adv. Immunol. 20, 125–195.
Simmons, R.L. and Russell, P. (1966) The histocompatibility antigens of fertilized mouse eggs and trophoblast. Ann. N.Y. Acad. Sci. 129, 35–45.
Simmons, R.L. and Russell, P.S. (1967) Xenogeneic antigens in mouse trophoblast. Transplantation, 5, 85–88.
Simmons, R.L., Lipschultz, M.L., Rios, A. and Ray, P.K. (1971) Failure of neuraminidase to unmask histocompatibility antigens on trophoblast. Nature New Biology, 231, 111–112.
Snell, G.D. (1968) The H-2 locus of the mouse: Observations and speculations concerning its comparative genetics and its polymorphism. Folia Biol. (Praha) 14, 335–358.
Snell, G.D. and Stevens, L.C. (1966) Early embryology. In: Biology of the Laboratory Mouse (Green, E. ed.) pp. 205–245, Blakiston, New York.
Snell, G.D. and Stimpfling, J. (1966) Genetics of tissue transplantation. In: Biology of the Laboratory Mouse (Green, E.L. ed.) pp. 457–472, Blakiston, New York.
Terasaki, P.I., Opelz, G. and Mickey, M.R. (1974) Histocompatibility and clinical kidney transplants. Transplantation Proceedings 6, 33–36.
Tuffrey, M., Bishun, N.P., and Barnes (1969) Porosity of the placenta to maternal cells in normally derived mice. Nature, 224, 701–704.
Voisin, G.A., Toullet, F., and D'Almeida, M. (1975) Characterization of spermatozoa auto-, iso- and alloantigens. Acta Endocrinol. 78, supplement 194, 173–222.
Vojtišková, M., Polačková, M. and Pokorná, Z. (1969) Histocompatibility antigens on mouse spermatozoa. Folia. Biol. (Praha) 15, 322–332.
Wachtel, S.S., Koo, G.C., Zuckerman, E., Hämmerling, U., Scheid, M.P. and Boyse, E.A. (1974) Serological cross reactivity between H-Y (male) antigens of mouse and man. Proc. Nat. Acad. Sci. U.S.A. 71, 1215–1218.
Wachtel, S.S., Koo, G.C. and Boyse, E.A. (1975a) Evolutionary conservation of H-Y (male) antigen. Nature, 254, 270–272.
Wachtel, S.S., Ohnu, S., Koo, G.C. and Boyse, E.A. (1975b) Possible role for H-Y antigen in the primary determination of sex. Nature, 257, 235–236.
Wild, A.E. (1974) Protein transport across the placenta. In: Transport at the Cellular Level, (Sleigh, M.A. and Jennings, D.H., eds.) pp. 521–546, Cambridge University Press, Cambridge.
Wimsatt, W.A. (1975) Some comparative aspects of implantation. Biol. Reprod. 12, 1–40.
Woodruff, M.F.A. (1958) Transplantation immunity and the immunological problem of pregnancy. Proc. Roy. Soc. B 148, 68–75.
Yanagisawa, K., Bennett, D., Boyse, E.A., Dunn, L.C., and Dimeo, A. (1974a) Serological identification of sperm antigens specified by lethal *t*-alleles in the mouse. Immunogenetics, 1, 57–67.
Yanagisawa, K., Pollard, D.R., Bennett, D., Dunn, L.C., and Boyse, E.A. (1974b) Transmission ratio distortion at the T-locus. Serological identification of two sperm populations in *t*-heterozygotes. Immunogenetics, 1, 91–96.

The transport of molecules across placental membranes 5

Richard K. MILLER, Thomas R. KOSZALKA and Robert L. BRENT

1. *Introduction*

In mammals the orderly development and growth of the various embryonic and fetal organ systems described elsewhere in this volume are ultimately dependent on the function of the extraembryonic membranes that form the placenta. In broad terms the placenta can be defined (Mossman, 1937) "as a union of fetal and maternal tissues for the purposes of physiological exchange". This definition encompasses the chorioallontoic placenta, the yolk sac placenta and the para-placental chorion. The functional importance of these various membranes in maternal-fetal exchange varies considerably between species and may further vary within the same species at different stages of gestation. Despite the remarkable degree of morphologic variation exhibited by placentae in different species, they share common functional properties in their adaptation to the transport of gases, nutrients and metabolites to and from the fetus, the elimination of fetal waste products and the elaboration of steroid and protein hormones.

Many questions have been raised concerning the regulatory role of the extraembryonic membranes in fetal development. For example, how do placentae regulate the movement of molecules, whether small or large, between the mother and the fetus? Which particular cells in placental membranes are involved in such transfer? Are there major differences in the transport capacities at different stages of gestation? Do all species transport similar materials by the same mechanism(s)? How sensitive are placental transport functions to alteration by xenobiotics, toxins or disease? If such alterations do occur what is their effect on fetal development? Is the placenta a site for primary teratogenic action? What is the contribution of altered placental function to retardation of fetal growth? This chapter cannot hope to provide an exhaustive discussion of each of these important questions. Rather we will focus our attention on the basic mechanisms by which materials are transported across the placenta since an understanding of transport processes is central to the study of each of the questions mentioned above. Most of the research reviewed in this chapter concerns the physiology of transport in the normal placenta. There is a growing awareness, however, that

G. Poste & G.L. Nicolson (eds.) The Cell Surface in Animal Embryogenesis and Development

alterations in placental transport function may be a significant factor in affecting the growth and development of the fetus. For this reason the final sections of this review will offer a brief outline of current knowledge of the role for placental pathology (including alterations in maternal blood supply) in fetal growth retardation and other disorders of the fetus.

Throughout pregnancy, the various placental membranes exhibit a fundamental feature of all biological membranes, namely, they show selective permeability to various materials. In the case of particulate materials, such as cells, the placenta severely restricts maternal-fetal transfer, a function aptly described as the "placental barrier". At the other end of the spectrum, however, the transfer of certain essential nutrients to the fetus is accelerated by specific transport enzymes and other carrier molecules present in the membranes of placental cells. It is also important to recognize that the observed transfer of material between the maternal and fetal circulations is a net expression of events involving a variable number of cell types interposed between the two circulations and cannot be simply assigned to a single cell type. Finally, it is difficult to generalize concerning transport processes in the placentae of different species. As will be discussed later there is a bewildering degree of interspecies variation in placental morphology and this is accompanied by equally marked variation in the transport of similar materials in different species. Yet further complications are imposed by the fact that even within a species the qualitative and quantitative transport capacities of the various extraembryonic membranes may change at different stages in gestation. Consequently, there is no single placenta which can be reliably used as a model to define structural or functional characteristics applicable to all species. It is presumed that at least some of the basic cellular activities involved in transplacental transport are similar in different species. However, similarity of function cannot be assumed without experimental study.

2. *Placental morphology*

2.1. General considerations

Details of the comparative anatomy of the placenta and the evolution of viviparity have been reviewed fully elsewhere (Bjorkman, 1970; Wynn, 1973; 1975; Mossman, 1974) and only a brief outline will be presented here.

As mammals have evolved the requirements for fetal nutrition have become more complicated. In the face of this ever increasing demand for a complex array of nutrients that rapidly exceeds the yolk storage capacity of the embryo, nature has equipped mammals with highly developed systems for maternal-fetal exchange in the form of the yolk sac placenta and the chorioallantoic placenta (Fig. 1).

The yolk sac (choriovitelline) placenta is the more primitive of the two types of placentation and is the principal means of fetal-maternal exchange in most marsupials, but it also coexists with the allantoic placenta in many Eutheria, such

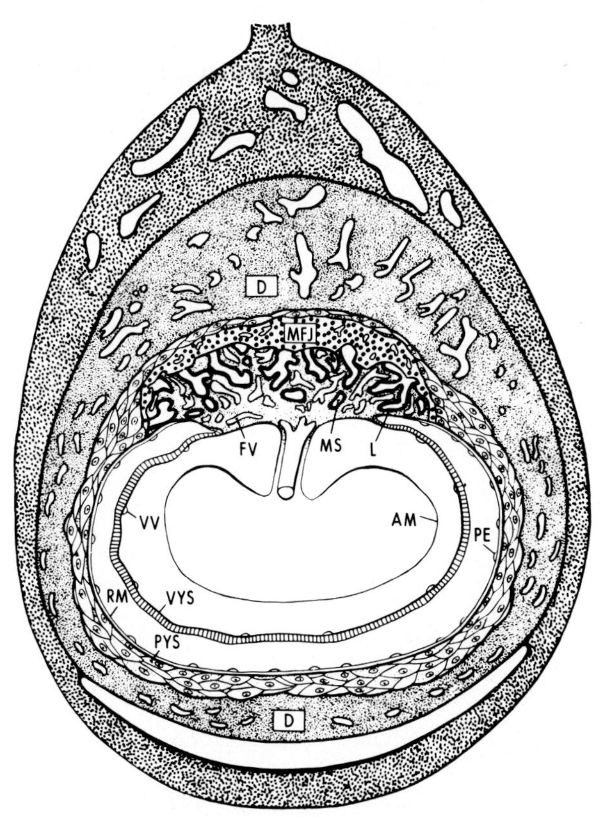

Fig. 1. Schematic representation of a cross-section of the feto-placental unit of the rat at day 11 of pregnancy (redrawn from Anderson, 1959).

The embryo is not depicted, but the umbilical vessels are shown and are connected to the chorioallantoic placenta which consists of the labryinth (L) and the maternal-fetal junction (MFJ). The labyrinth or region for maximal maternal-fetal exchange has both the fetal vessels (FV) and the maternal blood space (MS). The other extraembryonic membranes that surround the conceptus are the amnion (AM), the visceral yolk sac (VYS), Reichert's membrane (RM) and the parietal yolk sac (PYS). Surrounding these extraembryonic membranes are the maternal decidual tissues (D). The visceral yolk sac connects with the conceptus via the vitelline vessels (VV). This is the alternative circulation for maintaining the conceptus. On the embryonic side of the Reichert's membrane are the parietal endodermal cells (PE) which maintain Reichert's membrane until day 17–18 of gestation, when both the Reichert's membrane and the parietal yolk sac rupture and retract to the periphery of the chorioallantoic placenta.

as lagomorphs and rodents. In man true choriovitelline placentation is never well established because of the precocious development of the extraembryonic coelom which prevents contact of the yolk sac with the placenta. An entirely different form of yolk sac placentation occurs in certain mammals which develops by inversion of the germ layers. Such inverted yolk sac placentation occurs in rodents, lagomorphs, many bats and insectivores and also armadillos. In the formation of this type of placenta, mesoderm does not develop in the abembryonic hemisphere of the wall of the blastocyst so that the bilaminar omphalopleure remains as a very thin membrane in contact with the uterus. The highly vascular embryonic hemisphere of the yolk sac is then inserted into the abembryonic area with resulting apposition between its lining endoderm and the uterine mucosa over a wide area. The inverted yolk sac placenta is undoubtedly of considerable significance throughout most of pregnancy in many animals, notably in rodents. This consideration should be borne in mind in any extrapolation of data concerning placental transport processes derived from studies using these species.

The chorioallantoic placenta is the principal organ for maternal-fetal exchange in most higher mammals, including man. It is in the morphology of this type of placentation that interspecies variation is most obvious. The gross shape of this structure and the extent of its interaction with the uterine lining are determined by the initial distribution of villi which develop on the chorionic surface. In the horse and the pig, the distribution of villi over almost the entire chorionic surface produces a diffuse placenta. In the cow and the sheep, villi are localized in separate tufts which are distributed over the surface of the chorion to form a cotyledonary placenta. In carnivores the arrangement of chorionic villi into bands that run around the equator of the chorioallantois creates a zonary placenta. In man, apes, rodents and bats the chorioallantoic placenta is a single cake-shaped disc which results from the disappearance of villi from all but a circumscribed area on the chorion.

A variety of anatomical and histological criteria have been used in an attempt to develop a satisfactory classification for chorioallantoic placentae. Perhaps the most influential of the many proposed schemes has been the histological classification devised by Grosser (1909) and later modified by Mossman (1937) which is based on the intimacy and complexity of the zone of contact between the fetal chorion and the maternal tissues.

Grosser classified the chorioallantoic placenta of mammals into epitheliochorial, syndesmochorial, endothelialchorial and hemochorial depending upon the degree to which the maternal uterine tissues were eroded by foetal trophoblast. Grosser's classification placed considerable emphasis upon diffusion as a means of transporting material(s) across the placenta, with the obvious corollary that the number of cell layers separating maternal and fetal circulations would affect the ease with which material(s) could pass across the placenta. These assumptions formed the essential basis for the view that the placenta functioned as a protective "barrier" to restrict access of materials to the developing conceptus. The considerable influence of the "barrier" concept on opinions concerning transport mechanisms in the placenta is particularly obvious in the literature on the transmission of maternal antibodies to the fetus in utero. Until comparatively recently, many workers held to the view that species variation in the intrauterine transmission of antibodies to the fetus could be explained solely by the thickness of the placental barrier. The fact that antibody transmission occurred before birth in mammals with hemochorial placentae (rabbit, rat, guinea pig, man) was attributed to the thin placental barrier in these species while failure of antibodies to be transferred in species with epithelio- and syndesmochorial placentae (pig, horse, ungulates) was assumed to result from the presence of several layers of cells that provided an effective barrier to transport.

While such simplistic interpretations were perhaps not unreasonable at the time they were first proposed, it is now clear that the number of intervening cell layers between the maternal and fetal circulations has relatively little bearing on the transport of antibodies, proteins and many other classes of biologically active molecules. Rigid histological classifications neglect the dynamic transitions that occur within tissues. For example, capillaries may indent both the trophoblast and endometrium in an almost intra-epithelial location (Wynn, 1975). Thus, without changing the number of cellular layers the thickness of the "barrier" for diffusion is immediately reduced.

As we will discuss later the properties of the material to be transported and the surface properties of the various cells in the placental membranes appear to be of more importance in influencing placental transport. It seems likely that in the final analysis the process of transplacental transport will be defined in terms of the presence or absence of specific cell surface receptors, transport enzymes and other carrier molecules located in the plasma membranes of cells in the various extraembryonic membranes. The use of electronhistochemical and immunoelectronhistochemical techniques to study the plasma membrane properties of cells in placental membranes, including topographic mapping of the distribution of receptors and enzymes on the plasma membrane of individual cell types, will no doubt be undertaken within the next few years. It is anticipated that such studies will establish more meaningful correlations between the structural organization of placental membranes and their specific transport capacities.

In the remainder of this section we will present a brief outline of the histological and ultrastructural properties of rodent and human placentae since most of the data on transport mechanisms discussed in this article have been obtained from work on these species. Many of the structural features of these particular placentae are also represented in other species and descriptions of how particular structural features of placentation in man and rodents might be related to transport function may apply in general terms to other species. However, as mentioned earlier, detailed extrapolation is not without its dangers.

2.2. The rodent choriovitelline (yolk sac) placenta

The essential details of the morphogenesis of the yolk sac placenta in the rat have been outlined above and the stage of development of this structure by the 13–15th day of gestation is shown in Fig. 1, where the conceptus is shown surrounded by extraembryonic membranes including an outer wall, the parietal yolk sac (PYS in Fig. 1). The parietal yolk sac is embedded in decidual uterine tissue and is contiguous with the developing chorioallantoic placenta (Fig. 1). The parietal yolk sac is composed of two cell layers which are separated by a thick avascular basement membrane, or Reichert's membrane (Figs. 1 and 2). The outer layer of cells of the parietal yolk sac project into the uterine decidua and are made up of trophoblastic giant cells. The inner cell layer (parietal cells) of the

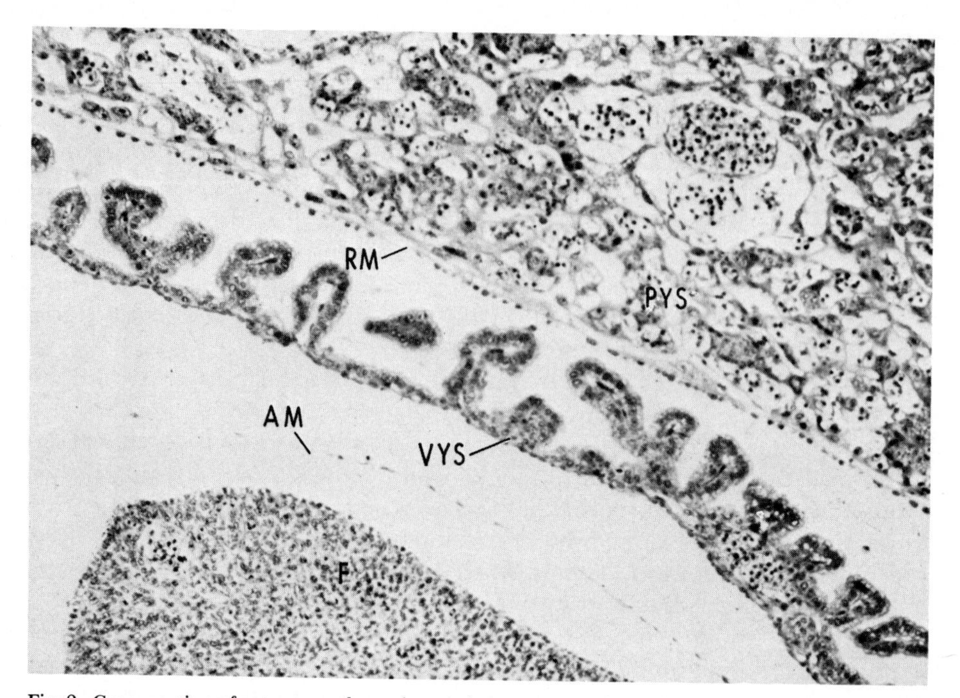

Fig. 2. Cross-section of rat extraembryonic membranes at a 14-day-old implantation site. F is a section of the fetal tissue. Surrounding the fetus is the amnion (AM), visceral yolk sac (VYS), parietal yolk sac (PYS), and Reichert's membrane (RM). Note the columnar epithelium of the visceral yolk sac. This membrane is very active in concentrating substances via active transport and pinocytosis. The parietal yolk sac consists of Reichert's membrane, endodermal parietal cells and giant trophoblastic cells as well as the maternal blood space. Along the fetal aspect of Reichert's membrane are the endodermal parietal cells. These cells synthesize the collagen necessary to produce Reichert's membrane. Magnification ×125.

PYS is contiguous with Reichert's membrane and adjacent to the yolk sac cavity. This cell layer is of endodermal origin. One of the known functions of the parietal cells of the PYS is to synthesize Reichert's membrane (Clark et al., 1975). The transport capabilities of Reichert's membrane and the parietal yolk sac have not been established. Reichert's membrane might, however, offer some diffusional limitation (Jollie, 1968).

At the junction of the chorioallantoic placenta and parietal yolk sac, Reichert's membrane is reflected onto the surface of the placenta; the parietal cells disappear and are replaced by a tall columnar epithelium. These cells are called visceral endodermal cells and constitute the outer layer of the visceral yolk sac (VYS in Fig. 1) which is continuous with the parietal yolk sac.

The choriovitelline membrane or visceral yolk sac consists of: (1) the visceral endodermal cells and their associated basement membrane; (2) mesenchyme and capillaries (including endothelial cells and their basement membrane); and (3) serosal cells and their associated basement membrane. Thus, the visceral yolk sac is composed of at least four different cell types and three basement membranes. The yolk sac completely envelopes the embryo which in turn is completely surrounded by yet a third, very thin (two cell thick), avascular membrane, the amnion.

The yolk sac serves an important nutrient and transport function for the rat embryo, especially during the early stages of gestation. On the 18th day of gestation the parietal yolk sac ruptures and retracts to the periphery of the chorioallantoic placenta. Thus, at term the extraembryonic membranes are the yolk-sac, amnion, and the chorioallantoic placenta in the rodent.

2.3. The human chorioallantoic placenta

The integral structure of the human chorioallantoic placenta is the villus (Fig. 3). The finger-like villi arise initially as avascular solid outgrowths from the chorion and are subsequently vascularized by the ingrowth of allantoic mesenchyme and vessels. With these villi, the surface area for maternal-fetal exchange is enormous. It has been estimated that at term the human chorioallantoic placenta has a surface area of 11 to 13 square meters. Thus, it is clear that the chorioallantoic placenta offers a major site of physiological exchange between mother and fetus. Even if the fetus is removed, the placenta will remain functional and continue to grow (Davies and Glasser, 1968; Panigel and Myers, 1972). Placental persistence under these conditions is related in part to the excellent maternal blood flow to this organ (approximately 20% of the cardiac output; Assali et al., 1968).

An extensive literature has developed on the histology and, more recently, the ultrastructure of the human chorioallantoic placenta (reviews, Panigel, 1975; Wynn, 1975).

The normal chorionic villus is completely covered by the syncytiotrophoblast which rests in part on an incomplete layer of cytotrophoblast cells and in part on a basement membrane (Fig. 4). The latter overlies the core of the villus which is composed of mesenchyme and the fetal capillaries (Fig. 4). We shall confine our remarks largely to the trophoblastic components.

The syncytiotrophoblast and the cytotrophoblast are cytologically very different (Fig. 4) yet both are of blastocystic ectodermal origin. The human syncytiotrophoblast is a true syncytium and is the single most active and variable component in the placenta. The syncytiotrophoblast displays a number of features common to epithelial cells involved in transport. The outer or maternal surface is covered with numerous microvilli (Figs. 3 and 4). These substantially increase the surface area available for exchange. Significant pinocytotic and/or micropinocytotic activity is also associated with this region and as we shall see later this may represent a major pathway for transplacental transport of certain proteins. The syncytiotrophoblast also has a well developed endoplasmic reticulum and Golgi apparatus, and these reflect the active synthetic and secretory activity of these cells. Secretory granules are commonly found in the syncytiotrophoblast but are rare in the cytotrophoblast. Numerous large mitochondria are also present which presumably supply the energy needed for transport and secretory functions (Hertig, 1968). Wislocki and Dempsey (1948) have demonstrated histochemically that estrogens and progesterone are synthesized in the

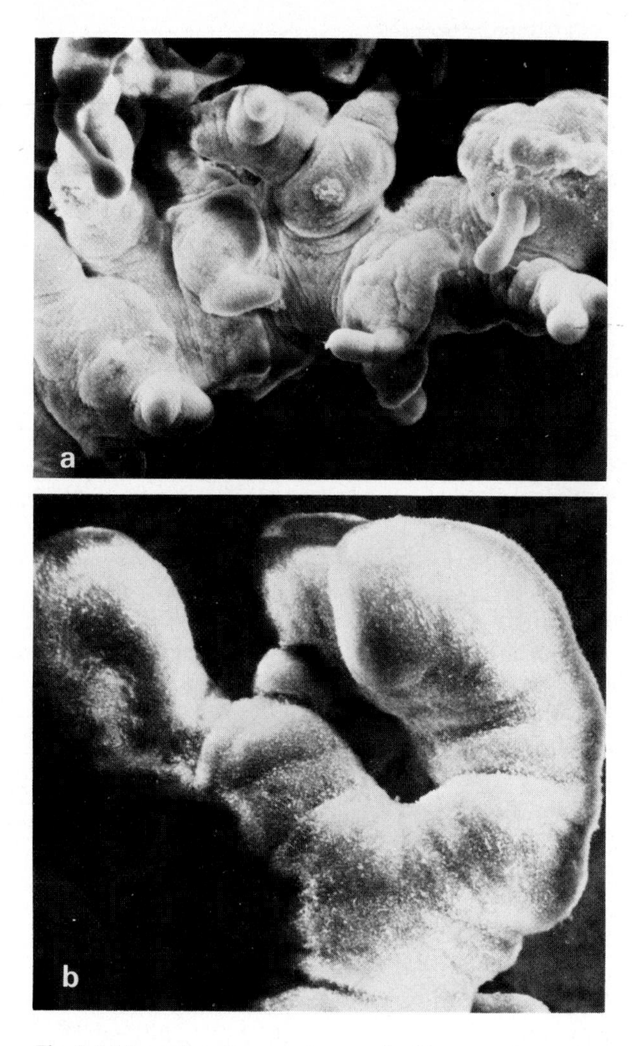

Fig. 3. (a) Scanning electronmicrograph of human placental villi at 10 to 14 weeks gestation. Note the syncytial sprouts at various stages of development projecting from the larger stem villi. Magnification ×340; (b) Scanning electronmicrograph of villi from 3(a), above, at higher magnification showing numerous microvilli which give the surface of the villus a soft velvety appearance. Magnification ×990. Both figures reproduced with permission from King and Menton (1975).

syncytium as are the chorionic gonadotropins (Wynn, 1968). Chorionic somatomammotropin (placental lactogen) has been similarly localized to syncytiotrophoblast (Sciarra et al., 1963). It is perhaps not unreasonable to suggest that the third postulated protein hormone of the placenta, chorionic thyrotropin, will also be eventually localized to the syncytiotrophoblast. The syncytial trophoblast may thus be regarded as the endocrinologically and morphologically mature form of trophoblast while the cytotrophoblast cells constitute the reserve, or stem, elements, whose main function is to form new syncytium.

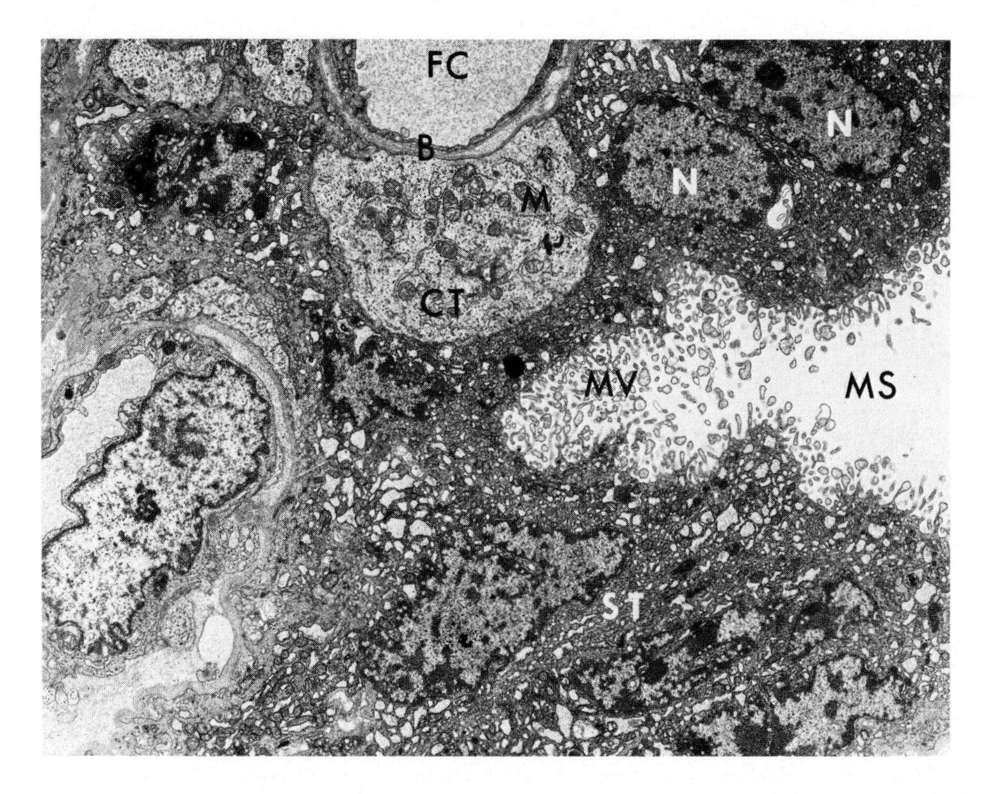

Fig. 4. Electronmicrograph of a cross-section of the human term placenta. MS is the maternal blood space. The placenta consists of two types of trophoblastic cells: the syncytiotrophoblast (ST) and the cytotrophoblast (CT). Note that in the syncytiotrophoblast there are numerous nuclei (N) as well as a prominent border of microvilli (MV) facing the maternal blood space. M are mitochondria; B is the basement membrane of the fetal capillary; FC is the fetal blood space or fetal capillary. Magnification ×7400.

Histochemical studies have further demonstrated the presence of enzymes such as ATPase (Okudaira et al., 1968) and alkaline phosphatase (Borst et al., 1973; Hulstaert et al., 1973) in association with the plasma membrane covering the microvilli on the maternal surface of the syncytiotrophoblast, which again suggests that it is this particular trophoblastic component which is primarily involved in transport of material across the placenta. ATPase activity in placental tissue has also been demonstrated biochemically. Shami and coworkers (Shami and Raddi, 1971, 1972; Shami et al., 1975) have demonstrated the presence of a Ca^{2+} or Mg^{2+} stimulated ATPase in guinea pig placentae and also in membrane vesicles prepared from this tissue. Miller and Berndt (1972) identified a Ca^{2+}-dependent ATPase in human placenta which is inhibited by ethacrynic acid in similar fashion to the enzyme preparation from guinea pig placenta. Miller and Berndt (1972) also demonstrated the presence of a Mg^{2+}-dependent Na^+-K^+ ATPase in human term placentae. As shown in Table 1 there is a marked difference between the Ca^{2+}-ATPase and the Na^+-K^+ ATPase from human

Table 1
ATPase activity in the human term placenta.

	Mg^{2+} ATPase	Ca^{2+} ATPase
	Hydrolysis	
Control (5 mM $MgCl_2$)	0.34 ± 0.14	
Control (5 mM $CaCl_2$)		0.27 ± 0.075
	% Δ from Controls	
No Na^+, K^+	-----	+11.8*
No Na^+	−14.7*	0.0
No K^+	−32.4*	0.0
Ouabain (0.1 mM)	−41.2*	0.0
Ethacrynic Acid (5 mM)	−76.5*	−74.1*
No Mg^{2+}	−94.1*	-----
No Ca^{2+}	-----	−92.6*
Blank	−95.6*	−95.6*

Reproduced with permission from Miller, R.K., and Berndt, W.O. (1973).

placentae. Only the Na^+-K^+ ATPase is inhibited by oubain or by removal of sodium or potassium ions, though both enzymes are inhibited by high concentrations of ethacrynic acid.

Many of the drug metabolizing enzymes such as aryl hydrocarbon hydroxylase are also known to be associated with the syncytiotrophoblast rather than with any of the other cell layers of the placenta (Juchau, 1973).

The other major trophoblastic cell type found in the human placenta is the cytotrophoblast or Langhans cell. By comparison to the syncytiotrophoblast these cells are structurally simple, with large nuclei and prominent nucleoli and contain only a very limited Golgi apparatus and endoplasmic reticulum. As mentioned, the importance of the cytotrophoblast layer is as a germinal zone for the syncytiotrophoblast. Although the cytotrophoblast layer is reduced and becomes discontinuous in the late stages of pregnancy, residual cytotrophoblast cells do persist until term.

The two trophoblastic layers described above are apposed to the trophoblast basement membrane which, in turn, overlies the mesenchymal core of the villus. The basement membrane appears as an electron dense substance in which fine filaments and occasional osmiophilic bodies can be seen. The thickness of the basement membrane increases as pregnancy progresses toward term, prompting the suggestion that it may act as a limiting factor in placental transport (Ashley, 1965).

Apart from the obvious ultrastructural specialization found in the placenta there are a number of other gross histological changes that occur during placental maturation that may be pertinent to transport processes.

The syncytial trophoblast, which is in direct contact with the maternal blood, is relatively thick in early placental development but becomes progressively thinner throughout gestation. During the same period the total thickness of the placental membrane may decrease from 0.025 mm to 0.002 mm. Apart from the thinning of the syncytiotrophoblast a number of other histological changes occur, all of which are probably consistent with an increased efficiency of transfer. These include: increasing subdivision of terminal villi thus increasing the ratio of villous surface to volume; reduction in the cytotrophoblast layer; and a reduction in the proportion of connective tissue elements in the villous core which allows closer contact between the fetal capillaries and the syncytiotrophoblast. At the same time the number of capillaries increases, the capillaries also widen and their endothelial lining becomes thinner. All of these changes would be expected to enhance the efficiency of maternal-fetal exchange. On the other hand, the thickening of the basement membranes of capillary endothelium and the trophoblast, obliteration of maternal and fetal vessels and the deposition of fibrinoid (see chapter 4 of this volume) as gestation progresses might well be expected to reduce the efficiency of transfer.

2.4. *The rodent chorioallantoic placenta*

The rodent chorioallantoic placenta is quite different from that of the human (Figs. 5 and 6). The chorioallantoic placenta in rodents has two well-defined regions: (1) the maternal-fetal function (MFJ in Fig. 1) and (2) the labyrinth (Figs.

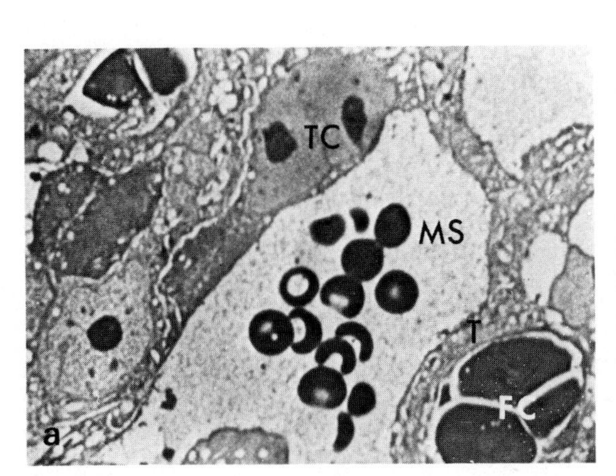

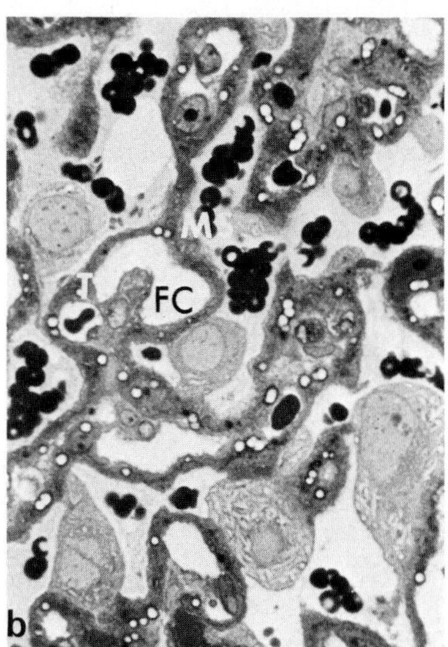

Fig. 5. (a) Cross-section of a 14 day chorioallantoic placenta of the rat showing the region for exchange of nutrients between mother and the conceptus. MS is the maternal blood space and FC is the fetal capillary or fetal blood space. Note the presence of large fetal red blood cells within the fetal capillary. T is the trophoblastic tissue separating the two blood compartments, and TC are the large, single trophoblastic cells. Magnification ×1200; (b) Cross-section of a near term chorioallantoic placenta of the rat at the labyrinth region of the placenta. Abbreviations are as in 5(a). Magnification ×600.

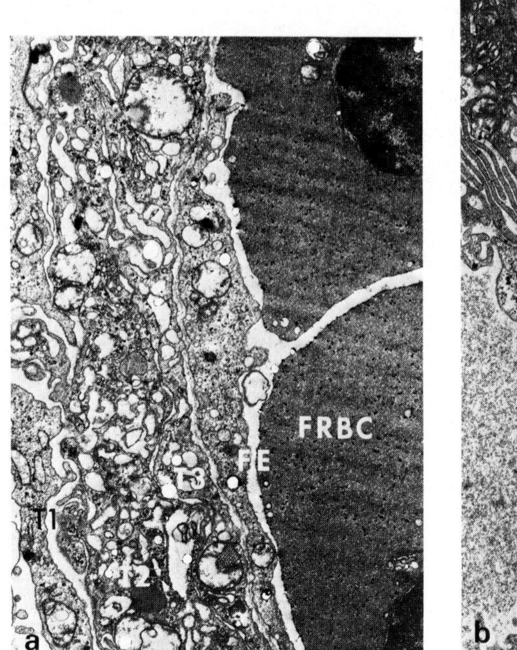

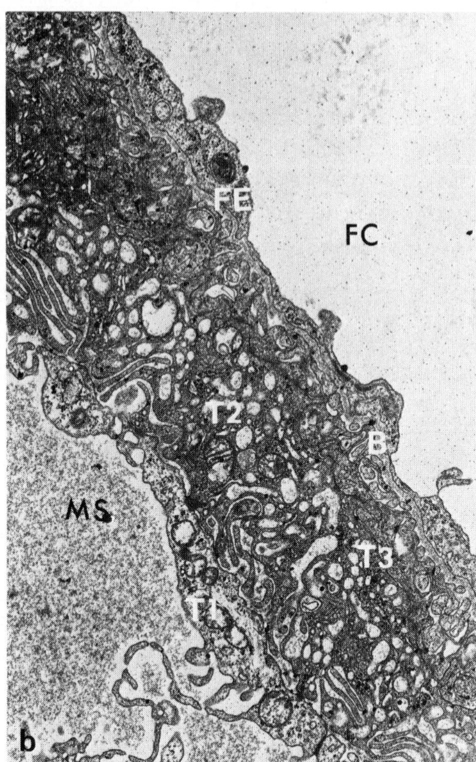

Fig. 6. (a) An electronmicroscopic section of the 14-day chorioallantoic placenta of the rat. Note the three trophoblastic layers separating the maternal blood space from the fetal blood space. These layers are T1, T2, and T3. Note the large fetal red blood cells (FRBC). A basement membrane and fetal endothelium (FE) are also present. Magnification ×9000. (b) An electronmicroscopic section of the term chorioallantoic placenta of the rat. The three trophoblastic layers (T1, T2, and T3) are present at term but there is an attenuation of the T1 layer when compared with the 14-day placenta and only T2 and T3 are prominent. B is the basement membrane. FE is the fetal endothelium of the blood vessel. FC is the fetal capillary space. MS is the maternal blood space. Magnification ×12 500.

1, 5, and 6). The maternal-fetal junction consists of small basophilic cells (spongioblasts), giant trophoblast cells and cells which contain large quantities of glycogen (all of fetal origin). In addition there are maternal blood channels. However, no exchange between mother and fetus occurs in this region and exchange occurs exclusively at the labyrinth. Also, rodents have three trophoblast layers in the fetal placenta separating the maternal and fetal circulation (Fig. 6b), rather than the two trophoblastic cell layers found in the human placenta. As gestation progresses the first trophoblastic layer in the rodent attenuates (Fig. 6b). Such differences are typical among species as well as within different gestational periods in the same species.

3. *Techniques for studying placental transport function*

3.1. In vivo human studies

The study of placental transport of molecules in man presents serious technical problems. The mother can be examined with relative ease, but the conceptus presents more difficulties due to its inaccessibility. In the past, most studies on placental transfer in humans have been restricted to the injection of a substance into the mother usually 30–60 minutes before delivery, after which both maternal and cord blood samples are collected at delivery, and qualitative and quantitative observations made concerning the ability of the drug to cross the placenta. One of the difficulties that arises using this method is that it is only possible to obtain one sample from the neonate, and it is a mixed arterio-venous blood sample. Thus, kinetic studies are impossible to carry out, and it is particularly difficult to establish a steady state.

Chinard et al. (1956) attempted to obtain more than one sample from the human umbilical cord after maternal glucose or fructose administration. Although an increase in fetal blood hexose concentrations was detected, still only two samples were obtained eleven minutes apart. Another technique was devised by Saling (1964) for measuring fetal acid-base balance. Instead of requiring a cesarean section for obtaining the fetal samples, he exposed the fetal head transvaginally after the fetal membranes had ruptured, and while the conceptus remained in utero, he obtained fetal blood by nicking fetal scalp vessels with a guarded knife tip. Asling et al. (1970) extended this technique to the study of the placental transfer of mepivacaine. Besides measuring drug concentration, these investigators were able to measure pCO_2, pH, pO_2 and heart rate. Again only two fetal blood samples were taken: one from the fetal scalp five minutes after the drug injection, and one from the umbilical cord at delivery. However, more samples could probably have been obtained during the interval between injection and delivery. Thus, with this technique, the placental transfer of many compounds can be examined in some detail at term. This promising approach should be exploited further by those interested in studying human placental transport in vivo.

Other studies have used the physiologic status of the fetus as an index of placental transport. For example, prolonged prothrombin times in babies born to mothers who have received oral anticoagulants, e.g., dicoumarol, have been interpreted as evidence that the drug had entered the conceptus (Kraus et al., 1949; Bloomfield, 1970). Recently it has been shown that electroencephalography can be used to detect alterations in fetal brain function after maternal injections of Meperidine or Carbocaine (Rosen et al., 1970). The symptoms of opium withdrawal in neonates are also becoming a more common clinical observation (Goodfriend et al., 1956; Stone et al., 1971). Even temperature measurements of neonates can perhaps be useful as an indicator of placental passage of compounds (Owen et al., 1972). Measurements of fetal heart rate have also been a popular parameter to assess movement of a compound across the placenta. For example, even though atropine was undetectable in human fetal blood, Johns (1965) was able to demonstrate drug-induced tachycardia in the conceptus following administration of atropine to the mother. However, the use of fetal responses to measure transfer must be qualified in that parameters other than the penetration of the compound into the placenta are responsible for the observed effect on the fetus. Such factors include the production of uterine contractions, alterations in uterine blood flow, metabolism of the compound to a more active metabolite or even the appearance of fetal distress.

More recently, transfer studies have been performed during early gestation to examine the distribution of substances such as antibiotics (Kauffman et al., 1975). However, the measurement of normal body constituents has not been extensively evaluated. Administration of materials to pregnant women immediately prior to abortion to determine whether the materials could be subsequently recovered in the aborted fetus have also been performed on occasion, but this approach is now limited by current guidelines concerning fetal research.

Most often a single injection of a compound is used to determine its placental permeability, but no single compartment is established as a constant. Therefore, rather sophisticated mathematical corrections must be made to correct for the different rates of disappearance of the substance from maternal blood and tissue, fetal blood and tissue, and placenta. If a substance in one of these compartments can be maintained at a constant concentration, then the placental transport characteristics can be established. Since constant infusion experiments have been performed in man (Glendening, 1965; Daubenfield et al., 1974; Kauffman et al., 1975) some data are available.

Thus, in human experimentation, the direct measurement of fetal blood is most feasible at term, but the assessment of the function of the fetal placental unit, whether at term or during early gestation is still technically difficult and is also accompanied by additional ethical problems when compared with the animal studies.

3.2. In vivo animal studies

Animal experiments offer greater technical freedom and can be subjected to better control than studies in man. The kinetics and characteristics of placental transport in laboratory animals are therefore better understood. It must be remembered, however, that there may be at least two functional placental transport systems in operation in lower animals, especially in the rodent. Noer and Mossman (1947) have described a technique to surgically isolate the yolk sac placenta of the rat which allows study of chorioallantoic placental transfer of substances for up to three days. By the fourth day the inextensible yolk sac membrane is too restrictive for the growing fetus and fetal death ensues. This technique has been further refined by Holsen and Wilson (1974) to study the in situ transport capabilities of the early yolk sac and placenta (day 12–14) for small molecules. Other techniques for the isolation of yolk sac near term have been useful to examine the transport of small molecules (Miller et al., 1974c). The same type of single sample experiments used in human studies can be used in animals where there are several conceptuses present in utero since the conceptuses may be removed individually (cf. Rosso, 1974). However, variations in blood flow and mechanical manipulation present experimental complications which may influence the outcome.

Recent procedures have been developed to obtain samples directly from the umbilical vessels of the rat fetus in situ as early as day 17 in pregnancy (Fantel, 1975). Such microtechniques allow direct measurement of venous and arterial blood levels and should be very useful in studying the transport of many different substances.

Perhaps the most complex and useful technique for the study of placental transport in vivo while maintaining all parameters as close to physiological as possible, including the concentration of any material that passes to the fetus, involves implantation of an indwelling catheter to sample both the amniotic fluid space and the fetal heart as performed by Almond et al. (1970). Large animals such as sheep and goats are obviously better for such experiments. Using this method, blood samples can be taken from the fetus and the mother, sometimes as often as ten times in six hours (Shoeman et al., 1972).

Other approaches to the study of placental transfer have involved externalizing the fetus, perfusing the umbilical vessels and placing chronic indwelling catheters directly into the fetal circulation. This approach enables consecutive blood samples to be obtained from the umbilical vessels as well as the uterine vessels after injections into either mother or fetus. These studies have been most useful in determining the kinetic parameters of various transport systems (Rudolph and Heymann, 1967a,b; Dawes, 1968; Power, 1972). These isolated in situ placental preparations eliminate the influence of the fetus on the placental transport capabilities, and by using the sheep or goat, a single cotyledon can be isolated without drastically compromising the fetus. One of the major limitations of this technique has been the need for large animals. Alexander et al. (1955) have taken these types of experiments one step further by removing the fetus and infusing solutions through the umbilical vessels. This technique allows for better control of the transfer process in that the effect of fetus on the placenta is completely eliminated. Also, serial blood samples can be obtained more easily and drug passage studied in either direction. More importantly, there is no involvement of other placental membranes such as the yolk sac. Money and Dancis (1960) have adapted these umbilical infusion techniques to small animals such as guinea pigs, thus allowing for easier laboratory investigation of the hemochorioplacenta. This technique has been further refined by Schröder et al. (1972a,b, 1975).

3.3. In vitro studies

Placentae (chorioallantoic or visceral yolk sac) can also be isolated and their functional properties examined under defined and controlled conditions in vitro. Importantly, there appears to be excellent agreement between results obtained in vitro and in vivo in studies of placental transfer mechanisms (review,

Miller and Berndt, 1975). Both in vitro and in vivo protocols are necessary to establish mechanisms of placental transport.

The in vitro techniques used for studying transport phenomena across placental membranes fall into four broad categories: (1) perfusion systems; (2) the Ussing cell; (3) tissue section or slices; and (4) membrane vesicle preparations. With these systems the movement of materials across the placenta, the concentration of materials within the placenta, and their metabolism by the placenta can each be examined under rigidly defined and controlled conditions.

3.3.1. Perfusion systems

This approach involves perfusion of the entire placenta or isolated sections of placenta. Near-term or term human (Dancis, 1975; Cedard, 1972, 1975) and guinea pig (Schröder et al., 1972) placentae have been used successfully in perfusion of the entire placental unit. Refinements in these techniques now permit perfusion of single human cotelydons (Schneider et al., 1972; Dancis, 1975). A solution is perfused through the umbilical vessels or other fetal vessels while a second solution bathes the entire maternal surface in a circulating pool. Alternatively, the second solution can be perfused through uterine vessels. Thus, flux measurements for a single molecule can be made across these membranes in relation to other molecules with known transport functions.

A number of substances (serotonin, bradykinin, and catecholamines) which alter blood flow have been studied in these systems and found to affect the passage of molecules (Ward and Auttieri, 1966, 1968; Morgan et al., 1972; Sherman and Gautieri, 1972). It should be emphasized that one of the major difficulties with the isolated perfusion technique is the requirement for a nearly perfectly delivered placenta with large fetal vessels and intact chorionic membranes, otherwise poor perfusion and leakage may occur. Further development of this system, by using the isolated cotyledon instead of the whole placenta (see above) has eliminated many of the technical problems associated with perfusion of the entire placenta.

Similar perfusion techniques can be utilized with the rodent yolk sac placenta in vitro. The visceral yolk sac can be isolated in the rat by placing a ligature between the chorioallantoic placenta and the visceral yolk sac. The visceral yolk sac-embryo unit is then incubated in a buffer solution containing the substance whose transport is to be measured (Netzloff et al., 1968; Kernis, 1969, 1971). The movement of the substance into both the visceral yolk sac and into the embryo is then monitored. Although the normal movement of materials into the embryo may be altered, perhaps due to reduction in oxygen tension, uptake of substances by the yolk sac appears to be unaffected and therefore provides a good indicator of transport function. This technique may prove to be particularly suitable for evaluating early embryonic function. Nonetheless, the system still requires further development and critical analysis.

More recently, tissue culture methods have been used to investigate visceral yolk sac function early in gestation (Miller and Runner, 1975). This system allows the investigator to explore and evaluate yolk sac function early in embryogenesis and to establish the effects of environmental agents on these transport parameters during the critical period of organogenesis.

3.3.2. Ussing cell measurements

The second category of in vitro experimentation involves measurement of fluxes in the Ussing chamber (Ussing, 1949). This chamber basically consists of two pools of bathing medium separated by a flat sheet of tissue. Although this technique was developed originally for transport studies in frog skin (Ussing, 1949), any suitable sheet of tissue can be used, e.g., visceral yolk sac (Deren et al., 1966), chorion (Battaglia et al., 1962; Moore et al., 1974; Seeds, 1967, 1970), amnion (Page et al., 1974) or chorioallantoic membrane (Moore et al., 1974). An excellent description of the technique and its potential applications has been provided by Clarkson and Lindemann (1969). The Ussing chamber is particularly useful for measuring the direction and rate of movement of materials across planar membranes (Stuart and Terepka, 1969; Knapowski et al., 1972; Seeds et al., 1973).

3.3.3. Section studies

A third in vitro approach to transport processes in placental membranes involves the use of slices or sections of these structures (Sybulski and Tremblay, 1967; Dancis et al., 1968; Longo et al., 1973; Smith et al., 1973; Miller and Berndt, 1974). This method consists of incubating thin tissue sections in a balanced salt solution under carefully controlled conditions of temperature, oxygen pressure, and pH. Uptake of specific substances can be monitored chemically or radiochemically in the sections after appropriate time intervals. In the same or parallel experiments, tissue oxygen consumption, inulin space and tissue electrolytes can also be measured. With these methods, a large number of variables can be studied using a single human placenta. Technical details of section preparation are presented in the various papers cited above.

The section technique, as well as the Ussing chamber, offer the investigator: good control over pH, pCO_2, pO_2, ion concentration, and hydrostatic pressure (Ussing Cell), and the use of chemical modifiers that may interact with tissue systems in vivo. The section technique can be a good index of the transport characteristics, but it cannot define the direction of transfer. Only cellular accumulation and efflux of a substance can be determined. Section studies must therefore be correlated with in vivo studies to determine the direction of transfer. Nevertheless, section studies offer opportunities for more exact definition of metabolic requirements, kinetic assessment of competitive and non-competitive substance interaction and ion requirements. Most importantly, the section technique enables the primary placental transport site, the chorioallantoic placenta, to be isolated and studied at different stages in gestation and in different species under identical conditions.

3.3.4. Membrane vesicle preparations

Recent refinements in technique have resulted in the development of methods for studying the movement of molecules across vesicles prepared from isolated cellular membranes derived from placental tissues. Although similar vesicle preparations have long been used for this purpose to study the properties of membranes from nerve and muscle, it is only within the last year that similar methods have been developed for placentae (Shami et al., 1974, 1975). A major difficulty associated with the technique concerns the problem of defining the cell type from which the membranes have been isolated. Despite this limitation it is anticipated that the future application of this method will be of considerable value in enlarging our understanding of placental transport processes.

4. *Mechanisms of placental transport*

4.1. *General aspects*

This section will examine the transfer characteristics of the various extraembryonic membranes in relationship to the penetration, binding, and metabolism of selected molecules.

The chorioallantoic placenta is perfused by two different blood supplies (maternal and fetal). Such a relationship creates a unique situation for the movement of compounds across the extraembryonic membranes. The movement of any molecule across the placental cell membrane is thus influenced by both maternal and fetal blood flow. Four factors are associated with the regulation of blood flow: (1) the basal motor activity of individual arteries; (2) the compression of arteries by myometrial contractions; (3) the back pressure from the intervillus space due to effects on venous drainage; and (4) the redistribution and shunting of blood flow from one region of the placenta to another. Even though the cell surface membranes are available for transfer, these extraplacental factors may alter inherent placental transfer mechanisms (see section 6).

Besides blood flow, other factors can affect the interpretation of data on placental transfer functions. The placenta is bounded by two protein-containing fluids which may exhibit quite different pH's. Such factors may dramatically modify the transfer of molecules by altering their binding affinity for various tissue or plasma components or by altering the degree of ionization of the molecules (see below).

The physico-chemical properties of a molecule exert considerable influence on its transport potential. Six basic properties are recognized as being important in this regard: (1) molecular weight; (2) lipid solubility; (3) degree of ionization; (4) ability to bind to tissue or plasma components; (5) susceptibility to chemical modification by the placenta (biotransformation); and (6) the rate of clearance of the molecule from the placenta and maternal-fetal tissues. Each of these will now be discussed in detail.

It has long been known that the *molecular weight* of a molecule correlates with its ability to penetrate biological membranes. As the molecular weight increases, especially above 1,000, the ability of a molecule to penetrate decreases dramatically. However, with every generalization, there are exceptions. The chorioallantoic and vitelline surface membranes do have the ability to selectively transfer large molecular weight proteins, notably gamma-globulins, from the mother to the fetus. The transfer characteristics of proteins will be examined later.

The *lipid solubility* of a molecule is also important in determining its ability to pass through biological membranes. A large partition coefficient for a molecule usually indicates that the molecule will easily penetrate through biological membranes. Placentae, in general, appear to conform to this concept. However, certain lipid soluble substances do penetrate through placental membranes at different rates than through membranes of other tissues, especially the central nervous system (Oh and Mirkin, 1971; Maickel and Snodgrass, 1973).

The degree of *ionization* of a molecule has been found to be a significant factor in determining permeability of a molecule through extraembryonic membranes. Penetration is graded, rather than an all-or-none relationship. Highly ionized substances (curare, para-aminohippuric acid, tetraethylammonium, pancuronium) (McNay et al., 1969; Spiers and Sim, 1972; Kivalo and Saarikoski, 1973) do penetrate through these membranes, though more slowly than uncharged species.

Binding of molecules to plasma and to tissues also has a profound influence on the resulting distribution of the molecules in the placental-conceptal unit. Binding of molecules to placental components eliminates the passage of a molecule from one blood pool to the other. Tissue binding may also create a sink or storage depot for certain molecules, which can begin to equilibrate with the blood when the blood levels of the compound are decreasing. The distribution of a compound within the blood and tissues is, of course, dependent upon the affinity of the particular binding sites for the molecule. Certain fetal tissues (liver, muscle, plasma) may have higher affinities for particular molecules than either the maternal tissue or the placental membranes. Thus, these substances may become concentrated to a higher level in fetal tissues than in the maternal tissue. Conversely, greater maternal plasma binding of a compound compared with the fetal plasma can significantly alter the amount of the compound available for passage into the conceptus (cf., Shoeman et al., 1972). A dramatic example of the influence of plasma and tissue binding on molecule distribution in the maternal-fetal unit has been observed in a study of the cholecystographic agent, iophenoxic acid. Eight to 10 year old children who had been exposed to this agent in utero have been found to have plasma drug concentrations similar to their mothers' (Shapiro, 1961; Carakushansky et al., 1969). The reason for the long equilibration and the equivalent plasma levels in the mother and progeny is that iophenoxic acid is bound to albumin with very high affinity (99.99%). The 0.01% of the drug which is free crosses the placenta and then binds to the fetal blood and other tissues. Due to both the high degree of fetal tissue and plasma binding as well as the placental transport capability, iophenoxic acid, which has a half-life of two and one-half years, is one of the unique "inheritable" drugs that may be passed in man through several generations and possibly many more in the rodent.

Differences in maternal and fetal plasma binding of molecules have also been noted for such substances as diphenylhydantoin (Shoeman et al., 1972), dieldrin (Eliason and Posner, 1971) and salicylates (Behrman and Battaglia, 1967). ^{14}C-Dieldrin was found to be concentrated in both the fetus and placenta of the rat when compared to maternal blood levels (Eliason and Posner, 1971). The concentration of this drug in the placenta is interesting in light of the high degree of binding to the plasma. Eliason and Posner (1971) further refined their investigations by measuring the effect of different plasma proteins on the passage of ^{14}C-Dieldrin across the guinea pig placenta in situ. Umbilical perfusion studies demonstrated that alpha-globulins maximally enhanced the transplacental movement of the label. Unfortunately, the influence of plasma proteins in the fetal circulation on placental levels of the labeled compounds was not reported.

Finally, *biotransformation* of molecules by placental, maternal and fetal tissues can modify the ability of a compound to pass from one compartment to another. Many of the factors outlined above can be readily altered by the addition of a hydroxyl group, removal of an amine group, conjugation, or complete metabolism of the compound. Many nutrients and xenobiotics can be metabolized by either the mother or the placenta resulting in the alteration of the distribution and the effect of the agent. The placenta contains aryl hydrocarbon hydroxylase, aromatases, other mixed function oxidases, as well as glucuronidase and sulfatase which metabolize both xenobiotics and hormones (Juchau, 1973). Certain of these metabolic processes can actually be induced by other xenobiotics such as polychlorinated compounds (Alvares and Kappas, 1975) or cigarette smoke (Nebert et al., 1969; Welch et al., 1969; Juchau, 1973). Thus, alterations in the biotransformation of molecules depending upon presence of inducers and inhibitors of metabolism in the mother or fetal-placental unit can create dramatic differences in the ability of the placenta to transfer molecules.

One of the more striking examples of the influence of biotransformation on transplacental movement of molecules is found in sheep. When glucose is administered intravenously to the mother, the fructose content of the fetal blood increases (Huggett et al., 1949; Tsoulos et al., 1971). Fructose is normally found in high concentrations in the fetal blood at birth but disappears within the first 24 to 48 hours (Cole and Hitchcock, 1946). Thus, glucose does not seem to cross from mother to fetus to any great extent; rather it is metabolized to fructose which then enters into the fetal blood compartment, except when a large load of glucose is administered.

Proteins (see section 5.4) and phospholipids (Brezenski et al., 1971; East et al., 1975) are other examples of substances that are degraded during passage from mother to fetus and their metabolites utilized by both the placenta and the conceptus.

In addition to the metabolism of nutrients, placental biotransformation of hormones such as the estrogens is important in the development of the conceptus. Maternal urinary levels of estriol are used as an index of the status of the human feto-placental unit. 16 alpha-OH-dihydroandrosterone is produced by the adrenals and the liver of the human fetus and is metabolized by the placenta and excreted into the maternal blood supply as estriol. A significant decrease in the maternal blood levels of estriols has been correlated with a functionally compromised fetus (Lundy et al., 1973).

Catecholamines have been observed to cross the extraembryonic membranes, but the rate of passage is very slow. The reason for limited passage of these agents is presumably due to the very high levels of monoamine oxidase (Morgan et al., 1972) and catechol-O-methyltransferase (Chen et al., 1974) contained in the placenta itself. Similarly, heparin is rapidly inactivated and metabolized by the mother and placenta, and thus does not affect the conceptus directly, whereas the coumarins do. The biotransformation of substances (whether nutrients, normal cellular products, or xenobiotics) can thus significantly influence their movement across extraembryonic membranes.

The *clearance* of molecules represents one of the most difficult parameters to evaluate experimentally in establishing the mechanisms of transport across extraembryonic membranes. Besides actual movement or penetration of material(s) through the extraembryonic membranes, the rate of presentation of these substances to the placenta must also be considered. This demands consideration of further complex functions such as the maternal tissue distribution of the molecules, rates of maternal excretion, and the blood flow characteristics of the uterine circulation. In addition, the final distribution within the conceptus can further hinder the interpretation of the data. Such difficulties arise as a result of fetal tissue distribution, utilization, and/or excretion of the molecules (cf. Kauffman et al., 1975). Finally, there is the inevitable complicating factor of species differences.

4.2. Specific transport mechanisms

4.2.1. Diffusion

This is a physico-chemical process by which molecules move across membranes in direct proportion to the electrochemical gradient(s) established across the membrane. Transfer is determined according to Fick's law which is defined as follows:

$$\text{Rate of diffusion} = \frac{KA(Cm - Ce)}{X}$$

where:

K = a diffusion constant characterizing the particular placental membrane in the species under study;
A = the area of the membrane over which diffusion occurs;
Cm = concentration of the unbound material in the maternal plasma;
Ce = concentration of the unbound material in the fetal blood; and
X = the thickness of the placental membrane.

Both the area and thickness of the placental membrane vary markedly, not only in different species, but also at different stages during development within the same animal. It seems likely that such changes will be of considerable functional significance in determining the efficiency of placental transport.

Oxygen, CO_2 and water are normally transferred in this manner, and it is thought that many compounds of small molecular weight and low ionic charge also cross by diffusion.

Antipyrine has been used as the standard for comparison of diffusion across the placenta membranes both in vivo and in vitro (Meschia et al., 1967; McNay et al., 1969). Numerous mathematical descriptions have been developed to account for diffusional processes and blood flow characteristics in the placenta in various species, and these should be consulted for full details (Meschia et al., 1967; Assali et al., 1968; Hill et al., 1972; Longo et al., 1972).

Simple diffusion always occurs down a concentration gradient (Fig. 7) and does not require metabolic energy. In Fig. 7 the flow of material from compartment 1 into a cell in the placenta and eventually into compartment 2 is shown to be

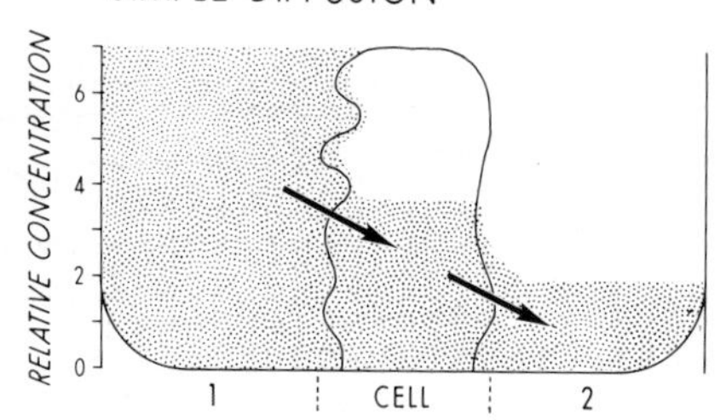

Fig. 7. Diagrammatic representation of movement of molecules across cells by simple diffusion. In the placenta, compartment 1 could be either the maternal blood space or the fetal blood space, while compartment 2 would be the reverse. The ordinate represents the relative concentration of a particular molecule in each compartment. Movement of molecules conforms to Fick's law (see text) and is down an electrochemical gradient.

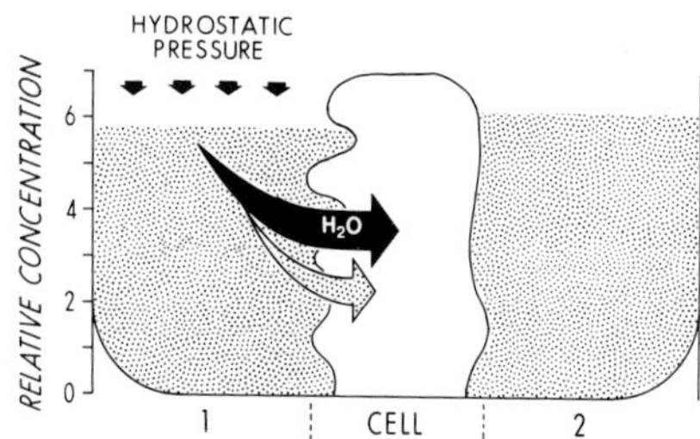

Fig. 8. Diagrammatic representation of the movement of molecules across cells by ultrafiltration. With a change in hydrostatic or osmotic pressure, gradients of pressure are produced which increase the movement of water across the cell from one compartment to another. The increased water flow will simultaneously transfer more organic solutes or ions from one compartment across the cell and into the other compartment than could normally be accounted for by simple diffusion. Thus the movement of molecules is dependent upon the rapid movement of water from one compartment through the cell into the other compartment.

down such a gradient and is dependent upon the membrane characteristics at the border of compartments 1 and 2. If hydrostatic or osmotic forces are applied to either of these compartments, the movement of water to equalize these forces can accelerate movement of an organic molecule or ion across the cell. This process is known as ultrafiltration (Fig. 8). Although ultrafiltration plays a well documented role in the movement of molecules across the membranes of the kidney glomerulus, the contribution of this process to placental transport is unknown. However, in view of the huge volumes of fluid passed between the mother and the conceptus, its role can by no means be excluded.

4.2.2. *Facilitated diffusion and active transport*

Specific proteins located in the cellular plasma membrane have been implicated in certain transport processes. Such proteins have been termed "carriers". In many biological systems the carrier molecules interact only with certain compounds having specific chemical characteristics (reviews, Stein, 1967; Christensen, 1975). For example, amino acids have affinities for specific carriers while sugars interact with different carrier molecules. In the case of amino acids, several different carriers have been identified for acidic, neutral and basic amino acids (see Christensen, 1975 and section 5.2).

Carrier-mediated transfer can be divided into two categories depending upon the energy requirements. The first does not require a direct energy input except to maintain the integrity of the cell membrane, and there is no concentration against an electrochemical gradient. This is termed facilitated diffusion (Fig. 9). The second type of carrier-mediated transfer involves transport of molecules against an electrochemical gradient and is strongly energy-dependent. This is termed active transport (Figs. 10 and 11).

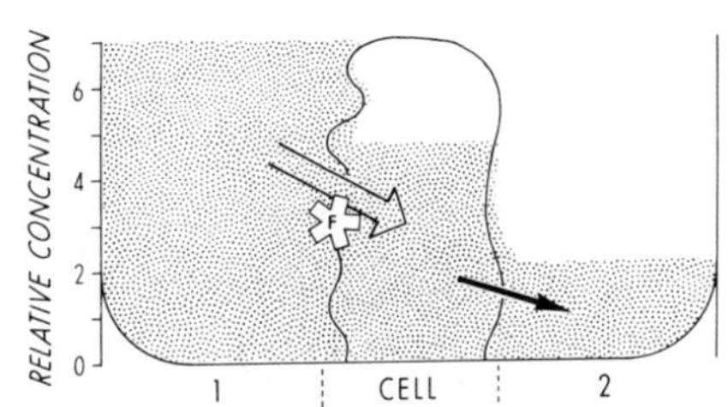

Fig. 9. Diagrammatic representation of the movement of molecules across cells by facilitated diffusion. Molecules are transferred into the cell via a specific membrane component or carrier (F). Movement of molecules from compartment 1 into the cell is quite rapid and exceeds the diffusional values based upon Fick's Law. Movement of molecules from the cell into compartment 2 then follows Fick's Law and is transferred from the cell into compartment 2 by a simple diffusional process. In the present representation where compartment 1 may be either maternal or fetal blood compartments. molecules are transferred more rapidly from compartment 1 into the cell, but they are not concentrated intracellularly and flow down their electrochemical gradient into compartment 2. No energy appears to be necessary to activate this carrier process except to the extent of maintaining the integrity of membrane function.

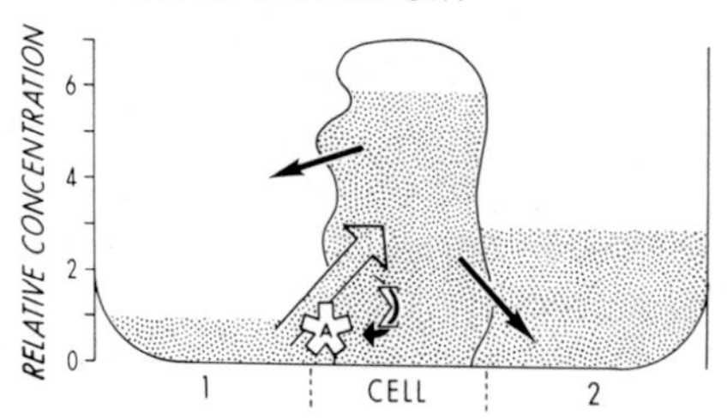

Fig. 10. Diagrammatic representation of the movement of molecules across cells by active transport. The molecules are transferred from compartment 1 into the cell by a mechanism which results in concentration of molecules within the cell. A concentration gradient is thus maintained between compartment 1 and compartment 2. A plasma membrane component or carrier (A) is specific for the movement of the particular molecule being transported. However, to maintain the higher concentration of molecules within the cell, energy is required to maintain carrier activity. Σ indicates the input of energy into the carrier mechanism. In addition, it should be noted that molecules can be moved from the cell into compartment 1 or compartment 2 and can also be occurring simultaneously by diffusional processes. However, the input of the carrier (A) can maintain a very high relative concentration of molecules within both the cell and compartment 2 in comparison to compartment 1. In this particular system, compartment 1 may be either the maternal or fetal compartment while compartment 2 would be the opposite compartment.

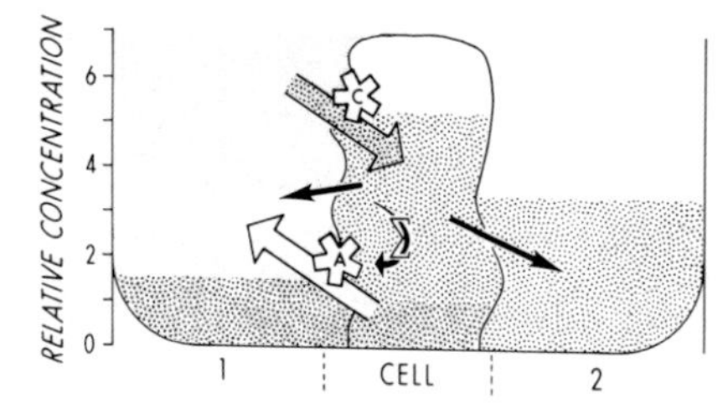

Fig. 11. Diagrammatic representation of the movement of molecules across cells by coupled transport. Two molecules are being transferred in this particular case: the coarser grains represent organic solutes such as amino acids or sugars; and the finer grains represent simple ions such as sodium, potassium, or other molecules. This model was proposed by Crane (1960) to account for the relationship between ion movement and organic solutes. Two carrier mechanisms (A and C) are involved in this process: the C-carrier which requires both molecules for maximal activity; and the A-carrier moves only the simple ions. The C-carrier or coupled carrier does not appear to require any additional energy input except to maintain the integrity of the cell, and thus the two molecules are moved into the cell down the concentration gradient of the ions (finer grains), which are found in much higher concentration in compartment 1. Thus the ion gradient determines the movement of the organic solute (coarse grains) into the cell. The A-carrier determines or maintains the ion gradient between compartment 1 and the cell by pumping the ions (fine grains) out of the cell. The A-carrier determines or maintains the ion gradient between compartment 1 and the cell by pumping the ions (fine grains) out of the cell. This process requires energy (Σ). Thus the organic solutes (coarse grains) can be concentrated inside the cell as well as in the compartment 2 by remaining within the cell while the coupled ion is transported out of the cell by the A-carrier. The organic solutes then diffuse down their gradients both into compartment 1 and compartment 2. If the process reached a steady state without the removal of the molecules by other organ systems (fetal tissues), then the organic solutes in the cell and in compartment 2 would become equal. The presence of this particular transport system in placental membranes has not been demonstrated definitively, and experiments to investigate its presence are in progress (see section 5.2).

The major point to be established is that the cells in the extraembryonic membranes contain specific plasma membrane components with high affinities for certain molecules. This characteristic can be distinguished by kinetic analyses. Since diffusional processes follow Fick's law, the following conditions indicate the presence of a carrier in transfer processes: (1) the appearance of a saturable transfer process at high substrate concentrations; (2) a faster rate of transfer than would be predicted on the basis of physico-chemical properties of the molecule alone; and (3) reduced transfer in the presence of similar molecules (competitive inhibition). To date such carrier-type transfer has been identified for sugars, amino acids, sodium and potassium, creatine, calcium, vitamin B12, and certain globulins in the extraembryonic membranes. In some instances the above mentioned substances have not satisfied *all* of the previously mentioned criteria for mediated transfer.

Carrier processes exhibit Michaelis-Menten type kinetics, and the V_{max} and K_m values can be established using the following equation:

$$F = \frac{S \cdot V_{max}}{K_m + S}$$

where:

F = the unidirectional flux
S = the substrate transferred
K_m = The Michaelis constant or the substrate concentration for half maximal unidirectional flux
V_{max} = maximal transport rate for the unidirectional flux.

When considering categories for transport of particular substances such as sugars or amino acids, it is important to remember that most of these substances are transported by more than one process. A considerable amount of molecular transfer can be provided by simple diffusion in addition to any carrier process. For example, it has been suggested for proteins by Gitlin (1974), and may be equally true for small molecules, that diffusional movement is a primary mechanism for the transfer of molecules through the extraembryonic membranes, and that the carrier processes serve only to supplement the basic diffusional processes. It may be conjectured that when considerable substrate is available, these carrier processes may not be of major importance. However, when the concepto-placental unit and/or mother are compromised by nutritional deficits or pathophysiological alterations, these carrier processes may become increasingly important mechanisms for sustaining the concepto-placental unit.

An additional factor which is pertinent to the various transport mechanisms discussed above and to simple diffusion concerns the amount of unstirred boundary layer space surrounding placental membranes. The unstirred layers are related to the potential chemical gradient for bulk movement of water or ions generated within the spaces located between and around cells. Few studies on placental transport have considered the importance of boundary layer effects. The difficulties encountered in attempting to define such parameters in placental membranes are considerable, and identification of any unstirred layer and its influence on transport is generally ignored. However, Page et al. (1974) using human term amnion in a modified-Ussing chamber determined a value for the influence of the unstirred layer and corrected the permeability coefficient for bulk water movement. Thus, in a placental membrane containing several cell types the importance of boundary layer effects may come to be better understood as techniques are refined to establish relevant transport coefficients and constants for transport across these membranes.

4.2.3. Pinocytosis

The ability of cells in placental membranes to form pinocytotic vesicles and thus engulf a small quantity of the surrounding extracellular fluid phase (together with any material which it may contain) offers a potential mechanism for transplacental transfer of a variety of materials (reviews, Wild, 1973, 1975). In its most conventional form pinocytosis merely serves as a pathway for the incorporation of material(s) into cells. Thus, in many cells internalization of one pinocytotic vesicle at the cell surface is followed by fusion of the vesicle with intracellular lysosomes (see Allison and Davies, 1974). However, in some cells electron microscopic studies have shown that pinocytotic vesicles can traverse the cytoplasm and fuse with the plasma membrane on the other side of the cell (Fig. 12). This phenomenon has been more fully documented in endothelial and mesothelial cells (Bruns and Palade, 1968; Casley-Smith and Chin, 1971;

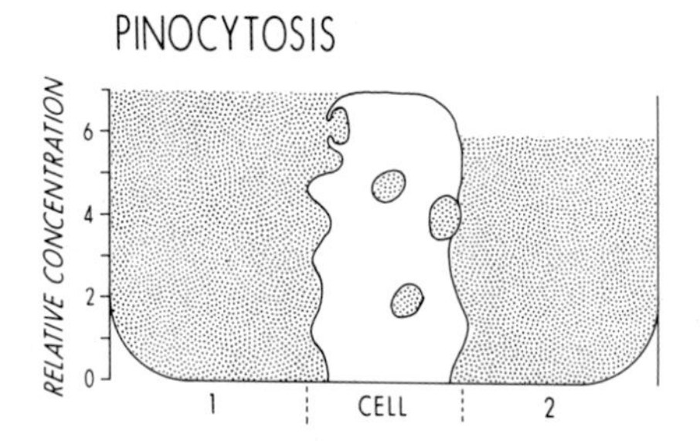

Fig. 12. Diagrammatic representation of the movement of molecules across cells by pinocytosis. Molecules are transferred from compartment 1 into the cell by vacuolization and internalization of a small segment of the cellular plasma membrane, thereby entrapping molecules adsorbed to the cell surface and molecules free in solution. Within the cell the pinocytotic vacuoles may fuse with lysosomes (not shown) resulting in exposure of the vacuole contents to lysosomal enzymes. There is some evidence to suggest that digestion of the vacuole contents may be confined to the free molecules and that substances bound to the membrane of the vacuole are protected from degradation. The vacuole may then undergo exocytotic fusion with the cellular plasma membrane on the opposite side of the cell, thus releasing the contents of the vacuole into compartment 2. A further possible pathway is that after its initial formation, the pinocytotic vacuole migrates to the opposite side of the cell to undergo fusion and release its contents without interacting with lysosomes.

Simionescu et al., 1973) where it is suggested to be an important pathway for the passage of large molecules through cellular barriers. There is a growing appreciation that a similar mechanism might be operating in the transfer of material from the maternal to the fetal side of the placenta. Wild (1973, 1975) has recently reviewed the evidence for pinocytotic activity in placental tissue and concluded that it may well be a major pathway for the absorption of proteins by the placenta. The evidence seems to be less clear, however, as to whether pinocytotic vesicles enable proteins or other materials to be transported across them by the type of process shown in Fig. 12. Further discussion of this topic will be presented in section 5.4.

4.2.4. Leakage via breaks in placental continuity

The presence of small defects or rents in the continuity of placental membranes has been identified as a relatively common occurrence in placentae from several species. Since such defects do not appear to develop in any predictable fashion, it seems unlikely that they provide anything other than a fortuitous pathway for the transfer of materials. Nonetheless, such openings would undoubtedly permit material to pass across the placenta (presumably in either direction). Indeed, such openings have been proposed as providing a means by which cells might pass between the maternal and fetal circulations with resulting risk of immune sensitization.

5. Transport of specific classes of materials across placentae

5.1. Carbohydrates

Sugars provide the major metabolic fuel for the developing fetus. Depending on the species, different sugars are used. In the ungulates (cows, sheep, goats) and the cetaceae (whales) fructose is the primary sugar found in the fetal blood, and the glycogen content of the placenta is minimal (Longo, 1972). In man, glucose is the principal sugar used by the fetus, and the glycogen level of the placenta is high (Longo, 1972), and only low concentrations of fructose are found in the blood.

The molecular weight and high polarity of sugars such as glucose suggest that their diffusion rate across the placenta would be slow, probably inadequate to meet fetal requirements. Widdus (1952) has established that glucose enters the fetus by facilitated diffusion involving a carrier molecule that is envisaged as shuttling back and forth across the cellular plasma membrane yet not requiring energy expenditure.

Placental transfer of sugars is stereospecific, a common characteristic of carrier mechanisms. Thus, D-xylose is transferred more rapidly than the L-form in man (Dancis et al., 1958), as are D-glucose and D-mannose when compared with L-glucose in the guinea pig (Schröder et al., 1975) (Fig. 13). In man, the rate of transfer for glucose across the placenta is quite rapid when compared with

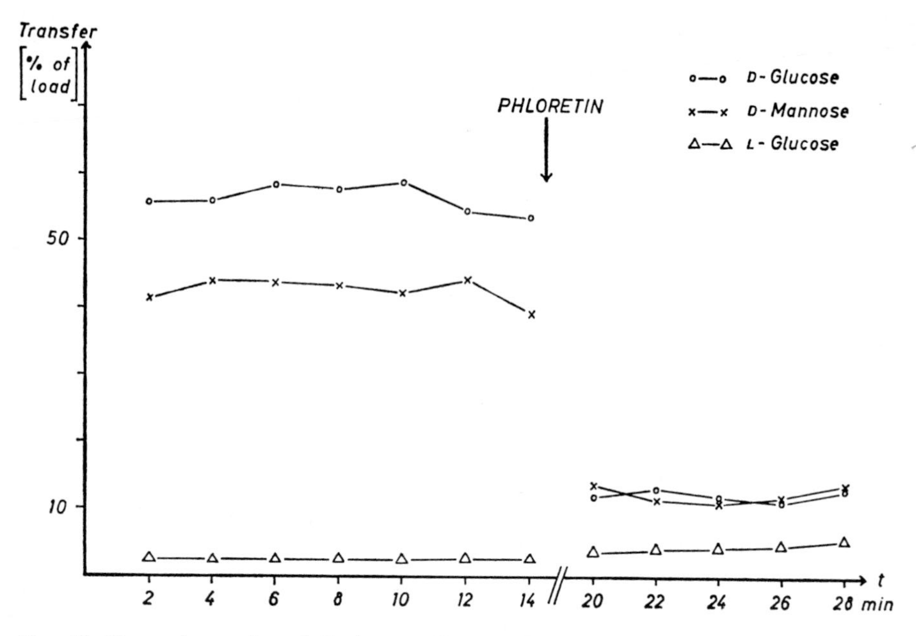

Fig. 13. Placental transfer of D-glucose (○———○), L-glucose (△———△) and D-mannose (×———×) in the perfused isolated guinea pig placenta. The ordinate represents the concentration of the sugars recovered in the fetal perfusate when the sugars are of maternal origin. The perfusion time in minutes is indicated on the abscissa. Addition of phloretin after 15 minutes results in a dramatic decrease in the movement of D-glucose and D-mannose, but movement of L-glucose remains at a low level. (Reproduced with permission from Schröder et al., 1975).

fructose (Holmberg et al., 1956), and L-glucuronic acid does not appear to cross the extraembryonic membranes in vitro (Folkart et al., 1960). The fetal blood levels of glucose are also consistently lower than the maternal venous levels (Zuspan et al., 1966). Part of this difference can probably be attributed to placental utilization of glucose. In vitro analyses have demonstrated that the placental membranes do not concentrate sugars (Battaglia et al., 1962; Deren et al., 1966; Padykula et al., 1966; Longo, unpublished observations).

The question of a saturable transport process in vivo has not been resolved. Cordero et al. (1970) demonstrated in a total of 32 women at term that when the maternal blood levels of glucose exceeded 450 mg %, the fetal blood glucose did not exceed a value much greater than 250 mg %. This observation was later confirmed by Oakley et al. (1972). On the other hand, Ely (1966) and Krauer et al. (1972) found saturation of glucose uptake in the near term guinea pig placenta. More recently, Chez et al. (1975), using two normal and four streptozoticin-treated premature Rhesus monkeys, could not detect saturation of uptake even when maternal levels exceeded 1,000 g%. Such differences among species will need to be resolved by using similar assay techniques and stages of gestation.

The use of non-metabolized sugars can be helpful in separating cellular transfer function from metabolism. 3-O-methylglucose is such a molecule and is not accumulated by the visceral yolk sac of either rabbit (Deren et al., 1966), guinea pig (Butt and Wilson, 1968) or rat (Padykula et al., 1966). However, some studies have demonstrated a specificity for sugar movement which suggests the involvement of a facilitated diffusional process. Furthermore, since the intracellular glucose concentration does not exceed that of the bathing medium (Battaglia et al., 1962), these data are consistent with sugar transfer occurring by facilitated diffusion. A possible exception to this generalization is found in the work of Holdsworth and Wilson (1967), who observed active accumulation of sugars by the visceral yolk sac of the chicken.

As with all carrier systems, the influence of counter transport by other molecules helps to identify specificity. Ely (1966) demonstrated the uphill movement of glucose across the placenta when the fetal side had high levels of galactose. Further specificity has been established by Schröder et al. (1975) using the in vitro guinea pig perfusion preparation. It is known that phlorizin inhibits the active transport of sugars by the intestine, while the aglycone of phlorizin, phloretin, reduces the facilitated transfer of sugars (Stein, 1967). Phloretin was found to inhibit D-glucose and D-mannose transfer across the placenta without decreasing L-glucose transfer (Fig. 13). Such specificity is similar to that demonstrated in the erythrocyte (Rosenberg and Wilbrandt, 1957). It should be noted, however, that Ely (1966) did not observe the phloretin inhibition, though this may be due to a concentration effect. Interestingly, the studies by Schröder et al. (1975) indicate a rather large diffusional component which may mask the facilitated diffusional process in certain preparations.

Even though insulin does not appear to stimulate amino acid uptake by the placenta (Miller and Berndt, 1975), there is still a controversy concerning the role of insulin in the metabolism and transport of sugars in the placenta in utero. Villee (1953a,b) and Litonjua et al. (1967) have observed stimulation of glucose uptake by insulin in vitro. In contrast, neither Battaglia et al. (1962) nor Szabo and Grimbaldi (1970) could reproduce this effect. A possible complication associated with all of these studies is the high insulinase activity of placental tissue itself (Posner, 1973). It is not clear just how much of an effect insulinase has on the insulin concentration in the bathing solution, but this might account for some of the reported differences.

Injections of insulin into the fetus in vivo increased the mean venous-arterial differences of glucose across the umbilical vessels in the sheep (Colwill et al., 1970). It was thought that such differences might be due to an insulin effect on placental transfer of glucose, even though the fetus appeared to be utilizing more glucose as shown by the marked decrease in arterial glucose levels. More recent studies in the sheep (Simmons et al., 1975) and the rat (Rabain and Picon, 1974) indicate that the primary action of insulin is not to control the transfer of glucose in the placentae, but rather to stimulate uptake of glucose by fetal tissues, which raises the possibility that the increased movement of glucose from mother to fetus resulted from falling fetal blood levels.

Placental transport of other carbohydrates has also been investigated. Sorbitol and mesinositol have been found to be in higher concentrations in the fetal than in the maternal blood (Campling and Nixon, 1954). However, the mechanism of transfer and metabolism of these molecules in the placenta are unknown.

5.2. Amino acids

The transfer of amino acids from the maternal circulation to the fetus has attracted considerable interest in view of the central importance of amino acids in the synthesis of proteins by the fetus. It is now well established that amino acids are transferred to the fetus by active transport against a substantial gradient.

Many studies have shown that the concentration of free amino acids in fetal blood is higher than in maternal blood (Christensen and Streicher, 1948; Crumpler et al., 1950; Clemetson and Churchman, 1954; Glendening et al., 1961; Ghadami and Pecora, 1964; Hayter et al., 1964; Knobil, 1965; Lindbland and Balderstan, 1967; Young and Prenton, 1969). In man, the normal fetal-to-maternal ratio varies (depending on the amino acid) from about 1.2 to 4.0, with a mean of about 1.8.

As in the case of sugars described in the preceding section, placental transfer of amino acids is stereospecific, with natural L-amino acids being transferred more rapidly than the D-form (Page et al., 1957; Kelly et al., 1964; Szabo and Grimbaldi, 1970).

Besides differences in transport between isomers of the same amino acids, Christensen and Streicher (1948) suggested that amino acids may be transported competitively. In their study, artificially high maternal blood levels of proline, histidine and methionine reduced fetal-maternal plasma ratios of glycine to one, whereas glutamate did not affect the glycine levels. Sybulski and Tremblay (1967) confirmed these earlier relationships in vitro using human placental sections. Both alanine and serine inhibited glycine uptake in the human term placenta but not phenylalanine, histidine, lysine and glutamate. Alpha-aminoisobutyric acid, a non-metabolized neutral amino acid, has also been studied in this in vitro system. Uptake of alpha-aminoisobutyric acid was depressed by glycine, and L-alanine, but not by D-alanine, lysine, glutamate, creatine or *N*-amidino alanine (Miller and Berndt, 1974 and unpublished observations). These last two compounds were tested because of evidence that, at least in the rat, creatine is concentrated in both the fetal blood and the extraembryonic membranes (Koszalka et al., 1972, 1975; Davis et al., 1973; Miller et al., 1974b). Schneider and Dancis (1974) studied the uptake of nine naturally occurring amino acids by human term placental sections over a two hour period. The acidic amino acids (aspartate and glutamate) exhibited the most rapid rate of uptake, while basic amino acids (arginine and lysine), neutral amino acids (alanine and glycine) and proline all exhibited a slower rate of uptake. These investigators also noted a rapid efflux of leucine from the placental sections which could account for the low concentration ratios observed for leucine and arginine. The efflux studies were continued over a long time period, and interpretation of the data is thus complicated by the possible effects of uptake processes. If, for example, the uptake processes for molecules like alpha-aminoisobutyric acid and lysine are sufficiently active to overwhelm the efflux process, then the dramatic differences in efflux rates between leucine and other amino acids may only be resolved by studying initial efflux rates with short incubations in the efflux solutions.

It appears from the above studies that there may be a number of different processes responsible for amino acid transport in mammals. Hill and Young (1973) have further refined the investigation of amino acid transport across the placenta by studying the patterns of free amino acids in the maternal blood, placenta and fetal perfusates using an in situ guinea pig placental perfusion preparation. This model system eliminates the influence of fetal metabolism and sequestration of the amino acids by the fetus. As seen in Table 2, all amino acids, both essential and non-essential, were transferred against a concentration gradient from mother to the fetus. Initially, the amino acid levels in the placental tissue were considerably higher than the levels found in the perfusate, though

Table 2
Free amino acid concentration in maternal and fetal plasmas, and in the reservoir after two hours closed circuit perfusion of the guinea pig placenta.

	Amino acid concentration (μM) (mean ± S.E.)				
Amino acid	Maternal plasma	Fetal plasma initial reservoir	Reservoir after 2 h perfusion	Fetal : maternal concentration ratios 1 h	2 h
Taurine	159 ± 53	139 ± 26	102 ± 12	0.9	0.6
Aspartate	14 ± 4	38 ± 8	27 ± 1	2.7	2.0
Threonine[a]	88 ± 34	397 ± 104	333 ± 27	4.5	3.8
Serine	81 ± 9	436 ± 88	958 ± 230	5.3	11.8
Glutamate	47 ± 6	153 ± 29	560 ± 109	3.2	12.0
Citrulline	11 ± 1	81 ± 20	347 ± 75	7.4	31.5
Glycine	225 ± 37	1142 ± 362	1704 ± 311	5.1	7.6
Alanine	132 ± 47	680 ± 112	1418 ± 121	5.2	10.8
α-Aminobutyrate	6 ± 2	17 ± 3	25 ± 6	2.8	4.0
Valine[a]	76 ± 18	273 ± 47	813 ± 54	3.5	10.7
Cystine	53 ± 29	89 ± 25	559 ± 141	1.7	10.6
Methionine[a]	12 ± 2	40 ± 11	117 ± 11	3.3	10.0
Isoleucine[a]	47 ± 7	146 ± 23	397 ± 27	3.1	8.5
Leucine[a]	69 ± 8	236 ± 37	721 ± 58	3.4	10.4
Tyrosine	39 ± 5	218 ± 43	367 ± 22	5.6	9.5
Phenylalanine[a]	33 ± 5	81 ± 12	372 ± 39	2.5	11.3
Ornithine	29 ± 2	62 ± 9	214 ± 23	2.1	7.3
Lysine[a]	70 ± 15	252 ± 60	1190 ± 118	3.6	17.0
Histidine	42 ± 5	124 ± 20	387 ± 27	2.9	9.2
Arginine	35 ± 6	151 ± 27	424 ± 58	4.3	12.2
n	5	6	5	---	----

[a]Essential for growth in young animals (Reproduced with permission from Hill, P.M.M. and Young, M., 1973).

Table 3
Concentration of individual free amino acids in placental tissue (μmole/kg wet weight) after perfusion of the guinea pig placenta.

Amino acid	Placental tissue mean ± s.e.	Ratio of placental levels to: Zero net transfer	Closed circuit reservoir	Plasmas: Maternal	Fetal
Taurine	910 ± 140	9.7	8.9	5.7	6.5
Aspartate	818 ± 154	32.7	30.3	60.6	21.5
Threonine	861 ± 157	2.1	2.6	9.8	2.2
Serine	1388 ± 58	2.9	1.5	17.1	3.2
Glutamate	3480 ± 238	16.3	6.2	74.7	22.7
Citrulline	137 ± 21	2.0	0.4	12.3	1.7
Glycine	3318 ± 226	5.3	2.0	14.7	2.9
Alanine	1641 ± 335	2.8	1.2	12.5	2.4
Valine	418 ± 23	1.5	0.5	5.5	1.5
Cystine	490 ± 18	1.7	0.9	9.3	5.5
Methionine	116 ± 17	2.4	1.0	9.9	2.9
Isoleucine	301 ± 13	1.5	0.8	6.4	2.1
Leucine	627 ± 131	1.5	0.9	9.1	2.7
Tyrosine	328 ± 51	1.9	0.9	8.5	1.5
Phenylalanine	707 ± 155	4.9	1.9	21.6	8.8
Ornithine	201 ± 32	2.4	0.9	6.9	3.2
Lysine	504 ± 80	1.5	0.4	7.2	2.0
Histidine	242 ± 28	1.6	0.6	5.8	1.9
Arginine	280 ± 54	1.8	0.7	1.6	1.9

Reproduced with permission from Hill, P.M.M. and Young, M. (1973).

when the perfusate was recirculated, the placental and perfusate levels of free amino acids were similar (Table 3). It would appear, therefore, that the placenta is the pumping station that concentrates the amino acids which must then diffuse into the fetal blood and finally reach equilibrium with the placental tissue depending upon the clearance factors present in the conceptus. One would not necessarily expect to see equilibration in vivo because of the high level of amino acid utilization by the developing fetus.

Dancis and Shafran (1958) suggested that a saturation phenomenon may exist in mammals for amino acid transport in vivo and Hill and Young (1973) demonstrated such a process. However, in vitro studies on human placental sections (Litonjua et al., 1967; Dancis et al., 1968) did not find a maximum uptake at low concentration (10^{-6} M) of alpha-aminoisobutyric acid. These observations suggest that another transfer mechanism such as diffusion may exist in addition to the active transport component. Indeed, Smith et al. (1973) and Miller and Berndt (1974) have recently found that the movement of alpha-aminoisobutyric acid involves at least two components.

The active component for transport has not been well defined in vivo. However, studies with placental sections in vitro have attempted to define some of the characteristics of this process. Litonjua et al. (1967) and Dancis et al. (1968)

examined amino acid uptake by placental sections and found that alpha-aminoisobutyric acid was accumulated by both human and guinea pig chorioallantoic placentae. Sybulski and Tremblay (1967) examined the transport and incorporation of ^{14}C-glycine into proteins in the human term placentae and found that placentae concentrated amino acids and that this required metabolic energy derived from both aerobic and anaerobic metabolism. Similarly, Longo et al. (1973) found that arsenate reduced the uptake of alpha-aminobutyric acid and that glycogen stores within human placentae were influential in maintaining the uptake process. In this laboratory oxidative inhibitors (e.g., nitrogen or dinitrophenol) and different metabolic substrates (e.g., acetate and glucose) have been used to test the characteristics of amino acid transport. As expected, the uptake of alpha-aminoisobutyric acid proceeded normally in an oxygen atmosphere when either acetate or glucose was present. However, in the presence of nitrogen or dinitrophenol, only glucose appeared to maintain uptake near normal levels (Miller and Berndt, 1974).

Naturally occurring amino acids such as glycine can be incorporated into proteins. Sybulski and Tremblay (1967) showed that puromycin, an inhibitor of protein synthesis, when added to placental sections at the time of incubation, did not significantly reduce glycine uptake by the human term placenta, even though the incorporation of glycine into proteins was decreased by 70%. This experiment suggests that the transport mechanism for amino acids in the normal human term placenta may not have an obligatory requirement for protein synthesis under normal incubation conditions. It should be noted, however, that the preincubation effect of enhanced uptake was influenced by the amount of protein synthesis in the tissue. The neutral amino acid transport system of the chorioallantoic placenta is dependent both on sodium and potassium ions (Miller and Berndt, 1974), as well as Mg^{2+}-dependent, $Na^{+}-K^{+}$-ATPase. This transport ATPase was identified in the human term placenta and found to be sensitive to ouabain (see section 2). Ouabain has in fact been found to inhibit amino acid transport in the placenta (Sybulski and Tremblay, 1967; Miller and Berndt, 1974).

As described previously, the transport of many organic compounds appears to require ATPase activity, and the $Na^{+}-K^{+}$-activated ATPase in particular. If sodium or potassium is removed from the incubation medium in vitro, there is a depression in the uptake of organic ions by the placenta. These observations are consistent with a dependence of organic ion transport on sodium and potassium ions. Two distinct transport processes could be involved: (1) the dependence of organic ion transport on the ion gradient (sodium), i.e., a coupled transport process such as suggested by Crane (1960); and (2) the dependence on metabolic energy input as represented by the transport ATPase and the electrical potential differences generated across the cellular membrane as proposed by Kimmich (1973). These differences in transport characteristics are quite important in relation to the actual transport mechanisms involved. With a coupled transport process the movement of the one substance (e.g., amino acids) involves the attachment to a carrier molecule which is moving sodium ions down their electrochemical gradient (Fig. 11). Once inside the cell, the amino acids are concentrated, while the sodium then is pumped back out of the cell with utilization of ATP. Thus the cell can increase its amino acid content while maintaining the sodium content and utilizing only the energy required for sodium transfer (Heinz, 1972). The direct linkage of metabolic energy to the transport of a particular substance such as an amino acid may be equally important. To date, however, no substantial evidence has been generated which distinguishes between either coupled transport or direct active transport of amino acids or any other substance in the extraembryonic membranes (Miller and Berndt, 1974).

This neutral amino acid transport in the human placenta is enhanced by preincubation at 37°C under in vitro conditions (Longo et al., 1973; Smith et al., 1973; Smith and Depper, 1974; Gusseck et al., 1975). Depending upon the length of preincubation, the kinetics of transport can be altered. Relatively short preincubation of human term placentae (30 to 45 minutes) altered the K_m of the transport process for α-aminoisobutyric acid. Longer preincubation (3 to 5 hours) produced increases in the V_{max} (Gusseck et al., 1975). Thus, these K_m changes do apparently reflect affinity differences in the carrier site, while the increases in the V_{max} reflect an increased number of carrier sites. These preincubation effects require both energy and protein synthesis (Smith and Depper, 1974; Gusseck et al., 1975). Similar findings have been noted for early human placentae (Miller et al., 1976b). More recently, Smith et al. (1975) reported that this enhancement is specific for the A type neutral amino acids, but creatine uptake is also enhanced by preincubation (Miller et al., 1976b). How specific and to what mechanisms (environmental, hormonal) can this enhancement be attributed, must yet be resolved; however, few other tissues (uterus – Riggs and Pan, 1972, and chicken heart – Gazzola et al., 1973) do demonstrate such enhancement.

Smith et al. (1973) and Miller and Berndt (1974) have found that the transport processes for amino acids in the placenta conform to Michaelis-Menten kinetics. The K_m for the uptake of alpha-aminoisobutyric acid by the human term placenta is 0.4–0.5 mM. The original observation concerning the kinetics of transport was found to be linear due to the inability to saturate the system. However, to obtain the above kinetic values a large diffusional component was either corrected mathematically (Smith et al., 1973) or determined experimentally by the reduction or elimination of the active component by: (1) removing sodium from the medium; (2) adding ouabain; or (3) adding dinitrophenol in the presence of acetate (Miller and Berndt, 1974). Data derived from 13 to 15-week-old human placentae demonstrated that this tissue can concentrate alpha-aminoisobutyric acid to levels two to three times as high as that found for the same tissue at term (Miller et al., 1976b).

Since a specific transport process seems to exist for neutral amino acids in placentae, the possible involvement of a transport carrier protein has been investigated. Meister (1973) has proposed that gamma-glutamyl transpeptidase is involved in the transport of amino acids from outside to inside the cell. This enzyme is a component of the gamma-glutamyl cycle which consists of five enzymes and requires three high energy phosphates for operation (Meister,

1973; Thompson and Meister, 1975). It is of interest therefore that Schulman et al. (1975) have recently reported the presence of gamma-glutamyl transpeptidase in human, mouse, rat, pig and horse placentae. Previous studies have also shown that the disease, 5-oxoprolinuria, is related to a defect in the cycle. This disease and an associated specific enzyme defect have now been discovered in a single human placenta (Van Der Werf et al., 1974). However, the relationship of this enzyme to transport is still speculative, particularly in view of the requirement for three high energy phosphates, the limited substrate specificity, and the lack of evidence for a sodium dependence for this process.

The transport of amino acids by placentae and extraembryonic membranes is widespread in the animal kingdom. The guinea pig (Butt and Wilson, 1968), rat (Miller et al., 1974a), rabbit (Miller and Berndt, 1972), and yolk sac of the chicken (Holdworth and Wilson, 1967) are all able to concentrate amino acids by active transport processes. There are conflicting data concerning the ability of the visceral yolk sac to accumulate and transport amino acids. Studies by Padykula et al. (1966) failed to show amino acid accumulation by the visceral yolk sac of the rat, although in this laboratory it was shown that alpha-aminoisobutyric acid was effectively concentrated by the visceral yolk sac from both the rat (Miller et al., 1974a,b) and the rabbit (Miller and Berndt, 1972). The difference in these results may be attributed to the short time period allowed for accumulation by Padykula's group. With longer incubation times, alpha-aminoisobutyric acid accumulated in the visceral yolk sac against a concentration gradient by an energy-dependent process which was specific for this neutral amino acid (Miller et al., 1974d). Also, comparison of amino acid uptake with sugar uptake (Padykula et al., 1966) showed that whereas sugars moved passively, the tissue levels of the valine were much larger than those of the sugar after 30 minutes of incubation in the visceral yolk sac preparation in vitro. These in vitro observations agreed with the observed accumulation of alpha-aminoisobutyric acid in the fetal blood in vivo as well as in the extraembryonic tissues of the pregnant rat (B.M. Davis, personal communication). Finally, Butt and Wilson (1968) using guinea pig visceral yolk sac sections found that both lysine and valine were concentrated in the tissue and that dinitrophenol (1 mM) reduced uptake by 42%. The same structure was also found to concentrate both lysine and valine to much higher levels early in gestation than that found at term. Thus, for both the chorioallantoic placenta and the visceral yolk sac, there appear to be transport processes which are specific for amino acids and are dependent on energy input.

5.3. Polypeptides

Polypeptides cross the placenta slowly, if at all. This conclusion has been derived largely from indirect evidence based on the apparent failure of several types of polypeptide hormones present in both maternal and fetal circulations to influence the other partner. Radioimmunoassay techniques provide a highly sensitive method for monitoring the presence of specific hormones. Using this technique it has been possible to actually demonstrate transplacental transfer of certain radiolabeled maternal endocrines such as insulin and various thyroid hormones (Myant, 1958; Gitlin et al., 1965a,b). However, there is a large body of evidence, albeit circumstantial, which suggests that even though some transfer may occur, it is insufficient to induce specific physiological response.

Maternal polypeptide hormones do not cross the placenta in sufficient amounts to influence the development of fetal organ systems in situations where the fetus has failed to produce the necessary hormone(s) for organ development and/or function. For example, anencephalic monsters have low circulating levels of growth hormone and hypoplastic adrenals due to lack of ACTH stimulation (for references see Dancis and Schneider, 1975), indicating that maternal pituitary hormones cannot compensate for fetal insufficiency. Similarly, athyreotic cretins have signs of thyroid deficiency at birth. The failure of maternal thyroid-stimulating hormone to cross the placenta suggested by this observation is similarly inferred from experimental studies where hypophysectomy in fetal sheep causes a fall in thyroid-stimulating hormone and failure of thyroid development (Thoburn and Hopkins, 1973). The relative impermeability of the human chorioallantoic placenta to polypeptide hormones is further demonstrated by the finding that chorionic gonadotropin and placental lactogen which are actually synthesized in the placenta and secreted into the maternal circulation are found in only very low levels in the fetal circulation (Kaplan and Grumbach, 1965).

These observations indicate that the fetal endocrine systems function autonomously and that the placenta provides an effective barrier to the transfer of active concentrations of hormones between mother and fetus. This barrier function is presumably necessary to ensure that the developing fetal organ systems are exposed to the correct amounts and appropriate balance of different hormones and are thus shielded from the risk of abnormal growth which might result from exposure to an aberrant hormonal regime (see section 5.6).

5.4. Proteins

Maternal plasma proteins, in spite of their large molecular size, are transferred to the fetus. There is, however, considerable selectivity in this process. For example, some maternal proteins with molecular weights in the 30,000 range, e.g., growth hormone (Gitlin et al., 1965b; King et al., 1971), ACTH (Miyakawa et al., 1974) and glucagon (Adam et al., 1972), do not appear to enter the fetal circulation to any significant extent while much larger macromolecules such as gamma-globulins (IgG) and polyvinylpyrrolidone (PVP) are transferred (Brambell et al., 1951; Faber et al., 1971; Kulangara and Schechtman, 1972; Sonada and Schlamowitz, 1972). The lack of importance of molecular size in determining placental transfer of proteins is reinforced by observations on the relative transport rates of intact IgG antibodies and fragments of IgG produced by proteolytic digestion. Univalent Fab pieces (molecular weight 50,000) produced by papain digestion of IgG are poorly transmitted compared with intact IgG (molecular weight 150,000) or Fc pieces (molecular weight 80,000) (Brambell et al., 1960). Similarly, observations on transplacental transfer of heavy (molecular weight 50–75,000) and light chains (molecular weight 22–24,000) produced from human IgG or IgM antibodies by cleavage of disulfide bonds have established that the former are readily transported to the rabbit fetus from the maternal circulation while the light chains show only a very low transfer capacity (Kaplan et al., 1965). In more recent studies Hemmings et al. (1975) have also shown that the charge of IgG molecules has little or no effect in determining their rate of transfer across the yolk sac placenta of the rabbit.

Studies on the transplacental transport of immunoglobulins have added the further dimension of specificity to the transport process in that transfer is highly selective for molecules from the same species (reviews, Gitlin, 1974, Wild, 1974, 1975). Selection also operates within classes of immunoglobulins and even subclasses (Wild, 1975).

Most of our knowledge concerning transplacental transport of proteins has come from studies on the transmission of maternal antibodies to the fetus. This subject has been recently reviewed in detail (Wild, 1973, 1974, 1975) and therefore only a brief outline will be given here. Discussion of the subject of antibody transfer from mother to fetus is complicated by the existence of considerable interspecies variation both in relation to the question of whether such transmission actually occurs and, if it does, which placental membranes are involved. A large body of research has shown that the major site for antibody transfer in rodents is the yolk sac placenta (Brambell et al., 1957; Brambell, 1970; Wild, 1973, 1974). In man, however, the yolk sac is a vestigial structure at the time that greatest antibody transfer is occurring (last trimester), and thus by default the chorioallantoic placenta has been implicated as the major site of transfer. Finally, in certain species, notably the ungulates, there appears to be little or no transfer of maternal antibodies to the fetus in utero, a deficiency that is compensated for by intestinal absorption of antibodies from the colostrum during the first few days of life.

Antibodies and other proteins injected into the uterine lumen of the maternal circulation can also gain entry to the fluids surrounding the fetus and be incorporated into the fetus from the latter. This is a consistent finding in the rabbit and certain rodents where entry of proteins into the exocoelomic fluid is rapid. From the exocoel, proteins can traverse the amnion and eventually reach the fetal gut after amniotic fluid is swallowed by the fetus. In the rabbit there is no evidence that antibodies are absorbed from the fetal gut (see Wild, 1974). On the other hand, the fetal rat gut absorbs antibodies and rapidly transports them to the fetal circulation (Brambell, 1970; Wild, 1974).

Dancis (1960) reported that radiolabeled albumin and immunoglobulins could be transferred across the placenta early in gestation, but the rate of transfer was low. These data agree with observations made in other studies with human (DuPam et al., 1958) and rodent placentae (Morphis and Gitlin, 1970) which demonstrated the presence of small amounts of maternal IgG in the fetus early in gestation. However, in both man (Gitlin and Basucci, 1969) and rodents (Morphis and Gitlin, 1970) the efficiency of the transfer process increased as gestation progresses (Fig. 14).

There is some disagreement in the literature concerning the type of kinetics that describe protein transfer across extraembryonic membranes. Sonada and Schlamowitz (1972) suggest that the transfer of both IgG and albumin across

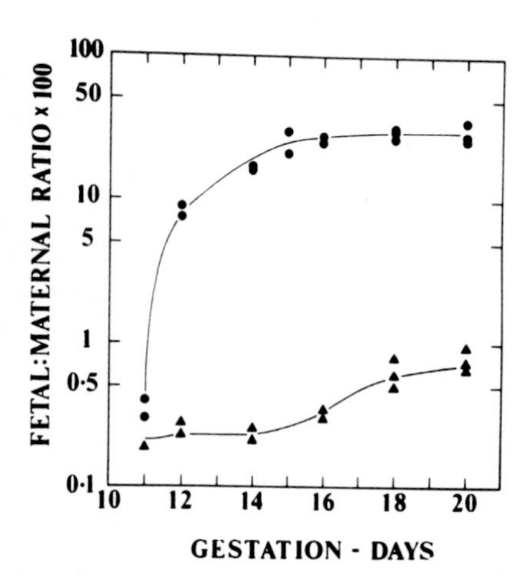

Fig. 14. Ratio of the concentration of human IgG (●——●) and human albumin (▲——▲) in fetal tissues and maternal serum 24 hours after intravenous injection of these agents into pregnant mice. Note that the ordinate is a log scale. (Reproduced with permission from Morphis and Gitlin, 1970.)

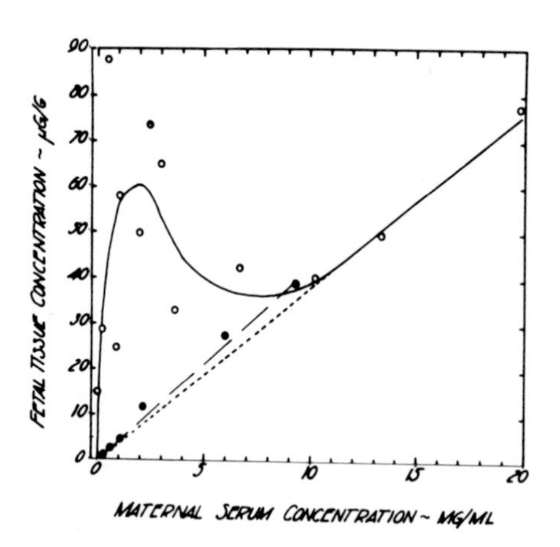

Fig. 15. The concentration of human IgG (○——○) and human albumin (●——●) in the fetus 24 hours after intravenous injection of these agents into pregnant mice. The short darkened line represents the extrapolation to 0 of the kinetic relation found between fetal and maternal concentration of IgG seen at higher maternal concentrations. (Reproduced with permission from Gitlin and Koch, 1968).

the placenta in utero obeys zero order kinetics. However, Gitlin has presented evidence which indicates that there are two possible mechanisms involved: (1) a first order process, most likely diffusional; and (2) a carrier-mediated process which is inhibited by high levels of IgG (Fig. 15). Perhaps the difficulties involved in establishing the mechanisms of transport in these different studies is related to the problem of accurately defining one transport process in the face of one or more coexisting transport systems. In addition, problems may arise from the sampling of the fetal tissue concentration rather than the actual level of circulating proteins.

Much of the current research on placental transport of antibodies and other proteins is concerned with evaluating the role of pinocytosis in the transfer of these materials. Interest in the role of pinocytosis in *both* transport and selection of proteins has developed largely from ideas first formulated by Brambell (see Brambell, 1970). The basic tenets of the "Brambell hypothesis" is that antibodies or other proteins first bind to appropriate receptors on the cell surface and are subsequently internalized within a pinocytotic vesicle. Within the vesicle the incorporated proteins are proposed as being protected from digestion by lysosomal enzymes by virtue of their binding to receptors. The final question of whether such vesicles can transit the cell and undergo exocytotic fusion at the opposite side of the cell (see Fig. 12) to discharge their contents intact is still not clear. There is at least some evidence for chemical modification of certain proteins within pinocytotic vesicles before they are released from the absorptive cell. For a full discussion of the histochemical and biochemical evidence in support of the role of pinocytosis in transplacental transport of proteins the reader is referred to the excellent and comprehensive reviews by Wild (1973, 1974). Most experimental observations on pinocytotic activity in placental membranes have been made on the yolk sac placenta of rodents (see Wild, 1973, 1974 and Williams et al., 1975a,b). Although pinocytosis has been demonstrated in trophoblastic cells in the chorioallantoic placenta of several species (for references see Wild, 1974), definitive evidence for transfer of materials across the entire placenta via this route has not yet been obtained.

As mentioned, an important element of transplacental transport of proteins concerns the high degree of selectivity exhibited by the transfer process. In the "Brambell hypothesis" this requirement is embodied in the proposed binding of proteins to specific receptors on the plasma membrane of the placental cells involved in absorption and transport. Substantial support for this proposal has been provided recently by Gitlin and Gitlin (1974) who demonstrated cell membrane receptors for the binding of specific proteins in mouse intestine, mouse placenta and in the human placenta. The Hofbauer cells which are present in the mesenchymal stroma of human placental villi have been associated with the IgG receptor. However, it is not yet clear whether these cells and their receptors are involved in immunoglobulin transport or if they carry the same placental receptors identified by Gitlin and Gitlin (1974). There is strong circumstantial evidence, reviewed by Wild (1975), that the receptors are associated with so-called coated micropinocytotic vesicles. Similar coated vesicles also appear to be involved in the selective absorption of proteins from the gut of newborn animals (see Rodewald, 1973) and other protein absorbing cells (see Wild, 1975). From an experimental standpoint, however, it would seem that any search for specific receptors in placental cells will be complicated by the marked degree of non-specific attachment of proteins to components in the so-called glycocalyx or cell coat which is present on the surface of cells both in the yolk sac and in the syncytiotrophoblast of the chorioallantoic placenta.

5.5. *Creatine*

Creatine, which together with phosphocreatine and ADP constitutes an important, reversible system for the rapid generation of ATP in muscle and neural tissue, is found normally in higher concentration in fetal blood than in the maternal blood of the rodent (Koszalka et al., 1972; 1975). Furthermore, constant infusion of ^{14}C-creatine into the maternal circulation of the rat under relative steady-state conditions maintained fetal plasma levels (Davis et al., 1973). Creatine is concentrated to higher levels in the extraembryonic membranes of the rat when compared either to fetal or to maternal blood and plasma levels, (Koszalka et al., 1972; 1975) (Table 4). It is proposed that to maintain the fetal-maternal gradient, creatine must first be concentrated to at least a specific minimal level in the extraembryonic membranes before being released into the fetal blood.

Other studies have investigated the directionality of creatine transport from mother to fetus and vice versa and the relative importance of the visceral yolk sac and chorioallantoic placenta of the rodent at term in maintaining the levels of creatine in the conceptus. There appears to be a unidirectional concentrative transport of ^{14}C-creatine from mother to fetus, which maintains placental and visceral yolk sac levels 4 to 5 times higher than when presented from the fetal side (Miller et al., 1974a). A similar relationship has also been found for alpha-aminoisobutyric acid. However, concentrations of creatine and urea in the

Table 4
Creatine content of maternal and fetal blood and extraembryonic membranes.

Creatine in whole blood (μg/ml)	
Maternal	Fetal
16.6 ± 6.6(6)[a]	61.2 ± 7.4 (33)

Creatine in placental unit (μg/g)		
Chorioallantoic placenta		Visceral yolk sac
Maternal-Fetal Junction	Labyrinth	
145 ± 14 (22)	127 ± 12 (22)	171 ± 31 (22)

[a]Values are means ± SD. Numbers in parentheses represent number of observations (reproduced with permission from Koszalka et al., 1975).

extraembryonic membranes were not found when these substances were presented from the maternal side (Miller et al., 1974a). When observing the passage of creatine from mother to fetus, it was found that removal of the visceral yolk sac did not alter the total amount of creatine transferred from mother to fetus indicating that the chorioallantoic placenta is the primary route for creatine movement into the fetal circulation (Miller et al., 1974a).

Using in vitro techniques, ^{14}C-creatine was found to be accumulated by human term placental slices (Miller et al., 1974b) and by rat placentae and visceral yolk sac (Miller et al., 1974d). The uptake process was found to have similar energy and ion requirements to that described earlier for the neutral amino acids. However, beta-guanidinopropionic acid reduced creatine uptake, while neutral amino acids such as L-alanine did not interact specifically. Other creatine analogs were also found to depress creatine uptake by these tissues.

As with the neutral amino acids, creatine transport by extraembryonic membranes is more rapid earlier in gestation. Furthermore, the total amount of material accumulated is greater. For example, the 13 to 18 week human placenta and the 12 to 14 day rat placenta and visceral yolk sac can accumulate creatine to levels 2 to 3 times greater than those seen at term (Miller et al., 1974b, 1974d, 1976b). Further studies have demonstrated that the transport process is saturable at high creatine levels. Similarly in vivo measurements of maternal and cord blood levels of creatine at term in man have shown that the fetal concentration of creatine is 41% higher than the maternal levels (Miller et al., 1976a). Thus creatine is actively transported from mother to fetus in both rodent and in man.

5.6. Lipids

Fetal fat is produced from free fatty acids transferred across the placenta and also by *de novo* synthesis by the fetus from carbohydrate and acetate. Free fatty acids from maternal plasma pass to the fetus by simple diffusion (see Longo, 1972; Dancis and Schneider, 1975). However, phospholipids from the maternal circulation are hydrolysed by the placenta before transfer and are resynthesized into phospholipids by the fetal liver (Brezenski et al., 1971; East et al., 1975). The amount of lipid made available to the fetus via this route appears to be small. Cholesterol appears to be transferred from the maternal to the fetal circulations without chemical modification though the transfer is both slow and limited (Pitkin et al., 1972). The cholesterol content of fetal blood remains significantly lower than that of maternal blood, presumably because the concentration of fetal β-lipoprotein to which it is bound is also present in lower amounts in the fetal circulation (see, Longo, 1972).

Both maternal and fetal blood contain relatively high levels of estrogens, progesterones and other steroids. Unlike the polypeptide hormones discussed earlier, several steroid hormones can pass rapidly across the placenta in either direction. Unconjugated estrogens are rapidly transferred, but conjugated estrogens show restricted transfer (Levitz and Dancis, 1963; Levitz et al., 1967). Progesterone and testosterone are also rapidly transferred across both the rodent and human placentae (for references see Longo, 1972; Dancis and Schneider, 1975).

The rapid transfer of steroid sex hormones across the placenta is also reflected in a number of clinical situations. Maternal virilizing tumors, such as arrhenoblastomas and possibly some adrenal tumors, cause virilization of genetic female fetuses, resulting in pseudohermaphroditism (Jones and Scott, 1968). Experimental treatment of pregnant monkeys (Biggs and Rose, 1947; Qazi and Thompson, 1972) with male sex hormones produces similar pseudohermaphroditism in female offspring. Masculinization of female human fetuses following administration of androgenic hormones for the treatment of breast cancer in pregnant women has also been reported (Grumbach and Ducharme, 1960). A major factor in the recent appearance of vaginal adenocarcinoma in a number of young women exposed in utero to diethylstilbestrol (Herbst et al., 1971) may be related to the rapid transplacental passage of diethylstilbestrol and/or its metabolites and their concentration as seen in the fetal blood of the mouse (Shah et al., 1975) and in both the fetal blood and reproductive organs of the rat (Miller, unpublished observations).

The transfer of bilirubin also conforms to the general principles outlined above for estrogens. Unconjugated bilirubin is rapidly transferred across the placenta in man and in monkeys, but the conjugated glucosiduronate is not (Schenker et al., 1964; Bashore et al., 1969; Lipstiz et al., 1973). The suppression of glucosiduronation in the human fetus just before birth facilitates the transplacental excretion of bilirubin from the fetus, and this explains why newborn infants affected with congenital hemolytic disease are not jaundiced.

5.7. *Electrolytes*

5.7.1. *Univalent cations and anions*

Simple inorganic ions have been implicated in the maintenance of cell function and integrity in many different organ systems. The extraembryonic membranes are no exception. To maintain the placental cells with an internal environment of low sodium and high potassium, the cell surface membrane functions are most important. Such fundamental requirements as an electrolyte pump with an associated energy supply, usually ATP, are necessary to establish appropriate ion gradients. However, besides the maintenance of their internal milieu, the movement of sodium and potassium ions across the placental membranes is of critical importance to the conceptus.

Electrolytes such as sodium, potassium, bicarbonate, and chloride ions, appear to determine the transplacental potential difference. The Nernst or Goldman-Field equations have been employed to determine the relationship between ion distribution and potential difference. Reported values for the potential difference across extraembryonic membranes in vivo range from the extremes of −133 millivolts in the goat (Meschia et al., 1958) to +15 millivolts in the rat (Mellor, 1969). Intermediate values have been observed in other species, for example, −35 millivolts in the sheep (Widdas, 1961) and −13 millivolts in the guinea pig (Mellor, 1969; Stulc et al., 1972). However, in both the rabbit (Mellor, 1969) and in man (Mellor et al., 1969), no potential difference has been observed across the placental membranes. These values are all expressed in relation to the charge on the fetus with respect to the mother.

Stulc et al. (1972) demonstrated that in the guinea pig the transplacental potential difference is sensitive to both hypoxia and to alterations in the potassium level of the fetal plasma. Recently, Fantel (1975) demonstrated that potassium movement across the placenta in the rat is in accordance with its electrochemical gradient. Thus, there does not appear to be any active transfer of this ion across the rat placenta. However, in both of these cases, the movement and distribution of electrolytes may be related to both a small active component and to a relatively large permeability of potassium in comparison to other ions across these fetal membranes. It would thus appear that the potential difference generated across the placentae is not due to potassium but may be the result of the ion pump utilizing sodium or chloride ions. This relationship has not been firmly established, however, and considerable controversy still exists in this particular area.

Since the initial studies by Flexner and Gelhorn (1942) and Flexner et al. (1948) using $^{22}Na^{+}$, most studies have measured transit times or rates of sodium passage from mother to fetus or vice-versa. In man, the maternal-to-fetal transfer rate for sodium is approximately 6 mg per gram of placenta per hour (Flexner and Gelhorn, 1942). This rate is rather constant for animals with hemochorial placentae (rabbit, rat, monkey, and guinea pig), though the rates decrease as follows in other species: man, rabbit > goat > swine. These differences in rates of transfer have been attributed to the thicknesses of the cellular layers between fetal and maternal blood in the various species (Flexner and Gelhorn, 1942). However, urea passage does not follow a similar pattern (Battaglia et al., 1962) which might be expected if the increasing number of cells served as a non-specific barrier to transit. Besides inter-species variation, Flexner et al. (1948) have observed a 70-fold increase in human placental permeability of sodium from 9 to 40 weeks of gestation followed by a rapid decrease in permeability near term.

Sodium exhibits differences in permeability characteristics when compared with tritiated water, antipyrine or urea. As noted previously, there are three major factors which influence transfer of substances across the placenta: (1) permeability of the membrane; (2) umbilical blood flow; and (3) uterine blood flow. Antipyrine and tritiated water are flow-limited, whereas urea is flow- and permeability-limited, while sodium appears to be permeability-limited (Battaglia et al., 1962). Yet in the perfused guinea pig placenta, a flow-dependent component for sodium has been observed. The slower the flow, the more sodium is transferred per unit volume. Nevertheless, it should be noted that at the faster flow rates, more sodium per unit time is transferred (Dancis and Money, 1960). More recently, Schneider et al. (1972), using the perfused human placenta, found that sodium moved quite rapidly from mother to fetus with transport characteristics similar to antipyrine. Other studies by Schröder et al.

(1972) have demonstrated a linear relationship between the transport of sodium and the perfusion rate. In addition, sodium was transferred in a similar manner in either direction with no indication of an active transport process, confirming previous observations. Yet the original observations of Flexner and Gelhorn (1942) and Flexner et al. (1948), later confirmed by Cox and Chalmers (1953a,b), indicated that there was a variable large sodium safety factor involved in the maintenance of the conceptus. This safety factor is defined as the ratio of the amount of sodium transferred to the fetus in excess of the amount of sodium utilized by the fetus. At 12 weeks of gestation in man the safety factor was 160, while at the 40th week it was 1130, which means that 1129 molecules of sodium must be returned to the mother for every molecule used by the fetus to maintain its environment. Since the transfer rate far exceeds the net transfer to the fetus, Flexner and Gelhorn (1942) assumed that sodium may be transferred in the forward and reverse direction at almost the same rate. However, recent studies reported in abstract form by Kelman and Twardock (1975) using the near term guinea pig, have found that while sodium is not actively transported from mother to fetus, potassium appears to be, even though they suggest potassium is not involved in maintaining the potential difference across these membranes. Thus, it has proven very difficult to determine whether an active or passive system exists in the placenta for sodium and/or potassium transport.

Crawford and McCance (1960) measured sodium fluxes and potential differences across the isolated swine chorioallantoic membrane in vitro using the Ussing chamber. They observed a small potential difference across the membrane favoring sodium transport from fetus to mother. This observation could account for the maintenance of fetal homeostasis by removing excess sodium transferred to the fetus. More recently, Stewart and Terepka (1969) and Moriaty and Hoghen (1970) described active sodium transport by the chick chorioallantoic membrane which exhibits a changing potential difference during development (−3 to 5 millivolts at 6 to 7 days; −40 millivolts at 17 days) as well as sodium concentration differences. Using human chorionic membranes, Knapowski et al. (1972) observed a potential difference as well as directional sodium transport. Vasopressin was shown to increase sodium transport from mother to fetus. Furthermore, Brame (1972) found that sodium moved rapidly from the amniotic fluid compartment to the maternal compartment when hypertonic saline solutions were injected into the amniotic cavity.

All of these in vitro studies indicate that there are probably two components involved in the movement of sodium across placental membranes: (1) a passive diffusional component; and (2) an active unidirectional component. Yet since these experiments were not performed using the chorionic frondosum, but rather the chorion (with the exception of the pig studies), these results reflect a more direct influence on amnion-maternal exchange rather than maternal-fetal exchange via the umbilical circulation. Since these membranes have similar origins, the existence of an active component must be seriously entertained.

The sodium content of the human placenta has been studied in vitro under conditions involving varying concentrations of different ions and in the presence of many different substances to determine the capability of the trophoblastic and placental tissues to maintain concentration differences. Berger and Van Hornstein (cited in Diem and Lenter, 1970) and Widdowson and Spray (1951) reported sodium and potassium values for human fresh term placentae to be approximately 100 m Eq/kg and 40 m Eq/kg, respectively. However, the tissues were washed in distilled water before the determinations were performed which may have caused loss of intracellular ions. A more recent study (Dawson et al., 1969) measured nine different cations and found that sodium, potassium, and calcium content were altered in placentae from toxemic patients. However, the sodium and potassium values in this study do not agree with other data (Widdowson and Spray, 1952; Berger and Van Hornstein cited in Diem and Lenter, 1970; Miller and Berndt, unpublished observations). The sodium content of these placentae was many fold higher, while the potassium content was two-thirds lower. Perhaps some of the differences can be attributed to the samples being frozen with no attempt having been made to wash out blood. It is doubtful, however, that differences in procedure would account for the extremely large sodium values. Yet within their experimental protocol, Dawson et al. (1969) observed a significant reduction in sodium content in toxemic placentae, whereas the potassium content was not significantly different from normal (Dawson et al., 1969). Zunker et al. (1965) obtained similar results for the potassium content of placentae from rats treated with oxidized cod liver oil and a diet deficient in vitamin E.

Studies in this laboratory have revealed that as might be expected, dinitrophenol (10^{-4} M), iodoacetamide (10^{-3} M), and ouabain (10^{-5} M) altered the ion content of the human placental sections, as did alterations in pH below 7 and above 8. Ouabain also alters the ion levels in the rabbit placentae (Miller, unpublished observations). These comments concerning the ion levels refer to total tissue ion content and give no information about free versus bound ions, or the possibility of compartmentalization. Thus, the reason for the relatively high sodium levels may be: (1) that the placenta has a different ion distribution from other epithelial tissues; or (2) that the placental tissue samples were not homogeneous and contained other cells or molecules either within or without the villus components which bound or compartmentalized the sodium.

It should be apparent from the above discussion that the questions concerning the mechanisms of movement for sodium and potassium, whether from mother to fetus or vice versa, have not been completely resolved. The in vitro data, as well as some in vivo findings, support the existence of an active transport process. However, other data suggest that sodium and potassium can penetrate equally well from mother to fetus and vice-versa. Again, it is emphasized that perhaps more than one component for transport may be involved, e.g., simple diffusion and active transport. Thus in certain instances, depending upon the experimental design, the diffusional elements may mask the active transport capability of these tissues.

5.7.2. Divalent cations

The movement of divalent cations, in particular calcium, has been linked to a specific transfer process in many tissues (Cooke and Robinson, 1971; Aviolo, 1972; Cittandini et al., 1973). Movement of calcium across cellular membranes plays a vital role in the maintenance of membrane permeability as well as membrane excitability. In the conceptus, calcium metabolism is of further importance, especially in the formation of bone, and blood coagulation products. Thus the movement of this ion through the extraembryonic membranes can be vital for the development of the conceptus.

Calcium is found in higher levels in the blood of the conceptus when compared to the blood of the mother (Mull, 1936; Crawford, 1965; Thalme, 1966; Delivoria-Papadopoulis et al., 1967; Armstrong et al., 1970; Tan and Raman, 1972; Pitkin, 1975). Such levels of calcium in the fetal blood do not appear to be related to the slightly larger calcium binding capacity of fetal serum compared to maternal serum (Twardock et al., 1971). Rather, it appears that ionized calcium makes up the major difference in total calcium between cord blood and maternal blood in man at term (Delivoria-Papadopoulis et al., 1967; Tan and Raman, 1972).

Species differences have been noted in the study of calcium movement across extraembryonic membranes. In the monkey, bidirectional movement of calcium was observed (MacDonald, 1965). However, in sheep only unidirectional transfer from mother to fetus has been reported (Braithwaite et al., 1972; Symonds et al., 1972). The bidirectional transport in the monkey is interesting when one considers the amount of calcium moved from mother to fetus in relation to the requirements for growth. Six to ten times the amount of calcium is transferred over and above what is needed for growth (MacDonald et al., 1965). The excess in movement of calcium from mother to fetus is similar to the observations noted above for sodium transfer.

Other studies have shown that if the calcium plasma level in the mother is increased, the movement of calcium is considerably different in different species. In the monkey, fetal calcium levels are elevated after increasing the levels of maternal calcium (Pitkin, 1975). In sheep, however, there does not appear to be a concomitant increase in fetal calcium following elevation of maternal calcium (Newman, 1957; Littledike et al., 1972). Such species differences have been postulated to be due to both active and diffusional components of the transfer processes. It has been calculated by Kornfeld and associates (see Pitkin, 1975) that the rate of active transport is similar in both species. The major difference appears to lie in the direction of the diffusional component of these species. Thus once again, the active component can be masked or altered depending upon the diffusional capabilities of these membranes.

To date little information has been obtained concerning the actual mechanism involved in the movement of calcium from mother to fetus, except for the identification of calcium ATPases in the guinea pig and human placentae. Recently, Shami et al. (1974, 1975), using placental plasma membrane vesicles, have begun to study the calcium uptake process in vitro. It was found that there was a direct relationship between ATP hydrolysis and calcium uptake in plasma membrane vesicles prepared from human placentae. If ATP was absent, uptake of calcium did not occur. Such a relationship may be due to the incorporation of phosphate into phospholipids which may interact with the calcium. However, this does not appear to be the case in the membrane vesicles. Rather, ATP is utilized directly as an energy donor. Thus the basic ground work has been laid for establishing the requirements for calcium transport across the placenta using in vitro techniques and infusion studies (Twardock and Austin, 1970).

The isolated chick chorioallantoic membrane has also been utilized to study both calcium transport and the relative effect of calcium on other transport processes (Coleman and Terepka, 1972; Garrison and Terepka, 1972a,b; Crooks and Simkiss, 1975). The chick membrane exhibits a sluggish active transport process which requires oxygen. In addition, the oxygen consumption of this tissue is dramatically increased in the presence of calcium. Other ions such as sodium appear to be necessary to maintain this transport process in the chicken.

Calcium thus appears to be transferred across the mammalian placenta from mother to fetus as well as across the avian chorioallantoic membranes from outside to inside by an energy-dependent active transport process. It is also recognized that diffusional processes can alter the movement of calcium in vivo as well as in vitro, but further studies will be needed to establish the relationships of specificity and other ion requirements for this particularly important transfer process.

The transfer of iron to the fetus has also received some attention, with most experimental studies being done in the rabbit. Iron circulates in maternal plasma partially bound to a specific protein, transferrin. Maternal iron is released from transferrin at the placenta by a metabolic process similar to that described for the reticulocyte (Laurall and Morgan, 1964; Larkin et al., 1970). Fetal requirements for iron are highest during the last third of pregnancy, and by term 90% of the iron turned over in the maternal plasma is directed toward the fetus (Bothwell et al., 1958). There is no transfer in the reverse direction. Maternal-to-fetal transfer is effected against a gradient.

5.8. Organic anions and cations

In tissues such as the kidney (Weiner and Mudge, 1964) and the central nervous system (Bierer and Heisey, 1972) there are specific transport sites for the movement of both organic anions and cations. Para-aminohippuric acid and tetraethylammonium have been used as prototype molecules for establishing the characteristics of these transport processes. Since the placenta also transfers many xenobiotics, it is necessary to ask whether there are similar specific transport carriers for organic cations and anions in placental membranes.

Using the constant infusion technique in sheep, McNay et al. (1969) demonstrated that both para-aminohippuric acid and tetraethylammonium cross these membranes with similar characteristics to those found for antipyrine, but at a slower rate. There also did not appear to be any differences in directionality associated with transfer from the mother to the conceptus or vice versa. Using the human chorion, Seeds et al. (1973) demonstrated that para-aminohippuric acid, as well as salicylate and tritiated water, penetrated through the chorion by means of large water-filled extracellular channels. In vitro studies using the section technique have established that neither para-aminohippuric acid nor tetraethylammonium are concentrated against a gradient by human term or early placentae, rabbit placentae or visceral yolk sac in the presence of such metabolic substrates as acetate, glucose, or succinate (Miller and Berndt, 1972 and unpublished observations). It would thus appear that carrier processes for organic anions or cations are not present in extraembryonic membranes of rodents, sheep or man. However, the in vitro studies with placental sections as well as the chorion studies may not accurately reflect in vivo transfer processes.

Few xenobiotic agents seem to be handled uniquely by the placental membranes, unless they are analogues of nutrients. Triamterene is such a compound and is a substituted pteridine. Furthermore, it is a diuretic which appears to be handled by a specific transport process in the extraembryonic membranes of sheep (McNay and Dayton, 1970) and the guinea pig (Dayton et al., 1972). Such specific transport of triamterene has been noted in the kidney (Lassen and Nielson, 1963; Wiebelhaus et al., 1965). Triamterene is another substance which appears to be transported by a non-diffusional process as well as by diffusion. The movement of triamterene across the sheep placenta has been compared with antipyrine, a molecule that passively penetrates through these membranes. Infusion studies have demonstrated that the movement of triamterene from mother to fetus far exceeded the rate of movement of antipyrine (McNay and Dayton, 1970). Further studies demonstrated that this movement could not be attributed to the high maternal binding capacity nor to the large amount of unbound, non-ionized triamterene in the fetal blood (McNay and Dayton, 1970; Dayton et al., 1972). Using guinea pigs and baboons, infusions of triamterene into the maternal blood also produced much lower levels in fetal blood; however, the placenta of the guinea pig appeared to concentrate triamterene to much higher levels than those found in the maternal plasma (Dayton et al., 1972).

Since triamterene is a base, one would think that its movement might be regulated by the organic base transport system as shown in other tissues. However, as mentioned above, there does not appear to be organic anion or cation transport systems in sheep, or possibly in man. Thus, there appears to be another transport process which concentrates triamterene in the maternal blood and placenta to much higher levels as seen in the fetal blood after fetal administration. The actual mechanism(s) involved in this transport process have not been defined. There may be an excretory mechanism for xenobiotics or purines or bases that is different from the organic cation transporting system, and it may have an important role in maintaining the fetal homeostatic environment.

Appreciation of the mechanisms determining whether particular xenobiotics will be able to cross the placenta and, equally importantly, the rate at which they cross, is also critical from the standpoint of assessing the teratogenic potential of the ever increasing number of xenobiotics with unknown biologic potential that are being introduced into the environment (see section 6.6).

5.9. *Nucleic acids*

The ability to synthesize nucleic acids from small molecules is widely distributed in cells, and the fetus appears to synthesize the bulk of its nucleic acids *de novo* (Dancis and Balis, 1954). Transplacental transfer of purine bases, nucleosides and nucleotides has been demonstrated (Hayashi et al., 1968), but its role in the fetal economy remains undefined.

5.10. *Vitamins*

5.10.1. *Water-soluble vitamins*

Group B and C vitamins are present in fetal blood in higher concentrations than in maternal blood. This suggests that these compounds are transferred to the fetus by an active transport mechanism(s). A potential complicating factor in experimental studies is that all water-soluble vitamins circulate in the blood in more than one form, making interpretation of transplacental kinetics uncertain.

The best data have been obtained with vitamin B_{12}. Vitamin B_{12} (cyanocobalamin) is a nucleotide, 5,6-dimethylbenziminazole linked at right angles to a 4 pyrrol ring, similar to porphyrin with a cobalt atom attached. A number of similar compounds, cobalamins, are found in nature. They differ by the kind of ligand attached to the cobalt atom (Halpern, 1974).

Since vitamin B_{12} is a coenzyme involved in DNA synthesis, a requirement for this substance in the developing fetus can be anticipated. Previous investigations have found that vitamin B_{12} is present at much higher levels in the fetal blood than in the maternal blood (Chow and Okuda, 1960; Lowenstein et al., 1960; Lubby et al., 1961). In fact, chorioallantoic placentae from man (Lubby et al., 1961), rabbits (Deren et al., 1966), rats (Padykula et al., 1966; Graber et al., 1971) and mice (Ullberg et al., 1967), as well as rodent visceral yolk sac (Deren et al., 1966; Padykula et al., 1966), concentrate vitamin B_{12} to even higher levels.

The absorption of free vitamin B_{12} from the intestinal contents involves a binding to a mucopolysaccharide (intrinsic factor) secreted by the parietal cells of the gastric mucosa (Shinton, 1972). This intrinsic factor facilitates the movement of vitamin B_{12} into the blood; it has not been identified in plasma (Shinton, 1972). Once in the plasma, vitamin B_{12} is bound to proteins, especially transcorrin I and II. Besides the intestine, other adult tissues transport vitamin B_{12}. Based on subcellular localization studies, the kidney (Newmark, 1972) and liver (Pletsch and Coffey, 1971) appear to incorporate vitamin B_{12} as a complex with proteins (transcorrin II) by pinocytosis.

An interesting observation by a number of different groups (Deren, 1966; Padykula et al., 1966; Ullberg et al., 1967; Graber et al., 1971) is the finding of a considerable time lag between injection of vitamin B_{12} into pregnant animals and its appearance in the fetal blood. During this time interval, whether in mouse, rat, or man, the placentae rapidly concentrated the vitamin B_{12} (Fig. 16).

Graber et al. (1971) demonstrated that trace amounts of vitamin B_{12} transferred to the conceptus per gram increased some ten-fold from day 10 to day 19 of gestation in the rat, whereas Padykula et al. (1966) observed that placental levels of vitamin B_{12} decreased as the visceral yolk sac levels increased. However, in all cases, the extraembryonic membranes concentrated vitamin B_{12} to rather high levels before any vitamin B_{12} appeared in the fetal tissues.

Intrinsic factor also alters vitamin B_{12} movement into the placentae. In vitro studies in rabbit (Deren et al., 1966) and rat (Padykula et al., 1966) yolk sac placentae have shown a 3- to 10-fold stimulation of vitamin B_{12} uptake. This stimulation by intrinsic factor is species specific. Only rabbit and hog intrinsic factor dramatically increase the vitamin B_{12} uptake by the rabbit visceral yolk sac, while in the visceral yolk sac of the rat, only rat intrinsic factor is effective. These findings are interesting in light of the observation that intrinsic factor has not been detected circulating in the blood. There are presently no data available on the role that transcorrin I and II might play in this transplacental movement. Meyer et al. (1974) have demonstrated that transcorrin II is the binding agent

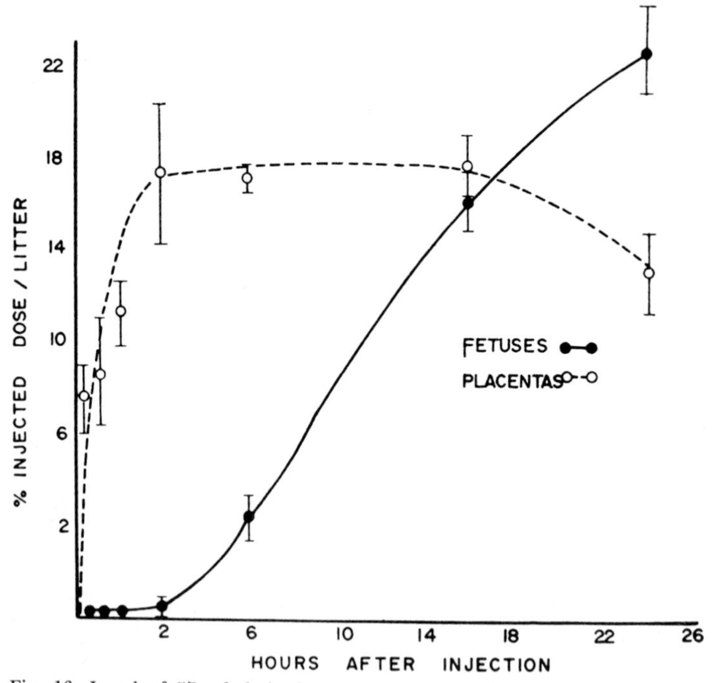

Fig. 16. Level of 57-cobalt in fetal and placental tissues after injection of ^{57}Co-vitamin B_{12} into pregnant rats at day 16 of gestation showing the considerable time delay before radiolabeled material was detectable in the fetus despite high levels of radiolabel in the placenta. Such differences are suggestive of compartmentalization of the injected material within the placenta and slow movement across the placenta to the fetus. (Reproduced with permission from Graber et al., 1971).

which facilitates the movement and concentration of vitamin B_{12} in 1210 ascites tumor cells. The question may be asked whether the free vitamin B_{12} moves by a different process than when associated with intrinsic factor or the transcorrins. No matter whether vitamin B_{12} is transferred alone or is co-transported in association with a protein, a concentrative process is involved.

5.10.2. *Fat-soluble vitamins*

The transfer of fat-soluble vitamins (groups A, D, E and K) is assumed to resemble that described earlier for lipids. Placental transfer is thus believed to occur by simple diffusion, and the levels of these compounds in the fetal blood are generally lower than in the maternal blood (see Longo, 1972).

5.11. *Gases*

The placenta has often been described as the fetal "lung". It is, however, far less efficient as an organ for gas exchange than the lung. The diffusion rate for gases per unit weight of placenta is approximately one-fiftieth that of lung. However, many of the fundamental principles which are pertinent to pulmonary gas exchange are also applicable to exchange in the placenta, except for the major difference that in the placenta exchange of gases occurs between two blood circulations.

The characteristics and requirements for respiratory gas exchange across human and animal placentae have been discussed at length in a recent symposium (see Longo and Bertels, 1972), and only a brief outline will be given here.

The movement of oxygen and carbon dioxide across the placenta is critically dependent upon the blood flow characteristics of both the maternal and fetal circulations (see Longo and Bertels, 1972). The respiratory gases O_2, CO_2 and the metabolically inert gases are presumed to cross the placenta by simple diffusion. Burns and Gurtner (1974) have suggested recently that a carrier mechanism involving cytochrome P-450 might play a role in facilitating the movement of oxygen across the placenta but Power and Loos (personal communication) could not detect a similar transport system in the rabbit placentae in vitro.

6. *Alterations in placental function: implications for transport and fetal development*

Although there have been many important advances in our understanding of the mechanisms of placental transport in the nature of placental disease, most studies dealing with placental transport have not utilized abnormal placentae. It has been burdensome enough to compare transport processes at different stages in one species and among many species. It is emphasized, however, that any discussion of alterations in the adequacies of maternal-fetal transfer must consider not only true placental tissues of fetal origin, but also the various maternal components, notably the maternal blood supply to the placenta. Most of our insight into the effects of a deranged supply line on fetal well-being has been derived from clinical studies in man, and this bias will be reflected in the literature reviewed in the following sections.

6.1. Placental disease

Detailed discussion of the pathology of the human placenta is beyond the scope of the present review, and the reader is referred to Benirschke and Driscoll (1967), Driscoll (1975) and Fox (1975). Despite the enormous literature on placental pathology, it has been extremely difficult to evaluate the effect of most types of placental disease on transport function. The correlation of specific morphologic alterations with metabolic abnormalities is therefore limited (Longo, 1972; Rushton, 1973). A number of investigators have attempted to experimentally induce placental disease in an effort to evaluate the effect of specific structural alterations on fetal development and viability (Brent and Franklin, 1960; Franklin and Brent, 1964; Wigglesworth, 1964; Hill et al., 1971; Myers et al., 1971; Panigel and Myers, 1972; Wallenberg et al., 1973). However, placental transport function was not examined in any of these studies.

The placenta differs from most other organs in that its pathology is largely quantitative rather than qualitative. A number of pathologic lesions such as infarction, hematoma, thrombus formation and fibrin deposition are found in placentae from normal pregnancies and their presence cannot strictly be considered pathological unless they are unduly numerous or large. Furthermore, many forms of placental pathology may interfere with function in only a portion of the placenta, while the remainder of the organ remains normal. Evaluation of the impact of such alterations on placental function must then take into account the remarkable reserve capacity of the placenta.

As mentioned earlier, there are a number of characteristic histologic changes that accompany normal aging of the human placenta (Table 5). Certain of these changes might be expected to reduce the efficiency of transport across the placenta, particularly in postmature pregnancies. Reduction in the surface area for exchange following focal separation of maternal and fetal tissues or infarct formation must certainly reduce the potential for transfer, or at least reduce the reserve capacity of the organ. Similarly, gross alterations such as deposition of fibrinoid, oedema and thickening of the basement membrane would be expected to change the diffusion characteristics of the placenta. It also seems likely that more subtle modification in placental composition may be occurring and could modify transport function. However, definitive evidence correlating specific histologic change with an alteration in transport function has yet to be obtained.

Finally, attempts to correlate placental disease with alterations in fetal growth

Table 5
Structural alterations accompanying aging of the human chorioallantoic placenta[a].

1. Increase in thickness of basal membrane of chorionic epithelium and decrease in diameter of villi
2. Decrease in villous epithelium (disappearance of cytotrophoblastic cells) and excess formation of syncytial knots
3. Progressive reduction in length of villi
4. Increase in free amino acids and in RNA, decrease in protein content
5. Increase in density of villous stroma (disappearance of most Hofbauer cells, sclerosis), decrease in capillaries and enhanced appearance of avascular villi – compensatory villous growth in peripheral parts of placenta
6. Increase in fibrotic foci
7. Increase in cyst formation
8. Calcification
9. Increase in degenerative change of decidual vessels (fibrinoid degeneration of intima)

[a]Modified from Vorherr (1975).

or viability are further complicated by the fact that many placental lesions have an extra-placental origin, for example, alteration in the maternal blood supply to the placenta.

6.2. Placental insufficiency and placental dysfunction

Such terms as placental dysfunction and placental insufficiency have been used widely in the clinical literature for many years and have been routinely offered as formal clinical diagnoses to account for fetal death or disease. However, the value of these terms is questionable since failure of the placenta to fulfill its multiple functions is rarely due to a primary fault within the placenta and is more commonly due to extraplacental causes which indirectly compromise placental function.

Gruenwald (1961, 1962, 1964) has written extensively about placental dysfunction, but recently (1974, 1975) he has had second thoughts about the accuracy of the term and he is no longer certain that there are any primary instances of placental dysfunction. In essence, he is saying that maternal conditions such as toxemia and hypertensive disease are the primary etiologies of placental dysfunction and that if we eliminate the maternal disease, we will markedly reduce the incidence of placental disease. Of course, he is correct in this concept. Secondly, he feels that too many investigators attribute a small infant to a small placenta (Gruenwald, 1974, 1975). Both Gruenwald's concepts as stated are biologically correct, but it also appears from the literature that many investigators

have long accepted these facts (see, for example, Brent and Jensh, 1967).
The term placental dysfunction as used in the remainder of this article does not imply primary placental disease. In many instances there may be no explanation for the existence of placental pathology. In fact, one can conjecture that placental pathology is secondary to fetal disease.

Although the growth curve for the placenta and fetus are somewhat different, there is relatively good correlation between placental and fetal size at term (Rosahn and Greene, 1936; Calkins, 1937). The growth rate of the placenta decreases as term approaches to a greater extent than does the fetal growth rate (Calkins, 1937; Hendricks, 1964; Tremblay et al., 1965). McLaren (1965) reported a finding that is quite commonly observed; namely, placental weight in rodents decreases on the last day of gestation (Table 6). The significance of the fact is not known, but it may account for some of the problems seen in postmature infants. Not only is there good evidence for a decrease in placental size at term, but also a decrease in placental function. When placental weight or function is decreased to the point of no reserve, small changes in placental weight and placental function produce marked changes in the birth weight (Pick, 1954; Kloosterman and Huidekoper, 1954; Hafez, 1963). Gruenwald believes that the effect of placental dysfunction is intimately related to the rapidity of its development. He refers to the chronic form of placental dysfunction as existing for weeks resulting in marked growth retardation and malnutrition (Gruenwald et al.,

Table 6
Placental weight on the 17th, 18th and 19th day of gestation in Q strain rats.

Number of implants surviving to mid-term	Number of pregnancies			Mean placental weight (mg)		
	17th day	18th day	19th day	17th day	18th day	19th day
6	0	1	1	—	105.0	121.7
7	0	3	1	—	117.1	122.9
8	0	1	8	—	102.5	97.3
9	0	4	5	—	113.6	100.0
10	3	7	4	114.5	103.6	93.6
11	0	9	6	—	111.1	100.6
12	1	8	5	96.7	106.0	95.3
13	1	4	4	100.7	109.2	89.8
14	3	2	3	111.4	107.5	91.7
15	1	1	3	104.0	127.3	83.6
16	0	1	0	—	110.0	—
17	0	0	1	—	—	91.9
Total	9	41	41			
	Mean litter size			12.4	10.9	10.8
	Mean placental weight			108.8	109.2	96.5

Reproduced with permission from McLaren (1965).

1963). The subacute form results from placental dysfunction of several days' duration resulting in malnourishment but not growth retardation (Gruenwald, 1962; Wagner, 1964). He places the infant who is postmature in this category (Gruenwald, 1962).

Maternal toxemia and hypertension, undernutrition and maternal vascular disease are some of the more common conditions associated with placental insufficiency (see, Rathburn, 1943; Kortenoever, 1950; McKiddie, 1950; Clifford et al., 1951; McKeown and Gibson, 1951; Clifford, 1954; Everitt, 1964; Gruenwald, 1964; Mead and Marcus, 1964; Iyengar, 1973; Schneider and Dancis, 1975; Vorherr, 1975). There have been few studies attempting to relate the biochemical pathology seen in toxemia to altered transport function or to intrauterine fetal growth retardation (IUGR). Iyengar (1973) reported that placentae from small-for-date pregnancies had lower nitrogen, DNA and RNA contents. Although the glycogen content was also lower, its rate of utilization was higher in the placentae from small-for-date infants. Most investigators seem to be more impressed with the placental pathology accompanying this condition. Hertig (1960) and Aherne and Dunhill (1966) reported that premature degeneration of the syncytiotrophoblast was common in placentae from toxemic mothers. Aherne and Dunhill (1966) also reported that in maternal hypertension and low birth weight infants the placental size, capillary surface area and villous surface area were all reduced (Table 7). Alvarez et al. (1972) and Fox (1975) reported that the number of syncytial sprouts and cytotrophoblast cells, as well as the thickness of the trophoblastic membrane, were greater in toxemic placentae than in normal ones. Hyperplasia of cytotrophoblastic cells has been proposed as

Table 7
Placental parameters and associated pathology in man[a].

Parameter	Normal[b]	Maternal hypertension[c]	Abnormally small infants[d]
Placental volume (ml)	488 (99)	363 (77)	350 (65)
Non-parenchyma (%)	20.8 (3.6)	21.6 (1.7)	29.0 (7.4)
Volume proportions of parenchyma			
Intervillous space	35.8 (3.2)	39.2 (5.7)	35.8 (4.9)
Chorionic villi	57.9 (5.7)	60.7 (6.1)	61.4 (6.2)
Fibrin	4.3 (2.1)	2.7 (1.0)	4.1 (2.0)
Villous surface area (m^2)	11.0 (1.3)	7.7 (1.6)	6.4 (1.4)
Capillary surface area (m^2)	12.2 (1.5)	10.3 (2.9)	6.8 (1.6)
Proportion of villus occupied by trophoblast	25.3 (1.9)	18.8 (1.0)	22.3 (3.7)

[a]Reproduced with permission from Aherne and Dunhill (1966).
[b]Mean values of 10 cases at 39–40 weeks' gestation.
[c]Mean values of 7 cases at 39–40 weeks' gestation.
[d]Mean values of 6 cases at 39–40 weeks' gestation.

occurring in response to reduced blood flow to the chorionic villi (Wigglesworth, 1962) and this conclusion has been reinforced by in vitro observations in which identical changes were induced in organ cultures of trophoblast by cultivation in an environment with a low oxygen tension (Fox, 1970). Cytotrophoblastic hyperplasia would thus appear to be a repair phenomenon that follows ischemic degeneration of the syncytiotrophoblast. Given the important role of the syncytiotrophoblast in placental transport and endocrine functions (see section 2), it is not unreasonable to conclude that ischemic degeneration of this component might have major consequences for transport of materials to the fetus. In addition, the overall thickening of the trophoblast membrane as a consequence of proliferation of cytotrophoblast would be expected to further hinder diffusional transport processes. Thus, many examples of so-called placental insufficiency may be due to circulatory disturbances that adversely affect blood flow through the utero-placental unit. This point will be discussed in detail in section 6.4.

6.3. Biochemical diseases of the placenta

Over the past 50 years a great deal has been learned about metabolic diseases in the human. Deficiencies or abnormalities affecting specific enzyme(s) have been identified as the etiologic factor in diseases affecting metabolism in a range of organ systems. The importance of enzyme variability and abnormality within the human placenta is only just beginning to be studied, and its possible contribution to placental metabolic disease or variability is not clear. Even within the same species the specific activity of some enzymes varies at different stages of gestation (Van Hein et al., 1974; Welsch, 1974). Edlow et al. (1971) studied the isozyme patterns of five enzymes involved in sugar metabolism and noted that one enzyme, lactate dehydrogenase, had one additional isozyme during the first trimester while the isozyme profiles of four other enzymes (glucose-6-phosphate dehydrogenase, 6-phosphogluconate dehydrogenase, β-glucuronidase and N-acetyl-β-glucosamidase) remained unchanged throughout pregnancy, though there was considerable variability in the specific activities of all five enzymes.

Enzyme activities may assist in understanding the functional role of certain placental metabolites. For instance, higher concentrations of placental choline acetyltransferase are found early in gestation, suggesting a function for acetylcholine beyond its proposed role in the delivery process (Sastry and Henderson, 1972; Welsch, 1974). Besides choline acetyltransferase, cholinesterase and acetylcholine are also present in the human placenta (Sastry et al., 1973).

Since high levels of exogenous acetylcholine do not have a significant vascular affect on the utero-placental unit, Harbison et al. (1975) proposed that this cholinergic system might regulate placental function, in particular its transport capabilities. These investigators found that in human placentae from spontaneous abortions, therapeutic abortions and normal deliveries, the levels of acetylcholine, choline acetyltransferase and acetylcholinesterase were highest between 16 and 20 weeks of gestation, and lowest at term and between 6 to 13 weeks.

These enzyme activities and acetylcholine content were shown to be temporally correlated with the ability of the human placenta to concentrate compounds under in vitro conditions. This study also demonstrated that the choline acetyltransferase levels in placentae from pre-eclampsic patients were one-third those found in normal placentae. Harbison et al. suggested that the low activity of the cholinergic system in placentae from pre-eclampsic patients may reflect a compromise of the placental transport system. Indeed, there is suggestive evidence that transport function may be reduced in the placentae from pre-eclampsic patients (Flexner et al., 1948; Cox and Chalmers, 1953a,b; Johnson and Clayton, 1957; Butterfield and O'Brien, 1963; Foley et al., 1967; Lindbland and Zetterstrom, 1968; Miller and Berndt, 1975). Harbison et al. (1975) have proposed a regulatory role for acetylcholine on membrane function in the placenta. The transport of molecules across the placenta is proposed as being facilitated by specific cholinergic-receptor interactions, and that transport is reduced when excess acetylcholine is released. This contrasts, however, with other studies using only hysterotomy, hysterectomy and normal term specimens (Miller et al., 1974b, 1976b).

Studies of placental sulfatase deficiency have also yielded interesting results and raise the important issue of whether specific enzyme deficiencies or abnormalities in the placenta can result in abortion or fetal pathology (France et al., 1973; France and Downey, 1974; Oakey et al., 1974), even though normal infants have been delivered from pregnancies in which the placenta is deficient in arylsulfatase. In one case it was determined that the enzyme deficiency was limited solely to the placenta, so it is conceivable that central enzyme deficiencies will not be manifest in either fetal or maternal organs (France and Downey, 1974).

Finally, variation in enzyme activity may be important in determining the ability of the placenta to chemically alter (biotransformation) materials derived from the maternal circulation. This is of considerable relevance to the question of whether teratogenic drugs or chemicals can gain access to the fetus (see section 6.5).

Besides enzyme variability and abnormality, it is possible that genetic and non-genetic factors may account for differences in other biochemical properties of the placenta, including membrane permeability and the number and types of plasma membrane receptor sites for particular classes of biologically active material.

6.4. Alterations in utero-placental circulation

Transfer of materials across the placenta involves exchange between two circulatory systems, that of the mother and that of the fetus. It is obvious, therefore, that changes in blood flow to or through the utero-placental unit will alter the kinetics of placental transport processes and that significant alteration in the utero-placental circulation may have profound consequences for the well being of the fetus (Brent and Franklin, 1960; Franklin and Brent, 1964).

Most attention has been given to the factors that reduce maternal (uterine) blood flow (reviews, Longo, 1972; Longo and Bartels, 1972; Adamsons and Myers, 1975). Reduction in maternal blood flow to the utero-placental unit not only decreases the nutrients available for transfer to the fetus but also may affect the function and viability of the placenta itself. Most examples of "placental insufficiency" (vide supra) appear to be associated with disorders of the maternal circulation to the utero-placental unit.

There are a variety of conditions affecting the mother that can alter the rate of blood flow through the intervillous space, and these fall into two general categories: those affecting maternal blood flow or pressure and those associated with abnormal maternal blood composition. Most of the more commonly encountered clinical conditions affecting the maternal circulation to the utero-placental unit are in the first category.

6.4.1. Alterations in blood supply to the utero-placental unit

Circulatory alterations may be acute or chronic. In the former, reduction in uterine blood flow is of considerable magnitude as, for example, in hypotensive circulatory collapse. The effect of such episodes on the fetus depends largely on their duration. Acute asphyxia of the fetus may lead to its death or, less commonly, brain injury. It is apparent, however, that substantial interference with the uterine circulation must occur to produce changes in the fetus of lasting consequence. Considerable changes in maternal arterial blood pressure can occur before circulation in the utero-placental unit is altered sufficiently to produce detectable change in oxygenation of the fetus (Myers, 1972). Similarly, experiments on primates have shown that reduction in maternal systemic blood pressure by as much as 50% does not change fetal oxygenation (Morishima et al., 1971). It is recognized, however, that a variety of anesthetics may induce complex hemodynamic effects that affect both maternal blood flow to the placenta and also directly affect the circulatory system of the fetus. Details are available in standard texts on obstetrical anesthesia.

Chronic changes in utero-placental circulation are more commonly encountered. These involve moderate reduction in uterine and/or placental blood flow of long duration and are frequently accompanied by growth retardation of the fetus. A strong association between impaired growth of the fetus and conditions known to decrease blood flow to the uterus has been established in epidemiological studies and by experimental studies in primates and rodents (see Adamsons and Myers, 1975).

From a clinical standpoint, conditions characterized by vasoconstriction within the uterine vasculature account for most examples in which the well being of the fetus is jeopardized because of inadequate circulation through the intervillous space. A wide variety of conditions can produce constriction of the uterine vasculature, but the more common causes are diseases such as pre-eclampsia, maternal hypertension and kidney disease.

Excessive adrenergic stimulation of the maternal vasculature will also reduce the perfusion pressure of maternal blood in the intervillous space. Experimental studies on primates have shown that activation of either alpha- or beta-adrenergic receptors or their blockade lead to impaired oxygenation of the fetus. Indeed, Adamsons et al. (1971) were able to induce fetal death by administration of large amounts of adrenergic agonists to pregnant rhesus monkeys.

In man, activation of the sympathetic nervous system of the mother can result both from endogenous release of catecholamines due to various maternal stress conditions or by incorporation of exogenous materials that are able to stimulate sympathetic ganglia or the adrenergic receptors of the surfaces of effector cells.

Alterations in uterine blood flow have been reported following the administration of a wide variety of drugs. Narcotic agents (Gautieri and Ciuchta, 1962; Klinge et al., 1966) and psychodelic drugs such as LSD, mescaline, bufotenine, psilocin and psilocybin (Juchau and Dyer, 1972) have been reported to alter placental blood flow in placental perfusion systems in vitro. Ergotamines have also been found to produce placental insufficiency by reducing uterine blood flow (Grauwiler and Schön, 1973; Leist and Grauwiler, 1973a,b, 1974). Acetylcholine has been reported to have a variable effect on uterine blood flow. Although a total increase in blood flow is observed following the injection of acetylcholine, it has been occasionally observed that the increased blood flow is via the myometrial arteries, while the blood flow through the placenta is actually decreased (Carter and Olin, 1973). Oxytocin has variable effects on blood flow in the perfused human placenta (Branda et al., 1973). Serotonin, in high doses, can produce abortion in the mouse and rat. This is believed to result from a reduction in blood flow due to the vasoconstrictive properties of this agent (Eliasson and Astrom, 1955; Juchau and Dyer, 1972), but other investigators have reported that serotonin also alters placental permeability (Robson and Sullivan, 1966; Honey et al., 1967). The role of histamine and serotonin in normal placental physiology has been clarified. Since both drugs are ordinarily present in physiological quantities, an essential physiological role may well exist. In vitro, serotonin (10^{-4} to 10^{-5} M) does not alter the uptake of amino acids into the human term placenta (Miller et al., 1973). Histamine is believed to play some role in maintaining vascular tone by acting as a vasodilator of placental vessels. Histamine appears to be metabolized in the maternal organism and fetus, and there are reports that the placenta can oxidatively deaminate histamine (Lindberg, 1973). A direct effect of many agents on placental transport function has not been investigated.

It is stressed, however, that the effect of vasoactive drugs is not always simple to interpret since many such drugs also reach the embryo. Certain drugs which affect the placental vasculature may simultaneously be affecting a number of tissues and biochemical processes in the mother, the placenta and the fetus (also see section 6.6, below).

In contrast to the marked alterations in maternal blood flow to the utero-placental unit produced by vasoactive agents, administration of similar drugs to the fetus has little or no effect on conductance in the umbilical circulation (see Adamsons and Myers, 1975).

Impairment of fetal circulation to the placenta is rare and is not generally considered as a contributory factor in fetal deprivation. Even gross malformations of the fetal heart and great vessels which are incompatible with postnatal survival may not impair circulation to the placenta and the somatic tissues of the fetus.

The fetal circulation is affected, however, by mechanical factors. Power and Longo (1973) have found that alterations in blood flow through the maternal vessels of the utero-placental unit can influence circulation in the fetal vasculature. Maternal hypertension and hypotension can thus affect blood flow within fetal placental vessels. Maternal conditions such as hypertension which impair circulation within the uterine vasculature may therefore simultaneously decrease umbilical blood flow and increase further the risk of impaired transport of oxygen and other nutrients to the fetus. Uterine contractions also reduce blood flow within the utero-placental unit (Ramsey et al., 1963; Borrell et al., 1965; Friess and Gobble, 1967) and may alter fetal placental circulation, especially as term approaches and contractions become more common.

Criteria for what constitutes a normal range for mechanical alterations in circulation within the utero-placental unit and the relationship of mechanical factors to other conditions that affect placental function have yet to be determined.

Circulation to the utero-placental unit is unusual in that there is no direct feedback between the tissues supplied (placental villi) and the sites where regulation of blood flow occurs (uterine spiral arteries). In most body tissues changes in blood flow occur in response to changes in the metabolic needs of the tissues supplied, and a mechanism for regulating blood flow resides locally within the tissue(s) supplied. The absence of local mechanism(s) for regulating blood flow within the utero-placental unit may be important for the reason that marked shifts in the maternal circulation will not occur. For example, fetal control over blood flow through the intervillous space might create problems for the mother since the fetus could then divert progressively more of the maternal cardiac output. The absence of local feedback control over blood flow to the utero-placental unit dictates, however, that circumstances in which maternal blood flow to the unit is reduced, cannot be compensated for and the placenta and the fetus will then be subject to deprivation, the severity of which reflects the extent of alteration in the maternal blood supply to the placenta.

6.4.2. Alterations in the composition of maternal blood

Maternal blood represents the major "environmental" exposure for the fetus and maternal homeostatic mechanisms thus provide the major protection for the fetus. In the absence of severe environmental stress or disease, the composition of the maternal blood is maintained within narrow limits, and the fetus is able to derive essential nutrients and remove its excretory products in a remarkably sheltered situation. The concept that the fetus constitutes a very successful "parasite" carries with it the idea that the fetus can obtain whatever it needs for normal growth and development even at the risk of causing maternal deprivation. If this were true, the status of the mother, short of catastrophic disease, would be expected to have little effect on the fetus. However, as we have seen in the preceding section, the fetus is by no means independent of maternal welfare. This conclusion is reinforced by observations on pregnancies in which maternal homeostasis is disturbed, resulting in alteration(s) of the composition of the maternal blood which, in turn, alters the supply of nutrients available to the fetus.

Under normal circumstances, the fetus derives adequate nutrients from the mother who derives her own supply from environmental sources and/or mobilization from her own tissues. Under conditions where maternal blood levels of specific nutrients cannot be maintained (e.g., lack of environmental source and/or depletion of mobilizable reserves) the fetus will then share in the malnutrition. The effect on the fetus will then depend on how essential particular nutrient(s) are to the fetus at the particular stage in its development when nutrient deprivation occurs. Animal studies have demonstrated that deficiencies of specific nutrients in the maternal diet can have marked effects on litter size, birth weight, incidence of congenital anomalies, and postnatal survival rates and growth patterns. Of particular significance is the finding that maternal malnutrition at critical periods may result in decreased brain weight, brain DNA and RNA content, alterations in myelogenesis and reduction in the number of neurons. Discussion of this important topic is beyond the scope of this article and the reader is referred to the comprehensive monograph by Wilson (1973).

As emphasized in the preceding section, alterations in the efficiency of oxygenation of the fetus exert considerable influence on the well being of the fetus and its growth. Apart from disturbances in fetal oxygenation resulting from circulatory alterations discussed above, changes in blood composition can also affect fetal oxygenation by influencing the oxygen carrying capacity of the blood or its oxygen content. The most common example is maternal anemia where a reduction in the red blood cell concentration reduces the oxygen content in the blood perfusing the intervillous space. However, the impact of maternal anemia on the fetus appears to be of limited effect, unless the anemia has reached extremes. This can be explained by the fact that in the presence of low hemoglobin concentrations, the pO_2 in the blood of the intervillous space is actually higher than when the same amount of oxygen is carried by a larger quantity of hemoglobin. Since the oxygen tension, as well as the quantity of oxygen, determines the amount transferred to the fetus, it is clear that maternal anemia carries less risk to the fetus than maternal hypoxemia. Furthermore, the viscosity of the maternal blood is decreased in anemia so that the blood flow through the intervillous space will, in fact, be increased at any given perfusion pressure.

6.5. Maternal factors

There are a large number of maternal factors which make an extremely important contribution to placental dysfunction or insufficiency. The maternal factors described in the literature range from biologically sound correlates such as maternal hypertensive disease discussed above to statistical correlates to less well defined factors such as socio-economic status. The role of maternal factors in fetal development have been extensively reviewed elsewhere (Brent and Jensh, 1967) and will only be mentioned briefly here. Furthermore, all maternal factors that lead to fetal problems are not necessarily mediated by placental insufficiency as even a secondary factor.

Fetal pathology and placental pathology may occur in a high incidence of pregnancies with a particular maternal disease (diabetes, phenylketonuria, sickle cell disease) or may only be recognized by analyzing large numbers of pregnancies because the effect is so minimal (smoking). In analyzing the literature dealing with maternal disease and maternal factors in pregnant women, it is obvious that most studies were undertaken to determine the impact of the disease on the pregnant woman and the viability of the offspring (Brent and Jensh, 1967). Few articles deal with the quality of the offspring and even less with the functional status of the pregnancy. Table 8 summarizes the various maternal factors reported to influence growth of the human fetus and their possible relationship to placental insufficiency.

We might focus on one maternal factor, namely smoking, as a prototype of our present understanding of a factor that is statistically correlated with fetotoxic effects but whose etiological relationship is controversial. The literature dealing with this matter is reviewed in the publication entitled "Smoking and Pregnancy" (U.S. Department of Health, Education and Welfare, 1973) and the article by Brent and Jensh (1967). Epidemiological studies indicate that smoking mothers have a higher incidence of fetal growth retardation, stillbirth, late fetal and neonatal deaths. Furthermore, the epidemiological evidence is such that

Table 8
Maternal disease and maternal factors that may or may not interfere with placental function in man.

Placental insufficiency	Possible placental insufficiency	Questionable placental insufficiency
Eclampsia	Small maternal heart volume	Maternal convulsive disorder
Toxemia	Small maternal size and stature	Contribution of prenatal care
Maternal hypertension		
Sickle cell disease	Increased maternal age and parity	Attempted abortion
Diabetes	Low socioeconomic class	Illegitimacy
Severe anemia	Smoking	Season
	Chronic alcoholism	Climate
	Drug addiction	Temperature
	Malnutrition	Amount of daylight
	Chronic maternal disease	

cigarette smoking appears to be etiologically related to the embryopathic results (U.S. Department of Health, Education and Welfare, 1973). It is of interest, however, that smoking mothers have a lower incidence of toxemia of pregnancy. This might indicate that the main effect of smoking is not on the placenta but directly on the developing fetus.

The mechanism of cigarette smoking induced pathology has been related to the following possibilities, either alone or in combination: decreased placental perfusion; toxic effect on the placenta; decreased maternal food intake; direct effect on the embryo; increased uterine activity; and interference with vitamin C and/or vitamin B metabolism. With such a complex list of potential etiologies one can understand why in vitro placental perfusion experiments or embryo culture experiments will not give a definitive answer. We know that nicotine can affect physiological activity of the fetus by affecting fetal breathing movements (Manning et al., 1975) and altering the fetal heart rate (Cloeren et al., 1974), but as yet, no experimental model can tell us the primary mechanism responsible for the increased occurrence of embryopathy in pregnancies of smoking mothers. Analysis of almost all so called maternal factors is beset with similar problems. In some instances an etiologic relationship can be demonstrated, but in most instances the exact mechanism remains elusive.

6.6. The placenta and teratogens

Most teratogenic or embryopathic agents also produce fetal growth retardation and a reduction in the size of the chorioallantoic placenta. The latter observation is not surprising since placental weight is directly correlated with fetal weight at term. It is thus often difficult to determine whether the reduction in placental size seen after exposure to teratogens is a primary or secondary effect. In most instances, it is secondary. The only exception to the general rule that the placenta is not often involved in teratogenesis concerns the inverted yolk sac placenta of rodents and rabbits. Trypan blue (Gillman et al., 1948; Beck and Lloyd, 1966; Lloyd and Beck, 1968) and some tissue antisera (Brent et al., 1961; Brent, 1966, 1967, 1971; Brent and Johnson, 1967; Jensen et al., 1975) have produced congenital malformations. Studies of antisera against yolk sac or placenta have also demonstrated a decrease in transport of macromolecules in vivo (Kobrin and Brent, 1973) or in vitro (Goetze et al., 1975). Few data are available concerning the effects of these antisera on the passage of small molecules. Perhaps there may be some selectivity of antisera action on large molecule transfer, e.g. on the pinocytotic processes. Although many cytotoxic drugs may interfere with placental function, their major effect appears to be on the cells in the embryo (Shepard, 1973).

The placenta does, however, play an important role in determining whether teratogenic agents can gain access to the embryo in sufficient concentration for teratogenic effects to occur. The concentration of a teratogen in the maternal blood is the major rate determining factor in placental transport, and it must be assumed that all drugs or chemicals present in the maternal circulation can cross the placenta by one mechanism or another. The critical question concerns the rate and the duration of transfer, since these will determine the final concentration of teratogen(s) to which embryonic tissues are exposed. Xenobiotics are assumed in most cases to cross the placenta by simple diffusion except where they are analogues of natural metabolites in which case they probably cross by the same mechanism as the natural substances they resemble.

As indicated throughout this article, the concentration of molecules in the maternal plasma is by no means the only determinant of the final dose of material which will reach the embryo or the fetus. For example, in the case of teratogenic drugs, many of these agents also affect the blood supply to the utero-placental unit. Finally, certain potential teratogens may be rendered harmless by chemical modification within the placenta (see Juchau, 1971, 1973; Juchau et al., 1974; Mirkin, 1974, 1975). Biotransformation of potentially embryopathic drugs and chemicals within the placenta must be better understood in order to more accurately predict the potential hazard (or lack of) to the fetus of new classes of chemicals which are constantly being introduced in our environment.

We also need to know more concerning the factors governing placental transfer of xenobiotics, particularly during the early stages of pregnancy when organogenesis is proceeding and the embryo is most susceptible to teratogenic insult. Although the large body of research reviewed in this chapter provides a useful framework of knowledge on placental transport on which to base further studies, much of the existing data are of little value in defining general principles for the transport of teratogens. The majority of studies on placental transport have been done in the latter stages of pregnancy after the period of highest teratogenic susceptibility has passed. It is also unfortunate that much of the work directly concerned with evaluating the teratogenicity of drugs or chemicals has been done with rodents and lagomorphs, since these species may have completely different placental transport activities from man due to the well developed yolk sac placenta in these species. The question of extrapolating data gained in these species to define teratologic hazard(s) in man must therefore be treated with considerable caution.

The phylogenetic relatedness of the simian primates to man has been advanced as a justification that they might be a more reliable test system for evaluating material(s) which could be teratogenic in man. Improvement in teratogenic testing technics, whether they be in vivo or in vitro will only develop as the mechanisms of teratogenesis are better understood. The transport studies reviewed in this chapter is one area of knowledge that must be expanded in order to establish the mechanisms of teratogenesis in man and lower species.

Acknowledgements

The authors wish to express their gratitude to Doctors Barbara M. Davis, Gordon Power, David Gusseck, William O. Berndt, Henry A. Thiede, and William Van Huysen for their critical review of this chapter.

The photomicrographs were prepared by Ms. Linda Biddle. The excellent typing assistance of Ms. Linda Mohan is gratefully acknowledged.

The work of the authors discussed in this chapter was supported in part by the following grants from the National Institutes of Health: AM13020, HD06360, HD370, HD630, and GRSG # RR-05403.

References

Adam, P.A.J., King, K.C., Schwartz, R. and Teramo, K., (1972) Human placental barrier to ^{125}I-glucagon early in gestation. J. Clin. Endocrin. Metabol. 34, 772–782.

Adamsons, K. and Myers, R.E. (1975) Circulation in the intervillous space; obstetrical considerations in fetal deprivation. In: The Placenta (Gruenwald, P., ed.) pp. 158–177, University Park Press, Baltimore.

Adamsons, K., Mueller-Heubach, E. and Myers, R.E. (1971) Production of fetal asphyxia in the rhesus monkey by administration of catecholamines to the mother. Am. J. Obstet. Gynecol. 109, 248–257.

Aherne, W. and Dunnill, M.S. (1966) Morphology of human placenta. Brit. Med. Bull. 22, 5–8.

Alexander, D.P., Andrews, R.D., Huggett, A. St. G., Nixon, D.A. and Widdas, W.F. (1955) The placental transfer of sugars in the sheep: the influence of concentration gradient upon the rates of hexose formation as shown in umbilical perfusion of the placenta. J. Physiol. 129, 367–383.

Allison, A.C. and Davies, P. (1974) Mechanisms of endocytosis and exocytosis. In: Transport at the Cellular Level (Sleich, M.A. and Jennings, D.H., eds.) pp. 419–446, Cambridge University Press, Cambridge.

Almond, C.H., Boulos, B.M., Davis, L.E. and MacKenzie, J.W. (1970) New surgical technique for studying placental transfer of drugs in vivo. J. Surg. Res. 10, 7–11.

Alvares, A.P. and Kappas, A. (1975) Induction of aryl hydrocarbon by polychlorinated biphenyls in the foeto-placental unit and neonatal livers during lactation. FEBS Letters 50, 172–174.

Alvarez, H., Medrano, C.V., Sala, M.A. and Benedetti, W.L. (1972) Trophoblast development gradient and its relationship to placental hemodynamics Part II: Study of fetal cotyledons from the toxemic placenta. Am. J. Obstet. Gynecol. 114, 873–878.

Anderson, J.W. (1959) The placental barrier to gamma-globulins in the rat. Am. J. Anat. 104, 403–430.

Armstrong, W.D., Singer, L. and Makowski, E.L. (1970) Placental transfer of fluoride and calcium. Am. J. Obstet. Gynecol. 107, 432–434.

Ashley, C.A. (1965) Study of the placenta with the electron microscope. Arch. Path. 80, 377–393.

Asling, J.H., Shnider, S.M., Margolis, A.J., Wilkinson, G.L., and Way, E.L. (1970) Para-cervical block anesthesia in obstetrics. Am. J. Obstet. Gynecol. 107, 626–634.

Assali, N.S., Dilts, Jr., P.V., Plentl, A.A., Kirschbaum, T.H. and Gross, S.J. (1968) Physiology of the placenta. In: Biology of Gestation (Assali, N.S., ed.) Vol. 1, pp. 185–289, Academic Press, New York.

Aviolo, L.V. (1972) Intestinal absorption of calcium, J.A.M.A., 219, 345–355.

Bashore, R.A., Smith, F. and Schenker, S. (1969) Placental transfer and disposition of bilirubin in the pregnant monkey. Am. J. Obstet. Gynecol. 103, 950–959.

Battaglia, F.C., Meschia, G., Blechner, J. and Barron, D.H. (1962) A method for the in vitro study of human chorion as a membrane system. Nature 196, 1061–1063.

Battaglia, F.C., Behrman, R.E., Meschia, G., Seeds, A.E. and Bruns, P.D. (1968) Clearance of inert molecules, Na and Cl ions across the primate placenta. Am. J. Obstet. Gynecol. 102, 1135–1141.

Beck, F. and Lloyd, J.B. (1966) The teratogenic effects of azo dyes. In: Advances in Teratology (Woollam, D.H.M., ed.) Vol. 1. pp. 131–193, Academic Press, New York.

Behrman, R.E. and Battaglia, F.C. (1967) Protein binding of human fetal and maternal plasmas to salicylate. J. Appl. Physiol. 22, 125–130.

Benirschke, K. and Driscoll, S.G. (1967) The Pathology of the Human Placenta. Springer-Verlag, New York.

Bierer, D.W. and Heisey, S.R. (1972) Organic anion transport in the maturing dog choroid plexus. Brain Res. 46, 113–119.

Biggs, R. and Rose, E. (1947) The familial incidence of adrenal hypertrophy and female pseudohermaphroditism. J. Obstet. Gynecol. Brit. Emp. 54, 369–374.

Björkman, N. (1970) An Atlas of Placental Fine Structure. Williams and Wilkins Co., Baltimore.

Bloomfield, D.K. (1970) Fetal deaths and malformations associated with the use of coumarin derivatives in pregnancy. Am. J. Obstet. Gynecol. 107, 883–888.

Borst, R.H., Kussather, E. and Schumann, R. (1973) Ultrastrukturelle Untersuchungen zuer Verteilung der alkalischen Phosphatase in Plazenten (materno-fetale Stromungseinheit) der menschlichen Plazenta. Arch. Gynäk. 215, 409–419.

Borrell, U., Fernstrom, I., Ohlson, L. and Wiqvist, N. (1965) Influence of uterine contractions on the uteroplacental blood flow at term. Am. J. Obstet. Gynecol. 93, 44–48.

Bothwell, T.H., Pribella, W.F., Mebust, W. and Finch, C.A. (1958) Iron metabolism in the pregnant rabbit: iron transport across the placenta. Am. J. Physiol. 193, 615–631.

Brambell, F.W.R. (1970) The Transmission of Passive Immunity From Mother to Young. North-Holland, Amsterdam.

Brambell, F.W.R., Hemmings, W.A. and Henderson, M. (1951) Antibodies and Embryos. The Athlone Press, London.

Brambell, F.W.R., Hemmings, W.A., Oakley, C.L. and Porter, R.R. (1960) The relative transmission of the fractions of papain hydrolyzed homologous gamma-globulin from the uterine cavity to the foetal circulation in the rabbit. Proc. R. Soc. Lond. 151, 478–489.

Brame, R.G. (1972) Quantitation of transport of sodium across the amnichorion. Am. J. Obstet. Gynecol. 113, 1085–1089.

Branda, L.A., Vaillancourt, P., and Kominkova, E. (1973) Effect of oxytocin on the perfused human placenta in vitro. Am. J. Obstet. Gynecol. 117, 1116–1125.

Brent, R.L. (1966) Immunologic aspects of developmental biology. In: Advances in Teratology (Woollam, D.H.M., ed.) Vol. I. pp. 82–129, Academic Press, New York.

Brent, R.L. (1967) The production of congenital malformations using tissue antisera. III. Placental antiserum. Proc. Soc. Exp. Biol. Med. 125, 1024–1029.

Brent, R.L. (1971) The effect of immune reactions on fetal development. In: Advances in the Biosciences (Raspe, G., ed.) Vol. VI. pp. 151–159, Pergamon Press, Oxford.

Brent, R.L. and Franklin, J.B. (1960) Uterine vascular clamping: New procedures for the study of congenital malformations. Science 132, 89–91.

Brent, R.L. and Jensh, R.J. (1967) Intrauterine growth retardation. In: Advances in Teratology (Woollam, D.H.M., eds.) Vol. II. pp. 107–169, Academic Press.

Brent, R.L. and A. Johnson (1967) The production of congenital malformations using tissue antisera. VI. Yolk sac. Fed. Proc. 26, 701–709.

Brent, R.L., Averich, E. and Drapiewski, V. (1961) Production of congenital malformations using tissue antisera. I. Kidney antisera. Proc. Soc. Exp. Biol. Med. 106, 523–526.

Brent, R.L., Johnson, A.J. and Jensen, M. (1971) The production of congenital malformations using tissue antisera. VII. Yolk sac antiserum. Teratology 4, 255–276.

Brezenski, J.J., Canazza, J. and Li, J. (1971) Role of placenta in fetal lipid metabolism. III. Formation of rabbit plasma phospholipids. Biochim. Biophys. Acta 239, 92–97.

Bruns, R.R. and Palade, G.E. (1968) Studies on blood capillaries. 1. General organization of blood capillaries in muscle. J. Cell Biol. 37, 244–276.

Burns, B. and Gurtner, G.H. (1974) A specific carrier for oxygen and carbon monoxide in the lung and placenta. Drug Metabol. Disp. 1, 374–379.

Butt, J.H. II, and Wilson, T.H. (1968) Development of sugar and amino acid transport by intestine and yolk sac of the guinea pig. Am. J. Physiol. 215, 1468–1477.

Butterfield, L.J., and O'Brien, D. (1963) The effect of maternal toxaemia and diabetes on transplacental gradients of free amino acids. Arch. Dis. Child. 38, 326–328.

Calkins, L.A. (1937) Placental variation: An analytical determination of its clinical importance. Am. J. Obstet. Gynecol. 33, 280.

Campling, J.D. and Nixon, D.A. (1954) The inositol content of foetal blood and foetal fluids. J. Physiol. 126, 71–80.

Carakushansky, G., Cardenas, L.E., and Gardner, L.I. (1969) Transplacental passage and persistence of iophenoxic acid (Teridax) in a child. Pediatrics 44, 1020–1021.

Carter, A.M. and Olin, T. (1973) Variation in the effect of acetylcholine on myometrial activity and maternal placental blood flow in the rabbit. J. Reprod. Fertil. 35, 73–80.

Casley-Smith, J.R. and Chin, J.C. (1971) The passage of cytoplasmic vesicles across endothelial and mesothelial cells. J. Microscopy 93, 167–189.

Cedard, L. (1972) Placental perfusion in vitro. Acta Endocrinol. suppl. 158, 331–343.

Cedard, L. and Alsot, E. (1975) Perfusion of placenta in vitro. In: Methods in Enzymology (Hardman, J.G. and O'Malley, B.W., eds) Vol. 39, pp. 244–252, Academic Press, New York.

Chen, C.H., Klein, D.C. and Robinson, J.C. (1974) Catechol-O-methyltransferase in rat placenta, human placenta, and choriocarcinoma. J. Reprod. Fertil. 39, 407–410.

Chez, R.A., Mintz, D.H., Reynolds, W.A. and Hutchinson, D.L. (1975) Maternal-fetal plasma glucose relationship in late monkey pregnancy. Am. J. Obstet. Gynecol. 121, 938–940.

Chinard, F.P., Danesino, V., Hartman, W.L., Huggett, A.St.G., Paul, W. and Reynolds, S.R.M. (1956) The Transmission of hexoses across the placenta in the human and the Rhesus monkey. J. Physiol. 132, 289–303.

Chow, B.F. and Okuda, K. (1960) Transfer of vitamins from mother to fetus. J.A.M.A. 172, 422–427.

Christensen, H.N. (1975) Biological Transport, 2nd edition, W.A. Benjamin, Inc. Reading, Massachusetts.

Christensen, H.N. and Streicher, J.A. (1948) Association between rapid growth and elevated cell concentrations of amino acids. I. In foetal tissues, J. Biol. Chem. 175, 95–100.

Cittadini, A., Scarpa, A. and Chance, B. (1973) Calcium transport in intact ehrlich ascites tumor cells. Biochim. Biophys. Acta 291, 246–259.

Clark, C.C., Tomichek, E.A., Koszalka, T.R., Minor, R.R. and Kefalides, N.A. (1975) The embryonic rat parietal yolk sac. The role of the parietal endoderm in the biosynthesis of basement membrane collagen and glycoprotein in vitro. J. Biol. Chem. 250, 5259–5267.

Clarkson, T.W. and Lindemann, B. (1969) Experiments on Na^+ transport of frog skin epithelium. In Laboratory Techniques in Membrane Biophysics, (Passow, H. and Stömpfli, R., eds.), pp 85–105. Springer-Verlag Berlin.

Clemetson, C.A.B. and Churchman, J. (1954) The placental transfer of amino acids in normal and toxaemic pregnancy. J. Obstet. Gynec. Brit. Commonw. 61, 364–371.

Clifford, S.H. (1954) Postmaturity with placental dysfunction. J. Pediat. 44, 1–13.

Clifford, S.H., Reid, D.E. and Worchester, J., (1951) Postmaturity. Am. J. Dis. Child. 82, 232.

Cloeren, S.E., Lippert, T.H. and Fridrich, R. (1974) The influence of cigarette smoking on fetal heart rate and uteroplacental blood volume. Arch. Gynäk. 216, 15–22.

Cole, S.W. and Hitchcock, M.W.S. (1946) Sugars in the foetal and maternal bloods of sheep. Biochem. J. 40.

Coleman, J.R. and Terepka, A.R. (1972) Fine structural changes associated with the onset of calcium, sodium and water transport by the chick chorioallantoic membrane. J. Membrane Biol. 7, 111–127.

Colwill, J.R., Davis, J.R., Meschia, G., Makowski, E.L., Beck, P. and Battaglia, F.C. (1970) Insulin-induced hypoglycemia in the ovine fetus in utero. Endocrinology 87, 710–715.

Cooke, W.J. and Robinson, J.D. (1971) Factors influencing calcium movements in rat brain slices. Am. J. Physiol. 221, 218–225.

Cordero, L., Yeh, S.Y., Grunt, J.A. and Anderson, G.G. (1970) Hypertonic glucose infusion during labor. Am. J. Obstet. Gynecol., 107, 295–302.

Cox, L.W. and Chalmers, T.A. (1953a) The transfer of sodium across the human placenta determined by Na^{24} tracer methods. J. Obstet. Gynecol. Brit. Commonw. 60, 203–213.

Cox, L.W. and Chalmers, T.A. (1953b) The effect of pre-eclampsia and toxaemia on the exchange of sodium in the body and the transfer of radius across the placenta measured by Na^{24} tracer methods. J. Obstet. Gynecol. Brit. Commonw. 60, 214–221.

Crane, R.K. (1960) Intestinal absorption of sugars. Physiol. Rev. 40, 789–825.

Crawford, J.D. and McCance, R.A. (1960) Sodium transport by the chorioallantoic membrane of the pig. J. Physiol. 151, 458–471.

Crawford, J.S. (1965) Maternal and cord blood at delivery. Biol. Neonate 8, 222–237.

Crumpler, H.R., Dent, C.E. and Lindan, O. (1950) The amino acid pattern in human foetal and maternal plasma at delivery. Biochem. J. 47, 223–227.

Crooks, R.J. and Simkiss, K. (1975) Calcium transport by the chick chorioallantois in vivo. Quart. J. Exptl. Physiol. 60, 55–63.

Dancis, J. (1975) Placental transfer studied in a perfusion system. In: Perinatal Pharmacology: Problems and Priorities (Dancis, J. and Hwang, J.C., eds.), pp. 101–108, Raven Press, New York.

Dancis, J. and Balis, M.E. (1954) The reutilization of nucleic acid catabolites. J. Biol. Chem. 207, 367–376.

Dancis, J. and Money, W.L. (1960) Transfer of sodium and iodo-antipyrine across guinea pig placenta with an in situ perfusion technique. Am. J. Obstet. Gynecol. 80, 215–220.

Dancis, J. and Schneider, H. (1975) Physiology: transfer and barrier function. In: The Placenta (Gruenwald, P., ed.), pp. 99–124. University Park Press, Baltimore.

Dancis, J., and Shafner, M. (1958) The origin of plasma proteins in the guinea pig fetus. J. Clin. Invest. 37, 1093–1096.

Dancis, J., Olsen, G. and Folkart, G. (1958) Transfer of histidine and xylose across the placenta and into the red blood cell and amniotic fluids. Am. J. Physiol. 194, 44–52.

Dancis, J., Money, W.L., Springer, D. and Levitz, M. (1968) Transport of amino acids by placenta. Am. J. Obstet. Gynecol. 101, 820–829.

Daubenfield, O., Modde, H. and Hirsch, H.A. (1974) Transfer of gentamicin to the foetus and the amniotic fluid during steady state in the mother. Arch. Gynäk. 217, 233–240.

Davies, J. and Glasser, S.R. (1968) Histological and fine structure observations on the placenta of the rat. Acta Anat. 69, 542–608.

Davis, B.M., Koszalka, T.R. and Miller, R.K. (1973) Transport of creatine from mother to fetus in the rat. Pharmacologist, 15, 199.

Dawes, G.S. (1968) Fetal and Neonatal Physiology. Year Book Publications, Chicago.

Dawson, E.B., Croft, H.A., Clark, R.R. and McGanity, W.J. (1969) Study of nine cation levels in term placentas. Am. J. Obstet. Gynecol. 103, 1144–1147.

Dayton, P.G., Pruitt, A.W., McNay, J.L. and Steinhorst, J. (1972) Studies with triamterene, a substituted pteridine. Unusual brain to plasma ratio in mammals. Neuropharm. 11, 435–446.

Delivoria-Papadopoulos, M., Battaglia, F.C., Bruns, P.D. and Meschia, G. (1967) Total, protein-bound, and ultrafiltrable calcium in maternal and fetal plasmas. Am. J. Physiol. 213, 363–366.

Deren, J.J., Padykula, H.A. and Wilson, T.H. (1966) Development of structure nd function in the mammalian yolk sac. II. Vitamin B_{12} uptake by rabbit yolk sacs, Dev. Biol. 13, 349–369.

Diem, K. and Lentir, C. (1970) Scientific Tables, 7th ed., Ciba-Geigy Ltd. Basle.

Driscoll, S.G. (1975) Placental manifestations of malformation and infection. In: The Placenta (Gruenwald, P., ed.) pp. 244–259, University Park Press, Baltimore.

Du Pam, R.M., Wenger, P., Koechli, S., Scheidegger, J.J. and Roux, J. (1958) Etude du passage de la γ-globuline marquée a travers le placenta humain. Clin. Chim. Acta 4, 110–115.

East, J.M., Chepenik, K.P. and Waite, B.M. (1975) Phospholipase A activities in rat placentas of 14 days gestation. Biochim. Biophys. Acta 388, 106–112.

Edlow, J.B., Huddleston, J.F., Lee, G., Peterson, W.F. and Robinson, J.C. (1971) Placental enzymes: specific activities and isoenzyme patterns during early and late gestation. Am. J. Obstet. Gynecol. 111, 360–364.

Eliason, B.C. and Posner, H.S. (1971) Placental passage of ^{14}C-dieldrin altered by gestational age and plasma proteins. Am. J. Obstet. Gynecol. 111, 925–929.

Eliasson, R. and Astrom, A. (1955) Pharmacological studies on the perfused human placenta. Acta Pharmacol. Toxicol. 11, 254.

Ely, P.A. (1966) The placental transfer of hexoses and polyols in the guinea pig, as shown by umbilical perfusion of the placenta. J. Physiol. 184, 255–271.

Everitt, G.C. (1964) Maternal undernutrition and retarded foetal development in merino sheep. Nature, 201, 1341–1342.

Faber, J.J., Green, T.J. and Long, L.R. (1971) Permeability of rabbit placenta to large molecules. Am. J. Physiol. 220, 688–693.

Fantel, A.G. (1975) Fetomaternal potassium relations in the fetal rat on the twentieth day of gestation. Pediat. Res. 9, 527–530.

Flexner, L.B. and Gellhorn, A. (1942) The comparative physiology of placental transfer. Am. J. Obstet. Gynecol. 43, 965–974.

Flexner, L.B., Cowie, D.B., Hellman, L.M., Wilde, W.S. and Vosburgh, G.J. (1948) The permeability of the human placenta to sodium in normal and abnormal pregnancies and the supply of Na^+ to the human fetus as determined with radioactive Na^+. Am. J. Obstet. Gynecol. 55, 469–480.

Foley, T.H., Holm, L.W., London, D.R. and Young, M. (1967) Preliminary observations on free amino acids in cow and calf plasma in prolonged gestation. Am. J. Obstet. Gynecol. 99, 1106–1109.

Folkart, G.R., Dancis, J. and Money, W.L. (1960) Transfer of carbohydrates across guinea pig placenta. Am. J. Obstet. Gynecol. 80, 221–224.

Fox, H. (1970) Effect of hypoxia on trophoblast in organ culture. Am. J. Obstet. Gynecol. 107, 1058–1063.

Fox, H. (1975) Morphological pathology of the placenta. In: The Placenta (Gruenwald, P., ed.) pp. 197–220, University Park Press, Baltimore.

France, J.T. and Downey, J.A. (1974) A study of arylsulfatase activity in children born of pregnancies affected with placental sulfatase deficiency. Biochem. Med. 10, 167–174.

France, J.T., Seddon, R.J. and Liggins, G.C. (1973) A study of a pregnancy with low estrogen production due to placental sulfatase deficiency. J. Clin. Endocrin. Metab. 36, 1–9.

Franklin, J.B. and Brent, R.L. (1964) The effect of uterine vascular clamping on the development of rat embryos three to fourteen days old. J. Morphol. 115, 273–290.

Garrison, J.C. and Terepka, A.R. (1972a) Calcium stimulated respiration and active calcium transport in the isolated chick chorioallantoic membrane. J. Membrane Biol. 7, 128–145.

Garrison, J.C. and Terepka, A.R. (1972b) The interrelationships between sodium ion, calcium transport and oxygen utilization in the isolated chick chorioallantoic membrane. J. Membrane Biol. 7, 146–163.

Gautieri, R.F. and Ciuchta, H.P. (1962) Effect of certain drugs on perfused human placenta. J. Pharm. Sci. 51, 55.

Gazzola, G.C., Franchi-Gazzola, R., Ronchi, R. and Guidotti, G.C. (1973) Regulation of amino acid transport in chick embryo heart cells. III. Formal identification of the A mediation as an adoptive transport system, Biochim. Biophys. Acta 311, 292–301.

Ghadami, H. and Pecora, P. (1964) Free amino acids of cord plasma as compared with maternal plasma during pregnancy. Pediatrics 33, 500–506.

Gillman, J., Gilbert, C., Gillman, T. and Spence, I. (1948) A preliminary report on hydrocephalus, spina bifida and other congenital anomalies in the rat produced by trypan blue. S. Afr. J. Med. Sci. 13, 47–90.

Gitlin, D. (1974) Protein transport across the placenta and protein turnover between amniotic fluid, maternal and fetal circulations. In: The Placenta Biological and Clinical Aspects (Moghissi, K.S., and Hafez, E.S.E., eds.), pp. 151–191, C.C. Thomas, Springfield, Ill.

Gitlin, D. and Biasucci, A. (1969) Development of γG, γA, γM, β1C/β1A, C'1 esterase inhibitor, ceruloplasmin, transferrin, hempexin, haptoglobin, fibrinogen, plasminogen, α_1-antitrypsin, orosomucoid, β-lipoprotein, α_2-macroglobulin and prealbumin in the human conceptus. J. Clin. Invest. 48, 1433–1449.

Gitlin, J.D. and Gitlin, D. (1974) Protein binding by specific receptors on human placenta, murine placenta, and suckling murine intestine in relation to protein transport across these tissues. J. Clin. Invest. 54, 1155–1166.

Gitlin, D. and Koch, C. (1968) On the mechanisms of maternofetal transfer of human albumin and γG globulin in the mouse. J. Clin. Invest. 47, 1204–1209.

Gitlin, D., Kumate, J. and Morales, C., (1965a) On the transport of insulin across the human placenta. Pediatrics 35, 65–69.

Gitlin, D., Kumate, J. and Morales, C. (1965b) Metabolism and maternofetal transfer of human growth hormone in the pregnant woman at term. J. Clin. Endocr. Metab. 25, 1599–1608.

Glendening, M.B. (1965) Transfer of infused D- and L-amino acids from mother to baby, Transcript of the Third Rochester Conference (Lund, C.J. and Thiede, H.A., eds.), pp. 280–297.

Glendening, M.B., Margolis, A.J., and Page, E.W. (1961) Amino acid concentrations in foetal and maternal plasma. Am. J. Obstet. Gynecol. 81, 591–595.

Goetze, T., Franke, H., Oswald, B., Schlag, B. and Goetze, E. (1975) Effects of goat anti-rat placenta IgG on the in vitro uptake of ^{125}I-labeled human serum albumin by the rat visceral yolk sac. Biol. Neonate, 27, 221–231.

Goldstein, A., Aronow, L. and Kalman, S.M. (1974) Principles of Drug Action. Hoeber Medical Division, New York.

Goodfriend, M.J., Shey, I.A. and Klein, M.D. (1956) The effects of maternal narcotic addiction on the newborn. Am. J. Obstet. Gynecol. 71, 29–36.

Graber, S.E., Scheffel, U., Hodkinson, B. and McIntyre, P.A. (1971) Placental transport of vitamin B_{12} in the pregnant rat. J. Clin. Invest. 50, 1000–1004.

Grauwiler, J. and Schon, H. (1973) Teratological experiments with ergotamine in mice, rats, and rabbits. Teratology, 7, 227–235.

Greenwald, P., Barlow, J.J., Nasca, P.C. and Burnett, W.S. (1971) Vaginal cancer after maternal treatment with synthetic estrogens. New Eng. J. Med. 285, 390–392.

Griess, F.C. and Gobble, F.L. (1967) Effect of sympathetic nerve stimulation on the uterine vascular bed. Am. J. Obstet. Gynecol. 97, 962–968.

Grosser, O. (1909) Vergleichende Anatomie und Entwicklungsgeschichte der Eihaute und der Placenta mit Besonderer Berucksichtigung des Menschen. Wilhelm Braumuller, Wien und Leipzig.

Gruenwald, P. (1961) Abnormalities of placental vascularity in relation to intrauterine deprivation and retardation of fetal growth; significance of avascular chorionic villi. New York State J. Med. 61, 1508–1513.

Gruenwald, P. (1962) Hypoxia of the fetus and newborn infant, Acute and Chronic Fetal Distress. Chicago Med. 64, 13–14.

Gruenwald, P. (1964) Infants and low birth weight among 5,000 deliveries. Pediatrics 34, 157–162.

Gruenwald, P. (1974) Placental insufficiency – a questionable concept. Arch. Dis. Child. 49, 915–916.

Gruenwald, P. (1975) ed., The Placenta, University Park Press, Baltimore.

Gruenwald, P., Dawkins, M. and Hepner, R. (1963) Chronic deprivation of the fetus. Sinai Hosp. J. (Balt.) 11, 51.

Grumbach, M.M. and Ducharme, J.R. (1960) The effect of androgens on fetal sexual development; androgen-induced female pseudohermaphroditism. Fertil. Steril. 11, 157–180.

Gusseck, D.J., Yuen, P. and Longo, L.D. (1975) Amino acid transport in placental slices, Mechanisms of increased accumulation by prolonged incubation, Biochim. Biophys. Acta 401, 278–284.

Hafez, E.S.E. (1963) Maternal influence on fetal size. Int. J. Fertil. 8, 547–553.

Halpern, J. (1974) Some aspects of organocobalt chemistry related to vitamin B_{12}. Ann. N.Y. Acad. Sci. 239, 2–21.

Harbison, R.D., Olubodewo, J., Dwivedi, C. and Sastry, B.V.R. (1975) Proposed role of the placental cholinergic system in the regulation of fetal growth and development. In: Basic and Therapeutic Aspects of Perinatal Pharmacology (Morselli, P.L., Garattini, S. and Sereni, F., eds.) pp. 107–120, Raven Press, New York.

Hayter, C.J., Hutchinson, E.A., Karvonen, M.J. and Young, M. (1964) Placental transport of aminoisobutyric acid in the unanesthetized guinea pig. J. Physiol. 175, 11–13.

Heinz, E. (1972) (ed.) Na^+-Linked Transport of Organic Solutes. Springer-Verlag, Berlin.

Hemmings, W.A., Jones, R.E. and Faulk, W.P. (1975) Transport across the rabbit foetal yolk sac of fractions of IgG from several mammalian species. Immunology 28, 411–418.

Hendricks, C.H. (1964) Patterns of fetal and placental growth in the second half of normal pregnancy. Obstet. Gynecol. 24, 357–365.

Herbst, A.L., Ulfelder, H. and Poskanzer, D.C. (1971) Adenocarcinoma of the vagina. New Eng. J. Med. 284, 878–881.

Hertig, A.T. (1968) Human Trophoblast. C.C. Thomas, Springfield, Ill.

Hill, D.E., Myers, R.E., Holt, A.B., Scott, R.E. and Cheek, D.B. (1971) Fetal growth retardation produced by experimental placental insufficiency in the rhesus monkey. II. Chemical composition of the brain, liver, muscle, and carcass. Biol. Neonate 19, 68–82.

Hill, E.P., Power, G.G. and Longo, L.D. (1972) A mathematical model of placental O_2 transfer with consideration of hemoglobin reaction rates. Am. J. Physiol. 222, 721–729.

Hill, P.M.M. and Young, M. (1973) Net placental transfer of free amino acids against varying concentrations. J. Physiol. 235, 409–422.

Hodari, A.A., Mariona, F.G., Houlihan, R.T. and Peralta, J. (1973) Creatinine transport in the maternal-fetal complex. Obstet. Gynec. 41, 47–55.

Holdsworth, C.D. and Wilson, T.H. (1967) Development of active sugar and amino acid transport in the yolk sac and intestine of chicken. Am. J. Physiol. 212, 233–239.

Holmberg, N.G., Kaplan, B., Karvonen, M.J., Lind, J. and Malm, M. (1956) Permeability of human placenta to glucose, fructose, and xylose. Acta Physiol. Scand. 36, 291–299.

Holsen, J.F., Jr. and Wilson, J.G. (1974) Changing placental roles of the rodent chorioallantoic and yolk sac placentae during organogenesis. Teratology, 9, a21.

Honey, D.P., Robson, J.M. and Sullivan, F.M. (1967) Mechanism of inhibitory action of 5-hydroxytryptamine on placental function. Am. J. Obstet. Gynecol. 99, 250–260.

Huggett, A. St. G., Warren, F.L. and Winterton, V.N. (1949) Origin and site of formation of fructose in the foetal sheep. Nature 164, 271.

Hultstaert, C.E., Torringa, J.L., Koudstaal, J., Hardonk, M.J. and Molenaar, I. (1973) The characteristic distribution of alkaline phosphatase in the full term human placenta. An electron cytochemical study. Gynecol. Invest. 4, 24–39.

Iyengar, L. (1973) Chemical composition of placenta in pregnancies with small-for-date infants. Am. J. Obstet. Gynecol. 116, 66–70.

Jensen, M., Koszalka, T.R. and Brent, R.L. (1975) Production of congenital malformations using tissue antisera. XV. Reichert's membrane and visceral yolk sac antisera. Dev. Biol. 42, 1–12.

Johns, A.H. (1965) Placental transfer of atropine and the effect on foetal heart rate. Brit. J. Anaes. 37, 57–60.

Johnson, T. and Clayton, C.G. (1957) Diffusion of radioactive radium in normotensive and pre-eclamptic pregnancies. Brit. Med. J. 1, 312–314.

Jollie, W.P. (1968) Changes in the fine structure of the parietal yolk sac of the rat placenta with increasing gestational age. Am. J. Anat. 122, 513–532.

Jones, H.W. and Scott, W.W. (1958) Hermaphroditism, Genital Anomalies and Related Endocrine Disorders. Williams and Wilkins, Baltimore.

Jost, A. (1973) Hormonal effects on fetal development: A survey. Clin. Pharm. Therap. 14, 714–720.

Jost, A. and Picon, L. (1970) Hormonal control of fetal development and metabolism. Adv. Metab. Dis. 4, 123–156.

Juchau, M.R. (1971) Human placental hydroxylation of 3,4-benzpyrene during early gestation and at term. Toxicol. Appl. Pharmacol. 18, 665–675.

Juchau, M.R. (1973) Placental metabolism in relation to toxicology. CRC Criticial Reviews in Toxicology 2, 125–158.

Juchau, M.R. and Dyer, D.C. (1972) Pharmacology of the placenta. Pediat. Clin. of North America 19, 65–79.

Juchau, M.R., Symms, K.G. and Zachariah, P.K. (1974) Drug-metabolizing enzymes in the placenta. In: Perinatal Pharmacology (Dancis, J. and Hwang, J.C., eds.) pp. 89–110, Raven Press, New York.

Kahn, J.B., Nicholson, D.B. and Assali, N.S. (1953) Placental transmission of thiobarbiturate in parturient women. Obstet. Gynecol. 1, 663–667.

Kaplan, S.L. and Grumbach, M.M. (1965) Serum chorionic "growth-hormone prolactin" and serum pituitary growth hormone in mother and fetus at term. J. Clin. Endocrinol. Metab. 25, 1370–1384.

Kaplan, K.C., Catsoulis, E.A. and Franklin, E.C. (1965) Materno-foetal transfer of human immune globulins and fragments in rabbits. Immunology 8, 354–359.

Kauffman, R.E., Morris, J.A. and Azarnoff, D.L. (1975) Placental transfer and fetal urinary excretion of gentamicin during constant rate maternal infusion. Pediat. Res. 9, 104–107.

Kelly, W.T., Hutchinson, D.H., Friedman, E.A. and Plentl, A.A. (1964) Placental transmission of tritium and carbon-14 labeled histidine enantiomorphs in primates. Am. J. Obstet. Gynecol. 89, 776–787.

Kelman, B.J. and Twardock, A.R. (1975) Calcium, sodium and potassium movements across the perfusing guinea pig placenta. Fed. Proc. 34, 317.

Kernis, M.M. (1971) Abnormal yolk sac function induced by chlorambucil. Experientia 27, 1329–1331.

Kernis, M.M. and Johnson, E.M. (1969) Effects of trypan blue and Niagara blue 2B on the in vitro absorption of ions by the rat visceral yolk sac. J. Embryol. Exp. Morph. 22, 115–125.

Kimmich, G.A. (1973) Coupling between Na^+ and sugar transport in the small intestine. Biochim. Biophys. Acta 300, 31–78.

King, B.F. and Menton, D.N. (1975) Scanning electron microscopy of human placental villi from early and late in gestation. Am. J. Obstet. Gynecol. 122, 824–828.

King, C.K., Adam, P.A.J., Schwartz, R. and Teramo, K. (1971) Human placental transfer of human growth hormone. Pediatrics. 48, 534–540.

Kivalo, I. and Saarikoski, S. (1973) Placental transmission and foetal uptake of ^{14}C-dimethyltubocurarine. Brit. J. Anaes. 44, 557–561.

Klinge, E., Mattila, M.J., Pentilla, O. and Jukarainer, E. (1966) Influence of drugs on vasoactive peptides and amines in perfused human placenta. Ann. Med. Exper. Fenn. 44, 369.

Kloosterman, C.J. and Huidekoper, B.L. (1954) Significance of placenta in "obstetrical mortality", study of 2000 births. Gynaecologia, 138, 529.

Knapowski, J., Feliks, M. and Adam, W. (1972) Badania nad aktywnym transportem sodu w blonach plodowych czlow ieka in vitro. Ginekologia Polska 43, 283–297.

Knobil, E. (1965) Placental amino acid transport in the Rhesus monkey, Transcript of the Third Rochester Conference (Lund, C.J., Thiede, H.A., eds.) 235–251.

Kobrin, L. and Brent, R.L. (1973) Effect of yolk sac antiserum on embryonic parameters of free and bound ^{14}C-leucine. Teratology 7, A-21.

Kortenoever, M.E. (1950) Pregnancy of long duration and post mature infant. Obstet. Gynecol. Survey 5, 812–814.

Koszalka, T.R., Jensh, R. and Brent, R.L. (1972) Creatine metabolism in the developing rat fetus. Comp. Biochem. Physiol. 41B, 217–229.

Koszalka, T.R., Jensh, R.P. and Brent, R.L. (1975) Placental transport of creatine in the rat. Proc. Soc. Exptl. Biol. Med. 148, 864–869.

Krauer, F.J., Joyce, J. and Young, M. (1972) The influence of high maternal plasma glucose levels and maternal blood flow on the placental transfer of glucose in the guinea pig. Diabetologia 9, 453–456.

Kraus, A.P., Perlow, S. and Singer, K., (1949) Danger of dicumarol treatment in pregnancy. J.A.M.A. 139, 758–762.

Kulangara, A.C. and Schechtman, A.M. (1972) Passage of heterologous serum proteins from mother into fetal compartments in the rabbit. Am. J. Physiol. 203, 1071–1080.

Larkin, E.C., Weintraub, L.R. and Crosby, W.H. (1970) Iron transport across rabbit allantoic placenta. Am. J. Physiol. 218, 7–16.

Lassen, J.B. and Nielsen, O.E. (1963) Investigations into the diuretic effect and elimination of triamterene. Acta Pharm. Toxicol. 20, 309–316.

Laurell, C.B. and Morgan, E. (1964) Iron exchange between transferrin and the placenta in the rat. Acta Physiol. Scand. 62, 271–305.

Leist, K.H. and Grauwiler, J. (1973a) Influence of the developmental stage on embryotoxicity following uterine vessel clamping in the rat. Teratology 8, 227–228.

Leist, K.H. and Grauwiler, J. (1973b) Transplacental passage of ^{3}H-ergotamine in the rat, and determination of the intraamniotic embryotoxicity of ergotamine. Experientia, 29, 764–767.
Leist, K.H. and Grauwiler, J. (1974a) Ergometrine and uteroplacental blood supply in pregnant rats. Teratology 10, 316–325.
Leist, K.H. and Grauwiler, J. (1974b) Fetal pathology in rats following uterine-vessel clamping on day 14 of gestation. Teratology 10, 55–67.
Levitz, M. and Dancis, J. (1963) Transfer of steroids between mother and fetus. Clin. Obstet. Gynecol. 6, 62–71.
Levitz, M., Condon, G.P., Dancis, J., Goebelsmann, V., Eriksson, G. and Diczfalusy, E. (1967) Transfer of estriol and estriol conjugates across the human placenta perfused in situ at midpregnancy. J. Clin. Endocrinol. Metabol. 27, 1723–1736.
Lindberg, S. (1963) ^{14}C-Histamine inactivation in vitro by human myometrial and placental tissues. Acta Obstet. Gynecol. Scand. 42, 27.
Lindbland, B.S. and Baldesten, H. (1967) The normal venous plasma free amino acid levels of non-pregnant women and of mother and child during delivery. Acta Paediat. Scand. 56, 37–48.
Lindbland, B.S. and Zetterstrom, R. (1968) The venous plasma free amino acid levels of mother and child during delivery. Acta Paediat. Scand. 57, 195–204.
Lipsitz, P.J., Flaxman, L.M., Tartow, L.R. and Malek, B.K. (1973) Maternal hyperbilirubinemia and the newborn. Am. J. Dis. Child. 126, 525–532.
Litonjua, A.D., Canlos, M., Soliman, J. and Paulino, D.Q. (1967) Uptake of alpha-aminoisobutyric acid in placental slices at term. Am. J. Obstet. Gynecol. 99, 242–246.
Littledike, E.T., Arnaud, C.D. and Whipp, S.C. (1972) Calcitonin secretion in ovine, porcine and bovine fetuses. Proc. Soc. Exp. Biol. Med. 139, 428–433.
Lloyd, J.B. and Beck, F. (1968) Evidence for a mechanism of teratogenesis involving inhibition of embryonic nutrition. Lab. Anim. Care 2, 171–180.
Longo, L.D. (1972) Disorders of placental transfer, In: Pathophysiology of Gestation (Assali, N., ed.) Vol. II, pp. 1–76, Academic Press, New York.
Longo, L.D. and Bartels, H. (1972) Respiratory gas exchange and blood flow in the placenta, U.S. Dept. of Health, Education and Welfare Publication No. (NIH) 73-361, Bethesda, Md.
Longo, L.D., Hill, E.P. and Power, G.G. (1972) Theoretical analysis of factors affecting placental O_2 transfer. Am. J. Physiol. 222, 730–737.
Longo, L.D., Yuen, P. and Gusseck, D.J. (1973) Anaerobic, glycogen-dependent transport of amino acids by the placenta. Nature 243, 531–533.
Lowenstein, L., Lalonde, M., Deschenes, E.B. and Shapiro, L. (1960) Vitamin B_{12} in pregnancy and the puerperium. J. Clin. Nutr. 8, 265–275.
Lubby, A.L., Cooperman, J.M., Stone, M.L. and Slobody, L.B. (1961) Physiology of Vitamin B_{12} in pregnancy, the placenta and the newborn. Am. J. Dis. Child. 102, 753–755.
Lundy, L.E., Wu, C-H and Lee, S.G. (1973) Estrogen assessments in the high risk pregnancy. Clin. Obstet. Gynecol. 16, 279–297.
MacDonald, N.S., Hutchinson, D.L., Hepler, M. and Flynn, E. (1965) Movement of calcium in both directions across the primate placenta. Proc. Soc. Exp. Biol. Med. 119, 476–481.
Maickel, R.P. and Snodgrass, W.R. (1973) Physico chemical factors in maternal-fetal distribution of drugs. Toxicol. Appl. Pharm. 26, 218–230.
Manning, F., Wyn Pugh, E. and Boddy, K. (1975) Effect of cigarette smoking on fetal breathing movements in normal pregnancies. Br. Med. J. i, 552–553.
Mayashi, T.T., Shin, D.H. and Wiand, S. (1968) Placental transfer of orotic acid, uridine and UMP.1. Comparison of acid-soluble and acid-unsoluble counts. Am. J. Obstet. Gynecol. 102, 1144–1149.
McKeown, T. and Gibson, J.R. (1951) Observations on all births (23, 970) in Birmingham, 1947. IV. Premature birth. Brit. Med. J. ii, 513–525.
McKiddie, J.M. (1950) Foetal mortality in postmature. J. Obstet. Gynecol. Brit. Emp. 56, 386.' Abstracted in Obstet. Gynec. Survey, 5, 44.
McLaren, A. (1965) Placental weight loss in late pregnancy. J. Reprod. Fertil. 9, 343–346.
McNay, J.L. and Dayton, P.G. (1970) Placental transfer of a substituted pteridine from fetus to mother. Science, 167, 988–990.
McNay, J.L., Fuller, E., Kishimoto, T., Malveaux, E. and Dayton, P.E. (1969) Lack of activity transfer

of ^{14}C-tetraethylammonium and para-aminohippuric acid by the term sheep placenta. Proc. Soc. Exp. Biol. Med. 131, 51–56.

Mead, P.B. and Marcus, S.L. (1964) Prolonged pregnancy. Am. J. Obstet. Gynecol. 89, 495–502.

Meister, A. (1973) On the enzymology of amino acid transport. Science, 180, 33–39.

Mellor, D.J. (1969) Potential differences between mother and fetus at different gestational ages in the rat, rabbit, and guinea pig. J. Physiol. 204, 395–405.

Mellor, D.J., Cockburn, F., Lees, M.M. and Blogden, A. (1969) Distribution of ions and electrical potential difference between mother and fetus in the human at term. J. Obstet. Gynecol. Brit. Commonw. 76, 993–998.

Meschia, G., Battaglia, F.C. and Bruns, P.D. (1967) Theoretical and experimental study of transplacental diffusion. J. Appl. Physiol. 22, 1171–1178.

Meschia, G., Wolkoff, A.S. and Barron, D.H. (1958) Difference in electrical potential across the placenta of goats. Proc. Nat. Acad. Sci. U.S.A. 44, 483–485.

Meyer, L.M., Gams, R.A., Ryel, E.M., Miller, I.E. and Kumar, S. (1974) Delivery of $^{57}Co-B_{12}$ to lymphoblasts derived from mice with transplanted 1210 ascites tumor cells by transcobalamine I, II, and III. Proc. Soc. Exptl. Biol. Med. 147, 679–680.

Miller, R.K. and Berndt, W.O. (1972) Active uptake of α-aminoisobutyrate (AIB) and passive uptake of p-aminohippurate (PAH) and tetraethylammonium by rabbit and human placentae. Fed. Proc. 31, 595.

Miller, R.K. and Berndt, W.O. (1973) Evidence for a Mg^{++} dependent $Na^{+} + K^{+}$ – activated ATPase and a Ca^{++} ATPase in human term placenta. Proc. Soc. Exptl. Biol. Med. 143, 118–122.

Miller, R.K. and Berndt, W.O. (1974) Characterization of neutral amino acid accumulation by human term placental slices. Am. J. Physiol. 227, 1236–1242.

Miller, R.K. and Berndt, W.O. (1975) Mechanisms of transport across the placenta; an in vitro approach. Life Sci. 16, 7–30.

Miller, R.K., Ferm, V.H. and Mudge, G.H. (1972) Placental transfer and tissue distribution of iophenoxic acid in the hamster. Am. J. Obstet. Gynecol. 114, 259–266.

Miller, R.K., Berndt, W.O., Koszalka, T.R. and Brent, R.L. (1973) The influence of serotonin and inhibitors of transport processes on human placental function in vitro. Teratology, 7, 23.

Miller, R.K., Davis, B.M., Brent, R.L. and Koszalka, T.R. (1974a) Directionality of ^{14}C-creatine transport from mother to fetus in the rat. Fed. Proc. 33, 223.

Miller, R.K., Davis, B.M., Brent, R.L. and Koszalka, T.R. (1974b) Transport of creatine in the human placenta. Pharmacologist, 16, 305.

Miller, R.K., Davis, B.M. and Koszalka, T.R. (1974c) Movement of substances into the feto-placental unit, National Workshop on Teratogenesis and Mutagenesis, Philadelphia.

Miller, R.K., Koszalka, T.R., Davis, B.M., Andrew, C.L. and Brent, R.L. (1974d) Placental transport of creatine and amino acids in the rat: An in vitro study. Teratology, 9, a28.

Miller, R.K., Reich, K.A., Fox, H.E., Davis, B.M., Brent, R.L. and Koszalka, T.R. (1976a) Creatine: transport by human placentae. Gynecol. Invest. 7, 36.

Miller, R.K., Reich, K.A. and Koszalka, T.R. (1976b) Transport of creatine and α-aminoisobutyric acid by the early human placenta. Fed. Proc. 35, 365.

Miller, S.A. and Runner, M.N. (1975) Differential permeability of murine visceral yolk sac to thymidine and to hydroxyurea. Dev. Biol. 45, 74–80.

Mirkin, B.L. (1973) Drug distribution in pregnancy. In: Fetal Pharmacology (Boreus, L., ed.) pp. 1–27, Raven Press, New York.

Mirkin, B.L. (1974) Fetal pharmacology. In: Modern Perinatal Medicine (Louis Gluck, ed.) Chapter 22. Year Book Medical Publishers, Chicago.

Mirkin, B.L. (1975) Perinatal pharmacology: Placental transfer, fetal localization, and neonatal disposition of drugs. Anesthesiology, 43, 156–170.

Miyakawa, I., Ikeda, I. and Maeyama, M. (1974) Transport of ACTH across human placenta. J. Clin. Endocr. Metab. 39, 440–442.

Money, W.L. and Dancis, J. (1960) Technique for the in situ study of placental transport in the pregnant guinea pig. Am. J. Obstet. Gynecol. 80, 209–214.

Moore, W.M.O., Ward, B.S. and Shields, R.A. (1974) Suitability of pig placenta for investigation of placental transfer. Am. J. Obstet. Gynecol. 120, 932–936.

Morgan, C.D., Sandler, M. and Panigel, M. (1972) Placental transfer of catecholamine in vitro and in vivo, Am. J. Obstet. Gynecol. 112, 1068–1075.

Moriaty, C.M. and Hoghen, C.A.M. (1970) Active Na^+ and Cl^- transport by the isolated chick chorioallantoic membrane. Biochim. Biophys. Acta 219, 463–470.

Morishima, H.O., Allen, I.H., Adamsons, K. and James, K.L. (1971) Anesthetic management for fetal surgery in the subhuman primate. Am. J. Obstet. Gynecol. 110, 926–935.

Morphis, L.G. and Gitlin, D. (1970) Maturation of the maternofoetal transport system for human γ-globulin in the mouse. Nature, 228, 573–575.

Mossman, H.W. (1937) Comparative morphogenesis of the fetal membranes and accessary uterine structures. Carnegie Inst. Wash. 262, 130–239.

Mossman, H.W. (1974) Structural changes in vertebrate fetal membranes associated with the adoption of viviparity. In: Obstetrics and Gynecology Annual (Wynn, R.M., ed.) vol. 3, pp. 7–32, Appleton-Century-Crofts, New York.

Mull, J.W. (1936) Variations in serum calcium and phosphorus during pregnancy. J. Clin. Invest. 15, 513–517.

Myant, N.B. (1958) Passage of thyroxine and tri-iodo-thyronine from mother to foetus in pregnant women. Clin. Sci. 17, 75–80.

Myers, R.E. (1972) Two patterns of perinatal brain damage and their conditions of occurrence. Am. J. Obstet. Gynecol. 112, 246–253.

Myers, R.E., Hill, D.E., Holt, A.B., Scott, R.E., Mellits, E.D. and Cheek, D.B. (1971) Fetal growth retardation produced by experimental placental insufficiency in the rhesus monkey. I. Body weight, organ size. Biol. Neonate 18, 379–394.

Nebert, D.W., Winker, J. and Gelboin, H.V. (1969) Aryl hydrocarbon hydroxylase activity in human placenta from cigarette smoking and non-smoking women. Cancer Res. 29, 1763–1767.

Netzloff, M.L., Chepenik, K.P., Johnson, E.M. and Kaplan, S. (1968) Respiration of rat embryos in culture. Life Sci. 7, 401–405.

Newman, R.L. (1957) Serum electrolytes in pregnancy, parturition and puerperium. Obstet. Gynecol. 10, 51–55.

Newmark, P.A. (1972) The mechanism of uptake of Vitamin B_{12} by the kidney of the rat in vivo. Biochim. Biophys. Acta 261, 85–93.

Noer, H.R. and Mossman, H.W. (1947) Surgical investigation of the function of the inverted yolk sac placenta in the rat. Anat. Rec. 98, 31–37.

Oakey, R.E., Cawood, M.L. and MacDonald, R.R. (1974) Biochemical and clinical observations in a pregnancy with placental sulphatase and other enzyme deficiencies. Clin. Endocrinol. 3, 131–148.

Oakley, N.W., Beard, R.W. and Turner, R.C. (1972), Effect of sustained maternal hyperglycaemia on the fetus in normal and diabetic pregnancies. Brit. Med. J. 1, 466–469.

Oh, Y. and Mirkin, B.L. (1971) Transfer of drugs into the central nervous system and across the placenta: A comparative study utilizing aminopyrine (A), diphenylhydantoin (D), sodium salicylate (S) and mecamylamine (M). Fed. Proc. 30, 2034–2037.

Okudaira, Y., Hashimoto, T., Hayakawa, K. and Suzuki, S. (1968) Localization of the adenosine triphosphatase activity in early human placenta. J. Electron Microscop. 17, 342–343.

Owen, J.R., Irani, S.F. and Blair, H.W. (1972) Effect of Diazepam administered to mothers during labour on temperature regulation of the neonate. Arch. Dis. Child. 47, 107–110.

Padykula, H.A., Deren, J.J. and Wilson, T.H. (1966) Development of structure and function in the mammalian yolk sac, III. The development of amino acid transport by rabbit yolk sac. Dev. Biol. 13, 370–384.

Page, E.W., Glendening, M.B., Margolis, A.J. and Harper, H.A. (1957) Transfer of D- and L-histidine across the human placenta. Am. J. Obstet. Gynecol. 73, 589–597.

Page, K.R., Abramovich, D.R. and Smith, M.R. (1974) The diffusion of tritiated water across isolated term human amnion. J. Membr. Biol. 18, 39–48.

Panigel, M. (1975) Electron microscopic studies on placental function and maternal fetal exchange. In: Modern Perinatal Medicine (Gluck, L., ed.) pp. 36–65, Year Book Medical Publishers, Chicago.

Panigel, M. and Myers, R.E. (1972) Histological and ultrastructural changes in rhesus monkey placenta following interruption of fetal placental circulation by fetectomy or interplacental umbilical vessel ligation. Acta Anat. 81, 481–506.

Paterson, P., Phillips, L., and Woods, C. (1967) Relationship between maternal and fetal blood glucose during labor. Am. J. Obstet. Gynecol. 98, 938–945.

Pick, W. (1954) Malnutrition of the newborn secondary to placental abnormalities. New Eng. J. Med. 250, 905–907.

Pitkin, R.M. (1975) Calcium metabolism in pregnancy: a review. Am. J. Obstet. Gynecol. 121, 724–737.

Pitkin, R.M., Connor, W.E. and Lin, D.S. (1972) Cholesterol metabolism and placental transfer in the pregnant Rhesus monkey. J. Clin. Invest. 51, 2584–2592.

Pletsch, Q. and Coffey, J.W. (1971) Intracellular distribution of radioactive vitamin B_{12} in rat liver. J. Biol. Chem. 246, 4619–4629.

Posner, B.I. (1973) Insulin metabolizing enzyme activities in human placental tissue. Diabetes, 22, 552–563.

Power, G.G. (1972) The placental sluice: maternal effects on the fetal circulation. In: Respiratory Gas Exchange and Blood Flow in the Placenta (Longo, L.D. and Bartels, H., eds.) pp. 191–205, Dept of HEW, Bethesda, Maryland.

Power, G.G. and Longo, L.D. (1973) Sluice flow in placenta: maternal vascular pressure effects on fetal circulation. Am. J. Physiol. 225, 1490–1496.

Qazi, Q.H. and Thompson, M.W. (1972) Genital changes in congenital virilizing adrenal hyperplasia. J. Pediat. 80, 653–654.

Quigley, H., Phillips, L.L. and McKay, D.G. (1965) Transport of radioactive sodium across the rat placenta in experimental toxemia of pregnancy. Am. J. Obstet. Gynecol. 91, 377–384.

Rabain, F. and Picon, L. (1974) Effect of insulin on the materno-fetal transfer of glucose in the rat. Horm. Metab. Res. 6, 376–380.

Ramsey, E.M., Corner, G.W. and Donner, M.W. (1963) Serial and cineradioangiographic visualization of maternal circulation in the primate (Hemochorial) placenta. Am. J. Obstet. Gynecol. 86, 213–219.

Rathbun, L.J. (1943) An analysis of 250 cases of postmaturity. Am. J. Obstet. Gynecol. 46, 278–285.

Riggs, T.R. and Pan, M.W. (1972) Transport of amino acids into oestrogen-primed uterus enhancement of the uptake by a preliminary incubation. Biochem. J. 128, 19–27.

Rodewald, R.B. (1973) Intestinal transport of antibodies in the newborn rat. J. Cell Biol. 58, 189–211.

Rosso, P. (1974) Maternal malnutrition and placental transfer of alpha-aminoisobutyric acid in the rat. Science 187, 648–649.

Rosahn, P.D. and Greene, H.S.N. (1936) The influence of intrauterine factors on the fetal weight of rabbits. J. Exp. Med. 63, 901–908.

Rosen, M., Scibetta, J.J. and Hochberg, C.J. (1970) Human fetal electroencephalogram. III. Pattern changes in presence of fetal heart rate alterations and after use of maternal medications. Obstet. Gynecol. 36, 132–140.

Rosenberg, T. and Wilbrandt, W. (1957) Strukturabhängigkeit der Hemmwirkung von Phlorizin und anderen Phloretinderivaten auf den Glukosetransport durch die Erythrocytenmembran. Helv. Physiol. Acta 15, 168–176.

Rudolph, A.M. and Heymann, M.A. (1967a) The circulation of the fetus in utero. Methods for studying distribution of blood flow, cardiac output and organ blood flow. Circul. Res. 21, 163–184.

Rudolph, A.M. and Heymann, M.A. (1967b) Validation of the antipyrine method for measuring fetal umbilical blood flow. Circul. Res. 21, 185–190.

Rushton, D.I. (1973) The placenta – An environmental problem. Brit. Med. J. 1, 344–348.

Saling, E. (1964) Mikroblutuntersuchungen am Feten. Ztschr. F. Geburtsh. Gynäk. 162, 56–75.

Sastry, B.V.R. and Henderson, G.I. (1972) Kinetic mechanisms of human placental choline acetyltransferase. Biochem. Pharmacol. 21, 787–802.

Sastry, B.V.R., Olubadewo, J. and Schmidt, D.E. (1973) Placental cholinergic system and occurrence of acetylcholine in placenta. Fed. Proc. 32, 742.

Schenker, S., Dawber, N.H. and Schmid, R. (1964) Bilirubin metabolism in the fetus. J. Clin. Invest. 43, 32–39.

Schneider, H. and Dancis, J. (1974) Amino acid transport in human placental slices. Am. J. Obstet. Gynecol. 120, 1092–1098.

Schneider, H., Panigel, M. and Dancis, J. (1972) Transfer across the perfused human placenta of antipyrine, sodium and leucine. Am. J. Obstet. Gynecol. 114, 822–828.

Schröder, H., Stalp, J. and Leichtweiss, H.P. (1972a) Untersuchungen zum Natriumtransport durch die isolierte Meerschweinchenplacenta. Pflügers Archiv Europ. J. Physiol. 332, Suppl. R21.

Schröder, H., Stolp, W. and Leichtweiss, H.P. (1972b) Measurements of Na^+ transport of the isolated, artificially perfused guinea pig placenta. Am. J. Obstet. Gynecol. 114, 51–57.

Schröder, H., Leichtweiss, H.P. and Madee, W. (1975) The transport of D-glucose, L-glucose, and D-mannose across the isolated guinea pig placenta. Pflügers Archiv Europ. J. Physiol. 356, 267–275.

Schulman, J.D., Coppola, P.T. and Clutterbuck-Jackson, S. (1975) γ-Glutamyltranspeptidase: a possible mediator of amino acid transport in human and other placenta. Gynecol. Invest. 6, 32–38.

Sciarra, J.J., Kaplan, S.L. and Grumbach, M.M. (1963) Localization of anti-human growth hormone serum within the human placenta: evidence for a human chorionic "growth hormone-prolactin". Nature, 199, 1005–1008.

Scott, J.M. and Henderson, A. (1972) Acute villous inflammation in the placenta following intrauterine transfusion. J. Clin. Path. 25, 872–875.

Seeds, A.E. (1967) Water transfer across the human amnion in response to osmotic gradients. Am. J. Obstet. Gynecol. 98, 568–571.

Seeds, A.E. (1970) Osmosis across term human placental membranes. Am. J. Physiol. 219, 551–554.

Seeds, A.E., Schrueffer, J.J., Reinhardt, J.A. and Garlid, K.D. (1973) Diffusion mechanisms across human placental tissue. Gynecol. Invest. 4, 31–37.

Shah, H.C. Gipson, S. and McLachlan, J.A. (1975) The fate of ^{14}C diethylstilbestrol in the pregnant mouse, Toxicol. Appl. Pharmacol. 33, 190.

Shami, Y. and Radde, I.C. (1971) Calcium-stimulated ATPase of guinea pig placenta. Biochim. Biophys. Acta 249, 345–352.

Shami, Y. and Radde, I.C. (1972) The effect of the Ca^{++}/Mg^{++} concentration ratio on placental (Ca^{++}–Mg^{++}) ATPase activity. Biochim. Biophys. Acta 255, 675–679.

Shami, Y., Messer, H.H. and Copp, D.H. (1974) Calcium binding to placental plasma membranes as measured by rate of diffuion in a flow dialysis system. Biochim. Biophys. Acta 339, 323–333.

Shami, Y., Messer, H.H. and Copp, D.N. (1975) Calcium uptake by placental plasma membrane vesicles. Biochim. Biophys. Acta 401, 256–264.

Shapiro, R. (1961) The effect of maternal ingestion of iophenoxic acid on the serum bound iodine of the progeny. New Eng. J. Med. 264, 378–381.

Shepard, T.H. (1973) Catalog of Teratogenic Agents. The Johns Hopkins University Press, Baltimore and London.

Sherman, W.T. and Gautieri, R.F. (1972) Effect of certain drugs on perfused human placenta X: norepinephrine release by bradykinin. J. Pharm. Sci. 61, 878–883.

Shinton, N.K. (1972) Vitamin B_{12} and folate metabolism. Brit. Med. J. 1 556–559.

Shoeman, D.W., Kauffman, R.E., Azarnoff, D.L. and Boulos, B.M. (1972) Placental transfer of diphenylhydantoin in the goat. Biochem. Pharm. 21, 1237–1243.

Simmons, M.A., Jones, M.D., Burd, L.I., Schreiner, R.L., Makowski, E.L., Meschia, G. and Battaglia, F.C. (1975) A direct effect of insulin upon placental glucose transport in vivo. Gynecol. Invest. 6, 31.

Simionescu, N., Simionescu, M. and Palade, G.E. (1973) Permeability of muscle capillaries to exogenous myoglobin. J. Cell Biol. 57, 424–452.

Smith, C.H., Adcock, E.W. III, Teasdale, F., Meschia, G. and Battaglia, F.G. (1973) Placental amino acid uptake: tissue preparation, kinetics, and preincubation effects. Am. J. Physiol. 224, 558–564.

Smith, C.H. and Depper, R. (1974) Placental amino acid uptake. II. Tissue incubation, fluid distribution and mechanisms of regulation, Pediat. Res. 8, 697–703.

Smith, C.H., Enders, R.H. and Judd, R.M. (1975) Placental transport systems for neutral amino acids, Pediat. Res. 9, 280.

Sonoda, S. and Schlamowitz, M. (1972) Kinetics and specificity of transfer of immunoglublin G and serum albumin across the rabbit yolk sac in utero. J. Immunol. 108, 807–818.

Spiers, I. and Sim, A.W. (1972) The placental transfer of pancuronium bromide. Brit. J. Anaesth. 44, 370–373.

Stein, W.D. (1967) The Movement of Molecules Across Cell Membranes. Academic Press, New York.

Stewart, M.E. and Terepka, A.R. (1969) Transport functions of the chick chorio-allantoic membrane. I. Normal histology and evidence for actine electrolyte transport from the allantoic membrane in vivo. Exp. Cell Res. 58, 93–106.

Stone, M.L., Salerno, L.J., Green, M. and Zelson, C. (1971) Narcotic addiction in pregnancy. Am. J. Obstet. Gynecol. 109, 716–723.

Štulc, J.J., Rietveld, W.J., Soeteman, D.W. and Versprille, A. (1972) The transplacental potential difference in guinea pigs. Biol. Neonate, 21, 130–147.

Sybulski, S. and Tremblay, P.C. (1967) Uptake and incorporation into protein of radioactive glycine by human placentas in vitro. Am. J. Obstet. Gynecol. 97, 1111–1118.

Symonds, H.W., Samson, B.F. and Twardock, A.R. (1972) The measurement of the transfer of calcium and phosphorus from foetus to dam in the sheep using whole body counter. Res. Vet. Sci. 13, 272–275.

Szabo, A.J. and Grimbaldi, R.D. (1970) The metabolism of placenta. Adv. Metabol. Dis. 4, 185–228.

Tan, M. and Raman, A. (1972) Maternal-fetal calcium relationships in man. Quart. J. Exp. Physiol. 57, 56–59.

Thalme, B. (1966) Electrolyte and acid-base balance in fetal and maternal blood. Acta Obstet. Gynecol. Scand. Suppl. 45(8), 1–118.

Thompson, G.A. and Meister, A. (1975) Utilization of L-cystine by the γ-glutamyl transpeptidase-γ-glutamyl cyclotransferase pathway. Proc. Nat. Acad. Sci. U.S.A. 72, 1985–1988.

Tremblay, P.C., Sybulski, S. and Maughan, G.B. (1965) Role of the placenta in fetal nutrition. Am. J. Obstet. Gynecol. 91, 597–605.

Tsoulos, N.G., Colwill, J.R., Battaglia, F.C., Makowski, E.L. and Meschia, G. (1971) Comparison of glucose, fructose, and O_2 uptakes by fetuses of fed and starved ewes. Am. J. Physiol. 221, 234–237.

Twardock, A.R. and Austin, M.K. (1970) Calcium transfer in perfused guinea pig placenta. Am. J. Physiol. 219, 540–545.

Twardock, A.R., Yung-Huei Kuo, E., Austin, M.K. and Hopkins, T.R. (1971) Protein binding of calcium and strontium in guinea pig maternal and fetal blood plasma. Am. J. Obstet. Gynecol. 110, 1008–1014.

Ullberg, S., Kristoffersson, H., Flodh, H. and Hanngren, A. (1967) Placental passage and fetal accumulation of labelled vitamin B_{12} in the mouse. Arch. Int. Pharmacodyn. 167, 431–449.

U.S. Department of Health, Education, and Welfare (1973) Public Health Service; Health Services and Mental Health Administration: Smoking and Pregnancy (Reprinted from The Health Consequences of Smoking, 1973). U.S. Government Printing Office, Washington, D.C.

Ussing, H.H. (1949) The active ion transport through the frog skin in the light of tracer studies. Acta Physiol. Scand. 17, 1–37.

Van der Werf, P., Stephani, R.A. and Meister, A. (1974) Accumulation of 5-oxoproline in mouse tissues after inhibition of 5-oxoprolinase and administration of amino acids; evidence for function of the γ-glutamyl cycle. Proc. Nat. Acad. Sci. U.S.A. 71, 1026–1029.

Van Hien, P., Kovacs, K. and Matkovics, B. (1974) Properties of enzymes I. Study of superoxide dismutase activity change in human placenta of different ages. Enzyme, 18, 341–347.

Villee, C.A. (1953a) The metabolism of human placenta in vitro. J. Biol. Chem. 205, 113–123.

Villee, C.A. (1953b) Regulation of blood glucose in the human fetus. J. Appl. Physiol. 5, 437–444.

Vorherr, H. (1975) Placental insufficiency in relation to post-term pregnancy and fetal postmaturity. Am. J. Obstet. Gynecol. 123, 67–103.

Wagner, M. (1964) Reply: letters to editor. J. Pediat. 64, 776–777.

Wallenburg, H.C.S., Hutchinson, D.L., Schuler, H.M., Stolte, L.A.M. and Janssens, J. (1973) The pathogenesis of placental infarction, II. An experimental study in the rhesus monkey placenta. Am. J. Obstet. Gynecol. 116, 841–846.

Ward, C.O. and Gautieri, R.F. (1966) Effect of certain drugs on perfused human placenta. VI. Serotonin antagonists. J. Pharm. Sci. 55, 474–478.

Ward, C.O. and Gautieri, R.F. (1968) Effect of certain drugs on perfused human placenta. VIII. J. Pharm. Sci. 57, 287–292.

Weiner, I.W. and Mudge, G.H. (1964) Renal tubular mechanisms for excretion of organic acids and bases. Am. J. Med. 36, 743–762.

Welch, R.M., Harrison, Y.E., Gomni, B.W., Poppers, P.J., Finster, M. and Cowney, A.H. (1969) Stimulatory effects of cigarette smoking on the hydroxylation of 3,4-benzpyrene and the N-demethylation of 3-methyl-4-monomethylaminoazobenzene by enzymes in human placenta, Clin. Pharm. Ther. 10, 100–110.

Welsch, F. (1974) Choline acetyltransferase of human placenta during the first trimester of pregnancy. Experientia, 30, 162–163.

Widdas, W.F. (1952) Inability of diffusion to account for placental glucose transfer in sheep and consideration of the kinetics of a possible carrier. J. Physiol. 118, 23–39.

Widdas, W.F. (1961) Transport mechanisms in the fetus. Brit. Med. Bull. 17, 107–111.

Widdowson, E.M. and Spray, C.M. (1951) Chemical development in utero. Arch. Dis. Childhood. 26, 205–214.

Wiebelhaus, V.D., Brennan, F.T., Maass, A.R., Sosnowski, G., Gressner, G., Weinstock, J., Azeff, M. and Jenkins, B. (1965) Effect of triamterene on the mechanism for urinary acidification. Pharmacology, 7, 165–173.

Wigglesworth, J.S. (1962) The Langhans layer in late pregnancy: a histological study of normal and abnormal cases. J. Obstet. Gynec. Brit. Commonw. 69, 355–359.

Wigglesworth, J.S. (1964) Experimental growth retardation in the foetal rat. J. Path. Bact. 88, 1–13.

Wild, A.E. (1970) Protein transmission across the rabbit foetal membranes. J. Embryol. Exp. Morphol. 24, 313–330.

Wild, A.E. (1973) Transport of immunoglobulins and other proteins from mother to young. In: Lysosomes in Biology and Pathology (Dingle, J.T., ed.) Vol. 3, pp. 169–215, North-Holland, Amsterdam.

Wild, A.E. (1974) Protein transport across the placenta. In: Transport at the Cellular Level (Sleigh, M.A. and Jennings, D.H., eds.) pp. 521–546, Cambridge University Press, Cambridge.

Wild, A.E. (1975) Role of the cell surface in selection during transport of proteins from mother to foetus and newly born. Phil. Trans. R. Soc. Lond. B271, 395–410.

Williams, K.E., Kidston, E.M., Beck, F. and Lloyd, J.B. (1975a) Quantitative studies of pinocytosis I. kinetics of uptake of (^{125}I) polyvinylpyrrolidone by rat yolk sac cultured in vitro. J. Cell Biol. 64, 113–122.

Williams, K.E., Kidston, E.M., Beck, F. and Lloyd, J.B. (1975b) Quantitative studies of pinocytosis II. kinetics of protein uptake and digestion by rat yolk sac cultured in vitro. J. Cell Biol. 64, 123–134.

Wilson, J.G. (1973) Environment and Birth Defects. Academic Press, New York.

Wislocki, G.B. and Dempsey, E.W. (1955) Electron microscopy of the placenta of the rat. Anat. Rec. 123, 33–63.

Wynn, R.M. (1968) Morphology of the placenta. In: Biology of Gestation (Assali, N.S., ed), Vol. I. pp. 93–152. Academic Press, New York.

Wynn, R.M. (1973) Fine structure of the placenta. In: Handbook of Physiology (Greep, R.O., ed.) Section 7: Endocrinology, Vol. II, Part 2, pp. 261–341, American Physiological Society, Washington, D.C.

Wynn, R.M. (1975) Fine structure of the placenta. In: The Placenta (Gruenwald, P., ed.) pp. 56–79, University Park Press, Baltimore.

Young, M. and Prenton, M.A. (1969) Maternal and fetal plasma amino acid concentrations during gestation and its retarded fetal growth. J. Obstet. Gynec. Brit. Commonw. 76, 333–344.

Zacks, S.I. and Wislocki, G.B. (1953) Placental esterases. Proc. Soc. Exp. Biol. Med. 84, 438–441.

Zunker, H.O., Phillips, L.L. and McKay, D.G. (1965) Potassium transport in kidney and placenta in normal pregnancy and in experimental toxemia. Am. J. Obstet. Gynecol. 91, 369–376.

Zuspan, F.P., Whaley, W.H., Nelson, G.H. and Ahlquist, R.P. (1966) Placental transfer of epinephrine. Am. J. Obstet. Gynecol. 95, 284–289.

On the mechanism of metazoan cell movements 6

J.P. TRINKAUS

1. Introduction

In studies of the cell movements of multicellular organisms most attention has been given to how cells move in culture. Until now this has been right and proper, for one can see more with greater ease in vitro and of course have more immediate and less ambiguous control of the cellular environment. But the central issue in the study of tissue cell movement is how cells move within the organism. This is a question of outstanding importance for a number of reasons. Cell movements are a universal feature of morphogenesis during embryonic development in Metazoa, in so far as we know, and in some plants as well. Cell movements have an important role in regeneration and repair. The mobilization of phagocytic and antibody-bearing cells at foci of inflammation is crucial in the defense of the organism against foreign bodies. And, finally, cell movements are an essential aspect of the spread of cancer.

Development has been divided into a *morphogenetic phase*, during which cells change shape or position and in doing so lay down the primitive body plan of the organism, molding in rough the form of the tissues and organs, and a *cytodifferentiation phase*, during which the individual cells acquire the special cellular and chemical features that characterize each particular histological state (Zwilling, 1968). This division is convenient, for in most instances the morphogenetic phase clearly precedes and lays the basis for the cytodifferentiation phase. Thus gastrulation precedes neurulation and neurulation precedes the differentiation of neurons. There are, however, exceptions to this rule, where the two may occur simultaneously. Good examples may be found in the formation of the oviduct (Wrenn, 1971), the lens (Yamada, 1967), the thyroid (Shain et al., 1972), and the pancreas (Rutter et al., 1968). Also, as Roux (1894) suggested years ago, it must be understood that a kind of morphogenetic cytodifferentiation is necessary at the beginning, in order for cells to engage in the shape changes and translocations that dominate the morphogenetic phase (Holtfreter, 1944; Trinkaus, 1963, 1973b; Gustafson and Wolpert, 1967; Coupé et al., 1973), and an understanding of the morphogenetic phase ultimately rests on an understanding of these early differentiations. Regardless of the usefulness of this

G. Poste & G.L. Nicolson (eds.) The Cell Surface in Animal Embryogenesis and Development

dichotomy, it is fair to state that the bulk of the work on developmental problems has centered on the cytodifferentiation phase, leaving somewhat to the side the equally important problem of how the cells got there in the first place. Thus, the mechanisms of gastrulation and organogenesis, in spite of their fundamental nature, remain among the least understood of all developmental problems, even though they have long been considered classic problems of embryology. I will, therefore, devote considerable attention to morphogenetic cell movements in this review, summarizing the state of our knowledge and emphasizing some of the investigations in most pressing demand.

Although, happily, most cells seem to remain in place once organogenesis and histogenesis are complete (but see p. 234), many kinds of cells can be stimulated to active mobility by removing part or all of an organ. Removal of a piece of skin, for example, initiates locomotory activity on the part of the cells at the new margins of the epidermis and the consequent epidermal spreading covers over the wound. When part of a limb is removed, cells at and near the point of amputation lose their former organization, epidermis spreads over the exposed surface and the inner stromal cells undergo an extensive rearrangement, which, along with active cell division and cytodifferentiation, results in regeneration and replacement of the lost parts. Individual neurons are famous for their ability to regenerate, and they can reinnervate the exact same peripheral (or central) end organ. Of course, in this case, it is only part of the cell, the extending axon, that is involved. The cell body, or soma, stays put. Finally, if differentiated tissues or parts of organs are removed and explanted in vitro under suitable culture conditions, cells will migrate actively from the explant out on the artificial substratum, demonstrating conclusively that virtually all tissue cells retain the capacity for locomotion.

The cancer problem presents one of the major reasons for interest in where and how cells move in vivo. For the extensive displacement of tumor cells when they become metastatic is apparently due in considerable part to locomotory activity by individual tumor cells. Spreading of tumor cells apparently occurs by two main methods: passive transport in the circulation and active spreading into adjacent tissues and organs. Passive movement in the blood and lymph is well-demonstrated (see p. 264); however, active spreading and invasion of individual cells and cell groups into adjacent tissues is less well understood, even though it seems to occur extensively in many neoplasms.

Unlike these various kinds of cells, where cell locomotion is associated with changes in the differentiation of cells, polymorphonuclear leukocytes, macrophages, lymphocytes and other like cells are and remain differentiated. Their mobility constitutes part of the defense against disease. Although it is one of the most spectacular of all cell movements in vivo, it is ill-understood. In addition to their important physiological function, these cells have a special interest; they are the only metazoan tissue cells known to be chemotactic (aside from certain spermatozoa). At least, they respond readily and decisively to chemotactic stimuli in vitro.

In all of these phenomena, the *occurrence* of cell movements is well-established.

But we know little of *mechanism*. And in the case of invasion by tumor cells we do not even know how much of the spread over short distances is due to active cell movement and how much to passive spread in the circulatory system or to continuing cell division.

It is convenient to consider the movements of cells within organisms under two rubrics: descriptive aspects and mechanism. Although convenient, this distinction is of course artificial. Careful description of what occurs normally can suggest mechanisms and a careful analysis of mechanism is bound to tell us more about the normal course of events. The kinds of descriptive questions to be asked are: do the cells actually translocate; when and where do these movements begin; when and where do they stop; what paths are followed; what is the rate of movement; does the rate vary, and, finally, what is their social behavior, i.e., do the cells move as individuals or collectively, as members of a cell stream or sheet?

The essential question under mechanism is, of course: just how do cells move in vivo; what is responsible for the changes in cell shape associated with movement; what forces cause them to commence and what causes them to stop; how are they motivated and guided in the course of their movements; how do they adhere to their substrata to gain traction for movement; what are the substrata? It should be stated bluntly at the outset that although we possess considerable knowledge of the descriptive aspects of certain cell movements, as during gastrulation, we have little understanding of mechanism. This is not surprising, since there has been little work on this aspect of cell movements in vivo, in contrast, at least in recent years, to the study of the mechanism of cell locomotion in vitro. Thus the purpose of this review, in so far as mechanism is concerned, is less to review what has been done, which is rather little, than to point out what needs to be done.

2. *Methods of studying the normal course of cell movements in vivo*

Although it may seem to many as belaboring the obvious, I feel it necessary to interject a word concerning methods of studying the normal course of cell movements within organisms. These are often casually referred to as "observational methods". In spite of several clear-headed articles on the matter through the years (most recently, Weston, 1967), one still perceives from time to time a tendency for preconceived notions to substitute for hardheaded evidence. Although inexcusable, this tendency is understandable. The problem of discerning with accuracy the extent and direction of movements of particular cells in the midst of other cells or intercellular materials presents a serious technical challenge, particularly in the case of opaque organisms. It is crucial, therefore, to understand the possibilities of the various methods available.

The classical method of descriptive embryology and pathology is tissue dissection and examination of sections of fixed material, in which the cells in question have not been marked. This approach and thin-sectioning electronmicroscopy (which is basically the same), may give some notion of the presence and extent of cell movements, particularly if the movement is massive and if the material is examined at carefully staged intervals. At best, however, important details will be lacking, or, worse, one will frequently be drawn toward totally erroneous conclusions (often by preconceived ideas) (see discussion by Campbell, 1974). This is especially likely where the movements of individual cells or small groups of cells are concerned. These reservations also apply to some extent to the study of fixed material in the scanning electron microscope (SEM). This powerful new method for examining the surfaces of cells can tell much about the extent and the manner of movement of cells in vivo, if the material is carefully staged in sequence. Indeed SEM is superior to old-fashioned histological methods and thin-sectioning electron microscopy in that it has the triple advantage of allowing at the same time high magnification, a view of a large part of the surface of cells, and increased depth of field. Nevertheless, the minute by minute details of changes in morphology, rate, and direction of movement of individual cells cannot be followed, since each new sequence, naturally, is of a new sample of cells.

In order to know with accuracy the course of cellular movements in fixed material, cells must be marked, generally grafted to an unmarked host, and followed stage by stage (Weston, 1967). A variety of natural markers have been used with success. The best of those currently available take advantage of the mononucleolate condition of a mutant of *Xenopus* (Blackler, 1962) and the different staining properties of the nuclei of Japanese quail and chick cells (Le Douarin, 1969). These techniques are most valuable because they give excellent resolution and are not limited to any particular kind of tissue cell. Other natural markers, which are limited to particular kinds of cells, and which have been useful for following the locomotory activities of particular cells include: alkaline phosphatase activity (Mintz, 1959), glycogen (Meyer, 1964; Steinberg, 1962), melanin (Townes and Holtfreter, 1955; Trinkaus and Lentz, 1964). The artificial cell marker par excellence is tritiated thymidine, introduced over 15 years ago (Hughes et al., 1958; Trinkaus and Gross, 1961) and used routinely ever since. It has the important advantage of labeling any dividing cell and its descendants. It has given conclusive results in situ, both by the technique of pulse-chase and in grafting experiments, where the cells in question are labeled and grafted to an unlabeled host, either to their normal site in order to observe normal cell movements (e.g. Weston, 1963; Chibon, 1966; Johnston, 1966) or elsewhere in an analysis of mechanism of movement (Weston and Butler, 1966). Markers such as these, which can be followed in fixed material, are essential if the cells to be traced reside in the interior of opaque organisms (the exact situation of most migratory cells, including in particular cancer cells).

If one wishes to follow cells on the surface of an organism, then one need not resort to fixation but may follow them in the living organism by heteroplastic grafting of pigmented cells to non-pigmented hosts (Harrison, 1904), by coloring them with a variety of vital dyes, as did Vogt in his at once pioneering and masterful studies of amphibian gastrulation (1929), or by placing adhering particles, such as carbon, on their surfaces (e.g., Spratt, 1946). One can even succeed in following cells in the interior of an organism by injecting particles and following their fate in sequential dissections (Ballard, 1966a,b,c).

Ideally, of course, the material of choice for studying cell movements in vivo is a transparent organism. With such material, cells can not only be followed by all of the techniques just mentioned but, in addition, much of their migratory behavior may be observed directly (Clark, 1912) and with cinemicrography (Speidel, 1935; Gustafson and Wolpert, 1967; Trinkaus, 1973a; Izzard, 1974; Bard and Hay, 1975). Recently, with the decrease in phase haloes provided by Nomarski interference optics, careful observations of all kinds are possible in vivo: second to second changes in cell shape and direction of movement, rate of locomotion, changing relations to other cells, to name a few. In short, the material par excellence for the study of cell movements in vivo is a transparent organism. If only all cell movements of interest occurred in such organisms!

Before leaving the subject of techniques of observation, it should be emphasized that the observer need not necessarily be wedded to such lucid material as Nature in her generosity has chosen to provide. Even in large opaque organisms ingenuity may create situations where direct observations of cell behavior may be made. The Clarks (1935) showed the way years ago in their employment of the rabbit ear chamber, a technique whose potentialities have hardly been realized.

3. The normal course of cell movements in vivo

It has been clear for some time that the social relations among moving cells vary greatly. Some cells appear to move completely as individuals, having no physical contact with other cells, at least during part of their migratory history. Others move in small cell clusters. Still others move in loose streams of cells, in which individual cells may be constantly shifting their contact relations with each other. Although not especially emphasized in the past (e.g. Trinkaus, 1965), as cell movements are studied more intimately, movement in streams is emerging as a more common method than previously realized. Cells that move in clusters and streams probably are also capable of translocating as individuals. Although at the extremes these categories are quite different, they obviously grade into one another frequently. It thus seems wise at this time to consider these three categories together.

Finally, cells may be in such close contact that they form dense tongues of cells or epithelioid sheets in which the individual cells cohere tightly and do not change position relative to one another during movement of the sheet. They are regimented to the movement of the mass. This appears to be a category of cell movement that is biologically valid and hence deserves separate treatment.

3.1. Movement of cells as individuals and in cell streams

Our knowledge of individual cell movement within organisms is very incomplete, without doubt for the simple reason that most organisms are opaque and movements of individual cells or even of small cell clusters might easily be overlooked. Moreover, wherever cell movements have been studied at all carefully in the more lucid systems, locomotion of individual cells and of cell clusters and streams are soon evident. This suggests that, were we able to see better, many new examples of such cell movements would be discovered. Indeed, I am convinced that movements of the sort I am about to describe are commonplace, at least during early development.

3.1.1. Normal cells

The best known normal movements of cells as individuals are those of macrophages and leukocytes.* Granted that much of the displacement of leukocytes is passive, in the flowing blood and lymph, nevertheless where they have been observed invading tissues they appear to do so as individuals (Marchesi, 1966, 1970; Wood et al., 1968). Less is known about the movements of macrophages (Jacoby, 1965).

During metazoan morphogenesis, the most famous, the most studied, and, indeed, the most spectacular movement of migratory cells in cell streams, and probably also as individuals, is the neural crest (Hörstadius, 1950; Chibon, 1974; Weston, 1970). Unfortunately, neural crest migration has never been observed directly, due to the large number of cells at this stage of development and to the opacity of amphibian and chick embryos, where it has been mainly studied. Hence, we do not really know whether cells actually ever migrate from the crest as separate individuals. The sparse scattering of crest cells labeled with tritiated thymidine in the ectoderm (Weston, 1963) and in the head mesenchyme (Johnston, 1966) is certainly suggestive of this. But, of course, one does not know how many unlabeled, hence undetectable, cells are moving with the labeled ones. When crest cells are taken from Japanese quail, however, one can be sure that all cells are labeled. Hence scattering of labeled cells in such an experiment clearly indicates individual cell movement (Le Douarin and Teillet, 1974). As crest cells move down the side of the neural tube to form eventually spinal ganglia and sympathetic ganglia they seem clearly to be in streams, as judged from the density of the population of labeled cells in radioautographs. The same may be said for the cells of the brain crest which form the visceral cartilages and head mesenchyme (Johnston, 1966).

Another reason for interest in the neural crest is that it provides one of the best examples of disaggregation and aggregation of cells as a morphogenetic mechanism. The early crest cells, which are clustered rather tightly together to form an axial ribbon on the dorso-lateral surface of the neural tube, disaggre-

*I am well aware that cell movements occur in many Metazoa besides those considered in this communication (in the Arthropoda, for example). Since there has been little or no study of them with marking methods, however, we have no accurate idea of their extent. In any event, this article is not intended to be a survey, wherein completeness is a primary objective. Rather it is an extended treatment of what we know and what we need to know about the mechanisms of cell movements in certain systems that for one reason or another are favorable for study and where there has been some analysis.

gate in an antero-posterior sequence to give rise to migratory cells. As these cells reach particular destinations, they may reaggregate to form rather dense cell masses, such as spinal and sympathetic ganglia and visceral cartilages.

Incidentally, pigment cells derived from the neural crest also provide one of the best examples of the capacity of certain adult normal tissue cells to migrate. If one grafts pigment free skin to a pigmented baby chick, melanoblasts from the surrounding skin will invade the graft and deposit pigment in the growing feather germs (Rawles, 1944). Likewise, melanoblasts from an adult chicken feather germ, when grafted to an embryo, will migrate in the wing bud, invade the forming feather germs and deposit melanin granules in their barb ridges (Trinkaus, 1948).

Primordial germ cells rank with the neural crest for the distances they migrate in certain species, such as the chick, and they migrate at least in part as individuals, since when identified in sections during their known period of migration, they are often clearly separated from other germ cells (see Willier, 1939). However, much of their long migration in the chick from the germinal crescent to the genital epithelium is by way of the early yolk sac circulatory system (Simon, 1960), a movement that is probably largely passive. Thus the migration of primordial germ cells in birds appears to begin by individual cell movement posteriad from the anterior germinal crescent, active invasion of the newly formed blood vessels of the yolk sac, passive transport by the circulating blood, and finally, after they somehow lodge in the genital epithelium, invasion of it to take their places in the differentiating gonad (Meyer, 1964). In amphibia, primordial germ cells lie deep in the floor of the blastocoel in the blastula stage and after gastrulation migrate in the endoderm to a dorsal position, whence they move into the mesoderm of the dorsal mesentery and, eventually, into the genital ridges (Blackler, 1958, 1962).

Other cells which in a certain sense engage in long-distance migration in vivo are neurons, both during embryogenesis and during regeneration in adults. To be sure, the soma or cell body of the neuron does not move, but the leading tip or "growth cone" of the advancing axon may travel extraordinarily long distances, particularly during the formation of the peripheral nervous system. Although nerve fibers often extend in bundles, they may also move as individuals. This "outgrowth" of the nerve axon is one of the oldest known cell movements of multicellular organisms. Postulated by His (1889) and Cajal (1890), on the basis of their histological observations, and demonstrated by Harrison in 1910 in tissue culture, it has been studied intensively ever since (see Weiss, 1945). This manner of axon extension in vivo was later studied by Speidel (1932, 1933) in the transparent tail of *Hyla*, in one of the first uses of cinemicrography to study cell movement, and found to be essentially the same as Harrison had found in culture. Axons were observed to advance with a leading active growth cone, branch, and make neuro-muscular junctions.

Although gastrulation is dominated in many forms by the spreading, folding and invagination of sheets of cells, some movements of gastrulation in all organisms that have been studied in detail are in the form of the migrations of individual cells and cell streams. Even in the Amphibia, where the movement of sheets dominates, invagination itself at its onset is characterized by the ingression of individual cells. Thus the so-called bottle cells sink in from the surface as separate cells and begin the process (Ruffini, 1925; Holtfreter, 1943). And this is not all. Quite recently, Løvtrup (1966, 1975) has presented evidence that Vogt's (1929) original fate map placing prospective mesoderm at the surface of the urodele early gastrula is incorrect. It is in the interior. Concurrently, Keller (1976) has found that in the anuran *Xenopus* the prospective mesoderm is not only in the interior but is not at all organized in a sheet. It is rather a somewhat loose aggregation of cells beneath the prospective endodermal and ectodermal sheet. It seems likely, therefore, that the prospective mesodermal cells in *Xenopus* move during gastrulation in a kind of broad cell stream.

Cells also move in a broad stream during avian gastrulation. Although cells move to the primitive streak in the epiblast sheet, once they reach the streak their behavior changes drastically. They aggregate in the form of an elongate axial wedge, in which the cells are closely in contact and assume a bottle shape (Balinsky and Walther, 1961; Granholm and Baker, 1970). Once inside, between the epiblast and the hypoblast, individual cells and small cell clusters break away from the central axial mass and move laterally to form the mesoblast (Balinsky and Walther, 1961; Trelstad et al., 1967; Hay, 1968). Although the movements of these cells have never been observed in the living organism, due to the opacity of the avian blastoderm, the fine-structural study of Trelstad et al. (1967) makes it clear that they move in a very loose cell stream, with many of them moving as individuals. There is no direct evidence for it, but one can easily *imagine* them constantly making and breaking contacts with each other and with the surfaces of the epiblast above and the hypoblast below, as they migrate outward. Somewhat later in development some of these new mesodermal cells aggregate in small clusters which move anteriorly to form the heart primordia (DeHaan, 1963).

In echinoderms, invagination is also presaged by the ingression of individual cells. Primary mesenchyme cells, derived from the micromeres, move into the blastocoel from the so-called vegetal plate (Driesch, 1896; Kinnander and Gustafson, 1960). But, as in the amphibians, these first cells to invade do not become part of the archenteron, even though they appear to lead the way. Rather, they move about as individuals on the inner surface of the ectodermal blastocoel wall and soon aggregate there to form a ring around the newly invaginating archenteron. From this ring two ventro-lateral branches extend toward the animal pole. Gustafson and Wolpert (1961) took advantage of the extreme transparency of certain sea urchin eggs to follow these movements in considerable detail. Later, after invagination has progressed 1/3 to 1/2 the distance toward the animal pole, another crop of mesenchyme cells appears at the tip of the archenteron – the so-called secondary mesenchyme cells. They too make connections with the ectodermal wall of the blastocoel, and, as they move about in a loose and constantly shifting group, invagination moves to completion.

Like echinoderm embryos, certain teleost eggs also possess a high degree of transparency. Hence we can affirm from direct observation of the activities of living cells that gastrulation in these embryos is dominated by the movement of cells as individuals and in streams. All of the so-called deep cells, which form the embryo proper, appear to move as individuals between the cell sheet of the enveloping layer above and the yolk syncytial layer below (Trinkaus, 1973a). Near the margin of the blastoderm, where they are concentrated in the germ ring, they move dorsad in a tight cell stream and converge toward the embryonic shield. Here, they adhere closely and begin embryo formation. While in the germ ring, however, each individual cell has an activity of its own; when filmed, the stream is revealed to be a seething mass of surface active cells, each of them contributing to the flow of the stream (Trinkaus, unpublished observations). In the area of the prospective yolk sac, deep cells move largely as individuals during gastrulation, although they are constantly aggregating with others to form small clusters, only to disaggregate again and move about further as individuals. Incidentally, the cinematographic studies of *Fundulus* deep cells confirm in part the marking studies of salmonid deep cells by Ballard (1973a,b,c), where the lesser transparency of the egg prohibits direct observation. In annual fish, deep cells go through a remarkable cycle of aggregation and disaggregation, moving in each phase largely as individuals (Wourms, 1972; Lesseps et al., 1975). Movements of individual cells continue after gastrulation in the teleost yolk sac. Stockard (1915) showed by direct observation that mesenchyme cells, propigment cells, and embryonic blood elements move about extensively, and aggregate, as they line up to form the yolk sac circulatory vessels and encase them with pigment cells.

In another transparent system, the tunicate stolen, there are also cells which engage actively in locomotion as individuals and they have provided excellent material for one of the most detailed studies of cell movement within an organism (Izzard, 1974). I will return to them later, when discussing mechanisms of cell locomotion.

Aside from the migrations of neural crest cells, the primordial germ cells, and the outgrowth of nerves, there are no other known long-distance morphogenetic cell migrations during post-gastrula embryonic development. There is, however, much apparent short-distance migratory activity on the part of individuals and groups of mesenchyme cells. The somites seem to form by the tight aggregation of loose axial mesenchyme lateral to the neural tube. Then, as the somite undergoes morphogenesis and the scleratome is formed, its mesial wall breaks open and disaggregates into cells which migrate around the embryonic spinal cord and notochord (Trelstad et al., 1967). Here they reaggregate and form the dense chondrogenic masses which will differentiate into vertebrae (see Hay, 1973). In the limb, in contrast, the prechondrogenic mass appears to rise as a result of proliferation (Zwilling, 1968), with cartilage being synthesized by cells which happen to lie in the center of the mass (Searls, 1967). Mesenchyme cells also migrate as individuals and in small cell clusters between the lens and the corneal stroma in the chick embryo to form the endothelial sheet separating the cornea from the anterior chamber (Bard et al., 1975; Nelson and Revel, 1975). In this system, the cell migrations are known in detail because they can be followed with both SEM and time-lapse cinemicrography, in spite of the fact that they occur in an opaque organism. Fortunately, the process continues in an apparently similar manner in organ culture.

Once cells have found their proper place in each organ system and cytodifferentiation is in full swing, it has been assumed that cell movement ceases. Indeed, it would seem that the integrity of the tissues and organs and of the organism itself depends on such immobility, and that any variance from it, as in the case of malignancy, could develop into a dangerous state of affairs. This presumed immobility was confirmed when Weston and Abercrombie (1967) joined pieces of embryonic liver and heart together, one of which was labeled, and found no cell intermingling. On the other hand, Wolff and Marin (1957) found that carmine particles introduced into embryonic chick liver fragments which presumably mark cells inside the fragment, moved around continuously. Wiseman and Steinberg (1973) have shown that thymidine labeled individual cells from chick embryonic heart, neural retina, and liver penetrate cell aggregates and intact pieces of the same organs when seeded on their surfaces. Armstrong and Armstrong (1973) have observed intermingling of mesenchyme cells when pieces of labeled and unlabeled mesonephros were joined in culture. In addition, Clark (1912) and Speidel (1933) observed that fibroblasts move within the tail of a transparent anuran tadpole. It thus is unquestionable that differentiated cells, albeit from embryos and larvae, can engage in locomotion within their own organs. It seems possible, therefore, that the stromal or fibroblast population of organs may normally translocate to a certain degree, particularly if disturbed. This suggests that these cells are not contact inhibiting in vivo (see p. 275). But how do we then explain the results of Weston and Abercrombie? Wiseman and Steinberg (1973) suggest that the trypsin treatment to which their cells were subjected may have rendered them more mobile. It is also possible that mobility in situ is confined to fibroblasts or mesenchymal cells and that under similar circumstances epithelial cells do not move, due perhaps to their firmer junctions (DiPasquale, 1975a). Liver, as used by Weston and Abercrombie, has much epithelial (parenchymal) tissue. But this does not explain the immobility of heart cells. Perhaps they do move, but so slowly (see Clarke, 1912) that their movement is overlooked in short term cultures.

Little is known about cell movement during regeneration, except that the limb regenerate has a local origin in vertebrates (Butler and O'Brien, 1942). There is much rearrangement of stromal cells inside and spreading of an epithelium to cover the wound surface. At various times, imaginary cells have been conjured up to explain the origin of the regenerate, especially in invertebrates. These cells, called neoblasts, have been assigned full pluripotency and legendary migratory capacities. But there is no good evidence for their existence. In any event, even if it be agreed that cells answering to their morphological description can be found, there is certainly no adequate evidence that they are actually pluripotent and (more important in the present context), that such cells engage in long-distance migrations (see Trinkaus, 1969, pp. 13–15).

In hydra, the cells and tissues are in continuous flux, as the animal undergoes continual growth and loss of tissues balanced by cell renewal. In addition to growth and differentiation, these processes involve cell migration, both as individuals and in sheets. Nematocysts, for example, migrate as individuals from the body column along the epidermal muscular layer into the tentacles, where they are mounted in batteries for capturing prey (Campbell, 1967). Although interstitial cells are usually considered to be a classic wandering cell, the evidence is mainly histological and therefore inadequate. Where there is evidence from cell labeling (Tardent and Morgenthaler, 1966), interstitial cells have been found to migrate; but their movements are much more limited than is generally supposed, both in terms of numbers of cells involved and distances traveled (Campbell, 1974).

The best analyzed morphogenetic movement by individual cells is found in the cellular slime molds. Here, individual cells, migrating on an inanimate substratum under the influence of a chemotactic attractant, form and join streams of cells moving centripetally toward a center. This aggregation phase of slime mold development is complete when all the cells in a given area have concentrated together in one center. Cell movement then occurs within the center, causing it to elongate into a slug, and continues, as shown by marking with vital dyes, while the slug moves about on the substratum.

Higher plants will not be considered in this chapter simply because they have evolved ways of undergoing morphogenesis without cell displacement. In this regard, in particular, they differ fundamentally from animals.

3.1.2. Cancer cells

In considering the movements of cancer cells, it should be emphasized at the outset that invasiveness is not abnormal per se. Normal tissue cells are also invasive at various times, especially during early development, and seem always able to move in tissue and organ culture.

Everyone appears to believe that when cancer cells become invasive they invade tissues adjacent to the original focus and others far away (after transport in the circulatory system), not only as cohesive tongues of many cells *but also as individual cells* (Willis, 1960). The evidence on which this belief is based is almost entirely observation of stained paraffin sections, the classical method of pathology. This method is of course limited as a way of studying a dynamic process such as cell movement, even under ideal circumstances. But, in justice, it should be recognized that in the case of tumors the circumstances are by no means ideal. It is virtually impossible to have a carefully timed series of observations. Nevertheless, the evidence seems clear and convincing that one of the ways in which cancers spread is by sending out whisps, streams, and tongues of cells into the interstices of the normal tissues adjacent to the growing cancer. The evidence is especially convincing where these extensions still possess anatomical continuity with the original tumor. Indeed, this may be the dominant form of cancer spread in tissues. Where supposed invasion by individual cells is concerned, however, the evidence is less convincing. If the tumor cells possess a distinctive cytology and if one observes a certain continuity, such as a line of them close to the tumor itself or close to massive invading wedges or tongues of tumor cells, then it seems possible that they got there by individual cell movement. However, it is also possible, for example, that they were simply left behind as the main tumor mass or the invading tongues retracted. If, on the other hand, the evidence rests solely on the discovery of isolated individual cells amongst normal tissue cells, it is not at all conclusive. Firstly, such cells may not be tumor cells. It is generally more difficult to identify cells as individuals in a tissue mass than those same cells in groups. Secondly, when anatomical continuity with the original tumor is lacking, we have no assurance that the cells were not carried there passively in capillaries. Wood et al. (1968) have tried to film cancer invasion in rabbit ear chambers by injecting tumor cells into blood vessels and observed cells penetrating capillary walls. However, they were unable to see them clearly after they entered the adjacent connective tissue, except that penetration through the capillary wall was followed by rapid formation of a tumor at that site. This, incidentally, argues against much migration through non-vascularized tissues.

What then can we say about invasion by individual cancer cells? In view of the evidence from fixed material and in view of the fact that many normal cells invade as individuals, especially during morphogenesis, it seems, on theoretical grounds, most *likely* that there is considerable invasive movement by tumor cells as individuals and in small clusters, as well as in cell streams and larger masses. What is required for proof is the indisputable kind of evidence we possess for normal tissue cells. To gain this, the same methods must be used, such as grafting labeled cancer cells to unlabeled normal sites and following them in closely timed sequences. This could be done both in organ culture (Pourreau-Schneider, 1961) and in vivo. Where possible, naturally, the direct approach of observing living tumor cells during the process of invasion would be most desirable. Perhaps tumor cells could be followed if injected directly into the tissues of transparent organisms, such as certain anuran larvae.

3.2. Movement of cells in sheets

3.2.1. Gastrulation

It has been said that the movement of cells in sheets is the dominant way that cells move in embryos (e.g. Trinkaus, 1965). Now, in the light of increased information, more importance should be given to the morphogenetic movements of individual cells and cell streams. However, the movements of sheets of cells remain the most obvious of all morphogenetic cell movements. In fact, such movements during gastrulation and neurulation are so massive as to involve a major part of the surface of the embryo, both externally and internally. And, in any case, it may be said without equivocation that the spreading, folding, and invagination of cell sheets that take place during the gastrulation and neurulation of some forms represent in their sweep and majesty one of the most wondrous of living phenomena. Because of this and because the work of gastrulation is no less than the laying down of the primitive body plan of the whole organism, these cell movements have excited the imagination of generations of embryologists.

The cell movements of gastrulation pose some of the oldest and least understood problems of development. Vertebrate gastrulation was first studied fully in the Amphibia (see Brachet, 1921; Vogt, 1929) and a pattern of movement emerged, which in its larger features occurs in most members of the higher vertebrates, though it varies somewhat in important detail from group to group. It should be emphasized, however, that gastrulation in other vertebrate groups does not necessarily follow the amphibian plan. This point needs emphasis because there has been an unfortunate tendency through the years to assume that all vertebrates gastrulate in basically the same way, hence follow the best-known plan, that of the Amphibia. This has on occasion led investigators to be satisfied with less evidence for their conclusions than they would have required had there been no preconceived idea. This proclivity no doubt has its origin in the time-hallowed belief that organisms are basically alike in their early development, in particular vertebrates (see Ballard, 1976). A good example of this is in the study of teleost gastrulation. Although the early post-Vogtian research on teleost development found them to gastrulate in a manner strikingly similar to that of the amphibians (Oppenheimer, 1936; Pasteels, 1936) meticulous further investigation, with better marking techniques, has shown the contrary to be so (Ballard, 1966a,b,c; 1968a,b; 1973a,b).

Another difficulty in the study of gastrulation is defining just what we mean by it, or when it begins or ends, especially since it differs in such important ways throughout the animal kingdom. For example, in amphibians, there is involution with the formation of an archenteron; in birds (and, presumably, mammals), there is ingression and no archenteron; in teleosts, there is neither involution, nor ingression, nor formation of an archenteron; and in echinoderms, there is invagination but apparently no involution. I do not wish to engage in a detailed discussion of this question; to do it justice would require too much space. I prefer to adopt part of the definition of Dalcq and Gerard (1935) and define gastrulation as that ensemble of processes during post-cleavage early development which when complete has brought the cells which will form the various organs to the places where those organs are to form. In essence, it is the embryo's way of laying down its primitive body plan.

Before continuing, I should make clear what I mean by *cell sheets* and define two terms which seem to have varied meanings in the literature, *invagination* and *involution.* Cells in sheets have two important and related characteristics. They are confluent and firmly adherent to one another with few or no apparent gaps between them, even at the level of fine-structure; and, in so far as we know, cells in sheets do not change position relative to one another. In addition, most cell sheets that engage in morphogenetic movements are cellular monolayers; but this does not always hold. I use the term invagination to mean the inpocketing of an unbroken sheet of cells, from a localized region, with no flow in from beyond. Cells may also move inside as individuals, i.e., by ingression, as in the primitive streak of avian embryos, and this is not to be confused with invagination. Involution is the flowing of a sheet of cells over the edge of an inpocketing or blastopore, in the case of gastrulation. By means of involution, cells that are some distance from the point of invagination move to the margin of the site of invagination, flow over it, and move inside.

Since the cell sheets involved in amphibian gastrulation are mainly all in one sheet, which constitutes the superficial cell layer(s) of the blastula (in urodeles at least), the precise coordination of movements is readily appreciated. Thus, as the prospective endoderm undergoes involution and invaginates to form the archenteron, it is replaced on the outer surface by the prospective ectoderm, which is spreading in epiboly. As the prospective endodermal sheet approaches the blastopore, it converges, apparently to squeeze into the relatively small blastopore. And, of course, once inside it diverges to form the broad archenteron. When invagination and involution are complete, and the archenteron fully formed, the epiboly of the prospective ectoderm is also complete and the entire gastrula is covered with ectoderm.

Neurulation in amphibian and chick embryos illustrates beautifully how a flat sheet of cells may roll up on itself and form a tube. In this case, the process takes place on the surface of the embryo, where all can see, and the tube formed is none other than the primordium of the brain and spinal cord of the whole organism. It is therefore not surprising that neurulation has been one of the most studied of all morphogenetic movements. I will discuss its mechanism later (p. 294).

In the echinoderm and in *Amphioxus,* simply to name two well-known representatives, cells move inside by the straightforward, classical invagination of a cell sheet, the vegetal plate (Conklin, 1932; Dan and Okazaki, 1956; Gustafson and Kinnander, 1956). Involution is apparently not involved, although the kind of marking of the surface which is necessary to be sure has apparently never been done.

Although the morphogenetic movements of gastrulation that are directly concerned with embryo formation in teleosts involve the movements of single cells and cell streams, the epiboly of the enveloping layer is one of the most spectacular examples of the spreading of a cell sheet. In the gastrula of *Fundulus,* for example, the epithelioid enveloping layer increases its surface area many fold during epiboly.

3.2.2. Organogenesis

Invagination is popularly associated with gastrulation, but in the vertebrates it occurs frequently after gastrulation during organogenesis (cf. Arey, 1954; Patten, 1968). Invagination is typically involved, for example, in the indentation of the optic vesicle, the lens placode, the olfactory placode, the otic placode, the stomodaeum, the proctodeum, the visceral clefts, and the formation of the Bowman's capsule. All of these movements seem to consist only of invagination, but since none has been investigated with careful marking experiments, we cannot say for sure whether involution is also involved.

The reverse of invagination – evagination – is a common method of forming diverticula from the tubes formed by gastrulation. Thus, the gut evaginates to form visceral pouches, the laryngo-tracheal groove, and the lung, the pancreatic and liver diverticula, to name the main ones. And the anterior hollow neural tube becomes brain by undergoing several evaginations, such as cerebral hemispheres, optic vesicles, mesencephalon and metencephalon. Evagination also is essential for the expansion of epithelial organs by branching. Typical examples of this are the salivary primordia (Bernfield et al., 1972) and the ureteric bud.

Recognition that these particular movements of cell sheets during organogenesis involve invaginations or evaginations comes from innumerable studies of serial sections of closely timed successive stages in development. So, there is no doubt that invagination or evagination, respectively, is taking place. What is not known is whether involution or evolution is involved. Since these morphogenetic movements take place discreetly in the inner confines of the embryo, the appropriate marking studies have not been done. However, they would be relatively easy to do for those invaginations that take place on the outer surface of the embryo. Perhaps such studies could be also performed in the case of the internal organs by culturing them in organ culture.

Movement of cells in sheets in the formation of organs and organisms is, of course, not confined to the vertebrates. Among the invertebrates perhaps the best studied of these movements are those of hydra (Shostak et al., 1965; Campbell, 1967). As the column of the organism expands, the ectoderm and endoderm both spread, sometimes together at the same rate and sometimes at different rates. This epithelial expansion apparently uses the mesoglea as a substratum (Shostak et al., 1965), a matter of some interest in the light of contemporary work on the use of intercellular materials as substrata for movement by vertebrate cells (see p. 274).

3.2.3. Wound closure

An ubiquitous example of the spreading of a cell sheet is the spreading of epithelia of all sorts to close wounds. This occurs in embryos, larvae, and adults throughout the Metazoa and on occasion may be so vast and rapid as to constitute a major cell movement. Indeed, response to wounding by spreading seems to be a general characteristic of epithelia. As Rand (1915) stated years ago "an epithelium will not tolerate free edge" and will generally spread quickly to restore its continuity. Characteristically, as soon as the free edges meet, movement of the sheet stops. In the sense that spreading of a cell sheet in wound closure restores the normal morphology of the organism, it is a morphogenetic movement. For the experimentalist, moreover, it offers a unique advantage that normal movements during morphogenesis lack. It can be started off and stopped almost at will.

4. *Cell movements in vitro*

4.1. *Mechanism of individual cell movements*

4.1.1. *Fibroblasts*

After this quick survey of cell movements within organisms, I wish now to discuss how cells move and what factors influence their locomotory activity and give direction to it. Although our primary concern in this review is the locomotion of cells in vivo, it is necessary first to summarize our understanding of cell locomotion in vitro. Most of what we know about tissue cell locomotion has been learned from its study under the artificial conditions of cell and tissue culture. I have no illusions that cell locomotion in culture necessarily proceeds in the same way as in vivo. Conditions in the two situations differ in important ways. The substrata used in culture to provide optimal optical conditions are usually plane surfaces of glass or plastic, hardly adequate substitutes for the substrata within organisms. Moreover, in order to observe cell surface activity in detail, precautions are taken to ensure that cells are observed without other cells on top. Such a monolayered condition is usually not found in vivo for individual moving cells or for cell groups, except for certain spreading epithelia. And, who knows how the various artificial culture media might alter cell locomotory properties? Nevertheless, when cells move, whether in vitro or in vivo, they can do so only by utilizing the locomotory machinery with which they are equipped. Thus, even though the precise use of this machinery may differ under different environmental conditions, its presence and some of its possible activities can probably best be discovered in artificial cell culture. This information is bound to give one an advantage in the study of cell movements in vivo. In the ensuing discussion, unless otherwise mentioned, the cells concerned will be fibroblast type cells, in both primary cultures from chick and mouse embryos and from established lines.

The locomotion of metazoan cells, whether fibroblasts or other cell types and whether in vitro or in vivo, is invariably accompanied by changes in cell form. Although it is tempting to explain these changes in terms of the laws of surface tension, as D'Arcy Thompson (1942) did in a most elegant way in his monumental treatise on form, this is inadmissible, mainly because the tension at the surface of cells is too low (see Danielli, 1945). How then are we to account for cell form changes? Clearly, there must be some kind of controlling cytoskeletal framework (Bonner, 1952). In modern terms, as everyone knows, this means microtubules and cortical microfilaments. And, in addition, since moving cells invariably adhere and conform to some extent to their substrata, the influence of such direct surface contacts on cell form is also not inconsiderable.

As fibroblasts engage in locomotion, they spread and flatten on the substratum, forming broad lamellipodia,* greatly increasing their apparent surface

*I shall avoid the terms "pseudopodium" and "amoeboid" in this paper, even though they are in common use, because they imply a similarity or even an identity with the locomotion of free-living

area. As they spread, the microvilli and blebs with which they are abundantly endowed in the rounded state diminish in number or disappear. Moreover, in certain cells, this occurs first in the marginal regions where the cells first spread. It seems likely, therefore, that blebs and microvilli represent membrane reserves which are available for the large increase in surface area that takes place when cells spread (Wolpert and Gingell, 1969; Follett and Goldman, 1970; Erickson and Trinkaus, 1976). Also, as a cell spreads, it adheres to the substratum, always near the margin of the spreading lamellipodia (Goodrich, 1924; Chambers and Fell, 1931; Curtis, 1964; Harris, 1973c; Lochner and Izzard, 1973; Revel et al., 1974) and perhaps lightly in non-marginal regions as well (Revel and Wolken, 1973; Lochner and Izzard, 1973). In so far as we know, spreading and adhesion to the substratum occur simultaneously. Rounded cells either are not adherent or adhere more lightly; they are more readily dislodged from the substratum. Flattened spread areas of cells are always adherent to the substratum.

Locomotion depends on advance of the lamellipodia over the substratum. If several lamellipodia spread in opposite directions, the cell as a whole may not move. If, however, one becomes larger, locomotion will generally take place in that direction. As the leading edge of this lamella advances, the cell will elongate in that direction, because the trailing edge of the cell remains adherent for a time. During its advance, the front end makes new adhesions to the substratum. Careful observation of moving fibroblasts with interference reflection microscopy has shown that individual points of adhesion (plaques) do not move. Instead, as old adhesions fade they are simultaneously replaced by new more forward adhesions formed ahead of the existing ones near the edge of the lamellipodium (Lochner and Izzard, 1973; Abercrombie and Dunn, 1975). These adhesions are apparently numerous and discrete and last from 10 to 25 seconds, depending on how fast the cell is moving. Cell spreading therefore seems to depend on the sequential formation of new plaques in front and the fading of old ones behind. The advance of the leading edge, however, is not at all steady; it consists of constantly alternating periods of small advances, retractions, and inactivity (Abercrombie et al., 1970a), which vary independently across the breadth of the lamellipodium. One point on the leading edge might be withdrawing while another is protruding. Forward advance of the lamella occurs in spite of these fluctuations because of the significantly greater frequency and duration of the protrusion phases. Since the protrusion phase usually lasts longer, the leading edge protrudes further than it retracts. The distance

---

amoebae. These terms are unwarranted when applied to most tissue cells, for, with few exceptions, their locomotion has either been shown not to be amoeboid or it has not been studied sufficiently to know. Loose use of these terms constitutes a substitution of terminology for knowledge and introduces additional confusion into an already confused literature. This is not to say that certain metazoan cells may not have amoeboid characteristics (see p. 273). But this is a subject for investigation that does not need the result prejudiced by terminology. For the present, therefore, I prefer to be phenomenological and use straightforward morphological terms like lamellipodium and filopodium that carry no other functional implication than that they are involved in locomotion.

involved in each protrusion is generally of the order of 5 μm. The speeds of protrusion and withdrawal are not significantly different.

Incidentally, independence of fluctuations of adjacent parts of the leading edge is consistent with local inhibition of ruffling (see p. 251) and suggests that both depend on local, compartmentalized factors, rather than freely diffusible ions.

When the leading edge has advanced some distance and the cell has become visibly elongate and oriented in the direction of locomotion, it becomes quite taut (Trinkaus et al., 1971), indicating that it is under much tension. Soon the tension will become so great as to pull the trailing edge away from the substratum. Immediately upon its detachment, the trailing edge retracts up to the point of the next adhesion. This may be due to active contraction of the cell, as is widely believed; but the possibility that it is wholly or in part due to passive elastic recoil of a stretched system has not been eliminated (Francis and Allen, 1971). Then, either the forward advance of the leading edge resumes, or the cell becomes quiescent for a while, or the cell margin forms a new lamellipodium which spreads in a new direction. If the new lamellipodium becomes larger than the old one, it will become a new leading edge and the cell itself will change direction. If both lamellipodia persist, have about the same size and spread in more or less opposite directions, the cell will cease moving, until one of them comes to dominate. When this happens, the other lamellipodium will become the trailing edge of the cell and will usually narrow as the cell elongates anew. The end result of these various activities is a jerky, apparently random kind of cell locomotion (Gail and Boone, 1970), with each successive leading edge crawling forward in its own fluctuating way, accompanied by much less frequent and always abrupt retractions of the trailing edge. This constantly varying behavior explains why, in spite of the forward thrust of the leading edge (as much as 5 μm/minute), the average speed of the whole cell over a longer period of time is quite low, of the order of 1 μm/minute (Abercrombie et al., 1970a).

The forward spreading of each lamellipodium is usually, but not always accompanied by ruffling (Abercrombie and Ambrose, 1958; Ambrose, 1961). Ruffles are local protrusions of the leading edge that uplift vertically from the substratum and often propagate back a certain distance on the upper cell surface (Ingram, 1969; Harris, 1969, 1973b). Since ruffling is usually associated with the locomotory advance of each lamella, it has been considered an essential part of the locomotory mechanism. This no longer appears to be so, however. Advance can occur without ruffling, both in a plasma clot or in fluid medium (Abercrombie et al., 1970b) and when cells crawl under a barrier, whether it be artificial (Heaysman, 1973) or other cells (Bell, 1976). It is probably significant that ruffles usually appear when there is a sudden change from withdrawal to protrusion. Normal cells will on occasion also form blebs at the leading edge which also propagate backward (Harris, 1973b). Their significance is not known.

When the trailing edge detaches from the substratum, it may leave very thin taut extensions of itself still attached to the glass or plastic. These are the so-called retraction fibers (Taylor and Robbins, 1963). As the cell moves along, these retraction fibers often break, leaving a tiny bit of the cell behind, still attached to the substratum. One of the main reasons for interest in retraction fibers is that their presence is positive proof that the cell was adhering at that point. It is not known whether adhesions of lamellae to the substratum are as punctate as suggested by the retraction fibers, but the observations of Lochner and Izzard (1973) (see p. 241) and SEM micrographs taken at a tilt (Revel et al., 1974) suggest that they are. In the SEM, lamellipodia appear to adhere at the edge by very small foot-like extensions from the under surface of lamellipodia, rather than over a broad front.

Another important feature of fibroblasts in culture is surface flow. When a moving cell encounters particles on the substratum it commonly picks them up and transports them backward on its upper surface (the surface away from the substratum) to the region of the nucleus (Abercrombie et al., 1970c; Harris and Dunn, 1972; Harris, 1973b). Particles will also be transported backward from the leading edge on the under surface a short distance, from 2–11 μm from the margin, apparently only as far back as the most forward adhesions of the lamella to the substratum (Harris and Dunn, 1973). These particles move a little faster than ruffles, about 3 μm per min. Although the interpretation of these results is controversial, i.e., particles may be inducing their own capping (de Petris and Raff, 1973), the view favored by these workers is that the particle movement reveals a hithertho unknown backward flow of cell surface, or of the parts of it that combine with the particle and that this flow is driven by the continuous addition of new membrane to the cell surface at the leading edge of the cell (see especially Harris, 1973b). Fibroblasts will also concentrate fluorescent *H-2* antibody bound to their surface into a cap in one area of the surface (Edidin and Weiss, 1972), giving conclusive additional evidence for surface flow. Moreover, the labeled antibody moves in a constant direction away from the cell's leading ruffled edge (see also de Petris and Raff, 1973). This concordance is marred, however, by the later discovery that not all fibroblasts cap (Edidin and Weiss, 1973). The reason for this is unclear, but it seems possible that it reflects differences in the mobility of membrane components among various cell lines. In any case, the capping of antibody by fibroblasts and lymphocytes is energy dependent (Taylor et al., 1971; Edidin and Weiss, 1972), suggesting a possible involvement of a submembrane contractile system. This important aspect of particle movement has not been studied.

Aside from its intrinsic general interest as additional evidence for the dynamic nature of the plasma membrane (Frye and Edidin, 1970; Singer, 1971; Singer and Nicolson, 1972), surface flow in moving fibroblasts has possibly important implications for two of the basic features of cell locomotion: change in cell form and cell traction against the substratum. One could conceive of the cell forming protrusions by moving membrane from one part of a cell to another and by inserting new membrane at the leading edge. Or, at least, insertion of new membrane could aid the process. And, if membrane flow is always backward, on the lower as well as the upper surface, then possibly the backward flowing adherent under surface could give the traction necessary for locomotion (Harris,

1973b). The important observation of Lochner and Izzard (1973) showing that points of adhesion to the substratum do not move is consistent with this hypothesis.

It must be emphasized at this point that, even though our knowledge of the detailed minute to minute locomotory behavior of fibroblasts in culture has expanded a great deal in the last five years or so, and even though in this period there have been a number of valiant efforts to understand the mechanism of this important process, we still cannot explain how cells move. It is therefore important to summarize briefly what has been learned and to pinpoint some of the areas of obscurity.

As it now appears, four of the important aspects of fibroblast spreading are: adhesion to the substratum; protrusion and withdrawal of the leading edge; increase in cell surface area, and cell elongation.

When a dissociated suspension of cells is plated onto a plane substratum, the rounded cells settle by gravity and begin to adhere to and spread on the bottom of the dish very soon after they contact it (Taylor, 1961; DiPasquale and Bell, 1974). No one knows exactly how soon this begins, because of the rudimentary state of our methods (for critical discussions, see Steinberg, 1964; Roth and Weston, 1967; Edwards and Campbell, 1971; Curtis, 1973; Walther et al., 1973). But, by the usual crude methods, such as seeing if the cells de-adhere when the culture dish is slightly agitated or turned on its side, fibroblasts begin adhering within minutes after contact with a glass or plastic substratum and most of them are quite firmly adherent at the end of an hour (Taylor, 1961). Coincident with this adhesion, the cell begins to flatten and spread on the substratum. Apparently cells make a number of close appositions with the plane substratum, in which the outer leaflet of the plasma membrane is separated from the substratum by less than 300 Å (Abercrombie et al., 1971; Heaysman and Pegrum, 1973; Bell, 1975; DiPasquale, 1975a). These points of adhesion, called plaques, are about a micrometer in length and breadth, and are found mostly near the edge of a lamellipodium. Since the location and dimensions of these plaques are roughly in accord with the observations with interference microscopy (Curtis, 1964; Lochner and Izzard, 1973), they are probably real, although the gap distances may well have been considerably distorted by fixation. Curtis (1964), for example, finds the gap to be of the order of 100 Å. There appear to be no specialized junctions or differentiations of the plasma membrane, such as one finds in tight and gap junctions. Looking at it naively, it is of interest that the junction between the cell surface and the glass or plastic is apparently an undifferentiated apposition and does not resemble the presumably firmer tight or gap junctions or desmosomes.

This is of much interest, for such an apposition could possibly provide sufficient adhesion for the traction required for spreading and yet, at the same time, be labile enough to fade quickly after the cell has made new more forward adhesions. Although this matter is poorly understood, it seems reasonable to assume that if cells adhere too firmly to the substratum, they might not be able to break their adhesions and thus would be immobilized, as, in fact, they are by polylysine (Erickson, unpublished observations). Likewise, if they adhere too weakly, they would de-adhere as the cell spreads and its tension increases, and immobilization would again be the result. The influence that different degrees of adhesion might have on cell locomotion has been nicely demonstrated by Carter (1965) and Harris (1973a). When cells are given a choice between more adhesive palladium and less adhesive cellulose acetate, they choose the palladium. This is not so much because they move better on the palladium – actually, they undergo locomotion on both substrata – but because they de-adhere more readily from the cellulose acetate.

An interesting and important feature of the plaques studied by Abercrombie et al. (1971) and Heaysman and Pegrum (1973) is the presence of a concentration of microfilaments in the cortical cytoplasm immediately beneath them. In a study which is a model for the kind of detailed attention such phenomena require, Heaysman and Pegrum (1971) also found microfilaments beneath cell to cell contacts (see also McNutt et al., 1973). They observed cells making contact in culture and then fixed them for thin sectioning at short intervals thereafter. When examined in the electron microscope these cells were found to have already made appositions with each other of the sort just described within 20 seconds of the first visible contact in the light microscope! Moreover, these quick contacts already have microfilaments concentrated just beneath them. This means that microfilaments mobilize or assemble at the point of contact within seconds after contact has been made! It is of course possible that cells make contact only where there are concentrations of microfilaments beneath the membrane. But this seems unlikely. If these microfilaments are contractile, their presence at a contact could explain the contraction that occurs in a fibroblast wherever it contacts another cell (Abercrombie, 1970; Abercrombie and Dunn, 1975). This has been termed contact retraction.

This evidence that cells can form appositions so quickly fits nicely the observation that as fibroblasts move they are constantly making new, more forward, adhesions to the substratum. The opposite theory, that adhesive appositions with the substratum may simply slide along as cell advances, is not completely outlandish, however, at least on theoretical grounds. It is not at all inconceivable that junctions and appositions might move, given the fluid properties of the lipid bilayer at physiological temperatures (Singer, 1971; Edidin, 1974). However, experimental evidence argues against this possibility. Point adhesions of lamellipodia to the substratum apparently do not move (Lochner and Izzard, 1973).

Since the entire lamellar region of a spreading cell is filled with a meshwork of microfilaments which apparently insert on the inner surface of the plasma membrane (Abercrombie et al., 1971; Spooner et al., 1971; Wessells et al., 1973; Goldman et al., 1973; DiPasquale, 1975a), everyone quite naturally believes that contractile activity involving these microfilaments is somehow involved in protrusion and withdrawal. This thought is supported by the finding that microfilaments concentrate beneath the plasma membrane just when and where it contacts the substratum (Abercrombie et al., 1971) and that where cells contact other cells they show contact retraction. Moreover, cytochalasin B, which inhibits microfilament contractility, freezes all activity of the leading edge, including ruffling (Carter, 1967; Wessells et al., 1971; Gail and Boone, 1971b) and Izzard and Izzard (1975) have shown that naked fibroblast cytoplasm is indeed contractile. This is as far as the relevant facts go, however. This has led to widespread conviction that contractile forces are involved in cell spreading. But we have no clear indication as to how. To fill this gap, Huxley (1973) has proposed a sliding filament model similar to that operating in muscle.

It is easy to speculate as to how contraction could cause withdrawal, by pulling on the plasma membrane and causing membrane to flow backward, but how could it cause protrusion? One possibility has come from altering the hydrostatic pressure of moving cells (Harris, 1973b; DiPasquale, 1975a). If the lowly permeable sugar, sorbitol, is added to the medium, the increased osmotic pressure causes a fall in the internal hydrostatic pressure of the cells. By filming cells under such treatment at high magnification, DiPasquale (1976) has shown that protrusion of the leading edge is reversibly inhibited by sorbitol in the medium. Since the internal hydrostatic pressure (turgor) of cells must be due to resistance exerted by the cell cortex, it is conceivable that contractile microfilaments control cell form changes in part by controlling cell turgor. Incidentally, although a certain amount of cytoplasmic flow must be involved in the spreading of cells from the rounded state and in fluctuations of the leading edge, it is not a dominant feature of fibroblast movement. There is no mass cytoplasmic flow, as, for example, in leukocytes (see p. 250).

Another possible factor at play in cell protrusion is rapid insertion of new surface membrane at the leading edge (with recycling of membrane from a sink near the nucleus). Abercrombie et al. (1970c) and Harris (1973b) suggest that this lies at the basis of rearward membrane flow, as shown by particle movement, and point out that this could also provide a mechanism for protrusion. In addition, it could explain ruffles, as a kind of buckling due to excess membrane formation. The greater frequency of ruffle formation during withdrawal (Abercrombie et al., 1970a) is consistent with this. The difficulty with the membrane insertion hypothesis of protrusion is that membrane insertion at the leading edge has not been established (see above).

One fly in the ointment has come from examining the cell surface with the SEM during spreading. Much of the surface of rounded, freshly seeded cells is in the form of abundant blebs and microvilli (Follett and Goldman, 1970; Porter et al., 1973). When BHK-21 fibroblasts begin to spread, these surface protuberances disappear, beginning near the cell margin in the newly forming lamellae. This raises the possibility that cell spreading, at least initially, does not require the formation of new membrane, but may utilize membrane already formed and stored in blebs and microvilli (Czarska and Grebecki, 1966; Wolpert and Gingell, 1969; Follett and Goldman, 1970; Erickson and Trinkaus, 1976). Moreover, when total surface area is estimated, taking microvilli into account, much of the surface of fully spread cells can be accounted for. If the surface membrane of rounded daughter cells were simply redistributed, it would provide the membrane required for the initial spreading of the cells. This suggests that some membrane of rounded cells flows toward the margin during cell spreading (Erickson and Trinkaus, 1976). Significantly, these microvilli are filled with microfilaments (Follett and Goldman, 1970; Szollosi, 1970; Betchaku and Trinkaus, 1974; Evans et al., 1974). Thus, submembranal contractile elements could well be involved in this redistribution.

Another possible contributing factor to cell spreading, which has received little attention, is deformation of the plasma membrane by molecules inserting differentially into the outer layer of the lipid bilayer (Sheetz and Singer, 1974). This would cause the membrane to evaginate and form protrusions. This has been demonstrated artificially with differentially inserting drugs in erythrocytes, but no physiological counterpart has yet been found.

A characteristic external feature of a fibroblast in full locomotion is its elongate form, oriented in the direction of locomotion, and a striking internal feature of such cells is the presence of abundant cytoplasmic microtubules oriented parallel to the direction of elongation of the cell. Because of the active role that microtubules appear to play in cell elongation in other systems (see below, p. 307), it has seemed likely that microtubules do the same in fibroblasts. Several laboratories have tested this hypothesis by treating elongate fibroblasts with colchicine* (Vasiliev et al., 1970; Gail and Boone, 1971a; Goldman, 1971). In most cases, this causes the cells to lose their elongate orientation and assume a rounded flattened, "pancake" morphology, with ruffling occurring more or less uniformly around the whole periphery. Since spreading in these treated cells occurs more or less equally around the periphery, the cells become immobilized. Sometimes cells treated with colchicine round up to a considerable degree and become more fluid, with exaggerated cytoplasmic flow (Harris and Bell, unpublished observations) somewhat like leukocytes whose locomotion, incidentally, is little affected by colchicine (Ramsey and Harris, 1972). All of this is consistent with the hypothesis that preservation of the elongate form of a fibroblast depends upon oriented microtubules. It does not tell us, however, whether microtubules actively promote cell elongation in the first place by their elongation and orientation, or whether they are oriented passively by the longitudinal tension in the cytoplasm of an actively elongating cell and then subsequently stabilize the elongate form. Goldman (1971) and Goldman and Knipe (1973) tackled this question by seeding rounded cells into medium containing colchicine. As expected, the cells attached to the substratum and spread on it but apparently did so more or less equally in all directions, which caused them to assume an epithelioid form. The finding that true epithelial cells possess very few microtubules (DiPasquale, 1975a) is consistent with this result. Significantly, when epithelial cells are separated from their sheet they bleb widely, do not spread and, of course, do not translocate

*Colchicine has been a most useful tool for studying developmental processes associated with microtubules because it causes disassembly of intact microtubules. It must be kept in mind, however, that colchicine has other effects that in certain circumstances may in their own way modify cell form and behavior (Bryan, 1974).

(Middleton, 1973; DiPasquale, 1975a). From all these results it appears that although microtubules are not necessary for spreading of fibroblasts, they are necessary in some way for their elongation and, therefore, for their locomotion. In corroboration of this, cyclic AMP, which enhances polymerization of tubulin subunits to form microtubules, causes fibroblasts and transformed cells to become elongate, with distinct narrow processes (Johnson et al., 1971; Hsie and Puck, 1971; DiPasquale et al., 1976) and increased numbers of microtubules (Porter et al., 1974).

Spreading appears to be the active feature of cell locomotion. However, were it not for retraction of the trailing edge, the cell as a whole would not displace. It has not been determined whether this retraction is due to active contraction or passive recoil of a stretched system. Unfortunately, treatments which inhibit contractility (such as cytochalasin B) inhibit it at the leading edge as well and thus prevent the forward spreading which ultimately causes detachment of the trailing edge. There is, however, some evidence favoring active contraction that has come in a round about way from study of elongating neurons. Ludueña and Wessells (1973) find that these cells, which lack the retraction phase of locomotion (nerve cell bodies do not move), also lack the longitudinally oriented sheath or cable microfilaments which characteristically extend the length of an elongated fibroblast. What is really needed, however, is a direct attack, in which the trailing edge of an elongate, stretched fibroblast is detached artificially in the presence of inhibitors of contraction such as cytochalasin B, and inhibitors of energy metabolism (on which active contraction depends). Neurons should also be detached to see if they retract.

Another aspect of retraction of the trailing edge that requires attention is the detachment process. Curtis (1964) observed that the edge detaches completely from the substratum, yet it is a common observation that when cells retract they often leave a part of themselves behind (Weiss and Lachmann, 1964; Poste et al., 1973; Terry and Culp, 1974). L. Weiss contends that the latter situation involving non-lethal cohesive rupture *within* the cell periphery is usual and normal in cellular separation and that adhesion of cells to, and their separation from, should not be treated as similar processes. This important matter obviously requires systematic investigation. It is of interest to note, however, that cells deposited onto substrata "coated" with materials deposited by previously attached cells appear to detect the deposited material(s) and show differences in their social and metabolic behavior compared with cells seeded onto unmodified substrata (Bolund et al., 1970; Culp, 1975; Weiss et al., 1975).

I will not deal with the possible chemical or physical basis of adhesion of cells to various substrata, artificial or natural. This is obviously a most important matter and the literature on it is large. Thus far, however, it is almost entirely inconclusive, in so far as valuable leads for cell movements are concerned. It does seem that sugar residues of the membrane glycoproteins are in some way involved, although it is not clear which ones and how (see reviews by Roseman, 1970; Kraemer, 1971; Roth, 1973; Curtis, 1973; Poste and Allison, 1973; Weiss, 1967).

A significant bit of evidence in support of this notion has come from study of agglutination of cells with certain lectins that combine with sugar groups at the cell surface. When ferritin labeled lectins are used, they are found to be concentrated at points of cell contact (Nicolson, 1972; Singer, 1974). This raises the possibility that upon the stimulus of contact, adhesive molecules move within the fluid lipid bilayer to the site of contact and by concentrating their combining groups make an adhesion possible. The observation that microfilaments are clustered in the cytoplasm beneath contacts between cells (Heaysman and Pegrum, 1973) supports the speculation that they perform the work of moving the glycoproteins within the bilayer to the site of contact.

4.1.2. Cancer cells

Although it is widely believed that transformed cells move differently from normal fibroblasts in vitro (e.g., Trinkaus, 1969, p. 22), this seems not to be so, at least for all transformed cells subjected to detailed study. Transformed cells may differ from fibroblasts in important ways, such as pattern of adhesion to the substratum, but the differences appear to be entirely quantitative (Leighton et al., 1959; Vasiliev and Gelfand, 1973; Bell, 1972, 1976; Harris, 1973c; Erickson, unpublished observations). In their morphology and manner of locomotion they appear to be either just like fibroblasts, or to differ from them only in degree. It therefore appears probable that all that I have been saying about the mechanism of locomotion of untransformed fibroblasts in vitro applies equally well to their transformed neoplastic counterparts.

4.1.3. Neurons

Neuronal extension is exceptional in that during advance of the leading edge the cell body remains fixed. In addition, however, the appearance of the advancing tip of a neurite is very different from that of a fibroblast or epithelial cell (Bray and Bunge, 1973; Wessells et al., 1973; Letourneau, 1975a,b). It lacks a broad, ruffling lamella and instead protrudes numerous constantly bending, rod-like, elongate extensions which frequently branch. They also may fuse with each other and retract into the cell (Spooner et al., 1974). These protrusions are referred to as microspikes, but they differ somewhat from those observed on fibroblasts and epithelial cells in that they often bend. Those on fibroblasts and epithelial cells, in contrast, act as if they were rigid rods, hinged at their base (Taylor and Robbins, 1963; DiPasquale, 1975a). Sometimes neurite microspikes flatten and spread momentarily on the substratum, but neurites never have a broad flattened area at the tip like a fibroblast. This highly motile, arborized tip is called a "growth cone," following the older literature in which the tip was first observed with the only means available at the time: in fixed sections (Cajal, 1890) and in the living with bright field microscopy (Harrison, 1910). Although this term is most misleading, it seems to have been solidified by upward of 75 years of usage. A neutral expression, like "extending tip", would be better.

An extending neurite adheres to the substratum in the region of its tip, often at the tips of microspikes. Whether there are other adhesions along the axon seems to be unknown. The microspikes, like the lamellae and microspikes of fibroblasts and epithelial cells, are filled with a meshwork of microfilaments

(Wessells et al., 1973). Also, like fibroblasts, when treated with cytochalasin B the protrusions are immobilized and retract, and the axon stops extending (Yamada et al., 1971). The axon itself possesses abundant microtubules oriented in its long axis and when these are disrupted by colchicine the axon retracts, indicating a dependence on microtubules for maintenance of the elongate state (Yamada et al., 1971). It is not at all understood how the tip of the axon extends over the substratum. Possibly it pulls itself along by adhesion of the tips of the microspikes to the substratum, followed by contraction of their microfilaments. It seems clear that the new membrane necessary for the spectacular extension of a neurite is added at the tip of the axon (Bray, 1970; Bray and Bunge, 1973). Marks on the axon always maintain the same position relative to the substratum and never move out toward the tip. Pfenninger and M.B. Bunge (1973) and Pfenninger and R.P. Bunge (1974) have recently provided evidence from freeze-fracture study of the tip that membrane vesicles in the cytoplasm may contribute to the increase in plasma membrane surface.

4.1.4. Leukocytes

The locomotion of polymorphonuclear keukocytes has particular relevance for the study of individual cell movement in vivo because of the amount of cytoplasmic flow involved (De Bruyn, 1944) (also see below p. 261). In this, of course, they are quite different from fibroblasts. Leukocytes and fibroblasts have an important similarity, however; they both thrust themselves forward by means of a flattened veil or lamellipodium (Robineaux, 1964). What happens is a rapid spreading of a broad, hyaline lamellipodium on one side of the leukocyte, followed by streaming of a large part of the granular cytoplasm into the lamellipodium, filling it and rounding off (Ramsey, 1972b, 1974). A new lamellipodium then forms, cytoplasm streams into it, and so on. In this manner, the bulk of the cell literally pours itself forward. As each new lamella forms and cytoplasm flows into it, the adhesions of the previous lamellipodium apparently remain and appear at the trailing end of the cell as a cluster of stretched retraction fibers. When these break, they form a small bump or "tail" at the rear end. The detailed adhesions of the lamellipodium to the substratum have not yet been studied. This streaming of cytoplasm into the lamellipodium is accompanied and perhaps caused by a contractile wave which is apparently generated in the cortical cytoplasm at the rear end of the cell and moves forward in a kind of peristalsis (Senda et al., 1975), like a "contractile ring", as described some years ago by Warren Lewis (1942) and De Bruyn (1946). Investigation of glycerinated leukocytes by Senda et al. (1975) has revealed a network of thick and thin filaments just inside the plasma membrane. The thin filaments complex with heavy meromyosin and are indubitably actin. The thick 150 Å filaments may be myosin. It seems highly likely, therefore, that these filaments provide the contractile force. If so, we may well be closer to an explanation of leukocyte locomotion than of the much more studied fibroblast. The observation that the speed of movement of leukocytes is markedly reduced by cytochalasin B is consistent with this interpretation (Zigmond and Hirsch, 1972; Ramsey and

Harris, 1972). Perhaps this is also related to their deformability (see Tickle and Trinkaus, 1973).

Incidentally, the average speed of leukocytes in culture is markedly higher than that of fibroblasts – about 10–12 μm per minute (Ramsey, 1972b), in contrast to about 1 μm per minute for fibroblasts (Abercrombie et al., 1970a). Perhaps this is related to the rapid cytoplasmic flow of leukocytes. Free-living amoebae also move quite fast, as do cells in vivo in which there is much cytoplasmic flow (see below, p. 264).

Polymorphonuclear leukocytes are of interest for another reason. They are chemotactic in vitro and, presumably, in vivo as well, where this sensitivity could have obvious physiological importance (Ramsey and Grant, 1974). Although it is not understood how the cells respond directionally to tiny differences in a gradient in concentration of an attractant, the mode of locomotion does not change under the influence of an attractant, nor does the speed. Chemotaxis is effected solely by a change in the direction of movement (Dixon and McCutcheon, 1936; Ramsey, 1972a, 1974). In human leukocytes, Ramsey (1972b, 1974) found that this directionality is achieved by cytoplasmic flow. Lamellipodia apparently extend on all sides of a cell, but cytoplasm tends to flow preferentially into lamellipodia that are on the side facing the attractant. In horse leukocytes, on the other hand, Zigmond (1974) observed that the frequency of "pseudopod" formation increases on the high concentration side of the cell. Conceivably, the wave of contraction generated near the trailing end of the cell is stimulated by a lower level of attractant concentration. Or perhaps the wave of contraction tends to move up the gradient in attractant concentration.

4.2. The social behavior of cells in vitro

4.2.1. Contact inhibition of cell movement

A form of in vitro cell behavior considered by many to provide a good model for the movement of both cell streams and certain cell sheets in vivo is contact inhibition of locomotion. This phenomenon has been recognized for some time (e.g. Loeb, 1921), but a rigorous quantitative foundation for it was first supplied by Abercrombie and Heaysman some twenty years ago in a series of brilliantly simple papers (1953, 1954). Inasmuch as contact inhibition has been thoroughly reviewed recently by Abercrombie (1970) and Martz and Steinberg (1973) and Harris (1974) there is no need for an exhaustive treatment here. In brief, when a fibroblast encounters another fibroblast in culture it either forms a lateral adhesion with it or moves under it. If they contact each other at their spreading lamellipodia they almost invariably cohere. When they cohere, ruffling is inhibited at the point of adhesion (Abercrombie and Ambrose, 1958; Trinkaus et al., 1971), the lamella contracts somewhat (Abercrombie and Dunn, 1975), and the cell stops moving. Such a contact inhibited cell will not move again until it forms a new lamellipodium elsewhere. As this new lamellipodium spreads, it will stretch the cell and eventually break the adhesion. Cell movement will then continue unimpeded until the cell meets another cell, etc. The most impressive triumphs of the principle of contact inhibition of movement are that it explains the directional migration of fibroblasts away from a solid explant in vitro to form the ever widening "zone of outgrowth" and the tendency of normal fibroblasts to form cell monolayers in culture (see description in Trinkaus, 1969, p. 22–25). It should be emphasized, however, that a fibroblast monolayer is somewhat of a steady state affair in which spaces form frequently between cells, only to be filled again as cells spread into them. Cells also underlap frequently. In a word, cells are more or less constantly milling about (Abercrombie and Heaysman, 1954; Bell, 1972; Martz and Steinberg, 1974).

Epithelial cells also show contact inhibition, both with other epithelial cells and with fibroblasts (Abercrombie and Middleton, 1968; DiPasquale, 1975a). As a consequence, they also spread only at their free margins and form cell monolayers with other epithelial cells. The feature that apparently distinguishes epithelial cells from fibroblasts in this regard is their tendency to form stronger and broader lateral adhesions. Thus they form the characteristic cohesive epithelial sheet, as opposed to the looser network of fibroblasts.

Contact inhibition appears to be due primarily if not entirely to two poorly understood features of cell social behavior in culture. Cells tend to adhere to other cells along their borders (Abercrombie, 1970), and tend not to spread on each other's upper surfaces (DiPasquale and Bell, 1974; Elsdale and Bard, 1974; Armstrong and Lackie, 1975; Vasiliev et al., 1975). This is very easy to demonstrate for the upper surface of an epithelial sheet, but it is apparently also true of fibroblasts (DiPasquale and Bell, 1974, 1975). Cells might also adhere more strongly to the substratum than to each other and thus not be able to leave the substratum when they contact each other (Abercrombie and Dunn, 1975). This would prevent overlapping (Abercrombie, 1961; Martz and Steinberg, 1973; Vesely and Weiss, 1973). It would not necessarily prevent ruffling, however. Moreover, cells have been found to prefer moving onto a less adhesive substratum to moving over each other (Harris, 1973a, 1974), an observation clearly not in accord with this "differential adhesion" model.

Fascinating though it is as an explanation of cell behavior in vitro, the possible significance of contact inhibition for the movement of cells in vivo was not widely appreciated until Abercrombie et al. (1957) observed that certain lines of sarcoma cells are not restricted in their movements by contact with normal fibroblasts. They appeared to move right over the fibroblasts. Since cancer cells move over normal tissue cells during invasion in vivo (i.e., they are not contact inhibited) and normal cells do not ordinarily move over normal cells in vivo, at least in adults (are contact inhibited), the contact behavior of cells in culture suddenly appeared to have relevance for cell behavior within organisms, in particular for invasiveness. Indeed, this apparent lack of contact inhibition by tumor cells in vitro quickly became an assay for neoplastic transformation of cells infected with oncogenic virus or treated with carcinogens or X-rays (Temin and Rubin, 1958; Vogt and Dulbecco, 1960; Sachs and Medina, 1961; Koprowski et al., 1962; Shein and Enders, 1962; Berwald and Sachs, 1965; Borek and Sachs, 1966; Rapp, 1974). Three criteria rapidly gained widespread acceptance as evidence of reduced contact inhibition of movement in vitro: the intermingling of cells; criss-crossing and nuclear overlap; and multilayering or piling up.

There are, however, certain difficulties with these criteria. Intermingling of cancer cells with normal cells in vitro may not be due to reduced contact inhibition, as is usually assumed. In studies, which have been largely overlooked, Barski and Belehradek (1965, 1968) observed tumor cells to invade an outgrowth of normal cells not by moving over them but by working their way through gaps in the outgrowth. They would not move over a coherent sheet of normal cells. As far as multilayering and cross-crossing are concerned, cells have rarely been observed at high magnification while engaged in the process. Observation of the end result has been considered sufficient, it being presumed that the cells reached this condition by overlapping, i.e., moving over the upper surfaces of other cells. Since the discovery that cells may crawl under each other without leaving the substratum (Boyde et al., 1969; Weston and Roth, 1969; Guelstein et al., 1973; Harris, 1973c) it has been clear that criss-crossing might result as well from underlapping as from overlapping. Criss-crossing from underlapping would tell nothing about contact inhibition, for, by moving under another cell, a cell avoids making a contact with its margin. The same reservation applies to the use of increased nuclear overlap as an indication of decreased contact inhibition. Nuclei could become overlapped as well from one cell moving under the other as from the reverse. As for multilayering or piling up, this could result from retraction of a cell sheet, which would have nothing to do with contact inhibition, or from continuing cell division, which would indicate reduced contact inhibition of cell division but not necessarily of cell movement.

With these reservations in mind, the contact behavior of two lines of transformed cells has recently been subjected to high magnification cinematographic analysis in our laboratory. Polyoma transformed mouse 3T3 cells (Bell, 1972, 1976) and polyoma transformed hamster BHK21 cells (Erickson, 1976) criss-cross to a high degree in culture, as shown by both cinemicrography and SEM. In all cases, however, when observed in the process, the presumed overlapping was found in fact to be due to one cell moving *under* another. No case of overlapping has been observed. In addition, when these cells do make adhesive contacts, both movement and ruffling are inhibited, just as with untransformed 3T3 and BHK21 cells. Moreover, and more to the point as far as invasiveness is concerned, each transformed line behaves in the same way when mixed with its untransformed counterpart. The conclusion from these studies is inescapable. These two transformed cell lines do not exhibit reduced contact inhibition of movement. They are different from the normal lines from which they were derived, to be sure, but not in terms of contact inhibiting behavior. They simply underlap more, apparently because they have fewer adhesions to the substratum, hence fewer impediments to underlapping. Although it would be premature to generalize on the basis of these results, two conclusions seem legitimate: (1) since it is clear that these two lines of transformed cells do not show reduced contact inhibition of cell locomotion, other transformed cells should be scrutinized equally closely; and (2) since transformed cells do not necessarily show reduced contact inhibition of movement, the value of the concept of contact inhibition for the study of cell invasiveness in vivo, whether of cancer cells in adults or of normal cells during morphogenesis, is thrown into question.

A recent study of contact interactions between polymorphonuclear leukocytes and fibroblasts (Armstrong and Lackie, 1975) has further complicated the picture. Although contact between these cells is not accompanied by diminution of ruffling of either cell, nevertheless they fail to crawl over each others' surfaces.

4.2.2. Contact guidance and cell alignment

It has often been proposed that the oriented movement of cells in culture along discontinuities of the substratum, first observed by Harrison (1914) and intensively studied and named contact guidance by Paul Weiss (e.g. 1945, 1961a), could serve as a model for directional movements. And, indeed, this notion has received support from many observations of cells moving along blood vessels, nerves, and other discontinuities in embryos (Speidel, 1935; Trinkaus, 1966). It needs to be emphasized, however, that although cells must become oriented (North–South) in order to move in a certain direction (North or South), the orientation does not in itself convey directionality. Something must be added, like contact inhibition or chemotaxis. Hence, the beautiful work on contact guidance of cells in culture has thus far been of little help in understanding morphogenetic movements, in spite of many optimistic allusions to it in this context (e.g. Trinkaus, 1965). Moreover, even the operation of contact guidance in vitro, in one of the classical examples of Weiss (1945), has recently been thrown into question. The tendency of neurites to extend preponderantly into the clot between two explants (the "two center effect"), has been shown not to occur if the nerve fibres move out individually and do not become fasciculated (Dunn, 1971). The nerve fibres leave an explant and move out in an overall radial fashion, like fibroblasts, because when a tip contacts another nerve fiber it is inhibited from extending further in that direction and subsequently extends in a new direction. Since a radial direction offers the lowest frequency of contacts with other fibers, nerves tend to extend radially.

Certain other kinds of cells, such as human fetal lung fibroblasts and BHK21 cells, elongate and orient toward each other in culture without any relation to discontinuities of the substratum and thus become aligned in parallel arrays. These arrays are of interest to an embryologist because they simulate the orderliness that characterizes cells arranged in tissues in vivo. Once these elongate cells have aligned themselves, however, they do not cease locomotion; they keep moving, constantly perfecting and reaffirming the parallel situation (Elsdale and Bard, 1972a). Thus active cell movement is necessary for both generating and maintaining the arrays. This is an excellent case of oriented control of cell shape and movement. It is not directional, however; the oriented cells move in one direction as well as the opposite. The elongate shape of these cells is apparently due to their abundant oriented microtubules. If seeded into medium containing colchicine, they attach and spread, but do not assume the elongate form (Goldman, 1971; Goldman and Knipe, 1973). Elongation in itself is not sufficient for alignment, however. Indeed, if the cells' adhesions to the substratum were confined to their ends, as is true of many elongate cells, these cells would tend to underlap extensively and criss-cross rather than align. Perhaps their adhesions are not confined to their ends. Erickson (unpublished observations) has checked this possibility by examining BHK21 cells with SEM at high magnification and found that they possess many tiny point adhesions to the substratum along their sides. If these prevented underlapping, these contact inhibiting elongate cells would have no choice but to line up in parallel array as their population density increases and they form monolayers.

Another feature of the behavior of fetal lung fibroblasts in culture that has probable significance for tissue formation in vivo is their tendency to form multilayers in which the successive layers of parallel arrayed cells are oriented orthogonally (Elsdale and Bard, 1972a). Since these cells apparently exhibit contact inhibition of movement when first plated out on glass or plastic, the explanation of this multilayering has special interest. It seems that the layers are partially separated by collagen, such that each layer does not really form completely on tops of other cells, but rather in part on a layer of collagen. In any case, when collagenase is added to the medium, orthogonal overgrowth is prevented, or, if orthogonal layers are already formed, they are disrupted. Because of the abundance of collagen in vivo, these results have important implications.

4.2.3. Directional cell movement

4.2.3.1. Contact inhibition. Since directionality is a cardinal feature of morphogenetic movements within organisms, control of the directionality of fibroblast movement from an explant deserves closer scrutiny. Although the overall movement of the cell population is away from the explant and therefore directional, movement of individual cells is directional only relative to the last cell that they contacted. They always move *away* from this cell into available free space on the glass or plastic substratum. This means that cell movement will more frequently be away from the explant than toward it, because there are more cells nearer the explant and more cell-free space away from it. However, individual cells can and do move toward the explant and in other directions as well. Because cells continue to move out of the explant into free space vacated by other cells that have moved further out and because cell division continues, the free-space near the explant quickly becomes occupied by cells, until they form a monolayer whose constituent cells are more or less inhibited from movement by contacts with other cells on all sides. Only cells at and near the periphery of this widening monolayer have substantial free space to move into and this is predominantly away from the monolayer, hence away from the explant. And, of course, as these cells move out, those behind them are no longer inhibited and they in turn can move out also. This directional movement thus has two important features: (1) cell movement is on the average directional, away from the explant, though in the more peripheral more sparsely populated regions of the zone of outgrowth, at any given moment, individual cells may not be moving directionally, relative to the explant; and (2) cell movement continues only peripherally. Cells near the explant, in an ever widening circumference, soon cease moving, due to contact inhibition, except for a small amount of milling about, and, of course, they inhibit further emigration of new cells from the explant. The whole cell mass thus moves only in the sense that the outer cells continually move away. It should be clear from this that contact inhibition of cell movement can give directionality to the movement of a population of cells only when the cell source is a single mass of many cells. If cells are dissociated and seeded randomly as a suspension, they will show contact inhibition as they collide and will tend to move away from each other, when they break away. But, there will be no overall directionality to their movement.

4.2.3.2. Chemotaxis. The chemotactic movement of a metazoan tissue cell such as the leukocyte toward an attractant would seem to be an ideal in vitro model for the directional movement of cell streams during animal morphogenesis. As in directional streams, chemotactic movement of leukocytes is directional in an overall sense, yet there is a certain amount of variable behavior on the part of individual cells (Ramsey, 1972a) since they change position relative to one another during movement. Also cells stop moving when they reach their destination. Another possible model is the chemotactically controlled streaming of slime mold cells toward a center. This is far and away the best analyzed example of directional cell movement and is clearly a morphogenetic movement in the best sense. Nevertheless, I will not consider either leukocyte or slime mold chemotaxis in this paper since both have been carefully reviewed recently (Bonner, 1971; Ramsey and Grant, 1974; Loomis, 1975).

Moreover, there is no undisputed evidence that chemotaxis operates during the normal morphogenetic movements of Metazoa. It may be objected, with good reason, that negative evidence here does not constitute disproof. However, since by and large science progresses best on the basis of probabilities, not improbabilities, it seems to me that until some good evidence appears for the operation of chemotaxis during morphogenetic cell movements or cancer invasion it is more profitable to look for other models.

4.2.3.3. Gradient in adhesiveness of the substratum. One of the signal achievements of cell culture has been to establish the importance of adhesiveness of the substratum in the control of cell behavior. Cells must adhere to their substratum in order to move: strongly enough to gain traction, but not so strongly that they become stuck in one place. Little wonder, therefore, that a gradient in differential adhesiveness should have much appeal as an in vitro model for giving cells directionality. The fundamental experiment on this matter is that of Carter (1965), in which he showed that cells will move up a gradient of increasing concentration of palladium, a substance to which they adhere and use as a substratum for locomotion. Carter believed that cells move up such a gradient directly, by a kind of passive wetting process. Harris (1973a), however, filmed the process and showed that this is not the mechanism. Cells actually move equally well on both palladium and on a substratum to which they adhere less strongly, such as cellulose acetate, but will detach more readily from the latter. Hence, they will tend to accumulate on a more adhesive substratum, not because they move onto it more readily, but because they detach more readily from a less adhesive one. Extending neurites also favor a more adhesive substratum, in this case polyornithine instead of palladium, and will move preferentially onto it (Letourneau, 1975a,b), possibly by the same mechanism.

This too has been an attractive model for explaining directional cell movements in vivo (e.g., DeHaan, 1963); but as yet no evidence has been forthcoming.

4.3. The spreading of epithelial sheets

An epithelium is defined as a tissue that lines a cavity or body surface, and epithelial sheets in vitro, in contrast to fibroblast sheets, are characteristically organized in tightly cohesive monolayers, or perhaps bilayers, in which individual cells adhere to their immediate neighbors over all or almost all of their lateral borders, right up to the margin of the sheet. Gaps (at the level of the light microscope) may appear between cells, but they are rare and quickly closed by spreading of cells on the margin of the gap. Cells do not change positions relative to one another. This is indicative of a contact inhibiting system. Except at these gaps, locomotory activity is confined to the marginal cells of the sheet (Holmes, 1914; Uhlenhuth, 1914; Matsumoto, 1918; Hitchcock, 1939; Danes, 1949) and when local portions of the margin spread more than the rest, the cells between such outgrowths come to be markedly stretched and tangentially oriented (Vaughan and Trinkaus, 1966). Because of this, the spreading of the entire sheet appears to depend on the activities of its marginal cells. Consistent with the observation that cells adhere firmly to the plane substratum only where they are actively spreading, adhesion of an epithelial sheet to the substratum is almost entirely by means of its marginal cells. Thus, when the margin of the sheet de-adheres, the entire sheet usually retracts immediately. Submarginal adhesions also occur (DiPasquale, 1975a), but these too seem always to be where cells are spreading, to fill a gap in the sheet.

The spreading cells at the margin of the sheet behave very much like fibroblasts, except that they are more tightly bound to the cells behind and on their sides and rarely break away (Vaughan and Trinkaus, 1966; Middleton, 1973; DiPasquale, 1975a). They possess a thin, fluctuating, ruffling lamella, which alternately protrudes, withdraws, or is stationary. Like fibroblasts (DiPasquale, (1975a, 1975b), translocation results from more time being devoted to protrusion (see p. 241). They may also spread in a rather bizarre manner without ruffling, which entails microspikes adhering to the substratum at their tips and cytoplasm spreading out in between them in a web-like fashion (DiPasquale, 1975a). As in fibroblasts, the lamellipodia and microspikes of these spreading epithelial cells are filled with a meshwork of 40–80 Å microfilaments. Moreover, spreading is inhibited by cytochalasin B (DiPasquale, 1975a). Because spreading and adhesion take place almost exclusively at the margin, the sheet is put under much tension as it spreads, and retracts greatly if detached.

Spreading epithelial cells show contact inhibition of cell movement. Epithelial sheets tend to be strongly monolayered (Middleton, 1972) and when epithelial cells contact other epithelial cells or fibroblasts and adhere, they inhibit each other's movement (Abercrombie and Middleton, 1968; DiPasquale, 1975a). Underlapping but not overlapping has been observed. Indeed, the upper surface of an epithelial sheet in culture does not support spreading by other cells (DiPasquale and Bell, 1974, 1975; Elsdale and Bard, 1974, 1975).

Epithelial cells in culture appear to differ from fibroblasts primarily in two ways: by their strong lateral adhesions and by their poor locomotion as individuals. The strong, extensive zonular adhesions which epithelial cells make with each other is their most distinctive feature, and, indeed, defines an "epithelium". Moreover, spreading of a sheet as a coordinated unit no doubt depends on these zonular adhesions. Although vertebrate epithelial cells are usually united by a variety of junctions and appositions, such as tight and gap junctions, 150–200 Å appositions, and desmosomes, it is the desmosomes that appear to be distinctive (DiPasquale, 1975). Fibroblasts and other kinds of tissue cells apparently never form them. Indeed, the strong intercellular attachments provided by desmosomes could well be the quality that makes epithelial cells form an "epithelium". In invertebrate epithelia, desmosomes are lacking, but their role is seemingly taken over by the separate junction.

The other striking feature of epithelial cells is their behavior when isolated from the sheet. If unattached to other cells, they become somewhat rounded, bleb wildly, have no particular orientation, do not spread on the substratum, and translocate little if at all. When they come into contact with the edge of the sheet, however, they join it and quickly spread on the substratum, like a typical marginal cell. They are clearly dependent on adhesion to other epithelial cells in order to spread (Middleton, 1973, 1976; DiPasquale, 1975a; see also Trinkaus, 1963). Epithelial cells thus appear to be dependent on the epithelial condition. They must have membership in a cell collective, the sheet, in order to spread like a normal epithelial cell. The fine-structural studies of DiPasquale (1975a) provide a probable explanation for this behavior. In contrast to fibroblasts, epithelial cells have very few microtubules. In view of the likely role of microtubules in the elongation of cells (see pp. 247–248), this paucity of microtubules could lie at the basis of the inability of isolated epithelial cells to spread into the elongate of stellate morphology necessary for locomotion. Indeed, the absence of microtubules in non-motile epithelial cells is added evidence of the importance of microtubules in fibroblast-type locomotion. The small number of microtubules, and the presence of desmosomes, could be the fundamental features of epithelial cells. It is conceivable, incidentally, that cells within an epithelial sheet manage without microtubules because they don't need them. Their extensive lateral adhesions to their neighbors suffice as a means of giving a definite form to the cells. As expected, colchicine has no effect on the spreading of an epithelial sheet during several hours of treatment (DiPasquale, 1975a).

5. *Modes of cell movement in vivo*

5.1. *Individual cells*

Our early understanding of how tissue cells, other than neuroblasts, undergo locomotion came largely from the pioneering efforts of the Lewises (1912, 1922) and the Clarks (1912, 1925), and their papers are still worth careful study. Since then, however, the in vitro approach of the Lewises has held sway and most investigations have been confined to observing cells in tissue culture. This is unfortunate, for cells do not necessarily behave within organisms as they do in culture. Indeed, cells do not even behave in the same way in vitro under different conditions. To put it simply, if we are to learn how cells move within organisms they must be studied there, where they can be observed in the particular environments that are their normal lot. This thought is obvious and elementary, but it needs emphasis. Our knowledge of how cells move in vivo is miniscule compared to what we know about their movements in vitro. To be sure, from time to time during the first half of the century there have been noble efforts to follow cell locomotion within organisms, but, with notable exceptions (such as Clark, 1912; Clark and Clark, 1925; and Speidel, 1933, 1935), they have told us little.

The observations of Clark (1912) and Clark and Clark (1925) on the movement of mesenchyme cells in the matrix of the transparent tail fin of an anuran larva deserve attention not only for their value as early explorations, but as well for the care and elegance of the observations. These cells are stellate or dendritic, possessing long thin branching processes emanating from a thick cell body. They translocate by moving cytoplasm from the thicker central part of the cell into certain of the long processes and withdrawing cytoplasm from processes on the other side. In this way, the main body of the cell is shifted from the retracting to the advancing processes. Except for this work and that of Speidel (1933, 1935), there has been no such detailed study since, until the last few years. What Harrison said in 1914 held once again, until quite recently. "Inexplicable as it may seem, very little of a definite nature has been added to our knowledge of this field since the period just cited" – in spite of the fact that the material has been available and the basic problems known all the time. There are, of course, bona fide technical impediments that have stood in the way of studies on cell locomotion in vivo, such as the opacity of most Metazoa, and these have no doubt held up advances in this area. But the main impediment has been lack of effort, based, I believe, on insufficient appreciation of the importance of the problem. Be that as it may, it is now obvious that the subject is ripe for renewed investigation with modern methods.

In modern times, cell locomotion has been studied in detail in vivo only in four organisms – sea urchin embryos, a fish embryo, the chick embryo, and a larval tunicate – and most of what one can say at this time is based on these studies. Hardly anything can be said of the locomotion of cancer cells in vivo. Even though sea urchin, fish, and chick embryos and larval tunicates are all widely separated phylogenetically and differ in many characteristics, their motile cells have some remarkable similarities.

5.1.1. Echinoderm mesenchyme cells

In both sea urchins and fish, contrary to general belief, cell movements begin prior to the onset of gastrulation. Primary mesenchyme cells in the echinoderm vegetal plate begin locomotory activity by resorbing the cilia on their outer surfaces, pulsating, and forming hemispherical protrusions called blebs (or zeiotic knobs) on their inner surfaces (Von Ubisch, 1937; Kinnander and Gustafson, 1960). Soon after, these cells lose their adhesions to the hyaline plasma layer and to each other and emerge into the blastocoel, where they creep about on the inner surface of the ectodermal blastocoel wall by means of thin, branched filopodia up to 30 μm long and about 0.5 μm or less in diameter (Gustafson and Wolpert, 1961; Gustafson, 1964). These filopodia attach to the blastocoel wall only at their tips. When the tip detaches (or fractures) the filopodium collapses. The point of attachment of these tips to the blastocoel wall is always at the junction between ectodermal cells, where the latter are cohering and thus are apparently more adhesive (Gustafson, 1963). Ultimately, the tips of several filopodia from different cells fuse and form a syncytial cable, in which the skeleton of the future pluteus will develop (Okazaki, 1960). As filopodia shorten, the cell is pulled toward the point of filopodial attachment. It has not been established whether the filopodia are extensions of blebs or are spun out separately, but Gustafson states (1963) that filopodia of secondary mesenchyme cells sometimes arise "from the pulsatory lobes themselves" (p. 578). They extend predominantly from one end of the cell, but others may form elsewhere on the cell surface. The presence of opposing filopodia may stretch the cell somewhat, but the predominant morphology of the cell body is rounded or pear-shaped.

The first phase of movement of these primary mesenchyme cells is a more or less radial dispersion away from the vegetal plate, with cells furthest from the center tending to move first (Gustafson and Wolpert, 1961), simulating contact inhibiting cells moving away from an explant. Some move half way to the animal pole. From this dispersed phase, cells gradually aggregate on the ectoderm to form a ring around the archenteron. It is not clear how they get there. It appears not to be a "directed" migration, however, as suggested in the title of the paper (1961), but more like a trapping of mesenchyme cells by a more adhesive region of the ectoderm, after random exploration by filopodia which constantly make and break their adhesions as they move about. The only part of these cells that adheres to other cells seems to be the tip of the filopodia. In this respect, primary mesenchyme cells are like fibroblasts in culture.

If mesenchyme cells are shifted to the animal pole by centrifugation, they nevertheless find their way back, apparently by random migration, and line up in the usual circle (Okazaki et al., 1962). In exogastrulae, the amount of ectoderm is greatly reduced; thus in order to adhere to ectoderm primary mesenchyme cells must travel a great distance, all the way to the animal pole. They do this in a remarkably *directed* way – almost in straight lines. There is at present no understanding of how they do this. In all of this, it seems clear that primary mesenchyme cells are not at all bothered by the necessity of crawling over ectodermal cells. Either they are not contact inhibited by ectoderm or they move over some intervening material. In the light of the studies of Elsdale and Bard (1972a), it would be interesting to know whether the two cell types are separated by extracellular material. Later, after invagination is well under way, secondary mesenchyme cells appear at the tip of the archenteron and they too move by means of filopodia (Dan and Okazaki, 1956; Gustafson and Kinnander, 1956; Gustafson and Wolpert, 1967) (see below, pp. 305–306).

5.1.2. Teleost deep cells

In the only other form in which the genesis of cell locomotion has been studied during early development, the teleost *Fundulus*, deep blastomeres within the blastoderm likewise begin locomotory surface activity by blebbing in mid-blastula, well before the beginning of gastrulation (Trinkaus, 1973a). These two instances suggest that cell movements may begin earlier than suspected in other forms as well, and by the same means. In *Fundulus*, the blebs have been shown to play an important role in locomotion, both in themselves and in the formation of other protuberances. At first, only a small proportion of the deep cells form blebs. But with time the number of cells blebbing increases steadily until by late blastula all or most deep cells are blebbing and the whole inner blastoderm is converted into a jostling mass. During this period, individual cells do not displace: each continues to bleb in place, and many show circus or limnicola movements (Holtfreter, 1947, fig. 3).

Cells begin to move in advanced blastulae. A cell forms a bleb, but, instead of retracting, the bleb extends, and cytoplasm flows into it, as indicated by a corresponding decrease in the size of the cell body. With this, the rounded cell body is displaced in the direction of the protruding bleb. Optical conditions in vivo have thus far precluded direct observation of cytoplasmic flow. Alternatively, a bleb elongates into a finger-like lobopodium, whose tip appears to adhere to the surface of another cell and spread lightly on it. Locomotion in such a case occurs either by the cell pouring into the lobopodium, as it were, or by the lobopodium shortening and pulling the cell body rapidly forward (or dislodging the cell to which it has adhered). This is a jerky, inch-worm like kind of locomotion. In some instances a lobopodium becomes stretched thin, and taut. In a word, it becomes a filopodium. Filopodia also form de novo, particularly later, during gastrulation. These act as do the filopodia of sea urchins – they adhere at the tip, are thin and straight, are often branched, and when they shorten they can move the cell body a considerable distance quite rapidly.

In contrast to sea urchin mesenchyme cells, *Fundulus* deep cells also form lamellipodia. These may be slight transitory flattenings of the leading edge of a bleb or short lobopodium, especially when they are first evident during blastula and early gastrula, or they may involve a considerable proportion of the cell. Lamellipodia appear to spread primarily on the under surface of the epithelioid enveloping layer or on the upper surface of the underlying yolk syncytial layer (periblast). In a few instances, undulations have been observed to form on the leading edge of lamellipodia and to propagate back on the free upper surface of the cell, like ruffles (Trinkaus, 1973b).

During all of these locomotory activities, the cell body remains rounded. Only in rare instances during blastula and early to middle gastrula stages have whole cells been observed to flatten, usually on the surface of the yolk syncytial layer. This occurs more frequently in late gastrula and after. Significantly, these flattened cells have not been observed to move. The typical modes of locomotion, by blebs, lobopodia, filopodia, and lamellipodia, continue throughout gastrulation. Although these several different modes of locomotion exist in the *Fundulus* blastoderm, we do not at this point have reason to believe that they represent different cell types. Cells have been observed to shift from one mode to another. An important feature of all these modes of locomotion is that the speed of movement is quite high, about 6–12 μm per minute, in contrast to fibroblasts but like leukocytes in vitro.

5.1.3. Tunicate tunic cells

In the transparent tunic of larvae of the tunicate *Botryllus*, there appear to be two kinds of locomotion: by means of rounded protrusions (blebs) into which there is protoplasmic flow, and by means of long, thin filopodia. In this case, these two kinds of locomotion appear to represent two different cell types. The former have not been studied in detail. The latter have been subjected to detailed study by Izzard (1974) and typically possess several radiating filopodia which adhere to the cuticle and by shortening move the cell about. Because of the relatively large number of competing filopodia, net displacement may be either extremely small, or considerable, if certain filopodia gain ascendancy. These filopodia seem to be like those of sea urchin mesenchyme in their relative length and thinness, but differ in their greater number per cell. Filopodia of both these organisms are longer and thinner and occur more frequently than those of *Fundulus* deep cells. Filopodia of *Botryllus* appear to form directly from the cell surface rather than from rounded protrusions.

5.1.4. Chick corneal mesenchyme

On the 4th day of chick embryonic development mesenchymal cells migrate as individuals into the swollen collagenous stroma between the cornea and the lens to form the cell sheet that will later differentiate into the endothelium separating the stroma from the anterior chamber of the eye. On the 6th day, fibroblasts, derived from the neural crest (Johnston et al., 1974), migrate into the stroma between the endothelium and the surface epithelium. These migrations are currently being subjected to intensive investigation with SEM, transmission electron microscopy and cinemicrography of living cells with Nomarski interference optics (Bard and Hay, 1975; Bard et al., 1975; Nelson and Revel, 1975). The study of living cells in situ at this relatively advanced stage of development is made possible by a happy circumstance; they continue their normal migratory movements in organ culture.

These cells have special interest because, in contrast to the other cells whose movements have been studied in vivo, they are engaged in definitive histogenesis at a relatively advanced stage of development. Thus, they might provide indications as to how genuine tissue cells migrate, as opposed to blastula and gastrula cells. Moreover, valid comparisons can be made of their movement in vivo with the much studied movement of fibroblasts in vitro.

In contrast to early embryonic cells, corneal endothelial cells have a somewhat flattened rather than a rounded cell body; but like some of them, they extend lamellipodia and numerous filopodia. Broad flattened non-ruffling lamellipodia along the cells' leading edges are best seen in SEM. They are not evident in the living, perhaps because optical limitations prevent visualization of such thin lamellae.

The fibroblasts that invade on the 6th day have spindle-shaped bodies and extend numerous branching filopodia (Bard and Hay, 1975). The spindle-shaped cell body could be due to tension exerted by firmly adhering filopodia extending from both the leading and trailing edge. The cell body is rounded only at mitosis. Blebs have not been seen. These dendritic cells pulled taut between opposing filopodia have a remarkable resemblance to fibroblasts in the anuran tail, also viewed in the light microscope (Clark, 1912). And, indeed, it appears that their locomotion is accomplished in the same way – flow of cytoplasm from the cell body into certain filopodia and not into others (Bard and Hay, 1975). Perhaps this is just how fibroblasts move within the organism.

In addition to observing these fibroblasts in situ, Bard and Hay took the important step of culturing them in vitro and made a significant discovery. On a plane substream of glass these cells behave like typical fibroblasts, forming broad lamellipodia, confirming the suggestion that the plane substratum of the typical tissue culture preparation has a substantial influence on cell form. More importantly, when cultured in a collagen gel (Elsdale and Bard, 1972b), these cells behave in essentially the same way as in vivo, assuming a spindle shape and extending long branching filopodia. This lends strong support to the suggestion that a collagen gel simulates the environment in which cells are moving within the organism. Incidentally, in a collagen gel, where details of cell behavior are more readily observed, these fibroblasts move by extending a long filopodial process, which stretches the cell, followed by detachment and quick retraction of the trailing edge.

Another observation of interest relates to the directional migration of these cells over the stroma eventually to achieve confluency. They appear to show contact inhibition of movement. That is, when a cell encounters another cell, its locomotion in that direction appears to cease and, when it begins again, it moves in another direction.

The presence of certain similar modes of cellular locomotion during the early development of such different organisms suggests that these may be representative of the ways in which early embryonic cells move generally in vivo. Because of a paucity of observations on other living embryos we are certainly in no position to generalize. However, it is of interest that the filopodial locomotion that Izzard (1974) analyzed in detail in the *Botryllus* tunic has also apparently been seen in other ascidians (Brien, 1930; Saint-Hilaire, 1931; Pérès, 1948). In the vertebrates, chick mesenchyme cells, known by marking studies to be emigrating from the primitive streak, have been examined in thin sections and found to have a highly dendritic form, with what appear to be filopodia (Trelstad et al., 1967). It should be pointed out, however, that in thin sections it would be hard to distinguish lamellipodia and filopodia. Blebbing or quasi-amoeboid locomotion, like that of *Fundulus* deep cells, has also been seen in annual fish embryos (Wourms, 1972) and apparently occurs as well in the tunicate tunic. Izzard (1974) and others (Seeliger, 1893; Saint-Hilaire, 1931; Cloney and Grimm, 1970) have found that presumptive tunic cells move by the eruption of blunt, hyaline protuberances (blebs). Incidentally, like deep cells (Trinkaus, 1973a), these cells have a high rate of locomotion, approximately 8 μm per minute (Izzard, 1974).

5.1.5. Tumor cells

It is well established that much of the translocation of cancer cells during metastasis is by passive transport in the circulatory system, both in blood vessels and lymphatics (for review, see Leighton, 1967). In spite of much study, however, the process is still poorly understood. Tumor emboli seem to favor lymph nodes and capillaries as points of exit from the circulation (Warren and Gates, 1936), but it is not known what causes them to lodge in certain loci and begin their invasion of adjacent tissues at that point. Nor is the process whereby tumor cells invade capillaries in the first place understood. All knowledge of this is based entirely on fixed material, either by chance observation in pathological material, or by following labeled cells.

Once cancer cells leave the circulation and invade solid tissues we know nothing about their mode or mechanism of locomotion. As mentioned, observations of transformed cells in vitro give little support to the belief that they move differently from their normal counterparts (see p. 249). Such differences as have been found appear to be mere quantitative variations from the norm. It may, of course, be that the "transformed" cells studied in vitro are not invasive in vivo and in some instances this is certainly a valid objection. What is really needed are more studies of cells known to be invasive taken directly from the organism and observed both in primary culture in vitro and in transplants in vivo.

To the best of my knowledge, the active locomotion of individual cancer cells within the living organism has been studied in only one laboratory (Wood, 1958; Wood et al., 1968). No details of the mode of cellular locomotion were uncovered in this study, but the method is of interest as a possible model for future investigation. Trypan blue stained skin carcinoma cells of the rabbit were injected into an artery next to a transparent rabbit ear chamber (Clark and Clark, 1935) and followed cinematographically as they entered the capillaries of the chamber, adhered to a capillary wall along with leukocytes, and later, following leukocytes, penetrated it to invade the adjacent connective tissue (Marchesi, 1970). Here, unfortunately, they quickly disappeared.

Two observations of Wood et al. (1968) are of interest. While still in capillaries individual carcinoma cells were observed to have a rounded contour (like sea urchin mesenchyme cells and *Fundulus* deep cells in vivo), but formed an apparently epithelioid "acinar pattern" when grouped together. The rate of movement of these rounded individual cancer cells, when adhering to the endothelial lining of capillaries, is quite high, 6–7 μm per minute, similar to that of the leukocytes filmed with them and, most interestingly, like that of rounded normal embryonic cells in vivo. Not surprisingly, they found that macrophages and fibroblasts move much more slowly.

It seems to me that this approach, injecting tumor cells into a relatively transparent system, perhaps transparent larval forms, where they can be followed cinematographically with Nomarski interference optics, combined, of course, with electron microscopy and other techniques, is the only one that is going to give us the kind of information we need, if we are one day to understand the modes of locomotion of individual tumor cells in vivo. In fact, given the significance of tumor cell invasion and metastasis it is extraordinary that research of this kind has not been attempted by others.

A dominant and persistent belief about the invasiveness of tumor cells is that they are less adhesive than normal tissue cells and hence tend to break away from the tumor mass and invade adjacent tissues or be wafted about by the circulatory system to distant organs. This is based mainly on Coman's (1944, 1960) observations that neoplastic cells are shaken apart or pulled apart more readily by microneedles than corresponding normal cells. The difficulty with this concept is the same as with all such measurements of quantitative differences in cell adhesiveness. We have no assurance that when cells are pulled apart the cell surfaces of the two cells are not torn, rather than cleanly separated (see p. 248). It could be argued, however, that it makes little difference, the point being that if cancer cells are more readily separated from one another, by whatever means, they will more readily slough from the original tumor mass and invade adjacent tissues or blood vessels.

5.2. Comparison with fibroblasts in vitro

Even though observation of individual embryonic cells in locomotion within organisms is still in its infancy, one important general feature is already evident. By and large these cells differ in an important way from the classical vertebrate fibroblast as we know it in vitro on a plane substratum. Typically, individual cells in motion in vivo are not highly flattened with thin ruffling lamellipodia. They appear to move more frequently by means of filopodia and blunt protrusions. And, even when they do form lamellipodia, as in *Fundulus*, the bulk of the cell body remains rounded. The key to this difference may well be the plane inanimate substratum on which cells are usually cultured in vitro. Harrison long ago pointed out in a classic paper (1914) that fibroblasts or mesenchyme cells will assume different shapes in vitro according to the mechanical conditions to which they are exposed. They are flat, if moving on a cover glass, and become stellate or dendritic, when moving in a fibrin network, or highly elongate, if moving on a stretched spider web. In their remarkable studies of mesenchyme cells in the living anuran tadpole, Clark (1912) and Clark and Clark (1925) showed these cells to be highly dendritic, possessing long thin processes we might call filopodia, when embedded in the fibrillar matrix of the tail fin. When they contact the wall of a capillary, however, they spread on this apparently flat surface and become highly flattened. Corneal fibroblasts may become highly flattened on a collagen mat (Bard et al., 1975; Nelson and Revel, 1975) and on glass (Bard and Hay, 1975) or become spindle-shaped, with long, thin processes when embedded in a collagen gel (Bard and Hay, 1975).

Another striking difference between individual cells that have been studied in vivo and fibroblasts and epithelial cells in vitro is the mobility of the cytoplasm. A frequent type of locomotion within *Fundulus* embryos and tunicate larvae is by means of blebs or lobopodia accompanied by much cytoplasmic flow. Although it now seems that there must be a certain small amount of cytoplasmic flow accompanying protrusion of the leading edge in fibroblasts and epithelial cells, these cells are never observed to move large masses of their cytoplasm during locomotion in vitro.

In this respect, a more apt comparison might be with polymorphonuclear leukocytes. They, of course, utilize extensive cytoplasmic flow constantly during their locomotion in vitro (Robineaux, 1964; Ramsey, 1972b), and presumably, in vivo as well. It is probably significant that leukocytes in vitro and cells moving by means of blebs in vivo in *Fundulus* (Trinkaus, 1973a) and *Botryllus* (Izzard, 1974) all move much faster than fibroblasts either in vitro or in vivo (Clark, 1912; Abercrombie et al., 1970a) – 8–10 μm/minute, as compared to 1 μm/minute for fibroblasts in vitro or even more slowly in vivo. From this, it would appear that bulk cytoplasmic flow makes for a higher rate of cell movement; but this is clearly not the whole story. *Fundulus* deep cells moving by means of filopodia or lamellipodia also move fast and primary mesenchymal cells of the sea urchin have been observed to thrust out filopodia at about the same high rate (Gustafson, 1964). Although the elevated rate of locomotion of these various cells is as yet unexplained, it is tempting to call attention to its association with roundedness of the cell body, in contrast to the highly flattened cell body of fibroblasts on a plane substratum. We know that in fibroblasts a flattened form is associated with firm adhesion to the substratum and a rounded form with less firm adhesion. By the same token, it seems probable that the rounded cell body of cells in vivo is evidence of less firm adhesion. Indeed, Gustafson (1964) states that the cell body of primary mesenchyme cells does not adhere at all. But we cannot be certain of this. In any event, it seems possible that the firm adhesion of the trailing edge of a fibroblast might impede locomotion, whereas, conversely, the more lightly adhering cell body of cells in vivo would offer less resistance to the actively protruding leading edge and hence would displace more readily. The very slow rate of locomotion of fibroblasts in vivo is also consistent with this idea. These highly stellate cells are anchored by adhering processes extending in all directions (Clark, 1912). From this we would predict that they would move only very slowly, if at all. Their rate of displacement was of the order of micrometers per day! But this might be abnormally low. As Clark remarked . . . "during most of the time over which the observations extended, the weather was unusually cold, . . ." (p. 369).

Bell and I (unpub.) have found that faster locomotion of *Fundulus* deep cells in vivo may indeed be due to lighter adhesion of the trailing edge. When these cells are cultured on glass in the absence of serum, they spread well, form ruffling lamellipodia, lose their rounded cell body, and remain stationary. In contrast, when serum is added to the medium in order to decrease adhesion to the substratum (Taylor, 1961), the cells form a single dominant lamellipodium, retain their rounded cell body, and engage in rapid locomotion at a speed of 6–8 μm/minute! By this means, therefore, we have apparently duplicated in vitro both the mode and the rate of deep cell locomotion in vivo.

In spite of these differences, there are sufficient similarities between cells thus far studied in vivo and in vitro to encourage the belief that they share certain basic mechanisms, even though this involves comparing cells from different stages of development. For example, lamellipodia of *Fundulus* deep cells have been observed to form undulations on the upper surface in vivo that propagate away from the leading edge like ruffles (Trinkaus, 1973b). As in vitro, these broad ruffles are only evident when the upper cell surface is free, i.e., exposed directly to fluid medium, in this case the fluid of the segmentation cavity. In the developing chick cornea, in contrast, mesenchyme cells with ruffles have been rarely seen (Bard et al., 1975; Nelson and Revel, 1975), even though some of the cells are highly flattened. Perhaps this is due to their lack of a free surface. These cells are embedded in the corneal stroma. When *Fundulus* deep cells pull apart in vivo they often form thin, taut retraction fibers, as do cells in vitro. Similarly, fibroblasts of the chick cornea have a spindle-shaped elongate form with thin extensions as do the same cells in vitro in collagen gels. Also, the leading edge of *Fundulus* deep cell extensions, whether they be lamellipodia, lobopodia or filopodia, can frequently be seen to flatten against their substrata, as do corneal fibroblasts. It has not yet been possible to observe minute by minute fluctuations of the leading edge of these cells, however, and this is required for accurate comparison. Much closer and more detailed observations in vivo are required. A major impediment, of course, is the unsatisfactory optical conditions in vivo, even in "transparent" eggs. However, observing this should not be insuperable. Filopodia with a thickness about 0.5 μm have been observed in sea urchin embryos (Gustafson and Wolpert, 1961).

It seems probable that a major cause of the different locomotory behavior of fibroblasts in vitro is the plane glass or plastic substratum on which they are usually cultured. This is ironic for the use of a transparent plane substratum is the major reason for the improved optical conditions in vitro. The observation of Elsdale and Bard (1972b) that when fibroblasts are cultured in a collagen gel, they do not form lamellipodia, but on the contrary assume an elongate in vivo-like morphology, with long thin filopodia, is completely in accord with this conclusion, at least in so far as fibroblasts tend to flatten on capillary walls. In the same vein, we have also found that *Fundulus* deep cells not only flatten on a relatively plane substratum (Trinkaus, 1973a) but adhere to it only at their margins, just as do fibroblasts in vitro. In SEM, deep cells can be seen adhering to the underlying yolk syncytial layer by a number of discrete "feet" all around their margins (Erickson and Trinkaus, unpublished observations), giving a picture very much like that seen by Revel et al. (1974) for fibroblasts in vitro. In transmission electron micrographs no contacts between the cell and the yolk syncytial layer are seen submarginally (Trinkaus and Lentz, 1967). Marginal cells of the epithelioid enveloping layer also adhere to the yolk syncytial layer all along their free margins, forming tight occluding junctions and close appositions exclusively in the most marginal region of each cell (Betchaku and Trinkaus, 1976), just like marginal cells of an epithelial sheet in culture (DiPasquale, 1975a).

5.3. *Neurons*

A cell that has recently been receiving careful attention in culture and in fine-structural studies, the neuron, has not been studied within organisms in any systematic way since the nineteen thirties. On the basis of observations on nerve outgrowth in the transparent tail of amphibian tadpoles, Speidel (1935) showed that advancing axons in vivo are quite similar to those studied earlier by Harrison (1910) in vitro. Both have "growth cones" with "filamentous processes" (microspikes) and when an axonal tip is about to retract "it draws in any filamentous processes and becomes quite rounded". Our contemporary eye would see this as de-adhesion of microspikes, on which the extended state depends. Also, he observed that on de-adhesion and retraction, the tip forms "small knoblike excrescences". We now know that formation of such knobs or blebs is a common sequel to de-adhesion of the edge of a fibroblast in vitro (Harris, 1973b). If Speidel, with limitations of equipment and none of our modern understanding of cell movement, but with superb observational talents, could see all this forty years ago, what more could be seen now! Clearly, studies like this need to be extended, with the advantage that the passage of time has granted.

6. *Mechanism of locomotion of individual cells and cell streams in vivo*

6.1. *Individual cells*

6.1.1. *Contractility and microfilaments*

There has been little analysis of the mechanism of cell locomotion in vivo. In summarizing what we do know, I think it best to consider first the matter of contractility. Even with our limited knowledge of the modes of cell movement in vivo, it is already evident that one of the most frequent ways cells do it is by extending a long protrusion, which adheres at its tip and then shortens, pulling the cell along. Everyone assumes that this is accomplished by active contraction, rather than passive recoil of a stretched elastic system, but in only one instance has proof been offered. Izzard (1974) took advantage of the fact that when filopodia of the tunic cells of *Botryllus* attach to the elastic cuticle they deform it to various degrees, giving a rough indication of changes in tension in the filopodia. If shortening is due to elastic recoil it should be accompanied by a decrease in tension. If, on the contrary, it is due to active contraction it should be accompanied by a maintenance or even an increase in tension. By careful observation of the tension exerted on the cuticle by shortening filopodia, Izzard found that the latter situation holds. The cuticle remains deformed as attached filopodia shorten. It seems clear, therefore, that in this instance cell movement is indeed due primarily to active contraction of certain adhering filopodia, accompanied by relaxation or detachment of opposing filopodia. Elastic recoil may occur, especially in the rounding up of a cell body, but it is not the major force. With this clear-cut demonstration that contractility plays a crucial role in the movement of these beautiful cells, the ground-work has been laid for a combined fine-structural and physiological analysis of the contractility and how it is controlled. Izzard's analysis naturally renders more probable the supposition that active contraction is involved wherever the shortening of a cell extension causes a cell to move. But proof of the sort he offers will be needed in each case. Moreover, it is not at all unlikely that elastic recoil is involved in some retractions, such as that of the trailing edge of a fibroblast when it detaches.

A case in point is the shortening of filopodia in the movement of sea urchin mesenchyme cells. Although they are assumed to contract actively and, in the case of secondary mesenchyme cells, do obviously pull out cones from the ectoderm wall (Gustafson and Kinnander, 1960), it has not been reported whether the tension continues or decreases as the filopodia shorten. However, Tilney and Gibbons (1969) noted the presence of 40–60 Å microfilaments in some filopodia and remark that filopodia which are attached and contain microfilaments are resistant to colchicine treatment but not to calcium-free sea water. There appears to have been no further study of this.

Fundulus deep cells also possess cortical microfilaments and they are located in the cortical cytoplasm just beneath the plasma membrane (Hogan and Trinkaus, in preparation). Thin cell extensions (filopodia or lamellipodia) are packed with microfilaments, as in other cells. When isolated blastoderms are treated with cytochalasin B which, among other effects, clumps microfilaments (Spooner et al., 1971; Miranda et al., 1974), dense clumps of microfilaments appear. Since cytochalasin B does not penetrate intact *Fundulus* eggs, we do not know its effect on deep cell locomotion. In fact, the investigation has not proceeded far enough at present to relate the presence of microfilaments to locomotory activity in deep cells (or any other individual cells in vivo), except by association. This must have a high priority in future investigation because of the probability that contractility plays an important role in cell locomotion. I must reemphasize, however, that in the locomotion of *Fundulus* deep cells we have not yet been able to distinguish between contractility and elastic recoil in the shortening of lobopodia and filopodia.

Microfilaments are also abundant in the long neck and microvillous extensions of the bottle cells of the amphibian blastopore (Baker, 1965) and in the neck of flask cells in the chick primitive streak (Balinsky and Walther, 1961). These important cells will be considered later in connection with their relationship to invagination (pp. 298 and 300).

6.1.2. Microtubules

In view of the widespread involvement of microtubules in determining the form of cells (see reviews by Tilney, 1968; Karfunkel, 1974; Spooner, 1974), it comes as no surprise that they are also present in migrating primary mesenchyme cells of the sea urchin, where their orientation, particularly in the filopodia, parallels the asymmetry of the cell (Gibbins et al., 1969). In the blastula, prior to the formation of filopodia, when presumably the cell is blebbing, the microtubules appear to radiate in all directions from the centrosphere. When primary mesenchyme cells are treated with colchicine or high hydrostatic pressure after filopodia have formed, the filopodia are reduced in number, but not eliminated, and the cells tend to spherulate (Tilney and Gibbins, 1969). However, if the syncytial cable is already formed, it persists. Although the authors propose that microtubules are important in the development of cell form but not in its maintenance, a role in maintenance does not seem to have been eliminated. Treatment with colchicine and high hydrostatic pressure not only prevents further development of elongate extensions but reverses some that are already there. It is also clear that other factors also operate, such as whether a filopodium is attached to another cell or not. However, there appears to have been no further study of the factors concerned in the generation of cell shape and in the operation of filopodia in this splendid material. It certainly deserves renewed attention, particularly in view of the important advances in our knowledge of the cytoskeleton that have taken place since 1969.

In contrast to sea urchin mesenchyme, motile *Fundulus* deep cells contain few cytoplasmic microtubules, although abundant microtubules are present in the spindle of dividing cells (Lentz and Trinkaus, 1967; Hogan and Trinkaus, unpublished observations). This is not surprising in blebbing cells and those engaged in a mode of locomotion where extensive cytoplasmic flow occurs. Such flow suggests that the cytoplasm has low viscosity, a condition which is consistent with small numbers or a lack of cytoplasmic microtubules. In view of this paucity of cytoplasmic microtubules we would not expect deep cell movement to be inhibited by colchicine. It isn't. In eggs treated with colchicine at the beginning of epiboly, deep cells continue locomotion (Trinkaus and Bell, unpublished observations). Significantly, however, the only form of locomotion that persists is that which involves massive cytoplasmic flow; locomotion by means of filopodia and lamellipodia is eliminated. Moreover, there is an increase in blebbing. Blebs often form on the top of blebs and thus build up long tortuous layered lobopodia. Since mitosis is blocked by colchicine, treated cells retain the relatively large size they possess at the beginning of gastrulation and, of course, do not increase in number.

6.1.3. Contact relations

Contact relations between moving cells and their substrata have not been studied during sea urchin gastrulation nor in the ascidian tunic. In the chick cornea, migrating endothelial cells make contact with collagen fibrils, but no specialized junctions are evident (Bard et al., 1975). In *Fundulus*, fine structural investigation of deep blastomeres has revealed a striking difference in the contact relations of non-motile and motile cells (Lentz and Trinkaus, 1967; Trinkaus and Lentz, 1967). In so far as can be determined from thin sections, non-motile deep blastomeres from early blastulae do not form junctions or appositions with each other and are usually widely separated. In contrast, motile cells from advanced blastulae and early gastrulae form 150–200 Å appositions wherever such a rounded protuberance contacts the surface of another cell. Since such appositions are probably adhesive (Farquhar and Palade, 1963), this confirms the evidence from films (Trinkaus, 1973a) that these cells begin adhering to each other when translocation begins. Gap junctions are also present between deep cells in early to mid-gastrula blastoderms, as confirmed by both transmission and freeze-fracture electron microscopy (Hogan and Trinkaus, in preparation). Neither tight junctions nor desmosomes have been seen. This is interesting because enveloping layer cells of *Fundulus*, which are epithelial and do not move over each other, possess both tight junctions and desmosomes. Thin sectioning has not been very helpful in study of the contacts of deep cells with the yolk syncytial layer. In transmission electron micrographs these cells seem merely to rest on the tips of microvilli on the surface of the yolk syncytial layer. To obtain more information on this important question, Erickson and I examined deep cells in the SEM. With this method, they were found to adhere to the yolk syncytial layer by means of numerous marginal feet and long filopodia. It therefore appears that these cells, like fibroblasts in culture adhere to their substratum largely or exclusively at small discrete points along their margins. Since the chance of sectioning through one of these feet is slim, the impression of non-adherence seen in transmission electronmicrographs (TEM) is explained. Deep cells also make contact with other deep cells by means of long filopodia, and by taut retraction fibers as well. In SEM the latter are often seen to terminate in a bulbous swelling, as do retraction fibers of fibroblasts on a glass or plastic substratum in culture (Erickson and Trinkaus, 1976).

6.1.4. Deformability

Because of the importance of blebs in *Fundulus* deep cell locomotion, and probably in the locomotion of sea urchin mesenchyme cells and "amoeboid" cells of ascidians as well, an understanding of bleb formation is basic to an understanding of locomotion in these kinds of cells. We have begun our investigation of blebbing by inducing blebs artificially and studying their characteristics (Tickle and Trinkaus, 1973). The method was to apply negative pressure to the cell surface in culture and draw out bleb-like or lobopodia-like protrusions (Mitchison and Swann, 1954; Weiss, 1966). When this was done (Tickle and Trinkaus, 1973), there was a decrease in the size of the cell body, just as in the formation of normal blebs and lobopodia (Trinkaus, 1973a), suggesting that there is cytoplasmic flow into the protrusions. Significantly, protrusions are more readily drawn from gastrula cells than from blastula cells. This finding correlates with the normally greater surface activity of gastrula cells and suggests that the normal basis of this greater activity is to be found in the cell surface and peripheral cytoplasm, rather than in internal forces (like, for example, increased internal hydrostatic pressure). Moreover, these correlations of experimental results with normal activity make it seem probable that we are dealing with the same properties in both and that therefore we have a legitimate experimental means of studying bleb formation. In continuing the analysis, the first question asked was whether surface deformation is due to local expansion or involves surface from the rest of the cell. To distinguish between these possibilities the cell surface was marked with particles of carbon and it was found that, as a protrusion forms, the carbon marks move toward the pipette (Tickle and Trinkaus, unpublished observations). This indicates that formation of a protrusion is not a local phenomenon but involves much more of the cell surface. Movement of particles argues against stretching and for surface flow. Carbon marks also move toward the site of formation of a normal bleb. Thus, it seems that surface flow is important also in formation of blebs, and perhaps other protuberances, and that gastrula cells could be more deformable because surface flow takes place more readily in them. A recent study of surface mobility in amphibian cells is consistent with this. Johnson and Smith (1976) have found that motile amphibian gastrula cells cap fluorescein labeled concanavalin A, whereas non-motile blastula cells do not. Clearly surface flow takes place more readily in gastrula cells.

The possibility that a sub-membranal contractile system is somehow involved in extensibility prompted some preliminary study of this. The results support this possibility. Deep cells are much more deformable in the presence of Ca^{2+}-free Holtfreter's solution, which interferes with cellular contractile activities. This increase in deformability with inhibition of contractility suggests a mechanism for the formation of blebs and other protrusions. All living cells appear to have a high degree of internal hydrostatic pressure which is constantly pressing against the cell surface. This causes cells to be spherical when in suspension. In such a state, any weakness in the contractile cortex of the cell would result in a protrusion forming at that locus. It was therefore with interest that we turned to thin-sectioning of the protrusions of *Fundulus* deep cells. We had already observed that rounded protrusions, presumably blebs, are quite empty of organelles, such as mitochondria and Golgi, confirming their hyaline appearance in the living (Trinkaus and Lentz, 1967). Further study (Betchaku and Trinkaus, unpublished observations) has revealed, in addition, that the microfilamentous cortex of such protrusions is distinctly thinner than that of adjacent parts of the cell. This observation is consistent with the hypothesis. A similar absence of organelles and a distinct thinning of the microfilamentous cortex has also been found in blebs of certain cell lines in vitro (L.B. Chen, unpub.; C.A. Erickson, unpub.), suggesting that this might be a wide-spread phenomenon. There is, however, an alternative hypothesis that invites investigation.

Conrad and Williams (1974) have presented evidence that microfilaments might function in bleb formation, as they do in cytokinesis, by constriction of a cortical band at the base of the bleb. A suggestive model for the retraction of a bleb has been provided in some remarkable studies of artificial blebs in *Physarum* (Wohlfarth-Botterman, 1964). Retraction of the bleb and flow of cytoplasm back into the cell is accompanied by the appearance of a dense layer of microfilaments in the bleb cortex.

The obvious resemblances of cells moving by means of blebs and lobopodia to free-living amoebae (such as *Amoeba proteus*) make it legitimate to ask whether or not we are in this instance at last dealing truly with a case of quasi-amoeboid movement in Metazoa (in contrast to that of fibroblasts and epithelial cells) (see above p. 241). In both deep cells and free-living amoebae, there is a quick bulging out of the cell surface to form a large rounded protuberance, accompanied by impressive cytoplasmic flow which often expands the protuberance and moves the whole cell along; and, in both cases, there is evidence that internal hydrostatic pressure, generated by a cortical contractile system, may be the motile force (Seravin, 1971; Wolpert and Gingell, 1968; Komnick et al., 1973). The relevant literature up to 1973 on the locomotion of free-living amoebae has been fully reviewed by Komnick et al. (1973). It now seems likely that this rich literature has important lessons for the study of certain motile cells during early metazoan development (and perhaps for certain aspects of leukocyte locomotion as well).

6.1.5. Starting

A basic question in the study of cell locomotion is how do cells start moving. In the very beginning of development, prior to normal cell movements, cells do not appear able yet to move. Thus, for them, the differentiation of motile capacity is an essential aspect of starting locomotion. The formation of blebs and their conversion into other cell extensions along with their adhesions to the available substrata, seems to be the way deep cells begin moving in *Fundulus*. Bleb formation also appears to be important in the onset of locomotion in sea urchin embryos. Perhaps local thinning of the contractile cortex of the cells is the fundamental step which permits a protrusion to form under the influence of the internal hydrostatic pressure of the cell.

Later in development, all tissue cells and cancer cells appear able to move under certain circumstances (in tissue culture, for example). Yet in general they do not. Capacity to move is not sufficient in itself to render a cell invasive, whether a normal tissue cell or a tumor cell. Stimuli are required. Such stimuli could come from a cell's environment.

Destruction of normal cells often accompanies cancer invasion and invasion of the endometrium by the trophoblast. Could this facilitate invasion?

With current interest in the developmental importance of the intercellular matrix, there has been frequent suggestion that changes in it could stimulate cell locomotion. Hyaluronic acid is a case in point. A large local increase in its concentration coincides with the onset of cell locomotion in a number of instances. Just as mesencephalic neural crest cells begin to move into cell-free space it becomes filled with glycosaminoglycans (GAGs) (acid mucopolysaccharides), of which hyaluronate is a major constituent (Pratt et al., 1975). In the eye, migration of crest derived fibroblasts into the corneal stroma is preceded and accompanied by a swelling of the stroma, with its orthogonally arrayed collagen fibrils (Trelstad and Coulombre, 1971) and an increase in hyaluronate (Toole and Trelstad, 1971; Trelstad et al., 1974), synthesized by the endothelium (Meier and Hay, 1973). Reducing GAG synthesis with 6-diazo-5-oxonorleucine (DON) leads to cleft palate, apparently because proliferation and migration of palatal fibroblasts is prevented (Pratt et al., 1973). Finally, migrating fibroblasts of the endocardial cushion of the heart traverse a cell-free space filled with GAGs, especially hyaluronate (the so-called cardiac jelly) (Manasek et al., 1973). Interference with hyaluronate synthesis prevents this migration (Fitzharris and Markwald, 1974).

Glycosaminoglycans have also been implicated in starting cell movement during the very earliest stages of development. In a provocative study of sulfated GAG synthesis during sea urchin development, Karp and Solursh (1974) made a suggestive discovery. The greatest incorporation of $^{35}SO_4$ into GAGs takes place at the very beginning of gastrulation, when primary mesenchyme cells are emerging from the vegetal plate. And, significantly, when normal embryos are raised in sulfate-free sea water, with a consequent drastic decrease in the ratio of their sulfated to nonsulfated polysaccharides, primary mesenchyme cells accumulate in the blastocoel but do not migrate.

Although these various correlations are provocative and amply sufficient to justify the intense current interest in the possibility that GAGs somehow make cell migration possible, the evidence is by no means clean (Manasek, 1975). The intercellular matrix often contains a number of GAGs, which in a given instance may be as important as hyaluronate. Moreover, collagen is always present, in one or another of its forms, at least in later embryogenesis. Inhibitors that prevent GAG synthesis, such as DON, are quite non-specific and very leaky; they by no means shut off all synthesis. But even if hyaluronate is usually the critical material, as is widely believed, its role in facilitating cell movement does not depend upon its sudden presence, as opposed to its previous absence. Hyaluronate is present in all these situations before cell movement begins. Perhaps a change in the state of hyaluronate is the critical event. In the cornea, in particular, the increase in hyaluronate at the time fibroblasts begin to invade is accompanied by enormous swelling of the stroma due to uptake and trapping of water by hyaluronate. This may separate the collagen layers (Trelstad and Coulombre, 1971) and conceivably create channels through which cells may move. In a word, it is simply not known just how collagen, hyaluronic acid and other glycosaminoglycans influence cell movement. To find this out, a first step would seem to be an in vitro attack, in which cell locomotory behavior is studied in and on substrata composed of different combinations of these obviously important intercellular materials (Pessac and Defendi, 1972; Terry and Culp, 1974), or by the use of inhibitors (Spooner and Conrad, 1975).

It has also been suggested that cells may begin migration by becoming contact inhibiting. Thus cells might suddenly leave the neural crest because they develop locomotory activity at their free edges, as do cells at the edge of an explant in culture. But once the lead cells have reached their destination and have stopped, as in the formation of a ganglion, what would keep those behind moving as before? It has also been suggested, on the basis of fine-structural studies, that chick mesoblast cells might leave the primitive streak to move out between the hyoblast and epiblast for the same reason (Trelstad et al., 1967; Hay, 1968). However, even though careful observations of fixed material may be suggestive, they are no more than that. Besides, we know that in some instances individual cell locomotion definitely begins *within* a cell mass, as in the blastoderm of *Fundulus.* Thus, intriguing though these suggestions may be, there is no hard evidence that contact inhibition is involved in starting cell locomotion in vivo.

A common feature of the onset of locomotion is that a number of cells begin to move simultaneously, or very soon after one another. As discussed, this could be due to an environment change common to all. It could also be due to contact and communication between cells, such as electrical or metabolic coupling. Since most embryonic cells seem to be coupled, this is not an implausible possibility (see chapter 8 in this volume by Sheridan). A way in which this might occur is discussed below (p. 280).

6.1.6. Stopping

There appear to be the two main ways of explaining cessation of cell movement when cells reach their appointed destinations: cell to cell interactions, and a change in the cells' non-cellular environment.

Contact inhibition of movement is an obvious model for stopping cells, once they contact each other. If, for example, cells are contact inhibiting but move exclusively over intercellular materials in vivo, then finally contacting the surfaces of other cells could stop their locomotion. As already emphasized (p. 255), the evidence is conflicting on whether cells in intact tissues will move over each other's surfaces. They certainly appear to do so in early embryogenesis. But even here the situation is not clear. Do cells actually move on the surfaces of other cells, or do they move over a thin layer of intervening intercellular material? Karp and Solursh (1974) observed rough material on the surface of primary mesenchyme cells of the sea urchin in the SEM and found that it disappears in sulfate-free sea water. On the other hand, some cells in some motile populations certainly make junctions and appositions with each other (Trinkaus and Lentz, 1967; Trelstad et al., 1967; Armstrong and Armstrong, 1973). And in the chick gastrula, where the basement lamina of the epiblast and hypoblast is incomplete and mesenchyme cells have a choice, as it were, between apparently bare cell surface and basement lamina, they choose both, forming connections with both. If then, cells can move over each other's surfaces in vivo under certain circumstances, why should they be inhibited by such contact under other circumstances? Careful investigation of this point is needed, in particular to see if there is a change in cell to cell contact relations when they cease migrating.

Other ways cells might stop each other's movement through interactions with each other are by coupling, with a signal passing from an immobilized cell to a moving one with which it is in contact, and by adhering specifically to other cells of the same kind (isotypic adhesion) or to cells of another kind (heterotypic adhesion). The model system for isotypic adhesion, of course, is tissue-specific cell sorting out or self-assembly in mixed cell aggregates. Dissociated cells of the same type, moving about within a cell aggregate, clearly have ways of recognizing each other and cohering when they make contact, whether by means of qualitatively-specific adhesions (Moscona, 1962; Roth, 1968, 1973; Balsamo and Lillien, 1975) or as a result of quantitative differences in degree of adhesiveness (Steinberg, 1964, 1970; Mostow, 1975). The legitimacy of using sorting out as a model for explaining isotypic cell adhesions in vivo received a strong boost recently from the experiments of Boucaut (1974a,b). Working with amphibian embryos, he injected labeled cells from the different germ layers of gastrulae into the blastocoel of blastulae and found that they tended to localize in the homologous regions of the developing host. Grafted endodermal cells, in particular, almost always localized in endoderm. Boucaut was thus able to show that these cells have the same affinities in vivo as in mixed aggregates in vitro (Townes and Holtfreter, 1955). This suggests that type-specific isotypic adhesion may really operate normally in vivo and play the role postulated for it. This sorting out model could be useful during normal development, however, only for cells which are more or less dispersed during their migrations, as is each cell type in a mixed aggregate. Thus, if migrating neural crest cells, for example, were differentiated as to type, their aggregation at their various destinations could in part be due to the same kind of factors that cause cells to sort out in mixed aggregates. The indications are, however, that they are pluripotent during migration and form their distinctive types only after arrival in their definitive positions (Weston and Butler, 1966; Le Douarin and Teillet, 1974) (also see below). This would seem to weigh against type-specific adhesion as a means of stopping a morphogenetic movement. It could be argued, however, that if cells differentiated toward the end of their migration, this could provide precisely the cell surface required for type specific adhesion and thus serve to stop migration at just the right moment.

Cells also adhere heterotypically in mixed aggregates, as in the restoration of organ structure, and Steinberg (1964) has provided a fascinating analysis of how this might help explain certain normal morphogenetic movements. The most spectacular examples of heterotypic adhesion, however, have come from studies of the specificity of nerve connections inside embryos (Gaze, 1970; Jacobson, 1970). When medial and lateral bundles of the optic nerves are cut and crossed, for example, they regenerate along their usual channels and reestablish proper connections in the optic tectum (Arora and Sperry, 1962; Sperry, 1965). This extraordinary specificity may be based in part on specific adhesion; labeled cells of the neural retina adhere preferentially in vitro to that region on the surface of the optic tectum with which they normally make connections (Barbera et al., 1973; Marchase et al., 1975). In view of results such as these, it seems likely indeed that specific heterotypic adhesion operates during normal morphogenesis in some instances to stop cell movements when the migratory cells have reached the right place. Such a mechanism could operate equally well for all kinds of morphogenetic cell movements, whether they take place as individuals, in cell streams, or in cohesive masses and cell sheets, and perhaps as well in certain metastases of tumor cells. A major aspect of the analysis of how the organism brings its morphogenetic movements to a halt must, therefore, concern itself with this fascinating phenomenon. Duplication of specific heteroadhesions in culture (Barbera et al., 1973; Marchase et al., 1975) could open the way for analysis in vitro.

The importance of the tissue environment in stopping cell movement is dramatically shown in three elegant experiments with neural crest cells. Weston and Butler (1966) found that when newly condensed neural crest was grafted into progressively older hosts, distal migration of labeled cells was progressively attenuated. In contrast, old neural crest grafted to young hosts gave rise to the full range of crest derivatives. More recently, Le Douarin and Teillet (1974) found that when they grafted the adrenomedullary region of the quail neural crest into the vagal region of the chick crest, the cells colonized the gut and differentiated into enteric ganglia, and when quail cephalic crest was transplanted to the adrenomedullary region of the chick, quail cells migrated into the suprarenal glands and differentiated into adrenomedullary cells. Similarly, Noden (1975) transplanted brachial neural crest to a cranial region and found that the transplanted cells mimicked the patterns of migration of the host region. From these results it may be concluded that both the pathways of migration and the locus at which neural crest cells stop migration is determined not by the cells themselves but by the environments they are exposed to during migration.

Just as hyaluronate might be an environmental factor facilitating migration, so a reduction in its concentration could cause migratory movements to stop. Such a correlation has been found in the migration of corneal fibroblasts. The concentration of hyaluronate is high during cell movement, but at the time cell movements cease and shortly thereafter there is a rapid decline in the incorporation of isotopic precursors into hyaluronate concomitant with increasing hyaluronidase activity (Toole and Trelstad, 1971).

Once cells cease migration and histogenesis begins, it is assumed that they remain largely in place, even though a small amount of displacement may occur (see above, p. 234). This topographic stability lies at the basis of the integrity of multicellular organisms and it is difficult to conceive of how tissues and organs could function normally without it. It has been thought to depend in large part on the kind of tissue-specific cell adhesions revealed by sorting out in mixed aggregates. The main evidence for this is that when these adhesions are disrupted and the resulting dissociated cells are mixed with other kinds of cells, they become migratory within the aggregate until they meet others of the same type. While these adhesions may well be crucial in immobilizing cells at first, the long-lasting, precisely ordered cell arrangements that characterize differentiated organs must depend more on the various junctions and appositions that soon form between cells. And they must also depend on the contact relations cells make with the abundant extracellular material (ECM) found in adult tissues and organs.

I have referred several times in this section to findings and concepts gained from study of cell self-assembly in mixed aggregates. This fascinating phenomenon, which is reviewed in detail in chapter 13 by Maslow, has certainly been important in directing our attention to the differentially adhesive properties of cells and the probable significance of these during morphogenesis. And it has led to a presently productive effort to isolate cell surface factors involved in specific cell adhesions. In spite of great expectations, however, and notwithstanding some brilliant research, work with sorting out has thus far been disappointing. It has contributed little of the kind of hard, detailed information about cell behavior that is required if we are to understand how cell movements are controlled within organisms. The problem seems to have been twofold: (1) the cells used in these studies are by and large derived from tissues and organs that never see each other during normal development; and (2) all takes place within the black box of the aggregate, where, hidden from the eye, the activities of the cells are prey to more or less whatever interpretations will logically unite the randomly distributed condition at the beginning with the sorted out condition at the end. Indeed, in this may well lie much of its seductiveness for many (see Mostow, 1975).

6.1.7. Directionality

Little is known about the factors involved in giving directionality to individual cell movements in vivo. Experiments such as that of Le Douarin and Teillet (1974), just cited, indicate that for some cells at least directionality is definitely imposed by the environment. Just how the environment communicates this information is not understood. Contact guidance, gradients in adhesiveness of the substratum, chemotaxis, and contact inhibition (pp. 251–257) have all been suggested many times, for obvious reasons; but there is no hard evidence from observations in vivo favoring any one of them. Nevertheless, it is common knowledge that migrating cells do seem to prefer certain routes, as along blood and lymph vessels, nerves, and, especially through and along connective tissues (see Trinkaus, 1965). Thus, Wolff and Schneider (1957) found that the preferred avenues of invasion of sarcoma cells into fragments of chick embryonic organs are through connective tissues, subepithelial spaces, and along connective tissue partitions in organs. These observations, incidentally, are consistent with the adage of pathologists that infiltrating tumor cells tend to follow "lines of least resistance" through organs. In human patients, this means that tumor cells move along lymph and blood vessels and through the interstices between tissues, such as coelomic spaces, cerebrospinal spaces, and epithelial cavities, "dissecting the tissues apart along planes of anatomical cleavage and occupying all available nooks and crannies amongst the tissue elements" (Willis, 1960, p. 153).

It seems possible, therefore, that for directional cell movements, just as for starting and stopping, the intercellular materials may play a crucial role. This optimistic statement is so general and tentative, however, that it really doesn't say very much. It would be more accurate to state that directionality, the most distinctive aspect of many movements of individual cells during morphogenesis, has thus far defied analysis (see next section for further discussion).

6.2. Cell streams

If we know little about how cells move as individuals within organsims, we know much less about how they move in cell streams. In particular, it is a complete mystery how cells achieve their directionality within the stream. Contact inhibition could conceivably play a role in the initial stages of movement in some instances, where peripheral cells move away from the main cellular mass to be followed by others, as in the neural crest and emigration of mesenchyme from the primitive streak. But the real difficulty is that some cell streams never have a leading edge. The germ ring of a teleost gastrula illustrates this well. It already rims the entire blastoderm when its dorsad streaming movements begin. This seems also to be true for the involuting mesoderm of amphibia.

In point of fact, in the teleost germ ring, the one cell stream that has been observed in the living organism (Trinkaus, 1973a), the movements of cells have thus far defied analysis. There are so many cells that the activities of individuals are obscured by the cell mass about them. Films reveal, nevertheless, that all cells, or virtually all, display a highly active surface and are apparently locomotory, but the extensive blebbing and changes in cell shape seem to occur in all directions. Thus, although the overall movement of the cell population is clearly directional, it appears that the movement of individual cells at any one time may not be, like birds in a flock.

The loose organization of other cell streams, such as chick and amphibian mesoblasts during gastrulation, and the presence of protuberances on the cells (Trelstad et al., 1967; Nakatsuji, 1974, 1975) also suggest a situation in which there is much individual cell locomotory activity. But because these cells have not been observed in the living organism, we can only conjecture as to whether this is so.

In the course of our studies of cell extensibility, Tickel and I (1976) have made a fascinating discovery that may have significance for the coordinated movements of cell masses. When *Fundulus* deep blastomeres are poked or nudged in culture, they tend to bleb on the other side of the cell. This suggests that the increase in blebbing activity that occurs normally during the blastula stage may be due to the increased probability of blebbing cells touching each other. Since blebs may function as organs of locomotion, one can also imagine this stimulation operating in cell streams, such as the germ ring. Cell movement forward might be enhanced by cells bumping into the rear of cells ahead of them. If so, this would eliminate the need of searching for more complicated controls.

To test whether this stimulus for blebbing could be transmitted to another cell, one cell of a doublet of two adhering cells was nudged. The result was that both cells blebbed, suggesting that the surface activity of adhering cells is coupled. This could provide a mechanism for the coordination of locomotory activity that occurs in directional cell movements, conceivably by the passage of ions. *Fundulus* deep cells have been shown to be electrically coupled (Bennett and Trinkaus, 1970).

Since our knowledge of the movement of tumor cells in streams, wedges, and tongues of cells is based entirely on static fixed material, observed mainly at the low magnification of the light microscope, we really know nothing about their surface activities, except that often the cells are more loosely associated than in normal tissues. We can therefore only make amateurish guesses as to what is going on. A good beginning would be to study some invading streams in fixed material with both transmission and scanning electron microscopy. But even this is especially difficult where cancer cells in vivo are concerned. In contrast to embryonic material, which can be staged accurately as to the time and place, one usually does not know about a tumor invasion until it is well under way. One could control this by implanting tumors and studying their invasive activities both in vivo and in organ culture in vitro (Wolff and Schneider, 1957).

Another kind of mass invasion by tongues and wedges of cells (or by a syncytium) occurs every time the trophoblast of a mammalian blastocyst penetrates the uterine endometrium during implantation. It is a true invasion, in that the cyto- or syncytio-trophoblast adheres to the endometrial epithelium, breaks through it and its basal lamina, and then moves into the congested stroma beneath. It differs from many neoplastic invasions, however, in that, characteristically, it is carefully controlled and limited in its extent. Implantation appears to be triggered by a lytic factor produced by the uterus (Mintz, 1970) and ceases once the blastocyst has buried itself in the endometrium. Unfortunately, in spite of the biological and medical importance of this fascinating phenomenon, whereby cells of one genotype normally and with impunity invade the tissues of another genotype, and in spite of a considerable amount of study (e.g., Lawn, 1969; Böving, 1971; Enders and Schafke, 1971; and Steer, 1971), implantation is little understood in so far as the mechanism of cell movement or of syncytial extension is concerned. This is no doubt due in large part to the fact that observations up until now have been confined to fixed material. The problem of understanding implantation is a truly challenging one. However, with modern techniques of cell labeling, the use of mutant strains, and with tissue and organ culture, it is surely not insoluble, especially since one can proceed on the basis of the fine morphological work that has already been done.

7. *Mechanism of spreading and folding of cell sheets in vivo*

7.1. *Spreading of sheets with a free edge*

7.1.1. *Wound closure*

Wound closure is to be distinguished from wound healing in that it involves only epithelial spreading and the events immediately associated with it. Wound healing, the term applied to the more or less complete restoration of the normal state, involves not only epithelial spreading, but much more in addition, such as regeneration of the dermis, glands, etc. In this review, I shall concern myself solely with epithelial wound closure.

Wound closure by epithelial cell sheets is an old subject and there has been much study of it through the years, presumably in part because of its obvious and direct clinical relevance. Nevertheless, we know little of its mechanism in so far as the locomotory activities of the cells are concerned. The mode of migrating epidermal cells during wound healing within organisms is simply not known. This is somewhat surprising in view of the relative accessibility of closing wounds, especially of the epidermis. Part of the reason is surely a certain preoccupation with mammalian material, particularly recently, where because of the requirement for a moist environment the living process is hidden from the eye by the scab or artificially applied dressings. As a consequence, much of our understanding has come from study of wound closure in water dwelling forms, especially amphibians, where observations on living epidermal cells can be made and where wound closure occurs with astonishing rapidity. A 1 mm^2 wound in the epidermis of *Necturus* closes in less than 2 hours at 18°C (Eycleshymer, 1907); thus marginal cells must be moving at an average rate of about 5 μm/minute. In frog tadpoles, the rate may be double this (Herrick, 1932). Moreover, the skin of a larval amphibian is a relatively simple system, consisting of a smooth epithelial sheet, 2–3 cells thick, resting on a tough, fibrous, cell-free basement lamina.

There seems to be general agreement that during wound closure an epithelium moves as a sheet, in which cells retain contact with each other, and that cell proliferation is not an important factor (e.g. Barfurth, 1891; Loeb and Strong, 1904; Arey, 1925). As T.H. Morgan stated in 1901, "the wound is covered not by individual cells wandering over the exposed surface, but by a steady advance of the smooth edge of the ectoderm toward a central point. . ." (p. 70). Moreover, like cells in an epithelial sheet in culture, all cells in the epidermis appear to be capable of movement. Wherever a wound is made, the marginal cells at the edge of the wound begin to move toward the center of the wound within a few minutes (Herrick, 1932). Indeed, as Lash (1955) has shown, the marginal cells are the first to move after wounding. But the cells back of the margin move also, as shown by observing the movement of epidermal gland cells as markers (Eycleshymer, 1907). When individual epidermal cells are artificially marked with carmine particles (Lash, 1955), non-marginal cells can be observed joining the advance at progressively later times, the further they are from the margin the later they join. This suggests that the non-marginal cells move passively, in response to active pull by the marginal cells, just as do non-marginal cells of an epithelial sheet spreading in vitro. The greater radial elongation of cells nearest the margin (Herrick, 1932) is consistent with this. Moreover, marked cells in the sheet are generally abruptly immobilized, when the marginal cells stop (Lash, 1955). In some instances, however, cells have been observed to pile up at the juncture when another sheet is encountered (Poynter, 1919; Herrick, 1932; Rand and Pierce, 1932; Chiakulas, 1952), as if submarginal cells continue to move after the marginal cells have stopped. Lash (1955) also observed this, but mainly when the sheet encountered an inanimate obstacle, such as tantalum, rather than living cells.

Regardless of what happens behind them, marginal cells appear to stop as soon as they contact the cells moving across the wound surface from the other side. This, of course, reminds one of contact inhibition of cell movement, as indeed does the commencement of movement by cells at the margin of the wound within minutes after they are provided with a free edge, by the wounding procedure. The reciprocal of this, as Weiss (1961b) has emphasized, is that the stationary state of cells in normal epithelia could be the result of inhibition brought about by contact on all sides with homologous neighbors. It is premature, however, to conclude at the present that contact inhibition is really at work. There have been no close observations of marginal cells in the process of making contact with other cells.

Although there is much vague reference in the literature to the "amoeboid" activity of epithelial sheets, little is known about the mechanism whereby the marginal cells perform their obviously important leadership function. In view of the obvious importance of ruffling leading lamellae in the spreading of the marginal cells of an epithelial sheet in culture, we might reasonably expect to find essentially the same process at work in wound closure. However, the detailed observations of the activities of the free margins of the marginal cells as they begin and cease spreading, which are necessary to determine whether this is so, have never been made. (Hudspeth's observation with Nomarski interference optics of "numerous microspike processes" extending from the free edge of a living marginal cell is consistent with what we would expect from our knowledge of spreading epithelial sheets in vitro, but is only a taste of the kind of observation that needs to be made.) Moreover, where we do have information from fixed material, it gives a confused picture. This is, as we know, not surprising, since when deductions concerning a dynamic process are made from fixed material, it is on the basis of the shape and position of cells killed at one instant in time. Thus in *Necturus*, the epidermis is observed in fixed material to be 6–7 cells thick at the margin of the closing wound (Eycleshymer, 1907), whereas in frog tadpoles the advancing epidermis thins out until it may be "only one or two cells in thickness" (Herrick, 1932). Perhaps in *Necturus* there was retraction of the edge at the time of fixation.

In closing wounds in mammalian skin, where there have been no observations in the living at all, the situation seems quite unsettled. Croft and Tarin (1970) observed a loosening of cellular contacts in the epidermis near the wound, with some loss of desmosomes, and small "pseudopods" projecting from the cells at the edge. Yet, on the other hand, they find the thickness of the wound epidermis to be greater than in the resting state. In both the pig and the mouse, Winter (1964) and Krawczyk (1971) report an elongation of cells at the advancing edge, with long lamellipodia or filopodia extending over the dermal surface appearing in the electron micrographs of Krawczyk. However, the cell sheet is not monolayered and both authors propose a kind of rolling of cells over one another, with the upper cells successively implanting on the wound surface. The obvious objection to this is that one cannot reason with confidence concerning modes of cell movement from fixed material alone. On the other hand, perhaps the mechanism of epidermal wound closure is different in amphibians and mammals. The much slower rate of epidermal advance in mammals [about 1 μm/minute in the pig (Winter, 1964)] is consistent with this possibility.

Our understanding of epithelial wound closure is obviously in an unsatisfactory state. Careful study of the activities of living cells is required. Where this has been done, one encounters a far more dynamic situation than one would suspect from observing fixed material alone. Hudspeth (1975) finds, for example, that epithelial cells move together and restore tight junctions, with concomitant restoration of transepithelial resistance, less than 30 minutes after removal of a single cell between them. In further investigation of wound closure, a number of questions are crying for answers. What is the detailed behavior of the leading edge? Do cells back of the wound margin ever separate? If they do, how do they close the gap? Where are the adhesions to the substratum? What are the fine-structural relations of cells to each other, near and away from the wound margin, and to the substrata over which they are moving? What happens when the opposing margins meet? Do cells pile up or is there a typical contact inhibiting situation? What role do elements of the cytoskeleton play? Some of the recent fine-structural studies of mammalian epidermal wound closure have provided some of the answers (see, Odland and Ross, 1968; Croft and Tarin, 1970; Krawczyk, 1971; among others), but because they have not been (and perhaps cannot be) combined with parallel studies of the living process, using time-lapse cinemicrography, cell marking, microdissection, and other such techniques, they are susceptible to misleading interpretations. In a word, what is really needed is the kind of multifaceted approach currently being utilized in vitro, to the extent that this is possible.

7.1.2. Epiboly of the Chick Yolk Sac

For some years, chick blastoderms were cultured with organ culture technique on plasma clots or agar in order to follow their morphogenetic movements (see, Spratt, 1946; Waddington, 1952). This worked reasonably well for following convergence of the epiblast, primitive streak formation and regression, and formation of the head process (but see Bellairs, 1971). However, the epibolic spreading of the area opaca, a normal accompaniment of these events in vivo, never occurred. The mystery was solved when New (1959), in an ingenious experiment, tried offering the blastoderm its normal in vivo substratum, the inner surface of the vitelline membrane. The upper surface of the leading edge of the blastoderm adhered readily to this membrane and outward spreading quickly ensued. Further investigation showed that adhesion and spreading by the upper surface of the blastoderm, i.e., the epiblast surface, is specific to it. If the blastoderm is placed with its under surface against the vitelline membrane, it does not adhere and eventually turns, so that its upper surface is once again in adhesive contact. It then spreads backward underneath itself, until the blastoderm is rolled up into a vesicle. This, incidentally, is proof positive that the blastoderm spreads as a result of the activity of its locomotory margin over a passive inanimate substratum, the vitelline membrane. Other evidence consistent with this conclusion is: (1) blastoderms expand only when their margins are spreading (New, 1959); and (2) the marginal region of a blastoderm will spread even though isolated from the rest of the blastoderm (Schlesinger, 1958). There seems little question that expansion on the inner surface of the vitelline membrane is precisely the normal mode of epiboly in ovo, but, because it takes place underneath the vitelline membrane, it is unfortunately utterly unavailable for investigation in situ.

Although the early chick blastoderm is hardly a simple epithelial sheet, nevertheless its manner of spreading is much like that of much simpler systems in vitro. It adheres to its substratum exclusively by a narrow marginal band of epiblast cells, and, if we consider the epiblast only, it spreads as a cohesive sheet. It has not been established whether the cells within the epiblast change positions relative to one another as they spread outward. In fixed material they appear to be tightly joined, at first by focal tight and close junctions by 18 hours of incubation (Trelstad et al., 1967) and by zonular tight junctions (zonulae occludentes) and desmosomes between one and two days of development (Bellairs, 1963), when the stretching is more intense. But this does not necessarily preclude a certain amount of independent cellular movement, with breaking and remaking of junctions. Schechtman (1942) inserted needles through marked areas of superficial layers of amphibian gastrulae, which at the time were spreading toward the blastopore, and found that the cells continue to move toward the blastopore willy nilly, around the needle. Perry (1975) (see also Sanders and Zalik, 1972) has since shown that these cells are joined by tight and close junctions and desmosomes.

It should be realized, however, that whereas the outward spreading of the epiblast of the area opaca over the yolk appears to resemble the centrifugal spreading of an epithelial sheet in vitro, the situation is complicated during gastrulation by the behavior of the proximal part of the same sheet. In the pellucid area, the epiblast sheet moves in the opposite direction, converging toward the primitive streak.

Although, like other spreading cell sheets, the epibolic spreading of the chick blastoderm depends on strong adhesion of its margin, unlike other spreading sheets, the only cells that will spread seemed at first to be those normally at the margin (New, 1959). Non-marginal cells did not spread, even when given a free edge. From this, it was concluded that normal marginal cells are intrinsically different. This would be an unusual situation, even for an in vivo system. Non-marginal cells of the teleost enveloping layer, for example, spread well over their normal substratum when given a free edge (p. 287). A suggestion that this may not be completely so for the chick comes from a study of Downie and Pegrum (1971), who find that non-marginal fragments can spread on the vitelline membrane under certain circumstances.

Another interesting feature of epibolic spreading of the chick blastoderm is the specificity of its substratum. Although the blastoderm margin will adhere to both the inner and outer surfaces of the vitelline membrane, it will spread only on the inner surface, its normal substratum (New, 1959). Perhaps it adheres too firmly to the outer surface and is thus immobilized, or too lightly and cannot gain traction. Bellairs (1963) examined the fine-structure and composition of the vitelline membrane and found it to be composed of two protein layers. With SEM (Bellairs et al., 1969), they appear to have a different fibrous structure. But neither the composition nor the structure gives a clue as to why the blastoderm spreads on one and not the other.

The locomotory activity of the marginal cells, as they spread, has been examined with both SEM and time-lapse cinemicrography by Bellairs et al. (1969) and, in spite of the difficult optical conditions (the light must pass through 12 μm of only partially transparent vitelline membrane), they observed that the leading edge is a very broad thin lamella ($\frac{1}{4}$–$\frac{1}{2}$ μm thick) which fluctuates. It extends forward, retracts and extends forward again, and forms ruffles, like a spreading epithelial sheet in culture. (The very long lamellae reported to be present at the cell margin in fixed material (Bellairs, 1963), were not observed in the living state.) It seems probable that what these investigators have observed in culture represents an approximation of what goes on in vivo, for in both cases the cells are spreading on the same substratum, the inner surface of the vitelline membrane. These observations have been recently confirmed by Downie and Pegrum (1971) in their observations of fixed material, who find that the blastoderm is just one cell thick at its margin. In addition, they assert that, in contrast to an epithelial sheet in vitro, the attached margin of the epiblast is about three cells wide. Immediately back of the margin, the epiblast is thickened into a trilayer where cells overlap each other (Bellairs, 1963; Downie and Pegrum, 1971). The evidence for adhesion to the vitelline membrane is based on proximity of the two layers to each other, as viewed in the electron microscope. Cells seem to be separated from the vitelline membrane by 60–70 Å, very much like cells in culture (Abercrombie et al., 1971; DiPasquale, 1975a). The quality of the micrographs is not sufficient to discern whether closer appositions are present. Both Bellairs (1963) and Downie and Pegrum (1971) noted projections of the plasma membrane into the vitelline membrane, which they call "attachment plaques". These plaques could conceivably help give the epiblast the grip on the substratum that it requires for locomotion. They could also prevent retraction of the epiblast sheet, as it becomes more and more stretched in epiboly.

Cells near the edge appear to be joined to each other by desmosomes and 150 Å appositions, with an occasional focal apposition of 60–70 Å. Desmosomes are more common in the region proximal to the edge.

Cortical microfilaments about 60 Å in diameter and oriented parallel to the direction of spreading are present just beneath the plasma membrane of the marginal cells where they contact the vitelline membrane. The attachment plaques also appear to contain them. This is similar to the concentration of microfilaments one finds in fibroblasts where they contact the substratum in vitro (Abercrombie et al., 1971). Microtubules are also present in marginal cells but they have no specific orientation, suggesting that they have little importance for the spreading process, as is the case for epithelial sheets in vitro (DiPasquale, 1975a, 1975b).

The fundamental discovery of New (1959) that the blastoderm spreads as a result of its adhesion to the inner surface of the vitelline membrane has clearly set the stage for further investigation of chick epiboly and we are now in a position to do much that needs to be done. On the matter of adhesion to the substratum, it would be of interest to know what kind of contacts epiblast cells make with the outer surface of the vitelline membrane, where they adhere but do not spread. Also, what are the surface activities and intercellular relations on this surface, as compared to on the inner surface, where the cells spread? Since it is unusual for a spreading sheet to adhere over a broad band of several cells at and near the margin, instead of exclusively by the most marginal cells, this should be checked by other methods, especially by micromanipulation of a living spreading edge. It would, of course, be helpful in such a study if one could see better what is going on as the leading edge spreads. Perhaps Nomarski optics would help. In addition, the activities of submarginal cells should be observed. In a self-contained cleidoic egg, the outer cell layer need not be as tight a system as in a teleost or amphibian egg, and the sheet may therefore not be so cohesive.

7.1.3. Teleost epiboly

In contrast to avian eggs, the eggs of certain teleosts offer a distinct advantage to the experimentalist; they are often highly transparent and therefore readily amenable to study in the living state. Principally for this reason, epiboly of the teleost *Fundulus* has become the most studied example of the spreading of a cell sheet with a free edge during morphogenesis. As a consequence, even though we do not yet understand certain crucial aspects of its mechanism, a number of points can be made.

The enveloping layer (EVL), which is the cell sheet that performs in epiboly, is an epithelial monolayer formed during cleavage and is derived from the surface cells of the blastoderm. It is attached marginally to the periblast or yolk syncytial layer (YSL), which underlies the blastoderm and separates it from the underlying yolk. The deep blastomeres, whose locomotory activity has already been discussed in detail, lie between the EVL and the YSL.

Epiboly consists of the spreading of the EVL and the YSL over a sphere of fluid yolk. They accomplish it together. Although the functional relationship between these two layers, one cellular and the other syncytial, during epibolic spreading is not completely understood, we now have a fair understanding of their structural relations.

Both layers participate in epiboly, but their spreading does not proceed in an equal fashion. At first, the YSL spreads well in advance of the EVL (phase I). But the latter soon catches up, spreading almost to the margin of the YSL. During the rest of epiboly (phase II), the two layers preserve this relationship, both spreading together, with the YSL a little in advance. The YSL appears to be an excellent living substratum for the spreading of the EVL cell sheet. Not only does it undergo epiboly with the EVL, but it has the capacity to continue and complete its epiboly independently of the EVL. When the latter is removed, YSL epiboly continues (Trinkaus, 1951). Moreover, the YSL will also serve as a substratum for the spreading of a detached blastoderm or one transplanted to it artificially from another egg (Devillers, 1952; Trinkaus, 1951, 1971). Significantly, such a transplanted blastoderm will not spread beyond the YSL onto the surface of the yolk, i.e., the yolk cytoplasmic layer (YCL). Clearly there is something special about the YSL surface. It was therefore of interest to discover that its surface is highly adhesive, both to cells and inanimate objects, in contrast to that of the YCL, which is very lowly adhesive. The manner of adhesion of the EVL to the YSL is poorly understood, as is true of cell to cell adhesions generally, but it has certain features in common with the adhesion of epithelial sheets to their substrata in culture. The EVL adheres to its YSL substratum exclusively by means of its most marginal cells (Trinkaus, 1951) and these marginal cells themselves adhere to the YSL exclusively in their most marginal regions (Betchaku and Trinkaus, unpublished observations).

Time lapse filming of normal epiboly reveals the EVL to be typically epithelioid (Trinkaus and Lentz, 1967; Trinkaus et al., unpublished observations). Individual submarginal cells show much surface activity, they contract and expand and undulate, and they may even appear to pull apart momentarily, with the gap being quickly filled by the surface active cells at the edge of the gap. Each cell thus apparently has the capacity to spread. This is confirmed when part of the blastoderm is removed. The new edges of the EVL invariably spread over the now exposed YSL until it is covered again. It is also confirmed in culture (Devillers et al., 1957). In spite of all this potential for locomotory activity, however, there is normally no displacement of individual cells in the EVL sheet before or during epiboly. Each keeps in its place, as good epithelial cells should, and the result is a stable cohesive sheet. Since during the immense spreading of epiboly there is relatively little cell division, the cell sheet at the end of the process is composed of not many more cells than at the beginning, a total of approximately 5000 (Betchaku and Trinkaus, 1976). The spreading of the EVL in epiboly thus involves the expansion of cells which are already largely there at the beginning. As a consequence, individual cells must expand considerably. This they do, and in the process increase their plane surface area about four-fold and become exceedingly thin. They average 0.35–0.7 μm thick in the non-nuclear region toward the end of epiboly. The great increase in surface area of the EVL and of its constituent cells is accompanied by a great increase in tension, as evidenced by the rapid retraction of the EVL sheet when its marginal attachment to the YSL is severed. In spite of this tension, however, the sheet remains intact, even in the most advanced stages of epiboly.

The low level of cell division, along with the immense increase in the area of each cell, suggests that cell division plays no essential role in epiboly. This conclusion is substantiated by treatment with colchicine. Epiboly continues even though mitosis is completely blocked (Kessel, 1960).

Fine-structural studies (Trinkaus and Lentz, 1967; Lentz and Trinkaus, 1971; Betchaku and Trinkaus, 1976) have revealed nothing extraordinarily new in the contacts that EVL cells make with each other. However, because we are dealing with a rapidly spreading sheet of cells in vivo they have a certain special interest. EVL cells make the standard kinds of junctions and appositions with each other that we have learned to associate with epithelia. In advanced epiboly, for example, when the EVL is under great tension, EVL cells are joined to each other by a circumferential apical junctional complex, composed of tight and close junctions, a more proximal circumferential desmosomal zone with abundant tonofilaments, gap junctions proximal to these, and, finally, extensive interdigitations of apposed plasma membranes separated by a space of about 200 Å. An interesting feature of the contacts between EVL cells is that they increase in complexity and extent as tension increases in the EVL during epiboly. Tight, gap and close junctions appear to be present all along. But desmosomes are present in blastulae only in an undifferentiated form and differentiate into typical desmosomes and increase in number slowly during the course of epiboly. Extensive interdigitations of apposed plasma membranes are present only in late stages of epiboly. This striking correlation of increase in differentiation, number and extent of junctions, all of which are known to serve an adhesive function, with the progress of epiboly suggests that they function to keep the cells together, in spite of the great stretching that occurs. In addition, the constant presence of an apical zonula occludens prevents leakage of ions and water and thus helps preserve the very low permeability of the EVL (Bennett and Trinkaus, 1970; Dunham et al., 1970).

The activity of marginal cells of the EVL during epiboly is of particular interest because in it we have a true example of the spreading of a free edge of an epithelium during normal morphogenesis. The marginal cells were filmed with Nomarski interference optics before and during epiboly. In late blastula, prior to epiboly, marginal cells appear perfectly quiescent, as expected; there is no fluctuating or undulatory activity of the leading edge (Trinkaus, Ramsey and Betchaku, unpublished observations). During phase I of epiboly, in contrast, when the margin of the EVL moves toward the margin of the external YSL, the EVL margin shows marked undulatory activity. These undulations appear to begin near the margin of the marginal cells, but one cannot be sure how near. The lamellar margin at this stage is so thin – 1 μm thick or less – that it is virtually impossible to see the edge with accuracy in the living egg. Although these undulations have not been observed to propagate backward, their very presence is suggestive of active locomotory activity on the part of these marginal cells, because of their resemblance to ruffles and the usual association of ruffles with active extension in vitro. However, in order to see if this is the case, other observations must be made. Besides, ruffling cells in culture do not need to ruffle in order to move. The thinness of the margin of these cells has also made it impossible to determine whether the leading edge shows fluctuating protrusions and retractions, as do spreading epithelial cells in culture (DiPasquale, 1975a).

After the margin of the EVL has progressed about $\frac{1}{5}$ the distance to the vegetal pole, and is almost at the margin of the YSL, and during the rest of epiboly (phase II), undulations are no longer evident. In so far as can be seen in films, the exceedingly thin marginal regions of these cells are completely quiescent. The edge of the enveloping layer moves as if guided by an unseen hand. It is of course possible that delicate fluctuations of the leading edge occur and are undetected in these films because of optical limitations. Be that as it may, we can only reason on the basis of what we see. And what we see suggests that the marginal cells of the EVL are inactive in so far as spreading is concerned during the bulk of epiboly and that therefore other forces must be at work. [Incidentally, at no time have marginal cells been seen to turn under the edge. Thus, in *Fundulus*, as in the trout (Ballard, 1966), there is no evidence for involution of cells at the margin of the blastoderm.]

It is apparent from these results that we must have further knowledge of the relation of the EVL to the YSL, if we are to understand how epiboly occurs. A first step was to observe the normal junctions between the two with transmission electron microscopy. This is no easy matter where one is dealing with an egg almost 2 mm in diameter and composed almost entirely of fluid yolk. T. Betchaku has accomplished this, however, with the following results (Betchaku and Trinkaus, in preparation). All contacts of marginal EVL cells with the YSL appear to be a mixture of tight and close junctions. There are no desmosomes, no interdigitations, and apparently no gap junctions. The presence of extensive tight junctions has recently been beautifully confirmed with freeze-fracture electron microscopy (Bennett and Gilula, 1974). In late blastula stage, prior to epibolic spreading, the contact is quite wide (12 to 14 μm). Significantly, when epiboly begins, the contact narrows considerably to 0.5–0.8 μm and remains so during all of phase I of epiboly. Then, as we enter phase II, when the EVL margin appears quiescent in films, the contact widens again and the leading edge of each EVL cell appears to be embedded in the YSL. In advanced epiboly, when the blastoderm is greatly expanded and there must be great tension on this marginal contact, the leading edge of the EVL appears to be thoroughly embedded in the YSL and in extensive contact, over a width of about 9 μm.

These fine-structural results are consistent with the picture revealed by filming and suggest a theory of epiboly. If active spreading over a substratum requires the constant making and breaking of junctions (Lochner and Izzard, 1973), the narrowing of the EVL–YSL contact in phase I is consistent with active advance of the marginal cells of the EVL over the YSL substratum and possibly is essential for starting the process. The undulatory activity near the leading edge at this time is also consistent with this possibility. In contrast, the gradual widening of the contact in phase II, coupled with quiescence of the marginal region of the marginal EVL cells, suggests that in this phase the EVL is not spreading actively over the YSL, but, on the contrary, is being pulled passively by the autonomously expanding YSL, to which it becomes ever more firmly joined. It is possibly significant that the rate of advance of the EVL margin changes from about 1.4 μm/minute in phase I to about 3.2 μm/minute in phase II.

SEM of the whole egg is confirmatory of this. Even though the surface of each EVL cell is thrown into folds adjacent to each junction with another EVL cell (as seems to be characteristic of epithelia generally), where each EVL cell joins the YSL its surface is smooth (Erickson and Trinkaus, unpublished). The surface of the external YSL, that narrow band of YSL just peripheral to the EVL margin, is smooth prior to and during the first phase of epiboly but as it narrows it becomes more and more folded. Could this folding be an active process due to a cortical contraction that causes the YSL to pull on the margin of the EVL? Examination of thin sections (Betchaku and Trinkaus, unpublished observations) shows these folds to be packed with 40–60 Å microfilaments. If these are contractile, as seems likely, then the contractile machinery for this postulated pulling activity is present.

If this tentative conclusion is correct, then the *Fundulus* egg presents us with a new kind of morphogenetic movement, one in which an epithelium spreads not because of the locomotory activity of its marginal cells, but because they are firmly attached to another layer which itself is actively spreading. The probable presence of such a relationship in *Fundulus* poses a general question as to whether displacement of certain cells in other systems may not likewise be passive and depend on their attachment to moving mass. Nerves and mounted nematocysts of hydra are a case in point. Apparently, they also move passively, carried along by the actively moving epithelial tissues to which they are intimately connected (Campbell, 1974) (see also p. 293).

If the spreading of the YSL is indeed the motive force in epiboly, an understanding of its expansion is crucial to understanding epiboly. Carbon marking of the YSL (Trinkaus, 1971) shows that it spreads in epiboly as a result of intrinsic expansion of its surface, i.e., it gains no surface from the YCL it replaces. In the light of the possible role that microvilli play as a source of reserve membrane in the spreading of fibroblasts in vitro (Wolpert and Gingell, 1969; Follett and Goldman, 1970; Erickson and Trinkaus, 1976), it was with much interest that we found that the internal YSL at the beginning of epiboly is covered with long microvilli, each of which is packed with microfilaments. Moreover, as epiboly progresses, these long microvilli are replaced by short microvilli (Betchaku and Trinkaus, 1974). Estimates show that the total surface area of the internal YSL of an early gastrula, taking the microvilli into account, is just as great as that of a late midgastrula. Thus the internal YSL possesses sufficient surface in its long microvilli at the beginning of epiboly to account for the spreading of more than half of epiboly, without the addition of new membrane. If there is in fact such a redistribution of surface, it seems possible that it could be mediated by the microfilaments that fill each villus, provided they are contractile and insert on the plasma membrane, both of which seems likely. In any case, the YSL as a whole is contractile. When treated with cytochalasin B, it relaxes and expands, and when treated with augmented Ca^{2+}, it contracts (Betchaku and Trinkaus, 1974). We have no notion as to how the YSL expands normally during the last part of epiboly. It seems certain, however, that much new surface membrane is added.

Incidentally, the lamellar cytoplasm of the marginal EVL cells is also packed with microfilaments, and these too are undoubtedly contractile; for, when a blastoderm is detached from the YSL, its margin contracts vigorously. This contraction is inhibited by Ca^{2+}-free Holtfreter's solution, EGTA, and cytochalasin B, and accelerated by augmented [Ca^{2+}]. Colchicine has no effect. As in the case of fibroblasts and epithelial cells in vitro, we have no good idea as to how microfilaments function in cell movement. Indeed, the matter is particularly puzzling in the case of marginal EVL cells, since it seems probable that much of their spreading is due not to their own activity but to the spreading substratum to which they are attached. However, the EVL is under much tension throughout epiboly, and perhaps its microfilaments are important in the generation of this, by a kind of isometric contraction.

As in epithelial wound closure, epiboly of both the EVL and the YSL stops when the yolk is entirely covered and the margins meet at the vegetal pole. Just before closure, EVL marginal cells become markedly elongated in the direction of the point of closure (Trinkaus, 1949), suggesting an accentuated pulling force. Conceivably, this is due to increased tug on the part of the YSL, or a draw-string effect, as the blastopore closes. In any case, when the EVL is detached from the YSL a short time before closure, YSL epiboly accelerates and the YSL closes its "blastopore" in a few minutes (Trinkaus, 1951). The cessation of EVL spreading without piling up of EVL cells when they meet at the very end of the process could indicate contact inhibition, if they are spreading actively, as in wound closure. Or, it could be due simply to fading of the pulling force, if the YSL is responsible for their spreading.

7.2. Spreading of cell sheets without a free edge

Although we have some comprehension of how a cell sheet with a free edge spreads in vitro and are approaching understanding of how such sheets spread in vivo, we have little inkling of how sheets lacking a free edge spread, either in vitro or in vivo. This is unfortunate, inasmuch as we have good reason to believe that the bulk of spreading by sheets of cells within organisms is by those lacking a free edge. The two best-known examples of such spreading are the movement of the presumptive endoderm toward the blastopore and its later expansion inside as the lining of the archenteron and the epiboly of the presumptive ectoderm of amphibian gastrulae and the convergence of the epiblast of the pellucid area of avian embryos (and presumably of other amniotes as well). But there are many other examples during development that are less well-known and, because they have not been studied with appropriate marking experiments, less sure. Here I would include, for example, the formation of the amnion, chorion, and allantois of amniotes, and the enormous surface expansions that take place as the coelom and its various mesenteries form. All the diverse invaginations and evaginations that occur during later organogenesis also probably involve some expansion of a cell sheet, but they will be considered later, along with other invaginations.

One can imagine a number of ways in which this mysterious spreading of sheets without edges might occur. In the first place, the sheet may be dragged along over the surface by the active ingression of some of its cells at a fixed locus, with the consequent tension and expansion being communicated to more distant cells of the sheet by their attachments and adhesions to each other (Revel et al.,

1973; Perry, 1975). Holtfreter (1943, 1944) has suggested that the bottle cells might do this in amphibian gastrulation and demonstrated that a graft of blastoporal cells will sink into an endodermal substratum, form bottle cells, and exert pull, as shown by deformation of cell shape, on the immediately surrounding endodermal cells at the surface. It is difficult to imagine, however, how the process would continue after the bottle cells are completely inside (see below, p. 300). Cells at the primitive streak of an avian embryo might conceivably exert a similar force. They assume a flask-shape, as they ingress (Balinsky and Walther, 1961; Granholm and Baker, 1970), but it seems not known whether they exert any pull on adjacent cells of the epiblast. Here the continual sinking in of cells at the streak could exert tension continuously as long as a streak is present. (It may be objected, of course, that such spreading as results from ingression is not strictly spreading without a free edge, particularly at the primitive streak, where as each cell disappears from the surface the cell behind it might momentarily be given a free edge).

A second possibility is that the sheet expands because of forces intrinsic to itself. Spemann suggests, for example, that spreading is an "inherent" property of the several parts of an early amphibian gastrula (1938, p. 103). Thus, when two ventral halves are grafted together, the prospective epidermis of both is thrown into ridges and folds (Spemann, 1931). Since there appears to have been no retraction of the margins of the explant, this is clear evidence of autonomous expansion of the ectoderm. Similarly, when the marginal zone is implanted near the animal pole it pushes out to form an excrescence (Spemann, 1931; see also Mangold, 1920) (However, the probable presence of prospective mesoderm in the graft may contribute to this protuberance). Similarly, in exogastrulae, even though the ectoderm is not underlain by mesoderm and endoderm, still it expands anyway, throwing itself into folds (Holtfreter, 1933). In both of these instances, it would be of interest to know just how much expansion occurs, and how this compares to the spreading that occurs during normal gastrulation. Certain other experiments, in which isolated pieces of ectoderm spread over mesoderm or in which endoderm spreads out through a hole in the roof of the archenteron (Holtfreter, 1944) and which purport to give evidence on the capacity of these germ layers for expansion, probably do not say anything about "inherent" spreading tendencies. Having been artificially provided with a free edge, both layers simply spread at their margins, as any epithelium would.

Just how in cellular terms a sheet without a free edge manages to spread is unknown, but four ways have been suggested: (1) addition of cells; (2) active spreading of its constituent cells; (3) rearrangement of its cells; and (4) passive translocation because of attachment to actively moving cells beneath.

Addition of cells could occur either by cell division or by insertion from beneath. Cell division seems not to be necessary for the cell movements of gastrulation (Kessel, 1960; Cooke, 1973). Even cellulation appears not to be necessary for much morphogenesis in *Chaetopterus* (Lillie, 1902). It might be quite important, however, in the spreading of cell sheets during later organogenesis. Insertion of cells from beneath has been suggested frequently and it is a common observation that as the ectoderm of an amphibian embryo expands to many times its original area during early development it comes to consist of a mosaic of dark and light cells, the latter apparently having inserted into it from beneath. Expansion as a result of insertion is therefore a real possibility, but it should be confirmed by cell to cell scrutiny. If it occurs frequently, it could go far to explain the expansion of an edgeless sheet.

The possibility that cells of a primordium might acquire inherent polarity at a precise time in development and elongate in a given direction has received support from two separate studies. Waddington (1941) found that isolated fragments of the notochord primordium constrict in their lateral dimensions and expand in an anteroposterior direction just as they do in situ, and Schechtman (1942) and Holtfreter (1944) showed that when the dorsal lip of the blastopore is grafted ectopically both the individual cells of the graft and the graft itself elongate along their anteroposterior axes. Although such a remarkable capacity for directional form change could account for some of the directional spreading of a sheet, it could not explain the extensive expansion and cell translocation that often occurs. Analysis of the tendency of cells to spread by observing cell behavior in culture has so far been discouraging. All embryonic cells tend to spread, more or less, on a glass substratum. Holtfreter (1947) found that ectodermal cells seem to spread more frequently than mesodermal or neural cells. The difference is, however, variable and seems not distinct enough to serve as a basis for further analysis.

Waddington (1962, p. 176) has pointed out that a certain limited amount of directional spreading could occur if cells simply rearranged themselves, lining up in an anteroposterior direction, perhaps by developing strong cell to cell attachment sites at their anterior and posterior ends. There seems to be no evidence for such a phenomenon in directionally spreading cell sheets, but this may be because no one has looked.

Finally, if a sheet lies on the surface of other cells that are engaged in active locomotion, it could conceivably be jostled along with them, having a ride, as it were, on their backs. Such a possibility comes to mind in the case of *Xenopus* gastrulation. Keller and Schoenwolf (1976) have shown that the presumptive mesodermal cells in these gastrulae are not arranged in a cohesive sheet but rather in a densely populated stream of cells, lying beneath and in contact with the presumptive endoderm and ectoderm at the beginning of gastrulation and, of course, more and more between them as involution proceeds. If the prospective mesodermal cells move toward the dorsal lip under their own steam, as apparently do cells in the germ ring of a teleost, and are attached to the cell layer peripheral to them, they could, indeed, exert directional force on it. It would be of interest to see what kinds of junctions prospective mesodermal and ectodermal cells make with each other before and during full gastrulation.

As stated at the outset of this section, we do not know how cell sheets without a free edge spread. I have tried to point out, however, that there are, nevertheless, several leads. These deserve to be followed up. In the pursuit of this analysis, however, it would be well to keep in mind the possibility that all of these and other forces not yet suggested may act in concert, each in itself being insufficient to bring about the whole process.

7.3. *Invagination and Folding*

The mechanisms of the invagination and folding of cell sheets have at long last been much studied recently, both in organ culture and in situ within the organism. In all cases, these deformations involve changes in the shape of individual cells within the sheet and since cells in sheets are closely joined together, changes in the form of cells are translated directly into coordinated changes in the form of the cell population. Thus if cells are constricted at their apical ends and become wedge-shaped, the previously flat sheet will acquire a concave curvature. Of all the various invaginations and foldings that take place during development the most intensively studied has been neurulation. Because of this, and because rolling up of a sheet of cells to form a hollow cylinder represents one of the basic ways in which cell sheets change their shape during morphogenesis, I have chosen to consider neurulation first.

7.3.1. *Neurulation*

The formation of the neural tube is the best studied of all morphogenetic movements for several reasons: its importance – it lays the basis for development of the central nervous system; its availability – it occurs right on the surface of the embryo; and its extent – in amphibians it may involve as much as half the surface ectoderm. Nonetheless, not so long ago one despaired at how frustrating it was "that the mechanism whereby this deceptively simple process occurs has until now resisted all analysis" (Trinkaus, 1969, p. 159). Happily, a number of investigations have since rendered this statement invalid. The basis for this was laid in four perceptive papers. The one that first really called attention to the possible importance of cytoskeletal elements in determining cell form and is, in fact, a landmark in the field, is that of Byers and Porter (1964). This paper showed that microtubules orient in the long axis of presumptive lens cells during the process of their elongation (see below p. 307). Waddington and Perry (1966) followed with the observation that elongating neural plate cells likewise possess abundant microtubules oriented in the long axis of each cell and, on the basis of this, suggested that microtubules are essential for cell elongation and hence for neural tube formation. The proposal that the microfilaments of the cytoskeleton could provide the contractile force for form changes in development came from Cloney's (1966) study of tail retraction during ascidian metamorphosis. As the tail retracts, the microfilament band of the epidermal cells thickens and the cell surface apical to the microfilaments buckles. Building on these observations of Cloney and on the ancient suggestion of Rhumbler (1902), followed by Lewis (1947) and Brachet (1960), that differential contraction of one side of the sheet could cause it to roll up, Baker and Schroeder (1967) observed bundles of 50 Å microfilaments in the apical cytoplasm of neural plate cells oriented parallel to the apical cell surface and proposed that these filaments are contractile and could be responsible for the apical narrowing in neural plate cells that occurs as the plate folds up to form the neural groove and tube. The subsequent researches that these papers stimulated have amply confirmed both hypotheses. The research on this subject has been carefully summarized in two recent reviews (Karfunkel, 1974; Spooner, 1974).

A number of lines of evidence have demonstrated that the forces necessary and sufficient for [illegible]rmation of neural folds and their approach to form the neural tube resid[illegible] neural ectoderm itself, once induction by the underlying chorda-me[illegible]ken place (Karfunkel, 1974). Thus, when neural plate cells are [illegible]heir elongate shape and even elongate further in culture (I[illegible]n the intact neural plate, elongation of cuboidal plate ce[illegible]lumnar cells, prior to neural fold formation, appears to dep[illegible] microtubules, just as predicted from morphological studies (W[illegible] Perry, 1966; Messier, 1969; Schroeder, 1970). Prior to cell elongation [illegible]bules are more or less randomly oriented. Consistent with this, they are also randomly oriented in flattened non-neural epidermal cells (Burnside, 1971). Coincident with elongation, however, they line up parallel to the axis of elongation in a striking way, and in some species increase in number. Moreover, when treated in situ with drugs that disrupt microtubules, such as vinblastine (Karfunkel, 1971) and colchicine (Löfberg and Jacobson, 1974) in amphibians and colchicine (Karfunkel, 1972) in the chick, the microtubules disappear and cells that were elongate prior to treatment lose their elongate shape and round up. Moreover, cells that would otherwise elongate fail to do so. Thus, when microtubules are disrupted, not only is cell elongation reversed, it is also prevented. As a consequence, microtubules seem to be necessary both for maintenance of the elongate state (see also Handel and Roth, 1971), and for its achievement in the first place. How the microtubules operate to effect this elongation is unknown for these cells (and for cells generally), but Burnside (1971) made an effort to get at the problem by counting microtubules at various locations along the apical-basal axis of neural plate cells of the newt *Taricha*, as they elongate. Not finding the differences in number predicted if microtubules were sliding past each other (see McIntosh et al., 1969) – the number of parallel microtubules does not differ at different apical-basal levels of the cell – she proposed that microtubules bring about cell elongation by displacement of cytoplasm toward the extending basal end of the cell. Although Messier (1972) provided support for this interpretation by showing that the migration of nuclei that occurs as the cells elongate is prevented when microtubules are disrupted by colchicine, it is in fact not known how microtubules cause elongation of these cells.

Once neural plate cells become elongate, the next step in their differentiation involves constriction of their apical ends, giving them a wedge-shape. Concomitant with this, the neural plate rolls up to form a tube. Rhumbler (1902) pointed out long ago that only a slight constriction of the apical ends of neural plate cells would be required to cause the cell sheet to roll up into a tube and that this could be due to active contraction. In a fine-structural study, Baker and Schroeder

(1967) provided impressive morphological evidence for just such a process. The circular band of cortical microfilaments immediately beneath the apical and lateral cell surface becomes increasingly dense as the cell apex constricts. Also, as the apex constricts the apical cell membrane becomes highly folded, presumably by buckling. Microfilaments with similar location and behavior have been observed during neurulation in other amphibians (Schroeder, 1970; Burnside, 1971; Karfunkel, 1971) and in the chick (Karfunkel, 1972).

This correlation of cytoplasmic structure with change in cell shape is arresting, and now, in view of abundant evidence that such cytoplasmic microfilaments are f-actin, are contractile (Pollard, 1972), and insert on the plasma membrane (Pollard and Korn, 1973; see also p. 245), it seems certain that the microfilaments of the neural plate are indeed contractile and could by their contraction cause constriction of the apical ends of neural plate cells and assumption of the wedge-shape. Unfortunately, however, we do not possess unequivocal direct evidence for this. The crux of the matter is that we lack an inhibitor which causes microfilament disassembly or in some uncomplicated way interferes with their function, as does colchicine for microtubules. The only recourse has been to use cytochalasin B, a drug known to inhibit contractility by unknown means. Possibly, it inhibits contraction by breaking the insertion of microfilaments on the plasma membrane (Miranda et al., 1974). In any case, when chick neurulae are treated with cytochalasin B, neural plate cells whose apices are constricted lose their apical constrictions (Karfunkel, 1972), and cells which would have become apically constricted in the absence of cytochalasin B fail to do so in its presence. Elongate cells treated with cytochalasin B retain their elongate form and their microtubules.

In apparent contradiction to this postulated role of the microfilaments, Jacobson (1962) inverted a neural plate and grafted it back upside down, with the result that it seemed to form a neural tube with normal orientation. When Karfunkel and Burnside repeated this experiment, however, the results were not in contradiction (see Karfunkel, 1974). Sections showed that the grafted neural plate formed an upside down neural tube, but since it was overgrown by the converging cut lateral margins of the host neural place it looked superficially like a normal neural tube with normal orientation. Burnside (1973) further tested the microfilament hypothesis by culturing isolated neural plates in Holtfreter's solution. Inasmuch as they curled in reverse and the curling was inhibited by cytochalasin B, it was of interest to see where their microfilaments were concentrated. Transmission electronmicrographs revealed a heavy concentration of microfilaments in reversed position, in what is normally basal cytoplasm, a result completely consistent with the hypothesis that contractile microfilaments are necessary for the rolling up of neural ectoderm.

In summary, although unambiguous proof is not yet at hand, there is strong evidence that neural plate cells constrict apically as a result of the contraction of apically concentrated, circumferential cortical microfilaments. This hypothesis should therefore be accepted as far and away the best working hypothesis available. The most convincing way to provide proof, of course, is to demonstrate the operation of a mechanism. This has not yet been done, but a lead as to how to go about it has come from Gingell's (1970) interesting manipulations of cortical contractility by the addition of Ca^{2+} ions. An approach such as this, combined with the application of lessons learned from current studies of non-muscle contractile proteins (see Pollard and Weihing, 1974), are clearly the next steps in the analysis of this fascinating and historically important developmental system.

An interesting accompaniment of the apical constriction of neural plate cells is the folding of the plasma membrane at the apical surface. Both transmission and scanning electron micrographs show it to be thrown into extensive folds and microvilli (Baker and Schroeder, 1967; Löfberg, 1974), especially where the cells are most constricted apically. Nearby unconstricted epidermal cells, in contrast, are relatively smooth, except, of course, for folds at their junctions with adjacent cells. It seems probable that these surface projections of neural groove cells represent excess plasma membrane that previously covered the much larger unconstricted apical surface, and that it has been thrown into folds and microvilli by the apical constriction of the cell. The circumferential apical tight junctions that unite all neural plate cells probably confine such excess membrane to the apical surface of the cell, i.e., prevent it from redistributing itself over the rest of the cell. If this interpretation is correct, and if isolated neural plate cells can be made to constrict apically, they should lack apical folds.

Another way in which cells might conceivably elongate and transform into a wedge-shape is by extending their lateral adhesions to each other – "zipping up", as it were (Gustafson and Wolpert, 1962, 1967). Everyone knows, for example, how a freshly seeded fibroblast extends and changes shape as it spreads adhesively on a plane substratum. If this occurred among cells of the neural plate, it could conceivably result in cell elongation. But if it were then to deform the elongate cells into a wedge-shape with normal orientation, the zipping up would have to be confined to the more apical lateral surfaces. This is an interesting hypothesis, but two observations weigh heavily against it: elongation of isolated neural plate cells (Holtfreter, 1947), and the appearance of folds and microvilli during apical constriction. If the wedge-shape resulted from cells extending their adhesions with each other, the "excess" apical surface membrane would have been used up laterally in new apical adhesions and junctions, instead of confined in folds and microvilli.

One of the grave difficulties in the study of primary neural induction has been lack of understanding of the chemical events that lie at the basis of the obvious morphological transformations. Now, with the research of the last several years, a partial chemical basis seems finally to have been established. The chordamesoderm must have its inductive effects at least in part by way of an influence on the cytoskeleton, with one of the first actions being a stimulation of assembly and orientation of microtubular subunits. The polarity of this orientation is no doubt related to the presence of the inductor on one side only. Concurrently, as microtubules align and the cell elongates, the inductor must somehow control the location and orientation of microfilaments at the apical end of the cell.

Following this, it must somehow stimulate these filaments to contract, with the consequent formation of a neural tube. The next steps in the analysis of neural induction are clearly to seek the stimuli emanating from the chordamesoderm which direct these cytoskeletal events. Two obvious thoughts are that the inductor might control the polymerization of microtubules by regulating the concentration of cyclic AMP in neural ectoderm (Johnson et al., 1972; Porter et al., 1974) and control the contraction of the microfilaments by regulating the uptake or release of Ca^{2+} ions (see Spooner, 1975). The rapidly growing literature on microtubule assembly and growth and on non-muscle contractility will doubtless provide other ideas for what is sure to be an exciting area of investigation.

Finally, our present understanding of neurulation illustrates admirably an important principle of morphogenesis, which I called attention to in the Introduction (p. 225). Cells must undergo cytodifferentiation in order to engage in morphogenetic movements. This principle applies generally, as in the surface activity and blebbing of sea urchin primary mesenchyme and *Fundulus* deep cells, but it is nowhere as well illustrated as in neurulation. In the neural ectoderm, this cytodifferentiation involves as a minimum the assembly and aligning of microtubules and the apical concentration, orientation, and contraction of microfilaments, with their consequent changes in cell form. It is only after these morphogenetic differentiations and the ensuing morphological transformations have taken place that the stage is set for the cytodifferentiation of new tissue cells, the neurons, and for the histogenesis of the central nervous system.

7.3.2. Gastrulation

7.3.2.1. Chick. The most impressive aspect of gastrulation is the movement of cells from the surface of the embryo into the interior. So much so, indeed, that in the minds of many gastrulation is just this. This is erroneous; these invasions may take place by one or a combination of three methods: ingression, invagination, and involution (see above, pp. 237–238).

In the chick blastoderm, cells ingress along an anteroposterior axis in the center of the area pellucida and by this not only move inside to form the mesoblast but in addition form a groove on the surface, the primitive groove, a mini, elongate invagination, which lasts only so long as the epiblast cells continue to sink in. It seems reasonable to assume, therefore, that they are responsible for it. It is not understood how these cells invade, but Balinsky and Walther in a paper remarkable for its insight at the time of its publication (1961!), described how the cells of the epiblast narrow apically and assume a flask shape as they invade the streak and noted that the apical neck region of these cells is filled with electron dense material. They postulated that this material is contractile and by this activity causes a narrowing of the cell apex, which forces the bulk of the cytoplasm downward, causing at the same time the adoption of the flask shape and ingression of the cells into the interior of the blastoderm. This material is no doubt microfilamentous and it seems reasonable to assume that it has the same function postulated for the apical microfilaments in neural ectoderm, both a narrowing of the apical part of the cell and an invagination of the cell layer, in this instance formation of the primitive groove.

Microtubules are also present in cells of the epiblast and are arranged parallel to the longitudinal axis of the cell (Sanders and Zalik, 1970). As these cells move into the primitive streak and elongate, they show another similarity with neural plate cells. The flask cells possess abundant, longitudinally oriented microtubules (Granholm and Baker, 1970) and a preliminary study by Granholm (1970) suggests that they are important in the maintenance of the elongate shape of these cells. Colchicine disrupts their microtubules, causes the cells to spherulate, and transforms the primitive groove into a linear mound. There is no experimental evidence as to whether the microtubules cause these cells to elongate in the first place.

Primitive streak cells differ from neural plate cells in one important respect. They break their attachments to other surface cells and slip down inside. This no doubt explains why invagination proceeds no further than the formation of a shallow groove. Because of this behavior, one might predict that epiblast cells lose their apical tight junctions, as they enter the streak. They do not. All streak cells whose apical surfaces are still at the surface of the streak possess a typical apical, circumferential, tight-close junctional complex, plus a more extended proximal region of close apposition (Granholm and Baker, 1970). Revel et al. (1973) report gap junctions as well. Predictably, the apical surface of each of these cells is highly convoluted. It would seem then that the crucial difference between cells of the primitive streak, which ingress and form only a shallow groove, and cells of the neural plate, which do not ingress and undergo complete invagination to form a tube, is the ingression of the streak cells. Presumably, streak cell ingression is facilitated by disappearance of their junctional attachments to their nearest neighbors, and Revel et al. (1973) present evidence that they do not completely disappear but remain as incomplete remnants on mesoblast cells that have just left the streak. Zonulae occludentes are no longer present, only isolated islands and clusters of strands (as seen in freeze fracture). Nothing is known about possible locomotory activity of the basal ends of these cells. The cells of the primitive streak are obviously begging for further study and models of what can and needs to be done are now available in abundance.

7.3.2.2. Amphibia. Ingression of superficial cells is also associated with invagination during amphibian gastrulation, at least in the beginning. The first cells invade at the site of the future blastopore and form the famous "bottle cells", which remain attached for a time to the surface of the gastrula at the apical tips of their very long, attenuated necks. Significantly, their invasion and elongation is accompanied by a surface indentation which marks the beginning of invagination (Ruffini, 1925; Holtfreter, 1943). Because of this association of events, it is hard to believe that bottle cells do not have something to do with the onset of invagination, as Rhumbler had already suggested back in 1902. However, experimental verification is necessary and Holtfreter (1944) provided some in one of his habitually clever experiments. When blastoporal lip is grafted in vitro to a substratum consisting of endoderm from the blastocoel, its cells sink into the endoderm, but remain attached to the surface by their apices. The result is that they attenuate, forming bottle cells, and the area of endoderm surface to which they are attached indents to form a small invagination. As a control, Holtfreter grafted a ball of endodermal cells from the floor of the blastocoel to the same substratum. These grafted cells also invade and become somewhat flask-shaped, but they do not retain their apical attachments and no invagination occurs.

From the old observations and this experiment, one can propose an hypothesis. Certain superficial cells of a late blastula, located where the blastopore will form, move actively into the interior. But they retain their apical adhesions to their neighbors, become greatly attenuated, and exert tension on the cells to which they are attached, pulling them inward. Thus invagination begins. The next question, of course, is how do they do it?

Apparently, the situation is basically the same as for invagination of the neural plate (Baker, 1965; Perry and Waddington, 1966). The long neck of each bottle cell is filled with microfilaments, particularly in the most apical region, and microtubules in the neck are oriented in its long axis. Moreover, at the apex of each bottle cell the plasma membrane is thrown into folds and long microvilli (Tarin, 1971b; Brick et al., 1974) and each bottle cell is attached to adjacent cells by circumferential tight and close junctions and interdigitating flanges. If the microtubules function to elongate the cells and the microfilaments to constrict them apically, this should in itself cause the superficial layer to become concave at that locus, thus beginning invagination. But invagination associated with still attached bottle cells continues beyond this point, which suggests that the rounded basal ends of the bottle cells actively invade the interior of the gastrula, pulling the apical ends and the surface of cells of the gastrula in behind them. Soon, however, the bottle cells lose their connections with the surface and merge into the endoderm of the floor of the blastocoel at the anterior end of the early archenteron, where they become incorporated into the pharyngeal epithelium. This postulated role of the bottle cells in the beginning phase of invagination seems reasonable, in view of their ultrastructure and what we know about other systems. However, for unexplained reasons, none of the postulated roles of the microfilaments and microtubules has yet been put to experimental test. And we know absolutely nothing about the likely invasive activity of the deep or basal ends of the bottle cells, except that they lack yolk platelets and hence have a somewhat hyaline look, like blebs of *Fundulus* deep cells (Nakatsuji, 1975). This is consistent with their having an active locomotory function. On the other hand, SEM of bottle cells reveals little of the surface activity that one associates with locomotion. Their basal ends are rounded and rarely form protuberances (Keller and Schoenwolf, 1976). Moreover, even if bottle cells do normally play a role in the beginning of archenteron formation, it appears that they are not indispensable for it. When they are removed prior to formation of the archenteron in *Xenopus* and *Bombina* it forms anyway (Cooke, 1975; Keller, unpublished observations). [For a summary of earlier notions of the "invasiveness" of bottle cells see Trinkaus (1965)].

In any event, the bottle cells could only account for the very beginning phase of invagination and not for the ensuing massive involution, as prospective endodermal and mesodermal cells pour over the blastoporal lips and move inside to form the archenteron. For this, Schechtman (1942) made certain observations that helped lay the basis for modern work. Although he placed emphasis on the correlative aspects of gastrulation, his experiments established quite elegantly that certain areas have an intrinsic capacity for their mass movements during gastrulation. The isolated dorsal lip, for example, has an autonomous capacity for expansion, demonstrating that its extension during gastrulation is brought about by forces within rather than from without. In addition, he showed that prospective head mesoderm will migrate across the inner surface of the prospective ectoderm in an explant, and that the actively moving prospective mesoderm and the adhering prospective endoderm spread together. Both of these observations, each in its own way, call attention to the need to pay attention to the activities of individual cells during involution.

This poses a serious technical problem in eggs as opaque as those of the Amphibia. A partial way around this has become available with scanning electron microscopy. The early results with this technique are revealing and, thus encouraging, but the problem of distinguishing fact from artifact in this material is extreme. Not only must one face the usual unknowns, such as shrinkage, but, in addition, in order to see what the cells in the interior look like the embryos must be broken open, with unknown consequences for the form of the cells. A partial control for this would be to open up embryos to expose the inner cells both before and after fixation and compare the results.

Nakatsuji (1974, 1975) has observed prospective endodermal and mesodermal cells of the roof of the archenteron as they spread anteriorly with both transmission and scanning electron microscopy and found that they form blebs, lobopodia, and, where they are attached to the inner surface of the blastocoel wall, lamellipodia. Karfunkel (1976) has studied cells at the leading edge of the ventral mesoderm and found similar blebbing behavior. The presence of such protrusions is consistent with these cells having locomotory activity as individuals.

The careful studies of Keller (1975, 1976) of normal gastrulation movements in *Xenopus* have laid the basis for an analysis of mechanism in this species that should relate closely to what is actually going on. There is now no doubt that what Løvtrup (1966, 1975) has proposed for the prospective mesoderm of urodeles definitely takes place in *Xenopus*. The prospective mesodermal cells are always beneath the surface, beneath the prospective endoderm prior to involution and between prospective endoderm and ectoderm, after involution. Moreover, they begin to move into the interior, in a kind of ingression, distinctly before the appearance of a blastoporal depression, confirming the results of Vogt (1929) for urodeles and of Nieuwkoop and Florschütz (1950) for *Xenopus*.

In an effort to understand the mechanism of involution, Keller and Schoenwolf (1976) have examined the form and contact relations of the cells of these spreading layers both before and during gastrulation in the SEM. Perhaps their most important discovery is that the arrangement of the prospective mesodermal cells differs drastically from that of the prospective ectodermal and endodermal cells. The latter have an epithelioid organization into cohesive layers, in which the cells are joined by a typical epithelial junctional complex of apical tight and close junctions and desmosomes (Sanders and Zalik, 1972; Perry, 1975). Perry also found them to be filled with actin-like and 10 nm microfilaments. The endoderm retains this organization after involution.

Before involution, the prospective mesodermal cells are columnar, with their long axes perpendicular to their direction of movement, and arranged quite regularly. In the zone of involution, however, they assume all sorts of shapes, from globular to vermiform, and are mixed together with no obvious order in their arrangement. After involution, the dorsal, chorda-mesodermal cells reassume the palisade arrangement possessed prior to involution. The prospective head, lateral, and ventral mesodermal cells, however, remain rounded or irregular in shape, have no orderly arrangement, and possess large rounded protrusions, fine extensions which might be filopodia or retraction fibers (or both), and lamellipodia.

These mesodermal cells clearly have the appearance of cells in active individual locomotion. Moreover, they are attached to both the overlying ectodermal cells and underlying endodermal cells. The appearance of these cells is highly significant, for the march of the mesodermal mantle in involution is led by the anterior, lateral and ventral cells. Putting all these observations together, a reasonable working hypothesis for involution (or at least spreading inside to enlarge the archenteron) in *Xenopus* would be the following (see Keller and Schoenwolf, 1976). Prospective head mesodermal cells become locomotory in the late blastula and move anteriorward inside the embryo as a cell stream on a substratum provided by the overlying prospective endodermal and ectodermal sheets, perhaps in the same way that deep cells of *Fundulus* utilize the overlying EVL sheet as a substratum. Because the prospective endodermal cells adhere to these moving prospective mesodermal cells, they are carried along on the backs, as it were, of the mesodermal cells, remaining in a sheet, however, because of their junctions. Because of the protrusions seen in the SEM, it seems reasonable to assume that these cells move like *Fundulus* deep cells (Trinkaus, 1973a) and that the stream of mesodermal cells moves very much like the stream of teleost deep cells in the germ ring, all cells in the stream being constantly in locomotion, and not like a fibroblast or epithelial sheet, in which the main locomotory activity is by the cells at the leading edge. As Keller and Schoenwolf (1976) point out, these protrusions are found on these prospective mesodermal cells along the full area of their involution and not just at their presumed "leading edge" adjacent to the blastocoel. If this hypothesis turns out to be essentially correct, we would have another example of an extensive sheet of cells spreading passively because of its attachment to an actively moving layer, like the EVL spreading because of its attachment to the spreading YSL (see p. 290). But it should be noted that if this hypothesis holds for the *Xenopus* gastrula, it would differ in an important way from *Fundulus* epiboly. Since the prospective mesodermal stream of *Xenopus* is attached to the endodermal sheet over its whole expanse, not just at its margin, each actively locomoting prospective mesodermal cell would be doing its part in carrying the attached endodermal sheet along. It might be objected that Holtfreter (1944) has shown the endoderm to have remarkable spreading properties of its own, for if a hole is made in the archenteron and ectoderm removed around it, the endoderm will spread out to cover the denuded areas. However, it seems probable that by artificially giving the endoderm free edge it responded like the epithelium that it is, and simply spread marginally until it encountered the free edge of another spreading epithelium, the ectoderm. It is therefore more like epithelial wound closure than the autonomous spreading of a germ layer that normally lacks a free edge.

There are various ways in which this hypothesis might be tested. If the prospective mesoderm is removed, there should be no endodermal involution. If ways are found, as I am sure they can be, of observing prospective mesodermal cells during involution, they should of course be in active locomotion, as individuals. And they should be inactive in the blastula, prior to their known time of ingression. It also may be possible to pick up some manifestations of these presumed changes in activity in culture, as we have been able to do in *Fundulus* (Trinkaus, 1963; Tickle and Trinkaus, 1973). Furthermore, since it appears that prospective mesodermal cells move in broad cell streams in vivo, it is possible that they stimulate each other's locomotory activity (see p. 279). This is the kind of matter that also might yield to an in vitro analysis. Finally, careful correlative transmission electronmicroscopic studies of junctions and cyto-skeleton should begin to give some structural basis for the locomotory changes observed.

An important aspect of study of the mechanism of amphibian gastrulation was begun by Holtfreter (1939), when he showed that germ layers have selective "affinity" for each other, ectoderm for ectoderm, endoderm for endoderm and mesoderm for mesoderm and that these affinities could play a role in the proper arrangements of the germ layers during gastrulation. When fragments of all three germ layers were combined, for example, a quasi embryo was reconsti-tuted, with ectoderm on the outside, endoderm forming a tube or vesicle inside, and mesoderm in between.

Steinberg (1964) has proposed that these affinities are based on quantitative differences in cell surface adhesiveness and that these differences in the adhesiveness of different prospective germ layers could direct their final disposition and therefore their movements relative to one another during gastrulation. The difficulty with such an approach is that it ignores certain details that are certainly crucial to the gastrulation process, for example, the different behavior of prospective dorsal mesoderm and of prospective anterior, ventral and lateral mesoderm. In a word, it has not proved to be a useful guide to the study of detailed cell movements during gastrulation. But it has had an heuristic effect in calling attention to the adhesive properties of cells as a possible factor controlling cell movements. Along these lines, Johnson (1970, 1972) has evidence, from a number of criteria, suggesting that changes in cell contact behavior, in particular adhesiveness, occur at the time gastrulation begins. The absence of these changes in hybrid frog embryos that do not gastrulate gives reason to suppose that these changes are necessary for gastrulation.

In another attempt to relate cell surface properties to their locomotory activity during gastrulation, Schaeffer et al. (1973) and Brick et al. (1974) have found that the electrophoretic mobility, and therefore, presumably, net surface charge density, of cells of the prospective mesoderm and anterior endoderm increases at the beginning of gastrulation, whereas that of the prospective ectoderm retains the lower value which all cells shared prior to gastrulation. This could mean that cells with increased surface charge repulse each other more (Dan, 1936; Bangham and Pethica, 1960; Pethica, 1961; Curtis, 1962) and thus have lower adhesiveness for each other. If cells were too adhesive to move, then a lowering of their adhesiveness could facilitate locomotion. If these results do indeed mean that these cells have reduced adhesiveness, then they are in apparent contradiction with the findings of Johnson (1970, 1972). It is not really possible at the present time to assess the significance of this. The two studies used different techniques and different cells and, in addition, both measured aspects of the cell surface whose significance for cell behavior is still ill understood. However, both of these investigations and those of Holtfreter and Steinberg point to aspects of the cell surface that are surely fundamental for cell movements. If in the future such studies are integrated closely with concomitant observations of particular cells known by marking experiments and direct observation to be engaged in locomotion in particular ways during the gastrulation process, they should contribute substantially to understanding finally what is going on.

Nor should we ignore the possibility that the ECM is important in these early gastrulation movements. There is nowhere near as much ECM between the cells of early embryos as there is frequently later on in development, but some is certainly present and could play the same role in controlling cell movements as has been postulated for later development. Kosher and Searls (1973), for instance, have found heavily sulfated GAG synthesis in the chorda-mesoderm of late gastrulae of *Rana*, and Tarin (1971a) found much metachromatic intercellular material between the loosely packed mesodermal cells of *Xenopus*. The synthesis of ECM by normal and hybrid gastrulae of *Rana* is currently being investigated by Johnson and he has found that hybrid embryos, which fail to gastrulate, produce reduced amounts of a high molecular weight, negatively charged material during the gastrula arrest period. Moreover, this material is confined to the cytoplasm in hybrid cells. In normal gastrulae, it moves into the extracellular spaces (unpublished observations).

7.3.2.3. Echinoderms and Amphioxus. In certain groups, such as echinoderms, the archenteron forms by a kind of textbook invagination, apparently without either ingression or involution. Ingression definitely does not contribute to the archenteron, although there is ingression of primary mesenchyme cells prior to invagination. Involution is universally assumed to be absent; but it must be admitted that the evidence against it is not rigorous. Surface cells peripheral to the locus of invagination have not been marked to determine whether they move to and over the blastoporal lips.

Many echinoderm gastrulae are particularly favorable material for the study of invagination, as well as for other morphogenetic movements, because of the transparency of the eggs, and a number of investigators have taken advantage of this feature through the years. This work has been thoroughly reviewed by Gustafson and Wolpert (1967) and discussed by myself (Trinkaus, 1969, pp. 136–145), and since there seems to have been no additional study of this beautiful material since, I feel free to refer the reader to these references and confine my remarks here mainly to suggestions as to what needs to be done.

The first phase of invagination clearly depends on forces intrinsic to the vegetal plate. Moore and Burt (1939) cut away the animal half and found that the vegetal half invaginated anyway, about 1/3 the distance to the animal pole. It is not understood how this happens. Gustafson and Wolpert (1967) have suggested that this is due to the combined loosening of adhesions of cells of the vegetal plate on their inner aspects, which has been observed and might cause this surface to expand, and a postulated confining ring of increased contact of cells of the blastocoel wall around the vegetal plate, which might restrict expansion to the vegetal plate. Thus confined, an expanding cell layer would have no choice but to buckle, in this case, inwardly, since it is on this side that the loosening occurs. In the light of research on other invaginations during the last years, this process needs careful study at the level of fine structure. The questions are obvious. Where are the microtubules and microfilaments and how are they oriented during succeeding phases? And what effect does modification of their activity have on invagination? What and where are the junctions and appositions between cells and how do they change during invagination? What, precisely, is the relation of cells of the vegetal plate to the hyaline plasma layer at the surface of the egg during invagination?

After this initial phase, secondary mesenchyme cells appear at the tip of the archenteron and make filopodial attachments to the blastocoel wall, predominantly in the area of the animal pole. With this, invagination continues to completion, with concomitant shortening of the filopodia. This age-old fundamental observation led Dan and Okazaki (1956) and Gustafson and Kinnander (1956) to propose independently that this second phase of invagination depends on pull by these "contracting" filopodia. Both the Japanese and Swedish schools have supported this hypothesis with much evidence (summarized in Gustafson and Wolpert, 1967) and it is the best hypothesis available. I have suggested elsewhere, however (Trinkaus, 1969), that several questions remain unanswered. The main problem is that none of the evidence eliminates another hypothesis, namely, that the archenteron continues to expand intrinsically, as during the first phase, but now becomes so extended that it needs anchoring to keep it from retracting and guidance to conduct it to precisely the correct place on the animal ectoderm. Moreover, in the light of later investigations of cell movement in vivo by means of shortening processes (Trinkaus, 1973a; Izzard, 1974), we must ask in addition whether the shortening of the filopodia is really due to active contraction or to elastic recoil (see p. 269). Incidentally, should the hypothesis of filopodial contractile pull prove valid, this would be another example of expansion of a sheet of cells by the activities of cells to which it is attached. The research of the Swedish and Japanese schools on invagination during echinoderm gastrulation have been most stimulating, but unfortunately have not provoked further investigation. Should added incentive be needed, the investigator might be reminded that this most favorable material abides in some of the most favorable locations.

Amphioxus is an excellent example of a form in which invagination during gastrulation seems to involve straightforward inpocketing of the prospective endoderm, uncomplicated by ingression, contractile filopodia, or involution (Conklin, 1932). It has not been established for sure, however, that there is not some involution. Indeed, Fig. 134–138 of Conklin (1932) suggest that perhaps there is a small amount of involution of prospective mesoderm. The blastula is a hollow ball whose wall is a cell monolayer. Gastrulation begins when the most vegetal cells of this wall elongate in an animal-vegetal direction to form the flattened vegetal or endodermal plate. Soon after its formation, this plate indents into the blastocoel in a classical invagination and soon contacts the inner wall of the blastocoel, the prospective ectoderm, in the animal half of the embryo. In the gastrula of *Amphioxus*, this happens quickly, for along with formation of the endodermal plate, the blastula flattens considerably in the animal-vegetal axis and the archenteron does not have far to go in order to reach the animal ectoderm. Conceivably, this eliminates a need for secondary mesenchyme filopodia. The archenteron of *Amphioxus* seems more like the archenteron at the end of the first phase of sea urchin invagination than that at the end of full invagination.

There has been no investigation of the mechanism of this elegant process in *Amphioxus* and I mention it here primarily in the hope of stimulating renewed interest in it and in the invagination of other forms in which gastrulation seems equally uncomplicated. What needs to be done is quite evident, from the example of vertebrate neurulation.

7.3.3. Organogenesis

Two of the ways in which cell sheets certainly move during organogenesis are invagination and branching. These are obvious from histological sections. In addition, there must be a considerable amount of epithelial spreading. But the extent of this is undetermined. Whether involution occurs is unknown.

7.3.3.1. Invagination. Invagination of flat epithelial sheets, as opposed to invagination in connection with branching, is commonplace during organogenesis, but it has been analyzed in only a few systems – neurulation, of course, and in the formation of the thyroid (Shain et al., 1972), the lens (Byers and Porter, 1964; Wrenn and Wessells, 1969), and the oviduct (Wrenn, 1971).

Because they have certain unique features of general interest for the mechanism of morphogenetic cell movements, I will concentrate on the latter two.

During invagination of the lens placode, after induction by the optic cup, cells change from a cuboidal form to an enormously elongate form only 3–4 μm thick and approximately 50 μm long. Byers and Porter (1964) showed that this elongation is accompanied by a spectacular orientation of long, straight microtubules in the long axis of each cell. These oriented microtubules persist as long as cells continue to elongate but disappear when elongation stops and the cells are about to form lens fibers. It certainly looks as if microtubules are responsible for the generation and maintenance of the elongate cell form. However, when treated with colcemid after the cells have become elongate, the elongate form is maintained, even though the microtubules have disappeared (Pearce and Zwaan, 1970; Piatigorsky et al., 1970). Microtubules are clearly not alone responsible for the maintenance of the elongate cell form. Perhaps once cells have elongated their form is stabilized by their adhesions to their closest neighbors. This seems quite likely. When lens cells are still cuboidal, they are widely separated, except apically, where they are tightly joined. During cell elongation, however, adjacent cells become more and more closely aligned, until finally the surfaces of each cell are in close apposition with those of its neighbors all along their lateral borders (Byers and Porter, 1964). This situation needs emphasis, for with current intense interest in the activity of cytoskeletal elements in the maintenance of cell form, the possibility that adhesions to other cells, by means of junctions and appositions, may be playing a role is often ignored. In so far as the process of cell elongation itself is concerned, in the light of results with other systems, it seems probable that the lengthening and orientation of the microtubules is an important, if not the only force at work. But there appears to be no experimental evidence on this. Since microfilaments are present in a typical band in the cortical cytoplasm at the apical end of each lens cell (Wrenn and Wessells, 1969), they may well be the contractile element that causes the apical narrowing of the cell and invagination, but at this point all we have is this anatomical correlation.

A neat example of how an inductor may have a morphogenetic effect such as invagination by way of an influence on the mobilization of microfilaments is provided in Wrenn's (1971) study of estrogenic stimulation of the morphogenesis of the chick oviduct. Epithelial cells of the immature oviduct contain microfilaments, to be sure, but they are scattered in bundles throughout the cytoplasm. After administration of estradiol, however, the microfilaments become concentrated in the apical cytoplasm. This is already evident 6 hours after administration of the hormone and by 24 hours and later large bundles of microfilaments may be seen oriented parallel to the apical surface. Concomitant with this apical concentration is a diminution in the number of microfilament bundles scattered generally in the cytoplasm and, significantly, the appearance of local invaginations of the luminal surface of the epithelium. When treated with cytochalasin B, the microfilament bundles become disorganized and the invaginations disappear.

The distinctive and hopeful feature of this system is that the microfilaments of the cell increase in number and become organized in the apical cytoplasm in response to a defined inductor, estradiol. However, the molecular events leading from the hormonal stimulus, which presumably acts initially at the level of the plasma membrane, to the eventual cytoplasmic response are as yet unknown.

7.3.3.2. Branching. The morphogenetic movements I have been discussing thus far have a common feature – they occur only once during development and with that the organ is established. This is not the case for certain other organs such as the salivary glands, the pancreas, and the lung, whose morphogenesis involves branching of an epithelial rudiment. The branching is repetitive and continues until the complex definitive morphology of the organ is established. Not surprisingly, the pattern of branching is characteristic for each organ primordium, and in each instance depends on an inductive interaction between the endodermally derived epithelium and its surrounding mesenchyme (see chapter by Saxén et al., in this volume). Since the discovery by Borghese (1950) and Grobstein (1953; see also 1967) that these systems will undergo morphogenesis in culture, they have been available for intensive investigation. I can be relatively brief in my discussion of the mechanism of branching because the subject has been fully reviewed elsewhere (Bernfield and Wessells, 1970; Spooner, 1973, 1974; Bernfield, 1973).

Branching morphogenesis characteristically takes place as follows. The primordium first arises as an epithelial bud; a salivary bud, for instance, protrudes from the oral epithelium. The bud forms a hollow stalk with a bulbous tip. An invagination which deepens into a cleft appears in the tip, separating it into two rounded lobes. Thus branching begins. A cleft appears in each lobe separating it into two lobules, which are in fact branches of the parental lobe. Deepening of the clefts and separation of each lobe into lobules is accompanied by cell division and growth of the new lobules. A cleft then appears in the tip of each new lobule, separating it in turn into two new branches, et cetera.

The branching system that has received the most attention has been the salivary gland, in particular the submandibular gland. Like other epithelial cells about to engage in foldings or invaginations, the cells of the salivary epithelial rudiment possess cortical bands of microfilaments. In the salivary rudiments, however, these bundles come to be concentrated in the cytoplasm of the basal end of the cells at the tip of each lobule, rather than the apical end. Since we are dealing with a tube covered with a basement lamina (Spooner and Wessells, 1972), the basal end of each cell is its outer surface. The cleft which divides the tip of each epithelial bud into two branches begins as a small invagination, formed by a narrowing of the basal ends of these cells, with accompanying folding of the cell surface. It is hypothesized that this invagination is accomplished by contraction of the microfilament bands (Bernfield and Wessells, 1970). Not unexpectedly, the process of cleft formation is blocked by cytochalasin B and clefts that are newly formed are reversed, with accompanying disorganization of the cortical bands of microfilaments (Spooner and Wessells, 1970).

Furthermore, when CB is removed, clearly defined apical cortical bands of microfilaments reappear accompanied by a renewal of cleft invagination. So, cleft formation seems indeed to be due to contraction of the basal ends of the cells, presumably in connection with their cortical, circumferentially oriented microfilament bands. The conclusion that these microfilaments are contractile is supported by the observation that they bind heavy meromyosin and are therefore undoubtedly actin (Spooner et al., 1973). Since nothing resembling myosin, either structurally or chemically, has yet been found, there is no structural basis as yet in these cells (nor generally in non-muscle cells) for an actomyosin contractile system like muscle.

It should be noted that older clefts, which are deeper, are not reversed by treatment with cytochalasin B, suggesting that their integrity depends on other factors. ECM has been implicated in this (see below), but perhaps the development of more extensive adhesions to adjacent cells is equally important.

Cytoplasmic microtubules, though present and mostly parallel to the long axis of the cell, seem not to be necessary for the maintenance of clefts (Bernfield and Wessells, 1970; Spooner and Wessells, 1972). In the presence of colchicine, all preexisting clefts remain, young as well as old, but no new clefts form. This latter result is somewhat puzzling, inasmuch as formation of new clefts is not inhibited if the colchicine is applied during recovery from CB treatment. After this, however, no more new clefts form. This contradiction does not appear to have been resolved. It seems likely that inhibition of cleft formation by colchicine is due not so much to an effect on cell form as to an inhibition of further growth of the salivary epithelial primordium, presumably because of inhibition of mitosis. Cell division thus seems necessary for the morphogenetic movements of branching morphogenesis, in contrast to other morphogenetic movements. Such a dependence upon cell division is not unexpected due to the continuing nature of the process and its constant need for additional cells.

In view of these studies, it seems established that branching morphogenesis depends basically on the dual processes of growth, supported by cell division, and cleft formation, due no doubt to contractile microfilaments. But what gives direction to the growth and what determines in which cells and where in those cells microfilaments will concentrate? There is now considerable interest in the possibility that ECM plays a determining role in all this. Both collagen and glycosaminoglycans are found at the interface between the epithelial primordium and the surrounding mesenchyme, organized both into a distinct basement lamina on the basal surfaces of the epithelial cells and looser surface associated material. In the branching lung primordium, collagen fibrils are highly ordered along the non-branching trachea, but randomly organized where branching occurs (Wessells, 1970), suggesting either that ordered collagen suppresses branching or that disordered collagen encourages it. The hypothesis that collagen is involved was tested by treating with collagenase (Grobstein and Cohen, 1965; Wessells and Cohen, 1968) and branching of buds of lung, uterus and salivary primordia were inhibited. It might be legitimately objected, however, that collagenases contain mucopolysaccharidases (Bernfield et al., 1972). Spooner

(1975) therefore treated primordia with certain inhibitors that block collagen synthesis and got the same effect. Just how collagen might be involved in the morphogenesis of these organs, however, is still a mystery. Bernfield and Wessells (1970) postulate that it may help stabilize the older clefts (which are insensitive to cytochalasin B) (but see Hay, 1973; Meier and Hay, 1974).

Glycosaminoglycans have been demonstrated by a number of criteria to be abundant all along the epithelial-mesenchymal interface in salivary primordia. Their rate of accumulation and rate of replacement after removal, however, is greater at the distal ends of the lobules, where branching occurs (Bernfield and Banerjee, 1972). Significantly, when GAGs are removed from the basal surface of the epithelial rudiment, branching morphogenesis ceases and does not recommence until GAGs reappear. Consistent with this, branching also stops when the epithelial rudiment is divested of its mesenchyme, if its GAGs are removed at the same time. If, however, the bulk of the GAGs is left intact when the mesenchyme is removed (Bernfield et al., 1972), normal morphogenesis ensues. And, when morphogenesis resumes, there is a greater accumulation of GAGs and more rapid cell division where new branching buds form.

It seems evident from these several correlations that surface associated glycosaminoglycans, and probably collagen and glycoproteins as well, are somehow involved in controlling branching morphogenesis of these epithelial lobules. The questions are: which and how? Some leads as to which GAGs are active come from enzyme digestion studies. Bernfield et al. (1972) treated epithelial rudiments with low concentrations of collagenase, and then exposed them to trypsin, hyaluronidase, chondroitinase AC and ABC, sialidase, and to the enzyme diluent Ca^{2+}, Mg^{2+}-free tyrodes, and then combined them with fresh mesenchyme. After these treatments, the only rudiments that continued morphogenesis were those subjected to the last two treatments. All the other treatments prevented branching morphogenesis. Since dermatan sulfate, keratan sulfate, heparan sulfate, and heparin are all resistant to both testicular hyaluronidase and chondroitinase AC, they do not appear to be involved. Bernfield et al. (1972) conclude, therefore, by elimination, that the most likely candidates for morphogenetically active glycosaminoglycans in the salivary primordium are the non-sulfated GAGs, chondroitin and hyaluronic acid, and the chondroitin sulfates.

Just how these glycosaminoglycans act in stimulating growth and cleft formation is a total mystery, as, indeed is the mode of action of these molecules in other morphogenetic systems as well. Bernfield and Banerjee (1972) speculate that GAG-protein complexes are involved in the initiation and localization of collagen fibrogenesis, with the resultant collagen fibers accumulating in the clefts and acting to stabilize the branched morphology as it develops. Another possible way GAGs could control morphogenesis is by binding Ca^{2+} ions. Ash et al. (1973) found that papaverine (which apparently inhibits calcium ion flux) and Ca^{2+}-free medium both inhibit salivary morphogenesis. Since glycosaminoglycans are negatively charged and could, therefore, bind calcium, they could possibly control the availability of calcium for the contractile process and thus control cleft formation (Spooner, 1975). These ideas are of course completely speculative and perhaps wrong. However, some morphogenetic action of these extra cellular materials must be found, for it seems certain that morphogenesis cannot proceed without them.

A special and fascinating feature of branching morphogenesis is its apparent dependence on cell division, in contrast to morphogenetic movements which occur at earlier stages of development. Since this is no doubt due to the repetitive nature of branching and the consequent need for more cells, it would be interesting to see if cell division is always an essential factor wherever morphogenetic movements are accompanied by growth.

8. Concluding remarks

Until recently, almost all studies of the locomotion of tissue cells have been of their movements in vitro. Although we do not yet know how cells move in vitro, we now have some good ideas, based on an expanding array of solid fact. Yet, even when we understand cell locomotion in culture, this may not tell us how cells move in vivo, where the cellular environments, in particular the substrata, are different. In order to learn about the locomotion of cells inside organisms, they must be studied there. There have been several attempts to study cell locomotion within organisms and, though all must be regarded as preliminary, considerable information has been gained. It is now undeniable that certain cells in certain organsisms will yield to modern techniques of observation and experiment and advance our understanding of this important matter. It is also clear that in vitro investigations are providing models that are eminently useful as guides for studies in vivo. There is thus substantial reason for optimism. Cell movement can be studied with profit in vivo, if one chooses one's cells carefully and takes advantage of lessons learned from investigations in vitro. What is obviously needed is a dual approach. Where possible, the same cells should be studied in their normal environment in situ and removed from it under various imposed conditions in culture, in both cases with full attention to modern advances in the chemical and physical properties of the plasma membrane and the structure and functioning of the cytoskeleton. It is only in this manner that we will approach an understanding of two of the fundamental problems of biology – how cells move directionally in the construction of organisms and invasively in the spread of cancer.

Acknowledgements

I am indebted to C.A. Erickson and R.E. Keller for valuable discussions and to Madeleine Trinkaus for inestimable aid with the manuscript. My current research, including the writing of this paper, is supported by grants from the National Science Foundation (BMS70-00610) and the National Institutes of Health (USPHS-HD 07137).

References

Abercrombie, M. (1961) The bases of the locomotory behaviour of fibroblasts. Exp. Cell Res. Suppl. 8, 188–198.

Abercrombie, M. (1970) Contact inhibition in tissue culture. In Vitro, 6, 128–142.

Abercrombie, M. and Ambrose, E.J. (1958) Interference microscope studies of cell contacts in tissue culture. Exp. Cell Res. 15, 332–345.

Abercrombie, M. and Dunn, G.A. (1975) Adhesions of fibroblasts to substratum during contact inhibition observed by interference reflection microscopy. Exp. Cell Res. 92, 57–62.

Abercrombie, M. and Heaysman, J.E.M. (1953) Observations on the social behaviour of cells in tissue culture. I. Speed of movement of chick heart fibroblasts in relation to their mutual contacts. Exp. Cell Res. 5, 111–131.

Abercrombie, M. and Heaysman, J.E.M. (1954) Observations on the social behaviour of cells in tissue culture. II. "Monolayering" of fibroblasts. Exp. Cell Res. 6, 293–306.

Abercrombie, M., Heaysman, J.E.M. and Karthauser, H.M. (1957) Social behaviour of cells in tissue culture. III. Mutual influence of sarcoma cells and fibroblasts. Exp. Cell Res. 13, 276–291.

Abercrombie, M., Heaysman, J.E.M. and Pegrum, S.M. (1970a) The locomotion of fibroblasts in culture. I. Movements of the leading edge. Exp. Cell Res. 59, 393–398.

Abercrombie, M., Heaysman, J.E.M. and Pegrum, S.M. (1970b) The locomotion of fibroblasts in culture. II. "Ruffling". Exp. Cell Res. 60, 437–444.

Abercrombie, M., Heaysman, J.E.M. and Pegrum, S.M. (1970c) The locomotion of fibroblasts in culture. III. Movements of particles on the dorsal surface of the leading lamella. Exp. Cell Res. 62, 389–398.

Abercrombie, M., Heaysman, J.E.M. and Pegrum, S.M. (1971) The locomotion of fibroblasts in culture. IV. Electron microscopy of the leading lamella. Exp. Cell Res. 67, 359–367.

Abercrombie, M. and Middleton, C.A. (1968) Epithelial-mesenchymal interactions affecting locomotion of cells in culture, In: Epithelial-Mesenchymal Interactions – 18th Hahnemann Symposium (Fleischmajer, R. and Billingham, R.E., eds.) pp. 56–63, The Williams and Wilkins Company, Baltimore.

Ambrose, E. (1961) The movement of fibrocytes. Exp. Cell Res. Suppl. 8, 54–73.

Arey, L.B. (1925) The method of repair in small wounds. Anat. Rec. 29, 345 (abstract).

Arey, L.B. (1954) Developmental Anatomy, 680 pp., W.B. Saunders Company, Philadelphia and London.

Armstrong, P.B. and Armstrong, M.T. (1973) Are cells in solid tissues immobile? Mesonephric mesenchyme studied in vitro. Dev. Biol. 35, 187–209.

Armstrong, P.B. and Lackie, J.M. (1975) Studies on intercellular invasion in vitro using rabbit peritoneal neutrophil granulocytes (PMNS) I. Role of contact inhibition of locomotion. J. Cell Biol. 65, 439–462.

Arora, H.L. and Sperry, R.W. (1962) Optic nerve regeneration after surgical cross-union of medial and lateral optic tracts. Am. Zool. 2, 389 (abstract 61).

Ash, J.F., Spooner, B.S. and Wessells, N.K. (1973) Effects of papaverine and calcium-free medium on salivary gland morphogenesis. Dev. Biol. 33, 463–469.

Baker, P.C. (1965) Fine structure and morphogenetic movements in the gastrula of the treefrog, *Hyla Regilla*. J. Cell Biol. 95–116.

Baker, P.C. and Schroeder, T.E. (1967) Cytoplasmic filaments and morphogenetic movements in the amphibian neural tube. Dev. Biol. 15, 432–450.

Balinsky, B.I. and Walther, H.H. (1961) The emigration of presumptive mesoblast from the primitive streak in the chick as studied with the electron microscope. Acta Embryol. Morphol. Exp. 4, 261–283.

Ballard, W.W. (1966a) The role of the cellular envelope in the morphogenetic movements of teleost embryos. J. Exp. Zool. 161, 193–200.

Ballard, W.W. (1966b) Origin of the hypoblast in *Salmo*. I. Does the blastodisc edge turn inward? J. Exp. Zool. 161, 201–210.

Ballard, W.W. (1966c) Origin of the hypoblast in *Salmo*. II. Outward movement of deep central cells. J. Exp. Zool. 161, 211–220.

Ballard, W.W. (1968) History of the hypoblast in *Salmo*. J. Exp. Zool. 168, 257–272.

Ballard, W.W. (1973a) Normal embryonic stages for salmonid fishes based on *Salmo gairdneri* Richardson and *Salvelinus fontinalis* (Mitchill). J. Exp. Zool. 184, 7–26.

Ballard, W.W. (1973b) Morphogenetic movements in *Salmo gairdneri*. Richardson. J. Exp. Zool. 184, 27–48.

Ballard, W.W. (1973c) A new fate map for *Salmo gairdneri*. J. Exp. Zool. 184, 49–74.

Ballard, W.W. (1976) Problems of gastrulation: real and verbal. BioScience, 26, 36–39.

Ballard, W.W. and Dodes, L.M. (1968) The morphogenetic movements at the lower surface of the blastodisc in salmonid embryos. J. Exp. Zool. 168, 67–84.

Balsamo, J. and Lilien, J. (1975) The binding of tissue-specific adhesive molecules to the cell surface. A molecular basis for specificity. Biochemistry, 14, 167–171.

Bangham, A.D. and Pethica, B.A. (1960) The adhesiveness of cells and the nature of the chemical groups at their surfaces. Proc. R. Soc. Edinburgh, Sect. B. 28, 43–50.

Barbera, A.J., Marchase, R.B. and Roth, S. (1973) Adhesive recognition and retinotectal specificity. Proc. Nat. Acad. Sci. U.S.A. 70, 2482–2486.

Bard, J.B.L. and Hay, E.D. (1975) The behavior of fibroblasts from the developing avian cornea. Morphology and movement in situ and in vitro. J. Cell Biol. 67, 400–418.

Bard, J.B.L., Hay, E.D. and Meller, S.M. (1975) Formation of the endothelium of the avian cornea: a study of cell movement in vivo. Dev. Biol. 42, 334–361.

Barfurth, D. (1891) Zur Regeneration der Gewebe. Arch. Microsk. Anat. 37, 406.

Barski, G. and Belehradek, J. (1965) Etude microcinématographique du mécanisme d'invasion cancéreuse en cultures de tissu normal associé aux cellules malignes. Exp. Cell Res. 37, 464–480.

Barski, G. and Belehradek, J. (1968) In vitro studies on tumor invasion. In: The Proliferation and Spread of Neoplastic Cells, pp. 511–531, Williams and Wilkins, Baltimore.

Bell, P.B., Jr. (1972) Criss-crossing, contact inhibition, and cell movement in cultures of normal and transformed 3T3 cells. J. Cell Biol. 55 (No. 2 Pt. 2), 16a.

Bell, P.B., Jr. (1976) Locomotory behavior, contact inhibition, and pattern formation of 3T3 and polyoma virus transformed 3T3 cells in culture. Submitted for publication.

Bellairs, R. (1963) Differentiation of the yolk sac of the chick studied by electron microscopy. J. Embryol. Exp. Morphol. 11, 201–225.

Bellairs, R. (1971) Developmental Processes in Higher Vertebrates, 366 pp. University of Miami Press, Coral Gables, Florida.

Bellairs, R., Boyde, A. and Heaysman, J.E.M. (1969) The relationship between the edge of the chick blastoderm and the vitelline membrane. Wilhelm Roux' Arch. Entwicklungsmech. Org. 163, 113–121.

Bennett, M.V.L. and Gilula, N.B. (1974) Membranes and junctions in developing *Fundulus* embryos. freeze-fracture and electrophysiology. J. Cell Biol. 63, 21a.

Bennett, M.V.L. and Trinkaus, J.P. (1970) Electrical coupling between embryonic cells by way of extracellular space and specialized junctions. J. Cell Biol. 44, 592–610.

Bernfield, M.R. (1973) Glycosaminoglycans and epithelial organ formation. Am. Zool. 13, 1067–1083.

Bernfield, M.R. and Banerjee, S.D. (1972) Acid mucopolysaccharide (glycosaminoglycan) at the epithelial-mesenchymal interface of mouse embryo salivary glands. J. Cell Biol. 52, 664–673.

Bernfield, M.R., Banerjee, S.D. and Cohen, R.H. (1972) Dependence of salivary epithelial morphology and branching morphogenesis upon acid mucopolysaccharide-protein (proteoglycan) at the epithelial surface. J. Cell Biol. 52, 674–689.

Bernfield, M.R. and Wessells, N.K. (1970) Intra- and extracellular control of epithelial morphogenesis. Dev. Biol. Suppl. 4, 195–249.

Berwald, Y. and Sachs, L. (1965) In vitro transformation of normal cells to tumor cells by carcinogenic hydrocarbons. J. Nat. Cancer Inst. 35, 641–661.

Betchaku, T. and Trinkaus, J.P. (1974) An in situ study of surface expansion during *Fundulus* epiboly. J. Cell Biol. 63, 24a.

Betchaku, T. and Trinkaus, J.P. (1976) A fine-structural analysis of epiboly in *Fundulus heteroclitus* (in preparation).

Blackler, A.W. (1958) Contribution to the study of germ-cells in the Anura. J. Embryol. Exp. Morphol. 6, 491–503.

Blackler, A.W. (1962) Transfer of primordial germ-cells between two subspecies of *Xenopus laevis*. J. Embryol. Exp. Morphol. 10, 641–651.

Bolund, L., Darzynkiewicz, Z. and Ringertz, N.R. (1970) Cell concentration and the staining properties of nuclear deoxyribonucleoprotein. Exp. Cell Res. 62, 76–89.

Bonner, J.T. (1952) Morphogenesis. pp. vi + 296. Princeton Univ. Press.

Bonner, J.T. (1971) Aggregation and differentiation in the cellular slime molds. Ann. Rev. Microbiol. 25, 75–92.

Borek, C. and Sachs, L. (1966) The difference in contact inhibition of cell replication between normal cells and cells transformed by different carcinogens. Proc. Nat. Acad. Sci. U.S.A. 56, 1705–1711.

Borghese, E. (1950) The development in vitro of the submandibular and sublingual glands of *Musculus*, J. Anat. 84, 287–302.

Boucaut, J.C. (1974a) Etude autoradiographique de la distribution de cellules embryonnaires isolées, transplantées dans le blastocèle chez *Pleurodeles waltlii* Michah. (Amphibien, Urodèle). Ann. Embryol. Morphog. 7, 7–50.

Boucaut, J.C. (1974b) Chimères intergénériques entre *Pleurodeles waltlii* Michah. et Ambystoma mexicanum Shaw. (Amphibiens, Urodèles). Ann. Embryol. Morphog. 7, 119–139.

Böving, B.G. (1971) Biomechanics of implantation. In: The Biology of the Blastocyst (Blandau, R.J., ed.) pp. 423–442, University of Chicago Press, Chicago, Ill.

Boyde, A., Grainger, F. and James, D.W. (1969) Scanning electron microscopic observations of chick embryo fibroblasts in vitro, with particular reference to the movement of cells under others. Z. Zellforsch. Mikrosc. Anat. 94, 46–55.

Brachet, A. (1921) Traité d'embryologie des vertébrés. 602 pp. Masson, Paris.

Brachet, J. (1960) The Biochemistry of Development. 320 pp. Pergamon Press, Oxford.

Bray, D. (1970) Surface movements during the growth of single explanted neurons. Proc. Nat. Acad. Sci. U.S.A. 65, 905–910.

Bray, D. and Bunge, M.B. (1973) The growth cone in neurite extension, In: Locomotion of Tissue Cells, Ciba Foundation Symposium 14 (new series) pp. 195–209, Elsevier, North-Holland, Amsterdam.

Brick, I., Schaeffer, B.E., Schaeffer, H.E. and Gennaro, J.F. Jr. (1974) Electrokinetic properties and morphologic characteristics of amphibian gastrula cells. Ann. N.Y. Acad. Sci. 238, 390–407.

Brien, P. (1930) Contribution à l'étude de la régénération naturelle et expérimentale chez les clavelinidae. Ann. Soc. R. Belg. 61, 19–112.

Bruyn, P.P.H. de (1944) Locomotion of blood cells in tissue cultures. Anat. Rec. 89, 43–63.

Bruyn, P.P.H. de (1946) The amoeboid movement of the mammalian leukocyte in tissue culture. Anat. Rec. 95, 177–192.

Bryan, J. (1974) Microtubules. BioScience 24, 701–711.

Burnside, B. (1971) Microtubules and microfilaments in newt neurulation. Dev. Biol. 26, 416–441.

Burnside, B. (1973) Microtubules and microfilaments in amphibian neurulation. Am. Zool. 13, 989–1006.

Butler, E.G. and O'Brien, J.P. (1942) Effects of localized X-radiation on regeneration of the urodele limb. Anat. Rec. 84, 407–413.

Byers, B. and Porter, K. (1964) Oriented microtubules in elongating cells of the developing lens rudiment after induction. Proc. Nat. Acad. Sci. U.S.A. 52, 1091–1099.

Cajal, S.R.Y. (1890) Sur l'origine et les ramifications des fibres nerveuses de la moelle embryonnaire. Anat. Anz. 5, 85–95 and 111–119.

Campbell, R.D. (1967) Tissue dynamics of steady state growth in *Hydra littoralis*. III. Behavior of specific cell types during tissue movements. J. Exp. Zool. 164, 379–391.

Campbell, R.D. (1974) Cell movements in *Hydra*. Amer. Zool. 14, 523–535.

Carter, S.B. (1965) Principles of cell motility: the direction of cell movement and cancer invasion. Nature 208, 1183–1187.

Carter, S.B. (1967) Effects of cytochalasins on mammalian cells. Nature (London) 213, 261–264.
Chambers, R. and Fell, H.B. (1931) Micro-operations on cells in tissue cultures. Proc. R. Soc. London, Ser. B, 109, 380–403.
Chiakulas, J.J. (1952) The role of tissue specificity in the healing of epithelial wounds. J. Exp. Zool. 121, 383–417.
Chibon, P. (1966) Analyse expérimentale de la régionalisation et des capacités morphogénétiques de la crête neurale chez l'Amphibien Urodèle *Pleurodeles waltlii* Michah. Mem. Soc. Zool. France 36, 1–107.
Chibon, P. (1974) Un système morphogénétique remarquable: La crête neurale des vertébrés. Ann. Biol. 13, 459–480.
Clark, E.R. (1912) Further observations on living growing lymphatics: their relation to the mesenchyme cells. Am. J. Anat. 13, 351–379.
Clark, E.R. and Clark, E.L. (1925) A. The development of adventitial (Rouget) cells on the blood capillaries of amphibian larvae. Am. J. Anat. 35, 329–264.
Clark, E.R. and Clark, E.L. (1935) Observations on changes in blood vascular endothelium in the living animal. Am. J. Anat. 57, 385–438.
Cloney, R.A. (1966) Cytoplasmic filaments and cell movements: epidermal cells during ascidian metamorphosis. J. Ultrastruct. Res. 14, 300–328.
Cloney, R.A. and Grimm, L. (1970) Transcellular emigration of blood cells during ascidian metamorphosis. Z. Zellforsch. Mikrosc. Anat. 107, 157–173.
Coman, D.R. (1944) Decreased mutual adhesiveness, a property of cells from squamous cell carcinomas. Cancer Res. 4, 625–629.
Coman, D.R. (1960) Reduction in cellular adhesiveness upon contact with a carcinogen. Cancer Res. 20, 1202–1204.
Conklin, E.G. (1932) The embryology of *Amphioxus*. J. Morphol. 54, 69–151.
Conrad, G.W. and Williams, D.C. (1974) Polar lobe formation and cytokinesis in fertilized eggs of *Ilyanassa obsoleta*. II. Large bleb formation caused by high concentrations of exogenous calcium ions. Dev. Biol. 37, 280–294.
Cooke, J. (1973) Properties of the primary organization field in the embryo of *Xenopus laevis*. IV. Pattern formation and regulation following early inhibition of mitosis. J. Embryol. Exp. Morphol. 30, 49–62.
Cooke, J. (1975) Local autonomy of gastrulation movements after dorsal lip removal in two anuran amphibians. J. Embryol. Exp. Morphol. 33, 147–157.
Coupé, R., Roubaud, P., Boulekbache, H. and Devillers, C. (1973) Etude de la dissociation cellulaire *in vitro* sur les premiers stades du développement embryonnaire de la Truite (*Salmo irideus* Gibb.). C.R. Acad. Sc. Paris 277, 353–356.
Croft, C.B. and Tarin, D. (1970) Ultrastructural studies of wound healing in mouse skin. I. Epithelial behaviour. J. Anat. 106, 66–77.
Culp, L.A. (1975) Topography of substrate-attached glycoproteins from normal and virus-transformed cells. Exp. Cell Res. 92, 467–477.
Curtis, A.S.G. (1962) Cell contact and adhesion. Biol. Rev. Cambridge Phil. Soc. 37, 82–129.
Curtis, A.S.G. (1964) The mechanism of adhesion of cells to glass. A study by interference reflection microscopy. J. Cell Biol. 20, 199–215.
Curtis, A.S.G. (1973) Cell Adhesion. In: Progress in Biophysics and Molecular Biology (Butler, A.S.G. and Noble, D., eds.) pp. 315–386, Pergamon Press, Oxford.
Czarska, L. and Grebecki, A. (1966) Membrane folding and plasma-membrane ratio in the movement and shape transformation in *Amoeba proteus*. Acta Protozool. 4, 201–240.
Dalcq, A. et Gérard, P. (1935) Brachet's Traité d'Embryologie des Vertébrés. 690 pp. Masson, Paris.
Dan, K. (1936) Electrokinetic studies of marine ova. III. Physiol. Zool. 9, 43–57.
Dan, K. and Okazaki, K. (1956) Cyto-embryological studies of sea urchins. III. Role of the secondary mesenchyme cells in the formation of the primitive gut in sea urchin larvae. Biol. Bull. (Woods Hole, Mass.) 110, 29–42.
Danes, B. (1949) Pulmonary epithelium of the newt *Triturus viridiscens*, studied in living cultures with cinematographic apparatus and phase contrast technique. J. Exp. Zool. 112, 417–447.

Danielli, J.F. (1945) Some reflections on the forms of simpler cells. In: Essays on Growth and Form (Le Gros Clark, W.E. and Medawar, P.B. eds.) pp. 296–308, Clarendon Press, Oxford.

DeHaan, R.L. (1963) Migration patterns of the precardiac mesoderm in the early chick embryo. Exp. Cell Res. 29, 554–560.

Devillers, Ch. (1952) Coordination des forces épiboliques dans la gastrulation de *Salmo*. Bull. Soc. Zool. France, 77, 304–309.

Devillers, Ch., Colas, J., Richard, L. (1957) Différentiation *in vitro* de blastoderms de truite (*Salmo irideus*) dépourvu de couche enveloppante. J. Embryol. Exp. Morphol. 5, 264–273.

DiPasquale, A. (1975a) Locomotory activity of epithelial cells in culture. Exp. Cell Res. 94, 191–215.

DiPasquale, A. (1975b) Locomotion of epithelial cells: factors involved in extension of the leading edge. Exp. Cell Res., 95, 425–439.

DiPasquale, A. and Bell, P.B., Jr. (1974) The upper cell surface: its inability to support active cell movement in culture. J. Cell Biol. 62, 198–214.

DiPasquale, A. and Bell, P.B., Jr. (1975) Comments on reported observations of cells spreading on the upper surfaces of other cells in culture. J. Cell Biol. 66, 216–218.

DiPasquale, A., McGuire, J., Moellmann, G. and Wasserman, S. (1976) Stimulation of microtubule assembly in Greene melanoma cells by dibutyryl cyclic adenosine monophosphate or cholera toxin. Submitted for publication.

Dixon, H.M. and McCutcheon, M. (1936) Chemotropism of leucocytes in relation to their rate of locomotion. Proc. Soc. Exp. Biol. Med. 34, 173–176.

Downie, J.R. and Pegrum, S.M. (1971) Organization of the chick blastoderm edge. J. Embryol. Exp. Morphol. 26, 623–635.

Driesch, H. (1896) Die taktische Reizbarkeit der Mesenchymzellen von *Echinus microtuberculatus*. Wilhelm Roux' Arch. Entwicklungsmech. Org. 3, 362–380.

Dunham, P.B., Cass, A., Trinkaus, J.P., and Bennett, M.V.L. (1970) Water permeability of *Fundulus* eggs. Biol. Bull. (Woods Hole, Mass.) 139, 420–421.

Dunn, G.A. (1971) Mutual contact inhibition of extension of chick sensory nerve fibres in vitro. J. Compar. Neur. 143, 491–507.

Eycleshymer, A.C. (1907) The closing of wounds in the larval *Necturus*. Am. J. Anat. 7, 317–325.

Edidin, M. (1974) Rotational and translational diffusion in membranes. Ann. Rev. Biophys. Bioeng. 3, 179–201.

Edidin, M. and Weiss, A. (1972) Antigen cap formation in cultured fibroblasts: a reflection of membrane fluidity and of cell motility. Proc. Nat. Acad. Sci. U.S.A. 69, 2456–2459.

Edidin, M. and Weiss, A. (1973) Restriction of antigen mobility in the plasma membranes of some cultured fibroblasts. In: Control of Proliferation of Animal Cells (Clarkson, B. and Baserga, R., eds.), pp. 213–219. Cold Spring Harbor Laboratory, New York.

Edwards, J.G. and Campbell, J.A. (1971) The aggregation of trypsinized BHK21 cells. J. Cell Sci. 8, 53–72.

Elsdale, T. and Bard, J. (1972a) Cellular interactions in mass cultures of human diploid fibroblasts. Nature, 236, 152–155.

Elsdale, T. and Bard, J. (1972b) Collagen substrata for studies on cell behavior. J. Cell. Biol. 54, 626–637.

Elsdale, T. and Bard, J. (1974) Cellular interactions in morphogenesis of epithelial-mesenchymal systems. J. Cell Biol. 63, 343–349.

Elsdale, T. and Bard, J. (1975) Is stickiness of the upper surface of an attached epithelium in culture an indicator of functional insufficiency? J. Cell Biol. 66, 218–219.

Enders, A.C. and Schlafke, S. (1971) Penetration of the uterine epithelium during implantation in the rabbit. Am. J. Anat. 132, 219–240.

Erickson, C.A. (1976) Both BHK and polyoma transformed BHK cells show contact inhibition of locomotion. J. Cell Biol., in press.

Erickson, C.A. and Trinkaus, J.P. (1976) Microvilli and blebs as sources of reserve surface membrane during cell spreading. Exp. Cell Res, 99, 375–384.

Evans, R.B., Morhenn, V., Jones, A.L. and Tompkins, G.M. (1974) Concomitant effects of insulin on surface membrane conformation and polysome profiles of serum-starved Balb/c 3T3 fibroblasts. J. Cell Biol. 61, 95–106.

Farquahar, M.G. and Palade, G.E. (1963) Junctional complexes in various epithelia. J. Cell Biol. 17, 375–412.

Fitzharris, T.P. and Markwald, R.R. (1974) Structural components of cardiac jelly. J. Cell Biol. 63, 101a.

Follett, E.A.C. and Goldman, R.D. (1970) The occurrence of microvilli during spreading and growth of BHK21/C13 fibroblasts. Exp. Cell Res. 59, 124–136.

Francis, D.W. and Allen, R.D. (1971) Induced birefringence as evidence of endoplasmic viscoelasticity in *Chaos carolinensis*. J. Mechanochem. Cell Motility, 1, 1–6.

Frye, L.D. and Edidin, M. (1970) The rapid intermixing of cell surface antigens after formation of mouse-human heterokaryons. J. Cell Sci. 7, 319–335.

Gail, M.H. and Boone, C.W. (1970) The locomotion of mouse fibroblasts in tissue culture. Biophys. J. 10, 980–993.

Gail, M.H. and Boone, C.W. (1971a) Effect of colcemid on fibroblast motility. Exp. Cell Res. 65, 221–227.

Gail, M.H. and Boone, C.W. (1971b) Cytochalasin effects on BALB/3T3 fibroblasts: dose dependent, reversible alteration of motility and cytoplasmic cleavage. Exp. Cell Res. 68, 226–228.

Gaze, R.M. (1970) The Formation of Nerve Connections. 288 pp. Academic Press, London, New York.

Gibbins, J.R., Tilney, L.G. and Porter, K.R. (1969) Microtubules in the formation and development of the primary mesenchyme in *Arbacia Punctulata*. I. The distribution of microtubules. J. Cell Biol. 41, 201–226.

Gingell, D. (1970) Contractile responses at the surface of an amphibian egg. J. Embryol. Exp. Morphol. 23, 583–609.

Goldman, R.D. (1971) The role of three cytoplasmic fibers in BHK-21 cell motility. I. Microtubules and the effects of colchicine. J. Cell Biol. 51, 752–762.

Goldman, R.D. and Knipe, D.M. (1973) The functions of cytoplasmic fibers in non-muscle cell motility. Cold Spring Harbor Symposium Quant. Biol. 37, 523.

Goldman, R.D., Berg, G., Bushnell, A., Chang, C.M., Dickerman, L., Hopkins, N., Miller, M.L., Pollack, R. and Wang, E. (1973) Fibrillar systems in cell motility. In: Locomotion of Tissue Cells, Ciba Foundation Symposium 14 (new series) pp. 83–107, Elsevier/North-Holland, Amsterdam.

Goodrich, H.B. (1924) Cell behavior in tissue cultures. Biol. Bull. (Woods Hole, Mass.) 46, 252–262.

Granholm, N.H. (1970) Effects of microtubular inhibiting agents on cell shape in the primitive streak of the avian embryo. Am. Zool. 10, 320 (abstract 315a).

Granholm, N.H. and Baker, J.R. (1970) Cytoplasmic microtubules and the mechanism of avian gastrulation. Dev. Biol. 23, 563–584.

Grobstein, C. (1953) Epithelio-mesenchymal specificity in the morphogenesis of mouse submandibular rudiments in vitro. J. Exp. Zool. 124, 383–404.

Grobstein, C. (1967) Mechanisms of organogenetic tissue interaction. Nat. Cancer Inst. Monogr. 26, 279–299.

Grobstein, C. and Cohen, J.H. (1965) Collagenase: effects on the morphogenesis of embryonic salivary epithelium in vitro. Science, 150, 626–628.

Guelstein, V.I., Ivanova, O.Y., Margolis, L.B., Vasiliev, J.M. and Gelfand, I.M. (1973) Contact inhibition of movement in the cultures of transformed cells. Proc. Nat. Acad. Sci. U.S.A. 70, 2011–2014.

Gustafson, T. (1963) Cellular mechanisms in the morphogenesis of the sea urchin embryo. Cell contacts within the ectoderm and between mesenchyme and ectoderm cells. Exp. Cell Res. 32, 570–589.

Gustafson, T. (1964) The role and activities of pseudopodia during morphogenesis of the sea urchin larva. In: Primitive Motile Systems in Cell Biology (Allen, R.D. and Kamiya, N., eds.), pp. 333–349. Academic Press, New York.

Gustafson, T. and Kinnander, H. (1956) Microaquaria for time-lapse cinematographic studies of morphogenesis in swimming larvae and observations on sea urchin gastrulation. Exp. Cell Res. 11, 36–51.
Gustafson, T. and Kinnander, H. (1960) Cellular mechanisms in morphogenesis of the sea urchin gastrula. The oral contact. Exp. Cell Res. 21, 361–373.
Gustafson, T. and Wolpert, L. (1961) Studies on the cellular basis of morphogenesis in the sea urchin embryo. Directed movements of primary mesenchyme cells in normal and vegetalized larvae. Exp. Cell Res. 24, 64–79.
Gustafson, T. and Wolpert, L. (1962) Cellular mechanisms in the morphogenesis of the sea urchin larva. Change in shape of cell sheets. Exp. Cell Res. 27, 260–279.
Gustafson, T. and Wolpert, L. (1967) Cellular movement and contact in sea urchin morphogenesis. Biol. Rev. Cambridge Philos. Soc. 42, 442–498.
Handel, M.A. and Roth, L.E. (1971) Cell shape and morphology of the neural tube: implications for microtubule function. Dev. Biol. 25, 78–95.
Harris, A. (1969) Initiation and propagation of the ruffle in fibroblast locomotion. J. Cell Biol. 43, 165a–166a.
Harris, A. (1973a) Behavior of cultured cells on substrata of variable adhesiveness. Exp. Cell Res. 77, 285–297.
Harris, A. (1973b) Cell surface movements related to cell locomotion. In: Locomotion of Tissue Cells. Ciba Foundation Symposium 14 (new series) pp. 3–26, Elsevier, North Holland, Amsterdam.
Harris, A. (1973c) Location of cellular adhesions to solid substrata. Dev. Biol. 35, 97–114.
Harris, A. (1974) Contact inhibition of cell locomotion. In: Cell Communication (Cox, R.P., ed.) pp. 147–185, John Wiley & Sons, New York.
Harris, A. and Dunn, G. (1972) Centripetal transport of attached particles on both surfaces of moving fibroblasts. Exp. Cell. Res. 73, 519–523.
Harrison, R.G. (1904) Experimentelle Untersuchungen über die Entwicklung der Sinnesorgane der Seitenlinie bei den Amphibien. Arch. Mikroskop. Anat. 63, 35–149.
Harrison, R.G. (1910) The outgrowth of the nerve fiber as a mode of protoplasmic movement. J. Exp. Zool. 9, 787–846.
Harrison, R.G. (1914) The reaction of embryonic cells to solid structure. J. Exp. Zool. 17, 521–544.
Hay, E. (1968) Organization and fine structure of epithelium and mesenchyme in the developing chick embryo, In: Epithelial-Mesenchymal Interactions – 18th Hahnemann Symposium (Fleischmajer, R. and Billingham, R.E., eds.) pp. 31–55, Williams & Wilkins Company, Baltimore.
Hay, E. (1973) Origin and role of collagen in the embryo. Am. Zool. 13, 1085–1107.
Heaysman, J.E.M. (1973) Comment. In: Locomotion of Tissue Cells, Ciba Foundation Symposium 14 (new series) p. 248, Elsevier, North-Holland, Amsterdam.
Heaysman, J.E.M. and Pegrum, S.M. (1973) Early contacts between fibroblasts. An ultrastructural study. Exp. Cell Res. 78, 71–78.
Herrick, E.H. (1932) Mechanism of movement of epidermis, especially its melanophores, in wound healing, and behavior of skin grafts in frog tadpoles. Biol. Bull. (Woods Hole, Mass.) 63, 271–286.
His, W. (1889) Die Neuroblasten und deren Enstehung in embryonalen. Mark. Arch. Anat. Physiol. Anat. Abtheil.
Hitchcock, H.B. (1939) The behavior of adult amphibian skin cultured in vivo and in vitro. J. Exp. Zool. 81, 299–329.
Hogan, J.C., Jr. and Trinkaus, J.P. (1976) Transmission and freeze-fracture electron microscopy of internal structure and junctions of *Fundulus* deep cells. In preparation.
Holmes, S.J. (1914) The behavior of the epidermis of amphibians when cultivated outside the body. J. Exp. Zool. 17, 281–295.
Holtfreter, J. (1933) Die totale Exogastrulation, eine Selbstablösung des Ektoderms vom Entomesoderm. Wilhelm Roux' Arch. Entwicklungsmech. Org. 129, 669–793.
Holtfreter, J. (1939) Gewäbeaffinität, ein Mittel der embryonalen Formbildung. Arch. Exp. Zellforsch. 23, 169–209.
Holtfreter, J. (1943) A study of the mechanics of gastrulation: Part I. J. Exp. Zool. 94, 261–318.
Holtfreter, J. (1944) A study of the mechanics of gastrulation: Part II. J. Exp. Zool. 95, 171–212.

Holtfreter, J. (1947) Observations on the migration, aggregation, and phagocytosis of embryonic cells. J. Morphol. 80, 25–55.

Holtfreter, J. (1948) Significance of the cell membrane in embryonic processes. Ann. N.Y. Acad. Sci. 49, 709–760.

Hörstadius, S. (1950) The Neural Crest, 111 pp. Cambridge Univ. Press, Cambridge, England.

Hsie, A.W. and Puck, T.T. (1971) Morphological transformation of Chinese hamster cells by dibutyryl adenosine cyclic 3′:5′-monophosphate and testosterone. Proc. Nat. Acad. Sci. U.S.A. 68, 358–361.

Hudspeth, A.J. (1975) Establishment of tight junctions between epithelial cells. Proc. Nat. Acad. Sci. U.S.A. 72, 2711–2713.

Hughes, W.L., Bond, V.P., Brecher, G., Cronkite, E.P., Painter, R.B., Quastler, H. and Sherman, F.G. (1958) Cellular proliferation in the mouse as revealed by autoradiography with tritiated thymidine. Proc. Nat. Acad. Sci. U.S.A. 44, 476–483.

Huxley, H.E. (1973) Muscular contraction and cell motility. Nature, 243, 445–449.

Ingram, V. (1969) A side view of moving fibroblasts. Nature, 222, 641–644.

Izzard, C.S. (1974) Contractile filopodia and in vivo cell movements in the tunic of the ascidian, *Botryllus schlosseri*. J. Cell Sci. 15, 513–535.

Izzard, C.S. and Izzard, S.L. (1975) Calcium regulation of the contractile state of isolated mammalian fibroblast cytoplasm. J. Cell Sci. 18, 241–256.

Jacobson, C.O. (1962) Cell migration in the neural plate and the process of neurulation in the axolotl larva. Zoologiska Bidrag Från Uppsala, 35, 433–449.

Jacobson, M. (1970) Developmental Neurobiology, 465 pp. Holt, Rinehart and Winston, New York.

Jacoby, F. (1965) Macrophages. In: Cells and Tissues in Culture, vol. 2 (Willmer, E.N., ed.) pp. 1–93, Academic Press, New York.

Johnson, G.S., Friedman, R.M. and Pastan, I. (1971) Restoration of several morphological characteristics of normal fibroblasts in sarcoma cells treated with adenosine-3′:5′-cyclic monophosphate and its derivatives. Proc. Nat. Acad. Sci. U.S.A. 68, 425–429.

Johnson, G.S., Morgan, W.D. and Pastan, I. (1972) Regulation of cell motility by cyclic AMP. Nature, 235, 54–56.

Johnson, K.E. (1970) The role of changes in cell contact behavior in amphibian gastrulation. J. Exp. Zool. 175, 391–427.

Johnson, K.E. (1972) The extent of cell contact and the relative frequency of small and large gaps between presumptive mesodermal cells in normal gastrulae of *Rana pipiens* and the arrested gastrulae of the *Rana pipiens* ♀ × *Rana catesbeiana* ♂ hybrid. J. Exp. Zool. 170, 227–238.

Johnson, K.E. and Smith, E.P. (1976) Studies on the binding of Concanavalin A to dissociated embryonic amphibian cells. Exp. Cell Res., in press.

Johnston, M.C. (1966) A radioautographic study of the migration and fate of cranial neural crest cells in the chick embryo. Anat. Rec. 156, 143–156.

Johnston, M.C., Bhakdinaronk, A. and Reid, Y.C. (1974) An expanded role of the neural crest in oral and pharyngeal development, in 4th Symp. on oral sensation and perception (Bosma, J.F. ed.) pp. 37–52, US Govt., Print. Off. Washington.

Karfunkel, P. (1971) The role of microtubules and microfilaments in neurulation in *Xenopus*. Dev. Biol. 25, 30–56.

Karfunkel, P. (1972) The activity of microtubules and microfilaments in neurulation in the chick. J. Exp. Zool. 181, 289–302.

Karfunkel, P. (1974) The mechanisms of neural tube formation. Int. Rev. Cytol. 38, 245–271.

Karfunkel, P. (1976) SEM analysis of mesodermal migration in amphibian gastrulation, in preparation.

Karp, G.C. and Solursh, M. (1974) Acid mucopolysaccharide metabolism, the cell surface and primary mesenchyme cell activity in the sea urchin embryo. Dev. Biol. 41, 110–123.

Keller, R.E. (1975) Vital dye mapping of the gastrula and neurula of *Xenopus laevis*. I. Prospective areas and morphogenetic movements of the superficial layer. Dev. Biol. 42, 222–241.

Keller, R.E. (1976) Vital dye mapping of the gastrula and neurula of *Xenopus laevis*. II. Prospective areas and morphogenetic movements of the deep layer. Dev. Biol., in press.

Keller, R.E. and Schoenwolf, G. (1976) A scanning electron microscopic study of cellular morphology, contact and arrangement in the gastrula of *Xenopus laevis* (submitted for publication).

Kessel, R.G. (1960) The role of cell division in gastrulation of *Fundulus heteroclitus*. Exp. Cell Res. 20, 277–282.

Kinnander, H. and Gustafson, T. (1960) Further studies on the cellular basis of gastrulation in the sea urchin larva. Exp. Cell Res. 19, 278–290.

Komnick, H., Stockem, W. and Wohlfarth-Bottermann, K.E. (1973) Cell motility: mechanisms in protoplasmic streaming and ameboid movement. Int. Rev. Cytol. 15, 169–249.

Koprowski, H., Pontén, J.A., Jensen, F., Raudin, R.G., Moorehead, P. and Saksela, E. (1962) Transformation of cultures of human tissue infected with simian virus SV40. J. Cell Comp. Physiol. 59, 281–292.

Kosher, R. and Searls, R. (1973) Sulfated mucopolysaccharide synthesis during the development of *Rana pipiens*. Dev. Biol. 32, 50–68.

Kraemer, P.M. (1971) Complex carbohydrates of animal cells: biochemistry and physiology of the cell periphery. Biomembranes, 1, 67–190.

Krawczyk, W.S. (1971) A pattern of epidermal cell migration during wound healing. J. Cell. Biol. 49, 247–263.

Lash, J.W. (1955) Studies on wound closure in urodeles. J. Exp. Zool. 128, 13–28.

Lawn, D.M. (1969) Uterine implantation of the mammalian ovum. Proc. Roy. Soc. Med. 62, 141–143.

Le Douarin, N. (1969) Particularités du noyau interphasique chez la caille japonaise (Coturnix Coturnix Japonica). Bull. Biol. 103, 435–452.

Le Douarin, N. and Teillet, M.A.M. (1974) Experimental analysis of the migration and differentiation of neuroblasts of the autonomic nervous system and of neurectodermal mesenchymal derivatives, using a biological cell marking technique. Dev. Biol. 41, 162–184.

Leighton, J. (1967) The Spread of Cancer. 208 pp. Academic Press, New York.

Leighton, J., Kalla, R.L., Kline, I. and Belkin, M. (1959) Pathogenesis of tumor invasion: I. Interaction between normal tissues and "transformed" cells in tissue culture. Cancer Res. 19, 23–27.

Lentz, T.L. and Trinkaus, J.P. (1967) A fine structural study of cytodifferentiation during cleavage, blastula, and gastrula stages of *Fundulus heteroclitus*. J. Cell Biol. 32, 121–138.

Lentz, T.L. and Trinkaus, J.P. (1971) Differentiation of the junctional complex of surface cells in the developing *Fundulus* blastoderm. J. Cell Biol. 48, 455–472.

Lesseps, R.J., Geurts van Kessel, A.H.M. and Denucé, J.M. (1975) Cell patterns and cell movements during early development of an annual fish, *Nothobranchius neumanni*. J. Exp. Zool. 193, 137–146.

Letourneau, P.C. (1975a) Possible roles for cell-to-substratum adhesion in neuronal morphogenesis. Dev. Biol. 44, 77–91.

Letourneau, P.C. (1975b) Cell-to-substratum adhesion and guidance of axonal elongation. Dev. Biol. 44, 92–101.

Lewis, M.R. and Lewis, W.H. (1912) Membrane formations from tissues transplanted into artificial media. Anat. Rec. 6, 195–205.

Lewis, W.H. (1922) The adhesive quality of cells. Anat. Rec. 23, 387–392.

Lewis, W.H. (1942) The relation of the viscosity changes of protoplasm to ameboid locomotion and cell division. In: The Structure of Protoplasm (Seifriz, W., ed.) pp. 163–197, Iowa State College Press, Ames Iowa.

Lewis, W.H. (1947) Mechanics of invagination. Anat. Rec. 97, 139–156.

Lillie, F.R. (1902) Differentiation without cleavage in the egg of the annelid *Chaetopterus pergamentaceus*. Wilhelm Roux' Arch. Entwicklungsmech. Org. 14, 477–499.

Lochner, L. and Izzard, C.S. (1973) Dynamic aspects of cell-substrate contact in fibroblast motility. J. Cell Biol. 59, 199a.

Loeb, L. (1921) Amoeboid movement, tissue formation and consistency of protoplasm. Am. J. Physiol. 56, 140–167.

Loeb, L. and Strong, R.M. (1904) On regeneration in the pigmented skin of the frog and on the character of the chromatophores. Am. J. Anat. 3, 275–283.

Löfberg, J. (1974) Apical surface topography of invaginating and noninvaginating cells. A scanning-

transmission study of amphibian neurulae. Dev. Biol. 36, 311–329.
Löfberg, J. and Jacobson, C.O. (1974) Effects of vinblastine sulphate, colchicine and guanosine triphosphate on cell morphogenesis during amphibian neurulation. Zoon 2, 85–98.
Loomis, W. (1975) Dictyostelium discoideum. A developmental system. Academic Press, New York.
Løvtrup, S. (1966) Morphogenesis in the amphibian embryo. Cell type distribution, germ layers, and fate maps. Acta Zool. (Stockholm) 47, 209–276.
Løvtrup, S. (1975) Fate maps and gastrulation in Amphibia – a critique of current views. Can. J. Zool. 53, 473–479.
Luduēna, M.A. and Wessells, N.K. (1973) Cell locomotion, nerve elongation, and microfilaments. Dev. Biol. 30, 427–440.
Manasek, F.J. (1975) The extracellular matrix: A dynamic component of the developing embryo. Curr. Top. Dev. Biol. 10, 35–102.
Manasek, F.J., Reid, M., Vinson, W. Seyer, J. and Johnson, R. (1973) Glycosaminoglycan synthesis by the early embryonic chick heart. Dev. Biol. 35, 332–348.
Mangold, O. (1920) Fragen der Regulation und Determination an umgeordneten Furchungsstadien und verschmolzenen Keimen von Triton. Wilhelm Roux' Arch. Entwicklungsmech. Org. 47, 250–301.
Marchase, R.B., Barbera, A.J. and Roth, S. (1975) A molecular approach to retinotectal specificity. In: Cell Patterning, Ciba Foundation Symposium 29 (new series) pp. 315–341, Elsevier/North-Holland, Amsterdam.
Marchesi, V.T. (1966) Mechanisms of cell migration and macromolecular transport across the walls of blood vessels. Gastroenterol., 51, 875–892.
Marchesi, V.T. (1970) Mechanisms of blood cell migration across blood vessel walls. In: Regulation of Hematopoiesis (Gordon, A.S., ed.) pp. 943–958. Appleton-Century-Crofts, New York.
Martinez-Palomo, A., Brailovsky, C. and Bernhard, W. (1969) Ultrastructural modification of the cell surface and intercellular contacts of some transformed cells, Cancer Res. 29, 925–937.
Martz, E. and Steinberg, M.S. (1973) Contact inhibition of what? An analytical review. J. Cell Physiol. 81, 25–38.
Martz, E. and Steinberg, M.S. (1974) Movement in a confluent 3T3 monolayer and the causes of contact inhibition of overlapping. J. Cell Sci. 15, 201–216.
Matsumoto, S. (1918) Contribution to the study of epithelial movement. The corneal epithelium of the frog in tissue culture. J. Exp. Zool. 26, 545–564.
McIntosh, J.R., Hepler, P.K. and Van Wie, D.G. (1969) Model for mitosis. Nature, 224, 659–663.
McNutt, S.N., Culp, L.A. and Black, P.H. (1971) Contact-inhibited revertant cell lines isolated from SV40-transformed cells. II. Ultrastructural study. J. Cell Biol. 50, 691–708.
Meier, S. and Hay, E.D. (1973) Synthesis of sulfated glycosaminoglycans by embryonic corneal epithelium. Dev. Biol. 35, 318–331.
Meier, S. and Hay, E.D. (1974) Control of corneal differentiation by extracellular materials. Collagen as a promoter and stabilizer of epithelial stroma production. Dev. Biol. 38, 249–270.
Messier, P.E. (1969) Effects of β-mercaptoethanol on the fine structure of the neural plate cells of the chick embryo. J. Embryol. Exp. Morphol. 21, 309–329.
Messier, P.E. (1972) The occurrence of nuclear migration under thiol treatment effective in inhibiting neurulation. J. Embryol. Exp. Morphol. 27, 577–584.
Meyer, D.B. (1964) The migration of primordial germ cells in the chick embryo. Dev. Biol. 10, 154–190.
Middleton, C.A. (1972) Contact inhibition of locomotion in cultures of pigmented retina epithelium. Exp. Cell Res. 70, 91–96.
Middleton, C.A. (1973) The control of epithelial cell locomotion in tissue culture. In: Locomotion of Tissue Cells, Ciba Foundation Symposium 14 (new series) pp. 251–270, Elsevier, North-Holland, Amsterdam.
Middleton, C.A. (1976) Contact-induced spreading is a new phenomena depending on cell contact. Nature, 259, 311–313.
Mintz, B. (1959) Continuity of the female germ cell line from embryo to adult. Arch. Anat. Microsc. Morphol. Exp. 48, Suppl. 155–172.

Mintz, B. (1970) Control of embryo implantation and survival. Adv. Biosci. 6, 317–342.
Miranda, A.F., Godman, G.C., Deitch, A.D. and Tanenbaum, S.W. (1974a) Action of cytochalasin D on cells of established lines. I. Early events. J. Cell Biol. 61, 481–500.
Miranda, A.F., Godman, G.C. and Tanenbaum, S.W. (1974b) Action of cytochalasin D on cells of established lines. II. Cortex and microfilaments. J. Cell Biol. 62, 406–423.
Mitchison, J.M. and Swann, M.M. (1954) The mechanical properties of the cell surface. I. The cell elastimeter. J. Exp. Biol. 31, 443–461.
Moore, A.R. and Burt, A.S. (1939) On the locus and nature of the forces causing gastrulation in the embryos of *Dendraster excentricus.* J. Exp. Zool. 82, 159–171.
Moscona, A.A. (1962) Analysis of cell recombinations in experimental synthesis of tissues in vitro. J. Cell. Comp. Physiol. Suppl. 1, 60, 65–80.
Mostow, G.D. (1975) (ed.) Mathematical models for cell rearrangement, 288 pp. Yale University Press, New Haven.
Nakatsuji, N. (1974) Studies on the gastrulation of amphibian embryos: pseudopodia in the gastrula of *Bufo bufo japonicus* and their significance to gastrulation. J. Embryol. Exp. Morphol. 32, 795–804.
Nakatsuji, N. (1975) Studies on the gastrulation of amphibian embryos: light and electron microscopic observation of a urodele *Cynops* pyrrhogaster. J. Embryol. Exp. Morphol. 34, 669–685.
Nelson, G.A. and Revel, J.P. (1975) Scanning electron microscopic study of cell movements in the corneal endothelium of the avian embryo. Dev. Biol. 42, 315–333.
New, D.A.T. (1959) Adhesive properties and expansion of the chick blastoderm. J. Exp. Embryol. Morphol. 7, 146–164.
Nicolson, G.L. (1972) Topography of membrane concanavalin A sites modified by proteolysis. Nature New Biology 239, 193–197.
Nieuwkoop, P.D. and Florschütz, P.A. (1950) Quelques caractères spéciaux de la gastrulation et de la neurulation de l'oeuf de *Xenopus laevis,* Daud. et de quelques autres Anoures. lère partie. Etude descriptive. Arch. Biol. 61, 113–150.
Noden, D.M. (1975) An analysis of the migratory behavior of avian cephalic neural crest cells. Dev. Biol. 42, 106–130.
Odland, G. and Ross, R. (1968) Human wound repair. I. Epidermal regeneration. J. Cell Biol. 39, 135–151.
Okazaki, K. (1960) Skeleton formation of sea urchin larvae. II. Organic matrix of the spicule. Embryologia 5, 283–320.
Okazaki, K., Fukushi, T. and Dan, K. (1962) Cyto-embryological studies of sea urchins. IV. Correlation between the shape of the ectodermal cells and the arrangement of the primary mesenchyme cells in sea urchin larvae. Acta. Embryol. Morphol. Exp. 5, 17–31.
Oppenheimer, J.M. (1936) Processes of localization in developing *Fundulus.* J. Exp. Zool. 73, 405–444.
Pasteels, J. (1936) Etudes sur la gastrulation des vertébrés méroblastiques. I. Téléostéens. Arch. Biol. 47, 205–308.
Patten, B.M. (1968) Human embryology, 651, pp., McGraw-Hill, New York.
Pearce, T.L. and Zwaan, J. (1970) A light and electron microscopic study of cell behavior and microtubules in the embryonic chicken lens using Colcemid. J. Embryol. Exp. Morphol. 23, 491–507.
Pérès, J.M. (1948) Recherches sur la genèse et la régénération de la tunique chez *Clavelina Lepadiformis* Muller. Arch. Anat. Microsc. Morphol. Exp. 37, 230–259.
Perry, M. (1975) Microfilaments in the external surface layer of the early amphibian embryo. J. Embryol. Exp. Morphol. 33, 127–146.
Perry, M. and Waddington, C.H. (1966) Ultrastructure of the blastopore cells in the newt. J. Embryol. Exp. Morphol. 15, 317–330.
Pessac, B. and Defendi, V. (1972) Cell aggregation: role of acid mucopolysaccharides. Science, 175, 898–900.
Pethica, B.A. (1961) The physical chemistry of cell adhesion. Exp. Cell Res. Suppl. 8, 123–140.
Petris, S. de and Raff, M.C. (1973). Fluidity of the plasma membrane and its implications for cell movement, In: Locomotion of Tissue Cells, Ciba Foundation Symposium 14 (new series) pp. 27–52. Elsevier/North-Holland, Amsterdam.

Pfenninger, K.H. and Bunge, M.B. (1973) Observations on plasmalemmal growth zones in developing neural tissue. J. Cell Biol. 59, 264a.

Pfenninger, K.H. and Bunge, R.P. (1974) Freeze-fracturing of nerve growth cones and young fibers. A study of developing plasma membrane. J. Cell Biol. 63, 180–196.

Piatigorsky, J., Webster, H. de F. and Craig, S.P. (1970) Ultrastructural and biochemical aspects of lens fiber formation in vivo and in vitro. J. Cell Biol. 47, 158a.

Pollard, T.D. (1972) Progress in understanding amoeboid movement at the molecular level. In: The Biology of Amoeba, (Jeon, K.W., ed.) pp. 291–317, Academic Press, New York.

Pollard, T.D. and Korn, E.D. (1973) Electron microscopic identification of actin associated with isolated amoeba plasma membranes. J. Biol. Chem. 248, 448–450.

Pollard, T.D. and Weihing, R.R. (1974) Actin and myosin and cell movement. CRC Crit. Rev. Biochem. 2, 1–65.

Porter, K., Prescott, D. and Frye, J. (1973) Changes in surface morphology of Chinese hamster ovary cells during the cell cycle. J. Cell Biol. 57, 815–836.

Porter, K., Puck, T.T., Hsie, A.W. and Kelley, D. (1974) An electron microscope study of the effects of dibutyryl cyclic AMP on Chinese hamster ovary cells. Cell, 2, 145–162.

Poste, G. and Allison, A.C. (1973) Membrane fusion. Biochim. Biophys. Acta 300, 421–465.

Poste, G., Greenham, L.W., Mallucci, L., Reeve, P. and Alexander, D.J. (1973) The study of cellular "microexudates" by ellipsometry and their relationship to the cell coat. Exp. Cell Res. 78, 303–313.

Pourreau-Schneider, N. (1961) Association d'un organe de poulet adulte avec un cancer de Mammifère en culture in vitro. C.R. Soc. Biol. 155, 1225–1227.

Poynter, C. (1919) Some observations on wound healing in the early embryo. Anat. Rec. 16, 1–23.

Pratt, R.M., Goggins, J.F., Wilk, A.L. and King, G.T.G. (1973) Acid mucopolysaccharide synthesis in the secondary palate of the developing rat at the time of rotation and fusion. Dev. Biol. 32, 230–237.

Pratt, R.M., Larsen, M.A. and Johnston, M.C. (1975) Migration of cranial neural crest cells in a cell-free hyaluronate-rich matrix. Dev. Biol. 44, 298–305.

Ramsey, W.S. (1972a) Analysis of individual leucocyte behavior during chemotaxis. Exp. Cell Res. 70, 129–139.

Ramsey, W.S. (1972b) Locomotion of human polymorphonuclear leucocytes. Exp. Cell Res. 72, 489–501.

Ramsey, W.S. (1974) Leucocyte locomotion and chemotaxis. Antibiot. Chemother. (Basel) 19, 179–190.

Ramsey, W.S. and Grant, L. (1974) Chemotaxis in The Inflammatory Process (Zweifach, B.W., Grant, L. and McCluskey, R.T., eds.) pp. 287–361, Academic Press, New York.

Ramsey, W.S. and Harris, A. (1972) Leucocyte locomotion and its inhibition by antimitotic drugs. Exp. Cell Res. 82, 262–270.

Rand, H.W. (1915) Wound closure in actinian tentacles with reference to the problem of organization. Wilhelm Roux' Arch. Entwicklungsmech. Org. 41, 159–214.

Rand, H.W. and Pierce, M.E. (1932) Skin grafting in frog tadpoles: local specificity of skin and behavior of epidermis. J. Exp. Zool. 62, 125–170.

Rapp, F. (1974) Virus-mediated transformation of mammalian cells, In: Developmental Aspects of Carcinogenesis and Immunity (King, T., ed.), Academic Press, New York.

Rawles, M.E. (1944) The migration of melanoblasts after hatching into pigment-free skin grafts of the common fowl. Physiol. Zool. 17, 167–183.

Revel, J.P., Hoch, P. and Ho, D. (1974) Adhesion of culture cells to their substratum. Exp. Cell Res. 84, 207–218.

Revel, J.P. and Wolken, K. (1973) Electronmicroscope investigations of the underside of cells in culture. Exp. Cell Res. 78, 1–14.

Revel, J.P., Yip, P. and Chang, L.L. (1973) Cell junctions in the early chick embryo – A freeze etch study. Dev. Biol. 35, 302–317.

Rhumbler, L. (1902) Zur Mechanik des Gastrulationsvorgänges, insbesondere der Invagination. Eine entwicklungsmechanische Studie. Wilhelm Roux' Arch. Entwicklungsmech. Org. 14, 401–476.

Robineaux, R. (1964) Movements of cells involved in inflammation and immunity. In: Primitive

Motile Systems in Cell Biology (Allen, R.D. and Kamiya, N., eds.) pp. 351–364. Academic Press, New York.

Roseman, S. (1970) The synthesis of complex carbohydrates by multiglycosyltransferase systems and their potential function in intercellular adhesion. Chem. Phys. Lipids 5, 270–297.

Roth, S. (1968) Studies on intercellular adhesive selectivity. Dev. Biol. 18, 602–631.

Roth, S. (1973) A Molecular model for cell interactions. Q. Rev. Biol. 48, 541–563.

Roth, S. and Weston, J.A. (1967) The measurement of intercellular adhesion. Proc. Nat. Acad. Sci. U.S.A. 58, 974–980.

Roux, W. (1894) Über den "Cytotrospismus" der Furchungzellen des Grasfrosches (*Rana fusca*). Arch. Entw. der Organismen, 1, 43–68.

Ruffini, A. (1925) Fisiogenia, 999 pp., Francesco Vallardi, Milano.

Rutter, W.J., Kemp, J.D., Bradshaw, W.S., Clark, W.R., Ronzio, R.A. and Sanders, T.G. (1968) Regulation of specific protein synthesis in cytodifferentiation. J. Cell Physiol. 72, Suppl. 1, 1–18.

Sachs, L. and Medina, D. (1961) In vitro transformation of normal cells by polyoma virus. Nature, 189, 457–458.

Saint-Hilaire, K. (1931) Morphogenetische Untersuchungen des Ascidienmantels. Zool. Jahrb. 54, 435–608.

Sanders, E.J. and Zalik, S.E. (1970) The occurence of microtubules in the pre-streak chick embryo. Protoplasma, 71, 203–208.

Sanders, E.J. and Zalik, S.E. (1972) The blastomere periphery of *Xenopus laevis*, with special reference to intercellular relationships. Wilhelm Roux' Arch. Entwicklungsmech. Org. 171, 181–194.

Schaeffer, B.E., Schaeffer, H.E. and Brick, I. (1973) Cell electrophoresis of amphibian blastula and gastrula cells: the relationship of surface charge and morphogenetic movement. Dev. Biol. 34, 66–76.

Schechtman, A.M. (1942) The mechanism of amphibian gastrulation. I. Gastrulation-promoting interactions between various regions of an anuran egg (*Hyla regilla*). Univ. Calif. Publ. Zool. 51, 1–40.

Schlesinger, A.G. (1958) The structural significance of the avian yolk in embryogenesis. J. Exp. Zool. 138, 223–258.

Schroeder, T.E. (1970) Neurulation in *Xenopus laevis*. An analysis and model based upon light and electron microscope. J. Embryol. Exp. Morphol. 23, 427–462.

Searls, R.L. (1967) The role of cell migration in the development of the embryonic chick limb bud. J. Exp. Zool. 166, 39–50.

Seeliger, O. (1893) Einige Beobachtungen über die Bildung des äusseren Mantels der Tunicaten. Z. Wiss. Zool. 56, 488–505.

Senda, N., Tamura, H., Shibata, N., Yoshitake, J., Kondo, K. and Tanaka, K. (1975) The mechanism of the movement of leukocytes. Exp. Cell Res. 91, 393–407.

Seravin, L.N. (1971) Mechanisms and coordination of cellular locomotion. Adv. Comp. Physiol. Biochem. 4, 37–111.

Shain, W.G., Hilfer, S.R. and Fonte, V.G. (1972) Early organogenesis of the embryonic chick thyroid. I. Morphology and biochemistry. Dev. Biol. 28, 202–218.

Sheetz, M.P. and Singer, S.J. (1974) Biological membranes as bilayer couples. A molecular mechanism of drug-erythrocyte interactions. Proc. Nat. Acad. Sci. U.S.A. 71, 4457–4461.

Shein, H.M. and Enders, J.F. (1962) Transformation induced by simian virus 40 in human renal cell cultures. I. Morphology and growth characteristics. Proc. Nat. Acad. Sci. U.S.A. 48, 1164–1172.

Sheridan, J. (1976) Cell coupling and cell communication during embryogenesis. In: The Cell Surface in Animal Embryogenesis and Development (Poste, G. and Nicolson, G.L., eds.) Ch. 8. North-Holland, Amsterdam.

Shostak, S., Patel, N.G. and Burnett, A.L. (1965) The role of the mesoglea in mass cell movements in hydra. Dev. Biol. 12, 434–450.

Simon, D. (1960) Contribution à l'étude de la circulation et du transport des gonocytes primaires dans les blastodermes d'oiseau cultivés in vitro. Arch. Anat. Microscop. Morphol. Exp. 49, 93–176.

Singer, S.J. (1971) The molecular organization of biological membranes. In: Structure and Function of Biological Membranes (Rothfield, L.I., ed.) p. 145–174. Academic Press, New York.

Singer, S.J. (1974) Molecular biology of cellular membranes with application to immunology. Adv. Immunol. 19, 1–66.

Singer, S.J. and Nicolson, G.L. (1972) The fluid mosaic model of the structure of cell membranes. Science, 175, 720–731.

Speidel, C.C. (1932) Studies of living nerves. I. The movements of individual sheath cells and nerve sprouts correlated with the process of myelin-sheath formation in amphibian larvae. J. Exp. Zool. 61, 279–331.

Speidel, C.C. (1933) Studies of living nerves. II. Activities of ameboid growth cones, sheath cells, and myelin segments, as revealed by prolonged observation of individual nerve fibers in frog tadpoles. Am. J. Anat. 52, 1–79.

Speidel, C.C. (1935) Studies of living nerves. III. Phenomena of nerve irritation and recovery, degeneration and repair. J. Comp. Neurol. 61, 1–80.

Spemann, H. (1931) Über den Anteil von Implantat und Wirtskeim und der Orientierung und Beschaffenheit der induzierte Embryonalanlage. Wilhelm Roux' Arch. Entwicklungsmech. Org. 123, 389–517.

Spemann, H. (1938) Embryonic Development and Induction, 401 pp., Yale University Press, New Haven.

Sperry, R.W. (1965) Embryogenesis of behavioral nerve nets. In: Organogenesis (DeHaan, R.L. and Ursprung, H., eds.) pp. 161–186, Holt, Rinehart and Winston, New York.

Spooner, B.S. (1973) Microfilaments, cell shape changes, and morphogenesis of salivary epithelium. Am. Zool. 13, 1007–1022.

Spooner, B.S. (1974) Morphogenesis of vertebrate organs. In: Concepts of Development (Lash, J. and Whittaker, J.R., eds), pp. 213–240, Sinauer Associates, Inc., Stamford, Connecticut.

Spooner, B.S. (1975) Microfilaments, microtubules, and extracellular materials in morphogenesis. BioScience 25, 440–451.

Spooner, B.S. and Conrad, G.W. (1975) The role of extracellular materials in cell movement. I. Inhibition of mucopolysaccharide synthesis does not stop ruffling membrane activity or cell movement. J. Cell Biol. 65, 286–297.

Spooner, B.S. and Wessells, N.K. (1970) Effects of cytochalasin B upon microfilaments involved in morphogenesis of salivary epithelium. Proc. Nat. Acad. Sci. U.S.A. 66, 360–364.

Spooner, B.S., Ash, J.F., Wrenn, J.T., Frater, R.B. and Wessells, N.K. (1973) Heavy meromyosin binding to microfilaments involved in cell and morphogenetic movements. Tissue Cell 5, 37–46.

Spooner, B.S., Yamada, K.M. and Wessells, N.K. (1971) Microfilaments and cell locomotion. J. Cell Biol. 49, 595–613.

Spooner, B.S., Luduẽna, M.A. and Wessells, N.K. (1974) Membrane fusion in the growth cone-microspike region of embryonic nerve cells undergoing axon elongation in cell culture. Tissue Cell 6, 399–409.

Spratt, N.T. (1946) Formation of the primitive streak in the explanted chick blastoderm marked with carbon particles. J. Exp. Zool. 103, 259–304.

Steer, H.W. (1971) Implantation of the rabbit blastocyst: the invasive phase. J. Anat. 110, 445–462.

Steinberg, M.S. (1962) Mechanism of tissue reconstruction by dissociated cells, II: Time-course of events. Science, 137, 762–763.

Steinberg, M.S. (1964) The problem of adhesive selectivity in cellular interactions. In: Cellular Membranes in Development (Locke, M., ed.) pp. 321–366, Academic Press, New York.

Steinberg, M.S. (1970) Does differential adhesion govern self-assembly processes in histogenesis? Equilibrium configurations and the emergence of a hierarchy among populations of embryonic cells. J. Exp. Zool. 173, 395–434.

Stockard, C.R. (1915) A study of wandering mesenchymal cells on the living yolk-sac and their developmental products: chromatophores, vascular endothelium and blood cells. Am. J. Anat. 18, 525–594.

Szollosi, D. (1970) Cortical cytoplasmic filaments of cleaving eggs: a structural element corresponding to the contractile ring. J. Cell Biol. 44, 192–209.

Tarin, D. (1971a) Histological features of neural induction in *Xenopus laevis.* J. Embryol. Exp. Morphol. 26, 543–570.

Tarin, D. (1971b) Scanning electron microscopical studies of the embryonic surface during gastrulation and neurulation in *Xenopus laevis.* J. Anat. 109, 535–547.

Taylor, A.C. (1961) Attachment and spreading of cells in culture. Exp. Cell Res. Suppl. 8, 154–173.

Taylor, A.C. and Robbins, E. (1963) Observations of microextensions from the surface of isolated vertebrate cells. Dev. Biol. 7, 660–673.

Taylor, R.B., Duffus, W.P.H., Raff, M.C. and Petris, S. de (1971) Redistribution and pinocytosis of lymphocyte surface immunoglobulin molecules induced by anti-immunoglobulin antibody. Nature New Biology 233, 225–229.

Temin, H.M. and Rubin, H. (1958) Characteristics of an assay for Rous sarcoma virus and Rous sarcoma cells in tissue culture. Virology 6, 669–688.

Terry, A.H. and Culp, L.A. (1974) Substrate-attached glycoproteins from normal and virus-transformed cells. Biochemistry 13, 414–425.

Thompson, D'Arcy W. (1942) On growth and form. pp. 1116. Cambridge Univ. Press, MacMillan, New York.

Tickle, C.A. and Trinkaus, J.P. (1973) Change in surface extensibility of *Fundulus* deep cells during early development. J. Cell Sci. 13, 721–726.

Tickle, C. and Trinkaus, J.P. (1976) Observations on nudging cells in culture. Nature, 261, 413.

Tilney, L.G. (1968) II. Ordering of subcellular units. The assembly of microtubules and their role in the development of cell form. Dev. Biol. Suppl. 2, 63–102.

Tilney, L.G. and Gibbins, J.R. (1969) Microtubules in the formation and development of the primary mesenchyme in *Arbacia Punctulata.* II. An experimental analysis of their role in development and maintenance of cell shape. J. Cell Biol. 41, 227–250.

Toole, B.P. and Trelstad, R.L. (1971) Hyaluronate production and removal during corneal development in the chick. Dev. Biol. 26, 28–35.

Townes, P.L. and Holtfreter, J. (1955) Directed movements and selective adhesion of embryonic amphibian cells. J. Exp. Zool. 128, 53–120.

Trelstad, R.L. and Coulombre, A.J. (1971) Morphogenesis of the collagenous stroma in the chick cornea. J. Cell Biol. 50, 840–858.

Trelstad, R.L., Hay, E.D. and Revel, J.P. (1967) Cell contact during early morphogenesis in the chick embryo. Dev. Biol. 16, 78–106.

Trelstad, R.L., Hayashi, K. and Toole, B.P. (1974) Epithelial collagens and glycosaminoglycans in the embryonic cornea. Macromolecular order and morphogenesis in the basement membrane. J. Cell Biol. 62, 815–830.

Trinkaus, J.P. (1948) Factors concerned in the response of melanoblasts to estrogen in the brown leghorn fowl. J. Exp. Zool. 109, 135–170.

Trinkaus, J.P. (1949) The surface gel layer of *Fundulus* eggs in relation to epiboly. Proc. Nat. Acad. Sci. U.S.A. 35, 218–225.

Trinkaus, J.P. (1951) A study of the mechanism of epiboly in the egg of *Fundulus heteroclitus.* J. Exp. Zool. 118, 269–320.

Trinkaus, J.P. (1963) The cellular basis of *Fundulus* epiboly. Adhesivity of blastula and gastrula cells in culture. Dev. Biol. 7, 513–532.

Trinkaus, J.P. (1965) Mechanism of morphogenetic movements. In: Organogenesis (DeHaan, R.L. and Ursprung, H., eds.) pp. 55–104, Holt, Rinehart and Winston, New York.

Trinkaus, J.P. (1966) Morphogenetic cell movements. In: Major Problems of Developmental Biology, The 25th Symposium of the Society for Developmental Biology, (Locke, M., ed.), pp. 125–176, Academic Press, New York.

Trinkaus, J.P. (1969) Cells into Organs: The Forces That Shape the Embryo, 253 pp., Prentice-Hall, Englewood Cliffs, New Jersey.

Trinkaus, J.P. (1971) Role of the periblast in *Fundulus* epiboly, (in Russian), Ontogenesis, 2, 401–405.

Trinkaus, J.P. (1973a) Surface activity and locomotion of *Fundulus* deep cells during blastula and gastrula stages. Dev. Biol. 30, 68–103.

Trinkaus, J.P. (1973b) Modes of cell locomotion in vivo. In: Locomotion of Tissue Cells, Ciba Foundation Symposium 14 (new series), pp. 233–249, Elsevier, North-Holland, Amsterdam.

Trinkaus, J.P. and Gross, M.C. (1961) The use of tritiated thymidine for marking migratory cells. Exp. Cell Res. 24, 52–57.

Trinkaus, J.P. and Lentz, T.L. (1964) Direct observation of type-specific segregation in mixed cell aggregates. Dev. Biol. 9, 115–136.

Trinkaus, J.P. and Lentz, T.L. (1967) Surface specializations of *Fundulus* cells and their relation to cell movements during gastrulation. J. Cell Biol. 32, 139–153.

Trinkaus, J.P., Betchaku, T. and Krulikowski, L.S. (1971) Local inhibition of ruffling during contact inhibition of cell movement. Exp. Cell Res. 64, 291–300.

Ubisch, L. Von (1937) Die Normale Skelettbildung bei *Echinocyamus pusillus* und *Psammechinus miliaris* und die Bedeutung dieser Vorgänge für die Analyse der Skelette von Keimblatt-Chimären. Z. Wiss. Zool. 149, 402–476.

Uhlenhuth, E. (1914) Cultivation of the skin epithelium of the adult frog, *Rana pipiens*. J. Exp. Med. 20, 614–635.

Vasiliev, J.M. and Gelfand, M. (1973) Interactions of normal and neoplastic fibroblasts with the substratum. In: Locomotion of Tissue Cells, Ciba Foundation Symposium 14 (new series) pp. 311–331, Elsevier, North Holland, Amsterdam.

Vasiliev, J.M., Gelfand, I.M., Domnina, L.V., Ivanova, O.Y., Komm, S.G., and Olshevskaja, L.V. (1970) Effect of Colcemid on the locomotory behavior of fibroblasts. J. Embryol. Exp. Morphol. 24, 625–640.

Vasiliev, J.M., Gelfand, I.M., Domnina, L.V., Zacharova, O.S. and Ljubimov, A.V. (1975) Contact inhibition of phagocytosis in epithelial sheets: alterations of cell surface properties induced by cell-cell contacts. Proc. Nat. Acad. Sci. U.S.A. 72, 719–722.

Vaughan, R.B. and Trinkaus, J.P. (1966) Movements of epithelial cell sheets in vitro. J. Cell Sci. 1, 407–413.

Vesely, P. and Weiss, R.A. (1973) Cell locomotion and contact inhibition of normal and neoplastic rat cells. Int. J. Cancer 11, 64–76.

Vogt, M. and Dulbecco, R. (1960) Virus-cell interaction with a tumor-producing virus. Proc. Nat. Acad. Sci. U.S.A. 46, 365–370.

Vogt, W. (1929) Gestaltungsanalyse am Amphibiankeim mit örtlicher Vitalfärbung. II. Teil. Gastrulation und Mesodermbildung bei Urodelen und Anuren. Wilhelm Roux' Arch. Entwicklungsmech. Org. 120, 384–706.

Waddington, C.H. (1941) Translocations of the organizer in the gastrula of *Discoglossus*. Proc. Zool. Soc. London 111, 189–198.

Waddington, C.H. (1952) The Epigenetics of Birds, 272 pp., Cambridge University Press, Cambridge.

Waddington, C.H. (1962) New Patterns in Genetics and Development, 271 pp., Columbia University Press, New York.

Waddington, C.H. and Perry, M.M. (1966) A note on the mechanisms of cell deformation in the neural folds of the amphibia. Exp. Cell Res. 41, 691–693.

Walther, B.T., Ohman, R. and Roseman, S. (1973) A quantitative assay for intercellular adhesion. Proc. Nat. Acad. Sci. U.S.A. 70, 1569–1573.

Warren, S. and Gates, O. (1936) The fate of intravenously injected tumor cells. Am. J. Cancer 27, 485–492.

Weiss, L. (1962) The mammalian tissue cell surface. In: The Structure and Function of the Membranes and Surfaces of Cells, Biochem. Soc. Sympos. No. 22 (Bell, D.J. and Grant, J.K., eds.) pp. 32–54, Cambridge University Press, Cambridge.

Weiss, L. (1966) Studies on cell deformability. II. Effects of some proteolytic enzymes. J. Cell Biol. 30, 39–43.

Weiss, L. and Lachmann, P.J. (1964) The origin of an antigenic zone surrounding HeLa cells cultured on glass. Exp. Cell Res. 36, 86–91.

Weiss, L., Poste, G., MacKearnin, A. and Willett, K. (1975) Growth of mammalian cells on substrates

coated with cellular microexudates. J. Cell Biol. 64, 135–145.
Weiss, P.A. (1945) Experiments on cell and axon orientation in vitro: the role of colloidal exudates in tissue organization. J. Exp. Zool. 100, 353–386.
Weiss, P.A. (1955) Nervous system (Neurogenesis). In: Analysis of Development (Willier, B.H., Weiss, P.A. and Hamburger, V., eds.) pp. 346–401, W.B. Saunders, Philadelphia.
Weiss, P.A. (1961a) Guiding principles in cell locomotion and aggregation. Exp. Cell Res. Suppl. 8, 260–281.
Weiss, P.A. (1961b) The biological foundations of wound repair. Harvey lectures, 55, 13–42.
Wessells, N.K. (1970) Mammalian lung development: interactions in formation and morphogenesis of tracheal buds. J. Exp. Zool. 175, 455–466.
Wessells, N.K. and Cohen, J.H. (1968) Effects of collagenase on developing epithelia in vitro: lung, ureteric bud, and pancreas. Dev. Biol. 18, 294–309.
Wessells, N.K., Spooner, B.S., Ash, J.F., Bradley, M.O., Ludueña, M.A., Taylor, E.L., Wrenn, J.T. and Yamada, K.M. (1971) Microfilaments in cellular and developmental processes. Science, 171, 135–153.
Wessells, N.K., Spooner, B. and Ludueña, M.A. (1973) Surface movements, microfilaments and cell locomotion. In: Locomotion of Tissue Cells, Ciba Foundation Symposium 14 (new series) pp. 53–82, Elsevier, North-Holland, Amsterdam.
Weston, J.A. (1963) A radioautographic analysis of the migration and localization of trunk neural crest in the chick. Dev. Biol. 6, 279–310.
Weston, J.A. (1967) Cell marking. In: Methods in Developmental Biology (Wilt, F.H. and Wessells, N.K., eds.) pp. 723–736. T.Y. Crowell, Boston.
Weston, J.A. (1970) The migration and differentiation of neural crest cells. Adv. Morphog. 8, 41–114.
Weston, J.A. and Abercrombie, M. (1967) Cell mobility in fused homo- and heteronomic tissue fragments. J. Exp. Zool. 164, 317–324.
Weston, J.A. and Butler, S.L. (1966) Temporal factors affecting localization of neural crest cells in the chicken embryo. Dev. Biol. 14, 246–266.
Weston, J.A. and Roth, S.A. (1969) Contact inhibition: behavioral manifestations of cellular adhesive properties in vitro. In: Cellular Recognition (Smith, R.T. and Good, R.A., eds.) pp. 159–167. Appleton-Century-Crofts, Educational Division, Meredith Corporation, New York.
Willier, B.H. (1939) The embryonic development of sex. In: Sex and Internal Secretions (Allen, E., ed.).
Willis, R.A. (1960) Pathology of Tumours, 1002 pp. Butterworths, London.
Winter, G.D. (1964) Movement of epidermal cells over the wound surface. In: Advances in Biology of Skin, Vol. V, (Montagna, W. and Billingham, R.E., eds.) The Macmillan Company, New York, pp. 113–127.
Wiseman, L.L. and Steinberg, M.S. (1973) The movement of single cells within solid tissue masses. Exp. Cell Res. 79, 468–471.
Wohlfarth-Botterman, K.E. (1964) Differentiations of the ground cytoplasm and their significance for the generation of the motive force of amoeboid movement. In: Primitive Motile Systems in Cell Biology (Allen, R.E. and Kamiya, N. eds.) pp. 79–109, Academic Press, New York.
Wolff, E. and Marin, L. (1957) Sur les mouvements des explants de foie embryonnaire cultivé in vitro. C.R. Acad. Sci. 244, 2745–2747.
Wolff, E. and Schneider, N. (1957) La culture d'un sarcome de souris sur des organes de poulet explantés in vitro. Arch. Anat. Micr. Morphol. Exp. 46, 173–198.
Wolpert, L. and Gingell, D. (1968) Cell surface membrane and amoeboid movement. In: Aspects of Cell Motility, pp. 169–198, Cambridge University Press, Cambridge.
Wolpert, L. and Gingell, D. (1969) The cell membrane and contact control. In: Ciba Found. Symposium on Homeostatic Regulators (Wolstenholme, G.E.W. and Knight, J. eds.) pp. 241–259, J. & A. Churchill Ltd., London.
Wood, S. (1958) Pathogenesis of metastasis formation observed in vivo in the rabbit ear chamber. Arch. Pathol. 66, 550–568.
Wood, S., Baker, R.R. and Marzocchi, B. (1968) In vivo studies of tumor behavior: locomotion of and interrelationships between normal cells and cancer cells. In: The Proliferation and Spread of

Neoplastic Cells (Twenty-first Annual Symposium on Fundamental Cancer Research, 1967) pp. 495–509, Williams and Wilkins, Baltimore, Maryland.

Wourms, J.P. (1972) The developmental biology of annual fishes. II. Naturally occurring dispersion and reaggregation of blastomeres during the development of annual fish eggs. J. Exp. Zool. 182, 169–200.

Wrenn, J.T. (1971) An analysis of tubular gland morphogenesis in chick oviduct. Dev. Biol. 26, 400–415.

Wrenn, J.T. and Wessells, N.K. (1969) An ultrastructural study of lens invagination in the mouse. J. Exp. Zool. 171, 359–368.

Yamada, K.M., Spooner, B.S. and Wessells, N.K. (1971) Ultrastructure and function of growth cones and axons of cultured nerve cells. J. Cell Biol. 49, 614–635.

Yamada, T. (1967) Cellular and subcellular events in Wolffian lens regeneration. Curr. Top. Dev. Biol. 2, 247–283.

Zigmond S. (1974) Mechanisms of sensing chemical gradients by polymorphonuclear leukocytes. Nature, 249, 450–452.

Zigmond, S.H. and Hirsch, J.G. (1973) Leukocyte locomotion and chemotaxis. J. Exp. Med. 137, 387–410.

Zwilling, E. (1968) Morphogenetic phases in development. In: The Emergence of Order in Developing Systems (Locke, M., ed.) Dev. Biol. Suppl. 2, 184–207. Academic Press, New York.

Inductive tissue interactions 7

Lauri SAXÉN, Marketta KARKINEN-JÄÄSKELÄINEN,
Eero LEHTONEN, Stig NORDLING
and Jorma WARTIOVAARA

1. Introduction

During embryogenesis, the cells of the organism differentiate in a synchronized, spatially and temporally controlled manner which implies the existence of a building plan, a general control mechanism functioning at various levels. Cytodifferentiation, the expression of different sets of genes in various cell types can theoretically be considered to operate through different regulative mechanisms: (1) *nuclear* regulation due to loss or amplification of genetic material; (2) *cytoplasmic* regulation operating via factors unevenly distributed in the egg cytoplasm and unequally inherited by the daughter cells; and (3) *extracellular* control systems ultimately regulating the reading of the genome. Examples of the first two mechanisms of cytodifferentiation are known in the animal kingdom, but are likely to be exceptions rather than the rule. Here we are concerned with the third alternative, an extracellular control system. Moreover, as determined cells form well-organized tissues and organs, synchronized morphogenesis implies communication between like and unlike cells.

There is convincing evidence that morphogenetic signalling systems are of at least two types: (1) organismal, humoral factors acting over long distances and operating on target cells in separate organ anlagen (morphogenetic hormones, growth-controlling factors); and (2) communication between cells and cell populations in close proximity. The latter, traditionally known as "embryonic induction", is said to take place "whenever in development two or more tissues of different history and properties become intimately associated and alteration of the developmental course of the interactants results" (Grobstein, 1956). Today, there is no real disagreement about the importance of these interactive events throughout development and in the regenerative processes in an adult organism (see, however, Holtzer, 1975). For recent monographs and reviews on the topic, the reader is referred to Grobstein (1967), Fleischmajer and Billingham (1968), Saxén and Kohonen (1969), Kratochwil (1972), Slavkin (1972), Wessells (1973) and Saxén (1975b).

Yet, despite more than 50 years of active research in the field, many fundamental questions remain unanswered. Hence any attempt at a synthetic

G. Poste & G.L. Nicolson (eds) The Cell Surface in Animal Embryogenesis and Development

treatment would be premature, and this review of research on inductive tissue interactions will rather be a presentation of generally accepted principles, with an outline of the current problems and hypotheses. The four basic topics to be discussed in this chapter are: (1) the sequential nature of inductive interactions, i.e. the continuity of inductive processes from the first steps of determination to the full development of the organism; (2) the specificity of induction, i.e. the changing relationship between truly "directive" interactions and "permissive" conditions guiding and maintaining differentiation; (3) the pathological implications of induction, i.e. failure of some interaction may result in abnormal cytodifferentiation, growth or morphogenesis (carcinogenesis, teratogenesis); and (4) the mechanism of inductive tissue interactions, i.e. the nature of the signal substances, their modes of transmission, and ultimately, the molecular basis of their action.

2. *Biology of inductive tissue interactions*

2.1. *Sequentiality of inductive tissue interactions*

2.1.1. *Introduction*

It has become evident that "embryonic induction" is not a single event programming the whole course of embryogenesis and its various steps of cytodifferentiation and morphogenesis. From the early morula stages onwards, the cells are exposed to a constantly changing micro-environment provided by the adjacent cells and tissues. Probably every developmental decision made by the cells until they ultimately attain the stable phenotype is guided by environmental stimuli, and in the adult organism such influences still control cell renewal. Each decision is followed by a new option. After loss of responsiveness (competence) to the previous inductive actions, the cells acquire a new "alertness" towards the next determinative and morphogenetic stimuli to which they are exposed.

Under certain conditions an early amphibian embryo (up to the young gastrula stage) can be divided into two halves, each giving rise to a complete embryo (p. 333). Thus the cells are still totipotent and capable of responding to all determinative stimuli. The first commitment known to take place in these embryos is determination into three germ layers, and this is guided by interactive processes between the presumptive ectodermal and endodermal cells leading to the formation of the mesodermal layer (Nieuwkoop, 1969, 1973). This stage is followed by inductive interactions between the three germ layers determining early organ anlagen. Subsequently a series of "secondary" inductions occur between the various tissue components within each organ.

In what follows, some serial inductive interactions have been chosen for somewhat more detailed description to illustrate this chain-like process controlling progressive differentiation.

2.1.2. Determination of the blastocyst

2.1.2.1. Determination of trophectodermal and inner cells. In experiments like those with amphibia cited above, a complete mammalian embryo is formed from a single blastomere. Rabbit and mouse embryos at the two-cell stage can be divided in two, and the resultant cells give rise to a complete embryo (Tarkowski, 1959; Seidel, 1960). In fact, any blastomere of the eight-cell rabbit egg can form a complete adult (Moore et al., 1968). When a mouse blastocyst is about to implant in the uterus it consists of approximately 64 cells, divided into two distinct types, inner cells and trophectodermal cells (see Graham, 1971). About thirteen of the cells are enclosed without contact with the exterior. They are surrounded by flat trophectodermal cells which form the blastocoel and face the external environment (Fig. 1). The two cell types become recognizable around the 16–32 cell stage, and are then no longer interconvertible. The small, round inner cells in their eccentric location in the blastocyst divide rapidly, with little sign of overt differentiation. If isolated, they are unable to implant in the uterus or evoke a decidual reaction. Grown alone in vitro, they are unable to pump fluid, and no blastocoel is formed. The trophoblast cells are tightly bound together, but if separated, the fragments do not reunite. After transfer to the uterus, they implant readily, and grown in vitro they pump fluid and form a blastocoel (see Gardner, 1972; Gardner and Papaioannou, 1975). Even small, isolated fragments form "miniblastocysts" (Sherman, 1975; see also Rossant, 1975a,b).

In the mammalian embryo both extrinsic and intrinsic factors have been suggested to cause early determination (see Graham, 1971; Hillman et al., 1972). The factors most often considered are the position of sperm penetration and the

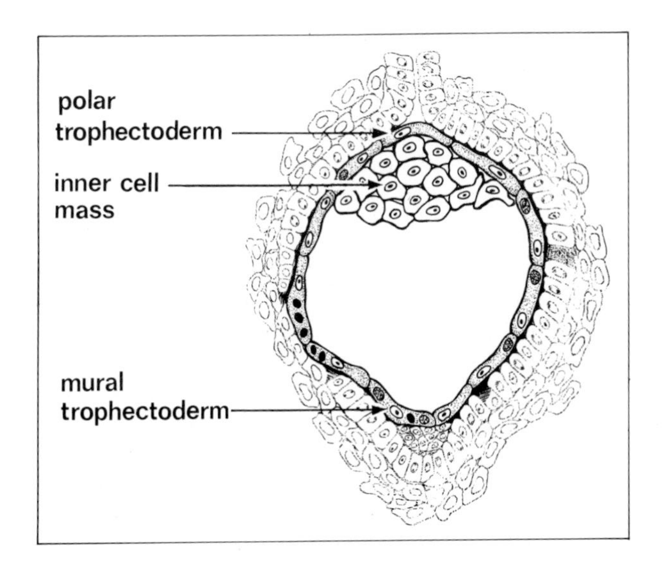

Fig. 1. Schematic drawing of a trophoblast.

micro-environment of the egg in the female reproductive tract. However, unfertilized eggs can be activated and will then form normal blastocysts with the two cell types both in vivo (Tarkowski et al., 1970) and in vitro (Graham, 1970). Thus the site of the penetration of the sperm is not essential for determination. Nor is any essential factor provided by the microenvironment of the egg, the female oviduct, as fertilized eggs form normal blastocysts in vitro. If then transferred to the uterus of a foster-mother, they develop into viable offspring (Whitten and Biggers, 1968). The factors responsible for early determination must thus be sought within the embryo itself.

Early differentiation in the cleaving egg could be determined by intrinsic mechanisms of three main types: (1) segregation of morphogenetic factors; (2) a reference point inside the embryo; and (3) the position of the cells. In the classic experiment with the amphibian egg mentioned above, the embryo could be divided into two blastomeres, each giving rise to a complete embryo (Endres, 1895). The experiment was only successful if the plane of bisection passed through the grey crescent region in the egg cytoplasm, dividing it into identical halves. The blastula and the gastrula can be similarly divided if the ligature is placed in the midline of a zygote (Spemann, 1901a, 1903). This crescent or cortical zone is a specialized part of the cytoplasm with morphogenetically important determinants (Curtis, 1965). It has been suggested, therefore, that future differentiation is already wholly determined in the cytoplasm of the fertilized egg and regulated through the uneven inheritance of cytoplasmic factors by the daughter cells (Dalcq, 1957). No such cytoplasmic factors have been identified in the mammalian embryo. It is not yet clear whether during cleavage the cells of an intact mammalian embryo always divide in certain planes. Thus the possibility of cytoplasmic determinants cannot be excluded in vivo. In vitro, however, the cells can be manipulated and their course of development can be regulated.

The second hypothesis resembles the first in implying a certain heterogeneity already present in the fertilized egg. It suggests that within the embryo there is a reference point, a source of gradients, and that future differentiation results from the different relations of the daughter cells to this point (see Wolpert, 1969 and chapter in this volume by McMahon and West). If such a reference point exists, the cells behave as if it were located in the centre of the embryo. Experimental evidence of the existence of such a reference point is still lacking.

The third hypothesis suggests that the determination of the cell is a consequence of its position within the embryo at the morula stage. Therefore the cells on the inside develop differently from the cells on the outside. The basic assumption was first made by Tarkowski and Wroblewska (1967), and had been implied before in the work of Mintz (1965). Graham and his associates (Graham, 1971; Hillman et al., 1972) obtained experimental evidence for this hypothesis by labelling embryos at the 4- or 8-cell stage with tritiated thymidine, and combining a labelled embryo with several unlabelled embryos to form a giant blastocyst. These experiments carried our understanding of development two steps further. Firstly, if any segregation of morphogenetic cytoplasmic factors does occur, it is reversible. Every cell from a 4-cell egg was able to form either an inner cell or a trophoblast cell. Moreover, the cells from 8-cell eggs were not distributed between the inner cell mass and the trophectoderm in the predestined proportions when combined with other embryos. Secondly, determination of the cells is entirely dependent on their position in relation to the other cells. If a labelled 4-cell embryo was completely enveloped with unlabelled embryos within 8 hours the cells of the labelled embryo had become irreversibly determined into inner cells.

This experiment showed that the inside-outside hypothesis is a sufficient explanation for early determination. However, as Graham (1971) has pointed out, early determination is not necessarily regulated in the same way in all mammals. For instance, the blastocyst of the Australian marsupial cat (*Dasyurus viverrinus*) is unilaminar and no group of cells is enclosed (see Graham op. cit.). The same may be true of several other yolky eggs. The physical arrangement of the cell in the marsupial cat does not exclude the possibility that cellular interaction is of importance in the early determination of the blastocyst.

2.1.2.2. Interdependence of trophectodermal and inner cells. In the mouse embryo, as mentioned above, the developmental heterogeneity between the two cell types, the inner cells and those of the trophectoderm is evident by the fourth day of development, and this diversity of the cell phenotype is believed to be brought about by a differential interrelationship of the cells in the morula (Graham, 1971; Hillman et al., 1972). A single inner cell from a 3 1/2-day embryo injected into the blastocoel of another embryo of the same age is always found in the inner cell mass of the host and injected trophectodermal cells form only derivatives of the trophectoderm (Gardner, 1971, 1974b). This proves that the determination of the two cell types is irreversible at this stage. From this stage on, each cell type follows its own developmental course. The trophectodermal cells, which constitute three-quarters of the blastocyst, establish intimate contact with the uterus. Mural trophectoderm adjacent to the blastocoel ceases to divide after the fifth day, but continues to replicate DNA and forms giant cells (Dickson, 1966; Sherman et al., 1972). In contrast, polar trophectoderm lying over the inner cell mass continues to divide and forms a conical mass of cells a few cell layers thick, the ectoplacental cone. At the periphery of the ectoplacental cone there are giant cells indistinguishable from the mural trophectodermal cells. The inner cells and the trophectodermal cells do not develop independently. On the contrary, proliferation of the cells in the ectoplacental cone is entirely dependent on the presence of the inner cells (Gardner et al., 1973; Gardner, 1974a; Ansell and Snow, 1975; Sherman, 1975). The inner cells can be removed by microsurgery (Gardner et al., 1973), or destroyed by treatment with tritiated thymidine (Ansell, 1975) or other chemicals lethal to rapidly dividing cells, such as BUdR, colcemid or cytosine arabinoside (Sherman, 1975). A trophoblast devoid of inner cells implants in the uterus and forms giant cells with onset of progesterone synthesis, but cell proliferation ceases (Sherman, 1975). Thus giant cell formation does not depend on the presence of the inner cells or on hatching (i.e. shedding off the zona pellucida). In the absence of inner cells, the trophoblast vesicle develops for only about 10 days, because new cells are not formed. These trophoblast vesicles never show any sign of inner cell formation. After reconstitution with inner cells injected into the trophoblast vesicle, the blastocyst develops normally (Gardner et al., 1973; Gardner, 1974b). The giant cells adjacent to the injected inner cells begin to divide again, and an ectoplacental cone is formed. The mitotic zone is only a few cell layers thick, and giant cell formation occurs at its periphery (Gardner and Papaioannou, 1975). In rat-mouse chimaeras, rat inner cells sustained the proliferation of mouse trophoblast cells equally well as mouse inner cells. Therefore Gardner and Papaioannou (1975) concluded that this stimulatory or inductive effect was not species-specific.

A fuller discussion of the biology of the trophoblast is given in chapter 3 of this volume by Sherman and Wudl.

2.1.3. Sequential induction of the central nervous system

2.1.3.1. Neural versus mesodermal induction. In the amphibian neuralization of the ectoderm and regionalization of the CNS are determined between the early gastrula and late neurula stages. A piece of presumptive neuroepithelium transplanted to the belly of another gastrula before invagination of the dorsal lip mesoderm develops according to its new surroundings and no extra neural structure is formed. This means that it is still undetermined. If the same operation is performed later at an early neurula stage, the transplant develops an extra neural plate, which shows that it was already determined when transplanted (Spemann, 1918). At a late neurula stage the regionality of the CNS has also been stabilized, as explants of neural plate cultured in vitro develop according to their original site (see Saxén and Toivonen, 1962; Toivonen, 1972).

The interactive events operating during this relatively short period of development are fairly well mapped out and provide a good example of sequential inductive processes. It is the invaginating dorsal lip mesoderm that provides the messages which guide the determination and differentiation of the neural plate. This relationship has repeatedly been demonstrated in implantation and transplantation experiments where the two tissues have been brought into contact. Regional differences exist in the inductive action of the mesoderm, and the regional characteristics of the ectoderm are built-in features of the inductor tissue (Fig. 2) (Ter Horst, 1948, Sala, 1955). Many tissues, also heterotypic ones, are inductively active. Their actions differ and mimic those of different regions of the normal inductors. Such artificial inductors can produce forebrain structures, caudal CNS derivatives or mesodermal structures only (Toivonen, 1940, 1954; Tiedemann and Tiedemann, 1959). Employment of such inductor tissues and their combinations resulted in the following working hypothesis (Toivonen and Saxén, 1955). Induction of the presumptive neural plate is brought about by two active principles, "neuralizing" and "mesodermalizing". A purely neuralizing action leads to the formation of neural structures of the most cranial part of the CNS (classic "archencephalic induction"), whereas the

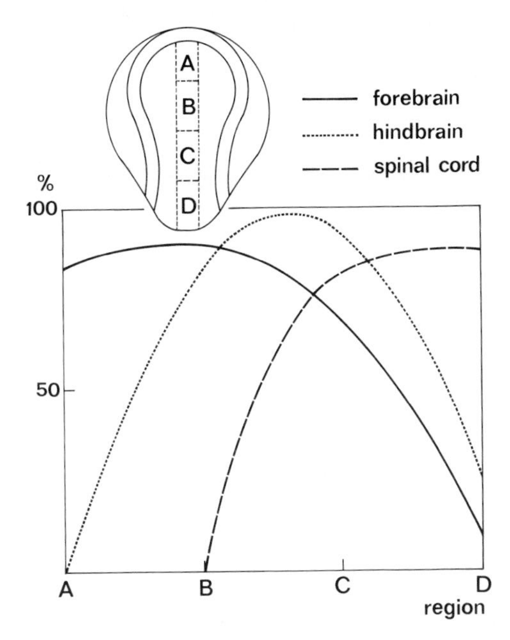

Fig. 2. Percentage induction of different regions of the CNS by various territories of the archenteron roof of an amphibian neurula (After Sala, 1955).

mesodermalizing action converts ectodermal cells into mesodermal structures, such as somites and notochord. Both principles in combination cause the regional characteristics of the CNS and the "induction" of caudal CNS derivatives (hindbrain, spinal cord). The hypothesis further suggests that gradients are formed, yielding different concentration ratios of the hypothetical inductor substances at different levels of the inductor.

The first step in testing this hypothesis was to mix artificial neuralizing and mesodermalizing inductors in different ratios and test the action of these mixtures on competent gastrula ectoderm. Pure neuralizing inductor produced forebrain vesicles. The higher the proportion of mesodermalizing inductor cells, the more caudal were the neural structures formed and the less frequent the forebrain derivatives (Fig. 3) (Saxén and Toivonen, 1961). Thus, various mixtures of the two artificial inductors closely mimicked the inductive actions of the different regions of the natural inductor.

2.1.3.2. Segregation of the central nervous system. To examine whether the two types of morphogenetic signals act simultaneously or sequentially on the target tissue, we designed the following experiment (Fig. 4) (Saxén et al., 1964). Isolated gastrula ectoderm was first exposed for 24 hours either to an artificial, purely neuralizing inductor or to a mainly mesodermalizing one (pure mesodermalizing inductors were not available). The inductor was then removed and the neuralized and mesodermalized ectoderms were cultured either separately or

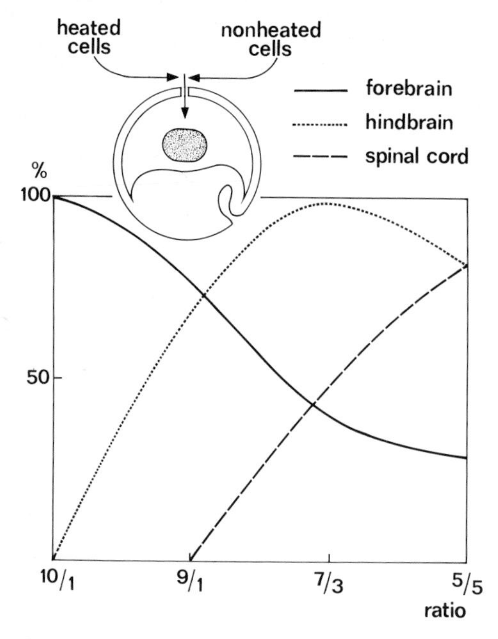

Fig. 3. Inductive action of different mixtures of heat-treated, neuralizing, and untreated, mesodermalizing HeLa cells on amphibian gastrula ectoderm. The ratio expresses neuralizing/mesodermalizing cells implanted into the blastocoel (After Saxén and Toivonen, 1961).

together. As expected, when cultured separately, the neuralized tissue gave rise exclusively to cranial neural derivatives (forebrain) and the mesodermalized cells to derivatives of axial mesoderm (notochord, muscle, etc.) with fragments of spinal cord. Very few of either type of explant contained hindbrain. When neuralized and mesodermalized cells were cultured together, nearly 75% of the explants gave rise to hindbrain. Such mixtures of ectodermal cells thus gave the same result as a mixture of neuralizing and mesodermalizing inductors acting on competent ectoderm. This indicates that induction of the central nervous system is a two-step process. First ectodermal cells are determined towards either neural or mesodermal cells, and then these two cell types interact and determine the regional characteristics of the CNS. This conclusion based on experiments with artificial inductors was subsequently tested in similar experiments with normally determined tissue components from the early neurula stages. Recombination of disaggregated cells from the anterior neural plate region (prospective forebrain) and from the caudal axial mesoderm resulted in the transformation of presumptive forebrain cells to caudal CNS derivatives (Toivonen and Saxén, 1966). Finally, the quantitative aspect of the original hypothesis was tested by mixing cells from the anterior neural plate and axial mesoderm in different ratios. A whole range of CNS structures developed in the aggregates and their regional characteristics were determined by the ratio of neural to mesodermal cells (Fig. 5) (Toivonen and Saxén, 1968). This lent further support for the existence of a gradient of inductors during normal development.

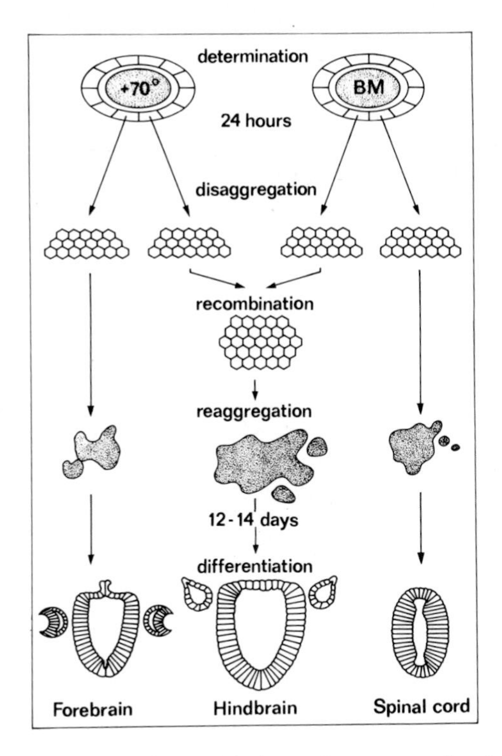

Fig. 4. Experimental design to demonstrate that determination of neural ectoderm and regionalization of the CNS are sequential events. Neural ectoderm was neuralized with heat-inactivated HeLa cells (+70°) or mesodermalized with guinea-pig bone-marrow (BM) (After Saxén et al., 1964).

In conclusion, a sequential interactive system regulates primary induction: in the initial phase, the prospective neural plate area becomes uniformly neuralized by an inductive stimulus from the invaginating dorsal lip mesoderm. During the second step, the mesodermal mantle induces the regional characteristics of the CNS.

2.1.4. Sequentiality in the induction of the lens

Spemann (1901b) based his classic concept of "lens induction" on the tissue interaction between the ectoderm and the optic bud derived from the neural tube. This interaction normally determines the size and position of the lens. It fairly soon became evident, however, that a lens was occasionally formed even in the absence of the optic bud (King, 1905; Spemann, 1912; Toivonen, 1940, 1945; Jacobson, 1958; Mizuno, 1970). It is now realized that the "lens-forming capacity" varies in extent in different amphibian species. In some it is present only in the presumptive lens ectoderm, but in others the whole head ectoderm is capable of forming a lens in the absence of the optic bud (see Twitty, 1955). These species differences have created much confusion and even the temperature during the experimental procedures affects the results.

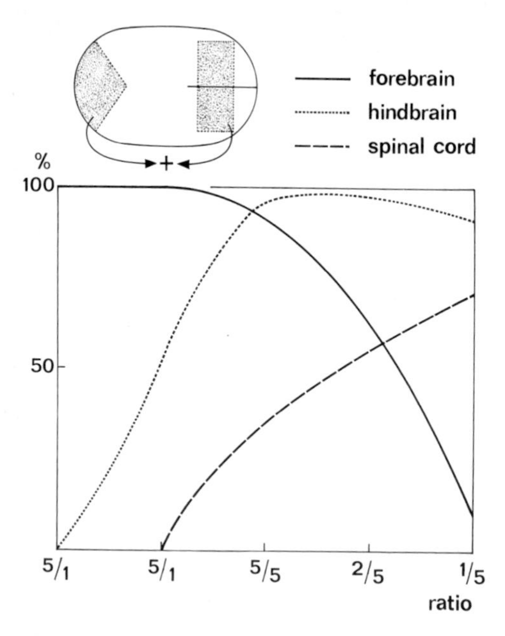

Fig. 5. Percentage differentiation of various CNS structures from reaggregates of anterior neural plate cells and cells from the caudal axial mesoderm mixed in different ratios (After Toivonen and Saxén, 1968).

The phenomenon suggests that the interaction of the optic bud and the presumptive lens ectoderm is only one link in a long chain of developmental events leading ultimately to the formation of a lens in the right proportions and position. The "predetermination" of the lens is initiated very early in embryonic development.

According to Jacobson (1958), presumptive lens ectoderm is first exposed to the inductive influence of pharyngeal endoderm during the early gastrula stages, and then to the inductive influence of heart mesoderm during the later gastrula stages as the mesodermal mantle advances towards the head (Fig. 6). Morphogenetic movements remove the presumptive lens ectoderm from the first two inductors, but if their influence is artificially prolonged, a lens is formed even in the absence of the optic bud. He also suggested that inductive influence of pharyngeal endoderm, heart mesoderm and optic bud are qualitatively similar, and additive, if acting simultaneously. The order of the inductors can be changed, and any one of them can be omitted provided that the influence of the other two is prolonged. The inductors and the responding presumptive lens ectoderm can also be taken from embryos of different ages. Moreover, the outcome is not altered even when stages of normal inductive interaction are skipped or repeated, provided that the presumptive lens ectoderm continuously is exposed to an inductor. The end result depends on the total time of exposure of the presumptive lens ectoderm to the inductive influences. This time is temperature-dependent, being longer at low temperatures. Jacobson (1966)

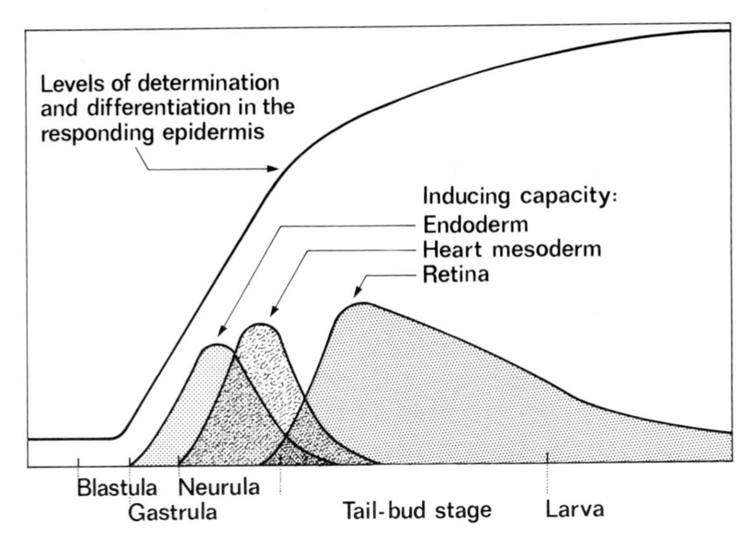

Fig. 6. Hypothesis on the effect on lens induction by three tissues to which presumptive lens ectoderm is exposed during normal development. Abscissa: developmental stage of the embryo. Ordinate: inductive capacity or level of determination (After Jacobson, 1966).

suggests that the "inductive capacity" of each inductor is distributed in the tissue in a gradient-like fashion. Lenses form in the area where the overlapping influences of different inductive signals are maximal. Even in the absence of an optic bud, lenses will form in their normal position, but they are somewhat smaller. Moreover, lenses degenerate if the inductors are removed, suggesting that the inductor has a sustaining effect, possibly continuing throughout life. Considerable inductive influence is also exerted on the ectoderm nearby, which is thus predisposed towards lens formation. Any slight added inductive stimulus will lead to lens formation in "ectopic" sites. Jacobson's experiments were made with the West Coast newt, but he suggests that in principle development is much the same in all vertebrates. The degree of predetermination of the ectoderm seems to be species-specific, varying in extent from a small area to the whole head or even to the whole body ectoderm (Harrison, 1920; Liedke, 1951; Jacobson, 1958; Mizuno, 1972).

2.1.5. Sequential epithelio-mesenchymal interactions

Apart from the integument, the best analyzed interactive processes are those occurring between the epithelial component and the mesenchymal stroma in various parenchymal organs, such as kidney, liver, pancreas, lung, thyroid (see Fleischmajer and Billingham, 1968). In most of these, progressive differentiation and stabilization of the components are guided by a series of inductive interactions.

2.1.5.1. Pancreas. A series of experiments by Rutter et al. (1964) and Wessells and Cohen (1967) on the isolation and separation of tissues has clearly shown the progressive developmental capacities in the pancreatic endodermal anlage and their dependence on exogenous stimuli. If the entire trunk region of a mouse embryo younger than the 7-somite stage is explanted in vitro, pancreatic acini are not formed. But explants between the 7- and 9-somite stages, consisting of the endodermal component and its mesenchymal stroma, continue to differentiate in vitro, if the two components remain in close association. Isolated from its stroma or combined with some heterologous mesenchyme the endoderm fails to form acini. At a slightly later stage, the dorsal endoderm achieves some degree of independence. Subsequent differentiation can then be supported by heterologous mesenchymes from other parenchymal organs. Around the 30-somite stage, the isolated endoderm becomes able to express its developmental capacities in the absence of any mesenchyme, provided that the medium is supplemented with certain cell-free supporting factors (p. 351).

2.1.5.2. Liver. Presumptive liver endoderm of the foregut region becomes programmed for hepatocyte differentiation as early as the 5-somite chick embryo (Le Douarin, 1964). This commitment depends on an intimate association with the hepatocardiac mesenchyme. The competence of the foregut cells to respond to the determinative mesenchymal stimulus is also acquired at about the 5-somite stage. Endoderm from earlier stages does not respond to the inductively active mesenchyme (Le Douarin, 1964, 1975).

The determined hepatocytes continue to be dependent on the adjacent liver mesenchyme. The sustaining influence provided by this tissue results in normal morphogenesis of the hepatic cords. When determined cells from embryos at the 15- to 26-somite stage were combined with various heterologous mesenchymes, the responses obtained were of two types. When combined with any derivative of the lateral plate mesenchyme (from which the liver mesenchyme is also derived), the endodermal hepatocytes underwent normal cytodifferentiation and morphogenesis, but when combined with cephalic or trunk mesenchyme they formed undifferentiated epithelial vesicles (Le Douarin, 1975).

Thus, liver development is another example which illustrates how organogenesis is guided by sequential interactive processes: in the first stage the presumptive hepatocytes are determined by the hepatocardiac mesenchyme, and in the lengthy second stage their morphogenesis, proliferation and functional activities are supported by continuous interaction with the liver mesenchyme.

2.1.5.3. Lung. The lung buds first become visible in a 25-somite mouse embryo, but their formation has been determined at a much earlier stage (Spooner and Wessells, 1970). When the presumptive pulmonary endoderm of embryos at the 2-somite stage is cultured with the surrounding mesenchyme, the explant forms buds. The result is the same when the endoderm is combined with certain heterotypic mesenchymes. In both cases, however, lung morphogenesis is arrested at this stage, and does not proceed to branching of the bronchial epithelium. Branching does occur if whole lung rudiments from older, 25-somite stage embryos are explanted or if the early presumptive endoderm is combined with bronchial mesenchyme from older embryos.

This two-step process, consisting of the "unspecific" initiation of bud formation and the strictly "specific" induction of branching in the bronchial epithelium, is in good accord with observations on normal lung development, as recently emphasized by Kratochwil (1972). In a study on the movements of the two components of the presumptive lung, endoderm and pulmonary mesenchyme, Rosenquist (1970) found that contact between them was not established until the 19-somite stage. As the initial bud-forming capacity could be demonstrated long before this stage (see above), it must have been induced by interaction with a non-lung mesenchyme. But the failure of these early explants to develop further shows that branching is dependent on contact with homotypic pulmonary mesenchyme.

2.1.5.4. Kidney. The metanephric kidney develops from two components, the branching ureter epithelium derived from the Wolffian duct and the mesenchymal metanephric blastema. Nothing is yet known about the early determinative events, but by the time of kidney morphogenesis the metanephric mesenchyme is already biased towards tubule formation, as no other embryonic mesenchyme can be induced to form tubules (Saxén, 1970a). The formation of secretory tubules in the blastema and the branching of the ureter bud are controlled by reciprocal interactions, and if they are separated development is arrested (Grobstein, 1953b; 1955a; Saxén et al., 1968; Saxén, 1971). Various embryonic tissues possess a capacity to induce tubules (Lombard and Grobstein, 1969; Unsworth and Grobstein, 1970), but attempts to mimic it with cell-free extracts or improved culture media have failed.

According to Gossens and Unsworth (1972), the triggering of secretory tubule formation is not sufficient to ensure continuous tubulogenesis: only small pretubular condensates developed if the contact between the metanephric mesenchyme and its inductor was broken after 30 hours of cultivation in a transfilter set-up, and on subcultivation of the single mesenchymes the condensates regressed within 6 days. Addition of uninduced kidney mesenchyme or heterologous mesenchyme stabilized tubule formation and led to elongation of the tubules. Interestingly enough, this second step could also be achieved with chick embryo extract added to the cultures. Gossens and Unsworth (1972) point out the apparent discrepancy between their results and the unpublished observations of Grobstein (see Grobstein, 1967) that 30 hours of contact is enough for subsequent tubule formation. They suggest that the discrepancy may be explained by the greater mesenchymal tissue mass and the higher embryo extract concentration used by the earlier workers. Using about 6 mesenchymes per explant but no embryo extract we have found that the minimum time of transfilter contact required for subsequent tubule formation is between 12 and 30 hours depending on the type of filter (Nordling et al., 1971; Wartiovaara et al., 1974). Gossens and Unsworth (1972) conclude that kidney tubule formation involves two steps: in the first step tubule formation is triggered and pretubular condensates formed, and in the second step development is stabilized. However, the stabilization may not be a genuine inductive event, but may only reflect nutritional or other unspecific effects in tissue cultures.

2.1.5.5. Thyroid. A stabilizing epithelio-mesenchymal interaction following an initial induction has also been suggested to occur between the epithelial secretory cells and the capsular mesenchyme in the thyroid gland (Hilfer and Hilfer, 1964; Hilfer, 1968). Isolated epithelial cells from 8-day chick embryos spread as a monolayer reaggregate and differentiate when combined with similarly treated capsular cells. Moreover, thyroid epithelial cells from a 16-day chick embryo with ultrastructural features characteristic of differentiation, still require the presence of the homologous mesenchymal cells to develop and maintain their follicular pattern. Fibroblasts from other embryonic organs do not have such a stabilizing influence. The author(s) conclude that an initial stage of inductive interaction between the epithelial and capsular cells is followed by a second-step interactive event responsible for stabilizing and maintaining the follicular pattern and epithelial specialization (Hilfer, 1968).

2.1.6. Homoiogenetic inductions

In certain experimental situations at least, it is possible to demonstrate a sequential interaction of a special type termed homoiogenetic. The term refers to a chain-like induction process in which one triggered cell can pass on the same message to other, uninduced cells. Deuchar (1971a) combined neuralized radioactively labelled cells from an amphibian gastrula with uninduced, unlabelled competent cells of the presumptive neuro-ectoderm. The neural structures which subsequently developed consisted of both labelled and unlabelled cells, which suggested that the neuralizing trigger was passed on from the neuralized cells to the originally uninduced cells. Using a biological nuclear marker (Le Douarin, 1974), Rasilo and Leikola (1976) found that when a piece of determined quail blastoderm was brought into contact with an undetermined chick blastoderm, the cells of the neural structures were derived from both quail and chick (Fig. 7).

Such a homoiogenetic induction has also been suggested to operate in chondrogenesis (Cooper, 1965), but it does not seem to form part of every inductive reaction. When chromosomal markers were used in the model system for induction of kidney tubules, no such second-step passage of the inductive message from cell to cell could be shown (Saxén and Saksela, 1971). In this particular system, the induction wave has been suggested to spread by cell migration (Saxén and Karkinen-Jääskeläinen, 1975).

2.1.7. Remarks

The examples presented above illustrate the progressive nature of cytodifferentiation and morphogenesis, and indicate that embryonic development is under the constant guidance of the microenvironment provided by the developing cells and tissues. Each developmental decision taken by the cells is apparently guided

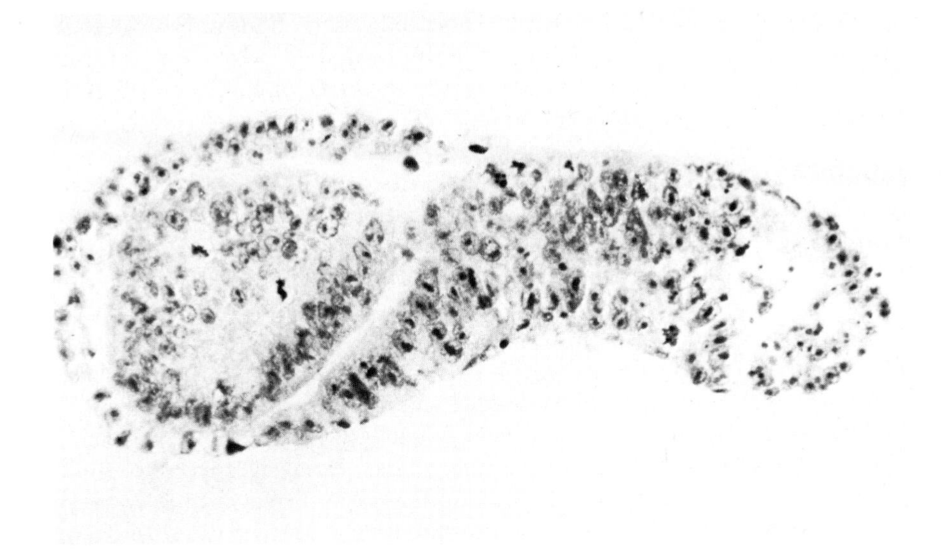

Fig. 7. Chick neural plate with neuralized quail cells. Note the large nuclear bodies of the quail cells (From Rasilo and Leikola, 1976).

by such exogenous influences. Once led into one pathway of development, the cells come under new environmental influences which either guide development one step further or support the differentiation already achieved. Most experiments are performed on short-term, isolated steps of this continuous process. The researcher is thus without sufficient knowledge of the preceding or subsequent events, the predetermined conditions of the interactants or the competence of the target cells which he has somewhat arbitrarily singled out for experimental analysis. It has to be remembered that each event studied is only a link in the chain of interactive processes, and an effort should always be made to find the right places for these pieces of information in the jig-saw puzzle of embryogenesis. Great care is also needed in the search for common denominators to various interactive situations, for their morphogenetic significance seems to vary from true determinative influences to mere supporting effects, as will be discussed in the next section.

2.2. Specificity of inductive tissue interactions

2.2.1. Introduction

The results of experiments with heterotypic tissue combinations in vitro (see below) have raised the question of the specificity of the inductive interactions. During normal development in vivo there is no possibility of nonspecific situations, because each interactive event must be strictly controlled both spatially and temporally if it is to lead to synchronous development of the interactants.

The question of specificity was raised by repeated observations indicating that the same morphogenetic result often could be obtained when one of the normally interacting tissues was replaced by a heterotypic one. As described in the previous section, tubule formation by metanephric mesenchyme can be triggered by a variety of tissues ranging from heterotypic mesenchymes to certain neoplastic tissues (Grobstein, 1955a; Unsworth and Grobstein, 1970; Auerbach, 1972). Branching and chemodifferentiation of pancreatic epithelium can be promoted by almost any heterotypic mesenchyme, provided that the epithelium has already reached a certain stage of determination (Rutter et al., 1967; Dieterlen-Lièvre and Hadorn, 1972). Thus, the morphogenetic signal required for normal development reflects merely a "permissive" condition which enables the target cells to express a pre-existing differentiative bias. The metanephric blastema is the only mesenchyme which responds to tubule inducers (Saxén, 1970a), and it has not been possible to induce pancreatic acini in heterotypic epithelia. Thus, cells must have received directive messages before coming in contact with the final inductor, and cells must have been instructed to choose one of the many differentiative pathways open to them during the early stages of embryogenesis.

Here a working hypothesis is presented about the relative importance of directive and permissive interactions during embryogenesis. This hypothesis will then be examined in the light of experimental data derived from model systems of inductive interactions. Assuming that differentiating cells (and cell populations) have a number of *developmental options*, it can be postulated that this number decreases progressively during development. With the gradual restriction of these options, the signals (environmental influences) guiding morphogenesis change from predominantly "directive" to predominantly "permissive" influences supporting the expression of the developmental options already taken by the cell (Fig. 8).

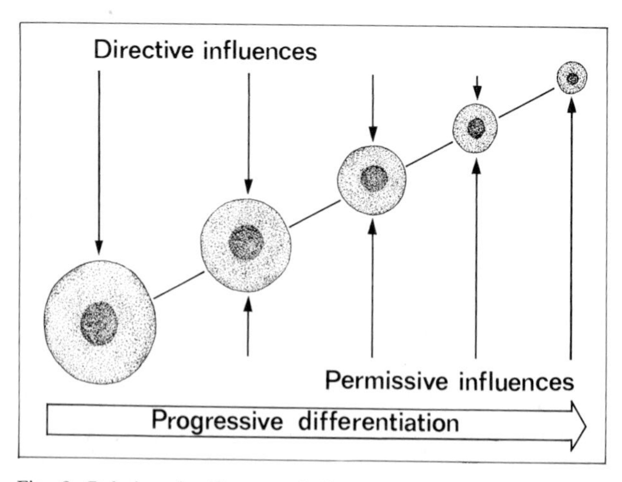

Fig. 8. Relative significance of directive and permissive interactive influences as differentiation progresses; a hypothesis. (See text for discussion.)

The experiments devised to test this simplified hypothesis are of two types: (1) experiments illustrating the extent to which a target cell population can be induced to differentiate by combining it with heterotypic tissues; and (2) experiments on the specific environmental requirements of a cell population separated from its normal counterpart. The first assumption in the hypothesis of Fig. 8, that there are a great number of options open to the cell during the early stages of development, has been verified repeatedly (pp. 332–335).

A later step of determination involving many options is the induction of the ectoderm during gastrulation and neurulation (reviews, Spemann, 1936; Holtfreter and Hamburger, 1955; Saxén and Toivonen, 1962; Deuchar, 1971b; Tiedemann, 1971; Nieuwkoop, 1973). A number of experiments using various inductors – tissues, fractions of such tissues or even pure chemicals – have shown that the cell population normally destined to become ectoderm can be converted into cells and organs of mesodermal, endodermal or mesectodermal type. Such results have generally been interpreted as demonstrating that the ectodermal cells are multipotent. The only argument raised against this view is that the target cell population (ectoderm) may already have undergone determination and consists of several categories, the various treatments selectively favouring the proliferation of one or more of these subpopulations while suppressing or killing the others (Ave et al., 1968). This notion was based on experiments in which the cells of the "competent" ectoderm could be separated electrophoretically into several subpopulations. With this technique, however, it was impossible to follow the subsequent development of these separated subpopulations, and some of our own results seem incompatible with this idea. For instance, the results illustrated in Fig. 5 indicate that a cell population already destined to become *neural* can be converted almost quantitatively into either forebrain, hindbrain or spinal cord structures without bringing in any new neural elements.

2.2.2. Directive influences in the development of the skin and its appendages

An example of cells with a long-persisting pluripotency and their requirement of directive morphogenetic influences is the integument, which can be converted experimentally into a variety of epidermal derivatives. Transplantation experiments in reptiles, birds and mammals have demonstrated two characteristics of the development of the epidermis and its appendages: (1) embryonic epidermis and the basal layer of the integument in more advanced stages possess a high degree of plasticity, and these cells retain several developmental options that can be unmasked experimentally; and (2) the dermal mesenchyme exerts a directive action on the epidermal target cells throughout development.

The two characteristics have been demonstrated in experiments within the same species where dermis from one region has been combined with epidermis from another region (homospecific/heterotypic) and in experiments where dermis from one species has been combined with epidermis from another species (heterospecific). One of the first experiments was performed by Sengel

(1958), who combined chick embryonic epidermis from the feather-bearing dorsal skin with mesenchyme from the tarsometatarsal region where the normal appendages are scales. The epidermis failed to produce feathers, but developed scales instead, thereby demonstrating the decisive influence of the heterotypic mesenchyme. This observation has subsequently been confirmed in several studies which have also indicated that the scale-inducing property of the tarsometatarsal mesenchyme develops rather late during development, after day 12 (Rawles, 1963).

Further homospecific/heterotypic experiments in birds and to a lesser degree in mammals have yielded convincing evidence of the directive role of the dermis in epidermal morphogenesis and in the morphogenesis and spatial distribution of feathers and hairs (Sengel et al., 1969; Dhouailly, 1970; Kollar, 1966, 1970; see also Kratochwil, 1972). Heterospecific recombination experiments have led Sengel's group to suggest that the dermal influence is sequential. First there is a message that initiates the formation of appendages, and then an influence that controls their morphogenesis. In the first step dermal influences trigger the formation of the early rudiments typical of each species. In the second step a more specific dermal influence determines the size and distribution of the appendages (Dhouailly and Sengel, 1972, 1973; Dhouailly, 1973, 1975; Sengel, 1975). An intra-class combination of epidermis and dermis from chick and duck results in normal development. The epidermis is able to respond to all the morphogenetic signals of the heterospecific dermis and expresses its own phenotypic characteristics under the direction of the latter. Also inter-class combination of epidermis and dermis from chick and mouse led to the onset of development of cutaneous appendages, which, however, remained immature. Obviously the epidermis was incapable of interpreting the heterospecific morphogenetic messages determining the size and distribution of the appendages. These observations led the authors to postulate that the class specificity is epidermis-dependent, whereas regional specificity is dermis-dependent. To test this hypothesis, they made extensive inter-class combinations of dermis and epidermis from lizard, chick and mouse.

In homospecific/heterotypic recombinations in the lizard, the type and distribution of the scales was determined by the dermal mesenchyme, as in previous experiments in the chick on the development of feathers (Dhouailly, 1975) (Fig. 9). In inter-class recombinations, chick tarsometatarsal dermis converted lizard epidermis into scales that were typically avian in distribution and shape, but had an abnormal, reptilian type of keratinization. All other heterospecific combinations produced in lizard epidermis immature scale buds that did not complete morphogenesis. The explants formed no feathers or hairs typical of the region or species of the dermal donor. In distribution and size, however, the immature scale buds corresponded roughly to the hair and feather buds of the species and region of the dermal mesenchyme. On the other hand, when lizard dermis was combined with heterospecific epidermis no morphogenesis occurred. Even the scale-forming epidermis of the tarso-metatarsal region of the chick failed to respond to the lizard dermis. Dhouailly concludes that the chick epidermis does

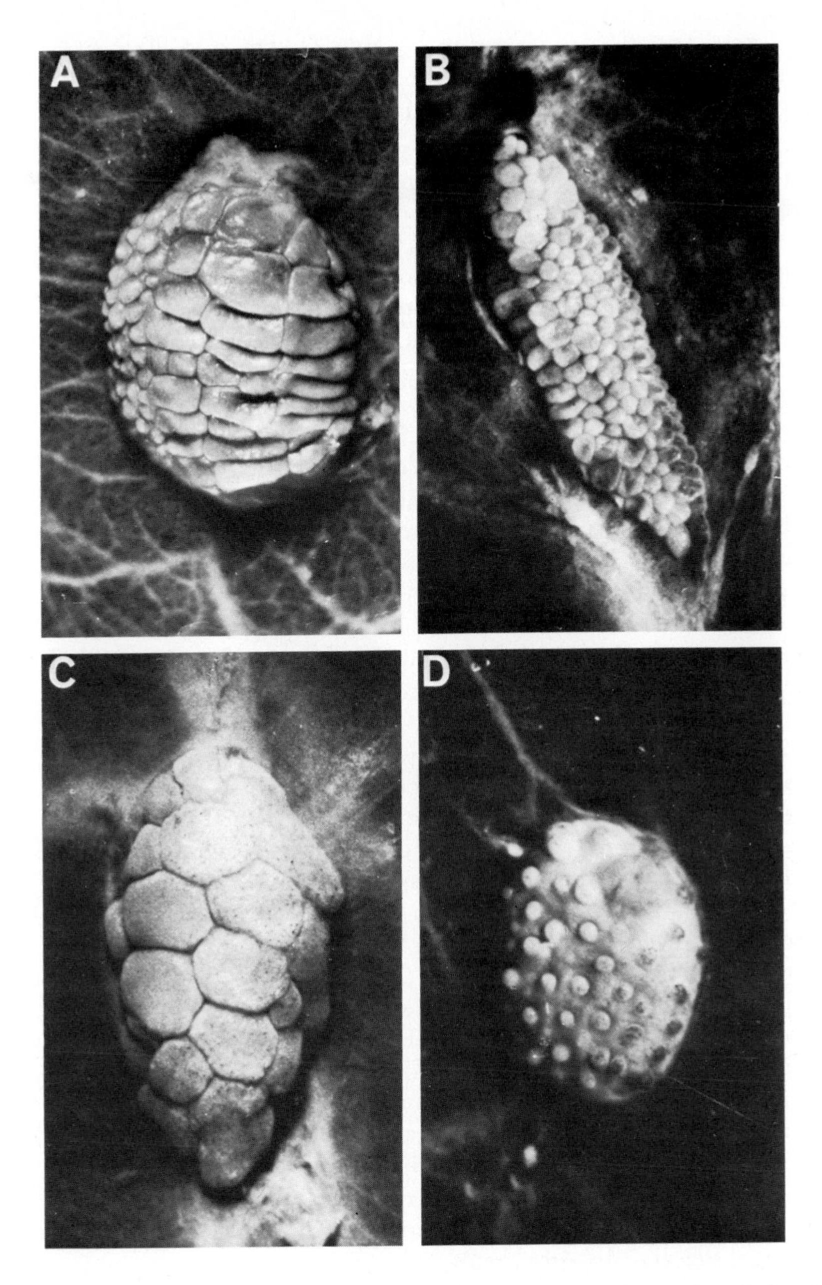

Fig. 9. Microphotographs of dermo-epidermal recombinations cultured for 8 days on chick chorioallantoic membrane (From Dhouailly, 1975). A. Dorsal epidermis and ventral dermis from a 23-day lizard embryo. Scales are of ventral type. B. Ventral epidermis and dorsal dermis from a 23-day lizard embryo. Scales are of dorsal type. C. Caudal epidermis from a 15-day lizard embryo and tarsometatarsal dermis from a 9-day chick embryo. Scales are of chick type. D. Dorsal epidermis from an 18-day lizard embryo and dorsal dermis from 7-day chick embryo. The small buds are arranged in a hexagonal pattern typical of feathers.

not primarily possess a scale-forming bias. This seems to be in good accord with Kratochwil's (1972) postulate that the primary, species-specific bias of chick epidermis is a feather-forming one. Dhouailly (1975) also concludes that the formation of scales in lizard epidermis is a class-specific property, whereas in the foot region of the chick it is "a secondary acquisition entirely dependent upon dermal properties". These results support the idea of a class-specific epidermal bias that can be expressed when triggered by "non-specific", heterotypic mesenchymal influences which determine the size and distribution of the appendages. Completion of their morphogenesis, again, seems to require specific directive influences from the homospecific mesenchyme.

Not only is the morphogenesis of the cutaneous appendages directed by dermal influences, but the entire embryonic epidermis can be modified by similar inductive actions of the mesenchyme. When the epidermis of young 5-day chick embryos is separated from the dermal stroma and combined experimentally with various heterotypic mesenchymes (gizzard, proventriculus, heart), the epidermal cells are converted into keratinizing squamous-type cells, into ciliated, cuboidal cells, or into mucus-producing ones, respectively (McLoughlin, 1961).

Even more striking deviations from the normal course of development of the epidermis have been reported after various heterotypic recombinations. When the plantar epidermis of a 14-day-old mouse embryo is combined with the mesenchymal dental papilla, the epithelial cells develop into enamel-secreting ameloblasts, whereas a reciprocal recombination does not differentiate into dental structures (Kollar and Baird, 1970). Epidermis of the young chick or duck is also profoundly affected when combined with dermal mesenchyme from the mammary region of the mouse; it develops invaginations bearing a close resemblance to the initial stages of morphogenesis of the mouse mammary gland (Propper, 1969, 1975).

Deviations from the normal development can be brought about in target cells even at advanced stages of differentiation. By the fifth day of incubation, the corneal epithelium of the chick is already differentiated, and a subepithelial, collagenous matrix has been deposited. Yet, when such epithelium is combined with heterotypic mesenchymes from more advanced embryos, the corneal cells can be induced to form both feathers and scales. An inter-class combination with mouse flank dermis will likewise lead to feather formation in the chick cornea (Coulombre and Coulombre, 1971). The results indicate that even after differentiating into corneal epithelium, the cells have retained some of their original developmental options and plasticity, and can still interpret certain morphogenetic messages released by other cells.

The directive influence of dermal mesenchyme on the overlying epidermal cells is not limited to the embryonic period, as shown by Billingham and Silvers (1967, 1968) in heterotypic dermo-epidermal recombinations in the guinea-pig. Epidermis and dermis of the sole of the foot, trunk and ear were reciprocally recombined and followed for more than 100 days to allow several renewals of the epidermis. In all instances the epidermis showed histological features characteristic of the region from which the dermis had been taken. For instance, pigmented epidermis from the ear, combined with dermal mesenchyme of the unpigmented sole, remained pigmented, but showed histological characteristics indistinguishable from the normal epidermis of the sole. Consequently, the authors concluded that even at these advanced stages the cells of the stratum germinativum are "equipotential" and capable of producing various types of epithelia. "This suggests that the conservation of these epidermal specificities is dependent upon the continuous action of specific, regionally distinctive, inductive stimuli derived from local populations of dermal fibroblasts" (Billingham and Silvers, 1968).

In conclusion, all the observations described here suggest that whereas epidermal cells have definite, genetically determined class-specific properties, and, hence, limited developmental options, their ultimate fate is controlled by dermal influences that show various degrees of specificity. Initiation of development, morphogenesis, and regional distribution of cutaneous appendages is controlled by directive influences from the dermal mesenchyme, and in more advanced stages the constantly renewing basal layer of the epidermis is similarly under the directive influence of the dermis. Any experimental interference with these interactive processes leads to abnormalities and disturbances of epidermal morphogenesis.

2.2.3. Permissive influences in the development of parenchymal organs

Parenchymal organs consist of parenchymal, endo- or mesodermal parenchymal cells and mesenchymatous stroma. Interactions between these components are to be expected and have, in fact, been demonstrated in all cases so far examined. The morphogenesis and chemodifferentiation of the epithelial component is controlled by the stromal mesenchyme, and, in many instances, the interaction is reciprocal. The specificity of these epithelio-mesenchymal interactions has been tested both in homospecific/heterotypic and in interspecific recombinations. Recent experience, however, has raised some doubts about the validity of these experiments, as the results may have been affected by the in vitro conditions in which most of them were performed (Lawson, 1974).

The rudiments used for testing the specificity of epithelio-mesenchymal interactions have mostly been glandular organs at rather advanced stages, and the epithelial components have already been determined and had lost most, if not all, of their developmental options. In embryos at the 30-somite stage, for example, the pancreatic epithelium no longer requires its mesenchymal stroma for normal branching and chemodifferentiation. The stroma can be replaced by almost any heterologous mesenchyme, by high levels of embryo extract, or by protein preparations promoting cell proliferation (Rutter et al., 1964, 1967, 1968; Wessells and Cohen, 1967; Fell and Grobstein, 1968; Pictet et al., 1974, 1975) (p. 342). Similarly, heterotypic mesenchyme can support the early development of tracheo-bronchial epithelium (Dameron, 1961, 1968) but not the typical branching of the bronchial tree (Spooner and Wessells, 1970). Again certain stages in liver morphogenesis can be induced with heterologous mesenchymes (Le Douarin, 1968, 1975) (p. 342).

The formation of kidney tubules in the metanephric mesenchyme, normally triggered by the ureteric epithelium, can be induced by a great variety of tissues that do not share any obvious feature: embryonic CNS; embryonic salivary epithelium; mesenchyme from embryonic salivary gland; head region; somites and long bones; and cells of a teratoid tumour (Grobstein, 1955a; Unsworth and Grobstein, 1970; Auerbach, 1972). As these various inductors do not divert the mesenchymal cells into any other course, and as the potent tubule inducers do not exert a similar action on any other embryonic mesenchyme tested, it is clear that the metanephric mesenchyme is already committed to tubule formation (Saxén, 1970a). Hence, the conclusion may be drawn that the cells of the metanephric blastema have lost their developmental options, but can only express their tubule-forming bias when properly triggered and sustained by non-specific permissive influences. This bias can remain covert throughout development. It will only be expressed when the metanephric cells are exposed to a suitable inductor. This latent potential has been demonstrated in certain neoplastic cells of metanephric origin (Wilm's tumour), which could be triggered to develop tubular elements when brought into contact with mouse brain in vitro (Crocker and Vernier, 1972).

Whereas all the target tissues listed above represent committed cells with a requirement for rather unspecific environmental permissive influences, other epithelio-mesenchymal interactions have been reported in which a more specific, homotypic effect is essential for normal development. Branching of the bronchial epithelium and early determination of both pancreatic and hepatic endoderm seem to require a homotypic mesenchymal stimulus. The target most extensively studied and referred to in connection with specific requirements is the epithelium of mouse embryonic salivary gland, which was originally thought to respond only to its homotypic mesenchyme (Grobstein, 1953a) or to the corresponding mesenchyme of the chick (Sherman, 1960).

The wide range of specificity in the epithelio-mesenchymal interactions, the extreme forms of which are the pancreatic and salivary mesenchymes, led Grobstein (1967) to suggest the existence of two mesenchymal factors: (1) a "mesenchyme common" (MC) factor prevents the spread of the epithelium in vitro and may cause the early development of the epithelium (formation of compact spheroid masses and "simple" cytodifferentiation); and (2) a "mesenchyme specific" (MS) factor is required for organized morphogenesis and may even "partially control the nature of the epithelial response". Grobstein (1967) also suggested that the MS factor may be more important early in development than in the later stages. The hypothesis is fully compatible with the results presented in the foregoing paragraph (skin), but may find less support from recent studies on epithelio-mesenchymal interactions in glandular organs.

Grobstein's (1967) hypothesis was based in part on results which suggested that in the interactions leading to salivary gland development the mesenchymal influence was highly specific (Grobstein, 1953a; Spooner and Wessells, 1972).

However, subsequent studies have indicated that even here heterotypic mesenchyme can mimic the effect of the homotypic salivary mesenchyme. Cunha (1972) reported that morphogenesis progressed when the salivary epithelium was combined with the mesenchyme of male secondary sex organs. Similarly, Lawson (1972, 1974) obtained morphogenesis and cytodifferentiation on combining both rat and mouse salivary epithelium with lung mesenchyme of the same species. She reported that twice the amount of lung mesenchyme was required as compared with experiments in which homotypic mesenchyme was used. She suggested that this, as well as the negative results of Grobstein (1953a) and Spooner and Wessels (1972), could be due to a different behavior of pulmonary and salivary mesenchymes grown on various non-nutrient substrates. Lung mesenchyme remained compact only on agar, where it also supported normal development of the epithelium. On Millipore filters lung mesenchyme spread rapidly and its "inductive" effect on the salivary epithelium was negligible. Probably the mesenchymal mass (or density) fell below the level required to support epithelial morphogenesis. Lung mesenchyme did not spread when cultured on Millipore filters, where it also exerted its morphogenetic influence on the epithelium. As the previous authors had used such non-nutrient substrates for cultivation of the lung mesenchyme, their negative results may well have been due to this spreading phenomenon. Consequently, no properties unique to the homotypic mesenchyme would be involved, and, as Lawson (1974) pointed out, the salivary model system does not necessarily call for a hypothesis involving mesenchyme-specific factor(s).

Directive mesenchymal influences have rarely been observed in epithelio-mesenchymal interactions, apart from those concerning epidermal differentiation (above). However, when combining embryonic mouse mammary epithelium with salivary mesenchyme, Kratochwil (1969) observed a dichotomous branching pattern and adenomere formation closely resembling the developmental processes in the salivary gland (Fig. 10). Similar directive influences have been reported by Bishop-Calame (1965) and David (1967, 1971) when ureteric and gastric epithelia were combined with various heterotypic mesenchymes. Occasionally, in their experiments, these epithelia acquired histological features typical of the normal counterpart of the heterotypic mesenchyme with which they were combined.

In conclusion, most of the epithelio-mesenchymal interactive situations tested involve rather advanced stages of differentiation, where the developmental options of the epithelial target cells are strictly limited. Consequently, the mesenchymal influence is primarily permissive in nature, and shows a high degree of non-specificity. Directive, more specific influences could be expected during the early development of parenchymal organs, when the cells chose between various developmental pathways.

2.2.4. Remarks

From the great variety of interactive situations reviewed, we may conclude that there is evidence to support the hypothesis illustrated in Fig. 8, and that

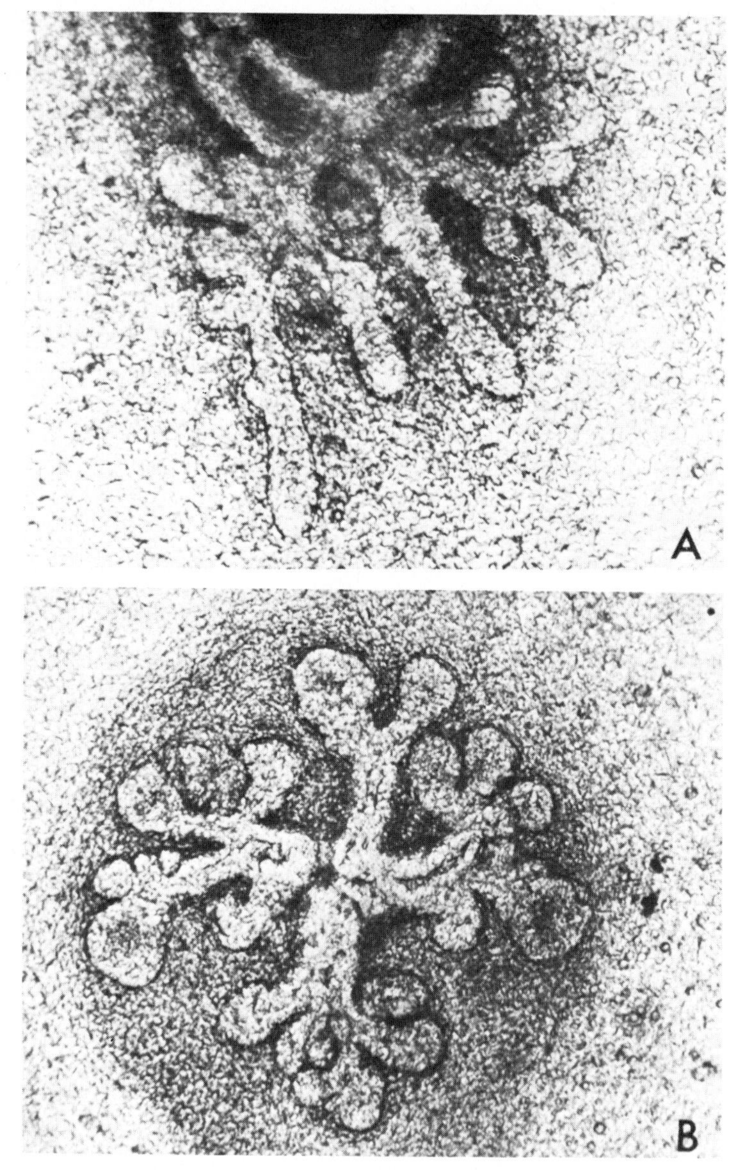

Fig. 10. Growth and branching of mouse mammary epithelium in homotypic (A) and salivary (B) mesenchyme. The latter shows dichotomous branching, which closely resembles salivary adenomere formation (From Kratochwil, 1969).

embryogenesis involves both "directive" and "permissive" influences. In normal development both are so highly specific as to ensure synchronized development, and there may not even be a clear-cut distinction between them. However, the importance of drawing this distinction becomes evident in discussions on the mechanism of morphogenetic interactions. Because of their different nature, there is no point in searching for common denominators. For example, the messages involved may vary from informative, directive molecules to merely nutritional or stimulating conditions.

2.3. Abnormal cell and tissue interactions

2.3.1. Introduction

The many examples reviewed in the previous paragraphs appear to justify the conclusion that inductive tissue interactions provide a central, vital guiding system throughout the course of embryogenesis, i.e. cytodifferentiation, proliferation and morphogenesis. Therefore, it seems justified to ask whether disturbances in normal developmental events could be due to abnormalities in these interactive processes. Would the absence or distortion of an interactive event lead to abortive or abnormal expression of the genetic potentialities of cells, to uncontrolled growth or to abnormal morphogenesis? The first two consequences bear a clear resemblance to neoplastic transformation of cells, and the latter alternative would lead to congenital defects at the organismal level. Many congenital defects can be traced back to genetically defective interactive processes or to exogenous interference with inductive events. In carcinogenesis mediated by abnormal interactions, the evidence is still insufficient but suggestive.

2.3.2. Tissue interactions and carcinogenesis

2.3.2.1. Experimental carcinogenesis of the skin. When methods for experimental induction of skin cancer became available in the nineteen thirties, it soon became apparent that an interaction between the dermis and the epidermis is necessary for the development of epithelial neoplasms (see Orr, 1963). In several experiments on this point the skin was painted with carcinogens, the epidermis removed from the painted area, and grafted on an untreated dermis (Billingham et al., 1951). An untreated epidermis was grafted on the painted area where, during the latent period that followed, the epidermis underwent regeneration and exfoliation with apparently normal multiplication of cells. No neoplastic features were seen either in the epidermis or in its appendages. Meanwhile, profound and progressive changes occurred in the dermal stroma. The normal, coarse network of interwoven collagen fibres was continuously replaced with sheets of fine non-refractible fibres arranged parallel to the skin surface (Orr, 1938). Orr used the term "permutation" for the stromal changes in the dermis at the site of the neoplasm. The term refers not only to the change in the collagen structure but also to local congestion, accumulation of mast cells and gradual fragmentation of elastic fibres. These changes seem to be non-specific, but some variation does occur between species. In the rabbit, accumulation of mucopolysaccharides is seen at the site of experimental neoplasia (Hamer and Marchant, 1957), but this is seldom seen in other species.

These studies suggest that the dermis plays a crucial role in the development of skin cancer. However, the possibility exists that hair-follicles and sebaceous glands left behind during the grafting operation might be the origin of the cancer. This was considered unlikely, because the hydrocarbon carcinogens always cause rapid and complete necrosis in the treated superficial dermis and especially in the hair-follicles and adnexal glands (see discussion by Orr in, Orr and Spencer, 1972). A contribution to this problem was made by Steinmüller (1971), who traced the origin of the neoplastic cells in inbred strains of mice and their F_1 hybrids. He repeated the experiments of Billingham et al. (1951), painting small areas of skin of the C57B1 × BALB/c hybrids with methylcholanthrene for 12 weeks. After another 2 weeks the epidermis and superficial dermis from the treated area were removed and replaced with trypsinized epidermis from unpainted parental BALB/c skin. When tumours developed, they were again transferred to similar F_1 hybrid mice or back to the BALB/c parent, which had been immunized against the other parental strain. The grafts grew well and the neoplastic changes progressed in the F_1 hybrids but not in immunized BALB/c mice. As the epithelium had to be histocompatible for successful transplantation the neoplastic cells were of F_1 hybrid origin. It was not clear whether the cells originated from proximal skin adnexes or from surrounding epidermis. In this connection, Orr (in Orr and Spencer, 1972) pointed out that a certain degree of caution is necessary in interpreting negative results of experiments with trypsin-treated, thin epidermal grafts, since failure may sometimes be due to technical difficulties.

The mechanism of tumour formation is entirely speculative. Local hypoxia and changes in pH due to vasoconstriction caused by carcinogens were suggested by Orr (1934, 1935, and 1937). As already mentioned, "permutation" of the dermis varies with the species, and it is difficult to make generalizations (Hamer and Marchant, 1957; Orr and Spencer, 1972). A possible exception might be the changes in the dermal collagen texture reported by Orr (1938). Tarin (1967), in an ultrastructural analysis, found marked thinning of the collagen fibrils during carcinogenesis in mouse skin. Mazzucco (1972) reported that a marked and continuous decrease in the amount of dermal collagen always precedes skin cancer formation in mice. This has also been observed in several other species. Tarin (1967) also showed that during experimental carcinogenesis the attachment of the epidermal cells to the basement membrane was disturbed. This could lead to loss of contact inhibition between the cells, with markedly increasing proliferation and a concomitant rise in collagenolytic activity (see discussion by Mazzucco, 1972). This may result in dermal changes, which could offer a new substrate for the epidermal cells and perhaps pave the way for neoplastic growth and invasion. In the light of the above-mentioned experiment, however, it seems apparent that the target site primarily affected is the dermis, which seems to regulate the subsequent development of the epidermis. This is not surprising, in view of the decisive role played by the dermis in the normal induction of the skin and epidermal appendages.

It has been suggested repeatedly that the epidermis may have inherent invasive properties, but is normally restrained by surrounding connective tissue. Epidermal cells from the adult epithelium when transferred to animal hosts after prolonged culture in vitro have been shown to cause tumour formation (Sanford et al., 1950). These findings led Tarin and Sturdee (1972) to analyze the behaviour of isolated epithelium in vivo, implanting it into the peritoneal cavity and anterior chamber of the eye. They found that, despite the viability of the epidermis, no sign of neoplastic growth resulted from removal of the dermal control. In certain other models where tumours have been induced experimentally, interruption of communication between cells has been suspected to be the factor causing neoplastic growth. Implants of solid plastic are tumorigenic (Johnston et al., 1972), as are pieces of Millipore filter implanted subcutaneously (Karp et al., 1973). The Millipore model has another peculiarity: tumorigenicity is inversely proportional to the pore size (Goldhaber, 1961; Karp et al., 1973). Filters with pore sizes of 0.22 μ or less were tumorigenic, whereas those with a larger pore size were not. In an inductive situation, cellular processes have been shown to pass through small pores with difficulty but through wider ones with ease (Wartiovaara et al., 1972, 1974; Lehtonen et al., 1975). Therefore it is possible that Millipore filters and pieces of solid plastic are tumorigenic because they disturb the communications between cells. The importance of cell-to-cell communication for normal tissue growth is also suggested by other findings. Absence of intercellular communication in cancer cells has been reported by Loewenstein (see Loewenstein, 1973) in some, but not all, solid epithelial tumours. Junctional coupling between such cells seems to be defective (Borek et al., 1969; Azarnia and Loewenstein, 1971), and endogenous nucleotides and their derivatives cannot pass from one cell to another (Azarnia et al., 1972; Azarnia and Loewenstein, 1973).

2.3.2.2. Virus-induced tumours. During the last decade Dawe and his associates have shown that epithelio-mesenchymal interactions are essential for the development of virus-induced tumours. Their model system consisted of a tumour induced in the mouse salivary gland by polyoma virus (PV). The tumour is transplantable, even when the transplant is infected in vitro. In culture, infected rudiments became transformed like those seen in vivo (see Dawe et al., 1968; Dawe, 1972). If the epithelium and mesenchyme of an infected rudiment are separated by trypsinization at an early stage before cultivation in vitro and then grown separately, neither tissue forms a tumour. But if the separated components are recombined, subsequent culture leads to neoplastic proliferation. Thus tumour formation is unaffected by the trypsin treatment, but depends on the presence of both mesenchyme and epithelium (Dawe et al., 1966b).

Experiments with salivary gland rudiments have shown that in vitro the epithelium can only complete normal morphogenesis when in contact with its own mesenchyme (p. 352). Correspondingly, if the salivary mesenchyme in the PV-infected recombinants was replaced by heterotypic mesenchyme, no transformation occurred (Dawe et al., 1966a). However, from the work of Lawson (1974) on the specificity of the mesenchymal requirement (p. 353), it appears that on suitable substrates other mesenchymes might also be compatible with tumour formation.

The stage of development of the salivary gland at the time of infection was important both in vitro and in vivo: very early rudiments failed to proliferate. This was seen only in rudiments infected around the time of overall morphogenetic development of the gland (Dawe et al., 1966b). To determine the origin of the tumour cells Dawe used recombinants in which one of the components carried a chromosomal marker. The neoplastic cells were karyotyped, and found to be of epithelial origin (Dawe et al., 1971).

In the odontogenetic epithelium PV induces a histologically characteristic ameloblastoma. The same response is seen if the tooth germ is infected and grown in vitro, but this tumour cannot be transplanted back into the animal host. If, however, the rudiment is transplanted immediately after infection, without cultivation, the transplant survives in the host and gives rise to ameloblastomas, but only if the strain LID/1 of PV has been used. In vitro the infected tooth germs develop a proliferative response, invading the underlying gelatin sponge after 24 days (Main, 1969). Even in uninfected control cultures separated odontogenic epithelium does not survive for more than 10 days in vitro. Infected mesenchyme survives for 50 days, showing viral damage, but, unlike whole rudiments, it gives no neoplastic response. If the mesenchyme was cultured for 24 days and fresh epithelium was then added, the latent period in the epithelium before tumour formation was shortened from the normal 24 days to 10 days. The result was similar if whole tooth germs were used instead of separated epithelium. Thus the epithelium was not programmed by its own mesenchyme, but by the precultivated mesenchyme (Main and Waheed, 1971).

Several speculations have been made as to how the mesenchyme contributes to the growth of the neoplasms induced by polyoma virus. Both "nutritive" and "permissive" effects have been suggested (see Main, 1972). In the last-mentioned experiment, however, precultivated mesenchyme programmed epithelium that lay beyond a mesenchyme of its own. This strongly suggests that the contribution is "inductive", on analogy with embryonic epithelio-mesenchymal interactions of considerable specificity.

2.3.2.3. Interactions between neoplastic and non-neoplastic tissues. Direct evidence of the importance of abnormal inductive interactions in carcinogenesis might be expected from combinations between neoplastic and non-neoplastic tissues. Readler and Lustig (1970) reported that peritumoral dermis from a skin cancer had a stronger growth-promoting effect on normal embryonic epidermis than mesenchyme from the lip and cheek or from a leucoplastic lesion. When mesenchymal stroma from the mammary gland of a normal mouse was combined with the epithelial component of a mammary adenocarcinoma tubules were formed, the DNA synthesis in the neoplastic cells declined, and gradually a mucopolysaccharide matrix appeared, which was not seen in the original tumour. The authors drew the conclusion that these changes, representing differentiation of the neoplastic cells, were initiated by agent(s) present in the normal mammary mesenchyme (DeCosse et al., 1973).

2.3.3. *Tissue interactions and teratogenesis*

2.3.3.1. Introduction. Abnormal development culminating in the birth of congenitally defective offspring may be expected to arise when inductive tissue interactions are deranged (see Zwilling, 1955; Saxén, 1970b, 1973, 1975a). Such derangements may be caused by: (1) changes in the inductive capacity of a tissue; (2) changes in the competence of the target cells; and (3) loss of contact between the interactants. These derangements may be the result of defective gene function or of epigenetic factors.

The part played by altered inductive interactions in abnormal development and production of congenital defects can be demonstrated in two ways. The interacting tissues can be exposed to physical or chemical treatment affecting one or both of them, or the development can be analyzed in mutant strains in which the type of defect suggests defective induction. In studies of the latter type, combinations between tissues of mutant and wild-type embryos have proved extremely useful ever since the pioneering work of Zwilling (1956a,b,c). In such experiments, it is possible to demonstrate and locate the primary defect in a particular interactive process.

2.3.3.2. Experimental interference with interactive processes. Since the early days of research on primary induction in amphibian gastrulae, developmental biologists have seen a wide variety of defects of the central nervous system resulting from accidental or intentional disruption of contact between the presumptive neural ectoderm and the underlying mesoderm. Such disruption can be achieved by total or partial prevention of invagination of the dorsal lip (inductor) tissue, or by surgical removal of the inductor or a part of it (see Mangold, 1961; Saxén and Toivonen, 1962). Another way of preventing close apposition of the interacting cells is to interpose non-permeable membranes, as shown by Brahma (1958) during primary induction, by McKeehan (1951) during induction of the lens, and by LeDouarin (1976) for the induction of the liver endoderm by its mesenchymal counterpart (p. 342).

A teratogenic treatment may change intercellular contact relationships. Bagg (1929) exposed pregnant mice to X-irradiation, performed a Caesarian section and examined the living embryos in situ. Haemorrhagic blebs were found in the distal parts of the embryonic limb-buds and these embryos developed different amputation defects with a varying degree of severity. The type and severity of the defect correlated with the site and size of the bleb previously recorded. Evidently the haemorrhagic blebs formed an obstacle between the interacting components, thus preventing limb induction at various stages and sites (Zwilling, 1969).

Several chemicals have been found to be teratogenic. Thalidomide seems to interfere with the induction of cartilage in avian and human nephric blastemas (Lash, 1963; Lash and Saxén, 1972), and nitrogen mustard treatment may upset the induction of limb-bud mesenchyme by the apical ectodermal ridge (Salzgeber, 1967). Recent experiments with various metabolic inhibitors have shown that kidney tubule induction and subsequent morphogenesis are sensitive to compounds that interfere with protein synthesis, whereas those acting at the transcriptional and/or translational level might not necessarily prevent induction. The interpretation of such results has to await a better understanding of the molecular basis of inductive tissue interactions.

2.3.3.3. Genetically determined defects in inductive interactions. Genetic defects can be thought to affect inductive tissue interactions at various levels. Disturbances in the growth of tissues or in the migration of cells may lead to abnormal cell relations and so to lack of the close appositions required during interactive processes. The metabolic pathway leading to the production and release of the signal substances by the inductor might be defective. A third possibility is that the target cells might not develop a proper responsiveness (competence) towards morphogenetic signals. Indeed, all these mechanisms have been demonstrated in mutant strains. Although we know the way in which the normal course of events is altered, we know almost nothing about the primary (molecular) genetic defect.

Disruption of relations between interacting cells should always lead to an abnormal inductive event, since it is a prerequisite for all inductive interactions that the interacting tissues are close to each other. Certain mutant genes seem to exert their action by causing such abnormal contact relations. In an anophthalmic mutant mouse, the inductor of the lens, the optic cup, is formed initially, but its growth seems to be retarded and it does not come close enough to the target tissue, the prospective lens epidermis (Chase and Chase, 1941; Chase, 1944). Consequently, no lens vesicle develops and the optic cup is covered with epidermis, producing the typical appearance of anophthalmia. In a recent re-examination of the embryogenesis in this strain of mice, Silver and Hughes (1974) arrived at a somewhat different conclusion. They showed that the mesenchymal cells between the interacting optic vesicle and the prospective lens epidermis are normally eliminated by morphogenetic cell death, but that in the mutant strain these cells fail to disappear. The authors concluded that these intervening cells may prevent the "flow of inductive substances" and thus lead to the ocular defects.

In the weaver-mutant mouse (*wv*), certain severe defects of the brain have also been traced back to defective cell interactions. During normal postnatal development, the granular cells of the cerebellum proliferate in its external layer and migrate from there to their final, adult position, where they mature. This migration is obviously guided by the fibres of the Bergman glial cells. In the *wv*

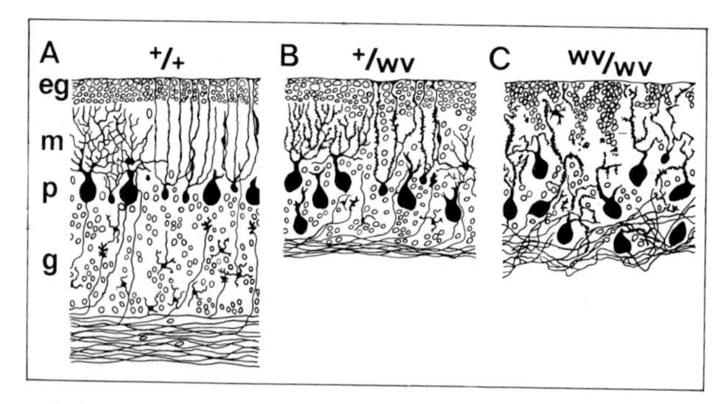

Fig. 11. Scheme of the postnatal architecture of the mouse cerebellum (A) and of the heterozygote (B) and homozygote (C) weaver mutant. eg: external granular layer, m: molecular layer, p: Purkinje cell layer, g: granular layer (After Rakic and Sidman, 1973).

mutant the number of Bergman cells is greatly reduced, and most of the migrating granular cells are thus deprived of support. As a result these cells fail to migrate, fail to mature and become degenerate (Fig. 11; Rakic and Sidman, 1973; Sidman, 1974).

A third example of mutant genes affecting cell and tissue relations is the work of Ede and his collaborators (Ede et al., 1974). The *talpid*^{3} mutant chick is characterized by limb malformations which the authors relate to abnormal adhesive properties of the mesenchymal cells in these embryos. The induction of the limb mesenchyme takes place at the interface between this mesenchyme and the apical ectodermal ridge. Here, the mesenchymal cells send out long filopodial processes, which establish contact with the basement membrane of the ectoderm (p. 376). In the *talpid*^{3} embryos, the filopodia are fewer and smaller, and consequently the area of contact between the ectoderm and mesenchyme seems to be reduced, which has been suggested to be the cause of the limb deformities. A detailed discussion of limb morphogenesis in normal chick embryos and in *talpid*^{3} mutants is given in Chapter 10 of this volume by Ede.

A defect in the inductor should naturally lead to failure of inductive interactions, but there are few well-established cases of such defects. According to the present, somewhat simplified hypothesis, development of the vertebrate limb is controlled by a sequence of inductive interactions. The proximo-distal programming of the mesenchymal cells in the bud is thought to be induced by the apical ectodermal ridge (AER), but this, in turn, is maintained by the mesenchymal cells, which produce a transmissible "maintenance factor" (MF) (Saunders, 1948; Zwilling, 1961, 1974; Saunders and Gasseling, 1963). According to Zwilling's (1974) hypothesis, the primary defect in the wingless mutant of the chick is a lack of the maintenance factor. The consequence is regression of the AER and blockade of limb induction. The results of recent work on a polydactylous mutant strain are in good accord with the above hypothesis. The limb bud of this mutant lacks the demarcating necrotic zones that normally determine its shape, and as a result it becomes abnormally wide. Consequently, the MF is distributed over a wider area than normal and results in an enlarged AER which, in turn, induces supernumerary digits (Hinchliffe and Ede, 1967; Hinchliffe and Thorogood, 1974).

A defect in target tissue has been reported more frequently. Two examples will illustrate genetic defects in which the primary lesion seems to be in the responsiveness of cells to an inductive signal. An eyeless mutant of the axolotl totally lacks optic vesicles. This defect was studied by Van Deusen (1973), who exposed presumptive neural ectoderm to the mesodermal inductor in combinations of wild-type and mutant tissues. A combination of mutant prechordal mesoderm with wild-type ectoderm led to the development of normal eyes, but not a combination of mutant ectoderm with wild-type mesoderm. Thus the defect seems to be in the responding tissue, the ectoderm.

The second example concerns the formation of the ectodermal scales on avian legs, a process known to be triggered by the underlying mesoderm (Sengel, 1958). A scaleless mutant exists, which lacks these epidermal derivatives. When mutant epidermis from the normally scale-forming area was combined with wild-type mesoderm, there were not even initial signs on scale formation, whereas the reciprocal combination yielded normal scales (Goetinck and Abbott, 1963; Sawyer and Abbott, 1972). Thus the conclusion is warranted that here, as in the previous example, the gene defect is expressed as an unresponsiveness of the target cell to inductive signals.

2.3.4. Remarks

The fragments of evidence presented here on the possible relation between inductive tissue interactions, on the one hand, and abnormal differentiation and morphogenesis, on the other, suggest that these interactive events offer a biologically sound and interesting basis for a new theory of carcinogenesis and teratogenesis. To construct this, we need a profound understanding of the mechanism of inductive tissue interactions, but, as will be discussed in the next section, such an understanding is still a distant goal.

3. Mechanisms of tissue interaction

3.1. Introduction

In the preceding section, some of the biological events, which take place during embryonic induction, have been analyzed. In the following, some of the mechanisms, which possibly operate in bringing about these morphogenetic events, will be discussed. This will include an analysis of the chemical characterization of signal substances, an analysis of how such signals are transmitted, as well as some speculations of the molecular models, which possibly operate.

3.2. Chemical characterization of signal substances

3.2.1. Introduction

A direct approach to the study of biologically active compounds would be to isolate them from the source tissue, to characterize them chemically, and to test their effects on the target cells. Such attempts have been made by developmental biologists since the early thirties, soon after the first observations suggested that "inductor substances" were transmissible. Few of these laborious studies have yielded chemically characterized, morphogenetically active substances. A brief account is given below of the different types of signal substances found, and some additional data on the chemistry of signal substances will be given in the discussion of "matrix interaction" and the various interactive mechanisms.

3.2.2. Inducing factors

3.2.2.1. RNA. As RNA is known to play a major role in the transmission of intracellular messages, it is not surprising that it has also been suggested to act as an inductor or messenger between cells. Even before RNA was recognized as an intracellular messenger, Brachet (1940, 1942) suggested that it might transmit morphogenetic signals. He compared induction to a virus infection and postulated intercellular transmission of RNA-rich particles. He based his hypothesis on the histochemical demonstration of RNA between the chordamesoderm and the overlying ectoderm, and suggested that there was a transfer of RNA between the two tissues. Since then, several other workers have presented data believed to support the view that RNA has an inductive function. From studies of staining properties, ultraviolet absorption, and the effect of RNase, McKeehan (1956) suggested that RNA may be transferred from the optic vesicle to the ectoderm during lens induction. With electron microscopy, Hunt (1961) showed that there consistently was an interepithelial space separating the optic vesicle and lens ectoderm. In the cytoplasm of both types of cells there were free electron dense particles, which he interpreted as ribonucleoprotein (RNP) and between the tissues there was a PAS-positive "cloud". During early lens induction, RNP particles were more abundant in the optic vesicle and later in the ectoderm. Hunt (1961) therefore agreed with the views of Spratt (1954), Niu (1956), Grobstein and Dalton (1957), and McKeehan (1958) that contact is not a constant relationship between inducer and induced. He also suggested that RNP was transferred from the optic vesicle to the presumptive lens ectoderm.

In the chordamesoderm-neurectoderm interphase in *Xenopus laevis* electron microscopy has revealed extracellular material which contains rinonucleoprotein particles (Kelley, 1969). This observation has been confirmed (Tarin, 1971, 1972, 1973) (Fig. 12). According to Tarin (1973), neither the RNA containing granules nor the glycosaminoglycan containing fibrils located extracellularly between the interaction tissues were essential mediators of the inductive stimulus. He suggests that either the passage of diffusible substances or direct contacts between the membranes are required for induction.

Another line of investigation has sought to establish the inductive effects of added RNA. Niu and his collaborators, who treated anterolateral fragments of chick blastoderm at the primitive streak stage with RNA isolated from heart and liver, found that only heart RNA caused the development of pulsating tissue (Niu and Mulherkar, 1970). Treatment of the heart RNA preparation with RNase abolished its heart-forming capacity. The same result was obtained when nuclear RNA from chicken hearts was added to post-nodal pieces of chick blastoderm at stage 4 (Niu and Deshpande, 1973). RNA from heart cell nuclei could be replaced by total RNA isolated from calf testicles, by poly-A-attached RNA (mRNA), or by mRNA isolated from either chick or calf hearts, but not by the small-molecular, filtrable fraction of testicular RNA devoid of mRNA (Lee and Niu, 1973; Niu et al., 1974). But as, unfortunately, no such experiments have so far been made in other systems, there is no way of knowing whether the inductive function of RNA has more general validity.

An inherent problem in such experiments is how to be sure that the responding tissue has not been predetermined and that the treatment is only permissive. Moreover, it is difficult to establish that the normal event and the one occurring in a

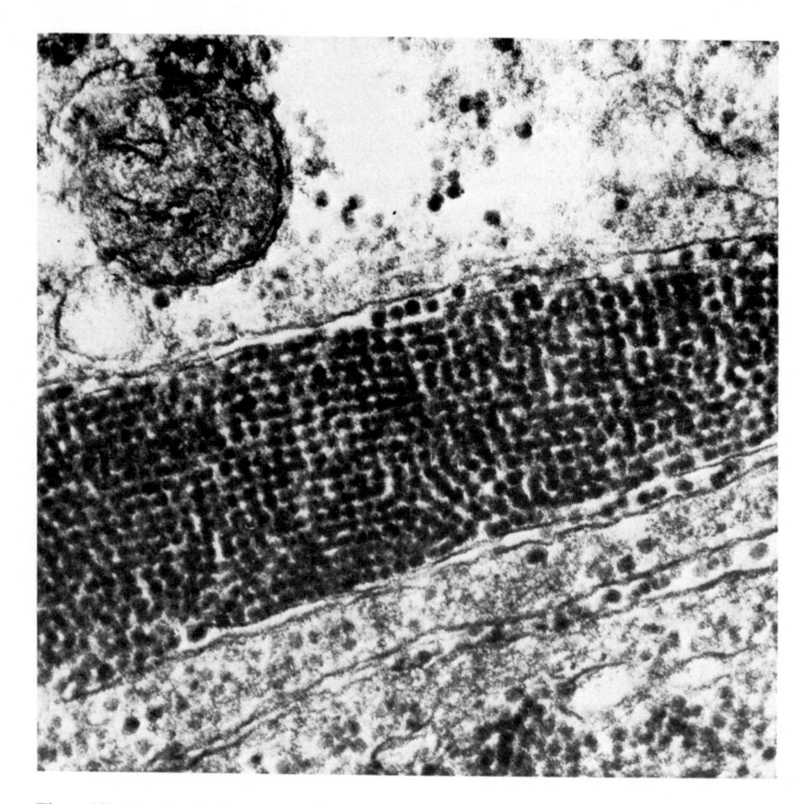

Fig. 12. Extracellular granules between an ectodermal and a mesodermal cell at the chordamesoderm-neurectodermal junction of a stage 12 *Xenopus laevis* embryo, ×80 000 (Courtesy of David Tarin).

highly artificial system in vitro are due to the same signal. Tissues in culture may utilize exogenous large-molecular RNA, but there is no evidence that this occurs in vivo. To study the passage of large nucleic acid molecules, Grainger and Wessells (1974) prelabelled lung mesenchyme with radioactive RNA or DNA precursors and cultivated it transfilter with the responding epithelium. In this system they were unable to detect transfer of either RNA or DNA. Although by no means excluding the possibility that RNA has an inductive capacity, this result casts some doubt on the general significance of responses achieved with RNA treatment.

The role of RNA has also been studied by Slavkin and his collaborators. By electron microscopy and biochemical studies these workers demonstrated that in the developing tooth the matrix between the epithelium and the mesenchyme is interspersed with vesicles containing RNA (Slavkin et al., 1969, 1970). Although the evidence was only circumstantial, they suggested that these RNA-containing vesicles might have a morphogenetic function (Slavkin and Croissant, 1973).

Lately, the likelihood that the RNA vesicles play an informative role has been somewhat diminished by the observation that at the time of morphogenesis odontoblastic cell processes and preameloblasts come in contact through holes in the basement lamina.

As an inducing molecule, RNA would satisfy the requirement that the inductive signal must have a large informative content. Temin (1971) has even suggested that through reverse transcriptase the inductive signal, RNA, may be incorporated into the genome, thus becoming truly instructive.

3.2.2.2. Mono- and divalent cations. The effect of mono- and divalent cations on the differentiation of presumptive neural ectoderm was studied extensively in *Rana pipiens* by Barth and Barth. These workers found that treatment for a few minutes in 0.1 M LiCl or cultivation for several hours in 0.005 M LiCl resulted in permanent transformation into pigment cells and mesenchyme (Barth and Barth, 1959). $CaCl_2$ or $MgSO_4$ treatment results in a differentiation into neural cells. Sequential treatment with $CaCl_2$ or $MgSO_4$ and LiCl gives rise to the formation of pigment cells even if the concentration of LiCl was too small to affect development alone (Barth and Barth, 1964). The authors interpreted these results as indicating a sequentiality of induction. In this system induction was reversible and if induction of pigment cells by LiCl was followed by a neural inductor, $CaCl_2$, neural cells and not pigment cells would develop. Treatment of the tissue with sucrose also affected its differentiation, but this effect was probably due to changes in the intracellular sodium concentration (Barth, 1966). The effects of Li^+ and sucrose were correlated with increased uptake of ^{22}Na (Barth and Barth, 1967). As the uptake of Na^+ and Ca^{2+} also increases during development in vivo, these authors suggested that normal embryonic induction is initiated by changes in intracellular cations (Barth and Barth, 1968). The effects of these ions have mainly been studied in primary induction, and generalizations would therefore be premature.

Lash and coworkers (1973) reported in a study on somite chondrogenesis that an increase in the concentration of potassium from 2.7 to 3.7 mM caused an increase in the incorporation of sulphur into chondroitin sulphate for the first 24 hours, followed by a decrease. If, after 24 hours, the explants were placed in a low-potassium medium, the rate of synthesis remained high. Potassium is also important for the normal development of human embryonic kidney (Crocker, 1973). With a potassium concentration of 3.25 to 5 mM, branching of the ureter bud was normal, but concentrations of 1.5 to 3 mM reduced branching, prevented nephron induction and caused dilatation of the ureter bud. These reports emphasize the importance of potassium for normal development, but they may not be genuine examples of induction.

3.2.2.3. Vegetalizing factor. A vegetalizing factor acting upon competent amphibian neuroectoderm was first isolated by Yamada (1962) from guinea-pig bone-marrow, and shown to be a protein. Tiedemann and his collaborators isolated a similar factor from 9-day-old chick embryos and characterized it chemically (Tiedemann and Tiedemann, 1959; Tiedemann, 1968, 1971, 1973). It is a protein, molecular weight 25,000–30,000, which induces competent gastrula ectoderm to form both mesodermal and endodermal structures. Another protein with similar biological activities has been isolated from young amphibian embryos (Faulhaber, 1972), but as only minute amounts can be recovered from such small embryos its chemical nature has been less well characterized.

Since gastrula ectoderm contains minor amounts of the vegetalizing factor in an apparently inactive form, a search was made for a substance that would specifically inhibit the inductor. Such an inhibitor was found in the $105,000 \times g$ supernatant of homogenates of 9-day chick embryos, which completely inhibited the action of the vegetalizing protein (Born et al., 1972).

The mode of action of such macromolecules has not yet been clarified, nor do we know whether they are involved in normal development. When the vegetalizing protein from chick embryos was tested on the chick, only a weak proliferation and a questionable inductive response were obtained (Deuchar, 1969).

3.2.2.4. Mesenchymal factor in pancreatic development. Mesenchymal factor (MF) acts upon isolated pancreatic epithelial cells, causing cell proliferation and morphogenesis. Normal morphogenesis and chemodifferentiation of the predetermined pancreatic epithelium requires merely a permissive mesenchymal influence (p. 351). Homotypic mesenchyme can be replaced by various heterotypic mesenchymes, by chick embryo extract, and by a particulate fraction of the latter (Rutter et al., 1964; Ronzio and Rutter, 1973). The factor was recently solubilized and concentrated by Amicon filtration, and shown to act on the surface of the pancreatic epithelial cells (Levine et al., 1973). The stimulation of DNA synthesis is dose-dependent, but a plateau is reached at a concentration of 2 mg/ml (Fig. 13). Since after covalent binding to Sepharose beads MF is still active, the authors suggest that there are intracellular second messengers. Dibutyryl cAMP and dibutyryl cGMP did not change MF activity, nor could they replace MF. Periodate oxidation abolished MF activity, but addition of dbcAMP restored it fully. The authors conclude that MF activity is cAMP-dependent. They suggest that the periodate-sensitive moiety may be the site which stimulates adenyl cyclase (Pictet et al., 1974, 1975).

3.2.2.5. Chondrogenic factors. In normal development, vertebral chondrogenesis requires inductor tissues. In vivo these have been shown to be the ventral spinal cord and the notochord (Holtzer and Detwiler, 1953; Strudel, 1953). Attempts to isolate and identify the chondrogenic factors involved have been hampered by the fact that even minor improvements in the culture conditions tend to stimulate chondrogenesis in somite cells (Ellison and Lash, 1971). Combinations of dissected pieces of notochord and somites result in rapid accumulation of cartilage matrix in the somites. If the extracellular matrix surrounding the notochord is removed with trypsin or with chondroitinases, its capacity to

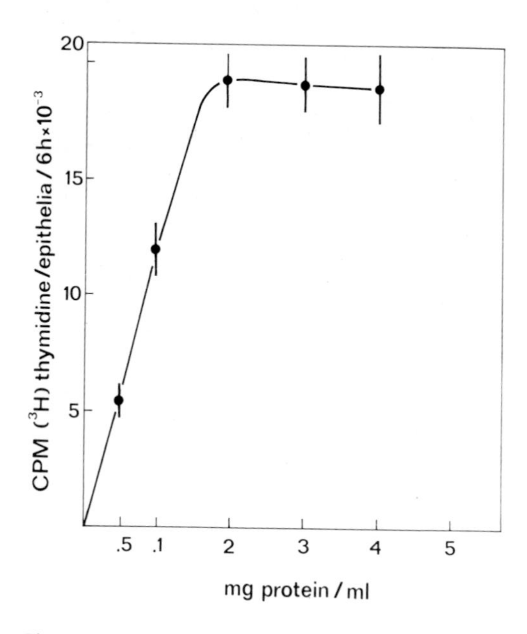

Fig. 13. Effect of mesenchymal factor isolated from chick embryo extract on the uptake of ³H-TdR by pancreatic epithelia from mouse embryos (After Pictet et al., 1975).

enhance chondrogenesis is impaired. Treatment of the matrix with collagenase, which does not remove perinotochordal glycosaminoglycans, has no effect. If pieces of trypsinized notochord are subcultured for 24 hours, proteoglycans are resynthesized, and their stimulatory effect upon somite chondrogenesis is restored (Kosher and Lash, 1975). The role of extracellular proteoglycans on chondrogenesis is further supported by the finding that exogenous, heterologous proteoglycans enhance cartilage formation of embryonic somites (Kosher et al., 1973).

3.2.3. Remarks

A variety of different substances ranging from simple metal ions to messenger RNA and more or less well-defined factors have an inductive capacity or at least influence development in certain systems. In no instance has the mechanism of action been shown, and the results are so far mainly descriptive. Recently, McMahon (1974, see also chapter 9 of this volume) has developed a unifying hypothesis: "Developmental processes are normal physiological functions and are guided by the same molecules that are used to regulate the physiological state of the adult" (p. 449). As regards extracellular substances that act as inductors, most of them are found in primary induction, and the majority of mammalian and avian induction systems require cell contacts with other cells or with an extracellular matrix.

3.3. *Transmission mechanisms in interactive processes*

3.3.1. *Introduction*

Since in most inductive interactions we know nothing of the nature of the signal substances, their distribution or mode of action, information about the mode of transmission of these signals is pertinent and might help us to solve the basic problem. This information is also vital in the search for the transmitter substances; should they be looked for in the extracellular space, on the cell surfaces or in the intracellular compartment? As long ago as 1955, Grobstein listed (1955b) three alternative transmission mechanisms: (1) long-range diffusion of signal substances; (2) action of extracellular matrix components; and (3) action mediated by cell contacts. Here a somewhat different operational distinction is suggested between the various possibilities (Fig. 14): (1) Transmission of signal substances across intercellular spaces, i.e. extracellular long-range diffusion (Fig. 14A). (2) Actions requiring contact between two cells or between a cell and an extracellular matrix: (a) interaction between complementary structures attached to the cell surfaces, to the extracellular matrix, or to both (Fig. 14B and

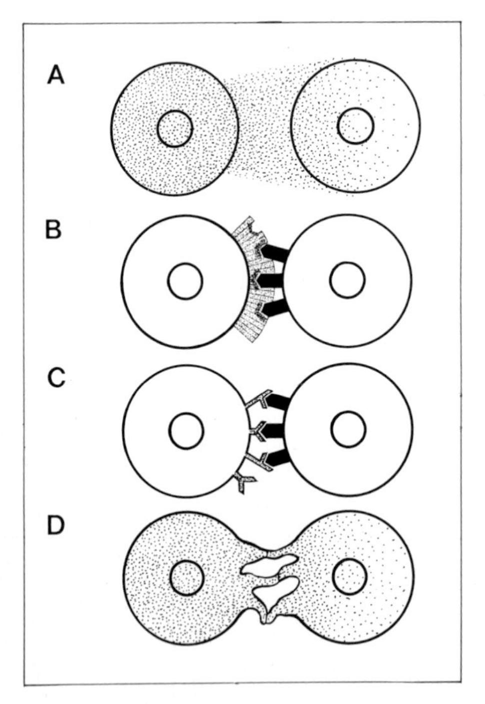

Fig. 14. Hypothesis on mechanisms of transmission of inductive signals. A. Extracellular diffusion. B. Matrix-mediated interaction. Complementary structures on the cell surface interact with the matrix. C. Cell surface mediated interaction. Complementary structures of the two cell surfaces interact. D. Cell junction mediated interaction. Direct intercytoplasmatic communication through membrane junctions.

C); and (b) transmission of intracellular molecules from one cell to another through membrane contacts such as gap junctions (Fig. 14D).

Our first alternative is identical with Grobstein's (1955b) first possibility, as it implies free movement of signal substances in the extracellular compartment. His "contact-mediated" mechanism has here been divided into two alternative modes of transmission, which differ in principle and which, as will be discussed later, would involve quite different molecules and pathways of communication. His matrix interaction where molecules with "restricted mobility" operate could belong to either of our alternatives depending on the degree of restriction. If the restriction is total, i.e. the signal molecules cannot move out of the matrix, contact between the matrix and the responding cell is required, and separation of the interactants by a small gap would then prevent induction. If the restriction in mobility of the signal substance is only partial, it is conceivable that the signal could leave the matrix. Then a small gap between the matrix and the tissue to be induced would not prevent induction. In this case the matrix itself would have no inductive action, it would just be coincidental to induction and might even slow it down. Matrix interaction as defined by Grobstein (1955b) would then correspond to extracellular long-range diffusion. In the following discussion we will confine the term matrix mediated interaction to those cases where contact between matrix and cell is required.

In the discussion of alternative transmission mechanisms, a great variety of directive and permissive influences have been lumped together under the term "inductive tissue interactions", although they do not necessarily operate through similar mechanisms. Therefore, there is no point in looking for factors common to the various interactive processes, but instead it would be worthwhile to test the validity of the hypotheses in the light of what we already know of various inductive interactions. Consequently, we will concentrate mainly on the problem of extracellular versus contact-mediated signal substances.

3.3.2. Extracellular transmission of molecules

3.3.2.1. Introduction. Extracellular transmission of molecules (diffusion) is defined here as a transfer of molecules released by the cells of the inductor to the target cells across an extracellular gap. This definition would include the classic concept of diffusible molecules (Bautzman et al., 1932; Holtfreter and Hamburger, 1955) but, as already discussed, not the hypothesis of transmissible matrix molecules with restricted mobility (Grobstein, 1955b). In many experiments, membrane filters with pores 0.1 to about 1 μm in diameter have been placed between the interacting tissues in the belief that this would exclude cell contacts. Usually, passage of inductive messages has been obtained (see Saxén and Kohonen, 1969; Saxén, 1972). The passage of the signal through the filter has generally been taken as evidence for extracellular transmission. Recently, it has been demonstrated that cytoplasmic processes can penetrate most filters, and that lack of demonstrable cell processes may be due to unsuitable fixation (p. 378). Therefore the earlier experiments have to be repeated by more modern methods and the earlier results re-evaluated. Some model systems have been reinvestigated for inductive interactions, but only in primary induction has it been possible to demonstrate with certainty that interaction between the tissues does not depend on contacts.

3.3.2.2. Primary embryonic induction. The original results of Saxén (1961, 1963) were based on filter experiments. When dorsal lip mesenchyme and presumptive neuroectoderm from amphibian gastrulae were separated by a Millipore filter with a nominal pore size of 0.8 μm, the ectodermal cells were frequently neuralized. Examination of the filters by electron microscopy did not reveal cytoplasmic material in the pores (Nyholm et al., 1962), and the authors concluded that the neuralizing effect was transmitted by a diffusible signal substance released by the normal inductor tissue. The same conclusion was reached by Gallera and his collaborators (Gallera, 1967; Gallera et al., 1968), who inserted Millipore filters between the ectoblast and the inductor (Hensen's node) in avian embryos. Induction was not prevented, and electron microscopy of the filters revealed no cytoplasmic contacts. Since recent findings in kidney tubule induction (p. 377) cast doubts on the validity of negative results obtained with earlier electron microscopic techniques, the matter was re-examined. Nucleopore filters were chosen because of their straight, regular-shaped pores and uniform structure as compared with the foamy Millipore material (Fig. 15; Cornell, 1969; England, 1969; Lehtonen et al., 1973). The filter assembly was similar to that used in previous studies (Saxén, 1961), with the interacting tissues separated by filters with different pore sizes (Toivonen et al., 1975). Through filters with pores of 0.1, 0.2 and 0.4 μm diameter neuralization occurred in about 80% of the cases. When the filters were examined using an improved method of fixation (p. 378), electron microscopy did not reveal penetration of cytoplasmic material, but occasionally a few granules in the pores. The granules resembled the extracellular material described by Tarin (1972, 1973) between the chorda-mesoderm and ectoderm in vivo. Since the presence of granules did not correlate with positive results, the authors took Tarin's view that this material was probably not the inductive signal. As transfilter cytoplasmic contacts were not detected, the conclusion was drawn that the neuralizing effect of dorsal lip mesoderm on ectoderm is mediated through extracellularly diffusible signal substances. This is not surprising, in view of the observation that neuralization can be produced by cell-free, soluble fractions (see Saxén and Toivonen, 1962).

Experiments with autoradiographic and immunofluorescence techniques have repeatedly demonstrated the passage of molecules from dorsal lip mesoderm to neuroectoderm (see Saxén, 1963; Toivonen et al., 1976). But the relevance of these observations is questionable, for Vainio et al. (1962) showed that both active and inactivated inductor tissues pass radioactive compounds to the ectoderm at the same rate. Therefore most, if not all, of the compounds transmitted are not concerned with the inductive signals.

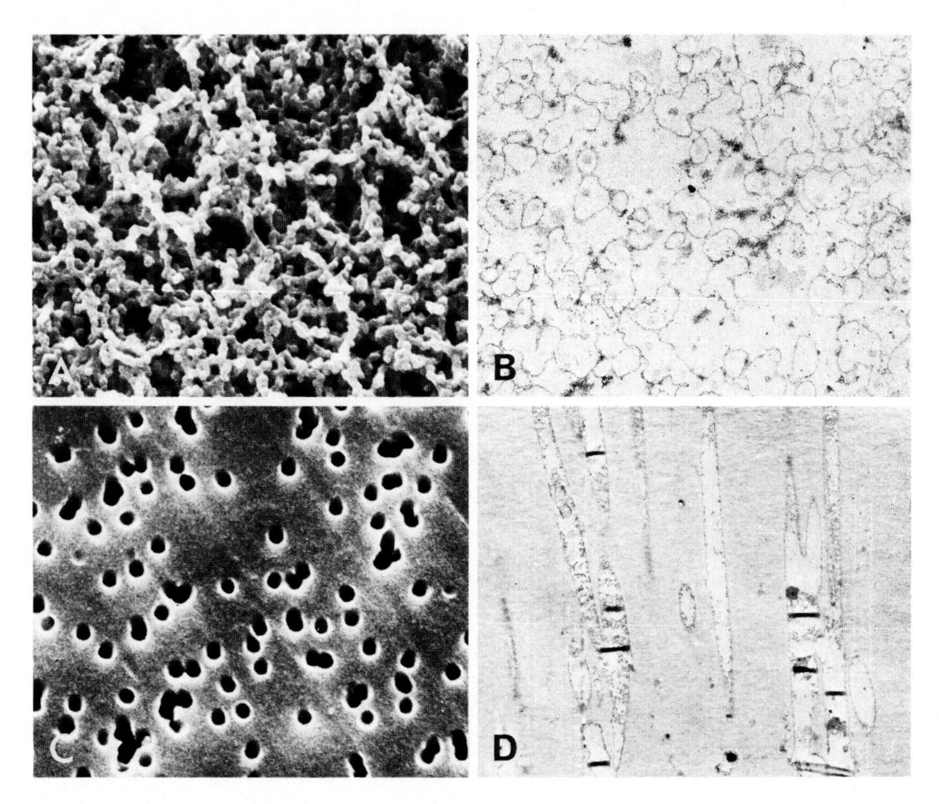

Fig. 15. Ultrastructure of membrane filters. A. Scanning electron micrograph of Millipore filter with a nominal pore size of 0.8 μm, ×4500. B. Transmission electron micrograph of filter in A, ×12,000 (From Lehtonen et al., 1973). C. Scanning electron micrograph of Nuclepore filter with a nominal pore size of 0.5 μm, ×4500. D. Transmission electron micrograph of filter in C, ×4500.

3.3.3. Cell-matrix interactions

3.3.3.1. Introduction. Since Grobstein (1955b) presented his hypothesis on transmission mechanisms (p. 368), interactions between the cell and its matrix (ECM) have been investigated extensively. His hypothesis of matrix interaction was based on two observations, that the inductive message could pass a porous filter about 20 μm thick, which by electron microscopy looked empty, and that the message could not cross a distance of 75 μm. The topic has recently been discussed at length at a symposium on "Extracellular Matrix Influences on Gene Expression" (Slavkin and Greulich, 1975). The chondrogenic factor in the perinotochordal ECM (p. 366) is a good example of ECM components with morphogenetic activities. Now some other examples of ECM components involved in inductive tissue interactions will be considered.

3.3.3.2. Skin. Differentiation of the epidermis in vitro requires the presence of a suitable substrate such as Millipore filter or collagen gel, which allows the cells to proliferate and inhibits keratinization by a mechanism of unknown nature (Wessells, 1964). In studies on the interaction between epidermis and mesenchyme, McLoughlin (1961, 1963) used two-layered, undifferentiated epidermis from a 5-day chick embryo (p. 350). In vivo this ectoderm was not yet capable of wound healing and did not show any sign of differentiation. But when combined in vitro with heart, limb bud, proventriculus, gizzard or other mesenchymes, it adopted a morphogenetic pattern varying from keratinization to mucus-producing ciliated epithelium. The first sign of orientation was the appearance of a PAS-positive basement membrane, which caused the nuclei of the epidermal cells nearby to line up, re-establishing the basal cell line. The more distant cells inserted long processes through the intervening tissue and made contact with the basement membrane, pulling themselves into the basal cell sheet, which then spread over the connective tissue in a characteristic manner. The spreading was most marked over the heart myoblasts and minimal over the limb mesenchyme. McLoughlin (1961) attributed the differences in the morphogenetic course and spreading pattern to "pre-existing differences in the intercellular material of various connective tissues". In the limb, for example, coarse collagen fibrils were associated with a matrix containing only neutral mucopolysaccharides, whereas the heart cultures contained fine collagen fibrils embedded in a matrix rich in acid mucopolysaccharides. Orientation was just as sharp with cell-free intercellular material of dermal origin as with connective tissues, and extended beyond the basal cell layer. Cotton threads, rayon or a cover glass had no such effect.

In vivo, the dermis forms a collagen lattice, which governs the orientation of the dermal cells. This lattice was first noted by Stuart and Moscona (1967), who studied the appearance of dermal papillae in the development of the feathers. The first sign of feather formation is an epidermal thickening, which is followed by condensation of cells in the dermis within 3–6 hours. This time is so short that it rules out cell division, the explanation suggested by Wessells (1965). Using colchicine, Stuart and Moscona (1967) showed that the condensates consisted mainly of cells which had migrated. In their movement, the cells followed a birefringent lattice, which first appeared in the middorsal line of the avian embryo and extended postero-laterally. The lattice divided the dermal cells into two populations: small, closely packed cells at the intersections and elongated cells extending from one dermal papilla to another. The orientation of the cells was thought to depend on the tendencies of fibroblasts, to migrate and to follow aligned substrates. A relative decrease in the number of fibres in parallel alignment or an increase in the number of intersections reduces the rate of cell migration (Fig. 16; Stuart et al., 1972).

It is thus hardly surprising that the cells should follow the lattice and form condensates in the intersections, but nothing is known of the events inducing the appearance of the lattice. As shown by recombination experiments with normal and scaleless mutant chicks, the defective tissue is not the dermis but the epidermis (Sengel and Abbot, 1962). The dermis lacks both the lattice and the dermal papillae necessary for feather formation, but collagen synthesis proceeds

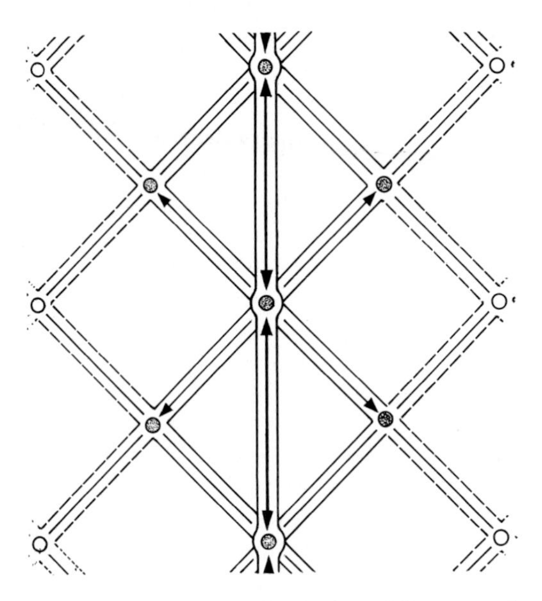

Fig. 16. Diagram of the skin in the saddle tract region of an embryonic chick to show the relative position of aligned fibrous material and dermal cells during development of papillae. The vertical structure in the diagram is the primary row, which contains major groups of aligned fibers and cells extending anterio-posteriorly between successively forming papillae (⊛). In the lateral region the major bands of aligned fibers and cells extend diagonally from the more mature medial papillae to the prospective lateral papillae (○). Note that papillae are formed where bands cross each other (After Stuart et al., 1972).

normally (Goetinck and Sekellick, 1970). Goetinck (1970) therefore suggested that the lattice is formed under epidermal influence.

Thus, although the genesis of the lattice is not yet understood, the model resembles that reported below in connection with glandular organs, in suggesting that the factor essential for some morphogenetic events is an organized intercellular matrix.

3.3.3.3. Glandular organs. The development of several glandular organs depends on epithelio-mesenchymal interactions between the branching epithelium and its mesenchymal stroma. In every instance reported, the interactants are separated by a basement membrane-like extracellular material or matrix, consisting of three major components: a large-molecular glycoprotein; a smaller glycoprotein with a low carbohydrate content; and a special type of collagen (Kefalides, 1968; Spiro and Fukushi, 1969). Experiments have been made to test the morphogenetic significance of these compounds, especially in the development of the embryonic salivary gland of the mouse. If salivary epithelium and its mesenchymal stroma are separated and cultured transfilter, the epithelium branches and differentiates. The discovery that treatment of the epithelium with collagenase prevented this morphogenesis led to the suggestion that collagen plays a morphogenetic role (Grobstein and Cohen, 1965; Wessells and Cohen,

1968). Later, however, Bernfield and Wessells (1970) reported that the collagenase preparations used in these studies were contaminated with mucopolysaccharidases. Mucopolysaccharides accumulate all over the epithelium, but especially around the branching points. After removal of collagen with low concentrations of purified collagenase, the mucopolysaccharides remain on the epithelium, which shows a normal branching pattern (Fig. 17). If extracellular mucopolysaccharides are removed with hyaluronidase, and the epithelium is recombined with its mesenchyme, branching and morphogenesis are seen only after a considerable lag period, which presumably reflects the time required for resynthesis (Bernfield et al., 1972). The authors conclude that the normal salivary gland morphogenesis is dependent on acid mucopolysaccharides in the epithelial basal lamina. The mode of action of this signal substance is unknown.

3.3.3.4. Lens. In the process of lens induction the optic vesicle and the head ectoderm come into intimate contact with each other, and are difficult to separate by microsurgery (McKeehan, 1951). But if, before they make contact, cellophane is inserted between the interacting tissues, the future development of the ectoderm into a lens is impaired (McKeehan, 1951). Induction was obtained if the optic vesicle was separated from the ectoderm by agar blocks 20 μm thick (McKeehan, 1958). By light microscopy, cytoplasmic processes could not be detected within the agar. Lens induction has been reported to occur through a Millipore filter with a thickness of 25 μm and a nominal pore size of 0.45 μm (Muthukkaruppan, 1965). These results have been taken to indicate that no cell contacts are formed between the two interacting tissues during the time of intimate adherence.

The interspace between the optic cup and the ectoderm has been studied repeatedly by electron microscopy in several species, with somewhat varying reports about its width, 0.08 to 6 μm (Cohen, 1961; Hunt, 1961; Weiss and Jackson, 1961; Hendrix and Zwaan, 1974, 1975). None of these authors have reported anything which would suggest the existence of cellular processes connecting the tissues. They all are of the opinion that during the interaction, in addition to a loose filamentous network embedded in the matrix, there are transverse filaments which span the interlayer and overlap the cell borders. Recently, Hendrix and Zwaan (1974, 1975) have reported that during this induction the amount of glycoproteins in the interspace increases markedly, and that the two epithelia are the source of the matrix materials. The matrix had a typical structure with globular aggregates and thin filaments. The authors propose that the macromolecules interact at different concentrations to form aggregates of various structure by a process of self-assembly. They furthermore suggest that changes in the density of the matrix leading to increased adhesion between lens rudiment and optic vesicle could restrict lateral spreading of lens cells and thereby influence morphogenesis.

3.3.3.5. Cornea. The differentiation of corneal epithelium depends on an inductive influence from the lens capsule (Hay and Revel, 1969). This may be an

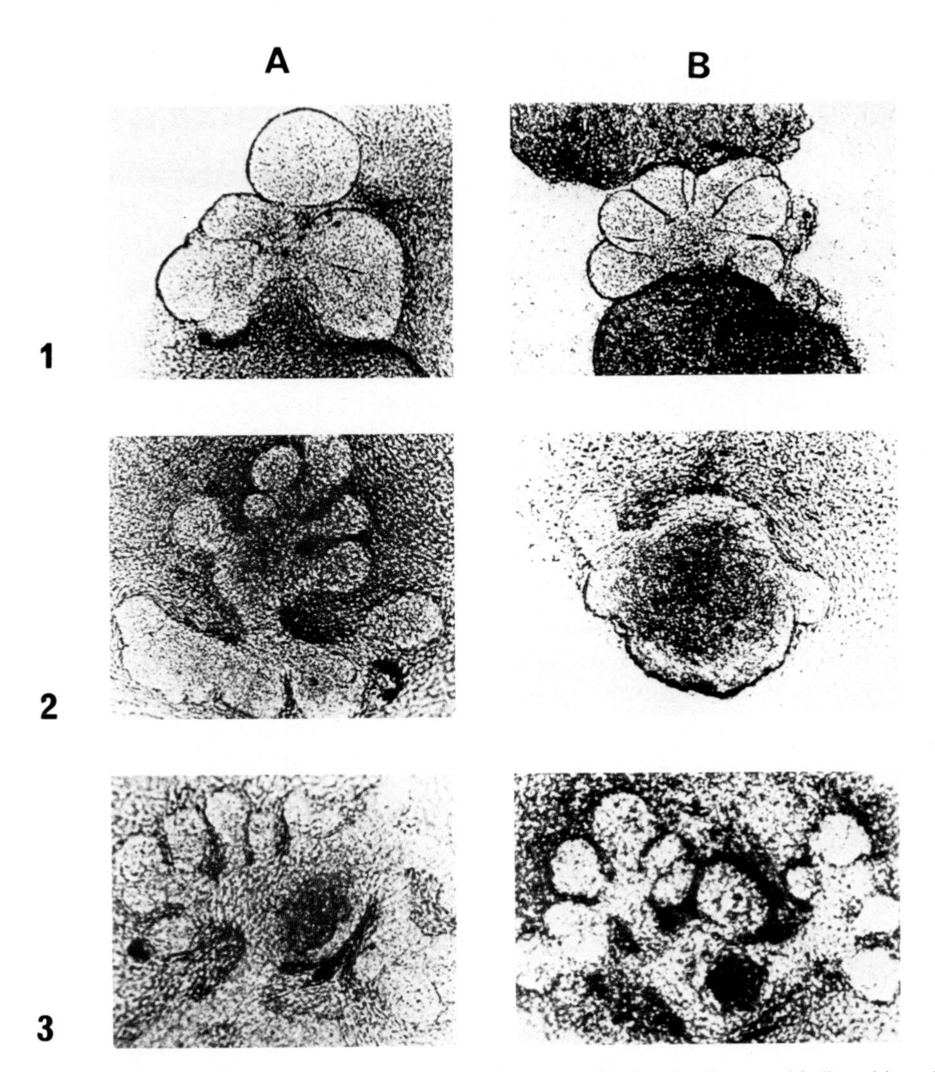

Fig. 17. Effect of enzyme treatment on the development of isolated salivary epithelia cultivated in combination with fresh salivary mesenchyme (Bernfield and Wessells, 1970). A. Collagenase, mucopolysaccharidase free, (0.6 mg/ml). B. Collagenase (0.6 mg/ml) + hyaluronidase (3 μg/ml) 1. at time of explantation; 2. at 24 hr of culture; and 3. at 48 hr of culture.

additional example of extracellular material which controls differentiation. Even NaOH-extracted lens capsule, cartilaginous collagen, collagen-rich vitreous humour and tendon collagen affect the corneal epithelium, enhancing its synthesis of an extracellular glycosaminoglycan matrix (Hay and Dodson, 1973; Meier and Hay, 1974). Recently, this interactive event was studied by interposing Nuclepore filters between the corneal epithelium and killed lens capsule (Meier and Hay, 1975). Corneal epithelial cells were shown to send cytoplasmic processes through the filter to make contact with the lens capsule. Synthesis of the epithelial glycosaminoglycan matrix was quantitatively dependent on the thickness of the filter and on the pore size (Fig. 18). The authors suggested that in this interactive system there is a direct interdependence between the surface of the corneal epithelial cells and the extracellular material of the lens capsule.

3.3.3.6. Limb bud. The ultrastructure of the ectodermal-mesenchymal region of the embryonic developing limb bud has been studied in several amphibians (Balinsky, 1972; Tarin and Sturdee, 1974; Kelley and Bluemink, 1974), in the chick (Jurand, 1965; Bérczy, 1966; Ede et al., 1974; also see chapter 10 of this volume), in the mouse (Jurand, 1965) and in man (Kelley, 1973).

All these studies have shown that the mesenchymal cells make contact with the basal lamina, but in no case was a mesenchymal cell process found to penetrate through the lamina. In a stage 22–26 chick, Jurand (1965) reported discontinuity of the basal lamina under the apical ectodermal ridge (AER) of the wing bud. This has been confirmed by Ede et al. (1974), but neither of these authors observed contacts between epithelium and mesenchyme. In both scanning and transmission EM mesenchymal cytoplasmic processes were found to be in close contact with the basal lamina of the ectoderm (Ede et al., 1974). Jurand's (1965) result in the chick was not confirmed in *Xenopus* by Tarin and Sturdee (1974), who suggested that the discontinuities in the basal lamina might be fixation artefacts.

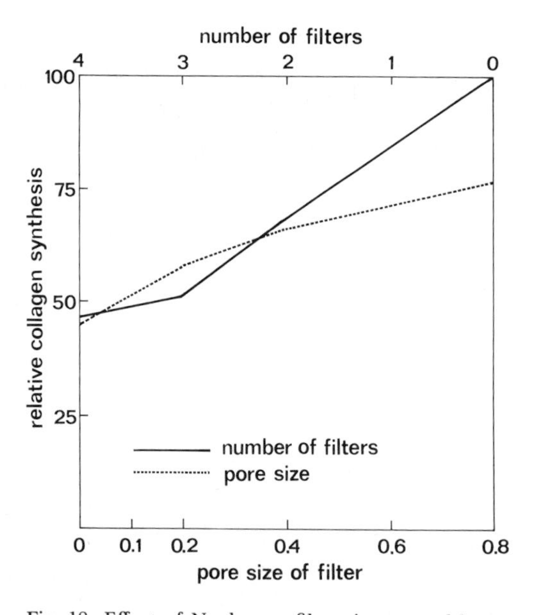

Fig. 18. Effect of Nuclepore filters interposed between corneal epithelium and lens capsule on the synthesis of epithelial glycosaminoglycans, expressed as the incorporation of radioactively labelled precursors. The incorporation, when the tissues were in direct contact, is taken as 100. Pore size "0" denotes effect of plain filter without underlying lens capsule (After Meier and Hay, 1975).

3.3.4. *Cell-mediated interactions*

3.3.4.1. Introduction. The possibility of cell contacts mediating developmental signals, originally proposed by Spemann (1912), has been revived in a modern form. At the highest magnification available to light microscopy, two cells were said to be in contact when no intercellular space could be seen. Because of the limited resolution of light microscopy, such cells could still be 200 nm apart. With the advent of electron microscopy the problem of what is meant by the expression cell contact became even more complex. Over what distance can cells still be said to be in contact? Does contact imply touching of membranes, or can cells be in contact through molecules at the cell periphery? How wide an intervening layer can there be of acid mucopolysaccharides or other negatively charged substances? The cell periphery (Weiss, 1969) may be regarded as an integral part of the cell. Accordingly, cells are considered to be in contact when the outer layers of the apposing trilaminar cell membranes are at most 10–20 nm apart. This is a slight modification of the definition of Weiss in 1958: "We shall consider a cell in contact with another body not only if the two surfaces are in direct apposition, but also if they are separated by a narrow space occupied by a molecular population whose free mobility is restrained". Obviously, molecules of the cell surfaces can form temporary links between the two surfaces when the surfaces are sufficiently close to each other and can thus transmit messages that could not be conveyed by extracellular diffusion of the same molecules.

3.3.4.2. Testing of contact hypothesis with transfilter experiments. The transfilter technique widely used in the study of inductive interactions (see p. 369) was designed to exclude cytoplasmic contacts between the interacting tissues and thus to distinguish between interactions involving membrane contacts and those involving diffusible inductive agents (Grobstein, 1953b). A large number of tissue interactions have since been studied by the transfilter technique (Table 1). In all these studies the successful induction obtained was taken as evidence of transmission mechanisms other than those based on actual cell contacts.

Grobstein and Dalton (1957) studied the transfilter induction of mouse kidney tubules by the dorsal spinal cord. They found that the inductive influence could pass Millipore filters with pore sizes down to 0.1 μm. Their filters were of different thicknesses. Thin 16–19 μm filters with a pore size of 0.1 μm permitted induction in all but one of 7 explants, but when the filter thickness was increased to 32–38 μm, kidney tubules were formed in only 3 cases out of 7. By electron microscopy, cytoplasmic material was seen deep within the filters with larger pores, but in those with 0.1 μm pores the cytoplasmic material extended into the filters only for 1–2 μm. The conclusion was that "the inductive activity in this system is not dependent upon cytoplasmic contact and hence resides in materials which are at least potentially extracytoplasmic", and that cytoplasmic penetration into filters with large pore sizes is co-incidental to induction.

Recently, Lehtonen et al. (1975) have reinvestigated kidney tubule induction through 0.1 μm pore size Millipore filters. In 6 cases out of 31, induction took

Table 1
Inductive tissue interactions operating through a filter membrane.

System	References
Epithelial-mesenchymal interactions	
Glandular organs	
Salivary gland	Grobstein (1953b)
Metanephric kidney	Grobstein (1953b, 1957)
Pancreas	Golosov and Grobstein (1962)
Thyroid gland	Hilfer (1968)
Lung	Taderera (1967)
Liver	LeDouarin (1976)
Limb bud	Saunders and Gasseling (1963)
Skin	Wessells (1962)
Tooth germ	Koch (1967)
Primary induction	
Neuralization of amphibian ectoderm	Saxén (1961, 1963)
Neuralization of chick ectoderm	Gallera (1967); Gallera et al. (1968)
Induction of cartilage	Lash et al. (1957); Cooper (1965)
Induction of lens	Muthukkaruppan (1965)
Interaction between corneal epithelium and lens capsule	Meier and Hay (1975)
Interaction between anterior pituitary and brain	Doskocil (1968)

place through filters 23–32 μm thick. In the positive cases the cytoplasmic processes penetrated deep into the filter pores. The discrepancy between these results and those of Grobstein and Dalton (1957) is explained by the difference in the methods of fixation. The chromeosmium fixative available in the 1950's did not preserve cytoplasmic processes within pores only 0.22 μm in diameter. In contrast, the more recent glutaraldehyde fixation was able to do this (Fig. 19). Even this fixation did not always preserve processes in small pores, the preservation depending on subtle modifications of the methods of fixation. Wartiovaara et al. (1974) used polycarbonate Nuclepore filters in similar experiments. In the tubule-forming explants, cytoplasmic processes made contacts with cells on the opposite side (Fig. 20). Little extracellular material was seen between the processes. The gap between the apposing cell membranes was often less than 10 nm, but specialized junctions were not identified. These results were taken to suggest that the most likely mechanism of communication in this transfilter induction is cellular contact, rather than long-range diffusion of signal substances (Wartiovaara et al., 1974). Additional evidence for the transfer of contact-mediated information has been obtained in transfilter experiments with another potent inductor, salivary mesenchyme (Lehtonen et al., 1975a). Induction by salivary mesenchyme required the use of Nuclepore filters with a pore size

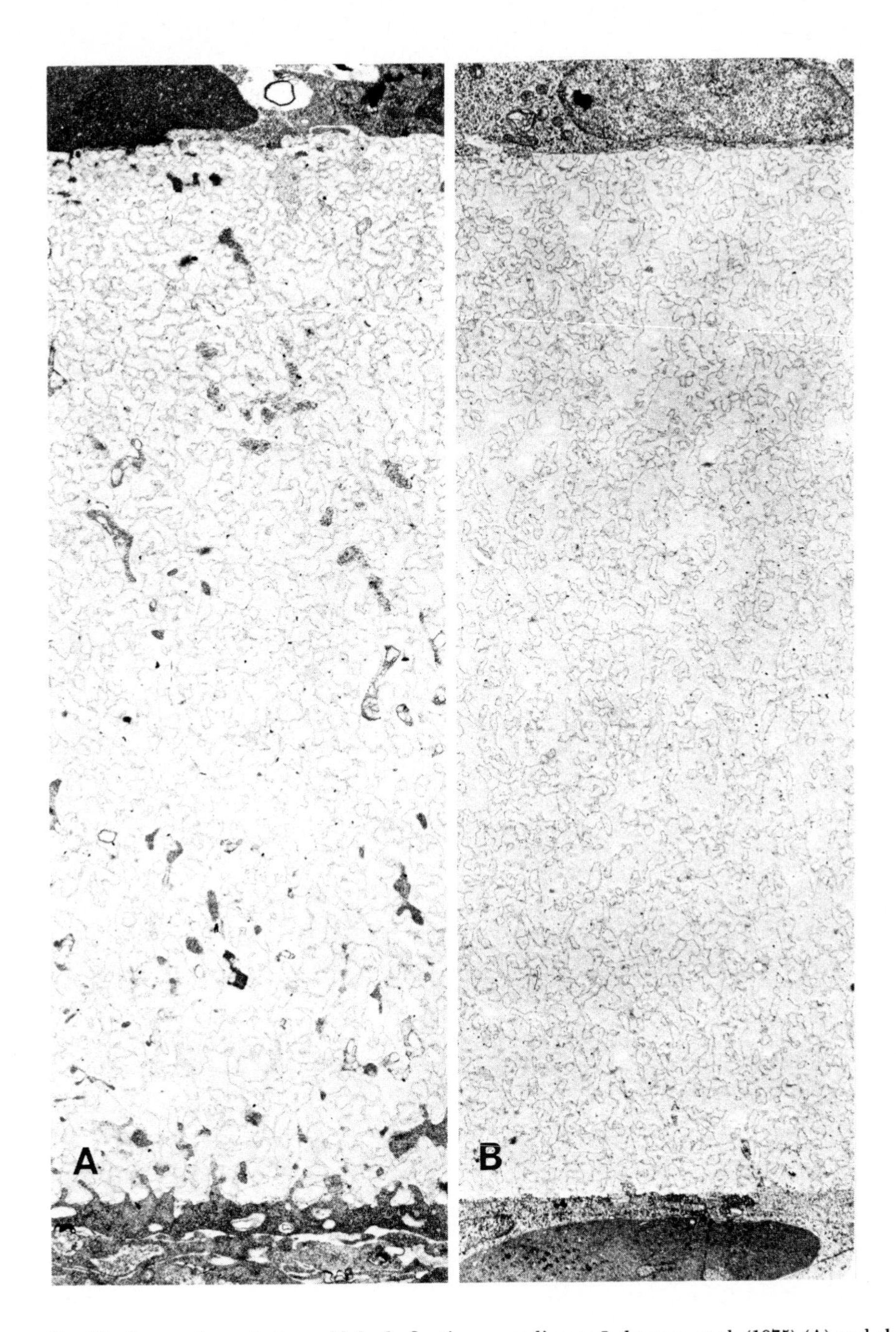

Fig. 19. Comparison of glutaraldehyde fixation according to Lehtonen et al. (1975) (A) and chrome osmium fixation according to Grobstein and Dalton (1957) (B) on the preservation of cytoplasmic processes within type 0.22 μm pore size Millipore filters in identical culture conditions. A × 13000, B × 9000 (From Lehtonen et al., 1975).

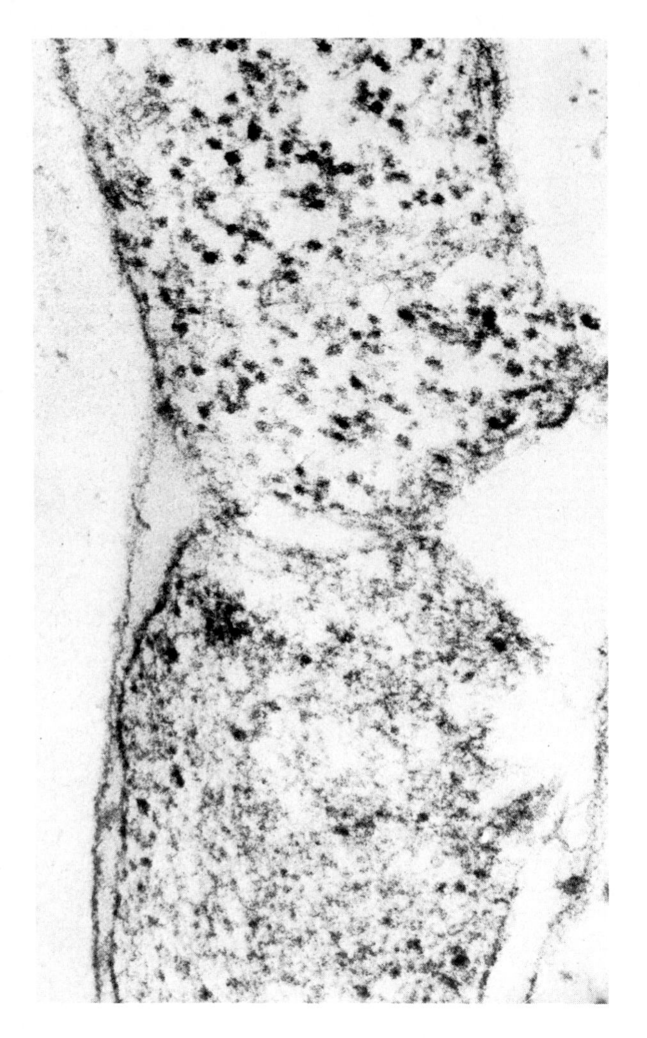

Fig. 20. Transmission electron micrograph of a transfilter explant on 0.6 μm pore size Nuclepore filter with kidney mesenchyme on top and spinal cord below the filter. Note the close apposition of the two tissues within the filter. Magnification ×60000.

of at least 1.0 μm as compared with the pore size of at least 0.1 μm when spinal cord was used as inductor. Electron microscopy revealed clear differences in the penetration of cytoplasmic processes from the two inductors through different filters. Again induction never occurred in transfilter experiments preventing cytoplasmic penetration.

3.3.4.3. Ultrastructural findings of in vivo contacts between interacting tissues

Early morphogenetic interactions. Little is known about the nature or mechanisms of the first determinative events in the early embryo. For example we do not know what triggers the proliferation of the trophectoderm in the blastocyst. A feature common to the early morphogenetic events seems to be the close contact between the cells involved (see Gardner and Papaioannou, 1975).

Early ultrastructural studies on rat blastocysts revealed several different kinds of cell-to-cell contacts at low magnification (Schlafke and Enders, 1967; Enders and Schlafke, 1967). More detailed studies with mouse blastocysts showed that specialized junctions exist between both like and unlike cells (Nadijcka and Hillman, 1974). Early in the blastocyst period which lasts altogether about 2 days, inner cells are connected with overlying trophectodermal cells by desmosomes and focal tight junctions. Later on, the number of junctions between unlike cells decreases, desmosomes especially become less frequent. Simultaneously, a basement membrane develops between the embryonic cell mass and the overlying trophectoderm. At a still later stage the basement membrane appears to be continuous except at some focal points, where tight junctions still attach the embryonic cell mass to the trophectoderm. No gap junctions were described in this study on the mouse blastocyst. Ducibella et al. (1975) using lanthanum as an electromicroscopic-tracer demonstrated gap and adhering junctions between cells of the inner cell mass and neighbouring trophectodermal cells in mouse blastocysts. In the same study they also used the freeze-fracture method and showed in rabbit blastocysts both gap and tight junctions between like cells. However, they did not comment on the nature of the contacts between the unlike cells of the blastocyst, i.e., cells of the inner cell mass and trophectodermal cells.

During the late blastocyst stages, a basal membrane underlies the embryonic trophoblast cells. It is during those stages that the inner cells are known to induce proliferation in the trophoblast cells (Gardner, 1972). The existence of this membrane might itself explain the sustaining effect on cell division by providing a favourable substrate for the basally located cells of the ectoplacental cone. It is unknown whether the basal membrane can be substituted by reconstituted collagen, by other substrates in vitro, or whether the presence of living cells is necessary.

In chick embryos at the 4-somite stage the cells of the primary mesenchyme are connected by close junctions (intercellular spaces less than 10 nm), and focal tight junctions, to the basal surfaces of the adjacent epithelia, the epiblast and the hypoblast (Trelstad et al., 1966, 1967). Junctions were also found between homotypic epiblast, mesoblast and hypoblast cells. Unfortunately, with the techniques then available, no distinction could be made between tight junctions and gap junctions. When cell contacts in early chick embryos were investigated by the freeze-etching technique (Revel et al., 1973), the "focal tight junctions" were found to be of both tight and gap types (p. 389). However, nothing was stated about the nature of the junctions between unlike cells, probably because of technical (orientation) difficulties. The existence of cell contacts between primary mesenchymal cells and hypoblast and epiblast has also been demonstrated in scanning electron microscopy (Fig. 21; Revel, 1974).

Liver. Junctions between unlike cells are not necessarily unique to primary inductive events in embryogenesis. Rifkind et al. (1969) studied the early

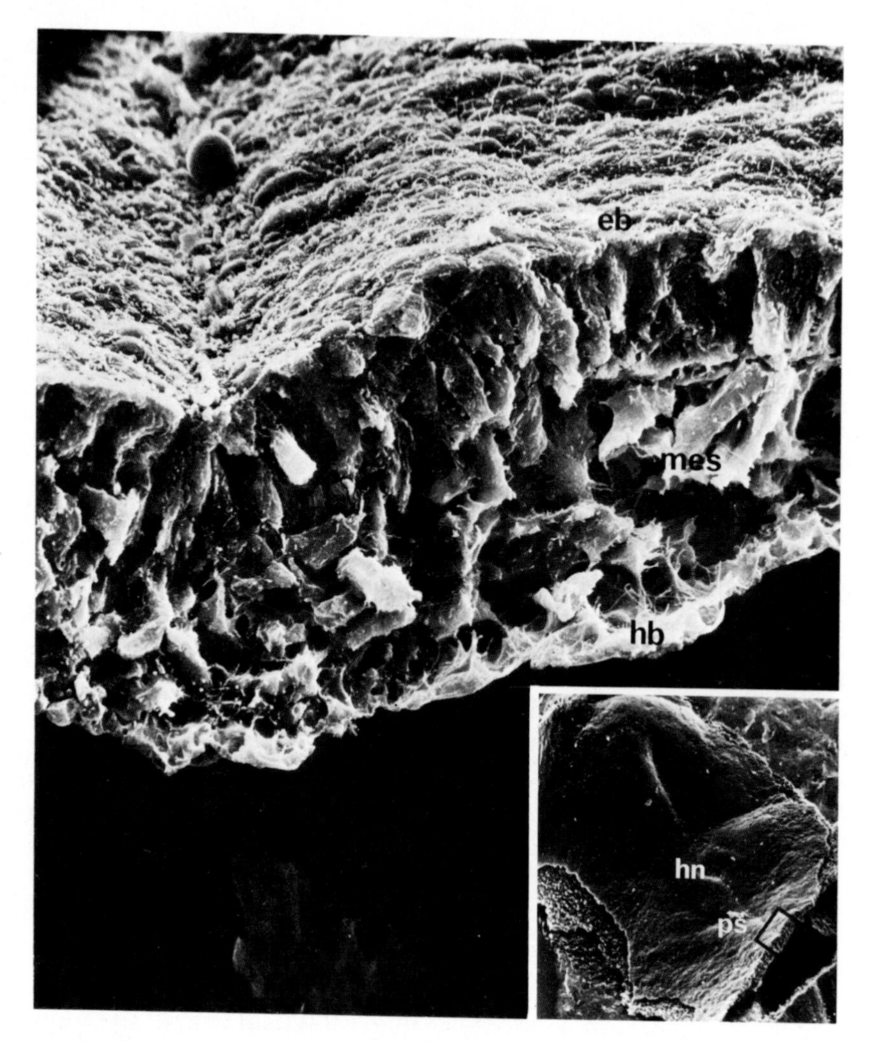

Fig. 21. Scanning electron micrograph of a critical point dried chick embryo, stage 5. The inset shows the whole embryo with the primitive streak (ps), Hensen's node (hn) and head process. The enclosed area is shown in the picture. Note the contacts made by the cells of the mesoblast (mes) with both hypoblast (hb) and epiblast (eb) inner surfaces (From Revel, 1974).

interaction between hepatic epithelium and mesenchyme, which leads to hepatocellular differentiation. No basement membrane was seen between epithelial cells and adjacent mesenchyme cells. Adherent junctions with a gap of about 15 nm between plasma membranes and desmosomes were established transiently across the epithelio-mesenchymal interface.

Tooth. In the cat, Pannese (1962) observed desmosomes and terminal bars between differentiated mesenchymal cells (odontoblasts), and undifferentiated epithelial cells (preameloblasts). Close contacts between odontoblasts and preameloblasts have also been reported in growing or differentiating teeth of the opossum (Lester, 1970), rat (Kallenbach, 1971) and cat (Silva and Kailis, 1972), but in these studies no specialized junctions were detected between the interacting cells. Kallenbach (1971) showed that during the differentiation of the rat incisor, odontoblast processes may penetrate up to 2 μm into the preameloblast layer, but always leave a gap at least 10 nm wide between the two cell types. Neither a basal lamina nor specialized contact sites were found between the apposing cell membranes. Kallenbach (1971) suggested that possibly the specialized junctions described by Pannese (1962) take place only between homotypic cells of the enamel organ.

Although reports on the nature of the contacts between interacting odontoblasts and preameloblasts are contradictory, it is evident that during tooth differentiation processes from odontoblasts make contact with the preameloblast cells through gaps in the basal lamina (Fig. 22; Slavkin and Bringas, 1976).

Duodenal mucosa. Mathan et al. (1972) have traced the development of the duodenal mucosa. 5–7 days before birth a well-defined basal lamina separates rat duodenal epithelium from mesenchymal cells. 3–4 days before birth the basal lamina is discontinuous in many areas. Epithelial cell processes project through the gaps in the basal lamina and form direct contacts with the mesenchymal cells. The epithelial and mesenchymal cell membranes are less than 10 nm apart at the contact sites. These epithelial cell processes exist in large numbers until 7–10 days after birth, and then decrease in both number and size until 3–4 weeks after birth. Thereafter the epithelio-mesenchymal interface resembles that seen in adult rats, with only occasional gaps in the basal lamina.

The appearance of epithelio-mesenchymal contacts in the rat duodenum coincides with a period of rapid growth and differentiation. Mathan et al. (1972) conclude that: "the presence of a large number of epithelio-mesenchymal contact sites during the period of rapid growth and differentiation of duodenal mucosa may reflect epithelio-mesenchymal cell interactions which may facilitate the maturation of the duodenal mucosa".

Submandibular gland. Epithelio-mesenchymal cell contacts have also been reported in the rat submandibular gland (Cutler and Chaudhry, 1973). These contacts, which were seen during the period of rapid branching and budding of the anlage (days 15 and 16), were made by either mesenchymal or epithelial cell extensions, which penetrated small gaps in the basal lamina (Fig. 23). Morphologically, the contact zones were of three different types, and the interspaces between the apposing cell membranes varied from 20 nm to 4 nm. Occasionally, the apposing membranes even seemed to show focal fusions which obliterated the gap between them. No septate or gap type junctions were seen.

Kidney. During kidney tubule induction in the developing mouse embryo cell contacts take place through gaps in the basal lamina at the inductively active tips of the branching ureter bud (Lehtonen, 1975). At these points the epithelial and mesenchymal cell membranes were mostly separated by 15 to 20 nm and in places by less than 10 nm. At a few focal points, the membranes of adjacent

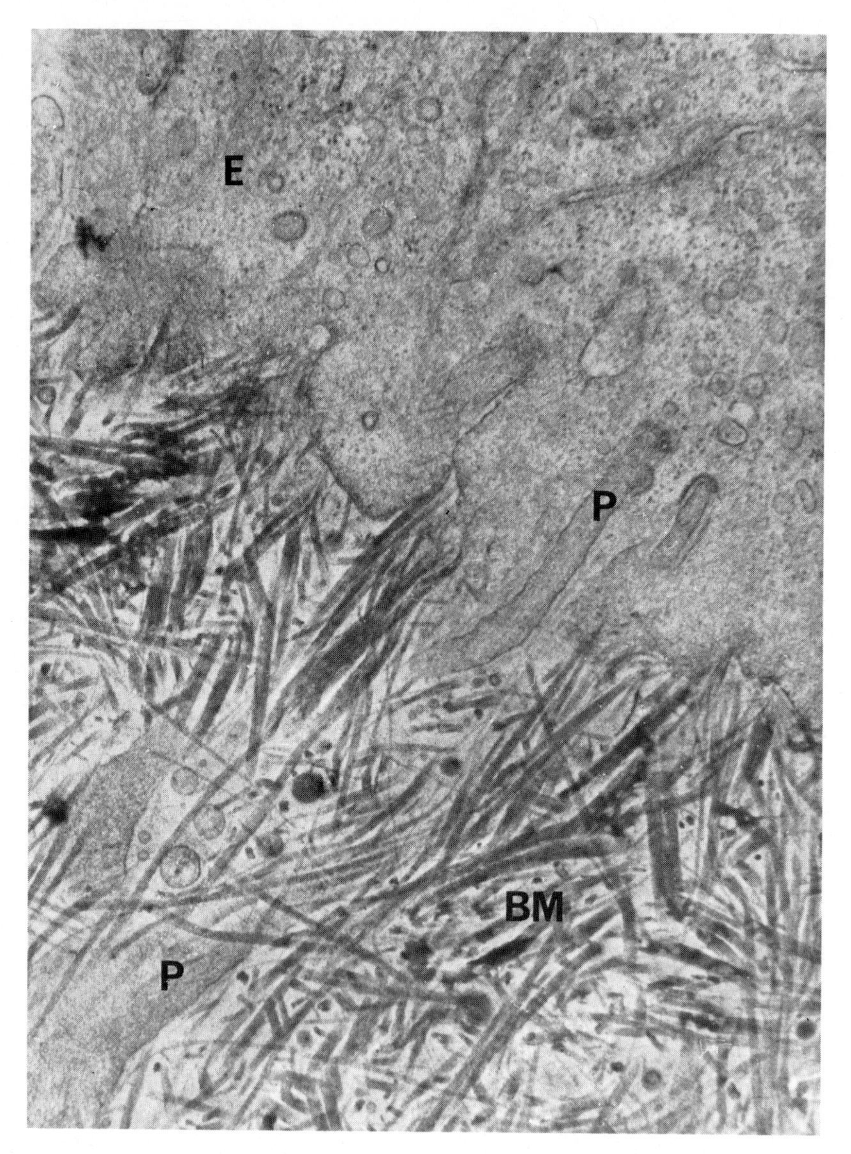

Fig. 22. Electron photomicrograph of the interface between preodontoblasts and preameloblasts during embryonic mouse tooth development. Note the penetration of preodontoblast cell processes (P) through basement membrane (BM) into clefts present on the undersurface of the epithelial cells (E). Magnification ×30000 (Courtesy of Harold Slavkin).

epithelial and mesenchymal cells seemed to fuse, but no specialized junctions could be identified at the epithelio-mesenchymal interface (Fig. 24).

3.3.5. Remarks

There are some well-documented examples of extracellular transmission of morphogenetic signals by diffusion. They seem all to be restricted to primary

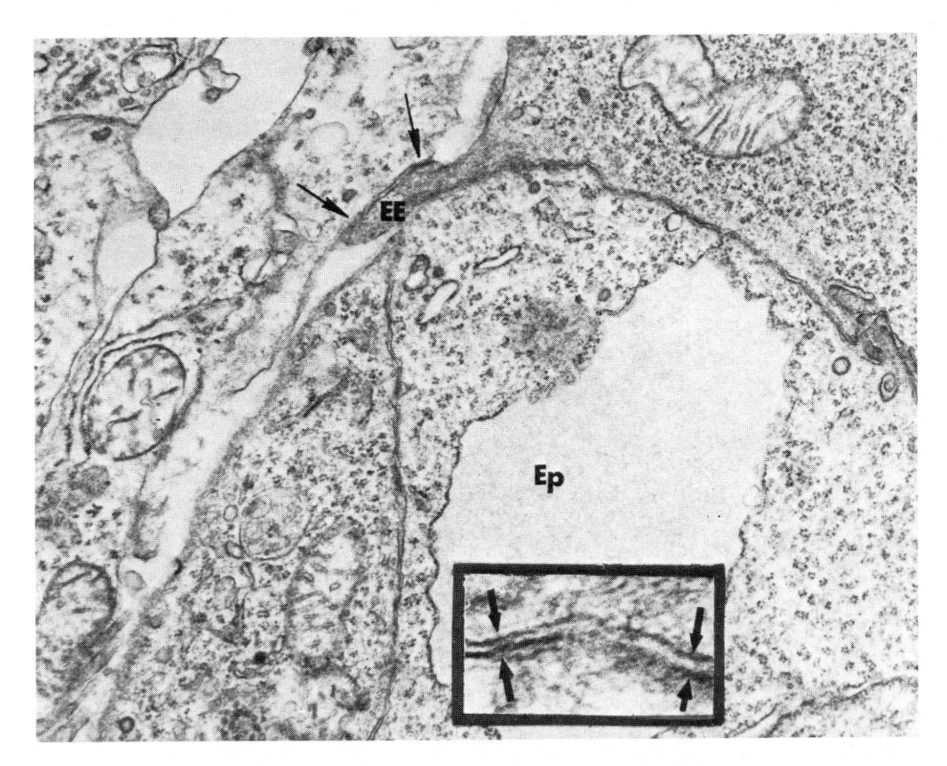

Fig. 23. Transmission electron micrograph of the submandibular gland from a 16-day-old rat embryo at the area of end bud epithelium (Ep). An epithelial extension (EE) makes a broad zone of contact with a neighbouring mesenchymal cell. Note broad dense zones of membrane fusion (arrows). Magnification × 21 000. Inset: Similar area of fusion after en bloc staining. The membranes are separated by a gap of approximately 4 nm (arrows). Magnification × 180000. (From Cutler and Chaudhry, 1973).

induction and they include the action of large molecules such as the vegetalizing protein isolated by Tiedemann (1973) as well as the action of monovalent ions as described by Barth and Barth (1968).

The examples of matrix mediated inductive actions are restricted to secondary inductive events in mammals and birds. Although salivary gland development, and probably also skin development, seem to depend on a morphogenetic action of an extracellular matrix on the responding cells, other inductive events currently believed to be matrix mediated may, in fact, be cell contact mediated. The demonstration of cell processes in a thick extracellular matrix may sometimes be technically difficult.

In an increasing number of inductive events cells and tissues appear to be in close apposition and may pass signals through thin basement membranes (with pores?) via cytoplasmic processes. This seems to be the case in many epithelio-mesenchymal interactions. It might be argued that the presence of contacting cell processes within an extracellular matrix does not prove that the cell contacts rather than the matrix are morphogenetically active. However, the occurrence

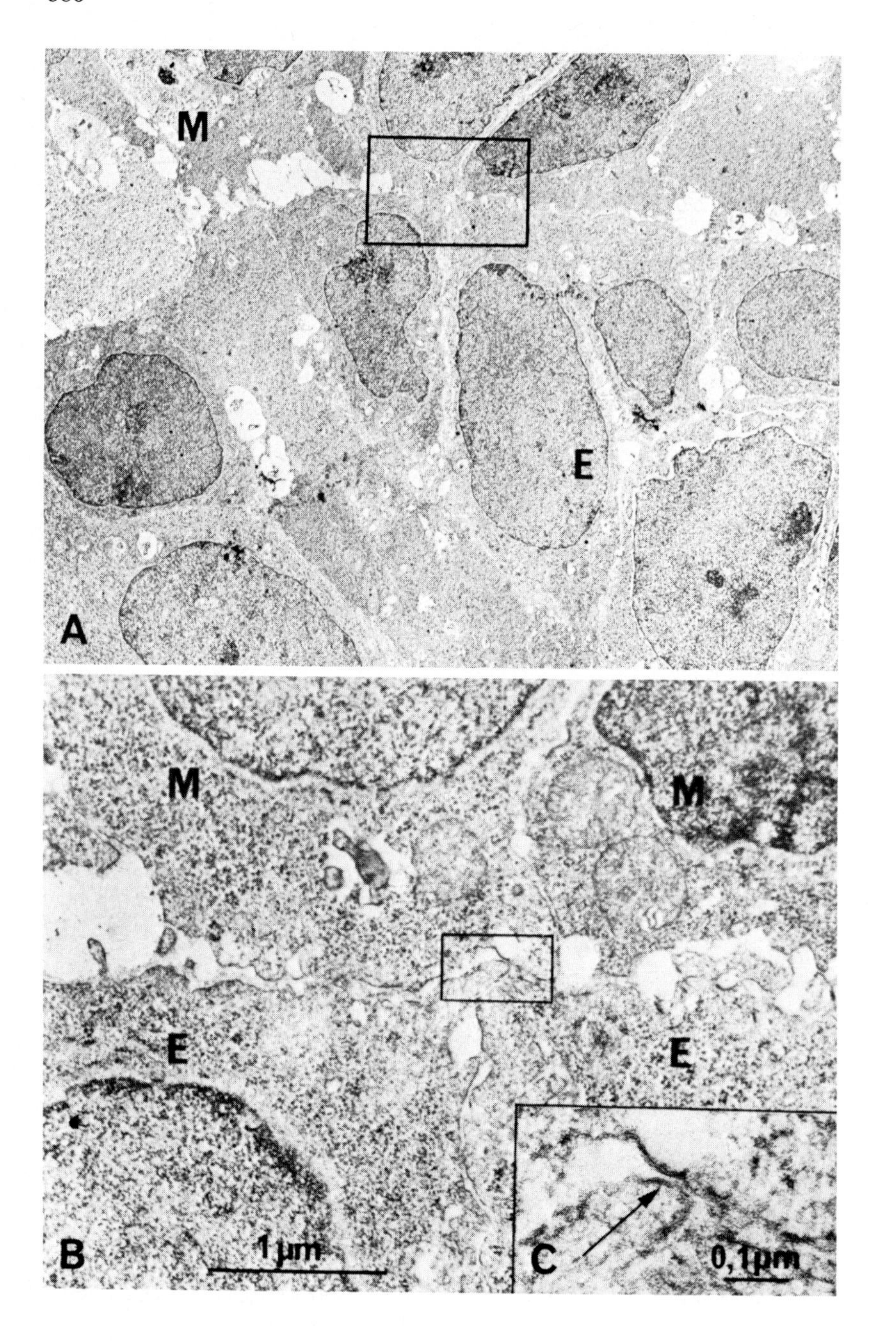

Fig. 24A-C. Transmission electron micrographs of a kidney from a 12-day mouse embryo. Three magnifications of the area of interaction between epithelial (E) and mesenchymal (M) cells. At places the membranes of the interacting cells are only about 10 nm apart (arrow). A × 6000, B × 30000, C × 110 000 (From Lehtonen, 1975).

of membrane contacts between interacting tissues makes it unnecessary to postulate extracellular transmission of active compounds, especially as the available data argue against extracellular diffusion in these cases. Before further speculations can be made on possible mechanisms for inductive interactions, cell relations during such events must be studied more thoroughly.

3.4. Molecular models for embryonic induction

3.4.1. Introduction

Hypotheses concerning the mechanisms by which inductive signals are transmitted (see p. 368) are based on two fundamentally different models of cell communication: (1) an interaction between complementary surface molecules, which might, but need not necessarily, involve a transfer of surface molecules; and (2) a transfer of molecules from the inducing to the responding cell, either through specialized contacts or through extracellular diffusion.

3.4.2. Interaction between complementary surface molecules

Good examples of how a molecule can trigger a cellular response without ever penetrating the cell membrane are the cAMP-mediated action of several hormones, or the lymphocyte-stimulating action of certain lectins. There is less evidence, however, for a cellular response triggered by contact with defined structures on another cell or on a large composite structure such as the extracellular matrix. The triggering of the cell-mediated immune response by histocompatibility antigens bound to the surface of another cell seems to be an example of cell interaction through complementary molecules.

As much as 30 years ago, Tyler (1946) and Weiss (1947) suggested that cells might have complementary surface structures, capable of interacting with each other in a lock-and-key fashion like antigens with antibodies. Such structures would form the basis of cell recognition, and the contacts made might even induce cells to undergo morphogenesis. This model would require the presence of a specific receptor on the cell surface triggered by a structure on another cell or on an extracellular matrix. Roseman (1970) advanced the hypothesis that cell adhesion takes place through enzyme-substrate complexes between cell surfaces. In a recent review on molecular models for cell interaction Roth (1973) reasoned that of the known systems of complementary molecules, antigen-antibody, sugar-lectin and enzyme-substrate, the last is the one most likely to be generally operative on cell surfaces (Fig. 25). The model assumes that some factor such as steric hindrance or the distance between enzyme and substrate sites prevents the reaction from taking place on the same cell. In cells of several types, embryonic retinal cells (Roth et al., 1971), human blood platelets (Bosmann, 1971; Jamieson et al., 1971), and cultured fibroblasts (Roth and White, 1972) recognition has been suggested to be due to the presence at the cell surfaces of glycosol transferases and their substrates. Such enzymes have also been reported on the surfaces of intestinal epithelial cells (Weiser, 1973) and cultured fibroblasts (Yogeeswaran et al., 1974). In all these studies the evidence for the presence of

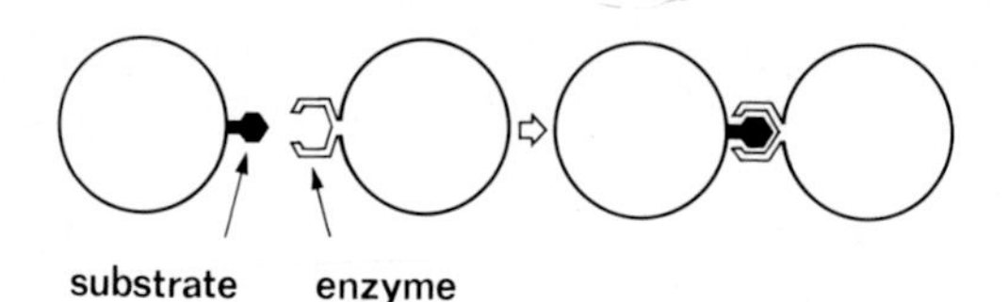

Fig. 25. Enzyme substrate model for cell interaction. The model assumes that the interacting cells become linked to each other through the formation of an enzyme-substrate type of molecular binding between surface associated group.

sugar transferases at the cell surface is indirect, as tests have concerned only the action of intact cells on cellular or other substrates. Autoradiography has shown that the products of these enzymes are located externally, but this is not conclusive evidence that the enzyme itself is located externally.

Recently, Evans (1974) has suggested another enzyme, nucleotide pyrophosphatase as a likely candidate for the control of cell interactions. He labelled plasma membrane vesicles and intact cells of liver parenchyma with ^{125}I, using the lactoperoxidase procedure with H_2O_2 generated by the glucose-glucose oxidase system (Hubbard and Cohn, 1972). This system labels only externally located proteins. Gel electrophoresis gave two major bands of glycosylated polypeptides with molecular weights of 100,000 and 130,000. The enzyme nucleotide pyrophosphatase, isolated as a sialoglycoprotein from liver plasma membranes, also has a molecular weight of 130,000. Since nucleotide pyrophosphatase is a marker of plasma membranes (Touster et al., 1970; Sanford and Rosenberg, 1972), whereas sugar transferases are located on the Golgi apparatus (Fleischer and Fleischer, 1970; Bergeron et al., 1973), he concluded that cell communication is more likely to be mediated by nucleotide pyrophosphatase than by sugar transferases. As yet, however, neither of these enzymes has been shown to play a role in morphogenetic events.

The glycosyltransferase hypothesis assumes that surface-associated or extracellular carbohydrate-containing material is present in embryonic tissues. The choice probably lies among hyaluronic acid, collagen and glycosaminoglycans. These substances all contain carbohydrates and have been shown to be involved in morphogenetic tissue interactions (p. 372 and p. 373).

3.4.3. Transfer of molecules between cells

The inductive message can involve the transmission of molecules from one cell to another, either by extracellular diffusion through an intercellular space or by intercellular transmission through specialized cell junctions. Extracellular diffusion seems to be operative mainly in primary induction (p. 370), and in this system available data are insufficient for molecular models. Evidence for intercellular transmission of molecules has come from studies on electrical coupling of cells and on metabolic cooperation between cells. Since Sheridan in Chapter 8 discusses cell coupling in communication, this will be dealt with only briefly here.

Electrical ionic coupling of cells means that there is less resistance between two cells than between the cell and its exterior. The lowered resistance is due to the passage of ions between the cells. Larger molecules may also pass, but there is some disagreement about the maximal size of molecules that can pass through the junctional sieve. Some have claimed that the junctions exclude molecules larger than about 1,000 daltons (see Sheridan, 1974), whereas in experiments on other cell types even RNA and protein have been reported to pass between cells in culture (Kolodny, 1971). This apparent transfer of substances of high molecular weight has been interpreted as a transfer of nucleotides between cells (see Sheridan, 1974).

It is generally believed that electric coupling of cells occurs through gap junctions (Pappas et al., 1971; Asada and Bennett, 1971; Bennett, 1973). The evidence is only circumstantial, however, and electrical coupling may also occur through junctions of other types (Rose, 1971). Coupling across the extracellular space has also been shown to occur in the developing teleost egg (Bennett and Trinkaus, 1970). Although gap junctions usually connect like cells, they may also join unlike cells (Brightman and Reese, 1969; Johnson et al., 1973).

The best evidence for the biological function of gap junctions has come from a cell-cell interaction termed metabolic cooperation (Subak-Sharpe et al., 1969). When cocultivated with normal cells genetically enzyme-deficient cells grew in a medium in which they did not proliferate when cultivated alone. The cooperation seemed to require direct physical contact between the normal and deficient cells, and according to Pitts (1971) the substrate rather than the enzyme moved from one cell to another. "This cooperation can also be demonstrated by the transfer of labelled nucleotides from one cell to another" (Fig. 26).

In tissue culture metabolic cooperation occurs only between cells which form gap junctions (Gilula et al., 1972). Numerous gap junctions were shown to exist between coupled fibroblastoid cells in culture, whereas none were found between non-coupled epitheloid cells (Hülser and Demsey, 1973). Azarnia et al. (1974) hybridized coupling human cells, which have gap junctions, with non-coupling mouse cells, devoid of gap junctions. Gap junctions and electrical coupling between the hybrid cells was observed.

An important finding is that gap junctions (low-resistance junctions) can form within minutes. It has also become evident that contacting cells may form low-resistance junctions for variable lengths of time (Loewenstein, 1967; Ito and Loewenstein, 1969; Bennett and Trinkaus, 1970; Sheridan, 1971). Cardiac myoblasts, grown in vitro, form close contacts, possibly gap junctions, within a few minutes to an hour after the establishment of initial contact. These cells attained synchrony of spontaneous contractions, which indicates that low-resistance junctions had been established (DeHaan and Hirakow, 1972).

Gap junctions have been demonstrated in embryonic tissues between ionically coupled cells (Fig. 27; Revel et al., 1973). However, there are no good examples of the developmental significance of gap junctions and electrical coupling between embryonic cells. In the squid embryo different tissues are electrically coupled to the yolk sac early in development, but later they become uncoupled. There seem to be differences in electrical coupling between adult and embryonic cells, for fluorescein will pass from one adult cell to another but not between two embryonic cells (Bennett et al., 1972).

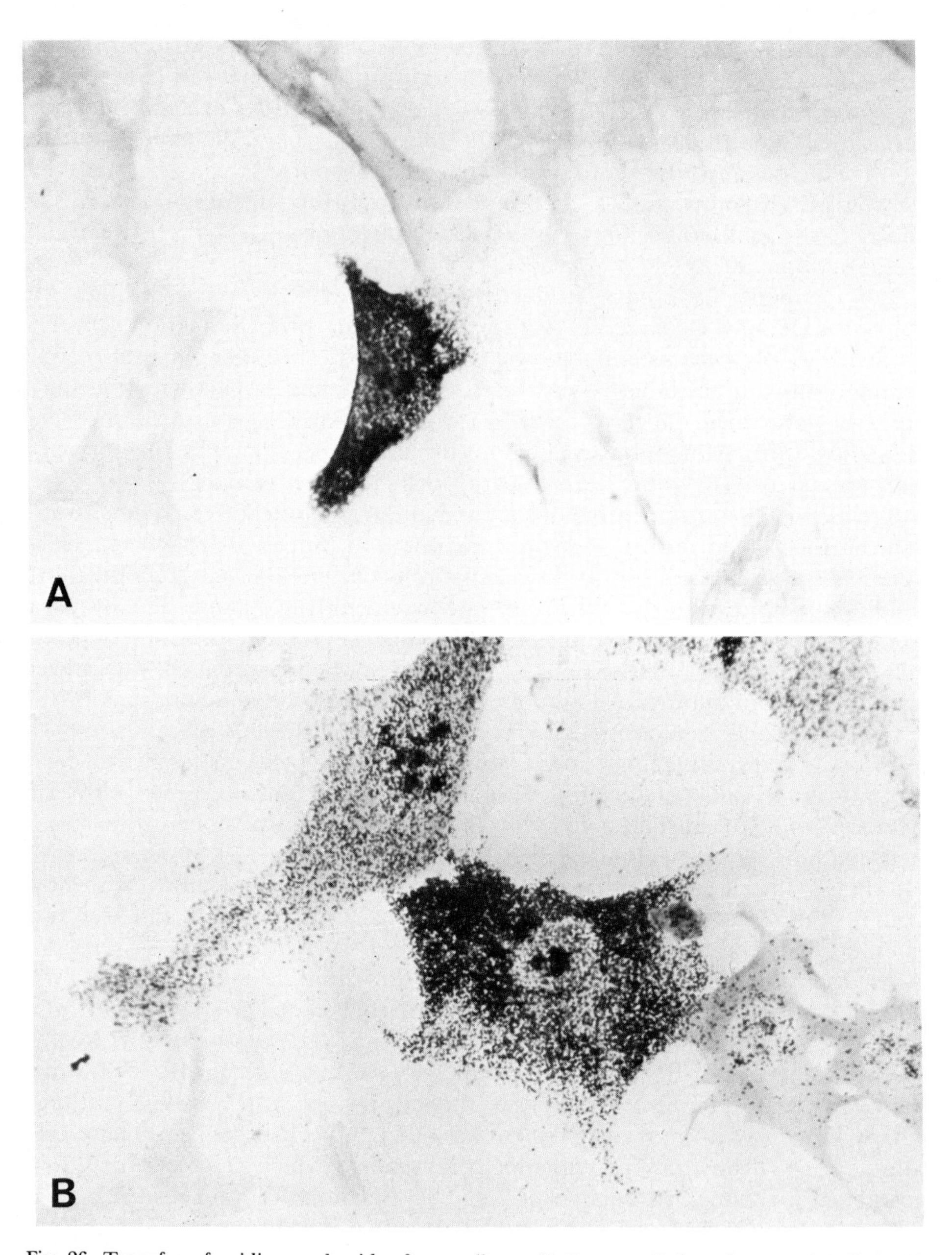

Fig. 26. Transfer of uridine nucleotides from cell to cell. Donor cells have been prelabelled with ^{3}H-uridine. A. Donor cell, HTC (hepatoma tissue cell line) is heavily labelled. The contacting recipient cell (BHK21/13) is unlabelled. HTC cells do not form junctions with BHK21/13 cells. B. Heavily labelled donor cells, BHK21/13 in contact with recipient BHK21/13 cells, which have taken up label. Junctions form between BHK21/13 cells. If after labelling donor cells are cultivated for 24 hr before being added to recipient cells no transfer of label occurs indicating that RNA cannot pass the junctions (Courtesy of M.E. Finbew, J.W. Simms and J.D. Pitts).

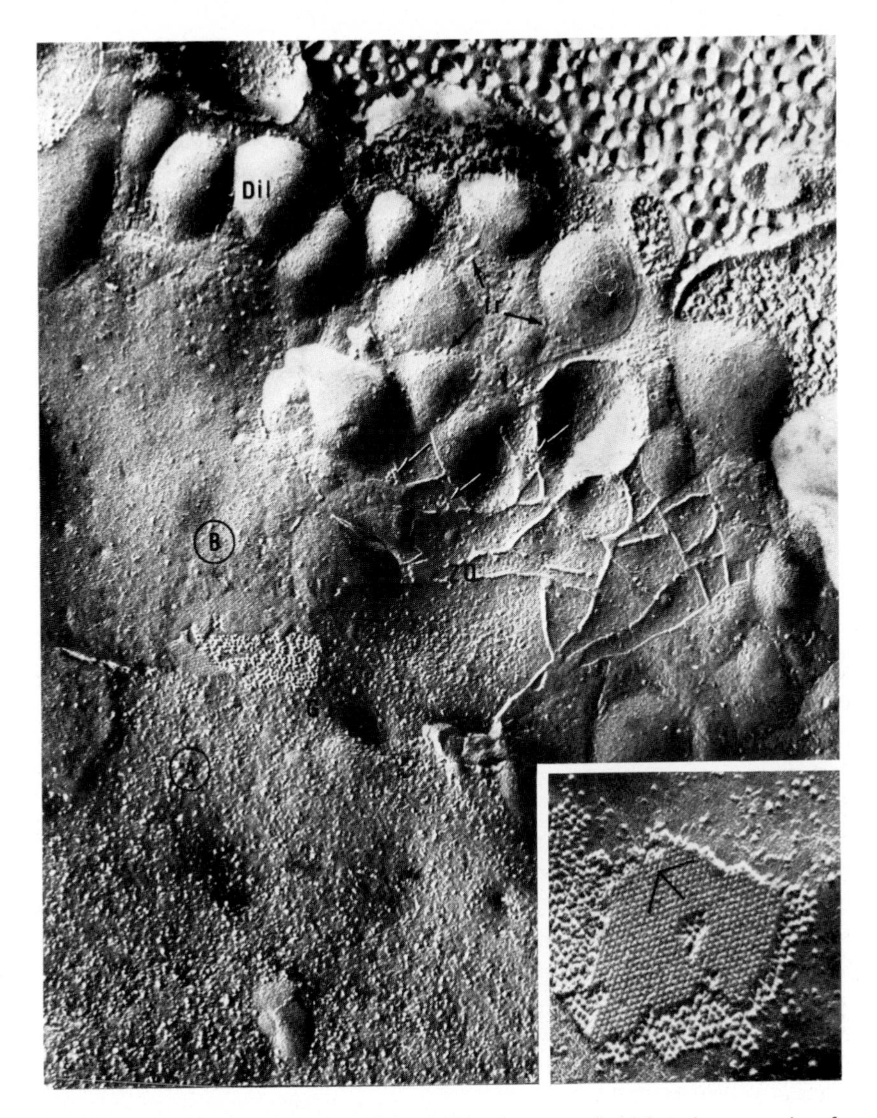

Fig. 27. The juxtaluminal portion of the epiblast in a stage 6 chick embryo seen in a freeze-cleaved sample. Interconnected strands of the zonula occludens (ZO) can be seen on A faces Ⓐ at the apex of the cells. The corresponding grooves on B faces Ⓑ are not seen very clearly because there are fragments of the ZO of the opposite cell still adhering to the grooves (fr) and because they are obscured by the dilations (Dil) of the junctional intercellular space. Near the ZO a gap junction (GJ) can be seen. The inset shows a higher magnification of another gap junction between cells of the epiblast. The close packing and orderliness of particles and the very precise distribution of pits is well illustrated. Magnification × 40000, inset × 75000. (From Revel et al., 1973).

The existence of low-resistance junctions during inductive events leads us to ask which types of signal molecules, if any, are transferred. The same question applies to extracellular diffusion. It has been believed that only substances with a high content of information can act as regulators of differentiation. Such a postulate would be necessary if, for its differentiation, the cell depended on receiving new information. However, the fertilized egg already contains all the information necessary for making of a new individual and differentiation seems to involve only triggering of the cell. Hence, inductors may be very simple substances, as suggested by Holtfreter (1938), Barth and Barth (1969, 1972) and by Wolpert (1971).

Recently, McMahon (1974) revived the notion of a simple inductor. He suggested that physiological and developmental regulatory mechanisms are similar and that developing tissues, like their fully differentiated counterparts, may be regulated by cyclic nucleotides, notably cAMP and cGMP neurotransmitters, but also by the intracellular concentration of inorganic ions. Studies on the effect of mono- and divalent cations on neural induction discussed earlier (p. 365) give some support for such ideas. A fuller discussion of this subject is given in the chapter by McMahon and West in this volume (p. 449).

3.4.4. Remarks

These two fundamentally different modes of interaction, interaction through complementary surface molecules and a transfer of molecules, seem to enclose all possible ways of molecular tissue interactions. Yet there is no good example that either of the two should operate. This is due to a lack of molecular data in morphogenetic interactions. It has even been suggested that the interaction between cells is not primarily based on chemical but on physical events. Gingell (1967) has suggested that changes in the electrostatic potential of cell surface membranes in close proximity might occur as a result of the approach of changed surfaces in electrolytic media. A change in surface potential could possibly alter the behaviour of a cell. Then mere proximity of cells could represent an inductive signal.

4. *Concluding remarks*

In his classic monograph, Hans Spemann (1936), the discoverer of embryonic induction, apologizes for the frequent use of expressions borrowed from psychologists, and hopes that these will soon be explained in physico-chemical terms. Now, forty years later, authors of a review on inductive tissue interactions feel very much like Spemann. Terms, such as "commitment", "decision", "communication" and many others, used throughout this chapter indicate that the problem is still at a phenomological level. While the findings of Spemann have been repeatedly confirmed and their biological significance conclusively shown, the molecular nature of inductive interactions remains poorly understood.

With an increasing understanding of the chemical basis of nucleo-cytoplasmic interactions and second messenger systems, the intracellular events of differentiation have been relatively well mapped out. The extracellular communication between two cells of different origin and developmental capacities is less well known and different communication mechanisms should be considered. Perhaps the long chain of studies on epithelio-mesenchymal interactions started in 1953 by Grobstein (1953a) and continued by several research groups will clarify some principles of intercellular communication. More attention should be focused not only on molecules in the extracellular matrix, but also on compounds at the cell surface and on the possibility of transmission of cytoplasmic signal substances. Such studies seem to constitute the main line of research in this field in the near future.

References

Ansell, J.D. (1975) The differentiation and development of mouse trophoblast, In: The Early Development of Mammals (Balls, M. and Wild, A.E., eds.), pp. 133–144, Cambridge University Press, Cambridge.

Ansell, J.D. and Snow, M.H.L. (1975) The development of trophoblast in vitro from blastocysts containing varying amounts of inner cell mass. J. Embryol. Exp. Morphol. 33, 177–185.

Asada, Y. and Bennett, M.V.L. (1971) Experimental alteration of coupling resistance at an electrotonic synapse. J. Cell Biol. 49, 159–172.

Auerbach, R. (1972) The use of tumors in the analysis of inductive tissue interactions. Develop. Biol. 28, 304–309.

Ave, K., Kawakami, I. and Sameshima, M. (1968) Studies on the heterogeneity of cell populations in amphibian presumptive epidermis, with reference to primary induction. Develop. Biol. 17, 617–626.

Azarnia, R., Larsen, W.J. and Loewenstein, W.R. (1974) The membrane junctions in communicating and noncommunicating cells, their hybrids, and segregants. Proc. Nat. Acad. Sci. U.S.A. 71, 880–884.

Azarnia, R. and Loewenstein, W.R. (1971) Intercellular communication and tissue growth. V. A cancer cell strain that fails to make permeable membrane junctions with normal cells. J. Membrane Biol. 6, 368–385.

Azarnia, R. and Loewenstein, W.R. (1973) Parallel correlation of cancerous growth and of a genetic defect of cell-to-cell communication. Nature, 241, 455–457.

Azarnia, R., Michalke, W. and Loewenstein, W.R. (1972) Intercellular communication and tissue growth. VI. Failure of exchange of endogeneous molecules between cancer cells with defective junctions and noncancerous cells. J. Membrane Biol. 10, 247–258.

Bagg, H.J. (1929) Hereditary abnormalities of the limbs, their origin and transmission. II. A morphological study with special reference to the etiology of clubfeet, syndactylism, hypodactylism, and congenital amputation of X-rayed mice. Am. J. Anat. 43, 167–219.

Balinsky, B.I. (1972) The fine structure of the amphibian limb bud. Acta Embryol. Morphol. Exp., Suppl., 455–470.

Barth, L.G. (1966) The role of sodium chloride in sequential induction of the presumptive epidermis of *Rana pipiens* gastrulae. Biol. Bull. 131, 415–426.

Barth, L.G. and Barth, L.J. (1959) Differentiation of cells of the *Rana pipiens* gastrula in unconditioned medium. J. Embryol. Exp. Morphol. 7, 210–222.

Barth, L.G. and Barth, L.J. (1964) Sequential induction of the presumptive epidermis of the *Rana pipiens* gastrula. Biol. Bull. 127, 413–427.

Barth, L.G. and Barth, L.J. (1967) The uptake of Na-22 during induction of presumptive epidermis cells of the *Rana pipiens* gastrula. Biol. Bull. 133, 495–501.

Barth, L.G. and Barth, L.J. (1968) The role of sodium chloride in the process of induction by lithium chloride in cells of the *Rana pipiens* gastrula. J. Embryol. Exp. Morphol. 19, 387–396.

Barth, L.G. and Barth, L.J. (1969) The sodium dependence of embryonic induction. Dev. Biol. 20, 236–262.

Barth, L.G. and Barth, L.J. (1972) ^{22}Sodium and ^{45}calcium uptake during embryonic induction in *Rana pipiens*. Dev. Biol. 28, 18–34.

Bautzmann, H., Holtfreter, J., Spemann, H. and Mangold, O. (1932) Versuche zur Analyse der Induktionsmittel in der Embryonalentwicklung. Naturwissenschaften 20, 971–974.

Bennett, M.V.L. (1973) Function of electrotonic junctions in embryonic and adult tissues. Fed. Proc. 32, 65–75.

Bennett, M.V.L. and Trinkaus, J.P. (1970) Electrical coupling between embryonic cells by way of extracellular space and specialized junctions. J. Cell Biol. 44, 592–610.

Bennett, M.V.L., Spira, M.E. and Pappas, G.D. (1972) Properties of electrotonic junctions between embryonic cells of *Fundulus*. Dev. Biol. 29, 419–435.

Bérczy, J. (1966) Zur Ultrastruktur der Extremitätenknospe. Anat. Entwicklungsgesch. 125, 295–315.

Bergeron, J.J.M., Ehrenreich, J.H., Siekevitz, P. and Palade, G.E. (1973) Golgi fractions prepared from rat liver homogenates. II. Biochemical characterization. J. Cell Biol. 59, 73–88.

Bernfield, M.R. and Wessells, N.K. (1970) Intra- and extracellular control of epithelial morphogenesis. In: Changing Syntheses in Development (Runner, M.N., ed.) pp. 195–249, Academic Press, New York.

Bernfield, M.R., Banerjee, S.D. and Cohn, R.H. (1972) Dependence of salivary epithelial morphology and branching morphogenesis upon acid mucopolysaccharide-protein (proteoglycan) at the epithelial surface. J. Cell Biol. 52, 674–689.

Billingham, R.E. and Silvers, W.K. (1967) Studies on the conservation of epidermal specificities of skin and certain mucosas in adult mammals. J. Exp. Med. 125, 429–446.

Billingham, R.E. and Silvers, W.K. (1968) Dermoepidermal interactions and epithelial specificity. In: Epithelial-Mesenchymal Interactions (Fleischmajer, R. and Billingham, R.E., eds.) pp. 252–266, Williams & Wilkins, Baltimore, Md.

Billingham, R.E., Orr, J.W. and Woodhouse, D.L. (1951) Transplantation of skin components chemical carcinogenesis with 20-methyl cholantrene. Brit. J. Cancer 5, 417–432.

Bishop-Calame, S. (1965) Etude d'associations hétérologues de l'uretère et de différents mésenchymes de l'embryon de poulet, par la technique des greffes chorio-allantoidiennes. J. Embryol. Exp. Morphol. 14, 247–253.

Borek, C., Higashino, S. and Loewenstein, W.R. (1969) Intercellular communication and tissue growth. IV. Conductance of membrane junctions of normal and cancerous cells in culture. J. Membrane Biol. 1, 274–283.

Born, J., Tiedemann, H. and Tiedemann, H. (1972) The mechanism of embryonic induction: Isolation of an inhibitor for the vegetalizing factor. Biochim. Biophys. Acta 279, 175–183.

Bosmann, H.B. (1971) Platelet adhesiveness and aggregation: the collagen: glycosyl, polypeptide: N-acetylgalactosaminyl- and glycoprotein: galactosyl transferases of human platelets. Biochem. Biophys. Res. Commun. 43, 1118–1124.

Brachet, J. (1940) Etude histochimique des protéins au cours de développement embryonnaire des Poissons, des Amphibiens et des Oiseaux. Arch. Biol. 51, 167–202.

Brachet, J. (1942) La localisation des acides pentose-nucléiques dans les tissues animaux et les oeufs d'Amphibiens en voie de développement. Arch. Biol. 53, 207–257.

Brahma, S.K. (1958) Experiments on the diffusibility of the Amphibian evocator. J. Embryol. Exp. Morphol. 6, 418–423.

Brightman, M.W. and Reese, T.S. (1969) Junctions between intimately apposed cell membranes in the vertebrate brain. J. Cell Biol. 40, 648–677.

Chase, H.B. (1944) Studies on an anophthalmic strain of mice. IV. A second major gene for anophthalmia. Genetics 29, 264–269.

Chase, H.B. and Chase, E.B. (1941) Studies on an anophthalmic strain of mice. I. Embryology of the eye region. J. Morphol. 68, 279–302.

Cohen, A.I. (1961) Electron microscopic observations of the developing mouse eye. I. Basement membranes during early development and lens formation. Dev. Biol. 3, 297–316.

Cooper, G.W. (1965) Induction of somite chondrogenesis by cartilage and notochord: A correlation between inductive activity and specific stages of cytodifferentiation. Dev. Biol. 12, 185–212.

Cornell, R. (1969) The use of Nucleopore filters in ultra-structural studies of cell cultures. Exp. Cell Res. 56, 156–158.

Coulombre, J.L. and Coulombre, A.J. (1971) Metaplastic induction of scales and feathers in the corneal anterior epithelium of the chick embryo. Dev. Biol. 25, 464–478.

Crocker, J.F.S. (1973) Human embryonic kidneys in organ culture: Abnormalities of development induced by decreased potassium. Science, 181, 1178–1179.

Crocker, J.F.S. and Vernier, R.L. (1972) Congenital nephroma of infancy: Induction of renal structures by organ culture. J. Pediat. 80, 69–73.

Cunha, G.R. (1972) Support of normal salivary gland morphogenesis by mesenchyme derived from accessory sexual glands of embryonic mice. Anat. Rec. 173, 205–212.

Curtis, A.S.G. (1965) Cortical inheritance in the amphibian *Xenopus laevis*: preliminary results. Arch. Biol. 76, 523–546.

Cutler, L.S. and Chaudhry, A.P. (1973) Intercellular contacts at the epithelial-mesenchymal interface during the prenatal development of the rat submandibular gland. Dev. Biol. 33, 229–240.

Dalcq, A.M. (1957) Introduction to General Embryology, Oxford University Press, Oxford.

Dameron, F. (1961) L'influence de divers mésenchymes sur la différenciation de l'épithélium pulmonaire de l'embryon de Poulet en culture in vitro. J. Embryol. Exp. Morphol. 9, 628–633.

Dameron, F. (1968) Etude expérimentale de l'organogenèse de poumon: nature et spécificité des interactions épitheliomésenchymateuses. J. Embryol. Exp. Morphol. 20, 151–167.

David, D. (1967) L'influence de divers mésenchymes sur la différenciation de l'épithélium gastrique du foetus de Lapin, en culture in vitro. C.R. Acad. Sci. 264, 1062–1065.

David, D. (1971) Influence de divers mésenchymes hétérologues sur l'évolution de l'épithélium gastrique du foetus de lapin. Action réciproque de l'épithélium sur le mésenchyme associé. Arch. Anat. Microsc. Morphol. Exp. 60, 421–442.

Dawe, C.J. (1972) Epithelial-mesenchymal interactions in relation to the genesis of polyoma virus-induced tumours of mouse salivary gland. In: Tissue Interactions in Carcinogenesis (Tarin, D., ed.) pp. 305–358, Academic Press, London.

Dawe, C.J., Main, J.H.P., Slatick, M.S. and Morgan, W.D. (1966a) Epigenetic factors in the neoplastic response to polyoma virus. In: Recent Results in Cancer Research, VI, Malignant Transformation by Viruses (Kirsten, W.H., ed.) pp. 20–33, Springer-Verlag, New York.

Dawe, C.J., Morgan, W.D. and Slatick, M.S. (1966b) Influence of epithelio-mesenchymal interactions on tumor induction by polyoma virus. Int. J. Cancer 1, 419–450.

Dawe, C.J., Morgan, W.D. and Slatick, M.S. (1968) Salivary gland neoplasms in the role of normal mesenchyme during salivary gland morphogenesis. In: Epithelio-Mesenchymal Interactions (Fleischmajer, R. and Billingham, R.E., eds.) pp. 295–312, Williams & Wilkins, Baltimore, Md.

Dawe, C.J., Whang-Peng, J., Morgan, W.D., Hearon, E.C. and Knutsen, T. (1971) Epithelial origin of salivary tumors in mice: evidence based on chromosome marked cells. Science, 171, 394–397.

DeCosse, J.J., Gossens, C.L. and Kuzma, J.F. (1973) Breast cancer: Induction of differentiation by embryonic tissue. Science, 181, 1057–1058.

DeHaan, R.L. and Hirakow, R. (1972) Synchronization of pulsation rates in isolated cardiac myocytes. Exp. Cell Res. 70, 214–220.

Deuchar, E.M. (1969) Effects of a mesoderm-inducing factor on early chick embryos. J. Embryol. Exp. Morphol. 22, 295–304.

Deuchar, E.M. (1971a) Transfer of the primary induction stimulus by small numbers of Amphibian ectoderm cells. Acta Embryol. Exp. 2, 93–101.

Deuchar, E.M. (1971b) *Xenopus laevis* and developmental biology. Biol. Rev. 47, 37–112.

Dhouailly, D. (1970) Déterminisme de la différenciation spécifique des plumes néoptiles et téléoptiles chez le Poulet et le Canard. J. Embryol. Exp. Morphol. 24, 73–94.

Dhouailly, D. (1973) Dermo-epidermal interactions between birds and mammals: differentiation of cutaneous appendages. J. Embryol. Exp. Morphol. 30, 587–603.

Dhouailly, D. (1975) Formation of cutaneous appendages in dermoepidermal recombinations between

reptiles, birds and mammals. Wilhelm Roux Arch. Entwicklungsmech. Organismen, 177, 323–332.
Dhouailly, D. and Sengel, P. (1972) La morphogenèse de la plume et du poil, étudiée par des associations hétérospécifiques de derme et d'épiderme entre le Poulet et la Souris. C.R. Acad. Sci. Ser. D275, 479–482.
Dhouailly, D. and Sengel, P. (1973) Interactions morphogènes entre l'épiderme de Reptile et le derme d'Oiseau ou de Mammifère. C.R. Acad. Sci. 277, 1221–1224.
Dickson, A.D. (1966) The form of mouse blastocyst. J. Anat. 100, 335–348.
Dieterlen-Lièvre, F. and Hadorn, H.B. (1972) Développement des enzymes exocrines dans les bourgeons pancréatiques chez l'embryon de poulet en présence de mésenchymes homologues et hétérologues. Wilhelm Roux Arch. Entwicklungsmech. Organismen, 170, 175–184.
Doskocil, M. (1968) Transfilter influence of brain tissue on the differentiation of the stomodeal epithelium in vitro. Ann. Med. Exp. Fenn. 46, 288–292.
Ducibella, Th., Albertini, D.F., Anderson, E. and Biggers, J.D. (1975) The preimplantation mammalian embryo: Characterization of intercellular junctions and their appearance during development. Dev. Biol. 45, 231–250.
Ede, D.A., Bellairs, R. and Bancroft, M. (1974) A scanning electron microscope study of the early limb-bud in normal and talpid[3] mutant chick embryos. J. Embryol. Exp. Morphol. 31, 761–785.
Ellison, M.L. and Lash, J.W. (1971) Environmental enhancement of in vitro chondrogenesis. Develop. Biol. 26, 486–496.
Enders, A.C. and Schlafke, S.J. (1967) A morphological analysis of the early implantation stages in the rat. Am. J. Anat. 120, 185–226.
Endres, H. (1895) Über Anstick- und Schnürversuche an Eiern von Triton taeniatus. Schles. Ges. Vaterländ. Kultur 73.
England, M.A. (1969) Millipore filters studied in isolation and in vitro by transmission electron microscopy and stereo-scanning electron microscopy. Exp. Cell Res. 54, 222–230.
Evans, W.H. (1974) Nucleotide pyrophosphatase: a sialoglycoprotein located on the hepatocyte surface. Nature, 250, 391–394.
Faulhaber, I. (1972) Die Induktionsleistung subzellularer Fraktionen aus der Gastrula von *Xenopus laevis*. Wilhelm Roux Arch. Entwicklungsmech. Organismen 171, 87–108.
Fell, P.E. and Grobstein, C. (1968) The influence of extra-epithelial factors on the growth of embryonic mouse pancreatic epithelium. Exp. Cell Res. 53, 301–304.
Fleischer, B. and Fleischer, S. (1970) Preparation and characterization of golgi membranes from rat liver. Biochim. Biophys. Acta, 219, 301–319.
Fleischmajer, R. and Billingham, R.E. (eds.) (1968) Epithelial-Mesenchymal Interactions, pp. 1–326, Williams & Wilkins, Baltimore, Md.
Gallera, J. (1967) L'induction neurogène chez les Oiseaux: passage du flux inducteur par le filtre millipore. Experientia 23, 461–464.
Gallera, J., Nicolet, G. and Baumann, M. (1968) Induction neurale chez les Oiseaux à travers un filtre millipore: étude au microscope optique et électronique. J. Embryol. Exp. Morphol. 19, 439–450.
Gardner, R.L. (1971) Manipulations on the blastocyst. Schering Symposium on 'Intrinsic and Extrinsic Factors in Early Mammalian Development' (Raspé, G., ed.) Advances in the Biosciences 6, 279–296, Pergamon Press, Viewig.
Gardner, R.L. (1972) An investigation of inner cell mass and trophoblast tissues following their isolation from the mouse blastocyst. J. Embryol. Exp. Morphol. 28, 279–312.
Gardner, R.L. (1974a) Microsurgical approaches to the study of early mammalian development, In: Birth Defects and Fetal Development, Endocrine and Metabolic Factors, the 7th Harold C. Mark Symposium (Moghissi, K.S., ed.) pp. 212–233, Charles C. Thomas, Springfield, Illinois.
Gardner, R.L. (1974b) Origin and properties of trophoblast. In: The Immunobiology of Trophoblast (Edwards, R.G., Howe, C.W.S. and Johnson, M.H., eds.) pp. 43–65, Cambridge University Press, Cambridge.
Gardner, R.L. and Papaioannou, V.E. (1975) Differentiation in the trophoectoderm and inner cell mass. In: The Early Development of Mammals (Balls, M. and Wild, A.E., eds.) pp. 107–132, Cambridge University Press, Cambridge.
Gardner, R.L., Papaioannou, V.E. and Barton, S. (1973) Origin of the ectoplacental cone and

secondary giant cells in mouse blastocysts reconstituted from isolated trophoblast and inner cell mass. J. Embryol. Exp. Morphol. 30, 561–572.

Gilula, N.B., Reeves, O.R. and Steinbach, A. (1972) Metabolic coupling, ionic coupling and cell contacts. Nature, 235, 262–265.

Gingell, D. (1967) Membrane surface potential in relation to a possible mechanism for intercellular interactions and cellular responses: A physical basis. J. Theoret. Biol. 17, 451–482.

Goetinck, P.F. (1970) Epithelial-mesenchymal interactions and collagen synthesis in the establishment of feather patterns in the chick embryo. J. Cell Biol. 47, 72a.

Goetinck, P.F. and Abbott, U.K. (1963) Tissue interaction in the scaleless mutant and the use of scaleless as an ectodermal marker in studies of normal limb differentiation. J. Exp. Zool. 154, 7–19.

Goetinck, P.F. and Sekellick, M.J. (1970) Early morphogenetic events in normal and mutant skin development in the chick embryo and their relationship to alkaline phosphatase activity. Dev. Biol. 21, 349–363.

Goldhaber, P. (1961) The influence of pore size on carcinogenicity of subcutaneously implanted Millipore filters. Proc. Am. Ass. Cancer Res. 3, 228–235.

Golosov, N. and Grobstein, C. (1962) Epithelio-mesenchymal interaction in pancreatic morphogenesis. Dev. Biol. 4, 242–255.

Gossens, C.L. and Unsworth, B.R. (1972) Evidence for a two-step mechanism operating during in vitro mouse kidney tubulogenesis. J. Embryol. Exp. Morphol. 28, 615–631.

Graham, C.F. (1970) Parthenogenetic mouse blastocysts. Nature, 226, 165–167.

Graham, C.F. (1971) The design of the mouse blastocyst. In: Control Mechanisms of Growth and Differentiation, XXV Symposia of the Society for Experimental Biology (Davies, D.D. and Balls, M., eds.) pp. 371–378, Cambridge University Press, Cambridge.

Grainger, R.M. and Wessells, N.K. (1974) Does RNA pass from mesenchyme to epithelium during an embryonic tissue interaction? Proc. Nat. Acad. Sci. U.S.A. 71, 4747–4751.

Grobstein, C. (1953a) Epithelio-mesenchymal specificity in the morphogenesis of mouse submandibular rudiments in vitro. J. Exp. Zool. 124, 383–413.

Grobstein, C. (1953b) Morphogenetic interaction between embryonic mouse tissues separated by a membrane filter. Nature, 172, 869–871.

Grobstein, C. (1955a) Inductive interaction in the development of the mouse metanephros. J. Exp. Zool. 130, 319–340.

Grobstein, C. (1955b) Tissue interaction in the morphogenesis of mouse embryonic rudiments in vitro. In: Aspects of Synthesis and Order in Growth (Rudnick, D., ed.) pp. 233–256, Princeton University Press, Princeton, N.J.

Grobstein, C. (1957) Some transmission characteristics of the tubule-inducing influence on mouse metanephrogenic mesenchyme. Exp. Cell Res. 13, 575–587.

Grobstein, C. (1967) Mechanisms of organogenetic tissue interaction. Nat. Cancer Inst. Monogr. 26, 279–299.

Grobstein, C. and Cohen, J. (1965) Collagenase: Effect on the morphogenesis of embryonic salivary epithelium in vitro. Science, 150, 626–628.

Grobstein, C. and Dalton, A.J. (1957) Kidney tubule induction in mouse metanephrogenic mesenchyme without cytoplasmic contact. J. Exp. Zool. 135, 57–73.

Hamer, D. and Marchant, J. (1957) Collagen and other constituents in the skin of normal, carcinogen-treated and castrated mice. Brit. J. Cancer 11, 445–451.

Harrison, R. (1920) Experiments on the lens in Amblystoma. Proc. Soc. Exp. Biol. Med. 17, 199–200.

Hay, E.D. and Dodson, J.W. (1973) Secretion of collagen by corneal epithelium. I. Morphology of the collagenous products produced by isolated epithelia grown on frozen-killed lens. J. Cell Biol. 57, 190–213.

Hay, E.D. and Revel, J.-P. (1969) Fine structure of the developing avian cornea. In: Monographs in Developmental Biology, Vol. 1 (Wolsky, A. and Chen, P.S., eds.) pp. 1–44, S. Karger, Basel.

Hendrix, R.W. and Zwaan, J. (1974) Changes in the glycoprotein concentration of the extracellular matrix between lens and optic vesicle associated with early lens differentiation. Differentiation, 2, 357–362.

Hendrix, R.W. and Zwaan, J. (1975) The matrix of the optic vesicle-presumptive lens interface during induction of the lens in the chicken embryo. J. Embryol. Exp. Morphol. 33, 1023–1049.

Hilfer, S.R. (1968) Cellular interactions in the genesis and maintenance of thyroid characteristics. In: Epithelial-Mesenchymal Interactions (Fleischmajer, R. and Billingham, R.E., eds.) pp. 177–199, Williams & Wilkins, Baltimore, Md.

Hilfer, S.R. and Hilfer, E.K. (1964) Influence of various mesodermal cell types on the fine structure of thyroid epithelial cells. J. Cell Biol. 23, 42A–42B.

Hillman, N., Sherman, M.I. and Graham, C. (1972) The effect of spatial arrangement on cell determination during mouse development. J. Embryol. Exp. Morphol. 28, 263–278.

Hinchliffe, J.R. and Ede, D.A. (1967) Limb development in the polydactylous talpid mutant of the fowl. J. Embryol. Exp. Morphol. 17, 385–404.

Hinchliffe, J.R. and Thorogood, P.V. (1974) Genetic inhibition of mesenchymal cell death and the development of form and skeletal pattern in the limbs of talpid^3 (ta^3) mutant chick embryos. J. Embryol. Exp. Morphol. 31, 747–760.

Holtfreter, J. (1938) Veränderung der Reaktionsweise im alternden isolierten Gastrulaektoderm. Wilhelm Roux Arch. Entwicklungsmech. Organismen 138, 163–196.

Holtfreter, J. and Hamburger, V. (1955) Embryogenesis: Progressive differentiation; Amphibians. In: Analysis of Development (Willier, B.H., Weiss, P.A. and Hamburger, V., eds.) pp. 230–296, W.B. Saunders, Philadelphia, Pa.

Holtzer, H. (1975) Discussion. In: Extracellular Matrix Influences on Gene Expression (Slavkin, H.C. and Greulich, R.C., eds.) pp. 570–573, Academic Press, New York.

Holtzer, H. and Detwiler, S.R. (1953) An experimental analysis of the development of the spinal column. III. Induction of skeletogenous cells. J. Exp. Zool. 123, 335–370.

Hubbard, A.L. and Cohn, Z.A. (1972) The enzymatic iodination of the red cell membrane. J. Cell Biol. 55, 390–405.

Hülser, D.F. and Demsey, A. (1973) Gap and low-resistance junctions between cells in culture. Z. Naturforsch. 28c, 603–606.

Hunt, H.H. (1961) A study of the fine structure of the optic vesicle and lens placode of the chick embryo during induction. Dev. Biol. 3, 175–209.

Ito, S. and Loewenstein, W.R. (1969) Ionic communication between early embryonic cells. Dev. Biol. 19, 228–243.

Jacobson, A.G. (1958) The roles of neural and non-neural tissues in lens induction. J. Exp. Zool. 139, 525–558.

Jacobson, A.G. (1966) Inductive processes in embryonic development. Science, 152, 25–34.

Jamieson, G.A., Urban, G.L. and Barber, A.J. (1971) Enzymatic basis for platelet: collagen adhesion as the primary step in haemostasis. Nature, New Biology 234, 5–7.

Johnson, R.G., Herman, W.S. and Preus, D.M. (1973) Homocellular and heterocellular gap junctions in *Limulus*: A thin-section and freeze-fracture study. J. Ultrastruct. Res. 43, 298–312.

Johnston, K.H., Ghobrial, H.K. and Buoen, L.C. (1972) Foreign body tumorigenesis in mice: Ultrastructure of preneoplastic tissue reactions. J. Nat. Cancer Inst. 49, 1311–1319.

Jurand, A. (1965) Ultrastructural aspects of early development of the fore-limb buds in the chick and the mouse. Proc. Roy. Soc. London Ser. B, 162, 387–405.

Kallenbach, E. (1971) Electron microscopy of the differentiating rat incisor ameloblast. J. Ultrastruct. Res. 35, 508–531.

Karp, R.D., Johnson, K.H., Buoen, L.C., Ghobrial, H.K.G., Brand, I. and Brand, K.G. (1973) Tumorigenesis by Millipore filters in mice: Histology and ultrastructure of tissue reactions as related to pore size. J. Nat. Cancer Inst. 51, 1275–1285.

Kefalides, N.A. (1968) Isolation and characterization of the collagen from glomerular basement membrane. Biochemistry 7, 3103–3112.

Kelley, R.O. (1969) An electron microscopic study of chordamesoderm-neurectoderm association in gastrulae of a toad, *Xenopus laevis*. J. Exp. Zool. 172, 153–180.

Kelley, R.O. (1973) Fine structure of the apical rim-mesenchyme complex during limb morphogenesis in man. J. Embryol. Exp. Morphol. 29, 117–131.

Kelley, R.O. and Bluemink, J.G. (1974) An ultrastructural analysis of cell and matrix differentiation during early limb development in *Xenopus laevis*. Dev. Biol. 37, 1–17.

King, H.D. (1905) Experimental studies on the eye of the frog embryo. Wilhelm Roux Arch. Entwicklungsmech. Organismen 19, 85–107.

Koch, W.E. (1967) In vitro differentiation of tooth rudiments of embryonic mice. I. Transfilter interaction of embryonic incisor tissues. J. Exp. Zool. 165, 155–170.

Kollar, E.J. (1966) An in vitro study of hair and vibrissae development in embryonic mouse skin. J. Invest. Dermatol. 46, 254–262.

Kollar, E.J. (1970) The induction of hair follicles by embryonic dermal papillae. J. Invest. Dermatol. 55, 374–378.

Kollar, E.J. and Baird, G.R. (1970) Tissue interaction in embryonic mouse tooth germs. II. The inductive role of the dental papilla. J. Embryol. Exp. Morphol. 24, 173–186.

Kolodny, G.M. (1971) Evidence for transfer of macromolecular RNA between mammalian cells in culture. Exp. Cell Res. 65, 313–324.

Kosher, R.A. and Lash, J.W. (1975) Notochordal stimulation of in vitro somite chondrogenesis before and after enzymatic removal of perinotochordal materials. Dev. Biol. 42, 362–378.

Kosher, R.A., Lash, J.W. and Minor, R.R. (1973) Environmental enhancement of in vitro chondrogenesis. IV. Stimulation of somite chondrogenesis by exogenous chondromucoprotein. Dev. Biol. 35, 210–220.

Kratochwil, K. (1969) Organ specificity in mesenchymal induction demonstrated in the embryonic development of the mammary gland of the mouse. Dev. Biol. 20, 46–71.

Kratochwil, K. (1972) Tissue interaction during embryonic development. In: Tissue Interactions and Carcinogenesis (Tarin, D., ed.) pp. 1–47, Academic Press, London.

Lash, J.W. (1963) Studies on the ability of embryonic mesonephros explants to form cartilage. Dev. Biol. 6, 219–232.

Lash, J.W. and Saxén, L. (1972) Human teratogenesis: In vitro studies on thalidomide inhibited chondrogenesis. Dev. Biol. 28, 61–70.

Lash, J., Holtzer, S. and Holtzer, H. (1957) An experimental analysis of the development of the spinal column. VI. Aspects of cartilage induction. Exp. Cell Res. 13, 292–303.

Lash, J.W., Rosene, K., Minor, R.R., Daniel, J.C. and Kosher, R.A. (1973) Environmental enhancement of in vitro chondrogenesis. III. The influence of external potassium ions and chondrogenic differentiation. Dev. Biol. 35, 370–375.

Lawson, K.A. (1972) The role of mesenchyme in the morphogenesis and functional differentiation of rat salivary epithelium. J. Embryol. Exp. Morphol. 27, 497–513.

Lawson, K.A. (1974) Mesenchyme specificity in rodent salivary gland development: the response of salivary epithelium to lung mesenchyme in vitro. J. Embryol. Exp. Morphol. 32, 469–493.

Le Douarin, N. (1964) Induction de l'endoderme préhépatique par le mésoderme de l'aire cardiaque chez l'embryon de poulet. J. Embryol. Exp. Morphol. 12, 651–664.

Le Douarin, N. (1968) Synthèse du glycogène dans les hépatocytes en voie de différenciation: role des mésenchymes homologue et hétérologues. Dev. Biol. 17, 101–114.

Le Douarin, N. (1974) Cell recognition based on natural morphological nuclear markers. Med. Biol. 52, 281–319.

Le Douarin, N. (1975) An experimental analysis of liver development. Med. Biol., 53, 427–455.

Lee, H. and Niu, M.C. (1973) Studies on biological potentiality of testis-RNA. I. Induction of axial structures in whole and excised chick blastoderms. In: The role of RNA in reproduction and development (Niu, M.C. and Segal, S., eds.) pp. 137–154, North-Holland, Amsterdam.

Lehtonen, E. (1975) Epithelio-mesenchymal interface during mouse kidney tubule induction in vivo. J. Embryol. Exp. Morphol. 34, 695–705.

Lehtonen, E., Nordling, S. and Wartiovaara, J. (1973) Permeability and structure of ethanol-sterilized nitrocellulose (Millipore) filters. Exp. Cell Res. 81, 169–174.

Lehtonen, E., Wartiovaara, J., Nordling, S. and Saxén, L. (1975) Demonstration of cytoplasmic processes in Millipore filters permitting kidney tubule induction. J. Embryol. Exp. Morphol. 33, 187–203.

Lester, K.S. (1970) On the nature of "fibrils" and tubules in developing enamel of the opossum, *Didelphis marsupialis.* J. Ultrastruct. Res. 30, 64–77.

Levine, S., Pictet, R. and Rutter, W.J. (1973) Control of cell proliferation and cytodifferentiation by a factor reacting with the cell surface. Nature, New Biology 246, 49–52.

Liedke, K.B. (1951) Lens competence in Amblystoma punctatum. J. Exp. Zool. 117, 573–591.

Loewenstein, W.R. (1967) On the genesis of cellular communication. Dev. Biol. 15, 503–520.

Loewenstein, W.R. (1973) Membrane junctions in growth and differentiation. Fed. Proc. 32, 60–64.

Lombard, M.-N. and Grobstein, C. (1969) Activity in various embryonic and postembryonic sources for induction of kidney tubules. Dev. Biol. 19, 41–51.

Main, J.H.P. (1969) Transformation of odontogenic epithelium by polyoma virus in vitro. J. Dent. Res. 48, 738–744.

Main, J.H.P. (1972) Epithelio-mesenchymal interactions in epithelial tumour induction. In: Cell Differentiation (Harris, R., Allin, P. and Viza, D., eds.) pp. 156–161, Munksgaard, Copenhagen.

Main, J.H.P. and Waheed, M.A. (1971) Epithelio-mesenchymal interactions in the proliferative response evoked by polyoma virus in odontogenic epithelium in vitro. J. Nat. Cancer Inst. 47, 711–726.

Mangold, O. (1961) Grundzüge der Entwicklungsphysiologie der Wirbeltiere mit besonderer Berücksichtigung der Missbildungen auf Grund experimenteller Arbeiten an Urodelen. Acta Genet. Med. Gemellol. 10, 1–49.

Mathan, M., Hermos, J.A. and Trier, J.S. (1972) Structural features of the epithelio-mesenchymal interface of rat duodenal mucosa during development. J. Cell Biol. 52, 577–588.

Mazzucco, K. (1972) The role of collagen in tissue interactions during carcinogenesis in mouse skin. In: Tissue Interactions in Carcinogenesis (Tarin, D., ed.) pp. 377–398, Academic Press, London.

McKeehan, M.S. (1951) Cytological aspects of embryonic lens induction in the chick. J. Exp. Zool. 117, 31–64.

McKeehan, M.S. (1956) The relative ribonucleic acid content of lens and retina during lens induction in the chick. Am. J. Anat. 99, 131–156.

McKeehan, M.S. (1958) Induction of portions of the chick lens without contact with the optic cup. Anat. Rec. 132, 297–305.

McLoughlin, C.B. (1961) The importance of mesenchymal factors in the differentiation in chick epidermis. II. Modification of epidermal differentiation by contact with different types of mesenchyme. J. Embryol. Exp. Morphol. 9, 385–409.

McLoughlin, C.B. (1963) Mesenchymal influences on epithelial differentiation. Symp. Soc. Exp. Biol. 17, 359–388.

McMahon, D. (1974) Chemical messengers in development: A hypothesis. Science, 185, 1012–1021.

Meier, S. and Hay, E.D. (1974) Control of corneal differentiation by extracellular materials. Collagen as a promoter and stabilizer of epithelial stroma production. Dev. Biol. 38, 249–270.

Meier, S. and Hay, E.D. (1975) Control of corneal differentiation in vitro by extracellular matrix. In: Extracellular Matrix Influences on Gene Expression (Slavkin, H.C. and Greulich, R.C., eds.) pp. 185–196, Academic Press, New York.

Mintz, B. (1965) Experimental genetic mosaicism in the mouse. In: Preimplantation Stages of Pregnancy (Wolstenholme, G.E.W. and O'Connor, M., eds.) pp. 194–207, Churchill, London.

Mizuno, M.T. (1970) Induction de christallin in vitro, chez le Poulet, en absence de la vésicule optique. C.R. Acad. Sci. 271, 2190–2197.

Mizuno, T. (1972) Lens differentiation in vitro in the absence of optic vesicle in the epiblast of chick blastoderm under the influence of skin dermis. J. Embryol. Exp. Morphol. 28, 117–132.

Moore, N.W., Adams, C.E. and Rowson, L.E.A. (1968) Developmental potential of single blastomers of the rabbit egg. J. Reprod. Fertil. 17, 527–531.

Muthukkaruppan, V. (1965) Inductive tissue interaction in the development of the mouse lens in vitro. J. Exp. Zool. 159, 269–288.

Nadijcka, M. and Hillman, N. (1974) Ultrastructural studies of the mouse blastocyst substages. J. Embryol. Exp. Morphol. 32, 675–695.

Nieuwkoop, P.D. (1969) The formation of the mesoderm in urodelan amphibians. I. Induction by the endoderm. Wilhelm Roux Arch. Entwicklungsmech. Organismen 162, 341–373.

Nieuwkoop, P.D. (1973) The "organization center" of the amphibian embryo: its origin, spatial organization, and morphogenetic action. Adv. Morphog. 10, 1–39.

Niu, M.C. (1956) New approaches to the problem of embryonic induction. In: Cellular Mechanisms in Differentiation and Growth (Rudnick, D., ed.) pp. 155–171, Princeton University Press, Princeton, N.J.

Niu, M.C. and Deshpande, A.K. (1973) The development of tubular heart in RNA-treated post-nodal pieces of chick blastoderm. J. Embryol. Exp. Morphol. 29, 485–501.

Niu, M.C. and Mulherkar, L. (1970) The role of exogenous heart-RNA in development of chick embryo in vitro. J. Embryol. Exp. Morphol. 24, 33–42.

Niu, M.C., Deshpande, A.K. and Niu, L.C. (1974) Poly(a)-attached RNA as activator in embryonic differentiation (38334). Proc. Soc. Exp. Biol. Med. 147, 318–322.

Nordling, S., Miettinen, H., Wartiovaara, J. and Saxén, L. (1971) Transmission and spread of embryonic induction. I. Temporal relationships in transfilter induction of kidney tubules in vitro. J. Embryol. Exp. Morphol. 26, 231–252.

Nyholm, M., Saxén, L., Toivonen, S. and Vainio, T. (1962) Electron microscopy of transfilter neural induction. Exp. Cell Res. 28, 209–212.

Orr, J.W. (1934) The influence of ischaemia on the development of tumours. Brit. J. Exp. Pathol. 15, 73–79.

Orr, J.W. (1935) The effect of interference with the vascular supply on the induction of dibenzanthracene tumours. Brit. J. Exp. Pathol. 16, 121–126.

Orr, J.W. (1937) The results of vital staining with phenol red during the progress of carcinogenesis in mice treated with tar, dibenzanthracene and benzpyrene. J. Pathol. Bacteriol. 44, 19–27.

Orr, J.W. (1938) The changes antecedent to tumour formation during the treatment of mouse skin with carcinogenic hydrocarbons. J. Pathol. Bacteriol. 46, 495–515.

Orr, J.W. (1963) The role of the stroma in epidermal carcinogenesis. Nat. Cancer Inst. Monogr. 10, 531–537.

Orr, J.W. and Spencer, W.K. (1972) Transplantation studies of the role of the stroma in epidermal carcinogenesis. In: Tissue Interactions in Carcinogenesis (Tarin, D., ed.) pp. 291–303, Academic Press, London.

Pannese, E. (1962) Observations on the ultrastructure of the enamel organ. III. Internal and external enamel epithelia. J. Ultrastruct. Res. 6, 186–204.

Pappas, G.D., Asada, Y. and Bennett, M.V.L. (1971) Morphological correlates of increased coupling resistance at an electronic synapse. J. Cell Biol. 49, 173–188.

Pictet, R.L., Filosa, S., Phelps, P. and Rutter, W.J. (1975) Control of DNA synthesis in the embryonic pancreas: Interaction of the mesenchymal factor and cyclic AMP. In: Extracellular Matrix Influences on Gene Expression (Slavkin, H.C. and Greulich, R.C., eds) pp. 531–540, Academic Press, New York.

Pictet, R., Levine, S., Filosa, S., Phelps, P. and Rutter, W.J. (1974) Control of cell proliferation and differentiation of embryonic rat pancreas by mesenchymal factor and cAMP. J. Cell Biol. 63, 270a.

Pitts, J.D. (1971) Molecular exchange and growth control in tissue culture. In: Ciba Foundation Symposium on Growth Control in Cell Cultures (Wolstenholme, G.E.W. and Knight, J., eds.) pp. 89–105, Churchill Livingstone, London.

Propper, A. (1969) Compétence de l'épiderme embryonnaire d'Oiseau vis-à-vis de l'inducteur mammaire mésenchymateux. C.R. Acad. Sci. 268, 1423–1426.

Propper, A. (1975) Epithelio-mesenchymal interactions between developing chick epidermis and rabbit embryo mammary mesenchyme. In: Extracellular Matrix Influences on Gene Expression (Slavkin, H.C. and Greulich, R.C., eds.) pp. 541–547, Academic Press, New York.

Rakic, P. and Sidman, R.L. (1973) Weaver mutant mouse cerebellum: Defective neuronal migration secondary to abnormality of Bergman glia. Proc. Nat. Acad. Sci. U.S.A. 70, 240–244.

Rasilo, M.-L. and Leikola, A. (1976) Neural induction by previously induced epiblast in avian embryo in vitro. Differentiation 5, 1–7.

Rawles, M.E. (1963) Tissue interactions in scale and feather development as studied in dermal-

epidermal recombinations. J. Embryol. Exp. Morphol. 11, 765–789.
Readler, P. and Lustig, E.S. (1970) Control of epithelial development in normal and pathological connective tissue from oral mucosa. Dev. Biol. 22, 84–95.
Revel, J.-P. (1974) Some aspects of cellular interactions in development. In: The Cell Surface in Development (Moscona, A.A., ed.) pp. 51–65, John Wiley & Sons, New York.
Revel, J.-P., Yip, P. and Chang, L.L. (1973) Cell junctions in the early chick embryo – a freeze etch study. Dev. Biol. 35, 302–317.
Rifkind, R.A., Chui, D. and Epler, H. (1969) An ultrastructural study of early morphogenetic events during the establishment of fetal hepatic erythropoiesis. J. Cell Biol. 40, 343–365.
Ronzio, R.A. and Rutter, W.A. (1973) Effects of partially purified factor from chick embryos on macromolecular synthesis of embryonic pancreatic epithelia. Dev. Biol. 30, 307–320.
Rose, B. (1971) Intercellular communication and some structural aspects of membrane junctions in a simple cell system. J. Membrane Biol. 5, 1–19.
Roseman, S. (1970) The synthesis of complex carbohydrates by multiglycosyltransferase systems and their potential function in intercellular adhesion. Chem. Phys. Lipids 5, 270–297.
Rosenquist, G.C. (1970) The origin and movement of prelung cells in the chick embryo as determined by radioautographic mapping. J. Embryol. Exp. Morphol. 24, 497–509.
Rossant, J. (1975a) Investigation of the determinative state of the mouse inner cell mass. I. Aggregation of isolated inner cell masses with morulae. J. Embryol. Exp. Morphol. 33, 979–990.
Rossant, J. (1975b) Investigation of the determinative state of the mouse inner cell mass. II. The fate of isolated inner cell masses transferred to the oviduct. J. Embryol. Exp. Morphol. 33, 991–1001.
Roth, S. (1973) A molecular model for cell interactions. Quart. Rev. Biol. 48, 541–563.
Roth, S. and White, D. (1972) Intercellular-contact and cell-surface galactosyl transferase activity. Proc. Nat. Acad. Sci. U.S.A. 69, 485–489.
Roth, S., McGuire, E.J. and Roseman, S. (1971) An assay for intercelular adhesive specificity. J. Cell Biol. 51, 525–535.
Rutter, W.J., Wessells, N.K. and Grobstein, C. (1964) Control of specific synthesis in the developing pancreas. Nat. Cancer Inst. Monogr. 13, 51–65.
Rutter, W.J., Ball, W.D., Bradshaw, W.S., Clark, W.R. and Sanders, T.G. (1967) Levels of regulation in cytodifferentiation. In: Morphological and Biochemical Aspects of Cytodifferentiation, Vol. 1 (Hagen, E., Wechsler, W. and Zilliken, F., eds.) pp. 110–124, Karger, Basel.
Rutter, W.J., Clark, W.R., Kemp, J.D., Bradhaw, W.S., Sanders, T.G. and Ball, W.D. (1968) Multiphasic regulation in cytodifferentiation. In: Epithelial-Mesenchymal Interactions (Fleischmajer, R. and Billingham, R.E., eds.) pp. 114–131, Williams & Wilkins, Baltimore, Md.
Sala, M. (1955) Distribution of activating and transforming influences in the archenteron roof during the induction of nervous system in Amphibians. Proc. Acad. Sci. Amsterdam, Ser. C, 58, 635–647.
Salzgeber, B. (1967) Sur l'étude expérimentale de la genèse de la phocomélie chez l'embryon de poulet. C.R. Acad. Sci. 264, 395–397.
Sanford, J.B. and Rosenberg, M.D. (1972) Nucleotide phosphohydrolase activities of the plasma membranes of embryonic chick liver cells. Biochim. Biophys. Acta 288, 333–346.
Sanford, K.K., Earle, W.R., Shelton, E., Schilling, E.L., Duchesne, E.M., Likely, G.D. and Becker, M.M. (1950) Production of malignancy in vitro. XII. Further transformations of mouse fibroblasts to sarcomatous cells. J. Nat. Cancer Inst. 11, 351–375.
Saunders, J.W., Jr. (1948) The proximo-distal sequence of origin of the parts of the chick wing and the role of the ectoderm. J. Exp. Zool. 108, 363–404.
Saunders, J.W., Jr. and Gasseling, M.T. (1963) Trans-filter propagation of apical ectoderm maintenance factor in the chick embryo wing bud. Dev. Biol. 7, 64–78.
Sawyer, R.H. and Abbott, U.K. (1972) Defective histogenesis in the anterior shank skin of the scaleless mutant. J. Exp. Zool. 181, 99–110.
Saxén, L. (1961) Transfilter neural induction of Amphibian ectoderm. Dev. Biol. 3, 140–152.
Saxén, L. (1963) The transmission of information during primary embryonic induction, in Biological Organization at Cellular and Supercellular Level (Harris, R.J.C., ed.) pp. 211–227, Academic Press, New York.

Saxén, L. (1970a) Failure to demonstrate tubule induction in a heterologous mesenchyme. Dev. Biol. 23, 511–523.

Saxén, L. (1970b) Defective regulatory mechanisms in teratogenesis. Int. J. Gynaecol. Obstet. 8, 798–804.

Saxén, L. (1971) Inductive interactions in kidney development, In: Control Mechanisms of Growth and Differentiation. XXV Symposia of the Society for Experimental Biology (Davies, D.D. and Balls, M., eds.) pp. 207–221, Cambridge University Press, Cambridge.

Saxén, L. (1972) Interactive mechanisms in morphogenesis. In: Tissue Interactions in Carcinogenesis (Tarin, D., ed.) pp. 49–80, Academic Press, London.

Saxén, L. (1973) Tissue interaction and teratogenesis, In: Pathobiology of Development (Perrin, E.V.D. and Finegold, M., eds.) pp. 31–51, Williams & Wilkins, Baltimore, Md.

Saxén, L. (1975a) Abnormal cell and tissue interactions. In: Handbook of Teratology (Wilson, J. and Fraser, C., eds.) Plenum Press, London (in press).

Saxén, L. (1975b) Embryonic induction. Clin. Obstet. Gynecol. 18, 149–175.

Saxén, L. and Karkinen-Jääskeläinen, M. (1975) Inductive interactions in morphogenesis. In: The Early Development of Mammals (Balls, M. and Wild, A., eds.) pp. 319–333, Cambridge University Press, Cambridge.

Saxén, L. and Kohonen, J. (1969) Inductive tissue interactions in vertebrate morphogenesis. Int. Rev. Exp. Pathol. 8, 57–128.

Saxén, L. and Saksela, E. (1971) Transmission and spread of embryonic induction. II. Exclusion of an assimilatory transmission mechanism in kidney tubule induction. Exp. Cell Res. 66, 369–377.

Saxén, L. and Toivonen, S. (1961) The two-gradient hypothesis in primary induction. The combined effect of two types of inductors mixed in different ratios. J. Embryol. Exp. Morphol. 9, 514–533.

Saxén, L. and Toivonen, S. (1962) Primary Embryonic Induction, pp. 1–271, Academic Press, London.

Saxén, L., Toivonen, S. and Vainio, T. (1964) Initial stimulus and subsequent interactions in embryonic induction. J. Embryol. Exp. Morphol. 12, 333–338.

Saxén, L., Koskimies, O., Lahti, A., Miettinen, H., Rapola, J. and Wartiovaara, J. (1968) Differentiation of kidney mesenchyme in an experimental model system. Adv. Morphog. 7, 251–293.

Schlafke, S.J. and Enders, A.C. (1967) Cytological changes during cleavage and blastocyst formation in the rat. J. Anat. 102, 13–32.

Seidel, F. (1960) Die Entwicklungsfähigkeiten isolierter Furchungszellen aus dem Ei des Kaninchens, *Oryctolagus cuniculus*. Wilhelm Roux Arch. Entwicklungsmech. Organismen 152, 43–130.

Sengel, P. (1958) Recherches expérimentales sur la différenciation des germes plumaires et du pigment de la peau de l'embryon de Poulet en culture in vitro. Ann. Sci. Nat. (Zool.) 20, 432–514.

Sengel, P. (1975) Feather pattern development. In: Cell Patterning, Ciba Foundation Symposium 29, pp. 51–70, Associated Scientific Publishers, Amsterdam.

Sengel, P. and Abbott, U.K. (1962) Comportement in vitro de l'épiderme et du derme d'embryon de poulet mutant "scaleless" en association avec le derme et l'épiderme d'embryon normal. C.R. Acad. Sci. 255, 1999–2000.

Sengel, P., Dhouailly, D. and Kieny, M. (1969) Aptitude des constituants cutanés de l'aptérie médio-ventrale à former des plumes. Dev. Biol. 19, 436–446.

Sheridan, J.D. (1971) Dye movement and low-resistance junctions between reaggregated embryonic cells. Dev. Biol. 26, 627–636.

Sheridan, J.D. (1974) Low-resistance junctions: Some functional considerations. In: The Cell Surface in Development (Moscona, A.A., ed.) pp. 187–206, John Wiley & Sons, New York.

Sherman, J.E. (1960) Description and experimental analysis of chick submandibular gland morphogenesis. Wis. Acad. Sci. Arts Lett. 49, 171–189.

Sherman, M.I. (1975) The role of cell-cell induction during early mouse embryogenesis. In: The Early Development of Mammals (Balls, M. and Wild, A.E., eds.) pp. 145–165, Cambridge University Press, Cambridge.

Sherman, M.I., McLaren, A. and Walker, P.M.B. (1972) Mechanism of accumulation of DNA in giant cells of mouse trophoblast. Nature, New Biology 238, 175–176.

Sidman, R.L. (1974) Contact interaction among developing mammalian brain cells. In: The Cell

Surface in Development (Moscona, A.A., ed.) pp. 221–253, John Wiley & Sons, New York.
Silva, D.G. and Kailis, D.G. (1972) Ultrastructural studies on the cervical loop and the development of the amelo-dentinal junction in the cat. Arch. Oral Biol. 17, 279–289.
Silver, J. and Hughes, A.F.W. (1974) The relationship between morphogenetic cell death and the development of congenital anophthalmia. J. Comp. Neurol. 157, 281–302.
Slavkin, H.C. (1972) Intercellular communication during odontogenesis. In: Developmental Aspects of Oral Biology (Slavkin, H.C. and Bavetta, L.A., eds.) pp. 165–199, Academic Press, London.
Slavkin, H.C. and Bringas, P. (1976) Epithelial-mesenchymal interactions during odontogenesis. IV. Morphological evidence for direct heterotypic cell-cell contacts. Develop. Biol. 50, 428–442.
Slavkin, H.C., Bringas, P. and Bavetta, L.A. (1969) Ribonucleic acid within the extracellular matrix during embryonic tooth formation. J. Cell. Physiol. 73, 179–190.
Slavkin, H.C. and Croissant, R. (1973) Intercellular communication during odontogenic epithelial-mesenchymal interactions: isolation of extracellular matrix vesicles containing RNA. In: The Role of RNA in Reproduction and Development (Niu, M.C. and Segal, S., eds.) pp. 247–258, North-Holland, Amsterdam.
Slavkin, H.C. and Greulich, R.G. (1975) (eds.) Extracellular Matrix Influences on Gene Expression, Academic Press, New York.
Slavkin, H.C., Flores, P., Bringas, P. and Bavetta, L.A. (1970) Epithelial-mesenchymal interactions during odontogenesis. I. Isolation of several intercellular matrix low molecular weight methylated RNAs. Dev. Biol. 23, 276–296.
Spemann, H. (1901a) Entwicklungsphysiologische Studien am Triton-Ei. I. Wilhelm Roux Arch. Entwicklungsmech. Organismen 12, 224–264.
Spemann, H. (1901b) Über Correlationen in der Entwicklung des Auges, Anat. Anzeiger 19, Ergänzungsheft, 61–79.
Spemann, H. (1903) Entwicklungsphysiologische Studien am Triton-Ei. III. Wilhelm Roux Arch. Entwicklungsmech. Organismen 16, 551–631.
Spemann, H. (1912) Zur Entwicklung des Wirbeltierauges. Zool. Jahrb. Abt. Allgem. Zool. Physiol. Tiere 32, 1–98.
Spemann, H. (1918) Über die Determination der ersten Organanlagen des Amphibienembryo. I-VI. Arch. Entwicklungsmech. Organismen 43, 448–555.
Spemann, H. (1936) Experimentelle Beiträge zu einer Theorie der Entwicklung, Springer-Verlag, Berlin.
Spiro, R.G. and Fukushi, S. (1969) The lens capsulae: Studies on the carbohydrate units. J. Biol. Chem. 244, 2049–2058.
Spooner, B.S. and Wessells, N.K. (1970) Mammalian lung development: Interactions in primordium formation and bronchial morphogenesis. J. Exp. Zool. 175, 445–454.
Spooner, B.S. and Wessells, N.K. (1972) An analysis of salivary gland morphogenesis: role of cytoplasmic microfilaments and microtubules. Dev. Biol. 27, 38–54.
Spratt, N.T., Jr. (1954) Physiological mechanisms in development. Physiol. Revs. 34, 1–24.
Steinmüller, D. (1971) Epidermal transplantation during chemical carcinogenesis. Cancer Res. 31, 2080–2084.
Strudel, G. (1953) L'influence morphogene du tube nerveux sur differenciation de la colonne vertebrale. C.R. Soc. Biol. 147, 132–133.
Stuart, E.S. and Moscona, A.A. (1967) Embryonic morphogenesis: Role of fibrous lattice in the development of feathers and feather patterns. Science, 157, 947–948.
Stuart, E.S., Garber, B. and Moscona, A.A. (1972) An analysis of feather germ formation in the embryo and in vitro, in normal development and in skin treated with hydrocortisone. J. Exp. Zool. 179, 97–118.
Subak-Sharpe, H., Bürk, R.R. and Pitts, J.D. (1969) Metabolic co-operation between biochemically marked mammalian cells in tissue culture. J. Cell Sci. 4, 353–367.
Taderera, J.V. (1967) Control of lung differentiation in vitro. Dev. Biol. 16, 489–512.
Tarin, D. (1967) Sequential electron microscopical study of experimental mouse skin carcinogenesis. Int. J. Cancer 2, 195–211.

Tarin, D. (1971) Histological features of neural induction in *Xenopus laevis*. J. Embryol. Exp. Morphol. 26, 543–570.

Tarin, D. (1972) Ultrastructural features of neural induction in *Xenopus laevis*. J. Anat. 111, 1–28.

Tarin, D. (1973) Histochemical and enzyme digestion studies on neural induction in *Xenopus laevis*. Differentiation, 1, 109–126.

Tarin, D. and Sturdee, A.P. (1974) Ultrastructural features of ectodermal-mesenchymal relationships in the developing limb of *Xenopus laevis*. J. Embryol. Exp. Morphol. 31, 287–303.

Tarkowski, A.K. (1959) Experiments on the development of isolated blastomers of mouse eggs. Nature, 184, 1286–1287.

Tarkowski, A.K. and Wroblewska, J. (1967) Development of blastomers of mouse eggs isolated at the 4- and 8-cell stage. J. Embryol. Exp. Morphol. 18, 155–180.

Tarkowski, A.K., Witkowska, A. and Nowicka, J. (1970) Experimental parthenogenesis in the mouse. Nature, 226, 162–165.

Temin, H.M. (1971) Guest editorial in issue No. 2, February (between p. 215 and 217). The protovirus hypothesis: Speculations on the significance of RNA-directed DNA synthesis for normal development and for carcinogenesis. J. Nat. Cancer Inst. 36, III-VII.

TerHorst, J. (1948) Differenzierungs- und Induktionsleistungen verschiedener Abschnitte der Medullarplatte und des Urdarmdaches von Triton im Kombinat. Wilhelm Roux Arch. Entwicklungsmech. Organismen 143, 275–303.

Tiedemann, H. (1968) Factors determining embryonic differentiation. J. Cell. Physiol. 72, Suppl. 1, 129–144.

Tiedemann, H. (1971) Extrinsic and intrinsic information transfer in early differentiation of amphibian embryos. In: Control Mechanisms of Growth and Differentiation (Davies, D.D. and Balls, M., eds.) pp. 223–234, Cambridge University Press, Cambridge.

Tiedemann, H. (1973) Pretranslational control in embryonic differentiation. In: Regulation of Transcription and Translation in Eukaryotes (Bautz, E.K.F., ed.) pp. 59–80, Springer-Verlag, Berlin.

Tiedemann, H. and Tiedemann, H. (1959) Versuche zur Gewinnung eines mesodermalen Induktionsstoffes aus Hühnerembryonen. Hoppe-Seyler's Z. Physiol. Chem. 314, 156–176.

Toivonen, S. (1940) Über die Leistungsspezifität der abnormen Induktoren im Implantatversuch bei Triton. Diss. Helsinki. Ann. Acad. Sci. Fenn. Ser. A IV, 55, No. 6, 1–150.

Toivonen, S. (1945) Zur Frage der Induktion selbständiger Linsen durch abnorme Induktoren im Implantatversuch bei Triton. Ann. Soc. Zool.-bot. Fenn. Vanamo 11, No. 3, 1–28.

Toivonen, S. (1954) The inducing action of the bone-marrow of the guinea-pig after alcohol and heat treatment in implantation and explantation experiments with embryos of *Triturus*. J. Embryol. Exp. Morphol. 2, 239–244.

Toivonen, S. (1972) Heterotypic tissue interactions in the segregation of the central nervous system (CNS). In: Cell Differentiation (Harris, R., Allin, P. and Viza, D., eds.) pp. 30–34, Munksgaard, Copenhagen.

Toivonen, S. and Saxén, L. (1955) The simultaneous inducing action of liver and bone-marrow of the guinea-pig in implantation and explantation experiments with embryos of *Triturus*. Exp. Cell Res., Suppl. 3, 346–357.

Toivonen, S. and Saxén, L. (1966) Late tissue interactions in the segregation of the central nervous system. Ann. Med. Exp. Biol. Fenn. 44, 128–130.

Toivonen, S. and Saxén, L. (1968) Morphogenetic interaction of presumptive neural and mesodermal cells mixed in different ratios. Science, 159, 539–540.

Toivonen, S., Tarin, D. and Saxén, L. (1976) Transmission of inductive signals during primary induction. Review. Differentiation 5, 49–55.

Toivonen, S., Tarin, D., Saxén, L., Tarin, P.J. and Wartiovaara, J. (1975) Transfilter studies on neural induction in the newt. Differentiation, 4, 1–7.

Touster, O., Aronson, N.N., Jr., Dulaney, J.T. and Hendrickson, H. (1970) Isolation of rat liver plasma membranes. Use of nucleotide pyrophosphatase and phosphodiesterase I as marker enzymes. J. Cell. Biol. 47, 604–618.

Trelstad, R.L., Hay, E.D. and Revel, J.-P. (1967) Cell contact during early morphogenesis in the chick embryo. Dev. Biol. 16, 78–106.

Trelstad, R.L., Revel, J.-P. and Hay, E.D. (1966) Tight junctions between cells in the early chick embryo as visualized within the electron microscope. J. Cell. Biol. 31, C6–C10.

Twitty, V.C. (1955) Eye. In: Analysis of Development (Willier, B.H., Weiss, P.A. and Hamburger, V., eds.) pp. 402–414, Saunders, Philadelphia, Pa.

Tyler, A. (1946) An autoantibody concept of cell structure growth, and differentiation. Growth 10, Symp. 6, 7–19.

Unsworth, B. and Grobstein, C. (1970) Induction of kidney tubules in mouse metanephrogenic mesenchyme by various embryonic mesenchymal tissues. Dev. Biol. 21, 547–556.

Vainio, T., Saxén, L., Toivonen, S. and Rapola, J. (1962) The transmission problem in primary embryonic induction. Exp. Cell Res. 27, 527–538.

Van Deusen, E. (1973) Experimental studies on a mutant gene (e) preventing the differentiation of eye and normal hypothalamus primordia in the axolotl. Dev. Biol. 34, 135–158.

Wartiovaara, J., Lehtonen, E., Nordling, S. and Saxén, L. (1972) Do membrane filters prevent cell contacts? Nature, 238, 407–408.

Wartiovaara, J., Nordling, S., Lehtonen, E. and Saxén, L. (1974) Transfilter induction of kidney tubules: Correlation with cytoplasmic penetration into Nucleopore filters. J. Embryol. Exp. Morphol. 31, 667–682.

Weiser, M.M. (1973) Intestinal epithelial cell surface membrane glycoprotein synthesis. II. Glycosyltransferases and endogenous acceptors of the undifferentiated cell surface membrane. J. Biol. Chem. 248, 2542–2548.

Weiss, L. (1969) The cell periphery. Int. Rev. Cytol. 26, 63–105.

Weiss, P. (1947) The problem of specificity in growth and development. Yale J. Biol. Med. 19, 235–278.

Weiss, P. (1958) Cell contact. Int. Rev. Cytol. 7, 391–423.

Weiss, P. and Jackson, S.F. (1961) Fine-structural changes associated with lens determination in the avian embryo. Dev. Biol. 3, 532–554.

Wessells, N.K. (1962) Tissue interactions during skin histodifferentiation. Dev. Biol. 4, 87–107.

Wessells, N.K. (1964) Substrate and nutrient effects upon epidermal basal cell orientation and proliferation. Proc. Nat. Acad. Sci. U.S.A. 52, 252–259.

Wessells, N.K. (1965) Morphology and proliferation during early feather development. Dev. Biol. 12, 131–153.

Wessells, N.K. (1973) Tissue interactions in development, In: Addison-Wesley Module, 9, pp. 1–43, Addison-Wesley Publ., Reading, Mass.

Wessells, N.K. and Cohen, J.H. (1967) Early pancreas organogenesis: morphogenesis, tissue interactions, and mass effects. Dev. Biol. 15, 237–270.

Wessells, N.K. and Cohen, J.H. (1968) Effects of collagenase on developing epithelia in vitro: lung, ureteric bud, and pancreas. Dev. Biol. 18, 294–309.

Whitten, W.K. and Biggers, J.D. (1968) Complete development in vitro of the preimplantation stages of the mouse embryo in a simple chemically defined medium. J. Reprod. Fertil. 17, 399–401.

Wolpert, L. (1969) Positional information and spatial pattern of cellular differentiation. J. Theor. Biol. 25, 1–48.

Wolpert, L. (1971) Positional information and pattern formation. Curr. Top. Dev. Biol. 6, 183–224.

Yamada, T. (1962) The inductive phenomenon as a tool for understanding the basic mechanism of differentiation. J. Cell. Comp. Physiol., Suppl. 1, 60, 49–64.

Yogeeswaran, G., Laine, R.A. and Hakomori, S. (1974) Mechanism of cell contact-dependent glycolipid synthesis: Further studies with glycolipid-glass complex. Biochem. Biophys. Res. Commun. 59, 591–599.

Zwilling, E. (1955) Teratogenesis. In: Analysis of Development (Willier, R.H., Weiss, P. and Hamburger, V., eds.) pp. 699–719, Saunders, Philadelphia, Pa.

Zwilling, E. (1956a) Interaction between limb bud ectoderm and mesoderm in the chick embryo. I. Axis establishment. J. Exp. Zool. 132, 157–171.

Zwilling, E. (1956b) Interaction between limb bud ectoderm and mesoderm in the chick embryo. IV. Experiments with a wingless mutant. J. Exp. Zool. 132, 241–253.
Zwilling, E. (1956c) Genetic mechanism in limb development. Cold Spring Harbor Symp. Quant. Biol. 21, 349–354.
Zwilling, E. (1961) Limb morphogenesis. Adv. Morphog. 1, 301–330.
Zwilling, E. (1969) Abnormal morphogenesis in limb development. In: Limb Development and Deformity: Problems of Evaluation and Rehabilitation (Swinyard, C.A., ed.) pp. 100–118, Charles C. Thomas, Springfield, Ill.
Zwilling, E. (1974) Effects of contact between mutant (wingless) limb buds and those of genetically normal chick embryos: confirmation of a hypothesis. Dev. Biol. 39, 37–48.

Cell coupling and cell communication during embryogenesis 8

J. D. SHERIDAN

1. Introduction

As soon as the egg divides, it must begin to coordinate first two, and ultimately myriad cells into a functional unit. This coordination depends on a number of mechanisms operating at many levels and certainly one of the more important of these is the cell surface. One surface-mediated interaction, cell coupling via low resistance junctions, has attracted special interest among both developmental and non-developmental biologists. Numerous physiological and ultrastructural studies have been made of cell coupling in embryos and have attempted to answer such questions as: Do embryonic junctions* have the same ultrastructure and permeability as their adult counterparts? How are junctions distributed in embryos? How are individual junctions assembled and how is their assembly regulated? What role, if any, do junctions play in development? The answers to these questions are still incomplete, and in some cases equivocal. The purpose of this review is to consider these partial answers, to compare, where appropriate, embryonic and non-embryonic junctions, and to suggest directions for future work.

2. General properties of "coupled" cells

Many general reviews about low resistance junctions are available (Loewenstein, 1966; Furshpan and Potter, 1968; Bennett, 1973; McNutt and Weinstein, 1973; Sheridan, 1974) so it is unnecessary to cover much historical background. Before considering embryonic junctions *per se*, however, I will first make some general comments about the properties of low resistance junctions in non-embryonic systems. Qualifications of some of these properties in embryonic systems will be discussed in depth later. (Throughout this article reviews and representative papers will be cited where appropriate.)

*For convenience, I will use "junctions" to refer to low resistance junctions unless specified otherwise.

G. Poste & G.L. Nicolson (eds.) The Cell Surface in Animal Embryogenesis and Development

2.1. *Electrical properties*

Low resistance junctions are points of intercellular apposition having an increased permeability to movement of small inorganic ions from cell to cell. (See Bennett, 1966, for a more quantitative or Sheridan, 1973, for a more qualitative approach.) Their high ionic permeability (= low resistance = high conductance) relative to nonjunctional membranes allows induced potential changes to be transferred from cell to cell, i.e. the junctions "electrically couple" cells. In general, the higher the ionic permeability of the junctions (or the lower its resistance) *relative to the permeability of non-junctional membrane* the more effectively the potential change is transferred. This is reflected in an increase in the coupling coefficient, i.e. the ratio of voltage change in one cell divided by the voltage change in the cell supplied with current. However, an increase in coupling coefficient can occur with a constant, or even decreased junctional permeability if the non-junctional permeability decreases sufficiently. Thus, other electrical data are necessary before changes in coupling coefficients can be ascribed solely to changes in junctional permeability.

The methods for estimating the ionic permeability of junctions are rather esoteric and are either suitable only for particular cell arrangements, e.g. where there is only one possible pathway for current to flow from cell to cell (Bennett, 1972) or, alternatively, for populations of cells that can reasonably be considered identical electrically (Siegenbeek et al., 1972). In the few studies of cell systems with suitable geometry, junctional ionic permeabilities (conductances) ranging from $10^{-4} - 10^{-8}$ mhos have been estimated (see Bennett et al., 1972; Siegenbeek et al., 1972).

The subunit structure of gap junctions, which are thought to be equivalent to low resistance junctions (see below), suggests that junctional permeability should be proportional to junctional area. If junctional area were known, it would thus be possible to carry the electrical analysis one step further by determining the degree of specialization of the junctional membranes. There have been no totally satisfactory direct comparisons of junctional area and junctional conductance. However, the few available experimental studies suggest specific conductances of the order of $10 - 10^{2}$ mhos/cm^{2} (Spira, 1971; Ito et al., 1974b) and these values also fit with theoretical calculations based on models of gap junctions (Bennett, 1974). Thus, the junctional membranes are 4–5 orders of magnitude more permeable to ions than non-junctional membranes.

2.2. *Permeability to larger molecules*

More general functional interest has been stimulated by the evidence that low resistance junctions are permeable to molecules larger than small inorganic ions (Loewenstein, 1966; Furshpan and Potter, 1968; Bennett, 1974). This evidence, which is only correlative, comes from two main types of experiments. The first type involves injecting a tracer molecule into one cell and watching its movement to adjacent cells. The fluorescent molecules, fluorescein (molecular weight 330) and procion yellow (molecular weight 650) have been particularly useful tracers (Payton et al., 1969; Bennett, 1974). They can be injected iontophoretically from micropipettes, can readily be detected at very low concentrations, have limited ability (essentially zero for procion yellow) to cross non-junctional membranes, and, in the case of procion yellow, can be fixed in the tissue and located in histological sections. Other molecules, such as sucrose (Bennett and Dunham, 1970) and dansylated amino acids (Johnson and Sheridan, 1971), have also been used.

In many experiments of this type, transfer of the small tracer has been correlated with electrical coupling: the tracers being exchanged when the cell are coupled electrically, but not when coupling is absent or experimentally disrupted.* In other experiments, however, cells fail to pass the tracers, and the implications of such experiments and their relevance to developmental systems is discussed in detail later. We need only comment here that there is no unequivocal evidence for a quantitative or qualitative discrepancy between permeability of junctions to small inorganic ions and to larger, tracer molecules.

An interesting variant of the injection approach has been devised for heart muscle (Weingart, 1974) and relies on the ability of cut muscle cells to remain open in Ca^{2+}-free solution but to reseal rapidly when normal Ca^{2+} is restored. A thin strand of purkinje or trabecular fibers is pulled through a diaphragm separating two compartments. One compartment contains Ca^{2+}-free solution, and in this compartment the muscle fibers are cut. A tracer, for example, procion yellow (Imanaga, 1974), is then added to the chamber and, after sufficient time elapses for uptake by the damaged fibers, the tracer is washed out of the extracellular space with Ca^{2+}-containing solution, allowing the cut fibers to seal. Then the tracer moves from cell to cell along the strand.

This method has been used to demonstrate movement of procion yellow and tetraethylammonium (Imanaga, 1974; Weingart, 1974). More recently, it has been reported that radiolabeled cyclic AMP (cAMP) is also transferred by heart cells in such a system (Tsien and Weingart, 1974), a result with profound functional implications as elaborated below in section 3.3. and elsewhere (Sheridan, 1974). However, only a brief report has appeared and it is not clear whether the cAMP moves prior to hydrolysis to 5′-AMP. In associated functional experiments, very high cAMP concentrations (100 mM) were used to produce rather small inotropic effects.

The second type of tracer experiment, with more direct functional implications, focusses on the transfer of metabolites between cells in contact in culture. This experiment also takes different forms. The earliest, and still most common, form involves culturing normal, wild-type cells with mutant cells unable to incorporate some exogenous radiolabelled precursor into cellular macromolecules (Subak-Sharpe, et al., 1966, 1969; Pitts, 1972). The wild type cells take up the precursor, e.g. hypoxanthine, convert it rapidly into small molecules

*This is an important control situation for it makes untenable the recent conclusion by van Venrooij et al. (1975) of fluorescein transfer by non-junctional routes.

retained by the cell membrane, e.g. nucleotides, and transfer these molecules to the mutant where they can be incorporated. Transfer is generally detected and quantitated by analyzing autoradiographs of the fixed and dried cells. In the case of wild-type and HGPRT-deficient mutant cells, transfer can also be measured by determining the total incorporation of exogenously derived nucleotides as a function of wild-type/mutant ratio (Sheridan et al., 1975a). Such experiments show not only that nucleotides are transferred but also that the cocultivation leads to alteration in the activity of enzymes of nucleotide metabolism in both wild-type and mutant cells.

An alternate approach is to prelabel the nucleotide pools of donor cells with uridine (Simms and Pitts, unpublished observations). Due chiefly to rapid turnover and recycling of heterogeneous RNA, the specific activity of the nucleotide pool remains high even after the cells are extensively washed. The labeled donors are then cocultivated with cold recipients and the transfer of nucleotides between the hot and cold cells is monitored autoradiographically.

In all the nucleotide experiments, transfer occurs between cells that are in direct contact and are known to be coupled electrically (and, where tested, transfer fluorescent tracers) but is absent when other forms of transfer are absent (Gilula et al., 1972).

Similar metabolic experiments have shown that glucose-6-phosphate, thymidine-X-P (Simms and Pitts, unpublished observations), and tetrahydrofolic acid (Finbow and Pitts, unpublished observations) are also transferred by junction-forming but not by non-junction forming cells in culture.

All of the tracer experiments suggest strongly, though circumstantially, that low resistance junctions act as rather non-selective sieves to pass a variety of small molecules. The upper cut-off in size of transferred molecules is not known. Pitts has shown that transfer of macromolecules does not occur in his culture experiments and in another study it was reported that Niagara Sky Blue (molecule weight 1000) is not exchanged by heart cells under conditions where procion yellow is transferred (Imanaga, 1974). Microperoxidase (molecular weight 1800) fails to cross the septal synapse in crayfish unless the cells are fixed (Bennett, 1974). Thus, a provisional upper limit of about 1000 daltons has been generally accepted as a working figure for the limit of junctional permeability.

2.3. Structure of low resistance junctions

Many lines of circumstantial evidence (see review by McNutt and Weinstein, 1973) suggest that gap junctions (described below) provide the structural basis for the transfer of small molecules observed in electrophysiological and tracer experiments.* (An important exception will be discussed below.)

*Although the evidence is strong that gap junctions mediate cell coupling, some coupling via other junctions, e.g. occluding or septate, should not be summarily dismissed. Nevertheless, it is unlikely these other junctions play significant roles in "communication", perhaps just indicating the minimal structure necessary for coupling to occur.

The most convincing observations are the following: (1) whenever cells capable of directly exchanging small molecules have been studied with freeze-fracture methods (the most sensitive for detecting small junctions), gap junctions have been found (this generalization cannot be made for any other type of junction); (2) when gap junctions are disrupted, as for example by hypertonic sucrose, transfer of small molecules is interrupted (Barr et al., 1965; Rose and Loewenstein 1975); (3) cells congenitally lacking gap junctions are unable to transfer small molecules (Gilula et al., 1972); and (4) the frequency of occurrence of gap junctions and distributions of the numbers of junctional particles/interface correlate well with the distribution of junctional permeabilities during junction formation (see below).

We have learned a great deal about the general structure of gap junctions and, more recently, even about their chemistry and molecular organization. Much of the basic information has been adequately reviewed elsewhere and only the highlights will be given here. The gap junction, or nexus, has the following characteristics (McNutt and Weinstein, 1973): (1) there is close membrane apposition and the extracellular space is narrowed to a 20–50 Å space; (2) the space is crossed by subunits which in negatively stained preparations produce a polygonal network observable in face-on views; and (3) there are two complementary sets of intramembranous subunits, also polygonally arranged and seen as 60–100 Å particles in freeze fracture replicas.

Freeze fracture studies have shown that gap junctions vary in the arrangement and packing of subunits (e.g. very loose, hexagonal, linear groups, domains), in their areas or number of subunits, and in the retention of subunits by the inner (P) or outer (E) membrane faces following fracture (McNutt and Weinstein, 1973; Sheridan and Johnson, 1975). The cause or significance of these variations is not clear. Some of the variations, e.g. retention of subunits by the E-face in arthropods, produce no obvious change in physiological properties (unless this is associated with the lability of junctional permeability in arthropods to altered cellular Ca^{2+}; see, Rose and Loewenstein, 1975). Other variations, such as tight versus loose packing of subunits, may be correlated with changes in junctional permeability (Perrachia, 1975), an exciting prospect if confirmed. It is also possible that some variations represent either different stages in assembly (or breakdown) of these structures or differences in the mechanisms underlying these processes (see below). Variation in junctional area or numbers of subunits is likely to change overall permeability, especially if each pair of complementary subunits contains an identical intercytoplasmic channel as most workers believe (Bennett, 1974).

Many of the inferences drawn from electron microscopic studies have been supported and extended by recent studies on the chemical constituents of the gap junctions and their molecular organization. Gap junctions can be isolated from liver (and less routinely from heart muscle) by virtue of their relative resistance to sarcosyl and enzymes such as collagenase and trypsin (Goodenough and Stoeckenius, 1972). The isolated junctions contain lipid (cholesterol and phospholipid) and only a few proteins, all of small molecular weight, with no detectable carbohydrate component or enzymatic activity.

The regularity of the packing of subunits in liver junctions has allowed X-ray diffraction analysis of pellets of gap junctions and optical diffraction analysis of negatively-strained gap junctions (Casper and Goodenough, unpublished observations). The two types of diffraction analysis give very comparable data concerning subunit size and packing order, both long and short-range. The optical diffraction pattern suggests that each subunit is itself composed of six particles, which fits with suggestions from electronmicroscopy and with the protein analysis. Particularly exciting evidence from the X-ray studies strongly suggests that the subunits in both apposed membranes are in register and that an aqueous channel passes from cell to cell down the middle of each complementary pair of subunits. The X-ray diffraction data thus provides some of the most direct support yet available for the model that has been offered to explain the permeability of gap junctions to small molecules.

These studies of low resistance and gap junctions in adult tissues therefore provide us with a basic functional and structural model to compare with evidence on the organization of these structures in embryonic systems.

3. *Embryonic cell junctions*

3.1. *Structure*

Studies of the structure of embryonic gap junctions and their distribution (see below) have until recently presented a somewhat confusing picture. This situation has a number of origins. First, embryos are generally difficult to prepare for high resolution microscopy. Second, even in well-fixed material, many (particularly earlier) studies have not distinguished gap from other close appositions, variously termed tight junctions, focal tight junctions or close junctions. Third, many of the observations on embryonic junctions have been made during studies initiated for other purposes and the necessary high resolution electronmicroscopy has often not been used. Very few studies on embryonic junctions *per se* have been made. Fourth, most of the electronmicroscopic studies have relied strictly on conventional thin section techniques, which provide minimal structural detail when gap junctions are large, and are even less useful when junctions are small as is often true in embryos.

The effects of these problems on our understanding of the distribution of gap junctions in embryos are substantial and are discussed below (3.3.). There are, however, enough studies of basic gap junction structure in embryos fixed successfully and investigated with freeze-fracture (Decker and Friend, 1974), or to lesser degree, lanthanum hydroxide (Ducibella et al., 1975) to provide at least some basis for comparison with adult junctions.

The results, however, are not especially surprising or illuminating. Gap junctions in embryos resemble those in adults: in conventional thin section there is the typical close membrane apposition with a 20–40 Å gap; colloidal lanthanum hydroxide infiltrates the gap, outlining intercellular subunits that can be seen in *en face* view and within each membrane aggregated intramembranous particles of 8–9 nm are found and which in most embryos are retained by the P-face in replicas.

The only striking difference from adult junctions are in numbers and organization of junctional particles. Embryonic junctions are often smaller with correspondingly fewer junctional particles. Also, the junctional particles are more frequently packed quite loosely or from unusual arrays. The smaller size of individual junctions may reflect a smaller total area of gap junctions per interface, but the necessary quantitation is lacking. It is likely that the looser packing and unusual arrays arise from faster turnover of embryonic junctions, i.e. from junctional formation or breakdown (see section 3.4. below). When the numbers of particles become very small (less than 10) or assume peculiar arrangements (e.g., linear strands) the question of definition becomes serious. As in some adult tissues such as retina (Raviola and Gilula, 1973), where the designation of intramembranous particle aggregates as gap junctions is somewhat arbitrary, corroborative physiological evidence for junctional transfer of small molecules is desirable but generally lacking. The need for caution in using the term "gap junction" is further emphasized by the existence of other arrays of particles (e.g. bladder, Kachadorian et al., 1975) that are unlikely to perform any junctional task.

Aside from these few qualifications, however, it appears that the structure of embryonic gap junctions presents no fundamental qualitative difference from that of adult junctions. This conclusion is important in evaluating suggestions from physiological studies that embryonic junctions have qualitative restrictions in permeability (see below).

3.2. Permeability of embryonic junctions

The permeability of embryonic junctions has been studied with the same types of methods applied to adult systems, though fewer embryonic systems have been tested. Each of these methods has elucidated a different aspect of permeability and is best discussed separately.

3.2.1. Ionic permeability

In adult tissues and other non-embryonic systems such as cell cultures in vitro, electrical coupling results unequivocally from specialized, low resistance junctions. In embryos, however, coupling arises from other sources as well. The study of these other sources and their distinction from junctional sources are important features of embryonic studies and should be discussed before we turn to low resistance junctions.

In whole embryos, as in other systems, coupling is determined by passing current into one cell and recording transmembrane potential changes inside a second cell with the reference electrode(s) in the extraembryonic medium. The presence of a substantial potential change in the second cell, by definition, shows that the cells are coupled. This means in essence that the resistance between the two intracellular electrodes is low relative to the resistance between the recording electrode and the reference electrode in the extraembryonic medium. The difference in relative resistances will occur whenever one or more of the following conditions hold: (1) the whole embryo is enveloped by a high resistance membrane; (2) the surface membrane of the cells is of high resistance, the remainder of the cell membrane is of normal or low resistance, and the intercellular space is isolated electrically from the extraembryonic space by an apical, occluding cell junction; (3) the surface membrane is of high resistance and the appositional membranes are of substantially lower resistance even if occluding junctions are absent; (4) the non-appositional membranes are of normal resistance and the entire appositional membranes are of very low resistance; and (5) all the membranes are of normal resistance except for specialized, junctional membranes of very low resistance (4 and 5 differ essentially in the area of membrane specialization and are difficult if not impossible to distinguish by electrophysiological measurements).

Conditions (1) or (2) above will lead to potential changes in the intercellular space of the same order as in the cells when a current pulse is passed either inside a cell or inside the intercellular space. Thus, the conditions can be detected or ruled out quite simply, as was done by Potter et al. (1966) in their classic study on early squid embryos. They excluded (1) and (2) by showing that coupling disappeared whenever the recording electrode was inside the embryo but not in a cell.

The situation in other embryos is different, however. In amphibians, during the cleavage stages, condition (3) appears to hold and during later stages, condition (2). Evidence for the presence of condition (3) during cleavage stages comes primarily from the work of Woodward (1968), de Laat and Bluemink (1974), and Slack and Warner (1973). They all conclude that the surface membrane of the egg has an unusually low ionic permeability (i.e., high resistance), whereas the new membrane produced during cleavage has a greater permeability, especially to K^+. As long as the cleaving egg remains within the vitelline membrane, most of the new membrane is hidden and is electrically isolated from the extraembryonic space by the high resistance, intercellular cleft. The new membrane, marked by its lack of associated pigment granules, can be artifically exposed to the surface either by removing the vitelline membrane, by treating with cytochalasin B, or by exposing the de-jellied egg to hypertonic medium. Once the new membrane is exposed, the overall ionic permeability of the cell membrane rises, becoming more K^+-selective, and the resting membrane potential becomes more negative. From a combined electronmicroscopic and electrophysiological analysis of these events, de Laat and Bluemink (1974) conclude that the cells in early cleavage stages should be electrically coupled (condition (3) holds) even in the absence of specialized junctions (condition (5)) or even of an abnormally low resistance of the entire appositional membrane (condition (4)).

In more advanced amphibian embryos the situation is complicated by the development of occluding junctions at the apices of the surface cells; thus condition (2) is established. In studies on morula and later stages, Ito and Loewenstein (1969) and Slack and Warner (1973) attempted to bypass the high resistance barrier (produced by the occluding junctions and the high surface membrane resistance) by splitting apart surface cells and thus producing an electrical shunt between the intraembryonic space and extraembryonic medium. The subsequent decrease in coupling coefficient (from .9–.6) was taken as evidence that the shunt was effective and the residual coupling as evidence for low resistance junctions. While this is a likely interpretation, a reduction in coupling by cell damage and some contribution by condition (3) were not ruled out. Nevertheless, even given these qualifications, there is strong evidence from these various experiments that some modification of the appositional membranes occurs and possibly involves specific junctions. This conclusion is further supported by the observation that coupling (and dye transfer, see below) develops between aggregated *Xenopus* (Sheridan, 1971) and newt (Ito et al., 1975a) cells and persists between isolated pairs of cells teased out from the inside of later *Xenopus* embryos (Slack and Warner, 1975).

Coupling via extracellular space occurs in other embryos as well. Bennett and Trinkaus (1970) analyzed in some detail the relative contributions of coupling via extracellular space and via low resistance junctions in fish embryos. They concluded that the enveloping cells at the surface are coupled in two ways: via the segmentation cavity, due to the high surface membrane resistance, occluding junctions, and lower resistance inner membranes, and via low resistance junctions, which resulted in greater coupling between two adjacent surface cells than between one surface cell and the segmentation cavity.

The possibility of coupling via extracellular space has been tested directly in other embryos (Sheridan, 1968). In chick embryos, for example, coupling was studied in isolated pieces of blastoderm with many potential shunts to bypass any surface barrier. Furthermore, in those studies it was often possible to advance an electrode into the intercellular space and show disappearance of coupling with other cells. The situation in the intact embryo, however, remains unclear.

Yet, the possible complications introduced by coupling via extracellular space need always be kept in mind in designing and interpreting experiments on whole embryos or even on certain isolated embryonic tissues. For example, in an early study (Loewenstein and Penn, 1967) of skin from amphibian larvae, the preparation was set up in a way that potentially retained a surface barrier. Thus, the data might have been influenced to an unknown degree by coupling via extracellular space and the conclusions should be viewed with this possible complication in mind.

As a group, these various studies show that coupling can occur in embryos via non-junctional pathways, but these cannot in most cases account quantitatively for the coupling seen. Except for the early cleavage stages, for which the extracellular route may be a sufficient explanation for coupling, specialized, low resistance junctions appear to be at least partially responsible.

Now that we have addressed and put into perspective the question of coupling via extracellular space, we can turn to the ionic permeability of low resistance junctions in embryos. Again, we can use low resistance junctions in non-embryonic systems for comparison but we are immediately confronted by another problem. What measure of ionic permeability is likely to provide the most meaningful comparison? As already noted, the degree of electrical coupling (i.e. coupling coefficient) is a feasible possibility. But this can give a misleading impression because it depends on non-junctional as well as junctional conductance or ionic permeability. This dual dependency is especially problematic in embryos, which, as mentioned, often have an exceptionally low non-junctional conductance (= high non-junctional resistance). As a result, the extremely high degree of coupling in early embryos, for example, is as much (or more) a reflection of non-junctional as junctional ionic permeability. Furthermore, as mentioned above, in the earliest embryos, low resistance junctions may not even be present.

The degree of coupling *per se* might be relevant, however, if the transfer of electrical potential changes (as distinct from bulk transfer of ions) is important functionally during development. There are characteristic changes in transmembrane potential in the early embryo (e.g., Slack et al., 1973), but it is unclear whether the potential changes themselves are important functionally or are merely the result of functionally important changes in ion concentrations. The distinction is important, for electrical coupling is a precise measure of the ability to transfer potential changes, but is not a direct measure of ability to transfer ions.

Just as in non-embryonic systems, we have few values for junctional conductances in embryos. The values we have come from four sources: (1) reaggregated *Fundulus* cells produce junctions whose ionic conductances vary from 10^{-6} to 10^{-5} mhos (Bennett and Trinkaus, 1970; Bennett et al., 1972); (2) reaggregated amphibian cells have ionic conductances of about 3×10^{-6} to 4×10^{-8} mhos (Ito et al., 1974b); (3) pairs of cells teased mechanically from *Xenopus* embryos have values ranging from 10^{-7} to 10^{-6} mhos (Slack and Warner, 1975); and (4) chick embryo cells, grown in primary culture have conductances estimated to be around 10^{-8} to 10^{-7} mhos (Seigenbeek et al., 1972). Besides these values, if the coupling conductance between cleavage stage *Xenopus* embryo cells were all due to junctions (which as we have seen is questionable) the value would be about 4×10^{-6} mhos (de Laat and Bluemink, 1974).

These values cover a wide range, and are not noticeably different from values from non-embryonic material. Yet, the value of such a comparison is questionable in the absence of a clear idea about function. Furthermore, the values are all obtained from artificial systems of uncertain relevance to the intact embryo.

As I have discussed in an earlier review (Sheridan, 1973), an argument can be made for representing the communicating ability of a junction by the ratio of junctional conductance (which is presumed proportional to junctional area) to cell volume. This argument is based on the premise that non-electrical communication most likely involves changes in total cellular concentration of small molecules transferred by junctions. If this premise is correct, the large size of many early embryonic cells, even with a "normal" junctional conductance, leads to a very small junctional conductance/volume ratio and thus poor relative communicating ability. (A related problem is discussed in the next section).

In embryos, however, it is possible that even small junctions (i.e., relative to cell volume) could permit sufficient transfer of small molecules to produce local changes even if too little were transferred to change appreciably the concentration in the whole cells, for example, movement of Ca^{2+} or cyclic nucleotides, which readily bind to cell constituents, might produce important local effects. (These possibilities will be discussed further in section 3.5.).

3.2.2. Permeability to tracer molecules

In a variety of non-embryonic systems, electrically coupled cells commonly pass other small tracers, such as fluorescein. However, in embryonic systems, dye transfer between coupled cells has been studied only in a few systems and in only one has dye transfer been demonstrated. It has been reported that fluorescein fails to transfer between cells in cleavage to blastula stage *Xenopus* embryos (Slack and Palmer, 1969), in *Fundulus* embryos (Bennett et al., 1972), and in *Asterias* embryos (Tupper and Saunders, 1972). The conclusion, often repeated in reviews (cf. Bennett, 1973), is that embryonic junctions are permeable to inorganic ions, but not to molecules the size of metabolites. This conclusion implies a qualitative difference between embryonic and non-embryonic junctions, and, if correct, has important functional implications. However, the support for this conclusion is not as clear as it seems and a detailed discussion is in order.

There are at least four possible a priori reasons for failure to detect transfer of dye molecules between electrically coupled cells: (1) coupling occurs via extracellular space as a result of one of the conditions discussed above; (2) the junctions have a low total ionic permeability (e.g. due to decreased area or number of subunits) relative to junctions mediating transfer in non-embryonic systems, so that the amount of dye transferred is below the limit of resolution; (3) the junctions have a total ionic permeability comparable to that in systems transferring tracers, but the cells are so large that the tracer never reaches a detectable concentration; and (4) the junctions have normal total ionic permeability, but decreased permeability to the small tracers (e.g. due to smaller diameter of channels in subunits which are increased in number). Reason (4) is the one most often quoted (cf. Bennett, 1974), but as we will see, in no case have all of the other possible reasons been excluded.

In starfish (*Asterias*) embryos, coupling is first detected at the 32 cells stage (Ashman et al., 1964; Tupper et al., 1970). The degree of coupling at that stage has not been accurately quantitated (because the voltage in one cell was recorded with the current passing electrode), but the degree of coupling is not especially high. Furthermore, without a measure of the non-junctional resistance, it is impossible to judge from the results how high an ionic permeability the junctions have. Thus, it is quite possible that the junctions also have a low ionic permeability (reason 2) and that the apparent lack of dye movement is due to the low sensitivity of the method for monitoring dye movement.

Related considerations apply to the *Xenopus* studies of Slack and Palmer (1969). First, as mentioned earlier, in early amphibian embryos, much of the coupling may occur via extracellular space and not involve junctions at all (reason 1). If so, the comparison with non-embryonic junctions is questionable. At best the high degree of coupling is probably misleading, and the junctions might be quite small with a low overall ionic as well as dye permeability (reason 2). These problems are compounded by the enormous size of the cells, which would make it difficult to detect small amounts of dye transfer (reason 3). Regarding detectability, it is significant that Slack and Palmer failed to detect fluorescein transfer during the 2-cell stage. Both de Laat and Bluemink (1974) and Sanders and Zalik (1972) report that a midpiece is retained until the beginning of second cleavage. If fluorescein transfer could not be detected with an intact cytoplasmic bridge it is even less likely to have been detected with a junction. Similar arguments might explain Baker and Warner's (1972) evidence that EGTA, injected into one cell, blocks cleavage in that cell but not in the adjacent one, implying that neither EGTA nor Ca^{2+} are transferred in sufficient amounts to affect the concentration in the adjacent cell.

The most convincing evidence that embryonic junctions have a low permeability to dyes relative to their permeability to ions comes from work by Bennett et al. (1972) on reaggregated *Fundulus* cells. Yet, even with these cells it is conceivable that transfer occurs, but is undetected due to the large size of the cells (reason 3). The ionic permeability of the *Fundulus* junctions was reported to be of the order of 10^{-6} mhos. In Novikoff hepatoma cells with junctions of comparable conductance, fluorescein transfer is first detected after about 20 sec. of continuously repeated injections (Johnson and Sheridan, 1972, and unpublished observations). The Novikoff cells, however, are about 15 μm in diameter, or roughly $\frac{1}{10}$ the diameter of the typical *Fundulus* cells used (see Fig. 9 in Bennett et al. 1972). If fluorescein were transferred between the *Fundulus* cells at the same rate as between the Novikoff cells, it would take about 100× as long to reach a detectable level in *Fundulus* as in Novikoff (i.e. fluorescence × (conc.) × (path length) × $1/r^2$). Thus, about 2000 seconds (= 33 minutes) would be required for transfer to be detected in the *Fundulus* system. The time discrepancy would be even greater if more time were required to get as high a concentration of fluorescein in the injected cell in the *Fundulus* experiment, or if fluorescein bound slowly to cytoplasmic constituents so that its effective free concentration were less than indicated by its fluorescence.

The only study to date demonstrating dye transfer between coupled embryonic cells was also carried out on reaggregated cells, but primarily from more advanced embryos (gastrulae and neurulae of *Xenopus*) (Sheridan, 1971). The results from those studies are consistent with the preceding arguments. First, the cells were generally smaller (less than 50 μm) than the earlier *Xenopus* cells and the *Fundulus* cells. Second, fluorescein was injected intermittently for the entire time of observation (up to about 10 minutes). Third, dye transfer was generally apparent only after about 5 minutes and in some cases no transfer was seen after

10 minutes of injection. Unfortunately in these studies, no estimate of junctional ionic permeability nor even of coupling coefficient was made, so it is difficult to conclude anything about the relation between junctional permeability to ions and fluorescein.

However, the experiments on reaggregated *Fundulus* and *Xenopus* cells are both of questionable relevance to the intact embryo. Not only are the reaggregation systems artificial, but the *Fundulus* experiments were done primarily (if not exclusively) on cells from the "enveloping layer", which is not part of the embryo proper, while the *Xenopus* cells were not always from clearly defined regions. Bennett et al. (1972) show a picture of dye transfer in vivo between deep blastomeres which will form embryonic tissues proper, but argue that this transfer was likely to be due to residual midpieces. The basis for this argument is unclear, however, and in fact it is difficult to understand why midpieces would be retained in the unusual pattern necessary to explain transfer to a few, but not all cells as in the picture. Transfer to some cells and not others would be more easily understood if they were joined by junctions of variable size and permeability. No dye injection experiments have yet been carried out on later *Xenopus* embryos, and the problems with the experiments on early embryos have been considered above.

As discussed above, there are no indications from electronmicroscopic studies that embryonic gap junctions differ qualitatively from adult junctions. Specifically there is no indication that the junctional particles in embryos are smaller as they might be if the putative channels had smaller diameters. Furthermore, there is no indication that embryonic junctions are especially large as would be needed for the junctions to have a "normal" ionic permeability, yet a restricted channel diameter. If the failure of fluorescein transfer in the *Fundulus* system were due to a decrease in a diameter of the channels through the subunits, then the specific junctional conductance would be decreased and thus the junctional area for a given conductance increased. For example, if the channel diameter were reduced to one half (from 20 Å to 10 Å) the cross-sectional area of the channel and the specific junctional conductance would both be decreased to one fourth. Instead of a 1 μm^2 junctional area for the 10^{-6} mhos, the predicted area would be 4 μm^2. In fact embryonic junctions tend to be smaller than adult junctions.

3.2.3. Permeability to nucleotides

As we have seen in section 3.2.1., there are other ways to test junctional permeability to small molecules. To date, no direct experiments utilizing these other methods to test junctional permeability in embryos have been published. However, there have been some intriguing reports that suggest junctional transfer of nucleotides can occur in embryonic systems.

All of these reports describe the transfer of radioactively labelled material from one group of embryonic cells, preincubated with [^{3}H]-uridine or [^{14}C]-uridine to another group of "cold" cells. In the first study, Kelley (1969) made "sandwiches" of "hot" mesoderm and "cold" ectoderm from *Xenopus* gastrulae and, from autoradiographs, concluded that there was transfer of label to the ectoderm after 4 hours culture. He suggested as one possibility that uridine-labelled nucleotides were transferred from the mesoderm to ectoderm.

In the second study, Grainger and Wessels (1974) preincubated chick embryonic lung mesoderm with [^{15}N] and [^{3}H]-ribonucleosides to produce heavy RNA labelled with [^{3}H]. The labelled mesoderm was incubated with unlabelled epithelial lung primordia across a Millipore filter. Following induction of epithelial differentiation, the RNA from mesoderm and epithelia was extracted, sedimented in cesium formate, and assayed for radioactivity. The results showed no detectable transfer of whole pieces of RNA (which would have been both heavy and labelled), but extensive transfer of the [^{3}H] label. A reasonable explanation is that rapid turnover of RNA (probably heterogeneous RNA) maintained a high level of [^{3}H]- and [^{15}N]-nucleotides in the pools of the mesoderm cells. Some of these labelled nucleotides transferred to the epithelia, and the [^{3}H] label was detectable whereas the [^{15}N] label was sufficiently dispersed throughout the RNA that no shift in density was seen in cesium formate gradients.

The mode of transfer is unclear. Millipore filters of the size used (.45 μm pore size) have been thought to preclude cell contact, but recent evidence suggests cells may infiltrate and even traverse the filters (see chapter in this book by Saxén et al., p. 377). If so, then cell contacts and even gap junctions might be produced, providing for transfer as they apparently do between various cultured cells (Gilula et al., 1972).

A third study was carried out on sea urchin embryos (Czihak and Horstadius, 1970). Embryos at the 16-cell stage were incubated with [^{14}C]-uridine. The radiolabelled micromeres were then incubated together with "cold" animal hemispheres from embryos arrested at the beginning of the 16-cell stage. After "several hours" of incubation, the associated groups of cells were fixed and studied autoradiographically. Transfer of label occurred somewhat variably, in which case the donor cells had chiefly nuclear labelling whereas the recipients had primarily cytoplasmic labelling. This pattern differs from that seen in in vitro studies with cultured cells prelabelled with 5-[^{3}H]-uridine (Simms and Pitts, unpublished observations). The difference can be attributed to the use of 2-[^{14}C]-uridine and the long coincubation times in the sea urchin experiments. The heavy nuclear labelling observed in the donor cells is explicable if a substantial proportion of the uridine had been converted to deoxythymidine nucleotides and then incorporated into DNA. The long culture period used would allow enough cell divisions to produce heavy nuclear labelling in donors. The cytoplasmic labelling found in recipient cells suggests that labelled uridine nucleotides are transferred, probably at the UTP level, which cannot be converted to dTMP and incorporation in the recipients is thus restricted to RNA.

Our discussion of the basic structural and permeability properties of gap and low resistance junctions in embryos has taken us through a number of controversial issues. In conclusion, however, there seem to be no unequivocal, qualitative differences between embryonic and non-embryonic junctions. There is, however, a possible quantitative difference, namely, a decrease in the ratio of junctional permeability (or area) to cell size, and this possible difference must be kept in mind when we discuss possible functions.

3.3. Distribution of low resistance junctions

The problem of distribution of low resistance junctions in embryonic systems is important both historically and conceptually. The earliest studies on embryos were motivated either explicitly or implicitly by the idea that the distribution of junctions might yield clues about their developmental signficance. Furthermore, early models for junctional involvement in the regulation of growth and differentiation were strongly dependent on changes in junctional distribution (Loewenstein, 1968).

The question of junctional distribution in embryos, like those of junctional structure and permeability, is incompletely answered. Few organisms have been studied, most of them vertebrates, and only certain stages of development have attracted much attention. We can see where the field stands, however, and what avenues remain for exploration, by considering three basic questions about junctional distribution: (1) when do junctions first appear during development?; (2) are junctions present between interacting cells during induction?; and (3) is junctional disappearance (or appearance) correlated with specific developmental stages or events?

(1) Electrophysiological and/or electronmicroscopic studies have been made of electrical coupling and junctions in a relatively small number of early embryos, and no clear pattern emerges. The blastomeres in early amphibian embryos (cleavage stages) are electrically coupled, but as we have seen it is not clear to what extent, if any, specialized junctions are involved. Furthermore, there is little unequivocal structural evidence for the presence of gap junctions in cleavage stage amphibian embryos. Sanders and Zalik (1972) first reported inability to find gap-like junctions between cells in the early stages, but later Singal and Sanders (1974) observed inter-blastomeric contacts resembling gap junctions. Freeze-fracture data will probably be necessary to resolve this problem.

Electrical coupling also appears during cleavage stages of tunicate embryos but there is no unequivocal evidence of either physiological or ultrastructural for specialized junctions at that stage (Takahashi et al., 1971).

Very early stages have also been investigated in certain marine invertebrates. As mentioned earlier, cells in starfish embryos reputedly do not form low resistance junctions until the 32-cell stage (Ashman et al., 1964; Tupper et al., 1970). However, the existence of weak coupling disrupted by the experimental procedure remains a possibility, especially in light of the very small sample size. Again, no electronmicroscopic data are available.

Gap junctions have been clearly demonstrated in freeze-fracture and lanthanum-treated material from early mammalian embryos (Ducibella et al.,

1975). The junctions are found as early as the 8-cell, morula stage, and thus represent the earliest appearance of unequivocal gap junctions in any embryo studies to date. Yet, there are no parallel physiological studies on mammalian embryos.

In embryos with meroblastic cleavage, all of the earlier cleavages are incomplete, so the question of junctions is somewhat academic and certainly not answerable with electrophysiological methods. Thus, of necessity, junctional studies in such embryos have focussed on later stages after which cells have become completely separated from their neighbors. In the squid embryo, coupling is found between blastoderm cells and the centrally-located yolk cell even after the blastoderm cells become segregated (Potter et al., 1966). In *Fundulus,* the enveloping cells are coupled to the yolk cell as well (Bennett and Trinkaus, 1970), but in this system much if not all of the coupling at this time is via extracellular space (see earlier discussion above).

The earliest stages studied in chick embryos were prior to primitive streak formation, well after the first cleavage occurred and after the blastodermal cells were very numerous and completely separated from neighbors. At this time coupling was extensive (unpublished observations), and electronmicroscopic studies (Revel et al., 1973) have shown frequent, small gap junctions between epiblast cells at about this same time.

In general these various studies leave open the possibility that low resistance (gap) junctions are present after first cleavage, but are only very small and perhaps labile in some embryos (e.g., amphibia, *Asterias, Fundulus*).

(2) The possibility that junctions occur between interacting cells during induction is an intriguing one and was a major stimulus for the studies of coupling in the early chick embryo and for many of the more recent studies on the developing neural axis in various amphibian embryos. The results of the coupling studies are equivocal. In the chick embryo, for example, coupling exists between essentially all of the presumptive tissues involved in the neural, notochordal axis, (e.g. notochord and neurectoderm) but it is not clear whether interacting cells are coupled directly or only indirectly via their common cellular connections with Hensen's node (Sheridan, 1966, 1968). Ultrastructural studies on the early chick embryo (Trelstad et al., 1967; Revel et al., 1973) do not provide enough information to distinguish between the indirect and direct coupling pathways, except to say that direct contact between interacting tissues, e.g. notochord and neurectoderm, anterior the Hensen's node region must be very infrequent if it occurs at all.

A similar situation appears to exist in amphibian gastrulae. Warner (1973) has reported that notochord cells are coupled electrically to overlying neural plate cells. In associated studies, however, she notes that current spreads a long distance throughout the neural plate and ectoderm cells at this stage, and it is therefore possible that the notochord-neural plate coupling is indirect. Again there are insufficient electronmicroscopic data to distinguish the possibilities.

It has been argued that direct connections between interacting cells, especially during so-called primary induction, are neither present nor necessary (cf. Meier and Hay, 1975). Evidence of their absence, however, is based strictly upon thin section methods, not freeze-fracture which would be necessary to detect small junctions. The lack of necessity of direct contact (via junction or otherwise) has been argued on the basis both of Grobstein's (1961) experiments showing passage of inductive signals across millipore filters and by the variety of non-cellular and non-specific exogenous stimuli, e.g. ionic changes, which can substitute for the inducing tissue as discussed elsewhere in this book by Saxén et al. (p. 362). However, induction can occur with a small number of competent, inducing cells (Deuchar, 1970), and it is hard to see how penetration of a few cell processes through the Millipore filters could ever be ruled out. The experiments with Nucleopore filters are more convincing, but again a few cell processes would be hard to exclude. Furthermore, the ability of seemingly non-specific stimuli to induce neurulation does not mean that the normal process lacks specificity.

In chapter 7, Dr. Saxén and his colleagues review this whole problem in greater detail. It should just be mentioned that the transfer of [^{3}H]-nucleotides from mesenchyme across a filter to ectoderm during induction, as reported by Grainger and Wessells (1974, and above), is best explained by direct cell contact and specialized junctions. Yet, this experiment is an artificial one as are many of those reported in the chapter by Dr. Saxén et al. and the relationship to the natural induction process is unclear. Some more general comments about induction systems are made below.

(3) Although electrical coupling is widespread in embryos beyond the blastula to gastrula stage (as are gap junctions in the few cases where definite studies are available), coupling must be lost between certain cells as they adopt their adult arrangement and differentiated state. For example, some junctions are certainly lost as the presumptive germ layers segment into different tissues. Potter et al. (1966) in their elegant study of the squid embryo reported the rather precipitous loss of coupling between the central yolk cell and nearly all other cells tested, including heart, otocyst, and retina, at a stage when those cells had already begun substantial differentiation (e.g. pigment was present in the retina, the heart was beating). Uncoupling between cells in different tissues, e.g. between heart and retina, was not directly tested, although the tentacle epidermis, which remained coupled to the yolk cell longer than other tissues, was clearly uncoupled from the other tissues. It is difficult, however, to draw any functional conclusions from the pattern of uncoupling in the squid embryo since it occurs in such a large number of tissues at once.

More interesting, and perhaps of greater developmental significance, are the cases in which cells in the same tissue lose their junctions. Most nerve cells are not electrically coupled or joined by gap junctions in the adult. Furthermore, nerves are not coupled to muscle or other target cells nor are skeletal muscles cells coupled to each other. Yet, the presumptive tissues from which these cells arise have extensive and presumably indiscriminate coupling. It is reasonable to expect that data concerning the timing of the selective loss of junctions between developing nerve or muscle cells might provide some insight into the junctional involvement, if any, in the control of differentiation.

Some interesting information on this issue has been obtained in a few studies on developing muscle and nerve cells. The situation with skeletal muscle is particularly interesting for there seem to be transient changes in junctional distribution as well as phylogenetic differences in the patterns of junctional loss. In early stages of somite formation in chick embryos (Sheridan, 1968) (and amphibians as well) intrasomite electrical coupling is extensive. The inevitable loss of coupling and gap junctions with muscle differentiation has not been followed in the chick, but in amphibians the pattern is complex and intriguing. According to Keeter et al. (1975), there is a transient appearance of gap junctions both within and between myotomes in stage 33–34 axolotl embryos. The intermyotome (and probably the intramyotome) gap junctions apparently form secondarily after the myotome cells begin their differentiation. The junctions then are lost, with the intermyotomal junctions persisting longer. Unpublished observations by Bennett (referred to by Keeter et al.) provide electrophysiological evidence of low resistance junctions. The intramyotomal gap junctions may be related to the transient gap junctions reported (Rash and Fambrough, 1972; Rash and Staehelin, 1974) to form before fusion of myoblasts.

Loss of coupling between developing skeletal muscle cells is not invariant in all embryos, however. In tunicates, for example, coupling, which is first seen at the cleavage stage, persists between skeletal muscle cells even after they develop their characteristic regenerative electrical responses (Takahashi et al. 1971). Presumably most if not all coupling is absent in adults, but even as late as the larval stage some cells remain coupled.

There is less information on the loss of coupling between developing nerve cells. The situation here is bound to be more complicated since some nerve cells in adults are well coupled, even though most are not. In a rather extensive survey of *Xenopus laevis* neural tube with thin section and lanthanum hydroxide methods, Hayes and Roberts (1975) demonstrate gap junctions between presumptive ventricular cells, but not between cells showing differentiated characteristics of nerve cells. However, as the authors point out, their methods do not exclude small junctions. This problem is another one which might only be resolved by freeze-fracture approaches and then only if differentiating nerve cells could be recognized in replicas.

Some recent interest has been generated by the brief report of Dixon and Cronly-Dillon (1972) that gap junctions are lost between central retina cells (and between pigmented and neural retina cells) in amphibian embryos coincident with the time that the ganglion cells become specified and begin to differentiate. Again the results were based on thin section studies, however, so that the true extent of junctional change is unknown. Nevertheless, some changes are likely and may have some role in the specification process (Jacobson and Hunt, 1973).

Although most neurones are destined to lose their gap junctions, there are two reports suggesting that transient reformation of gap junctions might be somehow involved in the early events of synapse formation. In the development of the visual system in *Daphnia*, as optic nerve fibers grow out toward the optic lamina, one fiber leads the rest and makes the first contact with lamina neuroblasts. This initial contact involves a complex wrapping of the incoming fiber terminal by the neuroblast. LoPresti et al. (1974) have recently shown that gap junctions are formed transiently between the optic fiber and the neuroblast, and suggest that these junctions provide for some kind of recognition process.

In a more artificial system, Fischbach (1972) has demonstrated the presence of weak electrical coupling between nerve and muscle cell in tissue culture. The coupling was rare, however, which may have indicated that coupling was transient or that it was undetectable due to the high axoplasmic resistance (coupling was tested between nerve somas and muscle cells). The intriguing possibility is that the coupling was associated in chemical synapse formation, which was supported by the finding that the coupled cells also were apparently connected by chemical synapses.

3.4. *Breakdown and formation of low resistance junctions*

Morphogenesis involves extensive changes in cellular arrangements, shape, and location. As a consequence, cells not only alter their area of contact with neighbors, but even change neighbors (Garrod and Steinberg, 1974). As these changes occur junctions must be broken down and reformed, and it is likely that these processes are carefully regulated. Thus, it is appropriate to turn next to the question of junctional breakdown and formation.

3.4.1. *Junctional breakdown*

We do not have a clear view of how the breakdown of gap junctions occurs in the embryo, or for that matter, in any system, but there seem to be two major possibilities. The first is that junctional particles in one membrane break their connections with complementary particles in the adjacent membrane. Then the disconnected aggregates of particles in the two membranes might dissociate into single particles or groups of particles. These in turn might be incorporated into new junctions that are formed elsewhere on the membrane. There is no compelling evidence for such a mechanism, and in fact the association of junctional particles across the extracellular space is resistant to most chemical treatments (cf. Goodenough and Revel, 1970), although gap junctions can be "unzippered" by hypertonic sucrose in some systems (Barr et al., 1965; Goodenough and Gilula, 1974). However, cells starting to migrate to other regions of the embryo (e.g. presumptive mesenchymal cells) (Revel et al., 1973) or rearranging within presumptive tissues (e.g. in neural plate) (Decker and Friend, 1974) often show variegated aggregates of particles which might represent breakdown of pre-existing junctions, but could also be sites of formation.

The second possibility is suggested by observations on artificially dissociated cells. When heart muscle (Muir, 1967), pancreas (Amsterdam and Jamieson, 1974), or clumps of Novikoff hepatoma cells (Johnson and Preus, unpublished observations) are dissociated with either enzymes (e.g. trypsin and collagenase) or EDTA, it appears that each junction is torn out of the membrane of one cell while remaining attached to the membrane of the adjacent cell. The cell losing the junction rapidly reseals (in the presence of suitable extracellular Ca^{2+}), while the other cell phagocytoses the junction. Gap junctions have been observed in intracellular vesicles and in various stages of degradation in dissociated cells, and in one system at least, the Novikoff hepatoma culture (Johnson et al., 1974), there are essentially no residual particle aggregates in the membranes of the freshly dissociated cells.

3.4.2. Junction formation

Rather more attention has been given to the formation of low resistance or gap junctions. There are now available a variety of data concerning the time course, the changes in permeability, the structural correlates, the metabolic requirements, and even possible mechanisms for regulation of junctional formation.

A consistent feature of junction is the rapidity with which electrical coupling is established. In the earliest study of junction formation, Loewenstein (1967) showed that dissociated sponge cells become electrically coupled within minutes after coming into contact. Similarly rapid onset of coupling has since been shown with reaggregating amphibian (Ito et al., 1974a) and fish (Bennett and Trinkaus, 1970) blastomeres. A particularly dramatic demonstration of the rapid development of coupling has come from studies of cardiac myoblasts in culture. These cells provide an especially useful model for studying junction formation for with them it is possible to monitor coupling without impaling the cells with microelectrodes. The myoblasts contract rhythmically in culture, each isolated cell showing an independent and generally different rate. However, shortly after two cells, beating at different rates, come into contact, the contractions become synchronized. Since the contractions are triggered by membrane action potentials, synchronization implies that the action potentials are transferred from cell to cell via low resistance pathways. Thus, the appearance of synchrony signals the formation of a low resistance junction. Because the cells are quite small and have very high membrane resistances (10^{10} ohms), they can be effectively coupled by relatively high junctional resistances, e.g. 10^8 ohms. This means that the development of synchrony is a very sensitive detector of junction initiation, but says rather less about any further development once a single efficient contact is made. De Haan and coworkers (Hirakow and De Haan, 1970; De Haan and Sachs, 1972) have made the most extensive use of this system and have shown that synchrony, i.e. junction formation, can occur within minutes after cell contact is made (as determined visually).

Even more rapid formation of junctions (within seconds) has been reported for amphibian (Ito et al., 1974a) and fish (Bennett and Spira, 1975) blastomeres that are first allowed to couple, then are pulled apart, and finally are manipulated back into contact. In the case of the amphibian cells and perhaps the fish cells as well, the cells are not completely separated but remain connected by fine "cytoplasmic strands". These strands are possibly attached via intact gap junctions which no longer can couple the cells because the longitudinal resistance of the thin strands is too high. When the cells are brought back into contact, the strands can retract and the junctions can couple the cells once again. Thus, the very rapid onset of coupling may have nothing to do with formation *per se.* (Another interpretation is discussed below.)

After coupling is first detected, there is a period during which coupling (as given by the coupling coefficient) increases gradually. This development of coupling has been followed continuously between pairs of reaggregating amphibian blastomeres (Ito et al., 1974b) as well as indirectly by sampling pairs of reaggregating hepatoma cells at different times of reaggregation (Johnson et al., 1974; Sheridan et al., 1975). The coupling coefficients and the input resistances have been used in both systems to estimate the value of junctional conductance (or resistance) at various times during formation.

Ito et al. (1974b) use a model which differs from that used by other workers and the differences warrant additional comment. The model shown in Fig. 3 of the paper by Ito et al. includes a "junctional insulation resistance", r_s, that in other models (e.g. Bennett, 1966; Sheridan, 1973) is assumed to be sufficiently large to be negligible. The rationale others have used to omit this resistance is complex, but is based ultimately on the fact that when two cells are already in contact and coupled, it is not possible to distinguish electrically between a model containing and one lacking a significantly small "insulation" resistance. Ito et al. include this resistance in order to fit with a structural model originally proposed to explain coupling thought to occur via septate junctions (Loewenstein, 1966) and subsequently applied, as here, to a quite different structure, the gap junction. In order to calculate the r_s term, however, Ito et al. are forced to assume that the non-junctional resistance of a cell does not change when it is placed into contact and begins to form a low resistance junction with another cell. This assumption is supported only by the fact that the non-junctional resistances differed by no more than 15% when they were measured in cells (the number of pairs not given) separated after formation had occurred. If, however, the non-junctional resistances were lower during the earlier stages of formation (e.g. 5–10 minutes after coupling is first detected), the remaining data would be easily accomodated by the alternative electrical mode that neglects r_s. This is in fact just the time when experimental manipulations, e.g. impalement by the V_1 electrode, might be producing their greatest effects. The distinction between the two models is at best clear only during the earliest phases of formation, for once the calculated r_s term becomes large relative to the calculated junctional resistance, r_c, the two models are electrically indistinguishable. That is, given the same non-junctional resistances, the input resistance and coupling coefficient within a few percent can be obtained with either model. The model with r_s, however, gives a junctional resistance about one-half that obtained with the other model. Furthermore, the models imply quite different formation mechanisms; the model with a substantial r_s term implies that the putative junctional subunits develop a high conductance *before* they link tightly with complementary particles across the extracellular space. The model neglecting r_s assumes the subunit conductance increases only *after* the linkage occurs. The ambiguity of the electrical basis for distinguishing the two models will probably only be resolved by appropriate ultrastructural studies. We have made some progress in this direction using reaggregating Novikoff hepatoma cells, and these studies are discussed below.

We have analyzed the development of coupling between Novikoff hepatoma cells (see Fig. 1), and have obtained estimates of the changes in junctional conductance during the first two hours of formation (Hammer and Sheridan, unpublished observations). Our conductance values are generally lower than Ito et al. report for comparable reaggregation times, which might reflect differences in the electrical models (we neglect the r_s term), in the species used, or simply in the area of apposition (the amphibian cells are 10–20 times larger in diameter). An important similarity is that both conductance estimates gradually increase with incubation time rather than show abrupt changes, although our estimates are for a whole population forming junctions asynchronously.

The development of junctional permeability to molecules larger than inorganic ions has been studied even less extensively. As discussed above (3.2.2.), reaggregated *Xenopus* cells from gastrulae and neurulae often can transfer fluorescein once they become electrically coupled (Sheridan, 1971). Freshly dissociated and replated tissue culture cells can transfer nucleotides in less than one hour after plating (Pitts, personal communication).

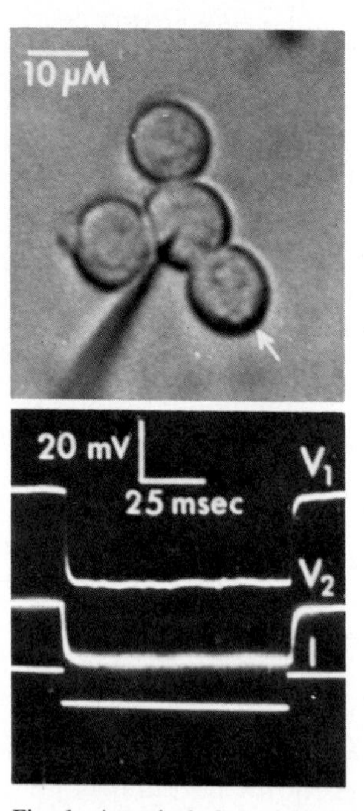

Fig. 1. A typical electrophysiological experiment on a cluster of reaggregated Novikoff hepatoma cells. The cells were dissociated with EDTA and reaggregated in the presence of Ca^{2+} for 120 min at 37°C and 5% CO_2. Electrical coupling was present between the central cell and all of the other cells, but the electrical records demonstrate coupling just to the cell indicated (arrow). The coupling coefficient (V_2/V_1) is 0.59 (unpublished work of M. Hammer and J. Sheridan). The current passed was 2.5×10^{-9} amphere.

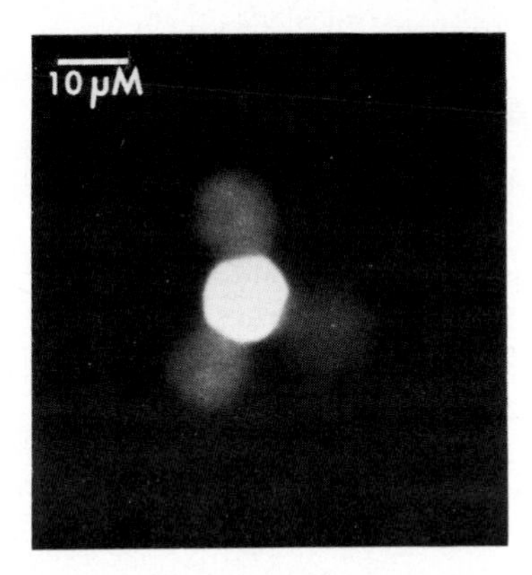

Fig. 2. The transfer of fluorescein between reaggregated Novikoff cells is shown after 60 minutes of reaggregation (37°C and 5% CO_2) and 5 minutes injection (intermittent, 90 msec every 200 msec). A 25 sec exposure was made on TriX film. All three peripheral cells were electrically coupling to the central, injected cell.

We have studied the development of the capability of reaggregated Novikoff cells to transfer fluorescein (Fig. 2) with some interesting, and not totally expected results: fluorescein only transfers between electrically coupled cells; not all coupled cells transfer fluorescein within the 5 minutes test period; cell pairs with higher coupling coefficients are more likely to transfer the dye, but many cells with large coupling coefficients cannot pass dye. The general parallel between electrical coupling and dye transfer is expected, but the lack of transfer between many relatively well coupled cells is not. A possible explanation is that the discrepancies arise from detectability problems due to variation in cell size (see above), but an intriguing possibility is that they occur when cells are coupled either via extracellular space as, for example, at formation plaques (see below) or via junctional subunits with reduced pore size. These are primarily structural issues and are discussed further below.

Study of the structural aspects of the formation of gap junctions has until recently been somewhat frustrating. The available techniques of thin section and even lanthanum-tracer are unsuitable for detecting small junctions, much less for characterizing them in detail. Freeze-fracture studies of systems involving rapid junctional development have proved eminently useful, however, and have recently provided some important insight into the process of junctional formation in a variety of systems (see, Revel et al., 1973; Decker and Friend, 1974; Johnson et al., 1974).

Nearly every one of these systems has unique characteristics, which is not surprising considering the fact that they involve cells with greatly different morphologies. However, certain fundamental similarities are emerging that are perhaps best exemplified by our observations on reaggregating Novikoff cells. This system has the advantage of providing a totally junction-free population of dissociated cells that forms only one junctional type, the gap junction (Johnson et al., 1974). Although some structural details remain unclear, we have suggested that the formation of gap junctions in this system occurs in the following manner:

(1) Initially the cells adhere over a relatively broad interface (although it is possible that the earliest contacts are made by numerous small microvilli). This adhesion appears to be necessary for gap junction formation to begin, but it can be distinguished experimentally from the formation process itself.

(2) The apposed cell membranes then interact over small areas producing "formation plaques", which are characterized by having a general deficiency of intramembraneous particles, a clustering of larger, 10–11 nm particles, and often a flattened appearance. The formation plaques occur in matched pairs that appear to develop simultaneously in the two apposed membranes while they are still separated by a large gap (Fig. 3).

(3) The gap separating apposed plaques next becomes narrower (2–4 nm), while the large particles begin to aggregate into tightly packed groupings that are indistinguishable from small gap junctions (Fig. 4).

(4) The small aggregates grow, both by fusion with other aggregates and by accretion of particles (Fig. 5).

(5) For a time, even aggregates are surrounded by extensive, particle-deficient membrane, the remnant of the plaque, but then this zone shrinks, ending up as the characteristic particle-free halo seen around most gap junctions between undissociated cells.

The most interesting feature of gap junction formation is the formation plaque, which also occurs consistently in a variety of embryos, such as amphibian neurulae (Decker and Friend, 1974) and somite-stage chicks (Revel et al., 1973) and in certain adult tissues, such as ovarian granulosa cells (Fletcher and Robertson, 1975) and even *Limulus* hepatopancreas (Johnson et al., 1974), stimulated to form junctions rapidly. Relatively particle-free membrane areas

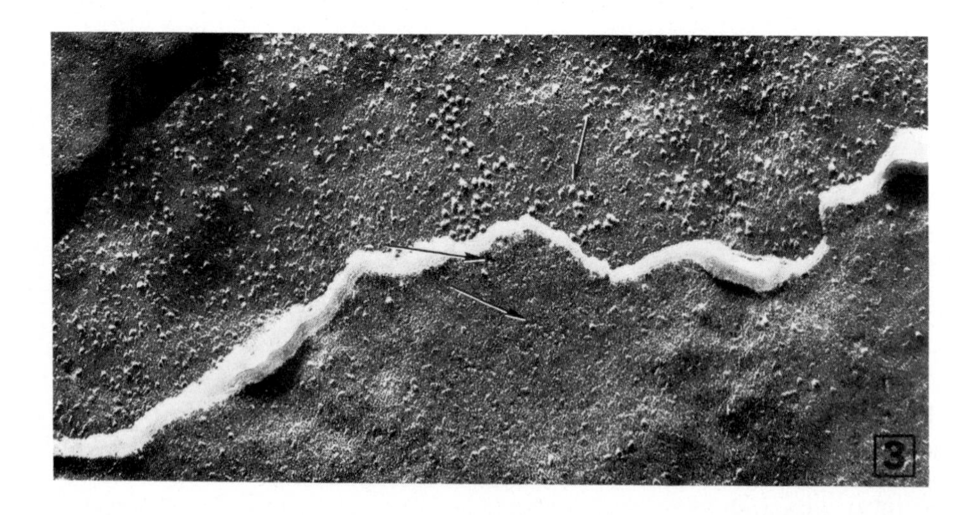

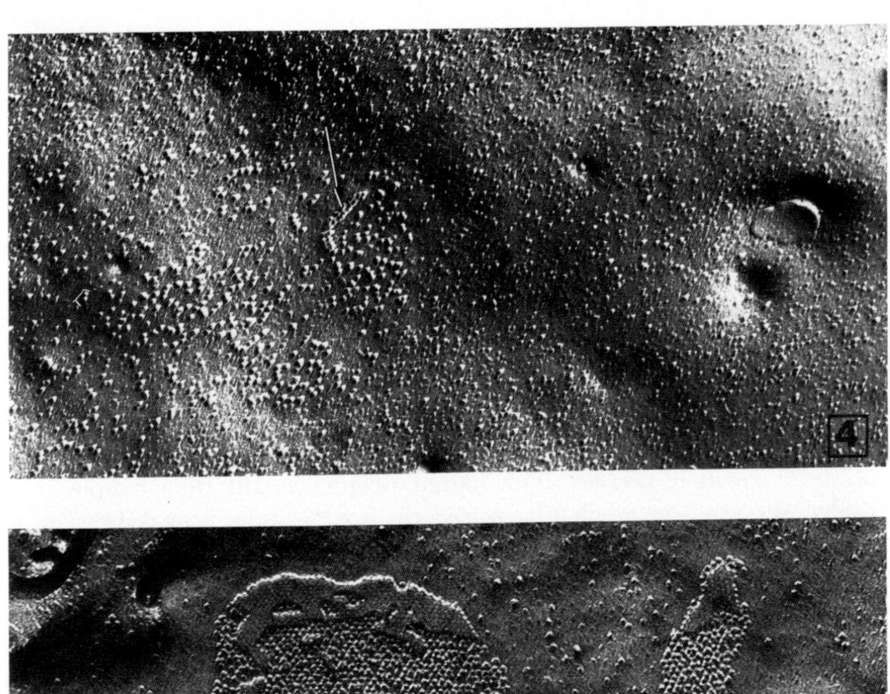

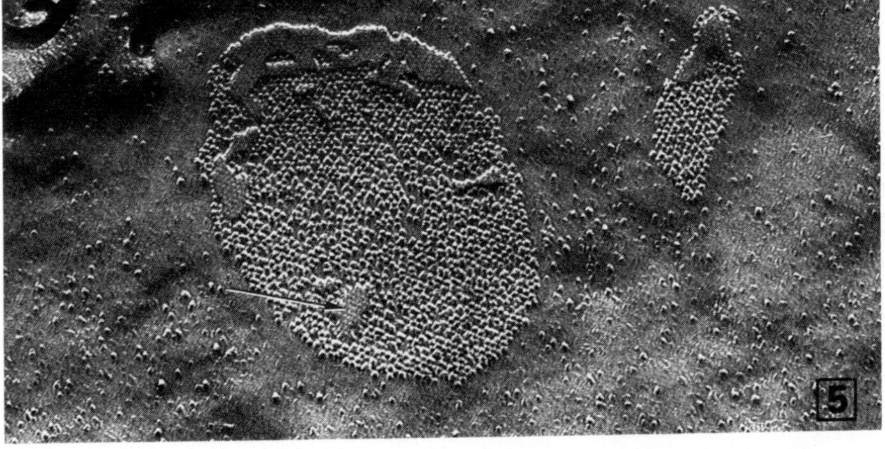

Fig. 3–5. Putative stages in gap junction formation between reaggregated Novikoff cells. Fig. 3. A "formation plaque" is shown with a relatively low overall density of particles on the P-face, a clustering of 10–11 nm particles (with corresponding parts on the E-face), and a flattening of the membrane (clearest on the E-face). The transition of the fracture plane from one membrane to the next shows another formation plaque in apposition across a relatively wide extracellular space. Fig. 4. In some plaques, the larger particles begin to aggregate (arrow) roughly hexagonally into rows or groups. Fig. 5. Further aggregation results in gap junctions, often multiple, with relatively particle-free adjacent membrane. Steps into the adjacent membrane expose E-face fragments with characteristic pits (arrow). Original micrographs at 100,000 X (unpublished work by R. Johnson, D. Preus, and R. Meyer).

resembling formation plaques, are also seen prior to formation of tight, or occluding, junctions in certain culture systems (Porvaznik and Johnson, unpublished observations) as well as formation of septate junctions in sea urchins (Gilula, 1973). The gap, tight, and septate junctions all involve specific, orderly arrangements of intramembranous particles and it is tempting to suggest that the particle-deficient zones are required for the intramembranous particles to associate into their characteristic arrays.

The formation plaques, with their loose clustering of large particles and their paired development across a wide extracellular space, raise some interesting questions about the development of the characteristic permeability of gap junctions. It is possible a priori that the junctional particles must aggregate before they become permeable to small molecules; that is, electrical coupling and dye transfer can occur only via aggregated particles. It is also possible, however, that the large, unaggregated particles in the formation plaques are also permeable to small molecules. The paired plaques might then couple cells electrically, though less efficiently, across the extracellular space. Our most recent quantitative structural data on reaggregating Novikoff cells indicate that the percentage of cell interfaces with aggregates is too low to account for the percentage of coupled cells found electrophysiologically, unless the percentage of interfaces with formation plaques is included. Thus, it is quite possible, if not likely, that weak coupling can occur in our system by way of formation plaques.

The percentage of cells transferring dye, however, is very close to the percentage of interfaces with aggregates suggesting that the aggregates are necessary for efficient transfer of molecules larger than small inorganic ions.

The suggested coupling by formation plaques and the restriction of dye transfer to particle aggregates have interesting implications for the interpretation of earlier studies on junctional formation between amphibian and fish embryonic cells. As discussed above, when coupled pairs of amphibian or fish cells are pulled apart and are then reassociated, coupling develops within seconds. It is an interesting possibility that this early coupling occurs by way of formation plaques that were present before the cells were pulled apart. Furthermore, if much of the coupling before the cells were dissociated occurred at formation plaques, the lack of fluorescein transfer (between fish cells) might also be explained.

The apparent transition from formation plaques to junctional aggregates and the implicit incorporation of the large plaque particles into junctions, suggests that some type of self-assembly process is involved. This suggestion is supported by our finding that neither protein synthesis nor ATP production (or cellular stores) is absolutely necessary for formation to occur, provided that any effects on initial cell adhesion are avoided by loose pelleting of the cells (Epstein et al., 1974). Protein synthesis has been shown by others to be unnecessary for development of synchrony between cardiac myoblasts in culture (Goshima, 1971) or for the ability of recently plated fibroblasts to transfer nucleotides (Pitts, unpublished observations). Dinitrophenol has been reported to block formation of electrical coupling between reaggregating newt embryo cells (Ito et al., 1974a), but in these studies ATP levels were not measured and the sample was too small to warrant any generalization.

If aggregation of junctional particles occurs by self-assembly, then the initiation of the process is a particularly critical event. Yet we know little about initiation except to say that cell adhesion is first necessary, and furthermore probably requires Ca^{2+} which may also play some direct role in the early stages of formation itself.

The proposed spontaneity of the aggregation of junctional particles also raises the issue of regulation: how do cells control formation? Recent evidence suggests that in some systems junctional formation can be promoted by certain steroid hormones (Merk and McNutt, 1972; Porvaznik and Johnson, unpublished observations), and is likely to involve induction of protein synthesis. The simplest explanation of these effects is that increased numbers of presumptive junctional particles are produced and this accelerates the self-assembly process. Another possibility is that some other membrane proteins involved in membrane "adhesion" are rate limiting and these are affected by this hormone.

Theoretically, another way of influencing junction formation would be to alter the membrane directly via for example changes in fluidity or the topography of membrane components.

We have investigated quite extensively the effects on formation of analogues of cyclic AMP and cyclic GMP which are known to affect such membrane-dependent cellular processes. We have found that cAMP promotes and cGMP retards gap junction formation as determined both ultrastructurally and physiologically (Sheridan et al., 1975b). The effects are transient, which may result from breakdown of the agents by phosphodiesterase in the serum, but they are dramatic enough to substantially modulate the ability of the population to communicate.

We do not know, of course, whether these effects are physiological. However, they are consistent with the idea that cAMP and increased junctional area act to stabilize growth, whereas cGMP and decreased junctional area act to destabilize or stimulate growth (see below).

3.5. Biological role of low resistance junctions

3.5.1. General comments

The investigation of low resistance junctions in embryonic systems has been motivated primarily by the conviction that they play some important and perhaps specific role in development (Furshpan and Potter, 1968; Loewenstein, 1968). Lacking hard experimental evidence for such a role, however, our ideas about the potential developmental significance of low resistance junctions rely heavily on speculative arguments. Nevertheless, some of these arguments can be quite forceful and they have the virtue of suggesting potentially fruitful directions for further work.

All of our current ideas about the function of low resistance junctions begin with the reasonable premise that they act as molecular sieves, allowing the passive diffusion of small molecules (of the order of 1000 daltons or less in molecular weight) from cell to cell while preventing the transfer of larger molecules. Net transfer of small molecules will occur, according to this premise, only when they are in different concentrations in the two cells or, since most natural molecules are charged, when there is an electrical potential difference between two cells. The transfer will always tend to break down the concentration and/or potential gradients, i.e. the cells will become more alike. However, this statement does not mean that the transferred molecules will always be equilibrated or that the potential differences will be abolished. That is, there is a dynamic aspect of junctional transfer. Furthermore, even if equilibrium is approached, the concentration or potential difference will not necessarily assume a simple arithmetic average of the initial values in the coupled cells. In any transfer, some cells will act as sources and others as sinks, and which group, if either, predominates depends on junctional factors, e.g. junctional size or overall permeability, as well as nonjunctional factors, e.g. volume or number of the "source" and "sink" cells or their respective metabolic capacities. Another way of putting this point is that, depending on the appropriate junctional and nonjunctional factors, transfer can either dampen a response by diluting out a critical change in small molecules or it can propagate the response by spreading the molecules throughout the population.

I have discussed in detail elsewhere (Sheridan, 1974) some possible ways these basic principles might apply to specific biological systems. Here I would like to elaborate on some of the ideas having more developmental significance and further to suggest some new possibilities.

3.5.2. Control of cell proliferation

Undoubtedly, one of the more important processes in development is cell proliferation. Most functional speculations about low resistance junctions have included some involvement in regulation of this process (Furshpan and Potter, 1968; Loewenstein, 1968; Socolar, 1973; Sheridan, 1974). In searching for a possible mechanism for this postulated function, our working premise about the selective permeability of low resistance junctions forces us to look for control molecules that are small enough to be transferred. The cyclic nucleotides, cAMP and cGMP, fit this description quite adequately, as I have previously discussed (Sheridan, 1971; 1974). Recently a few studies have provided new information about the changes in cyclic nucleotide levels during the cell cycle (Seifert and Rudland, 1974). These observations, in conjunction with the possible intercellular transfer of the cyclic nucleotides, have some interesting implications in terms of the regulation of cell proliferation. These implications are best discussed in terms of a model for regulation of cells growing as monolayers in culture. The primary aim of the model is to explain contact inhibition (or post-confluence inhibition) of cell growth, but it also serves to emphasize in a more general sense the possible functional properties of low resistance junctions.

"Contact inhibition of growth" refers to the inhibition of cell proliferation, with arrest in the early G1 phase of the cell cycle, that occurs when cells reach a tightly packed confluent arrangement in monolayer culture (see recent reviews by Harris, 1973; Martz and Steinberg, 1973). Previous attempts to explain this phenomenon in terms either of low resistance junctions or of cyclic nucleotide changes have failed to account for certain observations, namely, that the inhibition of any individual cell is not correlated with the number of neighbors it contacts (Martz and Steinberg, 1973), and that cyclic nucleotide levels do not correlate in a consistent way with the presence or absence of contact inhibiton (cf. Miller et al., 1975; vs. Sheppard, 1975).

My model is based on three premises:

(1) *cAMP and cGMP can be transferred from cell to cell by way of low resistance junctions.* This premise is consistent with the permeability of junctions to nucleotides and dyes of comparable size and charge. No direct unequivocal evidence for transfer of cyclic nucleotides is available, but suggestive findings have been obtained with strips of heart muscle (Tsien and Weingart, 1974, discussed above). Furthermore, intercellular transfer of cAMP may explain the apparent ability of antidiuretic hormone (ADH) to raise cAMP levels uniformly in the epithelial cells of the toad urinary bladder (Goodman et al., 1975) when only one cell type responds to ADH when isolated from the other epithelial cells (Scott et al., 1974).

(2) *The cGMP level increases in a pulse-like fashion during early G1 while there is a corresponding, transient decrease in cAMP levels. At other times during the cycle the cGMP levels remain constant at a low level while the cAMP levels are generally high (aside from a smaller transient decrease at the G1/S border).* This entire pattern has been seen in synchronized fibroblasts (Seifert and Rudland, 1974) and the cAMP changes have been observed in other systems (Sheppard, 1975).

(3) *The increase in cGMP (perhaps coupled with the decrease in cAMP) acts as a "trigger", necessary for the advance of the cell through G1 and the rest of the cycle. That is, once a critical threshold of cGMP is reached, the cell will proceed in the cycle.* A variety of conditions will cause cells to arrest in early G1. The point where the arrest occurs has been called the "restriction point" by Pardee (1974) and seems to be the same even though the conditions vary, e.g. confluency, low serum, certain amino acid deficiency. Rudland et al. (1974) showed that immediately after release of the restriction by the appropriate stimulus, e.g. serum or replacement of amino acids, a cGMP pulse is produced (also Moens et al., 1975; but see Miller et al., 1975). Also, abrupt and transient increases in cGMP are produced in lymphocytes stimulated to divide by PHA and other mitogens (Hadden et al., 1972). The cGMP pulse could be secondary to the cAMP decrease or to the change of some other cell property, e.g. divalent cation storage (Rubin, 1975) or ion transport (Rozengurt and Heppel, 1975). The model, while working best with a cGMP trigger, would also work if any other small molecule e.g. Ca^{2+} or Mg^{2+}, acted as the trigger.

In qualitative terms, the model operates quite simply. Consider a sparse culture of fibroblasts in which cell contact is rare. The cells will be growing asynchronously; thus each cell will be in a different phase of the cycle, with a relatively low proportion going through the cGMP-pulse (and cAMP decrease) at any time. As the cell density increases, more cells will come into contact and some will remain so long enough (15 to 60 minutes or so) for appreciable junctions to form. Some pairs of connected (and presumably "coupled") cells will have one cell in early G1, thus beginning a change in cGMP and cAMP. According to our first premise, cGMP and cAMP will move down their respective concentration gradients, cGMP going from the G1 cell to the other cell and cAMP in the opposite direction. The net effect will be that the early G1 cell will produce a smaller cGMP pulse and a smaller decrease in cAMP while the other cell will suffer a small increase in cGMP pulse and a small decrease in cAMP. If only two cells are involved in each interaction, it is unlikely that the changes in cyclic nucleotides will alter the subsequent behavior of the cells or of the population. Thus, the early G1 cell will still be "triggered" to proceed and the other cell will continue as well.

With further increase in cell density, however, the probability of an early G1 cell making contact and forming junctions with more than one other cell at the same time will increase. In any of these combinations of connected cells, most of the other cells will be outside the early G1 point and thus, irrespective of stage in the cycle, will have a low level of cGMP and a relatively high level of cAMP. These cells will be "sinks" for the cGMP coming from the early G1 cell (and sources of cAMP). If the number of non-early G1 cells is sufficiently large and the number and the extent of the junctions sufficiently great, the loss of cGMP from the early G1 cell (and perhaps its acquisition of cAMP) will prevent it from reaching threshold and thus the cell will arrest in early G1. The other cells, however, will continue their asynchronous progression through the cycle.

Clearly, with further increase in density, more early G1 cells will arrest until all cells are blocked. The cells will in effect "collect" at the "restriction point" in early G1 and the culture will be "contact inhibited".

The model is relatively simple and fully consistent with the observations (mentioned above) that have thwarted previous explanations of contact inhibition in terms of cell junctions and/or cyclic nucleotides. On the one hand, at the light microscope level, the early G1 cells in a clump cannot be recognized, nor is it possible to determine which cells in "contact" have formed extensive junctions. Thus, it is not surprising that there is no obvious correlation between number or even timing of contacts and the probability that a cell will continue to divide, even if, on the population basis, the junctions are essential to the arrest.

On the other hand, the key factor regarding the cyclic nucleotides is the transience of the critical changes, rather than the absolute levels, although these may be important. Thus, the inconsistency of the correlation of overall cAMP (Sheppard, 1975) and cGMP (Miller et al., 1975) levels with degree of confluence is not damaging to the model, but instead implies that *overall* cyclic nucleotide levels may play a less critical role in the inhibition.

The model can be extrapolated to in vivo growth regulation as well, although the support for the underlying premises is weaker. An interesting example (with general implications in terms of development) is the differentiation of stratified squamous epithelium. The basal cells, which comprise the chief mitotic population, have relatively small gap junctions (McNutt et al., 1971) and divide asynchronously at a regular rate. Once cells have divided, some move up into the next zone (intermediate) where they stop dividing and begin to differentiate. Part of this differentiation includes formation of large gap junctions (McNutt et al., 1971). Further outward migration, accompanied by differentiation into keratinocytes (in keratinized epithelia), culminates in cell death and desquamation.

According to my model, I would predict that the basal cells behave much like a just subconfluent population of cultured fibrolasts in which there is a significant number of newly divided cells producing cGMP pulses that are *not* damped out by transfer to neighbors; these *undamped* cells maintain the mitotic population. There are other daughter cells, however, that after division lose their cGMP to neighbors, fail to lower their cAMP, and thus arrest in G1 (or G0). These cells form the differentiating pool which supplies the intermediate zone. Once in this zone, differentiation progresses further while the cells become permanently arrested in G1 (G0). This explanation is strongly reminiscent of the "critical mitosis" idea since every cell beginning to differentiate has just finished a mitosis. However, some cells undergo many divisions before, by chance, forming the right number and extent of junctions with neighbors to become arrested in G1 (cf. Konigsberg and Buckley, 1974).

The unpredictability of the behavior of any particular cell is an important consequence of this model, and results in part from the variability in size and number of junctions in any population. Additional variability could arise, however, from differences in the threshold to cGMP changes or in the absolute size of the cGMP pulses. These variabilities introduce a randomness into the behavior of any individual cell, while allowing the response of the population to remain predictable in a statistical fashion.

Another consequence of the model with some developmental relevance is that *completely* synchronized cells should continue to cycle irrespective of the presence or absence of junctions. This is true because the cGMP-pulses (and cAMP decreases) occur in all cells at once so that no gradients exist and therefore there can be no net transfer. In many early embryos the cells cycles are extremely synchronized, and quite short (Dettlaff, 1964). The presence of junctions is questionable during the earliest stages, but even at later stages they would have little effect, according to the model, provided the cells remained synchronized.

The cells start to become asynchronous (at least in amphibian embryos) close to the time when morphogenetic movements begin and the stage is being set for initial inductive interactions. It is tempting to suggest that this coincidence is not fortuitous, but rather allows some cells to become slowed, or arrested, in G1 and therefore to begin differentiating. (These arguments are not meant to imply that mitotic arrest is necessary for differentiation, which is clearly not a valid generalization. Instead I am merely suggesting that differentiation is more likely to occur in a population of slowly growing cells).

Before leaving the topic of cell proliferation, some mention should be made of defects in regulation exemplified most dramatically by cancerous cells. Again the model I have suggested has some implications: cells can escape growth control either by being isolated from their neighbors during the critical early G1 period, or by producing an especially large cGMP pulse that cannot be dampened by transfer to neighbors. Effective isolation does not necessarily mean complete lack of junctions, but rather, according to the model, sufficiently small number (or size) *during* the early G1 period to minimize loss of cGMP. These ideas are not unique (Socolar, 1973; Sheridan, 1974), but they are more specific than previous suggestions. In particular they emphasize the potential importance of timing as well as quantitative extent of junctional formation.

3.5.3. Control of differentiation

In the previous section, I touched on differentiation only as a process occurring once cell proliferation has been arrested (or slowed). Here I want to touch briefly on two differentiative processes in which junction might play more specific roles. (Again the arguments are more illustrative rather than definitive).

The first process is induction, which I considered briefly in terms of junctional distribution and which is extensively discussed by Saxén et al. in another chapter. From these discussions we might conclude that junctions probably link cells within the induced and the inducing tissue and, in some specific interactions, e.g. in salivary gland morphogenesis, even between interacting tissues. If so, what do the junctions do? Again we are restricted to considering the transfer of small molecules, a choice which is rather different from the suggested involvement of RNA or proteins as inducers. Furthermore, we must account for the fact that induction appears to involve both "trigger-like" and amplified processes: the "trigger" is suggested by the evidence that once a cell makes a choice (is "committed") it may continue its differentiation irrespective of removal of the inducing tissues; the amplification is suggested by the evidence that induction tends to be "all or nothing" in a large part of the tissue and that just a few inducing cells are necessary for the response (Deuchar, 1970).

In one sense, low resistance junctions seem *by themselves* unlikely to provide for either triggering or amplification since they transfer molecules passively. However, if the molecules they transfer can act in these two ways the logical problem is solved. In fact the "trigger" mechanism merely requires a threshold, and the transfer of small molecules leading to a membrane change would be a reasonable possibility. The amplification, on the other hand, might occur by the activation of an enzyme or by stimulating a membrane transport system.

Thus, the answer to the question is that junctions could be involved in induction by transferring cyclic nucleotides (or perhaps divalent cations, such as Ca^{2+} or Mg^{2+}). The transfer could occur between the interacting tissues in which event the cyclic nucleotides or cations would be termed the "inducers". Alternatively (or additionally) the junctions might act to insure a uniform response of the induced tissue, especially where the inducer is associated with the extracellular matrix (Sheridan, 1974).

The second process involves the development of segmental patterns in the insect cuticle. This process is useful in illustrating the possible importance of quantitative variations in junctions. The process appears to depend on the transfer of "positional information" from cell to cell, and has been suggested to occur via low resistance junctions. This whole area is very complex and certain aspects are discussed elsewhere in this book. I wish just to make a few comments about one extensively studied system in which evidence for junctional involvement has been sought, but as yet not been found.

The critical problem is the following. There appears to be a gradient of positional information, running along the length of each segment of the insect epidermis and determining certain surface features of the cuticle, e.g. bristle arrangements. This gradient is repeated in an identical fashion in each segment since intersegmental exchange grafts of identical positions retain their original pattern whereas, either intrasegmental or intersegmental grafts from non-identical regions (e.g. anterior 1/3 to posterior 1/3) show an altered pattern. The molecular basis for the gradient is unknown, but Lawrence et al. (1972) suggest that it could depend on a gradient of a low molecular weight morphogen whose intracellular concentration specifies the position of each cell (see also Crick, 1970). These authors further suggest that each cell produces the morphogen, but also can be influenced by transfer of the morphogen to or from adjacent neighbors via cell junctions. The question then arises as to what happens at the intersegmental boundary where the low end of the gradient of one segment is abruptly succeeded by the high end of the next segment. If the morphogen could be transferred as effectively across the boundary as elsewhere, the abrupt change in gradient would appear impossible.

The simplest resolution of the problem that yet retains junctional involvement would be a lack of junctions between cells at the boundary. However, both electrophysiological (Warner and Lawrence, 1973; Caveney, 1974) and ultrastructural (Lawrence and Green, 1975) data indicate that junctions are present. Furthermore, the electrophysiological studies indicate that electrical coupling across the boundary is not obviously different from coupling within the segments. The electronmicroscopic study itself may provide an explanation, however. There are definite changes in cell shape at the boundary, i.e. the anterior cells of one segment differ from the posterior cells of the next. There may also be greater numbers or sizes of gap junctions between the anterior cells than between the posterior cells. Thus, there is a possible difference in junctional area/cell volume that could significantly alter the transfer of morphogens, but not the degree of electrical coupling (which depends on surface area rather than cell volume) (see Sheridan, 1973). A more quantitative analysis of gap junctions, perhaps with morphometric techniques, could help decide the issue. Also a study of gap junctions between transplanted and surrounding cells could give useful insight.

3.5.4. Homeostasis

Possible involvement of low resistance junctions in the control of cell growth and differentiation is particularly exciting. However, there are other interesting possibilities which might be generally termed homeostatic functions.

The first function under this heading, transfer of nutrients, may well be important in all embryos that have their internal store of nutrients (i.e. in yolk) distributed unequally (e.g. in fish or amphibians) or that rely on the environment (e.g. mammals). Experiments on the squid embryos (Potter et al., 1966) clearly demonstrate low resistance pathways from the yolk cell to all other cells at early stages. In the absence of blood sinuses or other features of a circulatory system, these pathways could insure that nutrients are distributed with a minimal loss of energy to transport systems, etc. Similar processes might occur in fish and birds. Even in amphibian embryos, the yolk is unequally distributed and thus junctions could serve to distribute nutrients at least prior to the development of the circulation. The early mammalian embryo must be nurtured by the mother before the placenta is developed and at this early time junctional transfer might play some role. In all these speculations we should keep in mind that nutrients are in their most useful form where they are small molecules well within the size range thought to penetrate low resistance junctions.

The second homeostatic function might involve ionic balance. It is clear that certain embryos, perhaps all vertebrate embryos, carefully regulate their intraembryonic ionic environment (cf. Slack et al., 1973). It is quite conceivable that junctional transfer of ions aids this process. The importance is not clear, but might provide a means for cells to share pumping loads and high energy compounds (e.g. ATP).

A non-embryonic system can be used to illustrate another potential involvement of junctions in ionic regulation. The toad urinary bladder is involved in water and salt balance. Retention of water and salt relies on active transport of Na^+ (and K^+) as well as passive reuptake of Na^+ and H_2O. These processes are regulated by a pituitary hormone, ADH, which acts primarily by increasing intracellular cAMP in the epithelial cells. The cAMP in turn presumably increases the permeability of the luminal membrane to Na^+ and H_2O, potentiating the reuptake. Recently it has been shown that the epithelium has two cell types, granular cells making up about 80% and mitochondrial-rich cells making up 20%. The granular cells are thought to be the major transporting cells (see Davies et al., 1974). However, when isolated, the granular cells do not respond to ADH by changes in cAMP whereas the mitochondrial-rich cells do (Scott et al., 1974). This apparent dilemma is clearly solved by suggesting that cAMP is transferred from the mitochondrial-rich cells to the granular cells. Support for this idea has been given by Goodman et al. (1975) who found uniform binding of fluorescent anti-cAMP antibody to all bladder epithelial cells after ADH administration. The major complication is that the ADH in vivo might be binding to granular cell receptors that are lost during isolation.

Possibility for an antagonistic action by some hormone(s) raising cGMP is suggested by the observation (Sheridan et al., unpublished observations) that 8 Br-cGMP at low concentration produces a decrease in short circuit current across frog skin.

The direct relevance of these various ideas to ion transport in a developing system is unclear, since hormone regulation probably develops late. Yet the principle of a junctional role in coordinating cellular responses is reasonable to consider.

4. *Concluding Remarks*

In this discussion I have attempted to illustrate and evaluate various aspects of cell coupling during development. Clearly many important issues remain unresolved. In particular we are still forced to speculate about the potential functional involvement of coupling in embryogenesis. Nevertheless, our understanding of the properties of low resistance junctions is increasing steadily and we are learning more about the mechanisms by which they are formed and modified. It is our hope that this basic information, along with further data on the distribution of junctions in embryos, will provide the necessary foundation for evaluating the various functional possibilities.

Acknowledgements

The author wishes to thank M. Hammer, R. Johnson, D. Preus, and R. Meyer for supplying unpublished material, and M. Lawrence for typing the manuscript. The author is an NCI Career Developmental Awardee; the unpublished work was supported in part by grants from the National Cancer Institute and the Heart and Lung Institute of the National Institutes of Health.

References

Amsterdam, A. and Jamieson, J.D. (1974) Studies on dispersed pancreatic exocrine cells. I. Dissociation technique and morphologic characteristics of separated cells. J. Cell Biol. 63, 1037–1056.

Ashman, R.F., Kanno, Y., and Loewenstein, W.R. (1964) The formation of a high resistance barrier in a dividing cell. Science 145, 604–605.

Baker, P.F. and Warner, A.E. (1972) Intracellular calcium and cell cleavage in early embryos of *Xenopus laevis*. J. Cell Biol. 53, 579–581.

Barr, L., Berger, W., and Dewey, M. (1965) Propagation of action potentials and the structure of the nexus in cardiac muscle. J. Gen. Physiol. 48, 797–823.

Bennett, M.V.L. (1966) Physiology of electrotonic junctions. Ann. N.Y. Acad. Sci. 137, 509–539.

Bennett, M.V.L. (1972) A comparison of electrically and chemically mediated transmission. In: Structure and Function of Synapses (Pappas, G.D. and Purpura, D.P., eds.) pp. 221–256, Raven Press, New York.

Bennett, M.V.L. (1973) Function of electrotonic junctions in embryonic and adult tissues. Fed. Proc. 31, 65–75.

Bennett, M.V.L. (1974) Permeability and structure of electrotonic junctions and intracellular movements of tracers. In: Intracellular Staining and Neurobiology (Kater, S.B. and Nicholson, C., eds.), pp. 115–134, Springer-Verlag, New York.

Bennett, M.V.L. and Dunham, P.B. (1970) Sucrose permeability of junctional membrane at an electrotonic synapse. Biophys. J. 10, 117a.

Bennett, M.V.L. and Spira, M.E. (1975) Rapid changes in electrotonic coupling between blastomeres of *Fundulus* eggs. J. Cell Biol. 67, 27a.

Bennett, M.V.L., Spira, M.E. and Pappas, G.D. (1972) Properties of electrotonic junctions between embryonic cells of *Fundulus*. Dev. Biol. 29, 419–435.

Bennett, M.V.L. and Trinkaus, J.P. (1970) Electrical coupling between embryonic cells by way of extracellular space and specialized junctions. J. Cell Biol. 44, 592–610.

Caveney, S. (1974) Intercellular communication in a positional field: movement of small ions between insect epidermal cells. Dev. Biol. 40, 311–322.

Crick, F. (1970) Diffusion in embryogenesis. Nature 225, 420–422.

Czihak, G., and Horstadius, S. (1970) Transplantation of RNA-labeled micromeres into animal halves of sea urchin embryos. A contribution to the problem of embryonic induction. Dev. Biol. 22, 15–30.

Decker, R. and Friend, D. (1974) Assembly of gap junctions during amphibian neurulation. J. Cell. Biol. 62, 32–47.

De Haan, R.L. and Sachs, H.G. (1972) Cell coupling in developing systems. In: Current Topics in Developmental Biology (Moscona, A.A. and Monroy, A., eds.), Vol. 7, pp. 193–228, Academic Press, New York.

de Laat, S.W. and Bluemink, J.G. (1974) New membrane formation during cytokinesis in normal and cytochalasin-B treated eggs of *Xenopus laevis*. II. Electrophysiological observation. J. Cell Biol. 60, 529–540.

Dettlaff, T.A. (1964) Cell division, duration of interkinetic states, and differentiation in early stages of embryonic development. Adv. Morphogenesis 3, 323–360.

Deuchar, E.M. (1970) Neural induction and differentiation with minimal numbers of cells. Dev. Biol. 22, 185–199.

Dixon, J.S. and Cronly-Dillon, J.R. (1972) The fine structure of the developing retina in *Xenopus laevis*. J. Embryol. Exp. Morph. 28, 659–666.

Ducibella, T., Albertini, D.F., Anderson, E. and Biggers, J.D. (1975) The preimplantation mammalian embryo: Characterization of intercellular junctions and their appearance during development. Dev. Biol. 45, 231–250.

Epstein, M. and Sheridan, J. (1974) Formation of low resistance junctions in the absence of protein synthesis and metabolic energy production. J. Cell Biol. 63, 95a.

Fischbach, G.D. (1972) Synapse formation between dissociated nerve and muscle cells in low density cell cultures. Dev. Biol. 28, 407–429.

Fletcher, W.H. and Robertson, J.D. (1975) Assembly of an "enclosed gap junction" by granulosa cells in the developing ovarian follicles of sexually immature rats. J. Cell Biol. 67, 116a.

Furshpan, E.J. and Potter, D.D. (1968) Low-resistance junctions between cells in embryos and tissue culture. In: Current Topics in Developmental Biology (Moscona, A.A. and Monroy, A., eds.) Vol. 3, pp. 95–127, Academic Press, New York.

Gilula, N.B., Reeves, O.R. and Steinbach, A. (1972) Metabolic coupling, ionic coupling and cell contacts. Nature 235, 262–265.

Gilula, N.B. (1973) Development of cell junctions. Am. Zoologist 13, 1109–1117.

Goodenough, D.A. and Gilula, N.B. (1974) The splitting of hepatocyte gap junctions and zonulae occludentes with hypertonic disaccharides. J. Cell Biol. 61, 575–590.

Goodenough, D.A. and Revel, J.P. (1970) A fine structure analysis of intercellular junctions in the mouse liver. J. Cell Biol. 45, 272–290.

Goodenough, D.A. and Stoeckenius, W. (1972) The isolation of mouse hepatocyte gap junctions. J. Cell Biol. 54, 646–656.

Goodman, D.B.P., Bloom, F.E., Battenberg, E.R., Rasmussen, H. and Davis, W.L. (1975) Immunofluorescent localization of cyclic AMP in toad urinary bladder: possible intercellular transfer. Science 188, 1023–1025.

Goshima, K. (1971) Synchronized beating of myocardial cells mediated by FL cells in monolayer culture and its inhibition by trypsin-treated FL cells. Exp. Cell Res. 65, 161–169.

Grainger, R.M. and Wessels, N.K. (1974) Does RNA pass from mesenchyme to epithelium during an embryonic tissue interaction? Proc. Nat. Acad. Sci. U.S.A. 71, 4747–4751.

Grobstein, C. (1961) Cell contact in relation to embryonic induction. Exp. Cell Res. 8 (Suppl) 234–245.

Hadden, J.W., Hadden, E.M., Haddox, M.K. and Goldberg, N.D. (1972) Guanosine 3′,5′-cyclic monophosphate: a possible intracellular mediator of mitogenic influences in lymphocytes. Proc. Nat. Acad. Sci. U.S.A. 69, 3024–3027.

Harris, A. (1974) Contact inhibition of cell locomotion. In: Cell Communication (Cox, R., ed.) pp. 147–185, John Wiley and Sons, New York.

Hay, E.D. (1968) Organization and fine structure of epithelium and mesenchyme in the developing chick embryo. In: Epithelial-Mesenchymal Interactions (Fleishmajer, R. and Billingham, R., eds.) pp. 31–35, Williams and Wilkins Co., Baltimore.

Hayes, B.P. and Roberts, A. (1973) Synaptic junction development in the spinal cord of an amphibian embryo: An electron microscope study. Z. Zellforsch 137, 251–269.

Hirakow, R. and De Haan, R.L. (1970) Synchronization and the formation of nexal junctions between isolated chick embryonic heart myocytes beating in culture. J. Cell Biol. 47, 88a.

Imanaga, I. (1974) Cell-to-cell diffusion of procion yellow in sheep and calf Purkinje fibers. J. Membrane Biol. 16, 381–388.

Ito, S. and Loewenstein, W.R. (1969) Ionic communication between early embryonic cells. Dev. Biol. 19, 228–243.

Ito, S., Sato, E., and Loewenstein, W.R. (1974a) Studies on the formation of a permeable cell membrane junction. I. Coupling under various conditions of membrane contact. Effects of colchicine, cytochalasin B., dinitrophenol. J. Membrane Biol. 19, 305–338.

Ito, S., Sato, E., and Loewenstein, W.R. (1974b) Studies on the formation of a permeable cell membrane junction. II. Evolving junctional conductance and junctional insulation. J. Membrane Biol. 19, 339–355.

Jacobson, M. and Hunt, R.K. (1973) The origins of nerve-cell specificity. Sci. Am. 228, 26–35.

Johnson, R., Hammer, M., Sheridan, J. and Revel, J.P. (1974) Gap junction formation between reaggregated Novikoff hepatoma cells. Proc. Nat. Acad. Sci. U.S.A. 71, 4536–4540.

Johnson, R., Herman, W. and Preus, D. (1973) Homocellular and heterocellular gap junctions in *Limulus*. J. Ultrastruct. Res. 43, 298–312.

Johnson, R.G. and Sheridan, J.D. (1971) Junctions between cancer cells in culture: Ultrastructure and permeability. Science 174, 717–719.

Kachadorian, W.A., Wade, J.B. and Discala, V.A. (1975) Vasopressin induced structural change in toad bladder luminal membrane. Science 190, 67–69.

Keeter, J.S., Pappas, G.D. and Model, P.G. (1975) Inter- and intramyotomal gap junctions in the *Axolotl* embryo. Dev. Biol. 45, 21–33.

Kelley, R.O. (1969) An electron microscopic study of chordamesoderm-neurectoderm association in gastrulae of a toad, *Xenopus laevis*. J. Exp. Zool. 172, 153–179.

Konigsberg, I.R. and Buckley, P.A. (1974) Regulation of the cell cycle and myogenesis by cell-medium interaction. In: Concepts of Development (Lash, J. and Whittaker, J.R., eds.) pp. 179–193, Sinauer Associates Inc., Stanford, Conn.

Lawrence, P.A., Crick, F.H.C. and Monro, M. (1972) A gradient of positional information in an insect, *Rhodnius*. J. Cell Sci. 11, 815–853.

Lawrence, P.A. and Green, S.M. (1975) The anatomy of a compartment border. The intersegmental boundary in *Oncopeltus*. J. Cell Biol. 65, 373–382.

Loewenstein, W.R. (1966) Permeability of membrane junctions. Ann. N.Y. Acad. Sci. 137, 441–472.

Loewenstein, W.R. (1967) On the genesis of cellular communication. Dev. Biol. 15, 503–520.

Loewenstein, W.R. (1968) Communication through cell junctions: Implications in growth control and differentiation. Dev. Biol. Suppl. 2, 151–183.

Loewenstein, W.R. and Penn, R.D. (1967) Intercellular communication and tissue growth. II. Tissue regeneration. J. Cell Biol. 33, 235–242.

Lopresti, V., Macagno, E.R. and Levinthal, C. (1974) Structure and development of neuronal connections in isogenic organisms transient gap junctions between growing optic axons and lamina neuroblasts. Proc. Nat. Acad. Sci. U.S.A. 71, 1099–1102.

Martz, E. and Steinberg, M.S. (1973) Contact inhibition of what? An analytical review. J. Cel. Physiol. 81, 25–37.

McNutt, N.S. and Weinstein, R.S. (1969) Carcinoma of the cervix: Deficiency of nexus intercellular junctions. Science, 165, 597–599.

McNutt, S. and Weinstein, R. (1973) Membrane ultrastructure at mammalian intercellular junctions. Prog. Biophys. Mol. Biol. 26, 45–101.

McNutt, N.S., Hershberg, R.A. and Weinstein, R.S. (1971) Further observations on the occurrence of nexuses in benign and malignant human cervical epithelium. J. Cell Biol. 51, 805–825.

Meier, S. and Hay, E.D. (1975) Stimulation of corneal differentiation by interaction between cell surface and extracellular matrix. I. Morphometric analysis of transfilter "induction". J. Cell Biol. 66, 275–291.

Merk, F.B. and McNutt, N.S. (1972) Nexus junctions between dividing and interphase granulosa cells of the rat ovary. J. Cell Biol. 55, 511–519.

Miller, Z., Lovelace, E., Gallo, M. and Pastan, I. (1975) Cyclic guanosine monophosphate and cellular growth. Science 190, 1213–1215.

Moens, W., Vokaer, A. and Kram, R. (1975) Cyclic AMP and cyclic GMP concentrations in serum- and density-restricted fibroblast cultures. Proc. Nat. Acad. Sci. U.S.A. 72, 1063–1067.

Muir, A.R. (1967) The effects of divalent cations on the ultrastructure of the perfused rat heart. J. Anat. 101, 239–261.

Pardee, A.B. (1964) Cell division and a hypothesis of cancer. Natl. Cancer Inst. Monograph 14, 7–20.

Pardee, A.B. (1974) A restriction point for control of normal animal cell proliferation. Proc. Nat. Acad. Sci. U.S.A. 71, 1286–1290.

Payton, B.W., Bennett, M.V.L. and Pappas, G.D. (1969) Permeability and structure of junctional membranes at an electrotonic synapse. Science, 166, 1641–1643.

Peracchia, C. and Fernandez-Jaimovich, M.E. (1975) Isolation of intramembrane particles from gap junctions. J. Cell Biol. 67, 330a.

Pitts, J.D. (1972) Direct interaction between animal cells. In: Cell Interactions (Silvestri, L.G., ed.). pp. 227–285, North-Holland, Amsterdam.

Potter, D.D., Furshpan, E.J. and Lennox, E.S. (1966) Connections between cells of the developing squid as revealed by electrophysiological methods. Proc. Nat. Acad. Sci. U.S.A. 55, 328–335.

Rash, J.E. and Fambrough, D. (1973) Ultrastructural and electrophysiological correlates of cell coupling and cytoplasmic fusion during myogenesis in vitro. Dev. Biol. 30, 166–186.

Rash, J.E. and Staehelin, L.A. (1974) Freeze-cleave demonstration of gap junctions between skeletal myogenic cells in vivo. Dev. Biol. 36, 455–461.

Raviola, E. and Gilula, N.B. (1973) Gap junctions between photoreceptor cells in the vertebrate retina. Proc. Nat. Acad. Sci. U.S.A. 70, 1677–1681.

Revel, J.P., Yip, P. and Chang, L.L. (1973) Cell junctions in the early chick embryo – a freeze etch study. Dev. Biol. 35, 302–317.

Rozengurt, E. and Heppel, L.A. (1975) Serum rapidly stimulates ouabain-sensitive $^{86}Rb^{+}$ influx in quiescent 3T3 cells. Proc. Nat. Acad. Sci. U.S.A. 72, 4492–4495.

Rubin, H. (1975) Central role for magnesium in coordinate control of metabolism and growth in animal cells. Proc. Nat. Acad. Sci. U.S.A. 72, 3551–3555.

Rudland, P.S., Seeley, M. and Seifert, W. (1974) Cyclic GMP and cyclic AMP levels in normal and transformed fibroblasts. Nature 251, 417–419.

Sanders, E.J. and Zalik, S.E. (1972) The blastomere periphery of *Xenopus laevis*, with special reference to intercellular relationships. Wilhelm Roux' Archiv. 171, 181–194.

Scott, W.N., Sapirstein, V.S. and Yoder, M.J. (1974) Partition of tissue functions in epithelia: Localization of enzymes in "Mitochondria-Rich" cells of toad urinary bladder. Science 184, 797–799.

Seifert, W.E. and Rudland, P.S. (1974) Possible involvement of cyclic GMP in growth control of cultured mouse cells. Nature, 248, 138–140.

Sheppard, J.R. (1975) Cyclic AMP and cell division. In: Molecular Pathology (Good, R.A., Day, S.B. and Yunis, J.J., eds.) pp. 405–439, C.C. Thomas, Springfield, Illinois.

Sheridan, J.D. (1966) Electrophysiological study of special connections between cells in the early chick embryo. J. Cell Biol. 31, c1–c5.

Sheridan, J.D. (1968) Electrophysiological evidence for low-resistance intercellular junctions in the early chick embryo. J. Cell Biol. 37, 650–659.

Sheridan, J.D. (1971) Dye movement and low resistance junctions between reaggregated embryonic cells. Dev. Biol. 26, 627–636.

Sheridan, J.D. (1973) Functional evaluation of low-resistance junctions: Influence of cell shape and size. Am. Zoologist, 13, 1119–1128.

Sheridan, J.D. (1974) Low resistance junctions: Some functional considerations. In: The Cell Surface in Development (Moscona, A.A., ed.) pp. 187–206, John Wiley and Sons, New York.

Sheridan, J.D. and Johnson, R.G. (1975) Cell functions and neoplasia. In: Molecular Pathology (Good, R., Day, S. and Yunis, J.J., eds.) pp. 354–378, C.C. Thomas, Springfield, Illinois.

Sheridan, J.D., Finbow, M. and Pitts, J.D. (1975a) Metabolic cooperation in culture: Possible involvement of junctional transfer in regulation of enzyme activities. J. Cell Biol. 67, 396a.

Sheridan, J.D., Hammer, M.G. and Johnson, R.G. (1975b) Cyclic nucleotide-induced changes in formation of low-resistance junctions. J. Cell Biol. 67, 395a.

Siegenbeck van Heukelom, J., van der Gon, J.J.D. and Prop. F.J.A. (1972) Model approaches for evaluation of cell coupling in monolayers. J. Memb. Biol. 7, 88–110.

Slack, C. and Palmer, J.P. (1969) The permeability of intercellular junctions in the early embryo of *Xenopus laevis*, studied with a fluorescent tracer. Exp. Cell Res. 55, 416–419.

Slack, C. and Warner, A.E. (1973) Intracellular and intercellular potentials in the early amphibian embryo. J. Physiol. 232, 313–330.

Slack, C. and Warner, A.E. (1975) Properties of surface and junctional membranes of embryonic cells isolated from blastula stages of *Xenopus laevis*. J. Physiol. 248, 97–120.

Slack, C., Warner, A.E. and Warren, R.L. (1973) The distribution of sodium and potassium in the early amphibian embryo. J. Physiol. 232, 297–312.

Socolar, S.J. (1973) Cell coupling in epithelia. Exp. Eye Res. 15, 693–698.

Spira, A.W. (1971) The nexus in the intercalated disc of the canine heart: Quantitative data for an estimation of its resistance. J. Ultrastruct. Res. 34, 409–425.

Subak-Sharpe, H., Bürk, R. and Pitts, J. (1966) Metabolic cooperation by cell-to-cell transfer between genetically different mammalian cells in tissue culture. Heredity, London, 21, 342–343.

Subak-Sharpe, H., Bürk, R.R., and Pitts, J.D. (1969) Metabolic co-operation betweeen biochemically marked mammalian cells in tissue culture. J. Cell Sci. 4, 353–367.

Takahashi, K., Shun-Ichi, M. and Kidokoro, Y. (1971) Development of excitability in embryonic muscle cell membranes in certain tunicates. Science, 171, 415–417.

Trelstad, R.L., Hay, E.D. and Revel, J.P. (1967) Cell contact during early morphogenesis in the chick embryo. Dev. Biol. 16, 78–106.

Tsien, R.W. and Weingart, R. (1974) Cyclic AMP: cell-to-cell movement and inotropic effect in ventricular muscle studied by a cut-end method. J. Physiol. 242, 95P–96P.

Tupper, J.T. and Saunders, J.W., Jr. (1972) Intercellular permeability in the early *Asterias* embryo. Dev. Biol. 27, 546–554.

Tupper, J., Saunders, J.W., Jr., and Edwards, C. (1970) The onset of electrical coupling between cells in the developing starfish embryo. J. Cell Biol. 46, 187–190.

van Venrooij, G.E.P.M., Hax, W.M.A., Schouten, V.J.A., Denier van der Gon, J.J. and van der Vorst, H.A. (1975) Absence of cell communication for fluorescein and dansylated amino acids in an electrotonic coupled cell system. Biochim. Biophys. Acta, 394, 620–632.

Warner, A.E. (1973) The electrical properties of the amphibian ectoderm during induction and early development of the nervous system. J. Physiol. 235, 267–286.

Warner, A.E. and Lawrence, P.A. (1973) Electrical coupling across developmental boundaries in insect epidermis. Nature, 245, 47–48.

Weingart, R. (1974) The permeability to tetraethylammonium ions of the surface membrane and the intercalated disks of sheep and calf myocardium. J. Physiol. 240, 741–762.

Woodward, D.J. (1968) Electrical signs of new membrane production during cleavage of *Rana pipiens* eggs. J. Gen. Physiol. 52, 509–531.

Transduction of positional information during development 9

Daniel McMAHON and Christopher WEST

1. Introduction

Experiments aimed at dissecting the phenomena of determination of position by cells during embryonic development have been pursued for at least two centuries. Work on positional determination reached a high point early in this century because of the efforts of Spemann, Mangold, P. Weiss and many others, but until very recently it has attracted diminishing attention from developmental biologists. During the last twenty five years in particular more attention has been directed toward understanding the problem of cellular differentiation in development and the mechanisms controlling gene expression in differentiated cells. Within the past few years, however, the question of positional information and its determination has once again begun to attract attention. Several factors have contributed to the renewed interest in this problem: a simple fascination with the dramatic process and results of position determination; a belief that insight into the mechanism of chemical transduction of positional information might contribute to a better understanding of gene regulation and cell differentiation; and possibly the intuition that the means for solving the problem of position determination might be at hand.

There appear to be three different mechanisms which are used to specify the position of a cell in an embryo. These are: (a) morphogenetic fields; (b) induction; and (c) mosaicism or cytoplasmic localization. Each of these will be examined later in this chapter. These phenomena differ both in their requirement for cellular interactions and in the nature of the cellular interactions involved. Cytoplasmic localization in its most extreme form does not require cellular interactions since substances which determine the developmental fate of the cells are localized in the cytoplasm and/or on the plasma membrane. In contrast, induction has a strict requirement for cell-to-cell interactions. Specifically, one group of cells acts upon another to induce a developmental response. However, the organization of morphogenetic fields is perhaps the most mysterious of the three and also the most difficult to study experimentally since within a field a group of cells behaves as an interacting and integrated system.

G. Poste & G.L. Nicolson (eds.) The Cell Surface in Animal Embryogenesis and Development

The expression of cellular genetic information in terms of positional determination and the subsequent development of pattern and form constitutes a formidable linkage between genetics, cellular differentiation and morphology. Clearly, a wide spectrum of research interests can be accommodated within such a broad-based subject. In this chapter, we have chosen to discuss what is known concerning the molecular biology of positional information and its determination. The large amount of work on the descriptive morphology of positional phenomena occurring in embryogenesis in different species will not be reviewed here, since it has been discussed fully elsewhere (Hadorn, 1966; Ursprung, 1966; Kühn, 1971; Bryant, 1974; Stocum, 1975). In addition to presenting an outline of current theories on the possible nature of the signals responsible for the specification of position, we will devote some attention to the question of how cells might interpret such signals. The interpretative process may be the key event in determining position since it is by this process that differing cellular responses will be established and thus, in turn, patterns produced. It seems likely that differences in the interpretative process in specific cell types result from differences in gene activity. Cellular responses to positional signals will thus be determined within a framework imposed by the existing program of gene expression in any particular cell. This introduces an attractive degree of simplicity to the determination of position since by using the same system of signals it becomes possible to generate a large number of different cellular responses merely by changing the cell's mechanism for signal interpretation. Since interpretation is presumably gene determined there is no difficulty in seeing how such diversity could be achieved. Furthermore, in any given situation the number of interpretative decisions that a cell can take may well be limited, in as much that the existing program of gene expression may restrict the response to a decision between only a few possibilities.

The experimental goals of molecular biologists who study position determination in development are three-fold. First, the mechanism(s) of chemical transduction of positional information must be understood. Second, the chemical nature of (and the temporal changes in) the interpretative system must be resolved. Finally, we can hope that there will be a simple underlying pattern for the nature of the chemical code. As we shall see in this article we are still a long way from realizing these goals.

2. *Classes of positional specification*

2.1. *Morphogenetic fields*

2.1.1. *General properties*

The concept of the morphogenetic field was initially developed by P. Weiss (1939). Morphogenetic fields have two distinguishing characteristics. The first is the negative property that they are refractory to experimental attempts to distinguish distinct groups of cells in the field. Although they give rise to at least two different kinds of cells it is not empirically possible to separate their constituent cells into a group which produces and one which reacts to a developmental stimulus. Secondly, the ratio of output cell types produced by the cells in a morphogenetic field is relatively independent of the absolute numbers of input cells.

2.1.2. Theories for position specification in a morphogenetic field

2.1.2.1. Gradients of diffusible morphogens. All of the theories which have been proposed for the specification of position in a morphogenetic field propose that a graded distribution of information about position is produced by some physical or biochemical process. The first such theory was initially proposed by Child (1928) and modified by Dalcq (1938) to explain specification of position by a regulative, or size invariant, mechanism. The "gradient theory" proposes that a gradient(s) of a diffusible morphogen(s) (i.e., a chemical which regulates the choice of developmental path by a cell) is established by diffusion. In its simplest formulation, cells at one end of a morphogenetic field are proposed to produce the morphogen which diffuses through the field to a sink where it is lost or destroyed. Modifications of the basic model which allow for size regulation are shown in Fig. 1. The first of these proposes that cells set the concentration of the

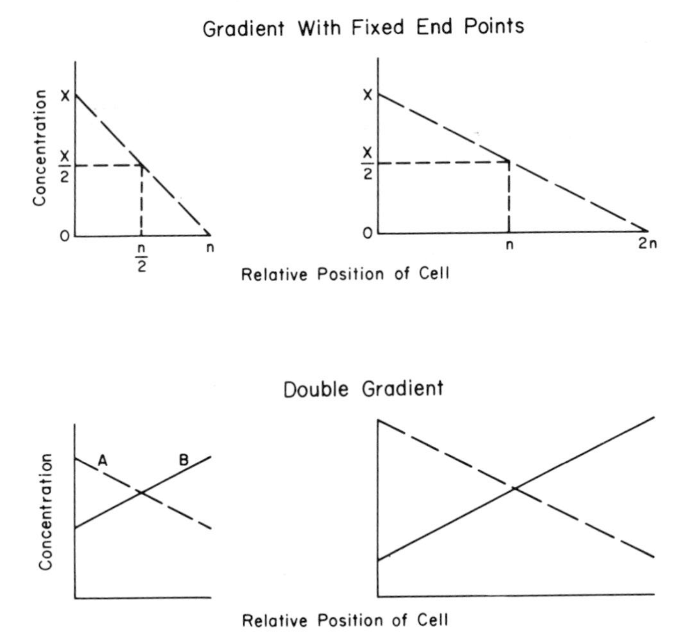

Fig. 1. Position determination using gradients of morphogens generated by diffusion. In the top panel a single gradient with fixed concentration of morphogen at each end is shown. The bottom panel illustrates a system which uses opposing gradients of two morphogens. See text for full discussion.

diffusible morphogen to a constant value at each end of the field. In Fig. 1 (top panel) the value at one end of the field is given as X and as zero at the other end. After a gradient has reached equilibrium, any specific concentration of morphogen between 0 and X specifies to a cell its relative position within the field. Thus a concentration of X/2 specifies a position in the middle of the field regardless of the number of cells assumed to be in the field. In the second situation, shown in the bottom panel of Fig. 1, a cell can determine its position by measuring the ratio between two different morphogens, A and B. Thus in Fig. 1 if a cell measured a ratio of $A/B > 1$ it would have determined that it was to the left of the center of the field and presumably could then act on this information. The significant features of this kind of theory are that the developmental choice of a cell is determined by the absolute or relative concentrations of a morphogen and that these concentrations are in turn determined by diffusion. Other variations of this idea can be developed (section 2.2.4). Crick (1970) has shown that a diffusion gradient could, in theory, be established in a morphogenetic field within the time available for determination if the morphogen was of relatively small (~500) molecular weight.

2.1.2.2. The phase-angle shift model. A very elegant model for position determination has been proposed by Goodwin and Cohen (1969). They propose that a group of cells emits periodic signals (which may be chemical, electrical, etc.) which travel through the field at different rates so that the signals become phase-shifted with respect to each other. The difference in phase becomes progressively greater for cells farther away from the signalling cells. This model is illustrated in Fig. 2. At time = 1 two coincident signals are being emitted. By time = 2 both signals have traveled into the field but one has traveled slightly farther than the other. This difference has increased further at t = 3. A new set of signals is also being generated at this time and they also begin to move into the field. Cells in this model are assumed to respond to the difference in phase angles (or alternatively time of arrival) of the signals. A linear gradient of phase angle differences is generated by this model which is then used to determine position. The example as posed is, however, too simple since it does not produce a regulative system. For example, imagine that the right half of the field is experimentally removed. Under these conditions the information provided to the cells in the left half is unaffected, and a third signal is thus necessary to make the system regulative.

2.1.2.3. The cell contact model. One of us has proposed a third model for position determination. Since dissatisfaction with some of the characteristics of the previous two models prompted this new model it is perhaps helpful to briefly mention some of the problems associated with the models discussed above. Both models generate linear gradients of positional information but development generates discontinuous cell types. In essence this leads to a requirement for some kind of amplification mechanism in cells that interpret the positional information in order that a small difference in positional informa-

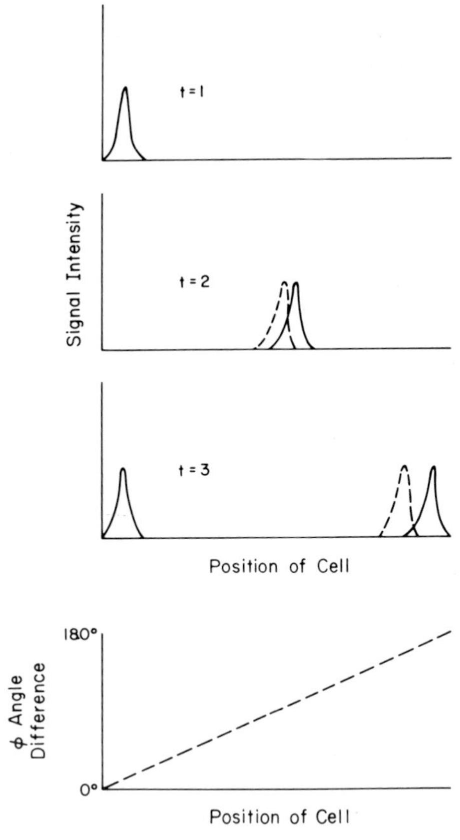

Fig. 2. Position determination using the phase-angle difference model. The top three panels show the progression through the field with time of two morphogenetic signals which move with different speeds. The difference in phase angle between the signals increases as a linear function of position in the field as shown in the bottom of the field. See text for full discussion.

tion can be used as a basis for switching cells from one path to another. While biochemical amplification mechanisms are relatively easy to formulate (for example, see Levine, 1966), the very requirement for one reduces the simplicity of the model. However, a considerably more serious problem underlies both theories. Each requires that a group(s) of cells acquire special properties, namely, that they become signal producers or destroyers. There is no basis within the theories for these properties of special groups of cells (although they are certainly a requirement for the mechanism). Once again, we are required to go outside the bounds of these theories to formulate a mechanism which produces these properties. Finally both theories are extremely difficult to falsify experimentally. This leads to both philosophical (Popper, 1959) and experimental difficulties.

The cell contact model (McMahon, 1973) is free from these problems. Cells are assumed to interact with one another via "contact sensing molecules" on their

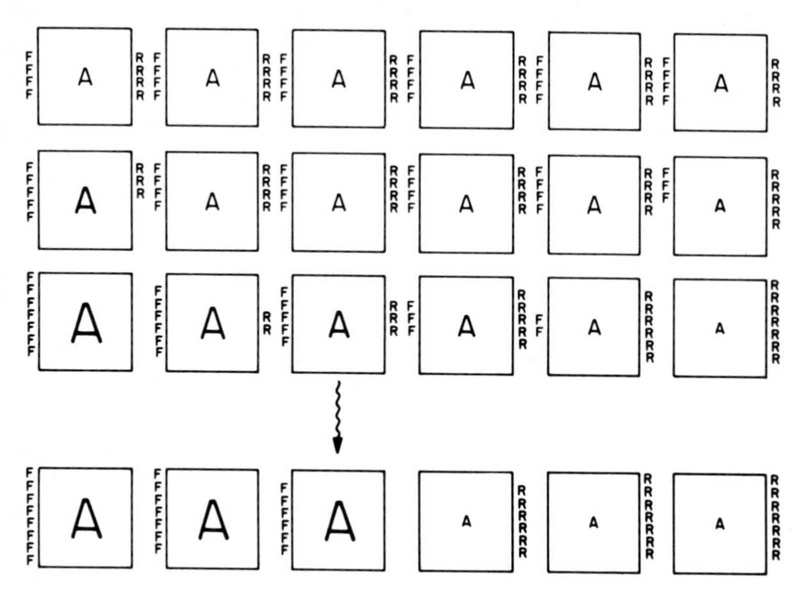

Fig. 3. Transmission of positional information in the cell-contact model. See text for full discussion.

plasma membranes. These respond to cell contact by regulating the intracellular concentration of a second messenger such as cAMP. Fig. 3 provides a qualitative illustration of the model. Each box is intended to represent a cell. Each row of six boxes represents a file of six cells in a morphogenetic field at a different time after position determination has begun. Each cell is postulated as having two complementary kinds of contact sensing molecules on it initially. One kind, F, when activated by contact with R lowers the concentration of an intracellular second messenger, A. Conversely when R is activated by contact with F on another cell it acts to increase the concentration of A. In addition the polarity which is characteristic of morphogenetic fields is assumed to be based upon a polarization of the contact sensing molecules in the cells. In the example given in Fig. 3 there are F's on the front of each cell and R's on the rear. Finally, the system is postulated to be regulated by negative feedback. Thus, as the concentration of A increases within a cell, the effectiveness or number of the molecules which increase the concentration of A (the R's) is also decreased, while those molecules which lead to a decrease in the concentration of A (the F's) are activated or increase in number. The inverse happens at the rear of the field. For convenience of discussion each half of the complementary system, F or R, is assumed to be a single molecule which affects both the concentration of A and stimulates its complementary molecule. This does not need to be so since F and R could each represent a group of molecules if each group were coordinately regulated.

The process of position specification in this model is triggered by a trivial physical fact. A cell must be at each end of a file of cells; so one cell must be at the front and another at the back. The F's on the first cell in a file are not activated by contact with another cell although its R's are activated. Therefore A is assumed to increase in the first cell. As A increases, the negative feedback intrinsic in the model drives down the number or activity of R's (which increase A) and drives up the F's (this attempted regulation is defeated since there are no R's to activate the F's of the first cell). As the R's (or R-associated system) decrease, the F's on the second cell are activated less and less. The concentration of A in this cell therefore increases and its environment becomes progressively like that of the front cell. The opposite events occur in the rear cells leading eventually to the situation presented in the last row of cells in Fig. 3, where two populations of cells exist with qualitatively different intracellular amounts of A. A quantitative solution of the equations which simulate this model for the pseudoplasmodium of *Dictyostelium discoideum,* in which A was assumed to be cAMP, lead to a similar result (McMahon, 1973). Some experimental tests of this model will be discussed later (Section 2.1.3.).

While the cell contact model was originally proposed for the psuedoplasmodium of *D. discoideum,* it can easily be applied to more complex organisms. It can be extended to fit a system exhibiting position determination in two dimensions by assuming another system of contact sensing molecules orthogonal to the first system on, for example, the medial and lateral edges of cells. If these controlled the distribution of another morphogen, B, and the system was independent of the first we could easily break a sheet of cells into four quadrants of defined (although not necessarily identical) size. If, however, the concentrations of the two morphogens interacted to produce their effects, then patterns of moderate complexity could readily be generated. A system like this can easily be extended into the third dimension, but this is probably not necessary since it is not clear that simultaneous determinations of position in three dimensions actually occur. The model also easily adapts to sequential temporal discriminations. Fields become independent of each other when they are separated by a noncellular gap or if abutting cells have an independent system of contact-sensing molecules. While in theory only two systems of contact-sensing molecules would be necessary in an embryo (if the groups of cells bearing them alternated), it seems likely that a number of sets of such molecules could exist, thus subdividing the developing organism into relatively small regions each with its own set of identifying molecules. One advantage that would be possessed by an organism with a relatively great number of systems of contact-sensing molecules over one with the theoretical minimum is that the multiple systems of molecules could serve as place markers for cells such as those of the neural crest which migrate from their place of origin to distant locations (Weston, 1970).

2.1.3. Specification of position in Dictyostelium discoideum

A fascinating example of position determination occurs in the pseudoplasmodium or slug of *D. discoideum.* Cells become committed to differentiate into either spore or stalk cells on the basis of their position in the pseudoplasmodium. These cells in the front third become the stalk of the resulting sorocarp and the cells of the two-thirds in the rear become spores (Raper, 1940; Bonner, 1967). The resulting proportion of stalk to spore cells in the sorocarp is approximately constant although the absolute numbers of cells in the psudoplasmodium can vary over four logs (Raper, 1941; Sussman, 1955). Bisected pseudoplasmodia can regulate to produce a normal proportion of stalk and spore cells (Raper, 1940).

Two models have been proposed as to how cells determine their position in the pseudoplasmodium. Ashworth (1971) has proposed that the slime sheath surrounding the pseudoplasmodium is altered as it moves over the cells and that the extent of this alteration provides cells with information about their position. Loomis (1975) has offered a variant of this model. He suggests that the slime sheath is thickened by addition of precursors as it passes over the pseudoplasmodium and it therefore becomes a barrier to the diffusion of a morphogen which determines cellular differentiation. Besides experiments which suggested that increasing amounts of N-acetylglucosamine were added to the sheath as it passed over the pseudoplasmodium, some support for this model was provided by dye-staining experiments which indicated that the slime sheath became a barrier to the diffusion of a tetrazolium dye as it passed along the pseudoplasmodium. However, other workers (Mine and Takeuchi, 1967) have shown that different tetrazolium dyes give radically different staining patterns. Some produce similar staining to that seen by Loomis, while others give an opposite pattern. (We are inclined to believe the pattern of staining is a function of the metabolism of the cells in the pseudoplasmodium.) In addition, the experiments describing the incorporation of N-acetylglucosamine into the sheath have recently been called into question (Watts and Treffry, 1975).

The other model applied to determination in the pseudoplasmodium is the cell contact model described above. Klaus and George (1974) produced small lesions in the pseudoplasmodium with a microbeam laser and looked to see whether the pseudoplasmodium might break into new pseudoplasmodia because of the cellular lesion since the cell contact model predicts that a continuous file of normal cells would be necessary for the organization of a single field and if this is broken the field will break down into independent fields. Klaus and George observed that two independent pseudoplasmodia were produced in one-third of the irradiations and they interpreted their results as in agreement with the cell contact model. Pan et al. (1974) have stained sectioned pseudoplasmodia with fluorescent antiserum against cAMP and showed that an abrupt discontinuity in the distribution of staining occurs during development. This result also agrees with a prediction of the cell contact model.

Finally, we (McMahon, 1975; McMahon et al., 1975) have examined the effects of purified plasma membrane on the biochemical development of separated pseudoplasmodial cells. The rationale for these experiments is the following. Commitment to differentiation does not occur and cannot be reversed in single cells (Gregg, 1965). Also, the dissociation of pseudoplasmodia into single cells disrupts their biochemical development which apparently does not proceed until the cells reaggregate (Loomis, 1969). The cell contact model suggests that much of this contact-mediated control is regulated by molecules on the plasma membrane. If this is true, then the addition of isolated plasma membranes might be expected to affect the course of biochemical development of the dissociated cells (assuming of course that the appropriate plasma membrane effector molecules are not affected by the membrane isolation technique). It is therefore of considerable interest that purified plasma membranes from pseudoplasmodial cells prevented morphogenesis when added to dissociated cells. Although the cells reaggregated they did not reform a pseudoplasmodium. Some major changes in the regulation of certain developmentally controlled enzymes were also detected in this system. Addition of isolated plasma membranes activated alkaline phosphatase in the dissociated cells and also prevented the normal developmental increases in glycogen phosphorylase and uridine diphosphoglucose pyrophosphorylase (McMahon et al., 1975). Activation of alkaline phosphatase by the added membranes did not require cellular protein synthesis and was not produced when log phase cells were treated with pseudoplasmodium membranes. These changes in enzyme profile were not produced when dissociated pseudoplasmodial cells were treated with membranes from a mutant, *agg 2,* which cannot develop, or with plasma membranes from sheep erythrocytes. Finally, activation of alkaline phosphatase and the disruption of morphogenesis occurred only when the dissociated cells and the added plasma membranes were in contact. Separation of cells from membranes by Millipore filter prevented the effects of the plasma membranes (McMahon, 1975). These results not only provide support for the idea that the plasma membrane is involved in communicating and transducing developmental information but they also provide a considerable incentive for characterizing the developmental regulation of plasma membrane components in *D. discoideum.*

2.2. Induction

Induction is a process in which the cellular interactions are of a more familiar nature, since one group of cells acts upon another. Induction is for this reason especially amenable to experimental analysis. A full discussion of the various experimental strategies employed in the study of induction phenomena and the range of stimuli that can provoke induction of competent cells is provided in chapter 7 of this volume by Saxén and his colleagues and we shall therefore restrict our comments to the role of induction in positional determination.

2.2.1. Postulated mechanisms for induction

2.2.1.1. Transfer of RNA. A mechanism which has received considerable attention in the past few years is the possibility that an informational macromolecule is transferred from inducing to induced cell. RNA has been particularly favored as a candidate in this regard (Niu, 1958; Czihak and Hörstadius, 1970). Niu (1958) has obtained data which indicates that specific induction can be produced with pure RNA in the embryonic chick. It is easy to envisage a variety of biochemical mechanisms by which the postulated inducing RNAs might act. They might be the mRNA for regulatory proteins, regulatory RNA themselves, or they might even be incorporated into the genomes of the recipient cells via reverse transcriptase. The recent results of Mishra et al. (1975) suggest that the last possibility should receive serious consideration. They demonstrated genetic transformation of *Neurospora crassa* treated with RNA prepared from another strain. If these results prove to be generally reproducible this will further enhance the attractiveness of the inducer RNA hypothesis.

However, a number of other experiments argue against a transfer of RNA from inducing to induced cells (Tarin, 1973; Spiegel and Rubinstein, 1972). Grainger and Wessels (1974) have used density labeling of nucleic acids followed by equilibrium density gradient centrifugation to show that no detectable RNA or DNA is transferred from inducing lung mesenchyme to induced lung epithelium. The sensitivity of their experiment was such that the transfer of 75 molecules of RNA to each induced cell would have been detected. The experiment does not therefore completely rule out the possibility that transferred RNA might produce the induction although it makes it less likely.

Experiments in which developing tissues are treated with nucleic acids or nucleases should be interpreted cautiously since both nucleases and nucleic acids may have effects other than those expected. For example, RNase (and other basic proteins) has been shown to activate adenylate cyclase in membranes treated with low concentrations and to inhibit it in plasma membranes treated with high concentrations (Wolff and Cook, 1975). As little as 10–20 μg/ml of RNase can significantly stimulate adenylate cyclase. Polyadenylic acid and poly A:U have also been shown to stimulate cyclic AMP synthesis in spleen cells (Winchurch et al., 1971). It seems reasonable therefore to suspect that adenylate cyclase and other membrane-associated enzymes may be affected when cells are treated with polyanions or polycations (Wolff and Cook, 1975; Sheetz and Singer, 1974).

2.2.1.2. The chemical messengers hypothesis. The chemical messengers hypothesis (McMahon, 1974) is quite different. Much of the information required for each developmental choice is assumed to be encoded in the inducible cells as a result of their previous history of development. Therefore the range of choices a cell can make at each step of position determination is assumed to be limited, perhaps even as small as two. Inducing cells are postulated as secreting inducers which are polypeptide hormones, neurotransmitters and other physiological messengers (hormones). These in turn affect the induced cells by triggering changes in intracellular chemical messengers such as cyclic nucleotides or inorganic ions which regulate metabolism and gene expression in the inducible cells. In this way the effects of inducers are assumed to regulate morphogenesis, division and differentiation of the induced cells. Rutter et al. (1973) have also suggested that cyclic nucleotides regulate proliferation of developing cells in the embryo.

2.2.2. *Cyclic AMP and the differentiation of vertebrate cells*

2.2.2.1. Limb bud cellular differentiation. Work has just begun on the involvement of cyclic nucleotides in development and the experimental picture is therefore not as complete as might be desired, though the results obtained so far emphasize the potential interest of continuing work in this area. A study of regenerating salamander limb has shown that an increase in the total content of cyclic AMP of the regenerating limb bud can be detected as regeneration progresses. If regeneration is prevented by hypophysectomy these changes do not occur (Sicard, 1975). Treatment of newt regeneration bastemas with dibutyryl cyclic AMP (optimum concentration = 10^{-5} M) has been shown to stimulate the incorporation of precursors into both protein and DNA (Babich and Foret, 1973; Foret and Babich, 1973). Sodium fluoride (10^{-2} M) also stimulates the incorporation of precursors and is more effective than dibutyryl cyclic AMP for stimulating incorporation into DNA. Although fluoride is a well known activator of adenylate cyclase in isolated membranes, it does not increase cyclic AMP levels in vivo. Thus this effect of fluoride could result from its known interference with various metabolic reactions (for example, the Krebs cycle).

Examination of the effects of cyclic AMP on development will probably prove more profitable when done with relatively simple systems. Zalin and her collaborators have examined the relationship between muscle formation in the embryonic chick limb and cyclic AMP. A pulsatile increase of cyclic AMP (10–15-fold for 1 hour) occurs about 6 hours before the initiation of myoblast fusion in vitro (Zalin and Montague, 1974). In vivo two smaller pulses of increased cyclic AMP concentration were detected in developing muscle (Zalin and Montague, 1975). It is not clear whether two pulses (rather than one) resulted because of asynchrony in cells in vivo or whether some other factor was responsible, since an attempt to correlate the cyclic AMP pulses with histological changes was not made. However, other measurements showed that DNA synthesis was inhibited when the pulses of increased cyclic AMP occurred. Finally, cultures of myoblasts can be caused to undergo premature fusion by treating them with prostaglandin E which produces a premature pulse of cyclic AMP in the cells (Zalin and Leaver, 1975). These studies emphasize the point that developmentally important changes in cyclic AMP may well occur in rather sharp and short bursts and may not necessarily require prolonged or sustained elevation of nucleotide levels. Only one paper has approached the possible relationship between cyclic AMP and commitment to a specific cellular differentiation in the limb bud. Solursh and Reiter (1975) treated undetermined (stage 24) chick limb bud mesodermal cells with dibutyryl cyclic AMP and theophylline. Their study was aimed at determining a possible relationship between inhibition of cell division and the differentiation of chondrocytes. Treatment with cyclic AMP stimulated the differentiation of chondrocytes and inhibited cell division (as indicated by DNA synthesis) though a variety of other agents that inhibited division did not promote chondrocyte differentiation. The appearance of chondrocytes in this system does not appear to result simply from differential cell survival. (Solursh, personal communication). Stimulation of the differentiation of chondrocytes from limb bud mesoderm by cyclic AMP was predicted on the basis of pharmacological data (McMahon, 1974).

2.2.2.2. Neural differentiation, ions and cyclic AMP. Neurons present a myriad of fascinating problems in developmental biology. Possessing the usual attributes of differentiated cells, unique morphologies and synthesis of specialized proteins, they are also frequently inducers and regulators of the development of other cells (Saxén and Toivonen, 1962) and their patterns of connectivity demonstrate mechanisms of positional discrimination of a high order (Hunt and Jacobson, 1974).

Neurons develop in many different circumstances in different organisms but their early development in amphibians has received the most investigation. Systematic study of neural differentiation in frogs dates back more than fifty years (Spemann, 1921). Barth (1965) and Barth and Barth (1963; 1968; 1969) have carefully examined the effects of inorganic cations on the induction of neurons from presumptive epidermis. They have shown that Li^+, Na^+ or Ca^{2+} are effective inducers, and that cells committed to become neurons can be induced again to differentiate into pigment cells by the same ions. A total of three types of neurons or pigment cells result from sequential inductions (Barth and Barth, 1963; Barth and Barth, 1967a).

Several lines of evidence support the hypothesis that ions are involved in the biochemistry of this induction. Isolated ectodermal cells are competent for these inductions only at the times of development at which they would normally occur in the embryo. The effect of the ionic inducer changes with time irrespective of whether the cells are left in the embryo or cultured in vitro (Barth and Barth, 1967a). This indicates that the "developmental clock" in the isolated cell is keeping accurate time. Changes in ionic flux accompany the normal induction of the neural plate in vivo (Barth and Barth, 1972) but in a hybrid embryo (*Rana sylvatica* and *Rana pipiens*) which does not gastrulate normally (which therefore prevents the induction in vivo), the changes in ion flux do not occur. However, if ectodermal cells are isolated from this hybrid they can be induced to differentiate into the neural series in vitro, indicating that the defect in the embryo results from a defect in induction rather than a defect in the developmental program of the embryo (Barth and Barth, 1967b).

The scope of this system has recently been extended by Taylor and his collaborators (Wahn et al., 1975a; Wahn et al., 1975b; Wahn et al., 1975c), who have shown that cyclic AMP and 8-bromo-cyclic AMP can induce the development of melanocytes and presumptive neurons and/or glia from presumptive epidermis of three different amphibians in vitro. The significance of these results is reinforced by the specificity of the effect, as shown by the lack of inducing effect of 2′,3′-AMP, 5′AMP, and dibutyryl cyclic GMP. However, butyrate induced the formation of neurons under some conditions. Hormones such as melanocyte stimulating hormone also stimulated neural differentiation. When whole embryos or explants of cells were treated with cAMP or dibutyryl cAMP no obvious morphological aberration was produced. The number of melanocytes was not increased though they did show earlier cytodifferentiation. Similar results were obtained when embryos or explants were treated with α or β MSH or ACTH. However, when large explants of cells were pulsed with dibutyryl cAMP and then dissociated into single cells, presumptive neurons and glia differentiated. This did not occur in similarly dissociated control cells which had not been pulsed with dibutyryl cAMP (Wahn, Lightbody, Ngo, Tchen, and Taylor, personal communication). Therefore, cell association seems to be able to suppress the inducing effect of cyclic AMP. Other agents are able to induce neural differentiation in the isolated system. These include acetylcholine and prostaglandins E_1, E_2 and F_2 (Wahn et al., personal communication). It would be interesting to know whether the effect of acetylcholine and the prostaglandins results in depolarization of the cells or activation of adenylate cyclase (Prasad, 1975). Similar neuralizing effects of cAMP have been reported using chick ectoderm (Bjerre, 1974).

Many questions occur when considering this work. For example, do the neurons and glia produced by in vitro induction possess all of the normal characteristics of these cells, such as excitable membranes and the ability to transmit an action potential? It would also be extremely interesting to know whether these cells are able to form synapses in vitro.

In order to demonstrate that cAMP is truly a component in the process of commitment to neural differentiation in vivo, it is necessary to show changes in cellular cAMP levels in those cells which become committed to neural differentiation at the time of induction. An experiment of Nanjundiah (1974), which took advantage of the chemotaxis of *D. discoideum* amoebae to cAMP, showed that the amoebae migrate to the general area of the future neural plate in the axolotl gastrula, suggesting that the neural plate area of the axolotl gastrula is probably leaking a high level of cAMP at this stage. This observation suggests that the cells of this region contain a high concentration of cAMP. However, the results of Reporter and Rosenquist (who made the first measurements of cAMP levels in embryos) indicate that, at the head process stage, the regions of the chick embryo which are fated to become neurons have a very low cAMP content (Reporter and Rosenquist, 1972). These results are not necessarily in conflict because it is reasonable to assume that levels of important morphogens will rapidly change during the course of development of a tissue. In this respect, the use in this system of the elegant fluorescent antibody technique for localizing cyclic AMP developed by Wedner and Steiner and their colleagues (Wedner et al., 1972; Ong et al., 1975) might prove helpful.

The spectrum of cells which different derivatives of cAMP produce is also interesting. A possible explanation for this result is that different derivatives of cAMP may have differing effects on protein kinases (Neelon and Birch, 1973; Granner et al., 1975). The ability of butyrate to produce neural differentiation occasionally is unexplained, but may be of more general significance since butyrate has recently been reported to be a very effective agent in eliciting cytodifferentiation of Friend-leukemia virus transformed cells (Leder and Leder, 1975).

The identity of the natural intercellular messenger for the induction of the neural plate continues to be a mystery. Antibodies against peptide hormones and prostaglandins provide a means to test whether either of these is involved in the natural induction. Peptide hormones are obviously very attractive candidates for inducers in this system and it might be profitable to test other neurophyseal hormones, especially vasotocin, for their effects on neural induction. It is especially appealing to consider that a hierarchy of peptidergic neurotransmitters exist which successively induce neural differentiations (McMahon, 1974).

2.2.3. Pattern regulation in Hydra

2.2.3.1. General properties. Hydra can be considered as two sheets of cells, separated by an acellular mesoglea, which have been folded into a cylinder 5 mm × 0.5 mm. Along its long axis, it is divided into several distinct morphological regions which are illustrated in Fig. 4. The end bearing the hypostome is called the distal end. Seven distinct morphological classes of cells are found in Hydra. Two of these, the epithelio-muscular cells and the gastric cells, are distributed uniformly along the longitudinal axis; the other five classes of cells are distributed as shown in Fig. 4. However, even among one class of cells such as the epithelial-muscular cells, regional specialization is also evident, as for example, in their lipid content (Fig. 4). Thus, on the basis of the morphology of the whole organism and the distribution and cytochemical properties of the constituent cells seven distinct regions can be identified (Fig. 4).

The morphology of Hydra is in some instances simply a function of the local cellular constitution, rather than requiring integrating factors acting on the whole organism. This was shown in a revealing experiment by Gierer et al. (1972). They dissociated Hydra into cells and then formed "sandwich" combinations of cells such as "head cells-foot cells-head cells" or "foot cells-head cells-foot cells". Sandwiches of the first type yielded a Hydra with heads at each end while the second type produced an organism with head structures (tentacles) in the middle. Nonetheless, the organism as generally constituted is governed by certain functions which tend to maintain and restore normal morphology. This system has properties characteristic of both a morphogenetic field and of an induction system. For example, the Hydra is regulative since parts cut from the organism are regenerated and normal morphology is maintained whether the individual is composed of 150,000 or 3000 cells (Bode et al., 1973). However, it is possible to induce formation of a secondary axis by grafting a head onto an individual. Structures found on the head (tentacles) can also be induced by extracts of Hydra and chemical treatment (Yasugi, 1974; Webster, 1967).

Several experiments have revealed that integrated individuals can be formed in the absence of some regions or cell types. A mutant discovered by Lenhoff (1965) produces individuals with a head at each end. This mutant indicates that the presence of the foot is not necessary for the formation of head or of buds. The opposite situation also occurs where the head can be transformed into a second foot by oligomycin (Hornbruch and Wolpert, 1975). These animals are able to bud, though two buds are formed suggesting that the budding region may be duplicated. Such animals indicate that a head is not absolutely necessary either for maintenance of the overall integrity of the organism or for budding, but could play a role in limiting the number of budding centers. Finally, treatment with agents which inhibit cell division (nitrogen mustard and UV irradiation) and destroy interstitial cells do not prevent pattern formation and regeneration (Briens and Reniers-Decoen, 1955; Diehl and Burnett, 1964).

Hydra responds to surgery such as amputation and implantations in characteristic ways that are presumed to be related to the normal process of pattern maintenance. The responses to surgery can be summarized as conforming to four general rules. When a piece of Hydra is removed the missing part reappears at the nearest available position to its original location. The larger the piece of tissue removed, the greater the amount of time required for its replacement (Weimer, 1928; Webster and Wolpert, 1966; Mookerjee and Bhattacharjee, 1967). A grafted part inhibits the regeneration of another copy of itself (MacWilliams and Kafatos, 1968; Hicklin et al., 1973). When grafts which create tissue duplications are made onto intact organisms, the likelihood that the graft will induce a secondary axis increases as the distance between the origin of the graft and its site of implantation increases. Otherwise the graft is resorbed. Removing the piece which is duplicated by the graft allows secondary axis induction at shorter distances (Webster, 1966; Browne, 1909).

2.2.3.2. A regional model for regulation via induction. We have constructed what we call the "regional" model of induction which we believe explains the basic properties of Hydra. The model is of the following form: (1) the cells characteristic of each region are induced by a morphogen which determines that region. Thus there may be seven morphogens in Hydra; and (2) an inducing cell normally secretes the morphogen characteristic of its region. However, secretion of the morphogen characteristic of an adjacent region only begins when the ratio of the ambient concentration of the local morphogen to that of the adjacent morphogen rises above a threshold.

Size invariance of the pattern is achieved in the following way. If the disc is amputated, there will be no disc morphogen. A decreased concentration will become evident first to peduncle cells at the wound surface and they will begin secreting basal disc morphogen. When there is sufficient basal disc morphogen, nearby cells will be induced in a way appropriate for a basal disc region, and recruitment of new cells to secrete basal disc morphogen will cease. Size regulation will have occurred. However, since the peduncle will now be small, a similar process of regulation will take place between peduncle and budding region cells. Regulation will sequentially progress through the rest of the animal.

The gradients of regeneration and secondary axis induction described above are natural consequences of this model. For instance, when a distal piece is excised, the amount of time required for hypostome regeneration will depend on the number of regions which must be sequentially reconstructed. Additionally, secondary axis induction is more likely to take place when the graft is

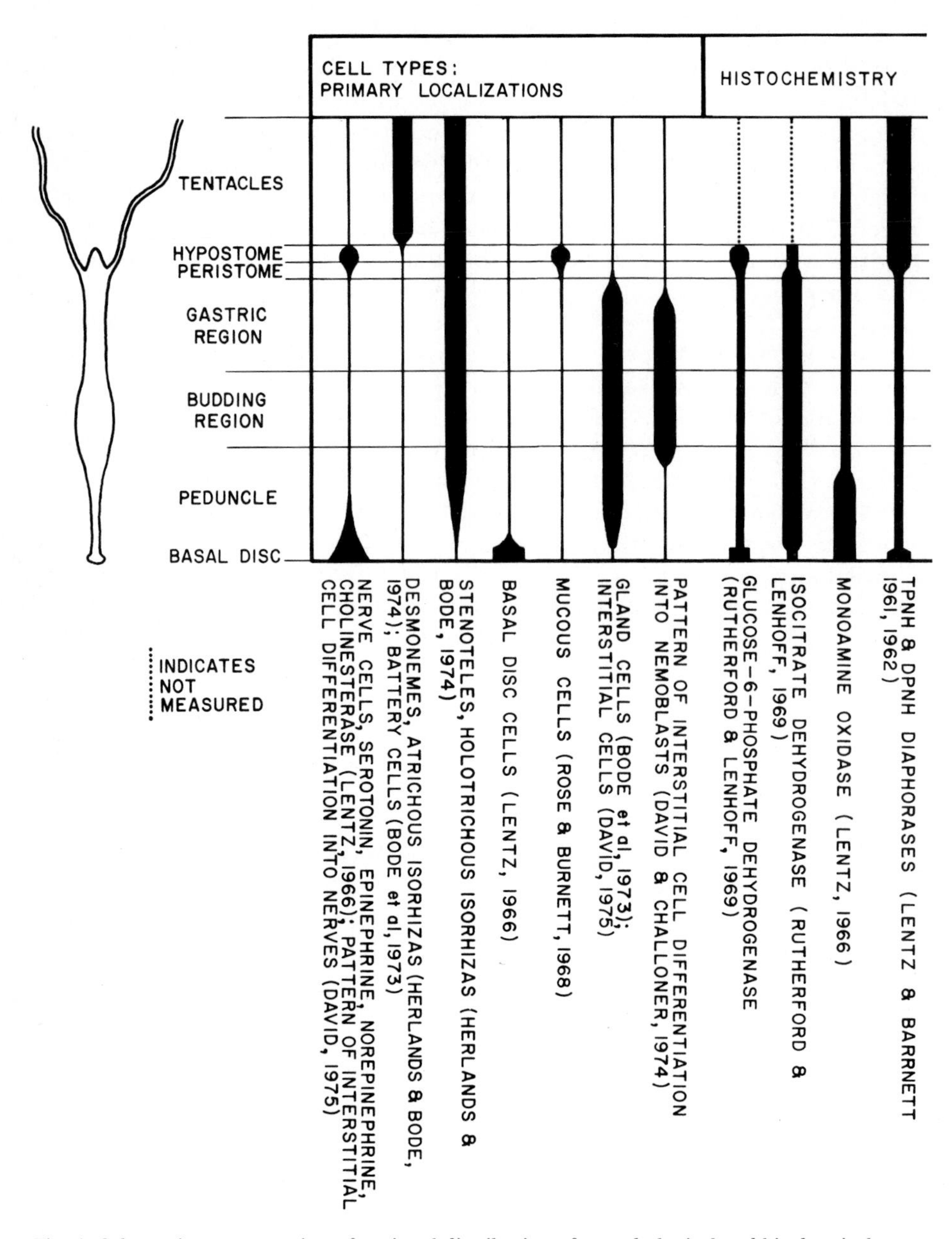

Fig. 4. Schematic representation of regional distribution of morphological and biochemical properties in Hydra. References to the original research are cited on the abscissa.

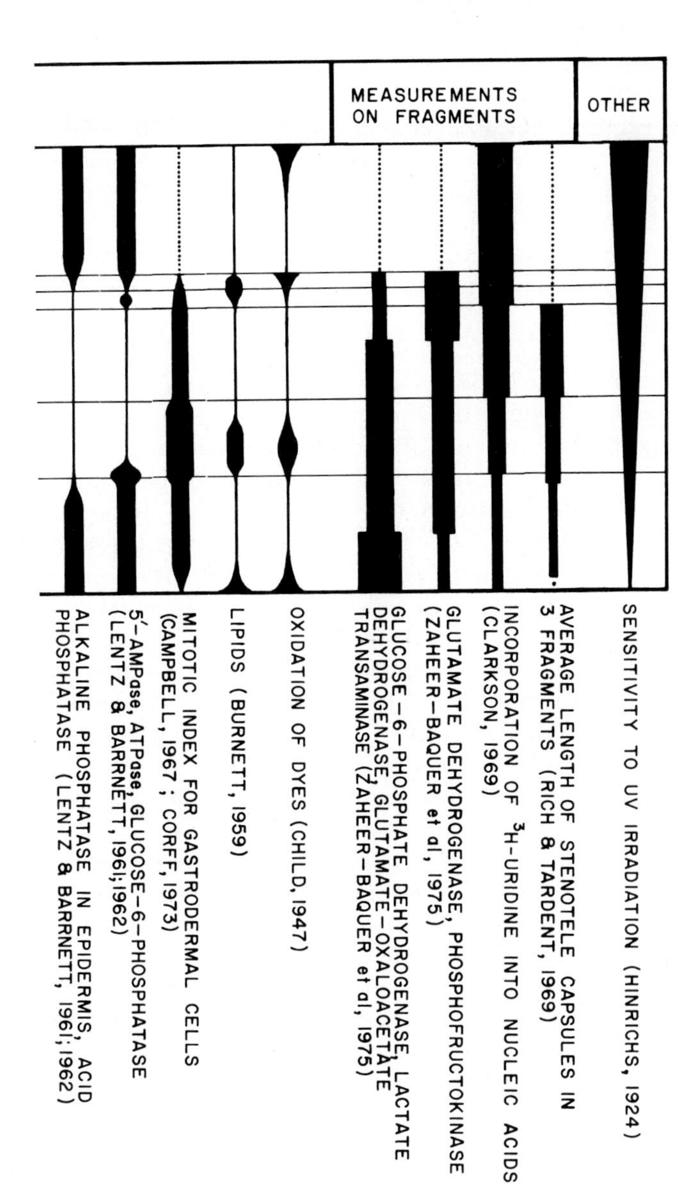
MEASUREMENTS ON FRAGMENTS
OTHER
SENSITIVITY TO UV IRRADIATION (HINRICHS, 1924)
AVERAGE LENGTH OF STENOTELE CAPSULES IN 3 FRAGMENTS (RICH & TARDENT, 1969)
INCORPORATION OF ³H-URIDINE INTO NUCLEIC ACIDS (CLARKSON, 1969)
GLUTAMATE DEHYDROGENASE, PHOSPHOFRUCTOKINASE (ZAHEER-BAQUER et al, 1975)
GLUCOSE-6-PHOSPHATE DEHYDROGENASE, LACTATE DEHYDROGENASE, GLUTAMATE-OXALOACETATE TRANSAMINASE (ZAHEER-BAQUER et al, 1975)
OXIDATION OF DYES (CHILD, 1947)
LIPIDS (BURNETT, 1959)
MITOTIC INDEX FOR GASTRODERMAL CELLS (CAMPBELL, 1967 ; CORFF, 1973)
5'-AMPase, ATPase, GLUCOSE-6-PHOSPHATASE (LENTZ & BARRNETT, 1961; 1962)
ALKALINE PHOSPHATASE IN EPIDERMIS, ACID PHOSPHATASE (LENTZ & BARRNETT, 1961; 1962)

implanted into a region which is not its original nearest neighbor, because the correct morphogen concentration there is already high. However, when the part duplicated by the graft is removed the morphogen concentration will become low and the region will be permitted to transdifferentiate new parts. Inhibition of disc regeneration by grafting another disc is also predicted by this model. A second disc implanted near the regenerating disc will inhibit recruitment of new cells to disc cells by raising the disc morphogen concentration too high. Finally, this model gives rise to the mosaic of biochemical and morphological properties in a convenient way.

2.2.3.3. Other models for regulation. Wolpert and his collaborators (1974) have extended ideas about gradients to Hydra. They postulate that cells initially learn their position from the local concentration of a diffusible molecule, S, and remember it in terms of a long term state, P. S is considered to be a linear gradient whose end points are set by a source in the hypostome and a sink in the basal disc. Whenever the concentration of S falls a given fraction below its original level remembered by P, as might happen for example at a site of surgery, a source is regenerated. This is assumed to happen most quickly in places where the concentration of S was originally the highest. In order to explain foot regeneration it is necessary to propose a source for another molecule at the proximal end. Since positional information is already specified, this gradient only serves a role in regeneration.

This model and others derived from it are very different from the regional model discussed in the preceding section because they propose that the patterning process is controlled by dominant regions or organizers located at one or both ends of the field. This hierarchy of control is not present in the regional model, where all the regions are codominant and interact on equal terms. The ends may have the misleading appearance of being dominant simply because they possess the most obvious morphological markers and because they are at the greatest distances from other parts of the animal.

There are several shortcomings to Wolpert's gradient theory, some of which also apply to other theories of this type. Firstly, the absence of a head or a foot (as found in certain mutants, and as a consequence of drug treatment) is difficult to reconcile with the loss of a source or a sink of positional information. Secondly, the model proposes mechanisms which may not be important for pattern formation and maintenance in intact organisms but apply only to regeneration. Thirdly, the theory fails to explain what governs the amount and territory of source and sink activity. Without regulation of these regions there will be problems in maintaining the proper gradient boundary conditions as a Hydra (or other tissue) grows and the source grows larger. Feedback mechanisms to accommodate this problem are complex. Finally, the model predicts that the tentacles would form a gradient of positional information analogous to that in the body column and respond to it like a body column. The regional model does not have these difficulties.

Newman (1974) has postulated the existence of two sources centered near the opposite ends of Hydra which secrete substances distributed in linear gradients.

When the head is excised the morphogen from the proximal source is expected to fall below a threshold (by leaking from the wound) and thus stimulate the distal source. As a result the distal end will recover its normal morphogen concentration. An interesting feature is that the hormonal product of one region stimulates the activity of another region when it falls below a certain level. This is an essential element of the regional model, although its form is slightly different. As the author acknowledges, there is no explanation for the distribution of the sources. The regional model solves this problem by making the different sources contiguous. Each source controls the size of its neighbor.

Gierer and Meinhardt (1972), building from the ideas of Turing (1952), have formulated equations which describe the nonlinear amplification of a prepattern of activator (A) and inhibitor (I) source concentrations. The place where the concentrations of the sources is high is selectively stimulated. This is accomplished through bimolecular activation of the source of A and monomolecular inhibition by A and I. As yet there are no known biochemical systems which act in this way. A is also lost slowly by diffusion. On the other hand, I stimulates its own production in a bimolecular reaction and inhibits it by a monomolecular mechanism. I is lost by diffusion. When these mechanisms act in concert a high concentration of A occurs at the highest point of source concentration and low concentrations occur elsewhere. This model generally agrees with the results concerning regeneration and inhibition of hypostomes as dominant centers, but offers no information for the patterning of other parts of Hydra. Additional models for pattern formation and maintenance have been reviewed elsewhere (Lesh-Laurie, 1973).

The properties of the regional model are quite different from the precepts of gradient theory. The regional hypothesis instead identifies each discrete cell community by a hormone and has these regions know their proper spatial relationships by allowing only particular pairwise associations.

2.2.3.4. The head morphogen. Lesh and Burnett (1964) and Lentz (1965) isolated a morphogen from Hydra which induces supernumerary heads and tentacles. More recently, Schaller (1973) and Schaller and Gierer (1973) have isolated a similar morphogen and have attempted to characterize it. Initial results indicate that it is a dialyzable molecule, probably a peptide, and is present in neurosecretory vesicles at the distal end of the animal. Schaller and Gierer demonstrated that it has a molecular weight of about 900.

The effect of crude extracts containing the morphogen on cell differentiation has been studied in a test system consisting of gastric annuli preloaded with high concentrations of interstitial cells. As well as inducing supernumerary tentacles (Lesh and Burnett, 1966), the morphogen also causes differentiation of mucous cells and nematoblasts (Lesh, 1970). Thus a morphogen which induces head morphology also seems to induce distal cells, a finding in general agreement with experiments which indicate that the morphogenesis of head structures may be a function of the underlying cell differentiation. This effect of the crude extract on cell differentiation is quite similar to the effect of high concentrations of sodium and calcium salts on whole animals, in which the gland cells turn into mucous cells and the interstitial cells all turn into nematoblasts (Macklin and Burnett, 1966). High concentrations of other ions do not have this effect. This raises the speculative possibility that the morphogen may exert its effect(s) by increasing the intracellular concentration of sodium and/or calcium. In this sense, the morphogen might be homologous to the peptides produced in the brains of vertebrates such as the octapeptide hormones vasopressin and oxytocin released from neurons terminating in the neurohypophysis of higher organisms. For example, in the toad bladder epithelium these hormones increase the permeability of the apical plasma membrane to sodium, which then enters the cells down a concentration gradient (Civan and Frazier, 1968).

In the vertebrate brain, the release of vasopressin is stimulated by acetylcholine and inhibited by norepinephrine (Barker, Crayton and Nicoll, 1971). Hydra head morphogen release may be under similar control. In Hydra, Lentz and Barnett (1963) have done work with neuropharmacological inhibitors which suggests that cholinergic functions are necessary for head formation. Head structure formation was stimulated by hemicholinium-3, decamethonium and a combination of eserine and methylcholine. These drugs caused multiple groups of tentacles to form in regenerating proximal halves. These treatments would be expected to temporarily increase the secretion of acetylcholine concentration or act as agonists. Hemicholinium blocks uptake of the transmitter, decamethonium depolarizes cholinergic postsynaptic membranes, and the eserine-methylcholine combination inhibits the hydrolysis of acetylcholine and serves as an agonist (see Goodman and Gilman, 1975).

A hypothetical scheme suggested by these findings is that acetylcholine might mediate head formation by triggering the release of a vasopressin-like hormone from peptidergic neurons, with the hormone acting to increase the permeability of susceptible cells to sodium and calcium. Cells with high intracellular sodium and calcium concentrations then form a head. On the other hand, the proximal portion of hydra may be induced by a hormone like prolactin which stimulates the Na^{+}–K^{+} stimulated, Mg^{2+} dependent ATPase and inhibits the permeability of plasma membranes to ions in other systems (Falconer and Rowe, 1975; Ensor and Ball, 1972).

2.3. Mosaic development

2.3.1. General properties

2.3.1.1. Relationship to regulation. Mosaic development is well-served by its name. The embryo or parts of the embryo appear to develop as a mosaic of cells which are irreversibly committed to their developmental fate (see Davidson, 1968; Kühn, 1971). Each cell's differentiation is an expression of its position in the embryo but has little or no dependence on interaction with other cells. Rather the cell's differentiation depends on the particular region of zygote cytoplasm and plasma membrane which it retains after cell division. Consequently, even the two cells which are produced by the first embryonic cleavage may have different fates and express these autonomously. The mosaicism of the embryo may be partially or entirely impressed on the egg itself (e.g., *Protophormia* and *Sciona*), or may result from the action of environmental stimuli on the egg (e.g., *Fucus*).

Mosaic development is not an aberrant or isolated form of positional specification but is integrated with other ways of specifying position. Phylogenetic groups of organisms whose embryos are predominantly mosaic are interspersed with others whose cellular position determination depends heavily on cell interactions (Kühn, 1971). The development of a single organism may involve both cellular interactions (field or induction processes) and mosaic determination. For example, the blastoderm of the *Drosophila melanogaster* embryo is a mosaic (Chan and Gehring, 1971; Bownes and Sang, 1974a,b), but later development of the imaginal discs (whose progenitor cells were determined in the blastoderm) is regulated by cellular interactions (Bryant, 1971; Nöthiger, 1972; Schubiger, 1971; Bryant, 1974). The development of the embryo of the leaf hopper, *Euscelis plebejus,* appears to be determined as a mosaic along the long axis of the egg and to be regulative across the transverse axis (Sander, 1971). The embryo of a frog depends heavily on cellular interactions in development but at least some of these (the invagination of the chordamesoderm) are determined by mosaicism of the plasma membrane and the associated cortical cytoplasm. This has been demonstrated in the very elegant experiments of Curtis (1962) who transplanted small pieces of plasma membrane (and associated cortex) between embryos. Finally an early experiment by Roux (1888) reveals the subtle boundary between experiments which demonstrate mosaicism and those which demonstrate regulation. Roux killed one of the two blastomeres of a two-celled frog embryo by pricking it with a hot needle. Only a half-embryo developed from the remaining blastomere. However, forty-five years later, Schmidt (1933) showed that an entire embryo could develop from the remaining blastomere if the remnant of the damaged cell was removed.

2.3.1.2. Specification of pattern after fertilization of the egg. Development of the zygote before cell division begins is frequently a time of considerable change which may be reflected in relatively subtle changes such as the development of a polarized current of ionic calcium, as in *Fucus,* (Robinson and Jaffe, 1975) or may be characterized by considerable movements of cytoplasm as in *Rana.* The final position of cytoplasmic components of the egg is often important in determining the subsequent course of development of cellular determination in mosaic organisms (Kühn, 1971). In some eggs cytoplasmic determinants can be moved by centrifugation, while in others the prepattern is very resistant to attempts to disturb it by centrifugation. The pattern of distribution of localized cytoplasmic constituents in the egg (such as ascorbic acid granules in the egg of *Aplysia*), can be moved by centrifugation but is restored when the eggs are removed from the centrifuge (Ries, 1939). In the fly, *Protophormia,* the pattern of larval cuticular segments is incomplete when the egg is laid as Herth and Sander (1973) have shown by separating the preblastoderm embryo into 2 independent segments. The complete pattern does not develop, however, until very near the time at which the blastoderm forms.

2.3.1.3. Experimental attempts to localize morphogens. Experimental analysis of the basis of action of compounds localized in mosaic embryo cytoplasm has just begun. One approach is to use mutants which lay defective eggs. Although a relatively large number of such mutants have been isolated in various organisms, only a few have been analyzed biochemically. Two mutants of *D. melanogaster* which lay defective eggs have been investigated in some detail. In one of these, *Deep Orange*, it has proved possible to rescue the defective eggs by the injection of cytoplasm from normal eggs. The identity of the component(s) present in normal egg cytoplasm is unknown (Garen and Gehring, 1972; Garen, personal communication). More progress has been obtained with the mutant, *rudimentary*, which can be rescued by injection of normal egg cytoplasm or by the injection of pyrimidines into the defective egg (Okada et al., 1974). The pyrimidine deficiency affects the development of different organs to different degrees, but the basis for this differential effect is unresolved.

Microinjections of cytoplasm have been used to analyze the origins of the primordial germ cells of *D. melanogaster.* When pole plasm, the portion of the egg cytoplasm which would normally be distributed to the primordial germ cells, is injected into an egg (even into an abnormal site in the egg) it allows the nuclei surrounded by it to differentiate into primordial germ cells (Illmensee and Mahowald, 1974; Warm, 1975). The identity of the component responsible for this determination is unknown but it has been suggested to be a species of RNA. RNA rich granules are found in the pole plasm and irradiation of the pole plasm with ultraviolet light prevents the differentiation of the primordial germ cells. The action spectrum for inactivation is that of a nucleoprotein (Illmensee and Mahowald, 1974; Warm, 1975).

2.3.2. Plasma membrane lipids and mosaicism

Little is known concerning the biochemical mechanism(s) of cytoplasmic localization. However, much of the thinking about this subject focuses on the possibility that the plasma membrane of the embryo may contain at least some of the information which specifies the fate of developing cells. In this section we shall discuss some possibilities for the role of lipids in this pre-pattern.

The lipid composition of the plasma membrane can be separated into relatively discrete phases in the plane of the membrane. This process in turn is potentially subject to cellular regulation by the local concentration of cations such as Ca^{2+} (Shimshick and McConnell, 1973; Ohnishi and Ito, 1974; Papahadjopoulos and Poste, 1975) and the regulation of the phosphorylation and dephosphorylation of phosphatidylinositol is likely to be quite important in situations of this sort (Buckley and Hawthorne, 1972). Regular arrays of lipids and proteins in membranes can also probably be produced by mutual interactions (cf. Henderson and Unwin, 1975). Lateral separation of discrete lipid phases could lead to at least local differentiations of the plasma membranes of developing cells which could in turn affect the physiology of developing cells in a number of ways. The most obvious of these are variations in permeability to inorganic ions (Watlington and Harlan, 1969) or nutrients (Cox et al., 1975) and alterations in the activity of enzymes associated with the plasma membrane (for review see Skou, 1975; Schwartz et al., 1975). Finally, the lipids of the plasma membrane may also regulate the assembly of intracellular structures. The assembly of the head of the bacteriophage T4 appears to depend upon some property of the *E. coli* inner membrane (Simon, 1972). The assembly of the head may be prevented by mutations of the host bacteria. One of these host mutants has a plasma membrane of unusually high viscosity which contains an altered array of fatty acids and phospholipids. The mutant allows assembly of the head at high temperatures and the nature of the host defects suggest that the viscosity of the lipid phase of the membrane may be important for the assembly of the bacteriophage head (Simon et al., 1975). It is easy to conceive of analogous relationships between the plasma membranes of eukaryotes and the assembly of microtubules and microfilaments of actin and myosin.

2.3.3. Actomyosin and patterns

The localization of any of a wide variety of proteins on the plasma membrane could also provide a plausible basis for mosaicism. However, we shall limit our discussion only to the localization of the proteins associated with actomyosin microfilaments. The experiments of Arnold and Williams-Arnold (1974) have shown that cytochalasin B is able to disrupt the "pre-pattern" of a cephalopod embryo, suggesting that actin-containing microfilaments might be involved in the positioning of the localized information in this embryo. Microfilaments are a cellular system which could provide a unique and highly flexible basis for both the stability of a pre-pattern and the ability to restore an experimentally disrupted pre-pattern. As discussed below, they also have the ability to specifically adsorb various intracellular proteins and changes in the association of such proteins with microfilaments might provide a system for modulating aspects of cell metabolism.

2.3.3.1. Localization of actin and myosin in vivo. The "muscle" proteins, actin, myosin, and tropomyosin, are found in a wide variety of eukaryotic cells in microfilaments located in the cortex and associated with the plasma membrane (Ishikawa et al., 1969; Mabuchi, 1973; Goldman, 1975; Lazarides, 1975; Röhlich, 1975). The disposition of these microfilaments differs among cell types (Goldman, 1975; Lazarides, 1975; Sanger, 1975), and is altered by a variety of stimuli including attachment of the cell to the substratum (Goldman, 1975). Actin, tropomyosin, and α-actinin are also associated with the perinuclear region of the cell (Lazarides, 1975) and with nuclear chromatin (Douvas et al., 1975) while actin is also found in the nucleolus (Sanger, 1975).

Actin, α-actinin, and tropomyosin seem to play separate but related roles in cells. α-actinin is localized at the focus of attachment of radiating nets of actin fibers, some of which also contain tropomyosin (Lazarides, 1975). In these sites it might fulfill a function analogous to that of the Z disks of muscles which contain α-actinin and tropomyosin and which provide points of attachment for actin fibers of opposite polarities (Schollmeyer et al., 1974; Pepe, 1975). The regularity of pattern and placement of these fibers suggests that a mosaic of "Z-line equivalents" perhaps containing α-actinin in the plasma membrane of the zygote could provide a molecular basis for the migration of nuclei to the cortex of the insect egg which occurs in blastoderm formation, for asymmetric cleavages, and for a variety of similar phenomena.

In addition it is easy to conceive a molecular basis for the changes in the pattern of a cleavage or of the distribution of localized material in the egg which often occurs during the development of embryos. This is postulated to result from the removal of "Z-line equivalents" which is triggered during the cell cycle. We suggest that this removal and possibly the exposure of new sites for actinin insertion might be controlled by an intracellular neutral protease regulated by a cellular chemical messenger (perhaps a combination of Ca^{2+} and cAMP). The Ca^{2+}-activated protease of myofibrils (Reddy et al., 1975) is a model for such an activity.

2.3.3.2. Membrane-associated cytoskeletal elements and topography of plasma membrane components. Evidence obtained in many laboratories over the past five years has shown that plasma membrane components are able to diffuse laterally within the plane of the membrane (reviews, Edelman, 1974; Edidin, 1974; Singer, 1974; Papahadjopoulos and Poste, 1975). The mobility of different membrane components appears to vary markedly. Phospholipids diffuse at a high rate, though this may be modified in specific regions or "domains" within the lipid bilayer by such phenomena as phase separation (vide supra). Certain integral membrane proteins and glycoproteins are also able to move laterally within the plasma membrane, though at rates slower than phospholipids while other integral proteins appear to be relatively immobile or "frozen" within the membrane. The differing mobilities of different classes of integral protein and glycoproteins within the plasma membrane, and even within specific regions of the cell surface, suggests that the cell may possess a control mechanism to limit the mobility of particular membrane components in order to maintain a specific topographic pattern(s) of molecules on the cell surface.

There is a growing body of evidence which suggests that the mobility of certain integral membrane proteins and glycoproteins may be controlled by their linkage to membrane-associated microfilaments and microtubules on the inner (cytoplasmic)face of the plasma membrane (reviews, Edelman, 1974; McConnell, 1975; Poste and Weiss, 1976). The major functional implication of this type of structural organization is that internal cytoskeletal elements such as microfilaments could exert *trans*-membrane control of the mobility of surface components on the outer face of the plasma membrane carried on integral membrane proteins. In addition, the fact that mobility of certain membrane components is restricted, often within specific regions of the cell surface, raises the interesting possibility that such restrictions might form the basis for stabilizing highly ordered topographic displays or "patterns" of molecular determinants on the cell surface. Such displays may well be highly specific, differing on different cell types, on different regions of the same cell and might also change during different stages of cellular activity. In short, such patterns, could be fundamental to the inherent specificity of cell surface organization. (See McMahon, 1973).

The existence of cell-specific or phase-specific topographic patterns of components on the cell surface could constitute a form of "molecular braille" for mediating various cell contact and recognition phenomena, including the transmission of positional information. Finally, the system of *trans*-membrane control of cell surface components by microfilaments and other internal cytoskeletal elements also permits topographic redistribution of surface components in response to various environmental stimuli. Reversible changes in the linkage of particular classes of plasma membrane components with membrane-associated microfilaments could occur in response to the binding of such agents as mitogens, serum factors, hormones, neurotransmitters etc. to the cell surface and thus provide a mechanism for rapid and reversible modulation of cell surface properties. While many aspects of these proposals are presently speculative, the potential implications of topographic redistribution of membrane components as a mechanism for altering cell surface properties are of sufficient importance to dictate that this topic has emerged as an area of intensive research activity.

2.3.3.3. Adsorption of "soluble" enzymes to intracellular microfilament networks. The placement and arrangement of an actomyosin network might also provide the basis for a much wider series of phenomena. Actin is associated with the plasma membrane of cells ranging from the amoebae of *D. discoideum* (Hoffman and McMahon, 1975) to the mammalian erythrocyte (Tilney and Detmers, 1975). Actin is now known to be the inhibitor of DNase I which has been found in the cytoplasm of many cells (Lazarides and Lindberg, 1974). Actin binds DNase I strongly and at the same time completely inhibits its activity. While it does not appear likely that DNase I is involved in the regulation of gene activity, this reaction may provide a model for a more general class of mutual interactions between proteins which bind to DNA and to actin.

The elegant work of Pette and his collaborators has provided some insight into the interaction between F-actin and soluble enzymes. In striated muscle, the enzymes glycogen phosphorylase, phosphoglucomutase, glucose-phosphate isomerase, triosephosphate dehydrogenase, and perhaps lactic dehydrogenase are associated with the I band which contains F-actin (Sigel and Pette, 1969). Several of these enzymes have also been shown to adsorb to pure F-actin or F-actin-troponin-tropomyosin complex in vitro (Clarke and Masters, 1975). The association of aldolase with actin has been studied in considerable detail and may provide a model for the association of other "soluble" enzymes with F-actin, etc. Aldolase can be removed from its binding site in the I band with 0.1 M phosphate buffer and then replaced with added aldolase in a KCl–$MgCl_2$ solution (Arnold et al., 1969). The properties of aldolase change when bound to actin, its V_{max} being increased by two and the K_m for FDP increased by 10-fold (Arnold and Pette, 1970). The increase in K_m may be sufficient to functionally inactivate the aldolase in vivo (Clarke and Masters, 1972). The binding is decreased by substrate and by ATP at physiological levels and is very sensitive to release by phosphate ion (Arnold and Pette, 1970). Some evidence suggests that the extent of binding of aldolase to the actin in vivo might be related to the functional state of the cell (Starlinger, 1967). In addition, the stability of aldolase may be affected by the binding. In vitro experiments show that the sensitivity of aldolase to trypsin or chymotrypsin degradation is enhanced by the presence of actin (Dedman et al., 1975). This sensitization to degradation is rather specific since bovine serum albumin does not replace actin in sensitizing aldolase nor does actin stimulate the proteolytic degradation of phosphofructokinase (Dedman et al., 1975) an enzyme which also appears to bind an actin.

Hexokinase also exhibits properties which are analogous to those described for aldolase. However, its site of binding appears to be distinctly different, sites on the outer membrane of the mitochondrion (Rose and Warms, 1966; Karpatkin, 1967; Wilson, 1968; Sigel and Pette, 1969; Tuttle and Wilson, 1970). Like aldolase the activity of the enzyme seems to be altered by binding and the equilibrium between bound and unbound enzyme seems to be controlled by the interplay of inorganic salt and metabolite concentrations (Rose and Warms, 1966; Wilson, 1968; Karpatkin, 1967; Tubble and Wilson, 1970).

If we assume that the pattern of intermediary metabolism in a cell may have an important influence on its differentiation (see Section 3) then selective adsorption and desorption of the enzymes of intermediary metabolism from surfaces such as microfilaments within the cell (perhaps in association with the assembly and disassembly of these structures) may have a role in regulating development. It is also conceivable that some hormones might exert their effects by modulating such a system.

The actomyosin microfilaments provide a surface for adsorption which has many advantages. They not only allow for a pattern of static placement of adsorbed components but also allow the possibility of movement in response to extracellular stimuli. For example, an increase in permeability of the plasma membrane to Ca^{2+} or a redistribution of molecules on the external surface of the plasma membrane could each provide a means for altering the functional activity of adsorbed enzymes by creating changes in the microfilament system.

An asymmetrical distribution of microfilaments within a cell would mean that different pathways of metabolism might predominate at different sites within the same cell. For example, glycolysis might predominate at one end of a cell and the hexose monophosphate shunt at the other. The asymmetric distribution of enzymes might be amplified upon cell division with a consequent asymmetric distribution of proteins to the progeny cells. Finally, the binding of adsorbed proteins to the microfilaments might be regulated by a variety of stimuli including changes in membrane permeability, covalent modification of the microfilaments or the adsorbed proteins (for example by phosphorylation), and by changes in the microfilament network induced by cell-to-cell contact or by binding of specific hormones, mitogens etc. to receptors on the cell surface.

Two systems in which these proposals on the association between contractile proteins and enzymes of intermediary metabolism might be profitably investigated are the sea urchin egg and mammalian cells transformed by tumor viruses.

Following fertilization, the earliest biochemical change in the sea urchin zygote is apparently an activation of carbohydrate metabolism, particularly the hexose monophosphate shunt (Isono, 1963), which is the predominant path of glucose oxidation in the early embryo (Krahl, 1956; Bäckström et al., 1960; Isono and Yasumasu, 1968). A number of metabolic intermediates rapidly change in concentration (see Giudice, 1973, for review). Triosephosphates, for example, increase by 10 to 20-fold within minutes (Aketa et al., 1964). Fructose-1,6-bisphosphate aldolase (Ishiwara, 1959), glucose-6-phosphase dehydrogenase (Isono, 1963) and glycogen phosphorylase (Giudice, 1973, p. 184) are each released from an insoluble fraction of the egg into the soluble fraction on fertilization. It is tempting to consider that fertilization which is known to trigger a number of changes in the microfilament-containing cortex such as the release of cortical granules (Runnström, 1966) might also release sequestered enzymes from actomyosin microfilaments and thereby trigger some of the biochemical changes which immediately follow fertilization. As discussed above, two of the released enzymes are known to be adsorbed to actin in other systems.

Mammalian cells transformed by simian virus 40 have been shown to contain a rather diffuse distribution of actin and myosin in their cytoplasm and lack the ordered bundles of microfilaments characteristic of untransformed 3T3 cells (Pollack et al., 1975). Neoplastic cells have long been known to have more active glycolysis than most normal cells (excepting nervous tissue). This problem which was initially pursued by Warburg and many others (see Weinhouse, 1955, for review) has become the subject of renewed investigation. Several investigators have shown that cells transformed by both RNA- and DNA-containing tumor viruses take up 2-deoxyglucose more rapidly and have more rapid glycolysis than similarly treated normal cells (Singh et al., 1974a). Pool size crossover-analysis of such cells indicates that hexokinase, phosphofructokinase and pyruvate kinase activities are greater in the transformed cells than in untransformed cells (Singh et al., 1974b). Analysis of temperature sensitive mutant-transformed cells indicates that the increase of enzyme activities, except for lactate dehydrogenase, is closely correlated with the transformed phenotype (Singh et al., 1974a). It may therefore be worthwhile to determine whether there is a causal relationship between the disturbance of the assembly of microfilaments, enhanced enzyme activities and the activation of glycolysis in transformed cells. The investigations discussed above have been pursued in different lines of cells so it is imperative that the same cell line be examined for metabolic alterations and disruption of microfilament bundles. Nevertheless, it is noteworthy that three of the enzymes whose activity is increased are phosphofructokinase (Sigel and Pette, 1969) and lactate dehydrogenase or pyruvate kinase (Sigel and Pette, 1969; Clarke and Masters, 1974), all of which are associated with actin in other cell types.

We are more inclined to believe that both effects (activation of glycolysis and disappearance of bundles of microfilaments) might result from an underlying alteration in cytoplasmic composition. One immediate possibility involves possible alterations in the cellular content of inorganic ions. The systems outlined above are not well enough defined to discuss in any detail, but an increase in the intracellular concentration of the cations together with inorganic phosphate might well influence the integrity of microfilament systems and thus, in turn, the activity of certain classes of enzymes.

2.3.4. Mosaicism – Experimental strategies

There are several experimental approaches that could be used to study the role of actin-containing microfilaments in mosaic specification. One is the use of inhibitors. Arnold and Williams-Arnold (1974) have already shown that cytochalasin B disrupts the pattern of information in cephalopod development. It might therefore be of interest to study the effect of a complementary inhibitor, phalloidin, on mosaicism. This inhibitor is produced by the fungus, *Amanita phalloides* and stimulates the formation of microfilaments containing actin, and stabilizes them once formed (Wieland and Govindan, 1974). In addition, it blocks the inhibition of DNase I by actin (Schäfer et al., 1975) so it might be of use in investigating the relationship between adsorption of proteins to microfilaments and development.

Injection of antibodies against actin and myosin into eggs is also an appealing strategy for examining the potential importance of the contractile proteins in mosaicism. This approach is open ended and need not be limited to these proteins. For example, antibodies raised against various plasma membrane components could be used. This method of approach is appealing because it identifies the specific components important in pattern specification, allows them to be purified by immunoadsorption, and also allows their localization within the egg.

3. Metabolic control of development

3.1. Metabolic messengers and coding

Morphogenetic fields, induction and mosaicism provide convenient categories in which to place the phenomena of position determination. However, as discussed here, and elsewhere (McMahon, 1974), it is appealing to consider the possibility that only superficial differences exist between these phenomena and that they rest upon similar biochemical foundations. We will therefore briefly discuss some general ideas regarding the possible nature and organization of the biochemical processes which regulate the differential activity of genes and other phenomena characteristic of development such as cellular movement and multiplication.

The differentiation of many simple eukaryotes is regulated by perturbation of metabolism. Differentiation of *D. discoideum* (Bonner, 1967), *Chlamydomonas reinhardi* (Sager and Granick, 1954), and various fungi (Hawker, 1957) is triggered by starvation for organic or inorganic nutrients and can be reversed or prevented by readdition to the medium of the critical components (Sager and Granick, 1954; Kates and Jones, 1964; Lee, 1972). The study of the simple differentiation of prokaryotes has yielded similar results (Freese et al., 1974; Wireman and Dworkin, 1975). It is possible that the basic mechanisms of development, conserved in evolution, rest upon simple metabolic signals (Child, 1928; McMahon, 1973; McMahon, 1974; Seglen, 1974; Tomkins, 1975).

3.1.1. Classes of metabolic messengers

Regulation of metabolism and possibly of development can be approached by considering both the general nature of the regulatory molecules and the probable nature and duration of their effects on macromolecules and the cell.

Two general classes of metabolic messengers can be recognized: metabolites per se (e.g., fructose-1,6,bisphosphate, phenylalanine, etc.); and molecules which we have previously called chemical messengers (e.g., inorganic ions, cyclic nucleotides, neurotransmitters, etc.). Individual metabolites seem to have effects only on a restricted array of macromolecules and to have effects which are rather limited in time. Since these molecules are inextricably interwoven into metabolism, the degree to which they can vary as metabolic signals is strictly limited. But this variation does not apply to chemical messengers which are derivatives of metabolites (and inorganic ions). They can act as indicators of the average state of a section of metabolism (e.g., carbohydrate metabolism, nitrogen metabolism, etc.) or of the state of the cell over a period of time. These subjects have been discussed more fully by Seglen (1974) and Tomkins (1975).

3.1.2. Levels of metabolic coding

It is useful to classify the effects of metabolic messengers according to the nature and persistence of their effects on macromolecules. Three potential levels of control can be considered. These differ not only according to the chemical nature of the regulatory event, but also according to the length of time the regulatory event may potentially affect the cell.

The first level of the hierarchy includes all interactions between metabolic messengers and macromolecules which do not result in covalent modification of the macromolecule. Gathered together in this class are such diverse processes as the interaction of inducers with repressors, allosteric regulation, competitive inhibition of enzymes, etc. Regulation at this level is generally remembered by the cell for only seconds or minutes (see Frieden, 1970) and therefore persists for the duration of the change in concentration of the messenger. In fairness it must be pointed out that even interactions such as these may generate metastable states, in the right environmental conditions, if a system with some positive feedback is involved. Thus *E. coli* may maintain the induced state of the *lac* operon while grown in a concentration of inducer which is below the threshold for induction of uninduced cells because of the efficiency of the β-galactoside permease for accumulating inducer in the cell (see Novick and Weiner, 1957, for discussion).

The second level of regulation is covalent modification of a macromolecule in response to change in concentration of a metabolite as chemical messenger. The addition of new functional groups to macromolecules by ATP, GTP, S-adenosylmethionine, NAD, and acetyl coenzyme A and their reversal are examples of this class (see Holzer and Duntze, 1971, for review).

Since the discovery of the metabolic importance of cyclic AMP by Rall and Sutherland (Sutherland and Rall, 1960; Rall and Sutherland, 1961) studies of protein phosphorylation have dominated the study of covalent modification of macromolecules. This system provides an archetype with which to consider the general possibilities for this level of regulation. Protein phosphorylation may be regulated by the supply of ATP (Siess and Wieland, 1975; Carlson and Kim, 1974; Schlender and Reiman, 1975), or by chemical messengers such as cyclic AMP or cyclic GMP (see McMahon, 1974, for review). No kinases activated by metabolites have been described, but the possibility that kinases are regulated by the concentration of intermediates of carbohydrate metabolism such as phosphoenolpyruvate should be considered seriously.

Another important regulatory mechanism may prove to be the selective removal of parts of macromolecules. Partial proteolysis could be such a mechanism. The work of Krebs and others has shown that the site of phosphorylation can be removed from several enzymes, regulated by phosphorylation, without inactivating the enzyme. This limited proteolysis is instead functionally equivalent to dephosphorylating the enzyme (Fischer et al., 1959; Nolan et al., 1964; Bergstrom et al., 1975). Phosphorylase kinase is an exception to this rule. A calcium-dependent protease cleaves the site of phosphorylation and irreversibly activates the phosphorylase kinase (Huston and Krebs, 1968). So in this case the effect of the proteolysis is equivalent to phosphorylating the enzyme. Other types of protein-associated regulation are affected by proteolysis. Mild proteolysis removes the K^+ regulatory site from fructose-1,6-biphosphatase but does not inactivate the enzyme (Colombo and Marcus, 1973).

Events accommodated within this second level of regulation can be considered as providing a short term biochemical memory. They can potentially affect cellular functions for hours after a triggering stimulus, though even if the modification is not reversed by a countervailing stimulus, the modified molecule will be degraded or diluted by an increase in cellular mass. Therefore it seems unlikely that a stimulus would be "remembered" for more than one or two cell generations.

Regulation of the activity of genes by such a mechanism has only been demonstrated clearly in one case. Zillig and his collaborators have shown that the transcription of the "early" genes of bacteriophage T7 is terminated when *E. coli* RNA polymerase is phosphorylated by a protein. The structural gene for the protein kinase is itself an early gene of the bacteriophage. The activity of the kinase is inhibited by high (1 mM) concentrations of cyclic AMP (Zillig et al., 1975). A "bootstrap" mechanism like that used by T7 for regulating the transcription of its early genes is an attractive mechanism for the regulation of gene activity in development of eukaryotes by metabolic messengers. Thus at successive stages of development, level two regulatory enzymes might be produced which, as a result of changes in the concentration of chemical messengers, would control the expression of later genes, including the structural genes which code for new enzymes that also operate at this second level (McMahon, 1974).

The kinds of regulation described above are quite familiar. However, we wish to postulate the existence of self-perpetuating modification of macromolecules, a process which we shall refer to as level three regulation. Although no such processes have yet been demonstrated to regulate either metabolism or development, we believe this kind of long-term biochemical memory will eventually be important at least in development.

There are two documented biochemical systems which exemplify a level three system. The *E. coli* modification-restriction system provides one model. One strand of methylated DNA in a double-stranded molecule promotes the methylation, rather than the endonucleolytic cleavage, of the other strand in the presence of S-adenosylmethionine and restriction endonuclease (Meselson and Yuan, 1968; Vovis et al., 1974). Therefore the lifetime of the methylated state of DNA is potentially infinite. However, the modification can be erased if DNA replicates in the absence of sufficient S-adenosylmethionine to promote methylation or in the absence of the enzyme. In fact, bacteriophage DNA is not efficiently modified when replicated in hosts whose supply of methionine is limited (Arber, 1965).

Type II protein kinase from cardiac muscle is an example of similar importance. The activated catalytic subunit of this kinase can phosphorylate the regulatory subunit of the enzyme. This phosphorylation increases the affinity of the catalytic subunit for cyclic AMP and decreases its affinity for the catalytic subunit (O. Rosen, personal communication). Both of these effects increase the probability that the enzyme will be dissociated and therefore activated. A similar mechanism which could lead to a self-perpetuating activated protein kinase (or other enzyme with analogous effects) is easy to envisage. After a single pulse of cyclic AMP at the appropriate time it could perpetuate its modified state through an unlimited number of generations and so remember the stimulus. The activated state could be potentially erased by a phosphoprotein phosphatase or by a reversal of the phosphorylation catalyzed by the kinase itself (Rosen and Erlichman, 1975; Rosen, personal communication).

Type three regulation provides a class of regulatory mechanisms which are potentially reversible though normally very stable. These generally seem to be the properties of the determined state.

3.1.3. Modification of macromolecules by methylation

Several metabolites which donate modifying groups to macromolecules have been mentioned above. One of these, S-adenosylmethionine, which has a wide range of acceptor macromolecules, provides an example of our current state of understanding of regulation via these simple modifications.

The cells of *E. coli* use methylation to label their DNA so that foreign DNA such as that of invading viruses, may be recognized and destroyed. The same enzyme may participate in either the methylation of DNA or its endonucleolytic cleavage as mentioned above. S-adenosylmethionine is required for both functions (Meselson and Yuan, 1968; Meselson et al., 1972; Yuan et al., 1975). Several investigators have recently proposed that similar modification-restriction mechanisms might produce selective silencing of eukaryotic DNA in development (Holliday and Pugh, 1975; Riggs, 1975; Sager and Kitchin, 1975). Two other functions of DNA methylation are possible. The expression of structural genes might be altered permanently by modifying regulatory genes by methylation. Methylation of the bases in a structural gene might provide a means of altering the coding properties of a structural gene itself. Alkylation of the bases via methylation could provide a mechanism of controlled mutagenesis of structural genes in development thereby altering the amino acid sequence of their protein products. The production of the variable regions of immunoglobulins might result from a somatic mutagenesis of this type. However, although the pattern of methylation of DNA changes during the development of the sea urchin embryo (Grippo et al., 1968), these changes have not been shown to be responsible for the regulation of gene expression during development.

A number of enzymes catalyze the methylation of proteins, including ribosomal and chromosomal proteins (Reporter, 1972; Kim, 1974; Cantoni, 1975) and polypeptide hormones (Diliberto and Axelrod, 1974). The pattern of histone methylation varies during the cell cycle suggesting these modifications may have a role in modifying chromosome structure or function (Borun et al., 1972). However there is still no unequivocal demonstration of a change in the function of a protein caused by its methylation.

The study of methylation of RNA has provided a great detail of information, including some which clearly indicates a functional role for the methylation. Helser, Davies and Dahlberg (1971) have shown that resistance of *E. coli* to the antibiotic kasugamycin is associated with a modification of the 16S ribosomal RNA. Specifically, two adenines in a hexanucleotide fragment of the 16S RNA are not methylated. Lai and Weisblum (1971) have shown that an acquired resistance to erythromycin by adaptation of *Staphylococcus aureus* is associated with the appearance of N^6-dimethyladenine in the 23S RNA. These experiments indicate that methylation of the bases of ribosomal RNA can be important in the function of bacterial ribosomes. The experiments of Reeves and Roth (1975) suggest that methylation of tRNA of *Salmonella typhimurium* may regulate the range of codons which it recognizes. Equivalent information is not available for eukaryotes.

The mRNAs of eukaryotes and of some of their viruses have recently been demonstrated to be methylated at a number of sites. Methylation of at least one site is clearly important for function of the mRNA. Many eukaryotic mRNAs are capped with a methylated guanine nucleotide attached to the 5′ end with reverse polarity (Perry and Kelley, 1974; Abraham et al., 1975; Furuichi et al., 1975; Levi and Shatkin, 1975; Perry et al., 1975). Methylation of this guanine has been shown to be dependent upon S-adenosylmethionine and to be a requirement for initiation of protein synthesis by at least some mRNAs (Both et al., 1975; Muthukrishnan et al., 1975). However this modification may not be required for the translation of all messengers since polio virus RNA apparently is not modified in this way (unpublished work cited in Nuss et al., 1975). The pattern of methylated bases varies among different mRNAs (Adams and Cory, 1975). For example globin mRNA (which is composed of the messengers for two different globin chains) has two different classes of 5′ termini (Perry and Scherer, 1975). It is obviously very attractive to consider the possibility that changes in the concentration of S-adenosylmethionine and changes in the spectrum of methylating enzymes can regulate the initiation of protein synthesis by different classes of mRNA during development.

3.1.4. Some other kinds of covalent modification

It is impossible to discuss the full array of metabolites which donate modifying groups to macromolecules in this limited space. However, two metabolites which may be important in gene regulation in eukaryotes can be briefly mentioned. One of these is NAD. Cellular levels of NAD could possibly be monitored (and used to regulate enzyme activity) continuously by molecules in the cell which are analogous to diphtheria toxin, an enzyme which attaches to receptors on the surface of a sensitive cell, is transferred to the cytoplasm, and which then transfers an adenosine diphosphoribose group from NAD (but not from NADH, NADP, or NADPH) to elongation factor-2 on the ribosome. This inactivates protein synthesis. Inactivation of protein synthesis is a rather drastic effect but modulation of the number of active ribosomes in the cell might be catalyzed by an analogous factor. The reaction catalyzed by diphtheria toxin is reversible, although cellular concentrations of NAD drive it to completion (see Collier, 1975, for review). An analogous reaction with a more favorable equilibrium constant could serve a regulatory role. A number of reactions are known in eukaryotic cells in which NAD acts as a donor of the adenosine diphosphoribose group to proteins (see Harris, 1973 and Anonymous, 1975 for reviews; Ueda et al., 1975; Smith and Stocken, 1975). Many of these reactions are not microscopically reversible but a thorough search for such reactions has not been made.

Acetyl-coenzyme A is also a well known donor of modifying groups to proteins (Johnson et al., 1973; Ruiz-Carrillo et al., 1975; Libby, 1973; Racey and Byvoet, 1972; Suria and Liew, 1974). However, practically nothing is known of the consequences of these modifications. Both the concentrations of NAD and of acetyl-coenzyme A in the cell can be modulated in a variety of ways including a shift of the metabolism of glucose from the hexose monophosphate shunt to glycolysis. Changes in the predominance of one path over the other might affect protein modifications which use these molecules as precursors via mass action.

It is interesting to consider the possibility that these other kinds of modification of macromolecules are regulated by their own set of chemical messengers. Such messengers have not been demonstrated, but, believing that they exist, we might assume that as cyclic AMP and cyclic GMP are related to ATP (the donor of modifying phosphate groups to proteins), these hypothetical chemical messengers might be derivative of the modifying metabolite. Thus for example spermine (or spermidine) or methylated adenines might be chemical messengers which regulate methylation. Biosynthetic products of acetate metabolism such as prostaglandins or steroids and related compounds, such as gibberellic acid or isopentenyl adenine, might regulate acetylation and so on. Alternatively, a molecule such as taurine might serve as a chemical messenger for acetylation.

3.2. *Chemical messengers and metabolites: synonymy and complementarity*

The suggestion that metabolism, mediated through chemical messengers and metabolites, regulates cell function and development may seem an attempt to propose a system of overwhelming complexity for the control of development. We do not believe that this is the case, but believe instead that an order underlies the regulation of metabolism and development. We believe that the basis of this order rests in a small number of relatively discrete metabolic states.

There may be two relationships in metabolic control which we propose to call synonymy and complementarity. Synonymy of chemical messengers and metabolites implies that these compounds have similar metabolic "meaning" (i.e., they tend to increase in concentration together and to exert similar effects on the cell). Complementarity of messengers implies that they, in general, have complementary effects on the cell. For example, the proposal of Goldberg et al. (1973) that cyclic AMP and cyclic GMP have opposing effects seems to be true in many, though not all, situations (Kneer et al., 1974).

Assuming that this is the case, groups of synonymous or complementary messengers can be assembled by considering the effects of potential chemical messengers on cell physiology and biochemistry and segregating them into classes on the basis of similarity or complementarity of their effects. Therefore we propose that cyclic GMP, K^+, Mg^{2+}, ATP, and GDP are associated in general with one state of the cell (which is generally anabolic or synthetic) whereas cyclic AMP, Na^+, Ca^{2+}, ADP, and GTP are associated with another state (which is generally catabolic). Within each group of synonyms above is one of a pair of complements. For example, the following may be pairs of complements: cyclic AMP and cyclic GMP; Na^+ and K^+; Ca^{2+} and Mg^{2+}; ADP and ATP; GTP and GDP.

4. *Concluding remarks*

In this review we have attempted to present a broad view of the possible biochemical mechanisms which might link the interpretation of a cell's position to the regulation of the course of its development. Therefore we have reviewed some classical opinions regarding the biochemistry of this process and have suggested, in addition, ideas which seem particularly attractive to us.

The suggestions we have offered can be briefly summarized. Modification of macromolecules may be responsible for the regulation of gene activity during development and self-perpetuating modified macromolecules (particularly enzymes) may provide the mechanism for "remembering" previous events in development. These modifications have been assumed to determine the course of the cell through a program of choices of genetic expression. This course of progress has, in turn, been postulated to be determined by the temporal changes in concentration of "metabolic messengers" (simple derivatives of common metabolites, inorganic ions and metabolites, such as ATP and S-adenosylmethionine which can donate modifying groups to macromolecules) within the cell. Finally the concentration of these metabolic messengers has been hypothesized to be regulated (as a function of the cell's position in an organism and its previous history of development) by hormones produced by neighboring cells, by cell contact or by the cytoplasmic localization of metabolites and metabolic enzymes.

References

Abraham, R., Rhodes, D.P. and Banerjee, A.K. (1975) The 5′ terminal structure of the methylated mRNA synthesized in vitro by vesicular stomatitis virus. Cell 5, 51–58.

Adams, J. and Cory, S. (1975) Modified nucleosides and bizarre 5′-termini in mouse myeloma mRNA. Nature 255, 28–32.

Aketa, K., Bianchetti, R., Marré, E. and Monroy, A. (1964) Hexose monophosphate levels as a limiting factor for respiration in unfertilized sea urchin eggs. Biochim. Biophys. Acta 86, 211–215.

Anonymous (1975) Seminar on Poly (ADP-Ribose) and ADP-ribosylation of protein. J. Biochem. 77, 1p–11p.

Arber, W. (1965) Host specificity of DNA produced by Escherichia coli V. The role of methionine in production of host specificity. J. Mol. Biol. 11, 247–256.

Arnold, H. and Pette, D. (1970) Binding of aldolase and triosephosphate dehydrogenase to F-actin and modification of catalytic properties of aldolase. Eur. J. Biochem. 15, 360–366.

Arnold, H., Nolte, J. and Pette, D. (1969) Quantitative and histochemical studies on the desorption and readsorption of aldolase in cross-striated muscle. J. Histochem. and Cytochem. 17, 314–320.

Arnold, J.M. and Williams-Arnold, L.D. (1974) Cortical-nuclear interactions in cephalopod development: cytochalasin B effects on the informational pattern in the cell surface. J. Embryol. Exp. Morphol. 31, 1–25.

Ashworth, J.M. (1971) Cell development in the cellular slime mould Dictyostelium discoideum. Symp. Soc. Exp. Biol. 25, 27–47.

Babich, G.L. and Foret, J.E. (1973) Effects of dibutyryl cyclic AMP and related compounds on newt limb regeneration blastemas in vivo. II. ^{14}C-leucine incorporation. Oncology 28, 89–95.

Bäckström, S., Hultin, K. and Hultin, T. (1960) Pathways of glucose metabolism in early sea urchin development. Exp. Cell Res. 19, 634–636.

Barker, J.L., Crayton, J.W. and Nicoll, R.A. (1971) Supraoptic neurosecretory cells: adrenergic and cholinergic sensitivity. Science 171, 208–210.

Barth, L.G. (1965) The nature of the action of ions as inductors. Biol. Bull. (Woods Hole) 129, 471–481.

Barth, L.G. and Barth, L.J. (1963) The relation between intensity of inductor and type of cellular differentiation of Rana pipiens presumptive epidermis. Biol. Bull. (Woods Hole) 124, 125–140.

Barth, L.G. and Barth, L.J. (1967a) Competence and sequential induction in presumptive epidermis of normal and hybrid frog gastrulae. Physiol. Zool. 40, 97–103.

Barth, L.G. and Barth, L.J. (1967b) The uptake of Na-22 during induction in presumptive epidermis cells of the Rana pipiens gastrula. Biol. Bull. (Woods Hole) 133, 495–501.

Barth, L.G. and Barth, L.J. (1968) The role of sodium chloride in the process of induction by lithium chloride in cells of the Rana pipiens gastrula. J. Embryol. Exp. Morphol. 19, 387–396.

Barth, L.G. and Barth, L.J. (1969) The sodium dependence of embryonic induction. Dev. Biol. 20, 236–262.

Barth, L.G. and Barth, L.J. (1972) ^{22}Sodium and ^{45}calcium uptake during embryonic induction in Rana pipiens. Dev. Biol. 28, 18–34.

Bergstrom, G., Ekman, P., Dahlquist, U., Humble, E. and Engström, L. (1975) Subtilisin-catalyzed removal of phosphorylated site of pig liver pyruvate kinase without inactivation of the enzyme. FEBS Lett. 56, 288–291.

Bjerre, B. (1974) A neuralizing influence of dibutyryl cyclic AMP on competent chick ectoderm. Experientia 30, 534–535.

Bode, H., Berking, S., David, C.N., Gierer, A., Schaller, H. and Trenkner, E. (1973) Quantitative analysis of cell types during growth and morphogenesis in Hydra. Wilhelm Roux' Arch. Entwicklungsmech. Organismen 171, 269–285.

Bonner, J.T. (1967) The Cellular Slime Molds. Princeton Univ. Press, Princeton, N.J.

Borun, T.W., Pearson, D. and Paik, W.K. (1972) Studies of histone methylation during the HeLa 5-3 cell cycle. J. Biol. Chem. 247, 4288–4298.

Both, G.W., Banerjee, A.K. and Shatkin, A.J. (1975) Methylation-dependent translation of viral messenger RNAs in vitro. Proc. Nat. Acad. Sci. U.S.A. 72, 1189–1193.

Bownes, M. and Sang, J.H. (1974a) Experimental manipulations of early Drosophila embryos. I. Adult and embryonic defects resulting from microcautery of nuclear multiplication and blastoderm stages. J. Embryol. Exp. Morphol. 32, 253–272.

Bownes, M. and Sang, J.H. (1974b) Experimental manipulations of early Drosophila embryos. II. Adult and embryonic defects resulting from the removal of blastoderm cells by pricking. J. Embryol. Exp. Morphol. 32, 273–285.

Briens, P. and Reniers-Decoen, M. (1955) La signification des cellules interstitielles des hydres d'eau douce et la problème de la réserve continuaire. Bull. Biol. Fr. Belg. 89, 258–325.

Browne, E.N. (1909) The production of new hydranths in Hydra by the insertion of small grafts. J. Exp. Zool. 8, 1–23.

Bryant, P.J. (1971) Regeneration and duplication following operations in situ on the imaginal discs of Drosophila melanogaster. Dev. Biol. 26, 637–651.

Bryant, P.J. (1974) Determination and pattern formation in the imaginal discs of Drosophila. Curr. Topics Dev. Biol. 8, 41–80.

Buckley, J.T. and Hawthorne, J.N. (1972) Erythrocyte membrane polyphosphoinositide metabolism and the regulation of calcium binding. J. Biol. Chem. 247, 7218–7223.

Burnett, A.L. (1959) Histophysiology of growth in Hydra. J. Exp. Zool. 140, 281–341.

Campbell, R.D. (1967) Tissue dynamics of steady state growth in Hydra littoralis. I. Patterns of cell division. Dev. Biol. 15, 487–502.

Cantoni, G.L. (1975) Biological methylation: selected aspects. Ann. Rev. Biochem. 44, 435–451.

Carlson, C.A. and Kim, K.-H. (1974) Regulation of hepatic acetyl coenzyme A carboxylase by phosphorylation and dephosphorylation. Arch. Biochem. Biophys. 164, 478–489.

Chan, L.-N. and Gehring, W. (1971) Determination of blastoderm cells in Drosophila melanogaster. Proc. Nat. Acad. Sci. U.S.A. 68, 2217–2221.

Child, C.M. (1928) The physiological gradients. Protoplasma 5, 447–476.

Child, C.M. (1947) Oxidation and reduction of indicators by Hydra. J. Exp. Zool. 104, 153–195.

Civan, M.M. and Frazier, H.S. (1968) The site of the stimulation action of vasopressin on sodium transport in toad bladder. J. Gen. Physiol. 51, 589–605.

Clarke, F.M. and Masters, C.J. (1972) On the reversible and selective adsorption of aldolase isoenzymes in rat brain. Arch. Biochem. Biophys. 153, 258–265.

Clarke, F.M. and Masters, C.J. (1974) On the association of glycolytic components in skeletal muscle extracts. Biochim. Biophys. Acta 358, 193–207.

Clarke, F.M. and Masters, C.J. (1975) On the association of glycolytic enzymes with structural proteins of skeletal muscle. Biochim. Biophys. Acta 314, 517–526.

Clarke, M., Schatten, G., Mazia, D. and Spudich, J.A. (1975) Visualization of actin fibers associated with the cell membrane in amoebae of Dictyostelium discoideum. Proc. Nat. Acad. Sci. U.S.A. 72, 1758–1762.

Clarkson, S.G. (1969) Nucleic acid and protein synthesis and pattern regulation in Hydra. Regional patterns of synthesis and changes in synthesis during hypostome formation. J. Embryol. Exp. Morphol. 21, 33–54.

Colombo, G. and Marcus, F. (1973) Activation of fructose, 1,6-diphosphatase by potassium ions. Loss upon conversion of "neutral" to "alkaline" form. J. Biol. Chem. 248, 2743–2745.

Corff, Sondra. (1973) Organismal growth and the contribution of cell proliferation to net growth and maintenance of form. In: Biology of Hydra (Burnett, A.L. ed.) pp. 346–393, Academic Press, New York.

Collier, R.J. (1975) Diphtheria toxin: mode of action and structure. Bacteriol. Rev. 39, 54–85.

Cox, G.S., Weissbach, H. and Kaback, H.R. (1975) Transportation and Escherichia coli fatty acid auxotroph. A novel case of catabolite repression. J. Biol. Chem. 250, 4542–4548.

Crick, F.H.C. (1970) Diffusion in embryogenesis. Nature 225, 420–422.

Curtis, A.S.G. (1962) Morphogenetic interactions before gastrulation in the amphibian Xenopus laevis—the cortical field. J. Embryol. Exp. Morphol. 10, 410–422.

Czihak, G. and Hörstadius, S. (1970) Transplantation of RNA-labeled micromeres into animal halves of sea urchin embryos. A contribution to the problem of embryonic induction. Dev. Biol. 22, 15–30.

Dalcq, C.M. (1938) Form and Causality in Early Development. Cambridge Univ. Press, Cambridge.

David, C.N. (1976) Stem cell differentiation in Hydra, in ASM Symposium Cell Differentiation and Communication (Dworkin, M. and Shapiro, L. eds.) in press.

David, C.N. and Challoner, D. (1974) Distribution of interstitial cells and differentiating nematocytes in nests in Hydra attenuata. Amer. Zool. 14, 537–542.

Davidson, E.H. (1968) Gene Activity in Early Development. Academic Press, New York.

Dedman, J.R., Payne, D.M. and Harris, B.G. (1975) Increased proteolytic susceptibility of aldolase induced by actin binding. Biochem. Biophys. Res. Comm. 65, 1170–1176.

Diehl, F.A. and Burnett, A.L. (1964) The role of interstitial cells in the maintenance of Hydra. I. Specific destruction of interstitial cells in normal, asexual, non-budding animals. J. Exp. Zool. 155, 253–260.

Diliberto, E.J., Jr. and Axelrod, J. (1974) Characterization and substrate specificity of a protein carboxymethylase in the pituitary gland. Proc. Nat. Acad. Sci. U.S.A. 71, 1701–1704.

Douvas, A.S., Harrington, C.A. and Bonner, J. (1975) Major nonhistone proteins of rat liver chromatin: preliminary identification of myosin, actin, tubulin and tropomyosin. Proc. Nat. Acad. Sci. U.S.A. 72, 3902–3906.

Edelman, G.M. (1974) Origins and mechanisms of specificity in clonal selection. In: Cellular Selection and Regulation in the Immune Response (Edelman, G.M., ed.) pp. 1–38. Raven Press, New York.

Edidin, M. (1974) Two-dimensional diffusion in membranes. In: Transport at the Cellular Level (Sleigh, M.A. and Jennings, D.H., eds.) pp. 1–14, Cambridge University Press, Cambridge.

Ensor, D.M. and Ball, J.N. (1972) Prolactin and osmoregulation in fishes. Fed. Proc. 31, 1615–1623.

Falconer, I.R. and Rowe, J.M. (1975) Possible mechanism for action of prolactin on mammary cell sodium transport. Nature 256, 327–328.

Fischer, I.H., Graves, D.J., Crittenden, E.R.S. and Krebs, E.G. (1959) Structure of the site phosphorylated in the phosphorylase b to a reaction. J. Biol. Chem. 234, 1698–1704.

Foret, J.E. and Babich, G.L. (1973) Effects of dibutyryl cyclic AMP and related compounds on newt limb regeneration blastemas in vitro. I. ^{3}H-thymidine incorporation. Oncology 28, 83–88.

Freese, E., Ichikawa, T., Oh, Y.K., Freese, E.B. and Prasad, C. (1974) Deficiencies or excesses of metabolites interfering with differentiation. Proc. Nat. Acad. Sci. U.S.A. 71, 4188–4193.

Frieden, C. (1970) Kinetic aspects of the regulation of metabolic processes. The hysteritic enzyme concept. J. Biol. Chem. 245, 5788–5799.

Furuichi, Y., Morgan, M., Shatkin, A.J., Jelinek, W., Salditt-Georgieff, M. and Darnell, J.E. (1975) Methylated, blocked 5′ termini in HeLa cell mRNA. Proc. Nat. Acad. Sci. U.S.A. 72, 1904–1908.

Garen, A. and Gehring, W. (1972) Repair of the lethal developmental defect in deep orange embryos of Drosophila by injecting egg cytoplasm. Proc. Nat. Acad. Sci. U.S.A. 69, 2982–2985.

Gierer, A. and Meinhardt, H. (1972) A theory of biological pattern formation. Kybernetik 12, 30–39.

Gierer, A., Berking, S., Bode, H., David, C.N., Flick, K., Hansmann, G., Schaller, H. and Trenkner, E. (1972) Regeneration of Hydra from reaggregated cells. Nature New Biology 239, 98–101.

Giudice, G. (1973) Developmental Biology of the Sea Urchin Embryo. Academic Press, New York.

Goldberg, N.D., O'Dea, R.F. and Haddox, M.K. (1973) Cyclic GMP. In: Advances in Cyclic Nucleotide Research (Greengard, P. and Robison, A., eds.) vol. 3, pp. 155–223. Raven Press, New York.

Goldman, R.D. (1975) The use of heavy meromyosin binding as an ultrastructural cytochemical method for localizing and determining the possible functions of actin-like microfilaments in nonmuscle cells. J. Histochem. Cytochem. 23, 529–542.

Goodman, L.S. and Gilman, A. (1975) The Pharmacological Basis of Therapeutics, 5th ed., Macmillan, London.

Goodwin, B. and Cohen, M.H. (1969) A phase-shift model for the spatial and temporal organization of developing systems. J. Theoret. Biol. 25, 49–107.

Grainger, R.M. and Wessells, N.K. (1974) Does RNA pass from mesenchyme to epithelium during an embryonic tissue interaction? Proc. Nat. Acad. Sci. U.S.A. 71, 4747–4751.

Granner, D.K., Sellers, L., Lee, A., Butters, C. and Kutina, L. (1975) A comparison of the uptake, metabolism, and action of cyclic adenine nucleotides in cultured hepatoma cells. Arch. Biochem. Biophys. 169, 601–615.

Gregg, J.H. (1965) Regulation in the cellular slime molds. Dev. Biol. 12, 377–393.

Grippo, P., Iaccarino, M., Parisi, E. and Scarano, E. (1968) Methylation of DNA in developing sea urchin embryos. J. Mol. Biol. 36, 195–208.

Grobstein, C. (1967) Mechanisms of organogenetic tissue interaction. Nat. Cancer Inst. Monogr. 26, 279–299.

Hadorn, E. (1966) Dynamics of determination. In: Major Problems in Developmental Biology (Locke, M., ed.) pp. 85–104. Academic Press, New York.

Harris, M. (1973) Poly(ADP-Ribose) an international symposium. Fogarty International Center Proceedings #26, National Institutes of Health, Bethesda, Maryland.

Hawker, L.E. (1957) The Physiology of Reproduction in Fungi. Cambridge University Press, Cambridge, England.

Helser, T.L., Davies, J.E. and Dahlberg, J.E. (1971) Change in methylation of 16S ribosomal RNA associated with mutation to kasugamycin resistance in Escherichia coli. Nature, 233, 12–14.

Henderson, R. and Unwin, P.N.T. (1975) Three-dimensional model of purple membrane obtained by electron microscopy. Nature, 257, 28–32.

Herlands, R.L. and Bode, H.R. (1974) The influence of tissue polarity on nematocyte migration in Hydra attenuata. Dev. Biol. 40, 323–339.

Herth, W. and Sander, K. (1973) Mode and timing of body pattern formation (regionalization) in the early embryonic development of cyclorrhaphic dipterans (Protophormia, Drosophila). Wilhelm Roux Arch. Entwicklungsmech. Organismen 172, 1–27.

Hicklin, J., Hornbruch, A., Wolpert, L. and Clarke, M. (1973) Positional information and pattern regulation in Hydra: The formation of boundary regions following axial grafts. J. Embryol. Exp. Morphol. 1. 30, 701–725.

Hinrichs, M.G. (1924) A demonstration of the axial gradient by means of photolysis. J. Exp. Zool. 41, 21–31.

Hoffman, S. and McMahon, D. (1976) The role of the plasma membrane in the development of Dictyostelium discoideum. I. Developmental and topographic analysis of polypeptide composition. Submitted for publication.

Holliday, R. and Pugh, J.E. (1975) DNA modification mechanisms and gene activity during development. Science, 187, 226–232.

Holzer, H. and Duntze, W. (1971) Metabolic regulation by chemical modification of enzymes. Ann. Rev. Biochem. 40, 345–377.

Hornbruch, A. and Wolpert, L. (1975) Polarity reversal in Hydra by oligomycin. J. Embryol. Exp. Morphol. 33, 845–852.

Hunt, R.K. and Jacobson, M. (1974) Neuronal specificity revisited. Curr. Topics Dev. Biol. 8, 203–259.

Huston, R.B. and Krebs, E.G. (1968) Activation of skeletal muscle phosphorylase kinase by Ca^{++}. II. Identification of the kinase activating factor as a proteolytic enzyme. Biochemistry, 7, 2116–2122.

Illmensee, K. and Mahowald, A.P. (1974) Transplantation of posterior pole plasm in Drosophila. Induction of germ cells at the anterior pole of the egg. Proc. Nat. Acad. Sci. U.S.A. 71, 1016–1020.

Ishihara, K. (1959) Release and activation of aldolase at fertilization in sea urchin eggs. J. Fac. Sci. Univ. Tokyo Sect. IV 8, 71–93.

Ishikawa, H., Bischoff, R. and Holtzer, H. (1969) Formation of arrowhead complexes with heavy meromyosin in a variety of cell types. J. Cell Biol. 43, 312–328.

Isono, N. and Yasumasu, I. (1968) Pathways of carbohydrate breakdown in sea urchin eggs. Exp. Cell Res. 50, 616–626.

Isono, N. (1963) Carbohydrate metabolism in sea urchin eggs. IV. Intracellular localization of enzymes of pentose phosphate cycle in unfertilized and fertilized eggs. J. Fac. Sci. Univ. Tokyo, Sect. 4. 10, 37–49.

Johnson, W.W., Wilhelm, J.A. and Hnilica, L.S. (1973) Nuclear basic protein acetylation during early sea urchin development. Biochim. Biophys. Acta, 295, 150–158.

Karpatkin, S. (1967) Soluble and particulate hexokinase of frog skeletal muscle. J. Biol. Chem. 242, 3525–3530.

Kates, J.R. and Jones, R.F. (1964) The control of gametic differentiation in liquid culture of Chlamydomonas. J. Cell Comp. Physiol. 63, 157–164.

Kim, S. (1974) S-adenosylmethionine protein carboxyl methyltransferase from erythrocyte. Arch. Biochem. Biophys. 616, 652–657.

Klaus, M. and George, R.P. (1974) Microdissection of developmental stages of the cellular slime mold, Dictyostelium discoideum, using a ruby laser. Dev. Biol. 39, 183–188.

Kneer, N.M., Bosch, A.L., Clark, M.G. and Lardy, H.A. (1974) Glucose inhibition of epinephrine stimulation of hepatic gluconeogenesis by blockade of the α-receptor function. Proc. Nat. Acad. Sci. U.S.A. 71, 4523–4527.

Krahl, M.E. (1956) Oxidative pathways for glucose in eggs of the sea urchin. Biochim. Biophys. Acta. 20, 27–32.

Krakow, J.S. and Pastan, I. (1973) Cyclic adenosine monophosphate receptor: loss of cAMP-dependent DNA binding activity after proteolysis in the presence of cyclic adenosine monophosphate. Proc. Nat. Acad. Sci. U.S.A. 70, 2529–2533.

Kühn, A. (1971) Lectures on developmental physiology. Translated by R. Milkman. Springer-Verlag, New York.

Lai, C.J. and Weisblum, B. (1971) Altered methylation of ribosomal RNA in an erythromycin-resistant strain of Staphylococcus aureus. Proc. Nat. Acad. Sci. U.S.A. 68, 856–860.

Lazarides, E. (1975) Immunofluorescence studies on the structure of actin filaments in tissue culture cells. J. Histochem. Cytochem. 23, 507–528.

Lazarides, E. and Lindberg, U. (1974) Actin is the naturally occurring inhibitor of deoxyribonuclease I. Proc. Nat. Acad. Sci. U.S.A. 71, 4742–4746.

Leder, A. and Leder, P. (1975) Butyric acid, a potent inducer of erythroid differentiation in cultured erythroleukemic cells. Cell 5, 319–322.

Lee, K.-C. (1972) Cell electrophoresis of the cellular slime mould Dictyostelium discoideum II. Relevance of the changes in cell surface charge density to cell aggregation and morphogenesis. J. Cell Sci. 10, 249–265.

Lenhoff, H.M. (1965) Cellular segregation and heterocytic dominance. Science, 148, 1105–1107.

Lentz, T.L. (1965) Hydra: induction of supernumerary heads by isolated neurosecretory granules. Science, 150, 633–635.

Lentz, T.L. (1966) The Cell Biology of Hydra. John Wiley and Sons, New York.

Lentz, T.L. and Barrnett, R.J. (1961) Enzyme histochemistry of Hydra. J. Exp. Zool. 147, 125–149.

Lentz, T.L. and Barrnett, R.J. (1962) Changes in the distribution of enzyme activity in the regenerating Hydra. J. Exp. Zool. 150, 103–117.

Lentz, T.L. and Barrnett, R.J. (1963) The role of the nervous system in regenerating Hydra: The effect of neuropharmacological agents. J. Exp. Zool. 154, 305–327.

Lesh, G.E. (1970) A role of inductive factors in interstitial cell differentiation in Hydra. J. Exp. Zool. 173, 371–382.

Lesh-Laurie, G.E. (1973) Expression and Maintenance of Organismic Polarity. In: Biology of Hydra. (Burnett, A.L., ed.) pp. 143–168, Academic Press, New York.

Lesh, G.E. and Burnett, A.L. (1964) Some biological properties of the polarizing factor in Hydra. Nature, 204, 492–493.

Lesh, G.E. and Burnett, A.L. (1966) An analysis of the chemical control of polarized form in Hydra. J. Exp. Zool. 163, 55–78.

Levi, S. and Shatkin, A.J. (1975) Methylated simian virus 40-specific RNA from nuclei and cytoplasm of infected BSC-1 cells. Proc. Nat. Acad. Sci. U.S.A. 72, 2012–2016.

Levine, S. (1966) Enzyme amplifier kinetics. Science, 152, 651–653.

Libby, P.R. (1973) Histone acetylation and hormone action. Early effects of aldosterone on histone acetylation in rat kidney. Biochem. J. 134, 907–912.

Loomis, W.F. (1969) Developmental regulation of alkaline phosphatase in Dictyostelium discoideum. J. Bacteriol. 100, 417–422.

Loomis, W.F. (1975) Polarity and pattern in Dictyostelium. In: Pattern Formation and Gene Regulation in Development (McMahon, D., and Fox, C.F., eds.) pp. 109–128. W.A. Benjamin, Palo Alto.

McConnell, H.M. (1975) Coupling between lateral and perpendicular motion in biological membranes. In: Functional Linkage in Biomolecular Systems (Schmitt, F.O., Schneider, D.M. and Crothers, D.M., eds.) pp. 123–131. Raven Press, New York.

McMahon, D. (1973) A cell contact model for position determination in development. Proc. Nat. Acad. Sci. U.S.A. 70, 2396–2400.

McMahon, D. (1974) Chemical messengers in development: a hypothesis. Science, 185, 1012–1021.

McMahon, D. (1975) Activation of alkaline phosphatase in D. discoideum by purified plasma membranes. Submitted for publication.

McMahon, D., Hoffman, S., Fry, W., and West, C. (1975) The involvement of the plasma membrane in the development of Dictyostelium discoideum. In: Pattern Formation and Gene Regulation in Development (McMahon, D. and Fox, C.F., eds.) pp. 60–75. W.A. Benjamin, Palo Alto.

Mabuchi, I. (1973) A myosin-like protein in the cortical layer of the sea urchin egg. J. Cell Biol. 59, 542–547.

Macklin, M. and Burnett, A.L. (1966) Control of differentiation by calcium and sodium ions in Hydra pseudoligactis. Exp. Cell Res. 44, 665–668.

MacWilliams, H.K. and Kafatos, F.C. (1968) Hydra viridis: inhibition by basal disk of basal disk differentiation. Science, 159, 1246–1247.

Mahowald, A.P. (1968) Polar granules of Drosophila II. Ultrastructural changes during early embryogenesis. J. Exp. Zool. 167, 237–262.

Martin, G.S., Venuta, S., Weber, M. and Rubin, H. (1971) Temperature-dependent alterations in sugar transport in cells infected by a temperature-sensitive mutant of Rous sarcoma virus. Proc. Nat. Acad. Sci. U.S.A. 68, 2739–2741.

Meselson, M. and Yuan, R. (1968) DNA restriction enzyme from E. coli. Nature, 217, 1110–1114.

Meselson, M., Yuan, R. and Heywood, J. (1972) Restriction and modification of DNA. Ann. Rev. Biochem. 41, 447–466.

Mine, H. and Takeuchi, I. (1967) Tetrazolium reduction in slime mould development. Annual Report of Biological Works, Faculty of Science, Osaka University, 15, 97–111.

Mishra, N.C., Niu, M.C. and Tatum, E.L. (1975) Induction by RNA of inositol independence in Neurospora crassa. Proc. Nat. Acad. Sci. U.S.A. 72, 642–645.

Mookerjee, S. and Bhattacharjee, A. (1967) Regeneration time at the different levels of Hydra. Wilhelm Roux Arch. Entwicklungsmech. Organismen 158, 301–314.

Muthukrishinan, S., Both, G.W., Furuichi, Y. and Slatkin, A.J. (1975) 5′-terminal 7-methylguanosine in eukaryotic mRNA is required for translation. Nature, 255, 33–37.

Nanjundiah, V. (1974) A differential chemotactic response of slime mold amoebae to regions of the early amphibian embryo. Exp. Cell Res. 86, 408–411.

Neelon, F.A. and Birch, B.M. (1973) Cyclic adenosine 3′:5′-monophosphate-dependent protein kinase. Interaction with butyrulated analogues of cyclic adenosine 3′:5′ monophosphate. J. Biol. Chem. 248, 8361–8365.

Newman, S.A. (1974) The interaction of the organizing regions in Hydra and its possible relation to the role of the cut end in regeneration. J. Embryol. Exp. Morphol. 31, 541–555.

Niu, M.C. (1958) Thymus ribonucleic acid and embryonic differentiation. Proc. Nat. Acad. Sci. U.S.A. 44, 1264–1274.

Nolan, C., Novoa, W.J., Krebs, E.G. and Fischer, E.H. (1964) Further studies of the site phosphorylated in the phosphorylase b to a reaction. Biochem. 3, 542–551.

Nöthiger, R. (1972) The larval development of imaginal discs. In: Results and Problems in Cell Differentiation, vol. 5 (Ursprung, H. and Nöthiger, R., eds.) pp. 1–34, Springer-Verlag, Berlin.

Novick, A. and Weiner, M. (1957) Enzyme induction as an all-or-none phenomenon. Proc. Nat. Acad. Sci. U.S.A. 43, 553–566.

Nuss, D.L., Furuichi, Y., Koch, G. and Shatkin, A.J. (1975) Detection in HeLa cell extracts of a 7-methyl guanosine specific enzyme activity that cleaves m^7GpppN^m. Cell 6, 21–27.

Ohnishi, Shun-Ichi and Ito, T. (1974) Calcium-induced phase separations in phosphatidylserine-phosphatidylcholine membranes. Biochemistry, 13, 881–887.

Okada, M., Kleinman, I.A. and Schneiderman, H.A. (1974) Repair of a genetically-caused defect in oogenesis in Drosophila melanogaster by transplantation of cytoplasm from wild-type eggs and by injection of pyrimidine nucleosides. Dev. Biol. 37, 55–62.

Ong, S.-H., Whitley, T.H., Stowe, N.W. and Steiner, A.L. (1975) Immunohistochemical localization of 3′:5′-cyclic AMP and 3′:5′-cyclic GMP in rat liver, intestine and testis. Proc. Nat. Acad. Sci. U.S.A. 72, 2022–2026.

Pan, P., Bonner, J.T., Wedner, H.J. and Parker, C.W. (1974) Immunofluorescence evidence for the distribution of cyclic AMP in cells and cell masses of the cellular slime molds. Proc. Nat. Acad. Sci. U.S.A. 71, 1623–1625.

Papahadjopoulos, D. and Poste, G. (1975) Calcium-induced phase separation and fusion in phospholipid membranes. Biophys. J. 15, 945–948.

Pepe, F. (1975) Structure of muscle filaments from immunohistochemical and ultrastructural studies. J. Histochem. Cytochem. 23, 543–562.

Perry, R.P. and Kelley, D.E. (1974) Existence of methylated messenger RNA in mouse L cells. Cell 1, 37–42.

Perry, R.P., Kelley, D.E., Friderci, K.H. and Rottman, F.M. (1975) Methylated constituents of heterogenous nuclear RNA: presence in blocked 5′ terminal structures. Cell 6, 13–19.

Perry, R.P. and Scherer, K. (1975) The methylated constituents of globin in mRNA. FEBS Lett. 57, 73–77.

Picard, J.J. (1975a) Xenopus laevis cement gland as an experimental model for embryonic differentiation I. In vitro stimulation of differentiation by ammonium chloride. J. Embryol. Exp. Morphol. 33, 957–967.

Picard, J.J. (1975b) Xenopus laevis cement gland as an experimental model for embryonic differentiation II. The competence of embryonic cells. J. Embryol. Exp. Morphol. 33, 969–978.

Pollack, R., Osborn, M. and Weber, K. (1975) Patterns of organization of actin and myosin in normal and transformed cultured cells. Proc. Nat. Acad. Sci. U.S.A. 72, 994–998.

Popper, K.R. (1959) The Logic of Scientific Discovery. Basic Books, New York.

Poste, G. and Weiss, L. (1976) Some considerations on cell surface alterations in malignancy. In: Fundamental Aspects of Metastasis (Weiss, L., ed.) pp. 25–47. North-Holland, Amsterdam.

Prasad, K.N. (1975) Differentiation of neuroblastoma cells in culture. Biol. Rev. Camb. Phil. Soc. 50, 129–265.

Racey, L.A. and Byvoet, P. (1972) Histone acetyltransferase in chromatin. Evidence for in vitro enzymatic transfer of acetate from acetyl-coenzyme A to histones. Exp. Cell Res. 64, 366–370.

Rall, T.W., Sutherland, E.W. (1961) The regulatory role of adenosine-3′,5′-phosphate. Cold Spring Harbor Symp. Quant. Biol. 26, 347–354.

Raper, K.B. (1940) Pseudoplasmodium formation and organization in Dictyostelium discoideum. J. Elisha Mitchell Sci. Soc. 56, 241–282.

Raper, K.B. (1941) Developmental pattern in simple slime molds. Growth, Third Growth Symposium 5, 41–76.

Reddy, M.K., Etlinger, J.D., Rabinowitz, M., Fischma, D.A. and Zak, R. (1975) Removal of Z lines and α-actinin from isolated myofibrils by a calcium-activated neutral protease. J. Biol. Chem. 250, 4278–4284.

Reeves, R.H. and Roth, J.R. (1975) Transfer ribonucleic acid methylase deficiency found in UGA suppressor strains. J. Bacteriol. 124, 332–340.

Reporter, M. (1972) Methylation of basic residues in structural proteins. Mech. Age. Dev. 1, 367–372.

Reporter, M. and Rosenquist, G.C. (1972) Adenosine 3′,5′-monophosphate: regional differences in chick embryos at the head process stage. Science, 178, 628–630.

Rich, F. and Tardent, P. (1969) Untersuchungen zur Nematocysten-Differenzierung bei Hydra attenuata Pall. Rev. Suisse Zool. 76, 779–787.

Ries, E. (1939) Histochemische Sonderungsprozesse während der frühen Embryonalentwicklung verschiederer wirbelloser Tiere. Archiv für Exp. Zellforsch. 22, 569–586.

Riggs, A.D. (1975) X inactivation, differentiation, and DNA methylation. Cytogenetics and Cell Genetics, in press.

Robinson, K.R. and Jaffe (1975) Polarizing fucoid eggs drive a calcium current through themselves. Science, 187, 70–72.

Röhlich, P. (1975) Membrane-associated actin filaments in the cortical cytoplasm of the rat mast cell. Exp. Cell Res. 93, 293–298.

Rose, I.A. and Warms, J.U.B. (1966) Mitochondrial hexokinase. Release, rebinding and location. J. Biol. Chem. 242, 1635–1645.

Rose, P.G. and Burnett, A.L. (1968) An electron microscopic and histochemical study of the secretory cells in Hydra viridis. Wilhelm Roux Arch. Entwicklungsmech. Organismen 161, 281–297.

Rosen, O.M. and Erlichman, J. (1975) Reversible autophosphorylation of a cyclic 3′:5′-AMP-dependent protein kinase from bovine cardiac muscle. J. Biol. Chem. 250, 7788–7794.

Roux, W. (1888) Beiträge zur Entwicklungsmechanik des Embryo 5. Ueber die künstliche Hervorbringung halber Embryonen durch Zerstörung eines der beiden ersten Furchungskugeln, sowie über Nachentwicklung (Post-generation) der fehlenden Köperhälfte. Virchow's Arch. path. Anat. Physiol. 64, 113–154, 246–291.

Ruiz-Carrillo, A., Wangh, L.J. and Allfrey, V.G. (1975) Processing of newly synthesized histone molecules. Science, 190, 117–128.

Runnström, J. (1966) The vitelline membrane and cortical particles in sea urchin eggs and their function in maturation and fertilization. Adv. Morphogenesis 5, 221–325.

Rutherford, C.L. and Lenhoff, H.M. (1969) Enzymes of glucose catabolism in Hydra. II. Application of microfluorometric analyses to patterns of enzyme localization. Arch. Biochem. Biophys. 133, 128–136.

Rutter, W.J., Pictet, R.L. and Morris, P.W. (1973) Toward molecular mechanisms of developmental processes. Ann. Rev. Biochem. 42, 601–646.

Sager, R. and Granick, S. (1954) Nutritional control of sexuality in Chlamydomonas reinhardi. J. Gen. Physiol. 37, 729–742.

Sager, R. and Kitchin, R. (1975) Selective silencing of eukaryotic DNA. Science, 189, 426–433.

Sander, K. (1971) Pattern formation in longitudinal halves of leaf hopper eggs (Homoptera) and some remarks on the definition of "embryonic regulation." Wilhelm Roux Arch. Entwicklungsmech. Organismen 167, 336–352.

Sanger, J.W. (1975) Presence of actin during chromosomal movement. Proc. Nat. Acad. Sci. U.S.A. 72, 2451–2455.

Saxén, L. (1972) Interactive mechanisms in morphogenesis. In: Tissue Interactions in Carcinogenesis (Tarin, D., ed.) pp. 49–80, Academic Press, London.

Saxén, L. and Toivonen, S. (1962) Primary Embryonic Induction. Logos Press, London.
Schäfer, A., DeVries, J.X., Faulstich, H. and Wieland, Th. (1975) Phalloidin counteracts the inhibitory effect of actin on deoxyribonuclease I. FEBS Lett. 57, 51–54.
Schaller, H.C. (1973) Isolation and characterization of a low molecular weight substance activating head and bud formation in Hydra. J. Embryol. Exp. Morphol. 29, 27–38.
Schaller, H.C. and Gierer, A. (1973) Distribution of the head activating substance in Hydra and its localization in membranous particles in nerve cells. J. Embryol. Exp. Morphol. 29, 39–52.
Schlender, K.K. and Reiman, E.M. (1975) Isolation of a glycogen synthetase I kinase that is independent of adenosine 3′:5′-monophosphate. Proc. Nat. Acad. Sci. U.S.A. 72, 2197–2201.
Schollmeyer, J.V., Goll, D.E., Stromer, M.H., Dayton, W. and Singh, I. (1974) Studies on the composition of the Z disk. J. Cell Biol. 63, 3039.
Schmidt, G.A. (1933) Schnürungs- und Durchschneidungsversuche am Amphibienkeim. Wilhelm Roux Arch. Entwicklungsmech. Organismen 129, 1044.
Schubiger, G. (1971) Regeneration, duplication and transdetermination in fragments of the leg disc of Drosophila melanogaster. Dev. Biol. 26, 277–295.
Schwartz, A., Lindenmayer, G.E. and Allen, J.C. (1975) The sodium-potassium adenosine triphosphatase: pharmacological, physiological and biochemical aspects. Pharmacol. Rev. 27, 4–134.
Seglen, P.O. (1974) Differones. Norwegian J. Zool. 22, Suppl. 1, 1–132.
Sheetz, M.P. and Singer, S.J. (1974) Biological membranes as bilayer couples. A molecular mechanism of drug-erythrocyte interactions. Proc. Nat. Acad. Sci. U.S.A. 71, 4457–4461.
Shimshick, E.J. and McConnell, H.M. (1973) Lateral phase separation in phospholipid membranes. Biochemistry, 12, 2351–2360.
Sicard, R.E. (1975) The effects of hypophysectomy upon the endogenous levels of cAMP during forelimb regeneration of adult newts (Notophthalamus viridescens). Wilhelm Roux Arch. Entwicklungsmech. Organismen 177, 159–162.
Siess, E.A. and Wieland, D.H. (1975) Regulation of pyruvate dehydrogenase interconversion in isolated hepatocytes by the mitochondrial ATP/ADP ratio. FEBS Lett. 52, 226–230.
Sigel, P. and Pette, D. (1969) Intracellular localization of glycogenolytic and glycolytic enzymes in white and red rabbit skeletal muscle. J. Histochem. Cytochem. 17, 225–237.
Simon, L.D. (1972) Infection of Escherichia coli by T2 and T4 bacteriophages as seen in the electron microscope: T4 head morphogenesis. Proc. Nat. Acad. Sci. U.S.A. 69, 907–911.
Simon, L.D., McLaughlin, T.J.M., Snover, D., Ou, J., Grisham, C., Loeb, M. (1975) E. coli membrane lipid alteration affecting T4 capsid morphogenesis. Nature, 256, 383–386.
Singer, S.J. (1974) Molecular biology of cellular membranes with applications to immunology. Adv. Immunol. 19, 1–66.
Singh, M., Singh, U.N., August, J.T. and Horecker, B.L. (1974a) Alterations in glucose metabolism in chick embryo cells transformed by Rous sarcoma virus. Transformation-specific changes in the activities of key enzymes of the glycolytic and hexose monophosphate shunt pathways. Arch. Biochem. Biophys. 165, 240–246.
Singh, U.N., Singh, M., August, J.T. and Horecker, B.L. (1974b) Alterations in glucose metabolism in chick-embryo cells transformed by Rous sarcoma virus: intracellular levels of glycolytic intermediates. Proc. Nat. Acad. Sci. U.S.A. 71, 4129–4132.
Skou, J.C. (1975) The ($Na^+ + K^+$) activated enzyme system and its relationship to transport of sodium and potassium. Quart. Rev. Biophys. 7, 401–434.
Smith, J.A. and Stocken, L.A. (1975) Chemical and metabolic properties of adenosine diphosphate ribose derivatives of nuclear proteins. Biochem. J. 147, 523–529.
Solursh, M. and Reiter, R.S. (1975) Determination of limb bud chondrocytes during a transient block of the cell cycle. Cell Differ. 4, 131–137.
Spemann, H. (1921) Die Erzeugung tierischer Chimaeren durch heteroplastische embryonale Transplantation zwischen Triton cristatus u. taeniatus. Wilhelm Roux Arch. Entwicklungsmech. Organismen 48, 533–570.
Spiegel, M. and Rubinstein, N.A. (1972) Synthesis of RNA by dissociated cells of the sea urchin embryo. Exp. Cell Res. 70, 423–430.
Starlinger, H. (1967) Uber die Bindung dei Muskelaldolase an grob-dispersen Partikeln in

Homogenaten erregter Muskeln. Hoppe-Seyler Z. Physiol. Chem. 348, 864–870.

Stocum, D.L. (1975) Outgrowth and pattern formation during limb ontogeny and regeneration. Differentiation, 3, 167–182.

Suria, D. and Liew, C.C. (1974) Isolation of nuclear acidic proteins from rat tissues. Characterization of acetylated liver nuclear acidic proteins. Biochem. J. 137, 355–362.

Sussman, M. (1955) "Fruity" and other mutants of the cellular slime mould Dictyostelium discoideum: a study of developmental abberation. J. Gen. Microbiol. 13, 295–309.

Sutherland, E.W. and Rall, T.W. (1960) The relation of adenosine-3′,5′-phosphate to the actions of catecholamines and other hormones. Pharmacol. Rev. 12, 265–299.

Tarin, D. (1973) Histochemical and enzyme digestion studies on neural induction in Xenopus laevis. Differentiation, 1, 109–126.

Tilney, L.G. and Detmers, P. (1975) Actin in erythrocyte ghosts and its association with spectrin. Evidence for nonfilamentous form of these two molecules in situ. J. Cell Biol. 66, 508–520.

Tomkins, G.M. (1975) The metabolic code. Science, 189, 760–763.

Turing, A.M. (1952) The chemical basis of morphogenesis. Phil. Trans. R. Soc. London B. 237, 37–72.

Tuttle, J.P. and Wilson, J.E. (1970) Rat brain hexokinase: A kinetic comparison of soluble and particulate forms. Biochim. Biophys. Acta 212, 185–188.

Ueda, K., Omachi, A., Kawaichi, M. and Hayaishi, O. (1975) Natural occurrence of poly (ADP-ribosyl) histones in rat liver. Proc. Nat. Acad. Sci. U.S.A. 72, 205–209.

Ursprung, H. (1966) The formation of patterns in development. In: Major Problems in Developmental Biology (Locke, M., ed.) pp. 177–216, Academic Press, New York.

Vovis, G.F., Horiuchi, K. and Zinder, N.D. (1974) Kinetics of methylation of DNA by a reduction endonuclease from Escherichia coli B. Proc. Nat. Acad. Sci. U.S.A. 71, 3810–3813.

Wahn, H.L., Lightbody, L.E., Tchen, T.T. and Taylor, J.D. (1975a) Induction of neural differentiation in cultures of amphibian undetermined presumptive epidermis by cyclic AMP derivatives. Science, 188, 366–369.

Wahn, H.L., Taylor, J.D., and Tchen, T.T. (1975b) Acceleration of amphibian embryonic melanophore development by MSH, dibutyryl-cAMP and theophylline. Proc. IX Int. Pigment Cell Conf., (Riley, V., ed.) Vol. 2, (in press).

Wahn, H.L., Taylor, J.P. and Tschen, T.T. (1975c) MSH, c-AMP, and embryonic melanophore development in amphibians. Dev. Biol., submitted for publication.

Warm, R. (1975) Restoration of the capacity to form pole cells in u.v.-irradiated Drosophila embryos. J. Embryol. Exp. Morphol. 33, 1003–1011.

Watlington, C.O. and Harlan, W.R., Jr. (1969) Ion transport and lipid content of isolated frog skin. Amer. J. Physiol. 217, 1004–1008.

Watts, D.J. and Treffry, T.E. (1975) Incorporation of N-acetylglucosamine into the slime sheath of the cellular slime mould Dictyostelium discoideum. FEBS Letters 52, 262–264.

Webster, G. (1967) Studies on pattern regulation in Hydra. IV. The effect of colcemide and puromycin on polarity and regulation. J. Embryol. Exp. Morphol. 18, 181–197.

Webster, G. and Wolpert, L. (1966) Studies on pattern regulation in Hydra. I. Regional differences in time required for hypostome determination. J. Embryol. Exp. Morphol. 16, 91–104.

Wedner, H.J., Hoffer, B.J., Battenberg, E., Steiner, A.L., Parker, C.W. and Bloom, F.E. (1972) A method for detecting intracellular cyclic adenosine monophosphate by immunofluorescence. J. Histochem. Cytochem. 20, 293–295.

Weimer, B.R. (1928) The physiological gradients in Hydra. I. Reconstitution and budding in relation to length of piece and body level in Pelmatohydra oligactis. Physiol. Rev. 1, 183–230.

Weinhouse, S. (1955) Oxidative metabolism of neoplastic tissues. Adv. Cancer Res. 3, 269–325.

Weiss, P. (1939) Principles of Development. A Text in Experimental Embryology. Holt, New York.

Weston, J.A. (1970) The migration and differentiation of neural crest cells. Adv. Morphogen. 8, 41–114.

Wieland, Th. and Govindan, U.M. (1974) Phallotoxins bind to actin. FEBS Lett. 36, 351–353.

Wilson, J.E. (1968) Brain hexokinase. A proposed relation between soluble-particulate distribution and activity in vivo. J. Biol. Chem. 243, 3640–3647.

Winchurch, R., Ishizuka, M., Webb, D. and Braun, W. (1971) Adenyl cyclase activity of spleen cells exposed to immunoenhancing synthetic oligo- and polynucleotides. J. Immunol. 106, 1399–1400.

Wireman, J.W. and Dworkin, M. (1975) Morphogenesis and developmental interactions in Myxobacteria. Science, 189, 516–523.

Wolff, J. and Cook, G.H. (1975) Charge effects in the activation of adenylate cyclase. J. Biol. Chem. 250, 6897–6903.

Wolpert, L., Hornbruch, A. and Clarke, M.R.B. (1974) Positional information and positional signalling in Hydra. Am. Zool. 14, 647–663.

Yasugi, S. (1974) Observations on supernumerary head formation induced by lithium chloride treatment in the regenerating Hydra, Pelmatohydra robusta. Dev. Growth and Diff. 16, 171–180.

Yuan, R., Bickle, T.A., Ebbers, W. and Brack, C. (1975) Multiple steps in DNA recognition by restriction endonuclease from E. coli K. Nature, 256, 556–560.

Zaheer-Baquer, N.Z., McLean, P., Nornbuch, A. and Wolpert, L. (1975) Positional information and pattern regulation in Hydra enzyme profiles. J. Embryol. Exp. Morphol. 33, 853–867.

Zalin, R.J. and Leaver, R. (1975) The effect of a transient increase in intracellular cyclic AMP upon muscle cell fusion. FEBS Lett. 53, 33–36.

Zalin, R.J. and Montague, W. (1974) Changes in adenylate cyclase, cyclic AMP and protein kinase levels in chick myoblasts, and their relationship to differentiation. Cell, 2, 103–108.

Zalin, R.J. and Montague, W. (1975) Changes in cyclic AMP, adenylate cyclase and protein kinase levels during the development of embryonic chick skeletal muscle. Exp. Cell Res. 93, 55–62.

Zillig, W., Fujiki, H., Blum, W., Janekovic, D., Schweiger, M., Rahmsdorf, H.-J., Ponta, H. and Hirsch-Kaufmann, M. (1975). In vivo and in vitro phosphorylation of DNA-dependent RNA polymerase of Escherichia coli by bacteriophage T7-induced protein kinase. Proc. Nat. Acad. Sci. U.S.A. 72, 2506.

Cell interactions in vertebrate limb development 10

D. A. EDE

1. Introduction

The limbs of the fowl embryo, on which most investigations in this field have been carried out, appear at 3 days as slight bulges on the flank, the wing buds anteriorly and the leg buds posteriorly. These limb buds, which are flattened in the dorsoventral (DV) plane, elongate and expand at the distal end to produce a paddle-shaped outgrowth over the next 4 days (Fig. 1a); thereafter their shapes become more complicated as the distinguishing features of the wings and legs become more marked.

How the development of the limb is achieved depends upon the activities and interactions of the cells of which the limb bud is made up and, ultimately, upon the genes controlling these activities. Consequently, one major approach to a thorough understanding of the whole system is through the study of genetic

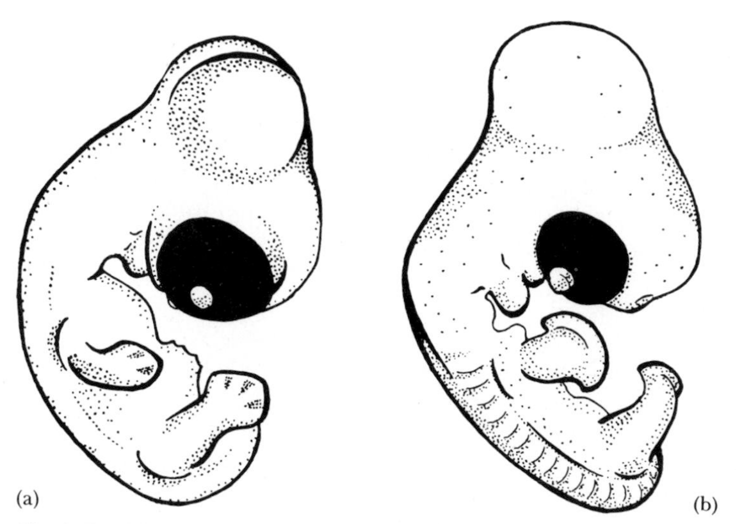

Fig. 1. Fowl embryo at 6 days of incubation: a. normal; b. *talpid*[3] mutant (redrawn from Ede and Kelly, 1964).

G. Poste & G.L. Nicolson (eds.) The Cell Surface in Animal Embryogenesis and Development

mutations which affect limb morphogenesis, especially if these mutations also produce detectable effects at the cellular level. One which has been studied in detail is the *talpid*^{3} mutant, produced by a recessive lethal gene, *ta*^{3}, (Ede and Kelly, 1964) which in homozygotes produces a dramatic distortion of limb development, with a fan-shaped rather than a paddle-shaped outgrowth (Fig. 1b). Homozygous *ta*^{3}/*ta*^{3} embryos invariably die, usually at 6–8 days of incubation, probably owing to breakdown of the general circulatory system, but parts grafted to normal embryos continue to develop and manifest their mutant characteristics (Hinchliffe and Ede, 1968).

2. *Overall patterns of development*

2.1. *Patterns of growth*

Patterns of growth for wing and leg buds in both normal and *talpid*^{3} embryos have been analysed by Wilby and Ede (1976) and are illustrated for development between 2 and 7 days, stages 18–30 according to the normal chick table (Hamburger and Hamilton, 1951), in Fig. 2. In Fig. 3 the position of the base line has been altered in order to reveal the growth patterns more clearly by maximizing the concentricity of the distal expansions, minimizing overlaps and aligning the major growth axes as far as possible. These drawings show that in each case a distinction can be made between two phases of growth in which first distal expansion and then proximal elongation is most prominent. Each limb shows an initial polarized expansion (stages 18–24) in which change in shape is confined to the distal tip, producing the paddle or modified paddle-shape, followed by a proximal elongation (stages 24–30) in which the developing paddle is displaced distally on an elongating stalk while the distal expansion becomes more uniform (stages 26–30). The differences between wing and leg, normal and abnormal, are essentially differences in the extent of each growth phase, e.g. *talpid*^{3} limbs are characterized by having a more uniform expansion and a less extensive proximal elongation, suggesting that there may be a constant control pattern, with small differences either in the degree to which cells respond to control signals or in the intensity of the signals in the different limb types. At all stages in its outgrowth the limb bud is flattened in the dorso-ventral axis.

2.2. *The limb skeleton*

As the limb bud develops, the pattern of cartilage elements which prefigures the limb skeleton emerges and can be observed by a number of techniques. Each element originates as a mesenchymal condensation which produces chondroitin sulphate, an acid mucopolysaccharide characteristic of cartilage which can be demonstrated histochemically using alcian blue or its synthesis detected autoradiographically by uptake of [^{35}S]-sulphate. Searls (1965) showed that sulphate incorporation is uniform throughout the limb bud up to stage 22 but that after

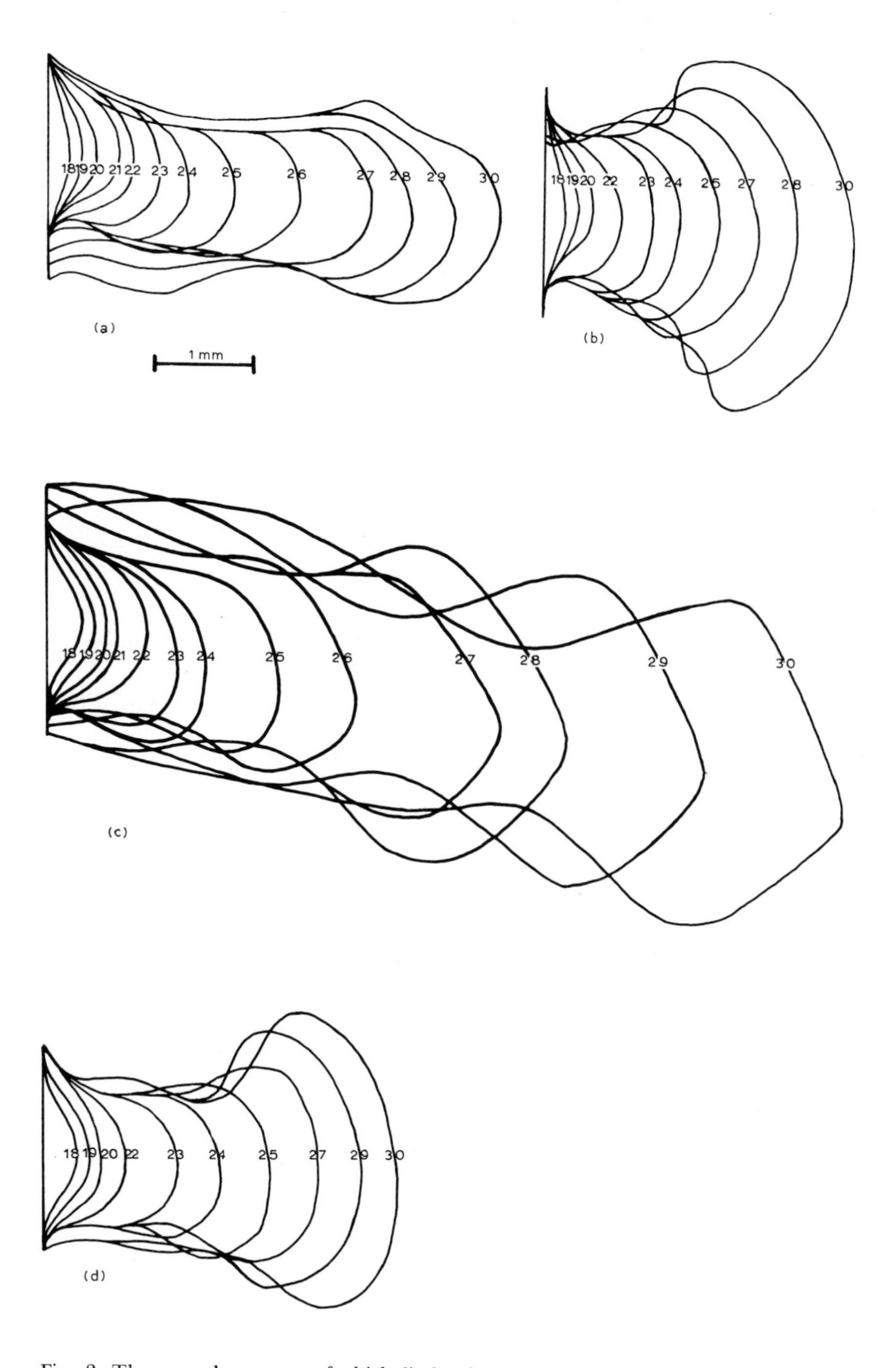

Fig. 2. The growth pattern of chick limbs drawn with a common baseline at the flank/limb intersection: a. normal wing; b. *talpid*[3] wing; c. normal leg, and d. *talpid*[3] leg. Numbers represent Hamburger-Hamilton stages (redrawn from Wilby and Ede, 1976).

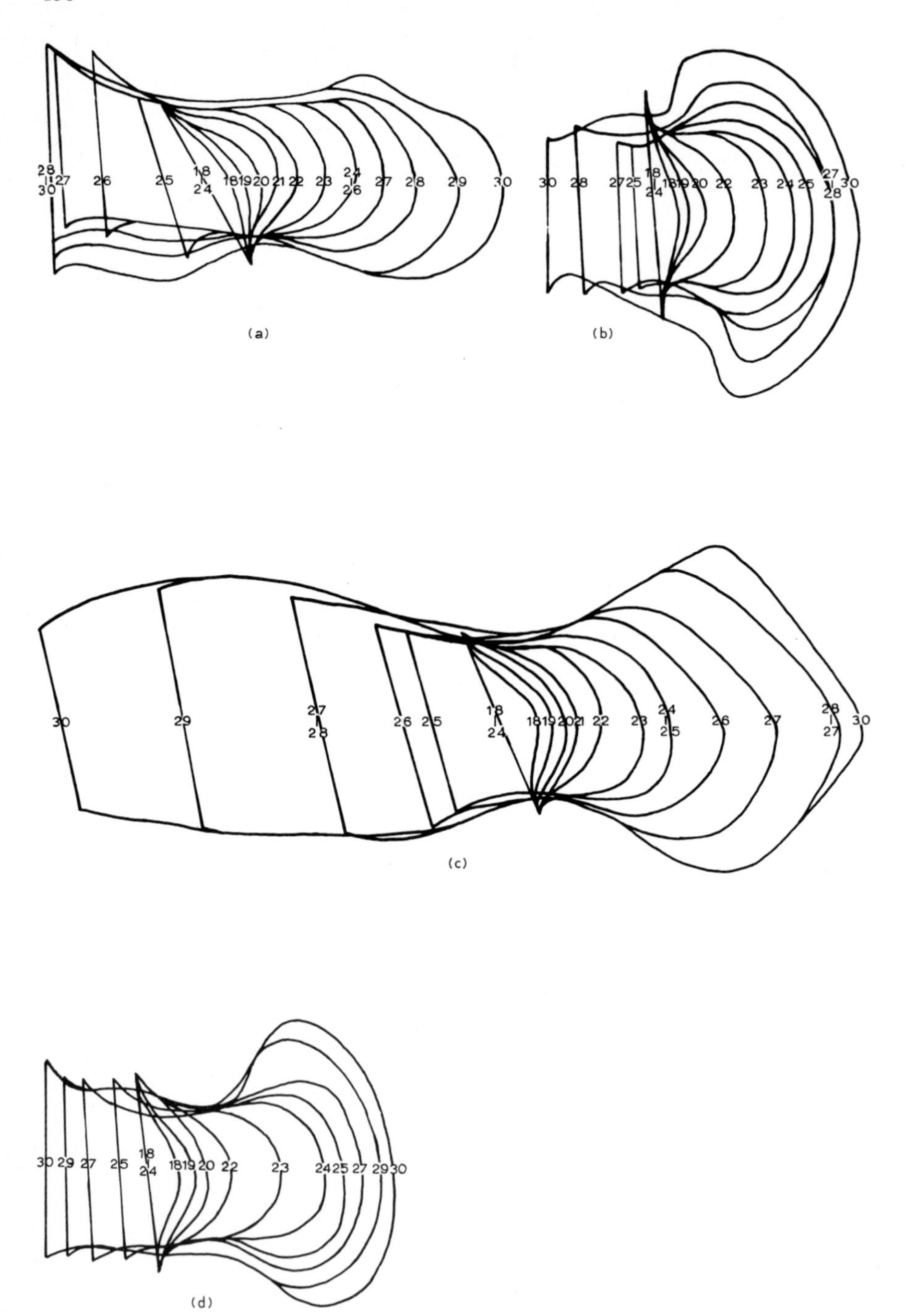

Fig. 3. The growth patterns of Fig. 2 drawn to emphasize the distinction between distal expansion and proximal elongation (redrawn from Ede and Wilby, 1976).

stage 24 it is high in the chondrogenic regions but very low elsewhere. Hinchliffe and Ede (1967) used the alcian blue technique in following skeletal development in normal and *talpid*[3] wings, and Hinchliffe and Thorogood (1974) used both techniques for the leg in each case. Development is essentially similar in fore and hind limb buds, with the most proximal elements appearing first and the most distal last, and division of each region (except the most proximal, which gives the single humerus or femur) into a number of elements in sequence along the anteroposterior (AP) axis, with more elements distally than proximally, to give the characteristic avian modification of the pentadactyl limb form. In the wing there is a proximal humerus, radius and ulna, a band of small carpal elements forming the wrist, a row of 4 metacarpals and, most distally, 4 digits, each with a characteristic number of phalanges.

The corresponding elements in the leg are the femur, tibia and fibula, tarsals, metatarsals and digits, whose emergence in the course of chondrogenesis is illustrated in Fig. 4 for stages 21–35, with the proximal parts truncated in

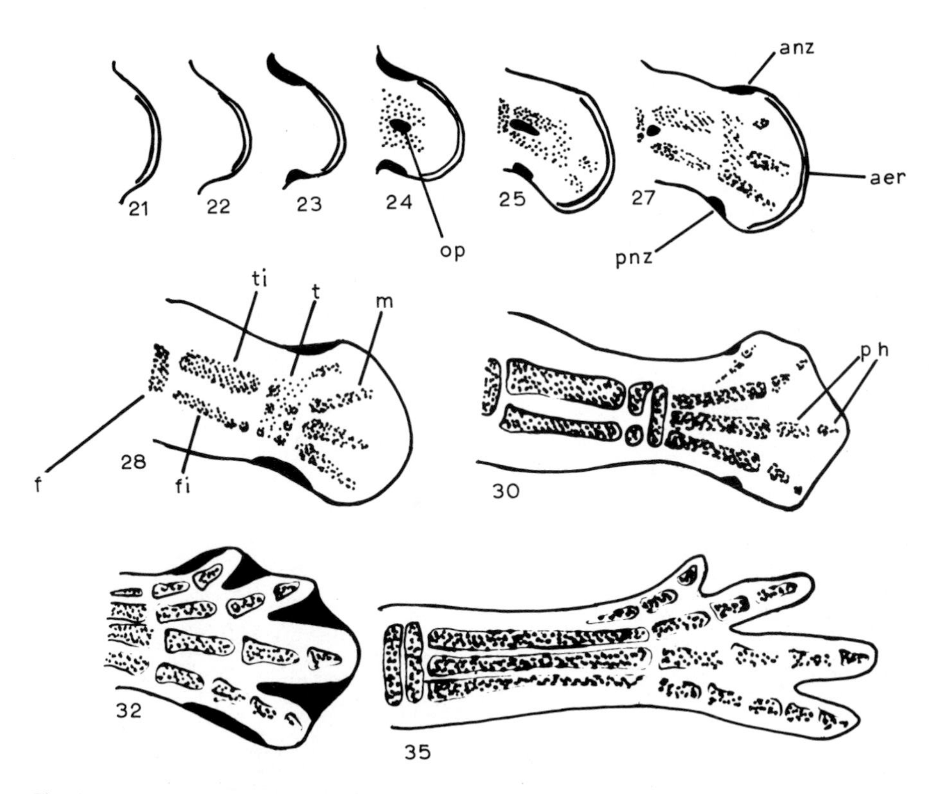

Fig. 4. Patterns of chondrogenesis and cell death in the normal chick hind limb (3.5–9 days; stages 21–35). Areas of cell death in solid black; precartilaginous condensations stippled; cartilaginous elements outlined. anz, anterior necrotic zone; aer, apical ectodermal ridge; fi, fibula; m, metatarsals; op, opaque patch; ph, phalanges; pnz, posterior necrotic zone; t, tarsals; ti, tibia (redrawn from Hinchliffe and Thorogood, 1974).

drawings of later stages. Up to Stage 22 uptake of radiolabeled sulphate is uniform throughout the limb bud; at stage 23 it is greater in the centre than at the periphery; at stage 24 the central chondrogenic region becomes Y-shaped, divided distally into tibia and fibula by a funnel-shaped area of low uptake which corresponds with a region of dying cells known as the opaque patch. At stages 25–26 the tibia and fibula are clearly established as separate elements and less well defined condensations representing the tarsals and the most posterior metatarsals have appeared distally. At stage 28 distinct tarsal elements are visible and all 4 metatarsals have appeared. By stage 30 the phalanges are appearing and all elements up to the metatarsals have sharply defined boundaries. By stage 32 all cartilages are clearly defined and the skeletal pattern is essentially complete.

Chondrogenesis in *talpid*[3] embryos, illustrated in Fig. 5, shows the same

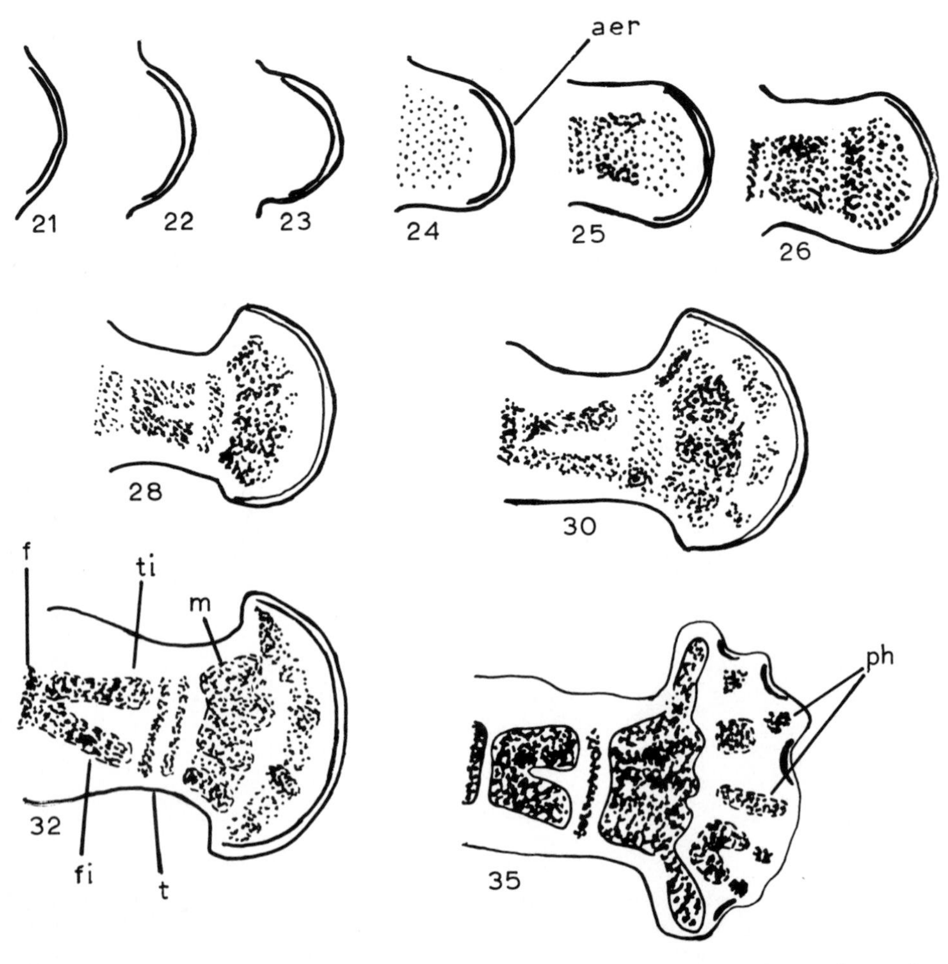

Fig. 5. Patterns of chondrogenesis and cell death in the *talpid*[3] chick hind limb. Conventions as for Fig. 4 (redrawn from Hinchliffe and Thorogood, 1974).

general pattern of emergence, but with some modifications. There is the same sequence of appearance of elements along the proximodistal (PD) axis, and a tendency for each region to break up into separate condensations, with more distally than proximally; but elements along the AP axis often fail to separate and there are more than the normal number of digits produced distally. The autoradiographic pattern is the same as in normal limb buds up to stage 24, but at this stage the central high-uptake region is poorly defined distally and there is no distal division into tibial and fibular condensations. In later stages the border between high and low activity is indistinct and shows a blurred gradation from the high activity representing the skeletal elements to the lower activity of much more extensive chondrogenic AP bands within which they develop. Thus there are no separate tarsal or metatarsal elements, the tibia and fibula are often fused and the 7–8 digits show varying degrees of fusion. The *talpid*^{3} pattern in the wing is essentially the same as in the leg, but with more fusion of adjacent elements.

2.3. Areas of cell death

Fig. 4 also illustrates another feature of the developing normal wing bud, which is easily made visible by staining the living embryo with Nile Blue sulphate or some other vital dye, namely the presence of highly localized areas of cell death. One of these, the opaque patch which separates the developing radial and ulnar condensations in the wing, and the tibial and fibular condensations in the leg, has been mentioned above. Dawd and Hinchliffe (1971) found that the opaque patch reaches its maximum extent at the time when the radius and ulna first appear and suggested a model accounting for the initial Y-shaped chondrogenic pattern based on the interaction of a central process of condensation with distal cell death and autolysis.

The patterns of cell death in avian limb morphogenesis have been reviewed by Hinchliffe (1974). One set of such regions with a clearly evident function comprises the interdigital necrotic zones (INZ). In the chick, cell death occurs in the interdigital tissue between all 4 digits beginning at Stage 31, becoming intense at stage 32 and terminating at stage 34, by which time the interdigital tissue has completely regressed, separating the digits. There appears to be no doubt that the role of the INZ is to shape the contours of the digits, since Saunders and Fallon (1966) showed that in a web-footed species, the duck, the INZs were extremely reduced in extent, and Menkes and Deleanu (1964) showed that suppressing the INZ in the chick embryo by treatment with Janus Green led to soft tissue syndactyly, i.e. effectively a webbed foot.

Much more puzzling are the anterior (ANZ) and posterior necrotic zones (PNZ) originally described by Saunders et al. (1962). Each consists of a well-defined area situated immediately beneath the ectodermal covering behind the expanded distal tip of the limb at the anterior or posterior margin, produced by a wave of cell death travelling distally as the bud grows in length, most obviously between stages 23–26. These authors proposed that the function of these zones might be to sculpt the characteristic limb contours, but the absence of these zones in some birds and mammals, e.g. Japanese Quail and the mouse, made this doubtful and Saunders (1966) showed that even in the chick, cell death in the PNZ could be prevented by grafting dorsal mesoderm in the prospective PNZ region, without any effect upon morphogenesis.

These regions of cell death are under genetic control; in the sex-linked *wingless* (*ws*) mutant of the fowl investigated by Hinchliffe and Ede (1973) the ANZ of the wing appears precociously and extends much beyond its normal boundary; in *talpid*^{3} mutants on the other hand cell death is repressed, leading to absence of the opaque patch and the ANZ and PNZ in both fore and hind limb buds (Hinchliffe and Ede, 1967; Hinchliffe and Thorogood, 1974).

3. *Tissue structure of the limb bud*

3.1. *The ectodermal covering*

The structure and ultrastructure of the limb ectoderm has been studied by Jurand (1965) and by Ede et al. (1974). The general ectoderm consists of two layers: an outer periderm and a deeper basal layer which rests upon a basal lamina (Figs. 6 and 7). The cells of the basal layer are columnar and arranged in a loose epithelium with large intercellular spaces separating them, except at the basal lamina where they are in contact with their neighbours all round; there are some other areas of contact distributed randomly over the main body of each cell. The much thinner periderm is composed of a single layer of flattened cells, orientated at right angles to the cells of the basal layer and in close contact with them. Each cell makes close contact with its neighbouring periderm cells around the whole of its periphery and the junctions between periderm cells show interlocking protruberances and indentations.

The structure of the ectoderm is uniform except at the distal edge of the limb bud, where it forms a thickened apical ectodermal ridge (AER), 25–35 μm high, displaced slightly towards the ventral side and rather higher posteriorly than anteriorly. Various aspects of its appearance in scanning electron microscope pictures and light microscope sections are illustrated in Figs. 8 and 10 (normal wing buds) and 9 and 11 (*talpid*^{3}). There is no apparent difference between normal and mutant in respect of this structure.

Amprino and Ambrosi (1973) have shown that the ectoderm slides in a P-D direction from the base towards the apex of the bud and that the origin, maintenance and gradual configurational changes of the AER depend, partly at least, on packing of the epithelial cells at the distal border, where the sliding fronts of the dorsal and ventral ectodermal sheets meet. They therefore regard the ridge, not as an anatomically distinct structure but simply as a band of the epithelial limb cover which, mainly on account of its location, acquires a compact structure and a peculiar physical denseness. The morphological observations of Ede et al. (1974) support this interpretation: the basal layer of the non-ridge ectoderm, with its columnar cells separated by wide gaps except at their base, forms in effect an extended caterpillar track which is adapted to flow smoothly over the limb contours. The thick periderm layer, with its closely interdigitating pavement-like cells, forms a flexible cover. The deformation of the basal cells at the collision line would give the appearance found in transverse sections of the AER, with tall wedge-shaped basal cells, arranged fan-wise, which retain their connection with the basal lamina though the nuclei become displaced at various levels. The periderm cells become rounded off and forced out of their previous cell contacts by the compression process, and many become necrotic.

However, this account may be inadequate, since in the mutant Eudiplopodia (Fraser and Abbott, 1971a,b) there is an additional AER produced on each limb bud, dorsal and slightly proximal to the primary one, and produced a day or more after it. This clearly makes the simple hypothesis of two colliding distal-moving sheets difficult to apply. This characteristic clearly implies some special property of the mutant ectoderm in its propensity to form the ridge structure: e.g. the AER will regenerate when removed from Eudiplopod leg buds as late as stage 20, whereas in normal limbs it cannot do so after 18, and in experimental conditions Eudiplopod limb ectoderm can form a ridge for several stages after normal ectoderm loses this capacity.

Searls and Zwilling (1964) first showed that the AER was capable of reorganizing itself after considerable disturbance and Errick and Saunders (1974) have shown that this may be taken to the extreme of turning the ectoderm inside-out over the limb bud mesoderm, with reorganization of the normal histology and polarity of non-ridge ectoderm within a few hours and of the AER after 3 days. The mechanism of this reorganization is unknown, but it must involve more complex cell interactions than the simple collision theory suggests.

3.2. The mesoderm

The cellular structure of the mesoderm has been studied by Ede et al. (1974), using light microscope and transmission and scanning electron microscope studies. In the limb bud prior to the appearance of precartilage condensations the mesenchyme cells are uniform in appearance and, generally, in distribution, though there is a tendency for looser packing in the central region of the bud in early stages. The cells are of characteristically irregular shape when seen in section (Fig. 12 and 13), with each nucleus about 4–8 μm in shape and occupying a large part of the cell body. The nuclei are overlain over much of their surface by a very thin layer of cytoplasm, with large cytoplasmic extensions in other regions, from which extend much finer tapering filopodia. These may be seen in sections in the large intercellular spaces and scanning electronmicrographs show that they are of considerable length, spanning the gaps between cells and travelling in some cases over the surface of cells, with fine arborizations at their tips. Each mesoderm cell is in contact with many others and examination of stereo-pairs shows that the contact is not confined to immediately adjacent cells, but many, by means of the long thread-like processes, extend over several cells.

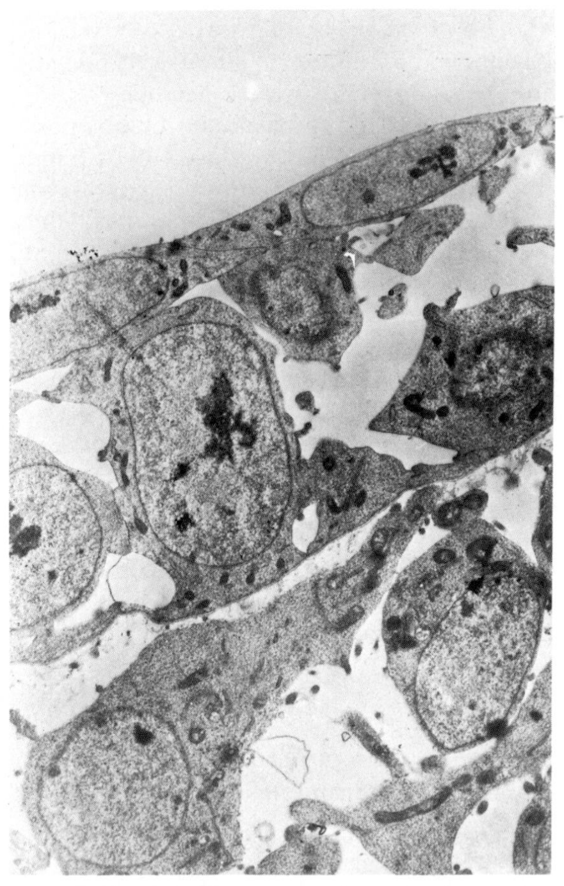

Fig. 6. Transmission electron microscope view of ectoderm and underlying mesenchyme just proximal to apical ectodermal ridge in normal chick limb bud, stage 25. (From Ede et al., 1974).

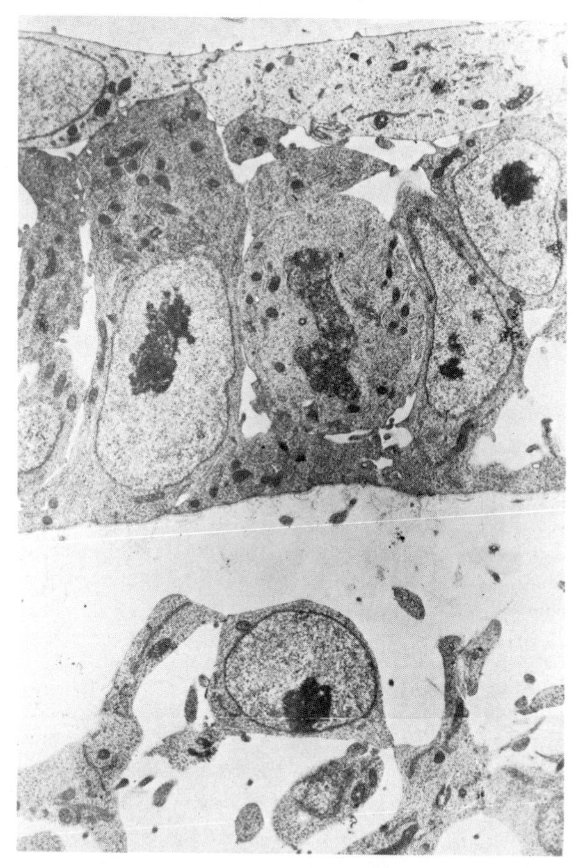

Fig. 7. Transmission electron microscope view of ectoderm and underlying mesenchyme just proximal to apical ectodermal ridge in *talpid*[3] chick limb bud, stage 25. (From Ede et al., 1974).

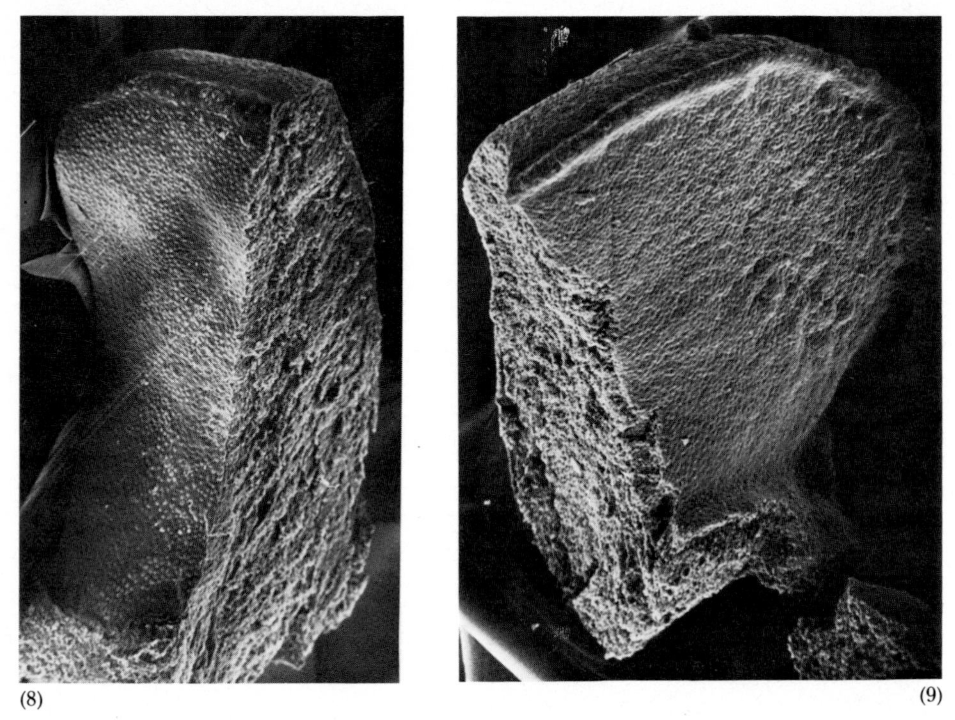

(8) (9)

Figs. 8 and 9. Scanning electron microscope views of wing buds of normal and *talpid*^{3} chick embryos, respectively, stage 25. Half of each limb bud has been cut away, the cut surface passing proximo-distally at right angles to the apical ectodermal ridge (from Ede et al., 1974).

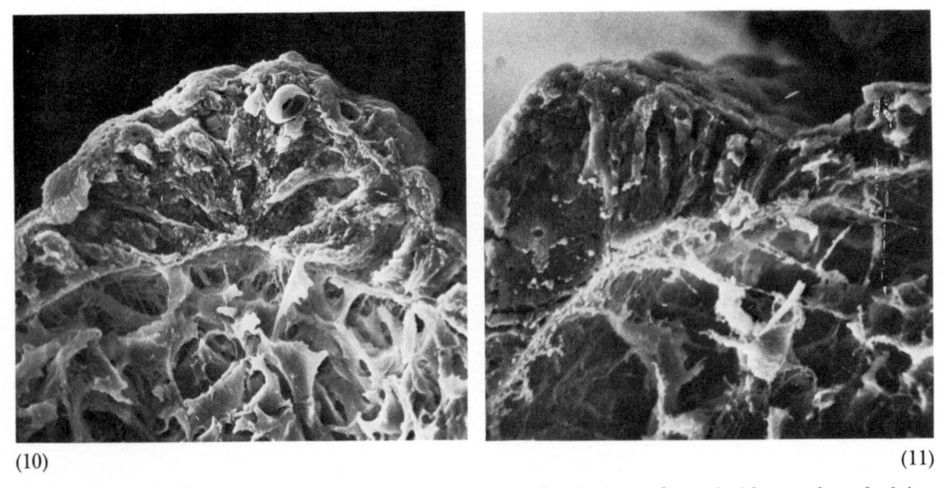

(10) (11)

Figs. 10 and 11. Scanning electron microscope views of apical ectodermal ridge and underlying mesoderm in a normal and a *talpid*^{3} chick embryo, respectively, stage 25 (from Ede et al., 1974).

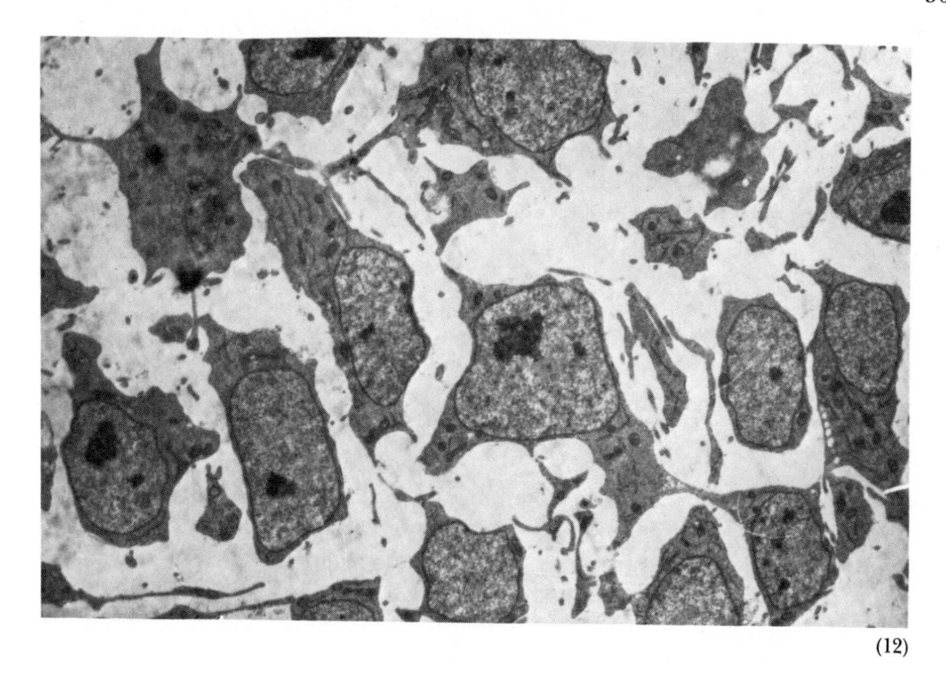

(12)

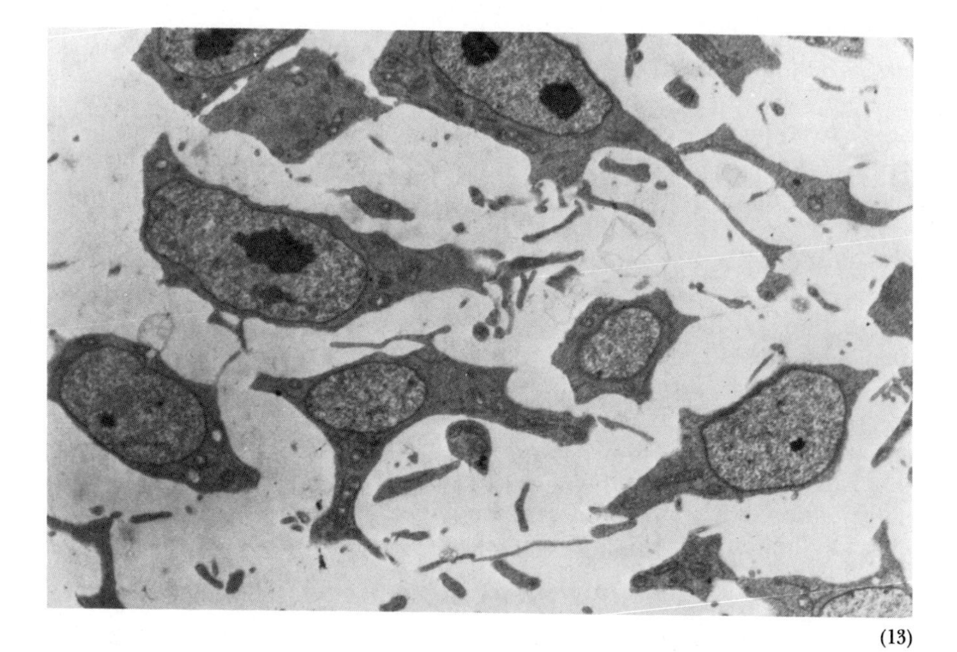

(13)

Figs. 12 and 13. Transmission electromicrographs of mesenchyme from normal and *talpid*[3] chick limb buds, respectively (from Ede et al., 1974).

Limb mesenchyme cells have sometimes been described as stellate but scanning electronmicrographs (Fig. 10 and 19) suggest that the mesenchyme cells tend to be flattened, and rather elongated, with more filopodia arising at each end than in the middle regions of the cell. This polarization of the mesenchyme cells is particularly evident immediately beneath the ectodermal layer where a tendency for cells to orientate at right angles to the interface produces a palisading effect.

Junctions between mesoderm cells consist almost entirely of tight junctions, with some intermediate junctions, but, unlike the basal layer of ectoderm cells, desmosomes are not present. Contact between ectoderm and mesoderm is made by filopodial extensions and more extensive lamellar extensions of the cytoplasm which appear to attach to the epidermal basal lamina. In normal embryos there is almost no gap between the basal lamina and the underlying ectoderm, but there is a gap of the order of 4–6 μm in depth immediately beneath the central cells of the AER. In *talpid*^{3} embryos there is a clear gap of up to 2 μm in all regions.

During the course of development the initial uniformity of the mesoderm cells is diminished as cell differentiation begins, to produce the distinctive histological structures associated with the skeleton, the muscles and the blood vessels. Most work has been done on skeletal development, especially on the early stages of chondrogenesis which leads to the establishment of the pattern of cartilage elements described above. This occurs through a process which has become known as "condensation" through the early work of Fell (1925) and Fell and Canti (1935), who described their origin through mesenchymal cells in these regions becoming more densely packed than elsewhere to form precartilage condensations which subsequently became transformed, through secretion of cartilage matrix, into the cartilages which will later become ossified to form the bones of the limb.

Each condensation is characterized not only by close packing of its cells but by the particular way in which they are arranged as seen in transverse sections (Fig. 14). Ede and Flint (1972) pointed out that this same pattern had been described by Anikin (1929) in developing amphibian limbs and that it was also characteristic of condensations arising in cultured limb mesenchyme cells in vitro. Within the condensation, cells are arranged concentrically and, except in the centre, are crescent shaped, with decreasing cell thickness and increasing radius of curvature towards the periphery. They suggested that this pattern arises through a cell or small group of cells at the condensation centre attracting peripheral neighbouring cells to crowd in on it, so that they would tend to wrap themselves around it, followed by this ring of cells attracting more peripheral cells, and so on. That this pattern is produced by cells crowding in on a central cell is demonstrated in the cell patterns formed in aggregation centres of *Dictyostelium minutum* (Gerisch, 1968), a slime mould in which the myxamoebae move into the centre from all directions rather than in streams as in better known species. The process has been studied at the ultrastructural level by Thorogood and Hinchliffe (1975) who confirmed the existence of early mesenchymal condensations

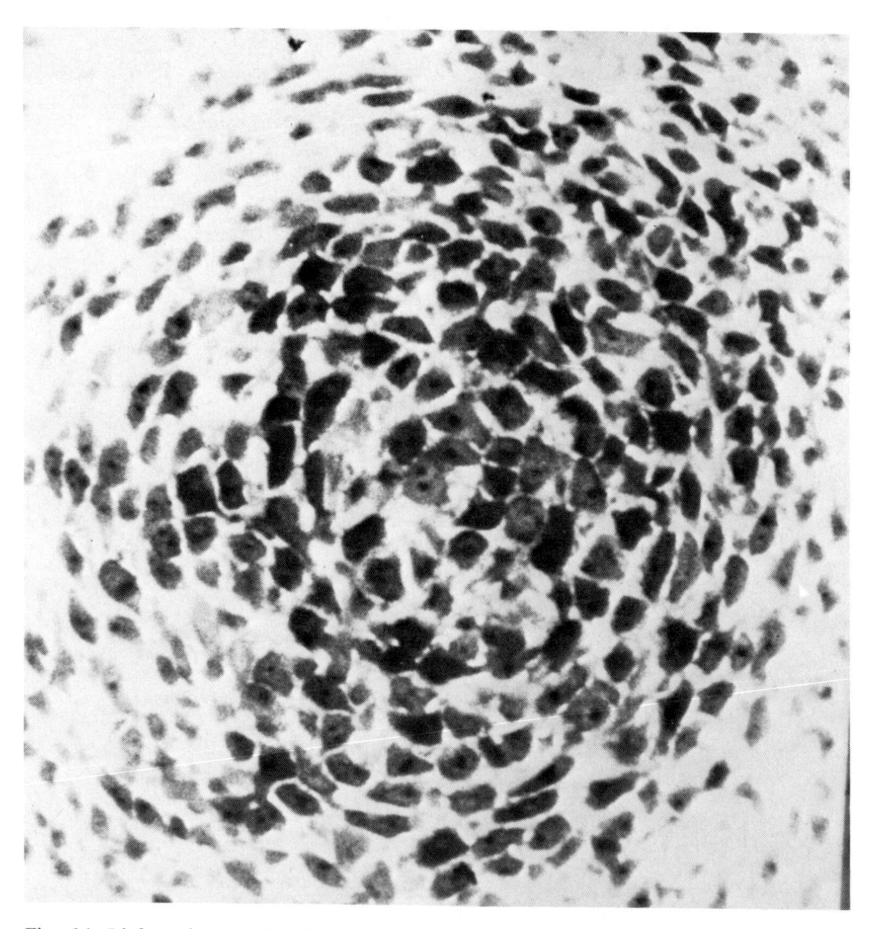

Fig. 14. Light micrograph of precartilage condensation in phalangeal region of normal chick embryo leg bud, stage 30; matrix secretion has begun. Transverse 2 μm section, stain toluidine blue.

composed of closely associated cells, and concluded in the absence of any evidence of differential mitosis being responsible, that they most probably arose through cell aggregation. Gould et al. (1972) have proposed a quite different hypothesis of cartilage condensation, based on the analysis of a series of sections through limbs at stage 24, from which they conclude that there is no condensation of cells in cartilage regions but rather a relative decrease in surrounding non-cartilage cell density. In this case the Anikin pattern could not arise through an aggregation process and Gould et al. (1974) have proposed that it is a consequence of cartilage matrix secretion, which forces randomly orientated cells into a concentric pattern by hydrostatic pressure. Searls et al. (1972) in an electron microscope study also failed to find true condensations of closely packed mesenchyme cells forming the initial skeletal blastema, but the investigation of Thorogood and Hinchliffe (1975) showed clearly (Fig. 15) that the

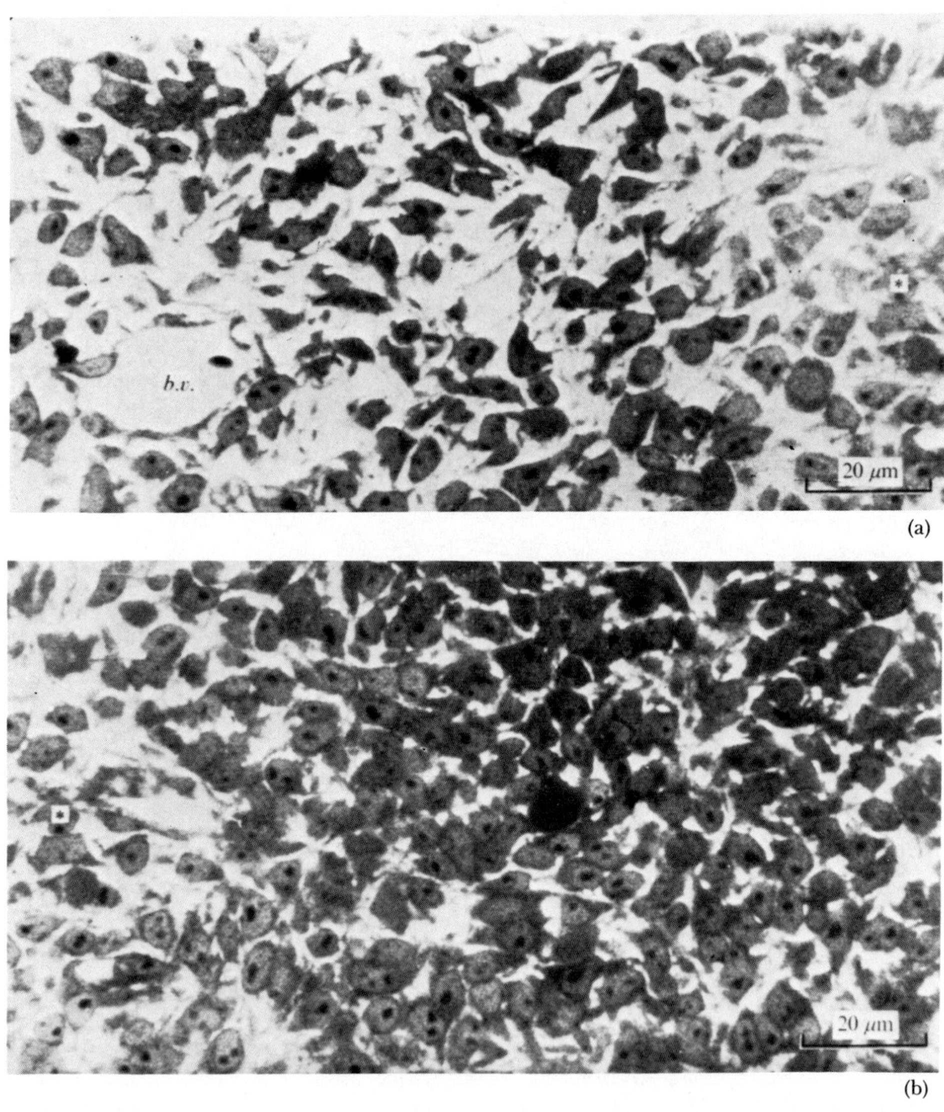

Fig. 15. Light micrograph of transverse 1 μm section of normal chick embryo leg bud, stage 26, in region of transition from undifferentiated loosely packed mesenchyme cells (a) to closely packed cells of the early precartilage condensation of the tibia (b); the point of overlap is marked by an asterisk. Stain, toluidine blue. bv, blood vessel (from Thorogood and Hinchliffe, 1975).

prechondrogenic cells of the leg bud were characterized by large areas of surface contact in the early stages of condensation formation, when there was an increase of about 60% in the cell packing density. The close association stage is a very transient one, since it is followed by secretion of the extracellular mucopolysaccharide matrix which forces the cells apart, and it may be that this is why close association has not been observed by others. Since the mechanism of condensation formation is closely connected with problems of the initiation of chondrogenesis, and with problems of pattern formation, it is important that these apparent contradictions should be resolved.

In *talpid*^{3} the process of condensation is affected, both in the embryo (Ede and Agerbak, 1968) and in vitro (Ede and Flint, 1972). The edges of the condensations are indistinct, the cells merging with those of the surrounding mesenchyme and neighbouring cartilage elements, e.g. the radius and ulna, and the metacarpals tend to merge and form a single band.

Events in the three chief areas of mesenchymal necrosis – anterior and posterior necrotic zones, the opaque patch, and the interdigital areas – have been described at tissue and cell levels by Hinchliffe and Thorogood (1974) for both normal and *talpid*^{3} mutant embryos. In normal limb buds mesenchyme in these areas is characterized by the presence of many densely staining, moribund and dead cells and cellular debris. This material is often in the form of phagocytic inclusions in the cytoplasm of large macrophage-like cells, as described in detail by Dawd and Hinchliffe (1971) in a cytological, cytochemical and electron microscope study of cells of the opaque patch. In *talpid*^{3} limb buds, on the other hand, where such cell necrosis is absent in the ANZ and PNZ and there is only occasionally a very reduced occurrence in the opaque patch and interdigital zones, the mesenchyme cells present a healthy and viable appearance. The histological appearance of cell death is normal in other mutants where its incidence is increased, such as *wingless* (*ws*) investigated by Hinchliffe and Ede (1973), but it appears precociously and extends beyond its normal boundaries. Somewhat similar extensions of normal cell death areas have been noted by Ede (1969) in the mutant *Ametapodia*, but whereas the extent in *wingless* is highly variable between different embryos and between right and left limb buds of the same embryo, in *Ametapodia* the pattern of cell death, as indicated by the effect it has upon the later development of the skeletal pattern, shows little variation among embryos and is highly symmetrical, suggesting that the process is under some sort of precise control.

The studies of Saunders and colleagues (Saunders et al., 1962; Saunders, 1966, Saunders and Fallon, 1966; Fallon and Saunders, 1968) on the PNZ have shown that cell death in this region is determined by a hierarchy of genetic, positional and temporal factors. Hinchliffe (1974) comments that cell death may therefore be thought of as the prospective fate of a group of competent cells, progressively determined by a sequence of morphogenetic signals. As noted above, the morphogenetic role of some of these regions, particularly the ANZ and PNZ remain obscure, and the fact that no cell death occurs in these regions in other embryos where the ultimate limb form is essentially the same, suggests that the cell death may be only an incidental effect or manifestation of a control process affecting some more developmentally significant event.

4. *Activity and interaction at the cell level*

4.1. Cell proliferation

The cellular activities which in concert produce the changes of form and pattern observed in limb development have been listed by Ede and Law (1969). These are mitosis, cell death, cell movements and orientations of any degree, and the changes which occur in the course of cell differentiation to give the various tissue types – changes in size and shape, and also in packing density through secretion of extracellular materials or other causes. The effects of these activities on the developing shape of the limb bud will also be affected by any constraints imposed by boundary membranes, e.g. on the mesoderm by the overlying ectoderm. Of these, rates and distribution of mitosis are the most important parameters producing the increasing mass of the limb and these have been investigated in the chick limb mesoderm by Cairns (1966), Hornbruch and Wolpert (1970), Janners and Searls (1970), Ede et al. (1975) and Thorogood and Hinchliffe (1975).

The existence of a mitotic gradient, with a high count distally, was first noted by Amprino (1965). It has been investigated by Hornbruch and Wolpert (1970), who showed that the overall level of mitosis dropped steeply between stages 18–23 and more gradually thereafter, and the existence of a gradient which was statistically significant from stage 25. They suggested that the gradient had no importance as a mechanism of limb outgrowth and that the elongation of the limb bud in its early stages was produced rather by the form of the overlying ectoderm, whose form would in turn be determined by its intercellular contacts, especially at the apical ectodermal ridge.

Ede et al. (1975) investigated the distribution of mitosis in the chick wing bud in more detail from stages 24–27 in the *talpid*^{3} mutant as well as in normal chicks. This confirmed the existence of an overall gradient at this time but revealed differences between the central and peripheral regions which relate to the presence of differentiating cartilage cells in the former. In normal embryos the gradient exists at all three stages in the central regions but there are great irregularities in the slope (Fig. 16a and c); in peripheral regions the gradient exists at the earlier stages, with a much smoother curve, but it has disappeared at stage 27 (Fig. 16e) when there is also a drop in the peripheral mitotic index from 4–9% to 1–4%. Janners and Searls (1970) showed that the gradient is not due to changes in length of the cell cycle. At stage 19 proximal cells are dividing rather more rapidly than distal cells (12.1 hour cycle against 13.6 hours) but that by stage 24 they have equalized at 13.2 hours. They have also shown that by stage 24, when mucopolysaccharide synthesis in connection with chondrogenesis is beginning, 75% of cells in the proximal region have stopped dividing, but only 25% in the subapical region. This is also the stage at which the precartilage condensation process begins, in which the chondrogenic regions become defined by an increased packing density, 30% over the surrounding mesenchyme according to Gould et al. (1972) and 60% according to Thorogood and Hinchliffe (1975). The process of cartilage differentiation, which commences with the

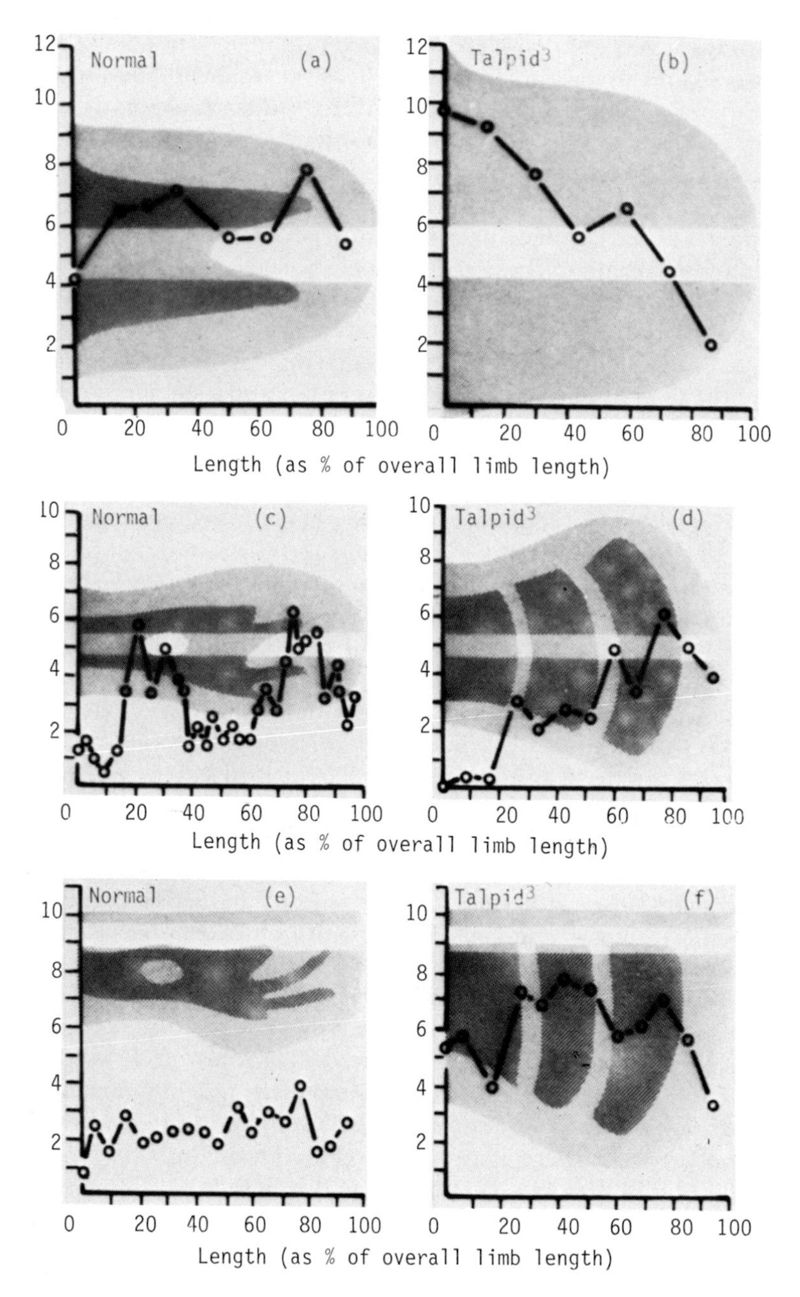

Fig. 16. Change of mitotic index with position in limb bud along proximo-distal axis in normal and *talpid*^{3} chick wing bud: a,b, stage 24, central region; c,d, stage 27, central region; e,f, anterior peripheral region, stage 27. The silhouette of the limb bud (light stippling) and of the chondrogenic regions (heavy stippling) is superimposed and the path along which mitoses were counted is shown as a light track; the outlines are distorted around this track which is drawn straight (redrawn from Ede et al., 1975).

proximal elements and spreads distally, is accompanied by both a decrease in the number of dividing cells and an increase in packing density, so that in the central region there is a gradient of each, with opposite directions and consequently an inverse relationship between the two. The irregularities in the slopes for the central region arise from the sequence of sampling sections passing through the complex alternation of chondrogenic and non-chondrogenic areas produced by the developing elements of the cartilage pattern and from temporal differences arising from the sequential origin of these skeletal elements, i.e., on entry into a cartilage region there is a sudden fall in mitotic index and on leaving it there is a rise. Additional irregularities in the slopes will be expected where the count passes through any area of cell death. Thus, at stages 24 and 25 there is a fall in the central counts at the point of divergence between radius and ulna, but not at stage 27; this corresponds to the region of cell death known as the "opaque patch" investigated by Hinchliffe and Thorogood (1974) which disappears by stage 27.

Summerbell and Wolpert (1972) established a relation between cell density and cell division in the early development of the chick wing and suggested that their finding that mitotic index is inversely proportional to cell density was important not only as being the first demonstration of density dependent growth control in vivo, but also provided a new mechanism for the control of growth and pattern in limb morphogenesis. For this it is assumed that there is a causal relationship such that increasing density inhibits mitosis; it is also assumed that the ectoderm is growing in such a way that free space becomes available between ectoderm and mesoderm, and faster at the distal tip. If the mesenchymal cells show contact inhibition of movement and tend to move into any free space, cell density will be reduced at the tip and lead to increased cell division there, thus establishing the proximo-distal gradient. The authors suggest that this model makes it possible to think of the AER as exerting its influence not through an inductive cellular interaction (see section 5.1.) but by controlling the growth of the ectoderm and thus mitosis in the underlying mesoderm. However, the statistically significant correlation between mitotic index and cell density is based upon the total pooled data, and much of this will relate to chondrogenic areas in which, for the reasons given above, an inverse relationship would be expected. Ede et al. (1975) found no significant correlation in the peripheral, i.e. non-chondrogenic regions, though a shallow mitotic gradient did exist at stage 25, suggesting that where chondrogenesis is not involved there may be no relationship. But there did appear to be a relationship at the non-chondrogenic tips, indicated as corresponding dips and rises on the graphs, which, whether or not it has the significance for outgrowth control that Summerbell and Wolpert suggest, deserves further investigation.

In *talpid*^{3} (Ede et al., 1975) at relatively late stages, (e.g. stage 27; Fig. 16d) the central mitotic counts follow the cartilage pattern closely. As in normal embryos, there is an overall inverse correlation between mitotic index and cell density in the central region which may again be explained as arising from the sequence of events in chondrogenesis. Cell death in the opaque patch has been shown by

Hinchliffe and Thorogood (1974) to be absent or markedly reduced, and there is correspondingly no dip at this point in the mitotic gradient at the region where the initial bifurcation between radius and ulna occurs in earlier stages (e.g. stage 24; Fig. 16b). At stage 24 there is a surprising inversion of the gradient found in all normal embryos and in *talpid*^{3} embryos at later stages, with a very high (up to 10%) mitotic index proximally. Jenner and Searls (1970) have shown that at stage 19 proximal cells are dividing rather more rapidly than distal cells (12.1 hour cycle against 13.6 hours) but that by stage 24 they are equalized at 13.2 hours. In *talpid*^{3} there is no information about cycle times, but the equalization has clearly not occurred and the divergence may have been exaggerated. In normal limbs the more rapid proliferation of proximal cells is off-set by the reduction in number of dividing cells as chondrogenesis begins but in *talpid*^{3} Hinchliffe and Ede (1968) found that chondrogenesis is retarded and at stage 24 a much higher proportion of central cells will still be dividing.

In peripheral regions in *talpid*^{3} there is a mitotic gradient with the lowest index proximally at stage 25 as in normal limbs, but whereas in normals there is an overall drop to very low levels at stage 27, in the mutant the level remains high (Fig. 16f). There is a peak of mitosis corresponding to the widest part of the limb paddle, and it may be that this contributes to the fan-shaped outgrowth of the mutant limb bud. This effect of the *talpid*^{3} gene in maintaining a high level of mitotic activity where a decline occurs normally has also been observed by Ede and Flint (1972) in cultured aggregates of limb bud mesenchyne cells, where mitotic activity in the non-chondrogenic periphery declined sharply in normal aggregates over 4h of culture but continued to increase in *talpid*^{3} aggregates. It therefore appears to be an autonomous effect of the gene on non-chondrogenic mesenchyme cells rather than a reaction to special conditions arising within the limb bud.

4.2. Cell death

Cell death in the limb bud is a most interesting phenomenon; its morphogenetic significance is by no means clear and it may be that its occurrence is something in the nature of a by-product of some more significant morphogenetic control process. It is programmed, presumably according to the position its cells are in at an early stage, but it can be re-programmed in certain situations. The pattern of cell death is under genetic control, and is modified in certain mutants.

The process of cell death and elimination in the anterior and posterior necrotic zones has been described by Hinchliffe and Ede (1973) in the limb bud of normal and *wingless* chick embryos. Within these areas, isolated mesenchymal cells deteriorate, die and are then ingested by neighbouring viable mesenchymal cells which thus become transformed into macrophages, first ingesting and then digesting further dead cells. This sequence has also been described in the opaque patch region of the central mesenchyme (Dawd and Hinchliffe, 1971). The necrotic regions are marked by high acid phosphatase activity. The enzyme is found in what appears to be viable mesenchymal cells, localized in lysosomes, but the most striking areas of intense activity are the macrophages which each contain several discrete areas of intense activity within the cytoplasm, and more diffuse activity associated with the dead cells in the digestive vacuoles. The intense acid phosphatase activity appears to indicate digestion of dead cells within macrophages by acid hydrolases rather than that cell death results from intracellular release of lysosomal enzymes. The process at cellular level is the same in the *wingless* mutant as in the normal embryo, but the anterior necrotic zone (ANZ) arises precociously, at stage 19 rather than stage 21 and comes to extend over the whole anterior, preradial, region of the limb bud. There is a pattern of cell death in the mutant, derived from the normal pattern apparently by simple extension of the ANZ, but there is considerable variation of expression between individuals and even between different limbs in the same individual. In another fowl mutant, *Ametapodia* (Ede, 1969), ANZ development is retarded, but the PNZ is enlarged and extends anteriorly through the metacarpal and metatarsal regions at stages 25–27. In this case the pattern of cell death, which leads to a highly distorted skeletal development, shows very little variation and limbs in opposite sides of the body are identical mirror-images. The control mechanism in normal embryos and *Ametapodia* mutants is precise, but imprecise in *wingless*.

Saunders and Fallon (1966) investigated the conditions under which cell death occurred in the limb and showed it to be programmed from the earliest appearance of the limb bud at stage 17, though no cells died or showed any sign of deleterious changes until stage 21 and the most intense necrosis appeared at stage 24. If the presumptive PNZ region at stage 17 (i.e. the stem cells from which the PNZ is derived) was transferred as a graft to the dorsal side of the embryo in the somite region, death occurred on schedule at stage 24. The cells which replace those removed from the PNZ region do not die, so it may be inferred that the "death clock" is set by some sequence of events prior to stage 17. However, the cells evidently can be reprogrammed so that they remain viable and undergo histodifferentiation, since presumptive PNZ tissues from stages 17–23 which are grafted to the dorsal side of the wing bud show no signs of necrosis and often produce an extra spur of cartilage. Presumptive PNZ tissues maintained in organ culture in vitro show necrosis on schedule, but this can be prevented by placing against them a piece of central wing or leg mesoderm from stage 24 embryos, even when separated by a Millipore filter, providing the PNZ tissue is extirpated before stage 21 and the filter pore size is not too small. A pore size of 0.45 μm allows effective interaction but with a pore size of 0.05 μm death occurs on schedule. After being maintained for 6 days in this type of culture, but not before, the PNZ cells remain healthy indefinitely in the absence of central mesoderm. It thus appears that after a prolonged period in contact with some factor the death clock can be turned off. Whether actual cell contact is necessary or not is uncertain, but Saunders and Fallon believe it is not.

Ede and Flint (1972) investigated the patterns of cell death in aggregates prepared from dissociated mesenchyme cells obtained from both normal and *talpid*^{3} mutant limb buds. In both cases the procedures caused a high level of cell death in the initial aggregates, but this level declined over the next 3 days in culture. Within the aggregates the patches of cells in the interior become chondrogenic, and in these regions cell death is reduced to a very low level; in the non-chondrogenic regions, the decline of cell death is less marked, and it follows that there is some special property of chondrogenic cells which leads either to more efficient mopping up of dead cells by macrophages or of less cell death among them.

In these aggregates the number of dead and dying cells is much less initially in *talpid*[3] than in normals; by 4 days there are no dead cells at all in *talpid*[3] pre-cartilage areas and though the decline is not so steep in non-cartilage areas, the number of dead cells here is still significantly less than in normal aggregates. *Talpid*[3] cells appear to be more viable in these conditions and this may be related to a cell property which also leads to the absence of necrotic zones in the intact limb, as described above. It appears that in the normal limb-bud in the fowl embryo and some but not all others, certain cell lineages are subject at a very early stage to factors which normally lead to cell death at a later stage, but that contact or near-contact with cells from some other region of the limb, or some mutant cell property, may prevent the implementation of this programme.

4.3. Cell movement

The occurrence of cell movement in the developing limb bud is in a different class from mitosis and cell death, since it cannot, with present techniques, be observed directly. But where there is a predominantly unidirectional outgrowth, with no obvious local distribution of mitosis to account for it and with mitotic figures orientated at random, it may be inferred that it must play some part. The difficulty is to establish its extent, and whether it involves active translational cell movement, or cells simply rotating around each other, or whether such movement as occurs is only passive, imposed by forces such as a constraining ectoderm. Searls (1967) and Stark and Searls (1973) have established that at least there is no cell dispersal and intermingling of cells, since labelled tissue fragments elongate but do not otherwise lose their original conformation when grafted into a limb site. This does not, as Fristrom and Fristrom (1975) have pointed out in the rather similar problem of evagination of imaginal disc evagination in insects, eliminate the possibility of cell movement, but requires that it should be limited and/or orderly.

Cell movement cannot be directly observed in the limb but the potential capacity of the mesenchymal cells for movement and their characteristic activity and interactions in movement can be investigated in monolayer culture. If cells from mutants exhibit abnormal behaviour and neighbour interactions in vivo, it may then be possible to relate this to the abnormalities in morphogenesis which distinguish the development of the mutant limb buds in vivo. This in vitro approach has been followed by Ede (1971, 1972), Ede and Wilby (1976) and Cairns (1975) in the fowl and by Elmer and Selleck (1975) and Kwasigroch and Kochhar (1975) in the mouse. Further cell properties related to movement, cell contact and adhesion, have been investigated in the formation of aggregates from suspensions of dissociated limb mesenchyme cells in culture (Ede and Agerbak, 1968; Ede and Flint, 1975a).

Ede and Flint (1975b) explanted mesenchyme fragments from early wing buds of normal and *talpid*[3] mutant chick embryos in plastic Petri dishes and studied the behaviour of individual cells as they moved out onto the plastic. The impression given by time-lapse ciné-films was that the mutant cells moved much more slowly than normal cells, and this observation was confirmed and refined by statistical analysis. Two parameters of cell movement were recorded: the distances moved over 100-second intervals and the length of time each cell spent at rest. It was shown that the average speed of movement over the whole path tracked for each cell, inclusive of time at rest, was significantly greater in normal than in *talpid*[3] cells, but there was no difference between distances moved over short time intervals of 100 seconds; the difference lay in the percentage of time spent at rest, which was very significantly higher in the mutant cells. This appeared to be related to differences in cell morphology, not recognisable in all cells but in the majority of them. Normal limb mesenchyme cells are typically elongated, with a ruffled membrane at the leading edge and attachments to the plastic and to other cells by long cytoplasmic filopodia, chiefly at the trailing edge. The attachments break as the cell advances and the cytoplasm contracts into the cell body. *Talpid*[3] cells are much more flattened than normals and when the cell is moving ruffled membranes are produced round much more of the cell periphery so that there is no obvious polarity; a few elongated filopodia are produced, but there are many more short spiky microvilli all round the cell periphery which attach to the plastic and to other cells. It is this absence of polarity and the all round attachment to the substrate that appears to immobilize the mutant cells for long periods.

These same differences in the morphology of the cell surface, with due allowances for the differences produced by the cells being in suspension and in three-dimensional aggregates rather than flattened on a plastic surface, account for the different patterns of aggregate size and shape produced by reaggregating suspensions of normal and *talpid*[3] limb mesenchyme cells in a gyratory shaker (Ede and Agerbak, 1968). Ede and Flint (1975a) showed by measurements of collision efficiency in a Couette viscometer that the mutant cells are more adhesive to each other than normal cells and by light and electron microscope studies on the process of aggregation in its early stages how this led to the observed differences in aggregate morphology. Aggregation in *talpid*[3] led to compact rounded structures consisting of cells tightly bound together, which having reached a certain size remained as discrete aggregates, whereas in normals there was from the beginning a looseness of arrangement and absence of organization in normal aggregates, which often began as strings of 3 or 4 cells rather than the spheroidal groups found in the mutant (Fig. 17 and 18). The surfaces of the resulting aggregates were irregular and rough, and collisions of aggregates lead to mutual adhesion and fusion to form a larger aggregate, producing ultimately a few large irregular aggregates instead of the many

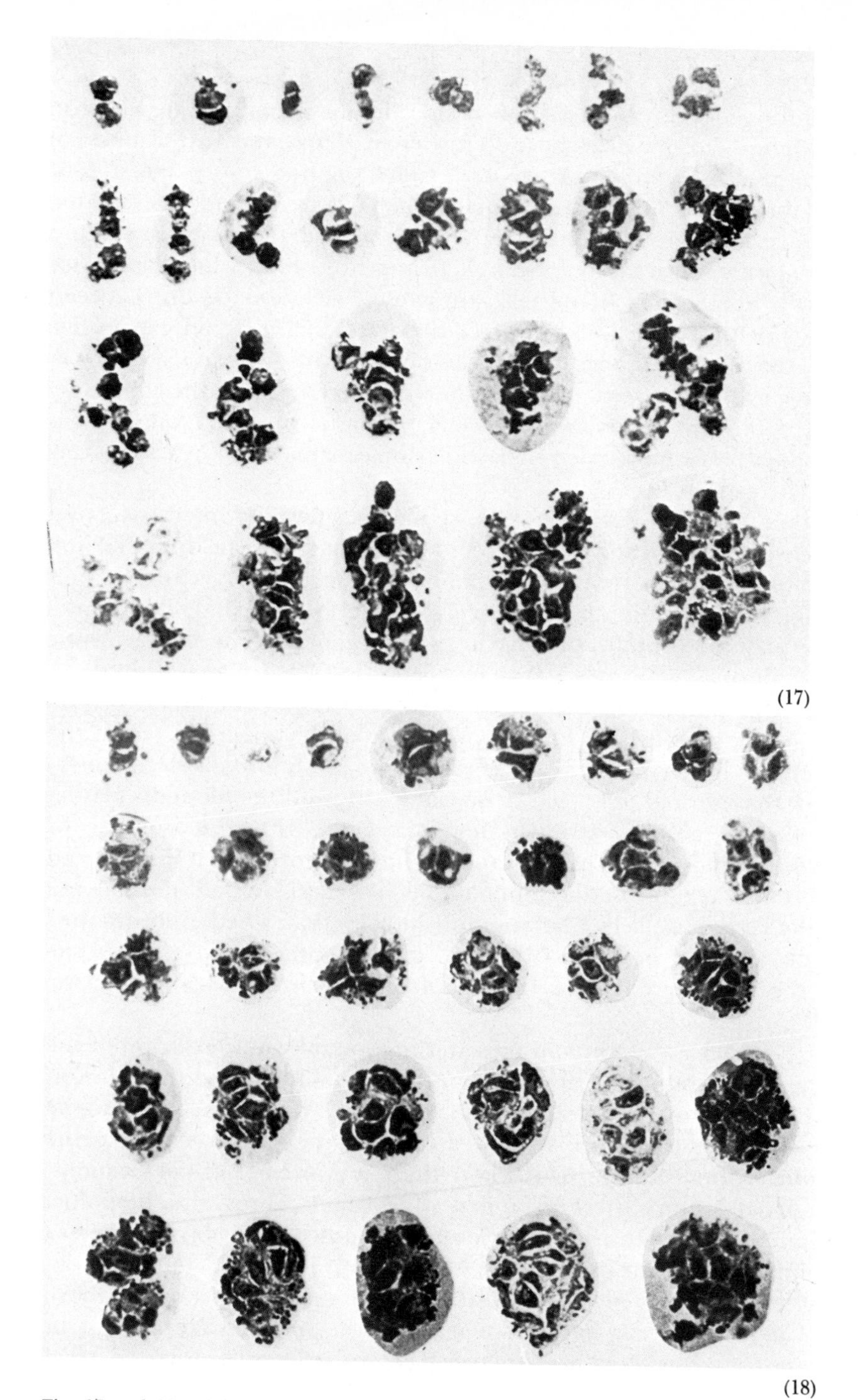

(17)

(18)

Fig. 17 and 18. Light micrographs of 1 μm sections of cells from normal and *talpid*[3] chick wing mesenchyme, respectively, reaggregated for 4 hours in rotation culture. The micrographs are arranged in order of increasing cell number and aggregate size in each case (from Ede and Flint, 1975a).

smaller and rounded aggregates found in the mutant material. The electron microscope studies showed that the differences in shape and in the internal arrangements of cells within normal and *talpid*^{3} aggregates is a result of differences in the extent of cell-to-cell contacts; *talpid*^{3} cells form more extended contacts and are more greatly deformed by them, and their aggregates are consequently more compact and spherical. It appeared that, when 2 cells are juxtaposed, whether normal or mutant, the closest cell contacts are between microvilli, i.e. very fine projections of the cell surface as well as by more extensive appositions of the cell membranes, and it seems likely that contact is made first at the tips of these cytoplasmic extensions, followed by extension of the contact by zipper action. As in the observations of cells on plastic substrata, *talpid*^{3} cells appear to produce many more microvilli, which are distributed more extensively over the cell surface.

In culture then, limb mesenchyme cells are highly active and interactive, and differences between normal and mutant cells lead to characteristic differences in the formation of aggregates from cell suspensions. The question arises whether corresponding characteristics appear in cells in the embryo and, if they do, whether they play any significant part in morphogenesis. The characteristic appearance of normal limb mesoderm has been described above. Ede et al. (1974) showed that in transmission electronmicrographs *talpid*^{3} limb mesoderm cells showed fewer of the long fine filopodial extensions which traverse the intercellular spaces in normal embryos (Fig. 12 and 13). Scanning electronmicrographs show that normal cells tend to be elongated, with the filopodia arising at each end. *Talpid*^{3} cells appear much more "ragged", with large numbers of shorter and thicker filopodia arising all round the edges of the cell (Fig. 19 and 20). The differences are shown diagrammatically in Fig. 21. In both normal and mutant cells the scanning electronmicrographs suggest that, whether or not they move, the cells are in a highly active state, for it is unlikely that either the filopodia or the lamellar edges of the cells from which they arise are static structures.

Interaction between ectoderm and mesoderm, especially in the region of the AER, is highly important in the development of the limb bud and the junction between them is therefore of special interest. In both the normal embryo and the *talpid*^{3} mutant the two layers are separated by the basal lamina except at the region immediately beneath the AER where there is a number of perforations. Beneath the basal lamina there is a mat of collagen fibres and filopodial extensions from the mesoderm cells are closely juxtaposed, probably attached to the basal lamina, and appear to be stretched taut. In *talpid*^{3} embryos the connexion is generally limited to the tips of the filopodia and, as noted above (section 3.2), there is a clear gap between ectoderm and mesoderm. But in normal embryos contact is often made between extensive lengths of mesodermal cells, suggesting that filopodia are establishing contact with the basal lamina, adhering to it and then extending the contact. The process may be similar to that proposed by Trelstad et al. (1967) based on their observations on primary mesenchyme cells in the early chick embryo, where "filopodia extending from

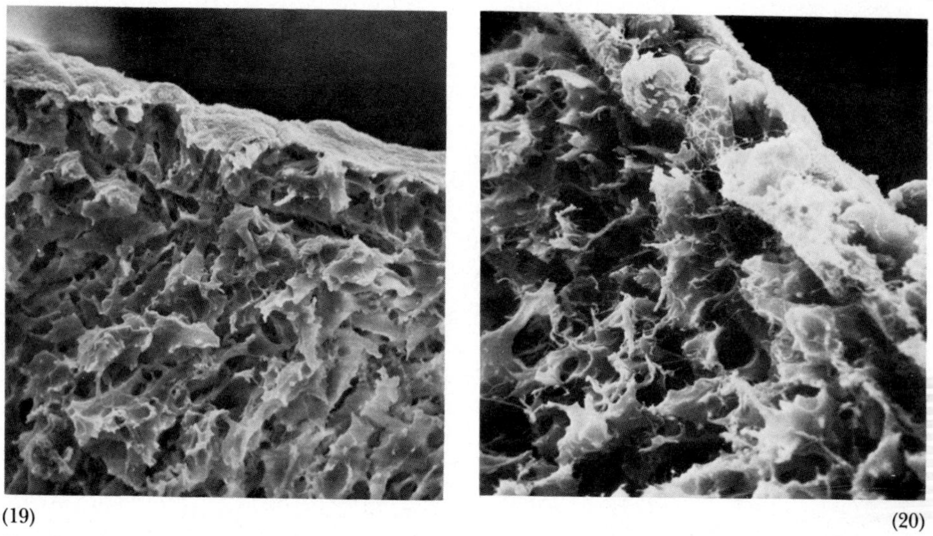

Fig. 19 and 20. Scanning electronmicroscope views of ectoderm and underlying wing-bud mesenchyme taken from regions just proximal to the apical ectodermal ridge of normal and *talpid*[3] embryos, respectively, stage 25 (from Ede et al., 1974).

Fig. 21. Diagrammatic representation of normal and *talpid*[3] wing-bud mesenchyme cells, partly in section (from Ede et al., 1974).

the advancing edge of the uppermost and lowermost mesenchymal cells seem to attach to the basement epithelial lamina (of the epiblast and hypoblast) as if they were feeling their way along the substratum". The distinct gap in *talpid*[3] limb buds may be due to the mesodermal cells adhering to each other more tenaciously than normal cells, for basically the same reason that they adhere more closely to each other in aggregates, and because, if they behave as they are observed to do in culture, they do not throw out so many of the very long filopodia which it appears may be necessary to make attachment to the ectodermal basal lamina. The situation is illustrated diagrammatically in Fig. 22.

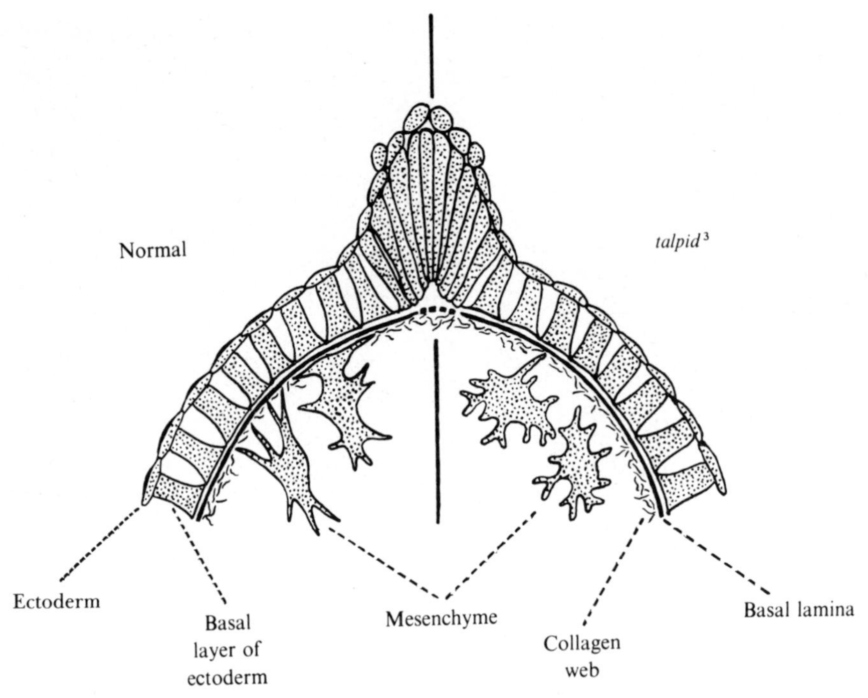

Fig. 22. Diagrammatic representation of the ectoderm and underlying mesoderm in the region of the AER, with a comparison of the ectoderm/mesoderm boundary in normal and *talpid*^{3} wing buds (from Ede et al., 1974).

All these observations suggest that the limb mesenchyme cells are in a state of considerable activity in the production of cytoplasmic processes, and at the ectoderm-mesoderm junction there is evidence that in normal embryos cells near the surface are attaching themselves to the basal lamina and hauling themselves towards it. In the interior of the limb bud at least some small movements must occur. Ede and Agerbak (1968) pointed out that in long bone rudiments the chondroblasts become aligned so that they are orientated at right angles to the long axis but that in *talpid*^{3} the chondroblasts are not aligned in this way; their inference was that some movement is necessary to produce the normal orientation and the inhibition of movement in *talpid*^{3} cells interferes with it. There are two other aspects of limb bud morphogenesis in which cell motility and cell to cell adhesion may play a part – in limb bud outgrowth and in the formation of precartilage condensations – and both of them are affected dramatically in the *talpid*^{3} mutant.

The increasing mass of the developing limb bud is produced by cell proliferation, but this cell activity in itself, even if there are local variations in mitosis, is insufficient to explain the particular mode of outgrowth which give it its distinctive elongated form. The work described above has shown that a proximo-distal gradient of mitosis does occur in the limb, at least from stage 24, but that it appears to be chiefly a by-product of the process of chondrogenesis.

Mitolo (1971) and Wilby and Ede (1976) have shown, using computer simulation techniques, that it is possible to obtain a limb-like outgrowth by means only of controlled mitotic distributions, but that the control would have to be complex and inflexible and that the gradients would have to be much steeper than those actually found in the developing limb. A much more simple and flexible control of outgrowth is obtainable if some distalward movement of the mesenchyme cells is introduced, and this would account for the "polarity of outgrowth" which Starke and Searls (1973) list together with three other cell activities – cell division, cell death and cell differentiation – as being necessary and sufficient to describe limb morphogenesis. Ede and Law (1969) have shown that computer simulation can generate a good approximation to normal limb bud outgrowth using a proximo-distal mitotic gradient together with some limited distalward cell movement and that reducing the movement produces a shape like that of the fan-shaped limb bud in the *talpid*[3] mutant. If distalward movement is essential for normal limb outgrowth, reduction of movement would produce a fan-shaped outgrowth; in *talpid*[3] the bud is fan-shaped and the observations reported above have shown that its mesenchyme cells do show restricted mobility in some circumstances. More elaborate and realistic simulations by Wilby and Ede (in preparation) suggest that the movement of cells around each other, probably following cell division, may be sufficient to produce the degree of outgrowth required. These considerations do not distinguish indisputably between active movements of the mesoderm cells and passive movements which are imposed on them by external constraints. Amprino (1965), though he points out that it is likely to be only one factor among others, and Hornbruch and Wolpert (1970) have emphasized the role the ectoderm might play, especially the thickened apical ridge, in moulding the growth of the underlying mesoderm. It is certainly the case that the isolated ectoderm, supported by the relatively rigid AER, does provide a jacket in which, in experiments, dissociated limb mesenchyme cells can be packed to produce a reconstituted limb which will go on to develop in a grafted site (Finch and Zwilling, 1971; Pautou, 1973). But isolated mesoderm can also maintain its shape and continues to do so in culture until cells begin to move actively away from it. In the course of evolution it is likely that ectodermal outgrowth and the growth of the underlying mesoderm have come to be well correlated, with interactions between them leading to a harmonious outgrowth, rather than one being entirely passive to the direction of the other. If the "collision" theory of AER formation described above is correct (section 3.1.) the ectoderm in the period when the AER is being formed must slide over a fairly rigid determined form, i.e. the underlying mesoderm. At a later stage the proposal by Ede et al. (1974) that mesodermal cells are attaching by filopodia to the ectodermal basal lamina and hauling themselves into close contact with it suggests that the ectoderm is providing a firm base for this activity. The mesenchymal cells are clearly neither immobilized nor behaving like passively flowing particles and the form of the outgrowing limb bud must be the result of a balance of forces exerted by all the cell activities discussed here together with the influence of the constraining ectoderm.

The experiments on cell movement and adhesion reported above have been made upon suspensions of pooled mesenchyme cells prepared from whole mesoderm of stage 24–26 embryos. Work by Cairns (1975) indicates that there is a detailed regional pattern of these cell properties even at very early stages of limb development. He isolated blocks of mesoderm about 0.1 mm in diameter from various regions of chick wing buds of stages 17–22 and cultured them individually in microtest plate wells and investigated both cell migration from the explants and cell death within the explants. The cells that migrated from blocks taken from a narrow distal band (up to 0.15 mm from the AER) were clearly different in morphology and behaviour from those coming from explants from any more proximal region. Cell migration occurred earlier from the proximal blocks; the proximal colonies so produced consisted largely of bipolar elongate cells which moved some distance away from and out of visible contact with neighbours, while distal colonies consisted only of stellate multipolar cells which remained in close contact with neighbouring cells. A more subtle regional variation was also observed in the antero-posterior axis in both cases. Work in progress by Bell and Ede confirms the existence of these regional differences in normal embryos but shows that in *talpid*^{3} embryos there is no distinction between distal and proximal colonies, cells in both resembling those of the distal colonies from normal embryos. This might suggest that in *talpid*^{3} cells of the narrow distal band are identical with normal cells of this region, but that the transformation to proximal type cells as the limb develops does not occur. But *talpid*^{3} and normal colonies may be distinguished by the presence of considerable debris indicating cell death in normal and its almost complete absence in mutant colonies. This is consistent with the findings by Ede and Flint (1972) reported above (section 3.2.) that *talpid*^{3} cells are more resistant to injury through suboptimal environmental conditions and suggests that there is a difference between the normal and mutant cells probably related to some cell surface property, which exists in all mesenchyme cells but which expresses itself as a behavioural difference only in the proximal cells.

Other disturbances of cell morphology and behaviour have been observed in mouse mutants, where again they are related to abnormal limb morphogenesis. Elmer and Selleck (1975) investigated this behaviour in vitro using mesenchyme cells from the *brachypod* (*pb*^{H}) mutant, which is characterized by skeletal malformations in the limb. They cultured cells from normal and *brachypod* embryos at 12 days, when no histological differentiation had occurred, and allowed them to grow at high densities over a 3 day culture period. In normal cultures, cells quickly became polarized and developed long tapering cytoplasmic projections, moving actively and coming together to form compact aggregates which quickly became chondrogenic. In contrast, mutant cultures formed large flattened stellate-shaped cells which produced more diffuse aggregates, in which chondrogenesis was retarded. The authors concluded that the brachypodism gene is acting at the site of the cell membrane which is causing an interference with cell functions peculiar to the developmental program of the skeletal limb elements.

Kwasigroch and Kochhar (1975) studied the in vitro behaviour of cells migrating from limb mesoderm fragments isolated from 12-day normal mouse embryos in experiments similar to those of Cairns on the chick described above, with strikingly similar results. In the mouse experiments, however, the division was made between the limb bud "shaft" and the paddle, so the mouse "distal" region includes a high proportion of cells which in the chick were shown to behave as "proximal". In the mouse the earliest cells to move out from both proximal and distal fragments were bipolar and in the proximal colonies most cells remained bipolar throughout the 3-day culture period and were predominantly orientated radially and migrated accordingly. Cells moving out from distal fragments changed within 2 days from being bipolar to stellate or multipolar in appearance; the cells showed no radial arrangement, the direction of migration was more random, and they were less mobile than proximal cells. When cells were subjected to excess vitamin A either added to the culture medium or treated in vitro before being removed to culture both shape and mobility were affected, both proximal and distal cells exhibiting the stellate shape and decreased rate of movement. Kochhar (1973) has shown that maternal hypervitaminosis produces very specific abnormalities in limb morphogenesis when retinoic (vitamin A) acid is administered to pregnant mice at carefully determined stages. Kwasigroch and Kochhar suggest that the effect of this compound may be to interfere with cell movements and interactions which are necessary for normal limb development, and in particular with the difference they observe between proximal and distal limb cells. If this were related to limb elongation it would account for the extreme shortening of the limbs found in a high proportion of treated embryos. They point out that a possible mode of action of vitamin A could be through its known effect on the integrity of cell surfaces, and thus on cell–cell interactions, by penetration into the lipid phase of their membranes.

The other aspect of limb morphogenesis in which cell movement may play an important part, that is in the origin of the precartilage condensations which prefigure the limb skeleton, is taken up in the discussion on chondrogenesis which follows in the next section.

4.4. Cell differentiation: chondrogenesis

By far the most striking and the most investigated aspects of cell differentiation in the limb bud are the events leading to the development of the cartilage skeleton, which is later transformed by ossification to bone, through chondrogenesis. Myogenesis is also important and involves an approximately equal number of cells; Gould et al. (1972) and Hilfer et al. (1973) have shown that distinct myogenic regions become apparent in the chick wing bud slightly earlier than chondrogenic ones, but the pattern of muscles is less well defined and myogenesis has therefore received less attention than chondrogenesis. Differentiation to form the connective tissue into which most of the rest of the mesoderm develops involves much less change in the morphology and biochemistry of the cells, which remain mesenchymal.

The most widely held view concerning cytodifferentiation originated with the work of Zwilling (1968) and Finch and Zwilling (1971) who proposed that development in the chick limb was divided into two phases: an initial phase in which the changes going on in the development of the limb were purely morphogenetic, establishing the characteristic form of the limb bud and determining the pattern of cytodifferentiation which ensues under normal circumstances and a second phase, beginning at about stage 25, when the fate of these cells become irreversibly stabilized and active cytodifferentiation occurs. This interpretation has been supported by the work of Medoff (1967), Searls (1967), Searls and Janners (1969), Schachter (1970), Caplan (1970, 1971), Ede and Flint (1972), Medoff and Zwilling (1972). All of this work, in various ways, appears to show that up to stage 25 cells in myogenic or chondrogenic regions, even when they are in the early stages of visible differentiation, are capable of becoming transformed in certain experimental situations in the limb or in vitro into cells of the other type, but that after stage 25 their fate is irrevocably fixed.

The sequence of events in chondrogenesis according to this view has been reviewed by Searls (1973) who distinguishes three major phases. In the first all mesoderm cells are phenotypically identical and all synthesize cartilage-specific (chondroitin sulphate) mucopolysaccharides. Medoff and Zwilling (1972) proposed that during this first phase all cells also synthesize myosin. In the second phase, beginning at stage 22, cells begin to be visibly differentiated; in the presumptive cartilage regions chondroitin sulphate synthesis is increased and myosin synthesis suppressed, and vice versa in the presumptive muscle regions. During this phase these changes are reversible. In the third phase, beginning at stage 25, the differentiation of the chondrogenic cells becomes visibly obvious through secretion of their characteristic intercellular metachromatic matrix and they become irreversibly stabilized as cartilage cells. Caplan (1970) has pointed out that this sudden and specific stabilization has the appearance of being an inductive event, and suggests that shifts in concentrations of substances of low molecular weight such as nicotinamide might be responsible for the switch; Caplan and Koutroupas (1973) postulate some sort of metabolic gradient situated radially across the limb which will have a controlling influence on differentiation.

A contrary view has been advanced by Dienstman et al. (1974) based upon cultures of chick leg bud mesoderm of stages 17–25 as monolayers over a period of 1–3 weeks in media which supported equally differentiation of muscle and cartilage. Clusters of myotubes and of chondrified cells were obtained and myogenic clones were made; mixed clones, containing both muscle and cartilage were never found. Only a small proportion of cells in any culture showed terminal myogenic or chondrogenic differentiation, but the percentage increased in cultures from older stages. The authors propose that there is no sudden stabilization event and that the mesoderm prior to stage 22 is not homogeneous; rather, it consists of a heterogeneous mixture of cell types. There are no mesenchymal cells that synthesize both cartilage and muscle specific substances and which may differentiate one or the other according to location, but there are common precursor mesoderm cells which produce distinct myogenic and chondrogenic lineages at least two generations before differentiated muscle and cartilage cells appear. Establishment of the lineages and terminal differentiation occurs asynchronously so that there is gradual development rather than sudden stabilization. On this view, production of cartilage or muscle in particular regions of the limb, or in particular culture conditions in vitro, would depend upon selection of one set of cell lineages rather than the other to survive and replicate. At present neither of these views is established conclusively, but the cell lineage interpretation of the experiments of Searls and Janners (1969) depends to some extent upon the occurrence of cell sorting and this has been shown not to occur in the experiments of Ede and Flint (1972) in aggregates consisting of mixed chondrogenic and non-chondrogenic cells from stage 26 limb buds, in which chondrogenesis occurs only centrally. Whichever view is taken, the visible manifestations of cartilage differentiation occur between stages 22 and 28, beginning with the appearance of the precartilage condensations of mesenchyme cells within the mesoderm, first described by Fell (1925) and Fell and Canti (1935), and believed by these authors to represent a close packing of the mesenchyme cells which would then move apart as the metachromatic intercellular matrix was established. More recent reports based upon electron microscope analysis of the chick wing bud have queried the existence of the early phase of intimate cell association (Gould et al., 1972, 1974; Searls et al., 1972) and proposed that the mesenchyme cells differentiate into chondroblasts and commence matrix secretion without going through the transient condensation stage. This work has been reviewed recently by Thorogood and Hinchliffe (1975) in the light of their own analysis of the condensation process during chondrogenesis in the chick hind limb bud, which concludes that there is a true condensation phase involving close cell to cell association.

All workers agree that there is an increase in cell density in the condensation regions (see above, section 4.1.) but there is disagreement regarding the extent of cell surface contact. According to Thorogood and Hinchliffe (1975) the sequence of events in the tibial rudiment is as follows. The undifferentiated mesenchyme consists of a loose cellular network in which intercellular contacts are at the tips of touching filopodia; they are almost always of the tight junction type and usually focal rather than extended. The first indication of a condensation is a localized increase in mesenchymal cell number, without at first any visible differentiation or formation of extensive areas of intercellular contact. The next stage is characterized by a considerable increase in cell packing density; the cells have rounded up and have few filopodia, with the surfaces of adjacent cells close together though not in direct contact (Fig. 15). The close packing is transient and is progressively lost as chondrogenesis proceeds and the cells become separated from each other. At the same time cartilage matrix components become detectable in the intercellular areas and it is the accumulating matrix which is presumably pushing the chondroblasts apart.

If, as this work appears to show, there is a real, though transient, condensation phase, this may represent a necessary prerequisite for the cell interactions which initiate chondrogenesis. Using chondrocytes derived from chick embryo somites Holtzer and co-workers have shown a clear relation between cellular adhesiveness and chondrogenesis in vitro. Abbot and Holtzer (1966) showed that a physical interaction through cell crowding was necessary to induce and maintain cartilage differentiation; clonal cultures of chondrocytes become strongly adherent to each other and began to synthesize chondroitin sulphate; absence of such interaction led to further DNA synthesis and cell proliferation. Holtzer et al. (1969) also showed that treatment with 5-bromodeoxyuridine caused the cells to lose their adhesiveness to each other and assume a fibroblast-like appearance, leading to suppression of chondrogenesis and inhibition of chrondoitin sulphate synthesis. Drews and Drews (1972, 1973) reached the same conclusion from their studies on cholinesterase activity in chick limb bud cells in vivo and in vitro.

The origin of the close packing leads us to resume discussion of the second aspect of cell movement in the limb bud deferred from section 4.3., i.e., its possible involvement in formation of the condensations. There is an increased cell packing density and no local increase in cell proliferation has been found to account for it, leaving cell movement as the most likely mechanism. Gould et al. (1972) suggested that the non-precartilage cells are dispersing while the precartilage cells remain steady, but Thorogood and Hinchliffe's (1975) observation of a 60% increase in cell packing together with a dramatic reduction of the intercellular spaces in the tibial region supports the suggestion of Ede and Aberbak (1968) and Ede (1971) that there is aggregative movement of cells onto a central cell or group of cells. This would account for the characteristic pattern of cells observed in transverse sections through condensations, first described by Anikin (1929) in amphibians but present in all vertebrates, with cells arranged concentrically and each one crescent-shaped, with decreased cell thickness and increasing radius of curvature towards the periphery (Fig. 14). Gould et al. (1974) believe that the Anikin pattern is produced by centrifugal pressure by secretion of matrix commencing with the central cells and this will certainly reinforce the pattern in the later stages of chondrogenesis. However, Thorogood and Hinchliffe's (1975) observations indicate that the close packing precedes matrix secretion and further observations are required to establish whether the pattern is established at that time. Either way, the pattern indicates that commencement of chondrogenesis is focal, at the centre of each skeletal rudiment. A similar centripetal aggregation of chondroblast cells was proposed by Holtfreter (1968) for in vitro formation of pharyngeal cartilage in the amphibian *Ambystoma*. In this case a cartilage inducer, the pharyngeal ectoderm, is necessary to initiate chondrogenesis, and when a fragment is placed in a dispersed culture of the appropriate mesectodermal cells, aggregation occurs, followed by matrix secretion within the aggregates. His hypothesis is that the primary "heterogenetic" induction which produces the aggregation is succeeded by a secondary "homoiotic" induction involving interactions between the cells of the condensation; the first produces new tendencies of motility and adhesiveness and the second leads to cytodifferentiation and matrix secretion. In the chick the initiator which produces the first phase is less well defined (see section 5.2 below), but a similar sequence may occur.

The hypothesis that cell movement and adhesion play an important part in chondrogenesis is supported by observations on *talpid*[3]. If centripetal cell crowding is a prerequisite of normal condensation and cartilage rudiment formation, increased adhesiveness and reduction of movement would be expected to lead to less well defined condensations and defective chondrogenesis, and this is precisely what is found in the mutant: the peripheral boundaries of condensations are indistinct, cells merging with those of the surrounding mesenchyme and with adjoining cartilage elements (Hinchliffe and Ede, 1967; Ede and Agerbak, 1968; Hinchliffe and Thorogood, 1974). Toole (1972) reached a similar conclusion in his work on hyaluronate turnover during chondrogenesis in the chick limb. He proposed that hyaluronate, which is synthesized in early limb bud stages but which drops rapidly in concentration during early stages of chondrogenesis, interferes with cell interactions leading to aggregation and consequently with matrix secretion, and that removal of hyaluronate, indicated by increased hyaluronidase activity when chondrogenesis begins, is necessary in order to allow the necessary freedom of cell movement. Toole et al. (1972) demonstrated a parallel effect of hyaluronate on chondrogenesis in vitro and proposed that its primary effect was likely to be upon the cell surface, acting there as a regulator, involved in controlling the timing of morphogenetic events requiring cell interactions; removal of hyaluronate by hyaluronidase would then result in cell interactions leading to "communication", i.e. in the condensation process, differentiation and immobilization in the correct place and sequence.

5. *Cell interactions and morphogenetic control mechanisms*

The most challenging problems in limb development are concerned with the integration of the activities and properties of the constituent cells described in preceding sections to give the harmoniously developing limb bud, with its characteristic three-dimensional overall form and internal skeletal structure, and which will have the potential of producing ultimately the precisely detailed anatomical complexity of the functional limb. There is at present a considerable amount of information available at the supracellular level (see review by Saunders, 1972) but at the cellular and molecular level almost all is conjectural. A number of hypothetical models have been devised but more experimental evidence is required.

In general, morphogenetic control mechanisms are of two sorts. In some systems, *regulation* is possible, such that the removal of any part is rapidly compensated for, with or without additional cell division. Other systems are characterized by *mosaic* development, such that, after varying development intervals, each part, down to the level of a single cell in some cases, can develop autonomously to its final state and no cognisance is taken of missing portions. These types of development obviously involve quite different degrees of cell interactions: regulative development must involve long range interactive control whereas mosaic development requires only the physical interaction of totally autonomous cells. Aspects of both types of development may occur in a single developmental system, and this appears to be the case in the limb. The control mechanism in the limb bud appears to differ along the three co-ordinate axes – proximo-distal, antero-posterior and dorso-ventral, and each of these will be discussed in turn.

5.1. The proximo-distal (P-D) axis

A real understanding of limb bud morphogenesis began with the work of Saunders (1948) and Zwilling (1949, 1955) leading to an appreciation of the critical importance of the apical ectoderm ridge in controlling events in the P-D axis (see review by Zwilling, 1961). The first step was the establishment of presumptive fate maps, showing that only those presumptive regions destined to form the most proximal parts of the limb exist in the earliest stages and that the rest are added in proximo-distal sequence as the bud grows outwards. The second step was to establish that this sequential process depends upon the presence of the AER space at the apex of the bud. These authors regarded the activity of the AER as an induction because grafting a supernumerary AER to the dorsal surface of a limb bud led to the formation of a supernumerary limb at this site (Saunders and Gasseling, 1968; Zwilling, 1956b,c). Zwilling (1956a) and Zwilling and Hansborough (1956) showed that interaction between ectoderm and underlying mesoderm was reciprocal, i.e., that the mesoderm was necessary to maintain the structure and activity of the AER, and suggested the existence of a maintenance factor which was produced by the mesoderm cells and which had this function. These concepts were further developed by Saunders and colleagues (Saunders et al., 1957, 1959; Saunders and Gasseling, 1965; Saunders and Reuss, 1974).

It is important to note that the activity of the AER controls both outgrowth of the limb bud and the sequential development of limb parts. Extirpation of the AER produces a limb in which those parts were missing whose territories, according to the presumptive fate maps, had not yet been laid down at the time of the experiment; outgrowth of the parts whose territories had been established completed their development (Saunders, 1948; Gasseling and Saunders, 1961). Summerbell (1974) extended and quantified these observations, showing that the level of extirpation was related extremely precisely to the length of the subsequent rudiment.

The question arose as to whether the effect of the level of extirpation reflected level-specific differences of the AER, i.e., whether the interaction with the underlying mesoderm involved an instruction signal. Experiments by Hampé (1959, 1960) appeared to show that this was the case, but subsequent very careful experiments by Pautou (1972) and Rubin and Saunders (1972) have demonstrated clearly that the reverse is true. These authors concluded that information for the proper sequence of limb levels is programmed not in the AER but in the responding mesoderm. The AER produces not a level-specific sequence of different inductive signals, but a constant signal that does not vary from level to level. Wolpert and colleagues (Summerbell et al., 1973; Wolpert et al., 1975) have proposed a model which has these properties. There is assumed to be a zone of mesoderm cells underlying the AER, called the progress zone and supposed to be (from the results of grafting experiments) about 300 μm deep in which the cells are all dividing and are not differentiated. The limits of the zone are determined, in some way, by interaction with the AER. These cells are characterized by some internal state which changes with time, called the positional value. As limb outgrowth proceeds, so some cells will be left behind by the advancing boundary of the zone, and having left the progress zone their positional value will cease to change and the subsequent fate of these cells is determined. Positional value has, in the author's terms, been transformed into positional information.

The change of state in the progress zone cells is autonomous, independent of cell interaction and must depend upon some sort of internal clock. Summerbell et al. (1973) have suggested it may be linked with mitosis, possibly in such a way that a change could occur at only one point in the mitotic cycle. But this, as Wolpert et al. (1975) point out, is incompatible with Summerbell's (1974) observations mentioned above, that require a continuous and highly specific change in positional value. This specificity would be difficult to account for on the basis of any sort of mitotic clock since mitosis is asynchronous, so that neighbouring cells on this, or indeed any sort of metabolic clock hypothesis, will have slightly different positional values. These authors propose that there may be local cellular interactions which will lead to an averaging among groups of cells. The only long range interaction involved is the signal emanating from the AER which specifies the zone within which the mesenchyme cells proliferate with change of positional value. Extirpation of the ridge destroys the progress zone and only those regions established by cells which have already left it can develop. There is at present no information about the nature of this signal. Errick and Saunders (1974) have shown that it is not dependent upon maintenance of the normal AER structure, since the limb bud continues to develop in an "inside-out" ectodermal jacket during the time when the AER is reorganizing its normal cellular structure. Furthermore, Cairns (1975) claims that the specific effect of ridge ectoderm on delaying cell death in cultured limb mesenchymal cells does not depend upon cellular contact. Both these results therefore suggest that a diffusible morphogen may be produced by AER cells.

The progress zone model is essentially one of mosaic development, with no capacity for regulation outside of the progress zone. Regulation has been claimed by Hampé (1959) and work by Kieny (1964a,b see discussion by Sengel in Wolpert et al., 1975) suggests that prospective distal material can be caused to change its fate and give rise to more proximal structures. This contradiction is at present unresolved.

5.2. The antero-posterior axis

In the course of grafting experiments on the posterior necrotic zone in the chick wing, Saunders and colleagues (see review by Saunders, 1972) discovered a clue to the solution to a very puzzling experimental result which led to a major advance in knowledge of limb morphogenesis control. The anomolous result was that rotation of the tip of an early wing bud through 180° led to the production of duplicated distal wing parts, with reversed symmetry anteriorly. Thus the digits in the normal wing are numbered II, III, IV in antero-posterior sequence; in the rotated tip the sequence produced is IV, III, II, II, III, IV (Amprino, 1968; Saunders and Gasseling, 1968).

These investigators discovered that this occurred because the transection was made through a small region of mesoderm cells, located almost coincidentally with the posterior necrotic zone, later named by Balcuns et al. (1970) the zone of polarizing activity (ZPA), which plays a key role in controlling development in this axis. The critical event in the rotation experiment was the transposition of part of the ZPA to an anterior position, and the same effect – producing a duplicated outgrowth with mirror symmetry–could be produced by grafting a portion of the ZPA alone. Insertion of the grafted ZPA at other sites, e.g. at the distal tip of the wing bud, also led to outgrowth of a supernumerary wing tip, and in all cases with the posterior side of the duplicated wing tip facing the graft site. Thus the ZPA has both an inductive activity in causing outgrowth, with concomitant elongation of the AER, and a polarizing activity in determining the antero-posterior sequence of wing parts. Its spatio temporal distribution has been investigated by MacCabe et al. (1973a) who found that at stages 17–22 the ZPA extended into both the wing bud and the posteriorly adjacent body wall; by stage 23 it lies entirely within the limb bud and by stage 27 it is found only near the apex of the bud in the hand-forming region. The ZPA graft may be taken from wing or leg, or indeed from homologous regions of other species, but the supernumerary structure is always of the kind (wing or leg) and species-specific character of the host bud. MacCabe and Saunders (1971) and MacCabe et al. (1973b) have also shown that limbs obtained by dissociating wing-bud mesoderm cells, packing them into an ectodermal jacket and grafting the reconstituted bud to a host embryo show no antero-posterior polarity, but that polarity can be imposed by placing an unfragmented piece of ZPA in the ectodermal jacket with the other mesoderm cells. The ZPA appears to lose its activity at the same time as the AER ceases to have the capacity to induce outgrowth, at stage 29.

Tickle et al. (1975) have proposed that control of development in the antero-posterior axis is based upon the existence of positional information provided by the different levels of some morphogen produced as a signal by the ZPA and set up as a concentration gradient by diffusion and breakdown through the labile mesoderm of the progress zone, the "nature of the digit being determined by its distance from the ZPA".

In the positional information type of theory, pattern determination is regarded as a two-stage process involving (1) an assignment of a specific positional value to each cell, followed by (2) the interpretation of this positional information in terms of molecular differentiation. According to this hypothesis the positional value (Wolpert et al., 1975) will determine whether a cell differentiates to form, say, cartilage or muscle, and also determines the future behaviour of the cells, such as their growth programme, which will produce humerus, radius, ulna or whatever is appropriate for that position. The cells in the progress zone of the limb bud are viewed as having two co-ordinates of position specified, one for the proximo-distal axis and the other for the antero-posterior axis, though the mechanisms are quite different and in the first case (see discussion in Wolpert et al., 1975) is strictly not a positional one. The first mechanism requires no signalling between the cells, except for the short-range interactions necessary to produce averaging of positional value; the second requires cell interaction leading to transmission of the positional signal. In both cases it is difficult to propose specific molecular models for the events involved in registering the positional information and translating it into the resulting changed cell states.

Wilby and Ede (1975) have proposed an alternative model for control of development in the antero-posterior axis which involves more specific though still hypothetical cellular interactions. The contrasting essential aspects of the models may be pointed out (see Fig. 23a and b). The positional information type of model uses a monotonically decreasing gradient of a diffusible morphogen to specify the positional value; thus in a simple case cells might interpret gradient levels between thresholds T1 and T2 as specifying cartilage development. The model becomes somewhat cumbersome when the pattern, as in the arrangement of the skeletal elements, is markedly periodic; to produce three bands of cartilage, chondrogenesis must be initiated between six threshold levels – T1–T2, T3 – T4, T5 – T6. This would require some biochemical switch mechanism sensitive to many triggering levels, calling for a very complex and sensitive, yet very stable, system. The alternative proposal is to use a periodic gradient and a

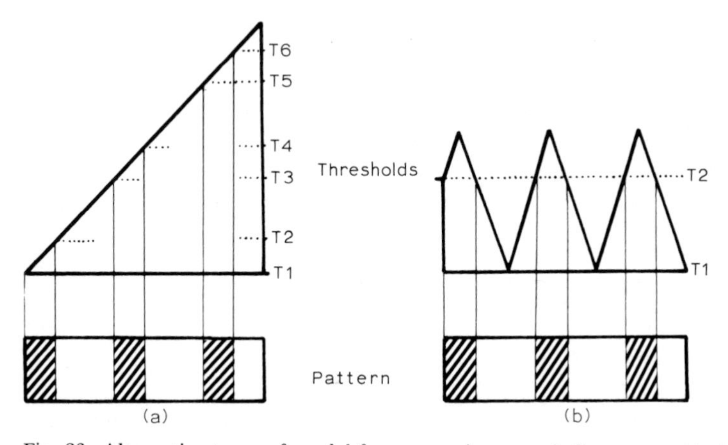

Fig. 23. Alternative types of model for generating a periodic pattern; a) via a monotonic gradient and multiple thresholds; and b) via a periodic gradient and a single threshold (redrawn from Wilby and Ede, 1975). See text for discussion.

single threshold for cartilage differentiation; such a system does not become more complex with an increasing number of elements, and extreme sensitivity and stability are not required since slight variations in concentration do not markedly alter the pattern. The stable periodic gradient can be set up if the cells have the following properties and interact in the following way: (1) cells are sensitive to their internal concentration of a freely diffusible morphogen M; (2) at concentrations of M below a lower threshold T1 cells are inactive; (3) at concentrations of M above T1, cells synthesize M; (4) at concentrations of M above a higher threshold T2 cells actively destroy M; and (5) the transformations "inactive to synthetic" and "synthetic to destructive" are *irreversible.*

The consequence of applying these rules in simple cell arrays and in the developing limb bud situation have been investigated using computer simulation techniques. In a single cell array (see Fig. 24) the rules result in the propagation of a constant velocity wave of M from an initiator region where cells are already synthesizing M. This initiator region is essential for the production of a repeatable pattern, since random initiation gives the same periodicity but variable position and polarity. Behind the wave front, the first gradient peak grows linearly by synthesis and then splits into two as the upper threshold is reached and destruction initiated. The interactions of synthesis and destruction and diffusion then cause the "trailing" peak to move backwards and downwards to stabilize between two areas of destruction and the "leading" peak to move

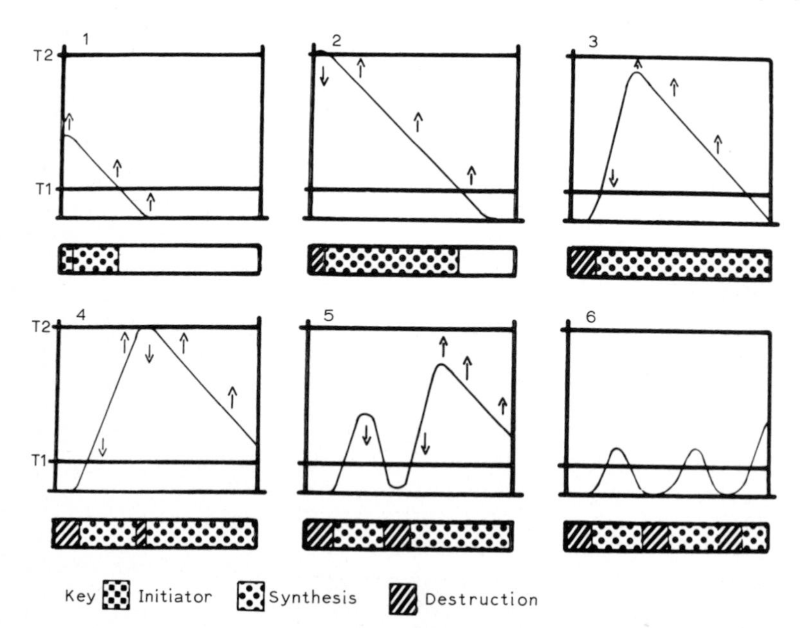

Fig. 24. Semi-diagrammatic representation of the periodic development of the Wilby-Ede waveform gradient model in one dimension. Arrows indicate direction of concentration changes: (1) induction of synthesis by initiator; (2) and (3) induction of destruction at second threshold; (4) and (5) development of second peak and destruction area; and (6) stable residual gradient (redrawn from Wilby and Ede, 1975). See text for discussion.

forwards and upwards to initiate a new area of destruction. The end result is a periodic pattern and a periodic residual gradient. Linking the initiation of destruction of M with the differentiation of cartilage allows patterning and differentiation to proceed simultaneously but the residual gradient is still available for any further patterning. The spread of the wave along the cell array is sufficient to specify a temporal developmental pattern and polarity since each cell will be switched on in strict order. This temporal pattern and polarity, which is in fact found in the developmental sequence of skeletal elements in the antero-posterior axis, might then control the programmes of growth and differentiation required to produce the different skeletal elements from the cartilage elements blocked out by the primary gradient pattern.

In two dimensions the pattern is more complex (Fig. 25) and consists of a curved wave front of synthesis spreading out from a line initiator, followed by point initiation of destruction. The areas of destruction, and therefore of cartilage synthesis, expand centrifugally, as we have seen that the cartilage elements in fact do (section 4.4). This is in striking contrast to any positional information model where, with a monotonic gradient, the interpretation must be either simultaneous across the rudiment or progressive from one side. In this simulation successive elements form at regular intervals, with a distinct polarity of size which results from the gradual steepening of the leading peak of M.

The model was applied within developing limb bud shapes, taking the ZPA as the initiator region and produced good approximations to the real skeletal pattern provided that the process was confined to a distal band of "cells" with no significant interactions with more proximal regions (Fig. 26a and b). This necessary restriction might be related to the existence of the progress zone, and/or to the existence of the distal band of cells characterized by the distinct behavioural and contact properties described above (section 4.3). The model requires a partitioning between this distal band and more proximal mesoderm which could most easily result from a change in the rate of diffusion of M from one to the other; changes in cell to cell contact might produce just this effect, but changes in cell membrane permeabilities might also be involved.

The model predicts that the initial cartilage pattern is a function of: (1) the overall limb shape; (2) the position of the initiator region; and (3) the rate of synthesis of M. The effects of changing the limb shape, and keeping other constants unchanged, may be tested by using the fan-shaped outline of the

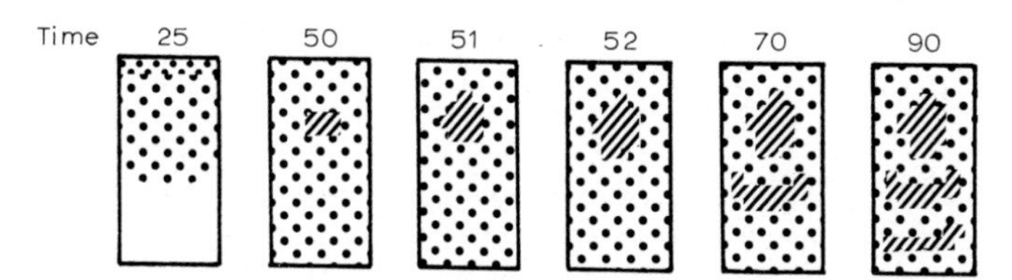

Fig. 25. Development of the model in two dimensions, showing shapes of differentiated areas. Times shown are computer steps. T25, spread of synthesis wave from initiator; T50–52, development of first destruction area; T70–90, final shape of subsequent areas (redrawn from Wilby and Ede, 1975).

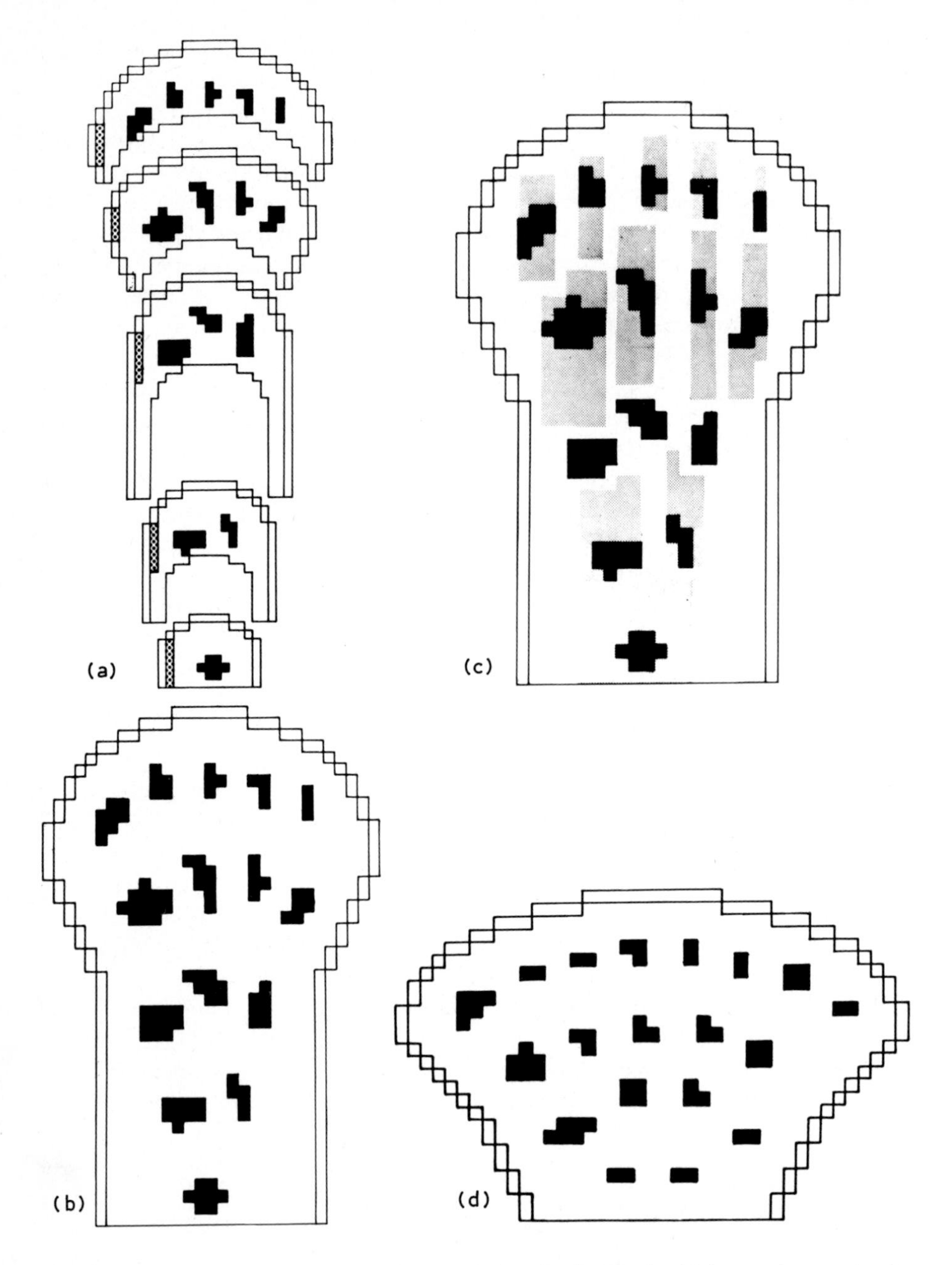

Fig. 26a–d. The wave-form gradient model applied to the limb bud, with interaction restricted to a narrow distal band, producing separate incremental growth patterns (a) which are combined to give the total pattern shown in (b). (c) shows the total pattern overlain by proposed growth areas in the chondrogenic regions, giving the final limb pattern; (d) is a simulated *talpid*[3]-like pattern produced by using the same constants as for a-c but within a fan-shaped outgrowth (redrawn from Wilby and Ede, 1975).

talpid[3] mutant limb bud and comparing the simulated cartilage pattern with that found in the real mutant (see Fig. 26d). The simulation produces a polydactylous pattern and a loss of the distinctive antero-posterior polarity of size and shape of the cartilage elements which are both characteristic of the mutant.

In both the positional information model of Tickle et al. (1975) and the wave-form gradient model of Wilby and Ede (1975) it has been supposed that the ZPA is active throughout the period during which the skeletal pattern is being established. Fallon and Crosby (1975), following up a suggestion by MacCabe et al. (1973a) that the ZPA played no part in normal development since it could be extirpated from stage 15 to stage 24 without any effect, or with minimal effects, upon normal limb development, appear to have confirmed that extirpation of the ZPA at stage 21 does not interfere with morphogenesis and shows that the ZPA is not re-established. Crosby and Fallon (1975) have also demonstrated the interesting fact that cells of the polarizing zone, when dissociated and co-aggregated with wing mesoderm, inhibit morphogenesis of that mesoderm when packed in ectodermal limb bud jackets, giving digit development very rarely, whereas mesoderm from which the ZPA has been extirpated before dissociation gives digit development in a much higher percentage of cases. Thus the ZPA when grafted as a unit, stimulates supernumerary limb outgrowth and imposes polarity, but when it is dissociated and interacts with the other wing mesenchyme as widely distributed cells it has the opposite effect of inhibiting outgrowth and morphogenesis. These recent discoveries regarding the ZPA call for much further investigation.

5.3. The dorso-ventral axis

Emphasis on the importance of the apical ectodermal ridge in accounts of limb morphogenesis may give the impression that the rest of the ectodermal covering has no significance other than as a confining jacket within which the development of the mesoderm proceeds. This is not the case, and interactions between mesoderm and non-ridge ectoderm have been discovered.

Stark and Searls (1974) showed that the degree of regulation following rotation through 180° proximo-distally of mesoderm in the prospective elbow region between stages 19 and 24 depended on whether the overlying dorsal and ventral ectoderm had also been rotated; regulation occurred more often and at later stages if the overlying ectoderm was not re-orientated. Work by Pautou and Kieny (1973) and MacCabe et al. (1974) has shown that one very important aspect of this interaction is in controlling the dorso-ventral polarity of the distal limb structures. In the more extensive report by MacCabe et al. mesodermal cores from leg buds of chick embryos of stages 19–22 were re-combined with corresponding ectodermal jackets, with antero-posterior and dorso-ventral axes arranged in all possible combinations, then allowed to develop as flank grafts in host embryos. Proximal parts of the resulting limb showed the original polarity of the mesodermal component and so also did distal (foot) parts, with one exception. This occurred whenever the dorso-ventral axis of the ectodermal component was reversed with respect to that of the mesoderm. In these cases the polarity of the resulting foot structures, including skeletal parts, followed the original polarity of the ectoderm. In this case the ectodermal component includes the AER and the possibility of some difference in activity between dorsal and ventral sides of the ridge is not excluded but, taken in conjunction with the results of Stark and Searls (1974) and other evidence, the authors conclude that the controlling differences more probably exist in the organization of the ectoderm proximal to the ridge on the dorsal and ventral sides of the bud.

6. *Concluding remarks*

In almost all investigations on limb morphogenesis interactions have been demonstrated at the tissue rather than the cellular level and the most important work for the future must be to bridge this gap in order to provide a satisfying explanation for these interactions and a solid foundation for analyzing the mechanisms by which they exert their control. The models which have been described above suggest areas in which investigations may be most fruitful. Foremost among these are the problems related to the definition and properties of the thin strip of mesenchyme immediately behind the apical ectodermal ridge. The special characteristics of cells in this region, their relation to each other, to cells of the AER ridge distally and to those of adjacent mesenchyme proximally call for intensive investigation. In this and other regions of the limb the changing nature of intercellular contacts, and the extent and nature of the communication between one cell and its neighbours will be essential for a full understanding of limb development. Exploring the relation between interactions at this level and genetic control mechanisms at the next level will call for the use of a range of mutant phenotypes in which these relationships are disturbed. Since the possibility of genetic work is so much greater in mice, and since techniques for experimental work on mammalian embryos have been much improved and extended, we expect some change of emphasis from studies on chick embryos to more work on developing limb buds in mouse embryos, and to chimaeric mixtures of cells and tissues of both as adumbrated in the exploratory study by Cairns (1965).

References

Abbott, J. and Holtzer, H. (1966) The loss of phenotypic traits by differentiated cells. III. The reversible behaviour of chondrocytes in primary cultures. J. Cell Biol. 28, 473–487.

Amprino, R. (1965) Aspects of limb morphogenesis in the chicken, In: Organogenesis (DeHaan, R. and Ursprung, H., eds.) pp. 225–281, Holt, Rinehart and Winston, New York.

Amprino, R. (1968) On the causality of the distal twinning of the chicken embryo wing. Arch. Biol. 79, 471–503.

Amprino, R. and Ambrosi, G. (1973) Experimental analysis of the chick embryo limb bud growth. Arch. Biol. 84, 35–86.

Anikin, A.W. (1929) Das morphogene Feld der Knorpelbildung. Wilhelm Roux Archiv. Entwicklungsmech. Organismen 114, 549–577.

Balcuns, A., Gasseling, M.T. and Saunders, J.W., Jr. (1970) Spatio-temporal distribution of a zone that controls antero-posterior polarity in the limb bud of the chick and other bird embryos. Am. Zool. (abstr.) 10, 323.

Cairns, J.M. (1965) Development of grafts from mouse embryos to the wing bud of the chick embryo. Dev. Biol. 12, 36–52.

Cairns, J.M. (1966) Cell generation times and growth of the chick wing bud. Am. Zool. 6, 3–8.

Cairns, J.M. (1975) The function of the apical ectodermal ridge and distinctive characteristics of adjacent distal mesoderm in the avian wing bud. J. Embryol. Exp. Morphol. 34, 155–169.

Caplan, A. (1970) Effects of the nicotinamide-sensitive teratogen 3-acetylpyridine on chick limb cells in culture. Exp. Cell Res. 62, 341–355.

Caplan, A. (1971) The teratogenic action of the nicotinamide analogues 3-acetylpyridine and 6-aminonicotinamide on developing chick embryos. J. Exp. Zool. 178, 351–358.

Caplan, A. and Koutroupas, S. (1973) The control of muscle and cartilage development in the chick limb: the role of differential vascularization. J. Embryol. Exp. Morphol. 29, 571–583.

Crosby, G.M. and Fallon, J.F. (1975) Inhibitory effect on limb morphogenesis by cells of the polarizing zone coaggregated with pre- or postaxial wing bud mesoderm. Dev. Biol. 46, 28–39.

Dawd, D.S. and Hinchliffe, J.R. (1971) Cell death in the "opaque patch" in the central mesenchyme of the developing chick limb: a cytological, cytochemical and electron microscopic analysis. J. Embryol. Exp. Morphol. 26, 401–424.

Dienstman, S.R., Biehl, J., Holtzer, S. and Holtzer, H. (1974) Myogenic and chondrogenic lineages in developing limb buds grown in vitro. Dev. Biol. 39, 83–95.

Drews, U. and Drews, U. (1972) Cholinesterase in der Extremitätenentwicklung des Huhnchens. I. Die Phasen der Cholinesterase Aktivität in der jungen Knospe und bei der Abgrenzung von Knorpel- und Muskelanlagen. Wilhelm Roux Archiv. Entwicklungsmech. Organismen 169, 70–86.

Drews, U. and Drews, U. (1973) Cholinesterase in der Extremitäten entwicklung des Hühnchens. II Fermentaktivitat und Bewegungsverhalten der präsumptiven Knorpelzellen in vitro. Wilhelm Roux Archiv. Entwicklungsmech. Organismen 173, 208–227.

Ede, D.A. (1969) Abnormal development at the cellular level in *talpid* and other mutants. In: The Fertility and Hatchability of the Hen's Egg (Carter, T.C. and Freeman, B.H., eds.) p. 199. Oliver and Boyd, Edinburgh.

Ede, D.A. (1971) Control of form and pattern in the vertebrate limb. In: Control Mechanisms of Growth and Differentiation, Symp. Soc. Exp. Biol. 25, 235–254.

Ede, D.A. (1972) Cell behaviour and embryonic development. Int. J. Neuroscience, 3, 165–174.

Ede, D.A. and Agerbak, G.S. (1968) Cell adhesion and movement in relation to the developing limb pattern in normal and *talpid*^{3} mutant chick embryos. J. Embryol. Exp. Morphol. 20, 81–100.

Ede, D.A. and Flint, O.P. (1972) Patterns of cell division, cell death and chondrogenesis in cultured aggregates of normal and *talpid*^{3} mutant chick limb mesenchyme cells. J. Embryol. Exp. Morphol. 27, 245–260.

Ede, D.A. and Flint, O.P. (1975a) Intercellular adhesion and formation of aggregates in normal and *talpid*^{3} mutant chick limb mesenchyme. J. Cell. Sci. 18, 97–111.

Ede, D.A. and Flint, O.P. (1975b) Cell movement and adhesion in the developing chick wing bud: studies on cultured mesenchyme cells from normal and *talpid*^{3} mutant embryos. J. Cell Sci. 18, 301–314.

Ede, D.A. and Kelly, W.A. (1964) Developmental abnormalities in the trunk and limbs of the *talpid*^{3} mutant of the fowl. J. Embryol. Exp. Morphol. 12, 339–356.

Ede, D.A. and Law, J.T. (1969) Computer simulation of vertebrate limb morphogenesis. Nature, 221, 244–248.

Ede, D.A. and Wilby, O.K. (1976) Analysis of cellular activities in the developing limb bud system. In: Automata, Languages, Development: At the Crossroads of Biology, Mathematics and Computer Science (Lindenmayer, A. and Rozenberg, G., eds.) North-Holland, Amsterdam (in press).

Ede, D.A., Bellairs, R. and Bancroft, M. (1974) A scanning electron microscope study of the early limb bud in normal and *talpid*^{3} mutant chick embryos. J. Embryol. Exp. Morph. 31, 761–785.

Ede, D.A., Flint, O.P. and Teague, P. (1975) Cell proliferation in the developing wing bud of normal and *talpid*[3] mutant chick embryos. J. Embryol. Exp. Morphol., in press.

Elmer, W.A. and Selleck, D.K. (1975) Effect of the brachypodism-H gene on chondrogenesis in vitro. J. Embryol. Exp. Morphol. 33, 371–386.

Errick, J.E. and Saunders, J.W., Jr. (1974) Effects of an "inside-out" limb-bud ectoderm on development of the avian limb. Dev. Biol. 41, 338–351.

Fallon, J.F. and Crosby, G.H. (1975) Normal development of the chick wing following removal of the polarizing zone. J. Exp. Zool. 193, 449–455.

Fallon, J.F. and Saunders, J.W., Jr. (1968) In vitro analysis of the control of cell death in a zone of prospective necrosis from the chick wing bud. Dev. Biol. 18, 553–570.

Fell, H.B. (1925) The histogenesis of cartilage and bones in the long bones of the embryonic fowl. J. Morphol. 40, 417–451.

Fell, H.B. and Canti, R.G. (1935) Experiments on the development in vitro of the avian knee joint. Proc. R. Soc. London Ser. B. 166, 316–351.

Finch, R. and Zwilling, E. (1971) Cultural stability of the morphogenetic properties of chick limb bud mesoderm. J. Exp. Zool. 176, 397–408.

Fraser, R.A. and Abbott, U.K. (1971a) Studies on limb morphogenesis. V. The expression of *eudiplopodia* and its experimental modification. J. Exp. Zool. 176, 219–236.

Fraser, R.A. and Abbott, U.K. (1971b) Studies on limb morphogenesis. VI. Experiments with early stages of the polydactylous mutant *eudiplopodia*. J. Exp. Zool. 176, 237–248.

Fristrom, D. and Fristrom, J.W. (1975) The mechanism of evagination of imaginal discs of *Drosophila melanogaster*. Dev. Biol. 43, 1–23.

Gasseling, M.T. and Saunders, J.W. Jr. (1961) Effects of the apical ectodermal ridge on growth of the versene-stripped chick limb bud. Dev. Biol. 3, 1–25.

Gerisch, G. (1968) Cell aggregation and differentiation in Dictyostelium. In: Current Topics in Developmental Biology (Moscona, A.A. and Monroy, A. eds.) vol. 3, pp. 157–197. Academic Press, New York.

Gould, R., Day, A. and Wolpert, L. (1972) Mesenchymal condensation and cell contact in early morphogenesis of the chick limb. Exp. Cell Res. 72, 325–336.

Gould, R.P., Selwood, L., Day, A. and Wolpert, L. (1974) The mechanism of cellular orientation during early cartilage formation in the chick limb and regenerating amphibian limb. Exp. Cell Res. 83, 287–296.

Hamburger, V. and Hamilton, H.L. (1951) A series of normal stages in the development of the chick embryo. J. Morphol. 88, 49–92.

Hampé, A. (1959) Contribution à l'étude du développement et de la régulation des déficiencies et des excédents dans la patte de l'embryon de poulet. Arch. Anat. Micros. Morphol. Exp. 48, 345–478.

Hampé, A. (1960) Sur l'induction et la compétence dans les relations entre l'epiblaste et le mesenchyme de la patte de Poulet. J. Embryol. Exp. Morphol. 8, 246–250.

Hilfer, S., Searls, R. and Fonte, V. (1973) An ultrastructural study of early myogenesis in the chick wing bud. Dev. Biol. 30, 374–391.

Hinchliffe, J.R. (1974) The patterns of cell death in chick limb morphogenesis. Libyan J. Sci. 4-A, 23–32.

Hinchliffe, J.R. and Ede, D.A. (1967) Limb development in the polydactylous *talpid*[3] mutant of the fowl. J. Embryol. Exp. Morphol. 17, 385–404.

Hinchliffe, J.R. and Ede, D.A. (1968) Abnormalities in bone and cartilage development in the *talpid*[3] mutant of the fowl. J. Embryol. Exp. Morphol. 19, 327–339.

Hinchliffe, J.R. and Ede, D.A. (1973) Cell death and the development of limb form and skeletal pattern in normal and *wingless* (*ws*) chick embryos. J. Embryol. Exp. Morphol. 30, 753–772.

Hinchliffe, J.R. and Thorogood, P.V. (1974) Genetic inhibition of mesenchymal cell death and the development of form and skeletal pattern in the limbs of *talpid*[3] (*ta*[3]) mutant chick embryos. J. Embryol. Exp. Morphol. 31, 747–760.

Holtfreter, J. (1968) Mesenchyme and epithelia in inductive and morphogenetic processes. In: Epithelial-Mesenchymal Interactions (Fleischmayer, P. and Billingham, R.E.., eds.) pp. 1–30, Williams and Wilkins, Baltimore, Md.

Holtzer, H., Bischoff, R. and Chacko, S. (1969) Activities of the cell surface during myogenesis and chondrogenesis. In: Cellular Recognition (Smith, R.T. and Good, R.A., eds.) pp. 139–149. North-Holland, Amsterdam.

Hornbruch, A. and Wolpert, L. (1970) Cell division in the early growth and morphogenesis of the chick limb. Nature 226, 764–766.

Janners, M. and Searls, R. (1970) Changes in rate of cellular proliferation during the differentiation of cartilage and muscle in the mesenchyme of the embryonic chick wing. Dev. Biol. 23, 136–165.

Jurand, A. (1965) Ultrastructural aspects of early development of the fore limb buds in the chick and the mouse. Proc. R. Soc. London Ser. B, 162, 387–405.

Kieny, M. (1964a) Etude du mécanisme de la regulation dans le developpement du bourgeon de membre de l'embryon de poulet. I. Regulation des excedents. Dev. Biol. 9, 197–229.

Kieny, M. (1964b) Etude du mécanisme de la regulation dans le developpement des bourgeon de membre de l'embryon de poulet. II. Regulation des deficiances dans les chimeres'aile-patte'et 'patte-aile', J. Embryol. Exp. Morphol. 12, 357–371.

Kochhar, D.M. (1973) Limb development in mouse embryos. I. Analysis of teratogenic effects of retinoic acid. Teratology I, 289–298.

Kwasigroch, T.E. and Kochhar, D.M. (1975) Locomotory behaviour of limb bud cells: Effect of excess vitamin A in vivo and in vitro. Exp. Cell Res. 95, 269–278.

MacCabe, J.A. and Saunders, J.W. Jr. (1971) The function of anteroposterior polarity in chick limbs developed from re-aggregated limb-bud mesoderm. Anat. Rec. (abstr.) 169, 372.

MacCabe, J.A., Errick, J. and Saunders, J.W., Jr. (1974) Ectodermal control of the dorsoventral axis of the leg bud of the chick embryo. Dev. Biol. 39, 69–82.

MacCabe, J.A., Gasseling, M.T. and Saunders, J.W. Jr. (1973a) Spatiotemporal distribution of mechanisms that control outgrowth and anteroposterior polarization of the limb bud in the chick embryo. Mech. Aging Dev. 2, 1–2.

MacCabe, J.A., Saunders, J.W. Jr., and Picket, M. (1973b) The control of the anteroposterior and dorsoventral axes in embryonic chick limbs constructed of dissociated and re-aggregated limb-bud mesoderm. Dev. Biol. 31, 323–335.

Medoff, J. (1967) Enzymatic events during cartilage differentiation in the chick embryonic limb bud. Dev. Biol. 16, 118–143.

Medoff, J. and Zwilling, E. (1972) Appearance of myosin in the chick limb bud. Dev. Biol. 28, 138–141.

Menkes, B. and Deleanu, M. (1964) Leg differentiation and experimental syndactyly in chick embryos. Rev. Roum. Embryol. Cytol. Ser. Embryol. 1, 69–77.

Mitolo, V. (1971) L'abozzo dell'ala nell'embrione di pollo: accrescimento e forma nei primi stadi di sviluppo. Boll. Soc. Ital. Biol. sper. 47, 634–637.

Pautou, M.-P. (1972) L'activité inductrice morphogène émanant de l'ectoderme apicale du bourgeon de patte chez les oiseaux n'a pas de qualification specifique. C.R. Acad. Sci., Paris Ser. D. 275, 1071–1074.

Pautou, M.-P. (1973) Analyse de la morphogènese du pied des oiseaux à l'aide de mélanges cellulaires interspecifiques. I. Etude morphologique. J. Embryol. Exp. Morphol. 29, 175–196.

Pautou, M.P. and Kieny, M. (1973) Interaction ecto-mésodermique dans l'établissement de la polarité dorso-ventrale du pied de l'embryon de poulet. C.R. Acad. Sci., Paris Ser. D. 277, 1225–1228.

Rubin, L. and Saunders, J.W., Jr. (1972) Ectodermal-mesodermal interactions in the growth of limb buds in the chick embryo: constancy and temporal limits of the ectodermal induction. Dev. Biol. 28, 94–112.

Saunders, J.W., Jr. (1948) The proximo-distal sequence of origin of the parts of the chick wing and the role of the ectoderm. J. Exp. Zool. 108, 363–403.

Saunders, J.W., Jr. (1966) Death in embryonic systems. Science, 154, 604–612.

Saunders, J.W., Jr. (1972) Developmental control of three-dimensional polarity in the avian limb. Ann. N.Y. Acad. Sci. 193, 29–42.

Saunders, J.W., Jr. and Fallon, J.F. (1966) Cell death in morphogenesis. In: Major Problems in Developmental Biology (ed. M. Locke) pp. 289–314, Academic Press, New York.

Saunders, J.W., Jr., and Gasseling, M.T. (1963) Trans-filter propagation of apical ectoderm maintenance factor in the chick embryo wing bud. Dev. Biol. 7, 64–78.

Saunders, J.W., Jr. and Gasseling, M.T. (1968) Ectodermal-mesenchymal interactions in the origin of limb symmetry. In: Epithelial-Mesenchymal Interactions (Fleischmajer, R. and Billingham, R.I., eds) pp. 78–97. Williams and Wilkins, Baltimore.

Saunders, J.W., Jr. and Reuss, C. (1974) Inductive and axial properties of prospective wing-bud mesoderm in the chick embryo. Dev. Biol. 38, 41–50.

Saunders, J.W., Jr., Cairns, J.M. and Gasseling, M.T. (1957) The role of the apical ridge of ectoderm in the differentiation of the morphological structure and inductive specificity of limb parts in the chick. J. Morphol. 101, 57–88.

Saunders, J.W., Jr., Gasseling, M.T. and Cairns, J.M. (1959) The differentiation of prospective thigh mesoderm grafted beneath the apical ectodermal ridge of the wing bud in the chick embryo. Dev. Biol. 1, 281–301.

Saunders, J.W., Jr., Gasseling, M.T. and Saunders, L.C. (1962) Cellular death in morphogenesis of the avian wing. Dev. Biol. 5, 147–178.

Schachter, L. (1970) Effect of conditioned media on differentiation in mass cultures of chick limb bud cells. Exp. Cell Res. 63, 19–32.

Searls, R.L. (1965) An autoradiographic study of the uptake of S^{35}-sulfate during the differentiation of limb bud cartilage. Dev. Biol. 11, 155–168.

Searls, R. (1967) The role of cell migration in the development of the embryonic chick limb bud. J. Exp. Zool. 166, 39–50.

Searls, R.L. (1973) Newer knowledge of chondrogenesis. Clin. Orthop. 96, 327–344.

Searls, R. and Janners, M. (1969) The stabilization of cartilage properties in the cartilage-forming mesenchyme of embryonic chick limb. J. Exp. Zool. 170, 365–376.

Searls, R.L. and Zwilling, E. (1964) Regeneration of the apical ectodermal ridge of the chick limb bud. Dev. Biol. 9, 38–55.

Searls, R.L., Hilfer, S.R. and Mirow, S.M. (1972) An ultrastructural study of early chondrogenesis in the chick wing bud. Dev. Biol. 28, 123–137.

Stark, R.J. and Searls, R.L. (1973) A description of chick wing bud development and a model of limb morphogenesis. Dev. Biol. 33, 138–158.

Stark, R.J. and Searls, R.L. (1974) The establishment of the cartilage pattern in the embryonic chick wing, and evidence for a role of the dorsal and ventral ectoderm in normal wing development. Dev. Biol. 38, 51–63.

Summerbell, D. (1974) A quantitative analysis of the effect of excision of the AER from the chick limb bud. J. Embryol. Exp. Morph. 32, 651–660.

Summerbell, D. and Wolpert, L. (1972) Cell density and cell division in the early morphogenesis of the chick wing. Nature New Biology 238, 24–26.

Summerbell, D., Lewis, J.H. and Wolpert, L. (1973) Positional information in chick limb morphogenesis. Nature, 244, 492–496.

Thorogood, P.V. and Hinchliffe, J.R. (1975) An analysis of the condensation process during chondrogenesis in the embryonic chick hind limb. J. Embryol. Exp. Morphol. 33, 581–606.

Tickle, C., Summerbell, D. and Wolpert, L. (1975) Positional signalling and specification of digits in chick limb morphogenesis. Nature, 254, 199–202.

Toole, B.P. (1972) Hyaluronate turnover during chondrogenesis in the developing chick limb and axial skeleton. Dev. Biol. 29, 321–329.

Toole, B.P., Jackson, G. and Gross, J. (1972) Hyaluronate in morphogenesis: inhibition of chondrogenesis in vitro. Proc. Nat. Acad. Sci. U.S.A. 69, 1284–1286.

Trelstad, R.L., Hay, E.D. and Revel, J.P. (1967) Cell contact during early morphogenesis in the chick embryo. Dev. Biol. 16, 78–106.

Wilby, O.K. and Ede, D.A. (1975) A model generating the pattern of cartilage skeletal elements in the embryonic chick limb. J. Theoret. Biol. 52, 199–217.

Wilby, O.K. and Ede, D.A. (1976) Computer simulation of vertebrate limb development: the effect of cell division control patterns. In: Automata, Languages, Development: At the Crossroads of Biology, Mathematics and Computer Science (Lindenmayer, A. and Rozenberg, G. eds.,) North-Holland, Amsterdam (in press).

Wolpert, C., Lewis, J. and Summerbell, D. (1975) Morphogenesis of the vertebrate limb. In: Cell Patterning. Ciba Foundation Symposium 29 (new series). Elsevier, Excerpta Medica, North-Holland, Amsterdam.

Zwilling, E. (1949) The role of epithelial components in the developmental origin of the "wingless" syndrome of chick embryos. J. Exp. Zool. 111, 175–187.

Zwilling, E. (1955) Ectoderm-mesoderm relationship in the development of the chick embryo limb bud. J. Exp. Zool. 128, 423–441.

Zwilling, E. (1956a) Interaction between limb bud ectoderm and mesoderm in the chick embryo. I. Axis establishment. J. Exp. Zool. 132, 157–172.

Zwilling, E. (1956b) Interaction between limb bud ectoderm and mesoderm in the chick embryo. II. Experimental limb duplication. J. Exp. Zool. 132, 173–187.

Zwilling, E. (1956c) Interaction between limb bud ectoderm and mesoderm in the chick embryo. IV. Experiments with a wingless mutant. J. Exp. Zool. 132, 241–254.

Zwilling, E. (1961) Limb morphogenesis. Adv. Morphogenesis 1, 301–330.

Zwilling, E. (1964) Development of fragmented and dissociated limb bud mesoderm. Dev. Biol. 9, 20–37.

Zwilling, E. (1968) Morphogenetic phases in development. Symp. Soc. Develop. Biol. 27, 184–207.

Zwilling, E. (1974) Effects of contact between mutant (*wingless*) limb buds and those of genetically normal chick embryos: confirmation of a hypothesis. Dev. Biol. 39, 37–48.

Zwilling, E. and Hansborough, L. (1956) Interaction between limb bud ectoderm and mesoderm in the chick embryo. III. Experiments with polydactylous limbs. J. Exp. Zool. 132, 219–239.

Heart development: interactions involved in cardiac morphogenesis 11

Francis J. MANASEK

1. Introduction

Early vertebrate cardiac development can be divided into several ontogenically sequential periods: (1) migration of precardiac cells; (2) cytodifferentiation; (3) primary morphogenesis; and (4) myocardial heterogeneity. Each of these somewhat arbitrary categories has a common denominator: successful completion involves a variety of cell and tissue interactions. Much as in any developing organ system, these interactions are complex but for the sake of simplicity we can consider them under three general headings: (1) cell-cell interactions involving like cells (homotypic); (2) cell-cell interactions involving dissimilar cells (heterotypic); and (3) cell-matrix interactions.

In succeeding sections these interactions will be described for each period of cardiac morphogenesis. Particular attention will be paid to studies in situ. This emphasis placed on investigation of intact hearts reflects an attempt to define events of organogenic significance by working with the organ itself. With an assurance of biological significance (to the intact organism) it may become more fruitful to investigate these events more critically at a lower level of integrative complexity (e.g., in cell culture) but with some confidence that such studies have relevance to mechanisms of organogenesis that operate in vivo. Clearly, only work on an intact organ developing in situ can provide, with assurance, a catalogue of ontogenic events with proper spatial and temporal correlation.

No attempt is made in this chapter to provide a comprehensive review of cardiogenesis. In addition to standard textbooks the reader is referred to reviews by DeHaan (1964) and Copenhaver (1955) for general overviews of cardiogenesis, and to Manasek (1975b,c) for reviews on special aspects of heart development and to O'Rahilly (1971) for a timetable of human development.

2. General considerations of embryonic cardiac development

Some basic salient features of the developing heart will be reviewed first in order to provide a general framework upon which to support detailed discussions. In

G. Poste & G.L. Nicolson (eds.) The Cell Surface in Animal Embryogenesis and Development

most vertebrates the heart follows a generally similar pattern of early embryonic development. Essentially, a tubular organ is formed first (see, Castro-Quezada et al., 1972; Argüello et al., 1975). The so-called tube is in reality a trough, or gutter-shaped structure since it initially is attached to the body by means of the dorsal mesocardium (Fig. 1). In hemichordates, such as the tunicata, this is the final form of the heart; it does not progress beyond the "tubular" stage. Histologically the tubular heart is a simple structure. Its outermost layer is a sheet of cells, the developing myocardium. This is histologically a true epithelium with specialized intercellular junctions (see Spira, 1971), a free apical surface and a basal surface that is associated with a basal lamina and extracellular matrix. The developing myocardium initially contains only developing muscle cells (Manasek, 1968). This is an important point because it means that all non-muscle cells are *added* to the myocardium later in development, rather than arising from progenitors that were part of the ontogenically primitive myocardium. The lumen is lined with a simple squamous endothelium, the endocardium (see Ojeda and Hurle, 1975). Thus, the young heart is an organ containing only epithelia. No mesenchymal cells participate in the histological organization of the heart until much later in embryonic life, (see, however, Virágh and Challice, 1973).

Another striking histological feature of the early heart is the relatively large extracellular compartment. The major portion of this compartment, a sleeve of ground substance interposed between myocardium and endocardium has traditionally been called cardiac jelly (Davis, 1924). The cardiac jelly contains the glycosaminoglycans, hyaluronate and chondroitin sulfate (Ortiz, 1958; Gessner and Boström, 1965; Gessner et al., 1965; Markwald and Adams Smith, 1972; Manasek, 1970a; 1973, 1975b; Manasek et al., 1973), type I-like collagen

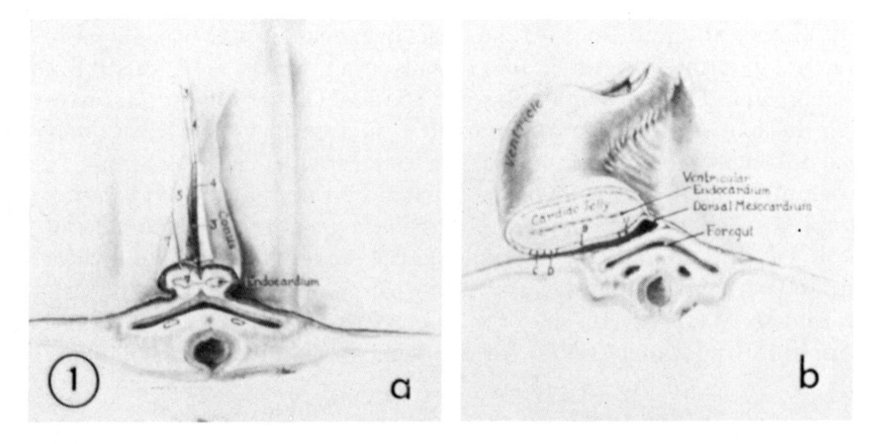

Fig. 1. Ventral views of (a) 10 somite chick embryo and (b) 16 somite chick embryo showing the internal structure and relationship of the heart to the rest of the embryo. The numbers in (a) indicate the origin of the cells, based on the grid numbers shown in Fig. 2. Figure from Rosenquist and DeHaan (1966) by permission of the Carnegie Institution of Washington.

(Johnson et al., 1974) and glycoproteins that are not yet fully characterized (Manasek, 1976b). It is abundantly clear that most, if not all, of the macromolecules of the cardiac jelly are synthesized by the developing myocardium (see discussions in Manasek 1973, 1975b). The production of extracellular macromolecules by the myocardium continues well after it has differentiated and is unequivocal cardiac muscle, but does gradually decrease as the heart matures. Throughout the early phases of heart formation, the cardiac jelly provides the immediate environment of the cells and tissues which comprise the developing organ.

Early in embryonic life the myocardium differentiates and its cells become functional cardiac myocytes. Irrespective of total length of the gestational period the heart begins to beat very shortly after the definitive embryonic mesoderm is formed. We may consider primitive streak stages of development as indicative of the approximate time mesoderm can be first identified unequivocally. Table 1 compares the time of the primitive streak stage to onset of cardiac function. The time period between these two events is invariably only a small fraction of total gestation. Clearly both biochemical and morphological events that lead to formation of a functional heart occur with great rapidity. Contractions appear to begin along the right side of the tubular heart (Patten and Kramer, 1933; see DeHaan, 1965 for discussion) but very rapidly the entire myocardium begins to contract. The onset of contraction is associated with the appearance of cross-banded myofibrils within the embryonic myocytes, and is correlated temporally with changes in heart shape, particularly the asymmetry introduced by the bending of the tubular heart to the right, a process called "looping".

The tubular heart has a single undivided lumen. Following onset of contractions, blood enters the venous end of the tube by means of a pair of veins, the coalesence of which delineates the early sinus venosus, and passes sequentially through the primitive atrium, primitive ventricle and out the truncus arteriosus from which it enters the aortic arches. Although undivided, the lumen is of

Table 1
Comparison of times of mesoderm formation (primitive streak stages), first heartbeats and gestation period.*

	Days after fertilization		
Animal	Primitive streak	Onset of heartbeat	Gestation period (days)
Frog	2–2.5	ca 5.5	ca 90
Chick	<1	1.5	21
Rat	8.5	9	22
Mouse	7	8	20
Rabbit	6.5–7	8	32
Man	15–16	22	ca 255

*Data taken from Sissman (1970).

unequal caliber along its length. Notable is the constriction located at the atrioventricular boundary. The lumen is narrowed at this site by an hypertrophy of mesenchymal cells derived from the endocardium. This tissue accumulates in mounds, called endocardial cushions, which protrude into the lumen and are the precursors of valves and portions of the interventricular septum (Patten et al., 1948). The tubular heart becomes convoluted, acquiring first a "C" shaped bend, convex to the right side. Bending continues, and at the same time internal remodeling occurs with the growth of septa; trabeculae carnae and valves (see DeHaan, 1967, 1968). The basic remodeling of the heart is completed early in development. For example by about day 5 the chick embryo's heart is anatomically a close minature of the mature heart (see de la Cruz et al., 1972).

While the architecture of the organ is undergoing change, its cellular composition changes also. The myocardium, initially a pure population of myocytes (Manasek, 1968, 1970b) becomes populated with fibroblasts, vascular endothelium, vascular smooth muscle, nerves, and a variety of other cell types. The outer cardiac surface becomes covered by the epicardium and a complex subepicardial layer develops (see Manasek, 1969).

We will first examine interactions leading to the formation of the tubular heart; then we shall consider interactions that result in increased histological complexity.

2.1. Migration of precardiac cells

Both premyocardial and preendocardial cells are identifiable long before they migrate to their destination and form the heart. Identification is not based on any unique cytological characteristics that distinguish them visibly from, say, renal progenitors. Indeed, at early stages most mesoderm looks alike even at the ultrastructural level. Specific tissue or organ progenitor cells are identifiable on the basis of their location with respect to well defined anatomical landmarks such as the neural tube, notochord ("head process"), the primitive pit or Hensen's node. Cells can therefore be characterized on the basis of the future behavior that is predicted for them. The most widely used method of determining the "fate" of cells occupying specific embryonic regions is to mark them and permit development to continue normally and to observe the ultimate location and phenotype of the cells previously marked. Marking could be accomplished through the use of vital dyes or, more elegantly, by use of labeled transplants (see Rosenquist, 1966). In the latter technique transplants of homologous regions from donor embryos which had their nuclei labeled previously with tritiated thymidine are grafted into unlabeled host embryos and the subsequent migration of labeled cells is determined by radioautographic analysis of serially sectioned embryos fixed at different post-graft times. Other methods include extirpation of specific regions and observing which structures were lacking after suitable development, or transplanting anatomically defined regions to a suitably "neutral" environment and observing what they developed into. In this way the "fates" of a large number of early embryonic regions have been mapped in some detail. Perhaps the most extensive fate maps are those of amphibians and chicks. Heart-forming regions in chicks are shown in Fig. 2.

The premyocardium is mesodermal (so-called splanchnic mesoderm) and forms in close association with the endoderm of the prospective foregut. Initially part of the lateral plate mesoderm which splits into two sheets (somatic and splanchnic), the precardiac mesoderm, is an epithelium. The precardiac cells are situated some distance from their ultimate site within the heart and must migrate

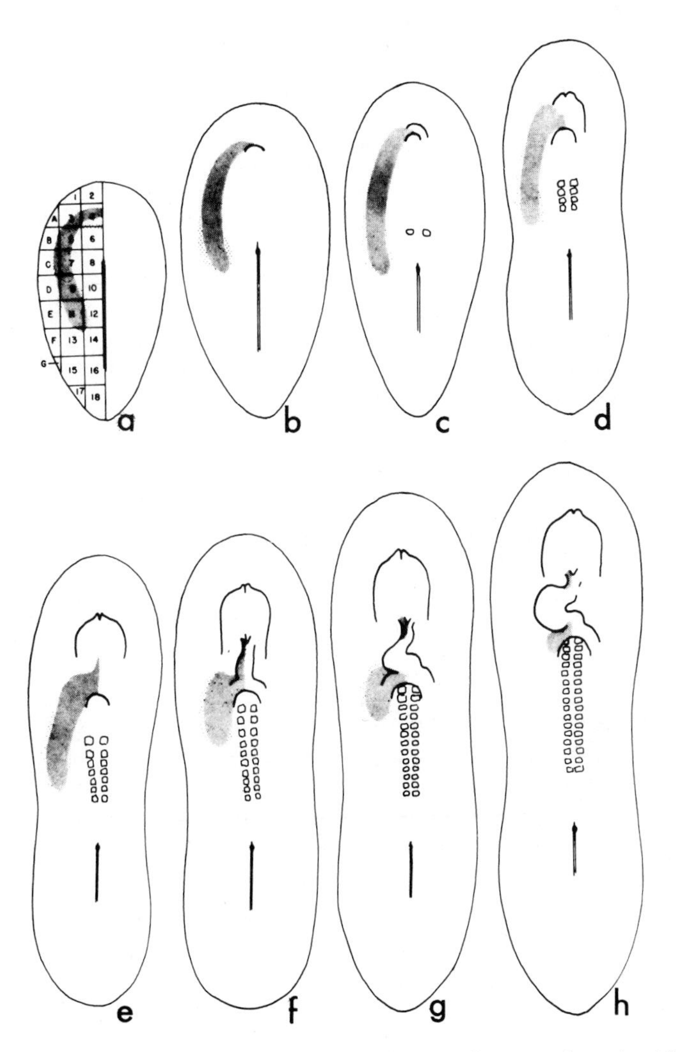

Fig. 2. The extent and distribution of precardiac mesoderm in chick embryo as determined by transplantation experiments in embryos ranging from Hamburger-Hamilton (1951) stages 5–12 (a to h, respectively). Densely hatched regions always contained pre-cardiac cells; lightly hatched did in some embryos only. Figure from Rosenquist and DeHaan (1966) by permission of the Carnegie Institution of Washington.

a considerable distance before they arrive at their definitive locations (see Wilens, 1955). All translocations of these cells are collective; in other words, precardiac cells are not free to migrate either individually or as clusters, but rather the entire sheet (epithelium) must move. The substituent cells retain their original position relative to each other within the epithelial layer. From the time the primary mesenchyme (see Hay, 1968 for classification) condenses to form the epithelial mesodermal lateral plate (a secondary epithelium according to Hay) cells that give rise to cardiac muscle and ultimately cardiac muscle itself are never isolated or mesenchymal, but always retain their epithelial associations. This is in distinct contrast to appendicular skeletal muscle (for discussion see Manasek, 1973) development, where myocytes are formed from freely wandering mesenchyme.

The evidence for migration of precardiac mesoderm as an intact epithelium is twofold: descriptive and experimental. Both electron and light microscope examination of precardiac mesoderm and the developing myocardium of the tubular heart stage has consistently failed, in the chick, to provide any evidence that these epithelial cells become mesenchymatous (Manasek, 1968). Throughout their migration and subsequent differentiation they are held together by apical junctional complexes and, subsequently, intercalated discs (see Spira, 1971). Experiments (Rosenquist and DeHaan, 1966; Stalsberg and DeHaan, 1969) in which ^{3}H-thymidine labeled grafts were followed (fate-mapping) showed conclusively that the labeled cells within the graft migrated together. They did not disperse and become separated from each other en route to the heart (Rosenquist and DeHaan, 1966). It is interesting to note that in contrast to premyocardial cells, the pre-endocardial cells migrated as individuals, or at best, small clusters. Labeled donor cells were found interspersed with unlabeled host cells suggesting that they were able to mix or reassociate more or less freely.

2.2. Mesoderm-endoderm interaction

The bilateral regions of pre-myocardial splanchnic mesoderm move toward the embryonic midline. Endoderm and pre-myocardial mesoderm form a sandwich of two dissimilar cell layers (endoderm and mesoderm) with a filling of extracellular matrix. (see Fig. 3). Collectively this sandwich is called the *splanchnopleure*. The basal surfaces of both endoderm and mesoderm come close together but electron microscope observation of these tissues in chick embryos has failed to reveal direct contact between opposing plasmalemmae (Manasek, 1975c). Thus, if any direct contact is made between endoderm and undifferentiated mesoderm such contacts must be small, scarce or restricted to an earlier stage, or all three.

Despite the fact that there may be no direct contact between endoderm and mesoderm these cell types interact, albeit over a distance. Precardiac mesoderm seemingly requires endoderm for successful myoblast-myocyte transformation. The dependence of cardiac muscle differentiation on endoderm has been documented most carefully in amphibia (see Copenhaver, 1955 and DeHaan, 1965 for reviews, and papers by Mangold, 1957; Jacobson, 1960, 1961; Jacobson

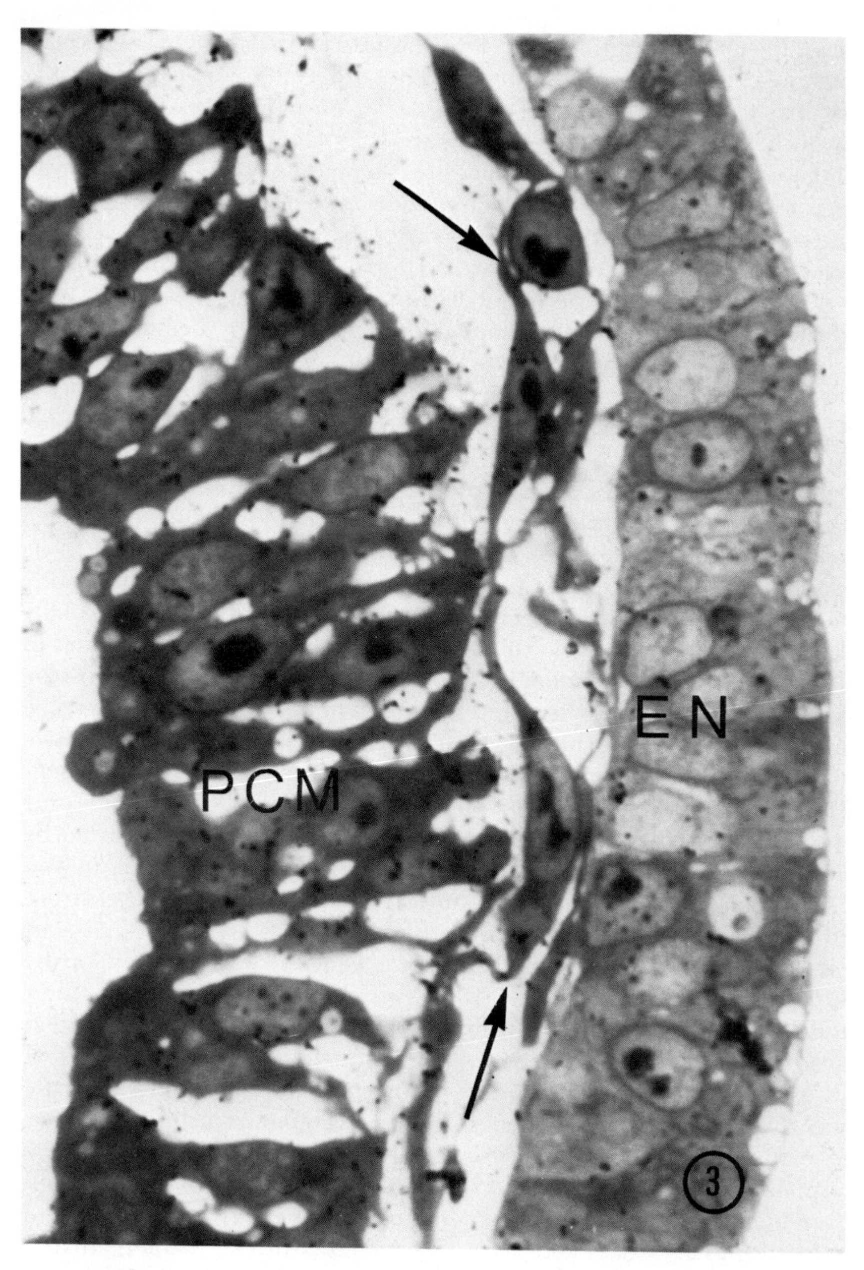

Fig. 3. Light microscope radioautograph showing a portion of the precardiac region of a stage 9 (7 somite) chick embryo labeled with $^{35}SO_4^=$. This is from a region of splanchnopleure that had already started folding toward the midline. The endoderm (EN) is therefore going to form foregut. Between the precardiac mesoderm (PCM) and endoderm are clusters of cells identified as pre-endocardium (arrows). As development progresses the mesoderm becomes more widely separated from both endoderm and endocardium, probably as a result of continued accumulation of extracellular matrix (cardiac jelly) synthesized largely by mesoderm. In this radioautograph most of the extracellular label is associated with mesoderm. ×2000.

and Duncan, 1968; Fullilove, 1970). In a large number of experiments it has been shown that anterior endoderm enhances the differentiation of precardiac mesoderm in vitro. Jacobson and Duncan (1968) have further shown that a fraction (which they call a "membrane fraction") of endoderm, obtained by passing homogenized endoderm through a Sephadex G-100 column, can substitute partially in vitro for the intact endoderm. This observation suggests that the active material is a large molecule since it apparently was recovered in the excluded volume of the column. Cardiac muscle development in amphibia, although it has been shown experimentally to require endoderm has not been examined morphologically to any great extent (see, however, Lemanski, 1973) and the normal, in situ relationship between endoderm and undifferentiated mesoderm has not been elucidated.

The factors that influence mesoderm in amphibians are far more complex than outlined above. Neural plate and anterior neural fold inhibits heart differentiation; ectoderm increases the frequency (percent explants differentiating) but not the rate (rapidity with which differentiation occurs). Only endoderm acts to increase both rate and frequency (Jacobson and Duncan, 1968). In its normal position in the developing embryo the mesoderm is subject to a combination of these factors at different times. This complex balance is virtually impossible to duplicate in vitro and this experimental limitation must be kept in mind. Interactions between endoderm and mesoderm do not appear to be obligatory for avian cardiac muscle differentiation. Phenotypic expression can be elicited in pre-cardiac mesoderm explanted shortly after it leaves the primitive streak and grown in culture in the absence of endoderm (Chacko and Rosenquist, 1974). Under these conditions phenotypic expression is dependent to some degree upon culture substrates. Similar observations had been made earlier by LeDouarin and his colleagues (Renaud and LeDouarin, 1968; LeDouarin and Renaud, 1969; LeDouarin, 1974). They showed that cultured avian precardiac mesoderm does not require endoderm to differentiate, but that the *extent* of phenotypic expression (maturation of cardiac myocytes) was retarded in the absence of endoderm.

Although avian precardiac mesoderm differentiates to some extent in vitro in the absence of endoderm the tissue does not complete normal histogenesis. If endoderm (or for that matter any other avian epithelium) was added to the culture the differentiating mesoderm formed a tubular structure that was similar, in many respects to a tubular heart. It is therefore possible to get a graded response from chick tissue and the greater the degree of cytodevelopment the greater the degree of organotypic morphology. It is interesting to consider the possibility that the latter depends upon the former; that a continued and normal sequence of myocardial cytodifferentiation and maturation is obligatory for normal organ shape. This had been proposed by Manasek et al., (1973) and is a point that will be discussed again later.

A few tentative conclusions can be drawn from the two systems most studied so far (chick and amphibian). Interactions between mesoderm and a dissimilar epithelium are seemingly necessary for normal myocardial development in both systems, except that cytodifferentiation in the avian system appears less dependent upon endoderm than in amphibian systems. The latter conclusion should be made with a degree of caution because of the in vitro nature of the experiments leading to them. Embryonic avian cells in culture require a more complex medium than do amphibian cells, largely because they lack the extensive intracellular yolk that is so characteristic of amphibian cells. Embryonic amphibium mesoderm used in studies of heart induction could therefore be grown in relatively simple media; in essence balanced salt solutions. On the other hand avian mesoderm was grown in more complex media on a variety of substrates. It is possible that some component of the avian medium substituted for endoderm or a factor normally elaborated by endoderm whereas in the simple balanced salt solution used in amphibian studies this artificial factor(s) was absent. The two culture systems are *not* analogous and comparisons between them may not be entirely valid. For example, it is not known if any of the constituents of the complex avian culture medium, if added to amphibian cultures, can initiate onset of cytodifferentiation.

If we examine the most recent experimental evidence bearing on amphibian heart induction more closely some interesting paradoxes become apparent. It is possible, although difficult, to remove completely the endoderm from embryos as young as the gastrula stage. If this is done (leaving neural progenitors in place) then hearts do not develop, an apparent confirmation of the demonstration in vitro that endoderm is necessary. Jacobson and Duncan (1968) were able to induce heart differentiation in 1 of 2 experimental embryos by inserting into the endoderm-less embryos pieces of gelatin into which they had incorporated conditioned medium produced by cultured anterior endoderm. In three cases where they used medium conditioned by whole endoderm (anterior plus posterior) no heart formation took place. Fifteen controls (Niu-Twitty medium in gelatin) were negative. Amano (1961), on the other hand, reported some heart induction in another amphibian species by using agar to replace surgically removed endoderm. Even if we ignore for the moment the statistical problems encountered in interpreting the experimental studies employing conditioned medium, the fact that non-specific "induction" (by agar) can occur makes this type of "replacement therapy" experiment difficult to interpret in principle. Non-specific induction may occur in culture experiments also. Jacobson and Duncan (1968) report that isolated cultured mesoderm differentiates in a small number of cases if grown in Niu-Twitty medium but not if grown in Holtfreter's solution. This could mean cells are healthier in Niu-Twitty, and hence more capable of expressing their slightly induced state, or that some substance in Niu-Twitty medium substitutes for the normal inducer (although not completely) or that mesoderm can modify or condition Niu-Twitty medium but not Holtfreter's solution. It is perhaps erroneous to consider Niu-Twitty the medium of choice because of its ability to elicit differentiation and there is no firm evidence that this differentiation represents an expression of a partly induced state (i.e. resulting from contact with the inductor only a short time prior to explantation).

The studies by Jacobson and Duncan (1968) showed that in hanging drop cultures anterior endoderm tissue was a potent inducer; cranial fold tissue a potent inhibitor. Sephadex G-100 excluded fraction material from *all* tissues tested (including cranial fold) *promoted* differentiation in hanging drop cultures over that seen in cultures containing mesoderm alone. Even the fraction obtained from anterior endoderm could not replace completely the activity of intact endoderm in inducing differentiation. Although it is possible that preparative procedures resulted in loss of some inducing ability in endoderm fractions it is difficult to see how tissue *inhibitory* effects could occur when intact cranial fold could yield an enhancing factor when prepared by similar techniques. Seemingly the efficacy of induction or inhibition is partly a function of state of assembly of the tissue. Intact tissues are more potent than fractions.

If we do not interpret these data with the presumption that the inducing tissue elaborates specific material that acts on mesoderm, then an alternative model of heart differentiation can be constructed that also fits the data. We do not need to consider induction or inhibition a unidirectional event. If we postulate that mesoderm is entirely capable of self-differentiation by means of creating its own unique environment (or field) then an inducing tissue is simply one that facilitates the production and maintenance of the field. This model implies that material is elaborated by the mesoderm which in turn results in its differentiation. Pre-cardiac mesoderm is known to elaborate matrix macromolecules (see Manasek, 1970b and also Figure 3) and recent studies by Gordon and Brice (1974a,b) suggest that factors influencing contractility are elaborated even by differentiated muscle in vitro. It need not be a single specific molecule that is involved in induction. Rather, differentiation may be a response to an environment that reflects an integration of individual molecular species. The creation of such a supramolecular organization would occur outside the cell, be dependent upon the physical chemical properties of the pre-existing environment and be entirely post-translational. Hence seemingly trivial differences in salt concentration and composition between Niu-Twitty and Holtfreter's medium could elicit different responses. It is possible, too, that disruption of the cranial fold simply physically changes its ability to effect creation of a suitable environment by mesoderm. The fact that in hanging drop culture (small volume) isolated mesoderm showed signs of differentiating in 7 days whereas in Petri dish culture (large volume) it took 3 weeks, also argues for a mesodermal source of factors necessary for differentiation. At the present time it is virtually impossible to predict the nature of these substances but it is tempting to suggest that their incorporation into a matrix is exceedingly important to subsequent development.

The answers to questions involving tissue interaction and heart induction are not yet clear. One of the problems is the fact that data obtained at different levels of organization are not entirely complementary and are difficult to fit into a comprehensive theory. Perhaps this suggests that different questions need be asked.

2.2.1. The mesodermal-endodermal interface

The nature of the interface between the endoderm and the precardiac splanchnic mesoderm during the time of mesodermal translocation and induction is of importance since this region probably mediates the interactions between these two dissimilar tissues. Furthermore, it has been suggested that endoderm acts as a substrate that contains directional information influencing the direction of migration of precardiac mesoderm (DeHaan, 1964; see also DeHaan, 1965).

The extracellular region between endoderm and precardiac mesoderm has not been subjected to critical chemical analysis for the period encompassing inductive interactions, myocardial migration or early pre-fusion stages of heart development. However, by inference as well as radioautographic evidence (Manasek, 1970a, see Fig. 3) we can assume that the extracellular regions contain hyaluronate and chondroitin sulfate as principal matrix glycosaminoglycans. Recently the presence of fucose-containing glycoproteins was discovered in this region, during stages shown in Fig. 4, by studies of synthesis using ^{3}H-fucose as labeled precursor (Manasek, 1976b). Most of these glycoproteins, as recovered after solubilization of tissue in SDS appear to be very large (>100,000) on the basis of their elution patterns from Agarose A-15 columns. From stages 8+ to 13 (Hamilton-Hamburger), newly synthesized fucose-containing glycoproteins elute from Agarose A-15 as molecules of progressively larger size, possibly indicating that these molecules aggregate more readily in older embryos than in younger. We presume that the apparent large size of these glycoproteins (most seem to range from about 100,000 to approximately 2×10^{6} as determined by molecular sieving) represent aggregates, although 6 M urea in the presence of SDS does not significantly decrease the apparent size of the labeled moieties (Manasek, unpublished observations). Alternatively they could be significantly smaller molecular weight molecules with a large Stokes radius. The resolution of

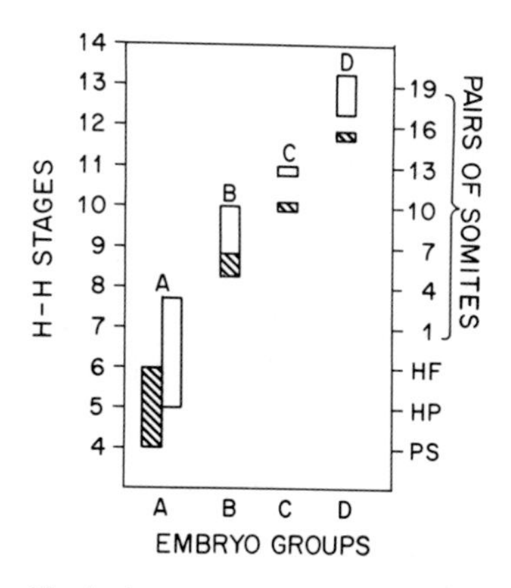

Fig. 4. Age range at start (hatched lines) and end (blank) of ^{3}H-fucose labeling experiments in intact, cultured embryos at various Hamburger-Hamilton (H-H) stages. Groups A, B, C correspond to graphs A, B, C in Fig. 5 and 6.

this problem awaits purification of the glycoproteins. It is interesting to note that the apparent increase in size of the labeled glycoprotein (presumed to represent aggregation) parallels the development of cardiac basal laminae, a relationship that will be discussed in more detail later.

The characterization of these molecules is not yet complete. At the present time work is underway in this laboratory in an attempt to accurately define the composition and structure of these glycoproteins. It is certain however, that in the developing heart, as elsewhere in the embryo (Manasek, 1975a,c) glycoproteins are widely distributed throughout the *extracellular matrix*, probably as structural components, and not restricted to cell surfaces. Furthermore, at least later in development there appears to be some homology between the Pronase released glycopeptides associated with heart cells and those in the true extracellular matrix (Manasek, 1976b).

The fact that glycoprotein synthesis could be demonstrated in early developing hearts raised some fundamental questions. Are glycoproteins, as a general class of molecule, restricted to heart-forming areas or are they present generally in early embryos? If they are present elsewhere, are there regional differences in molecular species? We furthermore were concerned with the distribution of these molecules. Are they principally cell-surface molecules or are they found in the matrix?

Synthesis of fucose-containing glycoproteins occurs throughout the chick embryo from as early as primitive streak stages. Embryos were explanted to New (1955) cultures and labeled with ^{3}H-fucose at the stages indicated in Fig. 4. Incubation continued for 6 hours in the presence of label and only normally developing embryos were collected. This meant that we could be assured that the material synthesized appeared in embryos that were normal by morphological criteria and presumably also by physiological criteria since the time course of their development was normal also. Axial regions were dissected and processed as described earlier (Manasek, 1976b). The data presented here show analyses of labeled molecules obtained from posterior axial regions (first somite to sinus rhomboidalis).

Fucose-containing molecules were synthesized throughout the stages examined. They elute from Agarose A-15 as shown in Fig. 5. With increasing developmental age (Fig. 5A to 5C) more label is recovered in the earlier eluting fractions. Peak I becomes prominent and Peak II elutes earlier. Peak III is largely unincorporated fucose and small molecular weight material. If labeled material is desalted first on Sephadex G-50 and only the excluded volume run through A-15, Peak III is not present. Prolonged incubation in high concentrations of SDS *does* alter the elution profile but only slightly.

The fucose labeled molecules are sensitive to Pronase digestion, further substantiating their identification as glycoproteins. This sensitivity can be demonstrated by means of their altered elution from Sephadex G-50 following exhaustive Pronase digestion (Fig. 6). An increase in label eluting between fractions 20 and 30 occurs with increasing age (Fig. 6). This could reflect a qualitative difference in glycopeptides or simply a quantitative difference.

Radioautography of ^{3}H-fucose labeled embryos has revealed the presence of incorporated label in extracellular matrix both in the heart (Manasek, 1975b) and in axial regions, particularly around the region of the developing sclerotome (Manasek, 1975c). The identification of glycoprotein in the extracellular compartment was also made in another way in an attempt to corroborate the radioautographic findings. Intact cultured chick embryos were labeled with ^{3}H-fucose and specific regions were collected and pooled. Cells and tissues were separated from matrix by means of mild digestion with testicular hyaluronidase. Most of the mass of extracellular matrix macromolecules in early embryos probably is glycosaminoglycan which is sensitive to hydrolysis by hyaluronidase. This enzyme reduces the viscosity and adhesivity of the matrix, permitting the embryo to literally "fall apart" when pipetted gently through the tip of a Pasteur pipette. Gentle centrifugation then separates cells and tissues from matrix components. Parenthetically, hyaluronidase is an excellent enzyme to use for the isolation and recovery of intact embryonic structures such as notochords or somites, but its efficacy declines rapidly as the matrix matures with the accumulation of collagen. Hyaluronidase is without direct effect on either collagen or glycoproteins. At the stages examined here, collagen is a minor component and this technique was used successfully to obtain glycoproteins from the matrix.

After solubilization in SDS both cell and matrix material were desalted on Sephadex G-50 to remove unincorporated fucose and material eluting in the excluded volume was digested exhaustively with Pronase and glycopeptides separated on Sephadex G-50 (Fig. 7). The presence of labeled material in the matrix fraction corroborates the radioautographic findings and a comparison of the glycopeptides obtained from cell and matrix fractions suggest close homology on the basis of this single criterion (elution from Sephadex G-50; see Fig. 7). Clearly, embryonic extracellular matrix contains glycoproteins, probably as structural components. They are not restricted to any single region of matrix but appear to be universal matrix components. An examination of Pronase released glycopeptides suggests ontogenic differences but their nature is still not clear, nor is it yet possible to make unequivocal statements about their possible differences in different regions. What is clear however is that models of cell-matrix interaction during ontogeny must consider glycoproteins as well as glycosaminoglycans and collagens as matrix macromolecules with possible biological activity. The apparent universality of these general classes of molecules makes it difficult to ascertain if any presumed biological specificity of action is a result of regional or temporal qualitative differences or is the result of quantitative differences among matrix macromolecules, the interactions of which impart unique properties to different matrices. This consideration of the integrative aspects of matrix has been discussed recently (Manasek, 1975c). These considerations are principally germane to the matrix of very young embryos prior to overt "connective tissue" differentiation. It is apparent too, that these questions cannot be answered by working only with a single system. For this reason we have begun to compare glycoproteins from a large number of developing tissues.

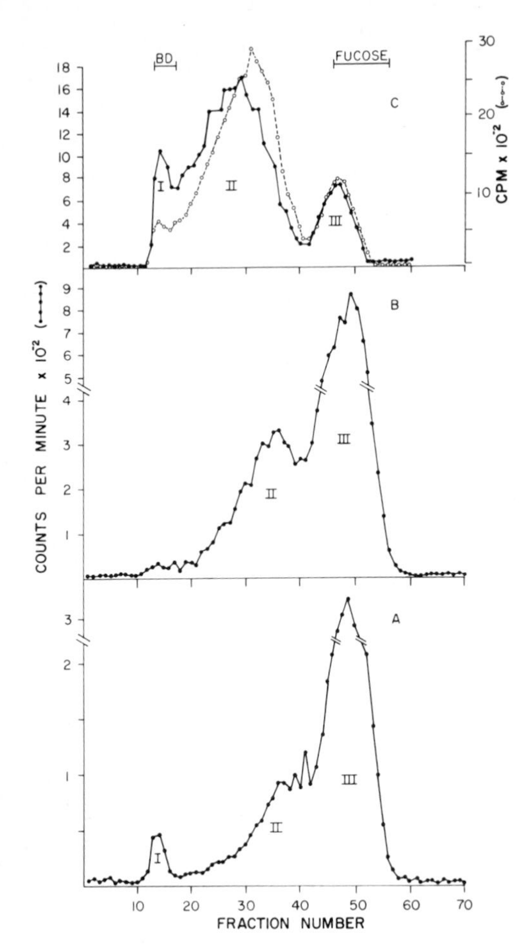

Fig. 5. Agarose A-15 separation of total fucose-labeled material obtained from embryonic trunks of age ranges A, B, C (see Fig. 4). Ten trunks (entire axial region) from each group, labeled for 6 hr, were used for each determination. Peak I is very large and seemingly more prominent in older (Fig. 5C; Group C) embryos than in younger (5A; B; Groups A, B respectively). Peak II also gains larger molecular weight material as embryonic age increases. Most of the material in Peak III represents monomeric fucose and small molecules but the A-15 does not resolve particularly well in this region. If material is desalted using Sephadex G-50 and the excluded volume run on A-15, Peak III is largely absent. Dashed line in Fig. 5C represents elution pattern after digestion with testicular hyaluronidase. This is identical to prolonged SDS treatment and implies that some of the large fucose-containing glycoproteins form aggregates, possibly with glycosaminoglycans. BD marks the region where blue dextran was recovered.

The findings, presented here and in earlier publications (Manasek, 1975a, b,c, 1976b) that fucose-containing glycoproteins are normal components of early embryonic extracellular matrices is in apparent contradiction to the recent findings of Pratt et al. (1975) who failed to demonstrate significant extracellular fucose-derived label in the region around neural crest. It should be emphasized, however, that it is very difficult to prove *absence* of synthesis by means of

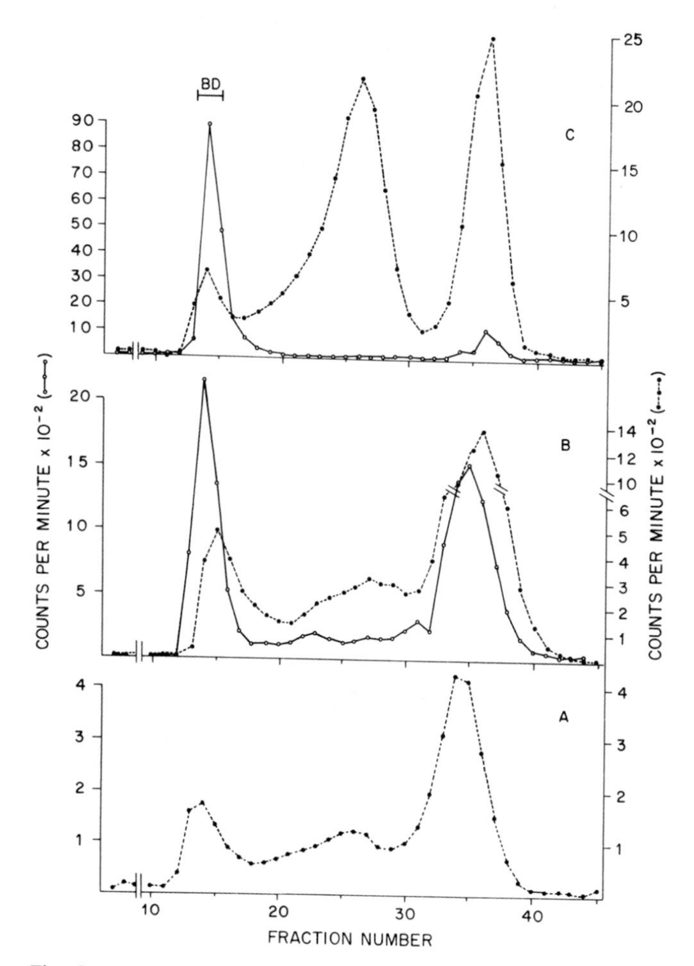

Fig. 6. Pronase digestion of the preparation described in Fig. 5 shows that the fucose labeled molecules are pronase-sensitive. The elution pattern of resulting glycopeptides (dashed line) from Sephadex G-50 is compared to that of solubilized, labeled molecules not exposed to pronase (solid line). Label in fractions 24–28 increases (relative to excluded label) with developmental age. In Group A, only sufficient material to examine glycopeptides was obtained. BD = blue dextran.

incorporation studies and the apparent discrepancy between the results may simply reflect some trivial experimental difference, such as in the amount of ^{3}H-fucose administered. If too little label is given, then radioautography would not be expected to demonstrate low specific activity end-product molecules but this should not be taken to mean an *absence* of synthesis. This problem has recently been discussed for other matrix components (Manasek, 1975c).

2.2.2. Basal laminae during early heart formation

Mature epithelia have a characteristic extracellular layer subjacent their basal surfaces called the *basal lamina*. Basal lamina is synonymous with the terms

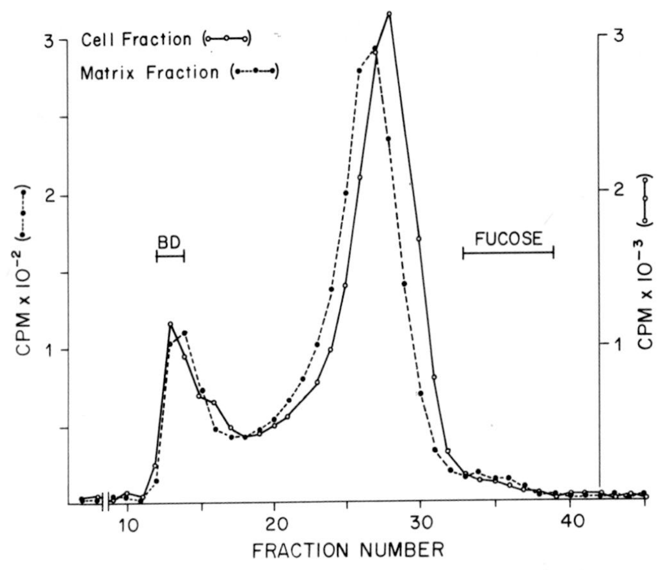

Fig. 7. Stage 14–16 embryos were labeled in culture with ^{3}H-fucose and their trunks separated into cell and matrix fractions by digestion with testicular hyaluronidase (Manasek, 1975c). Following exhaustive Pronase digestion, glycopeptides were separated on Sephadex G-50 as in Fig. 6. Note that the retarded peak is greater than that representing the excluded material. Prior to Pronase digestion the sample had been desalted on G-50 and the glycopeptides shown here are entirely from glycoproteins originally excluded by G-50.

basement lamina, basal lamella or basement lamella. The term basement membrane is often used synonymously by some authors. The basal lamina appears under developing epithelia at varying times during development. For example, as early as the time of laying the chick ectoderm has a demonstrable lamina (Low, 1967).

Although precardiac mesoderm is classified as an epithelium it does not have a lamina associated with its basal surface (Manasek, 1968). As seen with the electron microscope, its basal surface is largely bare. The endoderm, however, does have a basal lamina (Johnson et al., 1974) which is contacted by the precardiac splanchnic mesoderm (Manasek, 1975c). Thus, whatever interactions occur between these dissimilar tissues they probably are mediated by the endoderm basal lamina. Two possibilities present themselves. Either endodermal basal lamina components *themselves* are required for myocardial development or another endodermal product is involved, in which case this material would have to traverse the lamina and would, presumably, be a rather small molecule. Some basal laminae are known to contain collagen and glycoprotein. In particular those that are amenable to isolation and apparent purification have been examined compositionally. Renal glomerular, basal laminae and Descemet's membrane have been partially characterized and have been shown to contain both collagen and glycoprotein (see Kefalides, 1973). The collagen in these basal laminae is a unique non-fibrillar form seemingly restricted to basal laminae. It consists of three α1 chains that are unique because of their relatively high (as compared to other collagens) content of 3-hydroxyproline and high levels of covalently bound hexose (10–12%). These α1 chains have been termed α1 (IV) and the basal lamina collagen molecule can be written as $[\alpha 1\ (IV)]_3$ meaning that it consists of three α1 chains of Type IV. On the basis of such compositional analyses of isolated basal laminae it has been assumed that all laminae contain Type IV collagen and that the ontogenic appearance of morphologically recognizable basal laminae is synonymous with appearance of $[\alpha 1(IV)]_3$ collagen (Hay, 1973). To date there is no evidence that early embryonic basal laminae at the developmental stages where early inductive interactions are occurring, and consequently which interest us the most, contain any collagen as structural components, even Type IV collagen.

Endodermal basal lamina does contain a glycoprotein that can be labeled in situ with ^{3}H-fucose. This was demonstrated radioautographically (Manasek, 1976b). Since all the incorporated fucose was shown biochemically to be in glycoprotein the radioautographic localization of incorporated fucose could, with certainty, be considered to represent the histological location of glycoprotein.

Undifferentiated precardiac mesoderm therefore makes contact with a glycoprotein that probably was synthesized by a different tissue (endoderm). It is tempting to propose that this interaction between precardiac mesoderm and endodermal basal lamina glycoproteins is an obligatory step in myocardial development. Although not yet tested experimentally this hypothesis is consistent with several of the observations already discussed, viz., the importance of substrata in avian myocardial differentiation in vitro and the demonstrated requirement for endoderm for amphibian myocardial differentiation. The fact that a presumably large molecule(s) extracted from endoderm can substitute for intact tissue is also consistent with this hypothesis. In this respect it is intriguing that the fucose-containing glycoproteins isolated from early hearts (dissections that included endodermal material) are recovered as large molecules, either in aggregated or monomeric form.

2.2.2.1. Cardiac basal lamina formation. There are two basal laminae associated with early cardiogenesis: endodermal and myocardial. As already mentioned, the endodermal lamina is formed first, but based on the kinetics of incorporation of fucose label its glycoproteins are apparently turning over. It is proposed that elaboration of glycoproteins into a pre-existing solution (extracellular matrix) results in their precipitation because of steric exclusion from the solvent. The early cardiac extracellular matrix is a gel with high viscosity, a physical property noted quite early and which gave rise to the term "cardiac jelly" (Davis, 1924). The viscosity of this matrix is thought to result principally from the content of the glycosaminoglycan, hyaluronate (Manasek et al., 1973). Hyaluronate and chondroitin sulfate seem to be the principal glycosaminoglycans in embryonic cardiac matrix as determined biochemically. Solutions of these molecules have very interesting properties, among them being steric exclusion. Large, long chain polysaccharides have a large molecular domain from which other macromolecules are effectively excluded (Laurent, 1968). The exclusion of macromolecules from the solvent occupied by glycosaminoglycans would tend to decrease their solubility and result in precipitation. Thus glycoprotein secreted into a glycosaminoglycan matrix would not be expected to migrate or diffuse any great distance but a significant amount would precipitate close to the site of elaboration. This mechanism could explain in qualitative terms the biogenesis of basal laminae, since depending upon the glycosaminoglycan concentration precipitation could be relatively complete and rapid, giving rise to a discrete basal lamina.

This model requires the presence of a glycosaminoglycan-rich matrix *before* the synthesis and elaboration of lamina glycoprotein occurs. Certainly, in the case of the myocardium, its basal lamina is formed long after establishment of the glycosaminoglycan-rich cardiac jelly. Glycosaminoglycan synthesis has been demonstrated sufficiently early in development (Kosher and Searls, 1973; Manasek, 1976a) to suggest that in situ all basal laminae are formed only *after* a glycosaminoglycan matrix is present. At the present time we can express the steric exclusion model of lamina formation in qualitative terms only because there is no information available concerning absolute or relative amounts or concentrations of glycosaminoglycans in *any* developing extracellular matrix. Work is currently in progress to gather data that would permit quantitative modeling.

Radioautography shows that newly synthesized (hence labeled) macromolecules containing ^{3}H-fucose are concentrated in the region of the myocardial basal lamina (Fig. 8); fewer grains appear over the rest of the cardiac jelly. Based on biochemical analysis of total fucose labeled molecules it is certain that these silver grains delineate the location of glycoprotein. Although containing newly synthesized glycoprotein, the ontogenically new myocardial basal lamina is weakly anionic (Manasek, 1975b) and does not bind colloidal thorium, though the cardiac jelly per se contains regions that are strongly anionic by this criterion (Fig. 9). Seemingly there is a gradient of newly synthesized glycoprotein that is highest nearest the myocardium, while the reverse is true in terms of available negative charge. This region is entirely devoid of cells or cell processes so the materials being examined are true extracellular molecules.

Steric exclusion would not be expected to be complete. Nor would *all* excluded glycoproteins be expected to precipitate as basal lamina. Their large size would, however, prevent them from diffusing freely because of molecular sieve properties of glycosaminoglycan solutions. They would thus be immobilized partially in the matrix and their eventual distribution would probably depend more upon *translocation of the entire matrix rather than migration through it.* The relatively immobile glycoprotein could potentially lend specificity to a region of matrix. This highly speculative suggestion is an extension of the cell ligand hypothesis (for recent review see Moscona, 1974). Briefly this hypothesis predicts that cell recognition and selective cell adhesion resides in protein-carbohydrate surface molecules called ligands. If ligands on different cell surfaces are highly complementary this hypothesis predicts that recognition

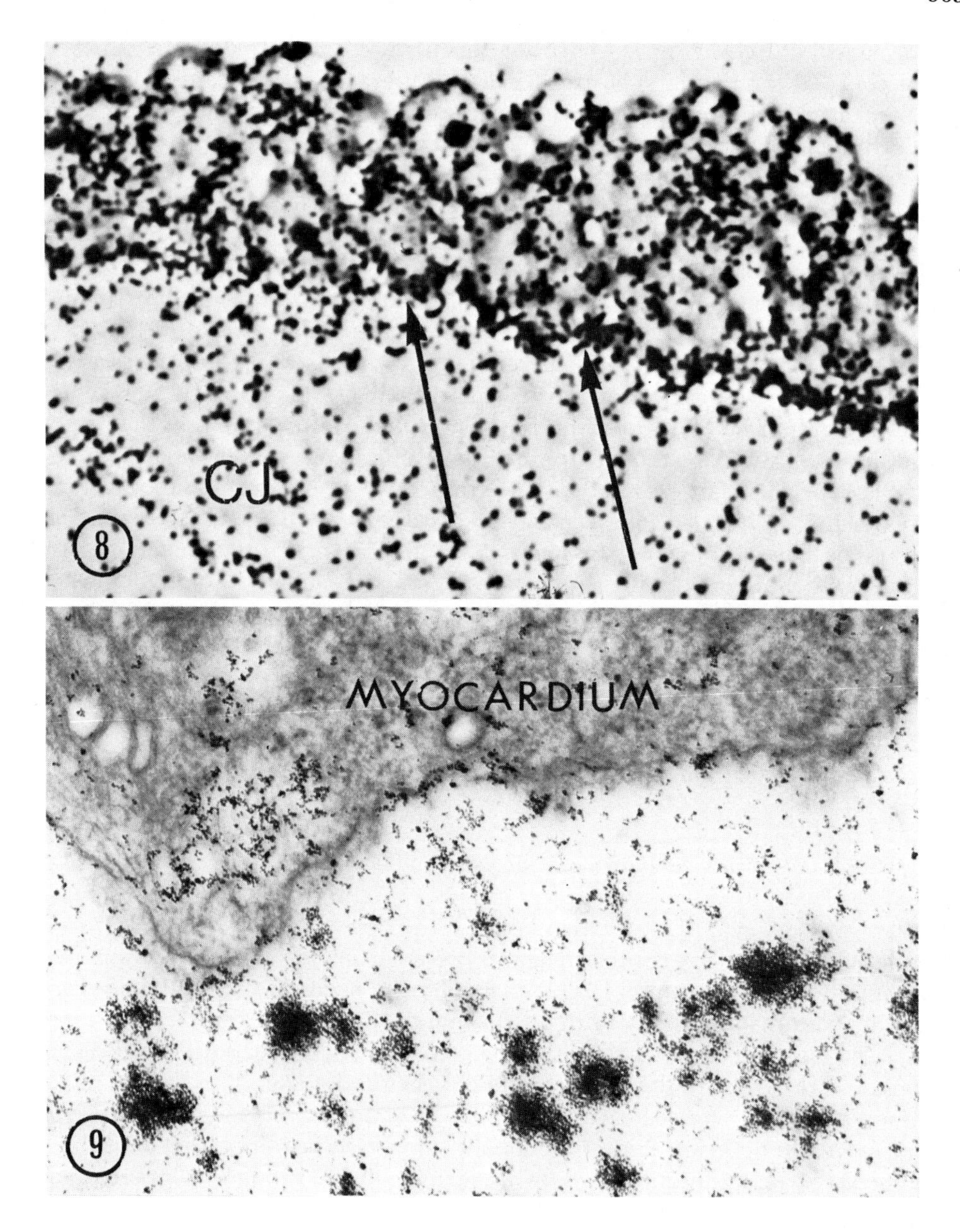

Fig. 8. Radioautograph of a portion of the myocardium and cardiac jelly of a group D embryo (see Fig. 4) labeled with ^{3}H-fucose and showing accumulation of incorporated label in the region of the myocardial basal lamina (arrows). The acellular cardiac jelly (CJ) is also labeled. ×1400.

Fig. 9. Colloidal thorium, used according to Revel (1964) demonstrates the location of anionic groups. The distribution of thorium does not coincide with the distribution of fucose label suggesting that the fucose labeled glycoproteins are not strongly anionic. Most of the material binding colloidal thorium is probably chondroitin sulfate. The basal lamina does not bind colloidal thorium although it is ruthenium red positive (Manasek, 1975b). ×35500.

occurs and adhesion results. Generally ligands are considered as cell surface molecules and the theory is based on cell-cell interactions that can be quantitatively measured under in vitro experimental conditions. As such it is a useful model upon which to base experiments designed to elucidate cell-cell interactions, but it does not readily offer testable predictions that could explain directed cell migration through a matrix as is the case with such a wide range of cells as the neural crest or the isolated pre-endocardial cells that migrate between endoderm and precardiac splanchnic mesoderm (Figs. 3 and 10). It is proposed that the glycoprotein normally resident within the matrix could create a cell or tissue-specific field that is recognized by transient cells, and therefore provide directional cues. This proposal is currently under intensive investigation in this laboratory where we are comparing the structure of cell and matrix glycoproteins.

Newly synthesized glycoprotein is not distributed uniformly within the heart. In addition to the obvious concentrations along ontogenically new basal laminae, (Fig. 8) particularly large accumulations are present in the dorsal mesocardium (Manasek, 1975b, 1976b). In the electron microscope these structures are similar in appearance to basal laminae (Johnson et al., 1974). They are prominent enough to be visible with the light microscope and light microscope radioautography confirms the presence of concentrations of fucose containing glycoprotein within these structures (Manasek, 1975b, 1976b). Clearly the matrix of the dorsal mesocardium is unique in the developing heart and it is tempting to suggest that the presence of these large, relatively immobile masses of glycoproteins in the matrix define spatially the embryonic midline and therefore provide positional information to migrating endocardium as well as myocardium (Manasek, 1976b).

2.3. *Myocardial cytodifferentiation and primary morphogenesis*

Cardiac muscle cytodifferentiation, defined as the acquisition of organized myofibrils by developing cardiac myocytes occurs in a near synchronized manner throughout the myocardium. In chicks, myofibrils become visible in the myocardium at around stage 9+ to 10 (8 to 10 somites). This event can be monitored readily by electron or polarized light microscopy. The appearance of fibrils in cells that were devoid of them a short while earlier denotes an *assembly* process; it does not necessarily correlate with the initiation of any translational or transcriptional event. Nonetheless it is a functionally important transition that delineates phenotypic expression and provides us with a readily monitored operational criterion of differentiation.

It is artificial and perhaps somewhat misleading to separate cytodifferentiation from precardiac cell migration since events that occur to myoblasts en route to the heart certainly are essential for cytodifferentiation. There is an overlap in time, and the events and interactions that occur during migration cannot yet be clearly separated.

Primary morphogenesis, the term used here to encompass all the early shape changes of the heart at the times that it is still a tubular organ (prior to

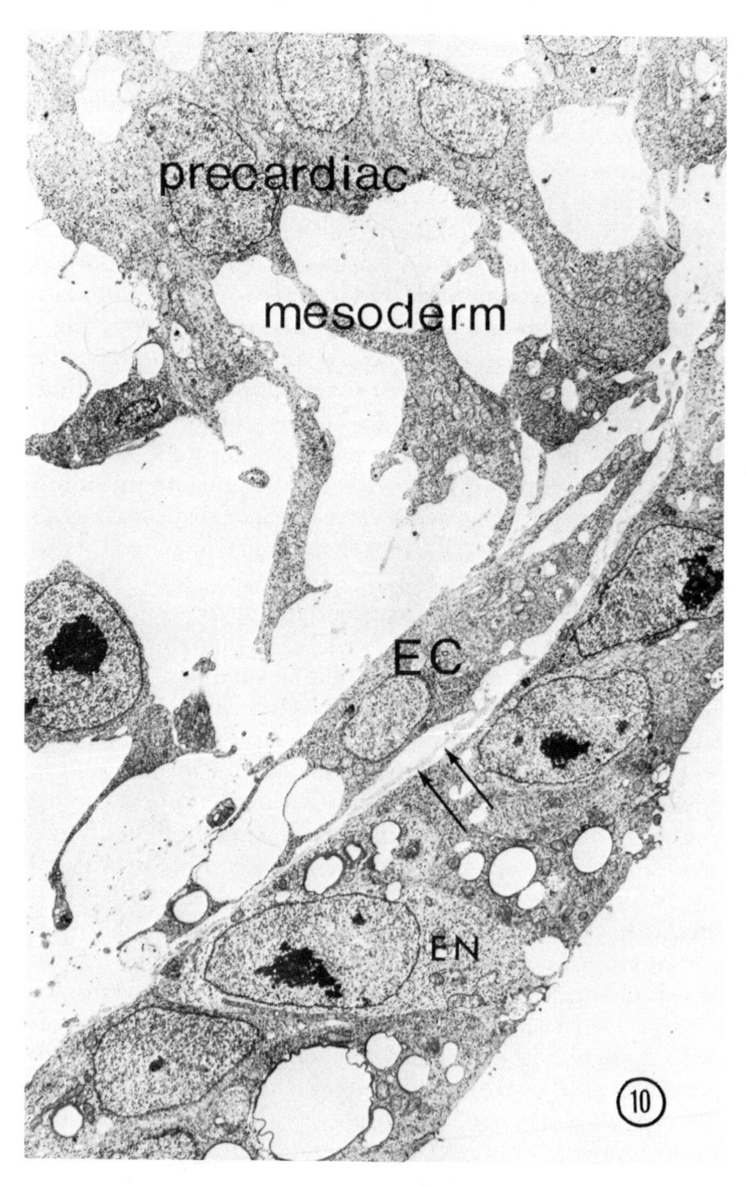

Fig. 10. Electronmicrograph of avian splanchnopleure at stage 8+(5 somites) embryo showing endoderm (EN) and mesoderm. In the precardiac region additional cells, presumably pre-endocardium (EC) are interposed between pre-cardiac mesoderm and endoderm. Neither pre-endocardium nor precardiac mesoderm has a basal lamina. A basal lamina is present along the endoderm and is most prominent along the region where dissimilar tissues are in close proximity (arrows). ×5060.

development of discrete chambers), is an integration of morphological events at the cellular level (for review see Stalsberg, 1970). The transformation of the midline tubular heart to a convoluted, looped structure (Fig. 11) is probably the immediate result of shape changes of individual cardiac muscle cells (Manasek, et al., 1972b). Briefly, cardiac myocytes occupying the prospective right side of the heart change from a roughly cuboidal shape to a more flattened, squamous shape (Fig. 12). This transformation is accompanied by a local increase in apical surface area. Since the myocytes do not change position relative to each other, the increased surface area cannot be accommodated by a reordering of the cells. Rather, the local change in apical surface area is accommodated by effecting a change in the shape of the organ: the heart bulges to the right. Thus, the intercellular relationships that define the epithelial character of the myocardium seem to be necessary for organ morphogenesis. Preliminary evidence obtained in this laboratory suggests that in the heart, the organization of the substrate (cardiac jelly) is an additional factor that is involved with morphogenesis. In this respect the interaction between myocardium and cardiac jelly may be analogous to that proposed for the lens rudiment and its underlying putative glycoproteins (Hendrix and Zwaan, 1974).

There are a number of morphological events occurring at the cellular level that correlate with cytodifferentiation. There is a striking change in the overall appearance of the myocardial layer. Prior to cytodifferentiation there were extensive intercellular spaces and the pre-myocytes were joined closely (by apical junctional complexes) only along their apical margins (see Spira, 1971). The lateral surfaces were widely separated and the basal surface was a poorly defined, irregular region devoid of basal lamina. Along with the appearance of myofibrils within the sarcoplasm there is a rapid increase in lateral cell bounding opposition with a concomitant decrease in intercellular space. The myocardial basal lamina begins to form at this time and becomes complete as the basal myocardial surface becomes a more regular surface (Manasek, 1968). It is tempting to propose that the basal lamina is, in some way, instrumental in effecting or stabilizing this change in myocardial architecture. These events are schematically shown in Fig. 13. In effect, the myocardial epithelium undergoes condensation along with a change in cell shape. This change is shown also to advantage in the light micrographs of frontal sections of a heart before and after looping (Fig. 12). These observations are consistent with the idea that the cell surface undergoes a developmental change, concomitant with differentiation, becoming more adhesive. This change in adhesivity would result in increased areas of cell contact. It has been suggested, based on other systems, that such histotypic organization is necessary for differentiation. For example in neural retina cells, glutamine synthetase, an enzyme whose presence is indicative of the differentiated state, is synthesized by neural retina in vitro only if histotypic cell aggregation occurs. If cells are grown in monolayers, glutamine synthetase is not induced even though the cells are in intimate contact with each other. Even in situ if retinotypic associations are prevented (in this case by BrdU) then differentiation does not occur (data reviewed in Moscona, 1974).

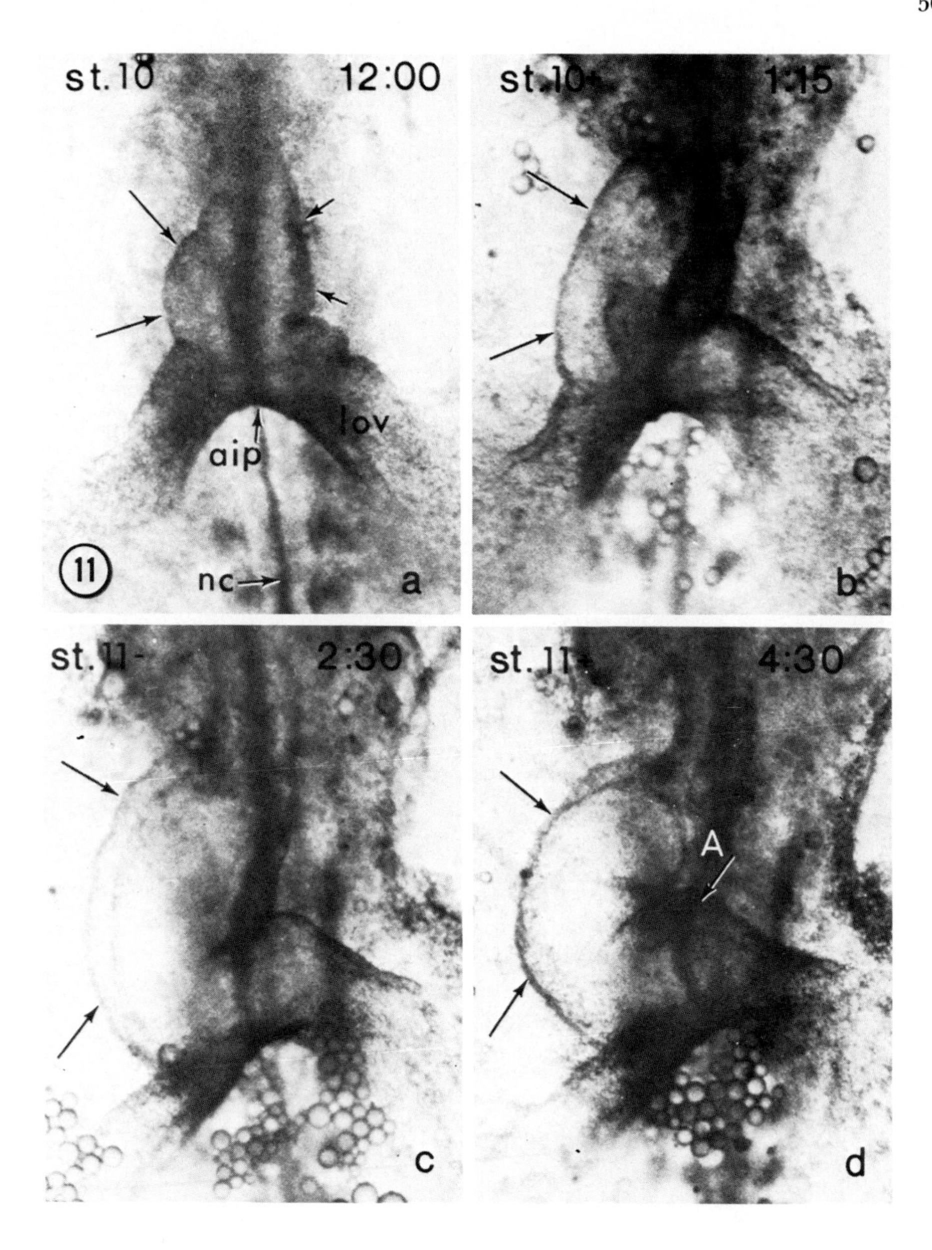

Fig. 11. Stages in the development of the normal "C"-shaped cardiac loop in the heart of a chick embryo grown in culture. The embryo's right side is on the left (ventral view). The notochord (nc) represents the embryonic midline; aip is the anterior intestinal portal. The total incubation time of 4.5 hr (12:00–4:30 in figures) was sufficient to result in the deformation of the midline tubular heart shown in a, (stage 10, 10 somites) to the typical looped structures seen in d. This is the time period that heartbeat is initiated, the myocardium develops myofibrils and blood flow begins. Figure from Manasek and Monroe (1972). ×80.

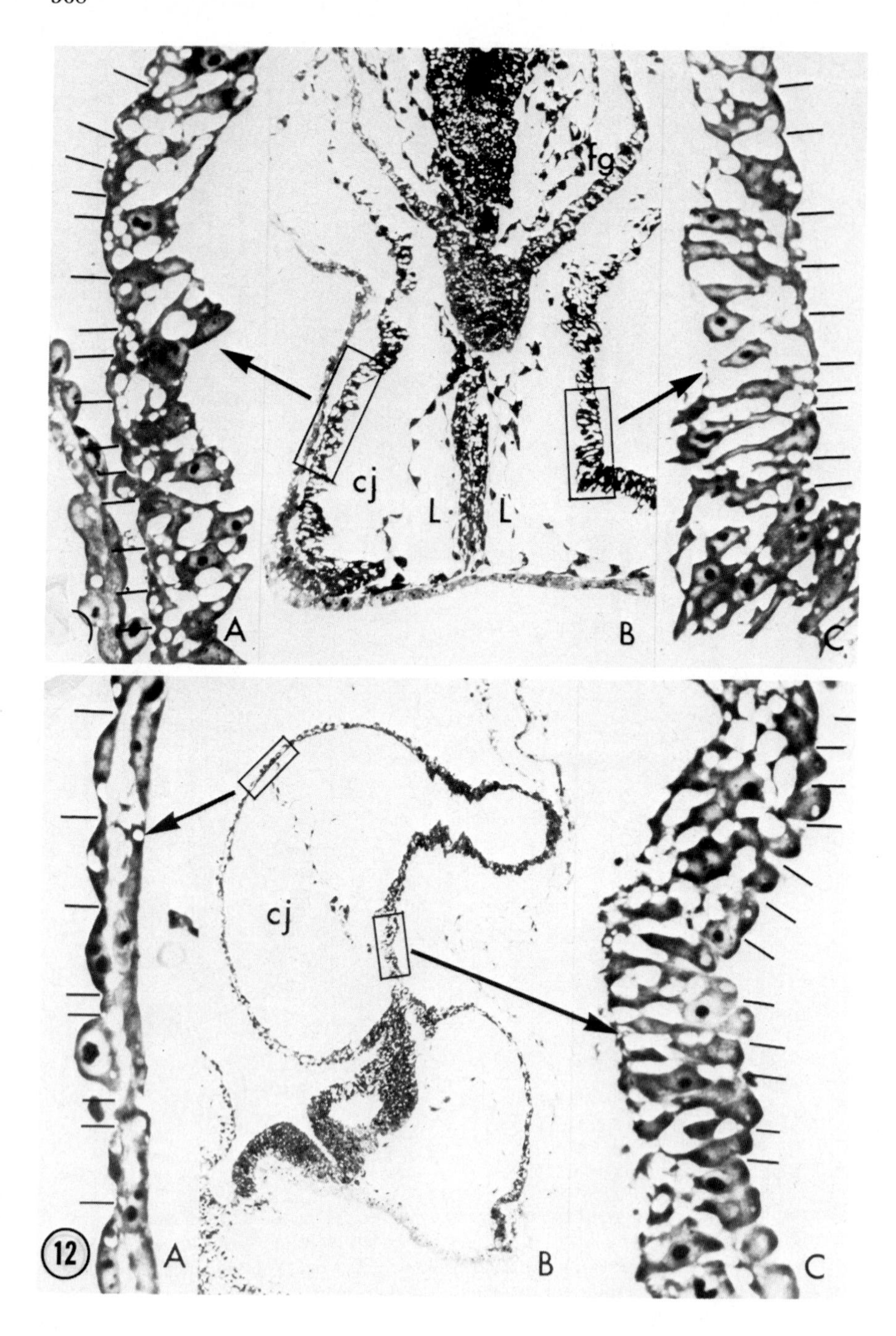

Fig. 12. Frontal sections (B) of the chick embryo heart before (top panel) and after looping (bottom panel) showing the differing histological features of the bulging right myocardial wall (A) and the concave left wall (B). Figure from Manasek et al. (1972b); A, ×940; B, ×76; and C, ×760.

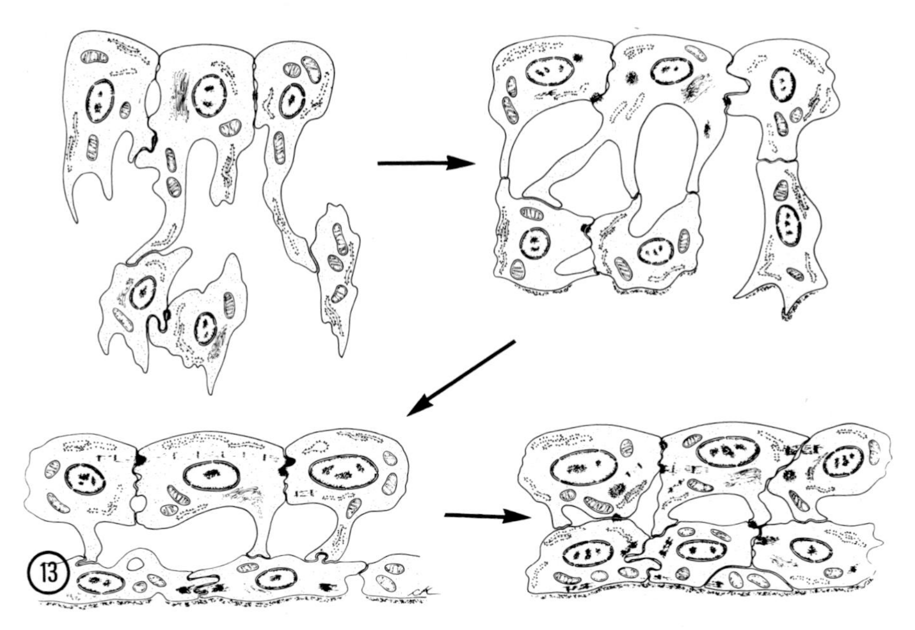

Fig. 13. Schematic representation of the changing myocardial wall architecture during the period of development ranging from 5 to 15 somites. This period encompasses myocardial phenotypic expression. In all cases the free outer (apical) surface is at the top and the basal cell layer bordering the cardiac jelly is at the bottom. The simultaneous development of the continuous basal surface; change in cell orientation; decrease in intercellular space and appearance of myofibrils and basal lamina are illustrated (drawing from Manasek, 1968).

3. *Experimental studies*

In the intact heart the normal sequence of myocardial cell shape changes as well as histotypic organization can be interrupted by administering the fungal metabolite, cytochalasin B. Cytochalasin B interferes with a large number of locomotory and morphogenetic events involving both single cells and developing organs. It is not my intent, nor is it possible to review here even a fraction of the significant literature on cytochalasin. However it seems likely that cytochalasin B does not grossly depress total protein synthesis (Estensen, 1971; Poste, 1972) nor does it appear to alter synthesis of extracellular glycosaminoglycans (Cohn et al., 1972). Cytochalasin B has several effects on cardiac myocytes. At high concentrations it dissociates myocytes grown in vitro, reduces the number of Na^+-carrying channels per unit area of membrane and disrupts myofibrillar architecture (Lieberman et al., 1973). When administered to normal embryos grown in culture, high concentrations of the drug also dissociate reversibly myofibrils in intact myocardia, hearts cease to beat (Manasek et al., 1972a) and the myocardium develops discontinuities as a result of cells separating under the action of cytochalasin (see Burnside and Manasek, 1972). These effects are both dose dependent and a function of embryonic age. Older hearts are more resistant to the effects of cytochalasin B. By judicious selection of cytochalasin concentration and embryonic age, its effects on the heart can, to some extent, be separated (Table 2). Of particular interest is the observation that low concentrations of cytochalasin have a profound effect when administered prior to myocardial differentiation and looping. In collaboration with Drs. Robert Waterman and M. Beth Burnside it was shown that the administration of cytochalasin B to embryos in New (1955) culture *prior* to myocardial condensation and differentiation has a number of effects at the cellular, tissue and organ level.

Table 2
Effects of cytochalasin B on intact hearts in different age embryos.*

Concentration of cytochalasin B	Embryo age	Incubation time	Effect
50 μg/ml	9 somites	several hours	no contractions; no fibrils form, no looping; cells separate
	11 somites	<45 min.	contractions stop; existing fibrils disrupt
		1 hour	cells separate; heart loses shape
	22–28 somites	>4 hours	contractions stop; existing fibrils intact; no new fibrils form; looping stops
10 μg/ml	11 somites	<2 hours	contractions continue; most fibrils intact; looping stops
		>2 hours	contractions stop; existing fibrils disrupt; looping stops
2 μg/ml	9 somites or younger	>6 hours	no contractions; no fibrils form; no looping
1 μg/ml	10–11 somites	>4–5 hours	weak contractions continue; existing fibrils intact; no new fibrils; looping stops
	12–13 somites	>4–5 hours	contractions continue; existing fibrils intact; no new fibrils; looping stops
	14–15 somites	>4–5 hours	contractions continue; existing fibrils intact; some new fibrils (?); looping sometimes continues

*Chick embryos were grown in New culture and various concentrations of cytochalasin B added to the medium according to standard techniques. Morphogenesis was followed in all cases photographically and representatives of each group prepared for histologic examination. Controls were done in all cases and were without effect.

3.1. Effect of cytochalasin B on the developing heart

A series of embryos, incubated for 5 hours with 2 μg/ml cytochalasin beginning at stage 10− (9 somites) was fixed and embedded in plastic. Controls, incubated in Tyrode's containing 1% DMSO were prepared similarly.

Frontal sections through the hearts of treated specimens (Fig. 14) demonstrated their overall symmetry (compare Fig. 14 to Fig. 12). In treated hearts both right and left walls are approximately the same thickness and consist of low columnar or cuboidal cells. In contrast, the right ventricular wall of looped hearts (such as shown in Fig. 12) invariably is thinner than the left (Manasek et al., 1972b). In treated embryos the dorsal mesocardium is very wide (Fig. 14) and the entire heart has a flattened appearance. The precardiac splanchnic mesoderm is attenuated and cell processes do not extend toward the foregut floor as they do normally (see Fig. 2 and 3 in Johnson et al., 1974).

3.1.1. Electron microscope studies

The myocardium normally becomes differentiated during the time periods used in this study. Myocardia of stage 10− embryos do not yet contain well formed myofibrils but within the ensuing few hours of development fibrillogenesis occurs and the cells become overtly differentiated (Manasek, 1968). Electron microscope observations of control embryos used in this study reveal a normal pattern of fibrillogenesis. Typical embryonic fibrils are seen in all myocardial cells examined in control embryos incubated with 1% DMSO. These fibrils are cross-striated and contain discrete, readily identifiable Z bands. Fibrils in various stages of assembly are also present (Fig. 15). The apical region of the developing myoepithelium demonstrates apical junctional complexes with small nexal junctions. These areas are frequently sites of developing intercalated discs (Fig. 16). Along the basal surface of the myocardium, a distinct basal lamina is formed.

Several ultrastructural aspects of the myocardium of cytochalasin-treated specimens remain normal. There is no evidence to suggest that any cells are undergoing agonal changes. With the dose used in the present study (2 μg/ml) desmosomes appear unaffected by cytochalasin B (Fig. 17). The general structure of apical junctional complexes remains unaltered and there is seemingly no diminution in number of punctate tight junctions. There is no evidence that cytochaslasin B at the concentration used in this study, overtly disrupts any pre-existing cell junctions in the embryonic heart (Figs. 18 and 19). In an earlier study it was clearly shown (Lieberman et al., 1973) that cytochalasin at concentrations of 10 μg/ml disrupts junctions between embryonic myocytes. Cytochalasin's effects on junctions seemingly demonstrated a dose-dependency.

Myofibrillar development is arrested in embryos exposed to cytochalasin B. Myofibrils consisting of repeating sarcomeres were never seen in any treated specimens. Seemingly, however, myofibrillar subunits are synthesized. Treated cells contain large numbers of thick and thin filaments, often with the appearance of being partially assembled into sarcomeres (Fig. 20). These groups of filaments are frequently associated with electron dense patches (Fig. 20) that have the appearance of early embryonic cardiac muscle Z bands, but more mature, highly ordered fibrils are absent. These morphological data suggest that the actual synthesis of myofibrillar precursors is not inhibited by cytochalasin B, but that their assembly into functional fibrils is. Polyribosomes, characteristic of developing muscle, are abundantly present in treated cells. The seeming integrity of these structures in conjunction with the accumulated myofibrillar material suggests that cytochalasin B does not disrupt polyribosome function.

Neither intercalated discs nor intact myofibrils are seen in treated hearts. It must be remembered that we added cytochalasin prior to the normal time of disc

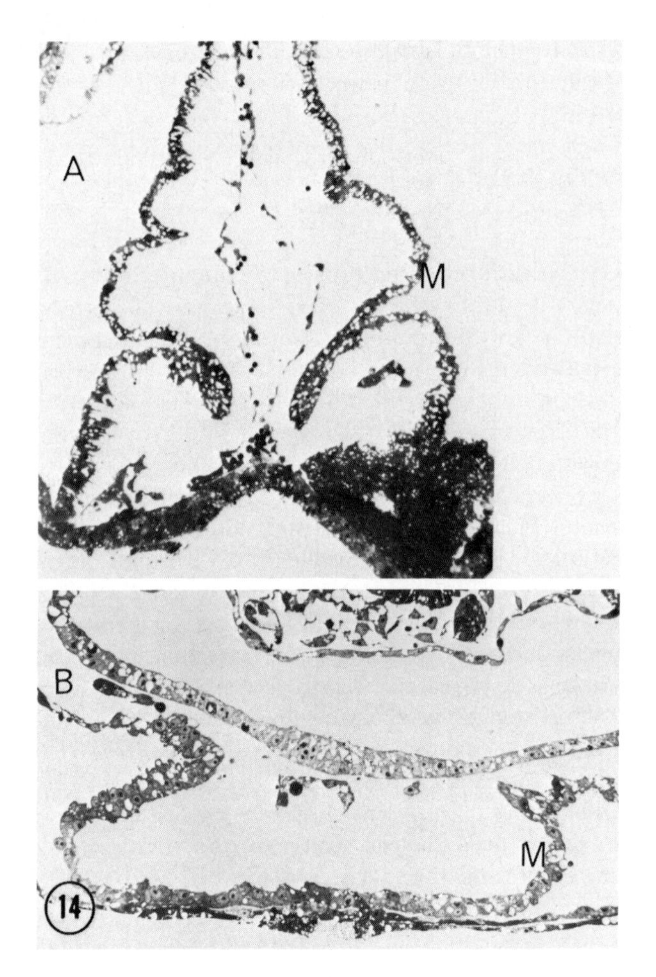

Fig. 14. Chick embryos (stage 10−; 9 somites) incubated in culture with 2 μg/ml cytochalasin B showing arrest of cardiac looping. After 5 hr frontal sections of plastic-embedded specimens show the marked symmetry and lack of differences in right and left myocardial wall thickness (compare to Fig. 12) typical of cytochalasin B-arrested hearts (A). Transverse sections (B) show that the dorsal mesocardium has failed to narrow in treated specimens. Compare to Fig. 1 and 3 in Johnson et al. (1974). The ultrastructural appearance of treated myocardium (M) are shown in subsequent electronmicrographs. Approximately ×170.

and fibril formation. Myofilaments, poorly organized into what appear to be myofibrillar segments do appear in treated hearts suggesting that the *synthesis* of *all* fibril components is not inhibited. However, none of these rudimentary incomplete fibrils was ever seen forming an intercalated disc. Thus, despite the fact that myofibrillar components are present they appear unable to make a functional association with the sarcolemma.

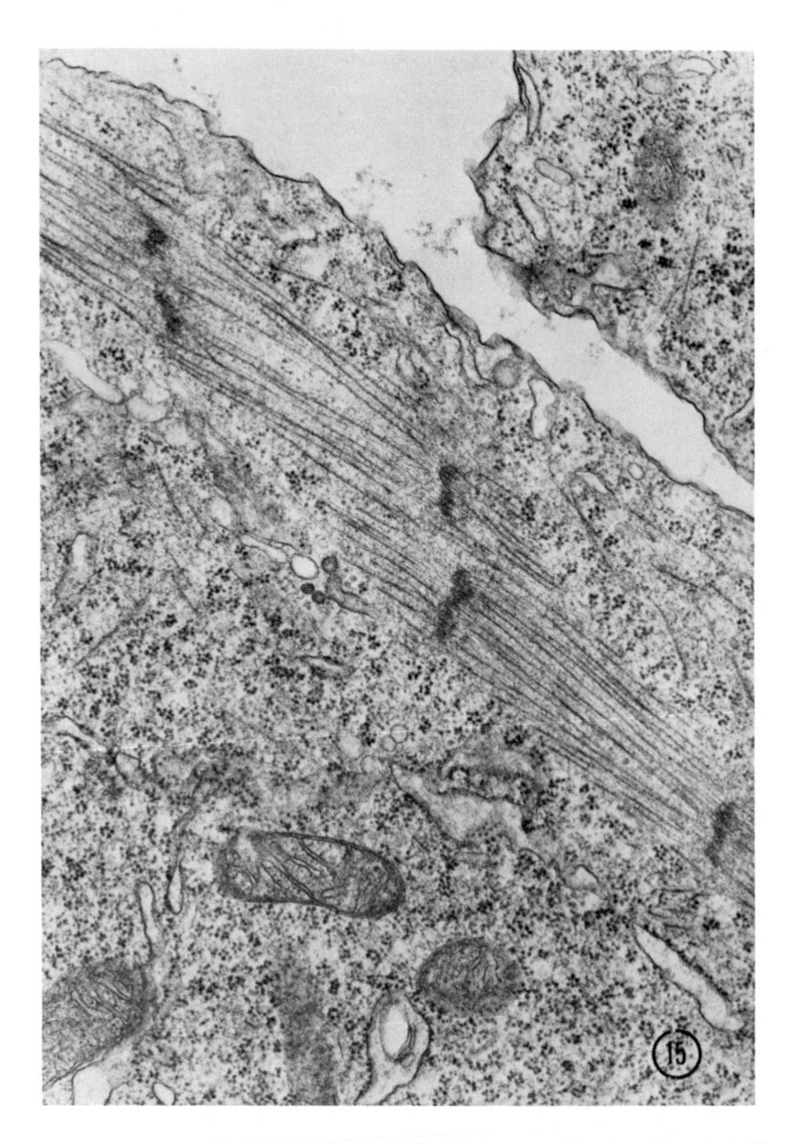

Fig. 15. Electronmicrograph of myocardium from control embryo (incubated in presence of DMSO) showing normal myocardial cytodifferentiation. Myofibrils, consisting of repeating sarcomeres are present in the cytoplasm. Figs. 15–20 show the myocardium of other control or cytochalasin B-treated (2 μg/ml) embryos incubated for 6 hr beginning at stage 9+. ×30000.

3.1.2. Primary morphogenesis is blocked by cytochalasin B

Cardiac looping is blocked by cytochalasin B (Fig. 21). The heart remains a midline structure if medium containing 2 μg/ml cytochalasin B is added to the cultured embryos prior to onset of normal looping. Hearts continue to increase in size in the presence of the drug but the profiles remain symmetrical and the ventricular bulge does not develop. In addition, hearts fail to establish normal

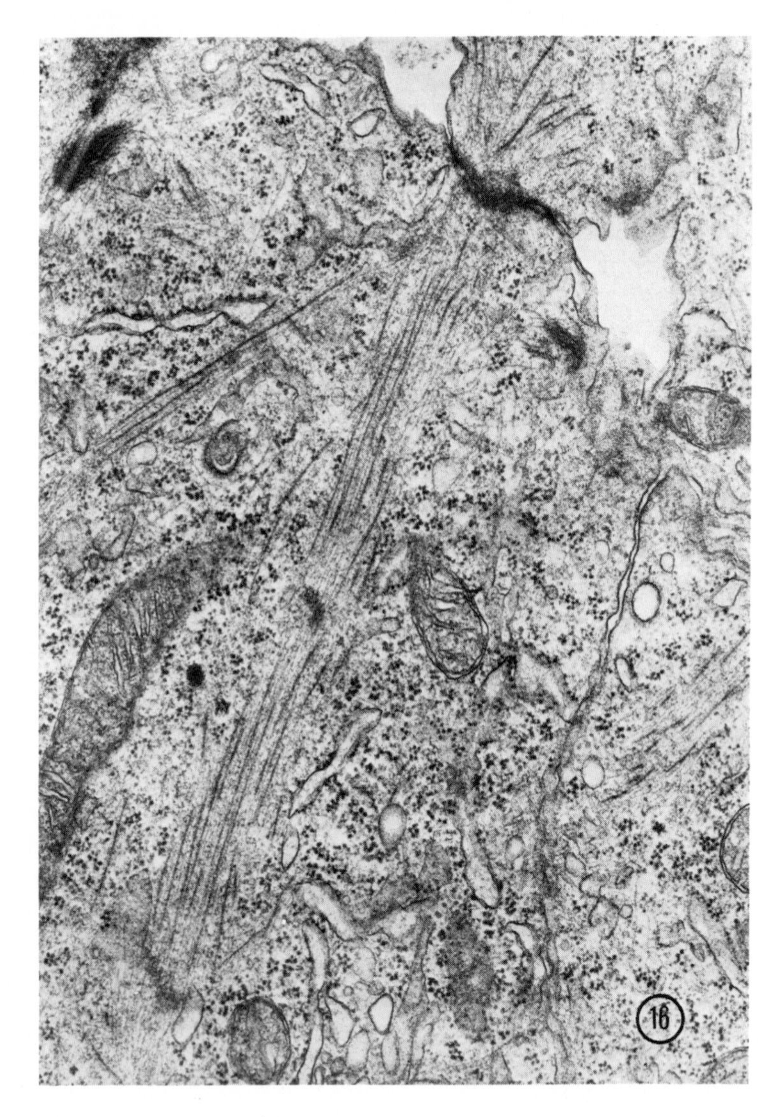

Fig. 16. At early stages of development, typical embryonic intercalated discs are formed at sites where developing fibrils insert into electron dense material associated with the sarcolemma. The intercalated disc seen in this control (DMSO-treated) is indistinguishable from normal. ×31 000.

spontaneous contractions in the presence of 2 μg/ml cytochalasin B and remain quiescent throughout the incubation period. Control embryos, grown in Tyrode's medium containing 1% DMSO undergo normal development. Hearts begin to beat and the normal sequence of shape change that characterizes looping, occurs within 5 hours of explanting (Fig. 22).

3.1.2.1. Iron oxide marking experiments. Following addition of cytochalasin B to cultured pre-loop embryos, pericardial membranes were removed and small

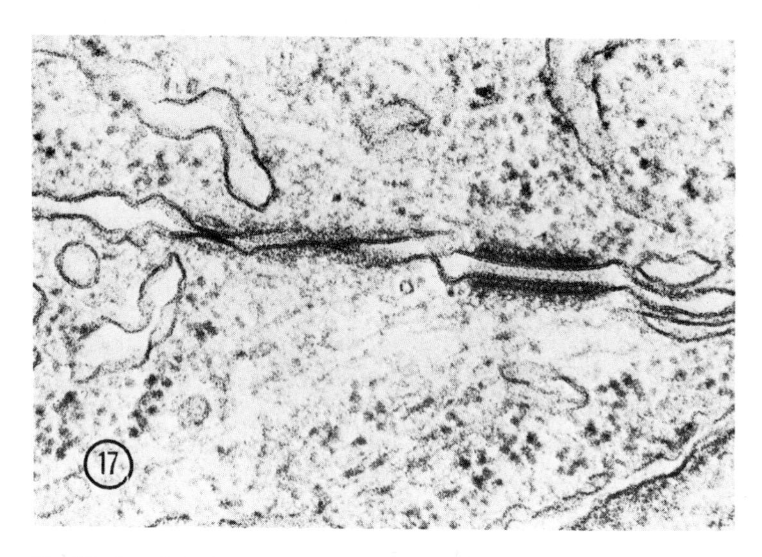

Fig. 17. Cytochalasin B at 2 μg/ml does not disrupt pre-existing desmosomes or nexal junctions as shown in this electronmicrograph from a treated heart. ×58000.

particles of iron oxide placed firmly on the bare myocardial surface. (For experimental details see Manasek et al., 1972). The position of these particles relative to each other can be measured precisely on photographs taken throughout the incubation period. Moreover, the position of particles relative to an extracardiac embryonic structure, such as the notochord (Fig. 23) can also be determined.

Although the heart increases in size during the incubation period (in this case $6\frac{1}{2}$ hours) the particles remain in approximately the same positions relative to each other and to the rest of the embryo. In no cases were they seen to move to the far right side as is the case during normal looping (Stalsberg and DeHaan, 1969; Manasek et al., 1972b). Iron oxide marking experiments thus show a striking failure of dextral rotation in cytochalasin-treated specimens.

Cytochalasin B thus has three principal effects on the early heart: (1) prevention of phenotypic expression; (2) inhibition of myocardial histogenesis; and (3) prevention of primary morphogenesis.

It is proposed that these three events are inextricably interrelated and that interference with myocardial histotypic development and subsequent prevention of cytodifferentiation disrupts organ morphogenesis. Perhaps the principal event which initiates the spectrum of cytochalasin-induced abnormalities is the prevention of histotypic development. Among its many biological effects (see comments by Holtzer and Sanger, 1972) cytochalasin B is thought to alter the adhesive properties of cells, which could explain (along with inhibition of locomotion) its effects on cell sorting in vitro (Armstrong and Parenti, 1972; Maslow and Mayhew, 1972, 1974; Steinberg and Wiseman, 1972) and also prevent, in situ, developmental changes that depend upon ontogenic changes in cell adhesion. Overton and Culver (1973) have proposed that the apparent reduction in adhesivity results from cytochalasin-induced surface immobility.

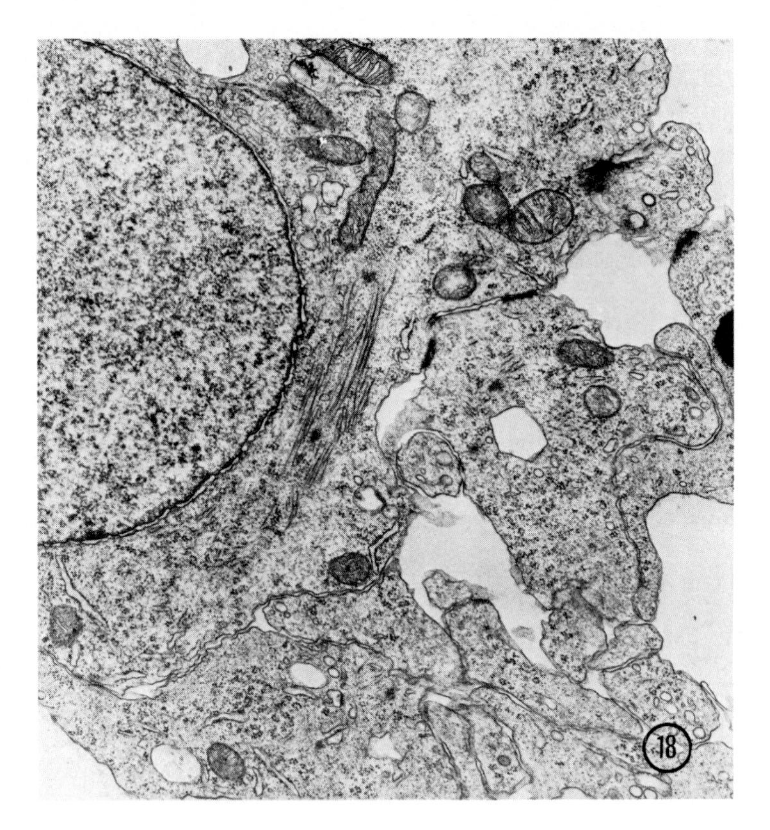

Fig. 18. Cytochalasin B prevents fibrillogenesis, but at sufficiently low doses (2 μg/ml) does not disrupt cell junctions as it does at higher doses (cf. Lieberman et al., 1973). This low power electron micrograph shows a portion of the myocardial wall of a stage 9+ embryo incubated with 2 μg/ml cytochalasin B, showing partly assembled myofibrils. The outer, apical surface is at the upper right. ×15500.

3.2. Morphogenesis and protein synthesis

Primary morphogenesis is dependent upon new protein synthesis. Introduction of cycloheximide, 250 μg/ml (a dose sufficient to block protein synthesis) to cultured chick embryos results in cessation of cardiac morphogenesis. If cycloheximide is added prior to myocardial cytodifferentiation and looping, the hearts do not loop (Fig. 24). If cycloheximide is added after looping has begun (Fig. 25) continued morphogenesis ceases. These studies, of course, do not identify the protein (or proteins) essential to morphogenesis but the rapidity with which morphogenesis ceases suggests that an extensive pool of the essential protein(s) does not exist.

Some indication that the synthesis of myospecific proteins are necessary for primary morphogenesis comes from the studies of Chacko and Joseph (1974) of the effects of 5-bromodeoxyuridine (BrdU) on early cardiac muscle cells. BrdU, a thymidine analogue, inhibits the expression of the differentiated state when

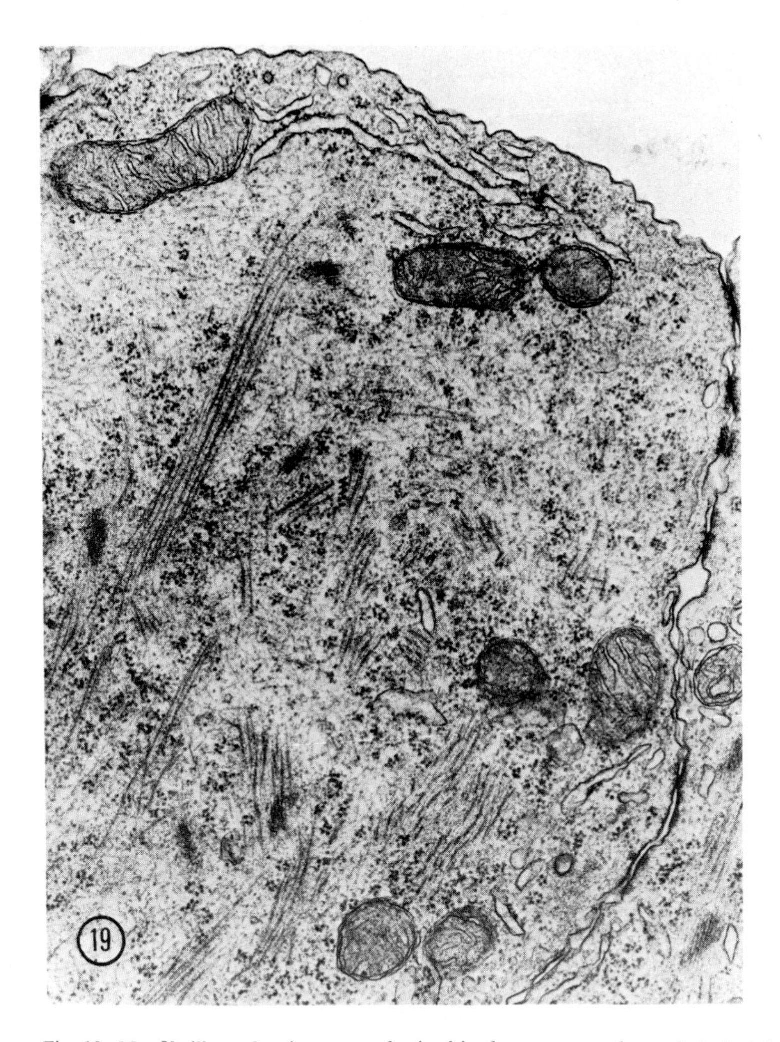

Fig. 19. Myofibrillar subunits are synthesized in the presence of cytochalasin B but do not assemble. The cytoplasm fills with partially assembled fibrils. Note the series of intact cell junctions. ×30 500.

incorporated into the DNA of undifferentiated cells (see Wilt and Anderson, 1972). In particular, skeletal myogenesis, defined on the basis of myoblast fusion and fibrillogenesis is inhibited if BrdU is incorporated into myoblasts (Stockdale et al., 1964; Coleman et al., 1969). Cardiac myoblasts (which normally do not fuse but rather differentiate as mononucleated cells) are also sensitive to BrdU (Chacko and Joseph, 1974) if the agent is administered sufficiently early in development. Sensitivity to BrdU is detected in cultured precardiac mesoderm cells derived from chick embryos younger than stage 7. Myofibrillogenesis (but not proliferation) is inhibited (Table 3). BrdU is without appreciable effect on differentiation in cells derived from embryos older than about stage 7. These cells continue to develop and express their phenotype. BrdU also has an effect

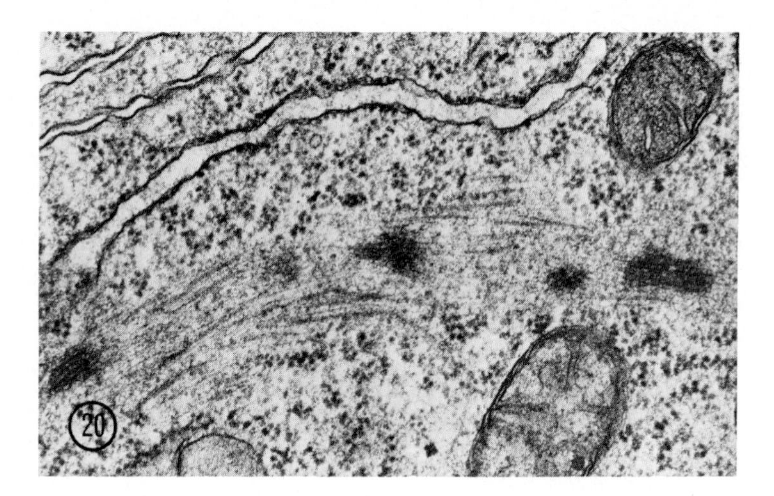

Fig. 20. In the presence of cytochalasin B, myofibrillar assembly does not progress normally. These structures, containing dense regions presumed to be nascent z-bands are typical of the point at which fibrillogenesis is arrested. These are similar to the structures shown by Holtzer et al. (1973). ×50 000.

Table 3
Effect of 5-bromodeoxyuridine on the differentiation of precardiac mesodermal explants at various stages*

Stage of the embryo	No. of explants cultured	No. of beating explants formed	Percent beating explants formed
6	100	1	1%
7	160	40	25%
8−	160	48	30%
8	170	76	45%
8+	200	106	53%
9	200	128	64%

*From Chacko and Joseph (1974).

on the intact developing heart. Chacko and Joseph noted a stage dependency of cardiogenesis in the presence of BrdU (Table 4). In this laboratory, BrdU was added to embryos grown in New culture to study its effects on development. Effects on cardiogenesis were seen when BrdU was administered as late as stage 9− (6 somites). Although fusion of right and left rudiments and the establishment of the midline tubular heart was not blocked, differentiation was retarded and looping was either retarded or abnormal (Fig. 26). Addition of BrdU to slightly older cultured embryos (stage 10−) did not block morphogenesis entirely but produced more or less subtle abnormalities (Fig. 27). In this case the ventricle bulged to the right but morphogenesis of the atrial region involving the left

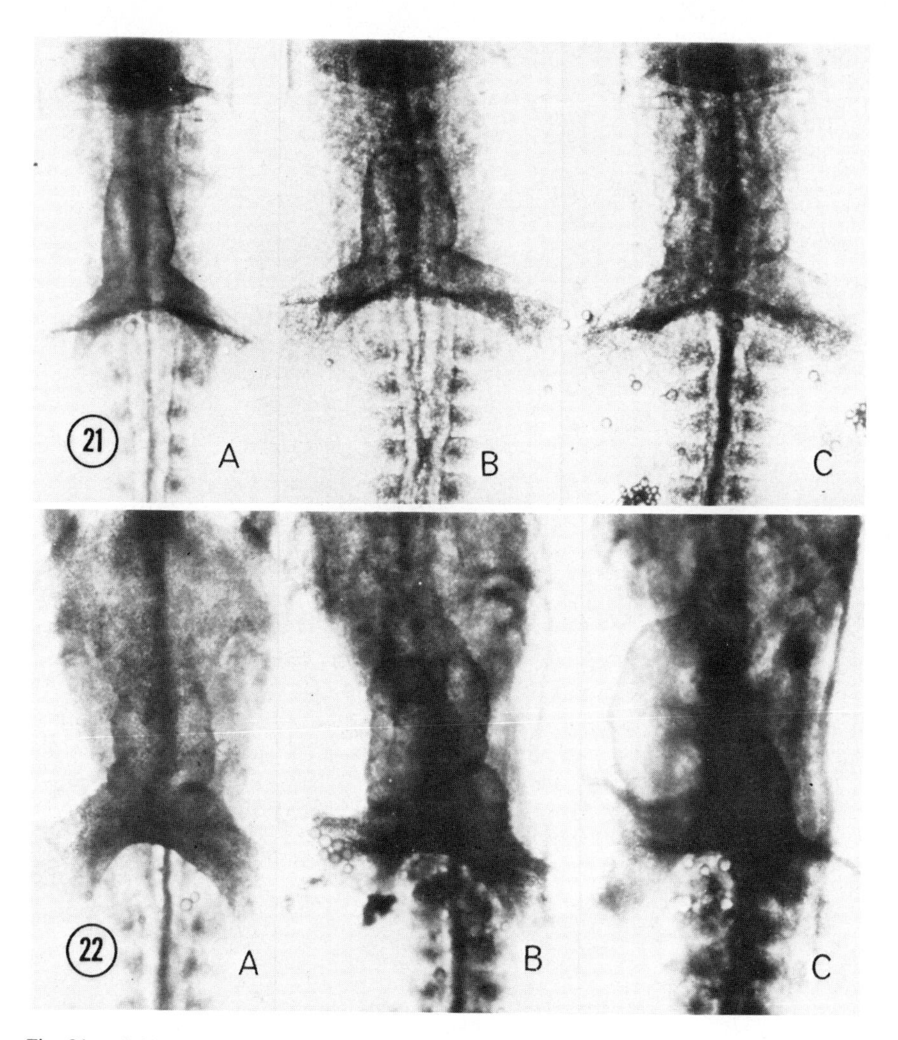

Fig. 21 and 22. Primary cardiac morphogenesis is sensitive to cytochalasin B. A stage 10− embryo explanted in medium containing 2 μg/ml cytochalasin B fails to develop a normal heart (Fig. 21A–C). The heart increases in size but remains a midline, symmetrical structure and does not loop. Control embryos, grown in identical medium without cytochalasin B show normal cardiac development (Fig. 22A–C). Approximately ×60.

omphalomesenteric vein was abnormal (compare Fig. 27b to Fig. 11). It is of interest to note that looping is normal in the presence of 2,4-dinitrophenol, an uncoupler of oxidative phosphorylation (Fig. 28). This is consistent with the proposals that early myocardium does not depend upon aerobic glycolysis (see review by DeHaan, 1965).

The effect of BrdU on myocardial cytodifferentiation was determined by examining the intact myocardium using polarized light microscopy. Although some evidence of fibrillogenesis was detected in those hearts exposed to BrdU as

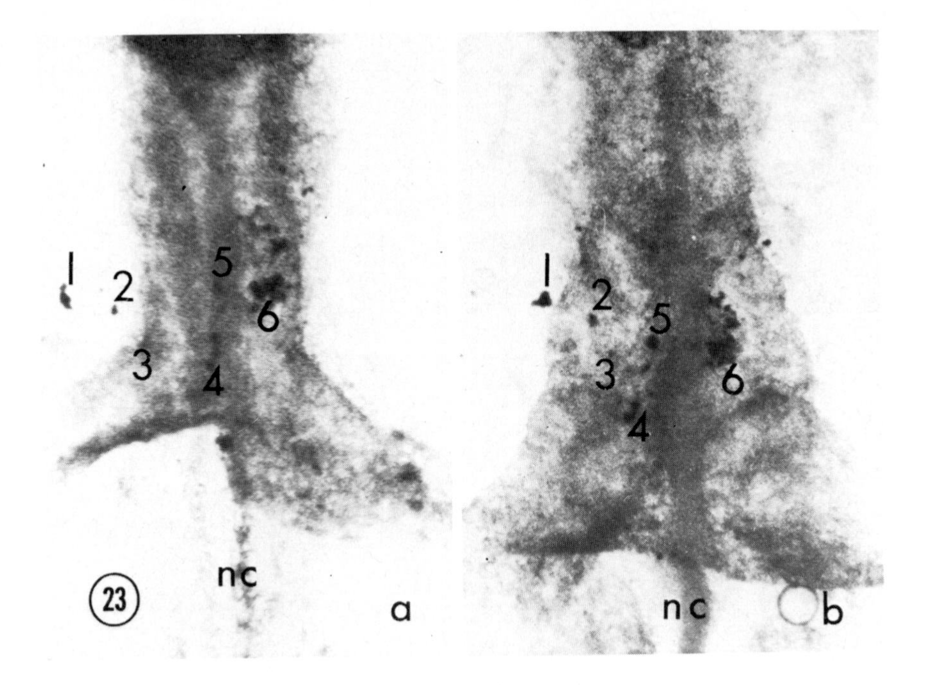

Fig. 23. Iron oxide particles were placed on the myocardial surface of stage 9+ embryos grown in culture exposed to medium containing 2 μg/ml cytochalasin B. (a). Six hours later (b) hearts had still not looped, and the myocardial surface had not undergone rotation, as shown by the failure of particles 4, 5, and 6 to change position relative to the notochord (nc). The particles 1, 2, 3 are not on the myocardial surface, but rather below it; the depth of field of the low power objective makes it seem as though they are on the same plane as 4, 5, 6. The technique used in this study is described in Manasek et al. (1972b). Approximately ×75.

Table 4
Effect of 5-bromodeoxyuridine (BrdU) on the development of heart in situ*

Stage at which BrdU added	Formation of the heart	Size of the heart formed	Ability to beat
5	No	—	—
6	No	—	—
7	Yes	Very small	Slow
8−	Yes	Very small	Slow
8	Yes	Small	Slow
8+	Yes	Normal	Normal
9	Yes	Normal	Normal

*From Chacko and Joseph (1974)

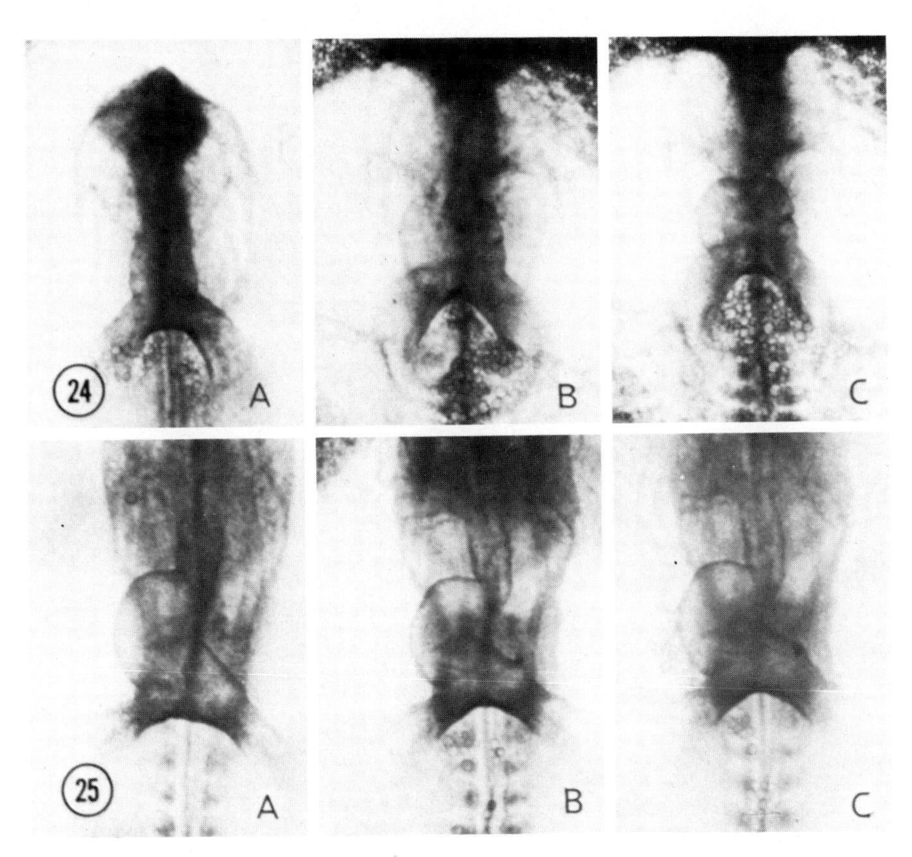

Fig. 24 and 25. Protein synthesis is required for primary heart morphogenesis. The presence of 250 μg/ml cycloheximide in the medium prevents development of a normal heart loop (Fig. 24A–C). The heart increases in size suggesting that migration of precardiac mesoderm into the rudiment is not stopped. The organ remains a midline, symmetrical structure (Fig. 24C). If a stage 11 (13 somite) embryo is explanted in the identical medium (Fig. 25A) cardiac morphogenesis ceases and the loop does not progress (Fig. 25B, C). Each sequence represent 6 hours of culture. Refer to Fig. 11 for normal development sequence.

late as stage 9−, the number of fibrils, as judged visually when compared to controls, was markedly reduced. In those hearts exposed to BrdU later in development a slight reduction in fibrillogenesis compared to controls might have resulted but it was not sufficiently great to be detected unequivocally by visual means alone. Collectively, these observations suggest that myocardial cells in situ respond to BrdU the same as in culture. Moreover, they again suggest a correlation between phenotypic expression (myofibrillogenesis) and cardiac morphogenesis. Protein synthesis is necessary (cycloheximide studies) but not sufficient (cytochalasin B studies) for looping. What is suggested is that differentiation must occur, myofibrillar proteins must be synthesized (BrdU studies) and moreover assembled into myofibrils (cytochalasin B studies) for events of cardiac morphogenesis to occur. It must be cautioned, however, that the precise

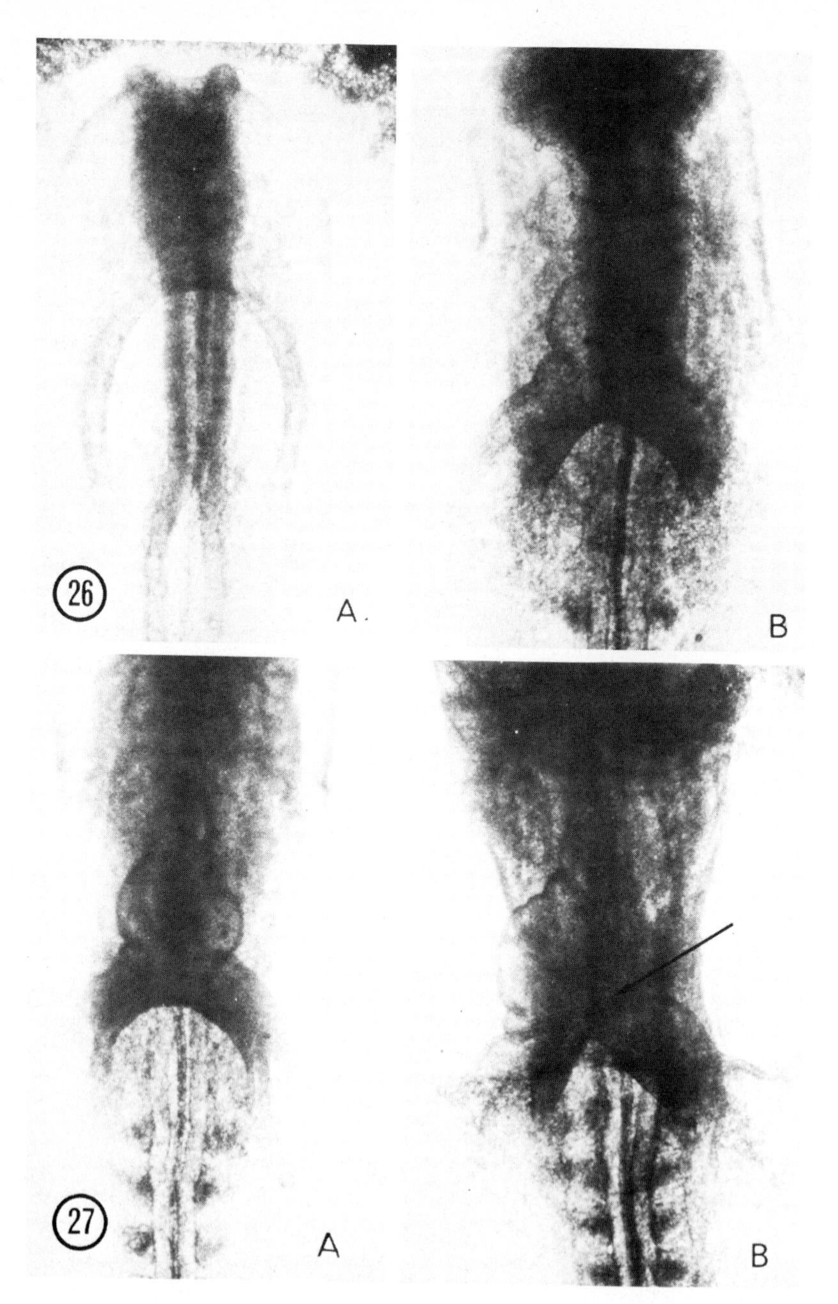

Fig. 26 and 27. Treatment of stage 9– (6 somites) embryos with BrdU (1×10^{-4} M) does not prevent development of the midline heart rudiment but it slows down its progress (Fig. 26A,B). When added at stage 10– (9 somites, Fig. 27A) BrdU does not prevent development of the loop. The heart develops its characteristic bulge to the right (Fig. 27B) but in many cases the overall morphology is not normal. Here, for example, the left omphalosmesenteric vein does not join the inflow side of the heart in normal fashion (arrow). It is interesting to note that the morphological defect involves tissue that was relatively undifferentiated at the time the embryo was exposed to BrdU and had not yet been incorporated into the myocardial wall (see fate map in Fig. 2). Approximately ×75.

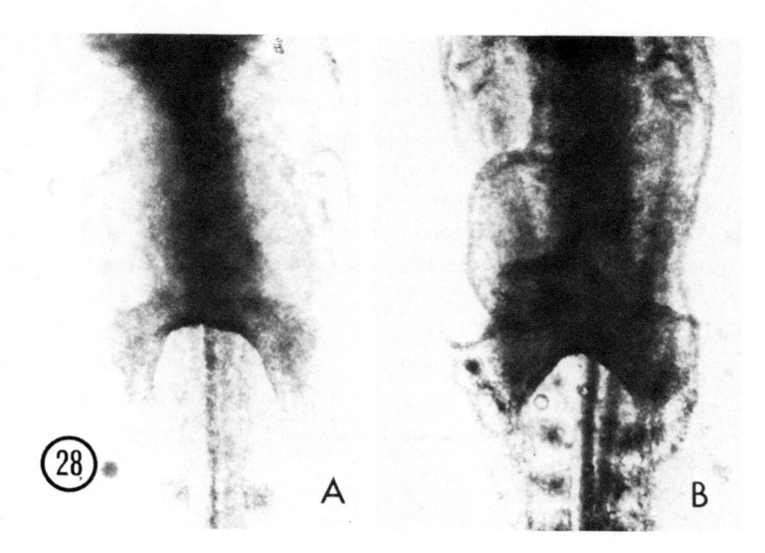

Fig. 28. Primary morphogenesis is not dependent upon aerobic glycolysis. A stage 9+ embryo (A) explanted into medium containing 1.5×10^{-4} M dinitrophenol developed a normal loop (B). Approximately ×75.

mechanism(s) by which BrdU and cytochalasin B exert their effects is not yet known and detailed appreciation of their primary effects are still sufficiently incomplete to rule out other mechanisms with absolute certainty. Nonetheless it can only be considered promising that primary cardiac morphogenesis can be manipulated in situ and these initial attempts suggest strongly that individual regulatory events can ultimately be discerned. Investigation solely at the cellular and subcellular level is by itself inadequate to achieve an understanding of the integrated events that mediate morphogenesis, but if combined with experimental and descriptive studies of the organ developing in situ should provide an increased understanding of organ morphogenesis.

4. *Histologic maturation of the heart wall*

The cardiac wall undergoes a number of striking changes during the heart's transition from a tubular peristaltic pump to a multichambered organ. During this time the distance between endocardium and myocardium decreases and these tissues interact; the myocardium itself acquires numerous non-muscle cell types and its outer surface becomes covered by epicardium (Kurkiwicz, 1909; Manasek, 1969; Lemanski, 1973; Virágh and Challice, 1973). Basically, the homogeneous all-myocyte myocardium becomes a complex heterogeneous tissue. This fact is overlooked often and most microscopic studies of myocardial development are concerned principally with myogenesis to the virtual exclusion of such components as blood vessels and true connective tissue cells.

Some of the various cell types found in the myocardium and their embryological origins are shown in Fig. 29. There is some degree of uncertainty about the origins of a number of these cells and little is known about the factors that initiate their entry into the heart. The transition from the histologic simplicity of the wall of the tubular heart to the complex architecture of the mature myocardium involves a number of interactions that are more amenable to descriptive inquiry than experimentation, partly because of difficulty encountered in attempting to manipulate experimentally a single parameter in such a complex system.

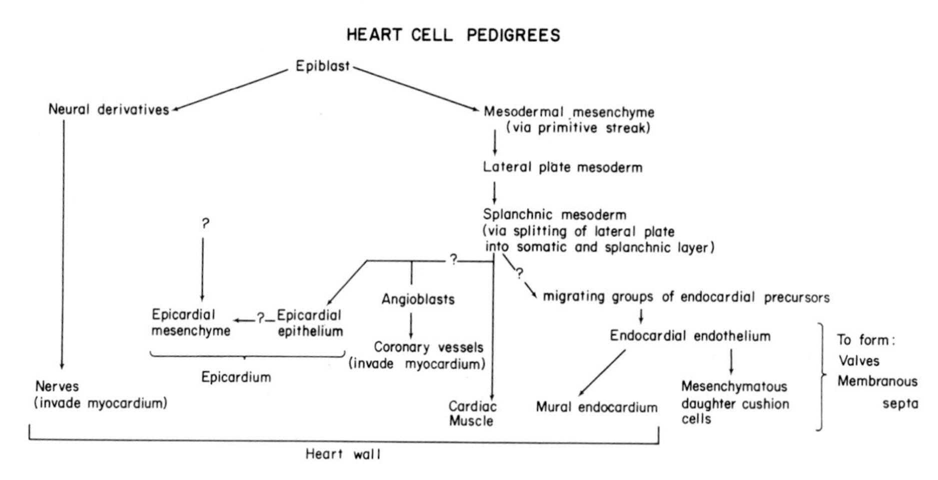

Fig. 29. Heart cell pedigrees.

4.1. Endocardial-myocardial interaction

Endocardium is probably derived from the lateral regions of precardiac splanchnic mesoderm. After condensation of the lateral plate mesoderm it splits into two layers, splanchnic and somatic, both of which are epithelial. Clusters of migrating cells separate themselves from the splanchnic mesoderm and move mediad between the more medial splanchnic mesoderm and the endoderm (Fig. 3 and 10). These endocardial forerunners again coalesce to form the intact, continuous endothelium of the heart, the endocardium (see Ojeda and Hurle, 1975). Operationally, two types of endocardial cells can be defined: those of the "endocardial cushion" region and those of the mural endocardium. Endocardial cells in the cushion region give rise to a population of mesenchymatous cells that populate the endocardial cushion connective tissue (Fig. 30). This event is not analogous to the dissolution of a primary or secondary epithelium to give rise to mesenchyme; rather the endocardium remains intact in the region of the developing cushion but its *progeny* are mesenchymal cells. Seemingly there is a change in surface characteristics of these daughter cells that enables them to

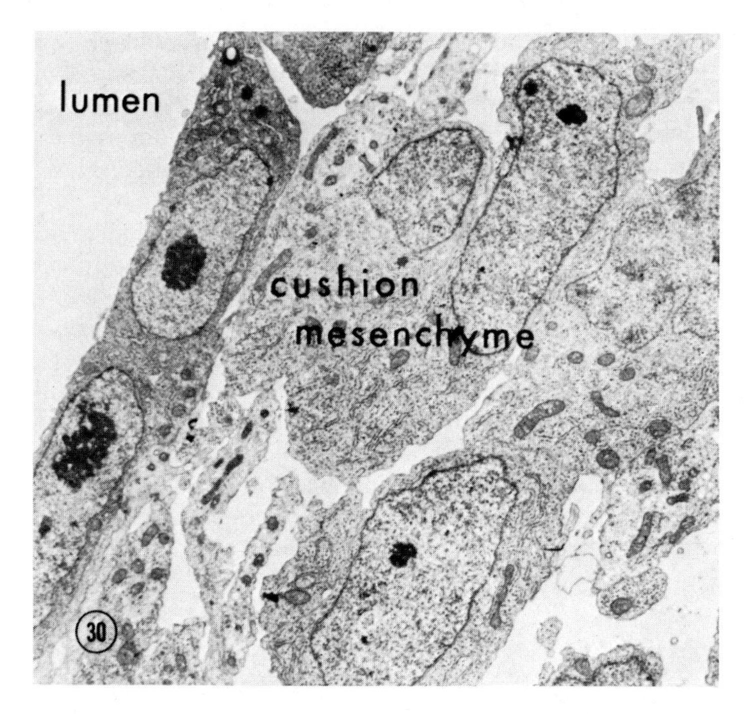

Fig. 30. Endocardium in the region of A-V canal proliferates to form the endocardial cushions which consist of mesenchyme and extracellular matrix. In this electronmicrograph the continuous endocardial endothelium bordering the lumen is seen along with its mesenchymal progeny. ×4500.

escape the endocardium whereas there is no change in the surfaces of the original endocardium since it remains an intact endothelium.

Differences have been noted between endocardium of cushion and non cushion areas and it has recently been proposed that mural and cushion endocardium represent divergent differentiation from a common endocardial precursor cell line (Markwald et al., 1975). There is no real evidence to support this view although it remains a possibility. Alternative possibilities are that the dichotomy between presumptive cushion and mural endocardium is manifest at the time the progenitors leave the splanchnic mesoderm and that these cells segregate into two groups as they are incorporated into the endocardium. This alternative is consistent with the observations of Rosenquist and DeHaan. These workers grafted ^{3}H-labeled splanchnopleure into labeled hosts and followed the migration of labeled cells radioautographically. Whereas label appeared in the myocardium in a predictable pattern (Rosenquist and DeHaan, 1966; Stalsberg and DeHaan, 1969) with individual labeled cells remaining together suggesting an *absence* of redistribution of myocardial cells the endocardium gave an entirely different picture. Discrete, small labeled grafts gave rise to labeled endocardial cells but these were distributed widely throughout the endocardium. Unlabeled cells were freely intermixed with labeled cells, suggesting that endocardial precursors migrate as individuals. Under these conditions they would be free to sort out and it is possible that cushion and mural endocardial progenitors segregate at this time. A third possibility is that no intrinsic difference exists between cushion and mural endocardium but that their behavior is different as a response to local differences in environment (Manasek, 1975b) particularly the cardiac matrix.

Endocardial cushion mesenchyme populates the developing cushion region and migrates centrifugally toward the myocardium. These cells approach, contact, but seemingly do not penetrate the myocardial basal lamina (Fig. 31). They are non-invasive. On the other hand, the mural endocardium, especially in the ventricular region is invasive in avian embryos. As the heart continues to become more convoluted and complex the cardiac jelly layer apparently becomes thinner, permitting close approximation between endocardium and myocardium. This interaction is not limited to contact, but rather the endocardium invades the myocardium (Henningsen and Schiebler, 1969; Manasek, 1970b). Endocardium does not become mesenchyme; rather this is an example of the endothelium intruding into the myocardium. Initially the endocardium

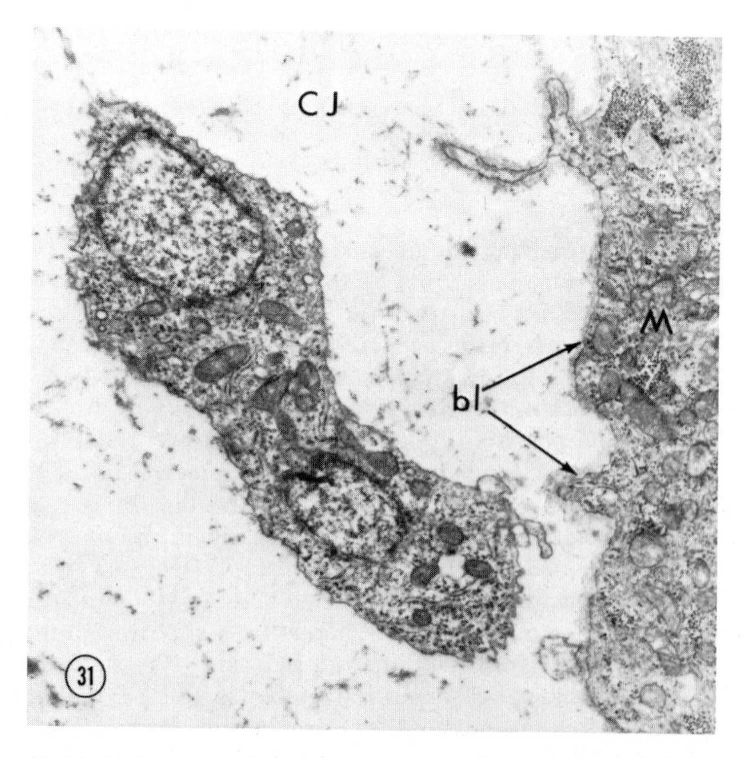

Fig. 31. Cushion mesenchyme migrates centrifigally through the cardiac jelly (CJ) until it contacts the intact basal lamina (bl) underlying the myocardium (M). These cushion cells seemingly do not cross the lamina and do not invade the developing myocardium. ×2000.

develops small outpocketings that contact the basal (innermost) layer of the myocardium. The previously intact layer of differentiated muscle cells separate at the point of closest approach of the endocardium which rapidly enters the gap. Broken or dead muscle cells have never been seen at these sites; rather it is clear that muscle cells separate as the result of loss of intercellular junctions.

After penetrating the innermost layer of muscle cells, the endocardium, in most cases, expands laterally. Throughout this process it retains a patent lumen that is continuous with the ventricular lumen and developing blood cells can be seen frequently in these endocardial outpocketings (Fig. 32).

Occasionally the endocardium continues to grow outward, penetrating even the outermost layer of myocardium. In these cases the endocardium penetrates completely the myocardial wall and only a single layer of endocardial cells is situated between the cardiac lumen and the pericardial region (Fig. 33). It is our impression that such complete penetrations of the myocardial wall by endocardium are common. In most instances it is likely that these endocardial herniations either regress spontaneously or are covered by restoration of myocardial continuity. It is tempting to propose that occasional massive endocardial herniations that do not regress spontaneously can result in Uhl's anomaly (Uhl, 1952). This congenital defect is characterized by absence of a portion of myocardium. In these regions endocardium is in close apposition to epicardium with some intervening connective tissue. It is not too difficult to visualize how such an anatomical relationship could be the result of an endocardial herniation during development, similar to, but more extensive than, that illustrated here.

4.2. Myocardial heterogeneity

The myocardium is subject to additional invasions by other cell types. Indeed, it is most likely that *all* of the non-muscle cells of the mature myocardium are added to it during ontogeny and do not arise from pre-existing precursor cells located within the myocardium of earlier stages. In other words the tissue heterogenity increases with development, and non-muscle cells (heterotypes) become dispersed through the myocardium. In addition to endocardium, the subepicardial layer also contributes to the non-myocyte population (Manasek, 1970b).

The signals that initiate the invasion of successive waves of different cell types are not known. Similarly factors that determine the locus of different cell types within the heart wall or the factors that prevent the sorting out of these heterogenous cells in vivo as they do in vitro are also unknown. There seemingly is a paradox when one considers the fact that many different cell types are contained within the heart wall, yet they remain dispersed throughout normal life. If, however, the myocardium is dissociated using enzymes such as collagenase and trypsin, and allowed to reaggregate in vitro, the cells segregate (Lieberman et al., 1972; Purdy et al., 1972; Shimada et al., 1974) in close agreement to Steinberg's models (Steinberg, 1970; Steinberg and Wiseman, 1972).

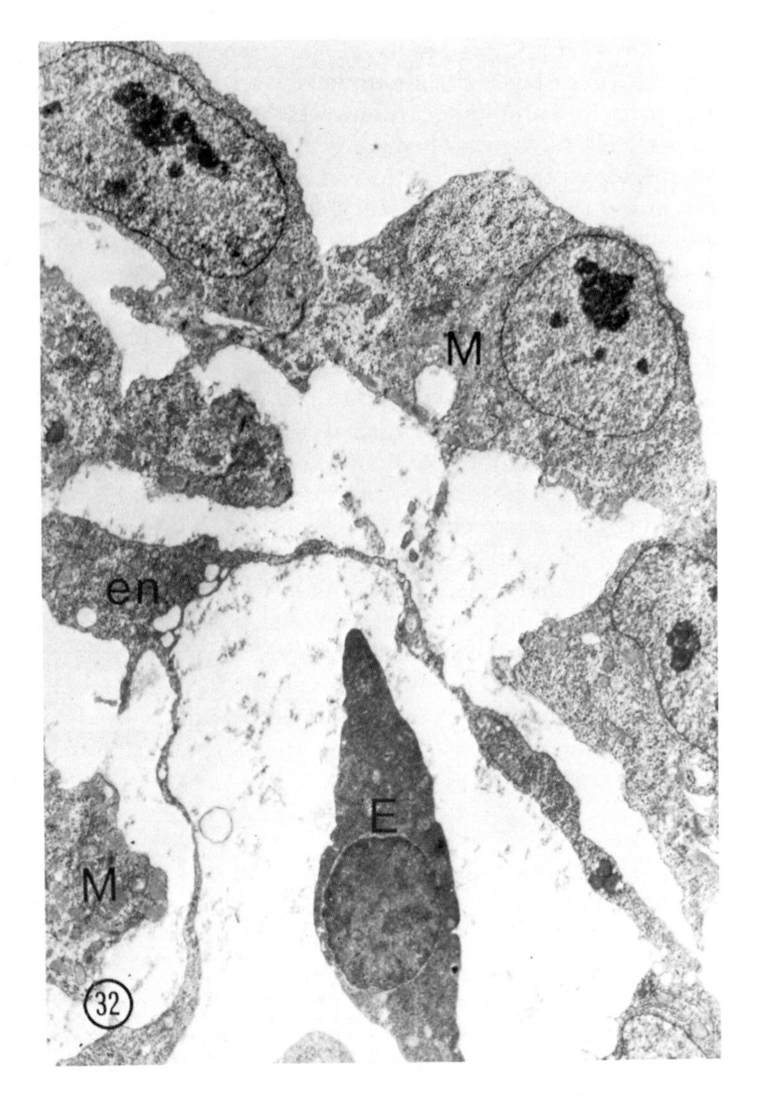

Fig. 32. Beginning at about day 3.5 of chick incubation the mural endocardium (en) invades the myocardium (M). These lumenal evaginations spread laterally and are often seen to contain developing erythrocytes (E). Prior to this "invasion" the ventricular myocardium contained only developing cardiac myocytes. ×5500.

The reasons for the behavioral differences between cells in situ and cells in vitro are not clear. Some inferential clues may be obtained from microscopic examination of the intact myocardium. In the heart, different cell types do not normally make extensive contacts with each other during ontogeny. For example, angioblasts most often are separated from developing muscle cells by a moderately electron dense material interposed between them (Manasek, 1971). Developing vascular smooth muscle also seemingly does not contact developing

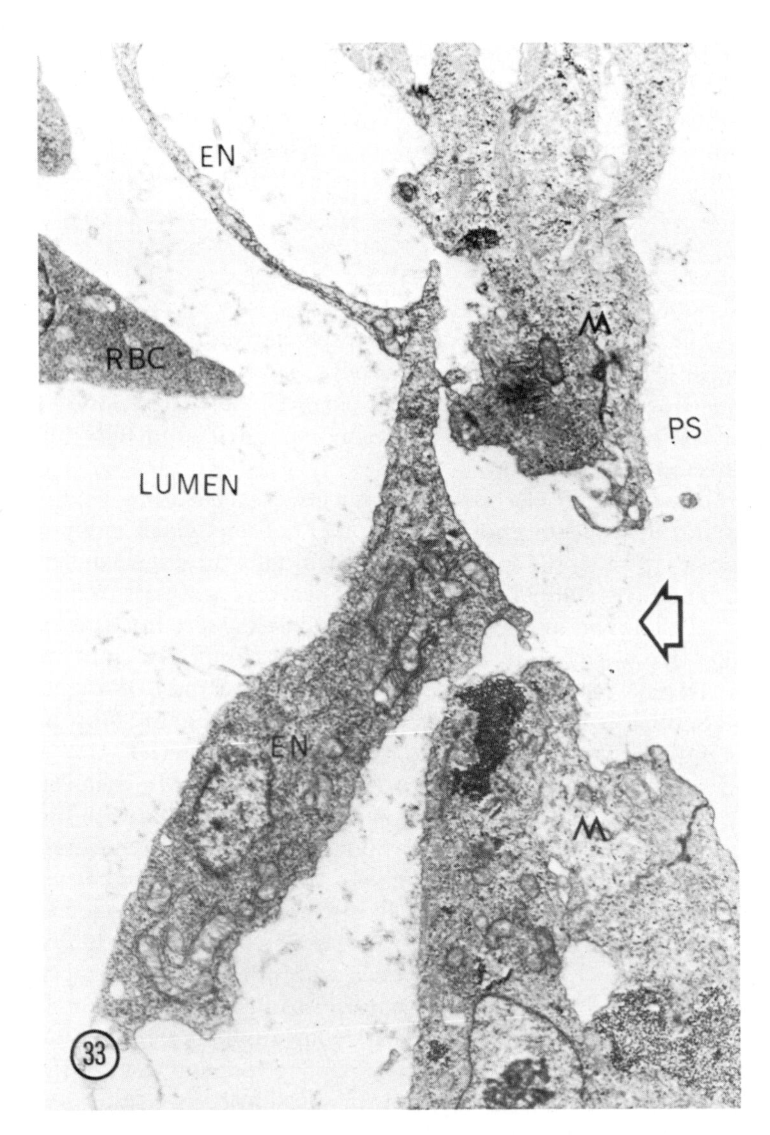

Fig. 33. During heart development (in this case in the 4th day of incubation) the mural endocardium (EN), formerly separated from myocardium (M) by cardiac jelly, invades the myocardium and not infrequently effects a perforation of the myocardium (arrow), coming in direct contact with the pericardial space (PS). ×6500.

cardiac muscle to any extent. Rather it is separated by extracellular material. In this laboratory we are currently examining the possibility that the extracellular material present in the normal heart may be involved directly in establishing and maintaining the complex distribution of different cell types within the myocardium. This is a particularly intriguing possibility because culture conditions can alter cell surface coats (Huet and Herzberg, 1973); level of sulfation of embryonic cardiac chondroitin sulfate (Manasek et al., 1973) and even the type of collagen synthesized by chondrocytes (Layman et al., 1972). Thus the environment (either the culture environment or the normal extracellular matrix) can regulate post translational events and, at least in some systems, gene expression which, in turn, influences the properties of the matrix (including the cell surface coat).

4.2.1. Embryonic myocardial glycoprotein

Because of the possibility that positional cues may be derived in part from glycoprotein present in cell surfaces, in the glycocalyx or in the dense matrix material which is a structural component of cardiac extracellular matrix, studies of glycoprotein synthesis in younger hearts are being extended to include the remainder of the developmental period.

We first attempted to ascertain if glycoproteins, as a general class of molecule, were synthesized by the heart throughout development. Intact chick embryo hearts were labeled with ^{3}H-fucose, solubilized in SDS, and the total labeled macromolecules examined. By solubilizing the entire heart we were assured of recovering all of the labeled macromolecules and our analyses were not limited to only those molecules that were readily extractable. Hearts from chick embryos of 3, 5, 7, 11, 14 and 18 days incubation were used. Aliquots of the SDS digest were passed through Sephadex G-50 to remove unincorporated fucose. Most of the macromolecules appeared in the excluded volume. This material (MW > 20,000) was collected and an aliquot passed through Agarose A-15. In material from 3, 5, and 7 day hearts a significant amount of label was recovered with the blue dextran marker but with increasing age proportionately less is recovered here relative to the more retarded fractions (Fig. 34). It is likely that the larger material is similar to the large molecular weight glycoprotein fraction detected in younger hearts (Manasek, 1976b) and which may represent an aggregate either of glycoprotein subunits or of glycoprotein complexed with other matrix components. The fucose labeled macromolecules are sensitive to Pronase digestion as shown by the alteration of their elution pattern from Sephadex G-50 (Fig. 35–38). These smaller degradation products, presumably glycopeptides, appear remarkably similar, suggesting close homology of the glycoproteins synthesized in hearts of these different ages (Fig. 35–38). Although this is consistent with the similarity of the elution pattern of the undigested glycoproteins on Agarose A-15 (Fig. 34), it must be remembered that comparisons based on Pronase released glycopeptides are relatively crude and imprecise. Nonetheless, throughout this age range there are two prominently labeled glycopeptide peaks: a large molecular size material eluting around fractions 12–16; and a more prominent peak eluting between fractions 25 and 30. There are also indications of an additional, poorly resolved peak coming off around fraction 20–22. Between 3 and 18 days there appears a progressive decrease in label associated with the largest glycopeptides (fractions 12–16) relative to those retarded by the Sephadex G-50 (compare Fig. 35 to Fig. 38). We are currently employing additional

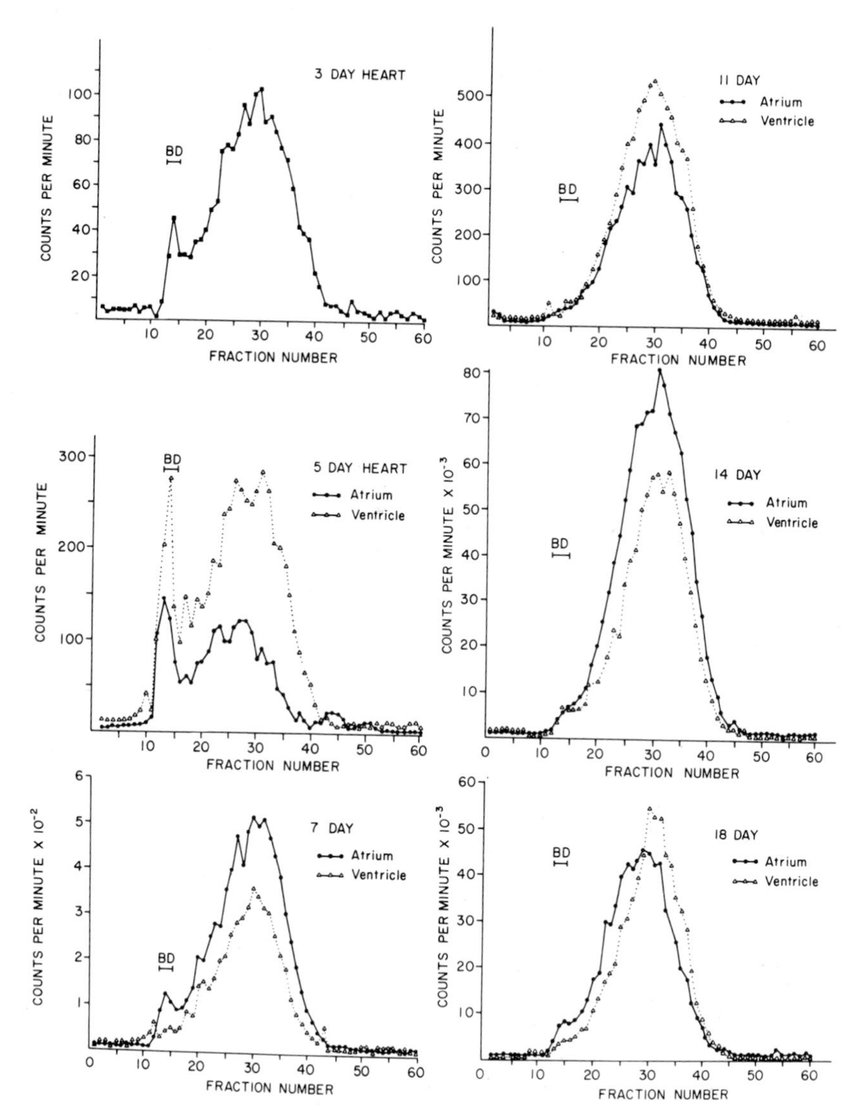

Fig. 34. Intact embryonic hearts were labeled for 6 hours in the presence of ^{3}H-fucose (200 μCi/ml). Ventricles and atria were separated except for 3 day old hearts which were left intact. After solubilizing (Manasek, 1976b) aliquots were desalted on Sephadex G-50 (0.9 × 50 cm) and the material recovered in the excluded volume chromatographed on Agarose A-15 as shown here. Note the age-dependent progressive decrease in relative amount of label co-eluting with blue dextran (BD). The retarded material elutes as a broad peak suggesting size heterogeneity. Counts per minute should not be compared between different ages since aliquot size was not corrected for differences in amount of starting tissue. In all instances the dotted line represents material obtained from ventricles, solid lines represent atrial material.

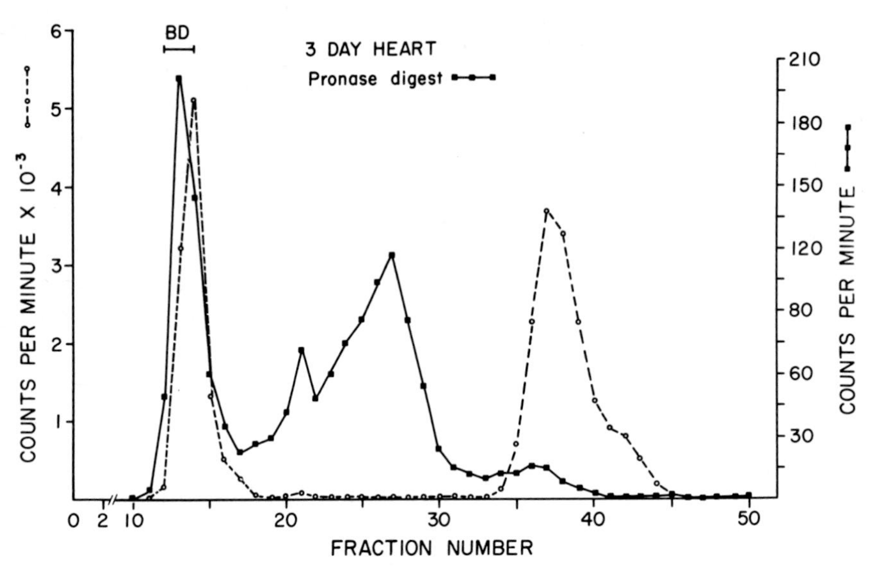

Fig. 35. Glycopeptides were characterized according to size on Sephadex G-50. Solubilized, labeled hearts were chromatographed on G-50 (dotted line) and the excluded volume (fractions 12–16) pooled. An aliquot was then digested exhaustively with Pronase and re-chromatographed on G-50 (solid line). Labeled molecules appear either with blue dextran (BD) or somewhat retarded (fractions 20–30). A small peak appears at about fraction 22.

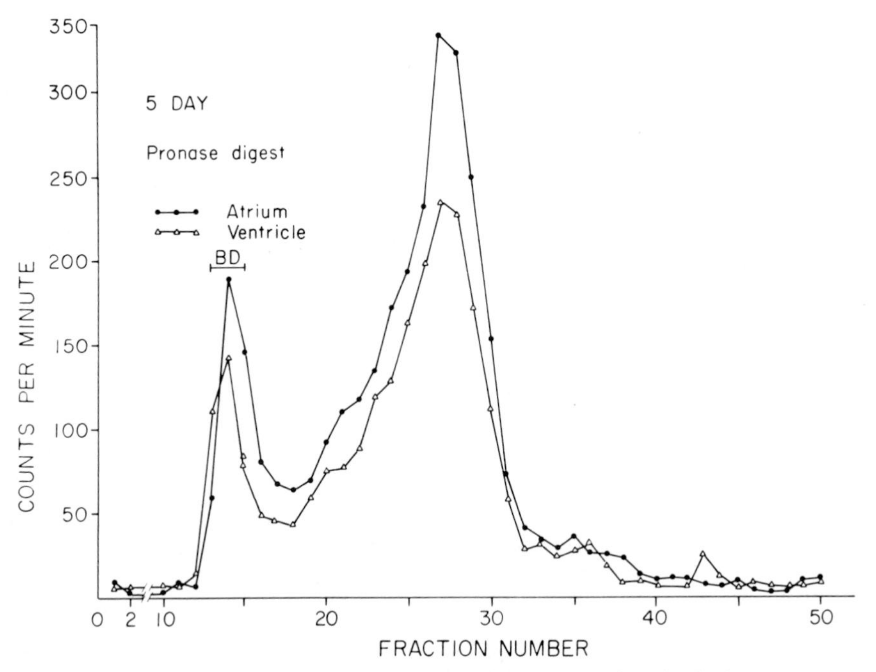

Fig. 36. Glycopeptides from 5 day atrium and ventricle, separated on Sephadex G-50 appear to be identical in terms of size distribution. Relatively more labeled molecules appear to be in the retarded fractions than in 3 day material (Fig. 35).

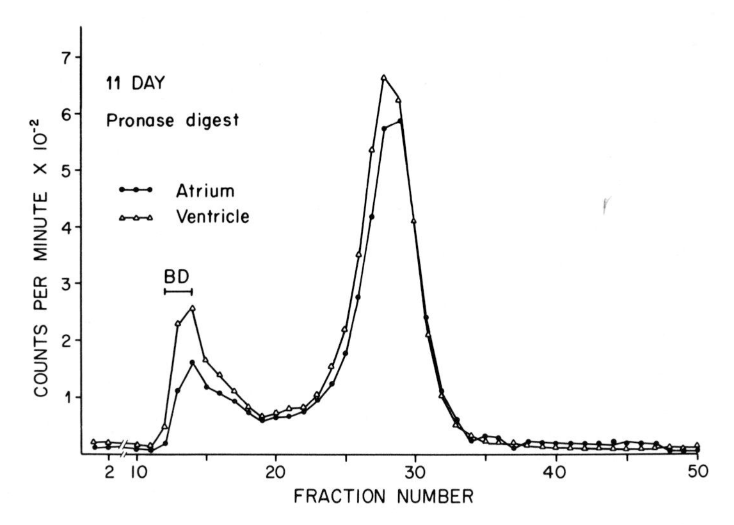

Fig. 37. Fucose-labeled glycopeptides from 11 day old embryonic atria and ventricles separated on Sephadex G-50.

techniques in an effort to establish more definitively ontogenic similarities or differences in fucose containing cardiac glycoproteins.

It must be emphasized that because we are examining only radioactive molecules we are dealing only with those recently synthesized (i.e., during the 6 hours the hearts were exposed to ^{3}H-fucose). The decrease in relative amount of label appearing in fractions co-eluting with Blue Dextran does not imply that aggregated forms are no longer *present*; it simply suggests that in older hearts newly synthesized glycoprotein does not appear in this form. It may take longer to form aggregates as a result of either changes in the rest of the extracellular matrix or it may represent a subtly different glycoprotein without this property.

Attempts to characterize electrophoretically either the glycoproteins or glycopeptides recovered by column chromatography have not yet been entirely successful. When subjected to acrylamide gel electrophoresis in the presence of SDS there is sufficient variability between presumably identical samples to make interpretation of results hazardous.

These studies of cardiac glycoprotein dealt with total labeled material. In the older stages described here we did not determine where the molecules were located. In younger stages it is clear that they are both cell and matrix associated (Manasek, 1976b) and it is likely that some remain as matrix components throughout development. Ontogenic changes in cardiac matrix, either as a result of quantitative or qualitative changes in synthesis or as an indirect result of matrix maturation resulting from extracellular molecule interactions are not yet sufficiently well understood to relate them directly to organogenic events. It is anticipated that continued studies of this complex system of glycosaminoglycans, collagens and non-collagenous glycoproteins will be particularly fruitful since it seems almost certain that this is a dynamic compartment of physiological

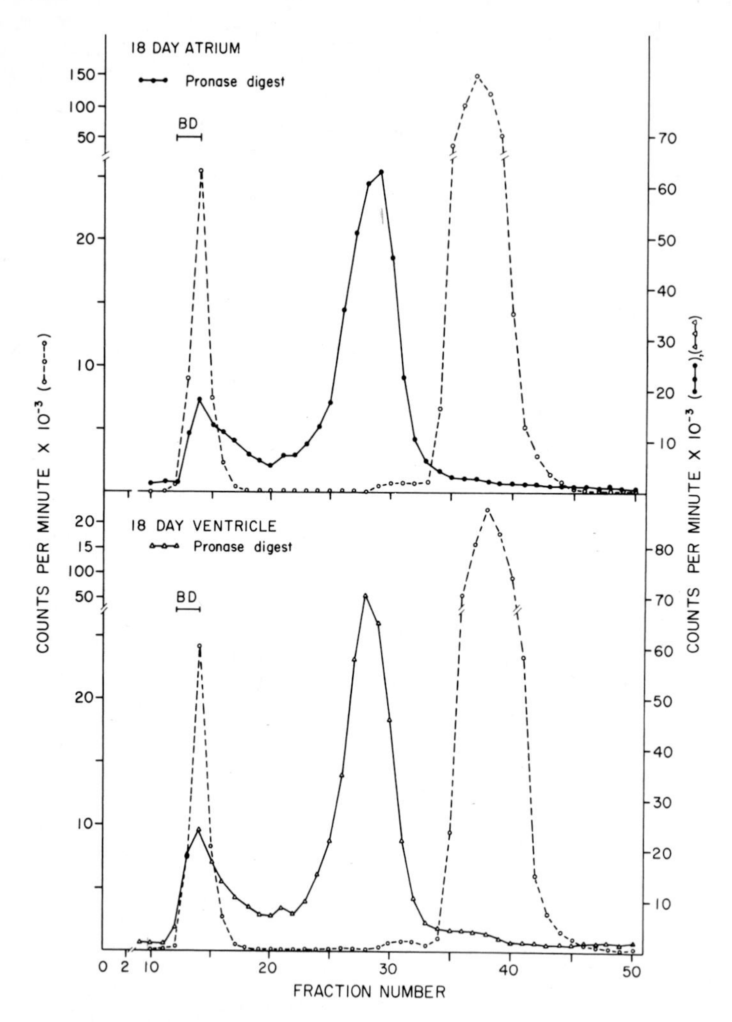

Fig. 38. Fucose-containing glycopeptides obtained from 18 day chick atria and ventricles appear to be similar to those synthesized from 5 days on. The major ontogenic differences seem to be in relative amount of label in different areas rather than developmental differences in size. The dotted lines represent recovery of label from G-50 prior to Pronase digestion. The material in the excluded volume was digested with Pronase and re-chromatographed (solid line) on G-50. Note the persistent small peak at about fraction 21–22.

importance that cannot be overlooked in attempts to understand the integration of the cells and tissues of the developing heart.

Acknowledgement

Original work reported here was supported by HL-13831 and by the Louis Block Fund, The University of Chicago. I thank Ms. J. Lacktis for her assistance.

References

Amano, H. (1961) Le rôle de l'entoblaste dans la formation du coeur chez l'urodèle. Comp. Rend. Soc. Biol. 155, 2218–2219.

Argüello, C., de la Cruz, M. and Sanchez Gòmez, C. (1975) Experimental study of the formation of the heart tube in the chick embryo. J. Embryol. Exp. Morphol. 33: 1–11.

Armstrong, P.B. and Parenti, D. (1972) Cell sorting in the presence of cytochalasin B. J. Cell Biol. 55, 542–553.

Burnside, B. and Manasek, F.J. (1972) Cytochalasin B: Problems in interpreting its effects on cells. Dev. Biol. 27: 443–444.

Castro-Quezada, A., Nadal-Ginard, B. and de la Cruz, M.V. (1972) Experimental study of the formation of the bulboventricular loop in the chick. J. Embryol. Exp. Morphol. 27, 623–637.

Chacko, S. and Joseph, X. (1974) The effect of 5-bromodeoxyuridine (BrdU) on cardiac muscle differentiation. Dev. Biol. 40, 340–354.

Chacko, S. and Rosenquist, G.C. (1974) Emergence of cardiac muscle cells from precardiac mesoderm. J. Cell Biol. 63, 55a.

Cohn, R.H., Banerjee, S.D., Shelton, E.R. and Bernfield, M.R. (1972) Cytochalasin B: Lack of effect on mucopolysaccharide synthesis and selective alterations in precursor uptake. Proc. Nat. Acad. Sci. U.S.A. 69, 2865–2869.

Coleman, J.R., Coleman, A.W. and Harline, E.J.H. (1969) A clonal study of the reversible inhibition of muscle differentiation by the halogenated thymidine analogue 5-bromodeoxyuridine. Dev. Biol. 19, 527–548.

Copenhaver, W.M. (1955) Heart, blood vessels, blood and endodermal derivatives. In: Analysis of development, (Willier, B., Weiss, P. and Hamburger, V. eds.) pp. 440–461, Saunders, Philadelphia.

de la Cruz, M.V., Muñoz-Armas, S. and Muñoz-Castellanos, L. (1972) Development of the chick heart. The Johns Hopkins University Press, Baltimore, Md.

Davis, C.L. (1924) The cardiac jelly of the chick embryo. Anat. Rec. 27, 201–202.

De Haan, R.L. (1964) Cell interactions and oriented movements during development. J. Exp. Zool. 157, 127–138.

De Haan, R.L. (1965) Morphogenesis of the vertebrate heart. In: Organogenesis, (De Haan, R. and Ursprung H., eds.) pp. 377–419. Holt, Rinehart and Winston, New York.

De Haan, R.L. (1967) Development of form in the embryonic heart. An experimental approach. Circulation, 35, 821–833.

De Haan, R.L. (1968) Emergence of form and function in the embryonic heart. Dev. Biol. Suppl. 2, 208–250.

Estensen, R.D. (1971) Cytochalasin B. I. Effects on cytokinesis of Novikoff hepatoma cells. Proc. Soc. Exp. Biol. Med. 136, 1256–1260.

Fullilove, S.L. (1970) Heart induction: Distribution of active factors in newt endoderm. J. Exp. Zool. 175, 323–326.

Gessner, I.H. and Boström, H. (1965) In vitro studies on ^{35}S-sulfate incorporation into the acid mucopolysaccharides of chick embryo cardiac jelly. J. Exp. Zool. 160, 283–290.

Gessner, I.H., Lorincz, A.E. and Boström, H. (1965) Acid mucopolysaccharide content of the cardiac jelly of the chick embryo. J. Exp. Zool. 160, 291–298.

Gordon, H.P. and Brice, M.A. (1974a) Intrinsic factors influencing the maintenance of contractile embryonic heart cells in vitro. I. Conditioned medium effects. Exp. Cell Res. 85, 303–310.

Gordon, H.P. and Brice, M.A. (1974b) Intrinsic factors influencing the maintenance of contractile embryonic heart cells in vitro. II. Biochemical analysis. Exp. Cell Res. 85, 311–319.

Hamburger, V. and Hamilton, H.L. (1951). A series of normal stages in the development of the chick embryo. J. Morphol. 88, 49–92.

Hay, E.D. (1968) Organization and fine structure of epithelium and mesenchyme in the developing chick embryo. In: Epithelial-mesenchyme interactions (Fleischmajer, R. and Billingham, R. eds.) pp. 31–55. The Williams and Wilkins Co., Baltimore, Md.

Hay, E.D. (1973) Origin and role of collagen in the embryo. Am. Zool. 13, 1085–1107.

Hendrix, R. and Zwaan, J. (1974) Changes in the glycoprotein concentration of the extracellular matrix between lens and optic vesicle associated with early lens differentiation. Differentiation, 2, 357–362.

Henningsen, B. and Schiebler, T.H. (1969) Zur Frühentwicklung der herzeigenen Strombahn. Elektronenmikroskopische Untersuchung an der Ratte. Z. Anat. Entwicklungsgesch, 130, 101–114.

Holtzer, H. and Sanger, J.W. (1972) Cytochalasin B: Microfilaments, cell movement and what else? Dev. Biol. 27, 444–446.

Holtzer, H., Sanger, J.W., Ishikawa, H. and Strahs, K. (1973) Selected topics in skeletal myogenesis. In: Cold Spring Harbor Symposia on Quantitative Biology XXXVII. pp. 549–566, Cold Spring Harbor Laboratory, New York.

Huet, Ch. and Herzberg, M. (1973) Effects of enzymes and EDTA on ruthenium red and concanavalin A labeling of the cell surface. J. Ultrastruct. Res. 42, 186–199.

Jacobson, A.G. (1960) Influence of ectoderm and endoderm on heart differentiation in the newt. Dev. Biol. 2, 138–154.

Jacobson, A.G. (1961) Heart determination in the newt. J. Exp. Zool. 146, 139–151.

Jacobson, A.G. and Duncan, J.T. (1968) Heart induction in salamanders. J. Exp. Zool. 167, 79–103.

Johnson, R.C., Manasek, F.J., Vinson, W.C. and Seyer, J. (1974) The biochemical and ultrastructural demonstration of collagen during early heart development. Dev. Biol. 36, 252–271.

Kalk, M. (1970) The organization of a tunicate heart. Tissue and Cell 2, 99–118.

Kefalides, N. (1973) Structure and biosynthesis of basement membranes. In: International Review of Connective Tissue Research, (Hall, D. and Jackson, D., eds.) pp. 63–104. Academic Press, New York.

Kosher, R.A. and Searls, R. (1973) Sulfated mucopolysaccharide synthesis during the development of *Rana pipiens.* Dev. Biol. 32, 50–68.

Kurkiewicz, T. (1909) O histogenezie mięśnia sercowego zwierzat kregowych. Bull. l'Acad. Sci. Cracovie, 148–191.

Laurent, T.C. (1968) The exclusion of macromolecules from polysaccharide media. In: The Chemical Physiology of the Mucopolysaccharides, (Quintarelli, G., ed.) pp. 153–168, Little Brown, Boston.

Layman, D.L., Sokoloff, L. and Miller, E.J. (1972) Collagen synthesis by articular chondrocytes in monolayer culture. Exp. Cell Res. 73, 107–112.

LeDouarin, G. (1974) Analyse expérimentale des premiers stades du développment cardiaque chez les vertébrés supérieurs. Ann. Biol. 13, 43–50.

LeDouarin, G. and Renaud, D. (1969) Etude morphologique et physiologique de la differenciation in vitro du mesoderme precardiaque de l'embryon de caille. Bull. Biol. 103, 453–468.

Lemanski, L.F. (1973) Heart development in the Mexican salamander, *Ambystoma mexicanum.* I. Gross anatomy, histology and histochemistry. J. Morphol. 139, 301–327.

Lieberman, M., Manasek, F.J., Sawanobori, T. and Johnson, E.A. (1973) Cytochalasin B: Its morphological and electrophysiological actions on synthetic strands of cardiac muscle. Dev. Biol. 31, 380–403.

Lieberman, M., Roggeveen, A.E., Purdy, J.E. and Johnson, E.A. (1972) Synthetic strands of cardiac muscle: Growth and physiological implication. Science, 175, 909–911.

Low, F.N. (1967) Developing boundary (basement) membranes in the chick embryo. Anat. Rec. 159, 231–238.

Manasek, F.J. (1968) Embryonic development of the heart. I. A light and electron microscopic study of myocardial development in the early chick embryo. J. Morphol. 125, 329–366.

Manasek, F.J. (1969) Embryonic development of the heart. II. Formation of the epicardium. J. Embryol. Exp. Morphol. 22, 333–348.

Manasek, F.J. (1970a) Sulfated extracellular matrix production in the embryonic heart and adjacent tissues. J. Exp. Zool. 174, 415–423.

Manasek, F.J. (1970b) Histogenesis of the embryonic myocardium. Am. J. Cardiol. 25, 149–168.

Manasek, F.J. (1971) The ultrastructure of embryonic myocardial blood vessels. Dev. Biol. 26, 42–54.

Manasek, F.J. (1973) Some comparative aspects of cardiac and skeletal myogenesis. In: Developmental Regulation. (Coward, S., ed.) pp. 193–218, Academic Press, New York.

Manasek, F.J. (1975a) Extracellular glycoproteins are synthesized during early development. Anat. Rec. 181, 420 (abst.).
Manasek, F.J. (1975b) The extracellular matrix of the early embryonic heart. In: Developmental and Physiological Correlates of Cardiac Muscle (Lieberman, M. and Sano, T., eds.) Raven Press, New York.
Manasek, F.J. (1975c) The extracellular matrix: A dynamic component of the developing embryo. In: Current topics in developmental biology. (Moscona, A.A. and Monroy, A., eds.). Academic Press. New York.
Manasek, F.J. (1976a) Macromolecules of the extracellular compartment of embryonic and mature hearts. Circ. Res. 38, 331–337.
Manasek, F.J. (1976b) Glycoprotein synthesis and tissue interaction during establishment of the functional embryonic chick heart. J. Mol. Cell Cardiol. 8, 389–402.
Manasek, F.J. and Monroe, R.G. (1972) Early cardiac morphogenesis is independent of function. Dev. Biol. 27, 584–588.
Manasek, F.J., Burnside, M.B. and Stroman, J. (1972a) The sensitivity of developing cardiac myofibrils to cytochalasin B. Proc. Nat. Acad. Sci. U.S.A. 69, 308–312.
Manasek, F.J., Burnside, M.B. and Waterman, R.L. (1972b) Myocardial cell shape changes as a mechanism of embryonic heart looping. Dev. Biol. 29, 349–371.
Manasek, F.J., Reid, M., Vinson, W., Seyer, J. and Johnson, R. (1973) Glycosaminoglycan synthesis by the early embryonic chick heart. Dev. Biol. 35, 332.
Mangold, O. (1957) Zur Analyse de Induktionsleistung des Entoderms der Neurula von Urodelen Naturwiss. 44, 289–290.
Markwald, R.R. and Adams Smith, W.N. (1972) Distribution of mucosubstances in the developing rat heart. J. Histochem. Cytochem. 20, 896–907.
Markwald, R.R., Fitzharris, T.P. and Adams Smith, W.N. (1975) Structural analysis of endocardial development. Dev. Biol. 42, 160–180.
Maslow, D.E. and Mayhew, E. (1972) Cytochalasin B prevents specific sorting of reaggregating embryonic cells. Science, 177, 281–282.
Maslow, D.E. and Mayhew, E. (1974) Histotypic cell aggregation in the presence of cytochalasin B. J. Cell Sci. 16, 651–663.
Moscona, A.A. (1974) Surface specification of embryonic cells: Lectin receptors, cell recognition and specific cell ligands. In: The Cell Surface in Development. (Moscona, A.A., ed.) pp. 67–99. John Wiley, New York.
New, D.A.T. (1955) A new technique for the cultivation of the chick embryo in vitro. J. Embryol. Exp. Morphol. 3, 320–331.
O'Rahilly, R. (1971) The timing and sequence of events in human cardiogenesis. Acta Anat. 79, 70–75.
Ojeda, J.L. and Hurle, J.M. (1975) Cell death during the formation of tubular heart of the chick embryo. J. Embryol. Exp. Morphol. 33, 523–534.
Oliphant, L.W. and Cloney, R.A. (1972) The ascidian myocardium: Sarcoplasmic reticulum and excitation-contraction coupling. Z. Zellforsch. 129, 395–412.
Ortiz, E.C. (1958) Estudio histoquimico de la gelatina cardiaca en el embryion de pollo. Arch. Inst. Cardiol. Mex. 28, 244–262.
Overton, J. and Culver, N. (1973) Desmosomes and their components after cell dissociation and reaggregation in the presence of cytochalasin B. J. Exp. Zool. 185, 341–356.
Patten, B.M. and Kramer, T.C. (1933) The initiation of contraction in the embryonic chick heart. Am. J. Anat. 53, 349–375.
Patten, B.M., Kramer, T.C. and Barry, A. (1948) Valvular action in the embryonic chick heart by localized apposition of endocardial masses. Anat. Rec. 102: 299–312.
Poste, G. (1972) Enucleation of mammalian cells by cytochalasin B. I. Characterization of anucleate cells. Exp. Cell Res. 73, 273–286.
Pratt, R.M., Larsen, M.A. and Johnston, M.C. (1975) Migration of neural crest cells in a cell-free hyaluronate-rich matrix. Dev. Biol. 44, 298–305.
Purdy, J.E., Lieberman, M., Roggeveen, A.E. and Kirk, R.G. (1972) Synthetic strands of cardiac muscle: Formation and ultrastructure. J. Cell Biol. 55, 563–578.

Rawles, M.E. (1943) The heart forming areas of the early chick blastoderm. Physiol. Zool. 16, 22–42.

Renaud, D. and Le Douarin, G. (1968) Influence de l'environnement tissulaire et des conditions de culture sur l'évolution du mésoderme précardiaque de l'embryon de poulet. Comp. Rend. Acad. Sci. Paris, 267, 431–434.

Revel, J.P. (1964) A stain for the ultrastructural localization of acid mucopolysaccharides. J. Microsc. 3, 535–544.

Rosenquist, G.C. (1966) A radioautographic study of labeled grafts in the chick blastoderm. Development from primitive streak to stage 12. Carnegie Inst. Wash. Contributions to Embryology, 263, 71–110.

Rosenquist, G.C. and De Haan, R.L. (1966) Migration of precardiac cells in the chick embryo: a radioautographic study. Carnegie Inst. Wash. Contributions to Embryology, 38, 111–121.

Shimada, Y., Moscona, A.A. and Fischman, D.A. (1974) Scanning electron microscopy of cell aggregation: Cardiac and mixed retina-cardiac cell suspensions. Dev. Biol. 36, 428–446.

Sissman, N.J. (1970) Developmental landmarks in cardiac morphogenesis: Comparative chronology. Am. J. Cardiol. 25, 141–148.

Spira, A.W. (1971) Cell junctions and their role in transmural diffusion in the embryonic chick heart. Z. Zellforsch. 120, 463–487.

Stalsberg, H. (1970) Mechanism of dextral looping of the embryonic heart. Am. J. Cardiol. 25, 265–271.

Stalsberg, H. and De Haan, R.L. (1969) The precardiac areas and formation of the tubular heart in the chick embryo. Dev. Biol. 19, 128–159.

Steinberg, M.S. (1970) Does differential adhesion govern self assembly processes in histogenesis? Equilibrium configuration and the emergence of a hierarchy among populations of embryonic cells. J. Exp. Zool. 173, 395–434.

Steinberg, M. and Wiseman, L.L. (1972) Do morphogenetic tissue rearrangements require active cell movements? J. Cell Biol. 55, 606–615.

Stockdale, F., Okazaki, K., Nameroff, M. and Holtzer, H. (1964) 5-Bromodeoxyuridine: Effect on myogenesis in vitro. Science, 146, 533.

Uhl, H.S.M. (1952) A previously undescribed congenital malformation of the heart: Almost total absence of the myocardium of the right ventricle. Bull. Johns Hopkins Hosp. 91, 197–209.

Virágh, S. and Challice, C.E. (1973) Origin and differentiation of cardiac muscle cells in the mouse. J. Ultrastruct. Res. 42, 1–24.

Wilens, S. (1955) The migration of heart mesoderm and associated areas in *Amblystoma punctatum*. J. Exp. Zool. 129, 579–602.

Wilt, F.H. and Anderson, M. (1972) The action of 5-bromodeoxyuridine on differentiation. Dev. Biol. 28, 443–447.

Development and differentiation of lymphocytes 12

Irving GOLDSCHNEIDER and Randall W. BARTON

1. Introduction

Adaptive immunological responsiveness is the property of a heterogeneous family of cells, the lymphocytes (immunocytes, antigen reactive cells). These cells have preformed membrane-bound receptors that are able to recognize specific molecular conformations called antigenic determinants (Wigzell, 1970; Paul, 1970; Basten et al., 1971; Roelants, 1972; Rutishauser and Edelman, 1972; Goldstein et al., 1973). Numerous clones of lymphocytes are present in the host, each of which can bind one, or at most a few, of the tens of thousands of antigenic determinants that exist. The process by which antigen receptors with such diverse specificities are generated is not known, but clearly either the genetic information for each specificity must be precoded (germ line theory) (Hood and Talmage, 1970; Premkumar et al., 1974) or variation must be introduced by mutation or genetic recombination into a part of the genome for the basic receptor molecule (somatic theories) (Smithies, 1967; Gally and Edelman, 1970; Jerne, 1971).

Interaction of antigens with compatible receptors stimulates lymphocytes to undergo blast transformation, clonal proliferation and functional differentiation, thereby triggering the effector limb of the immunological response (Gowans and McGregor, 1965; Kennedy et al., 1966; Shearer and Cudkowicz, 1969). This response may take two basic forms: direct interaction of lymphocytes or their surrogates with antigen (cell-mediated immunological reactions); and indirect interactions of lymphocytes with antigen through the secretion of antigen-specific immunoglobulin molecules which circulate in the body fluids (antibody-mediated immunological reactions). In higher vertebrates, each of these two forms of immunological response is the property of a separate population of lymphocytes: thymus-derived lymphocytes (T cells) producing cell-mediated reactions; and bursa- or bone marrow-derived lymphocytes (B cells) producing antibody-mediated reactions (see section 2). The objective of these immunological reactions is the elimination of the offending antigen, a condition known as immunity. However, like the proverbial double-edged

G. Poste & G.L. Nicolson (eds.) The Cell Surface in Animal Embryogenesis and Development

sword, immunological reactions can also redound to the detriment of the host, a condition known as allergy.

The subsequent encounter of the progeny of antigen-stimulated lymphocytes with the same antigen usually results in a quantitatively and qualitatively improved immunological response (booster or anamnestic response). This occurs partly as a result of prior proliferation and differentiation of specifically reactive cells, and partly as a result of the progressive selection by antigen of cells with average higher affinity receptors (Siskind and Benacerraf, 1969). Alternatively, the interaction of antigen with membrane-bound receptors may result in specific suppression of immunological responsiveness to a subsequent encounter with the same antigen, a phenomenon known as immunological tolerance (reviewed by Weigle, 1973). Such a mechanism normally allows the host to avoid the adverse consequences of immunological reactions directed by its own lymphocytes against its own antigens (autoallergy; autoimmunity).

The preceding outline of the cellular and molecular bases of adaptive immunological responsiveness generally fits the basic tenets of Burnet's clonal selection theory (1959).

Despite its complexity, the lymphocyte system lends itself to ontogenetic studies for several reasons, not the least of which is the fact that lymphocyte development and differentiation occurs in the adult as well as in the embryo. Lymphocytes are free-living cells in the body. They have characteristic migration pathways in blood and lymph and occupy different anatomical sites in lymphoid tissues during different stages in their development. As we will show, developmentally distinct subsets of lymphocytes differ according to size, buoyant density, electrophoretic mobility, antigenicity, immunological function, susceptibility to immunosuppressive agents, and the presence of various cell surface receptors for antigens, antibodies, complement components, hormones and lectins. As a result, lymphocytes can be isolated in large quantities at defined stages of development, or their development can be selectively blocked at different stages. Moreover, lymphocytes can be adoptively transferred to intermediate hosts or cultured in vitro, thus permitting study of their development under controlled conditions. Finally, being free-living cells, the surfaces of lymphocytes are available in their native state for detailed structural and functional studies.

In this chapter, the ontogeny of lymphocytes will be traced from the level of the lympho-hemopoietic stem cell to the appearance of immunologically competent antigen-reactive cells in the central and peripheral lymphoid tissues. Emphasis will be placed on the heterogeneity of lymphocyte populations and on the surface properties that characterize lymphocyte subsets. We will not formally address the differentiation that occurs when immunologically competent cells are activated by contact with antigens, antibodies or other ligands and stimulatory agents. Such differentiation occurs as part of the immunological (or pseudoimmunological) response and does not fall within the bounds of embryonic induction. However, it does illustrate that immunologically competent lymphocytes are not end stage cells, and in some circumstances may be considered stem cells of very restricted potential.

2. *Overview of lymphocyte phylogeny and ontogeny*

As described above, the hallmarks that distinguish immunological reactions from other types of cellular recognition phenomena are specificity and memory. These properties are evidenced by altered reactivity to a second challenge with a specific antigen, which usually occurs months or years later. By these criteria, only vertebrates engage in true immunological reactions. However, a fully developed immunological system did not suddenly appear in the vertebrates and ancestral forms of immunocytes and antigen-recognition molecules must exist in more primitive forms of animals (for recent reviews see Hildemann and Reddy, 1973; Marchalonis and Cone, 1973a,b; Burnet, 1974). For example, many invertebrates reject allografts and xenografts but accept isografts and autografts, implying a broad recognition of self and non-self. Although an antigen-recognition cell (i.e. primordial lymphocyte) has not been identified in such reactions, coelomocytes and hemocytes are likely prospects. These cells have been shown to be active in host defense mechanisms and foreign graft rejection in earthworms and lower chordates. Moreover, they may possess some degree of specificity and short term immunological memory in that they demonstrate accelerated second-set graft rejections after adoptive transfer to an intermediate host. Similarly, while contemporary immunoglobulins are not present in invertebrates, there is evidence that non-specific opsonins and perhaps "proto-immunoglobulins" may play a role in humoral immunity to infectious microorganisms. It is possible therefore that the origins of the immunological system will ultimately be traced to special adaptations of primitive cellular recognition phenomena in protozoans and early metazoans.

Integrated cellular systems for cell-mediated and antibody-mediated immunological reactions first appeared in the lower vertebrates (for recent reviews see Cooper, 1973; Du Pasquier, 1973; Marchalonis and Cone, 1973a). With the exception of the hagfish, all vertebrates examined so far have a thymus; and all above the level of the cyclostomes have abundant circulating lymphocytes and plasma cells. All produce at least one class of immunoglobulin. Cyclostomes and cartilagenous fish produce IgM only; most bony fishes and all amphibians and reptiles produce IgM and a 7S late-appearing immunoglobulin, possibly analogous to mammalian IgG; birds produce three classes of immunoglobulins (IgM, probably IgG and possibly IgA); and mammals produce five classes of immunoglobulins (IgM, IgG, IgA, IgD and IgE). Organized lymphoid areas (white pulp) are first seen in the spleens of bony fishes and organized bone marrow appears in amphibians. Primitive lymph nodes also appear in amphibians, but are well-developed only in mammals. As one might anticipate, the immunological repertoire, expressed in terms of the spectrum of antigens that can be recognized and the variety and vigor of immunological responses that can be generated, increases with increasing complexity of the lymphoid apparatus.

It has become evident in recent years that the lymphocyte system of higher vertebrates (birds and mammals) is composed of two developmentally and functionally distinct cell lines which arise from a common precursor in hemopoietic tissues (Roitt et al., 1969; Raff, 1973). One cell line gives rise to antibody-secreting plasma cells; the other to lymphocytes which initiate cell-mediated immunological reactions (e.g. delayed-type hypersensitivity reactions, homograft rejection, immunity to intracellular parasites). The existence of these two lines of lymphocytes was first suggested by observations of congenital immunological deficiency disorders in man (Bergsma and Good, 1968; Cooper et al., 1973), and by the effects of ablation of central lymphoid tissues on the lymphocyte systems of birds and rodents (Warner, Szenberg and Burnet, 1962; Cooper et al., 1966; Miller and Osoba, 1967; Waksman, Arnason and Jankovic, 1962).

Results obtained in these and in other investigations to be discussed later point to the scheme of lymphocyte ontogeny outlined in Fig. 1. Blood-borne lymphopoietic stem cells migrate from yolk sac, fetal liver, spleen and bone marrow to the thymus and bursa of Fabricius. The thymus arises as an epithelial invagination of the foregut (3rd and 4th branchial pouches); the bursa as an epithelial invagination of the hindgut. An epithelial analogue of the bursa of Fabricius has not been identified in mammals. Rather, the fetal liver and bone marrow appear to subserve the functions of the bursa in these species.

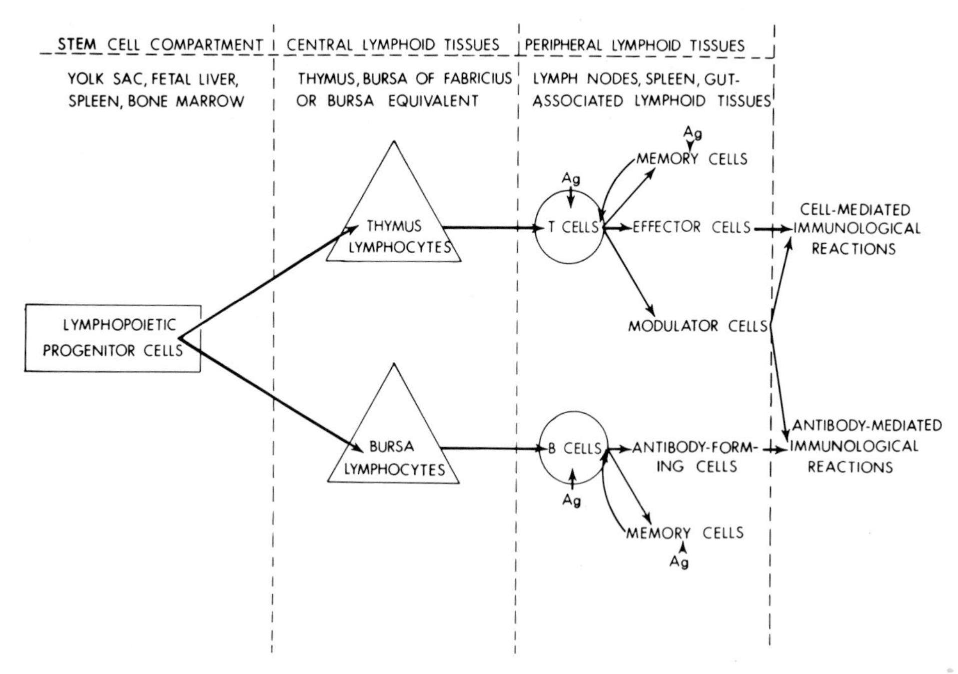

Fig. 1. Schematic representation of lymphocyte ontogeny. Two major lines of lymphocytes are depicted, T cells (thymus-derived lymphocytes) and B cells (bursa- or bone marrow-derived lymphocytes). Antigen-independent proliferation and differentiation occurs in stem cell compartment and central lymphoid tissues. Antigen-dependent proliferation and differentiation occurs in peripheral lymphoid tissues, as indicated (Ag). Memory cells serve as a reservoir of immunologically committed lymphocytes. See text section 2 for details.

Lymphopoietic stem cells which have reached the thymus, bursa of Fabricius or the mammalian "bursa equivalent" undergo rapid proliferation and differentiation to become immunologically competent lymphocytes, i.e. cells capable of binding antigen and initiating immunological reactions. This reflects the end of a complex and incompletely understood process by which myriad clones of lymphocytes are produced, each of which has receptors for a different and very restricted spectrum of antigenic determinants (generation of immunological diversity). Hormones which influence the proliferation and differentiation of thymus lymphocytes have been described and similar hormones may well exist for bursa lymphocytes as well.

After the acquisition of immunological competence, newly generated lymphocytes leave the central lymphoid tissues and migrate to discrete anatomical areas within the peripheral lymphoid tissues (spleen, lymph nodes, gut-associated lymphoid tissues). Lymphocytes which emigrate from the thymus are referred to as T cells, or thymus-derived cells. These constitute the line of cells that initiate cell-mediated immunological reactions. Lymphocytes which emigrate from the bursa of Fabricius or its mammalian equivalent are referred to as B cells, or bursa-derived or bone marrow-derived cells. These constitute the line of cells which initiate antibody-mediated immunological reactions. Some salient distinguishing characteristics of T and B cells are listed in Table 1.

As mentioned previously, T cells and B cells undergo further proliferation and differentiation after specific stimulation with antigens (clonal expansion) or non-specific stimulation with mitogens (polyclonal expansion). In the T cell system, subsets of cytotoxic, lymphokine-producing, modulator (amplifier, suppressor), helper and memory cells are produced. In the B cell system, antibody-secreting plasma cells and memory cells are formed. Synergy and antagony have been documented to occur between subsets of T cells in cell-mediated reactions and between T cells and B cells in antibody-mediated reactions. Such interactions between lymphocyte subsets usually involve the cooperation of a third cell type, the macrophage, which may be involved in the processing or presentation of antigen in an immunogenic form.

3. Lymphopoietic stem cells and progenitor cells

Cloning experiments using radiation-induced chromosomal markers have shown that cells of the erythrocytic, megakaryocytic, monocytic, granulocytic and lymphocytic series arise from a common pluripotent stem cell (Wu et al., 1968; Nowell et al., 1970). Such lympho-hemopoietic stem cells are present in the yolk sac of the embryo, in the liver and spleen of the fetus, and primarily in the bone marrow in the neonate and adult (Taylor, 1964; Goldschneider and McGregor, 1966; McGregor, 1968; Tyan, 1968; Tyan and Herzenberg, 1968; Moore and Metcalf, 1970). In severe combined immunodeficiency disorders of man, lymphopoiesis fails to occur, but hemopoiesis is normal. The defect appears to be due to the selective absence of lymphopoietic stem cells, inasmuch

Table 1
Distinguishing properties of T and B cells.

Parameters	T cells	B cells	References
1. Surface properties			
T cell-specific alloantigens and heteroantigens	+	–	(see section 5.3.1)
B cell-specific alloantigens and heteroantigens	–	+	(see section 5.3.1)
Readily detectable immunoglobulin	–[a]	+	(Raff, 1970; Rabellino et al., 1971; Vitetta et al., 1971)
Receptor for complement	–	+	(Bianco et al., 1970)
Receptor for immunoglobulin Fc	–[a]	+	(Basten et al., 1972a,b)
Receptor for sheep RBC's	+[b]	–	(Jondal et al., 1972)
More negatively charged	+	–	(Wioland et al., 1972)
More adherent to artificial surfaces	–	+	(Bianco et al., 1970)
Retained on antigen-coated columns	–	+	(Wigzell, 1970)
Retained on allogeneic monolayers	+	–	(Brondz, 1972)
2. Migration and anatomical distribution			
Germinal centers and lymphoid follicles	–	+	(Parrott and de Sousa, 1971; Howard, 1972; Goldschneider and McGregor, 1973)
Lymph node paracortex; spleen periarteriolar sheath; interfollicular areas in Peyer's patches	+	–	(Parrott and de Sousa, 1971; Howard, 1972; Goldschneider and McGregor, 1973)
Rapid recirculation from blood to lymph	+	–	(Goldschneider and McGregor, 1969; Howard, 1972)
3. Functional properties			
Cytotoxic lymphocyte precursors	+	–[c]	(Brunner and Cerottini, 1971)
Antibody-forming cell precursors	–	+	(Nossal et al., 1968)
Modulator cells in antibody- and cell-mediated immunological responses	+	–[d]	(Cantor and Asofsky, 1972; Gershon, 1974)
Helper cells in antibody responses to thymus-dependent antigens	+	–	(Miller et al., 1971)
Exhibit carrier specificity predominantly	+	–	(Mitchison et al., 1970; Katz and Benacerraf, 1972)
Exhibit hapten specificity predominantly	–	+	(Mitchison et al., 1970; Katz and Benacerraf, 1972)
Activated or tolerated at lower antigen concentrations	+	–	(Mitchison, 1971)
Proliferative response in mixed-lymphocyte reaction	+	–	(Johnston and Wilson, 1970; Vischer and Jaquet, 1972)
Proliferative response to PHA, Con-A	+	–[e]	(Greaves and Janossy, 1972)
Proliferative response to endotoxin	–	+	(Greaves and Janossy, 1972)
More sensitive to immunosuppressive agents	–	+	(Cohen and Claman, 1971; Cunningham and Sercarz, 1971; Turk and Poulter, 1972).

[a]May be present on activated T cells (see section 5.3.3).

[b]Primate T cells only.

[c]B cells may non-specifically kill antibody-coated target cells by binding to immunoglobulin Fc (see section 5.3.3).

[d]B cells may modulate immunological reactions indirectly via humoral antibodies or antigen-antibody complexes (MacLennan, 1972).

[e]Insolubilized PHA and Con-A can stimulate B cells (see section 5.3.3).

as it may be corrected by transplantation of bone marrow (Biggar et al., 1973). This implies that, in addition to pluripotent lympho-hemopoietic stem cells, separate populations of lymphopoietic and hemopoietic stem cells also exist.

Lympho-hemopoietic stem cells from prenatal and postnatal periods differ markedly in their physical characteristics and regenerative capacity. Stem cells in the embryo and fetus are large, low density cells with extensive proliferative capacities, whereas those in adults are small, high density cells with more limited proliferative capacities (Haskill and Moore, 1970; Metcalf and Moore, 1971). The morphological correlates of these stem cells in fetal hemopoietic and central lymphoid tissues are large, basophilic blast cells (Metcalf and Moore, 1971; Le Douarin and Joterau, 1975). In adult hemopoietic and regenerating thymus tissues the putative stem and progenitor cells appear as small lymphocyte-like cells, so called "lymphocyte-transitional cells" (Cudkowicz, Bennet and Shearer, 1964; Blackburn and Miller, 1967; Blomgren, 1969; Osmond et al., 1973; Yoffey, 1973). Although they constitute a rapidly dividing cell population, lymphocyte-transitional cells normally represent a small minority of nucleated cells in adult rodent bone marrow, perhaps because of their short transit time (3 days or less) (Harris, 1961; Everett and Caffrey, 1967; Rosse, 1971; Osmond and Nossal, 1974). However, they can be induced to proliferate extensively in response to irradiation (Harris and Kugler, 1967) or hypoxia (Yoffey et al., 1968) so as to constitute the major cell type in bone marrow. As anticipated, an increase in percentage of lymphocyte-like cells in bone marrow is accompanied by a comparable increase in lympho-hemopoietic stem cells. There is evidence to suggest that "adult-type" lympho-hemopoietic stem cells are lineal descendants of "embryonic-type" stem cells (Haskill and Moore, 1970; Moore and Metcalf, 1970; Moore et al., 1970).

Using a combination of density gradient centrifugation and filtration through a column of fine glass beads, we have isolated a stem cell-rich population of lymphocyte-like cells from rat bone marrow (Goldschneider, 1975b). These cells, which comprise approximately 7% of nucleated bone marrow cells, lack surface antigens characteristic of mature T and B cells. However, they bear a unique antigen, the bone marrow lymphocyte antigen (BMLA), which is not present on mature members of the lymphopoietic or hemopoietic cell systems. Antiserum to the bone marrow lymphocyte antigen abrogates all in vivo hemopoietic colony-forming potential and all lymphopoietic regenerating capacity of adult bone cell suspensions (Goldschneider, unpublished observations). It will therefore be of interest to determine if anti-BMLA serum can also abrogate the stem cell activity of embryonic hemopoietic tissues.

Pluripotent lympho-hemopoietic stem cells give rise to discrete progenitor cells for each of the formed elements of the blood through a complex series of proliferative and differentiative events (for a detailed discussion see Metcalf and Moore, 1971). Progenitor cells, unlike stem cells, are committed to specific pathways of cellular differentiation and lack the ability to replenish their own stock. They are generally smaller and more dense than stem cells and have different proliferative kinetics (Lajtha et al., 1969; Worton et al., 1969). Hemopoietic stem cells and progenitor cells in the mouse have also been shown to differ antigenically (van den Engh and Golub, 1974). Moreover, hemopoietic progenitor cells are able to form differentiated colonies in vitro, whereas hemopoietic stem cells only form differentiated colonies in vivo (Till and McCulloch, 1961; Pluznick and Sachs, 1965; Bradley and Metcalf, 1966; Metcalf and Moore, 1971). Unfortunately, neither lymphopoietic stem cells nor lymphopoietic progenitor cells form discrete colonies in vivo or in vitro.

Lafleur et al. (1972) and El-Arini and Osoba (1973) have separated the progenitors of T and B cells, respectively, from pluripotent stem cells in mouse bone marrow by a combination of velocity sedimentation and density centrifugation. As in the rat, these progenitor cells lack surface alloantigens characteristic of mature T and B cells (Schlesinger, 1972; Osmond and Nossal, 1974). Komuro and Boyse (1973) and Scheid et al. (1973) have induced the expression of thymocyte-specific antigens on the surface of antigenically "null" cells in mouse bone marrow by incubation of bone marrow cells in vitro with thymus extracts and non-specific activators of cyclic AMP. This suggests that these cells are genetically predetermined to become thymocytes. Moreover, these same cells were able to migrate to thymus (Komuro et al., 1975). These findings suggest, but do not prove, that separate progenitors for T and B cells exist, and that the role of the thymus (and presumably the bursa of Fabricius and its equivalent in mammals) is to direct the differentiation of these precommitted cells. A tentative scheme of lymphopoietic stem cell and progenitor cell development is shown in Fig. 2.

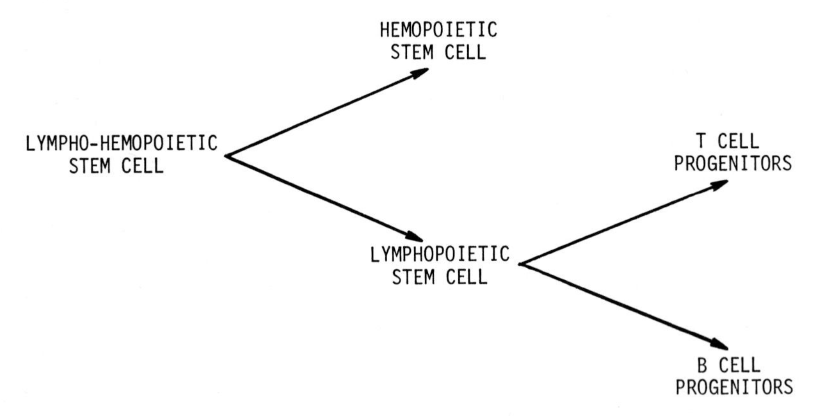

Fig. 2. Scheme of lymphopoietic progenitor cell development. See text section 3 for details.

4. *Central lymphoid tissues*

The central lymphoid tissues in birds and mammals consist of the thymus and the bursa of Fabricius or the bursa equivalent (bone marrow, fetal liver). These are the sites in which lymphopoietic progenitor cells, in the absence of exogenous antigenic stimulation, proliferate and differentiate to become T and B cells, which in turn populate the peripheral lymphoid tissues. The differentiative events fall into four major operationally-defined categories: (1) generation and/or expression of immunological diversity; (2) elimination or suppression of clones reactive with self-antigens ("forbidden clones"); (3) acquisition of immunological competence; and (4) acquisition of the ability to migrate to peripheral lymphoid tissues. In addition, lymphopoietic progenitor cells undergo marked changes in size, buoyant density, electrophoretic mobility, cell cycle, DNA polymerase activity, antigenicity, receptor activity, and sensitivity to immunosuppressive agents during their brief transit through the central lymphoid tissues. With the exception of the appearance of receptors for antigen, causal relationships between these acquired physical, biochemical and functional properties have not been established. As a result, multiple schemes for the differentiation of thymus lymphocytes (thymocytes) and bursa lymphocytes (bursa cells) have been proposed. Before discussing these schemes, it will be necessary to review some of the known properties of the central lymphoid tissues and of the cells that reside within them.

4.1. The thymus

4.1.1. Embryogenesis

The thymus develops as an invagination of epithelium from the 3rd and 4th pharyngeal pouches (Venzke, 1952; Hammond, 1954). It is first discernible on the 5th, 10th and 42nd day of embryonic development in the chicken, mouse and man, respectively (Venzke, 1952; Valdes-Dapena, 1957; Smith, 1965). Mesenchymal induction appears to be necessary for the growth and differentiation of the epithelial anlage, although the source of the undifferentiated mesenchyme is not critical (Auerbach, 1960). Induction can take place despite the interposition of a cell-impermeable Millipore membrane between the epithelium and the mesenchyme, suggesting that soluble factors are involved. Immigration of thymocyte progenitor cells into the thymus rudiment is also necessary for differentiation of the epithelial cells (see section 4.1.4).

Within a week after the arrival of the first thymocyte progenitors, the thymus loses contact with the pharyngeal pouches, becomes lymphoid in character, and assumes its adult form and anatomical location. In most mammals, the thymus appears as a bilobed organ in the superior mediastinum of the thoracic cavity, overlying the great vessels of the heart. However, in the guinea pig and in birds the thymus appears as a series of nodules along the internal jugular veins in the neck. This serves to graphically illustrate an elementary point that is frequently ignored in considerations of congenital or acquired thymic abnormalities, that animals have two thymuses. Thus, the "bilobed" thymus of most mammals, including man, is in fact, comprised of two developmentally, anatomically and functionally separate organs.

Histologically, the thymus is incompletely divided by septae into connecting lobules, each of which has a cortex and a medulla. The cortex contains approximately 90 percent of the total thymocytes, which are so tightly packed as to obscure the underlying epithelial parenchyma. Thymocytes are much more sparse in the medulla, thereby exposing the epithelial cells and abundant blood vessels in this region. Cortical and medullary thymocytes differ markedly in their physical, biochemical, antigenic and functional properties (see Table 2 and section 4.1.7). As a result, there is considerable controversy about their ontogenetic relationships and about their relationships to subsets of peripheral T cells (see section 4.1.8).

4.1.2. Thymocytopoiesis

It has been demonstrated by cross-transfusion studies in parabiotic chicken embryos (Moore and Owen, 1967a,b) and mice (Harris et al., 1964) that thymocytes are derived from blood-borne progenitor cells. Similar conclusions have been reached in studies involving explantation of thymuses at different stages of embryogenesis (Owen and Ritter, 1969); transplantation of thymuses into hosts having distinctive chromosomal, antigenic or nuclear markers (Harris and Ford, 1964; Metcalf and Wakonig-Vaartaja, 1964; Moore and Owen, 1967b; Owen and Ritter, 1969; LeDourin and Jotereau, 1975); parenteral injection of stem cell-rich populations into intact hosts (Ford and Micklem, 1963; also see section 3); and ablation of stem cell-rich hemopoietic tissues in embryos (Moore and Metcalf, 1970). The recent demonstration of an antigen (rat bone marrow lymphocyte antigen) that is present on both lympho-hemopoietic stem cells and on immature thymocytes further documents the extrathymic origin of thymocytes (Goldschneider, 1975b). Previous explantation studies (Auerbach, 1961) that had purported to show that thymocytes arose by in situ differentiation of thymus epithelial cells can be explained by the presence of progenitor cells which had migrated into the thymus anlage prior to the time of its explantation.

The thymus is ordinarily the first organ to become lymphoid in character during embryonic development. In the chicken, blood-borne progenitors begin to migrate into the epithelial anlage of the thymus between the 6th and 7th days of embryonic life; in the mouse, between the 11th and 12th days. These estimates were made by observing the development (or lack of development) of thymocytes in thymuses that had been explanted in vitro at different ages (Owen and Ritter, 1969). The arrival of thymocyte progenitors at the indicated time coincides with the appearance of large basophilic cells between the epithelial cells of the rudimentary thymus (Moore and Owen, 1967a,b; LeDourin and Jotereau, 1975). As discussed previously in section 3, it is not known whether separate progenitors exist for thymus and bursa cells, or whether specific commitment to a single line of development occurs after migration of a common lymphopoietic progenitor cell to thymus or bursa.

The first wave of thymocyte progenitors to migrate into the embryonic thymus undergoes little proliferation or differentiation for several days (Ball, 1963; LeDourin and Jotereau, 1975). At this stage of development, thymocyte progenitors form a fairly uniform population of large and medium size cells. Cortical and medullary zones of thymocytes are not evident. On or about the 16th day of embryogenesis in the mouse there is a burst of proliferative activity of thymocyte

Table 2
Distinguishing properties of cortical and medullary thymocytes.

Parameters	Cortical thymocytes	Medullary thymocytes	References
1. Physical and metabolic properties			
Buoyant density	high	low	(Colley et al., 1970b; Takiguchi Adler and Smith, 1971)
Cell size	small	medium	(Colley et al., 1970b; Takiguchi Adler and Smith, 1971)
Mitotic index	high	low	(Shortman and Jackson, 1974)
Terminal deoxynucleotidyl transferase	+	–	(Barton et al., 1975)
2. Surface properties			
Electrophoretic mobility	low	high	(Zeiller and Dolan, 1972)
Alloantigens:			
Thymus leukemia (TL) antigen	+	–	(Owen and Raff, 1970)
Thy-1 (θ), Ly-1, 2, 3 and 5	high	low	(Takiguchi et al., 1971; Lance et al., 1971)
Major histocompatibility antigens	low	high	(Konda et al., 1973; Order and Waksman, 1969)
Heteroantigens:			
Rat masked thymocyte antigen	–	+	(Goldschneider, 1975a)
Rat bone marrow lymphocyte antigen	+	–	(Goldschneider, 1975b)
3. Functional properties			
Sensitivity to cortisone and irradiation	high	low	(Blomgren and Andersson, 1971)
Responsive to phytohemagglutinin	–	+	(Stobo, 1972)
Responsive to concanavalin A	weak	strong	(Stobo, 1972)
Ability to recirculate from blood to lymph	–	+	(Lance et al., 1971)
Effector cells in cell-mediated immunological reactions	?	+	(Blomgren and Svedmyr, 1971)
Helper cells in antibody responses to thymus-dependent antigens	–	+	(Andersson and Blomgren, 1970)
Modulator cells in antibody- and cell-mediated immunological responses	+	–	(Cohen and Gershon, 1975)

progenitors. This is accompanied by a dramatic decrease in mean cell volume and a predominance of small thymocytes characteristic of the postnatal thymus. At this time thymocytes first express specific cell surface antigens (Schlesinger and Hurvitz, 1968a,b; Owen and Raff, 1970). Similar changes have been observed in the rat (Colley et al., 1970). Within one or two days, cortical and medullary zones become prominent, and immunologically competent cells first appear (Sosin et al., 1966; Schwarz, 1967; Kay, 1970; MacGillivray et al., 1970).

That the sudden change in cell size and function is due to the differentiation of large cells already in situ, rather than to recruitment of a population of small cells, is shown by explantation experiments. Mouse thymuses removed prior to day 16 and cultured in vitro or in cell-impermeable diffusion chambers in vivo were seen to undergo a similar shift in mean thymocyte volume as did intact thymuses (Owen and Ritter, 1969). Moreover, such explanted thymuses produced immunologically competent T cells (Ritter and Owen, cited in Metcalf and Moore, 1971, p. 242). It has also been demonstrated that the decrease in mean thymocyte volume is a consequence of cell division, rather than direct transformation of large to small cells (Borum, 1973).

During the time that the first wave of thymocyte progenitors is being "processed" by the thymus, the thymus is refractory to the immigration of new thymocyte progenitors (Moore and Owen, 1967b; LeDourin and Jotereau, 1975). However, after several days, it again becomes permissive, so that a new wave of progenitor cells arrives shortly before the time of birth or hatching. It is not known whether thymocyte progenitors continue to be processed in successive waves at 2 to 3 day intervals during postnatal life, or whether this is a peculiarity of the developing embryonic thymus. It may be of some significance that two other events coincide with the onset of thymocyte differentiation in the embryo: (1) vascularization of the thymus rudiment; (2) evidence of secretory activity by the thymus epithelial cells. These events will be discussed in more detail below (see sections 4.1.4 and 4.1.5).

The process of differentiation of thymocytes in the postnatal thymus takes approximately 3 days, during which time thymocyte progenitors undergo six to eight divisions (mean generation time approximately 9 hr) (Sainte-Marie and Leblond, 1964; Weissman, 1967; Michalke et al., 1969; Order and Waksman, 1969). Progressively smaller cells are produced as the mitotic wave progresses from the subcapsular region of the thymus cortex to the cortico-medullary junction.

The origins of medullary thymocytes are obscure. Radioautographic studies by Weissman (1967, 1973) and Fathman et al. (1975) who labelled cortical thymocytes by topical application or direct subcapsular intrathymic injection of [^{3}H]-thymidine, seem to indicate that medullary thymocytes are derived from cortical thymocytes. This notion has been challenged by the studies of Bryant (1971) and Shortman and Jackson (1974), who determined the incorporation of parenterally administered [^{3}H]-thymidine by thymocytes. These investigators concluded that the kinetics of proliferation of cortical and medullary thymocytes were incompatible with a precursor-progeny relationship, and that cortical and medullary thymocytes probably represented developmentally distinct cell populations. However, this evidence cannot exclude the possibility that a minor subset of cortical thymocytes gives rise to medullary thymocytes. Additional considerations of the ontogenetic relationship of cortical and medullary thymocytes are presented in section 4.1.8.

An intriguing aspect of thymocytopoiesis is its apparent autonomy from most immunological influences (Metcalf, 1963). This is in marked contrast to T cell proliferation and differentiation in peripheral lymphoid tissues. The following parameters have been found *not* to have an effect on the weight and growth of the thymus in the mouse: presence or absence of antigenic stimulation; removal or hypertransfusion of peripheral lymphocytes; partial thymectomy or transplantation of multiple thymus grafts. The last point is particularly suggestive of the fact that the primary proliferative stimulus for thymocytopoiesis must arise within the thymus itself. Thus, thymuses from infant mice continue their rapid growth when transplanted into adult mice whose own thymuses are undergoing age-related physiological involution. Moreover, the kinetics of thymocyte production in adult thymus grafts are characteristic of the strain and age of the donor of the thymus rather than of the host, even though host cells replace the donor thymocytes in the grafts (Metcalf et al., 1961).

The phenomenon of age-related involution of the thymus mentioned above suggests that thymocytopoiesis *is* affected by certain external, albeit non-immunological, influences. In several mammalian species that have been studied, thymus weight increases until puberty, after which it progressively declines to 50 percent or less of peak values (Metcalf, 1964; Solomon, 1971). By contrast, the mitotic index of thymocytes reaches its peak in late fetal life and early infancy, when it is more than 10 times as high as the mitotic index of peripheral lymphoid tissues. The mitotic index of thymocytes progressively decreases after infancy to reach levels by early adult life of approximately 50 percent of peak values. Thus, in prepubertal animals the absolute number of thymocytes increases at the same time that the rate of thymocyte proliferation decreases. However, after the onset of puberty, both the number of thymocytes and the rate of thymocyte proliferation decrease.

In addition to the onset of physiological involution at puberty, it has been widely observed that acute involution of the thymus may occur at any age under conditions of severe bodily stress (e.g. infections, trauma, burns) (cf. Boyd, 1932; Goldstein and MacKay, 1967; Gad and Clark, 1968). Under such conditions, the weight of the thymus and the number of thymocytes may decrease by more than 90 percent in a period of several days. Histologically, the cell loss is limited almost exclusively to the thymus cortex, where numerous pyknotic nuclei are evident during the early stages of the reaction.

The occurrence of age-dependent (physiological) and stress-induced (acute) involution of the thymus suggest that sex hormones and adrenal cortical hormones might exert regulatory influences on thymocytopoiesis. This has proved to be the case, although the exact mechanisms of action of these hormones on the thymus or on thymocytes and/or thymocyte progenitors are not known (for recent reviews of thymus-hormone interrelationships see Pierpaoli et al., 1970; Comsa, 1973). It has been known for many years that castration of male and female animals can prevent age-dependent involution of the thymus (Calzolari, 1898), and that atrophy of the thymus can be induced by injection of pituitary gonadotrophic hormones into normal, but not castrated, infant rats (Arvin and Allen, 1928). Administration of estradiol (Selye et al., 1936) or testosterone (Selye and Masson, 1939) inhibits the development of the thymus in the chicken embryo and also induces involution of the thymus in castrated guinea pigs (Cosma, 1953), apparently by altering the differentiation of the epithelial cells in the central lymphoid tissues (Szenberg, 1970). There is some evidence that sex hormones also may affect lymphopoiesis directly.

Adrenalectomy, like castration, inhibits age-dependent involution of the thymus (Jaffe, 1924). Combined adrenalectomy and castration has an even more pronounced effect (Marine et al., 1924). Mitotic indices increase markedly in adrenalectomized adult animals, and pyknotic indices decrease (Claesson, 1972; Shortman and Jackson, 1974). In prepubertal animals, adrenalectomy is followed by hyperplasia of thymocytes, with mitotic indices approaching values normally seen only in early postnatal and late fetal life. These effects are reversed when adrenalectomized animals are given adrenal glucocorticosteroids. It would appear, therefore, that the intrinsic rate of thymocytopoiesis set by the thymus is modulated throughout life by the physiological secretion of adrenal corticosteroids.

When administered in unphysiologically high concentrations, corticosteroids cause massive lysis of cortical thymocytes in vivo and in vitro (Dougherty, 1952; Ishidate and Metcalf, 1963), but do not affect immunologically competent medullary thymocytes (Blomgren and Andersson, 1971). The first step in this process appears to be binding of hormone to specific receptors in the thymocyte cytoplasm, followed by interaction of the hormone-receptor complex with the nucleus (Wira and Munck, 1974). However, it is not certain that thymocytolysis is the major mechanism by which endogenously produced corticosteroids affect thymocytopoiesis under physiological conditions or under conditions of stress. For example, cortisone has also been shown to alter the migratory properties of lymphocytes (Ernstrom, 1970; Cohen, 1972; Spry, 1972). Considering the normally rapid turnover of cortical thymocytes (almost 100% in 72 hr), modulation of progenitor cell inflow or thymocyte outflow could well mimic the effects of intrathymic cell death.

Other hormones which may influence thymocytopoiesis are pituitary growth hormone and thyroid hormone. Hypophysectomy causes premature involution of the thymus, whereas administration of growth hormone causes hypertrophy (Pierpaoli et al., 1970). The situation with respect to thyroid hormone is somewhat more complex. Whereas thyroidectomy causes thymus atrophy, administration of a low dose of thyroid hormone causes thymus hypertrophy, and administration of a high dose of thyroid hormone causes atrophy (Comsa, 1958). Surprisingly, there is evidence of an increased rate of thymocyte proliferation at both dose levels. This suggests that the atrophy observed after high doses of thyroid hormone is due to an increased rate of thymocyte migration to the peripheral lymphoid tissues. This may account, in part, for the lymphocytosis that is characteristically seen in clinical hyperthyroidism.

4.1.3. Thymus hormones

A number of "factors" have been extracted from thymus or from blood which are reputed to be hormones responsible for influencing the proliferation, differentiation and function of thymocyte progenitor cells, thymocytes and/or peripheral T cells (Luckey, 1973; White, 1975). At least 12 such factors have been described, nine of which have been partially characterized biochemically. All are proteins or polypeptides, with the exception of thymosterin which is a steroid of unknown structure. It is likely that there is a good deal of redundancy among the various factors isolated in different laboratories.

One of the first thymus factors to be described was the "lymphocytosis-stimulating factor" (LSF) of Metcalf (1956). This factor was obtained in saline extracts from the medullary regions of human or mouse thymuses, but not from the thymus cortex or other lymphoid tissues. As its name indicates, LSF was found to stimulate a transitory lymphocytosis in infant mice which reached its maximum 5 to 7 days after injection. Similar lymphocytosis-stimulating activity was demonstrated in the sera of mice and could be abolished by thymectomy.

Osoba and Miller (1965) showed that immunological competence was restored during pregnancy in thymectomized mice, apparently by humoral factors of fetal origin. A similar restoration of immunological competence was observed in thymectomized mice which had been implanted with thymus grafts in cell-impermeable diffusion chambers (Miller and Osoba, 1967; Stutman et al., 1970; Dardenne et al., 1974). More recently, Bach and Dardenne (1972) identified a hormonal factor that confers T cell properties on presumptive thymocyte progenitors. This factor is absent from the serum of thymectomized mice. "Thymus humoral factor" (THF), a low molecular weight peptide with comparable properties, has been isolated from mouse thymus by Trainin et al. (1966). THF appears to prevent wasting in neonatally thymectomized mice by enhancing lymphoid cell proliferation and by influencing the development of immunological competence by cells in bone marrow and peripheral lymphoid tissues.

The best characterized and most intensively studied thymus factor is thymosin, a homogeneous protein (MW 12,600) which was isolated from saline extracts of calf thymus by White and Goldstein (1970). Similar substances have subsequently been obtained from the thymuses of other species (Scheid et al., 1973). Thymosin has been demonstrated to restore immunological competence to neonatally thymectomized mice, to stimulate the growth of peripheral lymphoid tissues, and to induce the appearance of functional and antigenic T cell markers on bone marrow cells (Goldstein et al., 1972; Komuro and Boyse, 1973; Scheid et al., 1973). This last property is of special interest, in that it suggests that a specific thymus hormone can act selectively to induce the phenotypic expression of differentiated traits on genotypically precommitted thymocyte progenitors. Thus, in experiments with congenitally athymic *nude* mice (nu/nu), thymocyte progenitors from bone marrow were induced to express six sets of T cell-specific surface alloantigens and to functionally cooperate with B cells after incubation in vitro for 2 hr with thymosin (Scheid et al., 1973). T cell differentiation was also induced in vivo in congenitally athymic mice by intraperitoneal injection by thymosin.

It is important to note that, under physiological conditions, such antigenic and functional traits are only expressed on thymocyte progenitors *after* they have migrated to thymus (see section 4.1.2). This suggests that, if a thymic hormone normally acts to influence the differentiation of thymocyte progenitors, it does so at close range. It is of considerable interest, therefore, that thymosin has been demonstrated by immunofluorescence to be present within the cytoplasm of thymus epithelial cells (Mandi and Glant, 1973; also see section 4.1.4). However, there is also evidence that low concentrations of thymosin or a thymosin-like substance in blood may act at a distance to influence the further differentiation of a subset of immature T cells (T_1 cells; immediate post-thymic cells) in peripheral lymphoid tissues (Lonai et al., 1973; Stutman, 1975). Perhaps such cells require lower concentrations of hormone.

Thymopoietin (thymin), another polypeptide isolated from calf thymus, demonstrates similar functional properties to thymosin (Goldstein, 1975). However, thymopoietin, which may be a mixture of two closely related proteins, has a lower molecular weight (7,000) than thymosin, and is serologically distinct. Moreover, thymopoietin has the additional property of mediating the curare-like neuromuscular blockade found in myasthenia gravis. Immunofluorescence studies have localized thymopoietin to epithelial cells in the cortex or the medulla of the thymus, depending on the particular lot of antiserum that was used.

In addition to thymocytotropic factors that have been isolated from blood or thymus, a putative thymocytotropic factor has been isolated from mouse submaxillary salivary gland (Naughton et al., 1972). In low concentrations this substance stimulates blast transformation of thymocytes both in vivo and in vitro. In high concentrations it causes thymocyte lysis. The protein is an esteroprotease with a specificity for arginyl peptide bonds. Enzymes with similar biochemical properties have been identified as component parts of other cytotropic products of the submaxillary salivary gland, such as nerve growth factor and epithelial growth factor.

Several non-specific agents (e.g., poly A:U, endotoxin, cAMP), which have the common property of raising intracellular cAMP levels, have been shown to induce differentiation of thymocyte progenitors. This also appears to be the mechanism of action of "thymus humoral factor" (Kook and Trainin, 1975). Inasmuch as other substances that have the capacity to elevate intracellular cAMP concentrations probably exist in lymphoid and non-lymphoid tissues (Goldstein, 1975), it cannot necessarily be concluded that thymosin or any of the other thymocytotropic substances thus far isolated are important in the regulation of thymocyte proliferation and differentiation under physiological conditions. Such proof would require the ability to selectively interfere with thymocytopoiesis in vivo by specifically inactivating or depleting the putative thymocytotropic substance in question. Indeed, it has not yet been proved that any of the thymic factors are actually produced in the thymus. On balance, however, it would seem probable that the thymus directs thymocyte, and perhaps certain stages of peripheral T cell, differentiation at least in part through a hormonal mechanism.

4.1.4. Thymus epithelial cells

Thymus epithelial cells have been referred to traditionally as "epithelial-reticular cells", a contradiction in terms. They are true epithelial cells both ontogenetically (pharyngeal pouch epithelium) and morphologically (tonofibrils, desmosomes, formation of keratohyaline 'pearls'). Reticular cells, i.e. fibrocytes which produce reticulin, a variant of collagen, are found associated with blood vessels, and thus are particularly abundant in the inner medullary area of the thymus lobule. Thymus epithelial cells are also often confused with macrophages. The latter are found scattered throughout the thymus, particularly along cortical capillaries. The ultrastructural properties and anatomical arrangement of these cell types has been described in detail by Clark (1973).

Inasmuch as the thymus develops as an epithelial organ which is then infiltrated by thymocyte progenitor cells, it seems intuitively obvious that the thymus epithelial cells must play a central role in influencing the proliferation and differentiation of thymocytes. However, most of the evidence in support of this contention is indirect. For example, thymocyte progenitors fail to differentiate to T cells in man and mouse when the thymus epithelial anlage fails to develop (congenital thymus aplasia) or when it develops abnormally (congenital thymus hypoplasia or dysplasia) (Bergsma and Good, 1968; Pantelouris, 1968; DeSousa et al., 1969). Such conditions can be corrected by transplantation of normal thymus (Wortis et al., 1971; Biggar et al., 1973). Moreover, thymocytotropic factors appear to be produced by normal thymus tissue which has been depleted of thymocytes. This is the situation with thymus grafts that have been placed in cell impermeable diffusion chambers. Such grafts rapidly lose their complement of thymocytes, but are able to restore immunological competence to thymectomized hosts via humoral factors (section 4.1.3).

Other observations also suggest that the thymus epithelium forms the inductive microenvironment in which thymocytes differentiate. For example, thymocyte progenitor cells which enter the embryonic thymus do not divide or differentiate until the epithelial cells show signs of differentiation (Mandel, 1970). These changes include the formation of an elaborate network of cytoplasmic processes and evidence of secretory activity (rough endoplasmic reticulum, prominent Golgi zone, secretory-type vacuoles containing acidic glycoproteins) (Clark, 1968; Mandel, 1970). Electron micrographs show broad areas of contact between the cytoplasmic membranes of epithelial cell processes and developing thymocytes in the interstices formed by these processes. Some investigators feel that there is a non-random clustering of mitotically-active thymocytes in the vicinity of PAS-positive epithelial cells in the thymus cortex (Metcalf, 1963; Mandel, 1969). Such claims are difficult to evaluate given the ubiquity of epithelial processes in the thymus, and the possibility that some of the PAS-positive cells may be macrophages. Wekerle et al. (1973) and Waksal et al. (1975) have demonstrated that thymocytes can be induced to proliferate and undergo functional differentiation in vitro when cultured on a monolayer of thymus epithelial cells.

Another mechanism by which thymus epithelial cells might contribute to the microenvironment in which thymocytes differentiate is in the formation of the "blood-thymus barrier", which wholly or partially excludes many antigens from the thymus (Clark, 1964). This is discussed in section 4.1.5.

Not only do thymus epithelial cells, or their products, seem to be necessary for the differentiation of thymocyte progenitors, but thymocyte progenitors seem to be necessary for the differentiation of thymus epithelial cells. This is shown by a clinical syndrome known as severe combined immunological deficiency ('Swiss' form of agammaglobulinemia), in which neither T cells nor B cells form due to a congenital absence of lymphopoietic stem cells (Bergsma and Good, 1968). The thymus in patients with this disorder remains rudimentary. However, after a successful graft of bone marrow (containing lymphopoietic stem cells), the thymus epithelial cells differentiate, the thymus assumes its adult lymphoid character, and immunologically competent T cells are produced (Biggar et al., 1973). An analogous situation is seen experimentally in the failure of the embryonic thymus rudiment to differentiate further if it is explanted in vitro before thymus progenitor cells have had a chance to infiltrate the epithelial anlage (Owen and Ritter, 1969).

In certain species (mouse, guinea pig, man) some of the epithelial cells in the thymus medulla form Hassall's corpuscles, which are 25–100 μm diameter spheroidal aggregates of concentrically oriented epithelial cells (Mandel, 1968). Strikingly similar structures called "epithelial pearls" are characteristically found in well-differentiated squamous cell carcinomas, and may also be formed by non-neoplastic epithelial cells under a variety of pathological conditions. Like the "epithelial pearls" which they resemble so closely, Hassall's corpuscles undergo central necrosis, hyalinization and cystic degeneration as they mature (in other tissues such structures are called "keratohyaline pearls") (von Gaudecker and Schmale, 1974). Numerous putative functions have been ascribed to Hassall's corpuscles (e.g., production of thymus hormones, destruction of dead thymocytes, production or sequestration of immunoglobulins, uptake of circulating antigens). However, it is safe to say that none of these assertions has been proved. Indeed it may be that Hassall's corpuscles serve no special functions aside from those served by all other epithelial cells in the thymus medulla. The strongest evidence in favor of this view is the fact that the thymuses of closely related species may or may not have Hassall's corpuscles (e.g., mouse and rat, respectively). Also, the morphological and histochemical resemblance of Hassall's corpuscles to normal keratinizing stratified squamous epithelium suggests that they are formed as the result of an adaptive metaplastic process frequently seen when epithelial cells lose contact with body surface. Nonetheless, the formation of Hassall's corpuscles only after epithelial cell maturation has occurred, their absence in immunological deficiency syndromes associated with hypoplasia or dysplasia of the thymus, their restriction to the thymus medulla, and their intimate association with thin-walled arteriovenous shunts of the type commonly found in endocrine organs should counsel an open mind on the question of possible unique functions.

In addition to the formation of Hassall's corpuscles by some species, medullary and cortical epithelial cells differ in two other respects. First, medullary thymocytes develop large, complex, membrane-bound cystic structures which contain numerous microvilli and an amorphous PAS-positive material (Clark, 1968; Mandel, 1970). Cortical thymocytes, on the other hand, form small, simple cytoplasmic vesicles. If, as has been proposed, these vesicles are part of the cellular secretory apparatus, the morphological differences might reflect a difference in secretory products. Hence, it is of interest that antiserum to thymosin has been found to react with medullary epithelial cells only (Mandi and Glant, 1973), while of two antisera prepared to a mixture of thymopoietin I and II, one reacted selectively with medullary epithelial cells and the other with cortical epithelial cells (Goldstein, 1975). Moreover, Metcalf (1956) was only able to extract "lymphocytosis stimulating factor" from the medullary but not the cortical regions of thymus (see section 4.1.3). A second difference between epithelial cells in the medulla and the cortex of the thymus is the ability of some medullary epithelial cells to undergo proliferation and differentiation in postnatal life (Gad and Clark, 1968; Mandel, 1970). This is dramatically demonstrated during regeneration of the thymus after stress-induced acute involution, but is also evident in the "resting" thymus. It thus appears that the thymus medulla can act as a reservoir of immature epithelial cells throughout life.

4.1.5. The blood-thymus barrier

Few if any morphological changes occur in the thymus after parenteral administration of antigen. However, changes commonly associated with immunological responses in peripheral lymphoid tissues (e.g., formation of lymphoid follicles, germinal centers and plasma cells) are often seen in the thymus after direct injection of antigen (Marshall and White, 1961). These observations have been taken to imply that thymocytes can respond to antigens, but that a "blood-thymus barrier" must exist which normally prevents interaction of thymocytes with antigen. This interpretation is substantially correct, albeit for the wrong reasons.

First, medullary thymocytes and perhaps some cortical thymocytes *can* respond to antigenic stimulation (see section 4.1.7). However, the response is reflected by increased proliferation of immunologically competent thymocytes and differentiation to immunologically committed progeny. In contrast, plasma cellular responses which follow direct intrathymic injection of antigen are due to the recruitment of B cells from the blood, and may occur in any tissue in which antigens incite an inflammatory response. Second, a barrier *does* exist which inhibits the penetration of macromolecules from the blood into the thymus parenchyma, but this barrier is only present in the thymus cortex, and even there, it is incomplete (Raviola and Karnovsky, 1972). Thus, particulate and colloidal antigens (and other substances) readily gain access to medullary thymocytes, but only soluble antigens of low molecular weight gain access to cortical thymocytes.

These observations may help to explain why extremely small quantities of soluble antigens can induce immunological tolerance when injected parenterally, whereas large amounts of the same antigens in a particulate form are necessary to induce tolerance (Dresser and Mitchison, 1968). They also suggest an important physiological role for the blood-thymus barrier, namely, to prevent the wholesale induction of immunological tolerance to non-self antigens during critical stages of cortical thymocyte differentiation. Conversely, exposure of immunologically competent medullary thymocytes to non-self antigens might have a salutary effect by increasing the size of the responding clones and by expediting the release of specifically reactive cells to the peripheral lymphoid tissues. Thus, it is clear that some thymocytes are stimulated by the parenteral administration of antigen (Cohen et al., 1969). Also, there is a suggestion that increased numbers of thymocytes are mobilized to peripheral lymphoid tissues following such antigenic stimulation (Davies et al., 1966).

The anatomical basis for the blood-thymus barrier appears to rely mainly on the nature and arrangement of the blood vessels in the thymus, and secondarily on thymus macrophages and epithelial cells (Raviola and Karnovsky, 1972). Blood vessels with relatively "leaky" endothelial junctions (arterioles, postcapillary venules) are present almost exclusively in the medulla and the cortico-medullary junction. Conversely, capillaries with "tight" endothelial junctions are present almost exclusively in the cortex. These capillaries are surrounded by a wide connective tissue "space", which may connect with similar spaces in the inner medulla, and which any escaping antigens must traverse before reaching the parenchyma of the thymus cortex. Should antigen penetrate this far it would be hindered by epithelial cell processes and by macrophages, both of which are copiously applied to the reticular fibers of the perivascular spaces (Clark, 1964).

The same principles that govern the ingress of macromolecules into the thymus probably also apply to the ingress of thymocyte progenitors and the egress of thymocytes. In morphological studies at the light and electron microscopic levels, numerous cells can be seen to traverse the perivascular connective tissue spaces and to pass between endothelial cell junctions of blood vessels and lymphatics in the region of the medulla and cortico-medullary junction. Although the direction of migration of cells cannot be discerned from such studies, these are the anatomical regions in which the first lymphoid cells appear in the regenerating thymus. Moreover, large numbers of newly formed thymocytes have been recovered from veins and efferent lymphatics which drain the thymus, suggesting that at least some of the cells observed crossing the endothelium of medullary vessels were in the process of leaving the thymus. In contrast, few if any cells appear to traverse the endothelial cell junctions of capillaries in the thymus cortex, at least not in postnatal life.

The situation appears to be somewhat different in the immature thymus. There is a suggestion that blood vessels of the thymus in the fetal and neonatal period are more permeable to the passage of cells and antigens than in later life. Moreover, the first wave of thymocyte progenitors appears to move into the avascular epithelial anlage of the embryonic thymus by direct migration from the surrounding vascularized mesenchyme. It is possible that the different routes (non-vascular and vascular) followed by thymocyte progenitors in populating the thymus in fetal life account for the appearance of two successive waves of thymocytes during this period (see section 4.1.2).

4.1.6. Terminal deoxynucleotidyl transferase
Terminal deoxynucleotidyl transferase (TdT) is a unique DNA polymerizing enzyme which catalyzes the polymerization of monodeoxyribonucleotides in the absence of a template (Bollum, 1974). TdT has been found in high concentration in thymocytes of all species thus far tested (Chang, 1971) and in low concentrations in bone marrow cells (Coleman et al., 1974a,b; Bollum, 1975; Kung et al., 1975). It has also been found in peripheral leukocytes of patients with acute lymphoblastic and other forms of stem cell leukemia (Gallo, 1975) and in established tissue culture lines of neoplastic human (Srivastava and Minowada, 1973) and murine T cells (Bollum, personal communication). However, TdT has not been identified in the avian bursa of Fabricius (Chang, personal communication), in normal or neoplastic B cells or in normal peripheral T cells (Gallo, 1975). Neither does it appear to be present in non-lymphoid tissues (Chang, 1971).

Experiments using discontinuous BSA or Ficoll density gradients have indicated that TdT activity in the rat and mouse thymus is restricted to a population of moderately dense, cortisone-sensitive small thymocytes (Coleman et al., 1974a; Kung et al., 1975). In the rat, TdT-positive cells have been shown by antigenic and functional markers to comprise approximately 65% of cortical thymocytes (Barton et al., 1975). The remaining 35% of TdT-negative (or poor) cortical thymocytes form two distinct subsets of roughly equal proportions. One subset is composed of low density, cortisone-sensitive cells with high spontaneous [^{3}H]-thymidine incorporation; the other of high density, partially cortisone-resistant cells with low spontaneous [^{3}H]-thymidine incorporation. It is also likely that a fourth subset of cortical thymocytes exists that comprises less than 4 percent of the total population. This latter population has a very low buoyant density and high concentrations of histocompatibility antigens, suggesting that it contains the least mature members of the cortical thymocyte series (Order and Waksman, 1969). Unfortunately, it was not possible to determine whether the cells in this minor subset are TdT-positive or TdT-negative. Experiments in the mouse suggest that they may be TdT-positive (Kung et al., 1975).

Medullary thymocytes in the rat appear to be TdT-negative (Barton et al., 1975). This was demonstrated by selective lysis of medullary and cortical thymocytes with specific antisera and complement. Lysis of medullary thymocytes had no effect on total TdT levels, whereas lysis of cortical thymocytes decreased TdT levels proportionately to the percentage of killing.

The cellular origins of TdT activity in normal bone marrow are unknown. Inasmuch as leukocytes from stem cell leukemias are frequently TdT-positive, it seems reasonable to speculate that the enzyme is normally confined to lympho-hemopoietic stem cells and to lymphopoietic progenitor cells. This hypothesis is currently under investigation using the stem cell rich fraction of rat bone marrow lymphocytes discussed previously in section 3.

The physiological function of the terminal transferase enzyme in cortical thymocytes and bone marrow cells is unknown. Because of its unique functional properties and its peculiar cellular distribution, it has been postulated that TdT may play a role in the generation of immunological diversity in T cells (Baltimore, 1974; Bollum, 1974). The finding in the bursa of Fabricius of a DNA polymerizing enzyme with some of the functional attributes of TdT, but clearly distinguishable biochemically from TdT, has added further interest to this speculation (McCaffrey et al., 1974). The recent development of a highly specific antiserum to purified calf thymus TdT (Bollum, personal communication) should permit more exact studies of the cellular distribution, intracellular localization, biosynthesis and function of this intriguing enzyme.

4.1.7. Heterogeneity of thymocytes

It has long been known that a subset (5–10%) of thymocytes in the mouse and rat resembles mature T cells with respect to its ability to elicit lethal graft-versus-host reactions in genetically suitable hosts (Billingham and Silvers, 1961) and to recirculate from blood to lymph via postcapillary venules in lymph nodes (Goldschneider and McGregor, 1968a,b). Thymocytes in this functionally active subset are long-lived and home specifically to "thymus-dependent" areas of lymph node, spleen and Peyer's patches (Nossal, 1964; Murray and Woods, 1964; Parrot et al., 1966; Weissman, 1967; Goldschneider and McGregor, 1968a). That such thymocytes are not mature T cells which have migrated back to thymus is indicated by the following observations: (1) very few peripheral lymphocytes migrate to thymus in the adult animal (Gowans and Knight, 1964; Goldschneider and McGregor, 1968b); (2) immunological competence first appears in fetal or neonatal life among cells of the thymus, and the degree of immunological competence is the same in the thymuses from adults and newborn animals (Sosin et al., 1966; Schwarz, 1967; MacGillivray et al., 1970; Goldstein et al., 1971); (3) immunologically competent cells develop in explants of embryonic thymus in vitro or in diffusion chambers in vivo (Metcalf and Moore, 1971); and (4) immunologically competent thymocytes bear thymus-specific antigens that are absent from peripheral T cells (Colley et al., 1970a; Williams et al., 1971).

Subsequent studies, using sensitivity to cortisone as a means of selectively depleting cortical thymocytes (Dougherty, 1952; Ishidate and Metcalf, 1963) have shown that the immunologically competent subset of thymocytes is located in the thymus medulla. These cells are lower in average buoyant density than cortisone-sensitive thymocytes (Colley et al., 1970b; Takiguchi et al., 1971).

Cortisone-resistant medullary thymocytes have been found to have the following functional properties that are also ascribed to mature T cells: mixed lymphocyte reactivity (Weber, 1966); graft-versus-host reactivity (Blomgren and Andersson, 1969); ability to cooperate with B cells in humoral antibody responses (Andersson and Blomgren, 1970); generation of cytotoxic lymphocytes (Blomgren and Svedmyr, 1971); ability to recirculate from blood to lymph (Lance et al., 1971); responsiveness to phytohemagglutinin and concanavalin A (Blomgren and Svedmyr, 1971; Stobo, 1972); and formation of immune rosettes with sheep erythrocytes (Bach and Dardenne, 1972).

The preceding findings indicate that medullary thymocytes have cell surface receptors for antigens (see section 5.3.3.1), and are the immediate precursors of the population of immunologically competent, long-lived, cortisone-resistant, recirculating T cells in peripheral lymphoid tissues. The latter conclusion is strongly supported by the recent observation in the rat that medullary thymocytes bear a masked antigen (RMTA = rat masked thymocyte antigen) which is present in an unmasked form on T cells in the thymus-dependent zones of lymph node and spleen, but which is absent from cortical thymocytes (Goldschneider, 1975a). Significantly, RMTA can be unmasked by treating medullary thymocytes with neuraminidase or simply by culturing medullary thymocytes in vitro for several days (Goldschneider, unpublished observations). This latter observation, and the fact that RMTA is not masked on peripheral T cells, indicates that unmasking of RMTA is a physiological process that occurs shortly before or immediately after medullary thymocytes leave the thymus. It is not known whether physiological unmasking of RMTA is related to the loss of the rat thymus antigen (RTA), which also occurs at this time (Colley et al., 1970a).

Until very recently, the generally held presumption had been that cortisone-resistant thymocytes were the only immunologically competent cell types in the thymus. This is almost certainly not the case. Carefully conducted quantitative studies of graft-versus-host reactivity of thymocytes have shown that, while immunologically competent cells are markedly enriched by cortisone treatment, fully 50 percent of total graft-versus-host reactivity is lost by cortisone treatment (Tigelaar and Asofsky, 1973). Moreover, some cortical thymocytes have been found to have receptors for antigens (Lawrence et al., 1973) and to be responsive to concanavalin A stimulation (Stobo, 1972). In other studies, cortisone-sensitive thymocytes have been found to act as modulator cells (i.e. amplifier or suppressor cells) in graft-versus-host reactions (Tigelaar and Asofsky, 1973; Gershon et al., 1974), delayed-type hypersensitivity reactions (Ha et al., 1974), antibody responses of B cells to sheep erythrocytes and generation of cytotoxic lymphocytes (Cohen and Gershon, 1975).

Cortisone-sensitive T cells with modulator properties have also been identified in mouse and rat spleen (see section 5.4.1). These are rapidly dividing non-recirculating lymphocytes which are dependent upon the continued presence of the thymus for their survival or renewal. Antigenic analysis of rat thymocytes and peripheral T cells suggests that cortisone-sensitive T cells are derived from cortical thymocytes. Thus both cortical thymocytes and cortisone-sensitive T cells bear the rat bone marrow lymphocyte antigen (RBMLA) and the rat cortical thymocyte antigen (RCTA) (Goldschneider, 1975b; see section 5.3.1.2). Neither antigen is present on medullary thymocytes or on cortisone-resistant T cells.

Other differences between cortical and medullary thymocytes are listed in Table 2. The most notable of these in the mouse are qualitative and quantitative differences in the expression of cell surface alloantigens. These are discussed in detail in section 5.3.1.1.

In addition to heterogeneity between cortical and medullary thymocytes (macroheterogeneity), there is evidence that heterogeneity exists within populations of cortical and medullary thymocytes (microheterogeneity). It is not known whether such microheterogeneity reflects separate stages of differentiation of lineally-related thymocytes or parallel differentiation of thymocyte subsets having a common precursor (see section 4.1.8). Thus, at least four subsets of cortical thymocytes can be defined on the basis of differences in buoyant density, size, antigenicity, mitotic activity, viability and the presence of terminal deoxynucleotidyl transferase (Shortman and Jackson, 1974; Barton et al., 1975; Fathman et al., 1975; also see section 4.1.6). Similarly, populations of medullary thymocytes have been identified which differ in their respective abilities to exhibit mixed lymphocyte reactivity, graft-versus-host reactivity, and responsiveness to phytohemagglutinin (Colley et al., 1970b; Stobo, 1972). These properties appear sequentially in the neonatal thymus, denoting either step-wise maturation of a single subset of cells or sequential maturation of multiple subsets of cells.

4.1.8. Ontogenetic relationships of thymocyte subsets

The simplest scheme to attempt to explain the developmental relationships of thymocyte subsets is that proposed by Weissman (1973). In this scheme, cortical thymocytes, through a series of mitotic and differentiative events, give rise to medullary thymocytes, and these in turn give rise to peripheral T cells. The evidence in favor of this developmental pathway is that selective [^{3}H]-thymidine labelling of rapidly dividing cells in the thymus cortex results, after several mitotic cycles, in the appearance of labelled cortisone-resistant medullary thymocytes, and ultimately in labelled T cells in the peripheral lymphoid tissues. The most serious criticism of these experiments is the possible utilization or reutilization of the radioactive label by less rapidly dividing medullary thymocytes. Moreover, this scheme does not adequately account for the gross overproduction and more rapid turnover of cortical thymocytes relative to medullary thymocytes (Metcalf and Wiadrowski, 1966).

Shortman and Jackson (1974) compared the kinetics of proliferation of cortical and medullary thymocytes after parenteral administration of [^{3}H]-thymidine and concluded that these cells comprised two independent, self-generating populations of thymocytes. They postulated that cortical and medullary thymocytes arise from a common precursor, but that cortical thymocytes represent self-reactive clones of cells which have been stimulated to proliferate and to undergo sterile differentiation and death within the thymus. Medullary thymocytes, on the other hand, represent non-self-reactive clones of cells that are exported to peripheral lymphoid tissues. This scheme of thymocyte development also suffers certain deficiencies. Thus, intrathymic death can only account for a small fraction of the cortical thymocytes that are produced daily (Kindred, 1955; Sainte-Marie and Leblond, 1964; Michalke et al., 1969; Chanana et al., 1972); and massive numbers of rapidly dividing, cortisone-sensitive thymocytes (presumably cortical thymocytes) are known to be continually released from the thymus (Ernstrom and Sandberg, 1970; Chanana et al., 1971).

Bryant (1971) has also concluded that it is unlikely that cortical and medullary thymocytes have a parent-progeny relationship. He has postulated that cortical and medullary thymocytes may arise from separate progenitors, and that cortical thymocytes die almost immediately after release from the thymus.

None of the preceding schemes of thymocyte ontogeny allows for the possibility, indeed the probability, that some cortical thymocytes are functionally competent and give rise to a subset of peripheral T cells (see section 4.1.7). We have, therefore, proposed three variations on these schemes which are consistent with available data (Fig. 3). In the first variation, cortical thymocytes differentiate into a subset of T cell precursors and into medullary thymocytes. In the second variation, a common progenitor cell in thymus cortex gives rise to both cortical and medullary thymocytes. In the third variation, cortical and medullary thymocytes arise from separate progenitor cells. Other variations in which cortical thymocytes arise from medullary thymocytes or from a common progenitor in the thymus medulla are unlikely, inasmuch as antigens common to both thymocyte progenitors and cortical thymocytes (but not medullary thymocytes) have been identified (Goldschneider, 1975a), and inasmuch as the distribution of terminal deoxynucleotidyl transferase is probably restricted to thymocyte progenitors and cortical thymocytes (Barton et al., 1975). The final products of each of these provisional schemes of thymocyte development are a subset of modulator T cell precursors (cortical thymocytes) and a subset of effector T cell precursors (medullary thymocytes), each of which would migrate to peripheral lymphoid tissues. Clones of self-reactive cells in each subset would die within or without the thymus, or would survive in a tolerized state.

Whichever, if any, of the above schemes proves correct, the differentiation of thymocyte progenitors to thymocytes and of thymocytes to T cells involves major remodeling of the cell surface (see Table 2 and section 5.3). The progenitor cell → cortical thymocyte step involves de novo expression of TL, Thy-1 and Ly-1, 2, 3 and 5 alloantigens in the mouse, and the rat thymus antigen (RTA) in that species. This step is also accompanied in both species by a marked decrease in the concentration of major histocompatibility antigens. The cortical thymocyte → peripheral T cell step (or cortical thymocyte → medullary thymocyte step) involves the loss of the TL antigen, decreased expression of Thy-1 and Ly antigens, and increased expression of histocompatibility antigens. Terminal deoxynucleotidyl transferase activity is also lost at this stage. The progenitor cell → medullary thymocyte step involves de novo expression of the rat masked

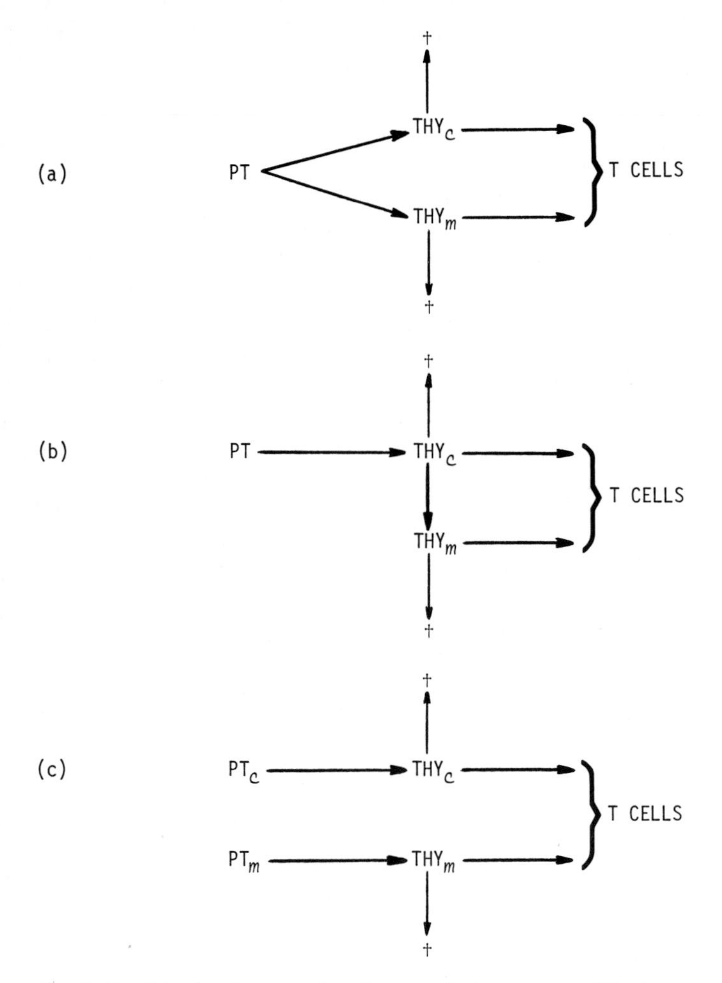

Fig. 3. Possible developmental pathways of thymocytes. PT, thymocyte progenitor cell; THY, thymocyte; c, cortical; m, medullary; †, death of self-reactive clones. See text section 4.1.8 for details.

thymocyte antigen (RMTA), which becomes unmasked during the medullary thymocyte → peripheral T cell step. The rat thymus antigen is also lost during this last step. If cortical thymocytes give rise to medullary thymocytes, the rat cortical thymocyte antigen (RCTA) and the rat bone marrow lymphocyte antigen (RBMLA) must be lost in the transition.

Another important event in the differentiation of progenitor cells to mature thymocytes is the expression of cell surface receptors for antigen. Such receptors are clearly present on mature medullary thymocytes, although their structure is a matter of great controversy (see section 5.3.3.1). Recently, antigen receptors have also been demonstrated on some cortical thymocytes (Lawrence et al., 1973).

The significance of the microheterogeneity that has been found within populations of cortical and medullary thymocytes is not known (see section 4.1.7). Some of the differences in physical, biochemical and functional properties undoubtedly represent stages in the differentiation of a single line of cells. However, others may reflect parallel differentiation of separate subsets of cells having a common precursor. This question is of central importance in attempting to determine the origins and ontogenetic relationships of functionally and antigenically distinct subsets of T cells in peripheral lymphoid tissues (see section 5.4.1).

4.2. Bursa of Fabricius

4.2.1. Formation of the epithelial anlage

The bursa of Fabricius arises between the 4th and 5th embryonic day as a focal proliferation and invagination of endodermal epithelial cells from the dorsal aspect of the primitive cloaca of the chicken. The bursa is an oval, saclike structure with many plicae or folds which project into the lumen. The epithelial lining of the bursa contains numerous lymphoid follicles, each of which is divided into a cortical and a medullary region by a layer of epithelial cells and associated basement membrane.

Lymphoid follicle development commences on day 12 or 13 within nodular foci of mitotically active epithelial cells (Ackerman and Knouff, 1959, 1964). These enlarging nodules of epithelial cells form the medullary regions of the developing follicles. Relatively few epithelial cells are present in the region of the developing follicle cortex. Near to the time of appearance of lymphoid cells in the bursal follicles the epithelial cells in the interior of the medulla undergo differentiation, becoming more stellate or reticular in appearance. The cytoplasmic processes of these cells interdigitate and are joined by desmosomes to form a supporting framework in the medulla (Ackerman, 1962; Ackerman and Knouff, 1964).

During follicle formation the epithelial nodules separate from the underlying tunica propria. Capillary loops are evident in the tunica propria near to the developing follicles, and vascular networks develop from these capillary loops. These networks surround the enlarging medullary area of each follicle, but no capillaries can be observed penetrating the medulla. Hence, only the cortex of the definitive follicle contains blood capillaries, some of which are in close association with the basement membrane encompassing the medulla (Ackerman and Knouff, 1959).

4.2.2. The bursa as a lymphoid organ

4.2.2.1. Embryonic and neonatal development. Studies employing chromosome markers in parabiotic chick embryos indicate that large, deeply basophilic progenitor cells of yolk sac origin migrate into the developing bursal medulla by day 13 or 14 of embryonic life (Moore and Owen, 1966). This is supported by the demonstration of a small percentage of yolk sac cells bearing bursal lymphocyte-specific cell surface antigens 4–5 days before lymphoid cells are observed in the bursa (Albini and Wick, 1975). The progenitor cells appear to enter the epithelial nodule by transit through the surrounding basement membrane (Tar et al., 1969).

Following progenitor cell appearance the bursal follicles rapidly acquire a lymphoid character. Within the enlarging follicles of 15-day embryos the mean generation time of follicular lymphocytes is 7–9 hours (Rubin et al., 1971). Although progenitor cells may continually enter the follicles (Tar et al., 1969), the marked increases in the number of bursal lymphocytes are probably the result of this rapid cell proliferation.

The synthesis of immunoglobulins also occurs soon after progenitor cell arrival. Immunoglobulin molecules (IgM) are detectable in bursal lymphoid follicles within one day after the appearance of yolk sac progenitor cells. The number of IgM-containing cells increases concomitantly with the increase in total lymphocytes within the enlarging medulla of the follicle (Kincade and Cooper, 1971). At this stage of development IgM is detected only in the bursa; no immunoglobulins are present in peripheral lymphoid tissues or in the circulation (Thorbecke et al., 1968; Kincade and Cooper, 1971; Tao-Wiedmann et al., 1975).

Shortly before hatching, the cortical region of bursal follicles begins to be populated with lymphoid cells. The sequential appearance of lymphocytes first in the medulla and then in the cortex is mimicked during regeneration of the bursa after X-irradiation (Cooper et al., 1972b). Autoradiography of the bursa after [^{3}H]-thymidine pulse labelling (Rubin et al., 1971) and morphological studies of the embryonic bursa (Ackerman and Knouff, 1964) have suggested that medullary lymphocytes may migrate through the basement membrane to the cortex.

Although cortical and medullary lymphocytes appear to be similar morphologically, immunofluorescent analysis of bursal tissue sections reveals that many medullary cells and few cortical cells contain detectable intracytoplasmic immunoglobulin (Thorbecke et al., 1968). However, virtually all lymphocytes in the bursa, both medullary and cortical, bear surface immunoglobulins (Kincade et al., 1971). Cooper et al. (1972b) have suggested that medullary lymphocytes synthesize immunoglobulins which are subsequently incorporated as antigen-receptors into the membranes of their progeny in the cortex. The putative, antigen-recognition cells in the cortex then emigrate to populate the B lymphocyte-specific areas in the peripheral lymphoid tissues. Although the relationship of medullary and cortical bursal lymphocytes has not been definitively established, competent antigen-recognition cells do arise within the bursal follicles (Waltenbaugh and Van Alten, 1974).

Five to 7 days after the initial appearance of IgM-containing cells within the medulla of bursal follicles, lymphoid cells possessing IgG appear (Thorbecke et al., 1968; Kincade and Cooper, 1971). The sequential appearance of IgM- and IgG-containing cells has also been observed in bursal organ cultures (Thorbecke et al., 1968). These observations suggest that the bursal IgG-containing cells may be derived from cells previously producing IgM. Such a possibility is supported by the findings that injection of 13-day embryos with heterologous antiserum to the heavy chain subunit of IgM suppresses the synthesis of IgG as well as IgM (Kincade et al., 1970). Early embryonic bursectomy results in agammaglobulinemia, but bursectomy 2 to 3 days before hatching abolishes production of IgG but not IgM (Cooper et al., 1969; Warner et al., 1969). Moreover, approximately 50% of the bursal cells from one month old chicks possess both IgG and IgM on their cell surfaces; whereas very few spleen cells do (Kincade and Cooper, 1971). Hence, a lineal conversion or "switchover" of IgM- to IgG-synthesizing cells, as opposed to a parallel differentiation of IgM- and IgG-synthesizing cells, seems to occur within the bursa (Cooper et al., 1972b).

Similarly, IgA-containing cells may be derived directly or indirectly from IgM-producing cells. Serum IgA is not detectable until 19 days after hatching, later than both IgM and IgG. Intraembryonic injection of antiserum to IgM, when combined with bursectomy at hatching, suppresses synthesis of IgA as well as IgG and IgM (Kincade and Cooper, 1973). Cooper et al. (1972b) have proposed that IgA-synthesizing cells may arise within the bursa from IgG-synthesizing cells (i.e. IgM → IgG → IgA). Alternatively, Martin and Leslie (1974) have suggested that IgA- and IgG-synthesizing bursa cells may arise by parallel differentiation of IgM-synthesizing cells (i.e. IgM → IgG and IgM → IgA).

It is important to note that bursal lymphoid development seems to proceed independently of antigen exposure. Normal bursal development occurs in chicks raised in a germ-free environment, and intravenous injection of antigens into chick embryos does not qualitatively or quantitatively alter bursal development (Kincade and Cooper, 1971).

4.2.2.2. B cell development in the adult chicken. Like the thymus, the bursa of Fabricius in most avian species undergoes physiological involution. In the White Leghorn chicken, for example, involution begins at approximately 3–6 weeks of age and is complete by 16 weeks (Glick, 1956, 1960). Toivanen et al. (1972b) have evidence suggesting that B cell progenitors exist in non-bursal locations in chickens in which bursal involution has occurred. Bone marrow and spleen cells from donors 10 weeks of age and older were capable of reconstituting humoral immunity in cyclophosphamide-treated, bursectomized chicks. The bone marrow and spleens of younger donors did not possess such capabilities. These "postbursal stem cells" (Toivanen and Toivanen, 1973) restore normal antibody production and normal splenic morphology 30–100 days after transplantation. However, unlike bursal cells from younger donors (Toivanen et al., 1972a) the "postbursal stem cells" have lost the capacity to migrate into the bursal rudiment and to restore normal bursal follicular morphology.

4.2.3. The bursa as an endocrine organ

Results of early experiments, in which bursal grafts in diffusion chambers appeared to restore specific antibody production in bursectomized recipients, suggested that a hormonal inducer of B cell differentiation existed (St. Pierre and Ackerman, 1965, 1966; Jankovic and Leskowitz, 1965). However, more recent studies have failed to corroborate these observations (Thompson and Cooper, 1971; Toivanen and Toivanen, 1973).

Support for a microenvironmental influence on bursal lymphopoiesis has also been presented (Moore and Owen, 1966; Toivanen et al., 1972a; Toivanen and Toivanen, 1973). These studies demonstrated that the bursal rudiment (epithelium) which remains after hormonal or cyclophosphamide treatment provides a microenvironment conducive to normal bursa lymphocyte development, and hence to B cell development. Bursa cells from early postnatal chicks (up to 2 weeks) were capable of reconstituting bursa lymphocyte-depleted chicks. Normal bursal and splenic morphology were restored, as was humoral immunity (Toivanen et al., 1972a; Toivanen and Toivanen, 1973). However, such reconstitution was not achieved if cyclophosphamide treatment was followed by surgical ablation of the bursal rudiment.

Bursal development can be modified by administration of certain hormones. Androgens can induce bursal involution or inhibit bursal development (Glick, 1964; Warner and Szenberg, 1964) in both male and female chick embryos. Moore and Owen (1966) suggested that this effect may be directed against the follicular epithelial cells, rather than against bursa lymphocytes or their progenitors. Involution of the bursal lymphoid follicles can also be induced by stress or by administration of corticosteroids (Glick, 1964; Warner and Szenberg, 1964). The mechanism of corticosteroid-mediated involution is not known.

4.3. Mammalian bursa-equivalent

4.3.1. Embryonic and neonatal B cell development

Mammals lack a homologue of the bursa of Fabricius, and considerable effort has gone into determining the site or sites in which antibody-forming cell precursors undergo differentiation. Such functional analogues of the bursa have been referred to as bursa-equivalent tissues. Essentially two groups of tissues have been suggested as bursal analogues in mammals, the gut-associated lymphoid tissues (GALT) and the hemopoietic tissues.

The proposal that the gut-associated lymphoid tissues (appendix and Peyer's patches) serve as the mammalian bursa-equivalent is based primarily on the results of extirpation experiments. Rabbits subjected to removal of gut-associated lymphoid tissues developed a persistent impairment of humoral immunity which was not corrected by injection of lympho-hemopoietic stem cells (Cooper et al., 1966a, 1968; Perey et al., 1968, 1970). No long term effects on cell-mediated immunity are observed in such animals (Cooper et al., 1966a). However, the role of the GALT as the mammalian bursa-equivalent must be questioned. Few lymphocytes and lymphoid follicles are contained in the gut-associated lymphoid tissues of neonatal rabbits (Thorbecke, 1960) and other rodents (Friedberg and Weissman, 1974). Also, development of the gut-associated tissues does not occur in germ-free animals, suggesting that antigenic stimulation is required. Friedberg and Weissman (1974) have determined that the level of proliferation in Peyer's patches of neonatal mice is insufficient to account for the rapid increase in the B cell population which occurs in the peripheral lymphoid tissues during neonatal development. More importantly, a survey of fetal and neonatal mouse tissues has revealed that immunoglobulin-bearing lymphocytes appear later in the gut-associated lymphoid tissues than in the spleen, liver or bone marrow (Nossal and Pike, 1973). These findings attenuate support for the gut-associated lymphoid tissues being the site of B cell generation in the mammal.

Recent studies have supported the hemopoietic tissues as the mammalian bursa-equivalent. Nossal and Pike (1973) first detected immunoglobulin-bearing lymphocytes in liver, spleen and bone marrow of fetal mice at day 16–17 of development. Similarly, Spear et al. (1973) observed B cells in fetal spleen at day 15–16, and Owen et al. (1974) demonstrated immunoglobulin-positive cells in liver and spleen at day 17. In addition, B lymphocytes in man are initially observed in the fetal liver at about the 9th week of gestation (Lawton et al., 1972). The proposition that fetal liver is a site of B lymphocyte generation in the mouse is supported by the observation that immunoglobulin-positive cells develop in vitro after 4 days culture of fetal liver from 14-day old embryos (Owen et al., 1974). Since no immunoglobulin-bearing cells were initially present in the 14-day liver explants, the studies indicate that the fetal liver not only contains B cell progenitors but also provides a proper microenvironment for their differentiation.

A variety of indirect evidence suggests that some degree of B cell maturation may occur in the murine spleen within the first 4 weeks of postnatal life. Thus, a substantial increase in the number of immunoglobulin-bearing and complement receptor lymphocytes is observed during this period (Spear et al., 1973; Gelfand et al., 1974; Sidman and Unanue, 1975). Also, the B cell population in the spleen begins to exhibit the magnitude and heterogeneity of antibody responses characteristic of adult mice (Goidl and Siskind, 1974; Spear and Edelman, 1974). It must be cautioned that these phenomena can also be explained by the migration of newly-formed B cells from bone marrow and/or liver to the spleen. A certain degree of B cell maturation may also occur in the spleen of the adult (see section 4.3.2).

4.3.2. B cell development in the adult

The generation of B cells in the bone marrow of adult rodents has been demonstrated in a number of studies. Although the majority of bone marrow small lymphocytes are non-dividing, they have a brief transit time and are replaced rapidly by newly-formed cells which arise in situ (Everett and Caffrey, 1967; Yoshida and Osmond, 1971; Brahim and Osmond, 1973; Osmond and Nossal, 1974; Rysser and Vassalli, 1974). The least mature of these rapidly dividing progenitor cells are referred to as "null" cells in that they lack the cell surface antigens of either T or B cells (Osmond and Nossal, 1974; Rysser and Vassalli, 1974). In vitro culture of null lymphocytes has revealed that with time an increasing proportion acquire cell surface immunoglobulins, i.e. they become

B cells. The surface Ig was noted to increase in density as the cells "matured" in culture. Lafleur et al. (1973) have demonstrated an intermediate stage of B cell development in mouse bone marrow. These "pre B" cells are large, low density cells which possess cell surface immunoglobulin, but which are not functionally competent. They can give rise to competent B cells in lethally irradiated host animals.

Although the bone marrow seems to be a source of B cell progenitors, it may not be an exclusive site for maturation of functional B lymphocytes. Indeed, B cell maturation has been demonstrated in adult mice whose bone marrow has been destroyed by ^{89}Sr (Phillips and Miller, 1974; Lawton et al., 1975). Also, Lafleur et al. (1973) found "pre B" cells in the spleen, but not in the lymph nodes, of adult mice. Rysser and Vassalli (1974) found that newly formed null bone marrow lymphocytes migrated primarily to the spleen when injected into syngeneic recipients. Within 20 hours after transfer, half of the null cells had acquired the characteristics of B cells.

The role of the spleen in B lymphocyte maturation in the rat has been presented by Strober (1975). He suggests that there is constant migration of immature B cells from the bone marrow to the spleen in the adult. The immature B cells differentiate into large lymphocytes bearing complement receptors and surface IgM. These large lymphocytes divide rapidly, producing progressively smaller and more dense progeny. A proportion of these small B cells lose complement receptors and surface IgM but gain other classes of surface immunoglobulins. The populations of these small, differentiated B cells are short-lived in the spleen; the cells are lost by death and by migration to other peripheral lymphoid tissues.

The preceding discussion has suggested that the bone marrow is a site of production of immature B cells which may subsequently migrate to the spleen to undergo further maturation. However, the inability of splenectomy (Strober, 1975) or bone marrow destruction (Phillips and Miller, 1974) to completely abrogate humoral immune development indicates that neither tissue is an exclusive site of B cell maturation nor is one tissue dependent on the other for B cell development. Hence, either bone marrow and spleen possess similar microenvironmental capabilities for B cell development or B cell differentiation can occur in additional, as yet unknown, sites within the adult mammal.

5. *Peripheral lymphoid tissues*

The peripheral (secondary) lymphoid tissues consist of lymph nodes, spleen, Peyer's patches in the terminal ileum, tonsillar and adenoidal tissues, and diffuse collections of lymphocytes in the submucosa and lamina propria of the gastrointestinal, respiratory and genitourinary tracts. They are populated by immunologically competent lymphocytes which migrate from the central (primary) lymphoid tissues during fetal and neonatal life. Although peripheral lymphocytes of both the T and B cell systems have enormous proliferative capacities when stimulated by antigen or other agents, they cannot sustain their populations indefinitely and must be replenished throughout postnatal life by newly formed cells from the central lymphoid tissues. Hence, congenital absence or perinatal removal of the thymus or bursa of Fabricius precludes population of the peripheral lymphoid tissues with T or B cells, respectively, and results in complete absence of cell-mediated or antibody-mediated immunological responsiveness. Extirpation of the thymus in adult animals results in a progressive depletion of peripheral T cells and loss of cell-mediated immunological competence over a period of months or years, depending on the species. However, combined thymectomy and irradiation (to destroy preexisting T cells) prevents regeneration of the T cell system even when lympho-hemopoietic stem cells are replaced by a bone marrow graft (see review by Miller and Osoba, 1967). The situation is somewhat different with respect to the B cell system, in which adult bursectomy and irradiation (or cyclophosphamide treatment) does not prevent the gradual regeneration of B cells. This suggests that some peripheral lymphoid tissues acquire the capacity to generate B cells in adult life, perhaps from progenitor cells that have already been processed by the central lymphoid tissues (see sections 4.2.2.2. and 4.3.2).

The peripheral lymphoid tissues serve three major functions. First, they are biological filters for antigens and other foreign or abnormal substances in blood, lymph and external secretions. Second, they are reservoirs of readily mobilizable populations of immunologically competent lymphocytes. And third, they provide the microenvironmental milieu for proliferation, differentiation and functional interaction of T cells, B cells and macrophages.

The present section will deal with the following aspects of T cell and B cell development in peripheral lymphoid tissues: anatomical distribution, migration pathways, surface properties, and cellular heterogeneity.

5.1. Anatomical distribution of lymphocytes

5.1.1. T cells

T cells occupy anatomically discrete areas in peripheral lymphoid tissues (Fig. 4). Studies in rats and mice involving thymus ablation or chronic drainage of lymphocytes from a thoracic duct fistula have shown that a population of long-lived, recirculating, thymus-dependent lymphocytes resides in the deep cortex (paracortex) of lymph nodes, in the periarteriolar sheath of spleen, and in the interfollicular areas of Peyer's patches (Waksman et al., 1962; Parrott et al., 1966; Goldschnieder and McGregor, 1968a). Depletion of T cells in these same areas has been observed in the peripheral lymphoid tissues of human beings and mice which have congenital absence of the thymus (de Sousa et al., 1969). Medullary thymocytes and T cells from thoracic duct lymph selectively migrate ("home") to these "thymus-dependent" areas of peripheral lymphoid tissues after intravenous injection (Parrott et al., 1966; Goldschnieder and McGregor, 1968a; Lance and Cooper, 1970; Howard et al., 1972). Moreover, some thymocytes that have been labelled in situ can be found several days later in these

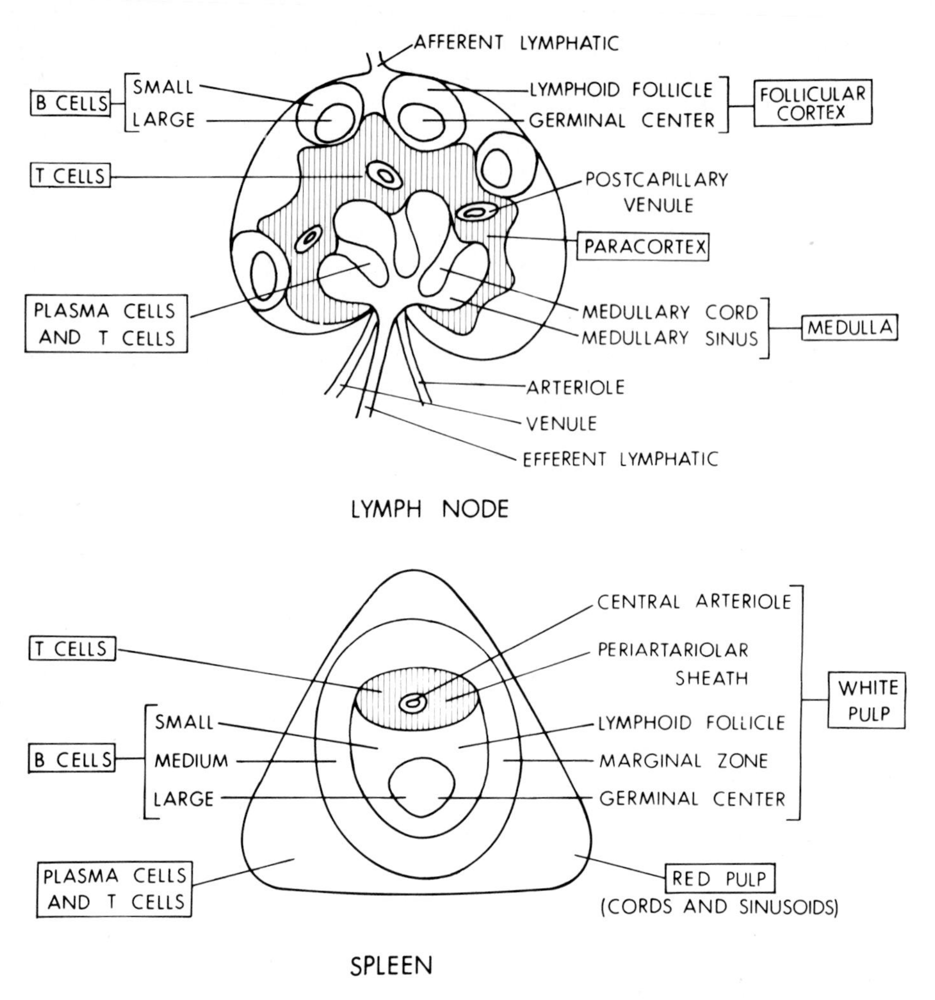

Fig. 4. Schematic representation of the anatomical distribution of T cells and B cells in mammalian lymph node and spleen. Histological and cytological correlates are designated on the right and left sides of the diagram, respectively. The hatched areas denote the location of recirculating T cells. See text section 5.1 for details.

thymus-dependent areas of lymph nodes, spleen and Peyer's patches (Nossal, 1964; Murray and Woods, 1964; Weissman, 1967; Linna, 1968).

Direct evidence that T cells normally reside in the thymus-dependent areas of lymph node and spleen has come from immunofluorescence studies using antisera specific for T cell surface antigens (Gutman and Weissman, 1972; Goldschneider and McGregor, 1973). In addition to mapping the anatomical distribution of T cells in frozen sections, these and other studies (Raff, 1971) have also determined the quantitative distribution of T cells in suspensions of cells from various tissues and body fluids. While there is some variation between species and strains, the relative proportions of T cells from each site are comparable, namely, thoracic duct lymph > lymph node > blood > spleen > Peyer's patches. A typical distribution of T cells in the rat is shown in Table 3.

Recent immunofluorescence studies have demonstrated two anatomically separate subsets of T cells in rat spleen (Goldschneider, 1975a,b). Using rabbit antisera specific for medullary thymocytes, a population of recirculating, cortisone-resistant T cells was identified in periarteriolar sheath of spleen white pulp, in thoracic duct lymph, and in paracortex of lymph node. A second subset of T cells was identified in the spleen. The cells in this latter subset are antigenically related to cortical thymocytes and are cortisone-sensitive. They appear to be distributed throughout red pulp of spleen as scattered clusters of small and medium size lymphocytes. The possible ontogenetic and functional import of T cell subsets in the rat and mouse will be discussed in section 5.4.1.

5.1.2. B cells

Histological studies of peripheral lymphoid tissues from mice and rats have revealed that discrete areas within peripheral lymphoid tissues remain populated with lymphocytes after neonatal thymectomy (Waksman et al., 1962; Parrott et al., 1966; Goldschneider and McGregor, 1968a). As shown in Fig. 4, these regions comprise the lymphoid follicles in the cortex of lymph nodes, the follicles and marginal zone areas in the white pulp of spleen, and the follicles in the gut-associated lymphoid tissues. Also, diffuse collections of plasma cells are found in the medulla of lymph nodes, the red pulp of spleen and the submucosa and lamina propria of the gastrointestinal, respiratory and genitourinary tracts. The presence of these "thymus-independent" areas has also been demonstrated in bursectomized chickens (Cooper et al., 1965, 1966b), humans with Bruton's agammaglobulinemia (Good, 1955) and congenitally athymic (nude) mice (De Sousa et al., 1969).

Table 3
Distribution of T and B lymphocytes in the rat.

	Percent of lymphocytes in each class[a]					
	thymus	lymph node	spleen	thoracic duct lymph	blood	bone marrow[b]
T cells	>99	81	66	87	61	13
B cells	<0.5	18	33	11	40	38

[a]Determined by indirect immunofluorescence using rabbit anti-rat T cell serum (Goldschneider and McGregor, 1973) to identify T cells and rabbit anti-rat Ig serum to identify B cells.

[b]Approximately half of the lymphocyte-like cells in bone marrow do not have T or B cell antigenic markers (Goldschneider and McGregor, 1973; Goldschneider, 1975b). These "null" cells are discussed in section 3.

The B cell nature of lymphoid follicles in the peripheral lymphoid tissues of mice and rats was confirmed by immunofluorescent analysis using antisera to cell surface immunoglobulin (Gutman and Weissman, 1972) and to B cell-specific heteroantigens (Goldschneider and McGregor, 1973). Using such antisera, Goldschneider and McGregor (1973) observed that, in addition to the small cells which populated the lymphoid follicles, the medium size lymphocytes in the marginal zone of rat spleen also were Ig^{+} and possessed B cell-specific antigens.

Antisera to immunoglobulins and to B cell-specific antigens have been used to determine the quantitative distribution of B lymphocytes in cell suspensions from a variety of tissues and body fluids (Raff, 1971; Gutman and Weissman, 1972; Goldschneider and McGregor, 1973). A representative distribution of B cells in the rat is presented in Table 3.

The specificity of these thymus-independent regions for B cells has been demonstrated by the selective homing of intravenously injected, radioactively labelled B cells. Following injection, labelled B cells migrate to and localize in the lymphoid follicles of lymph node and spleen and the lamina propria of the gastrointestinal tract in both mammalian and avian species (Howard, 1972; Gutman and Weissman, 1973; Sprent, 1973; DeKruyff et al., 1975; Durkin et al., 1975; Williams and Gowans, 1975) (see section 5.2.2.).

Peripheral lymphoid tissues in germ-free animals possess only primary follicles, which are composed of small B lymphocytes (Nieuwenhuis, 1971). Antigenic stimulation induces the formation of germinal centers within primary follicles. The germinal centers are composed of rapidly dividing medium and large lymphocytes which are enmeshed in the interdigitating processes of dendritic macrophages. The germinal center and its surrounding mantle or corona of lymphocytes (former primary follicle) are referred to as a secondary follicle. Various authors have concluded that the function of germinal centers is the generation of memory B cells, which are responsible for mediating an antibody response following subsequent antigen exposure (Wakefield and Thorbecke, 1968; Grobler et al., 1974; Nieuwenhuis and Keuning, 1974).

5.2. Migration pathways of lymphocytes

5.2.1. T cells

One of the striking properties of lymphocyte differentiation is the sequential acquisition (and loss) of the ability to selectively migrate via complex pathways to specific microenvironmental niches within central and peripheral lymphoid tissues. The recognition mechanisms which determine the directional migration of lymphocytes are unknown. However, the major pathways which lymphocytes, particularly T cells, follow in their migratory wanderings to and from lymphoid tissues have been traced. A composite scheme of the known or probable migration pathways of members of the T cell series is presented in Fig. 5. A discussion of these pathways is presented below.

The immunological functions of lymphocyte migrations through peripheral lymphoid tissues fall into three interrelated categories: immunological surveil-

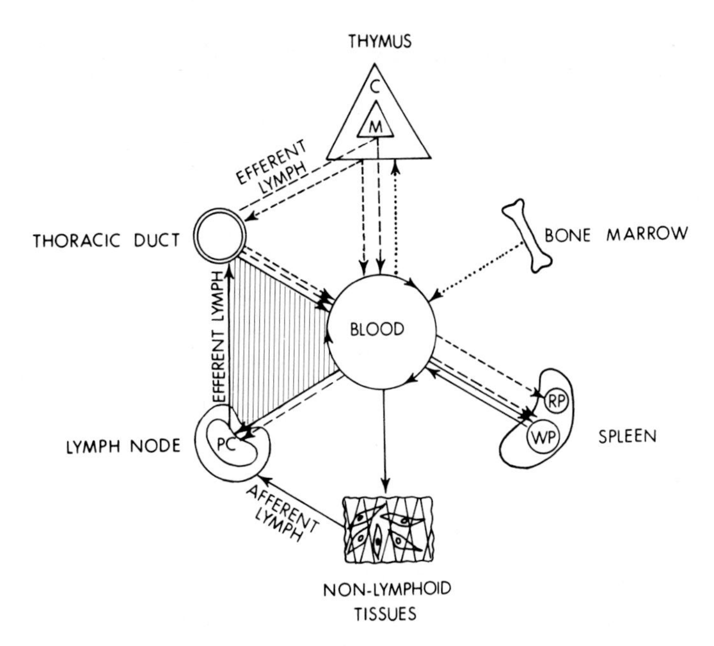

Fig. 5 Migration pathways of members of the T cell series. ...., thymocyte progenitor cells; ---, medullary thymocytes; -----, cortical thymocytes (probably pathway); ————, mature T cells; C, cortex; M, medulla; RP, red pulp; WP, white pulp; PC, paracortex; hatched area, major pathway of T cell recirculation from blood to lymph. See text section 5.2.1 for details.

lance; orchestration of immunological reactions; and dispersion of the immunological repertoire. Antigen-reactive cells constantly travel via blood and lymph through lymphoid and non-lymphoid tissues, thereby forming an economical and efficient surveillance system for antigens. Migration of lymphocytes also permits recruitment and short-range interactions of functionally distinct subsets of T cells, B cells and macrophages, thereby increasing the variety and effectiveness of immunological reactions. Lastly, migration of the progeny of antigen-stimulated lymphocytes (memory, modulator, helper and effector cells) constitutes a major mechanism whereby local immunization produces systemic immunity.

5.2.1.1. Migration of lymphopoietic progenitor cells to the thymus. Evidence has already been presented that thymocytes arise from migration of blood-borne progenitor cells into the thymus during embryonic and postnatal life (sections 3 and 4.1.2). In the embryo, thymocyte progenitors migrate from the blood into the perithymic mesenchyme, and thence into the epithelial anlage (Metcalf and Moore, 1971, p. 238; LeDourin and Jotereau, 1975). After the epithelial anlage has become vascularized, thymocyte progenitors enter the thymus directly, most probably by-passing between endothelial cell junctions of postcapillary venules in the region of the medulla and corticomedullary junction (Dukor et al., 1965; Blackburn and Miller, 1967). This is a highly selective migration pathway.

Neither hemopoietic cell precursors nor mature myeloid cells normally enter the thymus parenchyma (McKori et al., 1965; Micklem and Loutit, 1966). It is also a unidirectional pathway. Very few, if any, thymocytes are able to migrate back to thymus (Moore and Owen, 1967a,b; LeDouarin and Jotereau, 1975); and for the thymocyte progenitors which have already entered the thymus rapidly lose the ability to return to bone marrow (Order and Waksman, 1969). There is also some evidence that this may be a gated pathway, i.e. that "new" progenitor cells are not permitted to enter until "old" progenitor cells have been processed. This may explain the migration of at least two distinct waves of thymocyte progenitors with an intervening refractory period in the embryonic chicken and mouse thymus (Moore and Owen, 1967a,b; LeDouarin and Jotereau, 1975); and for the failure of the thymus of parabiotic animals to rapidly achieve random mixing of progenitors of donor and host origin (Harris et al., 1964).

5.2.1.2. Migration of thymocytes to peripheral lymphoid tissues. Several lines of evidence indicate that there is a unidirectional migration of thymocytes to peripheral lymphoid tissues throughout life. Direct intrathymic injection of [^{3}H]-thymidine has been followed by the appearance of labelled small lymphocytes in lymph nodes, spleen and Peyer's patches (Nossal, 1964; Murray and Woods, 1964; Weissman, 1967; Linna, 1968). Similarly, when a thymus is transplanted to an intermediate host, T lymphocytes of donor origin are found in peripheral lymphoid tissues within a few days. Simultaneously, the donor cell population in the thymus is replaced by thymocytes of host origin. Later, after passage through the donor thymus, mature T cells of host origin appear in the peripheral lymphoid tissues (Harris and Ford, 1964; Miller et al., 1965; Owen and Raff, 1970).

Cannulation of veins and lymphatics leading from the thymus in the guinea pig and calf has demonstrated that sufficient newly formed cells are released from the thymus each day to replace the population of lymphocytes in the blood approximately 4 times over (Ernstrom and Larsson, 1967; Cronkite and Chanana, 1970). Most of the emigrating thymocytes are rapidly dividing cells which home selectively to red pulp of spleen (Chanana et al., 1971). Long-lived thymocytes, on the other hand, tend to migrate to periarteriolar sheath of spleen and paracortex of lymph node. Similar homing patterns have been observed following intravenous injection of [^{3}H]-thymidine and [^{3}H]-uridine-labelled thymocytes into rats (Goldschneider and McGregor, 1968a). In the mouse, "spleen-seeking" and "lymph node-seeking" populations of thymocytes have been detected in serial transfers of thymocytes through intermediate hosts (Lance and Taub, 1969). The lymph node-seeking thymocytes are long-lived, cortisone-resistant cells which presumably arise from the thymus medulla (Andersson and Blomgren, 1970; Lance and Cooper, 1970). The spleen-seeking thymocytes are short-lived, cortisone-sensitive cells which probably arise from the thymus cortex (Zatz and Lance, 1970).

All of these observations are in keeping with the recent identification of two subsets of T cells in the rat: a cortisone-sensitive subset, possibly in spleen red pulp, and a cortisone-resistant subset in spleen white pulp and lymph node paracortex (Goldschneider, 1975a,b). These T cell subsets are related antigenically to cortical and medullary thymocytes, respectively. Thus, it appears that two migration pathways exist between thymus and peripheral lymphoid tissues, one for cortical thymocytes and one for medullary thymocytes.

Histological studies indicate that thymocytes leave the thymus by passing between endothelial cell junctions of postcapillary venules in the region of the medulla and corticomedullary junctions (Sainte-Marie and Leblond, 1964; Clark, 1973). The emigrating cells must first cross the perivascular connective tissue spaces and pass between the cytoplasmic processes of thymus epithelial cells and macrophages before reaching these blood vessels. That this is a unidirectional pathway is shown by the fact that few, if any, peripheral lymphocytes can migrate to the thymus, except in the neonatal period when thymic blood vessels may be more permeable (Gowans and Knight, 1964; Goldschneider and McGregor, 1968b). Increased permeability of thymic blood vessels in the neonatal period may also explain the release of less mature thymocytes than occurs in the adult (Weissman, 1967; Colley et al., 1970a).

Migration of thymocytes into spleen red pulp appears to involve the simple passage of cells from the blood sinuses to the splenic cords via open junctions between the lining endothelial cells (Goldschneider and McGregor, 1968a; Ford, 1969). Migration of thymocytes into white pulp of spleen and paracortex of lymph nodes is a more complex process, which is described below.

5.2.1.3. Recirculation of T cells from blood to lymph via postcapillary venules in lymph nodes. The phenomenon of recirculation of lymphocytes (primarily T cells) from blood to lymph was described by Gowans and Knight (1964) as the resolution of an apparent paradox involving the generation of thoracic duct lymphocytes in mammals. The paradox consisted of the observations that enormous numbers of lymphocytes were constantly transferred from the thoracic duct lymph to the blood, yet the rate of lymphopoiesis in the tissues drained by the thoracic duct was grossly inadequate to account for more than a small fraction of these lymphocytes. Moreover, most of the lymphocytes in thoracic duct lymph were long-lived cells (up to 100 days in the rat and up to 10 years in man) (Everett et al., 1964; Buckton et al., 1967).

In a series of ingenious experiments involving thoracic duct drainage and intravenous transfusion of radioactively labelled lymphocytes to intermediate hosts, Gowans and Knight (1964) demonstrated that at least 60 percent of lymphocytes in thoracic duct lymph continually recirculate from blood to lymph by crossing the endothelium of specialized postcapillary venules in the paracortex of lymph nodes (Fig. 4). These venules can be readily identified in histological sections by the presence of prominent cuboidal to columnar endothelial cells, which have abundant pyroninophilic (basophilic) cytoplasm and vesicular, apically oriented nuclei. Autoradiographic studies of adoptively transferred thoracic duct lymphocytes showed a rapid and selective "homing" of lymphocytes to postcapillary venules in lymph nodes (Gowans and Knight, 1964;

Goldschneider and McGregor, 1968a). Within 10 minutes of intravenous injection, large numbers of non-dividing small lymphocytes had marginated along the apical surface of the modified endothelial cells; within 20 minutes, they had begun to traverse the walls of the postcapillary venules; and within 30 minutes they had gained entrance to the perivenular tissues in the lymph node paracortex, thus completing the afferent arc of the recirculation pathway.

Early ultrastructural studies by Marchesi and Gowans (1964) indicated that recirculating lymphocytes were engulfed by the specialized endothelial cells which line postcapillary venules in lymph nodes, a process known as emperipolesis. Serial electronmicrographs suggested that such lymphocytes were transported transcytoplasmically in membrane-bound vesicles to the basilar side of the endothelial cell, from where they were discharged into the paracortex of the lymph node. This interpretation has recently been challenged by Schoefl (1972), who claims that recirculating lymphocytes pass through specialized junctions between adjacent endothelial cells. These junctions purportedly seal behind the migrating lymphocyte, thereby creating the appearance in certain planes of section of a lymphocyte within a membrane-bound vesicle. In either event, it is clear that the passage of lymphocytes across postcapillary venules in lymph nodes is a highly selective process by which the great majority of T cells, including those which have been newly released from the thymus medulla, gain access to lymph nodes. Only during conditions of local inflammation in lymph nodes do other cellular elements of the blood enter by this route. Moreover, this appears to be a unidirectional migration pathway. Migration of lymphocytes from lymph node to blood via postcapillary venules has not been documented.

The mechanism(s) by which T cells recognize postcapillary venule endothelial cells is obscure. The most likely possibilities are cell–cell recognition involving complementary surface receptors and/or the release of chemotactic factors in the vicinity of the postcapillary venules. A clue to the possible existence of receptors on lymphocytes is the reversible inhibitory effect of neuraminidase treatment on homing of lymphocytes to lymph node (Gesner et al., 1969). However, this effect may have a trivial explanation, such as temporary alteration of cell surface charge or temporary non-specific trapping of altered cells in the reticuloendothelial system of liver and spleen. There is no evidence for or against chemotaxis as a mechanism by which recirculating lymphocytes are attracted to postcapillary venules in lymph node. Nonetheless, it must be noted that lymphocytes are highly motile cells which can be induced to engage in chemotactic movement in the presence of a soluble factor (lymphokine) that is produced by activated lymphocytes (Ward et al., 1971).

Whatever the mechanism of lymphocyte-endothelial cell interaction, it is clear that the postcapillary venule endothelial cell is more than a passive participant. This is evidenced by its specialized morphology and by its reaction to the migration of lymphocytes into lymph node (Goldschneider and McGregor, 1968a). Thus, in rats which have been deprived of recirculating lymphocytes by neonatal thymectomy or by chronic drainage from a thoracic duct fistula, the cytoplasm of postcapillary venule endothelial cells becomes atrophic and loses its basophilia. The endothelial cells in lymph nodes from congenitally athymic mice and human beings are also atrophic. Conversely, the atrophic endothelial cells in T cell-deprived rats become hypertrophic and intensely basophilic within 24 hours after hypertransfusion with recirculating lymphocytes. The significance of these morphological and metabolic changes in relation to the traffic of lymphocytes through or between endothelial cells is not known.

The efferent arc of the recirculation pathway of T cells involves the active migration of lymphocytes from paracortex to medulla of lymph node and thence into the medullary sinuses. From here they are passively carried by efferent lymphatic vessels to the thoracic duct, which empties into the great veins in the thorax. Having reached the blood, the lymphocytes are free to migrate randomly to lymph nodes, thereby completing the cycle of recirculation.

The process of T cell recirculation is rapid, random and efficient. It has been estimated that 10–15 percent of lymphocytes in blood entering a lymph node are removed during a single passage through the node (Hall et al., 1967). In rats, it is not unusual to begin to recover T cells from the thoracic duct lymph 2 hours after intravenous injection (Gowans and Knight, 1964; Goldschneider and McGregor, 1968b). Complete mixing of donor and host lymphocytes in blood, lymph and lymphoid tissues usually occurs within 24 hours of adoptive transfer (Everett et al., 1964; Gowans and Knight, 1964). However, several caveats are in order. First, not all T cells are recirculating cells. Populations of thymocytes and peripheral T cells have been described in the mouse which preferentially migrate to spleen, but not to lymph node (Zatz and Lance, 1970). These cells are not depleted from spleen by thoracic duct drainage. They are rapidly dividing and cortisone sensitive. Their counterparts in the rat possibly reside in spleen red pulp and may be derived from cortical thymocytes (Goldschneider, 1975b). There is some evidence that such cells in spleen may acquire the property to recirculate from blood to lymph after they undergo antigen-induced differentiation (Cantor, 1972; Stutman, 1975). Conversely, a population of T cells has been described which loses its capacity to recirculate after antigenic stimulation (McGregor et al., 1971). These are large, rapidly-dividing T cells which participate in cell-mediated immunological responses. They have the peculiar propensity to migrate into inflammatory exudates, something that recirculating T cells are loath to do (Koster et al., 1971; Asherson and Allwood, 1972). However, the small, long-lived progeny of these non-recirculating T cells may once again join the pool of recirculating lymphocytes. Thus, the ability of T cells to recirculate from blood to lymph via postcapillary venules in lymph node is not an immutable property; rather, it appears to be restricted to certain stages of T cell differentiation.

A second caveat is that not all T cells which are present in thoracic duct lymph have passed through lymph node via postcapillary venules. Some have migrated to lymph node from blood by way of afferent lymphatics which drain non-lymphoid tissues (see section 5.2.1.5). Others have been newly formed in lymph node paracortex, i.e. they are the progeny of recirculating T cells (Hall et al., 1967). Both of these processes are markedly accelerated by antigenic stimulation.

A third caveat is that T (and B) cell accumulation in lymph nodes may not always be random. This is shown by three observations: (1) large lymphocytes tend to return to the analogous group of lymph nodes from which they were taken when transferred into normal or germ-free syngeneic hosts; (2) lymphocytes from antigen-primed animals tend to home to the lymph nodes draining the site of injection of the same antigen in an intermediate host; (3) lymphocytes with receptors for a particular antigen are selectively removed by passage through a host which contains that antigen (Griscelli et al., 1969; Ford and Atkins, 1971; Sprent et al., 1971; Rowley et al., 1972; McWilliams et al., 1975). The common factor in the last two (and perhaps the first) situations is the arrest or trapping by antigen of antigen-reactive lymphocytes in lymphoid tissues (and presumably in non-lymphoid tissues also), where they are stimulated to undergo blast transformation and clonal proliferation. Hence, the afferent arc of the recirculation pathway appears to be random, whereas the efferent arc in antigenically stimulated lymphoid tissues does not. However, the antigen-specific trap that is thus created gives the appearance of selective migration of lymphocytes to certain groups of lymph nodes. A mechanism that could explain the concomitant non-specific recruitment, or trapping, of T cells that also occurs in antigen-stimulated lymph nodes is the release by activated T cells of lymphokines which attract or retard other T cells (Hall and Morris, 1965; Ford, 1969; Ward et al., 1971; Zatz and Lance, 1971).

5.2.1.4. Migration of recirculating T cells through white pulp of spleen. Approximately one-half of the pool of recirculating T cells in the rat migrates through the spleen every 18 hours (Ford, 1969). The afferent arc of this specific migration pathway has been described in radioautographic studies of parenterally injected thymocytes and thoracic duct lymphocytes (Goldschneider and McGregor, 1968a). Within 10 minutes, small, non-dividing lymphocytes become concentrated in the concentrically arranged sinusoids in the marginal zone of splenic white pulp. By 20 minutes large numbers of labelled cells have entered the tissues of the marginal zone by passing through the loose endothelial cell junctions of these sinusoids. By 30 minutes some of the lymphocytes have migrated across the marginal zone into the loose connective tissue sheath that surrounds the central arteriole of the white pulp. Accumulation of labelled lymphocytes in the periarteriolar lymphoid sheath reaches its maximum at about 3 hours after the initial infusion of cells, and decreases thereafter until an equilibrium is reached. The modal transit time of migrating T cells in the spleen is approximately 5–7 hours.

Rapidly dividing thymocytes and T cells are excluded from the white pulp of spleen. They are shunted instead into sinusoids in the red pulp, from which some enter the red pulp cords. B cells and non-lymphocytic elements of the blood are also excluded from the periarteriolar lymphoid sheath.

Unlike the situation in lymph nodes, the endothelial cells between which T cells migrate to enter white pulp of spleen are not unusual morphologically; nor is there evidence of intimate contact between endothelial cells and the emigrating lymphocytes. This suggests that a chemotactic stimulus may be the major attractant of T cells to white pulp of spleen. Other evidence that the mechanism of homing to white pulp of spleen differs from that to lymph node is seen in the differential effects of neuraminidase and trypsin treatment on cells migrating to these organs (Gesner et al., 1969). Whereas both neuraminidase and trypsin inhibit migration of thoracic duct lymphocytes to lymph node, only neuraminidase blocks the migration to spleen white pulp.

Despite these differences, the same population of recirculating T cells appears to be involved in the migration to both lymph node paracortex and spleen white pulp. Both lymph node T cells and spleen white pulp T cells can be rapidly depleted by drainage of lymphocytes from a thoracic duct fistula; and T cells from spleen are readily able to cross the postcapillary venules of lymph node. This raises the question of the route and mechanism by which T cells leave the white pulp of spleen. That they readily do so is indicated by the ability to quantitatively recover white pulp T cells from the splenic veins, as well as from the thoracic duct (Goldschneider and McGregor, 1968a,b; Ford, 1969). Since the route to these veins is through the red pulp, it would appear that T cells must leave the periarteriolar sheath by reverse migration across the marginal zone. Mitchell (1973) has suggested that this may occur via a specialized set of bridging channels. There is no clue as to the stimulus that initiates T cell emigration from the spleen white pulp.

5.2.1.5. Circulation of T cells from blood to lymph via afferent lymphatics of lymph nodes. Lymphocytes which emigrate from blood vessels into connective tissue spaces of non-lymphoid organs and into body cavities are returned to local lymph nodes by way of afferent lymphatic vessels that empty into the subcapsular sinus of the lymph node (Morris, 1968). From here, the lymphocytes migrate passively through connecting lymphatic channels or actively through the substance of the lymph node to reach the efferent lymphatics in the lymph node medulla. As elsewhere, lymphocytes which encounter antigen or an inflammatory exudate in non-lymphoid tissue frequently are arrested in their migration, but their progeny may subsequently return to lymph node by afferent lymphatics. It has been estimated that approximately 3 percent of lymphocytes in thoracic duct lymph follow this migration pathway under normal conditions (Yoffey and Drinker, 1939). However, under conditions of increased traffic of lymphocytes through non-lymphoid organs (e.g. during an inflammatory reaction) this percentage may increase substantially (Smith et al., 1970).

5.2.2. B cells

As discussed in section 4, B cells in peripheral lymphoid tissues are derived from the bursa of Fabricius or its equivalent in mammals. Migration of bursal lymphocytes, which had been radioactively labelled either in situ or in vitro, to peripheral lymphoid tissues has been clearly demonstrated in avian species (Linna et al., 1968; Durkin et al., 1972; DeKruyff et al., 1975). This migration is depicted schematically in Fig. 6. Demonstration of similar B cell migration in

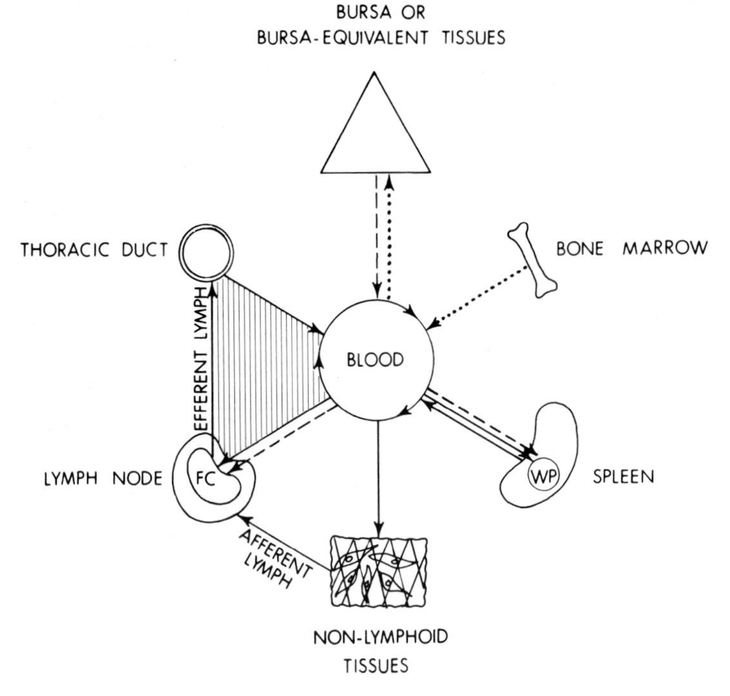

Fig. 6. Migration pathways of members of the B cell series. . . . . . , progenitors of bursa cells or their equivalent in mammals; ---, bursa cells or their equivalent; ————, memory B cells, WP, white pulp; FC, follicular cortex; hatched area, major pathway of B cell recirculation from blood to lymph. See text section 5.2.2 for details.

mammals has been hampered by the lack of a well defined bursal analogue. However, newly formed mouse bone marrow B lymphocytes or their immediate precursors have been reported to migrate to the spleen when injected intravenously into normal or lethally irradiated recipients (Lafleur et al., 1972; Rysser and Vassali, 1974). Nieuwenhuis and Keuning (1974) observed in rabbits that follicular structures in peripheral lymphoid tissues, which had been destroyed by X-irradiation, could be regenerated by cells migrating from bone marrow. In germ-free animals, the only apparent B cell-specific structures in peripheral lymphoid tissues are the primary lymphoid follicles (Nieuwenhuis, 1971). Hence, these structures probably develop independently of antigen exposure and they would seem to be populated by cells which are derived directly from bursa or bursa-equivalent tissues.

Weissman (1975) has observed that cell proliferation in primary follicles is rare; and, he suggests that population of primary follicles occurs via immigrant lymphocytes. Thus, it is significant that mature B cells from avian and mammalian species exhibit the ability to preferentially home to B cell-specific (thymus-independent) regions within peripheral lymphoid tissues (Parrott and DeSousa, 1971; Howard et al., 1972; Gutman and Weissman, 1973; Sprent, 1973;

DeKruyff et al., 1975; Durkin et al., 1975). Within one hour of intravenous injection, radioactively labelled peripheral B cells are initially detected around the postcapillary venules in the paracortex of lymph nodes and in the marginal sinus of white pulp of spleen. By 9–24 hours after injection, labelled B cells are found predominantly in primary follicles and in the lymphocyte corona or mantle of secondary lymphoid follicles (Gutman and Weissman, 1973; Durkin et al., 1975).

It has been suggested that cell surface characteristics determine specific homing properties of lymphocytes (Woodruff and Gesner, 1969; Parrott and DeSousa, 1971; Sprent, 1973; Durkin et al., 1975). Pretreatment of lymphocytes with agents which modify the cell surface, e.g., neuraminidase, does alter their migration patterns (Woodruff and Gesner, 1969; Gilette et al., 1972; Taub et al., 1972). Moreover, preincubation of B cells with anti-immunoglobulin impairs specific B cell homing, which suggests that surface immunoglobulin may play a role in B cell localization within peripheral lymphoid tissues (DeKruyff, Theis and Thorbecke, 1975; Durkin et al., 1975). However, cell-surface-modification experiments are difficult to interpret due to changes which may cause non-specific trapping of migrating lymphocytes by the reticulo-endothelial system.

Experiments in rats involving thoracic duct drainage have indicated that the majority of B cells in lymphoid follicles do not recirculate from blood to lymph (Goldschneider and McGregor, 1968a). Recently a population of B cells has been identified in the thoracic duct of mice (Sprent, 1973) and rats (Howard, 1972) which *does* recirculate, albeit, much more slowly than T cells (Howard, 1972; Sprent, 1973). Autoradiographic analysis of sections of peripheral lymphoid tissues revealed that these recirculating B cells traffic through the lymphocyte mantle or corona of secondary lymphoid follicles (Howard et al., 1972).

It is significant that many of the B cells which recirculate from blood to lymph in the rat have been identified as memory B cells (Strober, 1972; Strober and Dilley, 1973a,b). Intravenous injection of antigen-primed rat spleen cells into syngeneic hosts followed by thoracic duct drainage revealed that only memory B cells were capable of recirculating from blood to lymph; unprimed or virgin B cells did not exhibit the recirculation capability. Moreover, these memory B cells are relatively long-lived as compared to virgin B cells.

It has been suggested that memory B cells are derived from germinal centers following antigen stimulation (Wakefield and Thorbecke, 1968; Nieuwenhuis, 1971; Buerki et al., 1974; Grobler et al., 1974; Nieuwenuis and Keuning, 1974). Studies of lymphoid tissue regeneration in sublethally irradiated rabbits suggest that antigen-induced germinal centers may give rise to a migratory population(s) of B cells (Nieuwenhuis and Keuning, 1974; Nieuwenhuis et al., 1974). These authors believe that the germinal center-derived B cells migrate from the germinal centers through the lymphocyte corona of the secondary follicle to populate the surrounding marginal zone in the spleen and lymph nodes (marginal zone in lymph nodes = the 3–10 outer layers of cells in lymphoid follicles (van der Broeck, 1971)). These "marginal zone" cells might then become part of the memory B cell population which recirculates through the lymphocyte corona and marginal zone of lymphoid follicles in peripheral lymphoid tissues.

5.3. *Surface properties of lymphocytes*

5.3.1. *Lymphocyte-specific cell surface antigens*

A major advance in the study of T cell ontogeny and to a lesser extent of B cell ontogeny has been the detection of lymphocyte-specific cell surface antigens. Some of these antigens occur in multiple allelic forms (alloantigens) and can be detected by antisera (alloantisera) prepared in different inbred strains of the same animal species. This approach has generally been used in the mouse. Other lymphocyte-specific antigens (heteroantigens) which do not or are not known to exist in allelic forms can be detected by antisera (heteroantisera) prepared in members of different animal species. This approach has generally been used in the rat, rabbit, guinea pig, chicken and man. It is likely that some of these putative heteroantigens will ultimately prove to be alloantigens. In any event, both lymphocyte-specific alloantigens and heteroantigens are examples of "differentiation antigens," i.e. they are the products of selective gene expression which are restricted to a given cell type and which appear (and disappear) at defined stages in the development of that cell type (Boyse and Bennett, 1974). As such, they promise to provide important insights into the genetic and molecular events that underlie the biosynthesis, topographical display, and movement of molecules that are inserted into the plasma membrane. These topics have been reviewed elsewhere (Boyse and Bennett, 1974; Edidin, 1974; Singer, 1974). The emphasis here will be on the use of lymphocyte-specific antigens as markers to trace the major developmental stages of lymphocyte subpopulations.

5.3.1.1. T cell alloantigens. None of the T cell-specific alloantigens described so far appear to be present on the surface of lymphopoietic progenitor cells as tested by complement-mediated cytotoxicity. All are present in highest concentration on cortical thymocytes and in lower (or undetectable) concentrations on medullary thymocytes and peripheral T cells (Table 4). Their appearance on the surface of thymocyte progenitors can be induced within 2 hours, without the need for cell division, by thymosin and other substances that increase intracellular cAMP (see section 4.1.3). This may explain the sudden appearance of these alloantigens on the surface of cells in the mouse thymus on the 16th day of fetal life, the day on which differentiation of thymus epithelial cells and thymocytes is first observed (see section 4.1.2). That the thymus is normally responsible for inducing the expression of these antigens on T cell progenitors is shown by the failure of lymphocytes to display these antigens in congenitally thymusless or neonatally thymectomized mice. However, the continued expression of these antigens on recirculating T cells is not dependent on the continued presence of the thymus or on thymus humoral factors.

1. TL antigens. The TL (thymus-leukemia) antigen system of the mouse consists of at least 4 antigens (Tla.1, Tla.2, Tla.3 and Tla.4) which are coded by a complex locus adjacent to the D end of the major histocompatibility antigen complex on chromosome 17 (Boyse et al., 1968a). They are glycoproteins with molecular weights of 40,000 to 50,000 daltons (Muramatsu et al., 1973). The TL

Table 4
Distribution of alloantigens among members of the T lymphocyte series in the mouse.

	Relative concentration of alloantigens on cell surface[a]			
Antigen[b]	thymocyte progenitors	cortical thymocytes	medullary thymocytes	peripheral T cells[c]
H-2	+ + +	+	+ +	+ + +
TL	−	+ + +	−	−
Thy-1 (θ)	−	+ + +	+ +	+
Ly-1	−	+ + +	+ +	+
Ly-2/Ly-3	−	+ + +	+ +	+
Ly-5	−	+ + +	?	+

[a]Expressed in arbitrary units: (+ + +), high concentration; (+ +), intermediate concentration; (+), low concentration; (−), undetectable.
[b]See text section 5.3.1 for description of antigens and for references.
[c]Average values for unfractionated T cell populations. However, there is considerable qualitative and quantitative variation in the expression of these antigens among T cell subsets (see section 5.3.1 and 5.4.1).

antigens are normally present only on cortical thymocytes and are expressed in one of the following phenotypic displays: Tla.1, .2, .3; Tla.2; or Tla-negative. However, they may be expressed in one of the following abnormal phenotypes in T cell leukemias: Tla.1, .2, and Tla.1, .2, .4. In addition, TL-positive leukemia cells may appear in animals whose thymocytes are TL-negative (Boyse and Old, 1969). These observations suggest that the Tla complex consists of structural genes and regulator genes, the latter determining the expression (or non-expression) of the former. Moreover, in the normal host, derepression of the structural genes by the regulator genes is dependent upon the microenvironment of the thymus cortex, inasmuch as medullary thymocytes and peripheral T cells from TL-positive animals are invariably TL-negative.

Another interesting phenomenon involving TL antigens is that of antigenic modulation (Boyse et al., 1967). In this process TL-positive leukemia cells and TL-positive thymocytes become TL-negative when exposed in vivo or in vitro to anti-TL antigen serum. While this is an energy-dependent process (Old et al., 1968) it does not require cap formation, since it can also be induced by monovalent antibody fragments (Lamm et al., 1968). Rather it would appear to occur by pinocytosis of the antigen-antibody complexes, or by shedding of these complexes from the cell surface (Yu and Cohen, 1974). In animals which express more than one TL antigen, antiserum to any one of the antigens will usually cause modulation of the other antigens, suggesting that they are physically linked or at least closely approximated on the cell surface (Old et al., 1968). Antigenic modulation is a reversible process; removal of the source of antibody causes the cells to rapidly become TL-positive again.

The phenotypic expression of TL antigens effects the phenotypic expression of histocompatibility antigens, both in the normal state and during antibody-induced modulation of TL antigens. Thus, in homozygous TL-positive mice, concentrations of H-2D antigens are reduced approximately 50%, and in heterozygous TL-positive mice about 25% below levels in TL-negative mice (Boyse et al., 1968b). Conversely, increased expression of H-2D antigens accompanies the decreased concentration of TL antigens during antibody-induced modulation of the latter (Old et al., 1968). The mechanism underlying the reciprocal expression of TL and H-2D antigens is not known, although partial masking of the H-2D antigens by closely adjacent TL antigens would be the simplest explanation.

The suggestion has been made that the Tla locus may represent the integrated genome of a C type RNA leukemia virus whose expression is dependent upon certain genetic events in early T cell differentiation (Boyse et al., 1972). This suspicion is based on the similarly restricted distribution of the G_{IX} antigen to cortical thymocytes in mice which both harbor (but do not produce) the Gross leukemia virus genome and have the regulatory genes for the expression of the G_{IX} antigen (Stockert et al., 1971). All other lymphocytes as well as non-lymphocyte cell types are G_{IX}-negative unless they are induced to produce murine leukemia virus, whereupon they become G_{IX}-positive (even in the rat). As in the case of the TL antigens, one of the loci for the G_{IX} antigen is on the 9th linkage group (chromosome 17). Also, like TL antigens, the G_{IX} antigen can appear in leukemic cells of both G_{IX}-positive and G_{IX}-negative animals. Very recent biochemical studies indicate that the G_{IX} antigen is identical with the major envelope glycoprotein of the murine leukemia virus (Tung et al., 1975).

2. Ly antigens. Four sets of lymphocyte-specific (Ly) alloantigens have been identified on mouse thymocytes and peripheral T cells, but not on B cells (Boyse et al., 1968c, 1971b; Komuro et al., 1974). Each of these antigens is present in highest concentration on cortical thymocytes, in lower concentrations on medullary thymocytes, and in lowest concentrations on peripheral T cells (Aoki et al., 1969; Konda, et al., 1973; Schlesinger et al., 1973). The Ly antigens are not present on the progenitors of thymocytes (Schlesinger and Hurvitz, 1968b; Boyse and Old, 1969; Owen and Raff, 1970).

The Ly-1 locus, which codes for antigens Ly 1.1 and Ly 1.2, is present in linkage group XII (chromosome 19). The Ly-2 and Ly-3 loci, which code for antigens Ly 2.1, Ly 2.2 and Ly 3.1, Ly 3.2, respectively, are each located in linkage group XI (chromosome 6). Linkage studies have indicated that the Ly-2 and Ly-3 loci are closely linked or possibly identical loci. Serological studies indicate that both antigens are in close proximity on the cell surface. The location of the Ly-5 locus, which codes for antigens Ly 5.1 and Ly 5.2, is not yet known, but it is not linked with any of the other known T cell-specific alloantigens. None of the Ly alloantigens appears to be an histocompatibility antigen, although each may be linked to minor histocompatibility antigens.

Recently, both quantitative and qualitative differences in expression of Ly antigens has been found among functionally distinct subsets of T cells in mouse spleen (see section 5.4.1. for details). This suggests that these antigens may play a role in determining the types of immunological reactions in which a T cell may engage. In addition, the genetic locus for a specific peptide marker in the immunoglobulin light chain variable region has been found to be linked to or identical with the locus for expression of the Ly-3 antigen (Gottlieb, 1974).

3. Thy-1 antigens. The Thy-1 (theta; θ) antigen system is probably the best known of the "T cell-specific" alloantigens (Rief and Allen, 1964). It is not truly T cell specific, since it is also present on epidermal cells and in nervous tissues. However, it is not present on B cells, macrophages or any of the other formed blood elements. Like the Ly alloantigens, Thy-1 antigen occurs in two allelic forms (Thy-1.1 and Thy-1.2), is absent from thymocyte progenitor cells and is present in highest concentrations on cortical thymocytes and in progressively lower concentrations on medullary thymocytes and peripheral T cells. Also like the Ly antigens, the Th-1 antigen shows distinct quantitative differences in expression among T cell subsets (see section 5.4.1).

The Thy-1 locus is located in linkage group II (chromosome 9) of the mouse (Blankenhorn and Douglas, 1972; Itakura et al., 1972). An antigen which has a tissue distribution similar to that of the Thy-1 antigens and which cross-reacts serologically with the Thy-1.1 antigen has been described in the rat (Douglas, 1972; Peter et al., 1973; Acton et al., 1974). It is not known if the T cell-specific alloantigen described by Lubaroff (1973) in the rat is related to Thy-1 or any of the other T cell alloantigens in the mouse.

5.3.1.2. T cell heteroantigens. The major advantage of using heteroantisera to study cellular differentiation is the ability, at least theoretically, to raise antibodies against differentiation antigens on any cell type from any strain or individual in any species (cf. Goldschneider and Moscona, 1972). It is not surprising therefore that a variety of cell surface heteroantigens that are specific for members of the T cell series have been described in the rat and in other animal species, including man (Table 5). Although heteroantigens do not lend themselves to the types of genetic analysis possible with alloantigens, they have

Table 5
Distribution of heteroantigens among members of the T lymphocyte series in the rat.

	Anatomical location of positive lymphocytes[a]					
Antigen[b]	bone marrow	thymus cortex	thymus medulla	lymph node paracortex	spleen white pulp	spleen red pulp
RLTA	+	+	±	−	−	−
RTA	−	+	+	−	−	−
RTLA	−	+	+	+	+	+
RBMLA, RCTA	+	+	−	−		+
RMTA	−	−	+	+	+	−

[a](+), antigen positive; (−), antigen negative; (±), weakly antigen positive.
[b]See text section 5.3.1 for description of antigens and for references.

proved to be extremely useful markers for tracing the ontogeny of T cells. This is due to the fact that the expression of different heteroantigens has tended to be qualitatively restricted to different stages of T cell differentiation, thereby forming partially overlapping antigenic patterns which delineate probable developmental pathways. The major drawback to the widespread use of heteroantisera to study T cell ontogeny is the need for extensive cross-absorption to remove antibodies to species-specific and other irrelevant antigens. However, the unique reactivities of these antisera more than justifies the effort.

1. T lymphocyte antigens (TLA). Antigens specific for the surfaces of all thymocytes and peripheral T cells have been identified in the rat, mouse, chicken, guinea pig, rabbit and man (Shigeno et al., 1968; Forget et al., 1970; McArthur et al., 1971; Malchow et al., 1972; Shevach et al., 1972a; Aiuti and Wigzell, 1973; Goldschneider and McGregor, 1973). Antisera to such antigens are generally prepared against thymocytes or peripheral lymphocytes and absorbed with a variety of non-T cells including homologous erythrocytes, macrophages and normal or neoplastic B cells. It is not known whether TLA is a single antigen or several antigens, or whether TLA from one animal species cross-reacts serologically with TLA from other species. Where it has been sought TLA has not been found on thymocyte progenitors; neither has an antigen that is restricted to peripheral T cells been described.

2. Thymocyte-specific antigens (TA). Antigens specific for the surface of thymocytes, but not peripheral T cells or thymocyte progenitor cells, have been identified in the rat and the cow (Potworowski and Nairn, 1967; Colley, 1970a; Williams et al., 1971). One such antigen has been characterized as a homogenous protein (Bachvaroff et al., 1969). Antisera to thymocyte-specific antigens are usually prepared against normal thymocytes and absorbed with peripheral lymphocytes, as well as other cell types. Both cortical and medullary thymocytes are TA-positive.

At least two thymocyte-specific antigens exist in the rat, one of which is expressed during fetal and neonatal life only (Potworowski and Nairn, 1967), and one of which is expressed during both fetal and adult life (Colley et al., 1970a). These findings suggest that either the macro- and/or micro-environmental factors which influence thymocyte development differ in fetal and postnatal life or that the thymocyte progenitors themselves differ during these two periods. While the former possibility is undoubtedly true, the latter possibility also receives support from the different physical properties and proliferative potentialities or lympho-hemopoietic stem cells in fetal and adult life (see sections 3 and 4.1). Another possibility that is suggested by these observations is that the T cells which are formed during fetal and postnatal life may differ functionally. While there is no direct evidence to confirm or deny this speculation, it does appear that thymocytes are released to the peripheral lymphoid tissues in a less mature state in the fetus than in the adult (Weissman, 1967). Thus, numerous cells bearing thymus-specific antigens are present in peripheral lymphoid tissues during late fetal life (Colley et al., 1970b). Such antigens are not detected on thymus-derived lymphocytes in postnatal life, not even on newly released cells in lymphatics and venules which drain the thymus (Williams et al., 1971).

3. Rat masked thymocyte antigen (RMTA). This antigen is present in a masked form on medullary thymocytes and in an unmasked form on a subset of peripheral T cells (Goldschneider, 1975a). It is not present on cortical thymocytes, on thymocyte progenitor cells, or on a second subset of peripheral T cells. The RMTA can be unmasked by treatment of medullary thymocytes with neuraminidase. Treatment with trypsin, on the other hand, removes or destroys RMTA. Antiserum to RMTA is prepared against thoracic duct lymphocytes and absorbed with syngeneic erythrocytes, bone marrow cells and thymocytes to achieve specificity.

The fact that RMTA is found in a masked form on medullary thymocytes and in an unmasked form on a subset of peripheral T cells suggests that: (a) medullary thymocytes are the immediate precursors of this subset of T cells; and (b) unmasking of RMTA occurs either immediately before or immediately after medullary thymocytes are released to the periphery. A possible corollary of (b) is that unmasking of RMTA is necessary for the emigration of medullary thymocytes from the thymus, for the homing of these cells to peripheral lymphoid tissues, or for some other functional attribute of mature T cells. It is of interest in this regard, that RMTA becomes unmasked spontaneously after several days incubation of medullary thymocytes in vitro (Goldschneider, unpublished observation). Moreover, it has been observed, that the glycocalyx of medullary thymocytes is significantly thicker than that of peripheral T cells (Santer et al., 1973), raising the possibility that remodeling of the glycocalyx is involved in the unmasking process in vivo.

The subset of T cells which bears the rat masked thymocyte antigen is located in the paracortex of lymph node and the periarteriolar lymphoid sheath of spleen white pulp. It belongs to the pool of long-lived, immunologically competent T cells which continually recirculates from blood to lymph. A similar subset of T cells ("T_2" cells) has been described in the mouse (Cantor, 1972).

4. Mouse peripheral lymphocyte antigen (MPLA). This is an incompletely described antigen that appears to be restricted in distribution to mouse medullary thymocytes and peripheral T cells (Raff and Cantor, 1971). Unlike the rat masked thymocyte antigen, MPLA is not present in a masked form on medullary thymocytes, nor is it known whether it is restricted to recirculating T cells.

5. Rat bone marrow lymphocyte antigen (RBMLA). This antigen is present on thymocyte progenitors in bone marrow, on cortical thymocytes and on a subset of peripheral T cells (Goldschneider, 1975b). It is also present on hemopoietic stem cells and progenitors of B cells. The RBMLA is not present on medullary thymocytes, on the subset of cells which bears the rat masked thymocyte antigen, on B cells, or on mature hemopoietic cells.

Antiserum to RBMLA is prepared against a "null" population of bone marrow lymphocytes and is absorbed with syngeneic erythrocytes and peritoneal exudate cells. The "null" population of bone marrow lymphocytes (or "lymphocyte-like" cells) comprises approximately 40% of bone marrow lymphocytes, and lacks T cell-specific and B cell-specific surface antigens normally found on lymph node lymphocytes (Goldschneider and McGregor, 1973). As mentioned previously in section 3, the "null" population of bone marrow lymphocytes is analogous to the "lymphocyte-transitional" cells of Yoffey (1973) and appears to contain most if not all of the lympho-hemopoietic stem cells and progenitor cells in adult bone marrow.

The subset of T cells which bears the rat bone marrow lymphocyte antigen appears to reside in red pulp of spleen (Goldschneider, 1975b). Unlike medullary thymocytes and unlike T cells in lymph node and white pulp of spleen, the RBMLA-positive T cells are sensitive to cortisone. Although nothing is known about their biological functions, it is possible that RBMLA-positive T cells in the rat are equivalent to the T_1 cell subset in the mouse (Cantor, 1972) or with cortisone-sensitive suppressor T cells (Gershon et al., 1974; Cohen and Gershon, 1975) (see section 5.4.1).

The presence of the RBMLA on cortical thymocytes and on a subset of peripheral T cells suggests that cortical thymocytes are the immediate precursors of these T cells. This view is strengthened by the finding that medullary thymocytes and a second subset of T cells lack the bone marrow lymphocyte antigen and instead bear the masked thymocyte antigen. It is also supported by the observation that massive numbers of short-lived, cortisone-sensitive thymocytes (presumably cortical thymocytes) are continually released from the thymus, and that these cells selectively migrate to red pulp of spleen (see section 5.2.1.2). Thus it would appear that at least two lines of T cells exist in the rat, RBMLA-positive T cells derived from cortical thymocytes and RMTA-positive T cells derived from medullary thymocytes.

6. Rat cortical thymocyte antigen (RCTA). This antigen has a distribution similar to that of the rat bone marrow lymphocyte antigen, and may in fact be identical to RBMLA (Gregoire, Barton and Goldschneider, work in progress). However, it is detected by antiserum prepared against purified populations of cortical thymocytes rather than against the "null" population of bone marrow lymphocytes. It is possible, therefore, that RCTA will prove to have a more restricted range of distribution among bone marrow lymphocytes than does RBMLA, possibly only being present on lymphopoietic stem cells or progenitors of cortical thymocytes.

7. Rat low electrophoretic mobility thymocyte antigen (RLTA). This antigen is present in high concentrations on a "null" subset of bone marrow lymphocytes and on cortical thymocytes; in low concentration on medullary thymocytes; and probably not at all on peripheral T cells (Zeiller and Dolan, 1972). Antiserum to RLTA is prepared against a population of thymocytes which has been separated on the basis of its low electrophoretic mobility. Most of these cells are cortical thymocytes, but some may be medullary thymocytes. It is not clear what relationship, if any, RLTA bears to the rat cortical thymocyte antigen and the rat bone marrow lymphocyte antigen discussed above.

5.3.1.3. B cell alloantigens

1. Plasma cell antigen (PC-1). The PC-1 antigen is present on antibody-secreting cells (both normal and neoplastic plasma cells) of some strains of mice (Takahashi et al., 1970). PC-1 is also found in brain, liver and kidney tissue but it is not detectable in normal, unstimulated T cells, B cells and leukemic cells. The antigen is specified by a single mendelian dominant gene locus, Pca (plasma cell antigen), which is not closely linked with the H-2 region. The strain distribution of PC-1 does not parallel that of other cell surface alloantigens, which suggests that it is a distinct surface alloantigen.

2. Ly-4 antigen. The Ly-4 gene locus, originally described by Snell et al. (1973), is the only locus of the Ly series which codes for a B cell-specific alloantigen in the mouse (McKenzie and Snell, 1975; McKenzie, 1975). Although the Ly-4 locus has not been mapped, Snell et al. (1973) state that it is distinct from other loci which code for lymphocyte-specific cell membrane alloantigens. Only one antigenic specificity has been identified, Ly-4.2. This antigen is found on B cells (both virgin and memory) and on their antibody-secreting progeny in some strains of mice. However, few bone marrow lymphocytes possess Ly-4.2 determinants, which suggests that the antigen is not present on B cell progenitors. The Ly-4 antigen is not detectable on normal T cells or leukemias of T cell origin.

3. β antigen. This antigen was originally detected by the presence of B cell-specific antibodies in certain anti-H-2 antisera (Sachs and Cone, 1973). Subsequent immunizations between appropriate congenic strains of mice produced monospecific alloantisera to the β antigen. Genetic data indicate that the β antigen is determined by gene(s) in or to the left of the Ir-1 region of the H-2 complex (see section 5.3.2). The strain distribution and genetic location of the β antigen suggest that it is distinct from other known B cell-specific antigens (Sachs and Cone, 1973). In light of its link with the H-2 gene complex, it is possible that the β antigen may be associated with certain functions of the immune system (Sachs and Cone, 1973). Similar suggestions have been made for the I-region associated antigens (see below).

4. I-region associated (Ia) antigens. These cell membrane alloantigens are genetically determined by the I region of the H-2 complex in mice (see section 5.3.2). Reciprocal immunizations between congenic strains of mice which differ in the I region have enabled ten Ia antigenic specificities to be identified (Shreffler and David, 1975). Most of the anti-Ia antisera thus far studied detect primarily B cell-specific Ia antigens (Hämmerling et al., 1974; Hauptfield et al., 1974; Unanue et al., 1974; Delovitch and McDevitt, 1975). However, some anti-Ia sera also display reactivity to T cells (Götz, Reisfeld and Klein, 1973; Frelinger et al., 1974). Few bone marrow cells possess Ia antigens, which suggests that these antigens are not present on B cell progenitors (Hämmerling et al., 1974b). Also, Ia antigens have not been detected on a variety of murine tumor cells (Cullen et al., 1974; Delovitch and McDevitt, 1975).

The location of Ia genes in the major histocompatibility complex suggests they may be associated with immunological functions. Hämmerling et al. (1974a,b) postulate that the Ia antigens are involved in T and B cell collaboration. It is of interest that anti-Ia sera have been reported to bind to "factors" which purportedly mediate T and B cell cooperation (Amerding et al., 1974; Munro et al., 1974). Moreover Ia antisera have been shown to inhibit binding of immunoglobulin to Fc receptors on B cells. These latter receptors have also been suggested to mediate T-B cell interaction in the immune response (Dickler and Sachs, 1974) (also see section 5.3.3.2).

5.3.1.4. B cell heteroantigens

1. Immunoglobulin. Cell surface immunoglobulin (Ig) is the most commonly used antigenic marker to distinguish B and T cells (see Table 1). B cells generally bear large amounts of readily detectable Ig (10^4–10^5 molecules/cell) whereas T cells do not. However, Ig may be detected on T cells under certain circumstances (see section 5.3.3.1). The most common immunoglobulin class on B cells is IgM (monomeric), although IgG, IgA, IgD and IgE have also been identified. The biological function of cell surface immunoglobulin molecules as antigen receptors will be discussed in section 5.3.3.1.

2. Mouse-specific B lymphocyte antigen (MBLA). This is a species-specific surface antigen found on cells at all stages in the B cell series, including B cell progenitors in bone marrow and antibody-secreting plasma cells (Raff et al., 1971). Rysser and Vassalli (1974) have reported that B cell progenitors in mouse bone marrow acquire the MBLA prior to the development of detectable cell surface immunoglobulin. This indicates that commitment to B cell differentiation occurs early in lymphopoietic development. Antisera defining the MBLA were prepared in rabbits injected with lymph node cells from "B" mice. These are animals which have been reconstituted with B cell progenitors from fetal liver (Raff et al., 1971; Niederhuber and Möller, 1972). The ML2 heteroantigen described by Stout et al. (1975) has a similar distribution to the MBLA.

Another B cell-specific heteroantigen has been demonstrated by Zeiller and Pascher (1973) on an electrophoretically purified population of mouse spleen B cells. Antisera raised in rabbits to this B cell population reacted only with normal, resting B cells and plasma cells.

3. Rat-specific B cell antigen (RBLA). Antisera defining this antigen were prepared in rabbits against thoracic duct B cells from normal rats (Goldschneider and McGregor, 1973) or from rats which had been thymectomized, lethally irradiated and bone marrow reconstituted (Howard and Scott, 1974). Unlike the MBLA, the rat B cell antigen is present predominantly on mature, small B cells in lymphoid follicles and not on B cell precursors, germinal center cells, or plasma cells. It is possible that RBLA detects the population of recirculating memory B cells described earlier in section 5.2.2.

4. Th-B antigen. The Th-B antigen, described by Stout et al. (1975) in the mouse is shared by cortical thymocytes and B lymphocytes. The surface density of Th-B determinants is higher in 3-week-old mice than in 15-week-old mice. Also, the density of determinants is greater on bone marrow lymphocytes than splenic or lymph node B cells in 15-week-old mice. These observations led Stout et al. (1975) to suggest that Th-B is present predominantly on immature or precursor B cells. Moreover, the presence of Th-B on cortical thymocytes suggests that T and B cell progenitors share some cell surface determinants or that Th-B is present on a common progenitor of T and B cells. This is reminiscent of the rat bone marrow lymphocyte antigen (Goldschneider, 1975b) which is present on B and T cell precursors and cortical thymocytes (see section 5.3.1.2). Antisera defining the Th-B antigen were prepared in rabbits immunized with MOPC104E mouse plasmacytoma cells. The presence of the Th-B antigen on normal, murine plasma cells has not been reported.

5. Mouse-specific plasma cell antigen (MSPCA). This cell surface heteroantigen is present on antibody-secreting cells (both normal and neoplastic) but not on normal peripheral B cells and T cells (Takahashi et al., 1971). The MSPCA appears to be distinct from the PC-1 alloantigen. Brain, liver and kidney tissues are negative for the MSPCA but positive for PC-1. Heteroantisera specific for MSPCA were raised in rabbits against mouse myeloma MPC-67; similar results were obtained using MOPC104E (Takahashi et al., 1971).

5.3.2. Histocompatibility antigens

Histocompatibility antigens may be operationally defined as cell membrane alloantigens which determine compatibility or incompatibility of tissue grafts between members of the same species. More than 40 discrete genetic loci which code for histocompatibility antigens have been identified in the mouse. Of these, the most important is the major histocompatibility complex, the H-2 complex, which is part of the IXth linkage group (chromosome 17) (Table 6). Homologous major histocompatibility antigen loci have been found in man (HL-A complex) (Thorsby, 1974), and in other mammalian species (Ivanyi, 1970).

The genetics of the H-2 complex in the mouse is the best known of all surface membrane antigens in cell biology (see Klein, 1975 and Shreffler and David, 1975 for comprehensive reviews). The H-2 complex is extremely polymorphic and probably is comprised of several hundred genetic loci. Biochemical and structural studies on isolated H-2 antigens indicate that they are glycoproteins composed of approximately 90 percent protein and 10 percent carbohydrate. The antigenic determinants appear to reside in the protein moiety (Nathenson and Cullen, 1974). Using papain-solubilized human histocompatibility antigens, Strominger et al. (1974) have suggested that the basic structure of histocompatibility antigens is similar to that of the constant regions of immunoglobulin molecules. Thus, each antigenic unit is a dimer consisting of paired heavy and light chains. Both heavy and light chains have one intrachain disulfide bridge for each 11,000 to 12,000 MW of peptide. The light chains (MW 12,000 each) are similar or identical to β_2-microglobulin, a homologue of immunoglobulin constant region domains (see section 5.3.3.1). The heavy chains (MW 34,000 each) are linked by disulfide bridges. However, unlike immunoglobulin

Table 6
Products and functions of the major histocompatibility complex (H-2 complex) of the mouse.

	The major histocompatibility complex of the mouse[a]						
Marker loci:	H-2K	Ir-1A	Ir-1B	Ir-1C	Ss	H-2G	H-2D
Regions and subregions:	K	IA	IB	IC	S	G	D
		I (IA–IC)					
Products (major functions)							
1. Serologically-defined (SD) antigens (target antigens for cytotoxic T cells and humoral antibodies)	++	–	–	–	–	±	++
2. Lymphocyte-defined (LD) antigens (stimulating antigens in MLR and GVHR and in formation of cytotoxic T cells)	–	++	+	±	–	–	–
3. Immune response (Ir) gene products (control of immunological responses; possible antigen recognition sites on T cells)	–	++	+	±	–	–	–
4. I-region associated (Ia) antigens (may include LD antigens and Ir gene products, plus Fc receptors and receptors for T cell-B cell-macrophage interactions)	–	++	+	±	–	–	–
5. Control of serum complement levels	–	–	–	–	++	–	–

[a]The relative positions of the regions of the H-2 complex, which is a part of the IXth linkage group (chromosome 17), are indicated schematically. See text section 5.3.2 for details.

molecules, the light and heavy chains of histocompatibility antigens are associated by non-covalent bonds rather than by disulfide bridges.

Histocompatibility antigens are not unique to T or B cells, or for that matter to lymphocytes. They are expressed to a greater or lesser degree on all cell types (Edidin, 1972). However, the major histocompatibility complex and its products appear to play a crucially important role in all major aspects of T cell, and perhaps B cell, behavior (Table 6). Thus, products of the H-2 complex are responsible for stimulating allograft rejection, mixed lymphocyte reactions (MLR) and graft-versus-host reactions (GVHR) (Bach et al., 1972; Klein and Park, 1973; McKenzie and Snell, 1973; Klein et al., 1974). T cells are exceptionally reactive to histocompatibility as compared to non-histocompatibility antigens (Nisbet et al., 1969; Wilson and Nowell, 1970). The H-2 complex is also involved in the control of immunological responsiveness to a variety of antigens (Benacerraf and McDevitt, 1972), susceptibility to certain viral infections (Snell, 1968; Lilly, 1971), and in the cooperative interactions of T cells, B cells and macrophages (Katz et al., 1973; Rosenthal and Shevach, 1973). This implies that some of the products of the H-2 complex are involved in recognition phenomena. In addition, the H-2 complex controls the production of complement components, which function as humoral mediators of inflammation (Ferreira and Nussenzweig, 1975); and may control the expression of Fc receptors for cytophilic antibodies, and receptors for complement on B cells (Dickler and Sachs, 1974; Gelfand et al., 1974b) (also, see section 5.3.3.2). Finally, it has been postulated that the major histocompatibility antigens may aid in the somatic generation of immunological diversity by exerting selective pressure against self-reactive clones of developing lymphocytes and in favor of non-self-reactive clones (Jerne, 1971).

At least two and probably three classes of major histocompatibility complex gene products are displayed on the lymphocyte cell surface (Table 6): (1) serologically defined (SD) histocompatibility antigens against which humoral antibodies and cytotoxic ("killer") T cells react in graft rejection; (2) lymphocyte-defined (LD) antigens which stimulate T cells to differentiate into cytotoxic cells, and which enable T cells to cooperate with B cells in antibody responses; and (3) immune response (Ir) gene products, which control the recognition of many antigens (see below and section 5.3.3.1).

The approximate chromosomal locations of the genetic loci that control the products of the H-2 complex in the mouse have been mapped (Table 6). The serologically defined H-2 antigens appear to be controlled by two separate loci, H-2K and H-2D (Demant, 1973). These are duplicated genes which are positioned at opposite ends of the H-2 complex. As one might anticipate, there is extensive cross-reaction between the antigenic products of the H-2K and H-2D loci (Murphy and Shreffler, 1975). Both the lymphocyte-defined (LD) and the immune-response (Ir) loci are located in the I-region of the H-2 complex, immediately adjacent to the H-2K locus (Benacerraf and McDevitt, 1972; David and Shreffler, 1974). It is not yet clear whether there are separate LD and Ir loci or whether there is a single locus whose products are multifunctional. Recently, the development of congenic strains of mice that differ only in the I-region of the H-2 complex have made possible the development of antisera to I-region associated (Ia) cell surface antigens (Shreffler and David, 1975). These antigens are glycoproteins of approximately 30,000 MW (Cullen et al., 1974; Delovitch and McDevitt, 1975). With the aid of these antisera and congenic strains of mice it should be possible to resolve many of the questions about the genetics and functions of the I-region and its cell surface antigens (see Katz and Benacerraf, 1975).

Serologically defined histocompatibility antigens are present on all cell types, but their concentrations vary markedly, being highest on lymphocytes (Edidin, 1972). They appear to be randomly scattered over the cell surface (Parr and Oei, 1973), although they can be induced to undergo lateral translation in the membrane by cross-linking with antibodies (Davis, 1972; Raff and dePetris, 1973). Lymphocyte-defined antigens are also present on many cell types, but the superiority of allogenic lymphocytes as stimulators of T cells in mixed lymphocyte reactions suggests that marked quantitative and/or qualitative differences exist in the expression of these antigens on different cell types (Cerottini and Brunner, 1974). I-region associated (Ia) antigens are restricted in distribution to

B cells, T cells, macrophages, spermatocytes and epidermal cells (Hammerling et al., 1974a,b; Hauptfeld et al., 1974). Both qualitative and quantitative differences in Ia antigen expression occur on these different cell types, leaving open the possibility that the Ia antigens on T cells, B cells and macrophages may have special immunological functions (Frelinger et al., 1974; Gotze, 1974). It may be significant that most of the Ia antigens thus far identified have been on B cells and not on T cells (Sachs and Cone, 1973; Hämmerling et al., 1974a).

Among members of the T cell series, the serologically-defined (SD) histocompatibility antigens are present in high concentrations on thymocyte progenitors, on medullary thymocytes and on a subset of peripheral T cells (Aoki et al., 1969; Cantor, 1972; Schlesinger, 1972; Konda, Stockert and Smith, 1973). Serologically defined histocompatibility antigens are present in low concentrations on cortical thymocytes and on a second subset of peripheral T cells. Thus it appears that at all stages of T cell development, the concentrations of SD histocompatibility antigens and of thymocyte- and T cell-specific alloantigens are inversely related (Table 4). The genetic control mechanisms and the functional significance of these reciprocal shifts in T cell surface antigens, particularly the major surface remodeling that occurs on thymocytes, are unknown. Lymphocyte-defined and Ia antigens have not been detected on cortical thymocytes, but are present on medullary thymocytes and peripheral T cells (Frelinger et al., 1974; Gotze, 1974; Hauptfield et al., 1974).

5.3.3. Receptors and possible receptors

5.3.3.1. T cells

1. Antigen recognition molecules. There is abundant evidence that medullary thymocytes and mature T cells can be stimulated to undergo clonal proliferation and differentiation or be rendered immunologically tolerant by exposure to specific antigen. It is also clear that these responses are triggered by the interaction of antigens with receptors on the plasma membranes of T cells. Finally, it is evident that the capacity of lymphocytes in the T cell series to recognize antigens first appears in the thymus, primarily among medullary thymocytes, but also possibly among some cortical thymocytes. Yet, the nature of the antigen-binding molecules on the surface of thymocytes and T cells constitutes one of the most perplexing enigmas in immunobiology. This problem has been recently reviewed by Warner (1974), and will only be summarized here.

There are several reasons for the controversy about the nature of the antigen recognition molecules on T cells. First, unlike B cells, on which readily detectable membrane-bound immunoglobulin molecules bind large amounts of antigen, the number of receptors normally available to bind antigens on T cells is small (Wigzell, 1974) (estimates based on radioactive antigen binding studies indicate that 10^2 to 10^3 binding sites are present on T cells as compared to 10^4 to 10^5 sites on B cells) (Roelants, 1972). Second, antisera to known classes of immunoglobulins react weakly or not at all with T cells, positive reactions for light chains being more common than for heavy chains (Bankhurst and Warner, 1971; Rabellino et al., 1971; Raff, 1971; Nossal et al., 1972). Third, immunoglobulin molecules are readily extractable from the surfaces of B cells, whereas, with several notable exceptions, this does not appear to be true for T cells (see Marchalonis and Cone, 1973b; Greaves, 1975a,b for discussion). Fourth, T cells seem to "see" antigens differently than do B cells (Katz and Benacerraf, 1972). Thus T cells tend to bind larger and more complex antigenic determinants, to bind antigens less avidly, and to be stimulated or tolerized by lower doses of antigen than do B cells. Moreover, in T-B cell interactions, T cells generally react with the carrier portion of the antigen (usually a protein), whereas B cells generally react with the hapten portion of the same antigen (often a single antigenic determinant such as DNP). Fifth, T cells show a peculiar predisposition to react with products of the major histocompatibility complex (Nisbet et al., 1969; Wilson and Nowell, 1970), including those not readily detected by B cells (i.e. lymphocyte-defined antigens). Sixth, the ability of T cells to recognize and respond to many antigens is under the genetic control of the immune response (Ir) locus of the major histocompatibility complex (see section 5.3.2.), although this is probably also true to some extent for B cells. The Ir locus is not linked genetically to any of the loci which control the production of the classical immunoglobulin molecules (Herzenberg et al., 1968). And seventh, antigen binding by T cell receptors is readily blocked by antiserum to histocompatibility antigens, whereas antigen binding to B cells is not (Shevach et al., 1972b; Hämmerling and McDevitt, 1974).

The above observations suggest three major possibilities relative to the nature of the antigen-binding molecule on T cells: (a) the receptor is not an immunoglobulin molecule, but rather an unidentified product of the Ir locus (perhaps an Ia antigen); (b) the receptor is an undescribed class of immunoglobulin molecule (Ig "X" or Ig "T") which is produced by the Ir locus; and (c) there are two kinds of receptors, one of which is a classical immunoglobulin molecule and one of which is an unidentified product of the Ir locus (perhaps on Ia antigen). Simultaneous binding to both receptors would be necessary for activation of T cells by most antigens.

Inasmuch as the existence of a non-immunoglobulin antigen receptor that is a product of the Ir locus is entirely hypothetical at the moment, most investigative effort has gone into the search for plasma membrane-bound T cell immunoglobulin. Attempts have been made to demonstrate surface immunoglobulin on T cells by using anti-immunoglobulin sera to abrogate T cell function and by cytotoxic, immunofluorescence, electron-microscopic and biochemical techniques (see Warner, 1974). The conflicting conclusions from different laboratories using similar techniques can be summarized as follows: (a) there are no immunoglobulin molecules detectable on T cells; (b) immunoglobulin molecules are present in small quantities on most T cells; and (c) immunoglobulin molecules (monomeric IgM) are present in large quantities on most T cells.

While these positions would appear to be irreconcilable, there are several possible resolutions which favor the existence of substantial numbers of immunoglobulin (Ig) molecules on T cells. Greaves and Hogg (1971) have proposed that Ig molecules tend to be wholly or partially buried within the glycocalyx (Santer et al., 1972; Wioland et al., 1972) surrounding T cells, and therefore are not available to react with most anti-Ig sera or to bind large amounts of antigen. They have further postulated that such Ig might become "unmasked" during T cell activation. This view has received support from studies which show that Ig is more readily detectable on activated than on "resting" T cells (Biberfeld et al., 1971; Hellstrom et al., 1971; Jones and Roitt, 1972; Goldschneider and Cogen, 1973), and that longer stretches of heavy chain are available to be radiolabeled in the lactoperoxidase reaction on activated than on resting T cells (Marchalonis et al., 1972); and that activated T cells bind more antigen than do resting T cells (Roelants, 1972). However, other investigators have failed to verify some of these findings (Nossal et al., 1972; Vitetta et al., 1972).

Recently Cone (1975) has found evidence that subtle differences in the methods used in different laboratories to extract immunoglobulins from T cell surfaces may explain the diametrically opposed results that have been reported. In his experiments, the concentrations of detergent normally used to successfully extract surface Ig from B cells caused T cell surface Ig to associate with detergent micelles and to become unavailable for precipitation with anti-Ig sera. Based on this and other differences between T cell and B cell Ig, Cone has postulated that the antigen-recognition molecule on T cells represents a new class of immunoglobulin, "IgT", which cross-reacts serologically with IgM. This intriguing hypothesis remains to be confirmed.

It is not sufficient, of course, to simply demonstrate that Ig is present on T cells in order to establish its physiological role as an antigen receptor. It must be shown that it can specifically bind antigen and that it is produced by the T cell on which it is present. Cone et al. (1972) and Rollinghoff et al. (1973) have demonstrated that the putative IgT molecules isolated from antigen-stimulated T cells are able to specifically bind antigen in vitro. Feldmann et al. (1973) have demonstrated that putative immunoglobulin molecules spontaneously released by activated T cells in vitro are cytophilic for macrophages. Macrophages that are sensitized with IgT, can substitute for T cells in antigen-specific cooperative interactions with B cells. This suggests that the antigen receptor on T cells may be intimately associated with the products of the I locus of the H-2 complex which control T cell-macrophage and T cell-B cell interactions, or that it is one of the products of this locus (Katz and Benacerraf, 1975). Roelants et al. (1973) have demonstrated co-capping of antigen-binding molecules and Ig molecules on the surface of T cells, suggesting that they are identical. Moreover, they found that new antigen binding-sites and Ig molecules were regenerated within several hours after capping.

Perhaps the most convincing evidence that the antigen receptors on T cells are immunoglobulin-like molecules derives from the recent work of Binz and Wigzell (1975) in rats. These authors raised antisera in F_1 hybrid rats to purified T cells from one or the other parental strain of the hybrid combination. The antisera reacted selectively against donor parental strain T cells which had receptors for the histocompatibility alloantigens of the opposite parental strain.

The antisera also reacted with humoral antibodies that were raised against the identical histocompatibility alloantigens. These experiments indicate that the receptors on both T and B cells for the same histocompatibility antigens are similar or identical with respect to the structure of their antigen-binding sites, as judged by their shared idiotypes. Thus it may be concluded that at least the binding site on the T cell antigen receptor is similar if not identical to the variable portions of classical immunoglobulin light and/or heavy chains.

It must be cautioned that all of the experiments to date which indicate that the antigen receptor on T cells is an immunoglobulin molecule (or a portion thereof) may be susceptible to one or more of the following criticisms: (a) the observed activities of heterologous anti-Ig sera may be due to naturally-occurring antibodies to non-immunoglobulin membrane antigens; (b) "purified" T cell or thymocyte preparations may be contaminated with significant numbers of B cells; and (c) a significant proportion of T cells may have Fc receptors for cytophilic antibodies (see below). Although individual experiments may be guilty of these artifacts, it would appear that there is sufficient concurrence among the results of many different experimental approaches to justify a conclusion that membrane-bound, immunoglobulin-like molecules are produced by T cells and act as receptors for at least some antigens. This does not answer the questions relating to the amount of Ig on the surface of T cells; the orientation of the Ig in or on the plasma membrane; and the nature of the Ig molecule itself. If the antigen receptor proves to be a classical form of immunoglobulin molecule, it is likely that a separate set of modulating antigen receptors under the influence of the Ir genes will also be found. This possibility has been discussed by Feldmann (1973). If the antigen receptor proves to be a novel form of immunoglobulin molecule, it may be a product of the Ir genes. Evidence for an immunoglobulin-like structure of histocompatibility antigens has been presented by Strominger et al. (1974) (see sections 5.3.2. and 5.3.3.1). Recently, Katz and Benacerraf (1975) have outlined evidence for the existence of a trimolecular receptor complex which is responsible for the expression of immunological competence by T cells. According to their hypothesis, the trimolecular complex is under the control of the major histocompatibility complex and consists of: (a) an antigen receptor; (b) an I-region gene product that permits interaction of T cells with other lymphocytes and macrophages; and (c) an Ir-gene product that modulates or supplements the functions of the antigen receptor and the cell-interaction receptor.

2. Fc receptor. Using aggregated gamma globulin or antigen-antibody complexes, receptors for sites on the Fc portion of immunoglobulin heavy chains have been demonstrated on granulocytes, macrophages, platelets and B cells (Henson et al., 1972; Nussenzweig, 1974). Recently, Fc receptors have also been demonstrated on subsets of T cells. Stout and Herzenberg (1975) detected Fc receptors on approximately 23 percent of T cells in mouse lymph node and spleen and on approximately 10 percent of thymocytes. The majority of the F_c-positive (F_c^+) T cells were small lymphocytes which responded well to stimulation with concanavalin A (the F_c-negative (F_c^-) T cells did not), but which did not cooperate with B cells in antibody responses. Other investigators have not detected Fc receptors on small T cells, but have identified such receptors on approximately 50–60 percent of activated (i.e. large) T cells (Yoshida and Andersson, 1972; Andersson and Grey, 1974; Van Boxel and Rosenstreich, 1974). The reason for this discrepancy is not clear.

The ability of subsets of T cells to bind aggregated gamma globulin and antigen-antibody complexes at some stage in their development has important experimental and immunological implications. First, it indicates that the demonstration of Fc receptors is not as reliable a method to distinguish and separate T and B cells as was once assumed. Second, it indicates that the presence of immunoglobulin on the surface of a T cell does not necessarily indicate that the immunoglobulin was produced by that cell. Third, it suggests that antigen-antibody complexes may play a role in the specific or non-specific activation, suppression, recruitment or diversion of T lymphocytes during immunological reactions. As an example, antigen-antibody complexes are known to be able to inhibit cytotoxic T cells and thereby enhance tumor graft survival in sensitized hosts (Gorczynski et al., 1974). The mechanism for such inhibition is not known, but it could conceivably involve T cell Fc receptors.

3. C receptor. Receptors for complement components are present on the surfaces of granulocytes, macrophages, platelets and B cells (Henson et al., 1972; Nussenzweig, 1974). Complement receptors have not been identified on thymocytes or T cells.

4. β_2-microglobulin. β_2-microglobulin is a protein which is produced by and is present on the surface of a variety of cell types, including T and B cells (Poulik and Bloom, 1973). It was first discovered in the urine of human patients with renal tubular disorders, and is normally found in small quantities in serum, urine and cerebrospinal fluid (Berggard and Bearn, 1968). β_2-microglobulin consists of a single polypeptide chain of approximately 100 amino-acids (MW 11,600). It is of special interest from an immunological viewpoint because its amino acid sequence shows striking homology with the constant portions of immunoglobulin light and heavy chains, particularly with the C_{H3} region of IgG (Peterson et al., 1972). Although it does not cross-react serologically with classical Ig molecules, it has been postulated that β_2-microglobulin is a free Ig domain and that its gene is evolutionarily related to Ig precursor genes (Peterson et al., 1972; Smithies and Poulik, 1972). Inasmuch as the C_{H3} region is part of the Fc portion of the IgG molecule, it has been further postulated that β_2-microglobulin may share one or more of the receptor functions of the Fc fragment, such as binding and activation of complement components, and binding to Fc receptors on macrophages, B cells and possibly T cells.

A second observation of immunological interest is the apparent association of β_2-microglobulin with the major histocompatibility antigens (HL-A antigens) on the surface of human lymphocytes. Thus, solubilized HL-A antigens are precipitated by both anti-HL-A and anti-β_2-microglobulin sera, although the specificities of these two antisera are not cross-reactive (Peterson et al., 1974; Strominger et al., 1974). Similarly, when β_2-microglobulin is capped on the cell surface by exposure to the homologous antiserum, HL-A antigens are capped simultaneously (Poulik et al., 1973). However, not all β_2-microglobulin is co-capped by exposure to anti-HL-A serum (Neauport-Sautes et al., 1974). These results suggest that all serologically-defined HL-A antigens on the cell surface are associated with β_2-microglobulin, but not vice versa.

Treatment of HL-A antigens with papain produces two polypeptide fragments, the larger of which carries the HL-A serological activity and the smaller of which appears to be β_2-microglobulin or a molecule homologous to β_2-microglobulin (Nakamuro et al., 1973; Peterson et al., 1974). In the mouse, many H-2 antigen preparations contain a small fragment that may be homologous to human β_2-microglobulin. Based on these observations, Strominger et al. (1974) have postulated that association on the T cell surface of non-covalently bound β_2-microglobulin, serologically-defined histocompatibility antigen, and Ir gene product (Ia antigen?) may be homologous and ancestral to covalently bound light chain constant region, heavy chain constant region and variable regions (antigen combining sites) respectively, of classical immunoglobulin molecules.

β_2-microglobulin has also been found to be non-covalently associated with the TL antigens on cortical thymocytes (Ostberg et al., 1975; Vitetta et al., 1975b).

5. Receptors for sheep erythrocytes. Almost all thymocytes and peripheral T cells from human beings and other primates bear a receptor for sheep erythrocytes (SRBC) (Coombs et al., 1970; Lay et al., 1971; Froland, 1972; Jondal et al., 1972; Brown and Greaves, 1974). B cells apparently lack the SRBC-receptor. Consequently, the SRBC-receptor has become an extremely useful marker for identifying and isolating normal and neoplastic T cells by forming rosettes at 4°C ("E-rosettes") which are dissociable at 37°C. Cortical thymocytes form more stable E-rosettes than do medullary thymocytes and peripheral T cells, perhaps due to their lower negative surface charges (Galili and Schlesinger, 1975). The SRBC-receptor is destroyed (or removed) by exposure to trypsin or phospholipase A, but not to papain (Jondal et al., 1972; Chapel, 1973). It can be extracted in a functionally active state from the T cell surface by hypertonic KCl (Pyke et al., 1975). Nothing is known about its structure, relationship to other T cell-specific surface antigens, or physiological functions.

6. Receptors for hormones. Receptors for histamine, β-catecholamines and the E series prostaglandins have been identified on the surface of various leukocytes including thymocytes and peripheral T cells (Weinstein et al., 1973). Cells with one or more of these receptors can be removed by affinity chromatography on columns of hormone-coated Sepharose beads. Using histamine-coated beads, Shearer et al. (1974) were able to enhance plaque-forming cell responses of mouse spleen and bone marrow B cells, presumably by selectively retaining suppressor T cells and thymocytes on the column. Other authors have been able to show that cytotoxic T cells are selectively concentrated on histamine-coated columns (Simpson et al., 1973). It is important to note that, unlike insolubilized polypeptide hormones, which may alter cellular metabolism, insolubilized low molecular weight hormones act solely as specific ligands (Weinstein et al., 1973).

There are several pieces of evidence which suggest that in addition to their usual metabolic functions, endogenous β-adrenergic hormones may play a special role in the regulation of T cell differentiation and function. For example, histamine, β-catecholamines and prostaglandins have been shown to inhibit the cytotoxic activities of T cells against allogeneic target cells (Plaut et al., 1973a). Also, antigen-induced immunological unresponsiveness can be blocked by histamine and other agents that stimulate cyclic AMP (Mozes et al., 1972). Moreover, there is evidence that the numbers of histamine-binding sites increase after antigenic stimulation of T cells (Plaut et al., 1973b).

Receptors for polypeptide hormones are also present on thymocytes and peripheral lymphocytes. The receptor for insulin has been isolated and the kinetics and conditions of binding have been defined, as have some of the metabolic effects (Gavin et al., 1973). A receptor for growth hormone has also been described (Gavin et al., 1972).

7. Receptors for lectins. Lectins (hemagglutinins, phytomitogens) are proteins and glycoproteins which have the ability to specifically bind to cell surface glycoproteins, thereby causing agglutination and/or cell activation. The interaction of lectins with the cell surface can be competitively inhibited with simple sugars or glycosides. A number of lectins and several of the corresponding lectin-binding sites have been purified and chemically characterized (Lis and Sharon, 1973).

Lectins have been widely used as probes to study the nature of the lymphocyte surface and the mode of activation of lymphocytes by ligands (Moller, 1972; Edelman et al., 1974; Greaves, 1975a). The most commonly used lectins for this purpose have been phytohemagglutinin (PHA), concanavalin A (Con A), lentil mitogen and pokeweed mitogen (PWM). The former three lectins selectively stimulate T cells to undergo polyclonal blast transformation, cell division and functional differentiation. Pokeweed mitogen stimulates both T and B cells (Greaves and Janossy, 1972).

The failure of B cells to respond to stimulation with T cell "specific" lectins is not due to an absence of receptors for such lectins. Both T and B cells have approximately equal numbers of receptors on their surfaces for PHA, Con A and lentil lectin (Bauminger et al., 1972; Stobo et al., 1972). Moreover, B cells can be activated by PHA and Con A if the lectins are insolubilized by binding to Sepharose beads or by chemically crosslinking to plastic (Andersson et al., 1972b; Greaves and Bauminger, 1972). This is reminiscent of the stimulation of non-lymphoid cells by insolubilized hormones (Cuatrecasas, 1969). Also, B cells can be activated by soluble Con A if they are simultaneously exposed to soluble factors (lymphokines) that are produced by activated T cells (Andersson et al., 1973a; Rich and Pierce, 1973). These observations suggest that the threshold of activation by lectins is much higher for B cells than for T cells. This may be a general property of activation of B cells by ligands. It may help to explain why higher concentrations of antigens are needed to activate and to tolerize B cells than T cells, and why interaction with T cells is necessary for B cells to respond optimally to most antigens (Katz and Benacerraff, 1972).

The ability of members of the T cell series to respond to stimulation with lectins is a function of the state of cellular differentiation. Whereas receptors for a variety of lectins are present at the earliest stages of T cell ontogeny, only a minor subset of cortical thymocytes are able to respond to Con A (Stobo, 1972). Moreover, cortical thymocytes are unresponsive to PHA unless they are co-stimulated with lymphokines (Gery and Waksman, 1972). There is also marked variation among subsets of medullary thymocytes and peripheral T cells with respect to their ability to respond to different lectins (Colley et al., 1970b; Stobo et al., 1973). The use of lectins as probes of T cell heterogeneity is discussed in section 5.4.

5.3.3.2. B cells

1. Antigen recognition molecules. As originally postulated by Ehrlich (1900) and by Burnet (1959) the antigen receptors on antibody-forming cell precursors appear to be immunoglobulin molecules. Antisera directed against surface immunoglobulin molecules have been demonstrated to inhibit antigen binding to B cells; whereas antisera to non-immunoglobulin cell surface components have not (Walters and Wigzell, 1970; Unanue, 1971; Lawrence et al., 1973). Antigen-binding sites and cell surface immunoglobulin have been shown to redistribute concurrently on the surface of mouse spleen cells suggesting that they are identical molecules (Raff et al., 1973; Ashman, 1974). Moreover, Rolley and Marchalonis (1973) have been able to release intact immunoglobulin-antigen complexes from the surface of B cells using proteolytic enzymes. Although these results strongly suggest that cell surface immunoglobulins function as antigen receptors, the possibility that other non-immunoglobulin molecules (e.g. Ir gene products) may act in concert with immunoglobulin during antigen binding cannot be excluded.

Approximately 10^5 immunoglobulin molecules have been estimated to be present on the surface of B cells (Rabellino et al., 1970). These surface immunoglobulins appear to be the endogenous product of each B cell rather than passively adsorbed humoral antibodies. Thus, Unanue et al. (1972) have shown that B cells can replace their cell surface immunoglobulins when cultured in protein-free medium. Also, pulse-labelling experiments with radioactively labelled amino acids indicate that cell surface immunoglobulin is endogenously synthesized (Melchers and Anderson, 1973; Parkhouse, 1973). Finally, more than 95% of the immunoglobulin molecules on the surface of individual antigen-binding B cells are specific for one particular antigen (Raff et al., 1973). It seems improbable that such specificity could be attained other than by endogenous production of cell surface immunoglobulin. The fact that the antibody which is secreted by an antigen-stimulated B cell has the same specificity as the antigen receptors supports this proposition. This is important because most B cells also have receptors by which they can non-specifically bind the Fc fragments of certain classes of humoral antibodies (see below).

Many studies have examined the class distribution of cell surface immunoglobulin in a number of species (see Warner, 1974 for review). Although a wide range of results have been reported, IgM emerges as the predominant class of cell surface immunoglobulin. Both IgG and IgA are detected on varying proportions of lymphocytes; the proportion of IgA-bearing cells in lymph node and spleen is generally lower than that bearing IgG. Blood lymphocytes bearing cell surface IgD and IgE have been demonstrated in humans. However, the proportions of such cells are small in normal adults (Aisenberg and Block, 1972; Piessans et al., 1973).

The expression of cell surface IgD appears to be intimately involved with the development and differentiation of B cells. In humans, surface IgD is found on approximately 15% of the lymphocytes in umbilical cord blood (or 50% of the total cells bearing surface immunoglobulin) (Rowe et al., 1973a,b) and on 2–5% of peripheral blood lymphocytes in adults (Van Boxel et al., 1972; Rowe et al., 1973b). IgD is virtually undetectable in fetal serum and is present in low levels in adult serum (Spiegelberg, 1972). The observations that the proportion of IgD-bearing cells decreases in adults as compared to newborns, and that individual lymphocytes may bear both IgD and IgM having the same antigen-binding specificities have led to the suggestion that IgD is the first antigen receptor to appear on B cells during fetal life (Rowe et al., 1973b). According to this view, other classes of surface immunoglobulins would appear later in ontogeny.

More recently, an IgD-like molecule has been identified on the surface of mouse spleen cells (Melcher et al., 1974). However, IgD was not detected on spleen cells until 10–15 days after birth. Prior to that time only IgM was present. IgD was also present on the spleen cells of germfree mice (Vitetta et al., 1974a; Vitetta et al., 1975a). IgD was detected only on lymphocytes in peripheral lymphoid tissues where it comprised 85% of the surface immunoglobulin that was extractable from lymph node cells and 50% from spleen cells (Vitetta et al., 1974a; Vitetta et al., 1975a). Bone marrow lymphocytes from adult mice possessed no detectable IgD; only IgM was apparent. These observations coupled with those of Strober (1975) concerning B cell maturation in the spleen (see section 4.3.2) have led Vitetta et al. (1975a) to suggest that IgM is the "primordial" surface receptor in mice. In their view, the expression of IgD on maturing B cells occurs in the spleen prior to the emigration of putatively mature B cells to other peripheral lymphoid tissues.

2. Complement receptors. The complement system consists of 11 normal serum proteins which are serially activated upon interaction with antigen–antibody complexes to mediate diverse aspects of the inflammatory response. Receptors for complement components have been identified on lymphocytes from a variety of mammalian species. These receptors have been visually demonstrated by the binding of complement- and antibody-coated sheep erythrocytes (EAC) to the surface of lymphocytes. The formation of EAC rosettes can be inhibited by antiserum against the third component of complement, C3 (Eden et al., 1971); and, radioactively labelled, purified C3 will bind to lymphocytes (Bokisch and Theofilopoulos, 1973). In addition, Bokisch and Sobel (1974) have identified a receptor for the fourth component of complement on human B lymphocytes. These authors report that the C3 receptor and the C4 receptor cooperate in the binding of antigen-antibody-complement complexes to B cells.

Complement receptors on lymphocytes are considered to be markers for B cells, inasmuch as "complement receptor lymphocytes" (CRL) bear surface immunoglobulin (Bianco et al., 1970; Wigzell et al., 1973) and lack T cell-specific surface antigens (Bianco and Nussenzweig, 1971). In addition, neonatally thymectomized mice and congenitally athymic mice have increased proportions of complement receptor lymphocytes (Dukor et al., 1972; Möller, 1974).

However, not all putative B cells have complement receptors. Möller (1974) has observed that 30–40% of the lymphocytes in spleens of *nude* mice lack detectable complement receptors. Van Boxel et al. (1973) have also observed that the complement receptors are not present on all immunoglobulin-bearing cells in mouse spleen and lymph nodes. In neonatal mice, spleen cells bearing surface immunoglobulin are present at birth, whereas spleen cells bearing both complement receptors and immunoglobulin are not detected until 2 weeks after birth (Gelfand et al., 1974a). In addition, Möller (1974) has shown that B cells may not express detectable complement receptors following stimulation by mitogens or antigens. Hence, B cells lacking complement receptors may represent immature and/or activated B cells.

Gelfand et al. (1974b) have reported that the time of appearance during ontogeny of lymphocytes bearing complement receptors is under the control of two independent genes. One of these, CRL-1, is linked to the H-2 histocompatibility complex. Although CRL-1 is not a structural gene for the complement receptor, it does appear to control the timing of its expression. The presence of CRL-1 within the H-2 gene complex is of interest in light of the evidence that numerous immunological functions may be controlled within the H-2 region (see section 5.3.2).

Three overlapping functions have been ascribed to the complement receptors on B cells. First, complement receptors may aid in the localization of immune complexes within lymphoid follicles during the induction of immune responses (Dukor et al., 1970). Second, complement receptors may aid in cooperation between T and B cells. Pepys (1972) has suggested that bound antigens on T cells may fix complement. The resulting T cell-antigen-complement complexes may then interact with antigen and complement receptors on B cells to induce antibody synthesis. Pepys (1974) has supported this hypothesis by demonstrating that complement is required for induction of an antibody response to thymus-dependent antigens (i.e. antigens which require T and B cell cooperation for antibody production). Third, Dukor and Hartman (1973) have suggested that binding of complement, in addition to antigen binding, is necessary for B cell activation and initiation of antibody synthesis, the so-called "second signal" hypothesis.

3. Fc receptors. B cells are also able to bind antigen-antibody complexes and aggregated immunoglobulin (IgG mainly and IgM to a lesser extent) in the absence of complement (Basten et al., 1972a; Brown et al., 1970). Studies employing fragments of immunoglobulin, which are obtained by cleavage with papain, reveal that binding to B cells occurs via the Fc fragment of the immunoglobulin molecule, rather than by the antigen binding fragment (Fab) (Basten et al., 1972b; Paraskevas et al., 1972). Recently, B cells possessing the Fc receptor, but lacking complement receptors have been described (Parish, 1975). The Fc receptor is therefore distinct from antigen receptors and complement receptors (Basten et al., 1972a,b; Eden et al., 1973).

Mouse alloantisera against the H-2 complex have been shown to inhibit binding of immunoglobulin to the Fc receptor of B cells (Dickler and Sachs, 1974). Moreover, antisera against I region-associated (Ia) antigens of the H-2 complex also inhibited Fc receptor binding. As a result, Dickler and Sachs (1974) have suggested that initiation of T-B cell interaction in the immune response, which is controlled by the I region of the H-2 complex, may be mediated in part by the Fc receptor on B cells. It has also been postulated that suppression or feedback inhibition of antibody responses may occur via Fc receptor binding of antigen-antibody complexes to B cells (Basten et al., 1972a; Dickler and Sachs, 1974).

Perhaps the best documented function of Fc receptors on lymphocytes is in antibody-dependent cell-mediated cytotoxicity. Effector cells, which need not be primed against the target cell, bind to antibody-coated targets cells via Fc receptors and cause cell lysis. Aggregated immunoglobulin will inhibit the process, presumably by binding to the Fc receptors (van Boxel et al., 1974). Effector cells are not decreased in a lymphoid cell population depleted of T cells (van Boxel et al., 1972b). Hence, B lymphocytes (van Boxel et al., 1973) and antigenically "null" spleen lymphoid cells (Greenberg et al., 1973) have been suggested as the effectors of antibody-dependent cytotoxicity. However, macrophages and monocytes, which possess Fc receptors can also mediate such cytotoxicity (Perlmann and Perlmann, 1970; Allison, 1972; Dennert and Lennox, 1973). Recently, Cohen et al. (1975) have described a macrophage-like cell population and a lymphoid cell population in mouse spleen both of which are capable of mediating antibody-dependent cell-mediated cytoxicity.

5.4. *Heterogeneity of lymphocytes*

5.4.1. *T cells*

At least two subsets of T cells exist in the peripheral lymphoid tissues of mice and rats, and presumably other mammals as well. These subsets have been distinguished on the basis of differences in functions, anatomical distribution, migration pathways, surface antigens, susceptibility to immunosuppressive agents, responsiveness to phytomitogens, and adherence to artificial surfaces. Despite this wealth of descriptive information, little is known about the ontogenetic relationships of T cell subsets. Thus, it is not clear if T cell subsets constitute various stages of differentiation of a single cell line ($P \rightarrow T_1 \rightarrow T_2 \rightarrow T_3$, etc.); if they represent parallel differentiation of separate cell lines which arise from a common precursor ($P \rightarrow T_1 + T_2 + T_3$, etc.); or a combination of these possibilities, depending on the number of subsets that exist. Moreover, it is not clear in most instances whether antigenic stimulation is necessary for the functional differentiation of certain T cell subsets or merely for their clonal expansion.

The best described T cell subsets are the T_1 an T_2 cells of the mouse. These were first identified by Cantor and Asofsky (1972), who demonstrated that two different populations of parental strain T cells could act synergistically to initiate a graft-versus-host reaction in F_1 hybrids. The T_1 cell subset was found to be present in highest concentrations in spleen and thymus, whereas T_2 cells were present in highest concentrations in lymph node, thoracic duct lymph and blood. T_1 and T_2 cells were further distinguished in that the former were rapidly dividing, did not recirculate from blood to lymph, had moderately high concentrations of Thy-1 (theta) antigen, were sensitive to cortisone, did not participate in mixed lymphocyte reactions, and were markedly depleted within 2 weeks of adult thymectomy (Cantor, 1972). However, analysis of the synergistic interaction of these two cell populations suggested that antigen-stimulated T_1 cells could differentiate into T_2 cells in the presence of small numbers of activated T_2 cells. T_1 cells may therefore be analogous to the "immediate postthymic cell" described by Stutman (1975), which is thought to be an immature T cell of recent thymus origin which gives rise in the presence of circulating thymus hormone to fully competent T cells. Stutman and Good (1972) have further proposed that it is these competent T cells (putative T_2 cells) which are induced by antigenic stimulation to differentiate into a variety of memory and effector cell types (i.e. T_3, T_4, T_5 cells, etc.).

The existence of separate subsets of functionally distinct T cells is supported by the recent description of three subsets of Ly-antigen-positive T cells in mouse spleen (Cantor and Boyse, 1975a,b). Whereas mouse thymocytes bear the Ly 1,2 and 3 sets of alloantigens, populations of peripheral T cells were identified which displayed different phenotypic expressions of these antigens. An Ly-2,3^+ subset was found which contained the precursors of cytotoxic T cells; and an Ly-1^+ subset was found which cooperated with B cells in humoral antibody responses and with Ly-2,3^+ T cells in the generation of cytotoxic T cells. These studies suggest that the commitment to specific functions occurs among T cell subsets before, rather than as a consequence of, antigenic stimulation. No function has yet been ascribed to the Ly-1,2,3^+ T cell population, which is also present in mouse spleen. Ly-1,2,3^+ T cells are the only T cells present during the first week of life. They are progressively supplanted by Ly-1^+ and Ly-2,3^+ T cells, and are selectively depleted by thymectomy in the adult. They may therefore be the equivalent of T_1 cells and may represent the precursors of the Ly-1^+ and Ly-2,3^+ cells, which in turn may be subpopulations of T_2 cells. Alternatively, the Ly-1,2,3^+ cells may represent a line of T cells distinct from Ly-1^+ and Ly-2,3^+ cells.

Yet other subsets of T cells in the mouse have been identified according to their differential reactivities to phytomitogens and their differential susceptibilities to immunosuppressive agents. Stobo et al. (1973) have distinguished a recirculating population of radiosensitive, strongly Thy-1^+ (theta) T cells that is present in high concentration in lymph node; and a sessile population of radio-resistant, weakly THY-1^+ T cells that is present in high concentrations in spleen and bone marrow. The former subset responds equally well to stimulation with PHA and Con A, whereas the latter subset responds mainly to Con A. Bach and Dardenne (1973) showed that spontaneous (background) sheep RBC rosette-forming T cells from lymph node and spleen of non-immunized mice evidenced differential sensitivity to anti-Thy-1 serum, antilymphocyte serum and azathioprine in vitro. In each instance T-RFC from spleen were more sensitive than T-RFC from lymph node. Like T_1 cells, spleen T-RFC are depleted shortly after adult thymectomy and acquire the properties of lymph node T-RFC after antigenic stimulation. Another example of T cell heterogeneity is the differential effect of in vivo cortisone treatment on T cell functions (Segal et al., 1972). Gershon (1974) has shown that suppressor T cells in spleen may be derived from cortisone-sensitive thymocytes, whereas helper T cells are derived from cortisone-resistant thymocytes.

Suppressor T cells, which inhibit production by B cells and mitogen-induced blast transformation of T cells, have also been identified in rat spleen (Folch and Waksman, 1974). These suppressor cells are clearly different from the majority of T cells in that they adhere to glass wool and are responsive to cortisone treatment.

Other T cell subsets in the rat have been demonstrated with the aid of antisera to lymphocyte-specific heteroantigens (Goldschneider, 1975a,b; also see section 5.3.1). T cells in white pulp of spleen, paracortex of lymph node and thoracic duct lymph bear the rat masked thymocyte antigen (RMTA), whereas T cells which appear to reside in red pulp of spleen and in blood bear the rat bone marrow lymphocyte antigen (RBMLA). These antigens are also present on medullary thymocytes and cortical thymocytes, respectively. The $RBMLA^+$ T cell subset in the rat resembles the T_1 cell subset in the mouse in that it is non-recirculating and cortisone-sensitive. The $RMTA^+$ subset in the rat is analogous to the T_2 cell subset in the mouse in that it is recirculating and cortisone-resistant. It is of interest, therefore, that mouse T_1 and T_2 cells also resemble cortical and medullary thymocytes, respectively, with regard to suppressor, helper and effector cell properties.

The preceding observations suggest that T_1-type cells in the rat and mouse are derived from cortical thymocytes and that T_2-type cells are derived from medullary thymocytes. The demonstration that putative cortical and medullary thymocytes are normally released from the thymus and that they migrate to red and white pulp of spleen, respectively, supports this hypothesis (see section 5.2.1.2). However, a parallel development of T_1 and T_2 cells from cortical and medullary thymocytes does not exclude the possibilities that T_1 cells may act as precursors for T_2 cells, or that cortical thymocytes may act as precursors for medullary thymocytes. It is entirely possible that some cortical thymocytes may complete their maturation in the thymus, and that others may be released precociously and complete their maturation in the spleen. An hypothetical scheme that attempts to correlate the dichotomy of T cell development with the heterogeneity of T cell function is presented in Fig. 7.

5.4.2. B cells

At least four anatomically and presumably functionally distinct subsets of B cells exist in the peripheral lymphoid tissues (see Fig. 4): follicular cells, germinal

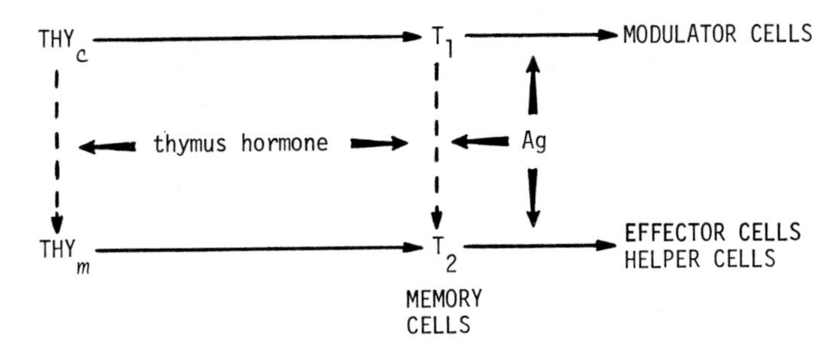

Fig. 7. Hypothetical scheme of T cell heterogeneity. THY_c, cortical thymocyte; THY_m, medullary thymocyte; T_1 and T_2, subsets of peripheral T cells; Ag, antigenic stimulation; ——, probable developmental pathway; -----, possible developmental pathway. See text section 5.4.1 for details.

center cells, marginal zone cells and plasma cells. The ontogenetic relationship between these major B cell subsets is unclear. The simplest scheme based on existing information is that bursa- or bone marrow-derived follicular B cells give rise to IgM-secreting plasma cells and to germinal center cells after primary antigenic stimulation. Germinal center cells in turn give rise to recirculating, memory B cells which reside in the marginal zones of lymphoid follicles, and to plasma cells which secrete antibody classes in addition to IgM. An alternative scheme would envision two separate lines of bursa- or bone marrow-derived cells, one involved in primary antibody responses and one mediating secondary (anamnestic, memory) antibody responses. The latter population might only be capable of responding to thymus-dependent antigens, whereas the former might respond to both thymus-dependent and thymus-independent antigens. The ensuing discussion will reflect on these two possible ontogenetic schemes.

Studies of antigen localization in primary lymphoid follicles in germ-free mice have revealed that antigen deposition occurs at sites of initiation of germinal center formation (Hanna et al., 1969). Similarly, Nossal et al. (1964) concluded that a causal relationship exists between antigen localization and germinal center development within primary follicles. These results suggest one of two possibilities concerning the derivation of germinal center cells: (1) germinal center cells arise from newly-formed B cells in central lymphoid tissues which migrate to sites of antigen localization; or (2) germinal center cells arise from follicular B cells. At the present time it is difficult to exclude either of these possibilities.

As mentioned previously (see section 5.2.2), germinal centers have been implicated in the generation of memory B cells. Wakefield and Thorbecke (1968) have shown that memory B cells in mice are generated in spleen white pulp only after germinal center formation has occurred. Similarly, autoradiographic studies of mouse lymph nodes following injection of tetanus toxoid indicate that germinal center formation precedes the memory, IgG, response (Buerki et al., 1974). Moreover, destruction of germinal centers by γ-irradiation results in the inhibition of the memory response to subsequent antigenic stimulation (Grobler et al., 1974).

The memory B cells which are generated in germinal centers appear to emigrate to other sites in peripheral lymphoid tissue. Hanna et al. (1968) have shown that in mice the capacity to elicit a memory response to a given antigen remains at maximum levels at a time when the antigen-induced germinal centers have become markedly reduced in size. Nieuwenhuis and Keuning (1974) have suggested that during germinal center formation in the rabbit, lymphocytes migrate from the germinal centers thrugh the lymphocyte corona of the secondary follicle to populate the surrounding marginal zone. These authors further suggest that these germinal center-derived cells comprise part of the recirculating memory B cell population.

It is of interest that virgin B cells can be stimulated to elicit a primary IgM antibody response in the absence of active germinal centers. Indeed, in germ-free mice, which contain no detectable germinal centers, a primary antibody response was observed prior to germinal center formation (Hanna and Hunter, 1971). Similar studies in conventionally raised rabbits revealed that the primary response to tetanus toxoid also occurs before germinal center formation (Buerki et al., 1974). Hence, IgM-secreting cells in the primary response do not appear to be derived from germinal center lymphocytes but, by exclusion, from virgin B cells in primary follicles or from an unidentified population of B cells.

In addition to the preceding differences in ontogeny and anatomical distribution, virgin B cells and memory B cells have been shown to differ in certain physiological and cell surface characteristics. Memory B cells, in contrast to virgin B cells, recirculate from blood to lymph, have a relatively slow turnover rate (>48 hours) (Strober and Dilley, 1973a), are less adherent to artificial surfaces (Schrader, 1974; Schlegel and Shortman, 1975) and exhibit a relatively low electrophoretic mobility (Schlegel et al., 1975).

A variety of indirect evidence has been presented which suggests that marginal zone lymphocytes may be the immediate precursors of antibody-forming plasma cells. Autoradiographic studies indicate that one of the initial cell populations to proliferate in the spleen following antigenic stimulation resides in the marginal zone (Nieuwenhuis, 1971; Pelc and Harris, 1973). Time course analysis reveals that [^{3}H]-thymidine-labelled marginal zone cells appear to migrate into the spleen red pulp where plasma cells are abundant. Significantly, Keuning and Bos (1967) have shown that during lymphoid regeneration following X-irradiation the return of antibody responsiveness correlates with the appearance of marginal zone lymphocytes. Assuming that plasma cell precursors do reside in the marginal zone, it is not yet clear whether these cells are involved in primary and/or memory antibody responses.

In addition to heterogeneity among B cells in peripheral lymphoid tissues, heterogeneity has also been demonstrated between B cells in bone marrow and spleen. Playfair and Purves (1971) have shown that a proportion of B cells from mouse bone marrow, B_1 cells, lack the capacity to act synergistically with T cells to elicit an antibody response in vitro. In contrast, the majority of spleen B cells, B_2 cells, possess this ability. Inasmuch as B and T cell cooperation has been suggested to be mediated via complement and/or Fc receptors (see section

5.3.3.2), the absence of such receptors on immature B cells in bone marrow may explain this phenomenon. This hypothesis is supported by the failure of complement receptor-bearing B lymphocytes to appear until 2 weeks after birth in the neonatal mouse (Gelfand et al., 1974a).

Other reports of B cell heterogeneity have also been correlated with B cell maturation. These reports indicate that B cells at different stages of development exhibit different patterns of responsiveness to mitogens (i.e. induction of DNA synthesis and/or immunoglobulin synthesis) (Gronowicz and Coutinho, 1975). Fetal liver cells and bone marrow cells can be induced to undergo DNA synthesis, but not immunoglobulin synthesis, by dextran sulphate (DxS), but purified protein derivative from tuberculin (PPD) does not stimulate these cells. However, spleen cells and peripheral lymph node cells can be activated by both DxS and PPD to synthesize immunoglobulin as well as DNA.

One of the central questions pertaining to B cell heterogeneity is the ontogenetic relationship of plasma cells which secrete different classes of immunoglobulins. At issue is the role of antigenic stimulation: does it induce a "switch" of immunoglobulin class in individual antibody-forming cell precursors or does it simply cause clonal expansion of antibody-forming cell precursors, each of which is committed to the production of a single immunoglobulin class?

The evidence in the chicken favors the view that the commitment to production of a single immunoglobulin class is an antigen-independent process which occurs as a result of a "switch" from IgM- to IgG- or IgA-secreting cells in the bursa of Fabricius (see section 4.2.2; also see reviews by Warner, 1974 and by Lawton and Cooper, 1974). Thus, bursectomy at different stages of perinatal development revealed that the initial B cells which emigrate from the bursa are committed to IgM synthesis only (Cooper et al., 1969; Ivanyi, 1973; Kincade et al., 1973). Similarly, injection of heterologous anti-IgM serum into chicks bursectomized after hatching resulted in suppression of IgM production but not IgG. Moreover, cells possessing both classes of immunoglobulin were rarely found in the spleen of normal chicks (Kincade and Cooper, 1971). It is possible, however, that this scheme may be altered by appropriate stimulation of peripheral B cells to allow a switch in production of heavy chain class by individual lymphocytes.

Evidence supporting antigen-dependent differentiation of antibody-forming cell precursors has been obtained in mice. Herrod and Warner (1972) reported that treatment of adult mouse spleen cells with heterologous antiserum to IgM suppressed both IgM and IgG production when the cells were transferred into irradiated congenic recipients. Similar experiments in vitro indicated that anti-IgM serum could suppress antigen-specific IgG, as well as IgM, production (Pierce et al., 1972a). In addition, treatment of antigen-primed mouse spleen cells with antiserum to IgM also resulted in suppression of specific IgA production (Pierce et al., 1972b). These studies support the idea that antigen exposure causes a "switch" in Ig synthesis by IgM-bearing peripheral B cells.

In contrast, Lawton and Cooper (1974) found that the spleens of germ-free and conventionally raised mice of the same strain had similar proportions of the different immunoglobulin classes of B cells. Similarly, fetal pigs, which are protected by the placenta from foreign antigens, had peripheral B cells which contained the different classes of immunoglobulins a month prior to birth (Binns et al., 1972). Pretreatment of mouse spleen cells with heterlogous antiserum against IgM was shown to suppress the synthesis of antigen-specific IgM antibodies, but not IgG antibodies, when these cells were transferred into irradiated recipients (Warner, 1972). Moreover, Kearney and Lawton (1975) demonstrated that cultures of mouse fetal liver and fetal spleen cells which were stimulated by bacterial lipopolysaccharide, a B cell mitogen, gave rise to cells which synthesized IgM, IgG and IgA. Prior to stimulation, such cells did not exhibit detectable surface immunoglobulin. These results support antigen-independent B cell development similar to that seen in the avian bursa of Fabricius.

In summary, there is good evidence in birds and in mammals that IgM-containing lymphocytes in central lymphoid tissues can give rise to IgG- and IgA-containing antibody-forming cell precursors in the absence of antigenic stimulation. There is also some evidence that a similar IgM → IgG and IgM → IgA "switch" can occur among antibody-forming cell precursors in peripheral lymphoid tissues as a result of antigenic stimulation. These are not necessarily mutually exclusive propositions (Fig. 8). However, given the complexity of the antibody response, it will be difficult to obtain proof of an antigen-dependent "switch" in Ig class within individual antibody-forming cell precursors. At the very least, such proof will require the demonstration of dual-producer cells, i.e. antibody-secreting cells containing two different immunoglobulin classes each of which has the same antigen binding specificity. Final proof will require the demonstration of an antigen-dependent "switch" in immunoglobulin class within clones of cells derived from single antibody-forming cell precursors. Until this fundamental problem is resolved, the question of the ontogenetic relationships of subsets of immunoglobulin-secreting cells will remain moot.

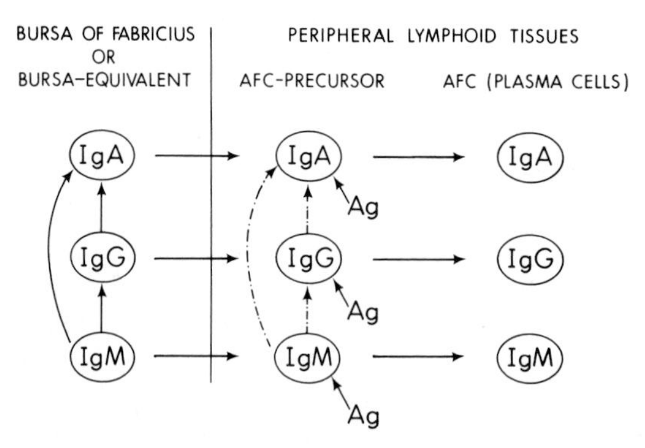

Fig. 8 Composite scheme of ontogeny of antibody-forming cells in central and peripheral lymphoid tissues. IgG, IgM, IgA, class of immunoglobulin present within individual cells; AFC, antibody-forming cell; Ag, antigenic stimulation; ——, probable developmental pathway; -----, possible developmental pathway.

References

Ackerman, G.A. (1962) Electron microscopy of the bursa of Fabricius of the embryonic chick with particular reference to the lympho-epithelial nodules. J. Cell Biol. 13, 127–146.

Ackerman, G.A. and Knauff, R.A. (1959) Lymphocytopoiesis in the bursa of Fabricius. Am. J. Anat. 104, 163–206.

Ackerman, G.A. and Knauff, R.A. (1964) Lymphocytopoietic activity in the bursa of Fabricius. In: The Thymus in Immunobiology (Good, R.A. and Gabrielson, E.E., eds.) pp. 123–149, Harper and Row, New York.

Acton, R.T., Morris, R.J. and Williams, A.F. (1974) Estimation of the amount and tissue distribution of rat Thy-1.1 antigen. Eur. J. Immunol. 4, 598–602.

Aisenberg, G.C. and Bloch, K. (1972) Immunoglobulins on the surface of neoplastic lymphocytes. N. Eng. J. Med. 287, 272–276.

Aiuti, F. and Wigzell, H. (1973) Function and distribution pattern of human T lymphocytes. I. Production of anti-T lymphocyte specific sera as estimated by cytotoxicity and elimination of function of lymphocytes. Clin. Exp. Immunol. 13, 171–181.

Albini, B. and Wick, G. (1975) Ontogeny of lymphoid cell surface determinants in the chicken. Int. Arch. Allergy Appl. Immunol. 48, 513–529.

Allison, A.C. (1972) Interactions of antibodies and effector cells in immunity against tumors. Ann. Inst. Pasteur 122, 619–631.

Anderson, C.L. and Grey, H.M. (1974) Receptors for aggregated IgG on mouse lymphocytes. Their presence on thymocytes, thymus-derived, and bone marrow-derived lymphocytes. J. Exp. Med. 139, 1175–1188.

Andersson, B. and Blomgren, H. (1970) Evidence for a small pool of immunocompetent cells in the mouse thymus. Its role in the humoral antibody response against sheep erythrocytes, bovine serum albumin, ovalbumin and NIP determinant. Cell. Immunol. 1, 362–371.

Andersson, J., Edelman, G.M., Möller, G. and Sjoberg, O. (1972) Activation of B lymphocytes by locally concentrated concanavalin A. Eur. J. Immunol. 2, 233–235.

Andersson, J., Möller, G. and Sjoberg, O. (1972) B lymphocytes can be stimulated by concanavalin A in the presence of humoral factors released by T cells. Eur. J. Immunol. 2, 99–101.

Aoki, T., Hämmerling, U., deHarven, E., Boyse, E.A. and Old, L.J. (1969) Antigenic structure of cell surfaces: An immunoferritin study of the occurrence and topography of H-2, θ and TL alloantigens on mouse cells. J. Exp. Med. 130, 979–1001.

Armerding, D., Sachs, D.H. and Katz, D.H. (1974) Activation of T and B lymphocytes in vitro. J. Exp. Med. 140, 1717–1722.

Arvin, G.C. and Allen, H.E. (1928) Variations in weight of thymus glands following administration of ovarian hormone and anterior hypophysis. Anat. Rec. 38, 39.

Asherson, G.L. and Allwood, G.G. (1972) Inflammatory lymphoid cells: Cells in immunized lymph node that move to sites of inflammation. Immunology, 22, 493–502.

Ashman, R.F. (1974) Concurrent redistribution of antigen receptors and surface immunoglobulin on antigen-binding cells. Immunol. 26, 539–548.

Auerbach, R. (1960) Morphogenetic interactions in the development of the mouse thymus gland. Dev. Biol. 2, 271–284.

Auerbach, R. (1961) Experimental analysis of the origin of cell types in the development of the mouse thymus. Dev. Biol. 3, 336–354.

Bach, J.F. and Dardenne, M. (1971) Etudes sur l'origine des cellules format les rosetes spontanées. C.R. Acad. Sci. 272, 1318–1319.

Bach, J.F. and Dardenne, M. (1972) Thymus dependency of rosette-forming cells: evidence for a circulating thymic hormone. Transplant. Proc. 4, 345–350.

Bach, J.F. and Dardenne, M. (1973) Antigen recognition by T lymphocytes. Evidence for two populations of thymus-derived, rosette-forming cells in spleen and lymph node. Cell. Immunol. 6, 394–406.

Bach, F.H., Widmer, M.B., Bach, M.L. and Klein, J. (1972) Serologically defined and lymphocyte-defined components of the major histocompatibility complex in the mouse. J. Exp. Med. 136, 1420–1444.

Bachvaroff, R., Galdiero, F. and Grabar, P. (1969) Anti-thymus cytotoxicity: Identification and isolation of rat thymus-specific membrane antigens and purification of the corresponding antibodies. J. Immunol. 103, 953–961.

Ball, W.D. (1963) A quantitative assessment of mouse thymus differentiation. Exp. Cell Res. 31, 82–88.

Baltimore, D. (1974) Is terminal deoxynucleotidyl transferase a somatic mutagen in lymphocytes? Nature, 248, 409–411.

Bankhurst, A.D. and Warner, N.C. (1971) Surface immunoglobulins on mouse lymphoid cells. J. Immunol. 107, 368–373.

Barton, R., Goldschneider, I. and Bollum, F.J. (1975) The distribution of terminal deoxynucleotidyl transferase (TdT) among subsets of thymocytes in the rat. J. Immunol., in press.

Basten, A., Miller, J.F.A.P., Warner, N.L. and Pye, J. (1971) Specific inactivation of thymus-derived (T) and non-thymus-derived (B) lymphocytes by ^{125}I-labelled antigen. Nature New Biology 231, 104–106.

Basten, A., Miller, J.F.A.P., Sprent, J. and Pye, J. (1972a) A receptor for antibody on B lymphocytes. I. Method of detection and functional significance. J. Exp. Med. 135, 610–626.

Basten, A., Warner, N.L. and Mandel, T. (1972b) A receptor for antibody on B lymphocytes. II. Immunochemical and electron microscopic characteristics. J. Exp. Med. 135, 627–642.

Bauminger, S., Greaves, M.F. and Janossy, G. (1972) Lymphocyte subpopulation-receptors for phytomitogens on T and B cells and activation by insoluble stimulants. Israel J. Med. Sci. 8, 640–641.

Benacerraf, B. and McDevitt, H.O. (1972) Histocompatibility-linked immune response genes. Science, 175, 273–279.

Berggard, I. and Bearn, A.G. (1968) Isolation and properties of a low molecular weight β2-globulin occurring in human biological fluids. J. Biol. Chem. 243, 4095–4103.

Bergsma, D. and Good, R.A. (eds.) Immunologic Deficiency Diseases in Man. The National Foundation/March of Dimes, White Plains, New York, 1968.

Bianco, C. and Nussenzweig, V. (1971) Theta-bearing and complement-receptor lymphocytes are distinct populations of cells. Science, 173, 154–156.

Bianco, C., Patrick, R. and Nussenzweig, V. (1970) A population of lymphocytes bearing a membrane receptor for antigen-antibody-complement complexes. I. Separation and characterization. J. Exp. Med. 132, 702–720.

Biberfeld, P., Biberfeld, G. and Perlmann, P. (1971) Surface immunoglobulin light chain determinants of normal and PHA-stimulated human blood lymphocytes studied by immunofluorescence and electron microscopy. Exp. Cell Res. 66, 177–189.

Biggar, W.D., Park, B.H. and Good, R.A. (1973) Immunologic reconstitution. Ann. Rev. Med. 24, 135–144.

Billingham, R.E. and Silvers, W.K. (1961) Quantitative studies on the ability of cells of different origins to induce tolerance of skin homografts and cause runt disease in neonatal mice. J. Exp. Zool. 146, 113–129.

Binns, R.M., Feinstein, A., Gurner, B.W. and Coombs, R.R.A. (1972) Immunoglobulin determinants on lymphocytes of adult, neonatal and foetal pigs. Nature New Biology 239, 114–116.

Binz, H and Wigzell, H. (1975) Shared idiotypic determinants on B and T cells reactive against the same antigenic determinants. I. Demonstration of similar or identical idiotypes and IgG molecules and T-cell receptors with specificity for the same alloantigens. J. Exp. Med. 142, 197–211.

Blackburn, W.R. and Miller, J.F.A.P. (1967) Electron microscopic studies of thymus graft regeneration and rejection. II. Syngeneic-irradiated grafts. Lab. Invest. 16, 833–846.

Blankenhorn, E.P. and Douglas, T.C. (1972) Location of the gene for theta antigen in the mouse. J. Hered. 63, 259–263.

Blomgren, H. (1969) The influence of the bone marrow on the repopulation of the thymus in X-irradiated mice. Exp. Cell Res. 58, 353–364.

Blomgren, H. and Andersson, B. (1969) Evidence for a small pool of immunocompetent cells in the mouse thymus. Exp. Cell Res. 57, 185–192.

Blomgren, H. and Andersson, B. (1971) Characteristics of the immunocompetent cells in the mouse thymus: Cell population changes during cortisone-induced atrophy and subsequent regeneration. Cell. Immunol. 1, 545–560.

Blomgren, H. and Svedmyr, E. (1971) In vitro stimulation of mouse thymus cells by PHA and allogenic cells. Cell. Immunol. 2, 285–299.

Bockman, D.E. and Cooper, M.D. (1973) Pinocytosis by epithelium associated with lymphoid follicles in the bursa of Fabricius, appendix, and Peyer's patches. An electron microscopic study. Am. J. Anat. 136, 455–478.

Bokisch, V.A. and Sobel, A.T. (1974) Receptor for the fourth component of complement on human B lymphocytes and cultured human lymphoblastoid cells. J. Exp. Med. 140, 1336–1347.

Bokisch, V.A. and Theofilopoulos, A.N. (1973) Receptor for native C3 on human lymphoblastoid cell lines. J. Immunol. 111, 300.

Bollum, F.J. (1974) Terminal deoxynucleotidyl transferase. In: The Enzymes (Boyer, P.D., ed.) 3rd. ed., pp. 145–171, Academic Press, New York.

Bollum, F.J. (1975) Terminal deoxynucleotidyl transferase (TdT) in dexamethasone-treated rat tissues. Fed. Proc. 34, 494.

Borum, K. (1973) Cell kinetics in mouse thymus studied by simultaneous use of ^{3}H-thymidine and colchicine. Cell Tissue Kinet. 6, 545–552.

Boyd, E. (1932) The weight of the thymus gland in health and in disease. Am. J. Dis. Child. 43, 1162–1214.

Boyse, E.A. and Bennett, D. (1974) Differentiation and the cell surface; illustrations from work with T cells and sperm. In: Cellular Selection and Regulation in the Immune Response (Edelman, G.M., ed.) pp. 155–176, Raven Press, New York.

Boyse, E.A. and Old, L.J. (1969) Some aspects of normal and abnormal cell surface genetics. Ann. Rev. Genetics 3, 269–290.

Boyse, E.A., Stockert, E. and Old, L.J. (1967) Modification of the antigenic structure of the cell membrane by thymus-leukemia (TL) antibody. Proc. Nat. Acad. Sci. U.S.A. 58, 954–957.

Boyse, E.A., Stockert, E. and Old, L.J. (1968a) Properties of 4 antigens specified by the Tla locus. Similarities and differences. In: Internat. Convocation on Immunology (Rose, N.R. and Milgram, F., eds.) pp. 353–357, S. Karger, Basel.

Boyse, E.A., Stockert, E. and Old, L.J. (1968b) Isoantigens of the H-2 and Tla loci of the mouse. Interactions affecting their representation on thymocytes. J. Exp. Med. 128, 85–95.

Boyse, E.A., Miyazawa, M., Aoki, T. and Old, L.J. (1968c) Two systems of lymphocyte isoantigens in the mouse. Proc. Roy. Soc. B170, 175–193.

Boyse, E.A., Old, L.J. and Stockert, E. (1971a) The relation of linkage group IX to leukemogenesis in the mouse. In: RNA Viruses and Host Genome in Oncogenesis (Emmelot, P. and Bentvelzen, P., eds.) pp. 171–185, North-Holland, Amsterdam.

Boyse, E.A., Itakura, K., Stockert, E., Iritani, C.A. and Miura, M. (1971b) A third locus specifying alloantigens expressed only on thymocytes and lymphocytes. Transplantation, 11, 351–353.

Bradley, T.R. and Metcalf, D. (1966) The Growth of Mouse Bone Marrow Cells in vitro. Aust. J. Exp. Biol. Med. Sci. 44, 287–300.

Brahim, F. and Osmond, D.G. (1973) The migration of lymphocytes from bone marrow to popliteal lymph nodes demonstrated by selective bone marrow labeling with ^{3}H-thymidine in vivo. Anat. Rec. 175, 737–746.

Brondz, B.D. (1972) Lymphocyte receptors and mechanisms of in vitro. Cell-mediated immune reactions. Transplant. Rev. 10, 112–151.

Brown, G. and Greaves, M.F. (1974) Cell surface markers for human T and B lymphocytes. Eur. J. Immunol. 4, 302–310.

Brown, J.C., deJesus, D.G., Holborrow, E.J. and Harris, F. (1970) Lymphocyte-mediated transport of aggregated human γ-globulin into germinal centre areas of normal mouse spleen. Nature, 228, 367–369.

Brunner, K.T. and Cerottini, J.-C. (1971) Cytotoxic lymphocytes as effector cells of cell-mediated immunity. In: Progress in Immunology (Amos, B., ed.) pp. 385–398, Academic Press, New York.

Bryant, B.J. (1971) Quantitative studies of thymocytokinetics. Adv. Exp. Med. Biol. 12, 103–112.

Buckton, K.E., Court-Brown, W.M. and Smith, P.G. (1967) Lymphocyte survival in men treated with X-rays for ankylosing spondylitis. Nature, 214, 470–473.

Buerki, H., Cottier, H., Hess, M.W., Laissue, J. and Stoner, R.D. (1974) Distinctive medullary and germinal center proliferative patterns in mouse lymph nodes after regional primary and secondary stimulation with tetanus toxoid. J. Immunol. 112, 1961–1970.

Burnet, F.M. (1959) The Clonal Selection Theory of Acquired Immunity. Cambridge University Press, Cambridge.

Burnet, F.M. (1974) Invertebrate precursors to immune responses. Contemp. Top. Immunobiol. 4, 13–24.

Calzolari, A. (1898) Recherces experimentales sur un rapport probable entre la fonction du thymus et celle testicules. Arch. Ital. Biol. (Turin) 30, 71–77.

Cantor, H. (1972) Two stages in development of lymphocytes. In: Cell Interactions (Silvestri, L.G., ed.) pp. 172–182, North-Holland, Amsterdam.

Cantor, H. and Asofsky, R. (1972) Synergy among lymphoid cells mediating the graft-versus-host response. III. Evidence for interaction between two classes of thymus-derived cells. J. Exp. Med. 135, 764–779.

Cantor, H. and Boyse, E.A. (1975a) Functional subclasses of T lymphocytes bearing different Ly antigens. I. The generation of functionally distinct T-cell subclasses is a differentiative process independent of antigen. J. Exp. Med. 141, 1376–1389.

Cantor, H. and Boyse, E.A. (1975b) Functional subclasses of T lymphocytes bearing different Ly antigens. II. Cooperation between subclasses of Ly^+ cells in the generation of killer activity. J. Exp. Med. 141, 1390–1399.

Cerottini, J.-C. and Brunner, K.T. (1974) Cell-mediated cytotoxicity, allograft rejection, and tumor immunity. Adv. Immunol. 18, 67–132.

Chapel, H.M. (1973) The effects of papain, trypsin, and phospholipase A on rosette formation. Transplantation, 15, 320–325.

Chanana, A.D., Cronkite, E.P., Joel, D.D., Williams, R.M. and Waksman, B.H. (1971) Migration of thymic lymphocytes: immunofluorescence and ^{3}HTdR labeling studies. Adv. Exp. Med. Biol. 12, 113–118.

Chanana, A.D., Joel, D.D., Schaedeli, J., Hess, M.W. and Cottier, H. (1973) Thymus cell migration: ^{3}HTdR-labeled and theta-positive cells in peripheral lymphoid tissues of newborn mice. Adv. Exp. Med. Biol. 29, 79–85.

Chang, L.M.S. (1971) Development of terminal deoxynucleotidyl transferase activity in embryonic calf thymus gland. Biochem. Biophys. Res. Commun. 44, 124–131.

Claesson, M.H. (1972) Diurnal variations in thymic lymphoid cell decay. Studies of intact, adrenalectomized, and adrenaline-treated mice. Acta Endocrinologica 70, 247–251.

Clark, S.L., Jr. (1964) The penetration of proteins and colloidal materials into the thymus from the blood stream. In: The Thymus, Wistar Inst. Monogr. No. 2 (Defendi, V. and Metcalf, D., eds.) pp. 9–32, Wistar Inst. Press, Philadelphia.

Clark, S.L., Jr. (1968) Incorporation of sulfate by the mouse thymus: its relation to secretion by medullary epithelial cells and to thymic lymphopoiesis. J. Exp. Med. 128, 927–957.

Clark, S.L., Jr. (1973) The intrathymic environment. Contemp. Top. Immunobiol. 2, 77–99.

Cohen, E.P., Majer, J. and Friedman, K. (1969) Responses to immunization in the thymus of the adult mouse. J. Exp. Med. 130, 1161–1174.

Cohen, J.J. (1972) Changes in lymphocyte circulation after hydrocortisone treatment. In: Cell Interactions (Silvestri, L. ed.) pp. 162–163, North-Holland, Amsterdam.

Cohen, J.J. and Claman, H.N. (1971) Thymus-marrow immunocompetence. V. Hydrocortisone-resistant cells and process in the hemolytic antibody response of mice. J. Exp. Med. 133, 1026–1034.

Cohen, P. and Gershon, R.K. (1975) The role of cortisone-sensitive thymocytes in DNA synthetic responses to antigen. Ann. N.Y. Acad. Sci. 249, 451–461.

Cohen, S.A., Ehrke, M.J. and Mihich, E. (1975) Mouse effector functions involved in the antibody-dependent cellular cytoxicity to xenogeneic erythrocytes. J. Immunol. 115, 1007–1012.
Coleman, M.S., Hutton, J.J. and Bollum, F.J. (1974a) Terminal deoxynucleotidyl transferase and DNA polymerase in classes of cells from rat thymus. Biochem. Biophys. Res. Commun. 58, 1104–1109.
Coleman, M.S., Hutton, J.J., DeSimone, P. and Bollum, F.J. (1974b) Terminal deoxynucleotidyl transferase in human leukemia. Proc. Nat. Acad. Sci. USA 71, 4404–4408.
Colley, D.G., Malakian, A. and Waksman, B.H. (1970a) Cellular differentiation in the thymus. II. Thymus-specific antigens in rat thymus and peripheral lymphoid cells. J. Immunol. 104, 585–592.
Colley, D.G., Shih Wu, A.Y. and Waksman, B.H. (1970b) Cellular differentiation in the thymus. III. Surface properties of rat thymus and lymph node cells separated on density gradients. J. Exp. Med. 132, 1107–1121.
Comsa, J. (1953) Nouvelles recherches sur les connexions physiologiques entre le thymus et les hormones sexuelles et le thymus chez le cobaye. Physiol. Comp. Oecol. 3, 128–134.
Comsa, J. (1958) Role of the thymus in the effect of thyroxin on leucopoiesis in guinea pigs. Acta. Endocrin. 27, 455–463.
Comsa, J. (1973) Hormonal interactions of the thymus. In: Thymic Hormones (Luckey, T.D., ed.) pp. 59–96, University Park Press, Baltimore, Md.
Cone, R.E. (1975) The T cell receptor problem. Progr. Allergy, in press.
Cone, R.E., Sprent, J. and Marchalonis, J.J. (1972) Antigen binding specificity of isolated cell surface immunoglobulin from thymus cells activated to histocompatibility antigens. Proc. Nat. Acad. Sci. U.S.A. 69, 2556–2560.
Coombs, R.R.A., Gurner, B.W., Wilson, A.B., Holm, C. and Lindgren, B. (1970) Rosette function between human lymphocytes and sheep red cells not involving immunoglobulin receptors. Int. Arch. Allergy Appl. Immunol. 39, 658–663.
Cooper, E.L. (1973) The thymus and lymphomyeloid system in poikilothermic vertebrates. Contemp. Top. Immunobiol. 2, 13–38.
Cooper, M.D., Peterson, R.D.A. and Good, R.A. (1965) Delineation of the thymic and bursal lymphoid systems in the chicken. Nature, 205, 143–146.
Cooper, M.D., Peterson, R.D.A., South, M.A. and Good, R.A. (1966) The function of the thymus system and the bursa system in the chicken. J. Exp. Med. 123, 75–102.
Cooper, M.D., Perey, D.Y., McKneally, M.F., Gabrielson, A.E., Sutherland, D.E.R. and Good, R.A. (1966) A mammalian equivalent of the avian bursa of Fabricius. Lancet I, 1388–1391.
Cooper, M.D., Perez, D.Y., Gabrielson, A.E., Sutherland, D.E.R., McKneally, M.F. and Good, R.A. (1968) Production of an antibody deficiency syndrome in rabbits by neonatal removal of organized intestinal lymphoid tissues. Int. Arch. Allergy Appl. Immunol. 33, 65–88.
Cooper, M.D., Cain, W.A., Van Alter, P.J. and Good, R.A. (1969) Development and function of the immunoglobulin-producing system: I. Effect of bursectomy at different stages of development on germinal centers, plasma cells, immunoglobulins and antibody production. Int. Arch. Allergy Appl. Immunol. 35, 242–252.
Cooper, M.D., Lawton, A.R. and Kincade, P.W. (1972a) A two-stage model for development of antibody-producing cells. Clin. Exp. Immunol. 11, 143–149.
Cooper, M.D., Lawton, A.R. and Kincade, P.W. (1972b) A developmental approach to the biological basis for antibody diversity. Contemp. Top. Immunobiol. 1, 33–68.
Cooper, M.D., Faulk, W.P., Fudenberg, H.H., Good, R.A., Hitzig, W., Kunkel, H., Rosen, F.S., Seligmann, M., Soothill, J. and Wedgwood, R.J. (1973) Classification of primary immunodeficiencies. New Eng. J. Med. 288, 966–967.
Cronkite, E.P. and Chanana, A.D. (1970) Lymphocytopoiesis. In: Formation and Destruction of Blood Cells (Greenwalt, T.J. and Jamieson, G.A., eds.) pp. 284–303, Lippincott, Philadelphia.
Cuatrecasas, P. (1969) Interaction of insulin with the cell membrane: The primary action of insulin. Proc. Nat. Acad. Sci. U.S.A. 63, 450–457.
Cudkowicz, G., Bennett, M. and Shearer, G.M. (1964) Pluripotent stem cell function of the mouse marrow "lymphocyte." Science, 144, 866–868.
Cullen, S.E., David, C.S., Shreffler, D.C. and Nathenson, S.G. (1974) Membrane molecules deter-

mined by the H-2 associated immune response region: isolation and some properties. Proc. Nat. Acad. Sci. U.S.A. 71, 648–652.

Cunningham, A.J. and Sercarz, E.E. (1971) The asynchronous development of immunobiological memory in helper (T) and precursor (B) cell lines. Eur. J. Immunol. 1, 413–421.

Dardenne, M., Papiernik, M., Bach, J.F. and Stutman, O. (1974) Studies on thymus products. III. Epithelial origin of the thymic hormone. Immunology, 27, 299–304.

David, C.S. and Shreffler, D.C. (1974) I region associated antigen system (Ia) of the mouse H-2 gene complex. Further definition of Ia specificities with restricted anti-Ia antisera. Transplantation, 18, 313–321.

Davies, A.J.S., Leuchars, E., Wallis, V. and Koller, P.C. (1966) The mitotic response of thymus-derived cells to antigenic stimulus. Transplantation, 4, 438–451.

Davis, W.C. (1972) H-2 antigen on cell membranes. An explanation for the alteration of distribution by indirect labeling techniques. Science, 175, 1006–1008.

De Kruyff, R.H., Durkin, H.G., Gilmour, D.G. and Thorbecke, G.J. (1975a) Migratory patterns of B lymphocytes. II. Fate of cells from central lymphoid organs in the chicken. Cell. Immunol. 16, 301–314.

De Kruyff, R.H., Gilmour, D.G. and Thorbecke, G.J. (1975b) Migratory patterns of B. lymphocytes. III. Inhibition of splenic follicular localization of transferred chicken bursa cells by preincubation with anti-Ig. J. Immunol., 114, 1700–1704.

Delovitch, T.L. and McDevitt, H.O. (1975) Isolation and characterization of murine Ia antigens. Immunogenetics, 2, 39–52.

Demant, P. (1973) H-2 gene complex and its role in alloimmune reactions. Transplant. Rev. 15, 162–200.

Dennert, G. and Lennox, E.S. (1973) Phagocytic cells as effectors in a cell-mediated immunity system. J. Immunol. 111, 1844–1854.

DeSousa, M.A.B., Parrott, D.M.V. and Pantelouris, E.M. (1969) The lymphoid tissues in mice with congenital aplasia of the thymus. Clin. Exp. Immunol. 4, 637–644.

Dickler, H.B. (1974) Studies of the human lymphocyte receptor for heat-aggregated or antigen-complexed immunoglobulin. J. Exp. Med. 140, 508–522.

Dickler, H.B. and Sachs, D.H. (1974) Evidence for identity or close association of the Fc receptor of B lymphocytes and alloantigens determined by the Ir region of the H-2 complex. J. Exp. Med. 140, 779–796.

Dougherty, T.F. (1952) Effects of hormones on lymphatic tissue. Physiol. Rev. 32, 339–340.

Douglas, T.C. (1972) Occurrence of a theta-like antigen in rats. J. Exp. Med. 136, 1054–1062.

Dresser, D.W. and Mitchison, N.A. (1968) The mechanism of immunological paralysis. Adv. Immunol. 8, 129–180.

Dukor, P. and Hartmann, K.U. (1973) Bound C3 as the second signal for B cell activation. Cell Immunol. 7, 349–356.

Dukor, P., Bianco, C. and Nussenzweig, V. (1970) Tissue localization of lymphocytes bearing a membrane receptor for antigen-antibody-complement complexes. Proc. Nat. Acad. Sci. U.S.A. 67, 991–997.

Dukor, P., Bianco, C. and Nussenzweig, V. (1971) Bone marrow origin of complement-receptor lymphocytes. Eur. J. Immunol. 1, 491–494.

Dukor, P., Miller, J.F.A.P., House, W. and Allman, V. (1965) Regeneration of thymus grafts. I. Histological and cytological aspects. Transplantation, 3, 639–668.

DuPasquier, L. (1973) Ontogeny of the immune response in cold blooded vertebrates. Curr. Topics Microbiol. Immunol. 61, 37–88.

Durkin, H.G., Caporale, L. and Thorbecke, G.J. (1975) Migratory patterns of B lymphocytes. I. Fate of cells from central and peripheral lymphoid organs in the rabbit and its selective alteration by anti-immunoglobulin. Cell. Immunol. 16, 285–300.

Durkin, H.G., Theis, G.A. and Thorbecke, G.J. (1972) Bursa of Fabricius as site of origin of germinal centre cells. Nature New Biology 235, 118–119.

Dwyer, J.M. and Warner, N.L. (1971) Antigen binding cells in embryonic chicken bursa and thymus. Nature New Biology 229, 210–211.

Edelman, G.M., Spear, P.G., Rutishauser, U. and Yahara, I. (1974) Receptor specificity and mitogenesis in lymphocyte populations. In: The Cell Surface in Development (Moscona, A.A., ed.) pp. 141–164, John Wiley and Sons, New York.

Eden, A., Bianco, C. and Nussenzweig, V. (1971) A population of lymphocytes bearing a membrane receptor for antigen-antibody-complement complexes. II. Specific isolation. Cell. Immunol. 2, 658–669.

Eden, A., Bianco, C. and Nussenzweig, V. (1973) Mechanism of binding of soluble immune complexes to lymphocytes. Cell. Immunol. 7, 459–473.

Edidin, M. (1972) The tissue distribution and cellular location of transplantation antigens. In: Transplantation Antigens (Kahan, B.D. and Reisfeld, R.A., eds.) pp. 125–140, Academic Press, New York.

Edidin, M. (1974) Arrangement and rearrangement of cell surface antigens in a fluid plasma membrane. In: Cellular Selection and Regulation in the Immune Response (Edelman, G.M., ed.) pp. 121–132, Raven Press, New York.

Ehrlich, P. (1900) On immunity with special reference to cell life. Proc. Roy. Soc. 66, 424–448.

El-Arini, M.O. and Osoba, D. (1973) Differentiation of thymus-derived cells from precursors in mouse bone marrow. J. Exp. Med. 137, 821–837.

Ernstrom, V. (1970) Hormonal influences on thymic release of lymphocytes into the blood. In: Hormones and the Immune Response, Ciba Foundation Study Group No. 36 (Wolstenholme, G.E.W. and Knight, J., eds.) pp. 53–59, Churchill, London.

Ernstrom, V. and Larsson, B. (1967) Export and import of lymphocytes in the thymus during steroid-induced involution and regeneration. Acta Path. Microbiol. Scand. 70, 371–384.

Ernstrom, V. and Sandberg, G. (1970) Quantitative relationship between release and intrathymic death of lymphocytes. Acta Path. Microbiol. Scand. 78, 362–363.

Everett, N.B. and Caffrey, R.W. (1967) Radioautographic studies of bone marrow small lymphocytes. In: The Lymphocyte in Immunology and Haemopoiesis (Yoffey, J.M., ed.) pp. 108–120, Edward Arnold, London.

Everett, N.B., Caffrey, R.W. and Rieke, W.O. (1964) Recirculation of lymphocytes. Ann. N.Y. Acad. Sci. 113, 887–897.

Fathman, C.G., Small, M., Herzenberg, L.A. and Weissman, I.L. (1975) Thymus cell maturation. II. Differentiation of three "mature" subclasses in vivo. Cell. Immunol. 15, 109–128.

Feldmann, M. (1973) Histocompatibility-linked immune-response genes. What do they do? Transplant. Proc. 5, 1803–1809.

Feldmann, M., Cone, R.E. and Marchalonis, J.J. (1973) Cell interactions in the immune response in vitro. VI. Mediation by T cell surface monomeric IgM. Cell. Immunol. 9, 1–11.

Ferreira, A. and Nussenzweig, V. (1975) Genetic linkage between serum levels of the third component of complement and the H-2 complex. J. Exp. Med. 141, 513–517.

Folch, H. and Waksman, B.H. (1974) The splenic suppressor cell. I. Activity of thymus-dependent adherent cells: Changes with age and stress. J. Immunol. 113, 127–139.

Ford, C.E. (1966) Traffic of lymphoid cells in the body. In: The Thymus, Ciba Symposium (Wolstenholme, G.E.W. and Porter, R., eds.) p. 131, Churchill, London.

Ford, C.E. and Micklem, H.S. (1963) The thymus and lymph nodes in radiation chimaeras. Lancet 1, 359–362.

Ford, W.L. (1969) The kinetics of lymphocyte recirculation within the rat spleen. Cell Tissue Kinet. 2, 171–191.

Ford, W.L. and Atkins, R.C. (1971) Specific unresponsiveness of recirculating lymphocytes after exposure to histocompatibility antigen in F_1 hybrid rats. Nature New Biology 234, 178–180.

Ford, W.L. and Gowans, J.L. (1969) The traffic of lymphocytes. Hematol. 6, 67–83.

Forget, A., Potworowski, E.F., Richer, G. and Borduas, A.G. (1970) Antigenic specificities of bursal and thymic lymphocytes in the chicken. Immunology, 19, 465–468.

Frelinger, J.A., Niederhuber, J.E., David, C.S. and Shreffler, D.C. (1974) Evidence for the expression of Ia (H-2 associated) antigens on thymus derived lymphocytes. J. Exp. Med. 140, 1273–1284.

Friedberg, S.H. and Weissman, I.L. (1974) Lymphoid tissue architecture. II. Ontogeny of peripheral T and B cells in mice: evidence against Peyer's patches as the site of generation of B cells. J. Immunol. 113, 1477–1492.

Froland, S.S. (1972) Binding of sheep erythrocytes to human lymphocytes. A probable marker of T lymphocytes. Scand. J. Immunol. 1, 269–280.

Gad, P. and Clark, S.L. (1968) Involution and regeneration of the thymus in mice induced by bacterial endotoxin and studied by quantitative histology and electron microscopy. Am. J. Anat. 122, 573–606.

Galili, U. and Schlesinger, M. (1975) Subpopulations of human thymus cells differing in their capacity to form stable E-rosettes and in their immunologic reactivity. J. Immunol. 115, 827–833.

Gallo, R.C. (1975) Terminal transferase and leukemia. New Eng. J. Med. 292, 804–805.

Gally, J.A. and Edelman, G.M. (1970) Somatic translocation of antibody genes. Nature, 227, 341–348.

Gavin, J.R., Archer, J.A., Lesniak, M.A., Gorden, P. and Roth, J. (1972) Hormone-receptor interactions in circulating cells: Studies in normal and pathologic states in man. J. Clin. Invest. 51, 35a.

Gavin, J.R., Gorden, P., Roth, J., Archer, J.A. and Buell, D.N. (1973) Characteristics of the human lymphocyte insulin receptor. J. Biol. Chem. 248, 2202–2207.

Gelfand, M.C., Elfenbein, G.J., Frank, M.M. and Paul, W.E. (1974a) Ontogeny of B lymphocytes. II. Relative rates of appearance of lymphocytes bearing surface immunoglobulin and complement receptors. J. Exp. Med. 139, 1125–1141.

Gelfand, M.C., Sachs, Lieberman, R. and Paul, W.E. (1974b) Ontogeny of B lymphocytes. III. Linkage of a gene controlling the rate of appearance of complement receptor lymphocytes. J. Exp. Med. 139, 1142–1153.

Gelfand, M.C., Asofsky, R. and Paul, W.E. (1974c) Ontogeny of B lymphocytes. I. In vitro appearance of Ig-bearing lymphocytes. Cell. Immunol. 14, 460–469.

Gershon, R.K. (1974) T cell control of antibody production. Contemp. Top. Immunobiol. 3, 1–40.

Gershon, R.K., Choen, P., Heuein, R. and Liebhaber, S.A. (1972) Suppressor T cells. J. Immunol. 108, 586–590.

Gershon, R.K., Kondo, K. and Lance, E.M. (1974) Immuno-regulatory role of spleen localizing thymocytes. J. Immunol. 112, 546–554.

Gery, I. and Waksman, B.H. (1972) Potentiation of the T lymphocyte response to mitogens. II. The cellular source of potentiating mediators. J. Exp. Med. 136, 143–155.

Gesner, B.M., Woodruff, J.J. and McCluskey, R.T. (1969) An autoradiographic study of the effect of neuraminidase or trypsin on transfused lymphocytes. Am. J. Pathol. 57, 215–230.

Gillette, R.W., McKenzie, G.O. and Swanson, M.H. (1973) Effect of concanavalin A on the homing of labeled T lymphocytes. J. Immunol. 111, 1902–1905.

Glick, B. (1956) The growth of the bursa of Fabricius in chickens. Poultry Sci. 35, 843–851.

Glick, B. (1960) Growth of the bursa of Fabricius and its relationship to the adrenal gland in the White Pekin duck, White Leghorn, outbred and inbred New Hampshire. Poultry Sci. 39, 130–139.

Glick, B. (1964) The bursa of Fabricius and the development of immunologic competence. In: The Thymus in Immunobiology (Good, R.A. and Gabrielson, A.E., eds.) pp. 343–358, Harper and Row, New York.

Goidl, E.A. and Siskind, G.W. (1974) Ontogeny of B-lymphocyte function. I. Restricted heterogeneity of the antibody response of B lymphocytes from neonatal and fetal mice. J. Exp. Med. 140, 1285–1302.

Goldschneider, I. (1974) Surface antigens and differentiation of thymus-dependent lymphocytes. In: The Cell Surface in Development (Moscona, A.A., ed.) pp. 165–185, Wiley-Interscience, New York.

Goldschneider, I. (1975a) Antigenic relationship between medullary thymocytes and a subpopulation of peripheral T cells in the rat: description of a masked antigen. Cell. Immunol. 16, 269–284.

Goldschneider, I. (1975b) Antigenic relationship between bone marrow lymphocytes, cortical thymocytes and a subpopulation of peripheral T cells in the rat: Description of a bone marrow lymphocyte antigen. Cell Immunol. (in press).

Goldschneider, I. and Cogen, R.B. (1973) Immunoglobulin molecules on the surface of activated T lymphocytes in the rat. J. Exp. Med. 138, 163–175.

Goldschneider, I. and McGregor, D.D. (1966) Development of immunologically competent cells in the rat. Nature, 212, 1433–1435.

Goldschneider, I. and McGregor, D.D. (1968a) Migration of lymphocytes and thymocytes in the rat.

I. The route of migration from blood to spleen and lymph nodes. J. Exp. Med., 127, 155–168.
Goldschneider, I., and McGregor, D.D. (1968b) Migration of lymphocytes and thymocytes in the rat. II. Circulation of lymphocytes and thymocytes from blood to lymph. Lab. Invest. 18, 397–406.
Goldschneider, I., and McGregor, D.D. (1973) Anatomical distribution of T and B lymphocytes in the rat: Development of lymphocyte-specific antisera. J. Exp. Med. 138, 1443–1465.
Goldschneider, I. and Moscona, A.A. (1972) Tissue-specific cell-surface antigens in embryonic cells. J. Cell. Biol. 53, 435–449.
Goldstein, A.L., Guha, A., Howe, M.L., and White, A. (1971) Ontogenesis of cell-mediated immunity in murine thymocytes and spleen cells and its acceleration by thymosin, a thymic hormone. J. Immunol. 106, 773–780.
Goldstein, A.L., Guha, A., Zatz, M.M., Hardy, M.A. and White, A. (1972) Purification and biological activity of thymosin, a hormone of the thymus gland. Proc. Nat. Acad. Sci. U.S.A. 69, 1800–1803.
Goldstein, G. (1975) The isolation of thymopoietin (thymin). Ann. N.Y. Acad. Sci. 249, 177–185.
Goldstein, G. and MacKay, I.R. (1967) The thymus in systemic lupus erythematosis: A quantitative histopathological analysis and comparison with stress involution. Brit. Med. J. 2, 475–478.
Goldstein, P., Blomgren, H. and Svedmyr, E.A. (1973) The extent of specific adsorption of cytotoxic educated thymus cells: evolution with time and number of injected cells. Cell. Immunol. 7, 213–221.
Good, R.A. (1955) Studies of agammaglobunemia. II. Failure of plasma cell formation in the bone marrow and lymph nodes of patients with agammaglobunemia. J. Lab. Clin. Med. 46, 167–181.
Gorczynski, R., Kontiainen, S., Mitchison, N.A. and Tigelaar, R.E. (1974) Antigen-antibody complexes as blocking factors on the T lymphocyte surface. In: Cellular Selection and Regulation in the Immune Response (Edelman, G.M., ed.) pp. 143–154, Raven Press, New York.
Gottlieb, P.D. (1974) Genetic correlation of a mouse light chain V-region marker with a thymocyte surface antigen. Immunogenetics, 1, 530.
Gotze, D. (1974) T(Iat)- and B(Iab)-cell alloantigens determined by the H-2 linked I region in mice. Immunogenetics, 1, 495–506.
Gotze, D., Reisfield, R.A. and Klein, J. (1973) Serological evidence for antigens controlled by the Ir region in mice. J. Exp. Med. 138, 1003–1008.
Gowans, J.L. and Knight, E.J. (1964) The route of recirculation of lymphocytes in the rat. Proc. Roy. Soc. London B158, 257–282.
Gowans, J.L. and McGregor, D.D. (1965) The immunological activities of lymphocytes. Prog. Allergy 9, 1–78.
Greaves, M. (1975) Antigen receptors on T lymphocytes: a solution in sight? Nature, 256, 92–93.
Greaves, M.F. (1975) Scratching the surface. In: Immune Recognition (Rosenthal, A.S., ed.) pp. 3–20, Academic Press, New York.
Greaves, M.F. and Bauminger, S. (1972) Activation of T and B lymphocytes by insoluble phytomitogens. Nature New Biology 235, 67–70.
Greaves, M.F. and Hogg, N.W. (1971) Antigen binding sites on mouse lymphoid cells. In: Cell Interactions and Receptor Antibodies in Immune Responses (Makela, O., Cross, A. and Kosunen, T.U., eds.) pp. 145–155, Academic Press, New York.
Greaves, M.F. and Janossy, G. (1972) Elicitation of selective T and B lymphocyte responses by cell surface binding ligands. Transplant. Rev. 11, 87–130.
Greaves, M.F., Owen, J.J.T. and Raff, M.C. (1973) T and B Lymphocytes, p. 185, American Elsevier, New York.
Greenberg, A.H., Hudson, L., Shen, L. and Roitt, I. (1973) Antibody-dependent cell-mediated cytotoxicity due to a "null" lymphoid cell. Nature New Biology 242, 111–113.
Griscelli, C., Vassalli, D. and McClusky, R.T. (1969) The distribution of large dividing lymph node cells in syngeneic recipient rats after intravenous injection. J. Exp. Med. 130, 1427–1451.
Grobler, P., Buerki, H., Cottier, H., Hess, M. W. and Stoner, R.D. (1974) Cellular bases for relative radioresistance of the antibody-forming system at advanced stages of the secondary response to tetanus toxoid in mice. J. Immunol. 112, 2154–2165.
Gronowicz, E. and Continho, A. (1975) Functional analysis of B cell heterogeneity. Transplant. Rev. 24, 3–40.

Grossi, C.E., Genta, V., Ferrarini, M. and Zaccheo, D. (1968) Localization of immunoglobulins in the developing chicken bursa of Fabricius. Rev. Franc. Étud. Clin. Biol. 13, 497–500.

Gutman, G.A. and Weissman, I.L. (1972) Lymphoid tissue architecture. Experimental analysis of the origin and distribution of T-cells and B-cells. Immunol. 23, 465–479.

Gutman, G. and Weissman, I.L. (1973) Homing properties of thymus-independent follicular lymphocytes. Transplantation 16, 621–629.

Ha, T.-Y., Waksman, B.H. and Treffers, H.P. (1974) The thymic suppressor cell. I. Separation of subpopulations with suppressor activity. J. Exp. Med. 139, 13–23.

Hall, J.G. and Morris, B. (1965) The immediate effect of antigens on the cell output of a lymph node. Brit. J. Exp. Pathol. 46, 450–454.

Hall, J.G., Morris, B., Moreno, G.D. and Bessis, M.C. (1967) The ultrastructure and function of the cells in lymph following antigenic stimulation. J. Exp. Med. 125, 9–110.

Hämmerling, G.J. and McDevitt, H.O. (1974) Antigen binding T and B lymphocytes. II. Studies on the inhibition of antigen binding to T and B cells by anti-immunoglobulin and anti-H-2 sera. J. Immunol. 112, 1734–1740.

Hämmerling, G.J., Deak, B.D., Mauve, G., Hämmerling, V. and McDevitt, H.O. (1974a). B lymphocyte alloantigens controlled by the I region of the major histocompatibility complex in mice. Immunogenetics. 1, 68–81.

Hämmerling, G.J., Mauve, G., Goldberg, E. and McDevitt, H.O. (1974b) Tissue distribution of Ia antigens. Ia on spermatozoa, macrophages, and epidermal cells. Immunogenetics, 1, 428–437.

Hammond, W.S. (1954) Origin of thymus in the chick embryo. J. Morphol. 95, 501–521.

Hanna, M.G. and Hunter, R.L. (1971) Localization of antigen and immune complexes, with special reference to germinal centers. Adv. Exp. Med. Biol. 12, 257–279.

Hanna, M.G., Nettesheim, P. and Walburg, H.E. (1969) A comparative study of the immune reaction in germfree and conventional mice. Adv. Exp. Med. Biol. 3, 237–248.

Hanna, M.G., Szakal, A.K. and Walburg, H.E. (1968) The relation of antigen and virus localization to the development and growth of lymphoid germinal centers. Adv. Exp. Med. Biol. 5, 149–165.

Harris, C. (1961) The lymphocyte-like cell in the marrow of rats. Blood, 8, 691–701.

Harris, J.E. and Ford, C.E. (1964) Cellular traffic of the thymus: Experiments with chromosome markers. Nature, 201, 884–885.

Harris, J.E., Ford, C.E., Barnes, D.W. and Evans, E.P. (1964) Cellular traffic of the thymus: experiments with chromosome markers. Evidence from parabiosis for an afferent stream of cells. Nature, 201, 886–887.

Harris, P.F. and Kugler, J.H. (1967) Transfusion of regenerating bone marrow into irradiated guinea pigs. In: The Lymphocyte in Immunology and Haemopoiesis (Yoffey, J.M., ed.) pp. 108–120, Edward Arnold, London.

Haskill, J.S. and Moore, M.A.S. (1970) Two dimensional cell separation: comparison of embryonic and adult haemopoietic stem cells. Nature, 226, 853–854.

Hauptfield, V., Hauptfield, M. and Klein, J. (1974) Tissue distribution of I region-associated antigens in the mouse. J. Immunol. 113, 181–188.

Hellström, V., Zeromski, J. and Perlmann, P. (1971) Immunoglobulin light chain determinants on unstimulated and stimulated human blood lymphocytes, assayed by indirect immunofluorescence. Immunology, 20, 1099–1111.

Henson, P.M., Johnson, H.B. and Speigelberg, H.C. (1972) The release of granule enzymes from human neutrophils stimulated by aggregated immunoglobulins of different classes and sublcasses. J. Immunol. 109, 1182–1192.

Herrod, H.G. and Warner, N.L. (1972) Inhibition by anti-μchain sera of the cellular transfer of antibody and immunoglobulin synthesis in mice. J. Immunol. 108, 1712–1717.

Herzenberg, L.A., McDevitt, H.O. and Herzenberg, L.A. (1968) Genetics of antibodies. Ann. Rev. Genetics 2, 209–244.

Hildemann, W.H. and Reddy, A.L. (1973) Phylogeny of immune responsiveness: marine invertebrates. Fed. Proc. 32, 2188–2194.

Hood, L. and Talmage, D.W. (1970) Mechanism of antibody diversity: Germ line basis for variability. Science, 168, 325–334.

Howard, J.C. (1972) The life-span and recirculation of marrow-derived small lymphocytes from the rat thoracic duct. J. Exp. Med. 135, 185–199.

Howard, J.C. and Scott, D.W. (1974) The identification of sera distinguishing marrow-derived and thymus-derived lymphocytes in the rat thoracic duct. Immunol. 27, 903–922.

Howard, J.C., Hunt, S. V. and Gowans, J.L. (1972) Identification of marrow-derived and thymus-derived small lymphocytes in the lymphoid tissue and thoracic duct of normal rats. J. Exp. Med. 135, 200–219.

Ishidate, M. and Metcalf, D. (1963) The pattern of lymphopoiesis in the mouse thymus after cortisone administration of adrenalectomy. Aust. J. Exp. Biol. 41, 637–649.

Itakura, K., Hutton, J.J., Boyse, E.A. and Old, L.J. (1972) Genetic linkage relationships of loci specifying differentiation alloantigens in the mouse. Transplantation, 13, 239–243.

Ivanyi, P. (1970) The major histocompatibility antigens in various species. Curr. Topics Microbiol. Immunol. 53, 1–90.

Ivanyi, J. (1973) Sequential recruitment of antibody class-committed B lymphocytes during otogeny. Eur. J. Immunol. 3, 789–793.

Jaffe, H.L. (1924) The influence of the suprarenal gland on the thymus. II. Direct evidence of regeneration of the involuted thymus following double suprarenalectomy. J. Exp. Med. 40, 619–625.

Jankovic, B.D. and Leskowitz, S. (1965) Restoration of antibody producing capacity in bursectomized chickens by bursal grafts in Millipore chambers. Proc. Soc. Exp. Biol. Med. 118, 1164–1166.

Janossy, G. and Greaves, M. (1975) Functional analysis of murine and human B lymphocyte subsets. Transplant. Rev. 24, 177–236.

Jerne, N.K. (1971) The somatic generation of immune recognition. Eur. J. Immunol. 1, 1–9.

Johnston, J.M. and Wilson, D.B. (1970) Origin of immunoreactive lymphocytes in rats. Cell. Immunol. 1, 430–444.

Jondal, M., Holm, G. and Wigzell, H. (1972) Surface markers on human T and B lymphocytes. I. A large population of lymphocytes forming non-immune rosettes with SRBC. J. Exp. Med. 136, 207–215.

Jones, G. and Roitt, I.M. (1972) Immunoglobulin determinants on lymphoid cells in culture. Cell. Immunol. 3, 478–492.

Kater, L. (1973) A note on Hassall's corpuscles. Contemp. Top. Immunobiol. 2, 101–109.

Katz, D.H. and Benacerraf, B. (1972) The regulatory influence of activated T cells on B cell responses to antigen. Adv. Immunol. 15, 1–94.

Katz, D.H. and Benacerraf, B. (1975) The function and interrelationships of T-cell receptors, Ir genes and other histocompatibility gene products. Transplant. Rev. 22, 175–195.

Katz, D.H., Hamaoka, T., Dorf, E.D., Maurer, P.H. and Benacerraf, B. (1973) Cell interactions between histoincompatible T and B lymphocytes. IV. Involvement of the immune response (Ir) gene in the control of lymphocyte interactions in responses controlled by the gene. J. Exp. Med. 138, 734–739.

Kay, H.E.M., Doe, J. and Hockley, A. (1970) Response of human foetal thymocytes to phytohemagglutinin (PHA). Immunology, 18, 393–396.

Kearney, J.F. and Lawton, A.R. (1975) B lymphocyte differentiation induced by lipopolysaccharide. II. Response of fetal lymphocytes. J. Immunol. 115, 677–681.

Kennedy, J.C., Till, J.E., Siminovich, L. and McCulloch, E.A. (1966) The proliferative capacity of antigen-sensitive precursors of hemolytic plaque-forming cells. J. Immunol. 96, 973–980.

Keuning, F.J. and Bos, W.H. (1967) Regeneration patterns of lymphoid follicles in the rabbit spleen after sublethal x-irradiation. In: Germinal Centers in Immune Responses (Cottier, H., Odartchenko, N., Schindler, R. and Congdon, C.C., eds.) pp. 250–258, Springer-Verlag, New York.

Kincade, P.W. and Cooper, M.D. (1971) Development and distribution of immunoglobulin-containing cells in the chicken. J. Immunol. 106, 371–382.

Kincade, P.W. and Cooper, M.D. (1973) Immunoglobulin A: Site and sequence of expression in developing chicks. Science, 179, 398–400.

Kincade, P.W., Lawton, A.R., Bockman, D.E. and Cooper, M.D. (1970) Suppression of immunoglobulin G synthesis as a result of antibody-mediated suppression of immunoglobulin M synthesis in chickens. Proc. Nat. Acad. Sci. USA 67, 1918–1925.

Kincade, P.W., Lawton, A.R. and Cooper, M.D. (1971) Restriction of surface immunoglobulin determinants to lymphocytes of the plasma cell line. J. Immunol. 106, 1421–1423.

Kincade, P.W., Self, K.S. and Cooper, M.D. (1973) Survival and function of bursa-derived cells in bursectomized chickens. Cell. Immunol. 8, 93–102.

Kindred, J.E. (1955) Quantitative studies on lymphoid tissues. Ann. N.Y. Acad. Sci. 59, 746–756.

Klein, J. (1973) List of congenic lines of mice. I. Lines with differences of alloantigen loci. Transplantation, 15, 137–153.

Klein, J. (1975) Biology of the Mouse Histocompatibility-2 Complex. Springer-Verlag, New York.

Klein, J. and Park, J.M. (1973) Graft-versus-host reaction across different regions of the H-2 complex of the mouse. J. Exp. Med. 137, 1213–1255.

Klein, J., Hauptfeld, M. and Hauptfeld, V. (1974) Evidence for a third, Ir-associated histocompatibility region in the H-2 complex of the mouse. Immunogenetics, 1, 45–56.

Komuro, K. and Boyse, E.A. (1973) Induction of T lymphocytes from precursor cells in vitro by a product of the thymus. J. Exp. Med. 138, 479–482.

Komuro, K., Goldstein, G. and Boyse, E.A. (1975) Thymus-repopulating capacity of cells that can be induced to differentiate to T cells in vitro. J. Immunol. 115, 195–198.

Komuro, K., Itakura, K., Boyse, E.A. and John, M. (1974) Ly-5; A new T-lymphocyte antigen system. Immunogenetics, 1, 452–456.

Konda, S., Stockert, E. and Smith, R.T. (1973) Immunologic properties of mouse thymus cells: membrane antigen patterns associated with various cell subpopulations. Cell. Immunol. 7, 275–289.

Kook, A.I. and Trainin, N. (1975) The control exerted by thymic hormone (THF) on cellular cAMP levels and immune reactivity of spleen cells in the MLC assay. J. Immunol. 115, 8–14.

Koster, F.T., McGregor, D.D. and MacKaness, G.B. (1971) The mediator of cellular immunity. II. Migration of immunologically committed lymphocytes into inflammatory exudates. J. Exp. Med. 133, 400–409.

Kung, P.C., Silverstone, A.E., McCaffrey, R.P. and Baltimore, D. (1975) Murine terminal deoxynucleotidyl transferase: cellular distribution and response to cortisone. J. Exp. Med. 141, 855–865.

Lafleur, L., Miller, R.G. and Phillips, R.A. (1972) A quantitative assay for the progenitors of bone marrow-associated lymphocytes. J. Exp. Med. 135, 1363–1374.

Lafleur, L., Miller, R.G. and Phillips, R.A. (1973) Restriction of specificity in the precursors of bone marrow-associated lymphocytes. J. Exp. Med. 137, 954–966.

Lajtha, L.G., Pozzi, L.U., Schofield, R. and Fox, M. (1969) Kinetic properties of hemopoietic stem cells. Cell Tissue Kinet. 2, 39–49.

Lamm, M.E., Boyse, E.A., Old, L.J., Lisowska-Bernstein, B. and Stockert, E. (1968) Modulation of TL (Thymus-Leukemia) antigens by Fab-fragments of TL antibody. J. Immunol. 101, 99–103.

Lance, E.M. and Cooper, S. (1970) Effects of cortisol and anti-lymphocyte serum on lymphoid populations. In: Hormones and the Immune Response. Ciba Foundation Study Group 36 (Wolstenholme, G.E.W. and Knight, J. eds.) pp. 73–95, Churchill, London.

Lance, E.M. and Taub, R.N. (1969) Segregation of lymphocyte populations through differential migration. Nature, 221, 841–843.

Lance, E., Cooper, S. and Boyse, E.A. (1971) Antigenic change and cell maturation in murine thymocytes. Cell. Immunol. 1, 536–544.

Lawrence, D.A., Spiegelberg, H.L. and Weigle, H.O. (1973) 2,4-dinitrophenyl receptors on mouse thymus and spleen cells. J. Exp. Med. 137, 470–482.

Lawton, A.R. and Cooper, M.D. (1974) Modification of B lymphocyte differentiation by anti-immunoglobulins. Contemp. Top. Immunobiol. 3, 193–226.

Lawton, A.R., Kincade, P.W. and Cooper, M.D. (1975) Sequential expression of germ line genes in development of immunoglobulin class diversity. Fed. Proc. 34, 33–39.

Lawton, A.R., Self, K.S., Rogal, S.A. and Cooper, M.D. (1972) Ontogeny of B-lymphocytes in the human fetus. Clin. Immunol. Immunopath. 1, 84–93.

Lay, W.H., Mendes, N.F., Bianco, C. and Nussenzweig, V. (1971) Binding of sheep red blood cells to a large population of human lymphocytes. Nature, 230, 531–532.

LeDouarin, N.M. and Jotereau, F.V. (1975) Tracing of cells of the avian thymus through embryonic life in interspecific chimeras. J. Exp. Med. 142, 17–40.

Lilly, F. (1971) H-2 membranes and viral leukemogenesis. In: Cellular Interactions in the Immune Response (Cohen, S., Cudkowicz, G. and McCluskey, R.T., eds.) pp. 103–108, Karger, Basel.

Linna, T.J. (1968) Cell migration from the thymus to other lymphoid organs in hamsters of different ages. Blood, 31, 727–746.

Linna, T.J., Brenning, T. and Hemmingsson, E. (1968) Lymphoid cell migration and germinal centers. Adv. Exp. Med. Biol. 5, 133–139.

Lis, H. and Sharon, N. (1973) The biochemistry of plant lectins (phytohemagglutinins). Ann. Rev. Biochem. 42, 541–574.

Lonai, P., Mogilner, B., Rotter, V. and Trainin, N. (1973) Studies on the effect of a thymic humoral factor on differentiation of thymus-derived lymphocytes. Eur. J. Immunol. 3, 21–26.

Lubaroff, D.M. (1973) An alloantigenic marker on rat thymus and thymus-derived cells. Transplant. Proc. 5, 115–118.

Luckey, T.D. (1973) Perspective of thymic hormones. In: Thymic Hormones (Luckey, T.D., ed.) pp. 275–314, University Park Press, Baltimore.

MacGillivray, M.H., Mayhew, B. and Rose, N.R. (1970) A comparison of the immunologic function of thymus cells of varying stages of maturation. Proc. Soc. Exp. Biol. Med. 133, 688–692.

MacLennan, I.C.M. (1972) Antibody in the induction and inhibition of lymphocyte cytotoxicity. Transplant. Rev. 13, 67–90.

Malchow, D., Droege, W. and Strominger, J.L. (1972) Solubilization and partial purification of lymphocyte specific antigens in the chicken. Eur. J. Immunol. 2, 30–35.

Mandel, T. (1968) The development and structure of Hassall's corpuscles in the guinea pig: a light and electron microscopic study. Z. Zellforsch. 89, 180–192.

Mandel, T. (1969) Epithelial cells and lymphopoiesis in the cortex of guinea-pig thymus. Aust. J. Exp. Biol. Med. Sci. 47, 153–155.

Mandel, T. (1970) Differentiation of epithelial cells in the mouse thymus. Z. Zellforsch. 106, 498–515.

Mandi, B. and Glant, T. (1973) Thymosin-producing cells of the thymus. Nature New Biology 246, 25.

Marchalonis, J.J. and Cone, R.E. (1973a) The phylogenetic emergence of vertebrate immunity. Aust. J. Exp. Biol. Med. Sci. 51, 461–488.

Marchalonis, J.J. and Cone, R.E. (1973b) Biochemical and biological characteristics of lymphocyte surface immunoglobulin. Transplant. Rev. 14, 3–49.

Marchalonis, J.J., Cone, R.E. and Atwell, J.L. (1972) Isolation and partial characterization of lymphocyte surface immunoglobulins. J. Exp. Med. 135, 956–971.

Marchesi, V.T. and Gowans, J.L. (1964) The migration of lymphocytes through the endothelium of venules in lymph nodes: an electronmicroscopic study. Proc. Roy. Soc. London, Ser. B. 159, 283–290.

Marine, D., Manley, O.T. and Bauman, E.J. (1924) The influence of thyroidectomy, gonadectomy, suprarenalectomy, and splenectomy on the thymus gland of rabbits. J. Exp. Med. 40, 429–443.

Marshall, A.H.E. and White, R.G. (1961) The immunological reactivity of the thymus. Brit. J. Exp. Pathol. 42, 379–385.

Martin, L.N. and Leslie, G.A. (1974) IgM-forming cells as the immediate precursor of IgA-producing cells during ontogeny of the immunoglobulin-producing system of the chicken. J. Immunol. 113, 120–126.

McArthur, W.P., Chapman, J. and Thorbecke, G.J. (1971) Immunocompetent cells of the chicken. I. Specific surface antigenic markers on bursa and thymus cells. J. Exp. Med. 134, 1036–1045.

McCaffrey, R., Smoler, D.F. and Baltimore, D. (1974) DNN polymerases in lymphoid cells. Haematol. Bluttrans. 14, 247–255.

McGregor, D.D. (1968) Bone marrow origin of immunologically competent lymphocytes in the rat. J. Exp. Med. 127, 953–966.

McGregor, D.D. (1969) Effect of tritiated thymidine and 5-bromodeoxyuridine on development of immunologically competent lymphocytes. Immunology, 16, 83–90.

McGregor, D.D., Koster, F.T. and Mackaness, G.B. (1971) The mediator of cellular immunity. I. The life-span and circulation dynamics of the immunologically committed lymphocyte. J. Exp. Med. 133, 389–399.

McKenzie, I.F.C. (1975) LY-4.2: a cell membrane alloantigen of murine B lymphocytes. II. Functional studies. J. Immunol. 114, 856–862.

McKenzie, I.F.C. and Snell, G.D. (1973) Comparative immunogenicity and enhanceability of individual H-2K and H-2D specificities of murine histocompatibility-2 complex. J. Exp. Med. 138, 259–277.

McKenzie, I.F.C. and Snell, G.D. (1975) LY-4.2: a cell membrane alloantigen of murine B lymphocytes. I. Population studies. J. Immunol. 114, 848–855.

McWilliams, M., Phillips-Quagliata, J.M. and Lamm, M.E. (1975) Characteristics of mesenteric lymph node cells homing to gut-associated lymphoid tissue in syngeneic mice. J. Immunol. 115, 54–58.

McKori, T., Chieco-Bianci, L. and Feldman, M. (1965) Production of clones of lymphoid cell populations. Nature, 203, 367–368.

Melcher, U., Vitetta, E.S., McWilliams, M., Lamm, M.E., Phillips-Quagliata, J.M. and Uhr, J.W. (1974) Cell surface immunoglobulin. X. Identification of an IgD-like molecule on the surface of murine splenocytes. J. Exp. Med. 140, 1427–1431.

Melchers, F. and Andersson, J. (1973) Synthesis, surface deposition and secretion of immunoglobulin M in bone marrow-derived lymphocytes before and after mitogenic stimulation. Transplant. Rev. 14, 76–130.

Metcalf, D. (1956) The thymic origin of the plasma lymphocytosis-stimulating factor. Brit. J. Cancer 10, 442–457.

Metcalf, D. (1963) The autonomous behavior of normal thymus grafts. Aust. J. Exp. Biol. Med. Sci. 41, Suppl. 437–447.

Metcalf, D. (1964) The thymus and lymphopoiesis. In: The Thymus in Immunobiology (Good, R.A. and Gabrielsen, A.E., eds.) pp. 150–182, Harper and Row, New York.

Metcalf, D. and Moore, M.A.S. (1971) Haemopoietic Cells. North-Holland Publishing Co., Amsterdam.

Metcalf, D. and Wakonig-Vaartaja (1964) Stem cell replacement in normal thymus grafts. Proc. Soc. Exp. Biol. Med. 115, 731–735.

Metcalf, D. and Wiadrowski, M. (1966) Autoradiographic analysis of lymphocyte proliferation in the thymus and in thymic lymphoma tissue. Cancer Res. 26, 483–491.

Metcalf, D., Sparrow, N., Nakamura, K. and Ishidate, M. (1961) The behaviour of thymus grafts in high and low leukemia strains of mice. Aust. J. Exp. Biol. Med. 39, 441–453.

Michalke, W.D., Hess, M.W., Riedwyl, H., Stoner, R.D. and Cottier, H. (1969) Thymic lymphopoiesis and cell loss in newborn mice. Blood J. Hematol. 33, 541–554.

Micklem, H.S. and Loutit, J.F. (1966) The radiation chimera-establishment and survival. In: Tissue Grafting and Radiation, pp. 77–118, Academic Press, London.

Miller, J.F.A.P. and Mitchell, G.F. (1969) Thymus and antigen reactive cells. Transplant. Rev. 1, 3–42.

Miller, J.F.A.P. and Osoba, D. (1967) Correct concepts of the immunological function of the thymus. Physiol. Rev. 47, 437–520.

Miller, J.F.A.P., Basten, A., Sprent, J. and Cheers, C. (1971) Interaction between lymphocytes in immune responses. Cell. Immunol. 2, 469–495.

Miller, J.F.A.P, Osoba, D. and Dukor, P. (1965) A humoral thymus mechanism responsible for immunologic maturation. Ann. N.Y. Acad. Sci. 124, 95–104.

Mitchison, N.A. (1971) The relative ability of T and B lymphocytes to see protein antigen. In: Cell Interactions and Receptor Antibodies in Immune Responses (Mäkelä, O., Cross, A. and Kosunen, T.U., eds.) pp. 249–260, Academic Press, New York.

Mitchison, N.A., Rajewsky, K. and Taylor, R.B. (1970) Cooperation of antigenic determinants and of cells in the induction of antibodies. In: Developmental Aspects of Antibody Formation and Structure (Sterzl J. and Riha, I., eds.) pp. 547–564, Academic Press, New York.

Möller, G. (1972) Lymphocyte activation by mitogens. Transplant. Rev. 11.

Möller, G. (1974) Effect of B-cell mitogens on lymphocyte subpopulations possessing C'3 and Fc receptors. J. Exp. Med. 139, 969–982.

Möller, G. and Svehag, S.-E. (1972) Specificity of lymphocyte-mediated cytotoxicity induced by in vitro antibody-coated target cells. Cell. Immunol. 4, 1–19.

Moore, M.A.S. and Metcalf, D. (1970) Ontogeny of the haemopoietic system: Yolk sac origin of in vivo and in vitro colony forming cells in the developing mouse embryo. Brit. J. Haematol. 18, 279–296.

Moore, M.A.S. and Owen, J.J.T. (1966) Experimental studies on the development of the bursa of Fabricius. Dev. Biol. 14, 40–51.

Moore, M.A.S. and Owen, J.J.T. (1967a) Chromosome marker studies on the irradiated chick embryo. Nature, 215, 1081–1082.

Moore, M.A.S. and Owen, J.J.T. (1967b) Experimental studies on the development of the thymus. J. Exp. Med. 126, 715–725.

Moore, M.A.S., McNeill, T.A. and Haskill, J.S. (1970) Density distribution analysis of in vivo and in vitro colony forming cells in developing fetal liver. J. Cell. Physiol. 75, 181–192.

Morris, B. (1968) Migration intratissulaire des lymphocytes du mouton. Nouv. Rev. Franc. Hémat. 8, 525–534.

Mosier, D.E. and Pierce, C.W. (1972) Functional maturation of thymic lymphocyte populations in vitro. J. Exp. Med. 136, 1484–1500.

Mozes, E., Shearer, G.M., Melman, K.L. and Bourne, H.R. (1972) In vitro correction of antigen-induced immune suppression: effects of poly (A): poly (U) and prostaglandin E_1. Cell Immunol. 9, 226–233.

Munro, A.J., Taussig, M.J., Campbell, R., Williams, H. and Lawson, Y. (1974) Antigen-specific T-cell factor in cell cooperation: physical properties and mapping in the left-hand (K) half of the H-2. J. Exp. Med. 140, 1579–1587.

Muramatsu, T., Nathenson, S.G., Boyse, E.A. and Old, L.J. (1973) Some biochemical properties of thymus leukemia antigens solubilized from cell membranes by papain digestion. J. Exp. Med. 137, 1256–1262.

Murphy, D.B. and Shreffler, D.C. (1975) Cross-reactivity between H-2K and H-2D products: I. Evidence for extensive and reciprocal serological cross-reactivity. J. Exp. Med. 141, 374–391.

Murray, R.G. and Woods, P.A. (1964) Studies on the fate of lymphocytes. III. The migration and metamorphosis of in situ labelled thymic lymphocytes. Anat. Rec. 150, 113–128.

Nakamuro, K., Tanigaki, N. and Pressman, D. (1973) Multiple common properties of human β_2-microglobulin and the common portion fragment derived from HL-A antigen molecules. Proc. Nat. Acad. Sci. U.S.A. 70, 2863–2865.

Nathenson, S.G. and Cullen, S.E. (1974) Biochemical properties and immunochemical relationships of mouse H-2 alloantigens. Biochim. Biophys. Acta. 344, 1–25.

Naughton, M.A., Geczy, C., Bender, V., Hoffman, H. and Hamilton, E. (1972) Esteropeptidase and thymotropic activity of a protein isolated from the mouse submaxillary gland. Biochim. Biophys. Acta. 263, 106–114.

Neauport-Sautes, C., Bismuth, A., Kourilsky, F.M. and Manuel, Y. (1974) Relationship between HL-A antigens and β_2-microglobulin as studied by immunofluorescence on the lymphocyte membrane. J. Exp. Med. 139, 957–968.

Niederhuber, J.E. and Moller, E. (1972) Antigenic markers on mouse lymphoid cells: the presence of MBLA on antibody-forming cells and antigen binding cells. Cell. Immunol. 3, 559–568.

Nieuwenhuis, P. (1971) On the origin and fate of immunologically competent cells. Wolters-Noordhoff, The Netherlands.

Nieuwenhuis, P. and Keuning, F.J. (1974) Germinal centers and the origin of the B-cell system. II. Germinal centers in the rabbit spleen and popliteal lymph nodes. Immunology, 26, 509–519.

Nieuwenhuis, P., van Nouhuijs, C.E., Eggens, J.H. and Keuning, F.J. (1974) Germinal centers and the origin of the B-cell system. I. Germinal centers in the rabbit appendix. Immunology, 26, 497–507.

Nisbet, N.W., Simonsen, M. and Zaleski, M. (1969) The frequency of antigen-sensitive cells in tissue transplantation: a commentary on clonal selection. J. Exp. Med. 129, 459–467.

Nossal, G.J.V. (1964) Studies on the rate of seeding of lymphocytes from the intact guinea pig thymus. Ann. N.Y. Acad. Sci. 120, 171–181.

Nossal, G.J.V. and Pike, B.L. (1973) Studies on the differentiation of B lymphocytes in the mouse. Immunology, 25, 33–45.

Nossal, G.J.V., Ada, G.L. and Austin, C.M. (1964) Antigens in immunity. IV. Cellular localization of ^{125}I- and ^{131}I-labelled flagella in lymph nodes. Aust. J. Exp. Biol. Med. Sci. 42, 311–330.

Nossal, G.J.V., Cunningham, A., Mitchell, G.F. and Miller, J.F.A.P. (1968) Cell to cell interaction in the immune response. III. Chromosomal marker analysis of single antibody-forming cells in reconstituted, irradiated and thymectomized mice. J. Exp. Med. 128, 839–854.

Nossal, G.J.V., Warner, N.L., Lewis, H. and Sprent, J. (1972) Quantitative features of a sandwich radioimmunolabeling technique for lymphocyte surface receptors. J. Exp. Med. 135, 405–428.

Nowell, P.C., Hirsch, B.E., Fox, D.H. and Wilson, D.B. (1970) Evidence for the existence of lympho-hematopoietic stem cells in the adult rat. J. Cell. Physiol. 75, 151–158.

Nussenzweig, V. (1974) Receptors for immune complexes on lymphocytes. Adv. Immunol. 19, 217–258.

Old, L.J., Stockert, E., Boyse, E.A. and Kim, J.H. (1968) Antigenic modulation. Loss of TL antigen from cells exposed to TL antibody. Study of the phenomenon in vitro. J. Exp. Med. 127, 523–539.

Order, S.E. and Waksman, B.H. (1969) Cellular differentiation in the thymus. Changes in size, antigenic character, and stem cell function of thymocytes during thymus repopulation following irradiation. Transplantation, 8, 783–800.

Osmond, D.G. and Nossal, G.J.V. (1974) Differentiation of lymphocytes in mouse bone marrow. II. Kinetics of maturation and renewal of antiglobulin-binding cells studied by double labeling. Cell. Immunol. 13, 132–145.

Osmond, D.G., Miller, S.C. and Yoshida, Y. (1973) Kinetic and haemopoietic properties of lymphoid cells in the bone marrow. In: Haemopoietic Stem Cells (Wolstenholme, G.E.W. and O'Conner, M. (ed.) pp. 131–156, Associated Scientific Publishers, Amsterdam, New York.

Osoba, D. (1965) Immune reactivity in mice thymectomized soon after birth: Normal response after pregnancy. Science, 147, 298–299.

Ostberg, L., Rask, L., Wigzell, H. and Peterson, P.A. (1975) Thymus leukemia antigen contains β_2-microglobulin. Nature, 253, 735–737.

Owen, J.J.T. and Raff, M.C. (1970) Studies on the differentiation of thymus-derived lymphocytes. J. Exp. Med. 132, 1216–1232.

Owen, J.J.T. and Ritter, M.A. (1969) Tissue interaction in the development of thymus lymphocytes. J. Exp. Med. 129, 431–437.

Owen, J.J.T., Cooper, M.D. and Raff, M.C. (1974) In vitro generation of B lymphocytes in mouse foetal liver, a mammalian bursa equivalent. Nature, 249, 361–363.

Pantelouris, E.M. (1968) Absence of the thymus in a mouse mutant. Nature, 217, 370–371.

Paraskeras, F., Lee, S.T., Orr, K.B. and Israels, L.G. (1972) A receptor for Fc on mouse B-lymphocytes. J. Immunol. 108, 1319–1327.

Parish, C.R. (1975) Separation and functional analysis of subpopulations of lymphocytes bearing complement and Fc receptors. Transplant. Rev. 25, 98–120.

Parkhouse, R.M.E. (1973) Assembly and secretion of immunoglobulin M (IgM) by plasma cells and lymphocytes. Transplant. Rev. 14, 131–144.

Parr, E.L. and Oei, J.S. (1973) Immobilization of membrane H-2 antigens by paraformaldehyde fixation. J. Cell. Biol. 59, 537–548.

Parrott, D.M.V. and de Sousa, M.A.B. (1971) Thymus-dependent and thymus-independent populations: origin, migratory patterns and lifespan. Clin. Exp. Immunol. 8, 663–684.

Parrott, D.M.V., de Sousa, M.A.B. and East, J. (1966) Thymus dependent areas in the lymphoid organs of neonatally thymectomized mice. J. Exp. Med. 123, 191–204.

Paul, W.E. (1970) Functional specificity of antigen-binding receptors of lymphocytes. Transplant. Rev. 5, 130–166.

Pelc, S.R. and Harris, G. (1973) Changes in DNA synthesis during immunization in mouse spleen as shown by autoradiographs with long exposure times. Adv. Exp. Med. Biol. 29, 683–691.

Pepys, M.B. (1972) Role of complement in induction of the allergic response. Nature New Biology 237, 157–159.

Pepys, M.B. (1974) Role of complement in induction of antibody production in vivo. Effect of cobra factor and other C3-reactive agents on thymus-dependent and thymus-independent antibody responses. J. Exp. Med. 140, 126–145.

Perey, D.Y.E., Cooper, M.D. and Good, R.A. (1968) Lymphoepithelial tissues of the intestine and differentiation of antibody production. Science, 161, 265–266.

Perey, D.Y.E., Frommel, D., Hong, R. and Good, R.A. (1970) The mammalian homologue of the avian bursa of Fabricius. II. Extirpation, lethal x-irradiation and reconstitution in rabbits. Effects on humoral immune responses, immunoglobulins and lymphoid tissues. Lab. Invest. 22, 212–227.

Perlmann, P. and Perlmann, H. (1970) Contactual lysis of antibody-coated chicken erthrocytes by purified lymphocytes. Cell. Immunol. 1, 300–315.

Peter, H., Clagett, J., Feldman, J.D. and Weigle, W.O. (1973) Rabbit antiserum to brain-associated thymus antigens of mouse and rat. I. Demonstration of antibodies cross-reacting to T cells of both species. J. Immunol. 110, 1077–1084.

Peterson, P.A., Cunningham, B.A., Berggard, I. and Edelman, G. (1972) β_2-microglobulin-A free immunoglobulin domain. Proc. Nat. Acad. Sci. U.S.A. 69, 1697–1701.

Peterson, P.A., Rask, L. and Lindblom, J.B. (1974) Highly purified papain-solubilized HL-A antigens contain β_2-microglobulin. Proc. Nat. Acad. Sci. U.S.A. 71, 35–39.

Phillips, R.A. and Miller, R.G. (1974) Marrow environment not required for differentiation of B lymphocytes. Nature, 251, 444–446.

Pierce, C.W., Solliday, S.M. and Asofsky, R. (1972a) Immune responses in vitro. IV. Suppression of primary γM, γG, and γA plaque-forming cell responses in mouse spleen cell cultures by class-specific antibody to mouse immunoglobulins. J. Exp. Med. 135, 675–697.

Pierce, C.W., Solliday, S.M. and Asofsky, R. (1972b) Immune responses in vitro. V. Suppression of γM, γG and γA plaque-forming cell responses in cultures of primed mouse spleen cells by class-specific antibody to mouse immunoglobulins. J. Exp. Med. 135, 698–710.

Pierpaoli, W., Fabris, N. and Sorkin, E. (1970) Developmental hormones and immunological maturation. In: Hormones in the Immune Response, Ciba Foundation Study Group No. 36 (Wolstenholme, G.E.W. and Knight, J., eds.) pp. 126–143, Churchill, London.

Piessans, W.F., Schur, P.H., Moloney, W.C. and Churchill, W.H. (1973) Lymphocyte surface immunoglobulins. Distribution and frequency in lymphoproliferative diseases. N. Eng. J. Med. 288, 176–180.

Plaut, M., Lichtenstein, L.M., Gillespie, E. and Henney, C.S. (1973a) Studies on the mechanism of lymphocyte mediated cytolysis. IV. Specificity of the histamine receptor on effector T cells. J. Immunol. 111, 389–394.

Plaut, M., Lichtenstein, L. M. and Henney, C.S. (1973b) Increase in histamine receptors on thymus-derived effector lymphocytes during the primary immune response to alloantigens. Nature, 244, 284–287.

Playfair, J.H.L. and Purves, E.C. (1971) Separate thymus dependent and thymus independent antibody forming cell precursors. Nature New Biology 231, 14–151.

Pluznik, D.H. and Sachs, L. (1965) The cloning of normal "mast" cells in tissue culture. J. Cell. Comp. Physiol. 66, 319–324.

Potworowski, E.F. and Nairn, R.C. (1967) Origin and fate of a thymocyte-specific antigen. Immunology, 13, 597–602.

Poulik, M.D., Bernoco, M., Bernoco, D. and Ceppellini, R. (1973) Aggregation of HL-A antigens at the lymphocyte surface induced by antiserum to β_2-microglobulin. Science, 182, 1352–1355.

Poulik, M.D. and Bloom, A.D. (1973) β_2-microglobulin production and secretion by lymphocytes in culture. J. Immunol. 110, 1430–1433.

Premkumar, E., Shoyab, M. and Williamson, A.R. (1974) Germ line basis for antibody diversity: Immunoglobulin V_H- and C_H-gene frequencies measured by DNA:RNA hybridization. Proc. Nat. Acad. Sci. U.S.A. 71, 99–103.

Pyke, K.W., Rawlings, G.A. and Gelfand, E.W. (1975) Isolation and characterization of the sheep erythrocyte receptor in man. J. Immunol. 115, 211–215.

Rabellino, E.S., Colon, S., Grey, H.M. and Unanue, E.R. (1971) Immunoglobulins on the surface of lymphocytes. I. Distribution and quantitation. J. Exp. Med. 133, 156–167.

Raff, M.C. (1970) Two distinct populations of peripheral lymphocytes in mice distinguishable by immunofluorescence. Immunology, 19, 637–650.

Raff, M.C. (1971) Surface antigenic markers for distinguishing T and B lymphocytes in mice. Transplant. Rev. 6, 52–80.

Raff, M.C. (1973) T and B lymphocytes and immune responses. Nature, 242, 19–23.

Raff, M.C. and Cantor, H. (1971) Subpopulations of thymus cells and thymus-derived lymphocytes. In: Progress in Immunology (Amos, B., ed.) pp. 83–93, Academic Press, New York.

Raff, M.C. and dePetris, S. (1973) Movement of lymphocyte surface antigens and receptors: The fluid nature of the lymphocyte plasma membrane and its immunological significance. Fed. Proc. 32, 48–54.

Raff, M.C., Feldman, M. and dePetris, S. (1973) Monospecificity of bone marrow-derived lymphocytes. J. Exp. Med. 137, 1024–1030.

Raff, M.C. Nase, S. and Mitchison, N.A. (1971) Mouse-specific B lymphocyte antigen (MBLA): a marker for thymus-independent lymphocytes. Nature, 230, 50–51.

Raviola, E. and Karnovsky, M.J. (1972) Evidence for a blood-thymus barrier using electron-opaque tracers. J. Exp. Med. 136, 466–498.

Reif, A.E. and Allen, J.M. (1964) The AKR thymic antigen and its distribution in leukemias and nervous tissues. J. Exp. Med. 120, 413–433.

Rich, R.R. and Pierce, C.W. (1973) Biological expressions of lymphocyte activation. I. Effects of phytomitogens on antibody synthesis in vitro. J. Exp. Med. 137, 205–223.

Roelants, G. (1972) Quantification of antigen specific T and B lymphocytes in mouse spleens. Nature New Biol. 236, 252–254.

Roelants, G., Forni, L. and Pernis, B. (1973) Blocking and redistribution ("capping") of antigen receptors on T and B lymphocytes by anti-immunoglobulin antibody. J. Exp. Med. 137, 1060–1077.

Roitt, I.M., Torrigiani, G., Greaves, M.F. and Brostoff, J. (1969) The cellular basis of immunological responses. Lancet ii, 367–371.

Rolley, R.T. and Marchalonis, J.J. (1973) Dynamics of receptor and antigen interaction at the lymphocyte. Transplant. Proc. 5, 71–74.

Rollinghoff, M., Wagner, H., Cone, R.E. and Marchalonis, J.J. (1973) Release of antigen-specific immunoglobulin from cytotoxic effector cells and syngeneic tumour immunity in vitro. Nature New Biology 243, 21–23.

Rosenthal, A.S. and Shevach, E.M. (1973) Function of macrophages in antigen recognition by guinea pig T lymphocytes. I. Requirement for histocompatible macrophages and lymphocytes. J. Exp. Med. 138, 1194–1212.

Rosse, C. (1971) Lymphocyte production and life-span in the bone marrow of the guinea pig. Blood 38, 372–377.

Rowe, D.S., Hug, K., Faulk, W.P., McCormick, J.N. and Gerber, H. (1973a) I_gD on the surface of peripheral blood lymphocytes of the human newborn. Nature New Biology 242, 155–157.

Rowe, D.S., Hug, K., Forni, L. and Pernis, B. (1973b) Immunoglobulin D as a lymphocyte receptor. J. Exp. Med. 138, 965–972.

Rowley, D.A., Gowans, J.L., Atkins, R.C., Ford, W.L. and Smith, M.E. (1972) The specific selection of recirculating lymphocytes by antigen in normal and preimmunized rats. J. Exp. Med. 136, 499–513.

Rubin, E., Cooper, M.D. and Krause, F.W. (1971) Kinetics of cellular proliferation in the bursa of Fabricius. Bacteriol. Proc. 7, 67.

Rutishauser, U. and Edelman, G.M. (1972) Binding of thymus- and bone marrow-derived lymphoid cells to antigen derivatized fibers. Proc. Nat. Acad. Sci. U.S.A. 69, 3774–3778.

Rysser, J.E. and Vassalli, P. (1974) Mouse bone marrow lymphocytes and their differentiation. J. Immunol. 113, 719–728.

Sachs, D.H. and Cone, J.L. (1973) A mouse B-cell alloantigen determined by gene(s) linked to the major histocompatibility complex. J. Exp. Med. 138, 1289–1304.

Sainte-Marie, G. and Leblond, C.P. (1964) Cytologic features and cellular immigration in the cortex and medulla of thymus in the young adult rat. Blood 23, 275–299.

Santer, V., Bankhurst, A.D. and Nossal, G.J.V. (1972) Ultrastructural distribution of surface immunoglobulin determinants on mouse lymphoid cells. Exp. Cell Res. 72, 377–386.

Santer, V., Cone, R.E. and Marchalonis, J.J. (1973) The glycoprotein surface coat on different classes of murine lymphocytes. Exp. Cell Res. 79, 404–416.

Scheid, M.P., Hoffmann, M.K., Komuro, K., Hämmerling, V., Abbott, J., Boyse, E.A., Cohen, G.H., Hooper, J.A., Schulot, R.S. and Goldstein, A.L. (1973) Differentiation of T cells induced by preparations from thymus and by non-thymic agents. J. Exp. Med. 138, 1027–1032.

Schlegel, R.A. and Shortman, K. (1975) Antigen-initiated B lymphocyte differentiation. IV. The adherence properties of antibody-forming cell progenitors from primed and unprimed mice. J. Immunol. 115, 94–99.

Schlegel, R.A., vonBoehmer, H. and Shortman, K. (1975) Antigen-initiated B lymphocyte differentiation. V. Electrophoretic reparation of different subpopulations of AFC progenitors for unprimed I_gM and memory I_gG responses to the NIP determinant. Cell. Immunol. 16, 203–217.

Schlesinger, M. (1972) Antigens of the thymus. Prog. Allergy 16, 214–299.

Schlesinger, M. and Hurvitz, D. (1968a) Differentiation of the thymus-leukemia (TL) antigen in the thymus of mouse embryos. Israel J. Med. Sci. 4, 1210–1215.

Schlesinger, M. and Hurvitz, D. (1968b) Serological analysis of thymus and spleen grafts. J. Exp. Med. 127, 1127–1137.

Schlesinger, M., Gottesfeld, S. and Korzash, Z. (1973) Thymus cell subpopulations separated on discontinuous BSA gradients: Antigenic properties and circulation capacity. Cell Immunol. 6, 49–58.

Schoefl, G.I. (1972) The migration of lymphocytes across the vascular endothelium in lymphoid tissue. J. Exp. Med. 136, 568–588.

Schrader, J.W. (1974) Evidence for the presence in unimmunized mice of two populations of bone marrow-derived B lymphocytes, defined by differences in adherence properties. Cell. Immunol. 10, 380–393.

Schwarz, R.M. (1967) Transformation of rat small lymphocytes with allogenic lymphoid cells. Am. J. Anat. 121, 559–570.

Segal, S., Cohen, I.R. and Feldman, M. (1972) Thymus-derived lymphocytes: Humoral and cellular reactions distinguished by hydrocortisone. Science, 175, 1126–1128.

Selye, H. and Masson, G. (1939) The effects of estrogens as modified by adrenal insufficiency. Endocrinology, 25, 211–215.

Selye, H., Harlow, C.M. and Collip, J.B. (1936) Auslösung der Alarmreaktion mit Follikelhormon. Endokrinologie, 18, 81–85.

Shearer, G.M. and Cudkowicz, G. (1969) Distinct events in the immune response elicited by transferred marrow and thymus cells. I. Antigen requirements and proliferation of thymic antigen-reactive cells. J. Exp. Med. 130, 1243–1261.

Shearer, G.M., Weinstein, Y. and Melman, K.L. (1974) Enhancement of immune response potential of mouse lymphoid cells fractionated over insolubilized conjugated histamine columns. J. Immunol. 113, 597–607.

Shevach, E., Green, I., Ellman, L. and Maillard, J. (1972) Heterologous antiserum to thymus-derived cells in the guinea pig. Nature New Biology 235, 19–21.

Shevach, E., Paul, W.E. and Green, I. (1972) Histocompatibility-linked immune response gene fraction in guinea pigs. Specific inhibition of antigen-induced lymphocyte proliferation by alloantisera. J. Exp. Med. 136, 1207–1221.

Shigeno, N., Hämmerling, V., Arpels, C., Boyse, E.A. and Old, L.J. (1968) Preparation of lymphocyte-specific antibody from anti-lymphocyte serum. Lancet ii, 320–323.

Shortman, K., Diener, E., Russell, P. and Armstrong, W.D. (1970) The role of nonlymphoid accessory cells in the immune response to different antigens. J. Exp. Med. 131, 461–482.

Shortman, K. and Jackson, H. (1974) The differentiation of T lymphocytes. I. Proliferation kinetics and interrelationships of subpopulations of mouse thymus cells. Cell. Immunol. 12, 230–255.

Shreffler, D. C. and David, C.S. (1975) The H-2 major histocompatibility complex and the I immune response region: Genetic variation, function and organization. Adv. Immunol. 20, 125–196.

Sidman, C.L. and Unanue, E.R. (1975) Development of B lymphocytes. I. Cell populations and a critical event during ontogeny. J. Immunol. 114, 1730–1735.

Simpson, E., Shearer, G.M., Weinstein, Y. and Melman, K.L. (1973) Induction of cytotoxic immune

responses in vitro to transplantation antigens: Effect of cell separation using histamine-coated sepharose beads. Fed. Proc. 32, 877.
Singer, S.J. (1974) Molecular biology of cellular membranes with applications to immunology. Adv. Immunol. 19, 1–66.
Siskind, G.W. and Benacerraf, B. (1969) Cell selection by antigen in the immune response. Adv. Immunol. 10, 1–50.
Smith, C. (1965) Studies on the thymus of the mammal. XIV. Histology and histochemistry of embryonic and early postnatal thymuses of C57BL/6 and AKR strain mice. Am. J. Anat. 116, 611–630.
Smith, J.B., McIntosh, G.H. and Morris, B. (1970) The traffic of cells through tissues: A study of peripheral lymph in sheep. J. Anat. 107, 87–100.
Smithies, O. (1967) The genetic basis of antibody variability. Cold Spring Harbor Symp. Quant. Biol. 37, 161–168.
Smithies, O. and Poulik, M.D. (1972) Initiation of protein synthesis of an unusual position at an immunoglobulin gene? Science, 175, 187–189.
Snell, G.D. (1968) The H-2 locus of the mouse: Observations and speculations concerning its comparative genetics and its polymorphism. Folia Biol. (Praha) 14, 335–358.
Snell, G.D., Cherry, M., McKenzie, I.F.C. and Bailey, D.W. (1973) LY-4, a new locus determining a lymphocyte cell-surface alloantigen in mice. Proc. Nat. Acad. Sci. U.S.A. 70, 1108–1111.
Solomon, J.B. (1971) Foetal and Neonatal Immunology, pp. 41–48, Elsevier, Amsterdam.
Sosin, H., Hilgard, H. and Martinez, C. (1966) The immunologic competence of mouse thymus cells measured by the graft vs. host spleen assay. J. Immunol. 96, 189–195.
Spear, P.G. and Edelman, G.M. (1974) Maturation of the humoral immune response in mice. J. Exp. Med. 139, 249–263.
Spear, P.G., Wang, A.-L., Rutishauser, U. and Edelman, G.M. (1973) Characterization of splenic lymphoid cells in fetal and newborn mice. J. Exp. Med. 138, 557–573.
Spiegelberg, H.L. (1972) γD immunoglobulin. Contemp. Top. Immunochem. 1, 165–180.
Sprent, J. (1973) Circulating T and B lymphocytes in the mouse. I. Migratory properties. Cell. Immunol. 7, 10–39.
Sprent, J., Miller, J.F.A.P. and Mitchell, G.F. (1971) Antigen-induced selective recruitment of circulating lymphocytes. Cell. Immunol. 2, 171–181.
Spry, C.J.F. (1972) Inhibition of lymphocyte recirculation by stress and corticotropin. Cell. Immunol. 4, 86–92.
Srivastava, B.I.S. and Minowada, J. (1973) Terminal deoxynucleotidyl transferase activity in a cell line (molt-4) derived from the peripheral blood of a patient with acute lymphoblastic leukemia. Biochem. Biophys. Res. Commun. 51, 529–535.
St. Pierre, R.L. and Ackerman, G.A. (1965) Bursa of Fabricius in chickens: possible humoral factor. Science, 147, 1307–1308.
St. Pierre, R.L. and Ackerman, G.A. (1966) Influence of bursa implantation upon lymphatic modules and plasma cells in spleens of bursectomized chickens. Proc. Soc. Exp. Biol. Med. 122, 1280–1282.
Stobo, J.D. (1972) Phytohemagglutinin and concanavalin A: Probes for murine 'T' cell activation and differentiation. Transplant. Rev. 11, 60–86.
Stobo, J.D., Paul, W.E. and Henney, C.S. (1973) Functional heterogeneity of murine lymphoid cells. IV. Allogeneic mixed lymphocyte reactivity and cytolytic activity as functions of distinct T cell subsets. J. Immunol. 110, 652–660.
Stobo, J.D., Rosenthal, A. and Paul, W.E. (1972) Functional heterogeneity of murine lymphoid cells. J. Immunol. 108, 1–17.
Stockert, E., Old, L.J. and Boyse, E.A. (1971) The G_{IX} system. A cell surface alloantigen associated with murine leukemia virus; implications regarding chromosomal integration of the viral genome. J. Exp. Med. 133, 1334–1335.
Stout, R.D. and Herzenberg, L.A. (1975) The Fc receptor on thymus-derived lymphocytes. I. Defection of a subpopulation of murine T lymphocytes bearing the Fc receptor. J. Exp. Med. 142, 611–621.
Stout, R.D., Yutoku, M., Grossberg, A., Pressman, D. and Herzenberg, L.A. (1975) A surface

membrane determinant shared by subpopulations of thymocytes and B lymphocytes. J. Immunol. 115, 508–512.

Strober, S. (1972) Initiation of antibody responses by different classes of lymphocytes. V. Fundamental changes in the physiological characteristics of virgin thymus-independent (B) lymphocytes and B memory cells. J. Exp. Med. 136, 851–871.

Strober, S. (1975) Maturation of B lymphocytes in the rat. II. Subpopulations of virgin B lymphocytes in the spleen and thoracic duct lymph. J. Immunol. 114, 877–885.

Strober, S. and Dilley, J. (1973a) Biological characteristics of T and B memory lymphocytes in the rat. J. Exp. Med. 137, 1275–1292.

Strober, S. and Dilley, J. (1973b) Maturation of B lymphocytes in the rat. I. Migration pattern, tissue distribution and turnover rate of unprimed and primed B lymphocytes involved in the adoptive antidinitrophenyl response. J. Exp. Med. 138, 1331–1344.

Strominger, J.L., Cresswell, P., Grey, H., Humphreys, R.H., Mann, D., McCune, J., Parham, P., Robb, R., Sanderson, A.R., Springer, T.A., Terhorst, C. and Turner, M.J. (1974) The immunoglobulin-like structure of human histocompatibility antigens. Transplant. Rev. 21, 126–143.

Stutman, O. (1975) Humoral thymic factors influencing postthymic cells. Ann. N.Y. Acad. Sci. 249, 89–105.

Stutman, O. and Good, R.A. (1972) Heterogeneity of lymphocyte populations. Rev. Eur. Etu. Clin. Biol. 17, 11–14.

Stutman, O., Yunis, E.J. and Good, R.A. (1970) Studies on thymus function. I. Cooperative effect of thymus function and lymphopoietic cells in restoration of neonatally thymectomized mice. J. Exp. Med. 132, 583–600.

Szenberg, A. (1970) Influence of testosterone in the primary lymphoid organs of the chicken. In: Hormones and the Immune Response. Ciba Foundation Study Group No. 36 (Wolstenholme, G.E.W. and Knight, J., eds.) pp. 42–45, Churchill, London.

Takahashi, T., Old, L.J. and Boyse, E.A. (1970) Surface alloantigens of plasma cells. J. Exp. Med. 131, 1325–1341.

Takahashi, T., Old, L.J., Hsu, C.-J. and Boyse, E.A. (1971) A new differentiation antigen on plasma cells. Eur. J. Immunol. 1, 478–482.

Takiguchi, T., Adler, W.H. and Smith, R.T. (1971) Identification of mouse thymus antigen recognition function in a minor, low-density, low θ cell subpopulation. Cell. Immunol. 2, 373–380.

Tao-Wiedmann, T.-W., Loor, F. and Hagg, L.-B. (1975) Development of surface immunoglobulins in the chicken. Immunology, 28, 821–830.

Tar, E., Olah, I. and Torö, I. (1969) Cell migration in the developing follicle of the bursa of Fabricius between cortex and medulla. Acta Biol. Acad. Sci. Hung. 20, 93–99.

Taub, R.N., Rosett, W., Adler, A. and Morse, S.I. (1972) Distribution of labelled lymph node cells in mice during the lymphocytosis induced by *Bordetella Pertussis*. J. Exp. Med. 136, 1581–1593.

Taylor, R.B. (1964) Pluripotential stem cells in mouse embryo liver. Brit. J. Exp. Pathol. 46, 376–383.

Thompson, J.H. and Cooper, M.D. (1971) Functional deficiency of autologous implants of the bursa of Fabricius in chickens. Transplantation, 11, 71–77.

Thorbecke, G.J. (1960) Gamma globulin and antibody formation in vitro. I. Gamma globulin formation in tissues from immature and normal adult rabbits. J. Exp. Med. 112, 279–292.

Thorbecke, G.J., Warner, N.L., Hochwald, G.M. and Ohanian, S.H. (1968) Immune globulin production by the bursa of Fabricius of young chickens. Immunology, 15, 123–134.

Thorsby, E. (1974) The human major histocompatibility system. Transplant. Rev. 18, 51–129.

Tigelaar, R.E. and Asofsky, R. (1973) Graft vs host reactivity of mouse thymocytes: Effect of cortisone pretreatment of donors. J. Immunol. 110, 567–574.

Till, J.E. and McCulloch, E.A. (1961) A direct measurement of the radiation sensitivity of normal mouse bone marrow cells. Radiat. Res. 14, 213–222.

Toivanen, P. and Toivanen, A. (1973) Bursal and postbursal stem cells in chicken. Functional characteristics. Eur. J. Immunol. 3, 585–595.

Toivanen, P., Toivanen, A. and Good, R.A. (1972a) Ontogeny of bursal function in chicken. I. Embryonic stem cell for humoral immunity. J. Immunol. 109, 1058–1070.

Toivanen, P., Toivanen, A., Linna, T.J. and Good, R.A. (1972b) Ontogeny of bursal function in chicken. II. Postembryonic stem cell for humoral immunity. J. Immunol. 109, 1071–1080.

Trainin, N., Bejerano, A., Strahilevitch, M., Goldring, D. and Small, M. (1966) A thymic factor preventing wasting and influencing lymphoiesis in mice. Israel J. Med. Sci. 2, 549–559.

Tung, J.-S., Vitetta, E.S., Fleissner, E. and Boyse, E.A. (1975) Biochemical evidence linking the G_{IX} thymocyte surface antigen to the $gp^{69/71}$ envelope glycoprotein of murine leukemia virus. J. Exp. Med. 141, 198–205.

Turk, J.L. and Poulter, L.W. (1972) Selective depletion of lymphoid tissue by cyclophosphamide. Clin. Exp. Immunol. 10, 285–296.

Tyan, M.L. (1968) Studies on the ontogeny of the mouse immune system. I. Cell-bound immunity. J. Immunol. 100, 535–542.

Tyan, M.L. and Herzenberg, L.A. (1968) Immunoglobulin production by embryonic tissues: Thymus independent. Proc. Soc. Exp. Biol. Med. 128, 952–954.

Unanue, E.R. (1971) Antigen-binding cells. I. Their identification and role in the immune response. J. Immunol. 107, 1168–1174.

Unanue, E.R., Dorf, M.E., David, C.S. and Benacerraf, B. (1974) The presence of I region-associated antigens on B cells in molecules distinct from immunoglobulin and H-2K and H-2D. Proc. Nat. Acad. Sci. U.S.A. 71, 5014–5016.

Unanue, E.R., Perkins, W.D. and Karnovsky, M.J. (1972) Endocytosis by lymphocytes of complexes of anti-Ig with membrane-bound Ig. J. Immunol. 108, 569–572.

Valdes-Dapena, M.A. (1957) An Atlas of Foetal and Neonatal Histology. J.B. Lippincott, Philadelphia.

van Boxel, J.A. and Rosenstreich, D.L. (1974) Binding of aggregated γ-globulin to activated T lymphocytes in the guinea pig. J. Exp. Med. 139, 1002–1012.

van Boxel, J.A., Paul, W.E., Frank, M.M. and Green, I. (1973) Antibody-dependent lymphoid cell-mediated cytotoxicity: role of lymphocytes bearing a receptor for complement. J. Immunol. 110, 1027–1036.

van Boxel, J.A., Paul, W.E., Terry, W.D. and Green, I. (1972a) IgD-bearing human lymphocytes. J. Immunol. 109, 648–651.

van Boxel, J.A., Stobo, J.D., Paul, W.E. and Green, I. (1972b) Antibody-dependent lymphoid cell-mediated cytotoxicity: no requirement for thymus-derived lymphocytes. Science, 175, 194–196.

Van der Broek, A.A. (1971) Immune Suppression and Histophysiology of the Immune Response. Groningen, The Netherlands.

Van den Engh, G.J. and E.S. Golub (1974) Antigenic differences between hemopoietic stem cells and myeloid progenitors. J. Exp. Med. 139, 1621–1627.

Venzke, W.G. (1952) Morphogenesis of the thymus of chicken embryos. Am. J. Vet. Res. 13, 395–404.

Vischer, T.L. and Jaquet, C. (1972) Effect of antibodies against immunoglobulins and the theta antigen on specific and non-specific stimulation of mouse spleen cells in vitro. Immunology, 22, 259–266.

Vitetta, E.S., Baur, S. and Uhr, J.W. (1971) Cell surface immunoglobulin. II. Isolation and characterization of immunoglobulin from mouse splenic lymphocytes. J. Exp. Med. 134, 242–264.

Vitetta, E.S., Bianco, C., Nussenzweig, V. and Uhr, J.W. (1972) Cell surface immunoglobulins. IV. Distribution among thymocytes, bone marrow cells, and their derived populations. J. Exp. Med. 136, 81–93.

Vitetta, E.S., Grundke-Iqbal, I., Holmes, K. and Uhr, J.W. (1974a) Cell surface immunoglobulin. VII. Synthesis, shedding and secretion of immunoglobulin by lymphoid cells of germ-free mice. J. Exp. Med. 139, 862–876.

Vitetta, E.S., Klein, J. and Uhr, J.W. (1974b) Partial characterization of Ia antigens from murine lymphoid cells. Immunogenetics, 1, 82–90.

Vitetta, E.S., Melcher, U., McWilliams, M., Lamm, M.E., Phillips-Quagliata, J.M. and Uhr, J.W. (1975a) Cell surface immunoglobulin. XI. The appearance of an IgD-like molecule on murine lymphoid cells during ontogeny. J. Exp. Med. 141, 206–215.

Vitetta, E. S., Uhr, J.W. and Boyse, E.A. (1975b) Association of β_2-microglobulin-like subunit with H-2 and TL alloantigens on murine thymocytes. J. Immunol. 114, 252–254.

von Gaudecker, B. and Schmale, E.-M. (1974) Similarities between Hassall's corpuscles of the human thymus and the epidermis. Cell Tissue Res. 151, 347–368.

Wakefield, J.D. and Thorbecke, G.J. (1968) Relationship of germinal centres in lymphoid tissue to immunological memory. I. Evidence for the formation of small lymphocytes upon transfer of primed splenic white pulp to syngeneic mice. J. Exp. Med. 128, 153–169.

Waksal, S.D., Cohen, I.R., Waksal, H.W., Wekerle, H., St. Pierre, R.L. and Feldman, M. (1975) Induction of T-cell differentiation in vitro by thymus epithelial cells. Ann. N.Y. Acad. Sci. 249, 492–498.

Waksman, B.H., Arnason, B.G. and Jankovic, B.D. (1962) Role of the thymus in immune reactions in rats. III. Changes in the lymphoid organs of thymectomized rats. J. Exp. Med. 116, 187–206.

Waltenbaugh, C.R. and VanAlten, P.J. (1974) The production of antibody by bursal lymphocytes. J. Immunol. 113, 1079–1084.

Walters, C.S. and Wigzell, H. (1970) Demonstration of heavy and light chain antigenic determinants on the cell-bound receptor for antigen. J. Exp. Med. 132, 1233–1249.

Ward, P.A., Offen, C.D. and Montgomery, J.R. (1971) Chemoattractants of leukocytes, with special reference to lymphocytes. Fed. Proc. 30, 17–21.

Warner, N.L. (1972) Surface immunoglobulins on lymphoid cells. Contemp. Top. Immunobiol. 1, 87–117.

Warner, N.L. (1974) Membrane immunoglobulins and antigen receptors on B and T lymphocytes. Adv. Immunol. 19, 67–216.

Warner, N.L. and Szenberg, A. (1964) The immunological function of the bursa of Fabricius in the chicken. Ann. Rev. Microbiol. 18, 253–268.

Warner, N.L., Szenberg, A. and Burnet, F.M. (1962) The immunological role of different lymphoid organs in the chicken. I. Dissociation of immunological responsiveness. Aust. J. Exp. Biol. Med. 116, 187–206.

Warner, N.C., Uhr, J.W., Thorbecke, G.J. and Ovary, Z. (1969) Immunoglobulins, antibodies and the bursa of Fabricius: induction of agammaglobulinemia and the loss of antibody-forming capacity by hormonal bursectomy. J. Immunol. 103, 1317–1330.

Weber, W.T. (1966) Difference between medullary and cortical thymic lymphocytes of the pig in their response to phytohemagglutinin. J. Cell. Physiol. 68, 117–126.

Weigle, W.O. (1973) Immunological Unresponsiveness. Adv. Immunol. 16, 61–122.

Weinstein, Y., Melmon, K.L., Bourne, H.R. and Sela, M. (1973) Specific leukocyte receptors for small endogenous hormones: detection by cell binding to solubilized hormone preparations. J. Clin. Invest. 52, 1349–1361.

Weissman, I.L. (1967) Thymus cell migration. J. Exp. Med. 126, 291–304.

Weissman, I.L. (1973) Thymus cell maturation. Studies on the origin of cortisone-resistant thymic lymphocytes. J. Exp. Med. 137, 504–510.

Weissman, I.L. (1975) Development and distribution of immunoglobulin-bearing cells in mice. Transplant. Rev. 24, 159–176.

Wekerle, H., Cohen, I.R. and Feldman, M. (1973) Thymus reticulum cell cultures confer T cell properties on spleen cells from thymus-deprived animals. Eur. J. Immunol. 3, 745–748.

White, A. (1975) Nature and biological activities of thymus hormones: Prospects for the future. Ann. N.Y. Acad. Sci. 249, 523–530.

White, A. and Goldstein, A.L. (1970) Thymosin, a thymic hormone influencing lymphoid cell immunological competence. In: Hormones in the Immune Response, Ciba Foundation Study Group No. 36. (Wolstenholme, G. E. W. and Knight, J., eds.) pp. 3–19, Churchill, London.

Wigzell, H. (1970) Specific fractionation of immunocompetent cells. Transplant. Rev. 5, 76–104.

Wigzell, H. (1974) On the relationship between cellular and humoral antibodies. Contemp. Top. Immunobiol. 3, 77–96.

Wigzell, H., Sundquist, K.G. and Yoshida, T.O. (1972) Separation of cells according to surface antigens by the use of antibody-coated columns. Fractionation of cells carrying immunoglobulins and blood group antigen. Scand. J. Immunol. 1, 75–87.

Williams, A.F. and Gowans, J.L. (1975) The presence of IgA on the surface of rat thoracic duct lymphocytes which contain internal IgA. J. Exp. Med. 141, 335–345.

In vitro analysis of surface specificity in embryonic cells 13

David E. MASLOW

1. Introduction

Embryogenesis in higher animals proceeds through a complex series of cellular interactions in which the cells of the embryo sort out and associate into specific multicellular groupings that give rise to tissues and organs. During this period individual cells and groups of cells continually change their relative positions and mutual adhesions. As described in other articles in this volume, organs such as the heart, the gonads, the adrenal cortex, the immune system, the hemopoietic tissues and various parts of the nervous system are formed from cells that originated at sites far away from their final location. These cells emigrate from their site of origin and migrate to their final location and reassociate there to establish the various tissue and organ primordia. It is now generally accepted that molecular events occurring at the cell surface are of considerable importance in the control of these morphogenetic processes. Of special significance in this context are the processes of cell recognition and selective cell adhesion. These particular facets of embryonic cell behavior are difficult to analyze in detail in the intact embryo and most of our present insight into the mechanisms of cell recognition and adhesion is derived from experiments done with model systems in vitro.

Several decades ago, Holtfreter dissociated tissues from early amphibian embryos and demonstrated how the resulting single cells were in certain cases able to reaggregate to form a semblance of the tissue from which they were derived. The introduction in the early nineteen fifties of effective dissociation techniques to prepare single cell suspension from the tissues and organs of mammals and birds was soon followed by the pioneering work of the Mosconas showing that trypsin-dissociated embryonic cells could be induced to form mixed cell aggregates within which the cells sorted out to establish specific cell associations with cells of their own type. Since 1952, and progressing at a growing pace, this experimental approach has made it possible to construct in vitro tissues from dispersed embryonic cells and thus analyze morphogenetic cell interactions under controlled conditions. The assumption implicit in the use of cell aggregation and tissue reconstruction techniques for studying the principles and

G. Poste & G.L. Nicolson (eds.) The Cell Surface in Animal Embryogenesis and Development

mechanisms underlying morphogenesis is that the interactions observed between particular cell types in vitro are for the most part similar to those occurring in vivo. The validity of this assumption has been demonstrated with varying degrees of success. In this article I will review some of the large body of work done on cell aggregation and tissue reconstruction in vitro and discuss how these techniques have contributed to our understanding of morphogenetic cell interactions in vivo.

Although such terms as "specificity" and "selectivity" have long been used in descriptions of morphogenetic cellular interactions, precise definition of these terms is difficult. In most instances, the term "specificity" has been used in a broad sense to denote the properties of a system in which the various interacting units are capable of *discrimination* and thus, at the same time, it implies selectivity of action and reaction. Viewed soberly, these terms add up to little more than a descriptive record. They tell us that cells can recognize matching and unmatching environments and cellular kinships and respond in discriminative fashion, though just how remains largely obscure. "Specificity" and "selectivity" are, however, eminently flexible terms and can be used in the same sense to simultaneously describe events occurring at different levels of biological organization ranging from the behavior of intact cells to the interaction between molecular species. Thus when referring to specificity in the cell recognition process and selectivity in the formation of cell adhesions we can also indicate that these responses reflect specificities residing in the cell surface of the interacting cells which, in turn, reflect specific patterns of macromolecular organization determined by a specific set of genetic information.

We face similar problems when defining cell adhesion. Despite the fundamental importance of cell adhesion and the large amount of work on the subject, there is no acceptable definition of this process. Operational definitions are legion. The major problem is that there is no reason to believe that each of the many methods used to study "adhesion" are measuring the same phenomenon. In most cases, individual working definitions have been used which refer to the formation of stable bonds between cells, or between cells and a substrate such as glass or plastic under the conditions used in the particular study. In this article, discussion of the adhesive process will be confined to a description of examples of selectivity in the adhesive interactions occurring between embryonic cells in mixed cell aggregates and an analysis of the possible mechanisms that underlie selectivity and which thus determine specificity. The equally complex problem of measuring the adhesive energy of cells to one another or to other surfaces will not be considered (for review see Curtis, 1973; Weiss, 1967).

2. *Analysis of sponge cell reaggregation*

Although this article will deal mainly with vertebrate cells, sponges warrant special mention since experiments done with these species have often served as models for similar studies using cells from vertebrate embryos. In addition to their relative ease of dissociation (Wilson, 1907), they are attractive organisms for the study of cellular interactions because they readily reform under appropriate conditions, perfect, albeit small sponges (Galtsoff, 1925). This may satisfy, in part, the justifiable criticisms concerning the applicability of in vitro experiments using vertebrate cells, particularly since similar results are obtained from analogous experiments using sponge and embryonic vertebrate systems (Jones, 1966b).

2.1. Specificity of reaggregation

The fundamental observation which provides the basis for most of the work on the specificity of cellular interactions in aggregating mixtures of cells is the finding that when dissociated cells from different species of sponges are mixed, the resulting reaggregated sponges consist almost exclusively of cells from the same species (Wilson, 1910; Galtsoff, 1923, 1925). Specificity of this type has been questioned in several reports (Curtis, 1962, 1970a; Sara et al. 1966a,b; Curtis and von de Vyver, 1971). However, in these particular studies sponge tissue was dissociated with EDTA whereas the original studies of Wilson and Galtsoff were done with dissociated cell populations obtained by pressing sponge tissue through bolting cloth. EDTA has been shown to have a toxic effect on vertebrate cells (Ball, 1966; Moscona and Moscona, 1967) and to destroy the activity of cell surface moieties necessary for aggregation, resulting in slower aggregate formation (Spiegel, 1955; Humphreys, 1970a). Such non-specific "injury" effects may result in the formation of non-stable heterotypic cell associations in a culture environment which does not present a constant shearing force as does the rotation culture vessel, although this would not account for those species which never exhibit specificity (vide infra).

The observations of Wilson and Galtsoff showing specificity of reaggregation have, however, been confirmed by many investigators using a variety of sponge species, but notably *Microciona prolifera* and *Haliciona occulata* (Humphreys et al., 1960a,b; Moscona, 1961a, 1968a; Humphreys, 1963, 1970a; McClay, 1971, 1974).

Specificity of aggregation is usually assayed between cells from sponge species of different colors so that the overall color of the aggregate can be used as a criterion for homogeneity. However, as Curtis (1970a,b) and Spiegel (1955) pointed out, the presence of only a few cells of one color in the mass of cells of the second color and species would not be readily detected. This was confirmed by McClay (1971) using labeled cell populations. To test if the presence of a few heterospecific cells is indicative of a lack of specificity as opposed to mere mechanical trapping, McClay recultured non-labeled aggregates of one cell type in suspensions of labeled cells of the same or a second type using the collecting aggregate technique of Roth and Weston (1967). Autoradiographs of sections of the aggregates showed that homotypic combinations resulted in internally mixed aggregates while heterotypic combinations showed few, if any, labeled cells at the surface, indicating the presence of a species recognition mechanism (Spiegel, 1954; McClay, 1971). Few heterotypic adhesions occurred after a recovery period of six hours, during which time contact is made between archiocytes and mucoid cells which appears to be necessary for specificity to be expressed (John et al., 1971).

The existence of this lag period may also explain the failure of Curtis (1970b) to detect any evidence of specificity of adhesion between cells from the species *Haliclona occulata* and *Halichondria panicea* which he studied using the Couette viscometer (Curtis, 1969). However, his data could be explained equally well by the need for cellular recovery from the dissociation process or the resynthesis of an aggregating factor removed by preparative washing. Sara et al. (1966a) report of mixed aggregates of sponge cells from a number of species of the class *Demospongiae* may be subject to similar criticism (even though their observations lasted 24 hours) since the mixed aggregates formed were quite small and the initial concentrations of cells in the experiments was sufficiently low to reduce the likelihood of contact between archeocytes and mucoid cells necessary for specificity (John et al., 1971). The low number of cells might also hinder the production of sufficient "factor" to promote specificity. Humphreys (1970a) found that more than 95% of cells in mixed cultures of *Microciona prolifera* and *Haliclona occulata* which he followed using time lapse photomicroscopy exhibited selectivity in the formation of stable associations. However, he cautioned that in every case except the two combinations he selected, sponge aggregation is not totally species specific. Thus, the earlier emphasis on the generality of the specific interactions between sponge cells has been wisely replaced with an interest in the mechanisms of those cases where they occur.

2.2. Cell surface molecules promoting adhesion

Sponge cells are also of interest since they provided the first evidence that specific cell surface molecules might play a role in the determination of adhesive specificity. Following the suggestion of Tyler (1947), Spiegel (1954) looked for the presence of macromolecules on the surface of sponge cells which could act in an antibody-antigen fashion to hold cells together. Antisera prepared from rabbits sensitized with cells from either of two species of sponge or a mixture of them, were tested for their effects on aggregation of the three types of cell suspensions. Homologous antisera inhibited aggregation, while heterologous antisera had little effect. While Spiegel's explanation that the antisera contained antibodies that were able to block species specific surface moieties that normally held like-cell surfaces together may not be in accord with current immunological theories, this approach at least served to focus attention on the possible role of a surface molecule in determining the specificity of cell interactions.

The major experimental approach adopted in an effort to establish the role of cell surface molecules in specific adhesion of sponge cells has involved dissociation of tissues by lengthy incubation in Ca^{2+} and Mg^{2+}-free sea water and harvesting an "organic factor" from the supernatant of the pelleted cells (Humphreys, 1963; Moscona, 1963). This factor, when stabilized with $CaCl_2$, has been shown to enhance the adhesion and aggregation of only the cell type from which it was derived (Humphreys, 1967). In some cases, the factor may also inhibit aggregation of some, but not all, heterotypic cells (Curtis and von de Vyver, 1971; McClay, 1974). Loewenstein (1967) confirmed the specificity of both cell adhesion and the effect of the isolated "factor" by demonstrating that intercellular communication developed only between cells of the same species and when "factor" from the same species was present.

While the early results of Humphreys and Moscona obtained with one or two sponge species have been confirmed in several laboratories, it is important to note that the production of an aggregation enhancement factor is not a characteristic of all species of sponge (Humphreys, 1970b). In addition, the specificity of the aggregation factor does not seem to be absolute. This was first demonstrated by MacLennan and Dodd (1967) who showed that a close taxonomic relationship decreases the specificity barrier. Similar responsiveness of sponge cells from the species aggregation factor produced by a closely related species has been reported by Turner and Burger (1973). It should be pointed out, however, that comparison of factor specificities should not really be done in assays which essentially measure and point out equilibrium conditions of aggregation. Comparison of aggregation kinetics, i.e. the initial velocities of aggregation in the presence of low amounts of aggregation factor would provide a more suitable and sensitive assay. Under these conditions species-specific differences in response to different aggregation factors would be detectable much earlier.

Chemically dissociated cells, unlike mechanically dissociated cells, do not aggregate at 4°C without the addition of aggregation "factor". This suggests that the chemical dissociation process removes the aggregation "factor" from its normal surface location, and that its resynthesis is inhibited at low temperatures (Humphreys, 1967). Moscona (1968a,b) has interpreted the function of the specific aggregation factor as an intercellular ligand. Alternatively, the factor may act to repair non-specific cellular damage caused by dissociation, or to inactivate an inhibitor of aggregation which is active only at low temperatures (Curtis and Greaves, 1965). These objections seem to be answered by the finding that unwashed cells with minimal damage ("factor" presumably not removed) show a greater enhancement of aggregation in the presence of the factor than "damaged cells", even at 24°C (Moscona, 1968a). Whether dissociated cells of any kind can be considered to be totally free from damage is unclear. The criticism dealing with potential sublethal cell damage is thus unresolved.

Attempts to isolate and characterize the aggregation factor have proceeded further in sponges than with cells from vertebrate embryos and the results are instructive if only in terms of the experimental approaches employed.

The chemical nature of the species-specific aggregation factor was first analyzed by indirect procedures. Margoliash et al. (1965) isolated a purified component capable of causing specific reaggregation of sponge cells which consisted largely of glycoproteins in units of approximately 25 Å diameter. The protein and sugar content of the factor were confirmed by Gasic and Galanti (1966), who also demonstrated inactivation of the factor by the proteolytic enzymes. They urged caution, however, before assuming that the isolated glycoprotein was the biologically active substance despite its apparent purity since secondary enzymatic effects on other cellular processes have not been ruled out.

To ensure that none of the active components of the factor were lost during purification, Henkart, Humphreys and Humphreys (1973) assayed for aggregation promoting activity at each step of their purification procedure. They determined that the factor contained protein and polysaccharide in about equal parts. It is of interest that the amino acid composition as determined by Margoliash et al. (1965) for *Microciona prolifera* and by Henkert et al. (1973) for *Microciona parthena* agree quite closely. Using a variety of techniques including sedimentation, light scattering, equilibrium dialysis and gel filtration and electron microscopy, Cauldwell et al. (1973) found the active component of the factor to be a large proteoglycan complex (molecular weight 20 million) consisting of a core and separable subunits extending radially up to 1600 Å. The differences in physical properties described by Margoliash et al. (1965) and by Cauldwell et al. (1973) may reflect improved preparative procedures available to the latter group. In a less rigorous isolation procedure, Muller and Zahn (1973) obtained evidence for an even larger (3000 Å length) particle with many radially arranged filaments. Whether the unusual radial molecules identified by Humphreys and colleagues and by Muller and Zahn are responsible for determining the specificity of sponge cell adhesion remains to be established.

It is clear, however, that some sponge cells exhibit specificity of reaggregation during the reformation of recognizable organisms from dissociated cells and that a "factor" can be isolated from some species of sponge which acts to enhance the aggregation of homologous cells and possibly even to inhibit aggregation of heterologous cells. The existence of specificity seems to depend not only on the species but also on the procedures used to dissociate the cells and the culture conditions. Despite the many elegant studies in this area, the concept of "specificity of adhesion" during aggregation remains undefined in at least two areas. One relates to the time of expression of specificity, either at the initial contact between cells or its later appearance in the final cellular configuration. The second question concerns the in vivo applicability of the in vitro results obtained with reaggregation "factor". Is the aggregation "factor", as isolated from some cells, an integral part of the cellular adhesive mechanism in vivo and does the in vitro response of cells to "factor" demonstrate an in vivo mechanism of specific cell behavior? Despite these unresolved questions, the success obtained with sponge cells suggests obvious avenues of approach for parallel studies on vertebrate cell aggregation factors.

3. *Analysis of vertebrate cell reaggregation*

3.1. *The effect of in vitro cultivation on cell surface properties*

The study of specific interactions among embryonic cells in vitro usually requires the dissociation of tissues into their component cells. Early studies using cells from amphibian, avian and mammalian tissues attempted to apply the dissociation procedures used successfully for sponges and other invertebrates, namely, incubation in Ca^{2+} free medium or changes of pH and mechanical agitation (Roux, 1894; Herbst, 1900). Holtfreter, for example, found that early embryonic amphibian tissues could be dissociated by brief incubation at pH 9.6–9.8 without preventing their normal development from proceeding independently of the rest of the embryo (Holtfreter, 1943a,c). It was soon realized, however, that specialized techniques were necessary to obtain large numbers of viable separate cells from most avian or mammalian tissues.

There are a myriad of techniques which can be used for cell dissociation. The book on tissue culture technique edited by Kruse and Patterson (1973) contains more than 40 short papers giving detailed methodology for the establishment and maintenance of primary cell cultures from a broad range of animal and plant tissues. These papers reflect the general reliance on the use of enzymes for tissue digestion to obtain dissociated cells. The articles by Rinaldini (1958, 1959), Weiss (1960), Moscona, Trowell and Willmer (1965) and Waymouth (1974) also provide valuable discussions of cell dissociation procedures and their effects on cells.

The major non-enzymatic method used in the dissociation of mammalian and avian tissues involves the use of chelating agents that bind the ions that are presumed to be neccessary for the maintenance of cell-to-cell adhesion. Rappaport and Howze (1966a,b), Rappaport (1966a,b), Gerschenson and Casanello (1968) and Casanello and Gerschenson (1970) have reported some success in dissociating rat liver cells and other tissues using tetraphenylboron, an agent which complexes K^{+} (Flaschka and Barnard, 1960). Tetraphenylboron has been shown however, to be of limited value in providing cells with normal metabolic patterns (Friedman and Epstein, 1967), ultrastructure (Harris and Leone, 1966) or high viability (Brown and Hilfer, 1968; Waymouth, 1974). The absence of recent reports on its use indicate that its utility is probably restricted to narrowly defined conditions.

Much greater use has been made of the Ca^{2+}-chelating agent, ethylenediaminetetraacetate (EDTA or Versene). The product of EDTA dissociation usually consists, however, of a mixture of single cells, clumps and debris (Anderson, 1953; Moscona et al., 1965). It is important to note that even the successful application of EDTA as a dissociating agent of higher vertebrate cells requires the use of some mechanical dispersion as well, and this is inevitably accompanied by varying degrees of cell trauma (Allen and Snow, 1970; Ball, 1966; Harris and Leone, 1966; Hilfer and Hilfer, 1966; Moscona and Moscona,

1967). EDTA has also been used in conjunction with trypsin to facilitate tissue disaggregation (Steinberg, 1967; Kemp et al. 1967 and many others).

EGTA (ethyleneglycol (2-aminoethylether)-N,N′-tetraacetic acid), which chelates Ca^{2+} but not Mg^{2+} (Schmid and Reilley, 1957) has not been as widely used even though it appears to have fewer detrimental side effects than EDTA (Waymouth, 1974). Kleinschuster and Moscona (1972) and Robb (1973) have reported its respective application in primary dissociation of embryonic neural retina and in the removal of cells from plastic. Another non-enzymatic technique involving the use of univalent antibodies to induce cell dissociation has been reported for slime molds (Bueg et al., 1971).

The proteolytic enzyme, trypsin, remains the most commonly used agent for tissue dissociation and production of large numbers of dissociated single cells. Most tissue dissociation procedures using trypsin are derived from the protocols established by Moscona (1952) and Moscona and Moscona (1952).

A major feature of enzymic tissue dissociation that is all too often ignored is that enzymes such as trypsin produce significant changes in the surface properties of cells. While such alterations may be of little significance or consequence for certain experiments, an appreciation of the surface alterations accompanying enzymic tissue dispersal is crucial if the dissociated cells are to be used for studies on cell adhesion. Although this would seem to be a case of stating the obvious, it is unfortunate, and indeed alarming, that the literature on cell adhesion and many other aspects of cell surface function abounds with instances where inadequate consideration has been given to this important point.

Among the reported effects of trypsin mediated cell disaggregation that may prove to be important in our ultimate understanding of cell interactions are: change in dry mass due to loss of surface materials (L. Weiss, 1958; Mallucci et al. 1972); changes in mechanical properties (L. Weiss, 1966); loss of surface microvilli (Edwards and Fogh, 1959; Dalen and Todd, 1971); loss of desmosomes (Overton, 1968; Fischman and Moscona, 1969); changes in electrophoretic mobility (Ponder, 1951; Engel, Pumper and Joseph, 1968; Kemp, 1969; Barnard et al., 1969; Maslow, 1970; Steinberg et al., 1973); reduced respiratory enzyme activity (Kellner et al., 1959); reduced adhesion to glass substrates (Easty et al., 1960; Hebb and Chu, 1960; Weiss and Kapes, 1966); and altered cell-to-cell adhesion (Weiss and Maslow, 1972). In addition, penetration of tryspin inside the cell (Hodges et al., 1973) may create a further range of alterations in cell function. Despite all these effects, cells remain viable during mild tryptic digestion (Rinaldini, 1959).

Trypsin is inadequate, however, for full disaggregation of some tissues, notably those rich in collagen for which collagenase has been used successfully (Hilfer, 1973; Lasfargues, 1973). Elastase has also been found effective for some tissues (Rinaldini, 1959; Levinson and Green, 1965; Phillips, 1972), as has pronase (Foley and Aftonomos, 1970; Gwatkin and Thomson, 1964; Gwatkin, 1973; Houba, 1967). Sanford (1974) reported that ultrasound, combined with trypsin and EDTA, enhanced the removal of strongly adhesive cells from glass.

While there is no reason to assume that isolated cells, regardless of their means of dissociation, interact with other cells in a manner comparable to the situation in vivo it is certainly preferable to attempt to overcome the immediate effects of the isolation procedure before proceeding with experimentation. Barnard et al. (1969) and Maslow (1970) used a Vibromixer apparatus (Ulrich and Moore, 1965), to examine the recovery of an electrokinetic steady-state by cells after trypsinization. They found that after 24 hours in vibromixer culture, trypsin-dissociated embryonic chick neural retina cells (Barnard et al., 1969) and liver cells (Maslow, 1970) reached a constant electrophoretic mobility, although the recovery patterns differed for the two cells types. A comparison of the aggregation kinetics of trypsinized and mechanically dissociated 6-day embryonic neural retina cells was made by Steinberg et al. (1973). This showed that cells prevented from aggregating by strong shearing forces showed a 30–35 minute lag before partial recovery from the tryspin procedure and a 70–80 minute lag for complete recovery to the level of the mechanically dissociated cells. The lag in the trypsinized cells was only minimally extended by the absence of serum. Kolodny (1972) reported a lag of similar duration with trypsinized mouse 3T3 cells.

The importance of allowing cells to recover from the effects of trypsin before using them in experiments is also illustrated by the results of Weiss and Maslow (1972). They compared the effect of cycloheximide on the aggregation of freshly trypsinized neural retina cells and similar cells that had been cultured for 24 hours in a Vibromixer apparatus, thus permitting recovery from the trypsinization. The ability of cycloheximide to inhibit aggregation of cells in a gyratory shaker culture was found to be greater in the freshly trypsinized cells than in the recovered cells. Similar results have also been obtained with puromycin (Maslow, unpublished data). Similarly, Demon et al. (1974) found that trypsinized HeLa cells aggregated differently after neuraminidase treatment than untrypsinized cells while Glaeser et al. (1968) showed that inhibitors of macromolecular synthesis were more effective on trypsin-dissociated cells than on EDTA-dissociated cells. The latter results do not, however, contradict the results of Kemp et al. (1967) who found that trypsin- and EDTA-dispersed cells aggregate similarly in the presence of puromycin. Rather Kemp's results suggest that the use of EDTA for tissue dissociation is also damaging to cells, and can affect their subsequent associative behavior.

If serum free medium is used after trypsinization of cells then even greater consideration must be given to possible problems resulting from the uptake or retention of active trypsin bound to the cell surface (Moscona et al., 1965; Poste, 1971; Rosenberg, 1960). The vigorous washing or prolonged maintenance in protein free salt solutions required to elute the bound enzyme may be detrimental to cell viability and behavior.

The possible role of the dissociation procedures in determining specific cell sorting patterns will be discussed further later in this article. The use of more purified trypsin preparations and various modifications to reduce the extent and severity of its effect on cells will be welcome, but the use of appropriate controls for any dissociative procedure is essential in all experimentation, and valid comparisons can be made only between results obtained using cells exposed to strictly comparable procedures.

3.2. Methods for the in vitro study of specificity in cell-to-cell adhesion

Most of our information on specificity in cell-to-cell adhesion has been derived from experiments performed in vitro and the bulk of this work has been done using a few standard techniques that promote the formation of cell aggregates. The striking consistency of aggregate formation under standard conditions dictates that changes in cell aggregation behavior can be used as a reliable and sensitive indicator of changes in the adhesive specificity of cells within the aggregate.

The essential justification for the use of aggregate formation in vitro as a model for normal development lies in the fact that normal tissue structures and cell-to-cell relationships are often found in aggregates formed from cultures of different cells derived from tissues that also interact in situ. For example, Ede and Agerbak (1968) and Ede and Flint (1975) have demonstrated an excellent correlation between the abnormal morphology of the chick *talpid*[3] mutant and the aggregate behavior of its cells in vitro.

3.2.1. Self aggregation in stationary culture

The first methods used to study the formation of multicellular groups from suspensions of isolated cells relied upon self-aggregation involving spontaneous histiotypic assembly of aggregates in stationary cultures (culture vessel does not move). In this technique cells come into contact through their own active movements or, as in hanging drop cultures, with the aid of gravity. One of the prime advantages of stationary culture is that it permits direct observation of the cells and their movements by time lapse microscopy without disturbing their growth (Capers, 1960; Trinkaus and Lentz, 1964; Mayhew and Blumenson, 1972; Garrod and Steinberg, 1973; Tolson et al., 1975).

The aggregation of cells in stationary cultures is subject to little precise control because of variation between seemingly identical cultures in such factors as the type of medium, pH, serum components, the physico-chemical properties of the substrate surface and distribution of cells on the substrate. The cells themselves also contribute to the formation of their own environment through the synthesis of cell products which they release onto the substrate or into the medium, or both. This material, which has been variously termed ground mat (P. Weiss, 1945), extracellular material (Moscona, 1960), microexudate (Rosenberg, 1960; Poste et al., 1973), conditioning factor (Rubin, 1966), microruptures (L. Weiss, 1961), exudate (Maslow and Weiss, 1972), or substrate attached material (Culp, 1974), represents modification of the environment by cultured cells, especially those growing in contact with a non-cellular substrate. The observation of Maslow and Weiss (1972) that some cells exhibit higher levels of exudation at the time of adhesion to a non-cellular substratum than to a cellular monolayer is also consistent with the concept of cellular modification of their environment, as are the observations of close correlation between the amount of microexudate deposited on the substrate and changes in the adhesive properties of the cells (Poste, 1971; Poste et al., 1973). The use of media in which cells have already grown (conditioned), or media containing specific components of embryo extract to promote the growth or differentiation of low concentrations of cells (Weiss et al., 1975; Wilde, 1961; Rubin 1966) or clones (Konigsberg, 1963; Konigsberg and Hauschka, 1966; Cahn and Cahn, 1966; Coon, 1966) may artificially produce the same result. These "conditioning" factors that enhance cell growth may well represent cell surface components (Rubin, 1966; Igarashi and Yaoi, 1975; Weiss et al., 1975).

The cellular material(s) exuded just prior to contact may also serve as a physico-chemically acceptable "glue" enabling the cell to overcome potential energy barriers to adhesion of cells (Weiss, 1970, 1973; Weiss and Neiders, 1971). It could satisfy this function if it consisted of fibrous molecules of either low radius of curvature or carrying a positive charge or no net change. Fibrillar, possible protein containing, material has been detected in the intercellular areas of electron micrographs of lanthanum stained reaggregates of trypsin dissociated embryonic chick cells (Overton, 1969; Khan and Overton, 1969) and the exudation of a ^{51}Cr-labelled proteinaceous material was found to occur at or near the time of adhesion of tumor cells to glass (Maslow and Weiss, 1972). In neither case was the material characterized.

The histotypic organization and differentiation of cells in stationary culture into morphologically and/or functionally identifiable tissue-like cell groups has been reported for: skeletal muscle (Holtzer, 1961; Konigsberg, 1963; Maslow, 1967, 1969); cardiac muscle (DeHaan, 1967; Gordon and Wilde, 1965; Gordon and Brice, 1974a,b,c,); liver (Kaighn and Prince, 1971, McLimans, 1969; Iype, 1971); lung (Grover, 1961; Douglas and Kaign, 1974); and various glands (Lee, 1971; Gailani et al., 1970; Spooner, 1970; Tashjian et al., 1970). Garrod and Steinberg (1973) and Tolson et al. (1975) observed sorting of cells from different tissues in monolayer cultures by time lapse microcinematography, and Moscona (1956) and Trinkaus and Groves (1955) obtained sorting and differentiation of the cells in mixed aggregates formed by depression slide culture and centrifugation, respectively.

The formation of skeletal muscle in vitro has been carefully studied, and is especially instructive concerning the expression and maintenance of specificity. A suspension of tryspin dissociated embryonic skeletal muscle consists of myoblasts and other nonmyogenic cells which can be distinguished by their morphology after adhesion (Konigsberg, 1963). Since skeletal muscle forms by the fusion of myoblasts to produce a syncytial state (Lash et al., 1957; Wilde, 1958, 1959; Capers, 1960; Cooper and Konigsberg, 1961; Yaffe and Feldman, 1965; and others) the nature of the cellular mechanism by which non-myoblastic cells are distinguished from myoblastic cells prior to fusion is of interest. It has been shown repeatedly that the fusion process is not species specific (Wilde, 1958, 1959; Yaffe and Feldman, 1965; Maslow, 1967, 1969) but permits myoblastic cells from mouse or rat to be incorporated into chick myotubes by fusion, albeit to a reduced extent (Shimada, 1968). One attempt to elucidate the mechanism underlying this phenomenon involved cultivation of freshly trypsin-dissociated embryonic mouse liver cells, with or without Actinomycin D treatment, together with dissociated chick skeletal muscle cells (Maslow, 1969). Only those liver cells treated with Actinomycin D were incorporated into the muscle fibers. It was suggested that this aberrant behavior resulted from the fact that the initial dissociation procedure damaged the normal liver cell surface, and that the Actinomycin D prevented the regeneration of normal surface components. The damaged liver cell was therefore not correctly recognized as "non-muscle" by the myoblasts and erroneously incorporated. Unfortunately, no data was provided in this study on the incorporation of liver cells treated with actinomycin after recovery from the trypsin procedure.

3.2.2. Rotation mediated aggregation

The bulk of experiments which have contributed to the study of specific adhesive interactions between cells in vitro, have used the rotation mediated cell aggregation system developed by Moscona (1961a). Although this procedure was designed originally to simply produce cell aggregates from single cell suspensions, it has since been modified to permit: (1) observations on the course of aggregation by cell counts of samples removed at known intervals (Curtis and Greaves, 1965; Orr and Roseman, 1969a,b); (2) turbidometric determinations (Kemp et al., 1967; Kemp, 1970); (3) collection of cells in suspension by aggregates or tissue fragments (Roth, 1968; Roth and Weston, 1967; Roth et al., 1971a; and (4) measurements of aggregate size, using either the average number of cells per aggregate (Armstrong, 1966; Steinberg, Armstrong and Granger, 1973) or the average diameter of the aggregate (Weiss and Maslow, 1972; Maslow and Mayhew, 1975). Techniques have also been devised for studying the aggregation of small numbers of cells based on the same principles (Henkart and Humphreys, 1970; Riddle, 1974).

In gyratory shaker cultures, the particles in suspension, whatever their means of dispersion or formation, are concentrated toward the vortex created by the spinning fluid. The formation of aggregates from suspensions of single cells is thus promoted by increasing the collision frequency of the cells in the more concentrated vortex. The increasing frequency of collision is countered by the shearing force produced by the moving medium, which tends to separate the cells and reduce the collision frequency. Thus, only those collisions which result in the cells becoming attached firmly enough to resist the shearing forces present will contribute to aggregate formation. Slower rotation speeds produce larger aggregates and faster speeds smaller ones and, if sufficiently fast, may largely prevent aggregate formation. Initial adhesions occur between individual cells in such systems but after a relatively short time, most cells are in aggregates and it is then the interaction of aggregates that is being observed. Under the same conditions different cells should produce aggregates of different sizes depending on their strengths of adhesion or resistance to separation, but other factors related to adhesion such as proportion of less adhesive cells within the population (cf. Moscona, 1965) or shape of the forming aggregates (Ede and Agerbak, 1968) may introduce modifications.

More consistent results are obtained using shaker cultures since many of the variables that cannot readily be controlled in stationary culture are eliminated (Moscona, 1961a, 1965). Orr and Roseman (1969a) obtained quantitative confirmation of the reproducibility of the shaker technique using Coulter Counter determinations of numbers of single cells remaining at different times after the initiation of culture. Nevertheless, because of different cell sizes and period of recovery from dissociation trauma, direct comparison of adhesive strengths cannot be made between different cell types (Moscona, 1961a; Roth, 1968). In addition, it should not be assumed that variation due to differences in media components and pH do not exist. No information about the role of cell movement in aggregate formation can be obtained based on results from shaker cultures, although some speculations about the importance of movement in the sorting of cells within the aggregate have been advanced (Steinberg, 1963a; Steinberg and Wiseman, 1972; Maslow and Mayhew, 1972, 1974).

In addition to the production of aggregates composed of a single cell type, the gyratory shaker technique can also be used to form mixed cell aggregates. Moscona (1961a) found that cells of two different tissue origins formed common aggregates in mixed gyratory shaker cultures, but grouped according to cell type, with one cell type occupying a generally peripheral position. Gyratory shaker mediated co-aggregation of heterotypic cell suspensions has been used repeatedly by Moscona and his co-workers in the analysis of adhesive interactions between cells from the integument (Garber and Moscona, 1964, 1967; Moscona and Garber, 1968; M. Moscona and Moscona, 1965; and Garber et al., 1968). In these cases, the pattern of cells and the extent of their sorting in the aggregates reflected the degree of homology of the tissues of origin. The sizes of the aggregates were found to be characteristic of the species and cell type used, with heterospecific aggregates exhibiting an intermediate size (Moscona, 1961a, 1962a). The characteristic sorting of different cell types within heterotypic aggregates has been described by numerous investigators and the literature on this subject will be discussed in more detail in section 4.

Experimental analysis of the adhesive interactions occurring between cells in mixed aggregates obviously demands that the different cell types can be identified reliably. Three general approaches have been used for this purpose. First, one cell can be labeled with a radioisotope (^{3}H-thymidine has been used widely) either by labeling in the intact embryo before tissue dissociation or immediately after dissociation (Trinkaus and Gross, 1961; Wiseman and Steinberg, 1973). The former is preferable since it eliminates the need for a culture period to achieve labeling of the cell population. Care must be taken, however, when dissociating tissues containing radiolabeled cells to ensure that labeled material released by lysed cells during the dissociation procedure does not contaminate unlabeled cells in the mixed culture. After incubation of labeled and unlabeled cells together, the resulting aggregates are fixed, sectioned and autoradiograms prepared to reveal the labeled cell population. This tedious procedure can, of course, be eliminated if a satisfactory cellular marker exists. For example, chromosomal markers are useful though they cannot be detected in sections of non-dividing cells. Pigment cells are ideal, but of limited availability (Trinkaus and Gross, 1961; Armstrong, 1971; Armstrong and Parenti, 1972).

Cells from different species can often be distinguished in culture based on differences in cell size and their histochemical staining properties. For example, mouse and chick cells can be distinguished in mixed cultures and this has been exploited in many investigations (Moscona, 1957; 1961a; Wolff, 1954; Wilde, 1958; Garber and Moscona, 1964; Yaffe and Feldman, 1965; Maslow, 1969; Burdick, 1970, 1972). The nuclei of mouse cells are larger and stain more darkly than chick nuclei with basic stains and with hematoxylin, though Holtzer (1961) has criticized the use of these staining differences to distinguish the identity of cells in cultures. Since the primary purpose of using cells derived from different organisms is to follow the behavior of cells of different tissue origin, it is also necessary to show that heterospecific mixtures of the same tissue type do not sort out from each other. The work of Moscona (1957, 1960, 1961a), Garber and Moscona (1964) and Garber et al. (1968) with many tissue combinations seemed to indicate that no sorting between tissues takes place based solely on species origin, but Burdick and Steinberg (1969) and Burdick (1970, 1972) have questioned the generality of this conclusion because of their inability to repeat some of the earlier results, as well as errors they point out in some of Moscona's reports. For some tissues, such as skin and neural retina, it should be noted that even the critics claim no sorting along species lines alone. Perhaps the wisest course of action now is to avoid the use of this attractive distinguishing feature, or at least to ensure that when cultures containing cells from different species and tissue origin are used that they are compared with control heterospecific cultures of the same tissue to establish that the latter do not give any indication of cell sorting after the same length of time in culture.

The final method used to delineate the relative positions of cells in mixed cell aggregates is based on differences in appearance following histochemical staining for specific cell constituents. For example, heart or liver cells, which are frequently co-aggregated with neural retina cells, can be distinguished from the retinal cells in mixed aggregates by their glycogen content detected by PAS staining (Steinberg, 1962a). This approach has the disadvantages of requiring some care in the histological preparation, and of being useful only in pairs of cell types that differ in a histochemically identifiable characteristic. It is also necessary that the cells not lose or change their staining properties as a result of the dissociation procedure or cultivation method. Fortunately, these criteria are usually met. Most of the results discussed below on specific cell sorting in mixed aggregates are based on observations obtained by use of appropriate staining techniques. In all sorting experiments, variation exists among aggregates in the same culture. Processing and sectioning may add additional variation to what is seen on the slide. Thus, it is not advisable to judge differences in sorting patterns based solely on photographs of single aggregates, nor to expect every aggregate to show the idealized sorting patterns.

3.2.3. The collecting aggregate technique

Roth and Weston (1967) and Roth (1968) have described a modification of the gyratory shaker aggregation technique that permits an evaluation of the probability of adhesion between cells and aggregates of the same or different types. Its major improvement over the reaggregation type experiment is that it reduces the number of variables which affect the rate of collision. Typically, radioactively labeled cell suspensions are cultured in gyratory shaker flasks together with either homotypic or heterotypic unlabeled aggregates of uniform size. After 4 to 6 hours, the aggregates and any labeled cells that have adhered are fixed and the number of radiolabeled cells adhering to the isotypic or heterotypic aggregates determined by counting the number of cells with radiolabeled nuclei from autoradiographs, or by direct liquid scintillation counting (Roth et al., 1971a). Using either procedure, the number of adherent cells has been found to vary with the size of the aggregate, the period of incubation, the concentration of cells in the suspension, and the temperature. Since these parameters can be kept uniform for the different cell aggregates, the finding that consistently higher numbers of cells adhere to homotypic aggregates compared to heterotypic aggregates, strongly indicates that specificity of adhesion can be expressed at the time of initial adhesions. This may, however, only reflect the behavior of the small percentage of cells which adhered to all the aggregates in a single culture flask.

McClay and Baker (1975) modified the collecting aggregate system by increasing the ratio of aggregates to cells in suspension and determined that even when the percentage of cells adhering to the aggregates was increased to approximately 50% of the initial cell population, specificity was still manifest. They also demonstrated, however, that there were subpopulations within each tissue that showed different patterns of adhesion and that differential trypsin recovery or the age of the embryo from which the cells were derived can perturb the adhesion patterns, especially during the first hour. Thus, this system, which minimizes many of the variables of the shaker technique, does not eliminate the possibility of a trypsin dissociation effect, particularly in view of Curtis' (1970c) data showing that even low levels of trypsin retained on the freshly dissociated cells are effective in reducing the adhesiveness of both homotypic and heterotypic aggregates. It is also of interest that fragments of tissues demonstrated the same selectivity as their respective aggregates (Roth, 1968).

These results do not contradict data which indicates that the initial adhesion between freshly dissociated cells in suspension is non-specific, since Roth indicated that the introduction of a heterotypic cell type into a homotypic suspension resulted in recruitment of heterotypic cells to the aggregates for up to 6–8 hours after dissociation. These mixed aggregates then attach to the preformed collecting aggregate through the homotypic cell (Roth, 1968). Although not discussed directly, the implication is that cells recover from their dissociation trauma after 6–8 hours and then from only isotypic adhesions. This would be in contradiction to unpublished results obtained by the present author showing that cells that had recovered from the dissociation procedure formed initially mixed aggregates which subsequently sorted. At some point these recovered cells must adhere to unlike cells for a randomly mixed aggregate to be formed, suggesting that the initial adhesions cannot be totally specific in the shaker system even with recovered cells.

McGuire (1972) attempted to resolve this inconsistency by proposing that the movement of the medium in a shaker culture provides greater resistance to the adhesion of single cells to large aggregates than adhesion between single cells or between aggregates. Thus, initial non-specific aggregation would occur in a single cell suspension even with recovered cells, but only homotypic adhesions are of sufficient strength to keep single cells attached to a preformed aggregate. It would appear, however, that *both* the aggregate and the surface of the isolated cell must be in the recovery lag period for non-specific adhesion to take place with collecting aggregates, so that differential recovery periods for different cell types may play a role in the appearance of specificity in this system, as suggested by Curtis (1970b). Experiments in which the cells in suspension and the aggregates share a 24 hour recovery period prior to mixing would be helpful in clarifying this point.

The collecting aggregate procedure, especially as modified by McClay and Baker (1975) using double cell labeling, provides quantitative data on the specificity of adhesion of cells to aggregates. However, even at the higher percentages of adherent cells, the cells found on the aggregate may still be a nonrepresentative fraction of the total population.

A method which provides a quantitative assay for specificity of adhesion of *all* cells in a population has been described by Walther et al. (1973). In this technique, the number of radiolabeled cells from a cell suspension that adhere to a monolayer of similar or different cells is determined by the amount of radiolabel recovered in association with the monolayer after various times. It can be assumed that all the cells will settle to the floor of the culture vessel and will thus have the opportunity to form adhesions with the monolayer. The monolayer is lysed after washing, and the total radioactivity associated with each monolayer is counted. This method is similar to that used by Maslow and Weiss (1972) but eliminates the need for tedious counting of labeled cells in autoradiographs. The variations in adhesive behavior observed using these techniques depend in part on the nature and prior history of the cells tested, thereby ruling out generalizations about specific cell adhesions. However, these methods may be useful for further quantitative analyses of the specificity of adhesion of some embryonic types, particularly under varying conditions. Gottlieb and Glaser (1975) have described a procedure to obtain monolayers almost immediately by the addition of the cell suspension to glass surfaces derivitized with glutaraldehyde-activated α-aminopropyl-triethoxysilane. This technique could markedly enhance the usefulness of monolayer-adhesion techniques.

3.2.4. Couette viscometer

The methods used by most investigators to study the specificity of interaction during the sorting process monitor the final relative positions of each cell type within the aggregate. Curtis (1969, 1970c) has developed a technique for studying the aggregation of cells which also provides a method of determining the absolute adhesiveness of the cells using a Couette viscometer. This consists of two concentric cylinders, the outer one rotated at known speeds around the freely suspended inner one. The cell suspension is placed in the approximately 1.5 mm space between them, and samples are removed at intervals for direct counting of the number of particles present. The average aggregate size in the viscometer, as in the gyratory shaker, depends on: (1) the adhesiveness of the cells; (2) the rate of aggregate formation through collisions of cells remaining permanent; and (3) the rate of break-up of aggregates due to shearing forces. As Curtis noted, an advantage of this system is the small variation in shear force due to the greater similarity in the velocity of medium flow. This contrasts with the gyratory flask system in which the variation in velocity, and thus shear force, is greater, producing a greater variation in aggregate size. Curtis claimed that the segregation of aggregates toward the vortex, away from the single cells concentrated at the periphery, is also reduced because of the narrow path of medium flow. The apparatus also permits control of differences in the viscosity of the medium, which would be reflected in a slowdown of the outer cylinder unless additional power is added.

In this system the aggregation rate is proportional to the collision rate if all the cells present are capable of adhering to each other. On the other hand, if two cell populations are present that can only form adhesions with cells of their own type, the rate of aggregation will be reduced. For an equal mixture of cells of the same adhesiveness that cannot form mixed adhesions, the rate of aggregation will be half that of a freely adhering population since half the collisions will not result in adhesions. Similarly, variation in absolute adhesiveness, or the degree of specific adhesion of one cell type, will be reflected in the rate of decrease in the number of particles in the aggregating cell populations and thus in the collision efficiency of the culture. It is assumed, without full justification, that the presence of two cell types will not result in some positive or negative interference in the aggregation process. It should be emphasized that, at best, this technique can only give information about the specificity of initial adhesions, and cannot be used to detect the appearance of specificity within the aggregates at a latter time. This is an important criticism since most evidence, as discussed below, supports the hypothesis that cell sorting patterns are not expressed at the time of formation of initial adhesions.

The rate of formation of aggregates was determined in Curtis' experiments (1969, 1970c; Curtis and Greaves, 1965) by manual hemocytometer counts of the total numbers of particles at intervals after the onset of culture. Attempts have been made subsequently to eliminate the tedium and potential for error associated with repeated manual counting by using electronic (Coulter) counters (Ball, 1966; Edwards and Campbell, 1971; Orr and Roseman, 1969a, b; McQuiddy and Lilien, 1971; Edwards, 1973) or a turbidometric assay method based on optical density changes (Jones, 1965; Cunningham and Hirst, 1967; Kemp, Jones, Cunningham and James, 1967; Kemp, 1970). These methods provide quantitative data, within their technological limits, on the rate of aggregation of cells, and even the size distribution of aggregates, but not on specific aggregation. Neither approach can distinguish cells of different types in aggregates formed in mixed cultures, even if of different sizes. Consequently, it cannot be determined if cells are aggregating only with similar cells. In addition, in both techniques cell death may simulate aggregation, since the counting mechanism depends on reduction in particle number.

4. *Some hypotheses to explain specific aggregation behavior*

4.1. *Specific aggregation factors*

The identification of "factors" that promoted specific adhesion and aggregation of sponge cells (section 2) has provided an obvious stimulus to search for similar factors in vertebrate cells.

The finding that inhibitors of cellular protein synthesis reduced the capacity of various avian and mammalian cells to form aggregates (Moscona, 1961b; Steinberg, 1962d; Curtis, 1963; Weiss, 1964; Ball, 1966; Moscona and Moscona, 1966; Richmond et al., 1968) prompted considerable speculation as to whether the inhibitors were blocking the synthesis of a cellular product that was responsible for aggregation, i.e. an aggregation "factor". However, Moscona showed that cell-free medium in which cells had been growing was also capable of producing a partially specific enhancement of aggregate size. Moscona proposed that this effect was due to a so called extracellular material (ECM) which was able to promote cell adhesion and aggregation by serving as a bridge between cells.

Steinberg (1963b) equated ECM with the stringy gel-like material present in suspensions of freshly trypsinized cells. This material dissolved in the presence of DNAse, indicating that a major structural component of the visible material probably consisted of nucleic acids liberated by dead cells during the tissue disaggregation procedure. However, the aggregation enhancement capacity of conditioned medium is not destroyed by DNAse incubation.

Enhancement of aggregation or adhesion by cell products present in the medium has also been shown by Wilde (1961, 1962), Lilien and Moscona (1967) Roth (1968) and Culp (1974). However, it is important to distinguish between non-specific enhancement of aggregation by conditioned medium, which may be due to a serum component bound by cells in the initial serum containing culture and released in the subsequent serum free culture (Lilien, 1968), and specific enhancement which is effective only on one cell type. Nir (1975 and personal communication) has estimated, based on calculations of Van der Waals attractive forces, that variations in the amounts of glycoprotein, protein and carbohydrate in the medium or bound to the cell surface resulting from the addition of a non-specific component could cause significant changes in the magnitude of attractive interactions between cells. Serum activation effects (Lilien, 1969; Balsamo and Lilien, 1974a) or the destruction of serum containing aggregation inhibitors (Curtis and Greaves, 1965) provide other possible mechanisms which could account for the non-specific enhancement. Rosenberg et al. (1969) found that a surface region fraction isolated from embryonic cells produced non-specific enhancement of aggregation of cells in vitro. The fraction was inactivated by heating, and was shown to bind only to the surfaces of living cells. The presence of fragments of surface material in media withdrawn from cultures cannot be ruled out, and takes an added significance in view of the recent demonstration of specific adhesion between isolated membranes and cells of the same type (Merrell and Glaser, 1973; Gottlieb et al., 1974).

Factors from cultured cells which act specifically to enhance reaggregation of homologous cells are obtained only in serum free cultures, and their effects on subsequent cell suspensions are masked in the presence of serum. Specific enhancement of aggregation by addition of medium containing cell-derived factor(s) has been described for neural retina (Moscona, 1962b; Lilien and Moscona, 1967; Lilien, 1968), brain (Garber and Moscona, 1967, 1972) and liver (Kuroda, 1968). However, Kuroda's report does not include the critical test for specificity of aggregation enhancing effect with heterologous cells. Details of the preparation of the aggregation enhancing factors vary, but those demonstrated to be specific involve growth of cells in two changes of serum-free medium for 1 or 2 days after growth for 2 and 3 days in the presence of serum. Dialysis of the cell free medium removes a non-specific aggregation enhancement component leaving a non-dialyzable constituent which displays an enhancement effect only on homologous cells. Characterization of aggregation enhancement factors from vertebrate cells has not proceeded as far as those found in sponge cells, but a relatively purified preparation of neural retina aggregating factor has been shown to be a glycoprotein of approximately 50,000 molecular weight (McClay and Moscona, 1974) and which lacks galactosyltransferase activity (Garfield et al., 1974). In addition, at least part of the specificity of binding of aggregation factors to cells is based on the type of carbohydrate moieties present in the complex glycoproteins on the cell surface (Balsamo and Lilien, 1975).

The aggregation enhancement effect of the cell product requires both uptake by (or association with) the cells and subsequent utilization by the cell (Moscona, 1963; Lilien, 1969). The existence of these two separate processes was demonstrated by dispersion of freshly dissociated neural retina cells into medium containing "factor" at 4°C, a temperature at which aggregates would not be formed. When returned to 37°C, typical enhancement of aggregation was observed. However, the medium, if separated from the cells after the incubation at 4°C and tested on another cell population, had no effect. These results indicate that while cells can bind the factor at 4°C, a further temperature sensitive, presumably biosynthetic process is needed for enhancement to be manifest (Lilien, 1968). Nevertheless, the specificity of the effect appears to be manifest at the uptake stage since repetition of these experiments using a liver cell suspension at 4°C resulted in no loss of "factor" from the medium when assayed at 37°C on neural retina cells.

Specificity in the binding of aggregation enhancing factors to the cell surface has also been demonstrated by the binding of radiolabeled neural retina and brain enhancement factors to heterologous cells only in the absence of serum, while the "factor" was bound to homologous cells even in the presence of serum (Balsamo and Lilien, 1974a). Localization of the factor at the cell surface of responding cells by specific antisera also suggested that the cell surface represents the site of action, though the need for a subsequent metabolic process emphasizes that its action is not necessarily limited to the surface. The obscure nature of the functional utilization of the factor is further shown by the observation that glutaraldehyde fixed cells bound to aggregation promoting factors can enhance aggregation of homologous living cells (Balsamo and Lilien, 1974b).

The experimental data obtained to date on aggregation promotion factors are of interest even though the enhancement effect requires an immediately preceding history of tryptic dissociation and cellular responsiveness is lost after even a short recovery period (Lilien, 1969; Balsamo and Lilien, 1974a). While the need for a trypsin perturbed cell surface lends some support to the hypothesis that the role of the factor is to replace or repair surface molecules involved in specific recognition removed or damaged by the exposure to trypsin, it also questions the functional relevance of a material that is apparently produced only in serum-free medium and which acts only on freshly trypsinized cells.

An initial approach toward extending these results to specificity in cell interactions in vivo has been made by Hausman and Moscona (1973) who attempted to correlate in vitro effects to molecular events at the transcriptional or translational level. However, it remains to be seen if their correlation between inhibition of aggregation, production of enhancement factor (but not the responsiveness to exogenously supplied homologous factor) and inhibition of glutamine synthetase synthesis is a causal one (Wiens and Moscona, 1972; Hausman and Moscona, 1973; Moscona and Wiens, 1975). Even if the production of the enhancement factor can be detected under the artificial conditions described above, evidence for a correlation between its production and an identifiable synthetic process lends support to the idea that the factor represents components of the cell surface or cell products normally involved in histiotypic interactions. Alternatively, the labile "third component" proposed by Balsamo and Lilien (1974b) may be the agent of real significance in vivo since its description would suggest a role for cells under physiological conditions (Fig. 1). It would also explain many of the temperature effects reported in connection with enhancement factors.

Pessac and Defendi (1972) have also suggested a three component system, two of which are receptors on each cell and the third an aggregation-promoting factor. All must be present for aggregation to occur, but a cell may produce the factor, the receptor, both or neither. The factor itself is capable of aggregating all cell types with receptors, so its usefulness as a determinant of specificity is limited. Weinbaum and Burger (1973) presented a similar model for sponge aggregation involving a released aggregation factor and a surface bound baseplate stabilized by calcium ions. Finally, it should not be overlooked that

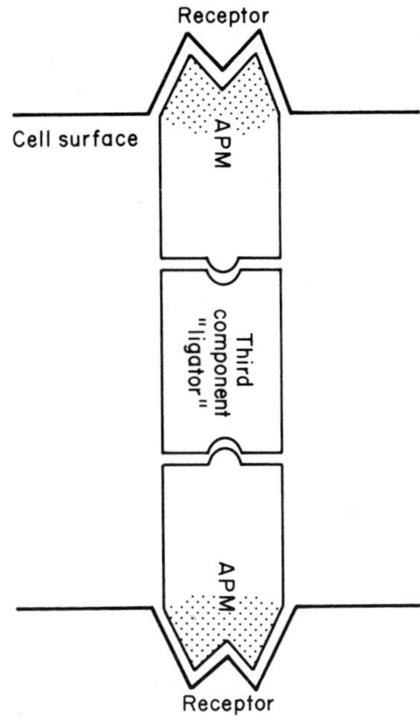

Fig. 1. Schematic representation of the three component model of specific cell adhesion showing specific aggregation promoting materials (APM) linked by a third component. The stippled areas on the APM's represent carbohydrates. Reproduced with permission from Balsamo and Lilien (1974b).

even if these factors are artifacts of the conditions under which they are harvested, their specific binding to cells in vitro may support the idea of molecular complementarity between cell surfaces, with cell surface moieties normally present serving as the complement to other surface molecules available for interactions only following trypsin exposure.

4.2. *Differential adhesion*

Although sorting according to tissue origin in suspensions of mixed cell types was studied by Moscona as early as 1952, most of his subsequent work has been concerned with other aspects of specificity, particularly the specific enhancement of aggregation of isotypic cells. The major work devoted to analyses of the sorting of cells in aggregates containing two cell types has come from Steinberg and his colleagues at Princeton.

The experimental results obtained by Steinberg and his group (to be discussed further) are interesting, but perhaps their greater contribution has been the introduction of concepts from the physical sciences and the demonstration of their possible applicability to cells. Steinberg applied thermodynamic free energy considerations to populations of two cell types which sort out from each other in mixed aggregates (Steinberg, 1962a,b,c, 1963a, 1964, 1970). The final equilibrium configuration of the cells within the aggregate was defined as that arrangement which reduces the free energy of the system to a minimum. This minimum is reached, according to Steinberg, when the total work done through adhesion in the system is raised to a maximum, which is a way of describing the situation when all the cells are so oriented that they adhere to each other with the greatest average tenacity. The "average" aspect is frequently overlooked by critics who question why the less adhesive phase should remain stable when adhesions between two different cell types is of greater strength (Jones and Morrison, 1969).

The differential adhesion hypothesis predicts that the less adhesive cells would form a stable outer layer when the average of the strengths of adhesion between the more adhesive cells and those between the less adhesive ones is greater than the strength of adhesion of a pair of cells of mixed types. This is expressed by the relationship:

$$\frac{Wa + Wb}{2} > W_{ab} \geq Wb$$

in which Wa, Wb and W_{ab} represent the works of adhesion of cells of the more adhesive type a, the less adhesive type b and mixed, respectively. When the average work of adhesion, $Wa + Wb/2$, is less than the adhesion between unlike cells, W_{ab}, a mixture of cells would be expected. When the work of adhesion of the less adhesive cell type, Wb, is greater than that of mixture, W_{ab}, separate aggregates would be formed, with appropriate gradations of envelopment for situations of near equality. The results expected from combinations of cells of differing relative cohesiveness are shown schematically in Fig. 2. From these derivations, Steinberg made a number of predictions about cell behavior (subsequently somewhat qualified), all of which were confirmed by experiments. In summary they state that if the more cohesive, internally segregating cell type is present as a very low percentage of the total cell population, cells of that type will not appear at the surface of the aggregate, but will not necessarily form one coalescent internal phase because of the reduced likelihood of them coming into contact. Furthermore, the mutual positions that cell types reach at equilibrium in different tissue pairings will assume a hierarchy in which a cell type that sorts externally to a second cell type will also sort externally to all cell types that sort internally to the second under standard conditions. Another proposal, also made by Townes and Holtfreter (1955), predicts that the equilibrium configuration of cell mixtures in aggregates will be the same as that for mixtures of the same intact tissue fragments. This is of some potential significance since it serves to give the aggregation of cells in suspension and the subsequent rearrangement of cells in aggregates a measure of relevance for possible events in vivo. However, as discussed by Curtis (1967), the fulfillment of these predictions does not necessarily prove Steinberg's hypothesis. The hierarchies of adhesiveness obtained by Steinberg's technique do not fully agree with those obtained using other methods, such as rate of aggregation and equilibrium aggregate size, possibly indicating the contribution of different aspects of cell behavior in aggregate formation under different conditions (Gershman, 1970; Hornby, 1973).

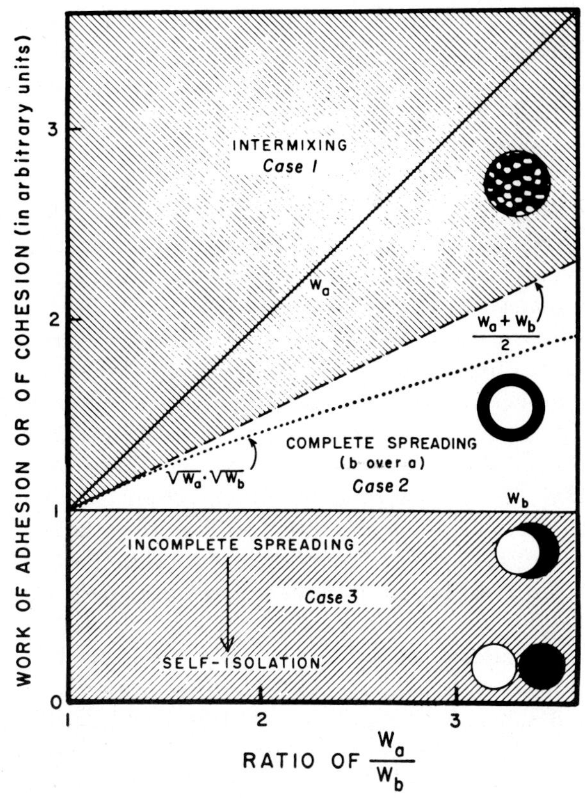

Fig. 2. Diagrammatic representation of the sorting patterns at equilibrium in coherent populations of two cell types of various cohesiveness according to the differential adhesion hypothesis. The work of cohesion of the less cohesive cell type (b), set equal to 1, is designated by line Wb; Wa denotes the work of cohesion of the more cohesive cells (a). The predicted pattern is revealed by the shading of the background at that point on the verticle line drawn from the appropriate abscissa value of Wa/Wb which corresponds to the work of adhesion of the two unlike cells as read on the ordinate. The dotted line, $\sqrt{Wa} \cdot \sqrt{Wb}$ indicates Wab for a system devoid of adhesive specificity. Reproduced with permission from Steinberg (1963a).

Exceptions to the hierarchal order of sorting have been shown for some combinations of chick embryonic heart and liver cells in which one cell population assumed the internal, discontinuous phase on some occasions and the external, continuous phase on others (Armstrong and Niederman, 1972; Wiseman et al., 1972). This variability, even though found in only some pairs of tissue types, must be explained within the framework of the differential adhesion hypothesis for the hypothesis to have credibility. Wiseman et al. (1972) presented data to indicate that time in culture, but not the extent of ontogenetic development of the embryo at the time of tissue excision, accounted for the variability of final sorting patterns following fusions of aggregate and tissue fragments of different sizes. The mixed aggregate experiments made by Gershman (1970) showing similar sorting patterns in combinations of 4–15 day embryonic chick tissues are also consistent with this conclusion. However, Lesseps (1973) and Lesseps and Brown (1974) found differences in the sorting of heart cells from embryos younger than 4 days, perhaps reflecting the morphogenetic development of the heart at that early stage. Unfortunately, Wiseman et al. (1972) did not report the envelopment behavior that followed the fusion of fragments of the same tissue and age in ovo, but differing only in period of incubation. Clearly the reported results with heart and liver combinations show that maintenance of cells in vitro alters the sorting patterns of these cell types, but data from homotypic combinations differing only in prior culture period would provide clear confirmation.

Maslow (1970) suggested that the variability of the recovery of liver cells from trypsinization, as monitored by recovery of their electrophoretic mobilities, may reflect the variability of the sorting behavior of liver tissues. However, Wiseman et al. (1972) have invoked the changes in the sorting patterns of heart and liver cells after periods in vitro as proof of the differential adhesion hypothesis while at the same time advancing the view that a time-dependent change in the behavior of the cells undergoing the rearrangement accounts for the variability of the sorting. Independent evidence, perhaps from some absolute measurement of adhesion, or at least from measurement of the rate of adhesion to a constant substrate, is clearly needed.

Phillips and Steinberg (1969) developed a modification of the sessile drop technique in an effort to determine absolute values for the adhesive properties of cells. This method involved calculations of the specific interfacial free energy of cell populations from the shape aggregate assumed during prolonged centrifugation. The assumption, made earlier by Holtfreter (1943a,b, 1944) and by Steinberg (1962c) is that cell populations, but not individual cells, in a mixed aggregate will behave similarly to a two phase system of mutually immiscible liquids such as water and oil, even though different mechanisms are clearly responsible. The results of long periods of centrifugation under fixed g-forces were that the aggregate assumed shapes characteristic of their cell type, whether the cells initially were in a spherical or flattened configuration. The relative shapes achieved by the aggregates agreed with the hierarchy of sorting in mixed aggregates with the most cohesive cells that more frequently sort internally also showing the greatest resistance to flattening. Phillips (personal communication) has also found that in mixed aggregates, each cell type assumed the equilibrium shape it had when cultured alone. These published conclusions, based on unpublished calculations, differ, however, from those of Curtis (1969, 1970b) based on his determinations of adhesive efficiencies.

While the determination of absolute values for the work of adhesion of cells in a population is of interest, cell heterogeneity would prevent the surface energy value from being more than an approximation for a specific cell. Noting this, Holtfreter suggested that the factors which cause instability in cell aggregates were of greater interest than those that seem to promote equilibrium configuration: "for it is these continuous deviations from an equilibrium emerging from the specific asymmetrical structure of living matter which are the essential feature of morphogenesis" (Holtfreter, 1943b).

Steinberg (1962e, 1964) has also proposed a stochastic model, later termed the site frequency model, to account for the differential adhesiveness advanced in his original hypothesis for cell sorting. This model, which has not been pursued, proposed that a quantitative difference in adhesive sites on the surface of cells could account for differential adhesion without the need for qualitative differences (Steinberg, 1958, 1962e, 1970). This conceptually attractive proposal suffers from a lack of evidence in its support, although there is no real reason to discard it. It is unclear, however, how tissue specific quantitative differences in surface molecules could be consistent with the tissue size related reversal in tissue sorting patterns reported by Wiseman, Steinberg and Phillips (1972).

4.3. Differences in periods of adhesivity

The changes in cell-to-cell contacts that must occur for the morphogenetic rearrangements characteristic of early embryological development to proceed, may result, at least in part, from changes in the relative adhesiveness of the cells, as proposed by Wiseman et al. (1972) and Armstrong and Niederman (1972). Since cultured tissue fragments also show changes in their in vitro behavior at ages when freshly excised tissues do not, the effect of the dissociation and culture procedures must also be considered in any analysis of sorting of different cells.

The comparison of the final sorting patterns and sizes of aggregates formed in rotation mediated cultures may reflect the adhesive stability or relative cellular adhesiveness of the two cell types, but does not give information on the probability of adhesion of two cells, either like or unlike. In addition, although both the shaker system and the fusion of tissue fragments permit assay of the equilibrium configuration they do not indicate when the specificity, if any, is expressed. Is specificity expressed at the initial formation of an adhesion between two cells or during the subsequent sorting of cells in the initially random aggregate? Curtis (1970c) has attempted to deal with this problem, and has developed an hypothesis to account for the appearance of sorting within aggregates based on his inability to detect specific cell adhesion.

Curtis proposes that the adhesive and/or the motile properties of cells change during the segregation period but the timing differs for different cell types. Thus one cell type becomes capable of forming stable specific adhesions before the other, thus dictating that homotypic adhesion will predominate. It is assumed that the trapping of cells at fixed positions within the aggregate by the formation of stable adhesions is initiated solely at the periphery of the aggregate although it is unclear why all of the cells of one type should not be trapped in their scattered positions at the same time, or at least that the trapping begin in scattered areas in the aggregate.

Curtis (1961) suggested that the trapping mechanism may result from viscosity changes in the periphery of cells at the surface of the aggregate, perhaps triggered by their being subject to reduced shear compared to that of the cells moving within the aggregate. No supporting data have been reported for cells in aggregates to confirm the differences in shear or viscosity that Curtis found for experimentally manipulated isolated cells. The results of Jones (1965) and Kemp and Jones (1970) showing that the addition of p-benzoquinone to aggregating muscle cells caused concentration-dependent inhibition in aggregation without effecting cell viability are consistent with Curtis' suggested mechanism. Quinone treated membranes have elevated surface viscosities, surface pressures, surface potentials and surface charge densities, all regarded as unfavorable to the formation of cell adhesions (Weiss, 1960).

The major support for the "timing" hypothesis comes from experiments using cells from embryonic amphibian tissue (Curtis, 1961) and sponges (Curtis, 1962a) in which one cell type is permitted to aggregate for 4 to 6 hours prior to the addition of the second and the final sorted pattern compared to that of control aggregates initiated with both tissues simultaneously. For both sponge and amphibian cells, the addition of one freshly dissociated cell type to the other after it had begun to aggregate resulted in a marked change in the final configuration of the sorted mixture. It would be of interest to determine if cultures in which one cell type, permitted to recover for 4 to 6 hours without aggregating, was added to a freshly dissociated second cell type, would behave similarly. For sponges the sorting may be dependent on differences in the aggregation rates of the cells from each species as expressed in isotypic cultures (Curtis, 1962a) but such differences have not been reported for amphibian cells.

These examples of temporal specificity have not been confirmed however, using chick or mammalian cells. We have shown, in fact, (Fig. 3) that chick embryonic heart and neural retina cells will sort in the same pattern after a 24 hour recovery period for one cell type just as in cultures initiated with freshly trypsinized cells. Similarly, Grover (1961) obtained normal sorting out with embryonic chick epithelial cells maintained in culture for 19 hours prior to the addition of the mesenchymal cells. Also, Maslow and Mayhew (1974) have found that the addition of the internally segregating freshly dissociated heart cells to neural retinal aggregates resulted in approximately half the aggregates showing the control sorted pattern, with the remainder of the aggregates containing only one cell type.

Using the Couette viscometer method, Curtis (1970c) has since extended his hypothesis to account for the results with aggregating suspensions of chick cells. In this case, a change in relative adhesiveness early in the recovery period following trypsinization is proposed as the mechanism that accounts for the final sorted configuration in mixed cell aggregates. According to this proposal, the initially more adhesive cell population begins to aggregate while the other cell type remains dissociated. The relative adhesiveness of the cells would gradually change. At some point, the previously less adhesive cell type, now more adhesive, would begin to aggregate onto the core formed by the first cell type, resulting in the observed concentric final pattern.

This proposal raises several questions. First, why should the newly adhesive cells, still in suspension, adhere to the aggregates of the now less adhesive cells in preference to forming isotypic aggregates? This is in direct contrast to the collecting aggregate experiments of Roth and Weston (1967) in which cells preferentially adhered to aggregates of the same type. Curtis attributed this

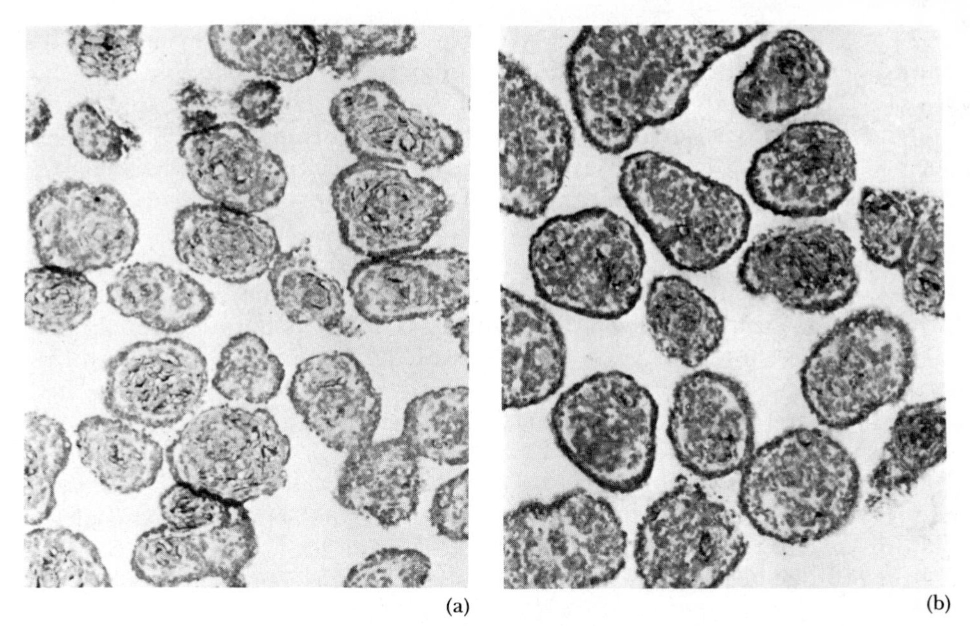

Fig. 3. Effect of prior recovery from trypsin mediated dissociation of one cell type on the sorting of aggregates formed in gyratory shaker culture of embryonic chick neural retina and heart cells. (a) Culture initiated with both cell types immediately after dissociation. X150. (b) Culture initiated with the neural retina cells after 24 hours in Vibromixer culture without aggregation and the heart cells freshly dissociated. X150. The PAS-negative staining retina cells are located peripherally to the PAS-positive staining heart cells in both cases, but the aggregates are larger when recovered cells are used (b). The heart cells, which are stained red, are incompletely represented by the dark areas in the black and white photographs.

seeming specificity of adhesion in collecting aggregate experiments to the difference in the immediacy of the trypsin exposure of the already formed aggregates and the newly trypsinized cells making heterotypic adhesions unlikely. However, in mixed cell suspensions, the random adhesions that would be expected during the cross over period of approximately equal adhesiveness would be unlikely to account for the universality of mixed but sorted aggregates, and the small number of homogeneous aggregates of the later aggregating cell type.

Secondly, the timing hypothesis proposes that the externally positioned cell type assumes that position because it becomes adhesive later than its co-aggregating cell type. This contrasts with the earlier hypothesis of Curtis (1961) which proposes that the externally aggregating cells in an initially mixed aggregate becomes trapped there because of earlier acquisition of the correct adhesive properties. Since the same cells would presumably have the same recovery pattern in both suspension culture and in preformed randomly mixed aggregates, these hypotheses would predict different final configurations for aggregates cultured in the two systems. In fact, the same pattern is obtained making it unlikely that a different mechanism acts in each case.

Finally, Curtis suggests that trypsin exposure plays a role in creating the impression of specific adhesion of cells to aggregates in the collecting aggregate system. While this possibility cannot be dismissed, it certainly cannot apply to the fragment fusion experiments of Roth (1968) and of Wiseman et al. (1972) in which the same final configurations are obtained as with cell suspensions. Similarly, while it is not unlikely that the carry over of trace amounts of trypsin into cultures of collecting aggregates could affect the subsequent appearance of specific adhesion, it is unclear why, as Curtis contends, it would result in an increase in the adhesion of cells to aggregates of the same type without the need for any physical mechanism to ensure the specificity of adhesion rather than resulting in greater variability of adhesion through disruption of the mechanisms that control its expression (Roth et al., 1971a). Loss of specificity of interaction between isolated membranes and homologous intact cells has been observed after trypsinization (Gottlieb, Merrell and Glaser, 1974).

Curtis (1970c) based his calculations of collision efficiency on data on the rates of loss of single cells and the formation of aggregates in Couette viscometer cultures of freshly dissociated neural retina and liver cells. The collision efficiency of the cells in these mixed cell suspensions was similar to that observed for isotypic cultures indicating no selectivity in the formation of cell adhesions. This absence of selectivity seems to contradict the portion of his hypothesis that proposes that the aggregation of cells occurs sequentially without the intermediate formation of a randomly mixed aggregate. On the other hand, plots of collision efficiencies of these cells at different periods of preincubation prior to the initiation of the aggregation promoting culture, as shown in Fig. 4, support

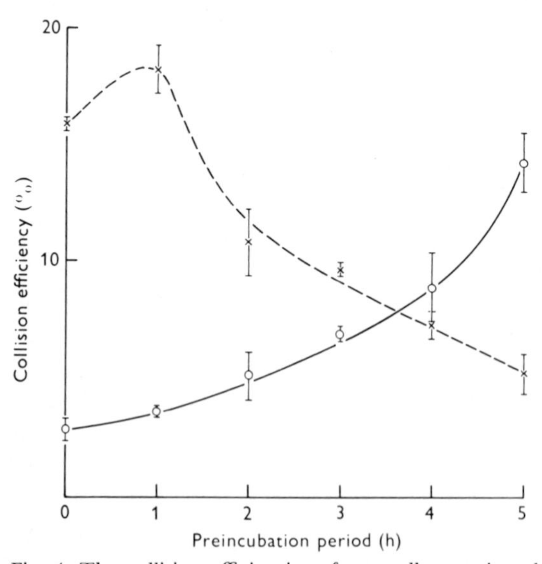

Fig. 4. The collision efficiencies of two cell types in relation to their period of preincubation in medium before reaggregation as measured by the Couette viscometer technique. The relative adhesiveness of the cell types appears to have reversed during the period in which reaggregation was prevented. Reproduced with permission from Curtis (1970c).

the predicted reversal of relative adhesiveness of the cells necessary for the sequential aggregation. As Curtis recognized, the critical data needed to support his hypothesis are: (1) an examination of the distribution of cells in aggregates at different stages in their formation; and (2) evidence that cell movement in aggregates is not important for the formation of the sorted state.

With reference to the first point, Armstrong (1971) examined the very early stages of aggregation of neural and pigmented retinal cells and found that whole living aggregates showed no sign of sorting of cells, unless Curtis' (1970b) rather tenuous definition of sorting as any cell in contact with at least one like cell is applied. Similarly, Shimada et al. (1974) found neural retina and heart cells intermixed in very early aggregates, although some clustering by cell type, possibly based on their marked size differences, was evident. Sheffield and Moscona (1969) and Sheffield (1970) also described adhesions among the different types of retinal cells in mixed aggregates as being initially non-specific, but after approximately two hours cell type specificity was expressed. Even aggregates of amphibian cells formed in stationary culture have a short initial period of random associations (Holtfreter, 1943c, 1944). The early onset of sorting by cell type in initially intermixed aggregates may thus account for the impression of initial sorting.

The importance of motility in the sorting of cells in mixed aggregates, is confirmed by the preferential formation of desmosomes between like cells in cytochalasin B treated aggregates that are incapable of sorting (Overton, 1973; 1974a,b; Overton and Culver, 1973). Since desmosome formation is indicative of specific cell adhesive behavior (Armstrong, 1970), the relative immobility of the cytochalasin B treated cells takes on added weight as the factor inhibiting full sorting. The need for cell movement during specific sorting in mixed aggregates has also been emphasized by Maslow and Mayhew (1974) and by Steinberg and Wiseman (1972), based on disruption of the sorting process by cytochalasin B. Thus, the available data concerning both the time of expression of specificity and the role of cell movement argue against the details of the Curtis hypothesis.

4.4. Oscillating adhesion sites

Curtis' hypothesis attempts to explain sorting of cells in mixed aggregates by the differences in adhesiveness of each cell at different times after exposure to the dissociation procedure without acknowledging the existence of specificity of adhesion between cells of different types or proposing a molecular mechanism to account for this temporal difference.

Jones and Morrison (1969) have attempted to explain the differences in adhesiveness that are responsible for the sorting phenomena observed in heterotypic cultures in terms of quantitative differences in the frequencies of availability of molecules on the cell surface that are capable of acting as linkage sites. Because of the ease of movement of different cell types relative to each other in a cohesive aggregate, they considered it unlikely that qualitatively different adhesive sites were responsible for differential adhesion. This would similarly rule out a separate mechanism for specific and non-specific adhesions, although different molecules could act to stabilize preferential adhesions once formed. Also, the possibility that different molecules, capable of interacting when in appropriate configurations, serve as the linkage sites cannot be excluded.

B.M. Jones (1966a; Jones and Morrison, 1969) proposed that the linkage sites on the surface of a cell oscillate between a position favorable to adhesion and one unfavorable to it. In this scheme, differential adhesiveness is envisaged as resulting from differences between the frequencies of oscillation of different cell types. The more closely the frequencies approached each other, the greater the opportunity for formation of bonds between the linkage sites and the firmer the attachment between the cells. The extent of sorting in a mixed aggregate would therefore reflect the degree of difference between the frequencies of oscillation of the cells used. To cite the extreme case, entirely separate aggregates would be formed when a wide discrepancy exists in the oscillation frequencies of the respective linkage sites. Figure 5 shows diagramatically how adhesive preferences could be determined by frequency of oscillation of linkage sites. In this example, a–c bonds would be favored over a–b or b–c bonds because of their close frequencies and more frequent periods for potential molecular interactions, but in all cases homotypic bonds would be preferred. Since this hypothesis assumes that adhesions between like cells are always more stable than those between unlike cells, the existence of sorted aggregates rather than separate ones indicates that the initially intermixed cells assume the configuration providing the most stable average adhesion of all the cells.

Changes in the metabolism of a cell could produce changes in its oscillation frequency, within characteristic tissue limits, with accompanying changes in the

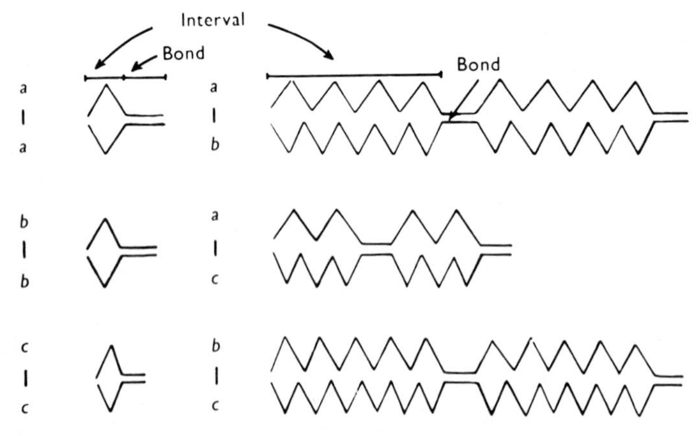

Fig. 5. Diagrammatic representation of the hypothesis of Jones and Morrison (1969) in which differences in the frequency of availability of apposed linkage sites are proposed as accounting for the preferential adhesion of like cells. In all cases, like cells could form bonds more readily than unlike cells, since their linkage sites will provide a more frequent opportunity of forming a bond. Reproduced with permission from Jones and Morrison (1969).

adhesive behavior of the cell. The reorganization of the early aggregate from its initial random configuration to a fully sorted condition occurs as a result of the substitution of stronger adhesions between like cells for weaker adhesions between unlike cells during the random movements of the cells. The initial period of non-specific aggregation can be explained by a delay in the expression of the cells' characteristic frequency during recovery from the dissociation trauma. Jones and Morrison (1969) have suggested a physico-chemical mechanism to support their hypothesis. It requires a surface molecule serving as the linkage and a means of oscillation. While there are many surface moieties that could serve as the linkage site, no evidence favoring one is provided.

The reports of actomyosin-like contractile proteins at the surface of cells other than muscle are of particular relevance to the suggestion of an oscillating mechanism. A possible role of cell surface contractile actomyosin in cell adhesion was presented by P.C.T. Jones (1966) and supported, but by no means proven, by the demonstrated enhancement of cell detachment (Jones, 1966a) and inhibition of aggregation by exogenous ATP (Jones and Kemp, 1970; Knight et al., 1966), presumably by affecting the movement of the linkage sites. In the last few years, considerable evidence has been obtained to indicate the presence of actin- and myosin-like molecules of non-muscle cells, both in association with the plasma membrane and in microfilaments (Pollard and Korn, 1973; Goldman et al., 1975; Wickus et al., 1975; Senda et al., 1975; also see reviews by Kemp, 1974 and Pollard and Weihing, 1974). It has been shown that antibodies directed against actomyosin can inhibit the aggregation of both muscle and liver cells from chick embryos (Jones et al., 1970), and that pretreatment of the anti-actomyosin antibodies with myosin abolished their ability to inhibit aggregation (Kemp et al., 1973a). These observations provide some support for the hypothesis that functional myosin is necessary for aggregation, perhaps in association with plasma membrane-associated actin (Kemp, 1974). The presence of actin and myosin-like components at the cell periphery, although consistent with an oscillatory mechanism, cannot be considered conclusive without further study of the other effects of anti-actomyosin antibodies and ATP on such factors as calcium ion uptake (Shlatz and Morinetti, 1972; Kemp, 1974) and cellular respiratory metabolism, both of which could markedly effect cell adhesive behavior.

4.5. Glycosyltransferases

All of the hypotheses of selectivity in cell adhesion propose the utilization of binding or linkage sites having a specific molecular composition, although the mechanisms by which these sites determine the cell's associative behavior are still largely conjectural. There are three types of molecular interactions involving cell surfaces that exhibit specificity: (1) antigen-antibody; (2) lectin-oligosaccharide; and (3) enzyme-substrate interactions. Roseman (1970, 1974) and Roth (1973) have proposed that the recognition of cells which results in specificity of adhesion is mediated via interactions between cell surface glycosyltransferases and cell surface glycosyl acceptors on separate cells. The two other types of interactions listed were discounted by them because the basic components are not a usual cell surface constituent. Their model, which also serves to explain cell separation is shown schematically in Fig. 6. The interaction between the enzyme and its substrate on the other cell accounts for the initial adhesion between the cells, but more permanent adhesions are achieved by other mechanisms such as the formation of desmosomes. Prior to the formation of these stabilizing factors, cell separation could occur by the completion of the reactions of each enzyme-substrate complex which would thus produce a modified substrate incapable of binding to the enzyme.

The biochemical background to this hypothesis, including discussion of glycoprotein structure and synthetic pathways, and the activity and subcellular localization of glycosyltransferases has been reviewed fully by Roseman (1970), Hughes (1973) and Kemp et al. (1973b) and only a brief outline will be presented here.

The general importance of carbohydrates in adhesion is suggested by the demonstrated needs for exogenous precursors of complex carbohydrates such as l-glutamine, d-glucosamine and d-mannosamine for tumor cell adhesion (Oppenheimer et al., 1969; Oppenheimer, 1973), though adhesion of some embryonic cells does not require them and may even be inhibited by their presence (Oppenheimer, 1973; Garber, 1963; Glaser et al., 1968; Balsamo and Lilien, 1975). Other simple sugars have also been shown to effect cell adhesion (Cox and Gesner, 1965), but the variability of the effect among the cell types tested tends to argue against a common mechanism.

The presence of glycosyltransferases on the surfaces of many cell types has been suggested by the ability of cells to carry out a transferase reaction without the substrate being transported into the cells (Roth et al., 1971a,b; Roth and White, 1972; Bosmann, 1971, 1972a,b; Weiser, 1973; McLean and Bosmann, 1975). The transferase activity remains with the cells after centrifugation, indicating binding, and isotypically labeled substrate is localized at the cell periphery. However this evidence does not prove the involvement of these enzymes in cell adhesion, especially since a large number of different kinds of transferases appear to be equally active at the cell surface (Patt and Grimes, 1974). The

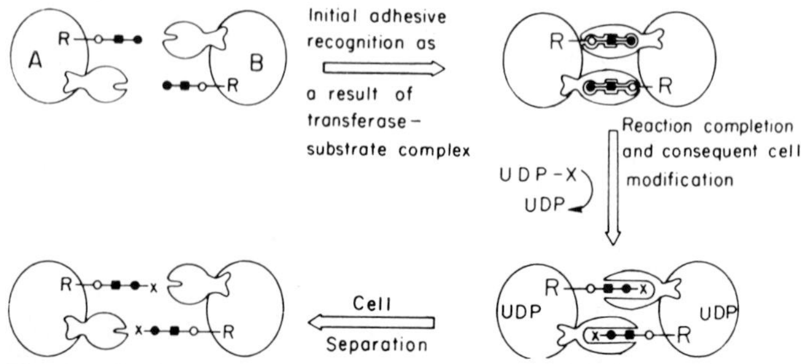

Fig. 6. Schematic representation of the role of proposed surface glycosyl transferases and their substrates in intercellular adhesive recognition. Reproduced with permission from Roth et al. (1971b).

possibility also exists that the enzymes localized on the surface may be contaminants released into the medium by lysed cells.

The model of Roth and Roseman would restrict interactions between enzymes and substrates on the same normal cell by localizing them sufficiently far apart, while the reduced adhesion displayed by normal cells during mitosis and by some malignant cells is proposed as resulting from changes in the proximity of the interacting molecules permitting self-glycosylation (Webb and Roth, 1974). The blocking of self-glycosylation may be the mechanism of action of teratoma cell adhesion factor (Oppenheimer, 1975) although Dorsey and Roth (1973) prefer to attribute the specificity of adhesion, rather than its enhancement to intercellular glycosyltransferase-substrate complex formation.

The direct evidence of glycosyltransferase-surface glycoprotein interactions serving as determinants of specificity of adhesion in embryonic cells is far from conclusive. Roth et al. (1971a) provide the best supporting data with their comparison of the number of adhesions to neural retina and liver aggregates from both homotypic and heterotypic cell suspensions in the presence or absence of β-galactosidases and other enzymes. The presence of β-galactosidases of different origins increased the adhesion of cells of both types to neural retina aggregates, and resulted in a decrease in the specificity of adhesion wth these aggregates. However, the enzymes had no effect on specific or non-specific adhesion to liver aggregates, weakening the case for a universal mechanism. In addition, a recent critique of the evidence in support of the localization and synthetic activity of cell surface glycosyltransferases emphasizes the present speculative nature of any hypothesis describing their activity in cell adhesion (Keenan and Morré, 1975).

4.6. Base pairing of surface polyribonucleotides

Another possible mechanism that could promote specificity in adhesive interactions is the complementary base-pair bonding found in nucleic acids. In a recent series of experiments, Mayhew (1974a; Juliano and Mayhew, 1972) found that exogenous polynucleotides can interact with and be detected at the surfaces of cells, and that these exogenous polynucleotides at the cell surface can participate in complementary reactions with unbound polynucleotides. Since RNA is known to be present as a normal component at many cell surfaces (Weiss and Mayhew, 1966; Bennett et al., 1969; Mayhew and Weiss, 1970), Mayhew (1974b) suggested that complementary base pairing of single stranded RNA on the surfaces of contacting cells might confer specificity on adhesions occurring between cells. Small nucleotide sequences could clearly carry sufficient information in the permutations of the four bases to ensure a large range of surface specificities. Orientation of the nucleotide sequences in close proximity could permit base pairing to occur, and cell movement would proceed by unzipping of the double strand. Cells having complementary polynucleotides affixed to their surfaces show increased adhesion to each other (Mayhew, 1975), demonstrating the feasibility of this hypothesis. However, further investigation using more complex polynucleotides is clearly required before firm conclusions can be reached. The question of how cellular orientation is achieved so as to result in the correct alignment of the binding polynucleotides remains unanswered. It also remains to be demonstrated that the RNA on cell surfaces exist in complementary sequences on similar cells, and that disruption or alteration of such surface RNA is accompanied by changes in the adhesive properties of cells.

4.7. Lectin binding sites

The ability of various lectins to bind to specific saccharide residues on the cell surface has dictated that these agents have emerged as useful probes of cell surface architecture (reviews, Sharon and Lis, 1972; Kemp et al., 1973b; Moscona, 1974). Lectins also offer a potential method for identifying the role of specific glycoproteins in cell adhesion and aggregation. Kleinshuster and Moscona (1972) found that EGTA-dispersed retinal cells harvested from young, 8 or 9-day chick embryos were readily agglutinated by low concentrations of concanavalin A (Con A) but similar retinal cell populations from 12 or 16 day embryos were less susceptible to agglutination and cells from 20 day embryos agglutinated only very poorly, if at all, even when exposed to massive doses of Con A. However, if these late fetal cells were lightly trypsinized they agglutinated as readily as early embryonic non-trypsinized cells. In contrast, wheat germ agglutinin agglutinated chick embryo cells only after trypsinization, regardless of their developmental age (Moscona, 1971). Although these observations provide convincing evidence that the surface properties of cells change during development, their relationship to cellular adhesive properties is unclear. Moscona (1974) has made the interesting suggestion that the developmental changes in cellular responsiveness to Con A might be correlated with changes in cell motility and reflect the role of the cell surface in cell movement. For example, in the early embryonic retina there are extensive morphogenetic movements. As morphogenesis is completed cell movements cease and this coincides with the gradual reduction in the agglutinability of the retinal cells. It is of interest to note that the highly motile gastrula cells of sea urchin embryos are also highly agglutinable by Con A (see, Moscona, 1974).

A more direct approach to evaluate the role of defined cell surface saccharide moieties in cell adhesion was made by Steinberg and Gepner (1973). These investigators attempted to block Con A receptor sites on the cell surface using a "univalent" preparation of Con A to determine whether the receptor sites were mediators of cell adhesion. This study, while potentially instructive, suffers from technical shortcomings associated with the method used to prepare "univalent" Con A. Steinberg and Gepner used chymotrypsin to digest Con A based on a procedure described by Burger and Noonan (1970). Apart from the lack of agglutinating activity in the final preparation, definitive evidence was not presented to support their claim that the lectin preparation consisted of univalent Con A. When mixed cell populations were treated with this preparation no differences were observed in the subsequent size of aggregates formed compared with untreated controls and the cell sorting pattern was also similar in both, though rearrangement was slower in the treated cultures. From this, Steinberg and Gepner concluded that Con A-binding sites (α-D-glucose-like pyranosides) were not involved in cell adhesion.

Evans and Jones (1974) have rightly criticized the enzyme procedure used by Steinberg and Gepner (1973) to produce a "univalent" Con A. They determined that incubation of native Con-A with a trypsin solution produced a mixture of agglutinating and non-agglutinating fractions. The presence of contaminating non-dissociated Con A may account for the full aggregation seen by Steinberg and Gepner. Using only the fraction that did not promote agglutination, Evans and Jones observed inhibition of aggregation of trypsin dissociated 9-day embryonic muscle cells, suggesting a common site for adhesion and Con-A binding. However, the adhesion and Con-A binding were shown to be separable, since 16-day embryonic cells were agglutinated by Con A without demonstrating spontaneous aggregation. In view of the various steric and charge effects accompanying lectin-carbohydrate interactions on the cell surface it is premature to assume that cells adhere via specific saccharides. Nonetheless, it seems likely that lectins can be used to provide further information on the cell surface molecules involved in adhesive interactions between cells so that final identification of the linkage or binding sites can be approached more accurately.

4.8. Surface charge distribution and the formation of probes

Attempts to define the mechanism of cell specificity in terms of the physicochemical behavior of colloids as generally described by Derjaguin and Landau (1941) and Verwey and Overbeek (1948) (DLVO theory) or quantum mechanical analyses of interactions between stable surfaces in non-physiologic environments have so far failed to provide sufficient insight to be of predictive value in analyzing adhesive interactions between living cells. In fact, the application of DLVO theory, which neglects all nonelectrostatic contributions to cell interactions, predicts the existence of net electrostatic repulsive forces between cells that would tend to prevent aggregation. This has led Weiss to propose that the protrusion of probes, as described originally by Bangham and Pethica (1960), and/or the extrusion of an intercellular "glue", as discussed, may serve to permit adhesion to occur (Weiss, 1970, 1971, 1973; Weiss and Neiders, 1971; Weiss and Harlos, 1972a, b; Maslow and Weiss, 1972).

The protrusion of probes with smaller radii of curvature than that of the cell offers a possible mechanism for promoting cell contact since its approach toward the cell surface would engender reduced electrostatic repulsion compared to the approach of the relatively flat surface of another cell (Bangham and Pethica, 1960; Weiss and Harlos, 1972a, b). Filopodia or microvilli satisfying this requirement have been seen repeatedly in electron micrographs of contacting cells (Lesseps, 1961, 1963; Fisher and Cooper, 1967; Pugh-Humphreys and Sinclair, 1970; Weiss and Subjeck, 1974a,b; Ede and Flint, 1975 and others) though the crucial question of whether the correlation is a causal one remains unclear.

The DLVO theory also fails to consider the heterogenous distribution of charge moieties on the cell surface (Danon et al., 1972; Weiss and Zeigel, 1972; Weiss et al., 1972; Weiss and Subjeck, 1974a). Non-uniform distribution of surface charges may well be of importance in determining aspects of the specificity of cell surface organization and would also be expected to influence the contact and adhesive interactions between cells. For example, Weiss and Subjeck (1974b) found that microvilli seem to preferentially align themselves with areas on the surface of a cell which they are approaching which have a low density of bound iron particles. It cannot be determined if the microvillus is attracted to the area or the area aligns itself and/or forms in response to the approaching microvillus.

5. *Concluding remarks*

The in vitro study of cellular adhesive specificities has resulted in the accumulation of considerable data from many experimental systems by numerous investigators and yet, as noted earlier, none of the hypotheses for cell-to-cell adhesive interactions proposed to date are fully consistent with the available experimental data.

Any model which attempts to explain the specificity of cell adhesion must accommodate the different levels of analysis followed by investigators in this field for it to have significance for living cells. The first level of analysis has involved characterization of the behavior of entire cells. It is difficult to conceive of complete embryological development without the capability of cells to adhere preferentially, although not necessarily to cells of the same type. The central issue in this phenomenological view of specificity is whether the appearance of specific cell interactions results from specificity of adhesion at the cellular level, or is merely a by-product of changes in the other aspects of the cell's behavior.

The second level of analysis is concerned with the determination and control of the specificity of adhesion. The differential adhesion hypothesis, as its proponents properly note, provides a model of cell behavior from this mechanistic viewpoint. Other useful contributions in this area include studies of the recovery of normal surface characteristics after exposure to dissociation trauma and observations on the dynamic changes in the surface structure of the cell that expose or conceal molecules necessary for adhesion during specific growth and/or developmental stages. Hypotheses suggesting the determination of specificity by quantitative rather than qualitative differences in surface moieties also fall in this category. Theoretical and mathematical simulations of specific cell behavior, which have not been reviewed at all in this paper, also deal with cells as mechanistic units without molecular involvement. While the current lack of orientation toward molecular interactions means that a full explanation of surface specificity will not result from these studies, they have undoubtedly contributed to our understanding of cell interactions and provided a necessary foundation for the future investigation of the molecular basis of cell adhesion.

Studies of cell and surface specificity at the molecular level show the least conflict in data and interpretation, perhaps because of their limited success to date. There seems to be a general acceptance of an important role for glycoproteins in cell adhesion, possibly serving directly as intercellular ligands or as linkage substrates for specific surface glycosyltransferases, though the exact mechanism remains to be established. The recent interesting findings indicating a possible relationship between heterogeneous distribution of surface charge and cell adhesion may also prove to be of importance.

Despite their different levels of approach, the various models of cell adhesion contain many common features. In this review I have selected, perhaps somewhat arbitrarily, data presented by the proponents of those hypotheses which seem to have general validity for surface specificity in cellular interactions in embryogenesis.

Areas in which additional information would be most useful in the elucidation of the mechanism of cell specificity in embryogenesis include the following: (1) detailed characterization of the various components present on the cell surface at specific stages in development and their topographic arrangement; (2) observations on contact and adhesive interactions between isolated plasma membranes and also possibly artificial membranes of defined composition. Some prototype experiments along these lines have been done already. Sussman and Boschwitz (1975) have shown that isolated membranes of slime mould cells display some of the specific adhesive properties exhibited by intact cells indicating the retention of at least some elements of specificity in a cell-free system. Similar adhesive specificity in isolated plasma membrane preparations has been demonstrated by Merrell and Glaser (1973) using chick cells; and (3) better utilization of cells from organisms with known genetic defects affecting early developmental processes, surface enzymes and transport activities and which may well be accompanied by changes in the adhesive properties of cells (see also chapter 9 of this volume).

Acknowledgements

The personal research cited in this paper was supported in part by Grant CA14370 from the National Institutes of Health. I am grateful to Drs. E. Mayhew, S. Nir and D. Stocum for useful discussions and comments on this manuscript.

References

Allen, A. and Snow, C. (1970) The effect of trypsin or ethylene-diaminetetraacetate on the surface of cells in tissue culture. Biochem. J. 117, 32p.

Anderson, N.G. (1953) The mass isolation of whole cells from rat liver. Science, 117, 627–628.

Armstrong, P.B. (1966) On the role of metal cations in cellular adhesion: effect on cell surface charge. J. Exp. Zool. 163, 99–110.

Armstrong, P.B. (1970) A fine structural study of adhesive cell junctions in heterotypic cell aggregates. J. Cell Biology 47, 197–210.

Armstrong, P.B. (1971) Light and electron microscope studies of cell sorting in combinations of chick embryo neural retina and retinal pigment epithelium. Wilhelm Roux Archiv Entwicklungsmech. Organismen 168, 125–141.

Armstrong, P.B. and Niederman, R. (1972) Reversal of tissue position after cell sorting. Dev. Biol. 28, 518–527.

Armstrong, P.B. and Parenti, D. (1972) Cell sorting in the presence of cytochalasin B. J. Cell. Biol. 55, 542–553.

Ball, W.D. (1966) Aggregation of dissociated embryonic chick cells at 3°C. Nature, 210, 1075–1076.

Balsamo, J. and Lilien, J. (1974a) Embryonic cell aggregation: kinetics and specificity of binding of enhancing factors. Proc. Nat. Acad. Sci. U.S.A. 71, 727–731.

Balsamo, J. and Lilien, J. (1974b) Functional identification of three components which mediate tissue-type specific embryonic cell adhesion. Nature, 251, 522–524.

Balsamo, J. and Lilien, J. (1975) The binding of tissue-specific adhesive molecules to the cell surface. A molecular basis for specificity. Biochem. 14, 167–171.

Bangham, A.D. and Pethica, B.A. (1960) The adhesiveness of cells and the nature of the chemical groups at their surfaces. Proc. Roy. Phys. Soc. Edinburgh 28, 43–50.

Barnard, P.J., Weiss, L. and Ratcliffe, T. (1969) Changes in the surface properties of embryonic chick neural retina cells after dissociation. Exp. Cell Res. 54, 293–301.

Bennett, M., Mayhew, E. and Weiss, L. (1969) RNA in the periphery of rapidly proliferating mouse lymphoid cells. J. Cell Physiol. 74, 183–190.

Bosmann, H.B. (1971) Platelet adhesiveness and aggregation: the collagen: glycosyl, polypeptide: N-acetylgalactosaminyl and glycoprotein: galactosyl transferase of human platelets. Biochem. Biophys. Res. Comm. 43, 1118–1124.

Bosmann, H.B. (1972a) Cell surface glycosyl transferases and acceptors in normal and RNA- and DNA-virus transformed fiborblasts. Biochem. Biophys. Res. Comm. 48, 523–529.

Bosmann, H.B. (1972b) Sialyl transferase activity in normal and RNA- and DNA-virus transformed cells utilizing desialyzed, trypsinized cell plasma membrane external surface glycoproteins as exogenous acceptors. Biochem. Biophys. Res. Comm. 49, 1256–1262.

Brown, J.M. and Hilfer, S.R. (1968) Effects of the dissociating agents collagenase, pronase and tetraphenylboron on embryonic chick thyroid and heart cells. J. Cell Biol. 39, 18a.

Beug, H., Gerisch, G. and Muller, E. (1971) Cell dissociation: Univalent antibodies as a possible alternative to proteolytic enzymes. Science, 173, 742–743.

Burdick, M.L. (1970) Cell sorting out according to species in aggregates containing mouse and chick embryonic limb mesoblast cells. J. Exp. Zool. 175, 375–368.

Burdick, M.L. (1972) Differences in the morphogenetic properties of mouse and chick embryonic liver cells. J. Exp. Zool. 180, 117–126.

Burdick, M.L. and Steinberg, M.S. (1969) Embryonic cell adhesiveness: Do species differences exist among warm-blooded vertebrates. Proc. Nat. Acad. Sci. U.S.A. 63, 1169–1173.

Burger, M.M. and Noonan, K.D. (1970) Restoration of normal growth by covering of agglutinin sites on tumor cell surface. Nature, 228, 512–515.

Cahn, R.D. and Cahn, M.B. (1966) Heritability of cellular differentiation: clonal growth and expression of differentiation in retinal pigment cells in vitro. Proc. Nat. Acad. Sci. U.S.A. 55, 106–114.

Capers, C.R. (1960) Multinucleation of skeletal muscle in vitro. J. Biophys. Biochem. Cytol. 7, 559–566.

Casanello, D.E. and Gerschenson, L.E. (1970) Some morphological and biochemical characteristics of isolated rat liver cells dissociated with sodium tetraphenylboron and cultured in suspension. Exp. Cell Res. 59, 283–290.

Cauldwell, C.B., Henkart, P. and Humphreys, T. (1973) Physical properties of sponge aggregation factor. A unique proteoglycan complex. Biochem. 12, 3051–3055.

Coon, H.G. (1966) Clonal stability and phenotypic expression of chick cartilage cells in vitro. Proc. Nat. Acad. Sci. U.S.A. 55, 66–73.

Cooper, W.G. and Konigsberg, I.R. (1961) Dynamics of myogenesis in vitro. Anat. Rec. 140, 195–206.

Cox, R.P. and Gesner, B.M. (1965) Effect of simple sugars on the morphology and growth pattern of mammalian cell cultures. Proc. Nat. Acad. Sci. U.S.A. 54, 1571–1579.

Culp, L.A. (1974) Substrate attached glycoproteins mediating adhesion of normal and virus transformed mouse fibroblasts. J. Cell Biol. 63, 71–83.
Cunningham, I. and Hirst, J.H.R. (1967) Analysis of a turbidimetric method for quantitatively estimating cell aggregation. Experientia 23, 693–695.
Curtis, A.S.G. (1961) Timing mechanisms in the specific adhesion of cells. Expt. Cell Res. Suppl. 8, 107–122.
Curtis, A.S.G. (1962) Pattern and mechanism in reaggregation of sponges. Nature, 196, 245–248.
Curtis, A.S.G. (1963) Effect of pH and temperature on cell reaggregation. Nature, 200, 1235–1236.
Curtis, A.S.G. (1967) The cell surface: Its molecular role in morphogenesis, Logos Press, London.
Curtis, A.S.G. (1969) The measurement of cell adhesiveness by an absolute method. J. Embryol. exp. Morph. 22, 305–325.
Curtis, A.S.G. (1970a) Re-examination of a supposed case of specific cell adhesion. Nature, 226, 260–261.
Curtis, A.S.G. (1970b) Problems and some solutions in the study of cellular aggregation. Symp. Zool. Soc. Lond., 25, 335–352.
Curtis, A.S.G. (1970c) On the occurrence of specific adhesion between cells. J. Embryol. exp. Morph. 23, 253–272.
Curtis, A.S.G. (1973) Cell adhesion. Prog. Biophys. Mol. Biol. 27, 315–384.
Curtis, A.S.G. and Greaves, M.F. (1965) The inhibition of cell aggregation by a pure serum protein. J. Embryol. Exp. Morphol. 13, 309–326.
Curtis, A.S.G. and van de Vyver, G. (1971) The control of cell adhesion in a morphogenetic system. J. Embryol. Exp. Morphol. 26, 295–312.
DeHaan, R.L. (1967) Regulation of spontaneous activity and growth of embryonic chick heart cells in tissue culture. Dev. Biol. 16, 216–249.
Dalen, H. and Todd, P.W. (1971) Surface morphology of trypsinized human cell in vitro. Exp. Cell Res. 66, 353–361.
Danon, D., Goldstein, L., Marikovsky, Y. and Skutelsky, E. (1972) Use of cationized ferritin as a label of negative charges on cell surfaces. J. Ultrastruct. Res. 38, 500–510.
Demon, J.J., Bruyneel, E.A. and Mareel, M.M. (1974) A study on the mechanism of intercellular adhesion. J. Cell Biol. 60, 641–652.
Derjaguin, B.V. and Landau, L.D. (1941) Theory of the stability of strongly charged lyophobic sols and the adhesion of strongly charged particles in solutions of electrolytes. Acta Physicochem. (U.S.S.R.) 14, 633–656.
Dorsey, J.K. and Roth, S.A. (1973) Adhesive specificity in normal and transformed mouse fibroblasts. Dev. Biol. 33, 249–256.
Douglas, W.H.J. and Kaighn, M.C. (1974) Clonal isolation of differentiated rat lung cells. In Vitro, 10, 230–237.
Easty, G.C., Easty, D.M. and Ambrose, E.J. (1960) Studies on cellular adhesiveness. Exp. Cell Res. 19, 539–548.
Ede, D.A. and Agerbak, G.S. (1968) Cell adhesion and movement in relation to the developing limb pattern in normal and Talpid[3] mutant chick embryos. J. Embryol. exp. Morph. 20, 81–100.
Ede, D.A. and Flint, O.P. (1975) Intercellular adhesion and formation of aggregates in normal and Talpid[3] mutant chick limb mesenchyme. J. Cell Sci. 18, 97–111.
Edwards, J. (1973) Intercellular Adhesion. In: New Techniques in Biophysics and Cell Biology (Pain, R.H. and Smith, B.J., eds) vol. 1, pp. 1–27, Wiley, London.
Edwards, J.G. and Campbell, J.A. (1971) The aggregation of trypsinized BHK21 cells. J. Cell Sci. 8, 53–72.
Edwards, G.A. and Fogh, J. (1959) Micromorphologic changes in human amnion cells during trypsinization. Cancer Res. 19, 608–611.
Engel, M.B., Pumper, R.W. and Joseph, N.R. (1968) Electrometric determination of surface charge of cultured mammalian cells. Proc. Soc. Exp. Biol. Med. 128, 990–996.
Evans, P.M. and Jones, B.M. (1974) Studies on cellular adhesion-aggregation. Consideration of involvement of concanavalin A receptors. Exp. Cell Res. 88, 56–62.
Fischman, D.A. and Moscona, A.A. (1969) An electron microscope study of in vitro dissociation and reaggregation of embryonic chick and mouse heart cells. J. Cell Biol. 43, 37a.

Fisher, H.W. and Cooper, T.W. (1967) Electron microscope studies of the microvilli of Hela cells. J. Cell Biol. 34, 569–576.

Flaschka, J. and Barnard, A.J. Jr. (1960) Tetraphenylboron (TPB) as an analytical reagent. Adv. Anal. Chem. Instr. 1, 1–117.

Foley, J.F. and Aftonomos, B. (1970) The use of pronase in tissue culture: a comparison with trypsin. J. Cell. Physiol. 75, 159–162.

Friedmann, T. and Epstein, C.J. (1967) The incorporation of (^{3}H) leucine into protein by tetraphenylboron- and citrate-dispersed rat line parenchymal cells. Biochim. Biophys. Acta 138, 622–624.

Gailani, S.D., Nussbaum, A., McDougall, W.J. and McLimans, W.F. (1970) Studies on hormone production by human fetal pituitary cell culture. Proc. Soc. Exp. Biol. Med. 134, 27–32.

Galtsoff, P.S. (1923) The amoeboid movement of dissociated sponge cells. Biol. Bull. 45, 153–161.

Galtsoff, P.S. (1925) Regeneration after dissociation (an experimental study on sponges). I. Behavior of dissociated cells of *microciona prolifera* under normal and altered conditions. J. Exp. Zool. 42, 183–221.

Garber, B. (1963) Inhibition of glucosamine of aggregation of dissociated embryonic cells. Dev. Biol. 7, 630–641.

Garber, B.B. and Moscona, A.A. (1964) Aggregation in vivo of dissociated cells. 1. Reconstruction of skin in the chorioallantoic membrane from suspensions of embryonic chick and mouse skin cells. J. Exp. Zool. 155, 179–202.

Garber, B.B. and Moscona, A.A. (1967) Suppression of feather morphogenesis in co-aggregates of cells from embryos of different ages. J. Exp. Zool. 164, 351–362.

Garber, B.B. and Moscona, A.A. (1972) Reconstruction of brain tissue from cell suspensions. II. Specific enhancement of aggregation of embryonic cerebral cells by supernatant from homologous cell cultures. Dev. Biol. 27, 235–243.

Garber, B., Kollar, E.J. and Moscona, A.A. (1968) Aggregation in vivo of dissociated cells. III. Effects of state of differentiation of cells on feather development in hybrid aggregates of embryonic mouse and chick cells. J. Exp. Zool. 168, 455–472.

Garfield, S., Hausman, R.E. and Moscona, A.A. (1974) Embryonic cell aggregation: Absence of galactosyltransferase activity in retina-specific cell-aggregating factor. Cell Diff. 3, 215–219.

Garrod, D.R. and Steinberg, M.S. (1973) Tissue-specific sorting-out in two dimensions in relation to contact inhibition of cell movement. Nature, 244, 568–569.

Gasic, G.J. and Galanti, N.L. (1966) Proteins and disulfide groups in the aggregation of dissociated cells of sea sponges. Science, 151, 203–205.

Gerschenson, L.E. and Casanello, D. (1968) Metabolism of rat liver cells cultured in suspension: Insulin and glucagon effects on glycogen level. Biochem. Biophys. Res. Comm. 33, 584–589.

Gershman, H. (1970) On the measurement of cell adhesiveness. J. Exp. Zool. 174, 391–406.

Gierer, A., Berking, S., Bode, H., David, C.N., Flick, K., Hansmann, G., Schaller, H. and Trenkner, E. (1972) Regeneration of hydra from reaggregated cells. Nature New Biology 239, 98–101.

Glaeser, R.M., Richmond, J.E. and Todd, P.W. (1968) Histotypic self-organization by trypsin-dissociated and EDTA-dissociated chick embryo cells. Exp. Cell Res. 52, 71–85.

Goldman, R.D., Lazarides, E., Pollack, R. and Weber, K. (1975) The distribution of actin in non-muscle cells. Exp. Cell Res. 90, 333–344.

Gordon, H.P. and Brice, M. (1974a) Intrinsic factors influencing the maintenance of contractile embryonic heart cells in vitro. I. The heart muscle conditioned medium effect. Exp. Cell Res. 85, 303–310.

Gordon, H.P. and Brice, M. (1974b) Intrinsic factors influencing the maintenance of contractile embryonic heart cells in vitro. II. Biochemical analysis of heart muscle conditioned medium. Exp. Cell Res. 85, 311–318.

Gordon, H.P. and Brice, M. (1974c) Intrinsic factors influencing the maintenance of contractile embryonic heart cells in vitro. III. The effect of inoculum level. Exp. Cell Res. 87, 409–412.

Gordon, H.P. and Wilde, C.E. Jr. (1965) "Conditioned" medium and heart muscle differentiation: Contrast between explants and disaggregated cells in chemically defined medium. Exp. Cell Res. 40, 438–442.

Gottlieb, D.I. and Glaser, L. (1975) A novel assay of neuronal cell adhesion. Biochem. Biophys. Res. Comm. 63, 815–821.

Gottlieb, D.I., Merrell, R. and Glaser, L. (1974) Temporal changes in embryonal cell surface recognition. Proc. Nat. Acad. Sci. U.S.A. 71, 1800–1802.

Grover, J.W. (1961) The enzymatic dissociation and reproducible reaggregation in vitro of 11-day embryonic chick lung. Dev. Biol. 3, 555–568.

Gwatkin, R.B.L. (1973) Pronase. In: Tissue Culture, Methods and Applications (Kruse, P.F., Jr. and Patterson, M.K., Jr., eds) pp. 3–5, Academic Press, New York.

Gwatkin, R.B.L. and Thompson, J.L. (1964) A new method for dispersing the cells of mammalian tissues. Nature, 201, 1242–1243.

Harris, C.C. and Leone, C.A. (1966) Some effects of EDTA and tetrophenylboron on the ultrastructure of mitochondria in mouse liver cells. J. Cell Biol. 28, 405–408.

Hausman, R.E. and Moscona, A.A. (1973) Cell surface interactions: differential inhibition by proflovine of embryonic cell aggregation and production of specific cell aggregation factor. Proc. Nat. Acad. Sci. U.S.A. 70, 3111–3114.

Hebb, C.R. and Chu, M.-Y.W. (1960) Reversible injury of L-strain mouse cells by trypsin. Exp. Cell Res. 20, 453–457.

Henkart, P. and Humphreys, T. (1970) Cell aggregation in small volumes on a gyratory shaker. Exp. Cell Res. 63, 224–227.

Henkart, P., Humphreys, S. and Humphreys, T. (1973) Characterization of sponge aggregation factor. A unique proteoglycan complex. Biochemistry, 12, 3045–3050.

Herbst, C. (1900) Über das Auseinandergehen im Furchungs und Gewebezellen in kalkfreiem Medium. Wilhelm Roux Archiv. Entwicklungsmech. Organismen 9, 424–463.

Hilfer, S.R. (1973) Collagenase treatment of chick heart and thyroid. In: Tissue Culture: Methods and Applications (Kruse, P.F. and Patterson, M.K., eds.) pp. 16–20, Academic Press, New York.

Hilfer, S.R. and Hilfer, E.K. (1966) Effects of dissociating agents on the fine structure of embryonic chick thyroid cells. J. Morphol. 119, 217–232.

Hodges, G.M., Livingston, D.C. and Franks, L.M. (1973) The localization of trypsin in cultured mammalian cells. J. Cell Sci. 12, 887–902.

Holtfreter, J. (1943a) Properties and functions of the surface coat in amphibian embryos. J. Exp. Zool. 93, 251–323.

Holtfreter, J. (1943b) A study of the mechanics of gastrulation; Part I. J. Exp. Zool. 94, 261–318.

Holtfreter, J. (1943c) Experimental studies on the development of the pronephros. Rev. Canad. de Biol. 3, 220–289.

Holtfreter, J. (1944) A study of the mechanics of gastrulation: Part II. J. Exp. Zool. 95, 171–212.

Holtzer, H. (1961) Aspects of chondrogenesis and myogenesis. In: Synthesis of Molecular and Cellular Structure (Rudnick, D. ed.) pp. 35–87, Ronald Press, New York.

Hornby, J.E. (1973) Measurements of cell adhesion. I. Quantitative studies of adhesion of embryonic chick cells. J. Embryol. Exp. Morphol. 30, 499–509.

Houba, V. (1967) The use of pronase for dispersing cells. Experientia, 23, 572.

Hughes, R.C. (1973) Glycoproteins as components of cellular membranes. Prog. Biophys. Mol. Biol. 26, 189–268.

Humphreys, T. (1963) Chemical dissolution and in vitro reconstruction of sponge cell adhesions. 1. Isolation and functional demonstration of components involved. Dev. Biol. 8, 27–47.

Humphreys, T. (1967) The cell surface and specific cell aggregation. In: The Specificity of Cell Surfaces (Davis, B.D. and Warren, L. eds.) pp. 195–210, Prentice-Hall, Englewood Cliffs.

Humphreys, T. (1970a) Species specific aggregation of dissociated sponge cells. Nature, 228, 685–686.

Humphreys, T. (1970b) Biochemical analysis of sponge cell aggregation. Symp. Zool. Soc. Lond. 25, 325–334.

Humphreys, T., Humphreys, S. and Moscona, A.A. (1960a) A procedure for obtaining completely dissociated sponge cells. Biol. Bull. 119, 294.

Humphreys, T., Humphreys, S. and Moscona, A.A. (1960b) Rotation mediated aggregation of dissociated sponge cells. Biol. Bull. 119, 295.

Igarashi, Y. and Yaoi, Y. (1975) Growth-enhancing protein obtained from cell surface of cultured fibroblasts. Nature, 254, 248–250.

Iype, P.T. (1971) Cultures of adult rat liver cells. 1. Establishment of monolayer cell-cultures from normal liver. J. Cell. Physiol. 78, 281–288.

John, H.A., Campo, M.S., Mackenzie, A.M. and Kemp, R.B. (1971) Role of different sponge cell types in species specific cell aggregation. Nature New Biology 230, 126–128.

Jones, B.M. (1965) Inhibitory effect of p-benzoquinone on the aggregation behavior of embryo-chick fibroblast cells. Nature, 205, 1280–1281.

Jones, B.M. (1966a) A unifying hypothesis of cell adhesion. Nature, 212, 362–365.

Jones, B.M. (1966b) Invertebrate tissue and organ culture in cell research. In: Cells and Tissues in Culture (Willmer, E.N., ed.), vol. 3, pp. 397–457, Academic Press, New York.

Jones, B.M. and Kemp, R.B. (1970) Aggregation and electrophoretic mobility studies on dissociated cells. II. Effects of ADP and ATP. Exp. Cell Res. 63, 301–308.

Jones, B.M. and Morrison, G.A. (1969) A molecular basis for indiscriminate and selective cell adhesion. J. Cell Sci. 4, 799–813.

Jones, B.M., Kemp, R.B. and Groschel-Stewart, U. (1970) Inhibition of cell aggregation by antibodies directed against actomyosin. Nature, 226, 261–262.

Jones, P.C.T. (1966) A contractile protein model for cell adhesion. Nature, 212, 365–369.

Juliano, R. and Mayhew, E. (1972) Interactions of polynucleotides with cultured mammalian cells. 1. Uptake of RNA by Ehrlich ascites carcinoma cells. Exp. Cell Res. 73, 3–12.

Kaighn, M.E. and Prince, A.M. (1971) Production of albumin and other serum proteins by clonal cultures of normal human liver. Proc. Natl. Acad. Sci. USA 68, 2396–2400.

Keenan, T.W. and Morré, D.J. (1975) Glycosyltransferases: Do they exist on the surface membrane of mammalian cells? FEBS Letters 55, 8–13.

Kellner, G., Broda, E., Suschny, O. and Rucker, W. (1959) Effects of trypsin treatment on tissue in culture, Exp. Cell Res. 18, 168–171.

Kemp, R.B. (1969) Studies on the effect of dissociating agents on the electrophoretic mobility and aggregative competence of embryonic chick muscle cells. Cytobios 2, 187–196.

Kemp, R.B. (1970) The effect of neuraminidase (3:2:1:18) on the aggregation of cells dissociated from embryonic chick muscle tissue. J. Cell Sci. 6, 751–766.

Kemp, R.B. (1974) Myosin-like proteins in the plasma membrane. In: Comparative Biochemistry and Physiology of Transport (Bolis, L., Block, K., Luria, S.E. and Lynen, F., eds.) 175–188, North-Holland, Amsterdam.

Kemp, R.B. and Jones, R.M. (1970) Aggregation and electrophoretic mobility studies on dissociated cells. 1. Effects of p-benzoquinone and tannic acid. Exp. Cell Res. 63, 293–300.

Kemp, R.B., Jones, B.M., Cunningham, I. and James, M.C.M. (1967) Quantitative investigation on the effect of puromycin on the aggregation of trypsin- and Versene-dissociated chick fibroblast cells. J. Cell Sci. 2, 323–340.

Kemp, R.B., Jones, R.M. and Groschel-Stewart, U. (1973a) Abolition by myosin and heavy meromyosin of the inhibitory effects of smooth-muscle actomyosin antibodies on cell aggregation in vitro. J. Cell Sci. 12, 631–639.

Kemp, R.B., Lloyd, C.W. and Cook, G.M.W. (1973b) Glycoproteins in cell adhesion. Prog. Surf. Memb. Sci. 7, 271–318.

Khan, T. and Overton, J. (1969) Staining of intercellular material in reaggregating chick liver and cartilage cells. J. Exp. Zool. 171, 161–174.

Kleinschuster, S.J. and Moscona, A.A. (1972) Interactions of embryonic and fetal neural retina cells with carbohydrate binding phytoagglutinins. Cell surface changes with differentiation. Exp. Cell Res. 70, 397–410.

Knight, V.A., Jones, B.M. and Jones, P.C.T. (1966) Inhibition of the aggregation of dissociated embryo-chick fibroblast cells by adenosine triphosphate. Nature, 210, 1008–1010.

Kolodny, G.M. (1972) Effect of various inhibitors on readhesion of trypsinized cells in culture. Exp. Cell Res. 70, 196–202.

Konigsberg, I.R. (1963) Clonal analysis of myogenesis. Science, 140, 1273–1284.

Konigsberg, I.R. and Hauschka, S.D. (1966) Cell interactions in the reproduction of cell type. In:

Reproduction: Molecular Subcellular and Cellular (Locke, M. ed.) pp. 243–289, Academic Press, New York.

Kruse, P.F. Jr. and Patterson, M.K., Jr. (1973) eds. Tissue Culture, Methods and Applications. Academic Press, New York.

Kuroda, Y. (1968) Preparation of an aggregation-promoting supernatant from embryonic chick liver cells. Exp. Cell Res. 49, 626–637.

Lasfargues, E.Y. (1973) Human mammary tumors. In: Tissue Culture: Methods and Applications (Kruse, P.F. and Patterson, M.K., eds.) pp. 45–50, Academic Press, New York.

Lash, J.W., Holtzer, H. and Swift, H. (1957) Regeneration of mature skeletal muscle. Anat. Rev. 128, 679–698.

Lee, H.H. (1971) Reaggregation and reorganization of juvenile chick testicular cells in vitro. Dev. Biol. 24, 322–334.

Lesseps, R. (1961) An electron microscopic study of dissociated embryonic cell surfaces. Am. Zool. 1, 458.

Lesseps, R.J. (1963) Cell surface projections: their role in the aggregation of embryonic chick cells as revealed by electron microscopy. J. Exp. Zool. 153, 171–182.

Lesseps, R.J. (1973) Developmental change in morphogenetic properties: embryonic chick heart tissue and cells segregate from other tissues in age-dependent patterns. J. Exp. Zool. 185, 159–168.

Lesseps, R.J. and Brown, S.A. (1974) Further evidence for a developmental change in morphogenetic properties of embryonic chick heart cells. J. Exp. Zool. 187, 261–266.

Levinson, C. and Green, J.W. (1965) Cellular injury resulting from tissue disaggregation. Exp. Cell Res. 39, 309–317.

Lilien, J.E. (1968) Specific enhancement of cell aggregation in vitro. Dev. Biol. 17, 657–678.

Lilien, J.E. (1969) Toward a molecular explanation for specific cell adhesion. In: Current Topics in Developmental Biology (Moscona, A.A. and Monroy, A., eds.) vol. 4, pp. 169–195, Academic Press, New York.

Lilien, J.E. and Moscona, A.A. (1967) Cell aggregation: Its enhancement by a supernatant from cultures of homologous cells. Science, 157, 70–72.

Lloyd, C.W. and Cook, G.M. (1974) On the mechanism of the increased aggregation by neuraminidase of 16C malignant rat dermal fibroblast in vitro. J. Cell Sci. 15, 575–590.

Loewenstein, W.R. (1967) On the genesis of cellular communication. Dev. Biol. 15, 503–520.

MacLennan, A.P. and Dodd, R.Y. (1967) Promoting activity of extracellular materials on sponge cell reaggregation. J. Embryol. Exp. Morphol. 17, 473–480.

Mallucci, L., Wells, V. and Young, M.R. (1972) Effect of trypsin on cell volume and mass. Nature New Biology 239, 53–55.

Margoliash, E., Schenck, J.R., Hargie, M.P., Burokos, S., Richter, W.R., Barlow, G.H. and Moscona, A.A. (1965) Characterization of specific cell aggregating materials from sponge cells. Biochem. Biophys. Res. Comm. 20, 383–388.

Maslow, D.E. (1967) The formation of multinucleated striated muscle. Am. Zool. 7, 751.

Maslow, D.E. (1969) Cell specificity in the formation of multinucleated striated muscle. Exp. Cell Res. 54, 381–390.

Maslow, D.E. (1970) Electrokinetic surfaces of trypsin-dissociated embryonic chick liver cells. Exp. Cell Res. 61, 266–270.

Maslow, D.E. and Mayhew, E. (1972) Cytochalasin B prevents specific sorting of reaggregating embryonic cells. Science, 177, 281–282.

Maslow, D.E. and Mayhew, E. (1974) Histotypic cell aggregation in the presence of cytochalasin B. J. Cell Sci. 16, 651–666.

Maslow, D.E. and Mayhew, E. (1975) Inhibition of embryonic cell aggregation by neoplastic cells. J. Nat. Cancer Inst. 54, 1097–1102.

Maslow, D.E. and Weiss, L. (1972) Cell exudation and cell adhesion. Exp. Cell Res. 71, 204–208.

Mayhew, E. (1974a) Interactions of polynucleotides with cultured mammalian cells. III. Initial binding at cell surfaces. Exp. Cell Res. 86, 87–94.

Mayhew, E. (1974b) Complementarily between cell surfaces. J. Theoret. Biol. 47, 483–484.

Mayhew, E. (1975) Interactions of polynucleotides with cultured mammalian cells. IV. Mediation of cell contact behavior. Exp. Cell Res., submitted.
Mayhew, E. and Blumenson, L.E. (1972) Aggregation of Burkitt lymphoma cells in stationary culture: Experimental and theoretical analysis. J. Cell Sci. 10, 749–758.
Mayhew, E. and Weiss, L. (1970) RNA in the cell periphery. In: Surface Chemistry of Biological Systems, pp. 191–208, Plenum Press, New York.
McClay, D.R. (1971) An autoradiographic analysis of the species specificity during sponge cell reaggregation. Biol. Bull. 141, 319–330.
McClay, D.R. (1974) Cell aggregation: properties of cell surface factors from five species of sponge. J. Exp. Zool. 188, 89–102.
McClay, D.R. and Baker, S.R. (1975) A kinetic study of embryonic cell adhesion. Dev. Biol. 43, 109–122.
McClay, D.R. and Moscona, A.A. (1974) Purification of the specific cell-aggregating factor from embryonic neural retina cells. Exp. Cell Res. 87, 438–443.
McGuire, E.J. (1972) Cell-cell recognition: Its role in intercellular adhesion. In: Proceedings of the Ninth Canadian Cancer Research Conference (Scholefield, P.G., ed.) pp. 123–139, Univ. Toronto Press, Toronto.
McLean, R.J. and Bosmann, H.B. (1975) Cell-cell interactions: enhancement of glycosyl transferase ectoenzyme systems during chlomydomonas genetic contact. Proc. Nat. Acad. Sci. U.S.A. 72, 310–313.
McLimans, W.F. (1969) Physiology of the cultured mammalian cell. In: Axenic Mammalian Cell Reactions (Tritsch, G.L., ed.) pp. 307–367, Dekker, New York.
McQuiddy, P. and Lilien, J. (1971) Sialic acid and cell aggregation. J. Cell Sci. 9, 823–833.
Merrell, R. and Glaser, L. (1973) Specific recognition of plasma membranes by embryonic cells. Proc. Nat. Acad. Sci. U.S.A. 70, 2794–2798.
Moscona, A. (1952) Cell suspensions from organ rudiments of chick embryos. Exp. Cell Res. 3, 535–539.
Moscona, A.A. (1956) Development of heterotypic combinations of dissociated embryonic chick cells. Proc. Soc. Exp. Biol. Med. 92, 410–416.
Moscona, A.A. (1957) The development in vitro of chimeric aggregates of dissociated embryonic chick and mouse cells. Proc. Nat. Acad. Sci. U.S.A. 43, 184–194.
Moscona, A.A. (1960) Patterns of and mechanisms of tissue reconstruction from dissociated cells. In: Developing Cell Systems and their Control (Rudnick, D., ed.) pp. 45–70, Ronald Press, New York.
Moscona, A.A. (1961a) Rotation mediated histogenetic aggregation of dissociated cells. Exp. Cell Res. 22, 455–475.
Moscona, A.A. (1961b) Effect of temperature on adhesion to glass and histogenetic cohesion of dissociated cells. Nature, 190, 408–409.
Moscona, A.A. (1962a) Analysis of cell recombinations in experimental synthesis of tissues in vitro. J. Cell. Comp. Physiol. 60, 65–79.
Moscona, A.A. (1962b) Synthesis of tissues in vitro from cells in suspension: Cellular and environmental factors. In: Biological Interactions in Normal and Neoplastic Growth (Brennan, M.J. and Simpson, W.L., eds.) pp. 113–126, Little Brown, Boston.
Moscona, A.A. (1963) Studies on cell aggregation: Demonstration of materials with selective cell-binding activity. Proc. Nat. Acad. Sci. U.S.A. 49, 742–747.
Moscona, A.A. (1965) Recombination of dissociated cells and the development of cell aggregates. In: Cells and Tissues In Culture (Willmer, E.N., ed.) Vol. 1, pp. 489–529, Academic Press, New York.
Moscona, A.A. (1968a) Cell aggregation: Properties of specific cell ligands and their role in the formation of multicellular systems. Dev. Biol. 18, 250–277.
Moscona, A.A. (1968b) Aggregation of sponge cells: Cell linking macromolecules and their role in the formation of multicellular systems. In Vitro 3, 13–21.
Moscona, A.A. (1971) Embryonic and neoplastic cell surfaces: Availability of receptors for concanavalin A wheat germ agglutinin. Science, 171, 905–907.
Moscona, A.A. (1974) Surface specification of embryonic cells: Lectin receptors, cell recognition and specific cell ligands. In: The Cell Surface in Development (Moscona, A.A., ed.) pp. 67–99, John Wiley, New York.

Moscona, A.A. and Garber, B.B. (1968) Reconstruction of skin from single cells and integumental differentiation in cell aggregates. In: Epithelial-Mesenchymal Interactions (Fleischmajer, R. and Billingham, R.E., eds.) pp. 230–243, Williams and Wilkins, Baltimore.

Moscona, A.A. and Moscona, H. (1952) The dissociation and aggregation of cells from organ rudiments of the early chick embryo. J. Anat. 86, 287–301.

Moscona, A.A. and Moscona, M.H. (1966) Aggregation of embryonic cells in a serum free medium and its inhibition at suboptimal temperatures. Exp. Cell Res. 41, 697–702.

Moscona, A.A. and Moscona, M.H. (1967) Comparison of aggregation of embryonic cells dissociated with trypsin or Versene. Exp. Cell Res. 45, 239–243.

Moscona, A.A., Trowell, O.A. and Willmer, E.N., (1965) Methods. In: Cells and Tissues in Culture (Willmer, E.N., ed.) Vol. 1, pp. 52–54, Academic Press, New York.

Moscona, A.A. and Wiens, A.W. (1975) Proflavine as a differential probe of gene expression: Inhibition of glutamine synthetase induction in embryonic retina. Dev. Biol. 44, 33–45.

Moscona, M.H. and Moscona, A.A. (1965) Control of differentiation in aggregates of embryonic skin cells: suppression of feather morphogenesis by cells from other tissues. Dev. Biol. 11, 402–423.

Muller, W.E.G. and Zahn, R.K. (1973) Purification and characterization of a species specific aggregation factor in sponges. Exp. Cell Res. 80, 95–104.

Nir, S. (1975) Long range intermolecular forces between macroscopic bodies; macroscopic and microscopic approaches. J. Theoret. Biol. 53, 83–100.

Oppenheimer, S.B. (1973) Utilization of L-glutamine in intercellular adhesion: Ascites tumor and embryonic cells. Exp. Cell Res. 77, 175–182.

Oppenheimer, S.B. (1975) Functional involvement of specific carbohydrate in teratoma cell adhesion factor. Exp. Cell Res. 92, 122–126.

Oppenheimer, S.B., Edidin, M., Orr, C.W. and Roseman, S. (1969) An L-glutamine requirement for intercellular adhesion. Proc. Nat. Acad. Sci. U.S.A. 63, 1395–1402.

Orr, C.W. and Roseman, S. (1969a) Intercellular adhesion. I. A quantitative assay for measuring the rate of adhesion. J. Membrane Biol. 1, 109–124.

Orr, C.W. and Roseman, S. (1969b) Intercellular adhesion. II. The purification and properties of a horse serum protein that promotes neural retina cell aggregation. J. Membrane Biol. 1, 125–143.

Overton, J. (1968) The fate of desmosomes in trypsinized tissue. J. Exp. Zool. 168, 203–214.

Overton, J. (1969) A fibrillar intercellular material between reaggregating embryonic chick cells. J. Cell Biol. 40, 136–143.

Overton, J. (1973) Experimental manipulation of desmosome formation. J. Cell Biol. 56, 636–646.

Overton, J. (1974a) Cell junctions and their development. Prog. Surf. Memb. Sci. 8, 161–208.

Overton, J. (1974b) Selective formation of desmosomes in chick cell reaggregates. Dev. Biol. 39, 210–255.

Overton, J. and Culver, N. (1973) Desmosomes and their components after cell dissociation and reaggregation in the presence of cytochalasin B. J. Exp. Zool. 185, 341–356.

Patt, L.M. and Grimes, W.J. (1974) Cell surface glycolipid and glycoprotein glycosyltransferases of normal and transformed cells. J. Biol. Chem. 249, 4157–4166.

Pessac, B. and Defendi, V. (1972) Evidence for distinct aggregation factors and receptors in cells. Nature New Biology 238, 13–15.

Pethica, B.A. (1961) The physical chemistry of cell adhesion. Exp. Cell Res., suppl. 8, 123–140.

Phillips, H.J. (1972) Dissociation of single cells from lung or kidney with elastase. In Vitro 8, 101–105.

Phillips, H.M. and Steinberg, M.S. (1969) Equilbrium measurements of embryonic chick cell adhesiveness. I. Shape equilibrium in centrifugal fields. Proc. Natl. Acad. Sci. U.S.A. 64, 121–127.

Pollard, T.D. and Korn, E.D. (1973) Electron microscopic identification of actin associated with isolated amoeba plasma membranes. J. Biol. Chem. 248, 448–450.

Pollard, T.D. and Weihung, R.R. (1974) Actin and myosin and cell movement. CRC Crit. Rev. Biochem. 2, 1–65.

Ponder, E. (1951) Effects produced by trypsin on certain properties of human red cells. Blood, 6, 350–356.

Poste, G. (1971) Tissue dissociation with proteolytic enzymes. Adsorption and activity of enzymes at the cell surface. Exp. Cell Res. 65, 359–367.

Poste, G., Greenham, L.W., Mallucci, L., Reeve, P. and Alexander, D.J. (1973) The study of cellular "microexudates" by ellipsometry and their relation to the cell coat. Exp. Cell Res. 78, 303–313.

Pugh-Humphreys, R.G.P. and Sinclair, W. (1970) Ultrastructural studies relating to the surface morphology of cultured cells. J. Cell Sci. 6, 477–484.

Rappaport, C. (1966a) Effect of temperature on dissociation of adult mouse liver with sodium tetraphenylboron. Proc. Soc. Exp. Biol. Med. 121, 1022–1025.

Rappaport, C. (1966b) Role of intercellular matrix in aggregation of mammalian cells. Proc. Soc. Exp. Biol. Med. 121, 1025–1028.

Rappaport, C. and Howze, G.B. (1966a) Dissociation of adult mouse liver by sodium tetraphenylboron, a potassium complexing agent. Proc. Soc. Exp. Biol. Med. 121, 1010–1016.

Rappaport, C. and Howze, G.B. (1966b) Further studies on the dissociation of adult mouse tissue. Proc. Soc. Exp. Biol. Med. 121, 1016–1021.

Richmond, J.E., Glaeser, R.M. and Todd, P. (1968) Protein synthesis and aggregation of embryonic cells. Exp. Cell Res. 52, 43–58.

Riddle, P.N. (1974) A microtube oscillation method for cellular reaggregation. Experientia, 30, 114–116.

Rinaldini, L.M. (1958) The isolation of living cells from animal tissues. Int. Rev. Cytol. 7, 587–647.

Rinaldini, L. (1959) An improved method for the isolation and quantitative cultivation of embryonic cells. Exp. Cell Res. 16, 477–505.

Robb, J.A. (1973) Microtest plates. B. Replica plating. In: Tissue Culture: Methods and Applications (Kruse, P.F., Jr. and Patterson, M.K., Jr., eds.) pp. 270–274, Academic Press, New York.

Roseman, S. (1970) The synthesis of complex carbohydrates by multiglycosyltransferase systems and their potential function in intercellular adhesion. Chem. Phys. Lipids 5, 270–297.

Roseman, S. (1974) Complex carbohydrates and intercellular adhesion. In: The Cell Surface in Development (Moscona, A.A., ed.) pp. 255–271, John Wiley, New York.

Rosenberg, M.D. (1960) Microexudates from cells grown in tissue culture. Biophys. J. 1, 137–159.

Rosenberg, M.D., Aufderheide, K. and Christianson, J. (1969) In vitro enhancement of cell clumping by surface region fractions. Exp. Cell Res. 57, 449–454.

Roth, S. (1968) Studies on intercellular adhesive selectivity. Dev. Biol. 18, 602–631.

Roth, S. (1973) A molecular model for cell interactions. Quart. Rev. Biol. 48, 541–563.

Roth, S.A. and Weston, J.A. (1967) The measurement of intercellular adhesion. Proc. Natl. Acad. Sci. U.S.A. 58, 974–980.

Roth, S.A. and White, D. (1972) Intercellular contact and cell-surface galactosyltransferase activity. Proc. Natl. Acad. Sci. U.S.A. 69, 485–489.

Roth, S., McGuire, E.J. and Roseman, S. (1971a) An assay for intercellular adhesive specificity. J. Cell Biol. 51, 525–535.

Roth, S.A., McGuire, E.J. and Roseman, S. (1971b) Evidence for cell-surface glycosyltransferases. Their potential role in cellular recognition. J. Cell Biol. 51, 536–547.

Roux, W. (1894) Uker den "cytotropismus" der Furchungszellen des Grasfrosches (Rana fusca). Wilhelm Roux Archiv. Entwicklungsmech. Organismen 1, 43–68.

Rubin, H. (1966) A substance in conditioned medium which enhances the growth of small numbers of chick embryo cells. Exp. Cell Res. 41, 138–148.

Sanford, W.C. (1974) A new method for dispersing strongly adhesive cells in tissue culture. In Vitro 10, 281–283.

Sara, M., Liaci, L. and Melone, N. (1966a) Bispecific cell aggregation in sponges. Nature, 210, 1167–1168.

Sara, M., Liaci, L. and Melone, N. (1966b) Mixed cell aggregation between sponges and the anthozoan *Anemonia sulcata.* Nature, 210, 1168–1169.

Schmid, R.W. and Reilley, C.N. (1957) New complexion for titration of calcium in the presence of magnesium. Anal. Chem. 29, 264–268.

Senda, N., Tamura, H., Shibita, N., Yoshitake, J., Kondo, K. and Tanaka, K. (1975) The mechanism of the movement of leucocytes. Exp. Cell Res. 91, 393–407.

Sharon, N. and Lis, H. (1972) Lectins: Cell agglutinating and sugar specific proteins. Science, 177, 949–959.

Sheffield, J.B. (1970) Studies on aggregation of embryonic cells: Initial cell adhesions and the formation of intercellular junctions. J. Morphol. 132, 245–264.

Sheffield, J.B. and Moscona, A.A. (1969) Early stages in the reaggregation of embryonic chick neural retina cells. Exp. Cell Res. 57, 462–466.

Shimada, Y. (1968) Suppression of myogenesis by heterotypic and heterospecific cells in monolayer cultures. Exp. Cell Res. 51, 564–578.

Shimada, Y., Moscona, A.A. and Fischman, D.A. (1974) Scanning electron microscopy of cell aggregation: cardiac and mixed retina-cardiac cell suspensions. Dev. Biol. 36, 428–446.

Shlatz, L. and Marinetti, G.V. (1972) Hormone-calcium interactions with the plasma membrane of rat liver cells. Science, 176, 175–177.

Spiegel, M. (1954) The role of specific surface antigens in cell adhesion. Part I. The reaggregation of sponge cells. Biol. Bull. 107, 130–148.

Spiegel, M. (1955) The reaggregation of dissociated sponge cells. Ann. N.Y. Acad. Sci. 60, 1056–1078.

Spooner, B.S. (1970) The expression of differentiation by chick embryo thyroid in cell culture. 1. Functional and fine structural stability in mass and clonal culture. J. Cell Physiol. 75, 33–47.

Steinberg, M.S. (1958) On the chemical bonds between animal cells. A mechanism for type-specific association. Am. Naturalist, 92, 65–81.

Steinberg, M.S. (1962a) On the mechanism of tissue reconstruction by dissociated cells. I. Population kinetics, differential adhesiveness, and the absence of directed migration. Proc. Nat. Acad. Sci. U.S.A. 48, 1577–1582.

Steinberg, M.S. (1962b) Mechanism of tissue reconstruction by dissociated cells. II. Time course of events. Science, 137, 762–763.

Steinberg, M.S. (1962c) On the mechanism of tissue reconstruction by dissociated cells. III. Free energy relations and the reorganization of fused heteronomic tissue fragments. Proc. Nat. Acad. Sci. U.S.A. 48, 1769–1776.

Steinberg, M.S. (1962d) The role of temperature in the control of aggregation of dissociated embryonic cells. Exp. Cell Res. 28, 1–10.

Steinberg, M.S. (1962e) Calcium complexing by embryonic cell surfaces: relation to intercellular adhesiveness. In: Biological Interactions in Normal and Neoplastic Growth (Brennan, M.J. and Simpson, W.L., eds.) pp. 127–140, Little Brown, Boston.

Steinberg, M.S. (1963a) Reconstruction of tissues by dissociated cells. Science, 141, 401–408.

Steinberg, M.S. (1963b) "ECM": Its nature, origin and function in cell aggregation. Exp. Cell Res. 30, 257–279.

Steinberg, M.S. (1964) The problem of adhesive selectivity in cellular interactions. In: Cellular Membranes in Development (Locke, M., ed.) pp. 321–366, Academic Press, New York.

Steinberg, M.S. (1967) Avian and mammalian cell dissociation. In: Methods in Developmental Biology (Wilt, F. and Wessells, N.K., eds.) pp. 565–572, Crowell, New York.

Steinberg, M.S. (1970) Does differential adhesion govern self-assembly processes in histogenesis? Equilbrium configuration and the energence of a hierarchy among populations of embryonic cells. J. Exp. Zool. 173, 395–434.

Steinberg, M.S., Armstrong, P.B. and Granger, R.E. (1973) On the recovery of adhesiveness by trypsin-dissociated cells. J. Membrane Biol. 13, 97–128.

Steinberg, M.S. and Gepner, I.A. (1973) Are concanavalin A receptor sites mediators of cell-cell adhesion. Nature New Biology 241, 249–251.

Steinberg, M.S. and Wiseman, L.L. (1972) Do morphogenetic tissue rearrangements require active cell movements? The reversible inhibition of cell sorting and tissue spreading by cytochalasin B. J. Cell Biol. 55, 606–615.

Sussman, M. and Boschwitz, C. (1975) Adhesive properties of cell ghosts derived from *Dictyostelium discoideum*. Dev. Biol. 44, 362–368.

Tashjian, A.H., Jr., Bancroft, F.C. and Levine, L. (1970) Production of both prolactin and growth hormones by clonal strains of rat pituitary tumor cells: Differential effects of hydrocortisone and tissue extracts. J. Cell Biol. 47, 61–70.

Tolson, N.D., Mayhew, E., Maslow, D.E. and Blumenson, L.E. (1975) Cell movement and the aggregation of embryonic cells: Experimental analysis and computer simulation. In preparation.

Townes, P.L. and Holtfreter, J. (1955) Directed movements and selective adhesion of embryonic amphibian cells. J. Exp. Zool. 128, 53–120.

Trinkaus, J.P. and Gross, M.C. (1961) The use of tritiated thymidine for marking migratory cells. Exp. Cell Res. 24, 52–57.

Trinkaus, J.P. and Groves, P.W. (1955) Differentiation in culture of mixed aggregates of dissociated tissue cells. Proc. Nat. Acad. Sci. U.S.A. 41, 787–795.

Trinkaus, J.P. and Lentz, J.P. (1964) Direct observation of type specific segregation in mixed cell aggregates. Dev. Biol. 9, 115–136.

Turner, R.S. and Burger, M.M. (1973) Involvement of a carbohydrate group in the active site for surface guided reassociation of animal cells. Nature, 244, 509–510.

Tyler, A. (1947) An auto-antibody concept of cell structure, growth and differentiation. Growth, 10, Suppl. 7–19.

Ulrich, K. and Moore, G.E. (1965) A vibrating mixer for agitation of suspension cultures of mammalian cells. Biotechnol. Bioeng. 7, 507–515.

Verwey, E.J.W. and Overbeek, J.T.G. (1948) Theory of the Stability of Lyophobic Colloids. Elsevier, Amsterdam.

Walther, B.T., Ohman, R. and Roseman, S. (1973) A quantitative assay for intercellular adhesion. Proc. Nat. Acad. Sci. U.S.A. 70, 1569–1573.

Waymouth, C. (1974) To disaggregate or not to disaggregate. Injury and cell disaggregation, transient or permanent? In Vitro 10, 97–111.

Webb, G.C. and Roth, S, S. (1974) Cell contact dependence of surface galactosyltransferase activity as a function of the cell cycle. J. Cell Biol. 63, 796–805.

Weinbaum, G. and Burger, M.M. (1973) Two component system for surface guided reassociation of animal cells. Nature, 244, 510–512.

Weiser, M.M. (1973) Intestinal epithelial cell surface membrane glycoprotein synthesis. II. Glycosyltransferases and endogenous acceptors of the undifferentiated cell surface membrane. J. Biol. Chem. 248, 2542–2548.

Weiss, L. (1958) The effects of trypsin on the size, viability and dry mass of sarcoma 37 cells. Exp. Cell Res. 14, 80–83.

Weiss, L. (1960) The adhesion of cells. Int. Rev. Cytol. 9, 187–225.

Weiss, L. (1961) Studies on cellular adhesion in tissue culture. IV. The alteration of substrata by cell surfaces. Exp. Cell Res. 25, 504–517.

Weiss, L. (1964) Studies on cellular adhesion in tissue culture. VII. Surface activity and cell detachment. Exp. Cell Res. 33, 277–288.

Weiss, L. (1966) Studies on cell deformability. II. Effects of some proteolytic enzymes. J. Cell Biol. 30, 39–43.

Weiss, L. (1967) The Cell Periphery, Metastasis and Other Contact Phenomena, North-Holland, Amsterdam.

Weiss, L. (1968) Studies on cell deformability V. Some effects of ribonuclease. J. Theoret. Biol. 18, 9–18.

Weiss, L. (1970) Cell contact phenomena. In: In Vitro: Advances in Tissue Culture (Waymouth, C., ed.) pp. 48–78, Williams and Wilkins, Baltimore.

Weiss, L. (1971) Biophysical aspects of initial cell interactions with solid surfaces. Fed. Proc. 30, 1649–1657.

Weiss, L. (1973) Contact between cancer and other cells. A biophysical approach. In: Chemotherapy of Cancer Dissemination and Metastasis (Garattini, S. and Franchi, G.) pp. 19–30, Raven Press, New York.

Weiss, L. (1974) Studies on cell adhesion in tissue culture. XV. Some effects of cycloheximide on cell detachment. Exptl. Cell Res. 86, 223–232.

Weiss, L. and Harlos, J.P. (1972a) Some speculations on the rate of adhesion of cells to coverslips. J. Theoret. Biol. 37, 169–179.

Weiss, L. and Harlos, J.P. (1972b) Short term interactions between cell surfaces. Prog. Surf. Sci. 1, 355–405.

Weiss, L. and Kapes, D.L. (1966) Observations on cell adhesion and separation following enzyme treatment. Exp. Cell Res. 41, 601–608.

Weiss, L. and Lachmann, P.J. (1964) The origin of an antigenic zone surrounding HeLa cells cultured on glass. Exp. Cell Res. 36, 86–91.

Weiss, L. and Maslow, D.E. (1972) Some effects of trypsin dissociation on the inhibition of reaggregation among embryonic chicken neural retina cells by cycloheximide. Dev. Biol. 29, 482–485.
Weiss, L. and Mayhew, E. (1966) The presence of ribonucleic acid within the peripheral zones of two types of mammalian cells. J. Cell Physiol. 68, 345–360.
Weiss, L. and Neiders, M.E. (1971) A biophysical approach to the adhesion of human gingival epithelial cells to tooth and glass surfaces. J. Periodont. Res. 6, 28–37.
Weiss, L. and Subjeck, J.R. (1974a) Interaction between the peripheries of Ehrlich ascites tumor cells as indicated by the binding of colloidal iron hydroxide particles. Int. J. Cancer 13, 143–150.
Weiss, L. and Subjeck, J.R. (1974b) The densities of colloidal iron hydroxide particles bound to microvilli and the spaces between them: Studies of glutaraldehyde-fixed Ehrlich ascites tumor cells. J. Cell Sci. 14, 215–223.
Weiss, L. and Zeigel, R. (1972) Heterogeneity of anionic sites at the electrokinetic surfaces of fixed Ehrlich ascites tumor cells. J. Theoret. Biol. 34, 21–27.
Weiss, L., Poste, G., MacKearnin, A. and Willett, K. (1975) Growth of mammalian cells on substrates coated with cellular microexudates. I. Effect on cell growth at low population densities. J. Cell Biol. 64, 135–145.
Weiss, L., Zeigel, R., Jung, O.S. and Bross, I.D.J. (1972) Binding of positively charged particles to glutaraldehyde-fixed human erythrocytes. Exp. Cell Res. 70, 57–64.
Weiss, P. (1945) Experiments on cell and axon orientation in vitro: The role of colloidal exudates in tissue organization. J. Exp. Zool. 100, 353–386.
Wickus, G., Gruenstein, E., Robbins, P.W. and Rich, A. (1975) Decrease in membrane-associated actin of fibroblasts after transformation by Rous sarcoma virus. Proc. Natl. Acad. Sci. U.S.A. 72, 746–749.
Wiens, A.W. and Moscona, A.A. (1972) Preferential inhibition by proflovin of the hormonal induction of glutamine synthetase in embryonic neural retina. Proc. Natl. Acad. Sci. U.S.A. 69, 1504–1507.
Wilde, C.E., Jr. (1958) The fusion of myoblasts; a morphogenetic mechanism in striated muscle differentiation. Anat. Rec. 132, 517–518.
Wilde, C.E., Jr. (1959) Differentiation in response to the biochemical environment. In: Cell, Organism and Milieu (Rudnick, D., ed.) pp. 3–45, Ronald Press, New York.
Wilde, C.E., Jr. (1961) The differentiation of vertebrate pigment cells. Adv. Morphogen. 1, 267–300.
Wilde, C.E., Jr. (1962) The role of cellular exudates in cell and tissue differentiation. In: Biological Interactions in Normal and Neoplastic Growth (Brennan, M.J. and Simpson, W.L., eds.) pp. 199–223, Little Brown, Boston.
Wilson, H.V. (1907) On some phenomena of coalescence and regeneration in sponges. J. Exp. Zool. 5, 245–258.
Wilson, H.V. (1910) Development of sponges from dissociated tissue cells. Bull. Bur. Fish 30, 1–30.
Wiseman, L.L. and Steinberg, M.S. (1973) The movement of single cells within solid tissue masses. Exp. Cell Res. 79, 468–471.
Wiseman, L.L., Steinberg, M.S. and Phillips, H.M. (1972) Experimental modulation of intercellular cohesiveness: Reversal of tissue assembly patterns. Dev. Biol. 28, 498–517.
Wolff, E. (1954) Potentialités et affinités des tissus, révélées par la culture in vitro d'organes en associations hétérogènes et xénoplastiques. Bull. Soc. Zool. France 79, 357–368.
Yaffe, D. and Feldman, M. (1965) The formation of hybrid multinucleated muscle fibers from myoblasts of different genetic origin. Dev. Biol. 11, 300–317.

Subject Index